PAUL L. BUTZER · ROLF J. NESSEL

FOURIER ANALYSIS AND APPROXIMATION

VOL. 1

MATHEMATISCHE REIHE
BAND 40

LEHRBÜCHER UND MONOGRAPHIEN
AUS DEM GEBIETE DER EXAKTEN WISSENSCHAFTEN

Fourier Analysis and Approximation

Vol. 1

One-Dimensional Theory

Paul L. Butzer
Professor of Mathematics

Rolf J. Nessel
Dozent of Mathematics

Rheinisch Westfälische Technische Hochschule Aachen

1971

BIRKHÄUSER VERLAG BASEL
UND STUTTGART

(LEHRBÜCHER UND MONOGRAPHIEN AUS DEM GEBIETE DER EXAKTEN WISSENSCHAFTEN, MATHEMATISCHE REIHE, BAND 40)
ISBN 978-3-0348-7450-2 ISBN 978-3-0348-7448-9 (eBook)
DOI 10.1007/978-3-0348-7448-9

United States Edition published by
ACADEMIC PRESS, INC.
111 Fifth Avenue, New York, New York 10003
(PURE AND APPLIED MATHEMATICS, A Series of Monographs and Textbooks, Volume 40–I)
ISBN 978-3-0348-7450-2 (Academic Press)

LIBRARY OF CONGRESS CATALOG CARD NUMBER: 77–145 668
AMS 1970 Subject Classification 41–01, 42–01, 26–01

WILLIAM CLOWES AND SONS LIMITED
LONDON, BECCLES AND COLCHESTER

to our parents

Preface

At the international conference on 'Harmonic Analysis and Integral Transforms', conducted by one of the authors at the Mathematical Research Institute in Oberwolfach (Black Forest) in August 1965, it was felt that there was a real need for a book on Fourier analysis stressing (i) parallel treatment of Fourier series and Fourier transforms from a transform point of view, (ii) treatment of Fourier transforms in $\mathsf{L}^p(\mathbb{R}^n)$-space not only for $p = 1$ and $p = 2$, (iii) classical solution of partial differential equations with completely rigorous proofs, (iv) theory of singular integrals of convolution type, (v) applications to approximation theory including saturation theory, (vi) multiplier theory, (vii) Hilbert transforms, Riesz fractional integrals, Bessel potentials, (viii) Fourier transform methods on locally compact groups. This study aims to consider these aspects, presenting a systematic treatment of Fourier analysis on the circle as well as on the infinite line, and of those areas of approximation theory which are in some way or other related thereto. A second volume is in preparation which goes beyond the one-dimensional theory presented here to cover the subject for functions of several variables.

Approximately a half of this first volume deals with the theories of Fourier series and of Fourier integrals from a transform point of view. This parallel treatment easily lends itself to an understanding of abstract harmonic analysis; the underlying classical theory is therefore presented in a form that is directed towards the case of arbitrary locally compact abelian groups, which are to be discussed in the second volume. The second half is concerned with the concepts making up the fundamental operation of Fourier analysis, namely convolution. Thus the leitmotiv of the approximation theoretic part is the theory of convolution integrals, the 'smoothing' of functions by such, and the study of the corresponding degree of approximation. Special as this approach may seem, it not only embraces many of the topics of the classical theory but also leads to significant new results, e.g., on summation processes of Fourier series, conjugate functions, fractional integration and differentiation, limiting behaviour of solutions of partial differential equations, and saturation theory.

On the other hand, no attempt is made to present an account of the theory of Fourier series or integrals per se, nor to prepare a book on classical approximation theory as such. Indeed, the theory of Fourier series is the central theme of the monumental treatises by A. Zygmund (1935, 1959) and N. K. Bari (1961). With respect to the theory of Fourier integrals we have aimed to bring portions of the fine treatises by E. C. Titchmarsh (1937) and S. Bochner–K. Chandrasekharan (1949) up to date, yet complementing them with Fourier transforms on the circle. Furthermore, a number of excellent books giving a broad coverage of approximation theory has appeared in the past years; for the classical ones we can single out N. I. ACHIESER (1947) and I. P. NATANSON (1949). In contrast, the present volume is meant to serve as a detailed introduction to each of these three major fields, emphasizing the underlying, unifying principles and culminating in saturation theory for convolution integrals.

Whereas many texts on approximation treat the matter only for continuous functions (and in L^p-space, if at all, separately), the present text handles it in the spaces $C_{2\pi}$ and $L^p_{2\pi}$, $1 \leq p < \infty$, simultaneously. The parallel treatment of periodic questions and those on the line, already mentioned in connection with Fourier transforms, is a characteristic feature of the entire material as presented in this volume. This exhibits the structures common to both theories (compare the treatment of Chapters 6 and 11, for example), usually discussed separately and independently. Whenever the material would be too analogous in statement or proof, emphasis is laid upon different methods of proof. However, the reader mainly interested in the periodic theory can proceed directly from Chapters 1 and 2 to Chapter 4 and from there to the relevant parts indicated in Chapter 6 and Sec. 7.1. He may then turn to Chapter 9, Sec. 10.1–10.4, Sec. 11.4–11.5, Sec. 12.1–12.2. On the other hand, Chapters 4 and 5, together with the basic material in Chapters 1 and 3 (Sec. 1.1–1.2, 1.4, 3.1–3.2) may serve as a short course on (classical) Fourier analysis; for selected applications one may then consult Chapters 6 and 7. As a matter of fact, Chapter 7 gives the first and best-known application of Fourier transform methods, namely to the solution of partial differential equations. In Chapters 10–13 these integral transform methods are developed and refined so as to handle profounder and more theoretical problems in approximation theory. A brief introductory course on classical approximation theory for periodic functions may be based on Chapters 1 and 2.

The present treatment is essentially self-contained; starting at an elementary level, the book progresses gradually but thoroughly to the advanced topics and to the frontiers of research in the field. Many of the results, especially those of Chapters 10–13, are presented here for the first time, at least in book-form. Although the presentation is completely rigorous from the mathematical point of view, the lowest possible level of abstraction has been selected without compromising accuracy. In many of the proofs intricate analysis is required. This we have carried out in detail not only since we believe it is more important to save the reader's time than to save paper, but because we believe firmly that the student reader should be able to follow each step of a proof. Despite the virtues of elegant brevity in the presentation of proofs, many recent texts have gone to the extreme of sacrificing understanding to the cost of all but the more expert in their fields. Although we have attempted to range both in depth and breadth, it remains inevitable that several themes have been slighted in a subject of rapidly increasing diversity and development. Presumably no apology is necessary for the fact that we have been guided in our selection by pursuing those topics which have caught our imagination; however, in the process we have attempted to illustrate a variety of analytical techniques.

With this step-by-step development the volume can be readily utilized by senior undergraduate students in mathematics, applied mathematics, and related fields such as mathematical physics. It is also hoped that the book will be useful as a reference for workers in the physical sciences. Indeed, the central theme is Fourier analysis and approximation, subjects of wide importance in many of the sciences. The principal prerequisites would be a solid course on advanced calculus as well as some working knowledge of elementary Lebesgue integration and functional analysis. To make the presentation self-contained these foundations are collected in a preliminary Chapter 0.

Following each section there is a total of approximately 550 problems, many consisting of several parts, ranging from fairly routine applications of the text material to others that extend the coverage of the book. Many of the more difficult ones are supplied with hints or with references to the pertinent literature. The results of problems are often used in subsequent sections. Each chapter ends with a section on 'Notes and Remarks'. These contain historical references and credits as well as detailed references to some 650 papers or books treating or supplementing specific results of a chapter. Important topics related to those treated but not included in the text are outlined here. Although we have tried our best to give everyone his full measure of credit, we apologize in advance for any oversights or inaccuracies in this regard. Here, as well as in the subject matter, we will appreciate any corrections suggested by the reader.

The second volume will deal with more abstract aspects of the material. Special emphasis is placed upon the theory in Euclidean n-space. Fourier transforms will be discussed in the setting of distribution theory, and a systematic account given of those parts of approximation theory concerned with functions of several variables. Included will be characterizations of saturation classes of singular integrals with radial or product kernels by Lipschitz conditions, Riesz transforms and fractional integrals, Bessel potentials, etc. by means of embedding theorems.

The material presented here first took form during several one-semester courses on Fourier series, on Fourier transforms, and on approximation theory given at the Technological University of Aachen during the past decade by one of the authors and assisted by the other. The third and final typewritten version was begun in the summer of 1966, as a joint effort of both authors. We have been especially fortunate with the assistance of several members of our team of collaborators. Dr. EBERHARD L. STARK read and checked the whole manuscript, gave helpful suggestions, edited every chapter, assisted in reading the proofs, and prepared the index. It is certain that without his patient and unstinting work nothing on the scale of the volume could have been completed. Drs. ERNST GÖRLICH and WALTER TREBELS gave valuable advice and criticism, read parts of the manuscript and set the authors straight on many a vital point; several portions of the manuscript were written in collaboration with Dr. TREBELS, including Chapter 11. Mr. FRIEDRICH ESSER assisted in reading the proofs. We are particularly indebted to our secretary Miss URSULA COMBACH who typed the final version cheerfully and with painstaking care; the earlier version had been capably typed by Mrs. KARIN KOCH and Mrs. DORIS EWERS.

We also wish to thank Mr. C. EINSELE of *Birkhäuser Verlag* for his patience, and the staff of *William Clowes and Sons Ltd.* for their skill and meticulous care in the production of this book.

Aachen, February, 1970 P. L. BUTZER and R. J. NESSEL

Contents

0

Preliminaries

In this chapter a number of fundamental concepts and results on real variable theory, Lebesgue integration, convolutions, functions of bounded variation, normed linear spaces, bounded linear operators, and functionals are listed, together with some of the conventions and terminology that will be adhered to throughout this book. For the convenience of the reader the results are set down in a form actually used in the text, and to that extent only. In order to refer to the concepts and propositions by number later on, some are presented in various different forms although unified formulations would be possible. Several of the more important facts are given with proof or also with explicit references; this will permit the reader to review the material.

The reader is also referred to the list of symbols.

0.1 Fundamentals on Lebesgue Integration

Let $\mathbb{R}$ be the set of all real numbers, also called the *real line* or *line group*, and $\mathbb{C}$ the set of all complex numbers, the complex plane. Unless specified otherwise, all functions $f, g, \ldots$ of this section are defined on $\mathbb{R}$ and assumed to be (complex-valued†) Lebesgue measurable (on $\mathbb{R}$). C denotes the set of all functions which are uniformly continuous and bounded on $\mathbb{R}$, endowed with the *norm*

(0.1.1) $$\|f\|_{\mathsf{C}} = \sup_{x \in \mathbb{R}} |f(x)|.$$

L^p is the set of functions which are Lebesgue integrable to the pth power over $\mathbb{R}$ if $1 \le p < \infty$, and essentially bounded (bounded almost everywhere) on $\mathbb{R}$ if $p = \infty$. For $f \in \mathsf{L}^p$

† It should be emphasized that, for the purposes of this text, the extension from real- to complex-valued functions is clear: A certain property holds for a complex-valued function if and only if it holds for the real part Re (f) as well as for the imaginary part Im (f).

$$\|f\|_p = \left\{\frac{1}{\sqrt{2\pi}} \int_{-\infty}^{\infty} |f(u)|^p \, du\right\}^{1/p} \tag{0.1.2}$$

if p is such that $1 \le p < \infty$, and in case $p = \infty$

$$\|f\|_\infty = \operatorname*{ess\,sup}_{x \in \mathbb{R}} |f(x)|. \tag{0.1.3}$$

Thus L^p consists precisely of those functions f for which the norm $\|f\|_p$ is a finite number. The constant factor $(1/\sqrt{2\pi})$ in (0.1.2) shall turn out to be convenient in Fourier analysis. In this connection, NL^1 is the set of those $f \in \mathsf{L}^1$ which are normalized by $\int_{-\infty}^{\infty} f(u)\,du = \sqrt{2\pi}$. If any confusion may occur, the more systematic notations $\mathsf{C}(\mathbb{R})$, $\mathsf{L}^p(\mathbb{R})$, $\|f\|_{\mathsf{L}^p}$, etc. are employed. Moreover, instead of $\|f\|$ the symbol $\|f(\circ)\|$ is also used, thus indicating the variable relative to which the norm is taken (e.g. $\|f(\circ + h)\|$). If f has an elementary representation, we write $\|f(x)\|$ (e.g. $\|(1 + x^2)^{-1}\|$).

Since $\mathbb{R}$ is noncompact, there are counterexamples (see Problem 5.2.1) proving that $f \in \mathsf{L}^p$ does not necessarily imply $f \in \mathsf{L}^q$ for any other value q, $1 \le q \le \infty$. However, if $f \in \mathsf{L}^p \cap \mathsf{L}^\infty$, an elementary estimate yields $f \in \mathsf{L}^q$ for every q with $p \le q \le \infty$. More generally, if $f \in \mathsf{L}^p \cap \mathsf{L}^q$ for some $1 \le p < q \le \infty$, then $f \in \mathsf{L}^s$ for every $p \le s \le q$. Indeed, setting

$$\mathbb{E}_1 = \{x \in \mathbb{R} \mid |f(x)| \le 1\}, \qquad \mathbb{E}_2 = \{x \in \mathbb{R} \mid |f(x)| > 1\},$$

one has $\mathbb{E}_1 \cup \mathbb{E}_2 = \mathbb{R}$, and for $q < \infty$

$$\begin{aligned}\sqrt{2\pi}\,\|f\|_s^s &= \int_{\mathbb{E}_1} |f(u)|^s \, du + \int_{\mathbb{E}_2} |f(u)|^s \, du \\ &\le \int_{\mathbb{E}_1} |f(u)|^p \, du + \int_{\mathbb{E}_2} |f(u)|^q \, du \le \sqrt{2\pi}(\|f\|_p^p + \|f\|_q^q).\end{aligned}$$

In this text, $\mathsf{X}(\mathbb{R})$ always denotes one of the spaces C or L^p, $1 \le p < \infty$. For $f, g \in \mathsf{X}(\mathbb{R})$ we write $f(x) = g(x)$ (a.e.) if equality holds for all $x \in \mathbb{R}$ in case $\mathsf{X}(\mathbb{R}) = \mathsf{C}$, and almost everywhere in case $\mathsf{X}(\mathbb{R}) = \mathsf{L}^p$, $1 \le p < \infty$, i.e., if $\|f - g\|_{\mathsf{X}(\mathbb{R})} = 0$. In this event we also write $f = g$ in $\mathsf{X}(\mathbb{R})$.

A real-valued function f is said to be *even, odd, positive* (non-negative), and *strictly* positive if $f(x) = f(-x)$, $f(x) = -f(-x)$, $f(x) \ge 0$, and $f(x) > 0$, respectively. It is called *monotonely increasing* (on $\mathbb{R}$) if $f(x_1) \le f(x_2)$ for $x_1 \le x_2$, *strictly* monotonely increasing if $f(x_1) < f(x_2)$ for $x_1 < x_2$, *monotonely decreasing* if $f(x_1) \ge f(x_2)$ for $x_1 \le x_2$. A sequence of functions $\{f_n(x)\}_{n=1}^\infty$ is said to be monotonely increasing if $f_n(x) \le f_{n+1}(x)$ for every x and $n = 1, 2, \ldots$.

Proposition 0.1.1. (***Fatou's lemma***). *Let* $\{f_n\}_{n=1}^\infty$ *be a sequence of positive functions on* $\mathbb{R}$. *If* $\liminf_{n\to\infty} f_n(x) = f(x)$ a.e., *then*

$$\int_{-\infty}^{\infty} f(u)\,du \le \liminf_{n\to\infty} \int_{-\infty}^{\infty} f_n(u)\,du.$$

Proposition 0.1.2. (***Lebesgue's monotone convergence theorem***). *Let* $\{f_n\}_{n=1}^\infty$ *be a monotonely increasing sequence of positive functions. If* $\lim_{n\to\infty} f_n(x) = f(x)$ a.e., *then*

$$\lim_{n\to\infty} \int_{-\infty}^{\infty} f_n(u)\,du = \int_{-\infty}^{\infty} f(u)\,du. \tag{0.1.4}$$

This result is also known as Beppo Levi's theorem, see also Prop. 0.3.3.

Proposition 0.1.3. (***Lebesgue's dominated convergence theorem***). *Let* $\{f_n\}_{n=1}^{\infty} \subset \mathsf{L}^1$, *and suppose that* $\lim_{n\to\infty} f_n(x) = f(x)$ a.e. *If there exists* $g \in \mathsf{L}^1$ *such that* $|f_n(x)| \le g(x)$ a.e. *for all n, then f belongs to* L^1, *and* (0.1.4) *holds.*

Proposition 0.1.4. (***Fubini's theorem***). *Let* $x, y \in \mathbb{R}$, *and* $f(x, y)$ *be a (complex-valued) function of two (real) variables defined and measurable on the two-dimensional Euclidean space* $\mathbb{R}^2$.

(i) *Suppose that* $f \in \mathsf{L}^1(\mathbb{R}^2)$, *i.e., the double integral* $\int_{-\infty}^{\infty}\int_{-\infty}^{\infty} f(x, y)\,dx\,dy$ *is absolutely convergent. Then, for almost all x,* $f(x, y)$ *is absolutely integrable over* $\mathbb{R}$ *with respect to the variable y, i.e.* $f(x, \circ) \in \mathsf{L}^1(\mathbb{R})$ a.e. *Moreover,* $\int_{-\infty}^{\infty} f(\circ, y)\,dy \in \mathsf{L}^1(\mathbb{R})$ *and*

$$\int_{-\infty}^{\infty}\left\{\int_{-\infty}^{\infty} f(x, y)\,dy\right\}dx = \int_{-\infty}^{\infty}\int_{-\infty}^{\infty} f(x, y)\,dx\,dy.$$

(ii) *Suppose that* $\int_{-\infty}^{\infty}\{\int_{-\infty}^{\infty} |f(x, y)|\,dy\}\,dx$ *exists as a finite number. Then* $f \in \mathsf{L}^1(\mathbb{R}^2)$ *and*

$$\int_{-\infty}^{\infty}\int_{-\infty}^{\infty} f(x, y)\,dx\,dy = \int_{-\infty}^{\infty}\left\{\int_{-\infty}^{\infty} f(x, y)\,dy\right\}dx = \int_{-\infty}^{\infty}\left\{\int_{-\infty}^{\infty} f(x, y)\,dx\right\}dy.$$

The second part of Prop. 0.1.4 is also associated with the name Tonelli–Hobson.

Proposition 0.1.5. (***Minkowski's inequality***). *Let* $f, g \in \mathsf{X}(\mathbb{R})$. *Then* $(f + g) \in \mathsf{X}(\mathbb{R})$ *and*

$$\|f + g\|_{\mathsf{X}(\mathbb{R})} \le \|f\|_{\mathsf{X}(\mathbb{R})} + \|g\|_{\mathsf{X}(\mathbb{R})}.$$

If p is such that $1 \le p \le \infty$, the *conjugate number* p' is defined through $(1/p) + (1/p') = 1$ in case $1 < p < \infty$, $p' = \infty$ if $p = 1$, and $p' = 1$ if $p = \infty$.

Proposition 0.1.6. (***Hölder's inequality***). *Let* $f \in \mathsf{L}^p$, $1 \le p \le \infty$, *and* $g \in \mathsf{L}^{p'}$. *Then* $fg \in \mathsf{L}^1$ *and* $\|fg\|_1 \le \|f\|_p \|g\|_{p'}$.

Let us emphasize that the assumption $f \in \mathsf{L}^p$, $1 \le p \le \infty$, means that f belongs to L^p for some fixed p with $1 \le p \le \infty$ (and not for all p of this interval). This will be the notation throughout this text. For $p = 2$ the assertion of Prop. 0.1.6. is also known as the inequality of Cauchy–Schwarz–Bunjakowski.

Proposition 0.1.7. (***Hölder–Minkowski inequality***). *Let* $f(x, y)$ *be defined and measurable on* $\mathbb{R}^2$. *If* $\|f(\circ, y)\|_{\mathsf{X}(\mathbb{R})} \in \mathsf{L}^1$, *then*

$$\left\|\int_{-\infty}^{\infty} f(\circ, y)\,dy\right\|_{\mathsf{X}(\mathbb{R})} \le \int_{-\infty}^{\infty} \|f(\circ, y)\|_{\mathsf{X}(\mathbb{R})}\,dy.$$

This is also known as the generalized Minkowski inequality (see HARDY–LITTLEWOOD–PÓLYA [1, p. 148]).

Proposition 0.1.8. (***Completeness of*** $\mathsf{X}(\mathbb{R})$). *Let* $\{f_n\}_{n=1}^{\infty} \subset \mathsf{X}(\mathbb{R})$ *be a Cauchy sequence in* $\mathsf{X}(\mathbb{R})$, *i.e.,* $\lim_{n,m\to\infty} \|f_n - f_m\|_{\mathsf{X}(\mathbb{R})} = 0$. *Then there exists* $f \in \mathsf{X}(\mathbb{R})$ *such that* $\lim_{n\to\infty} \|f_n - f\|_{\mathsf{X}(\mathbb{R})} = 0$. *The same assertion holds in* L^∞.

Let $f, f_1, f_2, \ldots \in \mathsf{L}^p$, $1 \le p \le \infty$. If $\lim_{n\to\infty} \|f_n - f\|_p = 0$, then f is called the *strong limit* or L^p*-limit* or *limit in the mean* of order p of the sequence $\{f_n\}_{n=1}^{\infty}$; one says that the sequence $\{f_n\}$ converges in L^p-norm or in the strong topology towards f, and writes $f = \text{s-lim}_{n\to\infty} f_n = \overset{(p)}{\text{l.i.m.}} f_n$. In C-space, strong convergence is uniform convergence.

Proposition 0.1.9. (***Continuity in the mean***). *If f belongs to $X(\mathbb{R})$, it follows that* $\lim_{h\to 0} \|f(\circ + h) - f(\circ)\|_{X(\mathbb{R})} = 0$.

The latter assertion does by no means hold if C is replaced by L^∞. Thus, considering strong convergence, we usually deal in this text with the spaces $X(\mathbb{R})$.

Proposition 0.1.10. (i) *Let f, $\{f_n\}_{n=1}^{\infty}$ be Lebesgue measurable functions such that $f_n \to f$ in measure, i.e., for each $\delta > 0$*

$$\lim_{n\to\infty} [\text{meas}\,\{x \mid |f(x) - f_n(x)| > \delta\}] = 0.$$

Then there exists a subsequence $\{f_{n_k}\}$ such that $\lim_{k\to\infty} f_{n_k}(x) = f(x)$ a.e.

(ii) *Let f, $\{f_n\}_{n=1}^{\infty}$ belong to L^p, $1 \le p < \infty$. If* $\lim_{n\to\infty} \|f - f_n\|_p = 0$, *then there exists a subsequence $\{f_{n_k}\}$ such that* $\lim_{k\to\infty} f_{n_k}(x) = f(x)$ a.e.

Proposition 0.1.11. *Let f, $\{f_n\}_{n=1}^{\infty}$ belong to L^p, $1 < p < \infty$. If there exists a constant M such that $\|f_n\|_p \le M$ for all $n = 1, 2, \ldots$, and if* $\lim_{n\to\infty} f_n(x) = f(x)$ a.e., *then for every $g \in L^{p'}$*

$$\lim_{n\to\infty} \int_{-\infty}^{\infty} f_n(x)g(x)\,dx = \int_{-\infty}^{\infty} f(x)g(x)\,dx.$$

0.2 Convolutions on the Line Group

Let f, g be two (complex-valued) functions defined and measurable on $\mathbb{R}$. The expression

$$(f * g)(x) = \frac{1}{\sqrt{2\pi}} \int_{-\infty}^{\infty} f(x - u)g(u)\,du \tag{0.2.1}$$

is called the *convolution* of f and g.

Proposition 0.2.1. *Let $f \in L^p$, $1 \le p \le \infty$, and $g \in L^{p'}$. Then $(f * g)(x)$ exists everywhere, belongs to C, and*

$$\|f * g\|_{C} \le \|f\|_p \|g\|_{p'}. \tag{0.2.2}$$

Moreover, if $1 < p < \infty$, then $f * g \in C_0$, i.e. $f * g \in C$ and $\lim_{|x|\to\infty} (f * g)(x) = 0$. *The same is true for $p = 1$ if, in addition, $g \in C_0$.*

Proof. Let $1 \le p < \infty$. Since $|(f * g)(x)| \le \|f\|_p \|g\|_{p'}$ by Hölder's inequality, the convolution $(f * g)(x)$ exists for every $x \in \mathbb{R}$. Furthermore,

$$|(f * g)(x + h) - (f * g)(x)| \le \|f(\circ + h) - f(\circ)\|_p \|g\|_{p'},$$

and therefore $f * g \in C$ by the continuity of f in the mean. Clearly, (0.2.2) holds. If $p = \infty$, the rôles of f and g may be interchanged.

Let $1 < p < \infty$. Given $\varepsilon > 0$, there exists a finite interval $-a \le u \le a$, such that

$$\frac{1}{\sqrt{2\pi}}\int_{|u|\ge a} |f(u)|^p\,du \le \varepsilon^p, \qquad \frac{1}{\sqrt{2\pi}}\int_{|u|\ge a} |g(u)|^{p'}\,du \le \varepsilon^{p'}.$$

If $x \in \mathbb{R}$ is such that $|x| > 2a$, then the interval $x - a \le u \le x + a$ is contained in the set $|u| > a$, and hence

$$\begin{aligned}|(f*g)(x)| &\le \frac{1}{\sqrt{2\pi}}\Big(\int_{-a}^{a} + \int_{|u|\ge a}\Big) |f(x-u)g(u)|\,du \\ &\le \Big(\frac{1}{\sqrt{2\pi}}\int_{-a}^{a} |f(x-u)|^p\,du\Big)^{1/p}\|g\|_{p'} + \|f\|_p\Big(\frac{1}{\sqrt{2\pi}}\int_{|u|\ge a}|g(u)|^{p'}\,du\Big)^{1/p'} \\ &\le \Big(\frac{1}{\sqrt{2\pi}}\int_{x-a}^{x+a} |f(u)|^p\,du\Big)^{1/p}\|g\|_{p'} + \|f\|_p\,\varepsilon \\ &\le \Big(\frac{1}{\sqrt{2\pi}}\int_{|u|\ge a} |f(u)|^p\,du\Big)^{1/p}\|g\|_{p'} + \|f\|_p\,\varepsilon \le (\|g\|_{p'} + \|f\|_p)\varepsilon.\end{aligned}$$

Thus $f*g$ vanishes at infinity, giving $f*g \in \mathsf{C}_0$. The same method of proof applies in case $p = 1$, $g \in \mathsf{C}_0$.

Proposition 0.2.2. *Let $f \in \mathsf{X}(\mathbb{R})$ and $g \in \mathsf{L}^1$. Then $(f*g)(x)$ exists* (a.e.) *as an absolutely convergent integral, $f*g \in \mathsf{X}(\mathbb{R})$, and*

$$\|f*g\|_{\mathsf{X}(\mathbb{R})} \le \|f\|_{\mathsf{X}(\mathbb{R})}\|g\|_1. \tag{0.2.3}$$

Proof. Let $\mathsf{X}(\mathbb{R}) = \mathsf{L}^p$, $1 \le p < \infty$. Since for almost all u

$$\frac{1}{\sqrt{2\pi}}\int_{-\infty}^{\infty} |f(x-u)|^p|g(u)|\,dx = |g(u)|\;\|f\|_p^p$$

which† belongs to L^1, it follows that

$$\frac{1}{\sqrt{2\pi}}\int_{-\infty}^{\infty} du\Big\{\frac{1}{\sqrt{2\pi}}\int_{-\infty}^{\infty}|f(x-u)|^p|g(u)|\,dx\Big\} = \|f\|_p^p\|g\|_1 \tag{0.2.4}$$

exists as a finite number. Therefore by Fubini's theorem

$$\frac{1}{\sqrt{2\pi}}\int_{-\infty}^{\infty} dx\Big\{\frac{1}{\sqrt{2\pi}}\int_{-\infty}^{\infty}|f(x-u)|^p|g(u)|\,du\Big\} \tag{0.2.5}$$

exists as well and is equal to $\|f\|_p^p\|g\|_1$. This implies that

$$\frac{1}{\sqrt{2\pi}}\int_{-\infty}^{\infty}|f(x-u)|^p|g(u)|\,du \tag{0.2.6}$$

exists for almost all $x \in \mathbb{R}$ and belongs to L^1. This proves the assertion for $p = 1$.

For $1 < p < \infty$ Hölder's inequality delivers

$$\begin{aligned}|(f*g)(x)| &\le \frac{1}{\sqrt{2\pi}}\int_{-\infty}^{\infty}|f(x-u)|\;|g(u)|^{1/p}|g(u)|^{1/p'}\,du \\ &\le \Big\{\frac{1}{\sqrt{2\pi}}\int_{-\infty}^{\infty}|f(x-u)|^p|g(u)|\,du\Big\}^{1/p}\Big\{\frac{1}{\sqrt{2\pi}}\int_{-\infty}^{\infty}|g(u)|\,du\Big\}^{1/p'}.\end{aligned}$$

† The fact that $|f(x-u)|^p|g(u)|$ is a measurable function on $\mathbb{R}^2$ is a rather delicate result of measure theory: see e.g. Hewitt–Stromberg [1, p. 396], Williamson [1, p. 65].

This shows by (0.2.6) that $(f * g)(x)$ exists almost everywhere as an absolutely convergent integral. Moreover, by (0.2.4) and (0.2.5)

$$\|f * g\|_p \le \|g\|_1^{1/p'}\left\{\frac{1}{\sqrt{2\pi}}\int_{-\infty}^{\infty} dx\left[\frac{1}{\sqrt{2\pi}}\int_{-\infty}^{\infty} |f(x-u)|^p|g(u)|\,du\right]\right\}^{1/p}$$

$$= \|g\|_1^{1/p'}\|f\|_p\|g\|_1^{1/p} = \|f\|_p\|g\|_1.$$

Finally, if $\mathsf{X}(\mathbb{R}) = \mathsf{C}$, it follows as in the proof of Prop. 0.2.1 that $(f * g)(x)$ exists for all x, belongs to C, and satisfies $\|f * g\|_{\mathsf{C}} \le \|f\|_{\mathsf{C}}\|g\|_1$.

If one of the above hypotheses is satisfied, it follows by an elementary substitution that convolving two functions f, g is a *commutative* operation, i.e. $(f * g)(x) = (g * f)(x)$ (a.e.). Obviously, convolution is *distributive*, i.e. $f * (g + h) = f * g + f * h$. Suppose now that f, g, h belong to L^1. Prop. 0.2.2 then implies $f * g \in \mathsf{L}^1$. Thus $(f * g) * h$ is well-defined as an element in L^1, and it follows by Fubini's theorem that convolution is also *associative*, i.e. $(f * g) * h = f * (g * h)$. The r-times convolution (*product*) $f * \cdots * f$ of $f \in \mathsf{L}^1$ with itself is denoted by $[f *]^r$. It exists as a function in L^1.

Convolution is an operation which leaves many of the structural properties of each of its members invariant (compare Sec. 1.1.2, Problems 5.1.1(iv), 6.3.5(vi)). Fundamental is the *translation-invariance* of convolutions. If T_a denotes the operation of translation by $a \in \mathbb{R}$, i.e. $(T_a f)(x) = f(x + a)$, this means that $T_a[f * g] = T_a f * g = f * T_a g$. As an important consequence, convolution is a smoothness increasing operation (compare Problems 3.1.4–3.1.6).

0.3 Further Sets of Functions and Sequences

Let $\mathbb{Z}$ be the set of all (positive and negative) integers, $\mathbb{P}$ the set of all positive integers including zero, and $\mathbb{N}$ the set of all naturals $1, 2, \ldots$. Let $r \in \mathbb{P}$. C^r denotes the class of all functions $f \in \mathsf{C}$ which are r-times differentiable on $\mathbb{R}$ and for which $f^{(j)} \in \mathsf{C}$ for each $1 \le j \le r$. Obviously, one sets $\mathsf{C}^0 = \mathsf{C}$ and $\mathsf{C}^\infty = \bigcap_{r=0}^{\infty} \mathsf{C}^r$. AC $(=\mathsf{AC}^0)$ is the set of all absolutely continuous functions f, i.e. $f(x)$ admits for every $x \in \mathbb{R}$ the representation $f(x) = \int_{-\infty}^{x} g(u)\,du$ for some $g \in \mathsf{L}^1$. Correspondingly, AC^{r-1} denotes the set of all functions which are $(r-1)$-times absolutely continuous, i.e. $f \in \mathsf{AC}^{r-1}$ means that there exists $g \in \mathsf{L}^1$ such that for every $x \in \mathbb{R}$

$$f(x) = \int_{-\infty}^{x} du_1 \int_{-\infty}^{u_1} du_2 \ldots \int_{-\infty}^{u_{r-1}} g(u_r)\,du_r, \tag{0.3.1}$$

each of the iterated integrals, possibly apart from the first, defining a function in L^1. The subscript 'o' stands for 'zero at infinity', whereas the subscript 'oo' means compact support. Thus, C_0^r denotes the set of those $f \in \mathsf{C}^r$ for which $\lim_{|x|\to\infty} f^{(j)}(x) = 0$ for each integer j with $0 \le j \le r$, C_{00}^r the set of those $f \in \mathsf{C}_0^r$ which have compact support. The subscript 'loc' stands for 'locally'. Thus C_{loc} is the set of all locally continuous functions, i.e. f is continuous on every finite interval. The class AC_{loc} is defined as the set of locally absolutely continuous functions, thus $f \in \mathsf{AC}_{\text{loc}}$ means that there exists

$g \in L^1_{loc}$ such that $f(x) - f(0) = \int_0^x g(u)\,du$ for every $x \in \mathbb{R}$. More generally, $f \in AC^{r-1}_{loc}$ means that there exists $g \in L^1_{loc}$ and constants $a_0, a_1, \ldots, a_{r-1}$ such that for every $x \in \mathbb{R}$

$$(0.3.2)\quad f(x) = a_0 + \int_0^x du_1\Big[a_1 + \int_0^{u_1}\Big[a_2 + \cdots + \int_0^{u_{r-2}} du_{r-1}\Big[a_{r-1} + \int_0^{u_{r-1}} g(u_r)\,du_r\Big]\ldots\Big]\Big]\dagger.$$

If $[a, b]$, (a, b), $[a, b)$, $(a, b]$ denote the sets of those x for which $a \le x \le b$, $a < x < b$, $a \le x < b$, $a < x \le b$, respectively (a, b may be infinite), then $X[a, b]$ is the set $C[a, b]$ of functions continuous on $[a, b]$ or $L^p[a, b]$, $1 \le p < \infty$, the set of functions Lebesgue integrable to the pth power over $[a, b]$.

A point $x \in \mathbb{R}$ is called a *D-point* of the function f if

$$(0.3.3)\qquad \int_0^h [f(x + u) - f(x)]\,du = o(h) \qquad (h \to 0).$$

D stands for differentiability; for, the D-points of an integrable f are precisely those points at which the indefinite integral of f is differentiable to the value $f(x)$. If

$$(0.3.4)\qquad \int_0^h |f(x + u) - f(x)|\,du = o(h) \qquad (h \to 0),$$

then x is called a *Lebesgue point* or *L-point* of f. Evidently, every L-point of f is a D-point of f but not conversely, and every point of continuity of an integrable f is an L-point. Moreover,

Proposition 0.3.1. *If $f \in L^1(a, b)$, then almost all points of the interval (a, b) are L-points, thus D-points, of f.*

Proposition 0.3.2. (***Mean value theorem***). *Let $\Delta_h^r f(x)$ be the rth (one-sided) difference of f at x (see Problem 1.5.2). If $f \in C^r_{loc}$, then $\lim_{h\to 0} h^{-r}\Delta_h^r f(x) = f^{(r)}(x)$ at every $x \in \mathbb{R}$, and in fact uniformly on every compact interval. If $f \in C^r$, then*

$$\lim_{h\to 0} \|h^{-r}\Delta_h^r f(\circ) - f^{(r)}(\circ)\|_C = 0.$$

Proposition 0.3.3. (***Theorem of B. Levi***). *Let $\{f_n\}_{n=1}^\infty \subset L^1(a, b)$. If $\sum_{n=1}^\infty \int_a^b |f_n(x)|\,dx < \infty$, then the series $\sum_{n=1}^\infty f_n(x)$ converges absolutely almost everywhere on (a, b) to a finite sum. If this sum is denoted by $f(x)$, then $f \in L^1(a, b)$ and*

$$\int_a^b f(x)\,dx = \sum_{n=1}^\infty \int_a^b f_n(x)\,dx.$$

The *characteristic function* $\kappa_{[a,b]}(x)$ of a (finite) interval $[a, b]$ is the function on $\mathbb{R}$ defined through

$$(0.3.5)\qquad \kappa_{[a,b]}(x) = \begin{cases} 1, & a \le x \le b \\ 0, & x \notin [a, b]. \end{cases}$$

† Obviously, the lower limit of integration in (0.3.2) may be replaced by any other real number; this only affects the value of the constants.

A *step function* is a finite linear combination of characteristic functions corresponding to (nonoverlapping) finite intervals. A function defined on some interval is called *simple* if it takes only a finite number of values (on measurable sets), and has compact support in case the interval does not have finite length. Thus the set of all simple functions on $\mathbb{R}$ is denoted by S_{00}.

Proposition 0.3.4. *The set of step functions is dense in* L^p *for* $1 \le p < \infty$, *i.e., given* $\varepsilon > 0$, *for every* $f \in \mathsf{L}^p$ *there exists a step function g such that* $\|f - g\|_p < \varepsilon$. *In particular,* S_{00} *is dense in* L^p, $1 \le p < \infty$. *Furthermore, if f has compact support, there exists a sequence* $\{g_n\}_{n=1}^{\infty} \subset \mathsf{S}_{00}$ *such that* $\lim_{n\to\infty} \|f - g_n\|_p = 0$, *and the support of* g_n *is contained in that of f, uniformly for all* $n \in \mathbb{N}$. *Finally,* C_{00} *is dense in* C_0 *and* L^p, $1 \le p < \infty$.

Let $\ldots c_{-1}, c_0, c_1, \ldots$, thus $\{c_k\}_{k=-\infty}^{\infty} \equiv c$, be a two-way sequence of complex numbers. Then l^p, more specifically $\mathsf{l}^p(\mathbb{Z})$, is the set of those sequences c for which either

$$\|c\|_p = \left\{\sum_{k=-\infty}^{\infty} |c_k|^p\right\}^{1/p} \qquad (1 \le p < \infty) \tag{0.3.6}$$

or

$$\|c\|_\infty = \sup_{k\in\mathbb{Z}} |c_k| \qquad (p = \infty) \tag{0.3.7}$$

is a finite number. Thus l^1 is the set of all those sequences c for which the series $\sum_{k=-\infty}^{\infty} c_k$ converges absolutely, l^∞ the set of all bounded sequences. The set of all $c \in \mathsf{l}^\infty$ for which $\lim_{|k|\to\infty} c_k = 0$ is denoted by l_0^∞. In many situations it is more convenient to look at a sequence $c = \{c_k\}_{k=-\infty}^{\infty}$ as a *function* defined on $\mathbb{Z}$. In this case we prefer the notation $c(k)$ instead of c_k. Then $\mathbb{Z}$ is often considered as a measure space in which each point has measure 1, and one has results corresponding to those of Sec. 0.1 for functions on $\mathbb{R}$. Thus for example

Proposition 0.3.5. (***Minkowski's inequality***). *Let* $c = \{c(k)\}_{k=-\infty}^{\infty}$, $d = \{d(k)\}_{k=-\infty}^{\infty}$ *be two functions of class* $\mathsf{l}^p(\mathbb{Z})$, $1 \le p \le \infty$. *Then* $(c + d) \in \mathsf{l}^p(\mathbb{Z})$ *and* $\|c + d\|_p \le \|c\|_p + \|d\|_p$, *i.e., for* $1 \le p < \infty$

$$\left\{\sum_{k=-\infty}^{\infty} |c(k) + d(k)|^p\right\}^{1/p} \le \left\{\sum_{k=-\infty}^{\infty} |c(k)|^p\right\}^{1/p} + \left\{\sum_{k=-\infty}^{\infty} |d(k)|^p\right\}^{1/p}.$$

Proposition 0.3.6. (***Hölder's inequality***). *Let* $c = \{c(k)\}_{k=-\infty}^{\infty}$, $d = \{d(k)\}_{k=-\infty}^{\infty}$ *be two functions such that* $c \in \mathsf{l}^p(\mathbb{Z})$, $1 \le p \le \infty$, *and* $d \in \mathsf{l}^{p'}(\mathbb{Z})$. *Then* $\{c(k)\, d(k)\}_{k=-\infty}^{\infty} \in \mathsf{l}^1(\mathbb{Z})$ *and* $\sum_{k=-\infty}^{\infty} |c(k)\, d(k)| \le \|c\|_p \|d\|_{p'}$.

For $p = 2$ the assertion of the latter proposition is also known as Bessel's inequality.

Similar definitions and results may be given when dealing with functions on $\mathbb{N}$, for example. Thus $\mathsf{l}^1(\mathbb{N})$, or in short l^1, is the set of all sequences $\{c_k\}_{k=1}^{\infty}$ for which the series $\sum_{k=1}^{\infty} c_k$ is absolutely convergent.

0.4 Periodic Functions and Their Convolution

A (complex-valued) function defined on $\mathbb{R}$ is said to be *periodic* if $f(x + 2\pi) = f(x)$ for all real x. We always deal with 2π-periodic functions. If f has period $\lambda > 0$, thus

$f(x+\lambda)=f(x)$ for all real x, f may be transposed into a 2π-periodic function by a linear substitution. Moreover, if f is defined on a *finite* interval $[a, b)$, a linear substitution yields a function defined on $[-\pi, \pi)$ which may then be extended to all of $\mathbb{R}$ by periodicity, i.e., by $f(x+2\pi)=f(x)$. Thus the normalization to 2π-periodic functions may be assumed without loss of generality, in particular, that $f(-\pi)=f(\pi)$. This implies that one may consider 2π-periodic functions to be defined on the circumference $\mathbb{T}$ of the unit circle in the complex plane, that is, on the *circle group* $\mathbb{T}=\{z\in\mathbb{C} \mid |z|=1\}$, and vice versa.

Let $C_{2\pi}$ denote the set of all periodic functions which are continuous for all x; one sets

$$\|f\|_{C_{2\pi}}=\sup_{|x|\le\pi}|f(x)|. \tag{0.4.1}$$

Obviously, $C_{2\pi}\subset C(\mathbb{R})$ and $\|f\|_{C_{2\pi}}=\|f\|_{C(\mathbb{R})}$ for every $f\in C_{2\pi}$ (compare (0.1.1)). The same remark holds for the class $L^\infty_{2\pi}$, the set of all periodic, Lebesgue measurable functions for which (compare (0.1.3))

$$\|f\|_{L^\infty_{2\pi}}=\operatorname*{ess\,sup}_{|x|\le\pi}|f(x)| \tag{0.4.2}$$

is finite. However, the situation is quite different for the class $L^p_{2\pi}$, $1\le p<\infty$, which consists of all periodic† functions, Lebesgue integrable to the pth power over $(-\pi, \pi)$ with

$$\|f\|_{L^p_{2\pi}}=\left\{\frac{1}{2\pi}\int_{-\pi}^{\pi}|f(u)|^p\,du\right\}^{1/p}. \tag{0.4.3}$$

Indeed, apart from the nullfunction, the intersection of $L^p(\mathbb{R})$ and $L^p_{2\pi}$ for $1\le p<\infty$ is void. Nevertheless, if no confusion may occur, also the shorter notation $\|f\|_p$ instead of $\|f\|_{L^p_{2\pi}}$ is employed. The fundamental difference between the classes $L^p(\mathbb{R})$ and $L^p_{2\pi}$ for $1\le p<\infty$ is that integration in (0.4.3) is extended over the finite interval $[-\pi, \pi]$, whereas in (0.1.2) it is extended over the whole (noncompact) real axis $\mathbb{R}$. As an immediate consequence one has that $L^p_{2\pi}\subset L^q_{2\pi}$ for all $p>q$. In particular, $L^p_{2\pi}\subset L^1_{2\pi}$ for all $p>1$. For this reason and the periodicity, the integral of f over an (arbitrary) interval of length 2π always has the same value for a function f belonging to one of the above classes, thus

$$\int_{-\pi}^{\pi}f(u)\,du=\int_{a-\pi}^{a+\pi}f(u)\,du \tag{0.4.4}$$

for any $a\in\mathbb{R}$. Indeed, setting $u=v+2\pi$, then $\int_{\pi}^{a+\pi}f(u)\,du=\int_{-\pi}^{a-\pi}f(v)\,dv$, and hence

$$\int_{a-\pi}^{a+\pi}f(u)\,du=\left(\int_{a-\pi}^{-\pi}+\int_{-\pi}^{\pi}+\int_{\pi}^{a+\pi}\right)f(u)\,du=\int_{-\pi}^{\pi}f(u)\,du.$$

Conversely, if (0.4.4) holds for some $f\in L^1_{\mathrm{loc}}$ for every $a\in\mathbb{R}$, then f is periodic, thus an element of $L^1_{2\pi}$. In the following, $X_{2\pi}$ always denotes one of the classes $C_{2\pi}$ or $L^p_{2\pi}$, $1\le p<\infty$.

† Since $f\in L^p_{2\pi}$ need only be defined almost everywhere, periodicity now means that $f(x+2\pi)=f(x)$ a.e. Note the different normalization in (0.1.2) and (0.4.3).

The basic facts on Lebesgue integration as stated in Prop. 0.1.1–0.1.11 are representative for any range of the interval of integration†. Thus, replacing $(-\infty, \infty)$ by the finite interval $(-\pi, \pi)$ yields corresponding theorems for periodic functions which are referred to by the same name. For example, *continuity in the mean* now refers to the fact that $\lim_{h\to 0} \|f(\circ + h) - f(\circ)\|_{X_{2\pi}} = 0$ for every $f \in X_{2\pi}$.

The same remarks hold for the analogs of Sec. 0.2. Indeed, for any two periodic functions f, g the *convolution* $f * g$ is defined through

$$(f * g)(x) = \frac{1}{2\pi} \int_{-\pi}^{\pi} f(x - u)g(u)\,du. \tag{0.4.5}$$

Obviously, the convolution is a periodic function if it exists.

Proposition 0.4.1. (i) *Let* $f \in L^p_{2\pi}$, $1 \le p \le \infty$, *and* $g \in L^{p'}_{2\pi}$. *Then* $(f * g)(x)$ *exists everywhere, belongs to* $C_{2\pi}$, *and* $\|f * g\|_{C_{2\pi}} \le \|f\|_p \|g\|_{p'}$.

(ii) *Let* $f \in X_{2\pi}$ *and* $g \in L^1_{2\pi}$. *Then* $(f * g)(x)$ *exists* (a.e.) *as an absolutely convergent integral,* $f * g \in X_{2\pi}$, *and* $\|f * g\|_{X_{2\pi}} \le \|f\|_{X_{2\pi}} \|g\|_1$.

The proof follows as in Sec. 0.2. Since $X_{2\pi} \subset L^1_{2\pi}$, the convolution $f * g$ of $f, g \in X_{2\pi}$ exists as a function in $X_{2\pi}$. Therefore, the convolution (0.4.5) is a commutative, distributive, and associative operation on $X_{2\pi}$; this follows by (0.4.4) and Fubini's theorem. Since there will be no danger of confusion with the nonperiodic case, the same notation $[f*]^r$ is employed for the r-times product $f * \cdots * f$ of $f \in X_{2\pi}$ with itself.

Concerning analogs of Sec. 0.3 for periodic functions, $C^r_{2\pi}$ denotes the set of all periodic functions which are r-times continuously differentiable. $AC_{2\pi}$ is the set of all absolutely continuous periodic functions, i.e. $f(x)$ is 2π-periodic and admits for every x the representation $f(x) - f(-\pi) = \int_{-\pi}^{x} g(u)\,du$ with some $g \in L^1_{2\pi}$; the periodicity of f then necessarily implies $\int_{-\pi}^{\pi} g(u)\,du = 0$. Results for 'loc'-classes of functions hold in particular for the corresponding classes of periodic functions. Indeed, the phrases '$f \in L^1_{2\pi}$' and '$f \in L^1_{loc}$ is 2π-periodic' have the same meaning. $S_{2\pi}$ is the class of all simple periodic functions, i.e. $f \in S_{2\pi}$ if f is 2π-periodic and takes only a finite number of values on measurable sets (over a period). Again $S_{2\pi}$ is dense in $X_{2\pi}$.

0.5 Functions of Bounded Variation on the Line Group

Let $BV = BV(\mathbb{R})$ be the set of all (complex-valued) functions $\mu, \nu, \ldots$ which are of bounded variation over $\mathbb{R}$ and *normalized* by‡ $\mu(-\infty) = \mu(-\infty + 0) = 0$,

† Indeed, one may formulate these facts for an arbitrary (finite or infinite) interval and even for more general measure spaces so as to cover, simultaneously, integration of functions, defined on $\mathbb{R}$, on $[a, b]$, on $\mathbb{Z}$, on the reals mod 2π, etc., with respect to Lebesgue measure or arbitrary measures (cf. Sec. 0.5, 0.6). However, we do not follow this unified approach since, in any case, we have to introduce the different notations.

‡ If μ is defined in a neighbourhood of x, then one sets $\mu(x + 0)$ $(\equiv \mu(x+)) = \lim_{t\to 0+} \mu(x + t)$, where $t \to 0+$ means that t is a positive quantity tending to zero. Correspondingly, one defines $\mu(x - 0) \equiv \mu(x-)$; thus $\mu(\infty - 0) = \lim_{x\to\infty} \mu(x)$.

$\mu(x) = [\mu(x+0) + \mu(x-0)]/2$ for $-\infty < x < \infty$, and $\mu(\infty) = \mu(\infty - 0)$. One sets (cf. NATANSON [6, p. 516 f])

$$\|\mu\|_{\mathsf{BV}} = \frac{1}{\sqrt{2\pi}} \int_{-\infty}^{\infty} |d\mu| \equiv \frac{1}{\sqrt{2\pi}} [\text{Var } \mu]_{-\infty}^{\infty}. \tag{0.5.1}$$

Of its subsets we mention NBV, the set of all $\mu \in \mathsf{BV}$ with $\int_{-\infty}^{\infty} d\mu = \sqrt{2\pi}$, and BV_0, the set of all $\mu \in \mathsf{BV}$ with $\int_{-\infty}^{\infty} d\mu = 0$. For $g \in \mathsf{L}^1$ the function μ defined by $\mu(x) = \int_{-\infty}^{x} \bar{g}(u)\, du$ is an element of BV, and $\|\mu\|_{\mathsf{BV}} = \|g\|_1$. Thus $\mathsf{AC} \subset \mathsf{BV}$.

Proposition 0.5.1. (***Jordan decomposition***). *A real-valued function μ belongs to* BV *if and only if there exist two functions $\mu_1, \mu_2 \in \mathsf{BV}$ which are positive and monotonely increasing on $\mathbb{R}$ such that $\mu(x) = \mu_1(x) - \mu_2(x)$ for all x and $\|\mu\|_{\mathsf{BV}} = \|\mu_1\|_{\mathsf{BV}} + \|\mu_2\|_{\mathsf{BV}}$. (For complex-valued μ these assertions hold for the real and imaginary part of μ, respectively.)*

In particular, a function $\mu \in \mathsf{BV}$ is bounded and has at most discontinuities of the first kind (finite jumps) which form a denumerable set Π_μ. For $\mu, \nu \in \mathsf{BV}$ let $\Pi_{\mu,\nu}$ be the set of all points x of the form $x = x_\mu + x_\nu$ with $x_\mu \in \Pi_\mu$, $x_\nu \in \Pi_\nu$, but whenever one of the sets Π_μ, Π_ν is empty, $\Pi_{\mu,\nu}$ is also assumed to be empty.

Let BV_{loc} be the set of all (complex-valued) functions which are locally of bounded variation, i.e., of bounded variation on every finite interval, and $\mathsf{BV}[a, b]$ the set of all those functions which are of bounded variation on a definite (finite) interval $[a, b]$. When speaking of functions of BV-classes, it is always assumed that these functions are normalized which includes that $\mu(x) = [\mu(x+0) + \mu(x-0)]/2$ for any such function μ and every point x in the interior of the interval of definition. When dealing with Stieltjes integrals, the normalization does not affect the value of the integral.

Concerning (proper and improper) Riemann–Stieltjes integrals, $\int_a^b f(u)\, d\mu(u)$ exists for $f \in \mathsf{C}[a, b]$, $\mu \in \mathsf{BV}[a, b]$ and

$$\int_a^b f(u)\, d\mu(u) = f(b)\mu(b) - f(a)\mu(a) - \int_a^b \mu(u)\, df(u). \tag{0.5.2}$$

This formula on *integration by parts* in particular shows that the integral $\int_a^b \mu(u)\, df(u)$ exists for every $\mu \in \mathsf{BV}[a, b]$, $f \in \mathsf{C}[a, b]$. Moreover, if $\mu, \nu \in \mathsf{BV}[a, b]$ have no common points of discontinuity, then each is Riemann–Stieltjes integrable with respect to the other from a to b. For these facts and the following proposition see in particular WIDDER [1, p. 3 ff; p. 248 ff].

Proposition 0.5.2. *Let $\mu, \nu \in \mathsf{BV}$. The* **convolution** *(or Stieltjes resultant) $\mu * d\nu$ as defined through (see also* (0.8.8))

$$(\mu * d\nu)(x) = \frac{1}{\sqrt{2\pi}} \int_{-\infty}^{\infty} \mu(x-u)\, d\nu(u) \tag{0.5.3}$$

exists for all $x \notin \Pi_{\mu,\nu}$ and

$$\lim_{x \to -\infty,\, x \notin \Pi_{\mu,\nu}} (\mu * d\nu)(x) = 0, \qquad \lim_{x \to \infty,\, x \notin \Pi_{\mu,\nu}} (\mu * d\nu)(x) = \mu(\infty)\nu(\infty)/\sqrt{2\pi}. \tag{0.5.4}$$

Moreover, $(\mu * d\nu)(x)$ can be defined on the set $\Pi_{\mu,\nu}$ so as to become an element of BV *satisfying*

$$\|\mu * d\nu\|_{\mathsf{BV}} \le \|\mu\|_{\mathsf{BV}}\|\nu\|_{\mathsf{BV}}. \tag{0.5.5}$$

In particular, $(\mu * d\nu)(x) = (\nu * d\mu)(x)$.

Proof. For x not in $\Pi_{\mu,\nu}$, $\mu(x-u)$ and $\nu(u)$ do not have points of discontinuity in common for all $u \in \mathbb{R}$. Hence the integrals

$$\int_M^N \mu(x-u)\,d\nu(u), \qquad \int_M^N \nu(x-u)\,d\mu(u)$$

exist for every $M, N \in \mathbb{R}$. Moreover, the limits for $M \to -\infty$, $N \to \infty$ exist since

$$\int_{-\infty}^{\infty} |\mu(x-u)|\,|d\nu(u)| < \infty, \qquad \int_{-\infty}^{\infty} |\nu(x-u)|\,|d\mu(u)| < \infty,$$

and integration by parts shows that for $x \notin \Pi_{\mu,\nu}$

$$\int_{-\infty}^{\infty} \mu(x-u)\,d\nu(u) = \int_{-\infty}^{\infty} \nu(x-u)\,d\mu(u), \tag{0.5.6}$$

since $\mu(-\infty) = \nu(-\infty) = 0$. Concerning (0.5.4), it suffices to prove the second relation. Since $\int_{-\infty}^{\infty} d\nu(u) = \nu(\infty)$ and

$$\left|(\mu * d\nu)(x) - \frac{\mu(\infty)\nu(\infty)}{\sqrt{2\pi}}\right| \le \frac{1}{\sqrt{2\pi}}\int_{-\infty}^{\infty} |\mu(x-u) - \mu(\infty)|\,|d\nu(u)|,$$

this would follow if

$$\lim_{x\to\infty,\, x\notin \Pi_{\mu,\nu}} \int_{-\infty}^{\infty} |\mu(x-u) - \mu(\infty)|\,|d\nu(u)| = 0.$$

Let $\varepsilon > 0$ be given. Since μ is bounded and $\nu \in \mathsf{BV}$, we may choose $N > 0$ such that the first of the following two inequalities holds for all $x \notin \Pi_{\mu,\nu}$:

$$\int_N^{\infty} |\mu(x-u) - \mu(\infty)|\,|d\nu(u)| < \varepsilon, \qquad \int_{-\infty}^{N} |\mu(x-u) - \mu(\infty)|\,|d\nu(u)| < \varepsilon\sqrt{2\pi}\,\|\nu\|_{\mathsf{BV}}.$$

Since $\mu(\infty) = \mu(\infty - 0)$, there exists $x_0 \in \mathbb{R}$ such that $|\mu(x-u) - \mu(\infty)| < \varepsilon$ for all $x \ge x_0$, $x \notin \Pi_{\mu,\nu}$, and $u \le N$; this implies the second relation for all $x \ge x_0$, $x \notin \Pi_{\mu,\nu}$, and thus (0.5.4). In order to define $(\mu * d\nu)(x)$ on the set $\Pi_{\mu,\nu}$, by the Jordan decomposition one may assume without loss of generality that μ and ν are monotonely increasing functions. Consequently, their convolution is monotonely increasing, and so $(\mu * d\nu)(x)$ has right- and left-hand limits at all points. Therefore one may define $(\mu * d\nu)(x)$ on the set $\Pi_{\mu,\nu}$ by

$$(\mu * d\nu)(x) = [(\mu * d\nu)(x+0) + (\mu * d\nu)(x-0)]/2. \tag{0.5.7}$$

Then, if $x_0 < x_1 < \cdots < x_n$ is an arbitrary set of points not in $\Pi_{\mu,\nu}$, one has

$$\sum_{k=1}^{n} |(\mu * d\nu)(x_k) - (\mu * d\nu)(x_{k-1})| \le \frac{1}{\sqrt{2\pi}}\int_{-\infty}^{\infty}\left(\sum_{k=1}^{n} |\mu(x_k - u) - \mu(x_{k-1} - u)|\right)|d\nu(u)|$$

$$\le \frac{1}{\sqrt{2\pi}}\int_{-\infty}^{\infty}\left(\sum_{k=1}^{n}\int_{x_{k-1}-u}^{x_k-u} |d\mu|\right)|d\nu(u)| \le \int_{-\infty}^{\infty} \|\mu\|_{\mathsf{BV}}\,|d\nu(u)| = \sqrt{2\pi}\,\|\mu\|_{\mathsf{BV}}\|\nu\|_{\mathsf{BV}}.$$

By virtue of the definition of $\mu * d\nu$ on the set $\Pi_{\mu,\nu}$ one sees by a limiting process that the above inequality holds for any set of points $\{x_k\}_{k=0}^n \subset \mathbb{R}$. Thus $\mu * dv$ is of bounded variation and satisfies (0.5.5). Indeed, in view of (0.5.4), (0.5.7), $\mu * dv$ is normalized so that it is a genuine element of BV. Finally, the commutativity of the convolution follows for all x by (0.5.6), (0.5.7), and a limiting process.

Proposition 0.5.3. *Every* $\mu \in \mathsf{BV}[a, b]$ *is differentiable almost everywhere on* $[a, b]$.

Proposition 0.5.4. *Let* $\mu \in \mathsf{BV}_{\text{loc}}$. *Suppose that* x *is a point where* $\mu'(x)$ *exists, and set* $\mu_x(u) = \mu(x + u) - \mu(x - u) - 2u\mu'(x)$. *Then* $[\text{Var}\,\mu_x(u)]_{u=0}^h = o(h)$, $h \to 0+$, *for almost all* x (*see* ZYGMUND [7I, p. 105]).

Concerning Lebesgue–Stieltjes integrals $\int f(u)\, d\mu(u)$, a list of properties might be given which would correspond to that of Sec. 0.1 for Lebesgue integrals, thus a Lebesgue dominated convergence theorem, a Fubini theorem, a Minkowski, Hölder or Hölder–Minkowski inequality. Therefore as a counterpart to Prop. 0.2.2

Proposition 0.5.5. *Let* $f \in \mathsf{X}(\mathbb{R})$, $\mu \in \mathsf{BV}$. *The* ***convolution*** $f * d\mu$ *as defined through*

$$(f * d\mu)(x) = \frac{1}{\sqrt{2\pi}} \int_{-\infty}^{\infty} f(x - u)\, d\mu(u) \tag{0.5.8}$$

exists (a.e.) *as an absolutely convergent integral,* $f * d\mu \in \mathsf{X}(\mathbb{R})$, *and*

$$\|f * d\mu\|_{\mathsf{X}(\mathbb{R})} \leq \|f\|_{\mathsf{X}(\mathbb{R})} \|\mu\|_{\mathsf{BV}}. \tag{0.5.9}$$

Proof. The case $\mathsf{X}(\mathbb{R}) = \mathsf{C}$ being obvious, let $\mathsf{X}(\mathbb{R}) = \mathsf{L}^p$, $1 \leq p < \infty$. It follows by Fubini's theorem that

$$\frac{1}{\sqrt{2\pi}} \int_{-\infty}^{\infty} dx \left\{ \frac{1}{\sqrt{2\pi}} \int_{-\infty}^{\infty} |f(x - u)|^p |d\mu(u)| \right\} = \|f\|_p^p \|\mu\|_{\mathsf{BV}}.$$

This implies that $\int_{-\infty}^{\infty} |f(x - u)|^p |d\mu(u)|$ exists for almost all x and belongs to L^1, proving the assertion for $p = 1$. If $1 < p < \infty$, by Hölder's inequality

$$\begin{aligned} |(f * d\mu)(x)| &\leq \frac{1}{\sqrt{2\pi}} \int_{-\infty}^{\infty} |f(x - u)|\, |d\mu(u)| \\ &\leq \left\{ \frac{1}{\sqrt{2\pi}} \int_{-\infty}^{\infty} |f(x - u)|^p |d\mu(u)| \right\}^{1/p} \left\{ \frac{1}{\sqrt{2\pi}} \int_{-\infty}^{\infty} |d\mu(u)| \right\}^{1/p'}, \end{aligned}$$

since the μ-measure of $\mathbb{R}$ is finite. Thus $(f * d\mu)(x)$ exists a.e., and the proof may be completed as for Prop. 0.2.2.

In the literature, Lebesgue–Stieltjes integrals are usually considered with respect to arbitrary Borel measures rather than for functions of bounded variation as is the case in this volume. But there is a close connection between Borel measures and BV-classes. Indeed, apart from a different normalization in the interior points (instead of $\mu(x) = [\mu(x + 0) + \mu(x - 0)]/2$ one assumes μ to be left-continuous, i.e. $\mu(x) = \mu(x - 0)$), there is a one-to-one correspondence between bounded Borel measures on $\mathbb{R}$ and functions $\mu \in \mathsf{BV}$, between Borel measures and functions $\mu \in \mathsf{BV}_{\text{loc}}$ with $\mu(0) = 0$, and between Borel measures on $[a, b]$ and functions $\mu \in \mathsf{BV}[a, b]$ with $\mu(a) = 0$.

0.6 The Class $\mathsf{BV}_{2\pi}$

Let $\mathsf{BV}_{2\pi}$ be the subset of those functions $\mu \in \mathsf{BV}_{\mathrm{loc}}$ for which for all x

$$\mu(x + 2\pi) = \mu(x) + [\mu(\pi) - \mu(-\pi)]. \tag{0.6.1}$$

In other words, $\mu \in \mathsf{BV}_{2\pi}$ if and only if μ is of bounded variation on every finite interval, is normalized for all x by $\mu(x) = [\mu(x + 0) + \mu(x - 0)]/2$, and satisfies (0.6.1). For $\mu \in \mathsf{BV}_{2\pi}$ one sets

$$\|\mu\|_{\mathsf{BV}_{2\pi}} = \frac{1}{2\pi} \int_{-\pi}^{\pi} |d\mu(u)| \equiv \frac{1}{2\pi} [\operatorname{Var} \mu(u)]_{u=-\pi}^{\pi}. \tag{0.6.2}$$

Every function μ which is defined and of bounded variation on $[-\pi, \pi]$ may be extended so as to become an element of $\mathsf{BV}_{2\pi}$ which is continuous in the points $(2k + 1)\pi$, $k \in \mathbb{Z}$. Indeed, without loss of generality (in particular, if dealing with Stieltjes integrals) one may first normalize μ on $[-\pi, \pi]$ by $\mu(-\pi) = \mu(-\pi + 0)$, $\mu(x) = [\mu(x + 0) + \mu(x - 0)]/2$ for $-\pi < x < \pi$, $\mu(\pi) = \mu(\pi - 0)$, and then extend μ from $[-\pi, \pi]$ to the whole real axis via (0.6.1).

A function $\mu \in \mathsf{BV}_{2\pi}$ is *not* necessarily periodic; it is periodic if and only if $\mu(-\pi) = \mu(\pi)$, i.e. if $\int_{-\pi}^{\pi} d\mu(u) = 0$. Thus $\mathsf{BV}_{2\pi}$ is to be distinguished from the set of those $\mu \in \mathsf{BV}_{\mathrm{loc}}$ which are periodic. However, the extension (0.6.1) is chosen in such a way that the integral $\int_{-\pi}^{\pi} f(u)\, d\mu(u)$ of a periodic function f with respect to $\mu \in \mathsf{BV}_{2\pi}$ may be evaluated over any other interval of length 2π; thus for $f \in \mathsf{C}_{2\pi}$, $\mu \in \mathsf{BV}_{2\pi}$, and any $a \in \mathbb{R}$

$$\int_{a}^{a+2\pi} f(u)\, d\mu(u) = \int_{-\pi}^{\pi} f(u)\, d\mu(u). \tag{0.6.3}$$

The particular case $f(x) \equiv \text{const}$ shows that (0.6.1) is also necessary for (0.6.3) to hold. If $\mu \in \mathsf{BV}_{2\pi}$, then the function $\mu(x) - (1/2\pi)[\mu(\pi) - \mu(-\pi)]x$ is periodic. For $f \in \mathsf{L}_{2\pi}^1$ the function μ as defined through $\mu(x) = \int_{-\pi}^{x} f(u)\, du$ is an absolutely continuous element of $\mathsf{BV}_{2\pi}$ with $\|\mu\|_{\mathsf{BV}_{2\pi}} = \|f\|_{\mathsf{L}_{2\pi}^1}$; indeed, by (0.4.4)

$$\mu(x + 2\pi) - \mu(x) = \int_{x}^{x+2\pi} f(u)\, du = \int_{-\pi}^{\pi} f(u)\, du.$$

Conversely, if for $f \in \mathsf{L}_{\mathrm{loc}}^1$ the function μ as defined through $\mu(x) = \int_{-\pi}^{x} f(u)\, du$ belongs to $\mathsf{BV}_{2\pi}$, then f is necessarily periodic, thus an element of $\mathsf{L}_{2\pi}^1$. For these facts as well as for the following propositions see ZYGMUND [7I, p. 11; p. 38f].

Proposition 0.6.1. *Let* $\mu \in \mathsf{BV}_{2\pi}$. *If* $\int_{-\pi}^{\pi} h(u)\, d\mu(u) = 0$ *for every* $h \in \mathsf{C}_{2\pi}$, *then* μ *is a constant (see also Prop.* 0.8.8).

Proposition 0.6.2. *Let* $\mu, \nu \in \mathsf{BV}_{2\pi}$. *The* ***convolution*** $\mu * d\nu$ *as defined through*

$$(\mu * d\nu)(x) = \frac{1}{2\pi} \int_{-\pi}^{\pi} \mu(x - u)\, d\nu(u) \tag{0.6.4}$$

exists for all $x \notin \Pi_{\mu,\nu}$ *as a Riemann–Stieltjes integral. Moreover,* $(\mu * d\nu)(x)$ *can be defined on the set* $\Pi_{\mu,\nu}$ *so as to become an element of* $\mathsf{BV}_{2\pi}$ *satisfying*

$$\|\mu * d\nu\|_{\mathsf{BV}_{2\pi}} \le \|\mu\|_{\mathsf{BV}_{2\pi}} \|\nu\|_{\mathsf{BV}_{2\pi}}. \tag{0.6.5}$$

The proof follows as for Prop. 0.5.2, the definition of $\mu * d\nu$ on the set $\Pi_{\mu,\nu}$ being again given by (0.5.7). In particular, $\mu * d\nu$ satisfies (0.6.1) since μ, ν have this property and hence

$$\begin{aligned}(\mu * d\nu)(x + 2\pi) - (\mu * d\nu)(x) &= (1/2\pi) \int_{-\pi}^{\pi} [\mu(x + 2\pi - u) - \mu(x - u)]\, d\nu(u) \\ &= (1/2\pi)[\mu(\pi) - \mu(-\pi)][\nu(\pi) - \nu(-\pi)];\end{aligned}$$

$\mu * d\nu$ is also correctly normalized since, for example,

$$\lim_{h\to 0+} (\mu * d\nu)(x + h) = \frac{1}{2\pi} \int_{-\pi}^{\pi} \Big[\lim_{h\to 0+} \mu(x + h - u)\Big]\, d\nu(u). \tag{0.6.6}$$

It is to be emphasized that the convolution $\mu * d\nu$ is *not* commutative. Indeed, by partial integration

$$(\mu * d\nu)(x) - (\nu * d\mu)(x) = (1/2\pi)[\mu(-\pi)\nu(\pi) - \mu(\pi)\nu(-\pi)]. \tag{0.6.7}$$

However, if $\mu, \nu \in \mathsf{BV}_{2\pi}$ would be additionally normalized by $\mu(-\pi) = \nu(-\pi) = 0$, commutativity would be realized, but the convolution of such functions does not necessarily satisfy $(\mu * d\nu)(-\pi) = 0$, i.e., the property $\mu(-\pi) = 0$ would not be invariant under convolution in $\mathsf{BV}_{2\pi}$.

As a counterpart to Prop. 0.5.5 one has

Proposition 0.6.3. *Let* $f \in \mathsf{X}_{2\pi}$, $\mu \in \mathsf{BV}_{2\pi}$. *The* ***convolution*** $f * d\mu$ *as defined through*

$$(f * d\mu)(x) = \frac{1}{2\pi} \int_{-\pi}^{\pi} f(x - u)\, d\mu(u) \tag{0.6.8}$$

exists (a.e.) *as an absolutely convergent integral,* $f * d\mu \in \mathsf{X}_{2\pi}$, *and*

$$\|f * d\mu\|_{\mathsf{X}_{2\pi}} \le \|f\|_{\mathsf{X}_{2\pi}} \|\mu\|_{\mathsf{BV}_{2\pi}}. \tag{0.6.9}$$

0.7 Normed Linear Spaces, Bounded Linear Operators

A real or complex *linear system* (vector space, manifold) is a set of (abstract) elements for which the operations of addition and of multiplication by real or complex numbers are defined such that these operations obey the usual rules of the algebra of vectors; in particular, a linear system contains a zero element which is denoted by θ. A set of n elements $f_1, \ldots, f_n$ of a linear system is said to be *linearly independent* if the equation $\sum_{k=1}^{n} \alpha_k f_k = 0$ (α_k scalars) implies $\alpha_1 = \alpha_2 = \cdots = \alpha_n = 0$; if the scalars α_k are not all zero, the set $\{f_k\}_{k=1}^{n}$ is linearly *dependent*. An expression of the form $\sum_{k=1}^{n} \alpha_k f_k$ is called a (finite) *linear combination.*

A linear system X is called a *normed linear space* if to each element (vector, point) $f \in \mathsf{X}$ there is associated a (unique) positive number $\|f\| = \|f\|_{\mathsf{X}}$, the *norm* of f, subject to the following conditions ($f, g \in X$, $\alpha \in \mathbb{C}$): (i) $\|f\| = 0$ if and only if $f = \theta$, (ii) $\|\alpha f\| = |\alpha|\,\|f\|$ (*homogeneity*), (iii) $\|f + g\| \le \|f\| + \|g\|$ (*triangular inequality*). A sequence $\{f_n\}_{n=1}^{\infty}$ of a normed linear space X is said to converge *strongly* or in X-*norm* or in the *strong topology* to $f \in X$, in notation $f = \text{s-lim}_{n\to\infty} f_n$, if $\lim_{n\to\infty} \|f - f_n\| = 0$. A subset B of X is said to be *closed* if B contains all its X-limit points, i.e., if for $\{f_n\} \subset B$ there exists $f \in X$ such that $\lim_{n\to\infty} \|f - f_n\| = 0$, then $f \in B$. The *closure* of an arbitrary set $B \subset X$ (in X-norm), denoted by $\bar{B} = \bar{B}^X$, is the smallest closed set containing B; thus $\bar{B}^X$ is the union of B and its X-limit points. The set $S_X(f_0, \rho) = \{f \in X \mid \|f - f_0\| \le \rho\}$ is called the (closed) *sphere* with center f_0 and radius ρ; thus $S_X(f_0, \rho) = \overline{S_X(f_0, \rho)}^X$. If $f_0 = \theta$, we abbreviate the notation and write $S_X(\rho)$. A set $B \subset X$ is called *bounded* if B is contained in some sphere, i.e., there exists a (finite) $M > 0$ such that $\|f\|_X \le M$ for all $f \in B$.

A sequence $\{f_n\}$ of a normed linear space X is called a (strong) *Cauchy sequence* in X if $\lim_{m,n\to\infty} \|f_m - f_n\| = 0$. The space X is said to be *complete* if every Cauchy sequence is convergent to some element in X, in other words, if $\lim_{m,n\to\infty} \|f_m - f_n\| = 0$ implies the *existence* of some $f \in X$ such that $\lim_{n\to\infty} \|f - f_n\| = 0$. A complete normed linear space is called a *Banach space*. Several examples of Banach spaces have already been introduced. Thus the spaces C and $C_{2\pi}$ of continuous functions are Banach spaces under the norms (0.1.1) and (0.4.1), respectively, strong convergence being equivalent to uniform convergence. Likewise, the Lebesgue spaces L^p or $L^p_{2\pi}$ are Banach spaces under the norm (0.1.2), (0.1.3) or (0.4.2), (0.4.3) with the usual convention that the elements of Lebesgue spaces are considered as equivalence classes consisting of those functions which are equal almost everywhere; the zero element is then the set of all functions f for which $f(x) = 0$ a.e. Further examples of Banach spaces are the classes l^p with norms given by (0.3.6), (0.3.7) and BV with norm (0.5.1). On the other hand, $BV_{2\pi}$ is not a normed linear space since $\|\mu\|_{BV_{2\pi}} = 0$ only implies $\mu(x) = \text{const}$. However, if one considers equivalence classes consisting of all functions in $BV_{2\pi}$ which differ only by a constant, then the set of these equivalence classes defines a Banach space $BV^0_{2\pi}$; in particular, each equivalence class contains a representative which is normalized at $-\pi$ to zero. In $BV^0_{2\pi}$ the convolution (0.6.4) is commutative (recall (0.6.7)).

A Banach space X is called a *Hilbert space* if the norm is induced by an inner product, i.e., to each pair $f, g \in X$ there is associated a (unique) complex number (f, g), the *inner product*, subject to the following conditions: (i) $(\alpha_1 f_1 + \alpha_2 f_2, g) = \alpha_1(f_1, g) + \alpha_2(f_2, g)$ for any scalars α_1, α_2, (ii) $(f, g) = \overline{(g, f)}$†, (iii) $(f, f) = \|f\|^2$. Examples of Hilbert spaces are given by L^2, $L^2_{2\pi}$ and l^2 with inner products

(0.7.1)
$$(f, g) = \frac{1}{\sqrt{2\pi}} \int_{-\infty}^{\infty} f(u)\overline{g(u)}\, du \qquad (f, g \in L^2),$$

(0.7.2)
$$(f, g) = \frac{1}{2\pi} \int_{-\pi}^{\pi} f(u)\overline{g(u)}\, du \qquad (f, g \in L^2_{2\pi}),$$

(0.7.3)
$$(f, g) = \sum_{k=-\infty}^{\infty} f_k \overline{g_k}. \qquad (f, g \in l^2).$$

† If c is a complex number, then $\bar{c}$ denotes the complex conjugate of c.

Let X and Y be two normed linear spaces which may be identical or distinct, and $\mathsf{D} \subseteq \mathsf{X}$ a subset. One says that T is an *operator* of D into Y if for each $f \in \mathsf{D}$ there is determined a unique $g \in \mathsf{Y}$, denoted by $g = Tf$ [or $g = T(f)$], called the *value* of T at f. The terms: operator, mapping, transformation are used synonymously. The set $\mathsf{D} = \mathsf{D}(T)$ is called the *domain* of T. If $D_1 \subseteq \mathsf{D}$, the set $\{T(f) \mid f \in \mathsf{D}_1\}$ is the *image* of D_1 under the mapping T and denoted by $\mathsf{T}(\mathsf{D}_1)$; in particular, T(D) is called the *range* of T. If one takes $\mathsf{D} = \mathsf{X}$, then T is said to be a mapping of X *into* Y if $\mathsf{T}(\mathsf{X}) \subseteq \mathsf{Y}$, and *onto* Y if $\mathsf{T}(\mathsf{X}) = \mathsf{Y}$. T is said to be *one-to-one* if $Tf_1 = Tf_2$ implies $f_1 = f_2$. If T is the mapping of X into itself defined by $Tf = f$ for all $f \in \mathsf{X}$, T is called the *identity operator* (of X) and denoted by I. The mapping T of X into Y is said to be *continuous* at the point $f_0 \in \mathsf{X}$ if to each $\varepsilon > 0$ there is a $\delta > 0$ such that $\|Tf - Tf_0\|_{\mathsf{Y}} < \varepsilon$ for all $f \in \mathsf{X}$ with $\|f - f_0\|_{\mathsf{X}} < \delta$. T is a continuous transformation of X into Y if T is continuous at every point of X. An operator T on some function space X is *positive* if $f(x) \geq 0$ for all $x \in \mathbb{R}$ implies $(Tf)(x) \geq 0$ for all $x \in \mathbb{R}$.

An operator T of X into Y is called *linear* if $T(\alpha_1 f_1 + \alpha_2 f_2) = \alpha_1 T(f_1) + \alpha_2 T(f_2)$ for all $f_1, f_2 \in \mathsf{X}$ and complex numbers α_1, α_2. T is called *bounded* if there exists a constant $M \geq 0$ such that $\|Tf\|_{\mathsf{Y}} \leq M\|f\|_{\mathsf{X}}$ for all $f \in \mathsf{X}$. The smallest possible value of M satisfying this inequality is said to be the *norm* or *bound* of T and denoted by $\|T\|$. From this definition it follows that

$$\|Tf\|_{\mathsf{Y}} \leq \|T\| \, \|f\|_{\mathsf{X}} \tag{0.7.4}$$

for every $f \in \mathsf{X}$, and

$$\|T\| = \sup_{f \neq \theta} \frac{\|Tf\|_{\mathsf{Y}}}{\|f\|_{\mathsf{X}}} = \sup_{f \neq \theta} \left\| T\left(\frac{f}{\|f\|_{\mathsf{X}}}\right) \right\|_{\mathsf{Y}} = \sup_{\|f\|_{\mathsf{X}} = 1} \|Tf\|_{\mathsf{Y}} = \sup_{\|f\|_{\mathsf{X}} \leq 1} \|Tf\|_{\mathsf{Y}}. \tag{0.7.5}$$

A linear operator T of X into Y is continuous if and only if T is continuous at a single point $f_0 \in \mathsf{X}$, or, if and only if T is bounded. The linear system of all bounded linear transformations of X into Y, endowed with norm (0.7.5), is again a normed linear space, denoted by [X, Y]. Consequently, the more precise notation $\|T\|_{[\mathsf{X},\mathsf{Y}]}$ is sometimes employed for the norm (0.7.5). If, in addition, Y is complete, thus a Banach space, so is [X, Y]. A sequence of operators $\{T_n\}_{n=1}^{\infty} \subset [\mathsf{X}, \mathsf{Y}]$ is said to converge *strongly* to the operator $T \in [\mathsf{X}, \mathsf{Y}]$ if $\lim_{n\to\infty} \|T_n f - Tf\|_{\mathsf{Y}} = 0$ for each $f \in \mathsf{X}$; $\{T_n\}$ is *strongly Cauchy convergent* (on X) if $\lim_{m,n\to\infty} \|T_m f - T_n f\|_{\mathsf{Y}} = 0$ for each $f \in \mathsf{X}$. If $\{T_n\}$ converges in the norm of [X, Y] towards T, i.e. $\lim_{n\to\infty} \|T_n - T\|_{[\mathsf{X},\mathsf{Y}]} = 0$, the sequence $\{T_n\}$ is said to converge *uniformly*, thus in the *uniform operator topology*, towards T; equivalently, $\lim_{n\to\infty} \|T_n f - Tf\|_{\mathsf{Y}} = 0$ uniformly for all $f \in \mathsf{X}$ with $\|f\|_{\mathsf{X}} \leq 1$. Evidently, uniform convergence of bounded linear operators implies strong convergence but not conversely. The transformation $T \in [\mathsf{X}, \mathsf{Y}]$ is said to be *isometric* if it preserves norms, i.e. $\|Tf\|_{\mathsf{Y}} = \|f\|_{\mathsf{X}}$ for every $f \in \mathsf{X}$; T defines a *contraction* if $\|Tf\|_{\mathsf{Y}} \leq \|f\|_{\mathsf{X}}$ for every $f \in \mathsf{X}$.

A subset A of a normed linear space X is said to be *dense* in X if to each $f \in \mathsf{X}$ and $\varepsilon > 0$ there exists $g \in \mathsf{A}$ such that $\|f - g\|_{\mathsf{X}} < \varepsilon$; A is said to be *fundamental* in X if the set of all finite linear combinations of elements of A is dense in X. A normed linear space X is called *separable* if it contains a denumerable dense subset. The spaces $\mathsf{X}_{2\pi}$ are separable (see Theorem 1.2.5) while $\mathsf{L}_{2\pi}^{\infty}$ is not.

Proposition 0.7.1. *Let* A *be a dense linear subset of a Banach space* X, *and suppose that* T_0 *is a bounded linear transformation of* A *into the Banach space* Y *with bound* $\|T_0\|_{[\mathsf{A},\mathsf{Y}]}$. *Then* T_0 *can be uniquely extended to a bounded linear transformation* T *of* X *into* Y *having the same bound, i.e.* $Tf = T_0 f$ *for all* $f \in \mathsf{A}$ *and* $\|T\|_{[\mathsf{X},\mathsf{Y}]} = \|T_0\|_{[\mathsf{A},\mathsf{Y}]}$.

Proposition 0.7.2. (***Uniform boundedness principle***). *Let* $\{T_n\}_{n=1}^{\infty}$ *be a sequence of bounded linear operators of the Banach space* X *into the normed linear space* Y. *If* $\{\|T_n f\|_{\mathsf{Y}}\}$ *is bounded for each* $f \in \mathsf{X}$ *separately, i.e., if for each* $f \in \mathsf{X}$ *there exists a constant* M_f *such that for all* $n \in \mathbb{N}$

$$\|T_n f\|_{\mathsf{Y}} \le M_f, \tag{0.7.6}$$

then the sequence $\{\|T_n\|_{[\mathsf{X},\mathsf{Y}]}\}$ *is bounded, i.e., there exists a constant* M *such that* $\|T_n f\|_{\mathsf{Y}} \le M\,\|f\|_{\mathsf{X}}$ *for all* $n \in \mathbb{N}$ *and* $f \in \mathsf{X}$.

Proof. Suppose that the sequence $\{\|T_n\|\}$ is not bounded. Replacing, if necessary, $\{\|T_n\|\}$ by a subsequence (which is also denoted by $\{\|T_n\|\}$), one may then assume that

$$\lim_{n\to\infty} \|T_n\| = \infty. \tag{0.7.7}$$

According to (0.7.5), there exists a sequence $\{f_n\}_{n=1}^{\infty} \subset \mathsf{X}$ such that for all n

$$\|T_n f_n\| > \tfrac{1}{2}\|T_n\|, \qquad \|f_n\| = 1, \tag{0.7.8}$$

and therefore by (0.7.7)

$$\lim_{n\to\infty} \|T_n f_n\| = \infty. \tag{0.7.9}$$

Now, let $\{\alpha_k\}_{k=1}^{\infty}$ be a sequence of positive numbers such that

$$\sum_{k=1}^{\infty} \alpha_k < \infty, \qquad \sum_{k=m+1}^{\infty} \alpha_k \le \alpha_m/5 \qquad (m \in \mathbb{N}) \tag{0.7.10}$$

(for instance, $\alpha_k = 6^{-k}$). By (0.7.7) one may find a positive integer n_1 such that $\alpha_1\|T_{n_1}\| > 5$. This determines f_{n_1} and therefore M_{n_1} of (0.7.6). By (0.7.7) again, one may choose n_2 with $n_2 > n_1$ such that $\alpha_2\|T_{n_2}\| > 5(\alpha_1 M_{n_1} + 2)$. Proceeding in this way, one arrives at a subsequence $\{T_{n_k}\} \subset \{T_n\}$ such that $n_{k+1} > n_k$ and

$$(1/5)\alpha_m\|T_{n_m}\| > \sum_{k=1}^{m-1} \alpha_k M_{n_k} + m \qquad (m \in \mathbb{N}). \tag{0.7.11}$$

For the corresponding subsequence $\{f_{n_k}\} \subset \{f_n\}$ set up $s_j = \sum_{k=1}^{j} \alpha_k f_{n_k}$. Then, since $\|f_n\| = 1$, for $i > j$

$$\|s_i - s_j\| = \Big\|\sum_{k=j+1}^{i} \alpha_k f_{n_k}\Big\| \le \sum_{k=j+1}^{i} \alpha_k,$$

which converges to zero for $i, j \to \infty$ by (0.7.10). Thus the elements $s_j \in \mathsf{X}$ form a Cauchy sequence, and since X is a Banach space, they converge to an element $g \in \mathsf{X}$. Therefore $g = \sum_{k=1}^{\infty} \alpha_k f_{n_k}$ is well-defined as an element in X, and

$$\begin{aligned}\|T_{n_m} g\| &= \Big\|T_{n_m}\Big(\sum_{k=1}^{m-1} \alpha_k f_{n_k}\Big) + T_{n_m}(\alpha_m f_{n_m}) + T_{n_m}\Big(\sum_{k=m+1}^{\infty} \alpha_k f_{n_k}\Big)\Big\| \\ &\ge \alpha_m\|T_{n_m}(f_{n_m})\| - \Big\|T_{n_m}\Big(\sum_{k=1}^{m-1} \alpha_k f_{n_k}\Big)\Big\| - \Big\|T_{n_m}\Big(\sum_{k=m+1}^{\infty} \alpha_k f_{n_k}\Big)\Big\|.\end{aligned}$$

Since by (0.7.6) and (0.7.11)

$$\left\| T_{n_m}\left(\sum_{k=1}^{m-1} \alpha_k f_{n_k}\right)\right\| \leq \sum_{k=1}^{m-1} \alpha_k \|T_{n_m} f_{n_k}\| \leq \sum_{k=1}^{m-1} \alpha_k M_{n_k} < (1/5)\alpha_m \|T_{n_m}\| - m$$

and by (0.7.8) and (0.7.10)

$$\left\| T_{n_m}\left(\sum_{k=m+1}^{\infty} \alpha_k f_{n_k}\right)\right\| \leq \sum_{k=m+1}^{\infty} \alpha_k \|T_{n_m}\| \, \|f_{n_k}\| \leq \|T_{n_m}\|\alpha_m/5,$$

it follows by (0.7.8) that for all positive integers m

$$\|T_{n_m} g\| \geq \alpha_m \|T_{n_m}\| \{\tfrac{1}{2} - \tfrac{1}{5} - \tfrac{1}{5}\} + m > m.$$

This implies $\lim_{m\to\infty} \|T_{n_m} g\| = \infty$, a contradiction to the assumption (0.7.6). This completes the proof (see also LORENTZ [3, p. 95 f]).

As an immediate application one has a theorem on the convergence of sequences of bounded linear operators; this theorem states in particular that the strong limit of a sequence of continuous linear transformations is continuous.

Proposition 0.7.3. (*Theorem of Banach–Steinhaus*). (a) *A sequence* $\{T_n\}_{n=1}^{\infty}$ *of bounded linear operators of a Banach space* X *into a Banach space* Y *converges strongly to a bounded linear operator* T *of* X *into* Y *if and only if there exists* (i) *a constant* $M > 0$ *such that* $\|T_n\|_{[X,Y]} \leq M$ *for all* $n \in \mathbb{N}$, *and* (ii) *a dense subset* A *of* X *such that the sequence* $\{T_n\}$ *is strongly Cauchy convergent on* A.

(b) *Let* X *be a Banach space and* $\{T_n\}_{n=1}^{\infty}$ *a sequence of bounded linear operators of* X *into itself. Then for each* $f \in X$

$$\lim_{n\to\infty} \|T_n f - f\|_X = 0 \tag{0.7.12}$$

if and only if there exists (i) *a constant* $M > 0$ *such that* $\|T_n\|_{[X,X]} \leq M$ *for all* $n \in \mathbb{N}$, *and* (ii) *a dense subset* A *of* X *such that* (0.7.12) *holds for each* $g \in A$.

Proof. To prove (a), let $\{T_n\} \subset [X, Y]$ converge strongly to $T \in [X, Y]$, i.e., for each $f \in X$, $\lim_{n\to\infty} \|T_n f - Tf\|_Y = 0$. Then the sequence $\{\|T_n f\|_Y\}$ is bounded for each $f \in X$, and thus $\|T_n\|_{[X,Y]} \leq M$ by Prop. 0.7.2. Moreover, $\lim_{m,n\to\infty} \|T_m f - T_n f\|_Y = 0$ for each $f \in X$, and $\{T_n\}$ is strongly Cauchy convergent on X, particularly on every subset of X. Conversely, let conditions (i) and (ii) be satisfied. Since A is dense in X, given $f \in X$ and $\varepsilon > 0$, there exists $g \in A$ such that $\|f - g\|_X < \varepsilon$. By the triangular inequality, the linearity of T_n, and by (i)

$$\begin{aligned} \|T_m f - T_n f\|_Y &\leq \|T_m(f - g)\|_Y + \|T_m g - T_n g\|_Y + \|T_n(f - g)\|_Y \\ &\leq 2M\varepsilon + \|T_m g - T_n g\|_Y. \end{aligned}$$

Thus, since $g \in A$, it follows by (ii) that $\lim_{m,n\to\infty} \|T_m f - T_n f\|_Y = 0$ for each $f \in X$, and $\{T_n\}$ is strongly Cauchy convergent on the whole space X. Since Y is complete, this implies the existence of an element $Tf \in Y$ such that $\lim_{n\to\infty} \|T_n f - Tf\|_Y = 0$ for each $f \in X$. Evidently, the operation T is linear. T is also bounded; for, since $\|T_n f\|_Y \leq M\|f\|_X$ uniformly for all $n \in \mathbb{N}$, it follows that $\|Tf\|_Y \leq M\|f\|_X$ for all $f \in X$. This establishes (a). The proof of the second version of the theorem of Banach–Steinhaus, namely (b), follows along the same lines.

Let X, Y be two normed linear spaces. The (cartesian) *product* X × Y is the set of all ordered pairs (f, g) with $f \in X$, $g \in Y$; thus $X \times Y = \{(f, g) \mid f \in X, g \in Y\}$. A linear transformation T of $D \subseteq X$ into Y is said to be *closed* if its *graph* $\{(f, Tf) \mid f \in D\}$ is a closed subspace of the *product space* $X \times Y$ as endowed with the norm $\|(f, g)\|_{X \times Y} = \|f\|_X + \|g\|_Y$, i.e., whenever for a sequence $\{f_n\} \subset D$ there exist $f \in X$ and $g \in Y$ such that $\lim_{n \to \infty} \|f - f_n\|_X = 0$ and $\lim_{n \to \infty} \|g - Tf_n\|_Y = 0$, then $f \in D$ and $Tf = g$. A bounded linear transformation of a closed domain $D \subseteq X$ into Y is closed.

Proposition 0.7.4. (*Closed graph theorem*). *A closed linear transformation of a Banach space* X *into a Banach space* Y *is bounded, thus continuous.*

The domain $D(T)$ of a closed linear operator T of a Banach space X into itself becomes a Banach space under the norm $\|f\|_{D(T)} = \|f\|_X + \|Tf\|_X$.

A (complex) Banach space X is said to be a (complex) *commutative Banach algebra* if to each pair f, g of elements in X there exists an element fg, called the *product*, such that (i) $\|fg\| \leq \|f\| \, \|g\|$ for every $f, g \in X$, and (ii) X becomes a commutative (complex) algebra under this multiplication on X, i.e., the *associative* law $f(gh) = (fg)h$, the *distributive* law $f(g + h) = fg + fh$, the *commutative* law $fg = gf$, and the relation $(\alpha f)g = \alpha(fg)$ hold for any $f, g, h \in X$, $\alpha \in \mathbb{C}$. If the algebra contains an element e such that $ef = f$ for every $f \in X$, then e is called the *unit* element of X. Examples of Banach algebras are given by the Banach spaces L^1, BV, $L^1_{2\pi}$, $BV^0_{2\pi}$ with convolution as multiplication (compare (0.2.3), (0.5.5), Prop. 0.4.1 (ii), (0.6.5), respectively). L^1, $L^1_{2\pi}$ are commutative Banach algebras *without* unit element (see Prop. 5.1.12, 4.1.4), BV, $BV^0_{2\pi}$ are those *with* unit element (see (5.3.3), (4.3.2); but note that $BV_{2\pi}$ is *not* a commutative Banach algebra, see (0.6.7)). The Banach spaces C, C_0, $C_{2\pi}$, L^∞, $L^\infty_{2\pi}$, l^∞, l^∞_0 are commutative Banach algebras if multiplication is defined through a pointwise product (e.g., if $f, g \in C$, then $(fg)(x) \equiv f(x)g(x)$; if $\{f_k\}_{k=-\infty}^{\infty}$, $\{g_k\}_{k=-\infty}^{\infty} \in l^\infty$, then $fg \equiv \{f_k g_k\}_{k=-\infty}^{\infty}$). The algebras C, $C_{2\pi}$, L^∞, $L^\infty_{2\pi}$, l^∞ do possess a unit element (namely $e(x) \equiv 1$ and $\{e_k\}_{k=-\infty}^{\infty}$ with $e_k \equiv 1$, respectively), whereas C_0, l^∞_0 do not. A further example of a not necessarily commutative Banach algebra is given by the space $[X, X]$ of all bounded linear operators of a Banach space X into itself, multiplication being defined through the composition $(TS)f = T(Sf)$ for $T, S \in [X, X]$, $f \in X$. This algebra has a unit element as given by the identity operator.

Speaking informally, two algebraic systems of the same nature are said to be *isomorphic* if there is a one-to-one mapping, called *isomorphism*, of one onto the other which preserves all relevant properties. For instance, two linear systems are isomorphic if there is a one-to-one *linear* mapping of one onto the other. An isomorphism T of a Hilbert space H_1 onto a Hilbert space H_2 is a one-to-one linear mapping of H_1 onto H_2 which also preserves inner products, i.e. $(Tf, Tg) = (f, g)$ for all $f, g \in H_1$; thus *Hilbert space isomorphisms* are always isometric mappings. Likewise, an (*algebra*) *isomorphism* T of a Banach algebra A_1 *into* a Banach algebra A_2 is a one-to-one linear mapping of A_1 onto the range of T in A_2 which preserves products, i.e. $T(fg) = (Tf)(Tg)$ for all $f, g \in A_1$.

0.8 Bounded Linear Functionals, Riesz Representation Theorems

Let X be a normed linear space. A bounded linear operator of X into (the Banach space) $\mathbb{R}$ or $\mathbb{C}$ of real or complex numbers is called a bounded linear *functional*. Functionals are denoted by $f^*, g^*, \ldots$, the value of f^* at $f \in X$ by $f^*(f)$ (which is a real or complex number), and the totality of all bounded linear functionals on X by

X^*, the *adjoint* or *dual* or *conjugate* of X. X^* becomes a Banach space under the norm (compare (0.7.4), (0.7.5))

$$\|f^*\| = \sup_{\|f\|=1} |f^*(f)| \qquad (f^* \in X^*). \tag{0.8.1}$$

Proposition 0.8.1. *Let X be a normed linear space. If $f \in X$ is such that $f^*(f) = 0$ for all $f^* \in X^*$, then $f = \theta$.*

Since the dual X^* of a normed linear space X is again a normed linear space (and even a Banach space), one may consider the dual of X^*, called the *second dual* of X and denoted by X^{**}. X^{**} is a Banach space. To each $f_0 \in X$ one may associate an element $f_0^{**} \in X^{**}$ via $f_0^{**}(f^*) = f^*(f_0)$, $f^* \in X^*$, which establishes a *natural mapping* $f_0 \to f_0^{**}$ of X onto a linear manifold X_0^{**} of X^{**}. This mapping is one-to-one and norm-preserving, and thus an (isometric) isomorphism of X onto X_0^{**}. In this sense, $X \subset X^{**}$. If $X = X^{**}$ under the natural mapping, then X is called *reflexive*. Examples of reflexive Banach spaces are given by $L_{2\pi}^p$ and L^p for $1 < p < \infty$. However, $C_{2\pi}$, $L_{2\pi}^1$, $L_{2\pi}^\infty$, C, L^1, L^∞ are not reflexive.

A sequence $\{f_n\}_{n=1}^\infty$ of elements of a normed linear space X is said to converge *weakly*, thus in the *weak topology*, to $f \in X$, in notation: $\text{w-lim}_{n\to\infty} f_n = f$, if $\lim_{n\to\infty} f^*(f_n) = f^*(f)$ for every $f^* \in X^*$.

Proposition 0.8.2. *Let X be a normed linear space and $\{f_n\}_{n=1}^\infty \subset X$, $f \in X$.* (i) *If the weak limit exists, it is unique.* (ii) *If $\{f_n\}$ converges strongly to f, it also converges weakly to f.* (iii) *If $\{f_n\}$ converges weakly to f, then $\|f\|_X \le \liminf_{n\to\infty} \|f_n\|_X$.*

Proposition 0.8.3. *A sequence $\{f_n\}_{n=1}^\infty$ of elements of a normed linear space X converges weakly to $f \in X$ if and only if there exists* (i) *a constant $M > 0$ such that $\|f_n\|_X \le M$ for all n, and* (ii) *a dense subset A of X^* such that $\lim_{n\to\infty} f^*(f_n) = f^*(f)$ for every $f^* \in A$.*

This is an immediate consequence of the theorem of Banach–Steinhaus.

Proposition 0.8.4. (***Weak compactness theorem***). *If X is a reflexive normed linear space, then each bounded sequence in X contains a weakly convergent subsequence.*

Analogously one may consider the weak topology of the dual space X^* of X. However, this topology turns out to be less useful than that obtained when restricting the functionals to $X \subset X^{**}$; the latter defines the *weak* topology* of X^*. Thus a sequence $\{f_n^*\}_{n=1}^\infty$ of bounded linear functionals on X converges *weakly** to $f^* \in X^*$, in notation: $\text{w*-lim}_{n\to\infty} f_n^* = f^*$, if $\lim_{n\to\infty} f_n^*(f) = f^*(f)$ for every $f \in X$; f^* is then called the *weak* limit* of this sequence. If X is reflexive, then the weak and weak* topology of X^* coincide. Applying the uniform boundedness principle, one immediately has

Proposition 0.8.5. *Let X be a Banach space and $\{f_n^*\}_{n=1}^\infty \subset X^*$, $f^* \in X^*$.* (i) *If the weak* limit exists, it is unique.* (ii) *If $\{f_n^*\}$ converges weakly* to f^*, then $\{\|f_n^*\|_{X^*}\}$ is bounded.* (iii) *If $\{f_n^*\}$ converges weakly* to f^*, then $\|f^*\|_{X^*} \le \liminf_{n\to\infty} \|f_n^*\|_{X^*}$.*

Proposition 0.8.6. *A sequence $\{f_n^*\}_{n=1}^\infty$ of bounded linear functionals on a Banach space X converges weakly* to $f^* \in X^*$ if and only if there exists* (i) *a constant $M > 0$ such that $\|f_n^*\|_{X^*} \le M$ for all n, and* (ii) *a dense subset A of X such that $\lim_{n\to\infty} f_n^*(f) = f^*(f)$ for every $f \in A$.*

For the following result (as well as for Prop. 0.8.4) one may consult TAYLOR [1, p. 209], ZAANEN [1, p. 155 ff].

Proposition 0.8.7. (***Weak* compactness theorem***). *If the normed linear space* X *is separable, then every sequence* $\{f_n^*\} \subset X^*$ *having uniformly bounded norms (i.e.* $\|f_n^*\|_{X^*} \le M$, *all n) contains a weakly* convergent subsequence.*

The foregoing implications of the Banach–Steinhaus theorem together with the above *selection principles* (Prop. 0.8.4, 0.8.7) are particularly useful in connection with analytic formulae giving a representation of all bounded linear functionals on a given space X. A list of four such results, unspecifically referred to as *(F.) Riesz representation theorem*, is added. It will be clear from the context, i.e., from the underlying Banach space, which result applies. Obviously, the integrals in the propositions to follow represent bounded linear functionals on the spaces in question.

Proposition 0.8.8. *If* f^* *is a bounded linear functional on* $C_{2\pi}$, *then there exists* $\mu \in BV_{2\pi}$ such *that*

$$f^*(f) = \frac{1}{2\pi}\int_{-\pi}^{\pi} f(u)\,\overline{d\mu(u)} \tag{0.8.2}$$

for every $f \in C_{2\pi}$. *The function* $\mu \in BV_{2\pi}$ *is uniquely determined by* f^* *up to a constant. Furthermore,*

$$\|f^*\| = \|\mu\|_{BV_{2\pi}} = \sup_{\|f\|_{C_{2\pi}}=1}\left|\frac{1}{2\pi}\int_{-\pi}^{\pi} f(u)\,\overline{d\mu(u)}\right|. \tag{0.8.3}$$

Thus $(C_{2\pi})^*$ *is isometrically isomorphic to* $BV_{2\pi}^0$ *(see also Prop.* 0.6.1).

Proposition 0.8.9. *To each bounded linear functional* f^* *on* $L_{2\pi}^p$, $1 \le p < \infty$, *there exists a unique* $g \in L_{2\pi}^{p'}$ *such that*

$$f^*(f) = \frac{1}{2\pi}\int_{-\pi}^{\pi} f(u)\overline{g(u)}\,du \tag{0.8.4}$$

for every $f \in L_{2\pi}^p$. *Moreover,*

$$\|f^*\| = \|g\|_{L_{2\pi}^{p'}} = \sup_{\|f\|_{L_{2\pi}^p}=1}\left|\frac{1}{2\pi}\int_{-\infty}^{\infty} f(u)\overline{g(u)}\,du\right|. \tag{0.8.5}$$

Thus $(L_{2\pi}^p)^*$ *is isometrically isomorphic to* $L_{2\pi}^{p'}$ *for* $1 \le p < \infty$. *In particular, if* $g \in L_{2\pi}^{p'}$ *is such that* $\int_{-\pi}^{\pi} f(u)g(u)\,du = 0$ *for every* $f \in L_{2\pi}^p$, *then* $g(u) = 0$ a.e.

Proposition 0.8.10. *If* f^* *is a bounded linear functional on* C_0, *then there exists a unique* $\mu \in BV$ *such that*

$$f^*(f) = \frac{1}{\sqrt{2\pi}}\int_{-\infty}^{\infty} f(u)\,\overline{d\mu(u)} \tag{0.8.6}$$

for every $f \in C_0$. *Moreover,*

$$\|f^*\| = \|\mu\|_{BV} = \sup_{\|f\|_{C_0}=1}\left|\frac{1}{\sqrt{2\pi}}\int_{-\infty}^{\infty} f(u)\,\overline{d\mu(u)}\right|. \tag{0.8.7}$$

Thus $(C_0)^*$ *is isometrically isomorphic to* BV. *In particular, if* $\mu \in BV$ *is such that* $\int_{-\infty}^{\infty} f(u)\,d\mu(u) = 0$ *for every* $f \in C_0$, *then* $\mu(u) = 0$ *for all* $u \in \mathbb{R}$.

Let $f \in \mathsf{C}_0$ and $\mu, \nu \in \mathsf{BV}$. Then the integral $\int_{-\infty}^{\infty} \int_{-\infty}^{\infty} f(x+u)\, d\mu(x)\, d\nu(u)$ defines a bounded linear functional on C_0. Therefore by the previous proposition there exists some $\xi \in \mathsf{BV}$ such that

$$(0.8.8) \qquad \frac{1}{\sqrt{2\pi}} \int_{-\infty}^{\infty} f(x)\, d\xi(x) = \frac{1}{2\pi} \int_{-\infty}^{\infty} \int_{-\infty}^{\infty} f(x+u)\, d\mu(x)\, d\nu(x) \qquad (f \in \mathsf{C}_0).$$

It turns out (cf. Problem 5.3.9) that $\xi(x) = (\mu * d\nu)(x)$; thus the right-hand side of (0.8.8) provides a further method of defining convolution of two functions $\mu, \nu \in \mathsf{BV}$ (compare (0.5.3)).

Proposition 0.8.11. *To each bounded linear functional f^* on L^p, $1 \le p < \infty$, there exists a unique $g \in \mathsf{L}^{p'}$ such that*

$$(0.8.9) \qquad f^*(f) = \frac{1}{\sqrt{2\pi}} \int_{-\infty}^{\infty} f(u)\overline{g(u)}\, du$$

for every $f \in \mathsf{L}^p$. Moreover,

$$(0.8.10) \qquad \|f^*\| = \|g\|_{\mathsf{L}^{p'}} = \sup_{\|f\|_{\mathsf{L}^p}=1} \left| \frac{1}{\sqrt{2\pi}} \int_{-\infty}^{\infty} f(u)\overline{g(u)}\, du \right|.$$

Thus $(\mathsf{L}^p)^*$ is isometrically isomorphic to $\mathsf{L}^{p'}$ for $1 \le p < \infty$. In particular, if $g \in \mathsf{L}^{p'}$ is such that $\int_{-\infty}^{\infty} f(u)g(u)\, du = 0$ for every $f \in \mathsf{L}^p$, then $g(u) = 0$ a.e.

As an application of Prop. 0.8.5(iii), 0.8.7, and the previous representation theorems one obtains, apart from Prop. 0.8.13(i), the following *selection principles*:

Proposition 0.8.12. (***Weak* compactness theorem for*** $\mathsf{L}^p_{2\pi}$, $1 < p \le \infty$). *Let $\{f_n\}_{n=1}^{\infty} \subset \mathsf{L}^p_{2\pi}$ for some $1 < p \le \infty$, and let there exist some constant $M > 0$ such that $\|f_n\|_p \le M$ for all $n \in \mathbb{N}$. Then there exists a subsequence $\{f_{n_k}\}$ and $g \in \mathsf{L}^p_{2\pi}$ such that for every $s \in \mathsf{L}^{p'}_{2\pi}$*

$$\lim_{k\to\infty} \int_{-\pi}^{\pi} f_{n_k}(u)s(u)\, du = \int_{-\pi}^{\pi} g(u)s(u)\, du.$$

Moreover, $\|g\|_p \le \liminf_{k\to\infty} \|f_{n_k}\|_p$.

Proposition 0.8.13. (i) (***Theorem of Helly***). *Let $\{\mu_n\}_{n=1}^{\infty} \subset \mathsf{BV}[a,b]$, and let there exist a constant $M > 0$ such that $|\mu_n(a)| \le M$ and $[\mathrm{Var}\ \mu_n]_a^b \le M$ for all $n \in \mathbb{N}$. Then there exists a subsequence $\{\mu_{n_k}\}$ and $\mu \in \mathsf{BV}[a,b]$ such that $\lim_{k\to\infty} \mu_{n_k}(x) = \mu(x)$ for all $x \in [a,b]$.*

(ii) (***Thereom of Helly–Bray for*** $\mathsf{BV}_{2\pi}$). *Let $\{\mu_n\} \subset \mathsf{BV}_{2\pi}$ and suppose that $\|\mu_n\|_{\mathsf{BV}_{2\pi}} \le M$ for all $n \in \mathbb{N}$. Then there exists a subsequence $\{\mu_{n_k}\}$ and $\mu \in \mathsf{BV}_{2\pi}$ such that for every $s \in \mathsf{C}_{2\pi}$*

$$\lim_{k\to\infty} \int_{-\pi}^{\pi} s(u)\, d\mu_{n_k}(u) = \int_{-\pi}^{\pi} s(u)\, d\mu(u).$$

Moreover, $\|\mu\|_{\mathsf{BV}_{2\pi}} \le \liminf_{k\to\infty} \|\mu_{n_k}\|_{\mathsf{BV}_{2\pi}}$.

Proposition 0.8.14. (***Weak* compactness theorem for*** L^p, $1 < p \le \infty$). *Let $\{f_n\}_{n=1}^{\infty} \subset \mathsf{L}^p$ for some $1 < p \le \infty$, and let there exist a constant $M > 0$ such that $\|f_n\|_p \le M$ for all $n \in \mathbb{N}$. Then there exists a subsequence $\{f_{n_k}\}$ and $g \in \mathsf{L}^p$ such that for every $s \in \mathsf{L}^{p'}$*

$$\lim_{k\to\infty} \int_{-\infty}^{\infty} f_{n_k}(u)s(u)\, du = \int_{-\infty}^{\infty} g(u)s(u)\, du.$$

Moreover, $\|g\|_p \le \liminf_{k\to\infty} \|f_{n_k}\|_p$.

Proposition 0.8.15. (***Theorem of Helly–Bray for*** BV). *Let $\{\mu_n\} \subset \mathsf{BV}$ and suppose that*

$\|\mu_n\|_{\mathsf{BV}} \le M$ *for all* $n \in \mathbb{N}$. *Then there exists a subsequence* $\{\mu_{n_k}\}$ *and* $\mu \in \mathsf{BV}$ *such that for every* $s \in \mathsf{C}_0$

$$\lim_{k\to\infty} \int_{-\infty}^{\infty} s(u)\, d\mu_{n_k}(u) = \int_{-\infty}^{\infty} s(u)\, d\mu(u).$$

Moreover, $\|\mu\|_{\mathsf{BV}} \le \liminf_{k\to\infty} \|\mu_{n_k}\|_{\mathsf{BV}}$.

The latter result is also referred to as the weak* compactness theorem for BV.

Finally a result concerning the boundedness of sequences of functions in L^p-spaces is given which indeed is an immediate consequence of the uniform boundedness principle (compare Prop. 0.8.5(ii), and the Riesz representation theorems (see also ZYGMUND [7I, p. 166]).

Proposition 0.8.16. *Let* $\{f_n\}_{n=1}^{\infty} \subset \mathsf{L}_{2\pi}^p$ *for some* $1 \le p \le \infty$, *and suppose that* $|\int_{-\pi}^{\pi} s(u) f_n(u)\, du|$ *is bounded in* $n \in \mathbb{N}$ *for each* $s \in \mathsf{L}_{2\pi}^{p'}$ *separately. Then there exists a constant* $M > 0$ *such that* $\|f_n\|_p \le M$ *for all* $n \in \mathbb{N}$.

A related result holds in L^p-space.

0.9 References

General references to the material of the whole chapter are HEWITT–STROMBERG [1], ROYDON [1], RUDIN [4]. Specifically, for the results on real variable theory and Lebesgue integration (Sec. 0.1–0.6) see ASPLUND–BUNGART [1], HALMOS [1], HILDEBRANDT [1], MCSHANE–BOTTS [1], MUNROE [1], TAYLOR [2], WILLIAMSON [1], ZAANEN [2]. For the basic facts on functional analysis (Sec. 0.7, 0.8) see BANACH [1], DUNFORD–SCHWARTZ [1], GARNIR–DE WILDE–SCHMETS [1], HALMOS [2], LJUSTERNIK–SOBOLEV [1], RIESZ–SZ.-NAGY [1], TAYLOR [1], YOSIDA [1], ZAANEN [1].

Part I

Approximation by Singular Integrals

One of the fundamental problems of analysis is to approximate a given function f in some sense or other by functions having certain properties, and generally, by functions which have 'better' properties than f. It is to be expected that the better-behaved functions are to be *constructed* from the given f by some smoothing operation on f itself. The approximation of f by singular convolution integrals is of special interest.

Given $f \in \mathsf{C}_{2\pi}$, a convolution integral of the type

$$(f * \chi_n)(x) \equiv (1/2\pi) \int_{-\pi}^{\pi} f(x-u)\chi_n(u)\, du$$

will be called a singular integral provided that the sequence $\{\chi_n(x)\}_{n=1}^{\infty}$ is a (periodic) kernel, specifically, $\chi_n \in \mathsf{L}_{2\pi}^1$ with $\int_{-\pi}^{\pi} \chi_n(u)\, du = 2\pi$ for each $n \in \mathbb{N}$. Such a sequence is said to be an approximate identity if, in addition, $\|\chi_n\|_1 \leq M$ for all n and $\lim_{n\to\infty} \int_{\delta \leq |u| \leq \pi} |\chi_n(u)|\, du = 0$ for each $0 < \delta < \pi$. If the functions χ_n happen to be positive, they usually have the familiar bell-shaped graph: the area under the curve $y = \chi_n(u)$ is equal to 2π, whereby, for increasing n, the peak at $u = 0$ becomes higher and narrower in such a way that the area under the curve near $u = 0$ comes out equal to 2π.

The name approximate identity is justified by the fact that for $f \in \mathsf{C}_{2\pi}$ the sequence $\{f * \chi_n\}$ tends uniformly to f for $n \to \infty$. Indeed, in view of the properties of $\{\chi_n(x)\}$, the magnitude of the convolution integral $f * \chi_n$ for large n essentially depends upon the value of its integrand near $u = 0$ (f being bounded). Since $f(x - u)$ is then near to $f(x)$ (f being continuous), $(f * \chi_n)(x)$ is roughly equal to $f(x)(1/2\pi) \int_{-\delta}^{\delta} \chi_n(u)\, du$, which tends to $f(x)$ for large n.

One of the important features of the convolution integral $f * \chi_n$ is that the 'best' properties of each of its factors are inherited by the product itself. This is due to the translation-invariance and commutativity of convolutions. Thus, if $f \in \mathsf{C}_{2\pi}$ is convolved with $\chi_n \in \mathsf{C}_{2\pi}$ for which $\chi_n^{(r)} \in \mathsf{C}_{2\pi}$, then the convolution product is r-times continuously differentiable.

Special as these convolution integrals may seem, they nevertheless subsume a large number of integrals of analysis; these occur in the theory of Fourier series, as solutions of partial differential equations, and in approximation theory. Each of these associations may be outlined briefly.

The theory of singular integrals is intimately connected with the theory of Fourier series. Thus the nth partial sum of the Fourier series of a function f may be written as a singular integral with the Dirichlet kernel, while the sequence of arithmetic means of these partial sums form a convolution integral with the Fejér kernel. It turns out that the Fejér kernel is an approximate identity, whereas the Dirichlet is not. The procedure of taking the first arithmetic means is known as Cesàro summation of the Fourier series of f. There are many other methods of summation, the kernels of which form approximate identities and which therefore sum Fourier series effectively. Some specific examples are provided by the Abel, Weierstrass, Riemann, and Riesz methods. Thus convolution is a vital concept in summability problems of the theory of divergent series.

A second connection is concerned with initial and boundary value problems in the theory of partial differential equations. Thus the solution of Dirichlet's boundary value problem for the unit disc is a singular integral having as approximate identity the Abel–Poisson kernel $\{p_r(x)\}$. It was FATOU who showed that if $f'(x_0)$ exists, then the derivative $(f * p_r)'(x_0)$ of the solution converges to $f'(x_0)$ when $r \to 1-$.

One of the vital problems of approximation theory is to estimate the error, or discrepancy, $(f * \chi_n)(x) - f(x)$ under varying hypothesis upon f and χ_n. As a rule, the smoother the function, the faster the error tends to zero for $n \to \infty$. These are direct theorems of approximation theory. Conversely, inverse theorems infer smoothness properties of f from the smallness of the error $(f * \chi_n)(x) - f(x)$. The smoothness properties upon f are usually given in terms of differentiability or Lipschitz properties. One speaks of an equivalence theorem in case the direct and inverse theorems are exact converses of each other.

These problems lead to the study of the classical theory of approximation of periodic functions by trigonometric polynomials, as associated with the names of D. JACKSON and S. N. BERNSTEIN. At this stage, the fundamental concept of best approximation of a periodic function by polynomials comes into play. The Jackson result is a direct theorem; the kernels that are used in its proofs are positive approximate identities.

Not only do singular convolution integrals deserve study on the circle group, but also on the line group. Given $f \in \mathsf{L}^p$, $1 \le p < \infty$, such an integral has the form

$$I(f; x; n) = \frac{1}{\sqrt{2\pi}} \int_{-\infty}^{\infty} f(x-u)\chi(u; n)\, du$$

with kernel $\{\chi(x; n)\}$ satisfying $\chi(\circ; n) \in \mathsf{L}^1$, $\int_{-\infty}^{\infty} \chi(u; n)\, du = \sqrt{2\pi}$ for each $n \in \mathbb{N}$. If, in addition, $\|\chi(\circ; n)\|_1 \le M$ for all n and $\lim_{n\to\infty} \int_{\delta \le |u|} |\chi(u; n)|\, du = 0$ for each $\delta > 0$, then $\{\chi(x; n)\}$ is called an approximate identity. However, it is important that detailed study can be restricted to the case $\chi(x; n) = n\chi(nx)$ with $\chi \in \mathsf{L}^1$, $\int_{-\infty}^{\infty} \chi(u)\, du = \sqrt{2\pi}$; in other words, $\chi(x; n)$ is generated by a function $\chi(x)$ of one variable through a simple scale change. These kernels are said to be of Fejér's type; correspondingly, the convolution integral

$$J(f; x; n) = \frac{n}{\sqrt{2\pi}} \int_{-\infty}^{\infty} f(x - u)\chi(nu)\, du.$$

It is basic that every kernel of Fejér's type is an approximate identity, and our remarks here are confined to them.

As for the circle group, the graph $y = n\chi(nu)$ for $n \to \infty$ reveals that $\{n\chi(nu)\}$ is a peaking kernel approaching the (Dirac) delta measure. Thus $J(f; x; n)$ tends in L^p-norm to f as $n \to \infty$, and one is interested in examining direct, inverse, and equivalence theorems for the approximation of f by $J(f; x; n)$. These problems again include a number of exciting questions. As an example we may mention the limiting behaviour of solutions of the heat equation for an infinite rod. Indeed, its elementary solution is Green's function $G(x, t) = (2t)^{-1/2} \exp\{-x^2/4t\}$, and, according to the generalized superposition principle, the general solution is then given by the convolution of $G(\circ, t)$ with f, f being the initial temperature distribution.

As we are dealing with integrals of convolution type, it is to be expected that Fourier transforms enter into the discussion. Indeed, the powerful techniques of Fourier analysis will be fully exploited. The Fourier transform of the (convolution) product $J(f; x; n)$ is the (pointwise) product of the transforms of f and $n\chi(n\,\circ)$, i.e. $f^\wedge(v)\chi^\wedge(v/n)$, thus separating the function f and the kernel. For large n the transform $\chi^\wedge(v/n)$ is close to $\chi^\wedge(0)$ which takes on the value 1 in view of the normalization of χ; on the other hand, the Fourier transform of the delta measure is equal to 1. Consequently, on the basis of the transform methods to be applied in our later approximation theoretical investigations, the behaviour of the transform $\chi^\wedge(v)$ near $v = 0$ will be the decisive one.

Furthermore, if f has period 2π and $\chi^\wedge(v) = 0$ for $|v| \geq 1$ (which implies that the transform of $n\chi(n\,\circ)$ vanishes for $|v| \geq n$), then $J(f; x; n)$ has Fourier coefficients which vanish for $|k| \geq n$, and is therefore a trigonometric polynomial of degree less than n. This gives the connection with the approximation of periodic functions by trigonometric polynomials. It is supplemented by the fact that every kernel χ of Fejér's type generates a periodic approximate identity via

$$\chi_n^*(x) = \sqrt{2\pi} \sum_{k=-\infty}^{\infty} n\chi(n(x + 2k\pi)).$$

For $f \in \mathsf{C}_{2\pi}$ it then follows that

$$\frac{1}{2\pi} \int_{-\pi}^{\pi} f(x - u)\chi_n^*(u)\, du = \frac{n}{\sqrt{2\pi}} \int_{-\infty}^{\infty} f(x - u)\chi(nu)\, du.$$

Since many important examples of periodic kernels are generated by some $\chi \in \mathsf{L}^1$, we can, for the corresponding singular integral of periodic functions, take our choice of either expression. The right-hand one has the advantage that the kernel is determined by scale change from one single generating function χ, whereas the periodic kernel $\{\chi_n^*(x)\}$ may have a more complicated functional dependence on the parameter n (though the Fourier coefficients of χ_n^* are given by $\chi^\wedge(k/n)$).

Having outlined some of the problems that are to be treated in Part I, we may conclude the introduction with a brief but more systematic outline of the individual chapters. Chapter 1 is exclusively concerned with singular integrals on the circle group, their fundamental properties, norm and pointwise convergence, direct approximation theorems, and asymptotic expansions. Included is a section on the classical theory of Fourier series. Chapter 2 is devoted to the theorems of Jackson and Bernstein for polynomials of best approximation. The basic properties of such polynomials, in particular their existence, are considered briefly. The chapter concludes with inverse theorems for convolution integrals which need not necessarily be polynomial summation processes nor have orders of approximation as good as polynomials of best approximation. Chapter 3 is devoted to a detailed study of singular integrals on the line group. In many respects, the treatment is parallel to that for the circle. On the other hand, it complements the periodic results by those of Sec. 3.1.2. The theory of Fourier transforms is only introduced in Part II and subsequently applied.

1

Singular Integrals of Periodic Functions

1.0 Introduction

Following our discussion of the purpose of the study of singular integrals in the introduction to Part I, the scope of this chapter may now be outlined briefly.

Sec. 1.1 deals with basic properties of singular integrals such as their convergence in the norms of the spaces $C_{2\pi}$, $L^p_{2\pi}$, $1 \leq p < \infty$. Convolution integrals of type $\chi_\rho * d\mu$, $\mu \in BV_{2\pi}$, are examined and a short discussion on strong and weak derivatives is included. In Sec. 1.2 the emphasis is upon the fundamental facts of trigonometric and Fourier series, summability of Fourier series, the Fejér and Abel–Poisson means. Summability of conjugate series is considered and Fourier–Stieltjes series are introduced. Sec. 1.3 is concerned with necessary and sufficient conditions assuring norm-convergence of the singular integral $I_\rho(f; x)$ towards f for all $f \in X_{2\pi}$. The fundamental Banach–Steinhaus theorem delivers such conditions in case of $C_{2\pi}$ or $L^1_{2\pi}$-space. For this purpose, the norms of convolution operators are determined in Sec. 1.3.1. In case the associated kernels are positive, an interesting theorem of Bohman–Korovkin (1952/53) states that norm-convergence takes place for all $f \in X_{2\pi}$ if and only if it does so for the two particular test functions $\cos x$ and $\sin x$.

In Sec. 1.4 pointwise convergence of singular integrals is studied, various hypotheses upon the kernel being taken into account; also the convergence almost everywhere of $\chi_\rho * d\mu$ is examined. An application to the Abel–Poisson integral leads to the theorem of Fatou.

Sec. 1.5 is concerned with questions on the order of approximation of periodic functions by positive singular integrals. Prop. 1.5.10 is a recast of the Bohman–Korovkin theorem in a quantitative form: the rapidity of convergence of $I_\rho(f; x)$ to $f(x)$ is estimated in terms of the rapidities of convergence of $I_\rho(\cos u; x)$ to $\cos x$, $I_\rho(\sin u; x)$ to $\sin x$. Subsection 1.5.4 is devoted to asymptotic expansions of positive singular integrals. Sec. 1.6 treats direct approximation theorems with applications to the integrals of Fejér–Korovkin and Fejér. There is also a short discussion on the best asymptotic or Nikolskiĭ constants for the measure of approximation. Sec. 1.7 is devoted to simple inverse small-o approximation theorems for singular integrals generated by row-finite θ-factors. Finally, there is a result by KOROVKIN on the critical order of approximation by positive integrals of this type.

1.1 Norm-Convergence and -Derivatives

1.1.1 Norm-Convergence

Let us begin with the definition of a kernel which is to generate our singular integrals.

Definition 1.1.1. *Let ρ be a parameter ranging over some set $\mathbb{A}$ which is either an interval (a, b) with $0 \le a < b \le +\infty$ or the set $\mathbb{N}$, and let ρ_0 be one of the points a, b or $+\infty$. A set of functions $\{\chi_\rho(x)\}$ will be called a (periodic)* ***kernel*** *if $\chi_\rho \in \mathsf{L}^1_{2\pi}$ for each $\rho \in \mathbb{A}$ and*

$$\int_{-\pi}^{\pi} \chi_\rho(u)\, du = 2\pi \qquad (\rho \in \mathbb{A}). \tag{1.1.1}$$

We call the kernel $\{\chi_\rho(x)\}$ ***real*** *if $\chi_\rho(x)$ is a real function,* ***bounded*** *if $\chi_\rho \in \mathsf{L}^\infty_{2\pi}$,* ***continuous*** *if $\chi_\rho(x) \in \mathsf{C}_{2\pi}$, and* ***absolutely continuous*** *if $\chi_\rho(x)$ is absolutely continuous for each $\rho \in \mathbb{A}$. A real kernel $\{\chi_\rho(x)\}$ is said to be* ***even*** *if $\chi_\rho(x) = \chi_\rho(-x)$ a.e.,* ***positive*** *if $\chi_\rho(x) \ge 0$ a.e. for each $\rho \in \mathbb{A}$.*

Instead of condition (1.1.1) one often assumes that

$$\lim_{\rho \to \rho_0} \int_{-\pi}^{\pi} \chi_\rho(u)\, du = 2\pi. \tag{1.1.2}$$

But there is no loss of generality to suppose that the functions $\chi_\rho(x)$ are normalized by (1.1.1) from the beginning.

Definition 1.1.2. *Let $f \in \mathsf{X}_{2\pi}$ and $\{\chi_\rho(x)\}$ be a kernel. Then we call an expression of the form*

$$I_\rho(f; x) = (f * \chi_\rho)(x) = \frac{1}{2\pi} \int_{-\pi}^{\pi} f(x - u) \chi_\rho(u)\, du \tag{1.1.3}$$

a (periodic) ***singular integral*** *(or convolution integral). We say that the singular integral is* ***positive, continuous*** *if the corresponding kernel is positive, continuous.*

Thus to each $f \in \mathsf{X}_{2\pi}$ we associate a set of functions $\{I_\rho(f; x)\}$ for which we have in view of Prop. 0.4.1

Proposition 1.1.3. *If $f \in \mathsf{X}_{2\pi}$ and $\{\chi_\rho(x)\}$ is a kernel, then $I_\rho(f; x) \in \mathsf{X}_{2\pi}$ for each $\rho \in \mathbb{A}$ and*

$$\|I_\rho(f; \circ)\|_{\mathsf{X}_{2\pi}} \le \|\chi_\rho\|_1 \|f\|_{\mathsf{X}_{2\pi}}. \tag{1.1.4}$$

Moreover, if the kernel is bounded, then $I_\rho(f; x)$ is a continuous function of x, thus belongs to $\mathsf{C}_{2\pi}$.

Setting $I_\rho(f; x) = [I_\rho f](x)$ for $\rho \in \mathbb{A}$, the integrals (1.1.3) define bounded linear transformations I_ρ of $X_{2\pi}$ into $X_{2\pi}$, determined by the kernel $\{\chi_\rho(x)\}$.

Of fundamental importance are theorems guaranteeing the convergence of $I_\rho(f; x)$ towards a given f as $\rho \to \rho_0$. In this respect we are mainly interested in theorems on norm-convergence, in other words, in theorems giving the strong convergence of the operators I_ρ towards the identity operator in $X_{2\pi}$. Assumptions assuring this fact lead one to the following definition of an approximate identity.

Definition 1.1.4. *A kernel $\{\chi_\rho(x)\}$ is called a (periodic)* ***approximate identity*** *if, with some constant $M > 0$,*

$$\|\chi_\rho\|_1 \le M \qquad (\rho \in \mathbb{A}), \tag{1.1.5}$$

$$\lim_{\rho\to\rho_0} \int_{\delta\le|u|\le\pi} |\chi_\rho(u)|\, du = 0 \qquad (0 < \delta < \pi). \tag{1.1.6}$$

We call an approximate identity ***even, positive, bounded*** *or* ***continuous*** *if the kernel is even, positive, bounded or continuous.*

Before proceeding to the main convergence theorem let us mention that, instead of (1.1.6), we shall often assume the property

$$\lim_{\rho\to\rho_0} \Big[\sup_{\delta\le|u|\le\pi} |\chi_\rho(u)|\Big] = 0 \qquad (0 < \delta < \pi), \tag{1.1.7}$$

which certainly implies (1.1.6). If the kernel is positive, then obviously (1.1.1) already implies (1.1.5) with $M = 1$.

Theorem 1.1.5. *If the kernel of the integral* (1.1.3) *is an approximate identity, then for every $f \in X_{2\pi}$*

$$\lim_{\rho\to\rho_0} \|I_\rho(f; \circ) - f(\circ)\|_{X_{2\pi}} = 0. \tag{1.1.8}$$

Proof. We first of all note that in view of (1.1.4) and (1.1.5) the singular integral (1.1.3) defines a set of uniformly bounded (with respect to $\rho \in \mathbb{A}$) operators on $X_{2\pi}$. Since by (1.1.1)

$$I_\rho(f; x) - f(x) = \frac{1}{2\pi}\int_{-\pi}^{\pi} [f(x-u) - f(x)]\chi_\rho(u)\, du, \tag{1.1.9}$$

we obtain, in case $X_{2\pi} = C_{2\pi}$, for any $0 < \delta < \pi$

$$\|I_\rho(f; \circ) - f(\circ)\|_{X_{2\pi}} \le \frac{1}{2\pi}\int_{-\pi}^{\pi} \|f(\circ - u) - f(\circ)\|_{X_{2\pi}} |\chi_\rho(u)|\, du \tag{1.1.10}$$

$$= \frac{1}{2\pi}\Bigg(\int_{|u|\le\delta} + \int_{\delta\le|u|\le\pi}\Bigg) \|f(\circ - u) - f(\circ)\|_{X_{2\pi}} |\chi_\rho(u)|\, du \equiv I_1 + I_2,$$

say. Since f is uniformly continuous, to each $\varepsilon > 0$ there is a $\delta > 0$ such that $\|f(\circ - u) - f(\circ)\|_{X_{2\pi}} \le \varepsilon$ for all $|u| \le \delta$. This implies $I_1 \le \varepsilon M$. Now take δ fixed. Then

$$I_2 \le 2\,\|f\|_{X_{2\pi}} \frac{1}{2\pi} \int_{\delta\le|u|\le\pi} |\chi_\rho(u)|\, du,$$

which tends to zero as $\rho \to \rho_0$ according to (1.1.6). Thus (1.1.8) holds in case $X_{2\pi} = C_{2\pi}$. If $X_{2\pi} = L^p_{2\pi}$, $1 \le p < \infty$, we proceed by the Hölder–Minkowski inequality to deduce (1.1.10). Since f is continuous in the mean (compare Sec. 0.4), the proof follows as before.

If we replace $X_{2\pi}$ by $L^\infty_{2\pi}$, (1.1.8) need no longer hold since f is not necessarily continuous in this norm. Then we have

Proposition 1.1.6. *Let $f \in L^\infty_{2\pi}$. If the kernel of the integral* (1.1.3) *is an approximate identity, then for every $s \in L^1_{2\pi}$*

$$\lim_{\rho\to\rho_0} \int_{-\pi}^{\pi} [I_\rho(f; x) - f(x)]s(x)\, dx = 0. \tag{1.1.11}$$

Proof. Again (1.1.9) holds, and hence by Fubini's theorem for every $s \in L^1_{2\pi}$ and $0 < \delta < \pi$

$$\left|\int_{-\pi}^{\pi} [I_\rho(f; x) - f(x)]s(x)\, dx\right| = \left|\frac{1}{2\pi}\int_{-\pi}^{\pi} \chi_\rho(u)\, du\left\{\int_{-\pi}^{\pi} [f(x-u) - f(x)]s(x)\, dx\right\}\right|$$
$$\le \frac{1}{2\pi}\left(\int_{|u|\le\delta} + \int_{\delta\le|u|\le\pi}\right)|\chi_\rho(u)|\, du \left|\int_{-\pi}^{\pi} f(x)[s(x+u) - s(x)]\, dx\right| \equiv I_1 + I_2,$$

say. But since $s \in L^1_{2\pi}$ is continuous in the mean, for each $\varepsilon > 0$ there is a δ such that $\|s(\circ + u) - s(\circ)\|_1 \le \varepsilon$ for all $|u| \le \delta$. Thus

$$I_1 \le M \sup_{|u|\le\delta} \left|\int_{-\pi}^{\pi} f(x)[s(x+u) - s(x)]\, dx\right| \le M\|f\|_\infty 2\pi\varepsilon.$$

Furthermore,

$$I_2 \le 2\, \|f\|_\infty \|s\|_1 \int_{\delta\le|u|\le\pi} |\chi_\rho(u)|\, du,$$

which proves (1.1.11) in view of (1.1.6).

Analogously to (1.1.3) one may also assign to each function $\mu \in BV_{2\pi}$ a singular integral

$$I_\rho(d\mu; x) = (\chi_\rho * d\mu)(x) = \frac{1}{2\pi}\int_{-\pi}^{\pi} \chi_\rho(x-u)\, d\mu(u) \tag{1.1.12}$$

for which the following proposition holds.

Proposition 1.1.7. *Let $\mu \in BV_{2\pi}$ and $\{\chi_\rho(x)\}$ be a kernel. Then $I_\rho(d\mu; x) \in L^1_{2\pi}$ for each $\rho \in \mathbb{A}$ and*

$$\|I(d\mu; \circ)\|_1 \le \|\chi_\rho\|_1 \|\mu\|_{BV_{2\pi}}. \tag{1.1.13}$$

If the kernel $\{\chi_\rho(x)\}$ is moreover an approximate identity, then for every $s \in C_{2\pi}$

$$\lim_{\rho\to\rho_0} \int_{-\pi}^{\pi} s(x) I_\rho(d\mu; x)\, dx = \int_{-\pi}^{\pi} s(x)\, d\mu(x). \tag{1.1.14}$$

Proof. (1.1.13) follows by Prop. 0.6.3. To establish (1.1.14) we have by Fubini's theorem

$$\int_{-\pi}^{\pi} s(x) I_\rho(d\mu; x)\, dx = \int_{-\pi}^{\pi} d\mu(u)\left\{\frac{1}{2\pi}\int_{-\pi}^{\pi} s(u+x)\chi_\rho(x)\, dx\right\}.$$

Since the expression in curly brackets tends uniformly to $s(u)$ by Theorem 1.1.5, (1.1.14) follows.

Let us conclude with an example. The (first) *integral means* (also called moving average or Steklov function, or last not least, singular integral of Riemann–Lebesgue) are defined by

$$A_h(f; x) = \frac{1}{h}\int_{x-(h/2)}^{x+(h/2)} f(u)\,du = \frac{1}{h}\int_{-h/2}^{h/2} f(x-u)\,du. \tag{1.1.15}$$

Here h is a positive parameter ranging over $\mathbb{A} = (0, 2\pi)$ and tending to $0+$. The integral (1.1.15) is of type (1.1.3). Indeed, setting $\rho = h$, the kernel $\{\chi_\rho(x)\}$ is given as the set of those 2π-periodic functions which, for each fixed h, coincide on $[-\pi, \pi)$ with $(2\pi/h)\kappa_{[-h/2,h/2]}(x)$ (recall definition (0.3.5) of characteristic functions). Obviously, this kernel is an even, positive, and bounded approximate identity. Therefore by Prop. 1.1.3, Theorem 1.1.5

Corollary 1.1.8. *Let $f \in \mathsf{X}_{2\pi}$. Then $A_h(f; x)$ belongs to $\mathsf{X}_{2\pi} \cap \mathsf{C}_{2\pi}$ such that $\|A_h(f; \circ)\|_{\mathsf{X}_{2\pi}} \le \|f\|_{\mathsf{X}_{2\pi}}$ for all $h \in (0, 2\pi)$ and*

$$\lim_{h\to 0+} \|A_h(f; \circ) - f(\circ)\|_{\mathsf{X}_{2\pi}} = 0.$$

Thus every integrable function f can be approximated in the mean arbitrary closely by continuous functions, namely by the $A_h(f; x)$. These are furthermore differentiable almost everywhere as stated by Prop. 0.3.1.

Further examples of singular integrals will be submitted during the course of the subsequent sections, see in particular Sec. 1.2.

1.1.2 Derivatives

While Theorem 1.1.5 was concerned with norm-convergence of singular integrals towards f, we shall now discuss approximation of derivatives. By the way, we shall illustrate our earlier remarks regarding convolutions as smoothness increasing operations. Roughly speaking, we shall show that if either factor is differentiable, so is the convolution product. To this end, let us first introduce some classes of functions.

In what follows, r always denotes a natural number. We set

$$\mathsf{W}^r_{\mathsf{X}_{2\pi}} = \begin{cases}\{f \in \mathsf{C}_{2\pi} \mid f \in \mathsf{C}^r_{2\pi}\}\\ \{f \in \mathsf{L}^p_{2\pi} \mid f = \phi \text{ a.e.}, \phi \in \mathsf{AC}^{r-1}_{2\pi}, \phi^{(r)} \in \mathsf{L}^p_{2\pi}\} \qquad (1 \le p < \infty).\end{cases} \tag{1.1.16}$$

Although the derivatives in this definition are to be taken in the pointwise sense, the fact that all derivatives occurring belong to the underlying space $\mathsf{X}_{2\pi}$ actually implies that the classes $\mathsf{W}^r_{\mathsf{X}_{2\pi}}$ may be precisely described through strong derivatives. To develop this point, we commence with

Definition 1.1.9. *If for $f \in \mathsf{X}_{2\pi}$ there exists $g \in \mathsf{X}_{2\pi}$ such that*

$$\lim_{h\to 0} \|h^{-1}[f(\circ + h) - f(\circ)] - g(\circ)\|_{\mathsf{X}_{2\pi}} = 0,$$

then g is called the uniform derivative of f if $X_{2\pi} = C_{2\pi}$, *and the derivative of f in the mean of order p if* $X_{2\pi} = L^p_{2\pi}$, $1 \le p < \infty$. *In short, we shall speak of these derivatives as the (first ordinary) derivative in* $X_{2\pi}$*-norm or* ***strong derivative***, *and denote g by* $D_s^{(1)}f$. *For any* $r \in \mathbb{N}$, *the rth strong*† *derivative of* $f \in X_{2\pi}$ *is then defined successively by* $D_s^{(r)}f = D_s^{(1)}(D_s^{(r-1)}f)$.

As an immediate consequence we state

Proposition 1.1.10. $f \in W^r_{X_{2\pi}}$ *implies the existence of the rth strong derivative* $D_s^{(r)}f$ *of f. If* $X_{2\pi} = C_{2\pi}$, *then* $(D_s^{(r)}f)(x) = f^{(r)}(x)$ *for all x, and if* $X_{2\pi} = L^p_{2\pi}$, $1 \le p < \infty$, *then* $(D_s^{(r)}f)(x) = \phi^{(r)}(x)$ a.e., *where* $\phi \in AC^{r-1}_{2\pi}$ *with* $\phi^{(r)} \in L^p_{2\pi}$ *is such that* $f(x) = \phi(x)$ a.e.

Proof. Let $X_{2\pi} = L^p_{2\pi}$, $1 \le p < \infty$. The assumption assures that for each $h \in \mathbb{R}$

$$\frac{f(x+h)-f(x)}{h} - \phi'(x) = \frac{1}{h}\int_0^h [\phi'(x+u) - \phi'(x)]\,du \quad \text{a.e.} \tag{1.1.17}$$

Therefore by the Hölder–Minkowski inequality and the continuity of ϕ' in the mean

$$\left\|\frac{f(\circ+h)-f(\circ)}{h} - \phi'(\circ)\right\|_p \le \int_0^1 \|\phi'(\circ+hu) - \phi'(\circ)\|_p\,du = o(1) \quad (h\to 0).$$

Thus the first strong derivative of f exists, in fact $D_s^{(1)}f = \phi'$. The proof is now completed by an obvious induction. Indeed, let $k < r$, and suppose that the kth strong derivative of f exists with $D_s^{(k)}f = \phi^{(k)}$. Since $\phi^{(k)}$ is absolutely continuous, for each $h \in \mathbb{R}$

$$\frac{(D_s^{(k)}f)(x+h) - (D_s^{(k)}f)(x)}{h} - \phi^{(k+1)}(x) = \frac{1}{h}\int_0^h [\phi^{(k+1)}(x+u) - \phi^{(k+1)}(x)]\,du \quad \text{a.e.}$$

This implies by the continuity of $\phi^{(k+1)}$ in $L^p_{2\pi}$-norm that

$$\left\|\frac{(D_s^{(k)}f)(\circ+h) - (D_s^{(k)}f)(\circ)}{h} - \phi^{(k+1)}(\circ)\right\|_p \le \int_0^1 \|\phi^{(k+1)}(\circ+hu) - \phi^{(k+1)}(\circ)\|_p\,du = o(1) \quad (h\to 0),$$

establishing the existence of the $(k+1)$th strong derivative of f and $D_s^{(k+1)}f = \phi^{(k+1)}$.

In $C_{2\pi}$-space the proof is similar, but more elementary.

In Sec. 10.1.3 we shall show that also the converse of Prop. 1.1.10 is valid. Surprisingly, this is even true for the (seemingly) more general concept of a weak derivative.

Definition 1.1.11. *If for* $f \in X_{2\pi}$ *there exists* $g \in X_{2\pi}$ *such that*

$$\lim_{h\to 0}\int_{-\pi}^{\pi} \frac{f(x+h)-f(x)}{h}\,d\eta(x) = \int_{-\pi}^{\pi} g(x)\,d\eta(x)$$

† Strong derivatives are denoted by $D_s^{(r)}f$ to distinguish them from usual pointwise derivatives $f^{(r)}$. The subscript s stands for strong.

for every $\eta \in \mathsf{BV}_{2\pi}$ *if* $\mathsf{X}_{2\pi} = \mathsf{C}_{2\pi}$, *and*

$$\lim_{h\to 0} \int_{-\pi}^{\pi} \frac{f(x+h)-f(x)}{h} s(x)\,dx = \int_{-\pi}^{\pi} g(x)s(x)\,dx$$

for every $s \in \mathsf{L}_{2\pi}^{p'}$ *if* $\mathsf{X}_{2\pi} = \mathsf{L}_{2\pi}^{p}$, $1 \le p < \infty$, *then g is called the (first ordinary)* ***weak derivative*** *of f, and denoted by* $D_w^{(1)}f$. *For any* $r \in \mathbb{N}$, *the rth weak derivative of* $f \in \mathsf{X}_{2\pi}$ *is defined successively by* $D_w^{(r)}f = D_w^{(1)}(D_w^{(r-1)}f)$.

As is to be expected, one has immediately

Proposition 1.1.12. *If the rth strong derivative of* $f \in \mathsf{X}_{2\pi}$ *exists, so does the rth weak derivative. In fact, we then have* $D_w^{(r)}f = D_s^{(r)}f$.

As a consequence of the preceding propositions we have that, if $f \in \mathsf{W}_{\mathsf{X}_{2\pi}}^r$, then the rth weak derivative of f exists and $(D_w^{(r)}f)(x) = f^{(r)}(x)$ if $\mathsf{X}_{2\pi} = \mathsf{C}_{2\pi}$, for example. For the converse see Sec. 10.1.3.

For purposes of illustration let us here state the following result which may be regarded as a fundamental theorem of the calculus involving strong and weak derivatives.

Proposition 1.1.13. *Let* $f \in \mathsf{X}_{2\pi}$. *If the first weak derivative of f exists and is zero, then* $f(x) =$ const (a.e.).

For a proof as well as for further results, including strong and weak generalized derivatives, see Sec. 10.1.

We now come to the actual problem of this subsection. However, instead of treating derivatives of the convolution of two arbitrary functions (to be found in Problem 1.1.8), we shall immediately turn to singular integrals (1.1.3), i.e., to convolutions for which one factor is a kernel.

Proposition 1.1.14. *Let* $f \in \mathsf{X}_{2\pi}$. *If the kernel* $\{\chi_\rho(x)\}$ *of the integral* $I_\rho(f; x)$ *is continuous such that* $\chi_\rho \in \mathsf{W}_{\mathsf{C}_{2\pi}}^r$ *for each* $\rho \in \mathbb{A}$, *then* $I_\rho(f; x)$ *is an r-times continuously differentiable function of x. In particular, for every x and* $\rho \in \mathbb{A}$

$$[I_\rho(f; \circ)]^{(r)}(x) = \frac{1}{2\pi} \int_{-\pi}^{\pi} f(x-u)\chi_\rho^{(r)}(u)\,du. \tag{1.1.18}$$

Proof. Let $r = 1$. Since $\chi_\rho \in \mathsf{C}_{2\pi}$, it follows that $\chi_\rho \in \mathsf{L}_{2\pi}^q$ for every $1 \le q \le \infty$. Therefore, for each $\rho \in \mathbb{A}$, $I_\rho(f; x)$ exists everywhere as a function in $\mathsf{C}_{2\pi}$ by Prop. 0.4.1. Moreover, since $\chi_\rho' \in \mathsf{C}_{2\pi}$, also $(f * \chi_\rho')(x)$ exists everywhere, and hence for arbitrary $h \in \mathbb{R}$

$$\frac{I_\rho(f; x+h) - I_\rho(f; x)}{h} - (f * \chi_\rho')(x)$$
$$= \frac{1}{2\pi} \int_{-\pi}^{\pi} f(u) \left[\frac{\chi_\rho(x+h-u) - \chi_\rho(x-u)}{h} - \chi_\rho'(x-u)\right] du.$$

Since $f \in \mathsf{X}_{2\pi}$ implies $f \in \mathsf{L}_{2\pi}^1$, we have

$$\left|\frac{I_\rho(f; x+h) - I_\rho(f; x)}{h} - \frac{1}{2\pi} \int_{-\pi}^{\pi} f(x-u)\chi_\rho'(u)\,du\right|$$
$$\le \|f\|_1 \left\| \frac{\chi_\rho(\circ + h) - \chi_\rho(\circ)}{h} - \chi_\rho'(\circ) \right\|_{\mathsf{C}_{2\pi}}$$

for every x. As $\chi_\rho \in \mathsf{W}^1_{\mathsf{C}_{2\pi}}$, the first uniform derivative of χ_ρ exists by Prop. 1.1.10. Thus it follows that $[I_\rho(f; \circ)]'(x)$ exists at every x and that (1.1.18) is valid for $r = 1$. If $r \in \mathbb{N}$ is arbitrary, the proof is completed by a straightforward induction.

Proposition 1.1.15. *Let $\{\chi_\rho(x)\}$ be the kernel of the integral $I_\rho(f; x)$. If $f \in \mathsf{W}^r_{\mathsf{C}_{2\pi}}$, then $I_\rho(f; x) \in \mathsf{W}^r_{\mathsf{C}_{2\pi}}$ for each $\rho \in \mathbb{A}$ and*

$$(1.1.19) \qquad [I_\rho(f; \circ)]^{(r)}(x) = I_\rho(f^{(r)}; x)$$

for every x. Moreover, if $\{\chi_\rho(x)\}$ is an approximate identity, then

$$(1.1.20) \qquad \lim_{\rho\to\rho_0} \|[I_\rho(f; \circ)]^{(r)}(\circ) - f^{(r)}(\circ)\|_{\mathsf{C}_{2\pi}} = 0.$$

Proof. Again we restrict ourselves to the case $r = 1$. We may proceed as in the proof of the preceding proposition to deduce

$$\left|\frac{I_\rho(f; x + h) - I_\rho(f; x)}{h} - (f' * \chi_\rho)(x)\right| \le \|\chi_\rho\|_1 \left\|\frac{f(\circ + h) - f(\circ)}{h} - f'(\circ)\right\|_{\mathsf{C}_{2\pi}}.$$

This proves (1.1.19), the rôles of f and χ_ρ being interchanged in comparison with (1.1.18). If $\{\chi_\rho(x)\}$ is an approximate identity, (1.1.20) now follows by Theorem 1.1.5 as applied to $f^{(r)}$.

Proposition 1.1.16. *Let the kernel $\{\chi_\rho(x)\}$ of the integral $I_\rho(f; x)$ be bounded. If $f \in \mathsf{W}^r_{\mathsf{L}^p_{2\pi}}$, $1 \le p < \infty$, then $I_\rho(f; x)$ is an r-times continuously differentiable function of x, and for every x and $\rho \in \mathbb{A}$*

$$(1.1.21) \qquad [I_\rho(f; \circ)]^{(r)}(x) = I_\rho(D_s^{(r)}f; x) = I_\rho(\phi^{(r)}; x),$$

where $\phi \in \mathsf{AC}^{r-1}_{2\pi}$ with $\phi^{(r)} \in \mathsf{L}^p_{2\pi}$ is such that $\phi(x) = f(x)$ a.e. Moreover, if $\{\chi_\rho(x)\}$ is an approximate identity, then

$$(1.1.22) \qquad \lim_{\rho\to\rho_0} \|[I_\rho(f; \circ)]^{(r)}(\circ) - (D_s^{(r)}f)(\circ)\|_p = 0.$$

Proof. Let $r = 1$. Since $\chi_\rho \in \mathsf{L}^\infty_{2\pi}$ implies $\chi_\rho \in \mathsf{L}^q_{2\pi}$ for every $1 \le q \le \infty$, the convolution $I_\rho(f; x)$ exists everywhere as a function in $\mathsf{C}_{2\pi}$ by Prop. 0.4.1. Furthermore, $D_s^{(1)}f$ exists by Prop. 1.1.10, and hence by Hölder's inequality

$$\left|\frac{I_\rho(f; x + h) - I_\rho(f; x)}{h} - I_\rho(D_s^{(1)}f; x)\right| \le \|\chi_\rho\|_{p'} \left\|\frac{f(\circ + h) - f(\circ)}{h} - (D_s^{(1)}f)(\circ)\right\|_p$$

for every x and $\rho \in \mathbb{A}$. Since the right-hand side tends to zero as $h \to 0$ by the definition of a strong derivative, (1.1.21) follows for $r = 1$. An obvious induction delivers the general case. Finally, Theorem 1.1.5 as applied to $D_s^{(r)}f$ gives (1.1.22).

Certainly, further results on derivatives of convolutions may be obtained by varying and even weakening the hypotheses (see also the Problems of this section). However, the above three propositions suffice in our later investigations on approximation by convolution integrals $I_\rho(f; x)$. They nevertheless illustrate the fact that a convolution is differentiable if either of the factors is differentiable. Here differentiability is actually taken in the strong sense. This enables one to use Hölder's inequality, thus avoiding the (sometimes more delicate) criteria for the interchange of integration and differentiation (with respect to a parameter) known from real variable theory. Prop. 1.1.14 is applicable to many of the examples of singular integrals that are to follow. Indeed, one reason for the interest in the approximation of f by specific integrals $I_\rho(f; x)$ is that the approximating functions $I_\rho(f; x)$ are smoother than f. This is achieved by the fact that many kernels belong to $C_{2\pi}^\infty$. Therefore we assumed χ_ρ to be continuous in Prop. 1.1.14. But the kernel of the integral means (1.1.15) is not continuous; here we have Problem 1.1.7(ii) which depends upon the particular structure of the integral $A_h(f; x)$. On the other hand, we may apply Prop. 1.1.15, 1.1.16, provided $f \in W_{X_{2\pi}}^r$.

Since the kernel of the integral means $A_h(f; x)$ is a bounded approximate identity, we have

Corollary 1.1.17. *Let $f \in W_{X_{2\pi}}^r$. Then $A_h(f; x)$ is an r-times continuously differentiable function of x and $[A_h(f; \circ)]^{(r)}(x) = A_h(D_s^{(r)}f; x)$ for every x and $h \in (0, 2\pi)$. Moreover,*

$$\lim_{h\to 0+} \|[A_h(f; \circ)]^r(\circ) - (D_s^{(r)}f)(\circ)\|_{X_{2\pi}} = 0.$$

Problems

1. (i) Let $\{\chi_\rho(x)\}$ be a kernel. Show that $\int_{-\pi}^{\pi} I_\rho(f; x)\, dx = \int_{-\pi}^{\pi} f(x)\, dx$ for every $f \in X_{2\pi}$ and $\rho \in \mathbb{A}$.
   (ii) Show that if $\{\chi_\rho(x)\}$ is a kernel [approximate identity], so is $\{\chi_\rho(-x)\}$.
2. Prove Theorem 1.1.5 for approximate identities satisfying (1.1.2) instead of (1.1.1) (compare also EDWARDS [1I, p. 60]).
3. Let $\{f_n\}_{n=1}^\infty$ be a sequence of functions in $X_{2\pi}$ which are bounded and continuous in $X_{2\pi}$-norm, uniformly with respect to $n \in \mathbb{N}$. If $\{\chi_\rho(x)\}$ is an approximate identity, show that $\lim_{\rho\to\rho_0} \|I_\rho(f_n; \circ) - f_n(\circ)\|_{X_{2\pi}} = 0$ uniformly for $n \in \mathbb{N}$.
4. (i) Let $\{\chi_\rho(x)\}$ be a kernel. Show that $\{(\chi_\rho * \chi_\rho)(x)\}$ is again a kernel, known as the *iterated kernel.*
   (ii) Let $\{\chi_\rho(x)\}$ be an approximate identity. Show that the iterated kernel again has the property (1.1.5). What can one say concerning (1.1.6)?
   (iii) Let the *iterated singular integral* be defined by

   $$I_\rho^2(f; x) = \frac{1}{2\pi}\int_{-\pi}^{\pi} f(x-u)(\chi_\rho * \chi_\rho)(u)\, du. \tag{1.1.23}$$

   If $\{\chi_\rho(x)\}$ is an approximate identity, show that $I_\rho^2(f; x) = I_\rho(I_\rho(f; \circ); x)$ (a.e.) and $\lim_{\rho\to\rho_0} \|I_\rho^2(f; \circ) - f(\circ)\|_{X_{2\pi}} = 0$ for every $f \in X_{2\pi}$. Extend to iterates of higher order. (Hint: Use the estimate

   $$\|I_\rho^2(f; \circ) - f(\circ)\|_{X_{2\pi}} \le \|I_\rho(I_\rho(f; \circ) - f(\circ); \circ)\|_{X_{2\pi}} + \|I_\rho(f; \circ) - f(\circ)\|_{X_{2\pi}})$$
5. Let the *rth integral means* $A_h^r(f; x)$ be defined successively by $A_h^r(f; x) = A_h(A_h^{r-1}(f; \circ); x)$. Show that $A_h^r(f; x)$ is of type (1.1.3) with kernel given by the r-times product $[(2\pi/h)\kappa_{[-h/2,h/2]}*]^r$. Furthermore,

   $$A_h^r(f; x) = h^{-r}\int_{-h/2}^{h/2}\cdots\int_{-h/2}^{h/2} f(x+u_1+\cdots+u_r)\, du_1\ldots du_r.$$

   Show that $A_h^r(f; x) \in C_{2\pi}$ for every $h \in (0, 2\pi)$ and $\lim_{h\to 0} \|A_h^r(f; \circ) - f(\circ)\|_{X_{2\pi}} = 0$ for every $f \in X_{2\pi}$ (and $r \in \mathbb{N}$).

6. Let f be defined in a neighbourhood of a point $x \in \mathbb{R}$. For (sufficiently small) $h \in \mathbb{R}$ the first *central difference* of f at x with respect to the increment h is defined by $\bar{\Delta}_h^1 f(x) = f(x + (h/2)) - f(x - (h/2))$, and the higher differences by $\bar{\Delta}_h^r f(x) = \bar{\Delta}_h^1 \bar{\Delta}_h^{r-1} f(x)$. Show (by induction) that

$$\bar{\Delta}_h^r f(x) = \sum_{k=0}^{r} (-1)^k \binom{r}{k} f\left(x + \left(\frac{r}{2} - k\right)h\right).$$

7. (i) Let $f \in X_{2\pi}$. Show that $\lim_{h\to 0} A_h^r(f; x) = f(x)$ almost everywhere, in particular at all points of continuity of f (for $r = 1$ compare with Prop. 0.3.1).
   (ii) Let $f \in X_{2\pi}$. Show that $A_h^r(f; x)$ has absolutely continuous derivatives of order $(r - 1)$, and $[A_h^r(f; \circ)]^{(r)}(x) = h^{-r} \bar{\Delta}_h^r f(x)$ (a.e.).
   (iii) Let $f \in W_{X_{2\pi}}^r$, and let $\phi \in AC_{2\pi}^{r-1}$ with $\phi^{(r)} \in X_{2\pi}$ be such that $\phi(x) = f(x)$ (a.e.). Show that

$$A_h^r(\phi^{(r)}; x) = A_h^r(D_s^{(r)} f; x) = [A_h^r(f; \circ)]^{(r)}(x) = h^{-r} \bar{\Delta}_h^r f(x) \quad \text{(a.e.)}.$$

   In particular, show that $A_h^2(f; x) \in W_{X_{2\pi}}^{r+2}$ and

$$[A_h^2(f; \circ)]^{(r+2)}(x) = h^{-2}[\phi^{(r)}(x + h) + \phi^{(r)}(x - h) - 2\phi^{(r)}(x)] \quad \text{(a.e.)}.$$

   (Hint: GRAVES [1, p. 254], TIMAN [2, p. 163 ff])

8. (i) Let $f \in W_{C_{2\pi}}^r$ and $g \in L_{2\pi}^1$. Show that $[f * g]^{(r)}(x) = (f^{(r)} * g)(x)$ for every x.
   (ii) Let $f \in W_{L_{2\pi}^p}^r$, $1 \le p < \infty$, and $g \in L_{2\pi}^{p'}$. Show that $(f * g)(x)$ is r-times continuously differentiable and that for every x

$$[f * g]^{(r)}(x) = (D_s^{(r)} f * g)(x) = (\phi^{(r)} * g)(x),$$

   where $\phi \in AC_{2\pi}^{r-1}$ with $\phi^{(r)} \in L_{2\pi}^p$ is such that $\phi(x) = f(x)$ a.e.
   (iii) Let $f \in W_{L_{2\pi}^p}^r$, $1 \le p < \infty$, and $g \in W_{L_{2\pi}^{p'}}^k$, $k \in \mathbb{N}$. Show that $(f * g)(x)$ is $(r + k)$-times continuously differentiable and that for every x

$$[f * g]^{(r+k)}(x) = (D_s^{(r)} f * g)^{(k)}(x) = (D_s^{(r)} f * D_s^{(k)} g)(x).$$

   (iv) Let $f \in W_{X_{2\pi}}^r$ and $g \in L_{2\pi}^1$. Show that the rth strong derivative (in $X_{2\pi}$) of $f * g$ exists and that $D_s^{(r)}[f * g] = (D_s^{(r)} f * g)$. (Theorem 10.1.12 indeed gives $f * g \in W_{X_{2\pi}}^r$, in other words, if $f \in W_{X_{2\pi}}^r$ and $g \in L_{2\pi}^1$, then also $f * g \in W_{X_{2\pi}}^r$.)
   (v) Let $f \in X_{2\pi}$ and $g \in W_{L_{2\pi}^1}^r$. Show that the rth strong derivative (in $X_{2\pi}$) of $f * g$ exists and that $D_s^{(r)}[f * g] = f * D_s^{(r)} g$, $D_s^{(r)} g$ being the rth $L_{2\pi}^1$-derivative (in fact $f * g \in W_{X_{2\pi}}^r$ by Theorem 10.1.12).

9. Let $f \in W_{X_{2\pi}}^r$, and let $\{\chi_\rho(x)\}$ be a continuous kernel such that $\chi_\rho \in W_{C_{2\pi}}^k$ for each $\rho \in \mathbb{A}$. Show that $I_\rho(f; x)$ is an $(r + k)$-times continuously differentiable function of x. Moreover, for every x and $\rho \in \mathbb{A}$

$$[I_\rho(f; \circ)]^{(r+k)}(x) = [I_\rho(D_s^{(r)} f; \circ)]^{(k)}(x) = (D_s^{(r)} f * \chi_\rho^{(k)})(x).$$

10. Apply Prop. 1.1.6, 1.1.7 to the integral means. Show in particular that if $\mu \in BV_{2\pi}$, then for every $s \in C_{2\pi}$

$$\lim_{h\to 0} \int_{-\pi}^{\pi} s(x) \frac{\mu(x + h) - \mu(x)}{h}\, dx = \int_{-\pi}^{\pi} s(x)\, d\mu(x).$$

11. (i) Let $\chi \in C_{2\pi}$ be such that $\chi(0) = 1$ and $0 \le \chi(x) < 1$ for all $x \in [-\pi, \pi]$, $x \ne 0$. Show that $\lim_{n\to\infty} (2\pi \|\chi^n\|_1)^{-1} \int_{-\delta}^{\delta} \chi^n(u)\, du = 1$ for each (fixed) $0 < \delta \le \pi$. Verify that $\{(\|\chi^n\|_1)^{-1} \chi^n(x)\}$ is a positive bounded approximate identity. (Hint: See also KOROVKIN [5, p. 19 f])
    (ii) Let the singular integral $I_n(f; x)$ be defined by ($n \in \mathbb{N}$)

$$I_n(f; x) = \frac{1}{2\pi \|\chi^n\|_1} \int_{-\pi}^{\pi} f(x - u) \chi^n(u)\, du,$$

    where χ is as in (i). Show that $\lim_{n\to\infty} \|I_n(f; \circ) - f(\circ)\|_{X_{2\pi}} = 0$ for every $f \in X_{2\pi}$.

(iii) As an example consider $\chi(x) = (1 + \cos x)/2$. (Hint: This leads to the singular integral of de La Vallée Poussin as treated in Sec. 2.5.2; see also KOROVKIN [5, p. 54], RUDIN [4, p. 89 f])

1.2 Summation of Fourier Series

1.2.1 Definitions

A *trigonometric polynomial* (or in short a *polynomial*) $t_n(x)$ of degree n, $n \in \mathbb{P}$, is an expression of the form

$$t_n(x) = \tfrac{1}{2}a_0 + \sum_{k=1}^{n} \{a_k \cos kx + b_k \sin kx\}, \tag{1.2.1}$$

where the *coefficients* a_k, b_k are arbitrary (complex) numbers independent of x. It will soon be clear why it is convenient to provide the constant term of (1.2.1) with the factor $\frac{1}{2}$. The set of all trigonometric polynomials of degree not exceeding n will be denoted by T_n. If $t_n \in \mathsf{T}_n$ and $(|a_n| + |b_n|) \neq 0$, we say that t_n is strictly of degree n.

Trigonometric series are series of the form

$$\tfrac{1}{2}a_0 + \sum_{k=1}^{\infty} \{a_k \cos kx + b_k \sin kx\} \tag{1.2.2}$$

with arbitrarily given coefficients a_k, b_k. The trigonometric series (1.2.2) may or may not converge. It converges at x if the sequence of *partial sums*

$$s_n(x) = \tfrac{1}{2}a_0 + \sum_{k=1}^{n} \{a_k \cos kx + b_k \sin kx\} \tag{1.2.3}$$

converges at x. In any case, if the coefficients $\{a_k, b_k\}$ are given, we may assign to them the trigonometric series (1.2.2), at least formally.

Applying Euler's formulae

$$\cos kx = \frac{e^{ikx} + e^{-ikx}}{2}, \qquad \sin kx = \frac{e^{ikx} - e^{-ikx}}{2i}, \tag{1.2.4}$$

we may write the nth partial sum (1.2.3) of (1.2.2) in the form

$$s_n(x) = \tfrac{1}{2}a_0 + \tfrac{1}{2} \sum_{k=1}^{n} \{(a_k - ib_k)\, e^{ikx} + (a_k + ib_k)\, e^{-ikx}\}.$$

This suggests that one defines a_k, b_k for $k \in \mathbb{Z}$ by the convention

$$a_{-k} = a_k, \qquad b_{-k} = -b_k, \qquad b_0 = 0 \qquad (k \in \mathbb{N}). \tag{1.2.5}$$

Then, if we set

$$c_k = \frac{a_k - ib_k}{2}, \tag{1.2.6}$$

we obtain for $s_n(x)$

$$s_n(x) = \sum_{k=-n}^{n} c_k\, e^{ikx}, \tag{1.2.7}$$

and thus the nth partial sum of the trigonometric series (1.2.2) corresponds to the nth *symmetric partial sum* of the series

$$\sum_{k=-\infty}^{\infty} c_k\, e^{ikx}. \tag{1.2.8}$$

Conversely, any series (1.2.8) with (complex) coefficients c_k can be written in the form (1.2.2) by setting $a_k = c_k + c_{-k}$ and $b_k = i(c_k - c_{-k})$. We call (1.2.2) the *real form* and (1.2.8) the *complex form* of a trigonometric series. The adjectives refer to the *real trigonometric system* $\{\cos kx, \sin kx\}_{k\in\mathbb{P}}$ and the *complex trigonometric system* $\{e^{ikx}\}_{k\in\mathbb{Z}}$. In addition, we may usually suppose that the coefficients in (1.2.1), (1.2.2) are real; if they are complex, the real and imaginary parts of (1.2.1), (1.2.2) can be taken separately. A complex trigonometric series (1.2.8) may be written in the form (1.2.2) with real coefficients a_k, b_k if and only if $c_{-k} = \overline{c_k}$ for all $k \in \mathbb{Z}$. Note that whenever we speak of convergence or summability of series (1.2.8) we shall always mean the limit, ordinary or generalized, of the symmetric partial sums (1.2.7).

Suppose now that the series (1.2.2) converges uniformly in x to a function f which is necessarily continuous. Then (see Problem 1.2.2) the coefficients a_k, b_k of (1.2.2) are uniquely determined by the sum f and given by

$$a_k = \frac{1}{\pi}\int_{-\pi}^{\pi} f(u)\cos ku\, du, \qquad b_k = \frac{1}{\pi}\int_{-\pi}^{\pi} f(u)\sin ku\, du. \tag{1.2.9}$$

On the other hand, the last formulae are meaningful for every $f \in \mathsf{L}^1_{2\pi}$. Thus we may evaluate coefficients a_k, b_k by (1.2.9) and form the corresponding trigonometric series (1.2.2) for any $f \in \mathsf{L}^1_{2\pi}$. Trigonometric series which are generated in this way are called Fourier series. The formal definition is given by

Definition 1.2.1. *For any $f \in \mathsf{L}^1_{2\pi}$, the numbers*

$$\left.\begin{aligned} f_c^\wedge(k) &= \frac{1}{\pi}\int_{-\pi}^{\pi} f(u)\cos ku\, du \\ f_s^\wedge(k) &= \frac{1}{\pi}\int_{-\pi}^{\pi} f(u)\sin ku\, du \end{aligned}\right\} \quad (k \in \mathbb{P}) \tag{1.2.10}$$

are called the ***real Fourier coefficients*** *of f, and the numbers*

$$f^\wedge(k) = \frac{1}{2\pi}\int_{-\pi}^{\pi} f(u)\, e^{-iku}\, du \qquad (k \in \mathbb{Z}) \tag{1.2.11}$$

the ***complex Fourier coefficients*** *of f. The corresponding trigonometric series*

$$f \sim \tfrac{1}{2} f_c^\wedge(0) + \sum_{k=1}^{\infty} \{f_c^\wedge(k)\cos kx + f_s^\wedge(k)\sin kx\}, \tag{1.2.12}$$

$$f \sim \sum_{k=-\infty}^{\infty} f^\wedge(k)\, e^{ikx} \tag{1.2.13}$$

are called the ***real*** *and* ***complex Fourier series*** *of f, and denoted by S[f]. Finally we write the* ***nth partial sum*** *of the Fourier series of f as*

$$(1.2.14)\quad S_n(f;x) = \tfrac{1}{2}f_c^\wedge(0) + \sum_{k=1}^{n}\{f_c^\wedge(k)\cos kx + f_s^\wedge(k)\sin kx\} = \sum_{k=-n}^{n} f^\wedge(k)\, e^{ikx}.$$

Note that the adjectives 'real' and 'complex' again refer to the trigonometric and exponential functions occurring. Thus real Fourier coefficients are complex numbers if f is complex-valued. The Fourier series of a function $f \in \mathsf{L}_{2\pi}^1$ is a trigonometric series but not conversely (see Problem 4.1.3). Therefore we changed the notation of the coefficients and wrote $f_c^\wedge(k), f_s^\wedge(k), f^\wedge(k)$ instead of a_k, b_k, c_k to distinguish between arbitrary coefficients and Fourier coefficients of a function f. Since the Fourier coefficients (1.2.10) and (1.2.11) in particular satisfy (1.2.5) and (1.2.6), the same notation $S[f]$ is justified for the real and complex Fourier series of f. A similar remark applies to (1.2.14). Finally, the sign $\sim$ in (1.2.12) or (1.2.13) denotes an equivalence relation and simply means that to each $f \in \mathsf{L}_{2\pi}^1$ we associate its Fourier series $S[f]$. This does not say anything about its convergence or representation for f.

Let us now indicate the connections between Fourier series and harmonic functions in order to develop the notion of a conjugate Fourier series. Let $z = \xi + i\eta = re^{ix}$ be a complex variable, and let $f \in \mathsf{L}_{2\pi}^1$ be a real-valued function (so that the real Fourier coefficients are indeed real numbers and $f^\wedge(-k) = \overline{f^\wedge(k)}$). Since the Fourier coefficients (real or complex) are bounded by $\|f\|_1$, the power series

$$(1.2.15)\qquad F(z) = f^\wedge(0) + 2\sum_{k=1}^{\infty} f^\wedge(k) z^k$$

defines a function $F(z) \equiv u(r, x) + iv(r, x)$, holomorphic in $|z| < 1$. Thus the real part

$$(1.2.16)\qquad u(r, x) = \tfrac{1}{2}f_c^\wedge(0) + \sum_{k=1}^{\infty} r^k\{f_c^\wedge(k)\cos kx + f_s^\wedge(k)\sin kx\}$$

and the imaginary part

$$(1.2.17)\qquad v(r, x) = \sum_{k=1}^{\infty} r^k\{f_c^\wedge(k)\sin kx - f_s^\wedge(k)\cos kx\}$$

of $F(z)$ are conjugate harmonic functions. Formally, $u(r, x)$ tends to the (real) Fourier series of f as $r \to 1-$. Concerning the conjugate harmonic function $v(r, x)$, its (formal) limit for $r \to 1-$ yields a series which we may call the conjugate Fourier series of f. This series may again be associated with f in the same way as its ordinary Fourier series.

More specifically we have

Definition 1.2.2. *Let* (1.2.12) *be the real Fourier series of a function* $f \in \mathsf{L}_{2\pi}^1$. *Then we call*

$$(1.2.18)\qquad \sum_{k=1}^{\infty}\{f_c^\wedge(k)\sin kx - f_s^\wedge(k)\cos kx\}$$

the ***conjugate*** *(or allied)* ***Fourier series*** *of f. If we consider* (1.2.12) *in its complex form* (1.2.13), *the conjugate Fourier series of f takes on the form*

$$(1.2.19)\qquad \sum_{k=-\infty}^{\infty}\{-i\,\operatorname{sgn} k\} f^\wedge(k)\, e^{ikx}.$$

With $S[f]$ denoting the Fourier series and $S_n(f;x)$ its nth partial sum, the corresponding notation for the conjugate Fourier series will be $S^\sim[f]$ and $S_n^\sim(f;x)$. Analogously, one may define ***conjugate trigonometric series*** *(to general trigonometric series (1.2.2) and (1.2.8)).*

Note that $S^{\sim\sim}[f] = -S[f] + f^\wedge(0)$, the equality only indicating that the (at least formal) series on both sides have the same coefficients. We shall return to this question in Chapter 9.

1.2.2 Dirichlet and Fejér Kernel

In order to study the convergence of the Fourier series associated with a function $f \in \mathsf{L}^1_{2\pi}$, we consider its partial sum. According to (1.2.10) we have

$$S_n(f;x) = \frac{1}{\pi}\int_{-\pi}^{\pi} f(u)\Big[\tfrac{1}{2} + \sum_{k=1}^{n} \{\cos ku \cos kx + \sin ku \sin kx\}\Big]\,du$$

$$= \frac{1}{2\pi}\int_{-\pi}^{\pi} f(u)\Big[1 + 2\sum_{k=1}^{n} \cos k(x-u)\Big]\,du.$$

As (see Problem 1.2.5)

$$(1.2.20)\quad D_n(x) \equiv 1 + 2\sum_{k=1}^{n} \cos kx = \begin{cases} \dfrac{\sin((2n+1)x/2)}{\sin(x/2)}, & x \neq 2j\pi \\ 2n+1, & x = 2j\pi \end{cases} \quad j \in \mathbb{Z},$$

we obtain

$$(1.2.21)\qquad S_n(f;x) = \frac{1}{2\pi}\int_{-\pi}^{\pi} f(x-u)D_n(u)\,du.$$

In view of (1.2.20) and Def. 1.1.1, the functions $D_n(x)$ define a kernel with parameter $\rho = n \in \mathbb{P}$ and $\rho_0 = \infty$. $\{D_n(x)\}$ is called the *Dirichlet kernel*, and the corresponding singular integral (1.2.21), i.e., the nth partial sum of the Fourier series of f, the *singular integral of Dirichlet*.

Thus Prop. 1.1.3 may be applied to the integral of Dirichlet. But it is crucial here that the Dirichlet kernel fails to be an approximate identity (and so Theorem 1.1.5 cannot be applied). This is shown by

Proposition 1.2.3. *If we define $L_n \equiv \|D_n\|_1$ as the* ***Lebesgue constants*** *(of the Fourier series), then*

$$(1.2.22)\qquad L_n = \frac{4}{\pi^2}\log n + O(1) \qquad (n \to \infty).$$

Proof. Obviously,

$$L_n = \frac{1}{\pi}\int_0^{\pi}\left|\frac{\sin((2n+1)u/2)}{\sin(u/2)}\right| du = \frac{1}{\pi}\int_0^{\pi}\left|\sin nu \cot\frac{u}{2} + \cos nu\right| du$$

$$= \frac{1}{\pi}\int_0^{\pi} |\sin nu| \cot\frac{u}{2}\,du + O(1).$$

Since the function $(1/u) - \cot u$ is bounded for $|u| \le \pi/2$, we have

$$\begin{aligned} L_n &= \frac{2}{\pi}\int_0^\pi \frac{|\sin nu|}{u}\,du + \frac{2}{\pi}\int_0^\pi |\sin nu|\left\{\frac{1}{2}\cot\frac{u}{2} - \frac{1}{u}\right\}du + O(1) \\ &= \frac{2}{\pi}\sum_{k=0}^{n-1}\int_{k\pi/n}^{(k+1)\pi/n} \frac{|\sin nu|}{u}\,du + O(1) \\ &= \frac{2}{\pi}\sum_{k=1}^{n-1}\int_0^{\pi/n} \frac{\sin nu}{(k\pi/n)+u}\,du + \frac{2}{\pi}\int_0^{\pi/n}\frac{\sin nu}{u}\,du + O(1) \\ &= \frac{2}{\pi}\int_0^{\pi/n} \sin nu\left\{\sum_{k=1}^{n-1}\frac{1}{(k\pi/n)+u}\right\}du + O(1). \end{aligned}$$

But in view of the inequality

$$\frac{n}{\pi}\left\{\frac{1}{2}+\frac{1}{3}+\cdots+\frac{1}{n}\right\} \le \sum_{k=1}^{n-1}\frac{1}{(k\pi/n)+u} \le \frac{n}{\pi}\left\{1+\frac{1}{2}+\cdots+\frac{1}{n-1}\right\},$$

valid for $0 \le u \le \pi/n$, it follows that

$$\sum_{k=1}^{n-1}\frac{1}{(k\pi/n)+u} = \frac{n}{\pi}\{\log n + O(1)\}.$$

The desired estimate (1.2.22) may now be obtained immediately.

The situation is very different if we replace the partial sums $S_n(f; x)$ of the Fourier series of f by the (first) arithmetic (or Fejér) means

$$\text{(1.2.23)}\qquad \sigma_n(f; x) \equiv \frac{1}{n+1}\sum_{k=0}^{n} S_k(f; x) = \frac{1}{2\pi}\int_{-\pi}^{\pi} f(x-u)\left\{\frac{1}{n+1}\sum_{k=0}^{n} D_k(u)\right\}du.$$

Making use of the representation (see Problem 1.2.6)

$$\text{(1.2.24)}\qquad \begin{aligned} F_n(x) &\equiv \frac{1}{n+1}\sum_{k=0}^{n} D_k(x) = 1 + 2\sum_{k=1}^{n}\left(1-\frac{k}{n+1}\right)\cos kx \\ &= \begin{cases} \dfrac{1}{n+1}\left[\dfrac{\sin((n+1)x/2)}{\sin(x/2)}\right]^2, & x \ne 2j\pi \\ n+1, & x = 2j\pi \end{cases} \quad j \in \mathbb{Z}, \end{aligned}$$

we have

$$\text{(1.2.25)}\qquad \sigma_n(f; x) = \frac{1}{2\pi}\int_{-\pi}^{\pi} f(x-u)F_n(u)\,du.$$

This is *Fejér's singular integral* belonging to f. $\{F_n(x)\}$ is called the *Fejér kernel*; it is an even, positive, and continuous kernel with parameter $\rho = n$ and $\rho_0 = \infty$. Moreover, $\{F_n(x)\}$ is an approximate identity satisfying (1.1.7) because for any fixed $0 < \delta < \pi$

$$\sup_{\delta \le |x| \le \pi} |F_n(x)| \le \frac{1}{(n+1)\sin^2(\delta/2)}.$$

Thus Theorem 1.1.5 may be applied to give

Corollary 1.2.4. *If* $f \in X_{2\pi}$, *then* $\|\sigma_n(f; \circ)\|_{X_{2\pi}} \le \|f\|_{X_{2\pi}}$ *and*

$$\text{(1.2.26)}\qquad \lim_{n\to\infty} \|\sigma_n(f; \circ) - f(\circ)\|_{X_{2\pi}} = 0.$$

1.2.3 Weierstrass Approximation Theorem

At this stage we shall give some applications of the results obtained so far. According to (1.2.23) the integral of Fejér may be rewritten in the form

$$(1.2.27)\quad \sigma_n(f;x) = \tfrac{1}{2}f_c^\wedge(0) + \sum_{k=1}^{n}\left(1 - \frac{k}{n+1}\right)\{f_c^\wedge(k)\cos kx + f_s^\wedge(k)\sin kx\}$$
$$= \sum_{k=-n}^{n}\left(1 - \frac{|k|}{n+1}\right) f^\wedge(k)\, e^{ikx}$$

which shows that it is a polynomial of degree n. Thus Cor. 1.2.4 contains the celebrated *Weierstrass approximation theorem* for periodic functions, namely

Theorem 1.2.5. *The set of all trigonometric polynomials forms a dense subset of* $X_{2\pi}$. *In particular, if* $f \in C_{2\pi}$, *then, given any* $\varepsilon > 0$, *there exists a natural number* $n = n(\varepsilon)$ *and a trigonometric polynomial* $t_n(x)$ *of degree n such that* $|f(x) - t_n(x)| < \varepsilon$ *uniformly for all x.*

Weierstrass' theorem is actually only an existence theorem as it states that there exist polynomials having the desired property. But in the proof of Theorem 1.2.5 via Cor. 1.2.4 we have at the same time *constructed* a well-defined sequence of polynomials, namely the Fejér means $\sigma_n(f;x)$ of the Fourier series of f.

Proposition 1.2.6. *The real trigonometric system is* ***closed*** *in* $X_{2\pi}$, *i.e., if* $f \in X_{2\pi}$, *then*

$$\int_{-\pi}^{\pi} f(u)\cos ku\, du = 0, \qquad \int_{-\pi}^{\pi} f(u)\sin ku\, du = 0 \qquad (k \in \mathbb{P})$$

implies $f(x) = 0$ (a.e.). *The same is true for the complex trigonometric system.*

Proof. The hypothesis means that all real Fourier coefficients $f_c^\wedge(k), f_s^\wedge(k)$ of $f \in X_{2\pi}$ vanish. Thus, by (1.2.27), $\sigma_n(f;x) \equiv 0$, from which $\|f\|_{X_{2\pi}} = 0$ by (1.2.26).

As an important consequence we state

Corollary 1.2.7. *A function* $f \in X_{2\pi}$ *is* ***uniquely*** *determined by its real (complex) Fourier coefficients. In other words, if the real (complex) Fourier coefficients of two functions of class* $X_{2\pi}$ *are equal for all* $k \in \mathbb{P}$ $(k \in \mathbb{Z})$, *then these functions are equal in* $X_{2\pi}$.

Thus a trigonometric series cannot be the Fourier series of more than one function in $X_{2\pi}$.

1.2.4 Summability of Fourier Series

We already saw that Theorem 1.1.5 does not assure norm-convergence of the partial sums of Fourier series of f towards f since the Dirichlet kernel does not constitute an approximate identity. But this theorem can be applied to the first arithmetic means of a Fourier series since they furnish a singular integral the kernel of which is an approximate identity. This suggests that we place more emphasis upon summability of Fourier series in order to produce norm-convergence in some generalized sense. We do not intend to treat summation of general sequences and series here but only consider the problem with respect to Fourier series in the following sense:

Let $\mathbb{A}$ be a parameter set as in Sec. 1.1 and $\rho \in \mathbb{A}$. $\{\theta_\rho(k)\}$ is called a *θ-factor* if, for each $\rho \in \mathbb{A}$, $\theta_\rho(k)$ is a real function on $\mathbb{Z}$ satisfying

$$\theta_\rho(k) \in \mathsf{l}^1, \qquad \theta_\rho(0) = 1, \qquad \theta_\rho(k) = \theta_\rho(-k). \tag{1.2.28}$$

Note that if $\mathbb{A} = \mathbb{P}$, $\rho = n$, $\rho_0 = \infty$, we may arrange the numbers $\theta_n(k)$ as an infinite matrix $\{\theta_n(k)\}$ with row-parameter n.

For $f \in \mathsf{L}^1_{2\pi}$ we may form the *θ-means*

$$U_\rho(f; x) = \tfrac{1}{2} f_c^\wedge(0) + \sum_{k=1}^{\infty} \theta_\rho(k)\{f_c^\wedge(k) \cos kx + f_s^\wedge(k) \sin kx\} \tag{1.2.29}$$

of the real Fourier series of f. Since $\theta_\rho \in \mathsf{l}^1$ and the Fourier coefficients are all bounded by $\|f\|_1$, the series on the right-hand side of (1.2.29) converges absolutely and uniformly in x, and thus defines a function $U_\rho(f; x) \in \mathsf{C}_{2\pi}$ for each $\rho \in \mathbb{A}$. If we begin with the complex form (1.2.13) of the Fourier series, we again arrive at the same $U_\rho(f; x)$, thus

$$U_\rho(f; x) = \sum_{k=-\infty}^{\infty} \theta_\rho(k) f^\wedge(k)\, e^{ikx}, \tag{1.2.30}$$

since θ_ρ is an even function on $\mathbb{Z}$. If we substitute (1.2.10) into (1.2.29), then

$$\begin{aligned} U_\rho(f; x) &= \frac{1}{2\pi} \int_{-\pi}^{\pi} f(x-u)\left\{1 + 2\sum_{k=1}^{\infty} \theta_\rho(k) \cos ku\right\} du \\ &= \frac{1}{2\pi} \int_{-\pi}^{\pi} f(x-u) C_\rho(u)\, du, \end{aligned} \tag{1.2.31}$$

where we have set

$$C_\rho(x) = 1 + 2\sum_{k=1}^{\infty} \theta_\rho(k) \cos kx, \tag{1.2.32}$$

the interchange of integration and summation being justified by the uniform convergence of the series. According to (1.2.28), the functions $C_\rho(x)$ define an even and continuous kernel. The assumption $\theta_\rho(0) = 1$ corresponds to (1.1.1); the weaker condition (1.1.2) would read $\lim_{\rho\to\rho_0} \theta_\rho(0) = 1$. The hypothesis that a θ-factor and thus the corresponding kernel $\{C_\rho(x)\}$ is even, reflects the fact that the convergence of a Fourier series in its, e.g., complex form is defined by the convergence of the symmetric partial sums (1.2.14), and it implies that (1.2.29) and (1.2.30) represent the same function for each $\rho \in \mathbb{A}$.

If the θ-means $U_\rho(f; x)$ of the Fourier series of $f \in \mathsf{L}^1_{2\pi}$ converge in some sense (pointwise, in norm, etc.) to a limit as $\rho \to \rho_0$, and if this limit coincides with the usual sum of the series in case the Fourier series converges in the ordinary sense, we call $\{\theta_\rho(k)\}$ a *convergence-factor* with respect to the limit notion under consideration. The θ-factor then defines a *summation process*; we say the Fourier series is *θ-summable* and call the limit its *θ-sum*.

As a first result we obtain by Theorem 1.1.5

Proposition 1.2.8. *Let $\{\theta_\rho(k)\}$ satisfy* (1.2.28) *such that the corresponding kernel $\{C_\rho(x)\}$ of* (1.2.32) *forms an approximate identity. Then for each $f \in X_{2\pi}$*

$$\lim_{\rho \to \rho_0} \|U_\rho(f; \circ) - f(\circ)\|_{X_{2\pi}} = 0, \tag{1.2.33}$$

i.e., the Fourier series of f is θ-summable to f in $X_{2\pi}$-norm.

As an example we take

$$\theta_r(k) = r^{|k|}, \quad \rho = r, \quad \mathbb{A} = [0, 1), \quad \rho_0 = 1. \tag{1.2.34}$$

Obviously, $\{r^{|k|}\}$ satisfies (1.2.28). It is called the *Abel–Poisson factor*. If we form the corresponding means (1.2.29), then we again arrive at (1.2.16). It will be convenient to redefine the function represented by the series (1.2.16) as $P_r(f; x)$, thus

$$\begin{aligned} P_r(f; x) &= \tfrac{1}{2} f_c^\wedge(0) + \sum_{k=1}^{\infty} r^k \{f_c^\wedge(k) \cos kx + f_s^\wedge(k) \sin kx\} \\ &= \sum_{k=-\infty}^{\infty} r^{|k|} f^\wedge(k)\, e^{ikx}. \end{aligned} \tag{1.2.35}$$

$P_r(f; x)$ are called the *Abel–Poisson means* of the Fourier series of f. On account of (1.2.31) we also have

$$P_r(f; x) = \frac{1}{2\pi} \int_{-\pi}^{\pi} f(x - u) p_r(u)\, du, \tag{1.2.36}$$

where in view of (1.2.32) and Problem 1.2.18(i) the kernel is given by

$$p_r(x) = 1 + 2 \sum_{k=1}^{\infty} r^k \cos kx = \frac{1 - r^2}{1 - 2r \cos x + r^2}. \tag{1.2.37}$$

$\{p_r(x)\}$ is called the *Abel–Poisson kernel*, and $P_r(f; x)$ the *singular integral of Abel–Poisson*. In Problem 1.2.18 we have collected some elementary facts concerning the Abel–Poisson kernel. $\{p_r(x)\}$ turns out to be even, positive, and continuous. Moreover, it is an approximate identity satisfying (1.1.7). Thus Prop. 1.2.8 yields

Corollary 1.2.9. *The Fourier series of $f \in X_{2\pi}$ is Abel–Poisson summable in $X_{2\pi}$-norm to f, i.e.,* $\lim_{r \to 1-} \|P_r(f; \circ) - f(\circ)\|_{X_{2\pi}} = 0$.

Let us return for a moment to (1.2.16). We already know that for real-valued $f \in X_{2\pi}$ the integral $P_r(f; x)$ is the real part of the function $F(z)$ of (1.2.15), holomorphic in $|z| < 1$. Therefore $P_r(f; x)$ is a harmonic function in the interior of the unit disc, that is, it satisfies Laplace's equation

$$\left\{\left(r \frac{\partial}{\partial r}\right)^2 + \frac{\partial^2}{\partial x^2}\right\} u(r, x) = 0 \qquad (-\pi \le x \le \pi, 0 \le r < 1) \tag{1.2.38}$$

if we take its polar-coordinate form, or

$$\left\{\frac{\partial^2}{\partial \xi^2} + \frac{\partial^2}{\partial \eta^2}\right\} u(\xi, \eta) = 0 \qquad (\xi^2 + \eta^2 < 1) \tag{1.2.39}$$

if we prefer cartesian coordinates and define $\xi = r \cos x$, $\eta = r \sin x$.

If we start off with an arbitrary function $f(x) \in X_{2\pi}$ and assume it to be defined on the unit circle $\{(r, x) \mid r = 1, -\pi \le x \le \pi\}$, then we may interpret Cor. 1.2.9 in the following sense:

Proposition 1.2.10. *The integral $P_r(f; x)$ of $f \in X_{2\pi}$ is a solution of Dirichlet's problem for the unit disc, the boundary value f being attained in $X_{2\pi}$-norm. In other words, $P_r(f; x)$ satisfies Laplace's equation in the interior of the unit disc and takes on the prescribed boundary values in the sense of Cor.* 1.2.9.

At this stage, the definition of the Dirichlet problem or first boundary value problem of potential theory for the unit disc has, in a sense, only informal character; it will be made more precise later on. In particular, it will be shown that the singular integral of Abel–Poisson gives the unique solution of the problem (Sec. 7.1.2).

1.2.5 Row-Finite θ-Factors

Let us now discuss θ-factors for which the parameter is discrete, i.e., $\mathbb{A} = \mathbb{P}$, $\rho = n$, $\rho_0 = \infty$, and for which the functions $\theta_n(k)$ have compact support on $\mathbb{Z}$ for each $n \in \mathbb{P}$. This means that the corresponding matrix $\{\theta_n(k)\}$ is row-finite. In this case (1.2.29) reduces to the finite sum

$$(1.2.40)\qquad U_n(f; x) = \tfrac{1}{2} f_c^\wedge(0) + \sum_{k=1}^{m(n)} \theta_n(k)\{f_c^\wedge(k) \cos kx + f_s^\wedge(k) \sin kx\}$$
$$= \sum_{k=-m(n)}^{m(n)} \theta_n(k) f^\wedge(k)\, e^{ikx},$$

where $m(n)$ increases with n on $\mathbb{P}$ and is such that $\theta_n(k) = 0$ for $|k| > m(n)$. We shall call such θ-factors *row-finite*. Of course, the first condition of (1.2.28) is now trivially satisfied, and the kernel (1.2.32) is the even trigonometric polynomial

$$(1.2.41)\qquad C_n(x) = 1 + 2 \sum_{k=1}^{m(n)} \theta_n(k) \cos kx.$$

The associated singular integral $U_n(f; x)$ of (1.2.40) then defines an operator which transforms the space $X_{2\pi}$ into the set $T_{m(n)}$ of polynomials of degree $m(n)$ at most. We call such operators (trigonometric) *polynomial operators* of degree $m(n)$.

We have already considered two examples of row-finite θ-factors. Thus

$$(1.2.42)\qquad \theta_n(k) = \begin{cases} 1, & |k| \le n \\ 0, & |k| > n \end{cases} \qquad m(n) = n,$$

$$(1.2.43)\qquad \theta_n(k) = \begin{cases} 1 - \dfrac{|k|}{n+1}, & |k| \le n \\ 0, & |k| > n \end{cases} \qquad m(n) = n$$

give the partial sums of the Fourier series and the Fejér means, respectively. Whereas the *Dirichlet factor* (1.2.42) corresponds to a kernel which fails to be an approximate identity, the *Fejér factor* (1.2.43) produces an approximate identity. For further examples of θ-factors we refer to the Problems.

1.2.6 Summability of Conjugate Series

If we apply a θ-factor to the conjugate Fourier series (1.2.18) of $f \in X_{2\pi}$, we obtain

$$(1.2.44)\qquad U_\rho^\sim(f; x) = \sum_{k=1}^{\infty} \theta_\rho(k)\{f_c^\wedge(k) \sin kx - f_s^\wedge(k) \cos kx\}.$$

According to (1.2.28), the series on the right-hand side converges absolutely and uniformly in x for each $\rho \in \mathbb{A}$ and thus defines a function of class $C_{2\pi}$ which we shall, corresponding to (1.2.29) and Def. 1.2.2, denote by $U_\rho^\sim(f; x)$. If we prefer the complex form (1.2.19) of the conjugate Fourier series of f, we obtain the same functions

$$U_\rho^\sim(f; x) = \sum_{k=-\infty}^{\infty} \theta_\rho(k)\{-i \operatorname{sgn} k\} f^\wedge(k)\, e^{ikx}. \tag{1.2.45}$$

Proceeding as in the case of ordinary Fourier series it follows that

$$\begin{aligned} U_\rho^\sim(f; x) &= \frac{1}{2\pi}\int_{-\pi}^{\pi} f(x-u)\Big\{2\sum_{k=1}^{\infty}\theta_\rho(k)\sin ku\Big\}\,du \\ &= \frac{1}{2\pi}\int_{-\pi}^{\pi} f(x-u) C_\rho^\sim(u)\,du, \end{aligned} \tag{1.2.46}$$

where we have set

$$C_\rho^\sim(x) = 2\sum_{k=1}^{\infty}\theta_\rho(k)\sin kx \qquad (\rho \in \mathbb{A}). \tag{1.2.47}$$

We call $C_\rho^\sim(x)$ the *conjugate function* of $C_\rho(x)$ and $U_\rho^\sim(f; x)$ the *conjugate* of the singular integral $U_\rho(f; x)$. Note that the functions $C_\rho^\sim(x)$ do not define a kernel since condition (1.1.1) is not satisfied. Indeed, $\int_{-\pi}^{\pi} C_\rho^\sim(u)\,du = 0$ for every $\rho \in \mathbb{A}$. Thus the summation of conjugate Fourier series possesses special features the discussion of which shall be left to Sec. 9.3.1. In Chapter 9 we shall also clear up the notion "conjugate" and "conjugate of" by giving a precise interpretation to this formal definition.

At this stage, let us substitute the factors of Dirichlet, Fejér, and Abel–Poisson into (1.2.47), obtaining

$$D_n^\sim(x) = 2\sum_{k=1}^{n}\sin kx = \cot(x/2) - \frac{\cos((2n+1)x/2)}{\sin(x/2)}, \tag{1.2.48}$$

$$F_n^\sim(x) = 2\sum_{k=1}^{n}\Big(1-\frac{k}{n+1}\Big)\sin kx = \cot(x/2) - \frac{\sin(n+1)x}{2(n+1)\sin^2(x/2)}, \tag{1.2.49}$$

$$p_r^\sim(x) = 2\sum_{k=1}^{\infty} r^k \sin kx = \frac{2r\sin x}{1-2r\cos x + r^2}. \tag{1.2.50}$$

Note that the conjugate of the singular integral of Dirichlet is nothing but the nth partial sum of the conjugate Fourier series of f, namely

$$S_n^\sim(f; x) = \frac{1}{2\pi}\int_{-\pi}^{\pi} f(x-u) D_n^\sim(u)\,du = \sum_{k=-n}^{n}\{-i \operatorname{sgn} k\} f^\wedge(k)\, e^{ikx}. \tag{1.2.51}$$

In the Abel–Poisson case the integral $P_r^\sim(f; x)$ is often denoted by $Q_r(f; x)$, i.e.,

$$Q_r(f; x) = \frac{1}{2\pi}\int_{-\pi}^{\pi} f(x-u) p_r^\sim(u)\,du, \tag{1.2.52}$$

which turns out for real-valued f to be the imaginary part of $F(z)$ of (1.2.15). Since $F(z)$ is holomorphic in $|z| < 1$, $Q_r(f; x)$ defines a harmonic function in the open unit disc. Moreover, $Q_r(f; x)$ is a conjugate harmonic to the integral $P_r(f; x)$ of Abel–Poisson, in the usual sense of the theory of functions of a complex variable.

1.2.7 Fourier–Stieltjes Series

We briefly introduce Fourier–Stieltjes series.

Definition 1.2.11. *For any $\mu \in \mathsf{BV}_{2\pi}$, the numbers*

$$\mu_c^\vee(k) = \frac{1}{\pi}\int_{-\pi}^{\pi} \cos ku \, d\mu(u), \qquad \mu_s^\vee(k) = \frac{1}{\pi}\int_{-\pi}^{\pi} \sin ku \, d\mu(u) \qquad (k \in \mathbb{P}) \tag{1.2.53}$$

are called the ***real Fourier–Stieltjes coefficients*** *of μ, and*

$$\mu^\vee(k) = \frac{1}{2\pi}\int_{-\pi}^{\pi} e^{-iku} \, d\mu(u) \qquad (k \in \mathbb{Z}) \tag{1.2.54}$$

the ***complex Fourier–Stieltjes coefficients*** *of μ. The corresponding trigonometric series*

$$d\mu \sim \tfrac{1}{2}\mu_c^\vee(0) + \sum_{k=1}^{\infty} \{\mu_c^\vee(k)\cos kx + \mu_s^\vee(k)\sin kx\}, \tag{1.2.55}$$

$$d\mu \sim \sum_{k=-\infty}^{\infty} \mu^\vee(k)\, e^{ikx} \tag{1.2.56}$$

are called the ***real*** *and* ***complex Fourier–Stieltjes series*** *of μ and denoted by $S[d\mu]$. The nth partial sum of the Fourier–Stieltjes series of μ is indicated by $S_n(d\mu; x)$. Conjugate Fourier–Stieltjes series are defined in an obvious way as for Fourier series.*

Concerning the meaning of the symbol $\sim$, the definition of convergence of Fourier–Stieltjes series, etc., the same remarks hold as for Fourier series. Note that the integrals in (1.2.53) and (1.2.54) are Riemann–Stieltjes integrals.

Proceeding just as for Fourier series, we obtain for the nth partial sum of the Fourier–Stieltjes series of $\mu \in \mathsf{BV}_{2\pi}$

$$S_n(d\mu; x) = \sum_{k=-n}^{n} \mu^\vee(k)\, e^{ikx} = \frac{1}{2\pi}\int_{-\pi}^{\pi} D_n(x-u)\, d\mu(u), \tag{1.2.57}$$

$\{D_n(x)\}$ being Dirichlet's kernel. Thus $S_n(d\mu; x)$ is a singular integral of type (1.1.12). Moreover, if we sum up Fourier–Stieltjes series by a θ-factor, we have (in correspondence with the notation in (1.1.12), (1.2.31))

$$U_\rho(d\mu; x) = \sum_{k=-\infty}^{\infty} \theta_\rho(k)\mu^\vee(k)\, e^{ikx} = \frac{1}{2\pi}\int_{-\pi}^{\pi} C_\rho(x-u)\, d\mu(u), \tag{1.2.58}$$

$\{C_\rho(x)\}$ being the kernel defined by (1.2.32). Thus Prop. 1.1.7 is applicable to the summation of Fourier–Stieltjes series and gives

Proposition 1.2.12. *Let $\{\theta_\rho(k)\}$ satisfy* (1.2.28) *such that the corresponding kernel $\{C_\rho(x)\}$ of* (1.2.32) *is an approximate identity. If $\mu \in \mathsf{BV}_{2\pi}$, then for every $s \in \mathsf{C}_{2\pi}$*

$$\lim_{\rho\to\rho_0} \int_{-\pi}^{\pi} s(x)U_\rho(d\mu; x)\, dx = \int_{-\pi}^{\pi} s(x)\, d\mu(x). \tag{1.2.59}$$

For a more detailed discussion of Fourier–Stieltjes series we refer to Sec. 4.3.

Problems

1. (i) (1.2.1) is called the *real* form of a trigonometric polynomial of degree n. Show that every $t_n \in \mathsf{T}_n$ may be written as $t_n(x) = \sum_{k=-n}^{n} c_k e^{ikx}$, the *complex* form of a trigonometric polynomial.
 (ii) Show that if $t_n \in \mathsf{T}_n$, $t_m \in \mathsf{T}_m$, then $t_n t_m \in \mathsf{T}_{n+m}$.
 (iii) Show that if $t_n \in \mathsf{T}_n$ is even, thus a cosine-polynomial, then t_n can be represented in the form $p_n(\cos x)$, where $p_n(x)$ is an algebraic polynomial of degree n, and conversely.
 (iv) If $t_n \in \mathsf{T}_n$ is even, show that $t_n(\arccos x)$ is an algebraic polynomial of degree n.
2. (i) Show that $\int_{-\pi}^{\pi} \cos mu \sin nu \, du = 0$,

$$\int_{-\pi}^{\pi} \cos mu \cos nu \, du = \begin{cases} \pi, & m = n \neq 0 \\ 2\pi, & m = n = 0 \\ 0, & m \neq n \end{cases} \qquad \int_{-\pi}^{\pi} \sin mu \sin nu \, du = \begin{cases} \pi, & m = n \neq 0 \\ 0, & m = n = 0 \\ 0, & m \neq n \end{cases}$$

 for $m, n \in \mathbb{P}$. Also verify that for $k, l \in \mathbb{Z}$

$$\int_{-\pi}^{\pi} e^{iku} e^{ilu} \, du = \begin{cases} 2\pi, & k = -l \\ 0, & k \neq -l. \end{cases}$$

 (ii) Suppose that the trigonometric series (1.2.2) converges uniformly in x to a function f. Show that the coefficients are determined by f through (1.2.9). (Hint: Insert the series (1.2.2) into $\int_{-\pi}^{\pi} f(u) \cos ku \, du$, interchange the order of summation and integration, and use (i))
 (iii) Suppose that the trigonometric series (1.2.8) converges uniformly in x to a function f. Show that the coefficients c_k are determined by f through $2\pi c_k = \int_{-\pi}^{\pi} f(u) e^{-iku} \, du$.
 (iv) Suppose that the trigonometric series (1.2.2) converges to a function $f(x)$ a.e., and the partial sums (1.2.3) remain bounded. Show that (1.2.9) is valid. (Hint: Lebesgue's dominated convergence theorem; see also ASPLUND–BUNGART [1, p. 424])
3. Let $f \in \mathsf{L}_{2\pi}^1$ and $c \in \mathbb{C}$. Show that

$$S_n(f; x) - c = \frac{1}{2\pi} \int_0^{\pi} [f(x + u) + f(x - u) - 2c] D_n(u) \, du.$$

 Thus the Fourier series of f converges at point x_0 to the value c if and only if one has $\lim_{n \to \infty} \int_0^{\pi} g(x_0, u) D_n(u) \, du = 0$, where $g(x_0, u) = f(x_0 + u) + f(x_0 - u) - 2c$.
4. Establish the following inequalities:
 (i) $|\sin x| \le |x|, \quad 1 - \cos x \le x^2/2 \qquad (x \in \mathbb{R})$,
 (ii) $\sin x \ge 2x/\pi, \quad 1 - \cos x \ge 2x^2/\pi^2 \qquad (0 \le x \le \pi/2)$,
 (iii) if $m, n \in \mathbb{N}$ with $m \le n$, then $|\sum_{k=m}^{n} e^{ikx}| \le |\sin(x/2)|^{-1}$ for all x with $x \neq 2\pi j$, $j \in \mathbb{Z}$. (Hint: Use the formula for the nth partial sum of a geometric series; see also Problem 1.2.5(i))
 (iv) Show that $(\pi x)^{-1} \sin \pi x \ge (1 + x^2)^{-1}(1 - x^2)$ for all $x \in \mathbb{R}$. (Hint: See REDHEFFER [1])
5. (i) Prove (1.2.20) and (1.2.48). (Hint: Either proceed by mathematical induction or use

$$\sum_{k=0}^{n} e^{ikx} = \frac{1 - e^{i(n+1)x}}{1 - e^{ix}} = \frac{1}{2} + \frac{\sin((2n + 1)x/2)}{2 \sin(x/2)} + i \frac{\cos(x/2) - \cos((2n + 1)x/2)}{2 \sin(x/2)} \Big)$$

 (ii) Show that $|D_n(x)| \le 1/\sin(\delta/2)$ uniformly for all x of $0 < \delta \le |x| \le \pi$ and $n \in \mathbb{N}$.
 (iii) Show that the Dirichlet kernel $\{D_n(x)\}$ cannot satisfy (1.1.7). (Hint: Consider the value of $D_n(x)$ at $x = \pi k/n$ for $k = 1, 2, \ldots, n$)

6. (i) Prove (1.2.24) and (1.2.49). (Hint: See the preceding Problem)
   (ii) Show that $F_n(x) \le 1/\sin(\delta/2)$ uniformly for all x of $0 < \delta \le |x| \le \pi$ and $n \in \mathbb{N}$.
   (iii) Show that $(n+1)F_n(x) = |\sum_{k=0}^{n} e^{ikx}|^2$.
   (iv) Show that $F_n(x) \le n+1$ for all x. (Hint: Use that $|D_n(x)| \le 2n+1$ for all x)
   (v) Show that $F_n(x) \le \pi^2 n$ for $|x| \le 1/n$, $n \ge 1$, and $F_n(x) \le \pi^2/(n+1)x^2$ for $1/n < |x| \le \pi$ (see Problem 1.2.4(ii)). Thus, setting $F_n^*(x) = 2\pi^2 n/(1+n^2x^2)$, then $F_n(x) \le F_n^*(x)$ for $|x| \le \pi$ and $n \ge 3$. Show that $\int_{-\pi}^{\pi} F_n^*(u)\,du \le 2\pi^3$ for all n.
7. (i) Let $a_0, a_1, a_2, \ldots$ and $b_0, b_1, b_2, \ldots$ be any complex numbers, and define $A_{-1} = 0$ and $A_k = \sum_{j=0}^{k} a_j$ for $k \ge 0$. Show that for $0 \le m \le n$
$$\sum_{k=m}^{n} a_k b_k = \sum_{k=m}^{n-1} A_k(b_k - b_{k+1}) - A_{m-1}b_m + A_n b_n.$$
   This formula is known as *Abel's transformation* or formula for partial summation.
   (ii) Show that $|\sum_{k=m}^{n} \lambda_k e^{ikx}| \le \lambda_m |\sin(x/2)|^{-1}$ for all $x \ne 2\pi j$, $j \in \mathbb{N}$, provided λ_k is real and $\lambda_m \ge \lambda_{m+1} \ge \cdots \ge \lambda_n \ge 0$. (Hint: Use (i) and Problem 1.2.4(iii))
   (iii) Let $\{\lambda_k\}_{k=0}^{\infty}$ be a sequence of positive real numbers which decrease monotonely to zero as $k \to \infty$. Show that the series $\sum_{k=0}^{\infty} \lambda_k \cos kx$ and $\sum_{k=1}^{\infty} \lambda_k \sin kx$ converge except possibly at $x = 2\pi j$, $j \in \mathbb{Z}$. The convergence is indeed uniform in every interval $[\delta, 2\pi - \delta]$, $0 < \delta < \pi$.
   (iv) Let $\{\lambda_k\}$ be as in (iii) and set
$$f(x) = (\lambda_0/2) + \sum_{k=1}^{\infty} \lambda_k \cos kx, \qquad g(x) = \sum_{k=1}^{\infty} \lambda_k \sin kx.$$
   As to f, show that if $f \in \mathsf{L}_{2\pi}^1$, then the series is the Fourier series of f; in other words, for $k \in \mathbb{P}$
$$\frac{1}{\pi}\int_{-\pi}^{\pi} f(u)\cos ku\,du = \lambda_k, \qquad \frac{1}{\pi}\int_{-\pi}^{\pi} f(u)\sin ku\,du = 0.$$
   State and prove similar assertions for g. (Hint: Hardy–Rogosinski [1, p. 32], Hewitt [1, p. 59]; compare also with Sec. 6.3.2, particularly Problems 6.3.2, 6.3.3)
8. (i) Let f be an integrable function over an interval (a, b). Show that
$$\lim_{\rho\to\infty}\int_a^b f(u)\cos\rho u\,du = 0 = \lim_{\rho\to\infty}\int_a^b f(u)\sin\rho u\,du.$$
   (Hint: Prove the above relations for characteristic functions of bounded intervals, proceed to step functions, and then use the fact that f may be approximated in the mean arbitrarily closely by step functions, see also Asplund–Bungart [1, p. 425])
   (ii) Let $f \in \mathsf{L}_{2\pi}^1$. Show that $\lim_{k\to\infty} f_c^\wedge(k) = \lim_{k\to\infty} f_s^\wedge(k) = \lim_{k\to\infty} f^\wedge(k) = 0$. (Hint: Use (i); see also Prop. 4.1.2)
9. (i) Show that the Fourier series of $f \in \mathsf{L}_{2\pi}^1$ converges at a given fixed point x_0 to the value c if and only if for some $0 < \delta < \pi$
$$\lim_{n\to\infty}\frac{1}{\pi}\int_0^{\delta}[f(x_0+u) + f(x_0-u) - 2c]\frac{\sin((2n+1)u/2)}{u}\,du = 0.$$
   This is known as *Riemann's principle of localization*. (Hint: Setting $g(x_0, u)$ as in Problem 1.2.3, then
$$\begin{aligned} S_n(f; x_0) - c &= \frac{1}{\pi}\int_0^{\delta} g(x_0,u)\frac{\sin((2n+1)u/2)}{u}\,du \\ &\quad + \frac{1}{\pi}\int_0^{\delta} g(x_0,u)\left[\frac{1}{2\sin(u/2)} - \frac{1}{u}\right]\sin((2n+1)u/2)\,du \\ &\quad + \frac{1}{2\pi}\int_{\delta}^{\pi}\frac{g(x_0,u)}{\sin(u/2)}\sin((2n+1)u/2)\,du \equiv I_1 + I_2 + I_3. \end{aligned}$$
   However, $\lim_{n\to\infty}(I_2 + I_3) = 0$ by Problem 1.2.8(i))

(ii) Let $f \in \mathsf{L}^1_{2\pi}$. If there is a point x_0 and $0 < \delta < \pi$ such that

$$\int_0^\delta \left| \frac{f(x_0 + u) + f(x_0 - u) - 2f(x_0)}{u} \right| du < \infty,$$

show that $\lim_{n \to \infty} S_n(f; x_0) = f(x_0)$. This is known as *Dini's condition* for the convergence of Fourier series. (Hint: Use (i) and Problem 1.2.8(i))

(iii) If $f \in \mathsf{L}^1_{2\pi}$ is differentiable at x_0, show that the Fourier series of f at x_0 converges to $f(x_0)$. (Hint: ASPLUND–BUNGART [1, pp. 430–431])

10. (i) Show that there is a constant $M > 0$ such that $|\int_a^b (\sin u/u)\, du| \le M$ for all a, $b \in \mathbb{R}$. (Hint: Use the *second mean value theorem*: If f is integrable and g positive and monotonely increasing on $[a, b]$, then there is a $\xi \in [a, b]$ such that $\int_a^b g(u)f(u)\, du = g(b -)\int_\xi^b f(u)\, du$)

(ii) Let $f \in \mathsf{L}^1_{2\pi}$ and suppose that f is of bounded variation in an interval $[x_0 - \delta, x_0 + \delta]$ around a point x_0. Show that the Fourier series of f converges at this point x_0 to the sum $[f(x_0 +) + f(x_0 -)]/2$. This is *Jordan's theorem*. (Hint: By Problems 1.2.3, 1.2.9(i) and the Jordan decomposition it suffices to show that for some $0 < \delta < \pi$

$$\lim_{n \to \infty} \frac{1}{\pi} \int_0^\delta \frac{\sin((2n + 1)u/2)}{u} g(u)\, du = 0,$$

where $g(x)$ is positive and monotonely increasing on $[0, \delta]$ with $g(0+) = 0$. To this end, by the second mean value theorem

$$\int_0^\delta g(u) \frac{\sin((2n + 1)u/2)}{u} du = g(\eta -) \int_\xi^\eta \frac{\sin((2n + 1)u/2)}{u} du + \int_\eta^\delta g(u) \frac{\sin((2n + 1)u/2)}{u} du.$$

Now apply (i) and Problem 1.2.8(i))

(iii) Let $f \in \mathsf{BV}_{loc}$ be 2π-periodic. Show that the Fourier series of f converges everywhere to the value $f(x)$. This is referred to as *Jordan's criterion*. (Hint: ASPLUND–BUNGART [1, pp. 432–436])

11. Let $f \in \mathsf{BV}_{loc}$ be 2π-periodic and denote the total variation of f over $[-\pi, \pi]$ by V.

(i) Show that $|f^\wedge(k)| \le V/(2\pi|k|)$ for every $k \ne 0$.

(ii) Show that $|S_n(f; x)| \le \|f\|_\infty + V$ for every x and $n \in \mathbb{N}$. (Hint: Show that

$$S_n(f; x) = \sigma_n(f; x) + \frac{1}{n + 1} \sum_{k=-n}^{n} |k| f^\wedge(k)\, e^{ikx}.$$

Now use (i) and Cor. 1.2.4, see also ACHIESER [2, p. 99])

(iii) Show that the partial sums of the Fourier series of f converge *boundedly* to f, i.e., $\lim_{n \to \infty} S_n(f; x) = f(x)$ a.e. and there is a constant M such that $|S_n(f; x)| \le M$ uniformly for all $n \in \mathbb{N}$.

12. Let f be a continuous function on the finite interval $[a, b]$ and $\varepsilon > 0$. Show that there exists a natural number $n = n(\varepsilon)$ and an algebraic polynomial $p_n(x)$ of degree n such that $|f(x) - p_n(x)| < \varepsilon$, uniformly for all $x \in [a, b]$. (Hint: Reduce by a linear substitution to the case $[-\pi/2, \pi/2]$, define f outside this interval so as to become a function in $\mathsf{C}_{2\pi}$, apply Theorem 1.2.5, and expand the trigonometric polynomial by Taylor's theorem into a power series; see also HARDY–ROGOSINSKI [1, p. 22])

13. (i) Let $f \in \mathsf{C}_{2\pi}$. If $f^\wedge(k) = 0$ for $|k| > n$, show that $f \in \mathsf{T}_n$.

(ii) Show that if the Fourier series of $f \in \mathsf{C}_{2\pi}$ converges uniformly in x, then it represents $f(x)$ at each x. (Hint: Use Cor. 1.2.7)

14. (i) If $\{\chi_\rho(x)\}$ is a kernel, show that $I_\rho(e^{in\circ}; x) = \chi_\rho^\wedge(n)\, e^{inx}$.

(ii) Let $\{\chi_\rho(x)\}$ be a kernel and t_n a complex trigonometric polynomial with coefficients c_k. Show that $I_\rho(t_n; x) = \sum_{k=-n}^{n} \chi_\rho^\wedge(k) c_k\, e^{ikx}$.

(iii) If t_n is given as in (ii), show that $S_m(t_n; x) = t_n(x)$ for $m \geq n$.
(iv) If $\{\chi_\rho(x)\}$ is an approximate identity, show that $\lim_{\rho\to\rho_0} \chi_\rho^\wedge(k) = 1$ for each $k \in \mathbb{Z}$. (Hint: Use Theorem 1.1.5)

15. (i) Let $f, g \in L^1_{2\pi}$. Show that if one of the factors f or g is a trigonometric polynomial of degree n, so is the convolution product $f * g$.
(ii) Let $\{\chi_n(x)\}$ be a continuous kernel with parameter $n \in \mathbb{N}$. The singular integral $I_n(f; x)$ is called *polynomial* if it maps $X_{2\pi}$ into T_n, thus if $I_n(f; x) \in T_n$ for every $f \in X_{2\pi}$. Show that a singular integral is polynomial if and only if $\chi_n \in T_n$. (Hint: If $I_n(f; x)$ is polynomial, show that $\chi_n^\wedge(k) = 0$ for $|k| > n$; use Problems 1.2.13, 1.2.14, see also BUTZER–NESSEL–SCHERER [1])

16. (i) The (first) *Cesàro means* of a sequence $\{a_n\}_{n=0}^\infty$ of complex numbers are defined as the sequence $\{A_n\}$, $A_n = (1/(n+1)) \sum_{k=0}^n a_k$. Show that if $\lim_{n\to\infty} a_n = a$, then also $\lim_{n\to\infty} A_n = a$.
(ii) Show that, at a point x_0, the Fourier series of a function $f \in C_{2\pi}$ either converges to $f(x_0)$ or diverges. (Hint: Use (i) and Cor. 1.2.4)

17. The *modified partial sums* $S_n^*(f; x)$ of the Fourier series of $f \in L^1_{2\pi}$ are defined (see ZYGMUND [7I, p. 50]) by

$$\begin{aligned} S_n^*(f; x) &= \tfrac{1}{2} f_c^\wedge(0) + \sum_{k=1}^{n-1} \{f_c^\wedge(k) \cos kx + f_s^\wedge(k) \sin kx\} \\ &\quad + \tfrac{1}{2}\{f_c^\wedge(n) \cos nx + f_s^\wedge(n) \sin nx\} \\ &= \sum_{k=-(n-1)}^{n-1} f^\wedge(k)\, e^{ikx} + \tfrac{1}{2}\{f^\wedge(-n)\, e^{-inx} + f^\wedge(n)\, e^{inx}\} \\ &= \tfrac{1}{2}[S_n(f; x) + S_{n-1}(f; x)]. \end{aligned}$$

(i) Show that $S_n^*(f; x) = (1/2\pi) \int_{-\pi}^{\pi} f(x-u) D_n^*(u)\, du$, where the *modified Dirichlet kernel* $\{D_n^*(x)\}$ is given through $D_n^*(x) = D_n(x) - \cos nx$.
(ii) Show that $D_n^*(x) = \cot(x/2) \sin nx$ for all $x \neq 2\pi j$, $j \in \mathbb{Z}$.
(iii) Show that $\lim_{n\to\infty} \|S_n(f; \circ) - S_n^*(f; \circ)\|_{C_{2\pi}} = 0$ for every $f \in X_{2\pi}$. (Hint: Use Problem 1.2.8(ii))

18. (i) Prove (1.2.37) and (1.2.50). (Hint: Use the fact that $1 + 2\sum_{k=1}^\infty z^k = (1+z)/(1-z)$ for $|z| < 1$, set $z = re^{ix}$, and consider the real and imaginary part of both sides)
(ii) Show that

$$p_r(x) = \frac{1-r^2}{(1-r)^2 + 2r(1-\cos x)} = \frac{1-r^2}{(1-r)^2 + 4r \sin^2(x/2)}.$$

Hence, for each $0 \leq r < 1$, $p_r(x)$ is an even, positive, and continuous function of x. In fact, $p_r(x), p_r^\sim(x) \in C^\infty_{2\pi}$ for each $0 \leq r < 1$. Show that $p_r'(x) \leq 0$ for $x \in [0, \pi]$, thus $p_r(x)$ is a monotonely decreasing function of x on $[0, \pi]$ for each $0 < r < 1$.
(iii) Show that $(1-r)/(1+r) \leq p_r(x) \leq (1+r)/(1-r)$, thus $p_r(x) \leq 2/(1-r)$ for all x and $0 \leq r < 1$. Furthermore, $p_r(x) \leq \pi^2(1-r)/2rx^2$ for $0 \leq x \leq \pi$ and $0 < r < 1$. Show that the kernel $\{p_r(x)\}$ satisfies (1.1.7).
(iv) Show that for all x and $0 \leq r < 1$

$$\sum_{k=1}^\infty \frac{1}{k} r^k \cos kx = -\tfrac{1}{2} \log(1 - 2r\cos x + r^2).$$

19. Let $f \in C_{2\pi}$ be continuously differentiable. Show that

(i) $\lim_{n\to\infty} \sigma_n'(f; x) = f'(x)$, (ii) $\lim_{r\to 1-} P_r'(f; x) = f'(x)$

uniformly for all x. (Hint: Use Prop. 1.1.15)

20. Let $\{\theta_n(k)\}$ be a row-finite θ-factor with $m(n) = n$. Setting $\theta_n(n+1) = 0$, suppose that, for each $n \in \mathbb{N}$, $\Delta^2\theta_n(k) \equiv \theta_n(k+2) - 2\theta_n(k+1) + \theta_n(k) \geq 0$ for all $k = 0, 1, \ldots, n-1$. Show that $\|C_n\|_1 \leq \frac{1}{2} + M(n+1)|\theta_n(n)|$, the kernel $\{C_n(x)\}$ being given by

(1.2.41). If $\theta_n(n) \geq 0$, one may take the constant M to be zero. (Hint: Apply Abel's transformation twice to $C_n(x)$; see also GERONIMUS [2] and Sec. 6.3.2).

21. For each $n \in \mathbb{N}$ let $t_n \in \mathsf{T}_n$ be such that t_n is positive and does not vanish identically. Suppose that to each $\varepsilon > 0$, $0 < \delta < \pi$ there corresponds an n_0 such that

$$\int_\delta^{2\pi-\delta} t_n(u)\,du < \varepsilon \int_0^{2\pi} t_n(u)\,du$$

for all $n \geq n_0$. Show that, for every $f \in \mathsf{C}_{2\pi}$,

$$\lim_{n\to\infty} (2\pi \|t_n\|_1)^{-1} \int_0^{2\pi} f(x-u)t_n(u)\,du = f(x)$$

uniformly for all x. (Hint: Theorem 1.1.5; see also PERRON [1])

22. (i) Show that the integral means of $f \in \mathsf{X}_{2\pi}$ admit the representation

$$A_h(f; x) = \sum_{k=-\infty}^{\infty} \frac{\sin(hk/2)}{(hk/2)} f^\wedge(k)\, e^{ikx},$$

the value at $k = 0$ being $f^\wedge(0)$. (Hint: Use Problems 1.1.7(ii), 1.2.10(iii))

(ii) Show that for the rth integral means of $f \in \mathsf{X}_{2\pi}$

$$A_h^r(f; x) = \sum_{k=-\infty}^{\infty} \left(\frac{\sin(hk/2)}{hk/2}\right)^r f^\wedge(k)\, e^{ikx},$$

the value at $k = 0$ being $f^\wedge(0)$. The right-hand side defines *Riemann's method* of summation of order r of the Fourier series of f. (Hint: For $r \geq 2$ one may apply Problem 1.2.13(ii))

1.3 Test Sets for Norm-Convergence

We return to questions concerning norm-convergence of the integral $I_\rho(f; x)$ towards f. In Theorem 1.1.5 we saw that one sufficient condition for convergence is that the corresponding kernel is an approximate identity. Another approach to this problem is the method of test functions. The problem here is to find particular sets of functions, so-called *test functions*, for which the integral $I_\rho(f; x)$ is defined such that, under suitable conditions upon the kernel, the norm-convergence for these test functions implies the norm-convergence for *all* functions of $\mathsf{X}_{2\pi}$.

1.3.1 Norms of Some Convolution Operators

By Prop. 1.1.3 the integral $I_\rho(f; x)$ forms a bounded linear transformation I_ρ on $\mathsf{X}_{2\pi}$ into $\mathsf{X}_{2\pi}$, the norm of which satisfies

$$\|I_\rho\|_{[\mathsf{X}_{2\pi},\mathsf{X}_{2\pi}]} \leq \|\chi_\rho\|_1 \qquad (\rho \in \mathbb{A}). \tag{1.3.1}$$

We are now interested in replacing the inequality in (1.3.1) by an equality and thus to give, for certain pairs of spaces $\mathsf{X}_{2\pi}$, an actual representation of the norm of the operator I_ρ.

Proposition 1.3.1. *Let $\{\chi_\rho(x)\}$ be a continuous kernel. Then for each $\rho \in \mathbb{A}$*

$$\|I_\rho\|_{[\mathsf{C}_{2\pi},\mathsf{C}_{2\pi}]} = \|\chi_\rho\|_1, \tag{1.3.2}$$

$$\|I_\rho\|_{[\mathsf{L}^p_{2\pi},\mathsf{C}_{2\pi}]} = \|\chi_\rho\|_{p'} \qquad (1 \leq p \leq \infty). \tag{1.3.3}$$

Proof. Let $\rho \in \mathbb{A}$ be fixed. Since $\{\chi_\rho(x)\}$ is a continuous kernel, the integral $I_\rho(f; x)$ defines a bounded linear transformation on $X_{2\pi}$ into $C_{2\pi}$ by Prop. 0.4.1. Therefore, for each $f \in X_{2\pi}$, there exists a point $x_f \in [-\pi, \pi]$ such that $\|I_\rho(f; \circ)\|_{C_{2\pi}} = |I_\rho(f; x_f)|$. This implies by (0.7.5)

$$\|I_\rho\|_{[X_{2\pi}, C_{2\pi}]} = \sup_{\|f\|_{X_{2\pi}}=1} |I_\rho(f; x_f)| = \| \sup_{\|f\|_{X_{2\pi}}=1} |I_\rho(f; \circ)| \|_{C_{2\pi}}.$$

Moreover, by Prop. 0.8.8, 0.8.9 and the periodicity of the functions in question, it follows that for each fixed $x \in [-\pi, \pi]$

$$\sup_{\|f\|_{X_{2\pi}}=1} |I_\rho(f; x)| = \sup_{\|f\|_{X_{2\pi}}=1} \left| \frac{1}{2\pi} \int_{-\pi}^{\pi} f(u)\chi_\rho(x-u)\, du \right| = \begin{cases} \|\chi_\rho\|_1, & X_{2\pi} = C_{2\pi} \\ \|\chi_\rho\|_{p'}, & X_{2\pi} = L^p_{2\pi}, \end{cases}$$

which proves (1.3.2) and (1.3.3) for $1 \le p < \infty$. The case $p = \infty$ follows similarly.

Proposition 1.3.2. *Let $\{\chi_\rho(x)\}$ be an even and continuous kernel. Then*

$$\|I_\rho\|_{[L^1_{2\pi}, X_{2\pi}]} = \|\chi_\rho\|_{X_{2\pi}} \qquad (\rho \in \mathbb{A}). \tag{1.3.4}$$

Proof. Let $1 \le p \le \infty$. Since the kernel is even, we have by Fubini's theorem for every $f \in L^p_{2\pi}$ and $h \in L^{p'}_{2\pi}$

$$\int_{-\pi}^{\pi} I_\rho(f; x)h(x)\, dx = \int_{-\pi}^{\pi} f(x)I_\rho(h; x)\, dx \qquad (\rho \in \mathbb{A}). \tag{1.3.5}$$

Using Prop. 0.8.8, 0.8.9 again, we obtain

$$\|I_\rho\|_{[L^1_{2\pi}, L^p_{2\pi}]} = \sup_{\|f\|_1=1} \|I_\rho(f; \circ)\|_p = \sup_{\|f\|_1=1} \sup_{\|h\|_{p'}=1} \left| \frac{1}{2\pi} \int_{-\pi}^{\pi} I_\rho(f; x)h(x)\, dx \right|$$
$$\le \sup_{\|f\|_1=1} \sup_{\|h\|_{p'}=1} \|f\|_1 \|I_\rho(h; \circ)\|_{C_{2\pi}} = \sup_{\|h\|_{p'}=1} \|I_\rho(h; \circ)\|_{C_{2\pi}} = \|\chi_\rho\|_p,$$

the last equality being true by (1.3.3). Consequently, we have for the one thing

$$\|I_\rho\|_{[L^1_{2\pi}, L^p_{2\pi}]} \le \|\chi_\rho\|_p \qquad (\rho \in \mathbb{A}), \tag{1.3.6}$$

and for another, by the same arguments,

$$\|\chi_\rho\|_p = \sup_{\|h\|_{p'}=1} \|I_\rho(h; \circ)\|_{C_{2\pi}} = \sup_{\|h\|_{p'}=1} \sup_{\|f\|_1=1} \left| \frac{1}{2\pi} \int_{-\pi}^{\pi} f(x)I_\rho(h; x)\, dx \right|$$
$$\le \sup_{\|h\|_{p'}=1} \sup_{\|f\|_1=1} \|h\|_{p'} \|I_\rho(f; \circ)\|_p = \sup_{\|f\|_1=1} \|I_\rho(f; \circ)\|_p = \|I_\rho\|_{[L^1_{2\pi}, L^p_{2\pi}]},$$

which together with (1.3.6) proves (1.3.4) for $X_{2\pi} = L^p_{2\pi}$, $1 \le p < \infty$. Since we have also shown that $\|I_\rho\|_{[L^1_{2\pi}, L^\infty_{2\pi}]} = \|\chi_\rho\|_{C_{2\pi}}$, and since $\|I_\rho(f; \circ)\|_\infty = \|I_\rho(f; \circ)\|_{C_{2\pi}}$ for every $f \in L^1_{2\pi}$, Prop. 1.3.2 is completely established.

1.3.2 Some Applications of the Theorem of Banach–Steinhaus

The theorem of Banach–Steinhaus, together with Theorem 1.2.5, gives a first result concerning sets of test functions for norm-convergence of $I_\rho(f; x)$ to f.

Proposition 1.3.3. *Let the kernel $\{\chi_\rho(x)\}$ of the singular integral $I_\rho(f; x)$ satisfy* (1.1.5). *If $\lim_{\rho\to\rho_0} \|I_\rho(h; \circ) - h(\circ)\|_{X_{2\pi}} = 0$ for all elements h of a set $A \subset X_{2\pi}$ which is dense in $X_{2\pi}$, then for every $f \in X_{2\pi}$*

$$\lim_{\rho\to\rho_0} \|I_\rho(f; \circ) - f(\circ)\|_{X_{2\pi}} = 0. \tag{1.3.7}$$

In particular, the real as well as the complex trigonometric system forms a test set for norm-convergence in $X_{2\pi}$.

In other words, if the kernel satisfies (1.1.5), and if (1.3.7) holds for each function $\cos kx$, $\sin kx$, $k \in \mathbb{P}$, then (1.3.7) holds for every $f \in X_{2\pi}$. Thus we have found a denumerable test set. We observe that this test set belongs to $X_{2\pi}$. But we emphasize that the functions of the test set need not necessarily belong to the space $X_{2\pi}$ but only to the domain of definition of the operators I_ρ.

As an example, let us consider the *singular integral of Rogosinski* defined for $f \in X_{2\pi}$ by

$$(1.3.8) \qquad B_n(f; x) = \frac{1}{2\pi} \int_{-\pi}^{\pi} f(x-u) b_n(u)\, du$$

with kernel given through

$$(1.3.9) \qquad b_n(x) = \frac{1}{2}\left[D_n\left(x + \frac{\pi}{2n+1}\right) + D_n\left(x - \frac{\pi}{2n+1}\right)\right],$$

$\{D_n(x)\}$ being Dirichlet's kernel. By Problem 1.3.2

$$(1.3.10) \qquad \int_{-\pi}^{\pi} |b_n(u)|\, du \le 4\pi^2 \qquad (n \in \mathbb{N}),$$

$$(1.3.11) \qquad b_n(x) = 1 + 2 \sum_{k=1}^{n} \cos \frac{k\pi}{2n+1} \cos kx.$$

Thus the Rogosinski kernel consists of cosine polynomials of degree n and satisfies (1.1.5); it is a kernel generated via (1.2.32) by the row-finite θ-factor $\theta_n(k) = \cos(k\pi/(2n+1))$ for $|k| \le n$, $= 0$ for $|k| > n$. Furthermore, by Problem 1.2.14(iii)

$$(1.3.12) \qquad B_n(t_m; x) = \frac{1}{2}\left\{t_m\left(x + \frac{\pi}{2n+1}\right) + t_m\left(x - \frac{\pi}{2n+1}\right)\right\}$$

for every trigonometric polynomial t_m of degree $m \le n$. Therefore

$$2\,\|B_n(t_m; \circ) - t_m(\circ)\|_{X_{2\pi}} \le \left\| t_m\left(\circ + \frac{\pi}{2n+1}\right) - t_m(\circ)\right\|_{X_{2\pi}} + \left\| t_m\left(\circ - \frac{\pi}{2n+1}\right) - t_m(\circ)\right\|_{X_{2\pi}}.$$

By the uniform continuity of the t_m, the right-hand side tends to zero as $n \to \infty$ for each t_m, i.e., the singular integral of Rogosinski satisfies the assumptions of Prop. 1.3.3, A being the set of all trigonometric polynomials. Therefore

Proposition 1.3.4. *For every $f \in X_{2\pi}$ the singular integral of Rogosinski converges in $X_{2\pi}$-norm to f as $n \to \infty$.*

It is to be noted that the result of Prop. 1.3.3, together with the uniform boundedness principle, even provides necessary and sufficient conditions such that (1.1.8) holds in $C_{2\pi}$- or $L^1_{2\pi}$-space. This is due to the fact that we have equality in relations (1.3.2) and (1.3.4). Indeed,

Theorem 1.3.5. *Let $\{\chi_\rho(x)\}$ be a continuous kernel. In order that* (1.3.7) *be valid in $C_{2\pi}$-norm for each $f \in C_{2\pi}$, it is necessary and sufficient that*

$$(1.3.13)\quad \begin{aligned} &\text{(i)}\ \|\chi_\rho\|_1 \le M && (\rho \in \mathbb{A}),\\ &\text{(ii)}\ \lim_{\rho\to\rho_0} \chi_\rho^\wedge(k) = 1 && (k \in \mathbb{Z}). \end{aligned}$$

The same assertion is valid in $L^1_{2\pi}$-space.

Proof. *Necessity.* Since (1.3.7) holds for every f of the Banach space $X_{2\pi} = C_{2\pi}$ or $L^1_{2\pi}$, Prop. 0.7.2 implies that the norms of the operators I_ρ are uniformly bounded, i.e., there is some constant $M > 0$ such that $\|I_\rho\|_{[X_{2\pi},X_{2\pi}]} \le M$ for all $\rho \in \mathbb{A}$. Thus (1.3.13)(i) holds by (1.3.2) or (1.3.4) (see also Problem 1.3.1(i)). Furthermore, if we apply (1.3.7) to the complex (or real) trigonometric system, we obtain (1.3.13)(ii) since for any $k \in \mathbb{Z}$

$$(1.3.14)\quad \begin{aligned} |\chi_\rho^\wedge(k) - 1| &= \|I_\rho(e^{iku};\circ) - e^{ik\circ}\|_{X_{2\pi}}\\ &\le \|I_\rho(\cos ku;\circ) - \cos k\circ\|_{X_{2\pi}} + \|I_\rho(\sin ku;\circ) - \sin k\circ\|_{X_{2\pi}}. \end{aligned}$$

Sufficiency. According to Prop. 1.3.3 we only have to verify (1.3.7) for all trigonometric polynomials. Since by Problem 1.2.14(ii)

$$(1.3.15)\qquad I_\rho(t_n;x) = \sum_{k=-n}^{n} \chi_\rho^\wedge(k) c_k\, e^{ikx}$$

for each $t_n \in T_n$ with coefficients c_k, this follows by (1.3.13)(ii), the sum being finite.

Concerning the spaces $L^p_{2\pi}$, $1 < p < \infty$, conditions (1.3.13) are sufficient but not necessary for (1.3.7) to be valid (see Problem 1.3.3(ii) and the remarks at the end of this subsection).

Theorem 1.3.5 may be applied in two ways. Since conditions (1.3.13) are sufficient, they may be used to demonstrate the convergence of singular integrals. Since they are necessary in $C_{2\pi}$ or $L^1_{2\pi}$, they may furthermore be used to establish the impossibility of convergence. For the first type of application consider again the integral of Rogosinski. Here (1.3.13)(i) is given by (1.3.10), whereas (ii) follows since $\lim_{n\to\infty} \cos(k\pi/(2n+1)) = 1$ for each $k \in \mathbb{Z}$. Thus Prop. 1.3.4 is a consequence of Theorem 1.3.5 (and Problem 1.3.3(ii)). As to the second type of application we have

Proposition 1.3.6. *There exists a function $f \in C_{2\pi}$ whose Fourier series does not converge uniformly.*

The proof follows immediately by Prop. 1.2.3 which shows that condition (1.3.13)(i) fails to hold for the singular integral of Dirichlet. Likewise one proves that $\lim_{n\to\infty} \|S_n(f;\circ) - f(\circ)\|_p = 0$ does not hold for every $f \in L^p_{2\pi}$ if $p = 1$. On the other hand, the latter assertion is valid if $1 < p < \infty$ (see Theorem 9.3.6). This would imply that conditions (1.3.13) are indeed not necessary for (1.3.7) to be valid in $L^p_{2\pi}$-space, $1 < p < \infty$.

1.3.3 Positive Kernels

As we have seen, the functions $\cos kx$, $\sin kx$, $k \in \mathbb{P}$, form a test set for the norm-convergence (1.3.7) of the integral $I_\rho(f; x)$ in case the kernel $\{\chi_\rho(x)\}$ satisfies (1.3.13)(i). Now it would be of interest to see whether this denumerable set of test functions can be reduced to a finite set. This will be possible if the kernel under consideration is (real and) positive.

Theorem 1.3.7. *Let the kernel* $\{\chi_\rho(x)\}$ *of the integral* (1.1.3) *be positive. The following four assertions are equivalent:*

(i) $\lim_{\rho\to\rho_0} \|I_\rho(f; \circ) - f(\circ)\|_{X_{2\pi}} = 0$ *for every* $f \in X_{2\pi}$,

(ii) $\lim_{\rho\to\rho_0} \|I_\rho(\cos u; \circ) - \cos \circ\|_{X_{2\pi}} = 0, \qquad \lim_{\rho\to\rho_0} \|I_\rho(\sin u; \circ) - \sin \circ\|_{X_{2\pi}} = 0,$

(iii) $\lim_{\rho\to\rho_0} I_\rho\left(\sin^2 \frac{u}{2}; 0\right) = 0,$

(iv) $\lim_{\rho\to\rho_0} \int_{\delta\le|u|\le\pi} \chi_\rho(u)\,du = 0 \qquad (0 < \delta < \pi).$

Proof. If (i) is satisfied, (ii) follows trivially since $\cos x$ and $\sin x$ belong to $X_{2\pi}$. Suppose (ii) is valid. In view of (1.1.1) we have for each $x \in \mathbb{R}$

$$(1.3.16) \qquad I_\rho\left(\sin^2 \frac{u}{2}; 0\right) = \frac{1}{2\pi}\int_{-\pi}^{\pi} \sin^2\left(\frac{x-u}{2}\right)\chi_\rho(x-u)\,du$$
$$= \tfrac{1}{2}\{1 - \cos x\, I_\rho(\cos u; x) - \sin x\, I_\rho(\sin u; x)\}.$$

Now the right-hand side converges in $X_{2\pi}$-norm to $\{1 - \cos^2 x - \sin^2 x\}/2 = 0$, and therefore, since the left-hand side is independent of x, relation (iii) follows. Since the kernel is positive, we have for each $0 < \delta < \pi$

$$I_\rho\left(\sin^2 \frac{u}{2}; 0\right) \ge \frac{1}{2\pi}\int_{\delta\le|u|\le\pi} \sin^2 \frac{u}{2}\chi_\rho(u)\,du \ge \frac{\sin^2(\delta/2)}{2\pi}\int_{\delta\le|u|\le\pi} \chi_\rho(u)\,du.$$

Thus (iii) implies (iv). Finally, if (iv) is satisfied, then the kernel $\{\chi_\rho(x)\}$ is an approximate identity, and the convergence in (i) follows by Theorem 1.1.5.

If we compare Theorem 1.3.7 with Theorem 1.1.5 we see that for positive kernels the convergence (1.1.8) necessarily implies that the kernel forms an approximate identity. In comparison with Prop. 1.3.3 the former theorem gives

Corollary 1.3.8. *For positive kernels the test set for the convergence of the integral* $I_\rho(f; x)$ *in* $X_{2\pi}$*-norm consists of the two functions* $\cos x$ *and* $\sin x$.

Note that the test set consists of the three functions 1, $\cos x$, $\sin x$ if the kernel satisfies (1.1.2) instead of (1.1.1) (see Problem 1.3.7).

A further generalization is possible.

Proposition 1.3.9. *Let the kernel* $\{\chi_\rho(x)\}$ *of the integral* (1.1.3) *be positive. Then* (1.3.7) *holds for each* $f \in X_{2\pi}$ *if and only if there exists a point* x_0 *such that*

$$(1.3.17) \qquad \lim_{\rho\to\rho_0} I_\rho(\cos u; x_0) = \cos x_0, \qquad \lim_{\rho\to\rho_0} I_\rho(\sin u; x_0) = \sin x_0.$$

Proof. Let (1.3.7) be satisfied. If $X_{2\pi} = C_{2\pi}$, norm-convergence means uniform convergence, and hence (1.3.17) follows trivially for every x_0. Let $X_{2\pi} = L^p_{2\pi}$, $1 \le p < \infty$. Then

$$\begin{aligned}|I_\rho(\cos u; 0) - 1| &= \left\{\frac{1}{2\pi}\int_{-\pi}^{\pi} |I_\rho(\cos u; 0) - 1|^p\,dx\right\}^{1/p}\\ &= \|\cos\circ\{I_\rho(\cos u;\circ) - \cos\circ\} + \sin\circ\{I_\rho(\sin u;\circ) - \sin\circ\}\|_p\\ &\le \|I_\rho(\cos u;\circ) - \cos\circ\|_p + \|I_\rho(\sin u;\circ) - \sin\circ\|_p,\\ |I_\rho(\sin u; 0)| &= \|\sin\circ\{I_\rho(\cos u;\circ) - \cos\circ\} - \cos\circ\{I_\rho(\sin u;\circ) - \sin\circ\}\|_p\\ &\le \|I_\rho(\cos u;\circ) - \cos\circ\|_p + \|I_\rho(\sin u;\circ) - \sin\circ\|_p.\end{aligned}$$

Therefore (1.3.17) holds for $x_0 = 0$. Furthermore, it can readily be shown that (1.3.17) holds for each real x_0.

Conversely, (1.3.16) for $x = x_0$ and (1.3.17) imply condition (iii) of Theorem 1.3.7, which completes the proof.

Thus in case of positive kernels the convergence of the integral $I_\rho(f; x)$ for only the two functions $\cos x$ and $\sin x$ at any one point x_0 implies the norm-convergence (1.3.7) for every $f \in X_{2\pi}$.

In connection with Theorem 1.3.5 we now have

Proposition 1.3.10. *Let the kernel* $\{\chi_\rho(x)\}$ *of the integral* (1.1.3) *be positive. The following assertions are equivalent:* (i) $\lim_{\rho\to\rho_0} \|I_\rho(f;\circ) - f(\circ)\|_{X_{2\pi}} = 0$ *for every* $f \in X_{2\pi}$, (ii) $\lim_{\rho\to\rho_0} \chi_\rho^\wedge(1) = 1$. *In this case,* $\lim_{\rho\to\rho_0} \chi_\rho^\wedge(k) = 1$ *for every* $k \in \mathbb{Z}$.

Proof. Obviously, (i) implies (ii) by setting $f(x) = e^{ix}$. Conversely, for $X_{2\pi} = C_{2\pi}$ it follows by (1.3.14) that $\lim_{\rho\to\rho_0} I_\rho(e^{iu}; x) = e^{ix}$ uniformly for all x; in other words, for any point x_0

$$\lim_{\rho\to\rho_0} [I_\rho(\cos u; x_0) + iI_\rho(\sin u; x_0)] = \cos x_0 + i\sin x_0.$$

Thus (1.3.17) is satisfied, and (i) follows for $C_{2\pi}$-space by Prop. 1.3.9. If $X_{2\pi} = L^p_{2\pi}$, $1 \le p < \infty$, then (1.3.14) delivers

$$\lim_{\rho\to\rho_0} \|[I_\rho(\cos u;\circ) - \cos\circ] + i[I_\rho(\sin u;\circ) - \sin\circ]\|_{X_{2\pi}} = 0.$$

This in turn implies that

$$\lim_{\rho\to\rho_0} \|I_\rho(\cos u;\circ) - \cos\circ\|_{X_{2\pi}} = 0, \qquad \lim_{\rho\to\rho_0} \|I_\rho(\sin u;\circ) - \sin\circ\|_{X_{2\pi}} = 0.$$

Now (i) follows by Theorem 1.3.7.

Therefore, for positive kernels, the test set for the convergence of the integral $I_\rho(f; x)$ in $X_{2\pi}$-norm consists of the one function e^{ix}, and the test condition is simply (ii) of the preceding proposition.

Problems

1. (i) Prove Prop. 1.3.2 for arbitrary (not necessarily even) continuous kernels $\{\chi_\rho(x)\}$. (Hint: Use Problem 1.1.1(ii))

(ii) Let $\mu \in \mathsf{BV}_{2\pi}$ and define an operator U of $\mathsf{C}_{2\pi}$ into $\mathsf{C}_{2\pi}$ through $Uf = f * d\mu$, $f \in \mathsf{C}_{2\pi}$. Show that $\|U\|_{[\mathsf{C}_{2\pi},\mathsf{C}_{2\pi}]} = \|\mu\|_{\mathsf{BV}_{2\pi}}$. (Hint: Modify the proof of Prop. 1.3.1)

2. (i) Prove (1.3.10) and show that $\|B_n(f;\circ)\|_{\mathsf{X}_{2\pi}} \le 2\pi \|f\|_{\mathsf{X}_{2\pi}}$ for every $f \in \mathsf{X}_{2\pi}$. (Hint: Use Problem 1.2.20 or NATANSON [8I, p. 217])
(ii) Prove (1.3.11).

3. Let $\{\chi_\rho(x)\}$ be the kernel of the singular integral $I_\rho(f; x)$.
(i) Show that the conditions $\|I_\rho\|_{[\mathsf{X}_{2\pi},\mathsf{X}_{2\pi}]} \le M$ for $\rho \in \mathbb{A}$ and (1.3.13)(ii) are necessary and sufficient for (1.3.7) to be valid for every $f \in \mathsf{X}_{2\pi}$.
(ii) Show that the conditions (1.3.13) are sufficient for (1.3.7) to be valid for every $f \in \mathsf{X}_{2\pi}$.
(iii) Show that condition (1.3.13)(ii) is necessary for (1.3.7) to be valid for every $f \in \mathsf{X}_{2\pi}$ (compare also with Problem 1.2.14(iv)).

4. Show that $\|S_n(f;\circ)\|_{\mathsf{C}_{2\pi}} = o(\log n)$, $n \to \infty$, for every $f \in \mathsf{C}_{2\pi}$. On the other hand, if $\lambda_n = o(\log n)$, $n \to \infty$, prove that there exists an $f \in \mathsf{C}_{2\pi}$ such that the sequence $\{S_n(f; 0)/\lambda_n\}$ is unbounded for $n \to \infty$. (Hint: Use Prop. 1.2.3 and the theorem of Banach–Steinhaus, see also RUDIN [4, p. 116])

5. (i) Let x_0 be any preassigned point. Show that there are functions $f \in \mathsf{C}_{2\pi}$ whose Fourier series are divergent at this point x_0. (Hint: For any x_0, $\{S_n(f; x_0)\}$ defines a sequence of bounded linear functionals S_n on $\mathsf{C}_{2\pi}$, the norms of which satisfy $\|S_n\| = L_n = O(\log n)$)
(ii) Show that $S_n(d\mu; x)$ does not converge for every x and $\mu \in \mathsf{BV}_{2\pi}$.

6. Let $\{\theta_\rho(k)\}$ be a θ-factor. Show that, for every $f \in \mathsf{C}_{2\pi}$, $\lim_{\rho\to\rho_0} \sum_{k=-\infty}^{\infty} \theta_\rho(k) f^\wedge(k) e^{ikx} = f(x)$ uniformly in x if and only if $\int_{-\pi}^{\pi} |1 + 2\sum_{k=1}^{\infty} \theta_\rho(k) \cos ku|\, du \le M$ for $\rho \in \mathbb{A}$ and $\lim_{\rho\to\rho_0} \theta_\rho(k) = 1$ for $k \in \mathbb{Z}$. Thus a necessary condition for uniform convergence for every $f \in \mathsf{C}_{2\pi}$ is that $|\theta_\rho(k)| \le M$ for all $\rho \in \mathbb{A}$ and $k \in \mathbb{Z}$. If, for example, $\{\theta_n(k)\}$ is a row-finite θ-factor with $m(n) = n$ (cf. (1.2.41)), then a further necessary condition for uniform convergence for every $f \in \mathsf{C}_{2\pi}$ is given by $|\sum_{k=1}^{n} [\theta_n(k)/(n + 1 - k)]| \le M_1$, independent of n. (Hint: See also BARI [1II, p. 2 ff] where also certain converses of the latter assertions are given, due to NIKOLSKIĬ [4], SZ.-NAGY [6]; compare also TELJAKOVSKIĬ [3])

7. Let $\{\chi_n(x)\}_{n=1}^{\infty}$ be a sequence of positive functions belonging to $\mathsf{L}_{2\pi}^1$ for each $n \in \mathbb{N}$, and let $I_n(f; x) = (1/2\pi)\int_{-\pi}^{\pi} f(x - u)\chi_n(u)\, du$. Show that the following assertions are equivalent:
(i) $\lim_{n\to\infty} I_n(f; x) = f(x)$ uniformly in x for each $f \in \mathsf{C}_{2\pi}$,
(ii) $\lim_{n\to\infty} I_n(1; x) = 1$, $\lim_{n\to\infty} I_n(\cos\circ; x) = \cos x$, $\lim_{n\to\infty} I_n(\sin\circ; x) = \sin x$ uniformly in x,
(iii) $\lim_{n\to\infty} (1/2\pi)\int_{|u|\le\delta} \chi_n(u)\, du = 1$, $\lim_{n\to\infty} \int_{\delta\le|u|\le\pi} \chi_n(u)\, du = 0$ for each $0 < \delta < \pi$.
State and prove counterparts in $\mathsf{L}_{2\pi}^p$-space. (Hint: Modify the proof of Theorem 1.3.7; see also KOROVKIN [5, p. 17 ff] in $\mathsf{C}_{2\pi}$-space and CURTIS [1], DZJADYK [2] in $\mathsf{L}_{2\pi}^p$-space)

8. Let $\{\theta_\rho(k)\}$ be a θ-factor such that $1 + 2\sum_{k=1}^{\infty} \theta_\rho(k) \cos kx \ge 0$ for all x and $\rho \in \mathbb{A}$. Show that $\lim_{\rho\to\rho_0} \|\sum_{k=-\infty}^{\infty} \theta_\rho(k) f^\wedge(k)\, e^{ik\circ} - f(\circ)\|_{\mathsf{X}_{2\pi}} = 0$ for every $f \in \mathsf{X}_{2\pi}$ if and only if $\lim_{\rho\to\rho_0} \theta_\rho(1) = 1$. In this case, $\lim_{\rho\to\rho_0} \theta_\rho(k) = 1$ for every $k \in \mathbb{Z}$.

9. (i) Show that the absolute term a of the even trigonometric polynomial of degree $2n - 2$

$$j_n(x) = \frac{3}{n(2n^2+1)} \left[\frac{\sin(nx/2)}{\sin(x/2)}\right]^4 = a + 2[\theta_{2n-2}(1)\cos x + \cdots + \theta_{2n-2}(2n-2)\cos(2n-2)x]$$

is equal to 1, and $\theta_{2n-2}(1) = 1 - (3/(2n^2 + 1))$. (Hint: NATANSON [8I, p. 79 ff])

(ii) Show that $\{j_n(x)\}_{n=1}^{\infty}$ is an even, positive, continuous approximate identity with parameter $n \in \mathbb{N}$, $n \to \infty$. (Hint: As to condition (1.1.6), use Problem 1.3.8, Theorem 1.3.7)

(iii) The integral

$$J_n(f; x) = \frac{1}{2\pi}\int_{-\pi}^{\pi} f(x - u) j_n(u)\, du$$

is called the *singular integral of Jackson*. Show that $\lim_{n\to\infty} \|J_n(f; \circ) - f(\circ)\|_{X_{2\pi}} = 0$ for every $f \in X_{2\pi}$.

10. The (special) *singular integral of Weierstrass* is defined for $f \in X_{2\pi}$ by

$$W_t(f; x) = \frac{1}{2\pi}\int_{-\pi}^{\pi} f(x - u)\theta_3(u, t)\, du.$$

$\theta_3(x, t)$ is *Jacobi's theta-function* as given for $x \in \mathbb{R}$, $t > 0$ by

$$\theta_3(x, t) = \sum_{k=-\infty}^{\infty} e^{-tk^2} e^{ikx} = 1 + 2\sum_{k=1}^{\infty} e^{-tk^2} \cos kx.$$

(i) Show that the series above is absolutely and uniformly convergent in x for each $t > 0$, and that $\{\theta_3(x, t)\}$ constitutes a kernel with parameter $t > 0$, $t \to 0+$.

(ii) Show that $\{\theta_3(x, t)\}$ is an even, positive, continuous approximate identity. (Hint: As to the positivity of $\theta_3(x, t)$ for each $t > 0$, either use the representation

$$\theta_3(x, t) = \prod_{k=1}^{\infty} (1 - e^{-2kt})(1 + 2\, e^{-(2k-1)t} \cos x + e^{-2(2k-1)t})$$

(see ERDÉLYI [1III, p. 177]) or use (3.1.36), (5.1.61). As to condition (1.1.6), use Problem 1.3.8, Theorem 1.3.7)

(iii) Show that $\lim_{t\to 0+} \| W_t(f; \circ) - f(\circ)\|_{X_{2\pi}} = 0$ for every $f \in X_{2\pi}$.

1.4 Pointwise Convergence

Up to the present we only considered norm-convergence of the singular integral $I_\rho(f; x)$ towards $f \in X_{2\pi}$ as $\rho \to \rho_0$. In trivial cases only does this give information about pointwise convergence. For instance, if $X_{2\pi} = C_{2\pi}$ and the kernel $\{\chi_\rho(x)\}$ is an approximate identity, then Theorem 1.1.5 implies $\lim_{\rho\to\rho_0} I_\rho(f; x) = f(x)$ uniformly and hence pointwise. A slight generalization is

Proposition 1.4.1. *Let $f \in X_{2\pi}$ and the kernel $\{\chi_\rho(x)\}$ of the integral $I_\rho(f; x)$ be an approximate identity satisfying* (1.1.7).

(a) *At every point x_0 of continuity of f,* $\lim_{\rho\to\rho_0} I_\rho(f; x_0) = f(x_0)$.

(b) *If f is continuous on $(a - \eta, b + \eta)$ for some $\eta > 0$, $a < b$, $a, b \in \mathbb{R}$, then* $\lim_{\rho\to\rho_0} I_\rho(f; x) = f(x)$ *uniformly on* $[a, b]$.

(c) *If the kernel $\{\chi_\rho(x)\}$ is, in addition, even, and x_0 is such that* $\lim_{h\to 0+} [f(x_0 + h) + f(x_0 - h)] = 2c$ *exists, then* $\lim_{\rho\to\rho_0} I_\rho(f; x_0) = c$.

The proof is essentially similar to that of Theorem 1.1.5 and left to Problem 1.4.1. Unfortunately, Prop. 1.4.1 only covers the situation when the point of convergence is a point of continuity of the function f or a point at which the one-sided limits of f exist. It thus applies to continuous functions or to those of bounded variation. This by no means solves the problem for Lebesgue spaces $L_{2\pi}^p$. In fact, we are then interested in theorems which assert pointwise convergence almost everywhere of $I_\rho(f; x)$ towards f as $\rho \to \rho_0$, a much more reasonable question since altering f on a set of measure zero does not alter the singular integral. In this respect we have

Proposition 1.4.2. *Let $f \in \mathsf{L}^1_{2\pi}$ and the kernel $\{\chi_\rho(x)\}$ of the integral $I_\rho(f; x)$ be an even, absolutely continuous approximate identity satisfying* (1.1.7) *and*

$$\frac{1}{2\pi}\int_0^\pi u\,|\chi'_\rho(u)|\,du \le M_1 \qquad (\rho \in \mathbb{A}). \tag{1.4.1}$$

Then at each point x for which

$$\int_0^h [f(x+u) + f(x-u) - 2f(x)]\,du = o(h) \qquad (h \to 0+), \tag{1.4.2}$$

thus for almost all x, we have

$$\lim_{\rho\to\rho_0} I_\rho(f; x) = f(x). \tag{1.4.3}$$

Proof. Since χ_ρ is even, we have by (1.1.1) for any $0 < \delta < \pi$

$$I_\rho(f; x) - f(x) = \frac{1}{2\pi}\left(\int_0^\delta + \int_\delta^\pi\right)[f(x+u) + f(x-u) - 2f(x)]\chi_\rho(u)\,du \equiv I_1 + I_2, \tag{1.4.4}$$

say. Setting

$$G(u) = \int_0^u [f(x+t) + f(x-t) - 2f(x)]\,dt, \tag{1.4.5}$$

then, according to (1.4.2), to each $\varepsilon > 0$ there exists a $\delta > 0$ such that $|G(u)| \le \varepsilon u$ for all $0 < u \le \delta$. We now fix this δ and estimate I_1 and I_2, respectively. By integration by parts we obtain

$$2\pi I_1 = G(\delta)\chi_\rho(\delta) - \int_0^\delta G(u)\chi'_\rho(u)\,du, \tag{1.4.6}$$

and therefore by (1.1.5) and (1.4.1)

$$|I_1| \le \frac{\varepsilon}{2\pi}\left[\delta\,|\chi_\rho(\delta)| + \int_0^\delta u\,|\chi'_\rho(u)|\,du\right] \le \varepsilon(M + 2M_1),$$

since one further integration by parts gives

$$\delta\,\chi_\rho(\delta) = \int_0^\delta \chi_\rho(u)\,du + \int_0^\delta u\chi'_\rho(u)\,du.$$

Regarding I_2, we have

$$|I_2| \le \sup_{\delta\le u\le\pi} |\chi_\rho(u)|\,\{2\,\|f\|_1 + |f(x)|\} \tag{1.4.7}$$

which in view of (1.1.7) tends to zero as $\rho \to \rho_0$. This proves (1.4.3) since by Prop. 0.3.1 almost all x satisfy (1.4.2).

It is useful to observe that if $\chi'_\rho(x)$ is of constant sign on $(0, \pi)$ and if $\chi_\rho(\pi)$ is bounded for $\rho \in \mathbb{A}$, then, in view of

$$\int_0^\pi u\chi'_\rho(u)\,du = \pi\chi_\rho(\pi) - \int_0^\pi \chi_\rho(u)\,du, \tag{1.4.8}$$

condition (1.4.1) is a consequence of (1.1.5). For example, by Problem 1.2.18 the Abel–Poisson kernel $\{p_r(x)\}$ is absolutely continuous such that $p'_r(x) \le 0$ on $[0, \pi]$ and

$p_r(\pi) = (1 - r)/(1 + r)$ for each $0 < r < 1$. Since this kernel forms an even approximate identity satisfying (1.1.7) we obtain for the singular integral $P_r(f; x)$ of Abel–Poisson

Corollary 1.4.3. *If* $f \in \mathsf{L}^p_{2\pi}$, $1 \le p \le \infty$, *then*

(1.4.9)
$$\lim_{r \to 1-} P_r(f; x) = f(x)$$

at each point x for which (1.4.2) *holds, consequently for almost all x.*

More generally, a constant sign of $\chi'_\rho(x)$ may be replaced by monotonicity. For such kernels we have

Proposition 1.4.4. *Let* $f \in \mathsf{L}^1_{2\pi}$ *and the kernel* $\{\chi_\rho(x)\}$ *of the integral* $I_\rho(f; x)$ *be an even, positive approximate identity satisfying*

(1.4.10) $\chi_\rho(x)$ *is monotonely decreasing on* $[0, \pi]$ *for each* $\rho \in \mathbb{A}$.

Then at each point x for which (1.4.2) *holds we have* (1.4.3).

Proof. Since χ_ρ is positive and monotonely decreasing, we have for each fixed $x \in (0, \pi]$

$$\int_{x/2}^{\pi} \chi_\rho(u)\, du \ge \int_{x/2}^{x} \chi_\rho(u)\, du \ge \frac{x}{2} \chi_\rho(x).$$

But according to (1.1.6) the left-hand side tends to zero as $\rho \to \rho_0$, and therefore

(1.4.11)
$$\lim_{\rho \to \rho_0} \chi_\rho(x) = 0 \qquad (0 < x \le \pi).$$

This, in particular, shows that (1.1.7) is satisfied.

Proceeding just as in the proof of Prop. 1.4.2, we obtain (1.4.4). If we again use (1.4.5) and integrate by parts (see (0.5.2)), we now have instead of (1.4.6)

$$2\pi I_1 = \int_0^\delta \chi_\rho(u)\, dG(u) = G(\delta)\chi_\rho(\delta) + \int_0^\delta G(u)\, d[-\chi_\rho(u)].$$

Hence

$$|I_1| \le \frac{\varepsilon}{2\pi}\left\{\delta\chi_\rho(\delta) + \int_0^\delta u\, d[-\chi_\rho(u)]\right\} = \frac{\varepsilon}{2\pi} \int_0^\delta \chi_\rho(u)\, du \le \varepsilon M.$$

Instead of (1.4.7) we have $|I_2| \le \chi_\rho(\delta)\{2\, \|f\|_1 + |f(x)|\}$, which, according to (1.4.11), tends to zero as $\rho \to \rho_0$.

We observe that condition (1.4.2) may equivalently be expressed by

$$\lim_{h \to 0+} (1/2h) \int_{-h}^{h} f(x + u)\, du = f(x).$$

Thus we may interpret the results of Prop. 1.4.2, 1.4.4 in the following way: If the particular singular integral $A_{2h}(f; x)$ of Riemann–Lebesgue (see (1.1.15)) converges at some point x to $f(x)$, then the general singular integral $I_\rho(f; x)$ converges at that point to $f(x)$.

On the other hand, there are some very important examples of singular integrals such as Fejér's singular integral, the kernels of which do not exactly satisfy the assumptions of Prop. 1.4.2 or 1.4.4, yet possess a majorant satisfying them. In this case we have

Proposition 1.4.5. *Let $f \in \mathsf{L}^1_{2\pi}$, and let the even kernel $\{\chi_\rho(x)\}$ of the integral $I_\rho(f; x)$ possess an absolutely continuous majorant $\{\chi^*_\rho(x)\}$ on $[0, \pi]$, thus $|\chi_\rho(x)| \le \chi^*_\rho(x)$ a.e. on $[0, \pi]$ for each $\rho \in \mathbb{A}$, which satisfies* (1.1.5), (1.1.7), *and* (1.4.1). *Then at each point x for which*

$$\int_0^h |f(x + u) + f(x - u) - 2f(x)|\, du = o(h) \qquad (h \to 0+) \tag{1.4.12}$$

we have

$$\lim_{\rho \to \rho_0} I_\rho(f; x) = f(x), \tag{1.4.13}$$

thus, almost everywhere.

Proposition 1.4.6. *Let $f \in \mathsf{L}^1_{2\pi}$, and let the even kernel $\{\chi_\rho(x)\}$ of the integral $I_\rho(f; x)$ possess a majorant $\{\chi^*_\rho(x)\}$ on $[0, \pi]$, satisfying* (1.1.5), (1.1.6), *and* (1.4.10). *Then condition* (1.4.12) *implies* (1.4.13).

If we replace the function G of (1.4.5) by

$$G^*(u) = \int_0^u |f(x + t) + f(x - t) - 2f(x)|\, dt, \tag{1.4.14}$$

then the proofs of the last two propositions are essentially those of Prop. 1.4.2 and 1.4.4, respectively, and left to the reader.

Let us consider the integral $\sigma_n(f; x)$ of Fejér defined by (1.2.25). According to Problem 1.2.6 Fejér's kernel $\{F_n(x)\}$ satisfies all the assumptions of Prop. 1.4.5 or 1.4.6, and hence

Corollary 1.4.7. *Let $f \in \mathsf{L}^p_{2\pi}$, $1 \le p \le \infty$. Then for almost all x*

$$\lim_{n \to \infty} \sigma_n(f; x) = f(x), \tag{1.4.15}$$

in particular at each x for which (1.4.12) *holds.*

Whereas the preceding results cover pointwise convergence of integrals $I_\rho(f; x)$ in case f belongs to $\mathsf{X}_{2\pi}$, the following propositions deal with convergence almost everywhere of $I_\rho(d\mu; x)$ defined in (1.1.12).

Proposition 1.4.8. *Let $\mu \in \mathsf{BV}_{2\pi}$, and let the kernel $\{\chi_\rho(x)\}$ of the integral $I_\rho(d\mu; x)$ be absolutely continuous and satisfy the assumptions of Prop.* 1.4.4. *Then at each point x for which $\mu'(x)$ exists,*

$$\lim_{\rho \to \rho_0} I_\rho(d\mu; x) = \mu'(x). \tag{1.4.16}$$

Proof. Obviously, the assertion is valid for the particular function $\mu(x) \equiv x$ by (1.1.1). Therefore it suffices to prove the proposition for $\mu(x) - \mu^\vee(0)x$, in other words, we may without loss of generality assume that μ is 2π-periodic. Then by partial integration

$$I_\rho(d\mu; x) = \frac{1}{2\pi}[\chi_\rho(x - u)\mu(u)]_{u=-\pi}^{\pi} + \frac{1}{2\pi}\int_{-\pi}^{\pi} \mu(u)\chi'_\rho(x - u)\, du = \frac{1}{2\pi}\int_{-\pi}^{\pi} \mu(x - u)\chi'_\rho(u)\, du.$$

Since the kernel is even, χ'_ρ is an odd function, and hence

$$I_\rho(d\mu; x) = \frac{\alpha_\rho}{2\pi}\int_0^\pi \frac{\mu(x + u) - \mu(x - u)}{\sin(u/2)}\, \Theta_\rho(u)\, du,$$

where we have set

$$\Theta_\rho(x) = -\frac{\sin(x/2)}{\alpha_\rho}\chi_\rho'(x), \qquad \alpha_\rho = \frac{1}{4\pi}\int_{-\pi}^{\pi}\chi_\rho(u)\cos(u/2)\,du - \frac{1}{\pi}\chi_\rho(\pi).$$

$\{\Theta_\rho(x)\}$ is a set† of even and positive functions on $[-\pi, \pi]$ which satisfy (1.1.1) and (1.1.6). Indeed, $\Theta_\rho(x)$ is even since the product of two odd functions is even. Furthermore, since χ_ρ satisfies (1.4.10), we have $\chi_\rho'(x) \le 0$ on $[0, \pi]$ and hence $\chi_\rho'(x) \ge 0$ on $[-\pi, 0]$. This implies $-\sin(x/2)\chi_\rho'(x) \ge 0$ on $[-\pi, \pi]$. Therefore by partial integration

$$\int_{-\pi}^{\pi}[-\sin(u/2)\chi_\rho'(u)]\,du = -\sin(u/2)\chi_\rho(u)|_{u=-\pi}^{\pi} + \tfrac{1}{2}\int_{-\pi}^{\pi}\chi_\rho(u)\cos(u/2)\,du = 2\pi\alpha_\rho.$$

Thus Θ_ρ satisfies (1.1.1). To verify (1.1.6), it follows that for each $0 < \delta < \pi$

$$\begin{aligned}\int_{\delta\le|u|\le\pi}|-\sin(u/2)\chi_\rho'(u)|\,du &= -2\int_{\delta}^{\pi}\sin(u/2)\chi_\rho'(u)\,du\\ &= -2\sin(u/2)\chi_\rho(u)\,|_{u=\delta}^{\pi} + \int_{\delta}^{\pi}\chi_\rho(u)\cos(u/2)\,du\\ &\le 2(\chi_\rho(\pi) + \chi_\rho(\delta)) + \int_{\delta}^{\pi}\chi_\rho(u)\,du = o(1) \qquad (\rho\to\rho_0)\end{aligned}$$

since χ_ρ satisfies (1.1.6) and (1.4.11). This proves (1.1.6) for $\{\Theta_\rho(x)\}$ since $\lim_{\rho\to\rho_0}\alpha_\rho = \frac{1}{2}$ as follows by (1.4.11) and an obvious modification of the proof of Prop. 1.4.1.

If μ is differentiable at x, then the function

$$G(u) = \frac{\mu(x+u) - \mu(x-u)}{2\sin(u/2)} - 2\mu'(x)$$

is continuous at $u = 0$ with $\lim_{u\to 0} G(u) = 0$ and bounded on $[0, \pi]$. Since

$$I_\rho(d\mu; x) - 2\alpha_\rho\mu'(x) = \frac{\alpha_\rho}{\pi}\int_0^{\pi}\left[\frac{\mu(x+u) - \mu(x-u)}{2\sin(u/2)} - 2\mu'(x)\right]\Theta_\rho(u)\,du,$$

it follows as in the proof of Theorem 1.1.5 that for $0 < \delta < \pi$

$$|I_\rho(d\mu; x) - 2\alpha_\rho\mu'(x)| \le \frac{\alpha_\rho}{\pi}\left(\int_0^{\delta} + \int_{\delta}^{\pi}\right)|G(u)|\,\Theta_\rho(u)\,du \equiv I_1 + I_2,$$

say. Given $\varepsilon > 0$, we may choose δ such that $|G(u)| < \varepsilon$ for $0 \le u \le \delta$. Then

$$I_1 \le \varepsilon\frac{\alpha_\rho}{\pi}\int_0^{\delta}\Theta_\rho(u)\,du \le \varepsilon\alpha_\rho, \qquad I_2 \le \|G\|_\infty\frac{\alpha_\rho}{\pi}\int_{\delta}^{\pi}\Theta_\rho(u)\,du,$$

which proves the assertion since $\lim_{\rho\to\rho_0}\alpha_\rho = \frac{1}{2}$ and since $\{\Theta_\rho(x)\}$ satisfies (1.1.6).

By Prop. 0.5.3 every $\mu \in \mathsf{BV}_{2\pi}$ is differentiable almost everywhere. Therefore for the Abel–Poisson integral

Corollary 1.4.9. *If $\mu \in \mathsf{BV}_{2\pi}$, then $\lim_{r\to 1-} P_r(d\mu; x) = \mu'(x)$ almost everywhere.*

As a counterpart to Prop. 1.4.6 we state

Proposition 1.4.10. *Let $\mu \in \mathsf{BV}_{2\pi}$, and let the kernel $\{\chi_\rho(x)\}$ of the integral $I_\rho(d\mu; x)$ satisfy the assumptions of Prop.* 1.4.6. *Then* (1.4.16) *holds almost everywhere.*

Proof. Since χ_ρ is an even function, we have

$$I_\rho(d\mu; x) = \frac{1}{2\pi}\int_0^{\pi}\chi_\rho(u)\,d[\mu(x+u) - \mu(x-u)],$$

† $\{\Theta_\rho(x)\}$ is not an approximate identity in the strict sense of our definition since Θ_ρ is not 2π-periodic.

and therefore by (1.1.1)

$$I_\rho(d\mu; x) - \mu'(x) = \frac{1}{2\pi}\int_0^\pi \chi_\rho(u)\, d[\mu(x+u) - \mu(x-u) - 2u\mu'(x)]$$

for each x where $\mu'(x)$ exists. Setting, for fixed x,

$$H(u) = \mu(x+u) - \mu(x-u) - 2u\mu'(x), \qquad V(t) = [\operatorname{Var} H(u)]_{u=0}^{t},$$

suppose that x is such that $H(u) = o(u)$, $u \to 0$, and $V(t) = o(t)$, $t \to 0+$; by Prop. 0.5.4 almost all x have these properties. For any $0 < \delta < \pi$

$$|I_\rho(d\mu; x) - \mu'(x)| \le \frac{1}{2\pi}\left(\int_0^\delta + \int_\delta^\pi\right)\chi_\rho^*(u)\, dV(u) \equiv I_1 + I_2,$$

say. Given $\varepsilon > 0$, let δ be such that $V(u) \le \varepsilon u$ for $0 \le u \le \delta$. If λ is such that $0 < \lambda < \delta$, then by partial integration and the monotonicity of χ_ρ^*

$$\begin{aligned}\int_\lambda^\delta \chi_\rho^*(u)\, dV(u) &\le \chi_\rho^*(\delta)V(\delta) + \chi_\rho^*(\lambda)V(\lambda) + \int_\lambda^\delta V(u)\, d[-\chi_\rho^*(u)]\\ &\le \varepsilon\left(\delta\chi_\rho^*(\delta) + \lambda\chi_\rho^*(\lambda) - \delta\chi_\rho^*(\delta) + \lambda\chi_\rho^*(\lambda) + \int_\lambda^\delta \chi_\rho^*(u)\, du\right)\\ &\le \varepsilon\left(2\int_0^\lambda \chi_\rho^*(u)\, du + \int_\lambda^\delta \chi_\rho^*(u)\, du\right) \le \varepsilon\, 2\pi\, \|\chi_\rho^*\|_1 \le \varepsilon\, 2\pi M,\end{aligned}$$

where the constant M is such that $\|\chi_\rho^*\|_1 \le M$ for all $\rho \in \mathbb{A}$. Since the above estimate is independent of λ, we conclude $I_1 \le \varepsilon M$. Furthermore

$$I_2 \le \chi_\rho^*(\delta)\frac{1}{2\pi}\int_\delta^\pi dV(u) \le \chi_\rho^*(\delta)(V(\pi)/2\pi),$$

which tends to zero as $\rho \to \rho_0$ since $\lim_{\rho\to\rho_0} \chi_\rho^*(\delta) = 0$ for each fixed δ (see (1.4.11)). This completes the proof.

The result of the last proposition may be applied to the integral of Fejér giving

Corollary 1.4.11. *If* $\mu \in \mathsf{BV}_{2\pi}$, *then* $\lim_{n\to\infty} \sigma_n(d\mu; x) = \mu'(x)$ *almost everywhere.*

Problems

1. Prove Prop. 1.4.1 and apply to the integral of Fejér. (Hint: Compare the estimate (1.4.7); see also Bari [1I, p. 133 f])
2. Let $f \in \mathsf{L}_{2\pi}^\infty$, and let the kernel $\{\chi_\rho(x)\}$ of the integral $I_\rho(f; x)$ satisfy the assumptions of Prop. 1.4.8 such that $\chi_\rho \in \mathsf{W}_{\mathsf{C}_{2\pi}}^1$ for each $\rho \in \mathbb{A}$. Suppose that $f(x)$ is differentiable at x_0. Show that $\lim_{\rho\to\rho_0} I_\rho'(f; x_0) = f'(x_0)$. As an application show that $\lim_{r\to 1-} P_r'(f; x_0) = f'(x_0)$; it is known as Fatou's theorem. (Hint: Use (1.1.18) and proceed as in the proof of Prop. 1.4.8; see also Zygmund [7I, p. 100], Timan [2, p. 135])
3. Let $f \in \mathsf{L}_{2\pi}^1$, and let the kernel $\{\chi_\rho(x)\}$ of the integral $I_\rho(f; x)$ satisfy the assumptions of Prop. 1.4.4. Show that at each point x for which
$$\int_0^h [f(x+u) + f(x-u) - 2f(x)]\, du = O(h^{1+\alpha}) \qquad (h \to 0+)$$
for some $\alpha > 0$, one has $|I_\rho(f; x) - f(x)| = O(m(\chi_\rho; \alpha))$, $\rho \to \rho_0$, where $m(\chi_\rho; \alpha)$ denotes the αth moment of χ_ρ (see (1.6.9)). (Hint: Show that $\chi_\rho(x) = O(m(\chi_\rho; \alpha))$, $\rho \to \rho_0$, for each x and proceed as in the proof of Prop. 1.4.4; see also Mamedov [5])
4. Let $\alpha > 0$. The (Cesàro) (C, α) *means* of the Fourier series of $f \in \mathsf{X}_{2\pi}$ are defined by
$$\sigma_{n,\alpha}(f; x) = \frac{1}{2\pi}\int_{-\pi}^{\pi} f(x-u)F_{n,\alpha}(u)\, du,$$

the kernel $\{F_{n,\alpha}(x)\}$ being given through

$$F_{n,\alpha}(x) = \sum_{k=-n}^{n} (A^{\alpha}_{n-|k|}/A^{\alpha}_{n})\, e^{ikx}, \qquad A^{\alpha}_{n} = \binom{n+\alpha}{n}.$$

Show that $\{F_{n,\alpha}(x)\}$ is an even approximate identity for each $\alpha > 0$. Thus $\sigma_{n,\alpha}(f; x)$ approximates every f in $X_{2\pi}$-norm as $n \to \infty$. Obviously, the $(C, 1)$ means are nothing but the Fejér means (1.2.25). Extend Cor. 1.4.7 to (C, α) means. (Hint: ZYGMUND [7I, p. 94])

1.5 Order of Approximation for Positive Singular Integrals

The first sections have included a discussion of the convergence of the integral $I_\rho(f; x)$ in $X_{2\pi}$-norm towards f. At this point we wish to study the rate of this convergence, in other words, the rate at which the positive numbers $\|I_\rho(f; \circ) - f(\circ)\|_{X_{2\pi}}$ tend to zero as $\rho \to \rho_0$. The order of approximation of f by $I_\rho(f; x)$ should depend upon the structural properties of the given function f. It is to be conjectured that stronger structural properties, that is, stronger continuity properties (including existence of derivatives, Lipschitz conditions on such derivatives, etc.), will be related to higher orders of approximation. Certainly, more refined properties of the kernels will also enter into consideration.

1.5.1 Modulus of Continuity and Lipschitz Classes

One measure for studying the structural properties of a function is given by the modulus of continuity for the space $X_{2\pi}$.

Definition 1.5.1. *For $f \in X_{2\pi}$ the **modulus of continuity** is defined for $\delta \geq 0$ by*

$$\omega(X_{2\pi}; f; \delta) = \sup_{|h| \leq \delta} \|f(\circ + h) - f(\circ)\|_{X_{2\pi}}.$$

Some elementary properties of $\omega(X_{2\pi}; f; \delta)$ are collected in the following

Lemma 1.5.2. *Let $f \in X_{2\pi}$.*

(i) *$\omega(X_{2\pi}; f; \delta)$ is a monotonely increasing function of δ, $\delta \geq 0$.*
(ii) *$\omega(X_{2\pi}; f; \lambda\delta) \leq (1 + \lambda)\omega(X_{2\pi}; f; \delta)$ for each $\lambda > 0$.*
(iii) *$\lim_{\delta \to 0+} \omega(X_{2\pi}; f; \delta) = 0$.*
(iv) *If $\omega(X_{2\pi}; f; \delta) = o(\delta)$ as $\delta \to 0+$, then f is constant* (a.e.).

In connection with even kernels we shall need the following generalization.

Definition 1.5.3. *For $f \in X_{2\pi}$ the **generalized modulus of continuity** is defined for $\delta \geq 0$ by*

$$\omega^*(X_{2\pi}; f; \delta) = \sup_{|h| \leq \delta} \|f(\circ + h) + f(\circ - h) - 2f(\circ)\|_{X_{2\pi}}.$$

Lemma 1.5.4. *Let $f \in X_{2\pi}$.*

(i) *$\omega^*(X_{2\pi}; f; \delta)$ is a monotonely increasing function of δ, $\delta \geq 0$.*
(ii) *$\omega^*(X_{2\pi}; f; \delta) \leq 2\omega(X_{2\pi}; f; \delta)$.*

(iii) $\omega^*(X_{2\pi}; f; \lambda\delta) \le (1+\lambda)^2 \omega^*(X_{2\pi}; f; \delta)$ *for each* $\lambda > 0$.
(iv) *If* $\omega^*(X_{2\pi}; f; \delta) = o(\delta^2)$ *as* $\delta \to 0+$, *then f is constant* (a.e.).

For the proofs of the preceding Lemmata one may consult Problem 1.5.1. Let us mention that $\omega(X_{2\pi}; f; \delta)$ and $\omega^*(X_{2\pi}; f; \delta)$ are sometimes referred to as the first and second modulus of continuity, respectively. See also Problem 1.5.3.

Definition 1.5.5. *A function $f \in X_{2\pi}$ is said to satisfy a **Lipschitz condition** of order α, $\alpha > 0$, in notation $f \in \mathrm{Lip}\,(X_{2\pi}; \alpha)$, if $\omega(X_{2\pi}; f; \delta) = O(\delta^\alpha)$. If $\omega(X_{2\pi}; f; \delta) = o(\delta^\alpha)$ as $\delta \to 0+$, we write $f \in \mathrm{lip}\,(X_{2\pi}; \alpha)$. For $r \in \mathbb{N}$, $\alpha > 0$ the class $W^{r,\alpha}_{X_{2\pi}}$ is defined as the set of those $f \in W^r_{X_{2\pi}}$ for which $\phi^{(r)} \in \mathrm{Lip}\,(X_{2\pi}; \alpha)$, where $\phi \in AC^{r-1}_{2\pi}$ with $\phi^{(r)} \in X_{2\pi}$ is such that $\phi(x) = f(x)$* (a.e.).

For $f \in \mathrm{Lip}\,(X_{2\pi}; \alpha)$ it is easily seen that there exists a constant M such that $\omega(X_{2\pi}; f; \delta) \le M\delta^\alpha$ for all $\delta > 0$. Consequently, $\mathrm{Lip}\,(X_{2\pi}; \alpha)_M$ denotes the set of those $f \in \mathrm{Lip}\,(X_{2\pi}; \alpha)$ for which $\omega(X_{2\pi}; f; \delta) \le M\delta^\alpha$ for all $\delta > 0$.

Definition 1.5.6. *A function $f \in X_{2\pi}$ satisfies a **generalized Lipschitz condition** of order α, $\alpha > 0$, in notation $f \in \mathrm{Lip}^*\,(X_{2\pi}; \alpha)$, if $\omega^*(X_{2\pi}; f; \delta) = O(\delta^\alpha)$ as $\delta \to 0+$. If $\omega^*(X_{2\pi}; f; \delta) = o(\delta^\alpha)$, we write $f \in \mathrm{lip}^*\,(X_{2\pi}; \alpha)$. For $r \in \mathbb{N}$, $\alpha > 0$ the class ${}^*W^{r,\alpha}_{X_{2\pi}}$ is defined as the set of those $f \in W^r_{X_{2\pi}}$ for which $\phi^{(r)} \in \mathrm{Lip}^*\,(X_{2\pi}; \alpha)$, where $\phi \in AC^{r-1}_{2\pi}$ with $\phi^{(r)} \in X_{2\pi}$ is such that $\phi(x) = f(x)$* (a.e.).

Without loss of generality we may restrict the order α in Def. 1.5.5 to $0 < \alpha \le 1$ and in Def. 1.5.6 to $0 < \alpha \le 2$, respectively, since $f \in \mathrm{lip}\,(X_{2\pi}; 1)$ as well as $f \in \mathrm{lip}^*\,(X_{2\pi}; 2)$ implies that f is a constant. This is Lemma 1.5.2(iv) and 1.5.4(iv), respectively. Moreover, it immediately follows from the definitions that $f \in \mathrm{Lip}\,(X_{2\pi}; \alpha)$ implies $f \in \mathrm{Lip}^*\,(X_{2\pi}; \alpha)$ for $0 < \alpha \le 1$. On the other hand, also the converse of the latter assertion is valid for $0 < \alpha < 1$. This is the result of Theorem 2.4.2. Thus the classes of functions $f \in X_{2\pi}$ which satisfy a Lipschitz or a generalized Lipschitz condition of order α are equivalent for $0 < \alpha < 1$. But if $f \in \mathrm{Lip}^*\,(X_{2\pi}; 1)$, then it is not necessarily true that $f \in \mathrm{Lip}\,(X_{2\pi}; 1)$. This follows by the example of Problem 1.5.1(iii).

Again it is an immediate consequence of the definition that for $f \in \mathrm{Lip}^*\,(X_{2\pi}; \alpha)$ there exists a constant M^* such that $\omega^*(X_{2\pi}, f; \delta) \le M^*\delta^\alpha$ for all $\delta > 0$. In this case we write $f \in \mathrm{Lip}^*\,(X_{2\pi}; \alpha)_{M^*}$, thus emphasizing the dependence upon the constant M^*. Obviously, $f \in \mathrm{Lip}\,(X_{2\pi}; \alpha)_M$ implies $f \in \mathrm{Lip}^*\,(X_{2\pi}; \alpha)_{2M}$.

1.5.2 Direct Approximation Theorems

We shall now return to the problem formulated in the beginning of this section. At first we shall consider singular integrals with (real) kernels which are even and positive. Moreover, it seems natural to assume that the integrals in question converge to f in $X_{2\pi}$-norm, i.e.,

$$\lim_{\rho\to\rho_0} \|I_\rho(f; \circ) - f(\circ)\|_{X_{2\pi}} = 0 \qquad (f \in X_{2\pi}). \tag{1.5.1}$$

Since by Prop. 1.3.10 the convergence (1.5.1) holds if and only if

$$\lim_{\rho\to\rho_0} (1 - \chi_\rho^\wedge(1)) = 0, \tag{1.5.2}$$

our considerations may be based on (1.5.2). Thus we expect that the discussion of the rate of convergence of (1.5.1) will be reduced, more or less, to a discussion of the rate of convergence of (1.5.2). In this section we shall only prove some *direct approximation theorems*, i.e., theorems which assert a certain order of magnitude of the approximation (1.5.1) provided the function f to be approximated satisfies certain structural properties.

Before proceeding to a basic lemma, we note that, since the kernel $\{\chi_\rho(x)\}$ is even, the complex Fourier coefficients $\chi_\rho\hat{}(k)$ of (1.2.11) are real. In fact,

$$\chi_\rho\hat{}(k) = \frac{1}{\pi}\int_0^\pi \chi_\rho(u)\cos ku\, du \qquad (k \in \mathbb{Z}), \tag{1.5.3}$$

and since the kernel is positive,

$$|\chi_\rho\hat{}(k)| \le \frac{1}{2\pi}\int_{-\pi}^\pi \chi_\rho(u)\, du = 1 \qquad (k \in \mathbb{Z},\ \rho \in \mathbb{A}). \tag{1.5.4}$$

Moreover, $\chi_\rho\hat{}(0) = 1$ for all $\rho \in \mathbb{A}$ by (1.1.1).

Lemma 1.5.7. *Let $\{\chi_\rho(x)\}$ be an even and positive kernel. Then*

$$\frac{1}{\pi}\int_0^\pi u^j\chi_\rho(u)\, du \le \left[\frac{\pi}{\sqrt{2}}\sqrt{1-\chi_\rho\hat{}(1)}\right]^j \qquad (j = 1, 2), \tag{1.5.5}$$

$$\frac{1}{\pi}\int_0^\pi u^j\chi_\rho(u)\, du \le \frac{\pi^2}{2}(1-\chi_\rho\hat{}(1))\left[\frac{\pi}{2}\sqrt{4-\frac{1-\chi_\rho\hat{}(2)}{1-\chi_\rho\hat{}(1)}}\right]^{j-2} \qquad (j = 3, 4). \tag{1.5.6}$$

Proof. To prove (1.5.5) for $j = 2$, we have by Problem 1.2.4(ii) and (1.5.3)

$$\frac{1}{\pi}\int_0^\pi u^2\chi_\rho(u)\, du \le \pi\int_0^\pi \sin^2\frac{u}{2}\chi_\rho(u)\, du = \frac{\pi^2}{2}(1-\chi_\rho\hat{}(1)).$$

For $j = 1$ the last result implies in view of Hölder's inequality

$$\frac{1}{\pi}\int_0^\pi u\chi_\rho(u)\, du \le \left\{\frac{1}{\pi}\int_0^\pi u^2\chi_\rho(u)\, du\right\}^{1/2}\left\{\frac{1}{\pi}\int_0^\pi \chi_\rho(u)\, du\right\}^{1/2} \le \frac{\pi}{\sqrt{2}}\sqrt{1-\chi_\rho\hat{}(1)}.$$

The proof of (1.5.6) runs along the same lines and is left to the reader.

Theorem 1.5.8. *Let $f \in X_{2\pi}$, and the kernel $\{\chi_\rho(x)\}$ of the singular integral $I_\rho(f; x)$ be even and positive. Then*

$$\|I_\rho(f; \circ) - f(\circ)\|_{X_{2\pi}} = O(\omega^*(X_{2\pi}; f; \sqrt{1-\chi_\rho\hat{}(1)})) \qquad (\rho \to \rho_0). \tag{1.5.7}$$

Proof. Since the kernel is even, we have

$$I_\rho(f; x) - f(x) = \frac{1}{2\pi}\int_0^\pi [f(x+u) + f(x-u) - 2f(x)]\chi_\rho(u)\, du, \tag{1.5.8}$$

and hence by the Hölder-Minkowski inequality

$$\|I_\rho(f; \circ) - f(\circ)\|_{X_{2\pi}} \le \frac{1}{2\pi}\int_0^\pi \|f(\circ+u) + f(\circ-u) - 2f(\circ)\|_{X_{2\pi}}\chi_\rho(u)\, du.$$

Therefore it follows by Lemmata 1.5.4(iii), 1.5.7 that for each $\lambda > 0$

$$(1.5.9)\qquad \begin{aligned}\|I_\rho(f;\circ) - f(\circ)\|_{X_{2\pi}} &\le \frac{1}{2\pi}\int_0^\pi \omega^*(X_{2\pi}; f; u)\chi_\rho(u)\,du \\ &\le \omega^*(X_{2\pi}; f; \lambda^{-1})\frac{1}{2\pi}\int_0^\pi (1+\lambda u)^2\chi_\rho(u)\,du \\ &\le \left\{\frac{1}{\sqrt{2}} + \frac{\lambda\pi}{2}\sqrt{1-\chi_\rho^\wedge(1)}\right\}^2 \omega^*(X_{2\pi}; f; \lambda^{-1}).\end{aligned}$$

This implies (1.5.7) by setting $\lambda = (1 - \chi_\rho^\wedge(1))^{-1/2}$.

Let us apply the last result to the singular integral $J_n(f; x)$ of Jackson as defined in Problem 1.3.9. Here $j_n^\wedge(1) = 1 - (3/(2n^2 + 1))$, and therefore

$$(1.5.10)\qquad 1 - j_n^\wedge(1) = O(n^{-2}) \qquad (n \to \infty).$$

Corollary 1.5.9. *If $f \in X_{2\pi}$, then*

$$\|J_n(f;\circ) - f(\circ)\|_{X_{2\pi}} = O(\omega^*(X_{2\pi}; f; n^{-1})) \qquad (n \to \infty).$$

Thus, if $f \in \mathsf{Lip}^*(X_{2\pi}; \alpha)$ for some $0 < \alpha \le 2$, then

$$(1.5.11)\qquad \|J_n(f;\circ) - f(\circ)\|_{X_{2\pi}} = O(n^{-\alpha}) \qquad (n \to \infty).$$

Thus, for the singular integral of Jackson the best possible order of approximation at this stage is $O(n^{-2})$; more refined properties upon f such as $f'' \in \mathsf{Lip}^*(X_{2\pi}; \alpha)$ do not imply better approximation. This depends neither on our methods of proof nor on the particular example, but turns out to be characteristic for positive polynomial operators. We shall return to this question in Sec. 1.7.

1.5.3 Method of Test Functions

It was shown in Prop. 1.3.9 that in case of positive singular integrals the test conditions (1.3.17) imply the convergence (1.3.7) for every $f \in X_{2\pi}$. Now the question naturally arises whether it is possible to strengthen (1.3.7) to a result concerning an order of approximation if the test conditions (1.3.17) include an order of approximation. The following proposition gives an affirmative answer.

Proposition 1.5.10. *Let the kernel $\{\chi_\rho(x)\}$ of the singular integral $I_\rho(f; x)$ be positive. If at some point x_0*

$$(1.5.12)\qquad \begin{aligned} I_\rho(\cos u; x_0) &= \cos x_0 - \beta_\rho(x_0), \qquad \lim_{\rho\to\rho_0} \beta_\rho(x_0) = 0, \\ I_\rho(\sin u; x_0) &= \sin x_0 - \gamma_\rho(x_0), \qquad \lim_{\rho\to\rho_0} \gamma_\rho(x_0) = 0,\end{aligned}$$

then, for each $f \in X_{2\pi}$,

$$(1.5.13)\qquad \|I_\rho(f;\circ) - f(\circ)\|_{X_{2\pi}} = O(\omega(X_{2\pi}; f; \sqrt{\beta_\rho(x_0)\cos x_0 + \gamma_\rho(x_0)\sin x_0})) \quad (\rho \to \rho_0).$$

Proof. As in the proof of Theorem 1.5.8 we obtain, for any $\lambda > 0$,

$$\begin{aligned}\|I_\rho(f;\circ) - f(\circ)\|_{X_{2\pi}} &\le \frac{1}{2\pi}\int_{-\pi}^{\pi} \|f(\circ - u) - f(\circ)\|_{X_{2\pi}}\chi_\rho(u)\,du \\ &\le \omega(X_{2\pi}; f; \lambda^{-1})\frac{1}{2\pi}\int_{-\pi}^{\pi}(1 + \lambda\,|u|)\chi_\rho(u)\,du \\ &\le \omega(X_{2\pi}; f; \lambda^{-1})\Big[1 + \lambda\Big\{\frac{1}{2\pi}\int_{-\pi}^{\pi} u^2\chi_\rho(u)\,du\Big\}^{1/2}\Big].\end{aligned}$$

But as in the proof of Lemma 1.5.7

$$\begin{aligned}\frac{1}{2\pi}\int_{-\pi}^{\pi} u^2\chi_\rho(u)\,du &\le \frac{\pi}{4}\int_{-\pi}^{\pi}(1 - \cos(x_0 - u))\chi_\rho(x_0 - u)\,du \\ &= \frac{\pi^2}{2}\{1 - \cos x_0 I_\rho(\cos u; x_0) - \sin x_0 I_\rho(\sin u; x_0)\} \\ &= \frac{\pi^2}{2}\{\beta_\rho(x_0)\cos x_0 + \gamma_\rho(x_0)\sin x_0\}.\end{aligned}$$

Therefore it follows that

$$\|I_\rho(f;\circ) - f(\circ)\|_{X_{2\pi}} \le \omega(X_{2\pi}; f; \lambda^{-1})\Big[1 + \frac{\lambda\pi}{\sqrt{2}}\sqrt{\beta_\rho(x_0)\cos x_0 + \gamma_\rho(x_0)\sin x_0}\Big],$$

which implies (1.5.13) by taking $\lambda^{-1} = \sqrt{\beta_\rho(x_0)\cos x_0 + \gamma_\rho(x_0)\sin x_0}$.

As an example we consider the singular integral (1.2.25) of Fejér. Here we have

$$\sigma_n(\cos u; x) = \cos x - \frac{1}{n+1}\cos x, \qquad \sigma_n(\sin u; x) = \sin x - \frac{1}{n+1}\sin x.$$

Hence (1.5.13) gives for every $f \in X_{2\pi}$

$$\|\sigma_n(f;\circ) - f(\circ)\|_{X_{2\pi}} = O(\omega(X_{2\pi}; f; n^{-1/2})) \qquad (n \to \infty);$$

see also Problem 1.5.10.

If the kernel $\{\chi_\rho(x)\}$ satisfying the assumptions of Prop. 1.5.10 is also even, then by (1.5.3)

$$\begin{aligned}I_\rho(\cos u; x) &= \cos x\,\chi_\rho^\wedge(1) = \cos x - (1 - \chi_\rho^\wedge(1))\cos x, \\ I_\rho(\sin u; x) &= \sin x\,\chi_\rho^\wedge(1) = \sin x - (1 - \chi_\rho^\wedge(1))\sin x,\end{aligned} \tag{1.5.14}$$

and therefore (1.5.13) implies

$$\|I_\rho(f;\circ) - f(\circ)\|_{X_{2\pi}} = O(\omega(X_{2\pi}; f; \sqrt{1 - \chi_\rho^\wedge(1)})) \qquad (\rho \to \rho_0);$$

thus we essentially obtain again (1.5.7) (see also Problem 1.5.8).

There is an interesting interpretation of Theorem 1.5.8 in connection with the method of test functions. By (1.5.11) we know that there are singular integrals $I_\rho(f; x)$ such that for every $f \in \mathrm{Lip}^*(X_{2\pi}; \alpha)$, $0 < \alpha \le 2$, we have

$$\|I_\rho(f;\circ) - f(\circ)\|_{X_{2\pi}} = O(\rho^{-\alpha}) \qquad (\rho \to \rho_0). \tag{1.5.15}$$

The problem that arises is to characterize such integrals. In this connection the method of test functions calls for a set of functions and for a condition such that if every function of the set satisfies the condition, then (1.5.15) holds for every $f \in \mathsf{Lip}^* (\mathsf{X}_{2\pi}; \alpha)$. Many important results may be interpreted in this way. Here we have

Proposition 1.5.11. *Let the kernel* $\{\chi_\rho(x)\}$ *of the singular integral* $I_\rho(f; x)$ *be even and positive. If*

$$1 - \chi_\rho^\wedge(1) = O(\rho^{-2}) \qquad (\rho \to \rho_0), \tag{1.5.16}$$

then (1.5.15) *holds for every* $f \in \mathsf{Lip}^* (\mathsf{X}_{2\pi}; \alpha)$, $0 < \alpha \leq 2$. *In other words, the test set for the assertion that* $f \in \mathsf{Lip}^* (\mathsf{X}_{2\pi}; \alpha)$ *implies* (1.5.15) *consists of the function* e^{ix} *only, and the test condition is*

$$\|I_\rho(e^{iu}; \circ) - e^{i\circ}\|_{\mathsf{X}_{2\pi}} = O(\rho^{-2}) \qquad (\rho \to \rho_0). \tag{1.5.17}$$

The proof follows immediately from (1.5.7) and the fact that

$$\|I_\rho(e^{iu}; \circ) - e^{i\circ}\|_{\mathsf{X}_{2\pi}} = 1 - \chi_\rho^\wedge(1). \tag{1.5.18}$$

We mention another important interpretation of Theorem 1.5.8. A sequence of bounded linear operators $T_n(f; x)$ which are polynomial of degree n such that for every $f \in \mathsf{X}_{2\pi}$

$$\|T_n(f; \circ) - f(\circ)\|_{\mathsf{X}_{2\pi}} = O(\omega^*(\mathsf{X}_{2\pi}; f; n^{-1})) \qquad (n \to \infty) \tag{1.5.19}$$

is called a Zygmund approximation sequence. We have

Proposition 1.5.12. *Let the kernel* $\{\chi_n(x)\}$ *of the singular integral* $I_n(f; x)$ *be even, continuous, and positive, and suppose that* $I_n(f; x)$ *is a polynomial operator of degree* n. *A necessary and sufficient condition that* $\{I_n(f; x)\}$ *be a Zygmund approximation sequence is that*

$$1 - \chi_n^\wedge(1) = O(n^{-2}) \qquad (n \to \infty). \tag{1.5.20}$$

Proof. We first of all note that in view of Problem 1.2.15 $\chi_n(x)$ is an even and positive polynomial of degree n. According to (1.5.18), the necessity of (1.5.20) follows immediately since $e^{ix} \in \mathsf{Lip}^* (\mathsf{X}_{2\pi}; 2)$. On the other hand, (1.5.7) implies the sufficiency.

For an example of a Zygmund approximation sequence we refer to Sec. 1.6.1.

1.5.4 Asymptotic Properties

We begin with a general theorem for positive singular integrals.

Theorem 1.5.13. *Let the kernel* $\{\chi_\rho(x)\}$ *of the singular integral* $I_\rho(f; x)$ *be positive and* $\zeta(x)$ *a function such that*

$$\zeta(x) \in \mathsf{C}_{2\pi}, \quad \zeta(x_0) = 0, \quad \zeta(x) > 0, \quad x \neq x_0, \quad -\pi \leq x, x_0 \leq \pi. \tag{1.5.21}$$

Suppose that for $h \in \mathsf{L}_{2\pi}^\infty$ *the limit*

$$\lim_{x \to x_0} h(x)/\zeta(x) = L \tag{1.5.22}$$

exists and is finite. Then (1.5.22) *implies*

$$\lim_{\rho\to\rho_0} I_\rho(h; x_0)/I_\rho(\zeta; x_0) = L \tag{1.5.23}$$

if and only if

$$\lim_{\rho\to\rho_0} \alpha_\rho(\delta)/I_\rho(\zeta; x_0) = 0 \tag{1.5.24}$$

holds for each $0 < \delta \le \pi$, *where*

$$\alpha_\rho(\delta) = (1/2\pi) \int_{\delta\le|u|\le\pi} \chi_\rho(u)\, du. \tag{1.5.25}$$

Proof. *Necessity.* We consider the particular function $h(x) = \zeta^2(x)$. Then, according to (1.5.21),

$$\lim_{x\to x_0} h(x)/\zeta(x) = \lim_{x\to x_0} \zeta(x) = 0.$$

Moreover, for this function

$$I_\rho(h; x_0) \ge \frac{1}{2\pi} \int_{\delta\le|u|\le\pi} \zeta^2(x_0 - u)\chi_\rho(u)\, du \ge m_\delta \alpha_\rho(\delta),$$

where $m_\delta \equiv \inf_{\delta\le|u|\le\pi} \zeta^2(x_0 - u) > 0$. It follows that

$$I_\rho(h; x_0)/I_\rho(\zeta; x_0) \ge m_\delta[\alpha_\rho(\delta)/I_\rho(\zeta; x_0)],$$

and thus (1.5.23) implies (1.5.24).

Sufficiency. According to (1.5.22), given any $\varepsilon > 0$ there is a $\delta > 0$ such that $|h(x_0 - u) - L\zeta(x_0 - u)| < \varepsilon\zeta(x_0 - u)$ for all $|u| \le \delta$. Since

$$I_\rho(h; x_0) - LI_\rho(\zeta; x_0)$$
$$= \frac{1}{2\pi}\left(\int_{|u|\le\delta} + \int_{\delta\le|u|\le\pi}\right)[h(x_0 - u) - L\zeta(x_0 - u)]\chi_\rho(u)\, du \equiv I_1 + I_2,$$

say, we therefore obtain

$$|I_1| \le \varepsilon I_\rho(\zeta; x_0), \qquad |I_2| \le \|h(\circ) - L\zeta(\circ)\|_\infty \alpha_\rho(\delta),$$

which establishes (1.5.23).

Before proceeding to the next result we give the following

Definition 1.5.14. *Let the function f be defined in a neighbourhood of the point x_0. If $\lim_{h\to 0} h^{-2}[f(x_0 + h) + f(x_0 - h) - 2f(x_0)]$ exists and is finite, this limit is called the* ***second Riemann derivative*** *of f at x_0; in notation: $f^{[2]}(x_0)$.*

For the general definition of a Riemann derivative as well as other generalized derivatives we refer to Sec. 5.1.4 (see also Problem 1.5.15).

Theorem 1.5.15. *Let the kernel $\{\chi_\rho(x)\}$ of the singular integral $I_\rho(f; x)$ be even and positive. If the functions $f, g \in \mathsf{L}^\infty_{2\pi}$ possess a second Riemann derivative at x_0, then (provided the denumerators are different from zero)*

$$\lim_{\rho\to\rho_0} \frac{I_\rho(f; x_0) - f(x_0)}{I_\rho(g; x_0) - g(x_0)} = \frac{f^{[2]}(x_0)}{g^{[2]}(x_0)} \tag{1.5.26}$$

if and only if

$$\lim_{\rho\to\rho_0} \frac{1 - \chi_\rho^\wedge(2)}{1 - \chi_\rho^\wedge(1)} = 4. \tag{1.5.27}$$

Proof. *Necessity.* If we set $g_k(x) = 1 - \cos kx$, $k = 1, 2$, then according to (1.5.3)

$$I_\rho(g_k; 0) = \frac{1}{2\pi}\int_{-\pi}^{\pi} (1 - \cos ku)\chi_\rho(u)\, du = 1 - \chi_\rho^\wedge(k).$$

Therefore, if (1.5.26) holds, we have

$$\lim_{\rho\to\rho_0} \frac{1 - \chi_\rho^\wedge(2)}{1 - \chi_\rho^\wedge(1)} = \lim_{\rho\to\rho_0} \frac{I_\rho(g_2; 0) - g_2(0)}{I_\rho(g_1; 0) - g_1(0)} = \frac{g_2''(0)}{g_1''(0)} = 4.$$

Sufficiency. If we set $2h(u) = f(x_0 + u) + f(x_0 - u) - 2f(x_0)$, then by (1.5.8)

$$I_\rho(f; x_0) - f(x_0) = \frac{1}{2\pi}\int_{-\pi}^{\pi} h(u)\chi_\rho(u)\, du = I_\rho(h; 0).$$

Since

$$\lim_{u\to 0} \frac{h(u)}{g_1(u)} = \lim_{u\to 0} \frac{f(x_0 + u) + f(x_0 - u) - 2f(x_0)}{2(1 - \cos u)} = f^{[2]}(x_0),$$

and since $\zeta(x) = g_1(x)$ obviously satisfies (1.5.21) for $x_0 = 0$, we obtain in virtue of Theorem 1.5.13 that

$$\lim_{\rho\to\rho_0} \frac{I_\rho(f; x_0) - f(x_0)}{I_\rho(g_1; 0)} = \lim_{\rho\to\rho_0} \frac{I_\rho(h; 0)}{I_\rho(g_1; 0)} = \lim_{u\to 0} \frac{h(u)}{g_1(u)} = f^{[2]}(x_0)$$

if and only if for each $0 < \delta < \pi$

$$\alpha_\rho(\delta) \equiv \frac{1}{2\pi} \int_{\delta\le|u|\le\pi} \chi_\rho(u)\, du = o(I_\rho(g_1; 0)) = o(1 - \chi_\rho^\wedge(1)) \qquad (\rho \to \rho_0).$$

But this is true since by (1.5.27)

$$\begin{aligned}\alpha_\rho(\delta) &\le \frac{1}{2\pi(1 - \cos\delta)^2} \int_{\delta\le|u|\le\pi} (1 - \cos u)^2 \chi_\rho(u)\, du \\ &\le \frac{1}{(1 - \cos\delta)^2}\left\{1 - 2\chi_\rho^\wedge(1) + \frac{1 + \chi_\rho^\wedge(2)}{2}\right\} \\ &= \frac{1 - \chi_\rho^\wedge(1)}{2(1 - \cos\delta)^2}\left\{4 - \frac{1 - \chi_\rho^\wedge(2)}{1 - \chi_\rho^\wedge(1)}\right\} = o(1 - \chi_\rho^\wedge(1)).\end{aligned}$$

Thus the proof is complete.

We remark that every singular integral which satisfies the assumptions of Theorem 1.5.15 admits the asymptotic expansion

$$I_\rho(f; x_0) - f(x_0) = (1 - \chi_\rho^\frown(1))f^{[2]}(x_0) + o(1 - \chi_\rho^\frown(1)) \qquad (\rho \to \rho_0). \tag{1.5.28}$$

To give a first example we consider the singular integral $W_t(f; x)$ of Weierstrass introduced in Problem 1.3.10. Since $[\theta_3(\circ; t)]^\frown = \exp\{-tk^2\}$, condition (1.5.27) is satisfied, and we obtain

Corollary 1.5.16. *If $f \in \mathsf{L}_{2\pi}^\infty$ has an ordinary second derivative at x_0, then for the integral of Weierstrass*

$$W_t(f; x_0) - f(x_0) = tf''(x_0) + o(t) \qquad (t \to 0+). \tag{1.5.29}$$

For further examples we refer to Problem 1.5.11. Unfortunately, there are very important singular integrals which do not possess the property (1.5.27). For instance we have for the Fejér kernel $\{F_n(x)\}$ of (1.2.24)

$$F_n^\frown(1) = 1 - \frac{1}{n+1}, \qquad F_n^\frown(2) = 1 - \frac{2}{n+1}, \qquad \lim_{n\to\infty} \frac{1 - F_n^\frown(2)}{1 - F_n^\frown(1)} = 2,$$

and for the Abel–Poisson kernel $\{p_r(x)\}$ of (1.2.37)

$$p_r^\frown(1) = r, \qquad p_r^\frown(2) = r^2, \qquad \lim_{r\to 1-} \frac{1 - p_r^\frown(2)}{1 - p_r^\frown(1)} = 2.$$

Thus Theorem 1.5.15 does not apply. But in comparison with Theorem 1.5.13 this only means that for those particular singular integrals $\zeta(x) = g_1(x) = 2\sin^2(x/2)$ is not a suitable choice for the function $\zeta(x)$. As is shown in Problem 1.5.12, $\zeta(x) = |\sin(x/2)|$ is now a suitable one.

Finally we mention that we may also apply the method of test functions in order to obtain general asymptotic expansions for positive singular integrals. Indeed, we have in continuation of Prop. 1.3.9 and 1.5.10

Proposition 1.5.17. *Let the kernel $\{\chi_\rho(x)\}$ of the singular integral $I_\rho(f; x)$ be positive. If at some point x_0 the representations*

$$\begin{aligned} I_\rho(\cos u; x_0) &= \cos x_0 + \eta_1(x_0)\,\varphi(\rho) + o(\varphi(\rho)), \\ & \qquad\qquad\qquad\qquad\qquad \lim_{\rho\to\rho_0} \varphi(\rho) = 0, \\ I_\rho(\sin u; x_0) &= \sin x_0 + \eta_2(x_0)\,\varphi(\rho) + o(\varphi(\rho)), \end{aligned} \tag{1.5.30}$$

hold, then for every $f \in \mathsf{C}_{2\pi}$ having a second derivative at x_0

$$\begin{aligned} \lim_{\rho\to\rho_0} [I_\rho(f; x_0) - f(x_0)]/\varphi(\rho) &= \{\eta_2(x_0)\cos x_0 - \eta_1(x_0)\sin x_0\}f'(x_0) \\ &\quad - \{\eta_1(x_0)\cos x_0 + \eta_2(x_0)\sin x_0\}f''(x_0), \end{aligned} \tag{1.5.31}$$

provided for each $0 < \delta < \pi$

$$\lim_{\rho\to\rho_0} (1/\varphi(\rho)) \int_{\delta\le|u|\le\pi} \sin^2\frac{u}{2}\,\chi_\rho(u)\,du = 0. \tag{1.5.32}$$

For the proof we refer to Problem 1.5.13. If the kernel $\{\chi_\rho(x)\}$ is also even, then $\varphi(\rho) = (1 - \chi_\rho^\wedge(1))^{-1}$, $\eta_1(x) = -\cos x$, $\eta_2(x) = -\sin x$ as follows by (1.5.14), and (1.5.32) reduces to (1.5.27). This again gives the expansion (1.5.28).

Problems

1. (i) Prove Lemma 1.5.2. (Hint: As to (ii) show that for any $\delta_1, \delta_2 > 0$
$$\omega(X_{2\pi}; f; \delta_1 + \delta_2) \le \omega(X_{2\pi}; f; \delta_1) + \omega(X_{2\pi}; f; \delta_2),$$
and thus $\omega(X_{2\pi}; f; n\delta) \le n\omega(X_{2\pi}; f; \delta)$ for any $n \in \mathbb{N}$. Now choose $n \in \mathbb{P}$ such that $n \le \lambda < n + 1$; see also NATANSON [8I, p. 75 ff], TIMAN [2, p. 96 ff]. As to (iv) use (ii) to show that $\omega(X_{2\pi}; f; \delta_2)/\delta_2 \le 2\omega(X_{2\pi}; f; \delta_1)/\delta_1$ for any $\delta_1 < \delta_2$, and thus $\lim_{\delta \to 0+} \omega(X_{2\pi}; f; \delta)/\delta > 0$ unless f is a constant (a.e.))
   (ii) Prove Lemma 1.5.4.
   (iii) Show that the function $f(x) = \sin x \log |\sin x|$ belongs to the class $\mathsf{Lip}^* (X_{2\pi}; 1)$ but not to the class $\mathsf{Lip} (X_{2\pi}; 1)$.
2. Let f be defined in a neighbourhood of a point $x \in \mathbb{R}$. For (sufficiently small) $h \in \mathbb{R}$ the first *one-sided difference* of f at x with respect to increment h is defined by $\Delta_h^1 f(x) = f(x + h) - f(x)$, and the higher differences by $\Delta_h^r f(x) = \Delta_h^1 \Delta_h^{r-1} f(x)$, $r \in \mathbb{N}$.
   (i) Show (by induction) that
$$\Delta_h^r f(x) = \sum_{k=0}^{r} (-1)^{r-k} \binom{r}{k} f(x + kh).$$
   (ii) If f has a bounded rth derivative in a neighbourhood of x, then (compare with Problem 1.1.7)
$$\Delta_h^r f(x) = \int_0^h \cdots \int_0^h f^{(r)}(x + u_1 + \cdots + u_r)\, du_1 \ldots du_r.$$
   (iii) Show (by induction) that for $n \in \mathbb{N}$
$$\Delta_{nh}^r f(x) = \sum_{k_1=0}^{n-1} \cdots \sum_{k_r=0}^{n-1} \Delta_h^r f(x + k_1 h + \cdots + k_r h).$$
3. For $f \in X_{2\pi}$ the *rth modulus of continuity* (smoothness) is defined for $\delta \ge 0$ by
$$\omega_r(X_{2\pi}; f; \delta) = \sup_{|h| \le \delta} \|\Delta_h^r f(\circ)\|_{X_{2\pi}}.$$
Show that the following properties hold for $\omega_r(X_{2\pi}; f; \delta)$:
   (i) $\omega_r(X_{2\pi}; f; \delta)$ is a monotonely increasing function of δ, $\delta \ge 0$.
   (ii) $\omega_r(X_{2\pi}; f; \delta) \le 2^{r-j}\omega_j(X_{2\pi}; f; \delta)$ for any $j \in \mathbb{N}$ with $j < r$. In particular, one has $\lim_{\delta \to 0+} \omega_r(X_{2\pi}; f; \delta) = 0$.
   (iii) $\omega_r(X_{2\pi}; f; \lambda\delta) \le (1 + \lambda)^r \omega_r(X_{2\pi}; f; \delta)$ for any $\lambda > 0$. (Hint: Use Problem 1.5.2(iii) to show that $\omega_r(X_{2\pi}; f; n\delta) \le n^r \omega_r(X_{2\pi}; f; \delta)$ for any $n \in \mathbb{N}$)
   (iv) If $f \in W_{X_{2\pi}}^r$ and $\phi \in AC_{2\pi}^{r-1}$ with $\phi^{(r)} \in X_{2\pi}$ is such that $\phi(x) = f(x)$ (a.e.), then $\omega_{j+r}(X_{2\pi}; f; \delta) \le \delta^r \omega_j(X_{2\pi}; \phi^{(r)}; \delta)$ for any $j \in \mathbb{N}$. In particular, if $f \in W_{X_{2\pi}}^1$, then $\omega(X_{2\pi}; f; \delta) \le \delta \|\phi'\|_{X_{2\pi}}$. (Hint: Use Problem 1.5.2(ii))
   (v) $\omega_r(X_{2\pi}; f; \delta_2)/\delta_2^r \le 2^r \omega_r(X_{2\pi}; f; \delta_1)/\delta_1^r$ for any $\delta_1 < \delta_2$. Thus, unless f is constant (a.e.) $\lim_{\delta \to 0+} \omega_r(X_{2\pi}; f; \delta)/\delta^r > 0$.

   For the proofs of these and further properties of $\omega_r(X_{2\pi}; f; \delta)$ see also TIMAN [2, p. 102 ff].
4. For $\alpha > 0$ the *Lipschitz class* $\mathsf{Lip}_r (X_{2\pi}; \alpha)$ is defined as the set of those functions $f \in X_{2\pi}$ for which $\omega_r(X_{2\pi}; f; \delta) = O(\delta^\alpha)$ as $\delta \to 0+$. The class $\mathsf{Lip}_r (X_{2\pi}; \alpha)_{M_r}$ consists of those $f \in \mathsf{Lip}_r (X_{2\pi}; \alpha)$ for which $\omega_r(X_{2\pi}; f; \delta) \le M_r \delta^\alpha$ for all $\delta > 0$, M_r being a given constant.

(i) Show that if $f \in \mathsf{Lip}_j(X_{2\pi}; \alpha)$ for some $j \in \mathbb{N}$, then $f \in \mathsf{Lip}_r(X_{2\pi}; \alpha)$ for every $r \in \mathbb{N}$ with $j < r$.

(ii) If $f \in \mathsf{Lip}_r(X_{2\pi}; \alpha)$, then there exists a constant M_r such that $\omega_r(X_{2\pi}; f; \delta) \le M_r \delta^\alpha$ for all $\delta > 0$.

(iii) Let $f \in X_{2\pi}$, $g \in L^1_{2\pi}$. Show that $\omega_r(X_{2\pi}; f * g; \delta) \le \|g\|_1 \omega_r(X_{2\pi}; f; \delta)$. In particular, $f \in \mathsf{Lip}_r(X_{2\pi}; \alpha)$ implies $f * g \in \mathsf{Lip}_r(X_{2\pi}; \alpha)$.

5. Show that for the second integral means $A_h^2(f; x)$ of $f \in X_{2\pi}$

$$\|A_h^2(f; \circ) - f(\circ)\|_{X_{2\pi}} \le \omega^*(X_{2\pi}; f; h) \qquad (h > 0).$$

(Hint: Use the fact that $A_h^2(f; x) = h^{-2} \int_0^h [f(x + u) + f(x - u)](h - u)\, du$)

6. Let $f \in X_{2\pi}$. Show that a necessary and sufficient condition for f to belong to $\mathsf{Lip}(X_{2\pi}; \alpha)$ is that the Fejér means $\sigma_n(f; x)$ belong to $\mathsf{Lip}(X_{2\pi}; \alpha)$, uniformly for $n \in \mathbb{N}$. Extend to general singular integrals. (Hint: As to the necessity use $\|\sigma_n(f; \circ)\|_{X_{2\pi}} \le \|f\|_{X_{2\pi}}$, for the sufficiency Cor. 1.4.7 and Fatou's lemma)

7. Let $f \in X_{2\pi}$ and the kernel $\{\chi_\rho(x)\}$ of the singular integral $I_\rho(f; x)$ be even and positive.

(i) If $f \in \mathsf{AC}_{2\pi}$ with $f' \in X_{2\pi}$, show that

$$\|I_\rho(f; \circ) - f(\circ)\|_{X_{2\pi}} = O(\sqrt{1 - \chi_\rho^\wedge(1)}\,\omega(X_{2\pi}; f'; \sqrt{1 - \chi_\rho^\wedge(1)})) \qquad (\rho \to \rho_0).$$

If $f \in \mathsf{AC}^1_{2\pi}$ with $f'' \in X_{2\pi}$, then $\|I_\rho(f; \circ) - f(\circ)\|_{X_{2\pi}} = O(1 - \chi_\rho^\wedge(1))$. (Hint: Use (1.5.7) and Problem 1.5.3(iv))

(ii) If $f \in \mathsf{AC}_{2\pi}$ with $f' \in X_{2\pi}$, show that for each $\lambda > 0$

$$\|I_\rho(f; \circ) - f(\circ)\|_{X_{2\pi}} \le \frac{\pi}{4}\{\sqrt{2}\,\sqrt{1 - \chi_\rho^\wedge(1)} + \lambda\pi(1 - \chi_\rho^\wedge(1))\}\omega(X_{2\pi}; f'; \lambda^{-1}).$$

(Hint: Use (1.5.8) and partial integration, and then proceed as in the proof of Theorem 1.5.8)

(iii) If $f \in \mathsf{AC}^1_{2\pi}$ with $f'' \in X_{2\pi}$, show that

$$\|I_\rho(f; \circ) - f(\circ)\|_{X_{2\pi}} \le \frac{\pi^2}{4}(1 - \chi_\rho^\wedge(1))\left\{\|f''\|_{X_{2\pi}} + \frac{3}{2}\omega^*\left(X_{2\pi}; f''; \sqrt{4 - \frac{1 - \chi_\rho^\wedge(2)}{1 - \chi_\rho^\wedge(1)}}\right)\right\}.$$

8. Let $f \in X_{2\pi}$ and the kernel $\{\chi_\rho(x)\}$ of the singular integral $I_\rho(f; x)$ be even and positive. Show that for each $\lambda > 0$

$$\|I_\rho(f; \circ) - f(\circ)\|_{X_{2\pi}} \le \left\{1 + \frac{\lambda\pi}{\sqrt{2}}\sqrt{1 - \chi_\rho^\wedge(1)}\right\}\omega(X_{2\pi}; f; \lambda^{-1}).$$

Thus in particular $\|I_\rho(f; \circ) - f(\circ)\|_{X_{2\pi}} = O(\omega(X_{2\pi}; f; \sqrt{1 - \chi_\rho^\wedge(1)}))$. If $\{\chi_\rho(x)\}$ is not even, $\chi_\rho^\wedge(1)$ is to be replaced by $\mathrm{Re}(\chi_\rho^\wedge(1)) = [\chi_\rho]_c^\wedge(1)/2$.

9. (i) Show that for the integral means $A_h(f; x)$ of $f \in X_{2\pi}$ one has

$$\|A_h(f; \circ) - f(\circ)\|_{X_{2\pi}} = O(\omega^*(X_{2\pi}; f; h)) \qquad (h \to 0+).$$

In particular, $f \in \mathsf{Lip}^*(X_{2\pi}; \alpha)$ implies $\|A_h(f; \circ) - f(\circ)\|_{X_{2\pi}} = O(h^\alpha)$. Formulate and prove counterparts for the rth integral means $A_h^r(f; x)$.

(ii) Show that for the singular integral $W_t(f; x)$ of Weierstrass of $f \in X_{2\pi}$ one has

$$\|W_t(f; \circ) - f(\circ)\|_{X_{2\pi}} = O(\omega^*(X_{2\pi}; f; \sqrt{t})) \qquad (t \to 0+).$$

In particular, $f \in \mathsf{Lip}^*(X_{2\pi}; \alpha)$ implies $\|W_t(f; \circ) - f(\circ)\|_{X_{2\pi}} = O(t^{\alpha/2})$.

10. (i) Show that for the singular integral $\sigma_n(f; x)$ of Fejér of $f \in X_{2\pi}$ one has

$$\|\sigma_n(f; \circ) - f(\circ)\|_{X_{2\pi}} = O(\omega^*(X_{2\pi}; f; n^{-1/2})) \qquad (n \to \infty).$$

In particular, $f \in \mathsf{Lip}^*(X_{2\pi}; \alpha)$ implies $\|\sigma_n(f; \circ) - f(\circ)\|_{X_{2\pi}} = O(n^{-\alpha/2})$.

(ii) Show that for the singular integral $P_r(f; x)$ of Abel–Poisson of $f \in X_{2\pi}$ one has

$$\|P_r(f; \circ) - f(\circ)\|_{X_{2\pi}} = O(\omega^*(X_{2\pi}; f; \sqrt{1-r})) \qquad (r \to 1-).$$

In particular, $f \in \mathsf{Lip}^*(X_{2\pi}; \alpha)$ implies $\|P_r(f; \circ) - f(\circ)\|_{X_{2\pi}} = O((1-r)^{\alpha/2})$. (Hint: Use Theorem 1.5.8; for improvements of the latter estimates see Cor. 1.6.5, Theorem 2.5.2)

11. Let $f \in L^\infty_{2\pi}$ have a second Riemann derivative at x_0. Show that

(i) $J_n(f; x_0) - f(x_0) = (3/2)n^{-2}f^{[2]}(x_0) + o(n^{-2}) \qquad (n \to \infty)$,
(ii) $A_h(f; x_0) - f(x_0) = (1/24)h^2 f^{[2]}(x_0) + o(h^2) \qquad (h \to 0+)$,

$J_n(f; x)$ and $A_h(f; x)$ being the singular integrals of Jackson and Riemann–Lebesgue, respectively. (Hint: Use Theorem 1.5.15, thus (1.5.28); to evaluate the second Fourier coefficient of the kernel, in the Jackson case see PETROV [1], MATSUOKA [3], for the integral means see Problem 1.2.22)

12. Let $f \in L^\infty_{2\pi}$ have a right-hand derivative $f'_+(x_0)$ and a left-hand derivative $f'_-(x_0)$ at x_0. Show that

(i) $\lim_{n\to\infty} (n/\log n)[\sigma_n(f; x_0) - f(x_0)] = (1/\pi)[f'_+(x_0) - f'_-(x_0)]$,
(ii) $\lim_{r\to 1-} ((1-r)\,|\log(1-r)|)^{-1}[P_r(f; x_0) - f(x_0)] = (1/\pi)[f'_+(x_0) - f'_-(x_0)]$.

(Hint: Use Theorem 1.5.13 with $\zeta(x) = |\sin(x/2)|$, see also NIKOLSKIĬ [3], NATANSON [8I, p. 163], KOROVKIN [5, p. 116] for (i) and MAMEDOV [9, p. 94] for (ii); $f'_+(x_0)$ is defined through $f'_+(x_0) = \lim_{h\to 0+}[f(x_0+h) - f(x_0)]/h$. For further asymptotic expansions see Cor. 9.2.9)

13. (i) Let $f \in C_{2\pi}$ have an ordinary second derivative at x_0. Show that

$$f(x) = f(x_0) + f'(x_0)\sin(x - x_0) + 2f''(x_0)\sin^2((x-x_0)/2) + \eta(x)\sin^2((x-x_0)/2),$$

where η is bounded with $\lim_{x\to x_0}\eta(x) = 0$. (Hint: Use L'Hospital's rule, see also NATANSON [8I, p. 212])

(ii) Prove Prop. 1.5.17. (Hint: Insert the expansion of (i) into $I_\rho(f; x_0) - f(x_0)$, see also MAMEDOV [2])

14. Let the kernel $\{\chi_\rho(x)\}$ of the singular integral $I_\rho(f; x)$ be even and positive and suppose that $\lim_{\rho\to\rho_0} m(\chi_\rho; 1) = 0$, where the αth (absolute) moment $m(\chi_\rho; \alpha)$ of χ_ρ is defined through (1.6.9).

(i) For $f \in X_{2\pi}$ show that $\|I_\rho(f; \circ) - f(\circ)\|_{X_{2\pi}} \le 2\omega(X_{2\pi}; f; m(\chi_\rho; 1))$. (Hint: See also Prop. 1.6.3)

(ii) For $f \in X_{2\pi}$ show that $\|\sigma_n(f; \circ) - f(\circ)\|_{X_{2\pi}} \le 2\omega(X_{2\pi}; f; \pi(1 + 2\log n)/4n)$. (Hint: Use Lemma 1.6.4)

(iii) Suppose that $\int_\delta^\pi \chi_\rho(u)\,du = o(m(\chi_\rho; 2))$, $\rho \to \rho_0$, for each $0 < \delta < \pi$. If $f \in C_{2\pi}$ has an ordinary second derivative at x_0, show that

$$I_\rho(f; x_0) - f(x_0) = (m(\chi_\rho; 2)/2)f''(x_0) + o(m(\chi_\rho; 2)) \qquad (\rho \to \rho_0).$$

Apply this to the singular integral of Jackson so as to obtain again the assertion of Problem 1.5.11(i) (with $f^{[2]}$ replaced by f''). (Hint: Use the expansion

$$f(x) = f(x_0) + f'(x_0)(x - x_0) + f''(x_0)(x-x_0)^2/2 + \eta(x)(x-x_0)^2,$$

where η is bounded with $\lim_{x\to x_0}\eta(x) = 0$, cf. Problem 1.5.13(i); see also NATANSON [4])

15. (i) Let the function f be defined in a neighbourhood of the point x_0. If

$$\lim_{h\to 0} h^{-1}[f(x_0 + (h/2)) - f(x_0 - (h/2))]$$

exists and is finite, this limit is called the *first Riemann derivative* of f at x_0; in notation: $f^{[1]}(x_0)$. Show that if the ordinary first derivative $f'(x_0)$ exists, so does $f^{[1]}(x_0)$ and $f^{[1]}(x_0) = f'(x_0)$.

(ii) Show that if the ordinary second derivative $f''(x_0)$ exists, so does $f^{[2]}(x_0)$, and $f^{[2]}(x_0) = f''(x_0)$.

16. Let $f \in \mathsf{L}^{\infty}_{2\pi}$, and the kernel $\{\chi_\rho(x)\}$ of the singular integral $I_\rho(f; x)$ be even and positive.

 (i) Suppose (see (1.6.9)) that $\lim_{\rho\to\rho_0} m(\chi_\rho; 1) = 0$ and $\int_\delta^\pi u\chi_\rho(u)\,du = o(m(\chi_\rho; 1))$, $\rho \to \rho_0$, for each $0 < \delta < \pi$. Show that if the one-sided derivatives $f'_+(x_0), f'_-(x_0)$ exist at x_0 (cf. Problem 1.5.12), then

$$I_\rho(f; x_0) - f(x_0) = (m(\chi_\rho; 1)/2)[f'_+(x_0) - f'_-(x_0)] + o(m(\chi_\rho; 1)) \qquad (\rho \to \rho_0).$$

 (Hint: Use (1.5.8) and the expansion

$$f(x_0 + u) + f(x_0 - u) - 2f(x_0) = [f'_+(x_0) - f'_-(x_0)]u + \eta(u)u,$$

 where η is bounded with $\lim_{u\to 0+} \eta(u) = 0$; see also TABERSKI [1])

 (ii) Establish again the assertions of Problem 1.5.12. (Hint: Compare also with Lemmata 1.6.4, 2.5.1)

 (iii) Suppose that $\lim_{\rho\to\rho_0} m(\chi_\rho; 2) = 0$ and $\int_\delta^\pi u^2\chi_\rho(u)\,du = o(m(\chi_\rho; 2))$, $\rho \to \rho_0$, for each $0 < \delta < \pi$. Show that if the second Riemann derivative $f^{[2]}(x_0)$ exists at x_0, then

$$\lim_{\rho\to\rho_0} [I_\rho(f; x_0) - f(x_0)]/m(\chi_\rho; 2) = f^{[2]}(x_0)/2.$$

17. Let the kernel $\{\chi_\rho(x)\}$ of the singular integral $I_\rho(f; x)$ be even and positive and let $m \in \mathbb{N}$. Then for each $f \in \mathsf{X}_{2\pi}$ with period $2\pi/m$

$$\|I_\rho(f; \circ) - f(\circ)\|_{\mathsf{X}_{2\pi}} = O(\omega^*(\mathsf{X}_{2\pi}; f; \sqrt{1 - \chi_\rho^\wedge(m)})) \qquad (\rho \to \rho_0).$$

 (Hint: See also DE VORE [2])

1.6 Further Direct Approximation Theorems, Nikolskiĭ Constants

We continue the investigations of the preceding section in order to consider direct approximation theorems for the singular integral $I_\rho(f; x)$. First we shall introduce the singular integral of Fejér–Korovkin and apply the results so far obtained. In discussing the integral of Fejér, it will be observed that the corresponding results are unsatisfactory and have to be completed by further direct theorems. The section concludes with a few selected results concerning best constants in asymptotic expansions for Lipschitz classes.

1.6.1 Singular Integral of Fejér–Korovkin

The *singular integral of Fejér–Korovkin* is defined for $f \in \mathsf{X}_{2\pi}$ by

$$K_n(f; x) = \frac{1}{2\pi}\int_{-\pi}^{\pi} f(x - u)k_n(u)\,du \tag{1.6.1}$$

with kernel $\{k_n(x)\}$ given by ($j \in \mathbb{Z}$)

$$k_n(x) = \begin{cases} \dfrac{2\sin^2(\pi/(n+2))}{n+2}\left[\dfrac{\cos((n+2)x/2)}{\cos(\pi/(n+2)) - \cos x}\right]^2, & x \neq \pm\dfrac{\pi}{n+2} + 2j\pi \\[2ex] (n+2)/2, & x = \pm\dfrac{\pi}{n+2} + 2j\pi \end{cases} \tag{1.6.2}$$

and discrete parameter $n \in \mathbb{P}$, $n \to \infty$. For each $n \in \mathbb{P}$, $k_n(x)$ is an even, positive trigonometric polynomial of degree n which may be represented as

$$k_n(x) = 1 + 2 \sum_{k=1}^{n} \theta_n(k) \cos kx \tag{1.6.3}$$

with θ-factor given for $1 \le k \le n$ by (cf. Problem 1.6.5)

$$\theta_n(k) = \frac{1}{2(n+2)\sin(\pi/(n+2))}\left[(n-k+3)\sin\frac{k+1}{n+2}\pi - (n-k+1)\sin\frac{k-1}{n+2}\pi\right]. \tag{1.6.4}$$

Thus $k_n^\wedge(1) \equiv \theta_n(1) = \cos(\pi/(n+2))$, and we note that in view of Problem 1.6.4 this value of the first coefficient already determines the kernel (1.6.2) uniquely.

Since $\{k_n(x)\}$ is a positive kernel and $\lim_{n\to\infty} k_n^\wedge(1) = 1$, it follows by Prop. 1.3.10 that $K_n(f; x)$ converges in $X_{2\pi}$-norm to f for every $f \in X_{2\pi}$. In order to apply the results of the last section, we observe that $1 - k_n^\wedge(1) = O(n^{-2})$ for $n \to \infty$. Thus the integral of Fejér–Korovkin has properties that are analogous to those of the integral $J_n(f; x)$ of Jackson (compare (1.5.10)). However, the degree of $J_n(f; x)$ is $(2n - 2)$ whereas the degree of $K_n(f; x)$ is at most n. As an immediate consequence of Theorem 1.5.8, in particular of (1.5.9), we note

Corollary 1.6.1. *If $f \in X_{2\pi}$, then*

$$\|K_n(f; \circ) - f(\circ)\|_{X_{2\pi}} \le 18\omega^*(X_{2\pi}; f; n^{-1}).$$

Thus, if $f \in \mathsf{Lip}^*(X_{2\pi}; \alpha)$ for some $0 < \alpha \le 2$, then

$$\|K_n(f; \circ) - f(\circ)\|_{X_{2\pi}} = O(n^{-\alpha}) \qquad (n \to \infty). \tag{1.6.5}$$

The kernel $\{k_n(x)\}$ of Fejér–Korovkin also satisfies (1.5.27). Indeed,

$$\frac{1 - k_n^\wedge(2)}{1 - k_n^\wedge(1)} = \frac{2(n+1)}{n+2}\left(1 + \cos\frac{\pi}{n+2}\right). \tag{1.6.6}$$

Therefore by Theorem 1.5.15

Corollary 1.6.2. *If $f \in \mathsf{L}_{2\pi}^\infty$ has an ordinary second derivative at x_0, then*

$$n^2[K_n(f; x_0) - f(x_0)] = \frac{\pi^2}{2} f''(x_0) + o(1) \qquad (n \to \infty). \tag{1.6.7}$$

1.6.2 Further Direct Approximation Theorems

In connection with Cor. 1.6.1 it is important that, in a certain sense, the results are best possible. In fact, if for example f is such that the approximation (1.6.5) holds, then, for $0 < \alpha < 2$, f belongs to $\mathsf{Lip}^*(X_{2\pi}; \alpha)$ as will follow in Sec. 2.4 (see Problem 2.4.6(i)). Thus for the integral of Fejér–Korovkin the order of approximation given by (1.6.5) cannot be improved for functions $f \in \mathsf{Lip}^*(X_{2\pi}; \alpha)$, $0 < \alpha < 2$.

For the singular integral of Fejér, the situation is quite different. Indeed, it follows by Theorem 1.5.8 that $f \in \mathsf{Lip}^*(X_{2\pi}; \alpha)$, $0 < \alpha < 1$, implies $\|\sigma_n(f; \circ) - f(\circ)\|_{X_{2\pi}} = O(n^{-\alpha/2})$ (see Problem 1.5.10). But this estimate is by no means best possible. For, as will be seen in the next corollary, $f \in \mathsf{Lip}^*(X_{2\pi}; \alpha)$, $0 < \alpha < 1$, even implies $\|\sigma_n(f; \circ) - f(\circ)\|_{X_{2\pi}} = O(n^{-\alpha})$, which again is best possible in the sense that the converse is also true (see Theorem 2.4.7).

To improve the order of approximation of Problem 1.5.10 we need the following result.

Proposition 1.6.3. *Suppose that the kernel $\{\chi_\rho(x)\}$ of the singular integral $I_\rho(f; x)$ is even. Then for every $f \in \mathsf{Lip}^*(X_{2\pi}; \alpha)$, $0 < \alpha \le 2$,*

$$\|I_\rho(f; \circ) - f(\circ)\|_{X_{2\pi}} = O(m(\chi_\rho; \alpha)) \qquad (\rho \to \rho_0), \tag{1.6.8}$$

where the αth ***absolute moment*** *$m(\chi_\rho; \alpha)$ of χ_ρ is defined by*

$$m(\chi_\rho; \alpha) = \frac{1}{2\pi}\int_{-\pi}^{\pi} |u|^\alpha |\chi_\rho(u)|\, du. \tag{1.6.9}$$

The proof is rather simple. Indeed, it follows as in the proof of Theorem 1.5.8 that

$$\|I_\rho(f; \circ) - f(\circ)\|_{X_{2\pi}} \le \frac{1}{2\pi}\int_0^\pi \omega^*(X_{2\pi}; f; u)\, |\chi_\rho(u)|\, du.$$

Thus in view of Problem 1.5.4(ii)

$$\|I_\rho(f; \circ) - f(\circ)\|_{X_{2\pi}} \le (M_2/2\pi)\int_0^\pi u^\alpha |\chi_\rho(u)|\, du,$$

which already establishes (1.6.8). Let us emphasize that Prop. 1.6.3 is valid for arbitrary kernels; the kernels need not be positive (or even, see Problem 1.6.7).

To apply (1.6.8) to particular singular integrals the moments (1.6.9) have to be estimated. It is to be expected that in this procedure the particular kernels must be estimated more refinedly. For the Fejér kernel a sharper version of Lemma 1.5.7 is valid.

Lemma 1.6.4. *Let $\{F_n(x)\}$ be the Fejér kernel. Then*

(i) $$\frac{1}{2\pi}\int_0^{1/n} u^\alpha F_n(u)\, du \le n^{-\alpha} \qquad (0 < \alpha \le 1),$$

(ii) $$\frac{1}{2\pi}\int_{1/n}^{\pi} u^\alpha F_n(u)\, du \le \begin{cases} \dfrac{\pi}{2}\dfrac{1}{1-\alpha}n^{-\alpha} & (0 < \alpha < 1) \\[2ex] \dfrac{\pi}{2}\dfrac{\log n + \log \pi}{n+1} & (\alpha = 1), \end{cases}$$

(iii) $$\frac{1}{2\pi}\int_0^{\pi} u^{1+\alpha} F_n(u)\, du \le \frac{\pi^2}{2\alpha} n^{-1} \qquad (0 < \alpha \le 1).$$

The proof follows from the inequalities (cf. Problem 1.2.4(ii))

$$u^\alpha F_n(u) \le \begin{cases} 2^{\alpha-2}\pi^2(n+1)^{1-\alpha} & (0 < u \le 1/n; 0 < \alpha \le 1) \\ \pi^2 u^{\alpha-2}/(n+1) & (0 < u \le \pi; 0 < \alpha \le 2). \end{cases} \tag{1.6.10}$$

Combining the results of Prop. 1.6.3 and Lemma 1.6.4 we obtain the following improvement of Problem 1.5.10.

Corollary 1.6.5. *For the singular integral $\sigma_n(f; x)$ of Fejér,*

$$f \in \mathsf{Lip}^* (\mathsf{X}_{2\pi}; \alpha) \Rightarrow \|\sigma_n(f; \circ) - f(\circ)\|_{\mathsf{X}_{2\pi}} = \begin{cases} O(n^{-\alpha}), & 0 < \alpha < 1 \\ O(n^{-1} \log n), & \alpha = 1 \\ O(n^{-1}), & 1 < \alpha \le 2. \end{cases}$$

It will be seen (Sec. 1.7, Problem 2.4.12) that even for $\alpha \ge 1$ the order of approximation in this corollary cannot be improved.

1.6.3 Nikolskiĭ Constants

As a consequence of Cor. 1.6.5 (see also the proof of Prop. 1.6.3) it follows that there exist constants C_α such that $f \in \mathsf{Lip}\,(\mathsf{X}_{2\pi}; \alpha)_M$ implies $\|\sigma_n(f; \circ) - f(\circ)\|_{\mathsf{X}_{2\pi}} \le C_\alpha M n^{-\alpha}$ for $0 < \alpha < 1$, $\le C_1 M n^{-1} \log n$ for $\alpha = 1$, the constant C_α being independent of f. In the proof of this fact we made no effort in obtaining the smallest such constant for every $f \in \mathsf{Lip}\,(\mathsf{X}_{2\pi}; \alpha)_M$. Here we shall give some basic results concerning this problem.

In this subsection we confine ourselves to the space $\mathsf{X}_{2\pi} = \mathsf{C}_{2\pi}$, and suppose throughout that the kernel $\{\chi_\rho(x)\}$ of the singular integral $I_\rho(f; x)$ is even and positive. The quantity

$$\Delta(\chi_\rho; \alpha) = \sup_{f \in \mathsf{Lip}\,(\mathsf{C}_{2\pi}; \alpha)_1} \|I_\rho(f; \circ) - f(\circ)\|_{\mathsf{C}_{2\pi}} \qquad (0 < \alpha \le 1) \tag{1.6.11}$$

is called the *measure of approximation* of the integral $I_\rho(f; x)$ for the class $\mathsf{Lip}\,(\mathsf{C}_{2\pi}; \alpha)_1$. Suppose that $\Delta(\chi_\rho; \alpha)$ admits an asymptotic expansion of the type

$$\Delta(\chi_\rho; \alpha) = N(\alpha)\varphi(\rho) + o(\varphi(\rho)) \qquad (\rho \to \rho_0), \tag{1.6.12}$$

where $N(\alpha)$ is a constant different from zero and $\varphi(\rho)$ is a positive function on $\mathbb{A}$ tending to zero as $\rho \to \rho_0$. Then the positive number $N(\alpha)$ is called the best asymptotic constant or *Nikolskiĭ constant* of the integral $I_\rho(f; x)$ for the class $\mathsf{Lip}\,(\mathsf{C}_{2\pi}; \alpha)_1$. As a first result we have (cf. (1.6.9)) that

$$\Delta(\chi_\rho; \alpha) = m(\chi_\rho; \alpha) \qquad (0 < \alpha \le 1). \tag{1.6.13}$$

Indeed, it follows as in the proof of Prop. 1.6.3 (cf. Problem 1.6.7) that for any $f \in \mathsf{Lip}\,(\mathsf{C}_{2\pi}; \alpha)_1$

$$\|I_\rho(f; \circ) - f(\circ)\|_{\mathsf{C}_{2\pi}} \le \frac{1}{2\pi} \int_{-\pi}^{\pi} |u|^\alpha \chi_\rho(u)\, du.$$

Therefore $\Delta(\chi_\rho; \alpha) \le m(\chi_\rho; \alpha)$. On the other hand, the 2π-periodic function $g_\alpha(x)$, $\alpha > 0$, which coincides with $|x|^\alpha$ on $[-\pi, \pi]$, belongs to $\mathsf{Lip}\,(\mathsf{C}_{2\pi}; \alpha)_1$ for $0 < \alpha \le 1$ (cf. Problem 1.6.9(i)). Therefore

$$\Delta(\chi_\rho; \alpha) \ge \|I_\rho(g_\alpha; \circ) - g_\alpha(\circ)\|_{\mathsf{C}_{2\pi}} \ge |I_\rho(g_\alpha; 0) - g_\alpha(0)| = m(\chi_\rho; \alpha),$$

which proves (1.6.13). We note that at the same time

$$\Delta(\chi_\rho; \alpha) = I_\rho(g_\alpha; 0) - g_\alpha(0) \qquad (0 < \alpha \le 1). \tag{1.6.14}$$

Relation (1.6.13) recasts the problem of determining the expansion (1.6.12) (and thus the Nikolskiĭ constant) for the integral under consideration to the corresponding one for the moment $m(\chi_\rho; \alpha)$. This, in particular in the formulation of (1.6.14), has the advantage that asymptotic expansions such as those discussed in Sec. 1.5.4 may be applied (see, for example, Problem 1.6.11(i)).

However, from this point of view, the results of Sec. 1.5.4, in particular Theorem 1.5.15, seem to be more appropriate for the following problem (the case $\alpha = 2$): If

$$\Delta^*(\chi_\rho; \alpha) = \sup_{f \in \mathsf{Lip}^*(\mathsf{C}_{2\pi}; \alpha)_2} \|I_\rho(f; \circ) - f(\circ)\|_{\mathsf{C}_{2\pi}} \qquad (0 < \alpha \le 2) \tag{1.6.15}$$

is the measure of approximation of the integral $I_\rho(f; x)$ for the class $\mathsf{Lip}^*(\mathsf{C}_{2\pi}; \alpha)_2$, to determine the Nikolskiĭ constant $N^*(\alpha)$ in the expansion (if it exists)

$$\Delta^*(\chi_\rho; \alpha) = N^*(\alpha)\varphi(\rho) + o(\varphi(\rho)) \qquad (\rho \to \rho_0). \tag{1.6.16}$$

As above it follows for $0 < \alpha \le 1$ that $\Delta^*(\chi_\rho; \alpha) = m(\chi_\rho; \alpha)$ (compare Problem 1.6.10). In case $1 < \alpha \le 2$, however, the function g_α does not belong to the class $\mathsf{Lip}^*(\mathsf{C}_{2\pi}; \alpha)_2$. Using $g_\alpha^*(x) = |2 \sin (x/2)|^\alpha$, one may analogously proceed to the inequality

$$\frac{1}{2\pi}\int_{-\pi}^{\pi} \left|2 \sin \frac{u}{2}\right|^\alpha \chi_\rho(u)\, du = I_\rho(g_\alpha^*; 0) - g_\alpha^*(0) \le \Delta^*(\chi_\rho; \alpha) \le m(\chi_\rho; \alpha), \tag{1.6.17}$$

now valid for all $0 < \alpha \le 2$ (see Problem 1.6.10).

Solutions of the above problems for arbitrary (fractional) values of α are often rather delicate. For the quite general class of so-called singular integrals of Fejér's type, however, an elementary solution is obtained in Sec. 3.3. For $\alpha = 2$ the following simple result is available.

Proposition 1.6.6. *Let the kernel* $\{\chi_\rho(x)\}$ *of the singular integral* $I_\rho(f; x)$ *be even and positive and satisfy* (1.5.27). *Then*

$$\Delta^*(\chi_\rho; 2) = 2(1 - \chi_\rho^\wedge(1)) + o(1 - \chi_\rho^\wedge(1)) \qquad (\rho \to \rho_0). \tag{1.6.18}$$

Proof. Since $\Delta^*(\chi_\rho; 2) \le m(\chi_\rho; 2)$ by the right-hand side of inequality (1.6.17), an application of Theorem 1.5.15 to $f = g_2$, $g = g_2^*$ at $x_0 = 0$ yields

$$\limsup_{\rho \to \rho_0} \frac{\Delta^*(\chi_\rho; 2)}{I_\rho([2 \sin (u/2)]^2; 0)} \le \lim_{\rho \to \rho_0} \frac{I_\rho(u^2; 0)}{I_\rho([2 \sin (u/2)]^2; 0)} = 1.$$

On the other hand, by the left-hand side of inequality (1.6.17)

$$\liminf_{\rho \to \rho_0} \frac{\Delta^*(\chi_\rho; 2)}{I_\rho([2 \sin (u/2)]^2; 0)} \ge 1,$$

and therefore

$$\lim_{\rho \to \rho_0} \frac{\Delta^*(\chi_\rho; 2)}{I_\rho([2 \sin (u/2)]^2; 0)} = 1.$$

But this is equivalent to (1.6.18) since by (1.5.3)

$$I_\rho([2\sin(u/2)]^2; 0) = \frac{1}{\pi}\int_{-\pi}^{\pi}(1-\cos u)\chi_\rho(u)\,du = 2(1-\chi_\rho^\wedge(1)).$$

For the singular integral of Fejér–Korovkin this result yields

Corollary 1.6.7. *If $\{k_n(x)\}$ is the kernel of Fejér–Korovkin, then*

$$\Delta^*(k_n; 2) = \pi^2 n^{-2} + o(n^{-2}) \qquad (n\to\infty).$$

Thus the Nikolskiĭ constant $N^*(2)$ has the value π^2.

Problems

1. (i) Show that any (real) $t_n \in \mathsf{T}_n$ with real coefficients can be represented in the form $t_n(x) = e^{-inx}p(e^{ix})$, where $p(z) = \sum_{j=0}^{2n} d_j z^j$ is an algebraic polynomial of degree $2n$ with the property that $p(z) = z^{2n}\bar{p}(1/z)$. Here $\bar{p}(z)$ denotes the polynomial $\bar{p}(z) = \sum_{j=0}^{2n} \overline{d_j} z^j$, $\overline{d_j}$ being the complex conjugate to d_j.
   (ii) If $p(z)$ denotes an algebraic polynomial of degree $2n$ for which $z^{2n}\bar{p}(1/z) = p(z)$, show that $e^{-inx}p(e^{ix})$ is a real trigonometric polynomial of degree n all of whose coefficients are real.
   (Hint: For this and the following four Problems see Fejér [1], Pólya–Szegö [1II, pp. 76–84])
2. Show that any (real) positive $t_n \in \mathsf{T}_n$ can be written in the form $t_n(x) = |h_n(e^{ix})|^2$, where $h_n(z)$ is an algebraic polynomial of degree n.
3. If $t_n(x)$ is a positive cosine-polynomial, show that it is always possible to find an algebraic polynomial $h_n(z)$ all of whose coefficients are real such that $t_n(x) = |h_n(e^{ix})|^2$.
4. If the even positive trigonometric polynomial $C_n(x)$ of degree n is given by $C_n(x) = 1 + 2\sum_{k=1}^{n}\theta_n(k)\cos kx$, show that (i) $C_n(x) \le n+1$, (ii) $|\theta_n(n)| \le \frac{1}{2}$, (iii) $|\theta_n(1)| \le \cos(\pi/(n+2))$. Furthermore, show that one has equality in (i) if and only if $C_n(x)$ is the Fejér kernel, in (ii) if and only if $C_n(x) = 1 + \cos nx$, in (iii) if and only if $C_n(x)$ is the Fejér–Korovkin kernel (with $\theta_n(1) = \cos(\pi/(n+2))$).
5. Show that the positive cosine-polynomial $C_n(x) = |\sum_{j=0}^{n} a_j e^{ixj}|^2/2A_n$ of degree n with $a_j = \sin((j+1)\pi/(n+2))$, $A_n = \sum_{j=0}^{n} a_j^2$, has the representation (1.6.2), and that $C_n^\wedge(k)$ is given by (1.6.4). (Hint: See also Korovkin [5, p. 75 f], Bohman [2], P. J. Davis [1, p. 337 f], Meinardus [1, p. 51 f], Mamedov [9, p. 65 f])
6. Let $t_n(x) = \sum_{k=-n}^{n} c_k e^{ikx}$, $c_k = \overline{c_{-k}}$, be any positive trigonometric polynomial and $t_n(x) \not\equiv 0$. Show that $|c_k| \le c_0\cos(\pi/([\![n/k]\!]+2))$ for $1 \le k \le n$. (Hint: See Szegö [1], Egerváry–Szász [1], Boas–Kac [1])
7. If $\{\chi_\rho(x)\}$ is the kernel of the singular integral $I_\rho(f; x)$, then for every $f \in \mathsf{Lip}(\mathsf{X}_{2\pi}; \alpha)$, $0 < \alpha \le 1$, one has $\|I_\rho(f; \circ) - f(\circ)\|_{\mathsf{X}_{2\pi}} = O(m(\chi_\rho; \alpha))$ as $\rho \to \rho_0$.
8. (i) Show that for the kernel $\{j_n(x)\}$ of Jackson (cf. Problem 1.3.9) $\Delta^*(j_n; 2) = 3n^{-2} + o(n^{-2})$ as $n\to\infty$.
   (ii) Show that for the kernel $\{\theta_3(x, t)\}$ of Weierstrass (cf. Problem 1.3.10) one has $\Delta^*(\theta_3(\circ, t); 2) = 2t + o(t)$ as $t \to 0+$.
   (iii) Show that for the kernel $\{\chi_h(x)\}$ of the integral means $A_h(f; x)$ (see (1.1.15), Problem 1.2.22; note that $\chi_h(x) = (2\pi/h)\kappa_{[-h/2, h/2]}(x)$ for $|x| \le \pi$) $\Delta^*(\chi_h; 2) = (1/12)h^2 + o(h^2)$ as $h \to 0+$.
   Thus for the Nikolskiĭ constants concerning the class $\mathsf{Lip}^*(\mathsf{C}_{2\pi}; 2)_2$ we have $N^*(2) = 3$, $= 2$, $= 1/12$, respectively.
9. (i) Let $g_\alpha(x)$ be the 2π-periodic function which coincides with $|x|^\alpha$ on $[-\pi, \pi]$. Show that $g_\alpha \in \mathsf{Lip}(\mathsf{C}_{2\pi}; \alpha)_1$ for $0 < \alpha \le 1$. (Hint: See e.g. Matsuoka [1])

(ii) Show that g_α as defined in (i) belongs to $\mathsf{Lip}^* (\mathsf{C}_{2\pi}; \alpha)_2$ for $0 < \alpha \le 1$, but does not for $1 < \alpha \le 2$. (Hint: BAUSOV [1])

(iii) Show that $g_\alpha^*(x) = |2 \sin (x/2)|^\alpha$ belongs to $\mathsf{Lip}^* (\mathsf{C}_{2\pi}; \alpha)_2$ for $0 < \alpha \le 2$. (Hint: TABERSKI [1], BAUSOV [1])

10. (i) Show that $\Delta^*(\chi_\rho; \alpha) = m(\chi_\rho; \alpha)$ for $0 < \alpha \le 1$. Thus $\Delta(\chi_\rho; \alpha) = \Delta^*(\chi_\rho; \alpha)$ for $0 < \alpha \le 1$. Prove (1.6.17). (Hint: Use the preceding Problem and proceed as in the proof of (1.6.13))

(ii) Show that $\Delta(\chi_h; \alpha) = [2^\alpha(1 + \alpha)]^{-1}h^\alpha$ for $0 < \alpha \le 1$, $\{\chi_h(x)\}$ being the kernel of the integral means. (Hint: Use (1.6.13).) Furthermore, show that $\Delta^*(\chi_h; \alpha) = [2^\alpha(1 + \alpha)]^{-1}h^\alpha + o(h^\alpha)$ for every $1 < \alpha \le 2$. (Hint: Use (1.6.17); compare with Problem 1.6.8(iii))

For a solution of the following Problems for fractional values of α one may also consult Sec. 3.3.

11. Let $\{F_n(x)\}$ be the Fejér kernel (1.2.24).

(i) Show that $\Delta(F_n; 1) = (2/\pi)n^{-1} \log n \,[1 + o(1)]$, $n \to \infty$. Thus $N(1) = 2/\pi$. (Hint: Use (1.6.14) and Problem 1.5.12; see also NATANSON [8I, p. 166])

(ii) Show that $\Delta(F_n; \alpha) = \{2\pi^{-1}(1 - \alpha)^{-1}\Gamma(\alpha) \sin (\alpha\pi/2)\}n^{-\alpha} + o(n^{-\alpha})$, $n \to \infty$, for $0 < \alpha < 1$. (Hint: NIKOLSKIĬ [1])

12. Show that for the Weierstrass kernel $\Delta^*(\theta_3 (\circ, t); \alpha) = \{2^\alpha\pi^{-1/2}\Gamma((1 + \alpha)/2)\}t^{\alpha/2} + o(t^{\alpha/2})$, $t \to 0+$, for $0 < \alpha \le 2$. (Hint: For $\alpha = 1$, 2 see KOROVKIN [4], also Problem 1.6.8(ii); for fractional α see BAUSOV [1])

13. (i) Show that for the Jackson kernel $\Delta^*(j_n; 1) = \{6\pi^{-1} \log 2\}n^{-1}[1 + o(1)]$, $n \to \infty$ (see also Problem 1.6.8(i)).

(ii) Show that for the Fejér–Korovkin kernel (see also Cor. 1.6.7)

$$\Delta^*(k_n; 1) = \left\{2 \int_0^\pi u^{-1} \sin u \, du - (4/\pi)\right\}n^{-1} + o(n^{-1}) \qquad (n \to \infty).$$

(Hint: NATANSON [4], PETROV [1, 2])

14. Show that for the Abel–Poisson kernel for $r \ge r_0 > 0$, $r \to 1-$,

$$\Delta(p_r; \alpha) = \begin{cases} [\cos (\alpha\pi/2)]^{-1}(1 - r)^\alpha + o((1 - r)^\alpha), & 0 < \alpha < 1 \\ (2/\pi)(1 - r) \,|\log (1 - r)|\, [1 + o(1)], & \alpha = 1. \end{cases}$$

(Hint: For $\alpha = 1$ use (1.6.14) and Problem 1.5.12, see NATANSON [3], TABERSKI [1] as well as SZ.-NAGY [4], TIMAN [1])

15. (i) Let the kernel $\{C_n(x)\}$ of the singular integral $U_n(f; x)$ be generated by a row-finite θ-factor with $m(n) = n$ (cf. (1.2.41)), and suppose that $\{C_n(x)\}$ is a positive kernel satisfying $\lim_{n\to\infty} \theta_n(1) = 1$. Let $b_n = \inf \Delta^*(C_n; 2)$, the infimum being taken over all kernels of the above type. Show that $\lim_{n\to\infty} n^2b_n = \pi^2$. Thus the Fejér–Korovkin kernel $\{k_n(x)\}$ is one example for which the above infimum is attained (see Cor. 1.6.7). (Hint: KOROVKIN [3])

(ii) Let the kernel $\{\chi_r(x)\}$ of the singular integral $I_r(f; x)$ be given by

$$\chi_r(x) = 1 + 2r \cos x + 2 \sum_{k=2}^\infty \theta_r(k) \cos kx \ge 0 \qquad (0 < r < 1).$$

Set $b_r = \inf \Delta^*(\chi_r; 2)$, the infimum being taken over all kernels of the above type. Show that $\lim_{r\to 1-} b_r/(1 - r) = 2$. Thus the Weierstrass kernel $\{\theta_3(x, t)\}$ with $t = \log (1/r)$ furnishes one example for which the above infimum is attained (see Problem 1.6.8(ii)). (Hint: KOROVKIN [4])

16. Let $\lambda(x)$ be a real-valued, continuous function on the interval [0, 1] such that $A_n \equiv \sum_{k=0}^n \lambda^2(k/n) > 0$ for each $n \in \mathbb{N}$. Set

$$l_n(x) = \frac{1}{A_n}\left| \sum_{k=0}^n \lambda\left(\frac{k}{n}\right) e^{ikx} \right|^2, \qquad \Phi_n(f; x) = \frac{1}{2\pi} \int_{-\pi}^\pi f(x - u)l_n(u) \, du.$$

(i) Show that $l_n(x) = 1 + 2\sum_{k=1}^{n} \theta_n(k) \cos kx$ with coefficients given by

$$\theta_n(k) = A_{k,n}/A_n, \qquad A_{k,n} = \sum_{j=0}^{n-k} \lambda\Big(\frac{j}{n}\Big)\lambda\Big(\frac{j+k}{n}\Big).$$

Thus l_n is an even and positive trigonometric polynomial of degree n at most.

(ii) Show that $\{l_n(x)\}$ is an approximate identity, thus $\lim_{n\to\infty} \|\Phi_n(f; \circ) - f(\circ)\|_{X_{2\pi}} = 0$ for every $f \in X_{2\pi}$. (Hint: Use Theorem 1.3.7, Prop. 1.3.10)

(iii) Show that the kernels of Fejér, Fejér–Korovkin (of degree $n - 2$) and Jackson (cf. (1.2.24), (1.6.2), Problem 1.3.9, respectively) furnish examples corresponding to $\lambda(x) = 1$ (cf. Problem 1.2.6(iii)), $\lambda(x) = \sin \pi x$ (cf. Problem 1.6.5), $\lambda(x) = 1 - |2x - 1|$, respectively.

(iv) Let there exist $b > 0$, $\delta > 0$ such that $\lambda^2(x) + \lambda^2(1 - x) \geq b$ for all $0 \leq x \leq \delta$. Show that with some $c > 0$

$$\Delta^*(l_n; 1) \geq c(\log n)/n + O(1/n) \qquad (n \to \infty).$$

(v) Let $\lambda \in \mathsf{BV}[0, 1]$ be such that $\lambda^2(0) + \lambda^2(1) > 0$. Show that

$$\lim_{n\to\infty} n(1 - l_n^\frown(k)) = k[\lambda^2(0) + \lambda^2(1)]/\Big\{2\int_0^1 \lambda^2(u)\,du\Big\},$$

$$\lim_{n\to\infty} (n/\log n)\Delta^*(l_n; 1) = [\lambda^2(0) + \lambda^2(1)]/\Big\{\pi\int_0^1 \lambda^2(u)\,du\Big\}.$$

Show furthermore that for $f \in \mathsf{L}_{2\pi}^\infty$ of Problems 1.5.12, 1.5.16(i) one has

$$\lim_{n\to\infty} \frac{n}{\log n}[\Phi_n(f; x) - f(x)] = \frac{f'_+(x) + f'_-(x)}{2}\,\frac{\lambda^2(0) + \lambda^2(1)}{\pi\int_0^1 \lambda^2(u)\,du}.$$

Apply to the singular integral of Fejér.

(vi) Let $\lambda(0) = \lambda(1) = 0$ and suppose that $\lambda(x)$ is piecewise continuously differentiable on $[0, 1]$. Show that

$$\lim_{n\to\infty} n^2(1 - l_n^\frown(k)) = (k^2/2)\Big\{\int_0^1 [\lambda'(u)]^2\,du\Big\}\Big/\Big\{\int_0^1 \lambda^2(u)\,du\Big\}.$$

Apply to the singular integrals of Fejér-Korovkin and Jackson. Examine (1.5.27) and Prop. 1.6.6.

(vii) Let $\lambda \in \mathsf{C}^2[0, 1]$. Show that for $0 < \alpha < 1$ and $n \to \infty$

$$\Delta^*(l_n; \alpha) = \frac{2\Gamma(\alpha)\sin(\alpha\pi/2)}{\pi\int_0^1 \lambda^2(u)\,du}\Big\{\int_0^1 x^{-\alpha}\Big[\int_0^{1-x} \lambda(x + u)\lambda'(u)\,du + \lambda(0)\lambda(x)\Big]dx\Big\}n^{-\alpha} + o(n^{-\alpha}).$$

If furthermore $\lambda(0) = \lambda(1) = 0$, then this expansion also holds for $1 \leq \alpha < 2$. For a result concerning $\alpha = 2$ compare (vi).

(Hint: As to (i)–(vi) see KOROVKIN [6], as to (vii) see BAUSOV [2, 3], the latter paper including the extension covering the Jackson integral)

1.7 Simple Inverse Approximation Theorems

In Sec. 1.5 and 1.6 we have in particular seen that if $f \in \mathsf{Lip}^*(X_{2\pi}; 2)$, then $\|\sigma_n(f; \circ) - f(\circ)\|_{X_{2\pi}} = O(n^{-1})$, $n \to \infty$. Moreover, this estimate cannot be improved by assuming the existence of additional derivatives of f. In fact, for the arbitrarily often differentiable function $f(x) = \exp\{ix\}$, $\sigma_n(f; x) = (n/(n + 1))f(x)$, and therefore

$\|\sigma_n(f;\circ) - f(\circ)\|_{X_{2\pi}} = 1/(n+1)$. On the other hand, if f is a constant, then obviously $\|\sigma_n(f;\circ) - f(\circ)\|_{X_{2\pi}} \equiv 0$, $n \in \mathbb{N}$, i.e., the constants are approximable by the Fejér means within any prescribed order. The question then arises whether the hypothesis $\|\sigma_n(f;\circ) - f(\circ)\|_{X_{2\pi}} = o(n^{-1})$, $n \to \infty$, would conversely imply that f is a constant. This is indeed the case and will be a consequence of

Theorem 1.7.1. *Let $f \in X_{2\pi}$, and the kernel $\{C_n(x)\}$ of the singular integral $U_n(f;x)$ be generated by a row-finite θ-factor, i.e., let* (*cf.* (1.2.41))

$$C_n(x) = 1 + 2\sum_{k=1}^{m(n)} \theta_n(k)\cos kx. \tag{1.7.1}$$

If there exists an $\alpha > 0$ such that for each fixed $k \in \mathbb{Z}$, $k \neq 0$,

$$\liminf_{n\to\infty} n^{\alpha}|C_n^{\wedge}(k) - 1| > 0, \tag{1.7.2}$$

then

$$\liminf_{n\to\infty} n^{\alpha}\|U_n(f;\circ) - f(\circ)\|_{X_{2\pi}} = 0 \tag{1.7.3}$$

implies that f is constant (a.e.).

Proof. First we note that for each $f \in X_{2\pi}$, $U_n(f;x)$ is a polynomial of degree $m(n)$ at most, given by (1.2.40). Moreover, by Problem 1.2.2(i) one has for each $k \in \mathbb{Z}$

$$[U_n(f;\circ)]^{\wedge}(k) = \sum_{j=-m(n)}^{m(n)} \theta_n(j) f^{\wedge}(j)\frac{1}{2\pi}\int_{-\pi}^{\pi} e^{i(j-k)x}\,dx = \begin{cases}\theta_n(k)f^{\wedge}(k), & |k| \le m(n)\\ 0, & |k| > m(n).\end{cases}$$

Therefore for $|k| \le m(n)$

$$\frac{1}{2\pi}\int_{-\pi}^{\pi} [U_n(f;x) - f(x)]\,e^{-ikx}\,dx = (\theta_n(k) - 1)f^{\wedge}(k),$$

and hence

$$|\theta_n(k) - 1|\,|f^{\wedge}(k)| \le \frac{1}{2\pi}\int_{-\pi}^{\pi} |U_n(f;x) - f(x)|\,dx. \tag{1.7.4}$$

In view of the hypotheses there exists a subsequence $\{n_j\}$, $\lim_{j\to\infty} n_j = \infty$, such that on the one hand $\lim_{j\to\infty} n_j^{\alpha}|\theta_{n_j}(k) - 1| = c_k > 0$ for each $k \in \mathbb{Z}$, $k \neq 0$, and on the other hand $\lim_{j\to\infty} n_j^{\alpha}\|U_{n_j}(f;\circ) - f(\circ)\|_1 = 0$. By (1.7.4) this implies $c_k|f^{\wedge}(k)| = 0$, and thus $f^{\wedge}(k) = 0$ for each $k \in \mathbb{Z}$, $k \neq 0$. Hence it follows by Cor. 1.2.7 that f is constant (a.e.).

For an application we return to the Fejér means $\sigma_n(f;x)$. By (1.2.43) the Fejér kernel is of the form (1.7.1) with $m(n) \equiv n$ and $\theta_n(k) = 1 - (|k|/(n+1))$. Moreover, it satisfies (1.7.2) with $\alpha = 1$. Therefore

Corollary 1.7.2. *If $f \in X_{2\pi}$ and*

$$\liminf_{n\to\infty} n\,\|\sigma_n(f;\circ) - f(\circ)\|_{X_{2\pi}} = 0, \tag{1.7.5}$$

then f is constant (a.e.).

In other words, in the spaces $X_{2\pi}$ the Fejér means $\sigma_n(f;x)$ never give an approximation to f with error $o(n^{-1})$ unless f is constant. This answers the question posed at the beginning. Furthermore, most examples of polynomial summation processes $U_n(f; x)$ of the Fourier series of f do possess the property (1.7.2) and thus admit an assertion analogous to Cor. 1.7.2 (cf. the Problems of this section). Hence, if one sums the Fourier series of $f \in X_{2\pi}$ by such processes, then on the one hand this produces convergence to every f and even approximation within a certain order which corresponds to the smoothness of f. But on the other hand, the order of approximation cannot be improved beyond the critical order $O(n^{-\alpha})$, $n \to \infty$, where α is given by (1.7.2). If this holds, the summation process $U_n(f; x)$ is said to be saturated, the relevant value α being the saturation index. We shall return to such questions in Chapter 12, give the formal definitions and deal with general singular integrals $I_\rho(f; x)$.

Here we mention a further interesting phenomenon. Let us suppose that $C_n(x)$ of (1.7.1) with $m(n) \equiv n$ is a positive function for each $n \in \mathbb{N}$. In view of Problem 1.6.4 $|\theta_n(1)| \le \cos(\pi/(n+2))$, and therefore the sequence $\{n^2|\theta_n(1) - 1|\}$ does not tend to zero as $n \to \infty$. Since $U_n(\cos u; x) = \theta_n(1)\cos x$ and correspondingly for $\sin x$, this implies (compare (1.5.14), (1.5.18))

Proposition 1.7.3. *Let the kernel $\{C_n(x)\}$ of the singular integral $U_n(f; x)$ be positive and given by* (1.7.1) *with $m(n) \equiv n$. Then the sequences $\{n^2\|U_n(\cos u; \circ) - \cos \circ\|_{X_{2\pi}}\}$ and $\{n^2\|U_n(\sin u; \circ) - \sin \circ\|_{X_{2\pi}}\}$ do not tend to zero as $n \to \infty$.*

Thus, even if f is an infinitely often differentiable function (as $\cos x$ is), the approximation of f by positive singular integrals cannot in general be better than $O(n^{-2})$, $n \to \infty$.

Moreover, we have in view of Problem 1.6.6 that if $C_n(x)$ of (1.7.1) with $m(n) \equiv n$ is a positive function for each $n \in \mathbb{N}$, then $|\theta_n(k)| \le \cos(\pi/([\![n/k]\!] + 2))$ for each $1 \le k \le n$. Therefore $\liminf_{n\to\infty} n^2|\theta_n(k) - 1| > 0$ for each $k \in \mathbb{Z}$, $k \ne 0$. This gives

Proposition 1.7.4. *If under the hypothesis of Prop.* 1.7.3

$$\liminf_{n\to\infty} n^2\|U_n(f; \circ) - f(\circ)\|_{X_{2\pi}} = 0$$

for some $f \in X_{2\pi}$, then f is constant (a.e.).

Problems

1. Let $f \in X_{2\pi}$ and $J_n(f; x)$ be the singular integral of Jackson. Show that f is constant (a.e.), provided $\liminf_{n\to\infty} n^2\|J_n(f; \circ) - f(\circ)\|_{X_{2\pi}} = 0$.
2. Let $f \in X_{2\pi}$ and $K_n(f; x)$ be the singular integral of Fejér–Korovkin. Show that f is constant (a.e.), provided $\liminf_{n\to\infty} n^2\|K_n(f; \circ) - f(\circ)\|_{X_{2\pi}} = 0$.
3. Let $f \in X_{2\pi}$ and $B_n(f; x)$ be the singular integral of Rogosinski. Show that f is constant (a.e.), provided $\liminf_{n\to\infty} n^2\|B_n(f; \circ) - f(\circ)\|_{X_{2\pi}} = 0$.
4. Let $f \in X_{2\pi}$ and the kernel $\{C_n(x)\}$ of the singular integral $U_n(f; x)$ be given by (1.7.1). If there exists an $\alpha > 0$ and $l \in \mathbb{N}$ such that for each fixed $k \in \mathbb{Z}$, $|k| > l$, condition (1.7.2) is satisfied, show that (1.7.3) implies that f is a polynomial of degree l at most.
5. Let $\{\chi_\rho(x)\}$ be the kernel of the singular integral $I_\rho(f; x)$, and let $\varphi(\rho)$ be any positive function on $\mathbb{A}$.

(i) Show that for each $k \in \mathbb{Z}$, $k \neq 0$, the following assertions are equivalent:

(a) One of the following two sequences does not tend to zero as $\rho \to \rho_0$:

$\{\varphi(\rho)\ \|I_\rho(\cos ku; \circ) - \cos k\circ\|_{X_{2\pi}}\}, \{\varphi(\rho)\ \|I_\rho(\sin ku; \circ) - \sin k\circ\|_{X_{2\pi}}\}$.

(b) The sequence $\{\varphi(\rho)\ |\chi_\rho^\wedge(k) - 1|\}$ does not tend to zero as $\rho \to \rho_0$.

(ii) If, furthermore, the kernel is even, show that for each $k \in \mathbb{Z}$, $k \neq 0$, the following assertions are equivalent:

(a) $\varphi(\rho)\ |\chi_\rho^\wedge(k) - 1|$ does not tend to zero as $\rho \to \rho_0$.
(b) $\varphi(\rho)\ \|I_\rho(\cos ku; \circ) - \cos k\circ\|_{X_{2\pi}}$ does not tend to zero as $\rho \to \rho_0$.
(c) $\varphi(\rho)\ \|I_\rho(\sin ku; \circ) - \sin k\circ\|_{X_{2\pi}}$ does not tend to zero as $\rho \to \rho_0$.

(Hint: See also Butzer–Nessel–Scherer [1])

1.8 Notes and Remarks

Sec. 1.1. An early and important paper on singular integrals is that of Lebesgue [2] (1909); an earlier but less general theory was given by Du Bois–Reymond and Dini. The work of Hahn [1] and Hobson [1] is also basic. Most of our results are to be found in Hardy–Rogosinski [1, pp. 53–67], and indeed in books on Fourier series such as Zygmund [7I, p. 84 ff], Edwards [1I, p. 59 ff]. For Theorem 1.1.5, Prop. 1.1.6, 1.1.7 see also Hoffman [1, pp. 17–20]. A sequence $\{\chi_n(x)\}_{n=1}^{\infty}$ defining an approximate identity (or unit) is sometimes referred to in the literature as a (Dirac) delta sequence. For the term 'approximate identity' compare also Prop. 4.1.4. For further general comments on singular integrals see the remarks to Sec. 3.1.

The term 'singular' was already used by Lebesgue [2], Hobson [2, Chapter 7] and de La Vallée Poussin [3]; it is also used by Titchmarsh [6, p. 30 ff], the Russian school, in particular by Natanson [6, pp. 309–359; 8], and by Hille–Phillips [1]. However, it is not standard and many authors have no specific name for it. Still others, especially the Chicago school of Zygmund, mean something very different by the word singular (see Calderón–Zygmund [1], Calderón [3]; the corresponding one-dimensional situation is treated in Part III and leads to conjugate functions and Hilbert transforms). In our concept a (convolution) integral is singular in the sense that the limit of $f * \chi_\rho$ for $\rho \to \rho_0$ delivers f; the individual $f * \chi_\rho$ may be very smooth. Therefore it would be more correct to call the whole set $\{I_\rho(f; x)\}$ of integrals (1.1.3) a singular integral, in correspondence with the definition of a kernel. Singular integrals need not necessarily be of convolution type. Indeed, there is a general theory dealing with integrals of type $\int_a^b f(u)\chi_n(x, u)\, du$ for which the reader is referred to Hobson [2, p. 422 ff], Natanson [6, p. 309 ff].

Strong and weak derivatives are considered in books on functional analysis, compare Ljusternik–Sobolew [1, p. 192 ff], Hille–Phillips [1, p. 58 ff]. Just as the usual pointwise derivative, they are defined successively by means of the derivatives of lower order (in contrast to the generalized derivatives considered in the pointwise sense in Sec. 5.1.4 and in the weak and strong topology in Sec. 10.1). For further comments, in particular for the connection with distributional derivatives, see the remarks to Sec. 3.1. Our results concerning derivatives of convolutions are standard topics in Fourier analysis, see e.g. Edwards [1I, p. 55 ff], Loomis [3, p. 165 f]; they followed immediately since we assumed strong differentiability. The same is true for relation (1.1.20) for which one may consult Edwards [1I, p. 60] or Natanson [8I, p. 216], for example. The corresponding pointwise assertions are less obvious, see Sec. 1.4.

Sec. 1.2. The material here constitutes the basic results of the classical theory of trigonometric and Fourier series to be found in any book on the subject. Our treatment is brief as ours is not intended as a text on Fourier series per se. We prefer to view Fourier coefficients from the standpoint of transform theory; this matter is dealt with systematically in Chapter 4. The standard references are also cited at the beginning of Sec. 4.4; see also the relevant chapters of texts on real variable theory, e.g. ASPLUND–BUNGART [1, Chapter 10]. In this line, the reader will note that the pointwise convergence theory of Fourier series has been relegated to the Problem-section. But this theory is covered in every book on Fourier series; let us just mention the excellent accounts in BARI [1], ZYGMUND [7].

For an elementary proof of an explicit formula for the numbers L_n of Prop. 1.2.3, due to FEJÉR, see CARLITZ [1]. In connection with the fact that the Dirichlet kernel does not constitute an approximate identity, KOREVAAR [1, p. 323] analysed its properties and considered 'delta sequences of Dirichlet type' from a general viewpoint.

The Weierstrass theorem has been the object of much study in recent years. M. H. STONE in 1937 treated very general approximation theorems concerning closed subalgebras of the Banach algebra (with pointwise operations) of continuous functions on any compact Hausdorff space. For theorems of Stone–Weierstrass-type compare HEWITT–STROMBERG [1, p, 90 ff], EDWARDS [3, Section 4.10]. The literature abounds with names for what we have called a closed system (Prop. 1.2.6); some speak of complete, fundamental or spanning systems.

Concerning summability, a number of books do not work with general θ-factors but with specific examples, such as Fejér or Abel, or with row-finite factors, thus with triangular matrices (see BARI [1II, p. 2 ff]). In most examples of interest (including those of this section) $\theta_\rho(k) = \theta(k/\rho)$, $\theta(x)$ being a definite function of (one) real variable x. The reasons for this become apparent in view of the results of Sec. 3.1.2, 5.1.5.

A fundamental theorem on linear methods of summation of sequences, the terms of which are general complex numbers, is due to O. TOEPLITZ (see HARDY [2], ZELLER [1] and the literature cited there): Given a matrix $A = (a_{jk})_{j,k=0}^{\infty}$, A associates with each sequence $\{s_k\}_{k=0}^{\infty}$ a sequence $\{\sigma_j\}_{j=0}^{\infty}$, defined by $\sigma_j = \sum_{k=0}^{\infty} a_{jk}s_k$, provided that these series converge. If $\sigma_j \to s$, the sequence $\{s_k\}$ is said to be *A-summable* to s; the σ_j are called the *A-means*. The theorem states that, for every convergent sequence $\{s_k\}$, $s_n \to s$ implies $\sigma_n \to s$ for $n \to \infty$ (thus A is *regular*) if and only if A is a Toeplitz matrix. Such matrices satisfy the three conditions: (i) $\lim_{j\to\infty} a_{jk} = 0$ $(k = 0, 1, \ldots)$, (ii) $\lim_{j\to\infty} \sum_{k=0}^{\infty} a_{jk} = 1$, (iii) $\sup_j \sum_{k=0}^{\infty} |a_{jk}| < \infty$. We are usually concerned with sequences $s_k = \sum_{l=0}^{k} c_l$ defined by partial sums of infinite series. The choices $a_{jk} = (j+1)^{-1}$ for $0 \le k \le j$, $= 0$ for $j < k$ and $a_{jk} = (1 - r_j)r_j^k$, $0 < r_j < 1$, $r_j \to 1-$, then lead to Cesàro and Abel summability of the series, respectively. In this general setting, a series summable Cesàro is summable Abel to the same sum, but not conversely. Thus many results involving Abel summability follow from corresponding theorems that treat Cesàro summability. Nevertheless, an independent study of the former is of interest, particularly when dealing with Fourier series of functions f. This is true, not only because such series may be Abel summable under weaker conditions on f than are necessary to guarantee their Cesàro summability, but also because the Abel means have special properties related to the theory of harmonic and analytic functions that are not enjoyed by the Cesàro means.

However, in contrast to the above Toeplitz approach, we commenced with the means (1.2.30). Thus we do not deal with A-means of the partial sums of the Fourier series (HARDY–ROGOSINSKI [1, p. 53 ff]), but insert the factors $\theta_\rho(k)$ directly into the series (ZYGMUND [7I, p. 84 ff]).

For an abstract approach to the Abel–Poisson integral see RUDIN [4, p. 109]. Concerning Problem 1.2.7(iv), although the series $(\lambda_0/2) + \sum_{k=1}^{\infty} \lambda_k \cos kx$ converges to a sum $f(x)$ for all $x \neq 2\pi j$, $j \in \mathbb{Z}$, it is not necessarily the Fourier series of f (see Problem 6.3.2(iii)). But if the sequence $\{\lambda_k\}$ is furthermore assumed to be convex on $\mathbb{P}$, then it follows that $f \in \mathsf{L}_{2\pi}^1$, and the above series is the Fourier series of f, see Sec. 6.3.2.

Sec. 1.3. For Prop. 1.3.1, 1.3.2 see GOES [1] and the literature cited there. As the operators in question are of convolution type, it was rather easy to determine their norm, Prop. 0.8.8,

0.8.9 on bounded linear functionals being applicable. Similar assertions are also available for more general kernels. If $\chi(x, u)$ is a continuous function of both variables x, u in the square $a \le x, u \le b$, and an operator T of $\mathsf{C}[a, b]$ into itself is defined by $(Tf)(x) = \int_a^b f(u)\chi(x, u)\,du$, then $\|T\| = \sup_{a \le x \le b} \int_a^b |\chi(x, u)|\,du$; compare KANTOROVICH–AKILOV [1, p. 108]. Prop. 1.3.3 seems to have been first stated explicitly by ORLICZ [1]. In the Russian literature the integral of Rogosinski is usually referred to as that of Bernstein, see NATANSON [8I, p. 217]. For Theorem 1.3.5 see also NATANSON [8III, p. 87 ff], KANTOROVICH–AKILOV [1, p. 255 ff].

Prop. 1.3.6 may be considered as a weak version of the famous theorem of DU BOIS-REYMOND (1872) as stated in Problem 1.3.5(i). This result has been extended in several directions. Thus there exist continuous functions whose Fourier series diverge on sets that are uncountable, of the second-category, and everywhere dense (see EDWARDS [1I, p. 155 ff], RUDIN [4, p. 101 ff]). In 1966 CARLESON [1] solved a problem posed by LUSIN in 1915. It states that the Fourier series of any function $f \in \mathsf{L}^2_{2\pi}$ converges pointwise almost everywhere to f. This result was extended by HUNT [1] to all $\mathsf{L}^p_{2\pi}$ with $p > 1$. In 1926 KOLMOGOROV showed that there exist integrable functions whose Fourier series diverge everywhere. For a proof see ZYGMUND [7I, p. 310 ff] or BARI [1I, p. 455 ff]. For a discussion from a general point of view, see KATZNELSON [1, p. 55 ff]. Other extensions are connected with the names of FABER, BERMAN, MARCINKIEWICZ, HARŠILADSE and LOZINSKIĬ (see Problems 2.4.13–2.4.17). Compare CHENEY [1, p. 210 ff], LORENTZ [3, p. 96 f].

For historical remarks on the integral of Jackson of Problem 1.3.9 and its many generalizations see e.g. GÖRLICH–STARK [2]. For remarks on the integral of Weierstrass of Problem 1.3.10 see BUTZER–GÖRLICH [2, p. 340, p. 366 f]. For comments to Sec. 1.3.3 see the Notes and Remarks to Sec. 1.5.

Sec. 1.4. Much of the material of this section was influenced by HARDY–ROGOSINSKI [1, p. 58 ff] and BOCHNER–CHANDRASEKHARAN [1, p. 13 ff]; see also the comments to Sec. 3.2. For Prop. 1.4.1 compare ZYGMUND [7I, p. 86]. Cor. 1.4.3, 1.4.7 are standard, the latter often called the theorem of Fejér–Lebesgue. Prop. 1.4.8 for the particular kernel of Abel–Poisson (then known as a theorem of Fatou) is to be found in ZYGMUND [7I, p. 100], HOFFMAN [1, p. 34 f]. Their proof looks simpler since they divide the central difference of $\mu(x)$ by $\sin x$ in which case $\{\Theta_\rho(x)\}$ constitutes an approximate identity. We divided by $\sin(x/2)$, there also being a singularity at π (compare (9.2.30)). Prop. 1.4.10 also does not seem to be stated explicitly for general kernels; for the (C, α) means it is given in ZYGMUND [7I, p. 105 f]. Problem 1.4.2, due to Fatou, may also be stated for nontangential limits; see BARI [1I, p. 152 ff], PRIVALOV [1].

At this stage it would be possible to study dominated convergence of singular integrals. An integral $I_\rho(f; x)$ with $f \in \mathsf{X}_{2\pi}$ converges *dominatedly* to f as $\rho \to \rho_0$ if $\lim_{\rho\to\rho_0} I_\rho(f; x) = f(x)$ a.e. and there exists some $g \in \mathsf{L}^1_{2\pi}$ such that $\sup_{\rho \in \mathbb{A}} |I_\rho(f; x)| \le g(x)$. We have seen (see Problems 1.2.10, 1.2.11) that if $f \in \mathsf{BV}_{\mathrm{loc}}$ is 2π-periodic, then the partial sums $S_n(f; x)$ converge *boundedly* to $f(x)$ as $n \to \infty$. More generally, if $\{\chi_\rho(x)\}$ is a kernel which is majorized by $\{\chi_\rho^*(x)\}$ satisfying (1.1.5) and (1.4.1) with constants M^* and M_1^*, respectively, then

$$\sup_{\rho \in \mathbb{A}} |I_\rho(f; x)| \le (M^* + 2M_1^*)\Theta(f; x).$$

The latter function, known as the Hardy–Littlewood majorant of f, is defined by

$$\Theta(f; x) = \sup_{0 < |t| \le \pi} \frac{1}{t} \int_x^{x+t} |f(u)|\,du \qquad (-\pi \le x \le \pi).$$

Let $f \in \mathsf{L}^p_{2\pi}$, $1 < p < \infty$. Then $\Theta(f; x) \in \mathsf{L}^p_{2\pi}$; in particular, the integral $\sigma_n(f; x)$ of Fejér converges dominatedly to f. Moreover, $\|\sup_{n\ge1} |\sigma_n(f; \circ)|\,\|_p \le A_p\|f\|_p$ for some constant A_p. This material receives excellent coverage in ZYGMUND [7I, p. 29 ff, 154 ff]; compare also K. L. PHILLIPS [1].

Sec. 1.5. At first some comments to Sec. 1.3.3. For continuous functions, the analog of Theorem 1.3.7 for approximating sums and its proof is due to BOHMAN [1]; his paper appeared on Aug. 8, 1952. The present theorem for singular integrals was given more than a year later by KOROVKIN [1] who was perhaps not aware of the BOHMAN paper. Statement (ii) is the essential new result; statements of type (iv) (plus (1.1.2)) may already be found in the work of BOREL, FRÉCHET and LEBESGUE. Theorems of type 1.3.7 gave rise to a considerable literature; they have also found their way into books on approximation, so, for instance, KOROVKIN [5, p. 1 ff], CHENEY [1, p. 123 ff], MEINARDUS [1, p. 6 ff], LORENTZ [3, p. 7 ff], MAMEDOV [9, p. 41 ff]. The theorem is often stated for positive linear operators (or functionals) which are not necessarily representable as integrals of convolution type. Theorem 1.3.7 for $L^p_{2\pi}$-functions is to be found in CURTIS [1], DZJADYK [2]. The useful term 'test set' is used in this connection by FREUD–KNAPOWSKI [1] and CURTIS [1].

Problems of the type considered in Sec. 1.5 were first discussed in any degree of generality by LEBESGUE [1] and DE LA VALLÉE POUSSIN [1] in 1908. Concerning parts (iv) of Lemmata 1.5.2, 1.5.4, a different proof by means of an elementary Fourier coefficient method is to be found in Sec. 10.1. For further results on moduli of continuity see TIMAN [2, p. 93 ff] and LORENTZ [3, p. 43 ff]. Lemma 1.5.7 and Theorem 1.5.8 (with ω^* replaced by ω) are taken from KOROVKIN [5, p. 72] (see Problem 1.5.8). This theorem as well as most theorems of this section may also be stated (in modified forms) for positive linear operators T_n. In this case, in the estimate (1.5.7) for example, the difference $1 - \chi^\wedge_\rho(1)$ is replaced by $\|(T_n[\sin^2 (u - \circ)/2](\circ)\|$; see MAMEDOV [1], SHISHA–MOND [1, 2]. Fourier coefficients not being available, the estimates are usually left in terms of moments of the kernels, compare MAMEDOV [6]. For Prop. 1.5.10 see also MAMEDOV [1]. In connection with Prop. 1.5.12, the word 'Zygmund approximation sequence' was coined by FREUD [1, 2].

Theorem 1.5.13 is due to KOROVKIN [2]; it is given there not only for trigonometric but also for algebraic approximation. In the latter case it generalizes a well-known result by E. V. VORONOVSKAJA on the asymptotic behaviour of Bernstein polynomials (see LORENTZ [1, p. 22]). For further, perhaps simpler results on asymptotic expansions of singular integrals compare Problems 1.5.14, 1.5.16. For Theorem 1.5.15 see KOROVKIN [3]; for Prop. 1.5.17 see MAMEDOV [2]. As a matter of fact, most of the results of this section, as well as many extensions of these, including algebraic approximation, are to be found in the book of MAMEDOV [9].

Apart from these few comments on approximation by positive operators it would be possible to cite a large number of rather recent papers connected with the present material, so, for example, TABERSKI [1], FREUD–KNAPOWSKI [1], SCHURER [1], FREUD [4], WULBERT [1], ŽUK [1], MÜLLER [1, 2], LEVIATAN–MÜLLER [1], and GÖRLICH–STARK [1, 2]. The last paper includes 140 references on the subject.

Summability of Fourier series of almost periodic functions and approximation for the relevant summation processes have been studied recently (see BREDIHINA [1, 2] and BUTZER–KEMPER [1]).

Sec. 1.6. The positive polynomial $k_n(x)$ for which $\theta_n(1) = \cos(\pi/(n + 2))$ was first considered by FEJÉR [1] in 1916, as PÓLYA–SZEGÖ [1II, Problems 40–42, 50–56] have remarked. FEJÉR showed that there exists exactly one positive, even trigonometric polynomial of degree n such that $\theta_n(1) = \cos(\pi/(n + 2))$ (Problem 1.6.4). PETROV [1, 2] states that this kernel was introduced by KOROVKIN without, however, citing any literature. On the other hand, KOROVKIN [3] uses this kernel and cites PÓLYA–SZEGÖ [1II]. Nevertheless it was KOROVKIN who first used the integral $K_n(f; x)$ explicitly and realized its importance in approximation (see proof of the Jackson theorem in Sec. 2.2). The matter was also considered by BOHMAN [2]; see also the literature given in Problem 1.6.5.

The results of Sec. 1.6.2 are standard topics in texts on approximation. Cor. 1.6.5 (with Lip instead of Lip*) seems to be due to BERNSTEIN [1] in $C_{2\pi}$-space and to SZÁSZ [2] in $L^p_{2\pi}$-space.

The work on best asymptotic constants was initiated by NIKOLSKIĬ [1] in 1940; he considered the Fejér means. The next important papers in the line are due to NIKOLSKIĬ [5], SZ.-NAGY [1, 2], and NATANSON [1, 2] (see Problem 2.5.9). All results of Sec. 1.6.3 up to

Prop. 1.6.6, which is due to MATSUOKA [2], are now standard and go back to NIKOLSKIĬ. Apart from the literature cited in the Problems, there are numerous papers treating various generalizations; see the fundamental paper of SZ.-NAGY [3] as well as LORCH [1], MAMEDOV [9], TELJAKOVSKIĬ [2, 4], BUTZER–STARK [1], STARK [1, 4], GÖRLICH–STARK [2] and the literature cited there; see also Sec. 3.3.

Sec. 1.7. Concerning Theorem 1.7.1 see Sec. 12.6. For Cor. 1.7.2 see also ZYGMUND [7I, p. 122]; in $C_{2\pi}$-space, for example, it may be strengthened to a pointwise version as follows. If $\liminf_{n\to\infty} n[\sigma_n(f; x) - f(x)] = 0$ for each real x, then f is identically constant. For such small-o pointwise results we refer to ANDRIENKO [1]; see Sec. 12.6. Prop. 1.7.3 is due to KOROVKIN [2], who actually stated it in more general terms: Let T_n be a positive linear polynomial operator of degree n, defined on $C_{2\pi}$. If $f_0(x) = 1, f_1(x) = \cos x, f_2(x) = \sin x$, then for at least one $i = 0, 1, 2$, $\|T_n f_i - f_i\| \neq o(n^{-2})$; compare LORENTZ [3, p. 94]. The question arises whether it is possible to obtain better approximation than $O(n^{-2})$ if the kernel of the convolution integral is allowed to have a finite number of changes of sign, so that it is a kernel of finite oscillation. This is indeed the case and a number of papers deal with such investigations. Thus BUTZER–NESSEL–SCHERER [1] considered even polynomial kernels $\{\chi_n(x)\}$ which possess $2s$ (s fixed, independent of n) changes of sign on $[-\pi, \pi]$. For these there exist smooth functions which cannot be approximated by the associated convolution integral with order better than $O(n^{-2s-2})$. Compare also KOROVKIN [7, 8] who considered this question in the setting of linear operators. For the actual construction of operators of the above type, for which orders of approximation better than $O(n^{-2})$ are in fact attained, see KOVALENKO [1], VINOGRADOVA [1], HOFF [1], BUTZER–STARK [2], and STARK [2, 3]. For an explicit proof of Prop. 1.7.4 independent of Problem 1.6.6 see CURTIS [1].

2

Theorems of Jackson and Bernstein for Polynomials of Best Approximation and for Singular Integrals

2.0 Introduction

This chapter is devoted to a brief study on *direct* theorems of D. JACKSON and *inverse* theorems of S. N. BERNSTEIN which play a fundamental rôle in approximation of periodic functions. These will be established not only for trigonometric polynomials of best approximation but theorems of these two types will also be studied for particular singular integrals.

Defining the best approximation of degree n of the function $f \in \mathsf{C}_{2\pi}$ by trigonometric polynomials t_n of degree n by $E_n(\mathsf{C}_{2\pi}; f) = \inf_{t_n \in \mathsf{T}_n} \|f - t_n\|_{\mathsf{C}_{2\pi}}$, the theorem of Weierstrass states simply that $E_n(\mathsf{C}_{2\pi}; f) \to 0$ as $n \to \infty$ for each $f \in \mathsf{C}_{2\pi}$. Yet it does not reveal how fast $E_n(\mathsf{C}_{2\pi}; f)$ tends to zero. The theorems under discussion relate the rapidity with which $E_n(\mathsf{C}_{2\pi}; f)$ approaches zero to the smoothness properties of f. Thus, the group of theorems due to Jackson states that the smoother the function, the faster $E_n(\mathsf{C}_{2\pi}; f)$ tends to zero. Conversely, the Bernstein theorems infer smoothness properties of f from the behaviour of $E_n(\mathsf{C}_{2\pi}; f)$. The smoothness properties of f are usually given in terms of its modulus of continuity, Lipschitz classes, and differentiability properties. More precisely, one Jackson theorem asserts that if the rth derivative $f^{(r)}$ belongs to $\mathsf{Lip}\,(\mathsf{C}_{2\pi}; \alpha)$ with $0 < \alpha \le 1$, then $E_n(\mathsf{C}_{2\pi}; f)$ must converge to zero with at least the rapidity of $n^{-r-\alpha}$. Conversely, if the sequence $n^{r+\alpha} E_n(\mathsf{C}_{2\pi}; f)$ is bounded, then $f^{(r)} \in \mathsf{Lip}\,(\mathsf{C}_{2\pi}; \alpha)$ in the case $0 < \alpha < 1$. If $\alpha = 1$, the class $\mathsf{Lip}\,(\mathsf{C}_{2\pi}; 1)$ must be replaced by $\mathsf{Lip}^*\,(\mathsf{C}_{2\pi}; 1)$ as A. ZYGMUND first observed. Indeed, the class $\{f \in \mathsf{C}_{2\pi} \mid \sup_{\delta > 0} \delta^{-1} \omega^*(\mathsf{C}_{2\pi}; f; \delta) < \infty\}$ is equal to the class $\{f \in \mathsf{C}_{2\pi} \mid \sup_{n \in \mathbb{N}} n E_n(\mathsf{C}_{2\pi}; f) < \infty\}$. It will be important to study the cases when the direct and inverse theorems match each other and are exact converses of one another. We then speak of an *equivalence* theorem.

The later part of the chapter, in contrast, deals with theorems of Jackson and Bernstein-type for singular integrals that are neither necessarily polynomial summation processes of the Fourier series of f nor have the same order of approximation to f as the polynomials of best approximation. Equivalence theorems for such integrals are also considered. A common feature of the chapter is the fact that the proofs of the inverse theorems for the polynomials of best approximation as well as for singular integrals are proved by a technique due to BERNSTEIN.

Sec. 2.1 is concerned with some basic properties of polynomials of best approximation, in particular with their existence. Sec. 2.2 is reserved to the group of theorems due to JACKSON, Sec. 2.3 to those of BERNSTEIN on best approximation by trigonometric polynomials. Sec. 2.4 contains some of the most useful typical applications of the fundamental theorems, so, for example, characterizations of Lipschitz classes, applications to the classical theory of Fourier series, and equivalence theorems for singular integrals such as those of Fejér and Rogosinski. Sec. 2.5 is devoted to direct and inverse theorems for singular integrals, in particular those of Abel–Poisson (not a polynomial summation process) and de La Vallée Poussin (having an approximation not as good as polynomials of best approximation).

2.1 Polynomials of Best Approximation

Let t_n be an arbitrary element of T_n. Given $f \in \mathsf{X}_{2\pi}$, the deviation $\delta(\mathsf{X}_{2\pi}; f; t_n)$ of f from t_n in $\mathsf{X}_{2\pi}$-metric is defined by

$$\delta(\mathsf{X}_{2\pi}; f; t_n) = \|f - t_n\|_{\mathsf{X}_{2\pi}}. \tag{2.1.1}$$

If we consider the set of numbers $\delta(\mathsf{X}_{2\pi}; f; t_n)$ for all $t_n \in \mathsf{T}_n$, then this set is bounded from below (as $\delta(\mathsf{X}_{2\pi}; f; t_n) \geq 0$) and thus possesses a greatest lower bound which will be denoted by

$$E_n(\mathsf{X}_{2\pi}; f) = \inf_{t_n \in \mathsf{T}_n} \delta(\mathsf{X}_{2\pi}; f; t_n) \tag{2.1.2}$$

and called the *best* (trigonometric) *approximation* of f by polynomials of degree n in $\mathsf{X}_{2\pi}$-space. If $f \in \mathsf{X}_{2\pi}$, then trivially

$$E_0(\mathsf{X}_{2\pi}; f) \geq E_1(\mathsf{X}_{2\pi}; f) \geq \cdots \geq E_n(\mathsf{X}_{2\pi}; f) \geq \cdots,$$

and the theorem of Weierstrass asserts that $\lim_{n\to\infty} E_n(\mathsf{X}_{2\pi}; f) = 0$.

The above terminology is clear if one can show that for $f \in \mathsf{X}_{2\pi}$ there actually *exists* an element in T_n for which the infimum (2.1.2) is attained. In this case, such a polynomial is called a *polynomial of best approximation* to f and is denoted by $t_n^* \equiv t_n^*(\mathsf{X}_{2\pi}; f)$, thus

$$E_n(\mathsf{X}_{2\pi}; f) = \|f - t_n^*(\mathsf{X}_{2\pi}; f)\|_{\mathsf{X}_{2\pi}}. \tag{2.1.3}$$

Theorem 2.1.1. *Given $f \in \mathsf{X}_{2\pi}$ and $n \in \mathbb{N}$, there exists a polynomial $t_n^*(\mathsf{X}_{2\pi}; f)$ of best approximation to f in $\mathsf{X}_{2\pi}$-space.*

Proof. In view of the definition of $E_n(\mathsf{X}_{2\pi};f)$ there exists a sequence of polynomials $\{t_{n,j}(x)\}_{j=1}^{\infty}$ of degree n, $t_{n,j}(x) = \sum_{k=-n}^{n} c_{k,j}\, e^{ikx}$, such that for each j

$$E_n(\mathsf{X}_{2\pi};f) \le \delta(\mathsf{X}_{2\pi};f;t_{n,j}) \le E_n(\mathsf{X}_{2\pi};f) + (1/j).$$

Since

$$\|t_{n,j}\|_{\mathsf{X}_{2\pi}} \le \|t_{n,j} - f\|_{\mathsf{X}_{2\pi}} + \|f\|_{\mathsf{X}_{2\pi}} \le E_n(\mathsf{X}_{2\pi};f) + \|f\|_{\mathsf{X}_{2\pi}} + (1/j),$$

the sequence $\{\|t_{n,j}\|_{\mathsf{X}_{2\pi}}\}$ is bounded for all $j \in \mathbb{N}$ by M, say. Let us assume for a moment that this in turn implies the boundedness of the coefficients $c_{k,j}$, i.e., there exists a constant M_1 such that $|c_{k,j}| \le M_1$ for all $|k| \le n$ and $j \in \mathbb{N}$. Then by the theorem of Bolzano–Weierstrass for complex numbers (and a finite diagonal process) there exists a subsequence $\{c_{k,j_l}\}_{l=1}^{\infty}$ which tends to a limit c_k^* (uniformly) for all $|k| \le n$. The corresponding subsequence $\{t_{n,j_l}\}$ then tends in $\mathsf{X}_{2\pi}$-norm to $t_n^*(x) = \sum_{k=-n}^{n} c_k^*\, e^{ikx}$ for $l \to \infty$. To show that $t_n^*(x)$ actually is a polynomial of best approximation we note that

$$\|f - t_n^*\|_{\mathsf{X}_{2\pi}} \le \|f - t_{n,j_l}\|_{\mathsf{X}_{2\pi}} + \|t_{n,j_l} - t_n^*\|_{\mathsf{X}_{2\pi}} \le E_n(\mathsf{X}_{2\pi};f) + (1/j_l) + \varepsilon_l,$$

where ε_l tends to zero as $l \to \infty$. This would give $\delta(\mathsf{X}_{2\pi};f;t_n^*) \le E_n(\mathsf{X}_{2\pi};f)$, and as the opposite inequality is obvious, the theorem would follow.

It remains to show that the coefficients of $t_{n,j}$ are bounded by M_1. Assume that a bound M_1 does not exist. Replacing, if necessary, $t_{n,j}$ by a subsequence (which we also denote by $t_{n,j}$) we may assume that $\max_{|k|\le n} |c_{k,j}|$ is different from zero, tends to infinity as $j \to \infty$, and that this maximum is attained for each $j \in \mathbb{N}$ for the same index k, say $k = n$. Put $p_{n,j}(x) = t_{n,j}(x)/c_{n,j}$. Then $\|p_{n,j}\|_{\mathsf{X}_{2\pi}} \le M/|c_{n,j}|$ which tends to zero as $j \to \infty$. Moreover, $p_{n,j}(x)$ is of the form

$$p_{n,j}(x) = \sum_{k=-n}^{n-1} c'_{k,j}\, e^{ikx} + e^{inx}$$

with $|c'_{k,j}| \le 1$ for all $|k| \le n$, $j \in \mathbb{N}$. Since the coefficients $c'_{k,j}$ are bounded, we may assume (replacing $p_{n,j}$ by a subsequence, if necessary) that they converge for each $|k| \le n$ to a limit c'_k with $c'_n = 1$ as $j \to \infty$. Then $\sum_{k=-n}^{n-1} c'_k\, e^{ikx} + e^{inx} = 0$, which is impossible since the functions $\exp\{ikx\}$, $|k| \le n$, are linearly independent (cf. Problem 1.2.2(i)). This completes the proof.

One could also establish the important property of uniqueness of the polynomial $t_n^*(\mathsf{X}_{2\pi};f)$ of best approximation. In $\mathsf{C}_{2\pi}$- and $\mathsf{L}_{2\pi}^1$-space this is due to the fact that the functions $\exp\{ikx\}$, $|k| \le n$, form a 'Chebychev system', whereas for $1 < p < \infty$ it is a consequence of the 'strict convexity' of the spaces $\mathsf{L}_{2\pi}^p$. In future we may then speak of $t_n^*(\mathsf{X}_{2\pi};f)$ as *the* polynomial of best approximation of degree n, and the use of the terminology $E_n(\mathsf{X}_{2\pi};f)$ being *the* best approximation to a given $f \in \mathsf{X}_{2\pi}$ by polynomials of degree n is then fully justified. We shall not enter into a thorough study of the properties of $t_n^*(\mathsf{X}_{2\pi};f)$, but refer the interested reader to the Notes and Remarks to this section.

Problems

1. Let $f \in \mathsf{X}_{2\pi}$ and λ be an arbitrary complex number. Show that $E_n(\mathsf{X}_{2\pi};\lambda f) = |\lambda|\, E_n(\mathsf{X}_{2\pi};f)$.
2. For every $f_1, f_2 \in \mathsf{X}_{2\pi}$ show that $E_n(\mathsf{X}_{2\pi};f_1 + f_2) \le E_n(\mathsf{X}_{2\pi};f_1) + E_n(\mathsf{X}_{2\pi};f_2)$.
3. For every $f_1, f_2 \in \mathsf{X}_{2\pi}$ show that $\psi(\lambda) = E_n(\mathsf{X}_{2\pi};f_1 + \lambda f_2)$ is a continuous function of λ. (Hint: NATANSON [8I, p. 110])
4. Let $f \in \mathsf{X}_{2\pi}$ and $t_n \in \mathsf{T}_n$. Then $E_n(\mathsf{X}_{2\pi};f + t_n) = E_n(\mathsf{X}_{2\pi};f)$.
5. Let $\{a_n\}_{n=0}^{\infty}$ be any sequence of positive numbers monotonely decreasing to zero as $n \to \infty$. Show that there exists $f \in \mathsf{X}_{2\pi}$ such that $E_n(\mathsf{X}_{2\pi};f) = a_n$ for every n. (Hint: NATANSON [8I, p. 109 ff], P. J. DAVIS [1, p. 332])

2.2 Theorems of Jackson

We state the *first theorem of Jackson* on the order of approximation of functions, having generalized modulus of continuity $\omega^*(X_{2\pi}; f; \delta)$, by means of (trigonometric) polynomials.

Theorem 2.2.1. *If $f \in X_{2\pi}$, then for each $n \in \mathbb{N}$*

$$E_n(X_{2\pi}; f) \le 18\omega^*(X_{2\pi}; f; n^{-1}).$$

Corollary 2.2.2. *If $f \in \mathsf{Lip}^*(X_{2\pi}; \alpha)$ for some $0 < \alpha \le 2$, then $E_n(X_{2\pi}; f) = O(n^{-\alpha})$ for $n \to \infty$.*

The proofs follow by an immediate application of Cor. 1.6.1. Indeed, the singular integral $K_n(f; x)$ of Fejér–Korovkin provides a polynomial of degree n satisfying the stated estimates which then in particular hold for $E_n(X_{2\pi}; f)$ as well.

The next problem is to generalize Theorem 2.2.1 to functions for which one or more derivatives exist. It is to be expected that the stronger the hypotheses upon f the more rapid the decrease of the quantity $E_n(X_{2\pi}; f)$ towards zero. This is indeed true, and we shall even establish the estimates for $E_n(X_{2\pi}; f)$ by actually *constructing* a polynomial of degree n having these properties. This will be achieved by introducing a singular integral which is built up from that of Fejér–Korovkin by taking a suitable linear combination of its iterates. But, in view of Prop. 1.7.4, the corresponding kernel cannot be positive. We state the so-called *second theorem of Jackson.*

Theorem 2.2.3. *Let $r \in \mathbb{N}$ and $f \in \mathsf{W}^r_{X_{2\pi}}$, i.e., let there exist a function $\phi \in \mathsf{AC}^{r-1}_{2\pi}$ with $\phi^{(r)} \in X_{2\pi}$ such that $f = \phi$ in $X_{2\pi}$. Then*

$$E_n(X_{2\pi}; f) \le \left(\frac{36}{n}\right)^r \|\phi^{(r)}\|_{X_{2\pi}} \qquad (n \in \mathbb{N}). \tag{2.2.1}$$

Moreover, for some constant C,

$$E_n(X_{2\pi}; f) \le Cn^{-r}\omega^*(X_{2\pi}; \phi^{(r)}; n^{-1}). \tag{2.2.2}$$

Proof. Let $K_n(f; x)$ be the singular integral of Fejér–Korovkin and K_n the bounded linear operator of $X_{2\pi}$ into T_n defined by $(K_n f)(x) = K_n(f; x)$. If we set $K^0 = I$ (the identity operator) and construct the powers of K_n (as usual) iteratively by $K_n^j = K_n^1(K_n^{j-1})$, $j \in \mathbb{N}$, then

$$(K_n - I)^r f = \sum_{j=0}^{r} (-1)^{r-j}\binom{r}{j} K_n^j f.$$

Putting $U_{r,n} = \sum_{j=1}^{r} (-1)^{j+1}\binom{r}{j} K_n^j$, then

$$U_{r,n} f - f = (-1)^{r-1}(K_n - I)^r f.$$

As K_n maps every $f \in X_{2\pi}$ into T_n, the same is true for every power of K_n, thus $U_{r,n} f \in \mathsf{T}_n$ for every $f \in X_{2\pi}$. $U_{r,n} f$ is again a singular integral of type (1.1.3) with kernel (compare Problem 1.1.4) $\chi_n(x) = \sum_{j=1}^{r} (-1)^{j+1}\binom{r}{j}[k_n *]^j$, k_n being the Fejér–Korovkin kernel.

To prove (2.2.1), we have in view of Cor. 1.6.1

$$E_n(X_{2\pi}; f) \le \|U_{r,n}f - f\|_{X_{2\pi}} = \|(K_n - I)[(K_n - I)^{r-1}\phi]\|_{X_{2\pi}}$$
$$\le 18\omega^*(X_{2\pi}; (K_n - I)^{r-1}\phi; n^{-1}).$$

Since $f \in W^r_{X_{2\pi}}$, it follows by Problem 1.5.3(ii), (iv) that

$$E_n(X_{2\pi}; f) \le \frac{36}{n} \|[(K_n - I)^{r-1}\phi]'\|_{X_{2\pi}}.$$

If $r = 1$, the proof is complete. If $r > 1$, then in view of Prop. 1.1.15, 1.1.16

$$[(K_n - I)^{r-1}\phi]'(x) = [(K_n - I)^{r-1}\phi'](x),$$

and therefore we may continue as above with

$$E_n(X_{2\pi}; f) \le \frac{36}{n} \|(K_n - I)[(K_n - I)^{r-2}\phi']\|_{X_{2\pi}}$$
$$\le \left(\frac{36}{n}\right)^2 \|[(K_n - I)^{r-2}\phi']'\|_{X_{2\pi}}.$$

By a repeated application of Prop. 1.1.15, 1.1.16, Cor. 1.6.1 and Problem 1.5.3 we have

$$E_n(X_{2\pi}; f) \le \left(\frac{36}{n}\right)^{r-1} \|(K_n - I)\phi^{(r-1)}\|_{X_{2\pi}},$$

and

$$E_n(X_{2\pi}; f) \le \left(\frac{36}{n}\right)^{r-1} 18\omega^*(X_{2\pi}; \phi^{(r-1)}; n^{-1}) \le \left(\frac{36}{n}\right)^{r} \|\phi^{(r)}\|_{X_{2\pi}}.$$

Now, (2.2.2) may be obtained from (2.2.1) by the following device. Recalling that $A_h^2(f; x)$ denotes the second integral mean of f (see Problem 1.1.5), we write $f(x) = A_h^2(f; x) + g(x)$. In view of Problem 1.1.7 it follows that $f \in W^r_{X_{2\pi}}$ implies $A_h^2(f; x) \in W^{r+2}_{X_{2\pi}}$ and

$$[A_h^2(f; \circ)]^{(r+2)}(x) = h^{-2}[\phi^{(r)}(x + h) + \phi^{(r)}(x - h) - 2\phi^{(r)}(x)].$$

Therefore

$$\|[A_h^2(f; \circ)]^{(r+2)}(\circ)\|_{X_{2\pi}} \le h^{-2}\omega^*(X_{2\pi}; \phi^{(r)}; h).$$

On the other hand, $g \in W^r_{X_{2\pi}}$, and since by Prop. 1.1.15, 1.1.16, and Problem 1.5.5

$$\|[A_h^2(\phi; \circ) - \phi(\circ)]^{(r)}\|_{X_{2\pi}} = \|A_h^2(\phi^{(r)}; \circ) - \phi^{(r)}(\circ)\|_{X_{2\pi}} \le \omega^*(X_{2\pi}; \phi^{(r)}; h),$$

we have by Problem 2.1.2 and by (2.2.1)

$$E_n(X_{2\pi}; f) \le E_n(X_{2\pi}; A_h^2(f; \circ)) + E_n(X_{2\pi}; g)$$
$$\le \left(\frac{36}{n}\right)^{r+2} h^{-2}\omega^*(X_{2\pi}; \phi^{(r)}; h) + \left(\frac{36}{n}\right)^{r} \omega^*(X_{2\pi}; \phi^{(r)}; h).$$

Finally, setting $h = 1/n$, $n \in \mathbb{N}$,

$$E_n(X_{2\pi}; f) \le ((36)^r(36^2 + 1))n^{-r}\omega^*(X_{2\pi}; \phi^{(r)}; n^{-1}),$$

and the proof is complete.

As an immediate consequence we state

Corollary 2.2.4. *If* $f \in {}^*W^{r,\alpha}_{X_{2\pi}}$ *for some* $r \in \mathbb{N}$ *and* $0 < \alpha \le 2$, *then* $E_n(X_{2\pi}; f) = O(n^{-r-\alpha})$ *for* $n \to \infty$.

Problems

1. Let $r \in \mathbb{N}$ and $f \in W^r_{X_{2\pi}}$, i.e., let there exist $\phi \in AC^{r-1}_{2\pi}$ with $\phi^{(r)} \in X_{2\pi}$ such that $f(x) = \phi(x)$ (a.e.). Show that there is some constant C such that
$$E_n(X_{2\pi}; f) \le Cn^{-r}\omega(X_{2\pi}; \phi^{(r)}; n^{-1}).$$
In particular, if $f \in W^{r,\alpha}_{X_{2\pi}}$, $0 < \alpha \le 1$, then $E_n(X_{2\pi}; f) = O(n^{-r-\alpha})$.
2. Let $f \in C_{2\pi}$ be r-times differentiable such that $f^{(r)} \in L^\infty_{2\pi}$. Show that $E_n(C_{2\pi}; f) \le (36/n)^r \|f^{(r)}\|_\infty$. Moreover, if $f^{(r)} \in C_{2\pi}$ is such that $f^{(r)} \in \mathsf{Lip}^*(C_{2\pi}; \alpha)$ for some $0 < \alpha \le 2$, then $E_n(C_{2\pi}; f) = O(n^{-r-\alpha})$.
3. Let $f \in X_{2\pi}$. If $f \in \mathsf{lip}(X_{2\pi}; \alpha)$ for some $0 < \alpha \le 1$, show that $E_n(X_{2\pi}; f) = o(n^{-\alpha})$.
4. For $f \in W^\infty_{X_{2\pi}}$ show that $\lim_{n\to\infty} n^\gamma E_n(X_{2\pi}; f) = 0$ for every $\gamma > 0$.

2.3 Theorems of Bernstein

It is the purpose of this section to establish results converse to those of the last section. In Cor. 2.2.4 it was shown that if f is r-times differentiable and $f^{(r)} \in \mathsf{Lip}(C_{2\pi}; \alpha)$, then $E_n(C_{2\pi}; f) = O(n^{-r-\alpha})$. The question then arises as to whether conversely the boundedness of the sequence $n^{r+\alpha}E_n(C_{2\pi}; f)$ implies the existence of $f^{(r)}$ such that $f^{(r)} \in \mathsf{Lip}(C_{2\pi}; \alpha)$. As we shall see, this is indeed the case; in other words, $E_n(C_{2\pi}; f)$ does not tend to zero more rapidly than $O(n^{-r-\alpha})$ in general, thus the estimate of Problem 2.2.1 cannot be improved for $0 < \alpha < 1$.

We begin by establishing the *inequality of Bernstein* on the order of growth of the derivatives of (trigonometric) polynomials. This inequality is fundamental for what is to follow.

Theorem 2.3.1. *If* t_n *is a polynomial of degree* n, *then*

$$\|t_n'\|_{X_{2\pi}} \le 2n\, \|t_n\|_{X_{2\pi}}. \tag{2.3.1}$$

Proof. We write (cf. Problem 1.2.14(iii))

$$t_n(x) = \frac{1}{2\pi}\int_{-\pi}^{\pi} t_n(u)\Big[1 + 2\sum_{k=1}^{n} \cos k(x-u)\Big]\, du.$$

Differentiation gives

$$t_n'(x) = \frac{1}{2\pi}\int_{-\pi}^{\pi} t_n(x+u)\Big[2\sum_{k=1}^{n} k \sin ku\Big]\, du.$$

To the term in brackets we may add the sum $2\sum_{k=1}^{n-1} k \sin(2n-k)u$ (all of its terms have degree greater than n) which does not change the value of the integral since $t_n \in T_n$. We then obtain

$$t_n'(x) = \frac{1}{2\pi}\int_{-\pi}^{\pi} t_n(x+u)\, 2n \sin nu\, F_{n-1}(u)\, du,$$

$\{F_n(x)\}$ being Fejér's kernel. Hence by the Hölder–Minkowski inequality

$$\|t_n'\|_{X_{2\pi}} \leq 2n \cdot \frac{1}{2\pi} \int_{-\pi}^{\pi} \|t_n(\circ + u)\|_{X_{2\pi}} F_{n-1}(u)\, du = 2n \|t_n\|_{X_{2\pi}},$$

and the proof is complete.

Corollary 2.3.2. *If $t_n \in T_n$, then for every $r \in \mathbb{N}$*

$$\|t_n^{(r)}\|_{X_{2\pi}} \leq (2n)^r \|t_n\|_{X_{2\pi}}. \tag{2.3.2}$$

We remark that by a more refined but nevertheless elementary argument one can show that the factor 2 in inequality (2.3.1) is superfluous. On the other hand, the resulting inequality $\|t_n'\|_{X_{2\pi}} \leq n \|t_n\|_{X_{2\pi}}$ cannot be improved as the example $t_n(x) = \sin nx$ shows.

Theorem 2.3.3. *If, for some $\alpha > 0$, $f \in X_{2\pi}$ can be approximated for each $n \in \mathbb{N}$ by a polynomial of degree n such that*

$$E_n(X_{2\pi}; f) = O(n^{-\alpha}) \qquad (n \to \infty), \tag{2.3.3}$$

then

$$\omega(X_{2\pi}; f; \delta) = \begin{cases} O(\delta^\alpha), & 0 < \alpha < 1 \\ O(\delta\,|\log \delta|), & \alpha = 1 \\ O(\delta), & \alpha > 1 \end{cases} \qquad (\delta \to 0+).$$

Proof. If t_n^* denotes the polynomial of best approximation to f, then $\|f - t_n^*\|_{X_{2\pi}} \leq Bn^{-\alpha}$ for some constant $B > 0$. We write $U_2(x) = t_{2^2}^*(x)$, $U_n(x) = t_{2^n}^*(x) - t_{2^{n-1}}^*(x)$, $n = 3, 4, \ldots$, $U_n(x)$ being a polynomial of degree 2^n. Then for $n = 3, 4, \ldots$

$$\|U_n\|_{X_{2\pi}} \leq \|t_{2^n}^* - f\|_{X_{2\pi}} + \|f - t_{2^{n-1}}^*\|_{X_{2\pi}} \leq (1 + 2^\alpha)B2^{-n\alpha},$$

and since $\|U_2\|_{X_{2\pi}} \leq B2^{-2\alpha} + \|f\|_{X_{2\pi}}$, we may choose a constant C such that for all $n = 2, 3, \ldots$

$$\|U_n\|_{X_{2\pi}} \leq C2^{-n\alpha}. \tag{2.3.4}$$

Now, $\sum_{k=2}^n U_k(x) = t_{2^n}^*(x)$ converges in $X_{2\pi}$-norm to f as $n \to \infty$, thus

$$\lim_{n\to\infty} \Big\| f - \sum_{k=2}^n U_k \Big\|_{X_{2\pi}} = 0. \tag{2.3.5}$$

Hence $\|f\|_{X_{2\pi}} \leq \sum_{k=2}^\infty \|U_k\|_{X_{2\pi}}$, and for any $h \in \mathbb{R}$ and integer $m \geq 2$

$$\|f(\circ + h) - f(\circ)\|_{X_{2\pi}} \leq \sum_{k=2}^m \|U_k(\circ + h) - U_k(\circ)\|_{X_{2\pi}} + 2 \sum_{k=m+1}^\infty \|U_k\|_{X_{2\pi}}.$$

Since U_n (a polynomial) is arbitrarily often differentiable, we have

$$U_k(x + h) - U_k(x) = \int_0^h U_k'(x + u)\, du.$$

Moreover, according to Bernstein's inequality

$$\|U_k'\|_{X_{2\pi}} \le 2\cdot 2^k \|U_k\|_{X_{2\pi}} \le 2C2^{k(1-\alpha)},$$

and thus

$$\|U_k(\circ + h) - U_k(\circ)\|_{X_{2\pi}} \le h\,\|U_k'\|_{X_{2\pi}} \le 2Ch2^{k(1-\alpha)}.$$

Collecting the results it follows that

$$\begin{aligned}\omega(X_{2\pi}; f; \delta) &\le 2C\Big\{\delta \sum_{k=2}^{m} 2^{k(1-\alpha)} + \sum_{k=m+1}^{\infty} 2^{-k\alpha}\Big\}\\ &\le 2C\delta \sum_{k=2}^{m} 2^{k(1-\alpha)} + \frac{2^{1-\alpha}C}{1-2^{-\alpha}} 2^{-m\alpha}.\end{aligned}$$

Supposing $0 < \delta < \frac{1}{2}$, we now choose $m \ge 2$ according to

$$2^{m-1} \le 1/\delta < 2^m. \tag{2.3.6}$$

Then with $C_1 = 2^{1-\alpha}C/(1-2^{-\alpha})$ we have

$$\omega(X_{2\pi}; f; \delta) \le 2C\delta \sum_{k=0}^{m} 2^{k(1-\alpha)} + C_1\delta^\alpha.$$

We now consider the different cases.

(i) If $0 < \alpha < 1$, then

$$\sum_{k=0}^{m} 2^{k(1-\alpha)} < \frac{2^{(m+1)(1-\alpha)}}{2^{1-\alpha} - 1},$$

and since $2^{m+1} \le 4/\delta$,

$$\omega(X_{2\pi}; f; \delta) \le 2C\delta \frac{4^{1-\alpha}}{2^{1-\alpha}-1}\, \delta^{\alpha-1} + C_1\delta^\alpha = O(\delta^\alpha).$$

(ii) If $\alpha = 1$, then $\omega(X_{2\pi}; f; \delta) \le 2C\delta(m+1) + C_1\delta$. But $2^{m+1} \le 4/\delta$ implies $m + 1 \le 2 + (|\log \delta|/\log 2)$, giving $\omega(X_{2\pi}; f; \delta) = O(\delta\,|\log \delta|)$.

(iii) If $\alpha > 1$, then

$$\omega(X_{2\pi}; f; \delta) \le \Big(\frac{2C}{1-2^{1-\alpha}} + C_1\delta^{\alpha-1}\Big)\delta = O(\delta).$$

Thereby the proof is complete.

Next we prove a theorem which establishes the existence of derivatives of a function $f \in X_{2\pi}$ if the sequence $\{E_n(X_{2\pi}; f)\}$ of its best approximations tends to zero sufficiently rapidly.

Theorem 2.3.4. *If, for some $r \in \mathbb{N}$ and $\alpha > 0$, $f \in X_{2\pi}$ can be approximated for each $n \in \mathbb{N}$ by a polynomial of degree n such that*

$$E_n(X_{2\pi}; f) = O(n^{-r-\alpha}) \qquad (n \to \infty),$$

then $f \in W^r_{X_{2\pi}}$. Moreover, if $\phi \in AC^{r-1}_{2\pi}$, $\phi^{(r)} \in X_{2\pi}$, is such that $\phi = f$ in $X_{2\pi}$, then

$$\omega(X_{2\pi}; \phi^{(r)}; \delta) = \begin{cases} O(\delta^\alpha), & 0 < \alpha < 1 \\ O(\delta\,|\log \delta|), & \alpha = 1 \\ O(\delta), & \alpha > 1 \end{cases} \qquad (\delta \to 0+).$$

Proof. By hypothesis $\|f - t_n^*\|_{X_{2\pi}} \le Bn^{-r-\alpha}$ for some constant $B > 0$. Introducing the polynomials $U_n(x)$ of degree 2^n as in the proof of Theorem 2.3.3, we now obtain instead of (2.3.4) that $\|U_n\|_{X_{2\pi}} \le C2^{-nr-n\alpha}$, $n = 2, 3, \ldots$, for some constant $C > 0$. The Bernstein inequality (2.3.2) implies that

$$\|U_n^{(j)}\|_{X_{2\pi}} \le 2^j 2^{jn} \|U_n\|_{X_{2\pi}} \le 2^j C 2^{-n(r-j)-n\alpha}. \tag{2.3.7}$$

Thus the series $\sum_{k=2}^{\infty} \|U_k^{(j)}\|_{X_{2\pi}}$ converge for each $j = 0, 1, \ldots, r$.

In case $X_{2\pi} = C_{2\pi}$ this implies that each series $\sum_{k=2}^{\infty} U_k^{(j)}(x)$ converges uniformly, and since in view of (2.3.5) $f(x) = \sum_{k=2}^{\infty} U_k(x)$, we have $f^{(r)} \in C_{2\pi}$ and

$$f^{(r)}(x) = \sum_{k=2}^{\infty} U_k^{(r)}(x).$$

If $X_{2\pi} = L_{2\pi}^p$, $1 \le p < \infty$, then $\sum_{k=2}^{\infty} \|U_k^{(j)}\|_{X_{2\pi}} < \infty$ implies that for each $j = 0, 1, \ldots, r$ the sequence of partial sums $\sum_{k=2}^{n} U_k^{(j)}(x)$ converges in $L_{2\pi}^p$-norm to a function $g_j \in L_{2\pi}^p$. Moreover (compare also Prop. 0.1.10), there exists a subsequence $\sum_{k=2}^{n_l} U_k^{(j)}(x)$ which converges almost everywhere to $g_j(x)$ for each $j = 0, 1, \ldots, r$. Suppose that x_0 is a point of convergence for each $j = 0, 1, \ldots, r$. Let us consider the following difference, taking $L_{2\pi}^p$-norms,

$$\begin{aligned}
&\left\| g_{j-1}(\circ) - g_{j-1}(x_0) - \int_{x_0}^{\circ} g_j(u)\,du \right\|_p \\
&\quad \le \left\| g_{j-1}(\circ) - \sum_{k=2}^{n_l} U_k^{(j-1)}(\circ) \right\|_p + \left| -g_{j-1}(x_0) + \sum_{k=2}^{n_l} U_k^{(j-1)}(x_0) \right| \\
&\qquad + \left\| \int_{x_0}^{\circ} \left[-g_j(u) + \sum_{k=2}^{n_l} U_k^{(j)}(u) \right] du \right\|_p .
\end{aligned}$$

Since the right-hand side tends to zero as $l \to \infty$, we have for each $j = 1, 2, \ldots, r$ that

$$g_{j-1}(x) - g_{j-1}(x_0) = \int_{x_0}^{x} g_j(u)\,du \quad \text{a.e.}$$

But, in view of (2.3.5), $g_0(x) = f(x)$ a.e. Therefore

$$\begin{aligned}
f(x) = f(x_0) &+ \int_{x_0}^{x} dx_1 \left[g_1(x_0) + \int_{x_0}^{x_1} dx_2 \left[g_2(x_0) + \cdots \right.\right. \\
&\left.\left. + \int_{x_0}^{x_{r-2}} dx_{r-1} \left[g_{r-1}(x_0) + \int_{x_0}^{x_{r-1}} g_r(x_r)\,dx_r \right] \ldots \right]\right] \quad \text{a.e.,}
\end{aligned} \tag{2.3.8}$$

and if we define ϕ by the right-hand side, then $f(x) = \phi(x)$ a.e., $\phi \in AC_{2\pi}^{r-1}$ and $\phi^{(r)} \in L_{2\pi}^p$. This, first of all, gives $f \in W_{X_{2\pi}}^r$ and

$$\lim_{n \to \infty} \left\| \phi^{(r)} - \sum_{k=2}^{n} U_k^{(r)} \right\|_{X_{2\pi}} = 0. \tag{2.3.9}$$

To complete the proof, we have in view of (2.3.7) and (2.3.9) that

$$\left\| \phi^{(r)} - \sum_{k=2}^{m} U_k^{(r)} \right\|_{X_{2\pi}} \le \sum_{k=m+1}^{\infty} \|U_k^{(r)}\|_{X_{2\pi}} \le 2^r C \sum_{k=m+1}^{\infty} 2^{-k\alpha} = C_2 2^{-m\alpha}$$

for every $m \in \mathbb{N}$, where C_2 is a certain constant. But since $\sum_{k=2}^{m} U_k^{(r)}(x) \in \mathsf{T}_{2^m}$, this gives $E_{2^m}(\mathsf{X}_{2\pi}; \phi^{(r)}) \le C_2 2^{-m\alpha}$. For each integer $n \ge 4$ we choose m such that $2^m \le n < 2^{m+1}$. Then

$$E_n(\mathsf{X}_{2\pi}; \phi^{(r)}) \le E_{2^m}(\mathsf{X}_{2\pi}; \phi^{(r)}) \le C_2 2^{-m\alpha} \le 2^\alpha C_2 n^{-\alpha}.$$

Thus $\phi^{(r)} \in \mathsf{X}_{2\pi}$ satisfies the assumptions of Bernstein's first Theorem 2.3.3, and therefore the proof is complete.

In connection with Theorem 2.3.3 we add the following result which will turn out to be very important for the characterization of the case $\alpha = 1$.

Theorem 2.3.5. *If, for some $\alpha > 0$, $f \in \mathsf{X}_{2\pi}$ can be approximated for each $n \in \mathbb{N}$ by a polynomial of degree n such that $E_n(\mathsf{X}_{2\pi}; f) = O(n^{-\alpha})$, then*

$$\omega^*(\mathsf{X}_{2\pi}; f; \delta) = \begin{cases} O(\delta^\alpha), & 0 < \alpha < 2 \\ O(\delta^2|\log \delta|), & \alpha = 2 \\ O(\delta^2), & \alpha > 2 \end{cases} \qquad (\delta \to 0+).$$

Proof. We proceed as in the proof of Theorem 2.3.3 and obtain (2.3.5). Now,

$$\|f(\circ + h) + f(\circ - h) - 2f(\circ)\|_{\mathsf{X}_{2\pi}} \le \sum_{k=2}^{m} \|U_k(\circ + h) + U_k(\circ - h) - 2U_k(\circ)\|_{\mathsf{X}_{2\pi}} + 4 \sum_{k=m+1}^{\infty} \|U_k\|_{\mathsf{X}_{2\pi}}.$$

As $U_k \in \mathsf{T}_{2^k}$, it follows that

$$U_k(x + h) + U_k(x - h) - 2U_k(x) = \int_{-h/2}^{h/2} \int_{-h/2}^{h/2} U_k''(x + u_1 + u_2)\, du_1\, du_2,$$

and thus by (2.3.4)

$$\|U_k(\circ + h) + U_k(\circ - h) - 2U_k(\circ)\|_{\mathsf{X}_{2\pi}} \le h^2 \|U_k''\|_{\mathsf{X}_{2\pi}} \le h^2 2^{2k+2} \|U_k\|_{\mathsf{X}_{2\pi}} \le 4Ch^2 2^{k(2-\alpha)}.$$

Therefore

$$\omega^*(\mathsf{X}_{2\pi}; f; \delta) \le 4C\Big\{\delta^2 \sum_{k=2}^{m} 2^{k(2-\alpha)} + \sum_{k=m+1}^{\infty} 2^{-k\alpha}\Big\}, \tag{2.3.10}$$

and if we choose m according to (2.3.6), then

$$\omega^*(\mathsf{X}_{2\pi}; f; \delta) \le 4C\delta^2 \sum_{k=2}^{m} 2^{k(2-\alpha)} + 2C_1\delta^\alpha.$$

Considering the cases as in the proof of Theorem 2.3.3, this finally yields the assertion.

Correspondingly, we have in connection with Theorem 2.3.4 the following

Theorem 2.3.6. *If, for some $r \in \mathbb{N}$ and $\alpha > 0$, $f \in \mathsf{X}_{2\pi}$ can be approximated for each $n \in \mathbb{N}$ by a polynomial of degree n such that $E_n(\mathsf{X}_{2\pi}; f) = O(n^{-r-\alpha})$, then $f \in \mathsf{W}^r_{\mathsf{X}_{2\pi}}$. Moreover, if $\phi \in \mathsf{AC}^{r-1}_{2\pi}$, $\phi^{(r)} \in \mathsf{X}_{2\pi}$, is such that $\phi = f$ in $\mathsf{X}_{2\pi}$, then*

$$\omega^*(\mathsf{X}_{2\pi}; \phi^{(r)}; \delta) = \begin{cases} O(\delta^\alpha), & 0 < \alpha < 2 \\ O(\delta^2|\log \delta|), & \alpha = 2 \\ O(\delta^2), & \alpha > 2 \end{cases} \qquad (\delta \to 0+).$$

The proof is left to Problem 2.3.2. The fundamental importance of the latter two theorems will be elucidated in the next section.

Problems

1. If, for some $\alpha > 0$, $f \in X_{2\pi}$ can be approximated for each $n \in \mathbb{N}$ by a polynomial of degree n such that $E_n(X_{2\pi}; f) = o(n^{-\alpha})$, $n \to \infty$, show that
$$\omega(X_{2\pi}; f; \delta) = \begin{cases} o(\delta^\alpha), & 0 < \alpha < 1 \\ o(\delta\,|\log \delta|), & \alpha = 1 \\ o(\delta), & \alpha > 1. \end{cases}$$
2. Prove Theorem 2.3.6.
3. If, for some $r \in \mathbb{N}$ and $\alpha > 0$, $f \in X_{2\pi}$ can be approximated for each $n \in \mathbb{N}$ by a polynomial of degree n such that $E_n(X_{2\pi}; f) = o(n^{-r-\alpha})$, $n \to \infty$, show that $f \in W^r_{X_{2\pi}}$. Moreover, if $\phi \in AC^{r-1}_{2\pi}$, $\phi^{(r)} \in X_{2\pi}$, is such that $\phi(x) = f(x)$ (a.e.), then
$$\omega^*(X_{2\pi}; \phi^{(r)}; \delta) = \begin{cases} o(\delta^\alpha), & 0 < \alpha < 2 \\ o(\delta^2\,|\log \delta|), & \alpha = 2 \\ o(\delta^2), & \alpha > 2. \end{cases}$$
(Hint: See also NATANSON [8I, p. 109], BUTZER–NESSEL [1])

2.4 Various Applications

Let us first summarize some of the results so far obtained in this chapter, and for the convenience of the reader state them in form of theorems.

Theorem 2.4.1. *Let* $f \in X_{2\pi}$.

(i) *For* $0 < \alpha < 1$, $E_n(X_{2\pi}; f) = O(n^{-\alpha}) \Leftrightarrow f \in \mathrm{Lip}\,(X_{2\pi}; \alpha)$,

(ii) $E_n(X_{2\pi}; f) = O(n^{-1}) \Leftrightarrow f \in \mathrm{Lip}^*\,(X_{2\pi}; 1)$.

To prove (i), the fact that $f \in \mathrm{Lip}\,(X_{2\pi}; \alpha)$ implies $E_n(X_{2\pi}; f) = O(n^{-\alpha})$ follows by Cor. 2.2.2, since in view of Problem 1.5.4 $f \in \mathrm{Lip}\,(X_{2\pi}; \alpha)$ yields $f \in \mathrm{Lip}^*\,(X_{2\pi}; \alpha)$. The converse is given by Theorem 2.3.3. Similarly, the equivalence of the statements of (ii) follows from Cor. 2.2.2 and Theorem 2.3.5.

In part (ii) of this theorem one cannot in general replace the class $\mathrm{Lip}^*\,(X_{2\pi}; 1)$ by $\mathrm{Lip}\,(X_{2\pi}; 1)$; indeed, if $f \in \mathrm{Lip}\,(X_{2\pi}; 1)$, then $E_n(X_{2\pi}; f) = O(n^{-1})$, but conversely, given the latter estimate, it does not necessarily follow that $f \in \mathrm{Lip}\,(X_{2\pi}; 1)$. This is verified by the example of Problem 2.4.2.

The question remains as to whether for functions $f \in \mathrm{Lip}\,(X_{2\pi}; 1)$ the quantity $E_n(X_{2\pi}; f)$ converges more rapidly to zero than the estimate $E_n(X_{2\pi}; f) = O(n^{-1})$ permits. By means of the example $f(x) = |\sin x|$ (cf. Problem 2.4.12(i)) it follows that the relation $\lim_{n\to\infty} nE_n(X_{2\pi}; f) = 0$ does not hold for every $f \in \mathrm{Lip}\,(X_{2\pi}; 1)$. Thus we are faced with the fact that the class $\mathrm{Lip}\,(X_{2\pi}; 1)$ is a true subclass of the class of function f for which $E_n(X_{2\pi}; f) = O(n^{-1})$, the latter being in turn contained in that class of functions f for which $\omega(X_{2\pi}; f; \delta) = O(\delta\,|\log \delta|)$, i.e.,

$$\mathrm{Lip}\,(X_{2\pi}; 1) \subset \{f \in X_{2\pi} \mid E_n(X_{2\pi}; f) = O(n^{-1})\} \subset \{f \in X_{2\pi} \mid \omega(X_{2\pi}; f; \delta) = O(\delta\,|\log \delta|)\}.$$

And part (ii) of Theorem 2.4.1 shows that the 'intermediate' class of functions is characterized as the class $\mathrm{Lip}^*\,(X_{2\pi}; 1)$. There seems to be no (cf. Sec. 2.6) simple characterization of the class $\mathrm{Lip}\,(X_{2\pi}; 1)$ expressed directly in terms of the order of best approximation. For very different characterizations of the class $\mathsf{Lip}\,(X_{2\pi};1)$ we refer to Sec. 10.2.

The results of this chapter may be used to prove certain nontrivial relations between Lipschitz classes. As an example we state

Theorem 2.4.2. *Let* $f \in X_{2\pi}$.

(i) *For* $0 < \alpha < 1$, $f \in \mathsf{Lip}^*\,(X_{2\pi};\alpha) \Leftrightarrow f \in \mathsf{Lip}\,(X_{2\pi};\alpha)$,

(ii) $$f \in \mathsf{Lip}^*\,(X_{2\pi};1) \Rightarrow \omega(X_{2\pi};f;\delta) = O(\delta\,|\log\delta|).$$

Proof. If $f \in \mathsf{Lip}\,(X_{2\pi};\alpha)$, then obviously $f \in \mathsf{Lip}^*\,(X_{2\pi};\alpha)$. Conversely, if $f \in \mathsf{Lip}^*\,(X_{2\pi};\alpha)$, then $E_n(X_{2\pi};f) = O(n^{-\alpha})$ by Cor. 2.2.2 which implies $f \in \mathsf{Lip}\,(X_{2\pi};\alpha)$ by Theorem 2.3.3 if $0 < \alpha < 1$. This proves (i). Regarding (ii), if $f \in \mathsf{Lip}^*\,(X_{2\pi};1)$, then $E_n(X_{2\pi};f) = O(n^{-1})$ by Cor. 2.2.2, and thus $\omega(X_{2\pi};f;\delta) = O(\delta\,|\log\delta|)$ by Theorem 2.3.3. In particular, $f \in \mathsf{Lip}^*\,(X_{2\pi};1)$ implies $f \in \mathsf{Lip}\,(X_{2\pi};\alpha)$ for every $0 < \alpha < 1$.

Next we consider certain applications of the theorems on best approximation to the theory of Fourier series.

Theorem 2.4.3. *Let $S_n(f;x)$ be the nth partial sum of the Fourier series of $f \in X_{2\pi}$, and $E_n(X_{2\pi};f)$ the best approximation of f by polynomials of degree n. There is a constant $c > 0$ such that*

$$\|S_n(f;\circ) - f(\circ)\|_{X_{2\pi}} \le c(1 + \log n)E_n(X_{2\pi};f). \tag{2.4.1}$$

Proof. Again, letting t_n^* denote the polynomial of best approximation for f, then $S_n(t_n^*;x) = t_n^*(x)$ by Problem 1.2.14(iii). Therefore we have by (1.1.4)

$$\begin{aligned}\|S_n(f;\circ) - f(\circ)\|_{X_{2\pi}} &\le \|S_n(f - t_n^*;\circ)\|_{X_{2\pi}} + \|t_n^* - f\|_{X_{2\pi}}\\ &\le \|D_n\|_1\|f - t_n^*\|_{X_{2\pi}} + \|f - t_n^*\|_{X_{2\pi}},\end{aligned}$$

which in view of Prop. 1.2.3 implies (2.4.1).

The inequality (2.4.1) shows that the approximation of f in $X_{2\pi}$-norm by the nth partial sum of its Fourier series differs from the best approximation $E_n(X_{2\pi};f)$ at most by the factor $O(\log n)$. Though it is not true that the $S_n(f;x)$ converge in $X_{2\pi}$-norm for *every* $f \in X_{2\pi}$, they approximate f as close as desired provided f is sufficiently smooth. Thus the nth partial sums of the Fourier series of f behave very differently to the Fejér means $\sigma_n(f;x)$, for example; these never approximate f better than $O(n^{-1})$ though they converge for *every* $f \in X_{2\pi}$ in $X_{2\pi}$-norm. Hence the order of approximation of f by the partial sums $S_n(f;x)$ is not limited by Theorem 1.7.1-type restrictions. Moreover, if one confines oneself to the spaces $L^p_{2\pi}$, $1 < p < \infty$, then it is even true that the partial sums $S_n(f;x)$ approximate *every* f in $L^p_{2\pi}$-norm and, in fact, with the order of best approximation; see Theorem 9.3.6, Prop. 9.3.8.

Corollary 2.4.4. *The Fourier series of a function $f \in C_{2\pi}$ is uniformly convergent to f if $\lim_{n\to\infty} E_n(C_{2\pi};f)\log n = 0$.*

Corollary 2.4.5. *If $f \in C_{2\pi}$ has generalized modulus of continuity $\omega^*(C_{2\pi};f;\delta)$ such that $\lim_{\delta\to 0+}\omega^*(C_{2\pi};f;\delta)\,|\log\delta| = 0$, then the Fourier series of f converges uniformly to f.*

Proof. According to Theorem 2.2.1 we have

$$E_n(\mathsf{C}_{2\pi};f)\log n = O(\log n\ \omega^*(\mathsf{C}_{2\pi};f;n^{-1})).$$

The hypothesis then implies that $E_n(\mathsf{C}_{2\pi};f)\log n = o(1)$, $n\to\infty$, and therefore Cor. 2.4.4 gives the assertion.

As an application of Cor. 2.2.4 we state

Corollary 2.4.6. *Let* $f\in\mathsf{C}_{2\pi}$ *be such that* $f^{(r)}(x)$ *exists and belongs to* $\mathsf{Lip}^*(\mathsf{C}_{2\pi};\alpha)$, $0<\alpha\le 2$. *Then*

$$\|S_n(f;\circ)-f(\circ)\|_{\mathsf{C}_{2\pi}} = O(n^{-r-\alpha}\log n) \qquad (n\to\infty).$$

For an extension in case of $\mathsf{L}^p_{2\pi}$-approximation, $1<p<\infty$, compare Cor. 9.3.9. Next we consider applications to the Fejér means $\sigma_n(f;x)$ of the Fourier series of f.

Theorem 2.4.7. *Let* $f\in\mathsf{X}_{2\pi}$.

(i) *For* $0<\alpha<1$, $\|\sigma_n(f;\circ)-f(\circ)\|_{\mathsf{X}_{2\pi}} = O(n^{-\alpha}) \Leftrightarrow f\in\mathsf{Lip}(\mathsf{X}_{2\pi};\alpha)$,

(ii) $\|\sigma_n(f;\circ)-f(\circ)\|_{\mathsf{X}_{2\pi}} = O(n^{-1}) \Rightarrow f\in\mathsf{Lip}^*(\mathsf{X}_{2\pi};1)$.

Proof. To prove (i), if $\|\sigma_n(f;\circ)-f(\circ)\|_{\mathsf{X}_{2\pi}} = O(n^{-\alpha})$, then in particular $E_n(\mathsf{X}_{2\pi};f) = O(n^{-\alpha})$, and thus $f\in\mathsf{Lip}(\mathsf{X}_{2\pi};\alpha)$ by Theorem 2.3.3. The converse assertion is given by Cor. 1.6.5. Concerning (ii), the hypothesis implies $E_n(\mathsf{X}_{2\pi};f) = O(n^{-1})$, and thus the assertion follows by Theorem 2.3.5.

Complementary to (ii), if $f\in\mathsf{Lip}^*(\mathsf{X}_{2\pi};1)$, then $\|\sigma_n(f;\circ)-f(\circ)\|_{\mathsf{X}_{2\pi}} = O(n^{-1}\log n)$ by Cor. 1.6.5, the latter estimate being best possible as the example of Problem 2.4.12(ii) shows (see also Cor. 9.3.10). Thus the results so far obtained do not enable one to characterize the class of functions $f\in\mathsf{X}_{2\pi}$ for which $\|\sigma_n(f;\circ)-f(\circ)\|_{\mathsf{X}_{2\pi}} = O(n^{-1})$. This will be considered in Sec. 12.2, completing our discussion on approximation by Fejér means; for, in view of Cor. 1.7.2, $\|\sigma_n(f;\circ)-f(\circ)\|_{\mathsf{X}_{2\pi}} = o(n^{-1})$ already implies that f must be constant.

As a final application, let us observe that it is also possible to use the theory of best approximation in order to establish direct approximation theorems for particular singular integrals. Apart from Theorem 2.4.3 we state

Theorem 2.4.8. *If* $B_n(f;x)$ *is the singular integral of Rogosinski of* $f\in\mathsf{X}_{2\pi}$, *then*

$$\|B_n(f;\circ)-f(\circ)\|_{\mathsf{X}_{2\pi}} \le (2\pi+1)E_n(\mathsf{X}_{2\pi};f) + 2\omega^*(\mathsf{X}_{2\pi};f;n^{-1}). \tag{2.4.2}$$

Proof. Let t_n^* be the polynomial of best approximation to f in $\mathsf{X}_{2\pi}$. By the Minkowski inequality

$$\|B_n(f;\circ)-f(\circ)\|_{\mathsf{X}_{2\pi}} \le \|B_n(f;\circ)-B_n(t_n^*;\circ)\|_{\mathsf{X}_{2\pi}} + \|B_n(t_n^*;\circ)-f(\circ)\|_{\mathsf{X}_{2\pi}}.$$

According to (1.3.10) it follows that

$$\|B_n(f;\circ)-B_n(t_n^*;\circ)\|_{\mathsf{X}_{2\pi}} \le \|b_n\|_1\|f-t_n^*\|_{\mathsf{X}_{2\pi}} \le 2\pi E_n(\mathsf{X}_{2\pi};f).$$

Moreover, (1.3.12) gives that

$$B_n(t_n^*; x) = \frac{1}{2}\left[t_n^*\left(x + \frac{\pi}{2n+1}\right) + t_n^*\left(x - \frac{\pi}{2n+1}\right)\right] \equiv \Theta_n(x),$$

say, and since

$$\left\| \Theta_n(\circ) - \frac{1}{2}\left[f\left(\circ + \frac{\pi}{2n+1}\right) + f\left(\circ - \frac{\pi}{2n+1}\right)\right] \right\|_{X_{2\pi}} \leq E_n(X_{2\pi}; f),$$

we obtain that

$$\|B_n(t_n^*; \circ) - f(\circ)\|_{X_{2\pi}} \leq E_n(X_{2\pi}; f) + \tfrac{1}{2}\omega^*\left(X_{2\pi}; f; \frac{\pi}{2n+1}\right),$$

which proves (2.4.2).

Theorem 2.4.9. *Let $B_n(f; x)$ be the singular integral of Rogosinski of $f \in X_{2\pi}$.*

(i) *For* $0 < \alpha < 2, f \in \mathsf{Lip}^*(X_{2\pi}; \alpha) \Leftrightarrow \|B_n(f; \circ) - f(\circ)\|_{X_{2\pi}} = O(n^{-\alpha})$,

(ii) $f \in \mathsf{Lip}^*(X_{2\pi}; 2) \Rightarrow \|B_n(f; \circ) - f(\circ)\|_{X_{2\pi}} = O(n^{-2})$,

(iii) $f = \text{const} \Leftrightarrow \|B_n(f; \circ) - f(\circ)\|_{X_{2\pi}} = o(n^{-2})$.

The proof is an immediate consequence of Theorems 2.2.1, 2.4.8, 2.3.5, and Problem 1.7.3.

Concerning the still missing opposite direction in part (ii) we observe that, if e.g. $X_{2\pi} = C_{2\pi}$, then $\|B_n(f; \circ) - f(\circ)\|_{C_{2\pi}} = O(n^{-2})$ implies $\omega^*(C_{2\pi}; f; \delta) = O(\delta^2|\log \delta|)$ if we use Theorem 2.3.5, and $f' \in \mathsf{Lip}^*(C_{2\pi}; 1)$, if we use Theorem 2.3.6 (cf. also Problem 2.4.10). In Sec. 12.2 we shall show that $\|B_n(f; \circ) - f(\circ)\|_{X_{2\pi}} = O(n^{-2})$ actually implies $f \in \mathsf{Lip}^*(X_{2\pi}; 2)$. This would complete the discussion concerning approximation by the singular integral of Rogosinski.

Problems

1. Let $f \in X_{2\pi}$ and t_n be any element of T_n. Show that there exists a constant B such that
$$\|S_n(f; \circ) - f(\circ)\|_{X_{2\pi}} \leq B(1 + \log n)\, \|f - t_n\|_{X_{2\pi}}.$$
2. Show that $f(x) = \sum_{k=1}^{\infty} k^{-2} \sin kx$ is an example of a function for which $E_n(C_{2\pi}; f) = O(n^{-1})$, but which does not belong to $\mathsf{Lip}(C_{2\pi}; 1)$. (Hint: Natanson [8I, p. 101 f])
3. Show that a function $f \in X_{2\pi}$ belongs to $W^{\infty}_{X_{2\pi}}$ if and only if $\lim_{n\to\infty} n^r E_n(X_{2\pi}; f) = 0$ for each $r \in \mathbb{N}$. (Hint: See also Natanson (8I, p. 106])
4. For $f \in C_{2\pi}$ and some $r \in \mathbb{P}$ show that
   (i) $E_n(C_{2\pi}; f) = o(n^{-r-\alpha}) \Leftrightarrow f^{(r)} \in \mathsf{lip}(C_{2\pi}; \alpha)$ $\qquad (0 < \alpha < 1)$,
   (ii) $E_n(C_{2\pi}; f) = o(n^{-r-1}) \Leftrightarrow f^{(r)} \in \mathsf{lip}^*(C_{2\pi}; 1)$.
5. Let $\Gamma_n(\alpha) = \sup_{f \in \mathsf{Lip}(X_{2\pi}; \alpha)_1} E_n(X_{2\pi}; f)$ for $0 < \alpha \leq 1$. Show that $K(\alpha)n^{-\alpha} \leq \Gamma_n(\alpha) \leq 12n^{-\alpha}$ for all $n \in \mathbb{N}$ where $K(\alpha)$ is a positive quantity depending only upon α. (Hint: Natanson [8I, p. 118])
6. Let $K_n(f; x)$ be the singular integral of Fejér–Korovkin. Show that
   (i) for $0 < \alpha < 2, f \in \mathsf{Lip}^*(X_{2\pi}; \alpha) \Leftrightarrow \|K_n(f; \circ) - f(\circ)\|_{X_{2\pi}} = O(n^{-\alpha})$,
   (ii) $f \in \mathsf{Lip}^*(X_{2\pi}; 2) \Rightarrow \|K_n(f; \circ) - f(\circ)\|_{X_{2\pi}} = O(n^{-2})$,
   (iii) $f = \text{const} \Leftrightarrow \|K_n(f; \circ) - f(\circ)\|_{X_{2\pi}} = o(n^{-2})$.

The same assertions hold if one replaces $K_n(f; x)$ by the singular integral $J_n(f; x)$ of Jackson.

7. For $f \in X_{2\pi}$ show that
   (i) $E_n(X_{2\pi}; f) = O(n^{-r-\alpha}) \Leftrightarrow f \in W^{r,\alpha}_{X_{2\pi}}$ $\quad (0 < \alpha < 1)$,
   (ii) $E_n(X_{2\pi}; f) = O(n^{-r-1}) \Leftrightarrow f \in {}^*W^{r,1}_{X_{2\pi}}$.
8. For $f \in X_{2\pi}$ and any $r \in \mathbb{N}$ show for $0 < \alpha < 1$ that $f \in W^{r,\alpha}_{X_{2\pi}}$ if and only if $f \in {}^*W^{r,\alpha}_{X_{2\pi}}$.
9. Show that $E_n(X_{2\pi}; f) = O(n^{-r-\alpha})$ for some $r \in \mathbb{N}$ and $0 < \alpha < 2$ if and only if $f \in {}^*W^{r,\alpha}_{X_{2\pi}}$. Discuss the case $\alpha = 2$.
10. Let $f \in X_{2\pi}$. Show that a necessary and sufficient condition such that $f \in \mathrm{Lip}^*(X_{2\pi}; \alpha)$ for some $1 < \alpha < 2$ is that $f \in W^{1,\alpha-1}_{X_{2\pi}}$, i.e., there exists ϕ such that $f(x) = \phi(x)$ (a.e.) and $\phi \in AC_{2\pi}$, $\phi' \in \mathrm{Lip}(X_{2\pi}; \alpha - 1)$. Furthermore, $f \in \mathrm{Lip}^*(X_{2\pi}; 2)$ implies $\phi' \in \mathrm{Lip}^*(X_{2\pi}; 1)$.
11. The *de La Vallée Poussin sums* $\tau_{2n-1}(f; x)$ of $f \in X_{2\pi}$ are defined by

$$\tau_{2n-1}(f; x) = n^{-1} \sum_{k=n}^{2n-1} S_k(f; x) = 2\sigma_{2n-1}(f; x) - \sigma_{n-1}(f; x),$$

where $S_n(f; x)$ and $\sigma_n(f; x)$ are the singular integrals of Dirichlet and Fejér, respectively. Prove that the sums $\tau_{2n-1}(f; x)$ define a summation process of the Fourier series of f corresponding to the row-finite θ-factor

$$\theta_{2n-1}(k) = \begin{cases} 1, & 1 \le |k| \le n \\ 2 - (|k|/n), & n + 1 \le |k| \le 2n - 1. \end{cases}$$

Thus $\tau_{2n-1}(f; x)$ is a singular integral with kernel

$$1 + 2\sum_{k=1}^{n} \cos kx + 2\sum_{k=n+1}^{2n-1} \left(2 - \frac{k}{n}\right) \cos kx \equiv D_n(x) + 2\sum_{k=n+1}^{2n-1} \left(2 - \frac{k}{n}\right) \cos kx.$$

Show that this kernel may be represented as $\{(1 + 2\cos nx)F_{n-1}(x)\}$. Furthermore,
   (i) $\tau_{2n-1}(t_m; x) = t_m(x)$ for every polynomial t_m of degree $m \le n$,
   (ii) $\|\tau_{2n-1}(f; \circ)\|_{X_{2\pi}} \le 3\,\|f\|_{X_{2\pi}}$,
   (iii) $\|\tau_{2n-1}(f; \circ) - f(\circ)\|_{X_{2\pi}} \le 4E_n(X_{2\pi}; f)$.
   (Hint: Natanson [8I, p. 167 ff], Zygmund [7I, p. 115 ff])
12. (i) Show that $f(x) = |\sin x|$ is an example of a function for which $f \in \mathrm{Lip}(C_{2\pi}; 1)$ but $E_n(C_{2\pi}; f) > 1/4\pi n$. (Hint: Lorentz [3, p. 93])
   (ii) Show that $f(x) = \sum_{k=1}^{\infty} k^{-2} \cos kx$ is an example of a function for which $f \in \mathrm{Lip}(C_{2\pi}; 1)$ but $\|\sigma_n(f; \circ) - f(\circ)\|_{C_{2\pi}} \ge (n + 1)^{-1}(1 + \log(n + 1))$.
   (iii) Show that $f(x) = \sum_{k=1}^{\infty} k^{-1} \sin kx$ belongs to $\mathrm{Lip}(L^1_{2\pi}; 1)$ but $\|S_n(f; \circ) - f(\circ)\|_1 \ne o(n^{-1} \log n)$. (Hint: Quade [1])
13. A bounded linear operator P of the Banach space X into X is called a *projection* if $P^2 = P$, i.e., $P(Pf) = Pf$ for every $f \in X$. A projection P is said to be *onto* $G \subseteq X$ if G is the range of P. Show that if P is a projection onto G, then $Pf = f$ for every $f \in G$.
14. Let P be a projection of $C_{2\pi}$ onto T_n.
   (i) If $f_u(x) \equiv f(x + u)$, show that for every $f \in C_{2\pi}$

$$\frac{1}{2\pi} \int_{-\pi}^{\pi} (Pf_u)(x - u)\, du = S_n(f; x),$$

   $S_n(f; x)$ being the singular integral of Dirichlet. (Hint: See Cheney [1, p. 210 ff], Lorentz [3, p. 96 ff])
   (ii) If S_n denotes the Fourier series projection of $C_{2\pi}$ onto T_n, i.e., $(S_n f)(x) \equiv S_n(f; x)$, show that $\|S_n\| \le \|P\|$. Thus the Fourier series projection is *minimal* among all projections of $C_{2\pi}$ onto T_n.

(iii) Show that every P which is representable as a convolution integral necessarily coincides with the Fourier series projection S_n. (Hint: Suppose that there exists some (continuous) $\chi \in \mathsf{L}_{2\pi}^1$ such that $(Pf)(x) = (f * \chi)(x)$ for every $f \in \mathsf{C}_{2\pi}$, and apply Problems 1.2.15 (ii), 2.4.13)

15. For each $n \in \mathbb{N}$, let P_n be a projection of $\mathsf{C}_{2\pi}$ onto T_n.
    (i) Show that $\lim_{n\to\infty} \|P_n\| = \infty$. (Hint: Use Prop. 1.2.3, 1.3.1, and the preceding Problem)
    (ii) Show that there exists $f \in \mathsf{C}_{2\pi}$ such that $\lim_{n\to\infty} \|P_n f\|_{\mathsf{C}_{2\pi}} = \infty$. (Hint: Use (i) and the uniform boundedness principle)
16. Show that there does not exist a sequence of bounded linear operators L_n on $\mathsf{C}_{2\pi}$ which are polynomial of degree n such that $\|L_n f - f\|_{\mathsf{C}_{2\pi}} = O(E_n(\mathsf{C}_{2\pi}; f))$ for every $f \in \mathsf{C}_{2\pi}$. (Hint: Suppose, conversely, that there exists such a sequence. Then $L_n t_n = t_n$ for every $t_n \in \mathsf{T}_n$, and hence L_n would be a projection onto T_n. Now construct a contradiction to Problem 2.4.15(ii))
17. State and prove the counterparts of the previous three Problems in $\mathsf{L}_{2\pi}^1$-space. However, in $\mathsf{L}_{2\pi}^p$-space, $1 < p < \infty$, the situation is quite different; see Prop. 9.3.8.
18. Let $f \in \mathsf{X}_{2\pi}$. Show that $\|\sigma_n(f; \circ) - f(\circ)\|_{\mathsf{X}_{2\pi}} \le (12/(n+1)) \sum_{k=1}^{n+1} E_k(\mathsf{X}_{2\pi}; f)$. (Hint: STEČKIN [3])

2.5 Approximation Theorems for Singular Integrals

In this section we continue the program begun in the foregoing section, this time to establish equivalence theorems on approximation by singular integrals. The methods of Sec. 2.4 are effective only for those integrals which define a polynomial summation process of the Fourier series of f and whose order of approximation to f is the same as the order of best approximation. Nevertheless, one can also extend these arguments (above all, also those of Sec. 2.3) to singular integrals which are not of this particular type. This section is devoted to two representative examples.

2.5.1 Singular Integral of Abel–Poisson

The integral of Abel–Poisson as defined in Sec. 1.2.4 furnishes a summation process of the Fourier series of f which is not of polynomial type. To establish (optimal) direct approximation theorems we proceed as in the case of the Fejér means (cf. Sec. 1.6.2) and first of all have

Lemma 2.5.1. *Let $\{p_r(x)\}$ be the Abel–Poisson kernel. Then*

$$\text{(i)}\quad \frac{1}{2\pi}\int_0^{1-r} u^\alpha p_r(u)\,du \le (1-r)^\alpha \qquad (0 < \alpha \le 1),$$

$$\text{(ii)}\quad \frac{1}{2\pi}\int_{1-r}^{\pi} u^\alpha p_r(u)\,du \le \begin{cases} \dfrac{\pi}{4(1-\alpha)r}(1-r)^\alpha & (0 < \alpha < 1) \\[2ex] \dfrac{\pi}{4r}(1-r)\log\dfrac{\pi}{1-r} & (\alpha = 1), \end{cases}$$

$$\text{(iii)}\quad \frac{1}{2\pi}\int_0^{\pi} u^{1+\alpha} p_r(u)\,du \le \frac{\pi^{1+\alpha}}{4\alpha r}(1-r) \qquad (0 < \alpha \le 1).$$

The proof follows by elementary computations using Problem 1.2.18(iii). An application of Prop. 1.6.3 is

Theorem 2.5.2. *If $P_r(f; x)$ is the singular integral of Abel–Poisson, then for $r \to 1-$*

$$f \in \mathsf{Lip}^*(\mathsf{X}_{2\pi}; \alpha) \Rightarrow \|P_r(f; \circ) - f(\circ)\|_{\mathsf{X}_{2\pi}} = \begin{cases} O((1-r)^\alpha), & 0 < \alpha < 1 \\ O((1-r)\,|\log(1-r)|), & \alpha = 1 \\ O(1-r), & 1 < \alpha \le 2. \end{cases}$$

We now consider the inverse problem for this singular integral. If we wish to adapt the methods of proof of Sec. 2.3, we first of all have to look for something in place of Bernstein's inequality.

Lemma 2.5.3. *For every $f \in \mathsf{X}_{2\pi}$ the singular integral $P_r(f; x)$ of Abel–Poisson has a continuous second derivative satisfying*

$$\|P_r''(f; \circ)\|_{\mathsf{X}_{2\pi}} \le \frac{4}{(1-r)^2}\|f\|_{\mathsf{X}_{2\pi}} \qquad (0 < r < 1). \tag{2.5.1}$$

Indeed, for each $0 < r < 1$ the Abel–Poisson kernel $\{p_r(x)\}$ of (1.2.37) is arbitrarily often differentiable with respect to x, and hence $P_r(f; x) \in \mathsf{C}_{2\pi}^\infty$ for every $f \in \mathsf{X}_{2\pi}$ by Prop. 1.1.14. Therefore

$$\|P_r''(f; \circ)\|_{\mathsf{X}_{2\pi}} \le \left\{\frac{1}{2\pi}\int_{-\pi}^{\pi} |p_r''(u)|\, du\right\} \|f\|_{\mathsf{X}_{2\pi}},$$

and the estimate (2.5.1) follows by an elementary calculation (cf. Problem 2.5.2).

Theorem 2.5.4. *If, for some $0 < \alpha \le 1$, $f \in \mathsf{X}_{2\pi}$ can be approximated by the singular integral of Abel–Poisson such that*

$$\|P_r(f; \circ) - f(\circ)\|_{\mathsf{X}_{2\pi}} = O((1-r)^\alpha) \qquad (r \to 1-),$$

then $f \in \mathsf{Lip}^*(\mathsf{X}_{2\pi}; \alpha)$.

Proof. We proceed as in the proofs of Theorems 2.3.3 and 2.3.5. By hypothesis there exists a constant $B > 0$ such that $\|P_r(f; \circ) - f(\circ)\|_{\mathsf{X}_{2\pi}} \le B(1-r)^\alpha$ for all $0 < r < 1$. If we set $1 - r_n = 2^{-n}$ and $U_2(x) = P_{r_2}(f; x)$, $U_n(x) = P_{r_n}(f; x) - P_{r_{n-1}}(f; x)$ for $n = 3, 4, \ldots$, then

$$\|P_{r_n}(f; \circ) - f(\circ)\|_{\mathsf{X}_{2\pi}} \le B2^{-n\alpha} \tag{2.5.2}$$

and

$$\|U_n\|_{\mathsf{X}_{2\pi}} \le \|P_{r_n}(f; \circ) - f(\circ)\|_{\mathsf{X}_{2\pi}} + \|P_{r_{n-1}}(f; \circ) - f(\circ)\|_{\mathsf{X}_{2\pi}} \le (1 + 2^\alpha)B2^{-n\alpha} \tag{2.5.3}$$

for $n = 3, 4, \ldots$. By Cor. 1.2.9, $\sum_{k=2}^n U_k(x) = P_{r_n}(f; x)$ converges in $\mathsf{X}_{2\pi}$-norm to f as $n \to \infty$, thus

$$\lim_{n\to\infty} \|f - \sum_{k=2}^n U_k\|_{\mathsf{X}_{2\pi}} = 0 \tag{2.5.4}$$

and $\|f\|_{\mathsf{X}_{2\pi}} \le \sum_{k=2}^\infty \|U_k\|_{\mathsf{X}_{2\pi}}$. Hence for every $h \in \mathbb{R}$ and integer $m \ge 2$

$$\|f(\circ + h) + f(\circ - h) - 2f(\circ)\|_{\mathsf{X}_{2\pi}} \le \sum_{k=2}^m \|U_k(\circ + h) + U_k(\circ - h) - 2U_k(\circ)\|_{\mathsf{X}_{2\pi}} + 4\sum_{k=m+1}^\infty \|U_k\|_{\mathsf{X}_{2\pi}}. \tag{2.5.5}$$

In view of Lemma 2.5.3, $P_r(f; x)$ is twice continuously differentiable for every $f \in X_{2\pi}$. Therefore

$$U_k(x + h) + U_k(x - h) - 2U_k(x) = \int_{-h/2}^{h/2} \int_{-h/2}^{h/2} U_k''(x + u_1 + u_2)\, du_1\, du_2,$$

and consequently

$$\|U_k(\circ + h) + U_k(\circ - h) - 2U_k(\circ)\|_{X_{2\pi}} \le h^2 \|U_k''\|_{X_{2\pi}}. \tag{2.5.6}$$

Since the convolution of $L^1_{2\pi}$-functions is a commutative and associative operation (by Fubini's theorem), we have $P_{r_k}(P_{r_{k-1}}(f; \circ); x) = P_{r_{k-1}}(P_{r_k}(f; \circ); x)$ and so the identity $(k = 3, 4, \ldots)$

$$U_k(x) = P_{r_k}(f(\circ) - P_{r_{k-1}}(f; \circ); x) - P_{r_{k-1}}(f(\circ) - P_{r_k}(f; \circ); x).$$

Now we apply (2.5.1) and (2.5.2) to deduce

$$\begin{aligned} \|U_k''\|_{X_{2\pi}} &\le \|P_{r_k}''(f(\circ) - P_{r_{k-1}}(f; \circ); \circ)\|_{X_{2\pi}} + \|P_{r_{k-1}}''(f(\circ) - P_{r_k}(f; \circ); \circ)\|_{X_{2\pi}} \\ &\le 4\cdot 2^{2k}\|f(\circ) - P_{r_{k-1}}(f; \circ)\|_{X_{2\pi}} + 2^{2k}\|f(\circ) - P_{r_k}(f; \circ)\|_{X_{2\pi}} \\ &\le 4\cdot 2^{2k}B2^{-\alpha(k-1)} + 2^{2k}B2^{-k\alpha} \le 9B2^{k(2-\alpha)}. \end{aligned} \tag{2.5.7}$$

Therefore it follows by (2.5.3), (2.5.5), and (2.5.6) that

$$\omega^*(X_{2\pi}; f; \delta) \le 12B\Big\{\delta^2 \sum_{k=2}^{m} 2^{k(2-\alpha)} + \sum_{k=m+1}^{\infty} 2^{-k\alpha}\Big\}. \tag{2.5.8}$$

Thus we have the same estimate (apart from the constant factor) for the generalized modulus of continuity of f as in (2.3.10), and we may complete the proof in similar fashion.

Corollary 2.5.5. *Let* $f \in X_{2\pi}$.

(i) *For* $0 < \alpha < 1$, $\|P_r(f; \circ) - f(\circ)\|_{X_{2\pi}} = O((1 - r)^\alpha) \Leftrightarrow f \in \mathrm{Lip}\,(X_{2\pi}; \alpha)$,

(ii) $\|P_r(f; \circ) - f(\circ)\|_{X_{2\pi}} = O(1 - r) \Rightarrow f \in \mathrm{Lip}^*\,(X_{2\pi}; 1)$.

This is an immediate consequence of Theorems 2.4.2, 2.5.2, and 2.5.4.

Cor. 2.5.5 is an equivalence theorem for the Abel–Poisson summation method in case $0 < \alpha < 1$. To solve the case $\alpha = 1$ completely, the method of proof in Theorem 2.5.4 does not seem to be sufficiently powerful. Indeed, $f \in \mathrm{Lip}^*\,(X_{2\pi}; 1)$ implies by Theorem 2.5.2 $\|P_r(f; \circ) - f(\circ)\|_{X_{2\pi}} = O((1 - r)\,|\log(1 - r)|)$, and this estimate cannot be improved in general as the example of Problem 2.5.1 shows. In Sec. 12.2 we shall see that the set of functions $f \in X_{2\pi}$ for which $\|P_r(f; \circ) - f(\circ)\|_{X_{2\pi}} = O(1 - r)$ is actually a proper subset of $\mathrm{Lip}^*\,(X_{2\pi}; 1)$. This would complete our discussion on the order of approximation by the integral of Abel–Poisson since an approximation $\|P_r(f; \circ) - f(\circ)\|_{X_{2\pi}} = O((1 - r)^\alpha)$ for $\alpha > 1$ is possible only for the trivial case of a constant function f. Indeed,

Proposition 2.5.6. *If* $f \in X_{2\pi}$ *and*

$$\liminf_{r\to 1-} \frac{1}{1 - r}\|P_r(f; \circ) - f(\circ)\|_{X_{2\pi}} = 0, \tag{2.5.9}$$

then f is constant (a.e.).

Proof. By definition,

$$P_r(f; x) = \sum_{k=-\infty}^{\infty} r^{|k|} f^\wedge(k)\, e^{ikx},$$

and since the series is uniformly convergent for each $0 < r < 1$, we have by Problem 1.2.2(i) for any $l \in \mathbb{Z}$

$$\frac{1}{2\pi}\int_{-\pi}^{\pi} P_r(f; x)\, e^{-ilx}\, dx = \sum_{k=-\infty}^{\infty} r^{|k|} f^\wedge(k) \frac{1}{2\pi}\int_{-\pi}^{\pi} e^{i(k-l)x}\, dx = r^{|l|} f^\wedge(l).$$

Therefore

$$\frac{1}{2\pi}\int_{-\pi}^{\pi} [P_r(f; x) - f(x)]\, e^{-ilx}\, dx = [r^{|l|} - 1] f^\wedge(l),$$

and hence for each $0 < r < 1$ and $l \in \mathbb{Z}$

$$[1 - r^{|l|}]\, |f^\wedge(l)| \le \|P_r(f; \circ) - f(\circ)\|_{X_{2\pi}}. \tag{2.5.10}$$

Now, on the one hand

$$\lim_{r \to 1-} \frac{1 - r^{|l|}}{1 - r} = |l|,$$

whilst on the other hand there exists by hypothesis a sequence $\{r_j\}$ with $r_j \to 1-$ as $j \to \infty$ such that

$$\lim_{j \to \infty} \frac{1}{1 - r_j} \|P_{r_j}(f; \circ) - f(\circ)\|_{X_{2\pi}} = 0.$$

This implies by (2.5.10) that $|l|\, |f^\wedge(l)| = 0$ for every $l \in \mathbb{Z}$ and thus $f^\wedge(l) = 0$ for every $l \in \mathbb{Z}$, $l \neq 0$. Finally f is constant (a.e.) by Cor. 1.2.7.

2.5.2 Singular Integral of de La Vallée Poussin

The *singular integral of de La Vallée Poussin* is defined for $f \in X_{2\pi}$ by (cf. also Problem 1.1.11)

$$V_n(f; x) = \frac{1}{2\pi}\int_{-\pi}^{\pi} f(x - u) v_n(u)\, du \tag{2.5.11}$$

with kernel $\{v_n(x)\}$ given by

$$v_n(x) = \frac{(n!)^2}{(2n)!}\left(2 \cos \frac{x}{2}\right)^{2n} \tag{2.5.12}$$

and discrete parameter $n \in \mathbb{P}$, $n \to \infty$. For each $n \in \mathbb{P}$, $v_n(x)$ is an even, positive trigonometric polynomial of degree n which may be represented as

$$v_n(x) = 1 + 2 \sum_{k=1}^{n} \theta_n(k) \cos kx, \qquad \theta_n(k) = \frac{(n!)^2}{(n-k)!(n+k)!}. \tag{2.5.13}$$

Thus the singular integral of de La Vallée Poussin associated with $f \in X_{2\pi}$ defines a polynomial summation process of the Fourier series of f given by

$$V_n(f; x) = \sum_{k=-n}^{n} \frac{(n!)^2}{(n-k)!(n+k)!} f^\wedge(k)\, e^{ikx}. \tag{2.5.14}$$

Indeed, the kernel $\{v_n(x)\}$ being positive with $\lim_{n\to\infty} \theta_n(1) = 1$, we may apply Prop. 1.3.10 to deduce

Proposition 2.5.7. *For every $f \in X_{2\pi}$, $\|V_n(f; \circ)\|_{X_{2\pi}} \le \|f\|_{X_{2\pi}}$ and*

$$\lim_{n \to \infty} \|V_n(f; \circ) - f(\circ)\|_{X_{2\pi}} = 0. \tag{2.5.15}$$

Regarding the approximation behaviour of $V_n(f; x)$ we note that

$$1 - v_n^\wedge(1) \equiv 1 - \theta_n(1) = \frac{1}{n+1},$$

and so in view of Theorem 1.5.8

Proposition 2.5.8. *For every $f \in X_{2\pi}$*

$$\|V_n(f; \circ) - f(\circ)\|_{X_{2\pi}} = O(\omega^*(X_{2\pi}; f; n^{-1/2})) \qquad (n \to \infty).$$

In particular, $f \in \mathsf{Lip}^*(X_{2\pi}; \alpha)$ for some $0 < \alpha \le 2$ implies

(2.5.16) $$\|V_n(f; \circ) - f(\circ)\|_{X_{2\pi}} = O(n^{-\alpha/2}) \qquad (n \to \infty).$$

The example $f(x) = |\sin x|^\alpha$ of Problem 2.5.6 shows that the estimate (2.5.16) cannot be improved in general. Thus the integral $V_n(f; x)$ of de La Vallée Poussin does not have such a good order of approximation to $f \in \mathsf{Lip}^*(X_{2\pi}; \alpha)$ as the order of best approximation which would be $O(n^{-\alpha})$. In consequence, we cannot apply the results of Sec. 2.3 to establish the converse of Prop. 2.5.8 since these would only infer that (2.5.16) implies $f \in \mathsf{Lip}^*(X_{2\pi}; \alpha/2)$.

Nevertheless, to obtain optimal results we first prove the following lemma which, in some sense, may be regarded as an improvement of the Bernstein inequality in case the polynomials are given by $V_n(f; x)$.

Lemma 2.5.9. *For every $f \in X_{2\pi}$*

(2.5.17) $$\|V_n''(f; \circ)\|_{X_{2\pi}} \le n \|f\|_{X_{2\pi}}.$$

Proof. Since

(2.5.18) $$v_n^\wedge(k) = \begin{cases} \dfrac{(n!)^2}{(n-k)!(n+k)!}, & |k| \le n \\ 0, & |k| > n, \end{cases}$$

it follows by direct computation that

(2.5.19) $$k^2 v_n^\wedge(k) = n^2[v_n^\wedge(k) - v_{n-1}^\wedge(k)] \qquad (n \in \mathbb{N}, k \in \mathbb{Z}),$$

(2.5.20) $$n^2[v_{n-1}(x) - v_n(x)] = \frac{n^2}{2} v_{n-1}(x)(1 - \cos x) - \frac{n}{2} v_n(x).$$

The integral $V_n(f; x)$, being a polynomial of degree n, is arbitrarily often differentiable. Therefore, for every $f \in X_{2\pi}$

(2.5.21) $$V_n''(f; x) = -n^2[V_n(f; x) - V_{n-1}(f; x)]$$

by (2.5.14) and (2.5.19), and thus by (2.5.20)

$$\begin{aligned} \|V_n''(f; \circ)\|_{X_{2\pi}} &= \left\| \frac{1}{2\pi} \int_{-\pi}^{\pi} f(\circ - u) \left[\frac{n^2}{2} v_{n-1}(u)(1 - \cos u) - \frac{n}{2} v_n(u)\right] du \right\|_{X_{2\pi}} \\ &\le \left\{\frac{n^2}{2} \|v_{n-1}(\circ)(1 - \cos \circ)\|_1 + \frac{n}{2} \|v_n\|_1\right\} \|f\|_{X_{2\pi}} \\ &= \left\{\frac{n^2}{2} [1 - v_{n-1}^\wedge(1)] + \frac{n}{2}\right\} \|f\|_{X_{2\pi}} = n \|f\|_{X_{2\pi}}. \end{aligned}$$

This establishes the lemma.

Theorem 2.5.10. *If, for some* $0 < \alpha \leq 1$, $f \in X_{2\pi}$ *can be approximated by the integral of de La Vallée Poussin such that* $\|V_n(f; \circ) - f(\circ)\|_{X_{2\pi}} = O(n^{-\alpha})$, *then*

$$\omega^*(X_{2\pi}; f; \delta) = \begin{cases} O(\delta^{2\alpha}), & 0 < \alpha < 1 \\ O(\delta^2 |\log \delta|), & \alpha = 1. \end{cases}$$

Proof. We shall follow the lines of the proof of Theorem 2.5.4. Indeed, by hypothesis $\|V_n(f; \circ) - f(\circ)\|_{X_{2\pi}} \leq Bn^{-\alpha}$ for some constant $B > 0$. Setting $U_2(x) = V_{2^2}(f, x)$, $U_n(x) = V_{2^n}(f; x) - V_{2^{n-1}}(f; x)$ for $n = 3, 4, \ldots$, then $U_n(x)$ is a polynomial of degree 2^n at most satisfying the estimate (2.5.3). Furthermore, the estimates (2.5.5) and (2.5.6) are valid as well. Since

$$U_k(x) = V_{2^k}(f(\circ) - V_{2^{k-1}}(f; \circ); x) - V_{2^{k-1}}(f(\circ) - V_{2^k}(f; \circ); x),$$

an application of Lemma 2.5.9 yields

$$\begin{aligned} \|U_k''\|_{X_{2\pi}} &\leq 2^k \|f(\circ) - V_{2^{k-1}}(f; \circ)\|_{X_{2\pi}} + 2^{k-1}\|f(\circ) - V_{2^k}(f; \circ)\|_{X_{2\pi}} \\ &\leq 2^k B 2^{-(k-1)\alpha} + 2^{k-1} B 2^{-k\alpha} \leq 3B 2^{k(1-\alpha)}. \end{aligned}$$

Therefore the generalized modulus of continuity of f admits the estimate

$$\omega^*(X_{2\pi}; f; \delta) \leq 12B\left\{\delta^2 \sum_{k=2}^{m} 2^{k(1-\alpha)} + \sum_{k=m+1}^{\infty} 2^{-k\alpha}\right\}.$$

If we suppose $0 < \delta < \frac{1}{2}$ and choose $m \geq 2$ according to

$$2^{m-1} \leq 1/\delta^2 < 2^m, \tag{2.5.22}$$

then we may complete the proof by considering the cases as in the proof of Theorem 2.3.3.

As a consequence of Prop. 2.5.8 and Theorem 2.5.10 we formulate

Corollary 2.5.11. *Let* $f \in X_{2\pi}$.

(i) *For* $0 < \alpha < 2$, $f \in \text{Lip}^*(X_{2\pi}; \alpha) \Leftrightarrow \|V_n(f; \circ) - f(\circ)\|_{X_{2\pi}} = O(n^{-\alpha/2})$,

(ii) $f \in \text{Lip}^*(X_{2\pi}; 2) \Rightarrow \|V_n(f; \circ) - f(\circ)\|_{X_{2\pi}} = O(n^{-1})$.

In Sec. 12.2 we shall show that also the converse to (ii) is valid. This would again complete our investigations concerning the integral of de La Vallée Poussin because of

Proposition 2.5.12. *If* $f \in X_{2\pi}$ *and*

$$\liminf_{n\to\infty} n\, \|V_n(f; \circ) - f(\circ)\|_{X_{2\pi}} = 0, \tag{2.5.23}$$

then f *is constant* (a.e.).

The proof follows by Theorem 1.7.1 since

$$\lim_{n\to\infty} n\, |v_n^\wedge(k) - 1| = \lim_{n\to\infty} n\left(1 - \frac{(n!)^2}{(n-k)!(n+k)!}\right) = k^2 \qquad (k \in \mathbb{Z}), \tag{2.5.24}$$

and so condition (1.7.2) is satisfied by the kernel $\{v_n(x)\}$ with $\alpha = 1$.

The question finally arises whether it would somehow be possible to extract a general method of proof of the inverse approximation theorems (as applied to the representative examples) so as to cover the general singular integral $I_\rho(f; x)$. This is indeed possible and is to be carried out in Problem 6.4.3 (and Sec. 3.5) for so-called singular integrals of Fejér's type (these are sufficiently general). However, in the case of general integrals $I_\rho(f; x)$, an inequality of Bernstein-type has to be postulated which would include Lemmata 2.5.3 and 2.5.9 as particular cases.

Problems

1. Show that the function $f(x) = \sum_{k=1}^{\infty} k^{-2} \cos kx$ belongs to $\mathsf{Lip}\,(\mathsf{C}_{2\pi}; 1)$, but $\|P_r(f; \circ) - f(\circ)\|_{\mathsf{C}_{2\pi}} \geq r^{-1}(1 - r)\,|\log(1 - r)|$ for $0 < r < 1$ (cf. Problem 2.4.12(ii)). Thus the estimate of Theorem 2.5.2 for $\alpha = 1$ cannot be improved in general. (Hint: See also BUTZER–BERENS [1, p. 123])
2. If $\{p_r(x)\}$ is the Abel–Poisson kernel, show that $\|p_r'\|_1 \leq (4/\pi)(1 - r)^{-1}$ as well as $\|p_r''\|_1 \leq 4(1 - r)^{-2}$ for $0 < r < 1$. (Hint: See also BERENS [1, pp. 33, 36], BUTZER–BERENS [1, p. 123])
3. For some $0 < \alpha \leq 1$, let $f \in \mathsf{X}_{2\pi}$ be approximable by the singular integral of Abel–Poisson such that $\|P_r(f; \circ) - f(\circ)\|_{\mathsf{X}_{2\pi}} = O((1 - r)^\alpha)$ as $r \to 1-$. Show that
$$\omega(\mathsf{X}_{2\pi}; f; \delta) = \begin{cases} O(\delta^\alpha), & 0 < \alpha < 1 \\ O(\delta\,|\log \delta|), & \alpha = 1. \end{cases}$$
4. Show that $(1/2\pi)\int_{-\pi}^{\pi}(2\cos(u/2))^{2n}\,du = (2n)!/(n!)^2$. Furthermore, establish (2.5.13) and (2.5.14). (Hint: See also NATANSON [8I, p. 9 ff, p. 222 f])
5. (i) Let $f \in \mathsf{X}_{2\pi}$. Show that $\lim_{n\to\infty} V_n(f; x_0) = f(x_0)$ at each point x_0 of continuity of f. (Hint: Use Prop. 1.4.1)
   (ii) Let $f \in \mathsf{X}_{2\pi}$. Show that $\lim_{n\to\infty} V_n(f; x) = f(x)$ a.e. (Hint: Use Prop. 1.4.4)
6. Show that $f(x) = |\sin x|^\alpha$ belongs to $\mathsf{Lip}\,(\mathsf{C}_{2\pi}; \alpha)$, but there exists a constant C such that $\|V_n(f; \circ) - f(\circ)\|_{\mathsf{C}_{2\pi}} \geq Cn^{-\alpha/2}$. (Hint: Compare with Problem 1.6.9(iii), see NATANSON [8I, p. 208])
7. Let $f \in \mathsf{L}_{2\pi}^{\infty}$ have an ordinary second derivative at x_0. Show that $V_n(f; x_0) - f(x_0) = n^{-1}f''(x_0) + o(n^{-1})$, $n \to \infty$. (Hint: Use Theorem 1.5.15, thus (1.5.28))
8. (i) Let $f \in \mathsf{C}_{2\pi}$ be continuously differentiable. Show that $\lim_{n\to\infty} V_n'(f; x) = f'(x)$ uniformly for all x. (Hint: Use Prop. 1.1.15)
   (ii) Let $f \in \mathsf{L}_{2\pi}^{\infty}$, and suppose that f is differentiable at x_0. Show that $\lim_{n\to\infty} V_n'(f; x_0) = f'(x_0)$. (Hint: Use (1.1.18) and proceed as in the proof of Prop. 1.4.8; see also NATANSON [8I, p. 215])
9. (i) Show that $\Delta^*(v_n; 2) = 2n^{-1} + o(n^{-1})$. Thus, for the Nikolskiĭ constant, $N^*(2) = 2$. (Hint: Use Prop. 1.6.6)
   (ii) Show that for each $0 < \alpha \leq 1$
$$\Delta(v_n; \alpha) = \{2^\alpha \pi^{-1/2}\Gamma((1 + \alpha)/2)\}n^{-\alpha/2} + o(n^{-\alpha/2}) \qquad (n \to \infty).$$
   (Hint: NATANSON [2], for $\alpha = 1$ see also [8I, p. 210])
   (iii) Show that the same formula as in (ii) for $\Delta(v_n; \alpha)$ holds for $\Delta^*(v_n; \alpha)$ for every $0 < \alpha \leq 2$. (Hint: TABERSKI [1])
10. Show that the geometric mean of a positive cosine-polynomial $C_n(x) = \lambda_0 + \sum_{k=1}^{n} \lambda_k \cos kx$, not identically zero, is given by
$$\exp\left\{\frac{1}{2\pi}\int_{-\pi}^{\pi} \log C_n(u)\,du\right\} = |h_n(0)|^2 = h_n^2(0),$$
where $h_n(z)$ is an algebraic polynomial in z having real coefficients such that $C_n(x) = |h_n(e^{ix})|^2$. (Hint: Apply Problem 1.6.3, see PÓLYA–SZEGÖ [1II, p. 84])
11. If $C_n(x)$ of the previous Problem has geometric mean equal to one, show that: (i) $C_n(x) \leq 2^{2n}$; equality holds if and only if $C_n(x) = 2^{2n}\cos^{2n}((x - x_0)/2)$, thus for the kernel of de La Vallée Poussin (apart from a constant). (ii) The arithmetic mean of $C_n(x)$ is given by $\lambda_0 = (1/2\pi)\int_{-\pi}^{\pi} C_n(u)\,du \leq \binom{2n}{n}$; when does equality hold in this case? (iii) $\lambda_k \leq 2\binom{2n}{n+k}$, $k = 1, 2, \ldots, n$; when does equality hold in this case?
12. Let $W_t(f; x)$ be the singular integral of Weierstrass of $f \in \mathsf{X}_{2\pi}$. Show that (compare Problem 1.5.9(ii))
$$\text{for } 0 < \alpha < 2,\ f \in \mathsf{Lip}^*(\mathsf{X}_{2\pi}; \alpha) \Leftrightarrow \|W_t(f; \circ) - f(\circ)\|_{\mathsf{X}_{2\pi}} = O(t^{\alpha/2}), \quad t \to 0+,$$
$$f \in \mathsf{Lip}^*(\mathsf{X}_{2\pi}; 2) \Rightarrow \|W_t(f; \circ) - f(\circ)\|_{\mathsf{X}_{2\pi}} = O(t), \quad t \to 0+.$$

2.6 Notes and Remarks

The presentation in this chapter is brief since this is not intended to be a text on approximation theory *per se*. The theorems of Sec. 2.1–2.4 are standard and do not represent the best possible results currently known: only that degree of generality is considered that is relevant to its later use in the text. The most useful general references here are LORENTZ [3] and TIMAN [2], where the leitmotiv is actually that of the order of approximation by polynomials of best approximation. One may also consult the classical treatises of ACHIESER [2] and NATANSON [8]. The results of Sec. 2.5 are not standard in texts.

Sec. 2.1. The existence theorem for best approximations (Theorem 2.1.1) was first established in the doctoral dissertation of KIRCHBERGER [1] in 1902 and also by BOREL [1] in 1905. P. L. CHEBYCHEV was perhaps one of the first (1857) to concern himself with polynomials of best approximation and their *characterization*, using the mathematical techniques of those days. The concept $E_n(\mathsf{X}_{2\pi}; f)$ is best treated from the point of view of the theory of normed linear spaces. If X is a normed linear space, and $g_1, \dots, g_n$ are n linearly independent given vectors in X, the best approximation of $f \in \mathsf{X}$ by linear combinations ('polynomials') $\sum_{k=1}^{n} \alpha_k g_k$ (α_k being scalars) is $E_n(f) = \min_{\alpha_k, 1 \le k \le n} \|f - \sum_{k=1}^{n} \alpha_k g_k\|$. Then the *existence* of polynomials of best approximation to f is a simple consequence of the fact that a continuous function defined on a compact set achieves its infimum.

But the question when such a polynomial of best approximation is *unique* is much more involved. A general reference to the unicity problem is CHENEY [1, Chapter 1; 3] and the literature cited there. Thus a sufficient condition which assures uniqueness is the strict convexity of X. X is called *strictly convex* if $\|f_1\| \le r$, $\|f\|_2 \le r$ imply $\|f_1 + f_2\| < 2r$ unless $f_1 = f_2$. Although the spaces $\mathsf{L}^p_{2\pi}$, $1 < p < \infty$, are strictly convex, this is not true for $p = 1$ nor for the space $\mathsf{C}_{2\pi}$. These spaces must therefore be treated by their own methods.

An important problem here is the characterization of the polynomial of best approximation. One such result in case $\{g_1, \dots, g_n\}$ is a system of elements of $\mathsf{C}_{2\pi}$ satisfying the Haar condition (often termed a Chebychev system) is the Borel–Young or Chebychev alternation theorem in case of uniform approximation. The system $\{g_1, \dots, g_n\}$ satisfies the *Haar condition* if each nontrivial polynomial $\sum_{k=1}^{n} \alpha_k g_k$ has at most $n - 1$ distinct roots (in any interval of length 2π). In this respect, Haar's unicity theorem states that the polynomial $\sum_{k=1}^{n} \alpha_k g_k$ of best approximation to $f \in \mathsf{C}_{2\pi}$ in the *uniform* norm is unique for all choices of f if and only if $\{g_1, \dots, g_n\}$ forms a Chebychev system.

The unicity problem in the case of best approximation in $\mathsf{L}^1_{2\pi}$-*norm* is solved by an important theorem of JACKSON: If $\{g_1, \dots, g_n\}$ is a Chebychev system in $\mathsf{C}_{2\pi}$, then each $f \in \mathsf{C}_{2\pi}$ possesses a unique polynomial of best approximation in $\mathsf{L}^1_{2\pi}$-norm, see CHENEY [1, p. 218 ff]. Other references: HAVINSON [1], KRIPKE–RIVLIN [1], PTÁK [1] and RIVLIN–SHAPIRO [1]. For the unicity and characterization problem in the setting of arbitrary Banach spaces we refer to PHELPS [1], GARKAVI [1] and to the many papers and monograph of SINGER [1–4]; see also AUMANN [1] as well as BUTZER–GÖRLICH–SCHERER [1].

The theory of *nonlinear* approximation, in particular on the existence, characterization and unicity of nonlinear best approximations, has been the subject of mathematical investigations only during the last few years. Here we must mention the names of MOTZKIN [2], RICE [1II], MEINARDUS [1, pp. 131–188], BROSOWSKI [1], as well as the literature cited in the last text. However, the special case of *rational* approximation is a much older discipline, see e.g. CHENEY [1, Chapter 5].

Sec. 2.2. Theorem 2.2.1 (in $\mathsf{C}_{2\pi}$-space) for the ordinary modulus of continuity was proved by JACKSON [1] in 1911 in his dissertation (written under E. LANDAU at Göttingen). These investigations were stimulated by earlier work of LEBESGUE [1] and DE LA VALLÉE POUSSIN [1] in 1908. It seems that KOROVKIN [5] was the first to use the polynomials $K_n(f; x)$ explicitly in proving the first Jackson theorem (JACKSON himself had employed the polynomials $J_n(f; x)$). The proof of Theorem 2.2.3, the second Jackson theorem, is based upon an interesting nonstandard method of NATANSON [5] which fits in well with our approach based upon the theory of *convolution integrals*. The proof of (2.2.2) is taken from ZYGMUND [3; 7I, p. 117]. The fact that one can replace the ordinary modulus of continuity by the generalized one (cf. Theorem 2.2.1 and the estimate (2.2.2)) was first noted by ZYGMUND [2] in 1945. Best approximation in $L^p_{2\pi}$-space, including the Jackson theorems, was first studied systematically by QUADE [1] (see also the text of TIMAN [2]). For various improvements of the Jackson theorems, determination of the best possible constants occurring, and various connected results see e.g. the references in CHENEY [1, p. 230]. For more recent papers see SHISHA [1], ROULIER [1], DE VORE [1], LORENTZ–ZELLER [1].

Sec. 2.3. The theorems here (in $C_{2\pi}$-space) are from the doctoral dissertation (1912) of BERNSTEIN [1]. The proof of Bernstein's inequality (2.3.1) given is due to F. RIESZ [1]; for the refined estimate $\|t_n'\|_{X_{2\pi}} \le n\|t_n\|_{X_{2\pi}}$ compare NATANSON [8I, p. 90 f]. The method of proof of Theorems 2.3.3, 2.3.4 is particularly interesting. To appreciate BERNSTEIN's argument (cf. H. S. SHAPIRO [1, p. 78]) it is worth while to note what a straightforward approach would yield. If $E_n(C_{2\pi}; f) = O(n^{-\alpha})$, then the polynomials t_n^* (of best approximation to f) are uniformly bounded, and thus by Bernstein's inequality $\|t_n^{*\prime}\|_{C_{2\pi}} = O(n)$. Therefore it follows for $h > 0$ that

$$\|f(\circ + h) - f(\circ)\|_{C_{2\pi}} \le \|t_n^*(\circ + h) - t_n^*(\circ)\|_{C_{2\pi}} + 2E_n(C_{2\pi}; f) \le C(nh + n^{-\alpha}) = O(h^{\alpha/(1+\alpha)})$$

if we choose n around $h^{-1/(1+\alpha)}$. This shows $f \in \mathrm{Lip}\,(C_{2\pi}; \alpha/(1 + \alpha))$ rather than $\mathrm{Lip}(C_{2\pi}; \alpha)$. The reason for this imperfect result is that the estimate $\|t_n^{*\prime}\|_{C_{2\pi}} = O(n)$ is too crude. The extra information that $\{t_n^*\}$ converges with a certain speed must be exploited to produce a sharper estimate for $t_n^{*\prime}$. In fact, we obtained $\|t_n^{*\prime}\|_{C_{2\pi}} = O(n^{1-\alpha})$.

The Bernstein method has been adapted to other situations, too, in particular to various singular integrals, see Sec. 2.5, 3.5, 13.3. Theorem 2.3.5 is due to ZYGMUND [2] (1945). We owe to Zygmund the discovery of the fact that $f \in \mathrm{Lip}^*\,(X_{2\pi}; 1)$ if and only if $E_n(X_{2\pi}; f) = O(n^{-1})$. He was also the first to recognize the importance of the function class $\mathrm{Lip}^*\,(X_{2\pi}; 1)$ in various branches of Fourier analysis. The Bernstein theorems in $L^p_{2\pi}$-space were first proved by QUADE [1]; see also TIMAN [2, p. 334 f].

These results have been developed further by many authors. There is an elegant generalization by DE LA VALLÉE POUSSIN [3, pp. 53–58] (extended to higher moduli of continuity by BUTZER–NESSEL [1]) as well as an extension of a different type by S. B. STEČKIN which has given rise to a large number of investigations on the subject. See e.g. LORENTZ [3, pp. 58–63], TIMAN [2, pp. 344–49]. For generalizations of the theorems of Jackson and Bernstein as well as of certain results of M. ZAMANSKY and S. B. STEČKIN (and their converses) to arbitrary Banach spaces (in the setting of the theory of intermediate spaces) the reader is referred to BUTZER–SCHERER [1–4].

Sec. 2.4. The applications presented here are standard, see e.g. NATANSON [8], GOLOMB [1]. Theorem 2.4.3 is proved in LEBESGUE [2]. Condition $\lim_{\delta\to 0+} \omega^*(C_{2\pi}; f; \delta)\,|\log \delta| = 0$ is known as the generalized Dini–Lipschitz condition. For Theorem 2.4.8 see ROGOSINSKI [1, 2], BERNSTEIN [3] or NATANSON [8I, p. 217 ff]; for part (iii) of Theorem 2.4.9 see ZYGMUND [3]. Problem 2.4.11 is due to DE LA VALLÉE POUSSIN [3, p. 34]. Concerning Problem 2.4.14(ii), CHENEY–HOBBY–MORRIS–SCHURER–WULBERT [1, 2], CHENEY–PRICE [1] have recently shown that the Fourier series projection is the *only* minimal projection of $C_{2\pi}$ onto T_n. $\mathrm{Lip}(C_{2\pi}; 1)$ cannot be characterized in terms of $E_n(C_{2\pi}; f)$; see SCHERER [1].

Let R_n be the operator which assigns to each $f \in C_{2\pi}$ its polynomial of best approximation, thus $R_n f = t_n^*(C_{2\pi}; f)$. For each $n \in \mathbb{N}$, R_n is an operator of $C_{2\pi}$ onto T_n with the property that $R_n t_n = t_n$ for all $t_n \in T_n$. Hence $R_n^2 = R_n$. Moreover, $\lim_{n\to\infty} \|R_n f - f\|_{C_{2\pi}} = 0$ by Weierstrass' theorem. The operator R_n is in fact continuous but not linear; see, for example, CHENEY [1, p. 210]. It is therefore natural to ask whether a *linear* operator can exist having the above properties. In view of Sec. 1.7 this operator cannot be positive. Indeed, the result of HARŠILADSE–LOZINSKIĬ of Problems 2.4.15–2.4.16 shows that no such operator does exist. In $L^1_{2\pi}$-space the situation is quite similar. However, in $L^p_{2\pi}$-space, $1 < p < \infty$, there does exist a sequence of bounded linear operators of $L^p_{2\pi}$ onto T_n, namely the partial sums of the Fourier series of f, whose order of approximation to a given f is indeed that of the best approximation $E_n(L^p_{2\pi}; f)$; see Prop. 9.3.8.

Sec. 2.5. Concerning Theorem 2.5.2 see ANGHELUTZA [1], also SALEM–ZYGMUND [1], NATANSON [3], TABERSKI [1], or BUTZER–BERENS [1, p. 122]. Theorem 2.5.4 was first established for the Abel–Poisson integral on the unit n-sphere for continuous functions by DU PLESSIS [2]. In the form given the result was rediscovered by BERENS [1] in his doctoral thesis. The method of proof of the theorem is that of BERNSTEIN (cf. Theorem 2.3.3). Prop. 2.5.6 is due to ZYGMUND [3].

The integral (2.5.11) was introduced by DE LA VALLÉE POUSSIN [2] in 1908. Further papers on the subject are NATANSON [2], BUTZER [1], TABERSKI [1], MATSUOKA [1]; see also NATANSON [8I, p. 206 ff]. Prop. 2.5.8 is due to NATANSON [1]. For Lemma 2.5.9 see BERENS [3, p. 57]; the reader may note the important rôle played by the identity (2.5.19) in the proof. Once such identities (cf. the treatment in BUTZER–PAWELKE [1]) have been established for a general class of singular integrals, a complete approximation theory may be built up for such a class. One identity of this type is given by the semi-group property (cf. Sec. 13.4.2). The proof of Theorem 2.5.10 in this form does not seem to be given elsewhere; see also BERENS [3, p. 56 ff]. For Prop. 2.5.12 see BUTZER [1].

There is also an abstract approach to the material of this section. Indeed, results of the type given by Cor. 2.5.5, 2.5.11 may be established for a general class of approximation processes on Banach spaces. Such a process (cf. Def. 12.0.1) is defined by a family of commutative, bounded linear operators $\{T_\rho\}_{\rho>0}$ on a Banach space X to itself which approximate the identity strongly as $\rho \to \infty$, and satisfy a Jackson-type inequality

$$\|T_\rho f - f\|_X \le C_1 \rho^{-\sigma} \|f\|_Y \qquad (f \in Y) \tag{2.6.1}$$

as well as a Bernstein-type inequality

$$\|T_\rho f\|_Y \le C_2 \rho^{\sigma} \|f\|_X \qquad (f \in X) \tag{2.6.2}$$

for a suitable Banach subspace Y of X and exponent $\sigma > 0$. In particular, 'polynomial' operators may be studied. In this respect, we refer to BUTZER–SCHERER [2, 3, 5] as well as to their monograph [1, p. 73 ff]. These results are also given in the setting of the theory of intermediate spaces.

3

Singular Integrals on the Line Group

3.0 Introduction

The material of this chapter is in many points a straightforward adaptation of the periodic counterparts of Chapter 1. This is particularly true for Sec. 3.1 which corresponds to Sec. 1.1, 1.3, thus treating convergence in the norms of the spaces $X(\mathbb{R})$. Special emphasis is placed upon the study of singular integrals of Fejér's type. In Sec. 3.1.2 to each approximate identity on the real line a periodic approximate identity is associated via (3.1.28), (3.1.55), respectively. Important examples of singular integrals such as those of Fejér, Gauss–Weierstrass, and Cauchy–Poisson are introduced. Sec. 3.2 deals with pointwise convergence of convolution integrals, the results correspond to those of Sec. 1.4. Sec. 3.3 is concerned with nonperiodic counterparts of Sec. 1.5, thus with questions on the order of approximation by positive singular integrals on the real line. The method of test functions is touched upon and certain asymptotic expansions are given. Furthermore, Nikolskiĭ constants for periodic singular integrals of Fejér's type with respect to Lipschitz classes are determined; these complete the results of Sec. 1.6.3 in the fractional case. Sec. 3.4 deals with direct approximation theorems for singular integrals, the kernels of which need not necessarily be positive. In case the order of approximation is $O(\rho^{-\alpha})$, $0 < \alpha \leqslant 2$, the results correspond to those of Sec. 1.6. For applications of these concerning higher order approximation we refer to Sec. 6.4 where certain periodic counterparts are also formulated. In Sec. 3.5 inverse approximation theorems for singular integrals of Fejér's type are given. The proofs follow by a direct adaptation of Bernstein's idea, already familiar from Sec. 2.3, 2.5. In Sec. 3.6 some aspects concerning shape preserving properties of approximation processes are discussed. For a certain class of functions f it is shown that the approximation of f by the Gauss–Weierstrass integral is monotone if and only if f is convex. The concept of variation diminishing kernels is introduced. The main result here is that a kernel is variation diminishing if and only if it is totally positive. For counterparts of Sec. 1.2 concerning the classical theory of Fourier series the reader is referred to Chapter 5.

3.1 Norm-Convergence

3.1.1 Definitions and Fundamental Properties

Just as for periodic singular integrals in Sec. 1.1 we begin with

Definition 3.1.1. *Let ρ be a positive parameter tending to infinity. A set of functions $\{\chi(x; \rho)\}$ will be called a nonperiodic kernel or a **kernel** on the real line if $\chi(\circ; \rho) \in \mathsf{L}^1$ for each $\rho > 0$ and*

$$\int_{-\infty}^{\infty} \chi(u; \rho)\, du = \sqrt{2\pi} \qquad (\rho > 0). \tag{3.1.1}$$

*A kernel $\{\chi(x; \rho)\}$ will be said to be **real, bounded, continuous** or **absolutely continuous** if $\chi(x; \rho)$ is a real, bounded, continuous or absolutely continuous function of x for each $\rho > 0$. A real kernel $\{\chi(x; \rho)\}$ is **even** or **positive** if $\chi(x; \rho) = \chi(-x; \rho)$ or $\chi(x; \rho) \geq 0$ a.e. for each $\rho > 0$.*

We shall usually just speak of a kernel, whether it is periodic or not. The distinction will be apparent either from the context or the different notations $\{\chi_\rho(x)\}$ and $\{\chi(x; \rho)\}$. Instead of (3.1.1) sometimes the condition

$$\lim_{\rho \to \infty} \int_{-\infty}^{\infty} \chi(u; \rho)\, du = \sqrt{2\pi} \tag{3.1.2}$$

is used in the literature. But there is no loss of generality since we may always normalize a kernel by (3.1.1). The normalization (3.1.1), which is different to that of (1.1.1), is rather convenient as will be shown in Chapter 5.

Definition 3.1.2. *For $f \in \mathsf{X}(\mathbb{R})$ the convolution*

$$I(f; x; \rho) = \frac{1}{\sqrt{2\pi}} \int_{-\infty}^{\infty} f(x - u)\chi(u; \rho)\, du \tag{3.1.3}$$

*defines a **singular integral** generated by the kernel $\{\chi(x; \rho)\}$. The singular integral is said to be **positive** or **continuous** if the kernel is positive or continuous.*

As an immediate consequence of Prop. 0.2.2 we state

Proposition 3.1.3. *Let $f \in \mathsf{X}(\mathbb{R})$ and $\{\chi(x; \rho)\}$ be a kernel. For each $\rho > 0$, it follows that $I(f; \circ; \rho) \in \mathsf{X}(\mathbb{R})$ and*

$$\|I(f; \circ; \rho)\|_{\mathsf{X}(\mathbb{R})} \leq \|\chi(\circ; \rho)\|_1 \|f\|_{\mathsf{X}(\mathbb{R})}. \tag{3.1.4}$$

With $[I(\rho)f](x) \equiv I(f; x; \rho)$, the integral (3.1.3) defines a bounded linear transformation $I(\rho)$ of $\mathsf{X}(\mathbb{R})$ into itself for each $\rho > 0$.

In order to produce convergence of the integral $I(f; x; \rho)$ towards f as $\rho \to \infty$ we introduce the notion of an approximate identity.

Definition 3.1.4. *A kernel $\{\chi(x; \rho)\}$ is called an **approximate identity** (on the real line) if there is some constant $M > 0$ with*

$$\|\chi(\circ; \rho)\|_1 \leq M \qquad (\rho > 0), \tag{3.1.5}$$

$$\lim_{\rho\to\infty} \int_{\delta \le |u|} |\chi(u;\rho)|\,du = 0 \qquad (\delta > 0). \tag{3.1.6}$$

Apart from (3.1.6) we sometimes use

$$\lim_{\rho\to\infty} \Big[\sup_{\delta\le|u|} |\chi(u;\rho)|\Big] = 0 \qquad (\delta > 0). \tag{3.1.7}$$

In contrast to the periodic case (see (1.1.6) and (1.1.7)) condition (3.1.7) does not imply (3.1.6). Another new feature is that in most of the examples the dependence of the kernel upon the parameter $\rho > 0$ takes the special form $\chi(x;\rho) = \rho\chi(\rho x)$. For this case it is easy to see that

Lemma 3.1.5. *$\{\rho\chi(\rho x)\}$ defines an approximate identity for every $\chi \in \mathsf{NL}^1$.*

For such kernels, which are said to be *kernels of Fejér's type*, the singular integral (3.1.3) is denoted by

$$J(f;x;\rho) = \frac{\rho}{\sqrt{2\pi}} \int_{-\infty}^{\infty} f(x-u)\chi(\rho u)\,du \tag{3.1.8}$$

and called a *singular integral of Fejér's type.* Since every $\chi \in \mathsf{NL}^1$ generates a kernel (and even an approximate identity) via $\{\rho\chi(\rho x)\}$, we shall abbreviate the notation and also call χ itself a kernel.

Concerning convergence of singular integrals we have

Theorem 3.1.6. *If the kernel $\{\chi(x;\rho)\}$ of the singular integral* (3.1.3) *is an approximate identity, then*

$$\lim_{\rho\to\infty} \|I(f;\circ;\rho) - f(\circ)\|_{\mathsf{X}(\mathbb{R})} = 0 \tag{3.1.9}$$

for every $f \in \mathsf{X}(\mathbb{R})$. If $f \in \mathsf{L}^\infty$, then for each $h \in \mathsf{L}^1$

$$\lim_{\rho\to\infty} \int_{-\infty}^{\infty} [I(f;x;\rho) - f(x)]h(x)\,dx = 0. \tag{3.1.10}$$

The proof is similar to that of the corresponding results on periodic functions and left to the reader. The same is true for the following

Proposition 3.1.7. *Let the kernel $\{\chi(x;\rho)\}$ of the singular integral* (3.1.3) *satisfy* (3.1.5). *If $\lim_{\rho\to\infty} \|I(h;\circ;\rho) - h(\circ)\|_{\mathsf{X}(\mathbb{R})} = 0$ for all elements h of a set $\mathsf{A} \subset \mathsf{X}(\mathbb{R})$ which is dense in $\mathsf{X}(\mathbb{R})$, then for every $f \in \mathsf{X}(\mathbb{R})$*

$$\lim_{\rho\to\infty} \|I(f;\circ;\rho) - f(\circ)\|_{\mathsf{X}(\mathbb{R})} = 0. \tag{3.1.11}$$

The problem now is to find a suitable dense subset A of $\mathsf{X}(\mathbb{R})$. To this end we have

Proposition 3.1.8. *For each $\alpha > 0$ the functions of the form $p(x)\exp\{-\alpha x^2\}$, where $p(x)$ is any algebraic polynomial, are dense in the spaces C_0 and L^p, $1 \le p < \infty$.*

For a proof one may consult Problems 3.1.18–3.1.20.

In order to specialize the last proposition to a more convenient form we introduce the *Hermite polynomials*

(3.1.12) $$h_n(x) = (-1)^n e^{x^2} \frac{d^n}{dx^n} \{e^{-x^2}\} \qquad (n \in \mathbb{P})$$

and the *Hermite functions*

(3.1.13) $$H_n(x) = h_n(x)\, e^{-x^2/2} \qquad (n \in \mathbb{P}),$$

the elementary properties of which are left to Problem 3.1.3. In terms of the Hermite functions, Prop. 3.1.8 now reads (see Problem 3.1.3(iv))

Corollary 3.1.9. *The Hermite functions form a **fundamental** set in the spaces C_0 and L^p, $1 \le p < \infty$, i.e., the linear manifold generated by the Hermite functions is dense in these spaces.*

3.1.2 Singular Integral of Fejér

Let us consider the *singular integral of Fejér*

(3.1.14) $$\sigma(f; x; \rho) = \frac{2}{\pi\rho} \int_{-\infty}^{\infty} f(x-u) \frac{\sin^2(\rho u/2)}{u^2}\, du$$

with parameter $\rho > 0$. Putting

(3.1.15) $$F(x) = \frac{1}{\sqrt{2\pi}} \left[\frac{\sin(x/2)}{x/2}\right]^2,$$

F is an even, positive function belonging to $\mathsf{C}_0 \cap \mathsf{L}^1$, thus to L^p for every $1 \le p \le \infty$. We may write the integral (3.1.14) as

(3.1.16) $$\sigma(f; x; \rho) = \frac{\rho}{\sqrt{2\pi}} \int_{-\infty}^{\infty} f(x-u) F(\rho u)\, du$$

which thus defines a singular integral of type (3.1.8). Indeed, F belongs to NL^1 since $\int_{-\infty}^{\infty} u^{-2} \sin^2 u\, du = \pi$ (compare Problem 5.2.8). According to Prop. 0.2.1, 3.1.3, Lemma 3.1.5, and Theorem 3.1.6 we have

Corollary 3.1.10. *For $f \in \mathsf{X}(\mathbb{R})$ the singular integral of Fejér exists for all $x \in \mathbb{R}$ and $\rho > 0$ and defines a function in $\mathsf{X}(\mathbb{R}) \cap \mathsf{C}$ satisfying*

(3.1.17) $$\|\sigma(f; \circ; \rho)\|_{\mathsf{X}(\mathbb{R})} \le \|f\|_{\mathsf{X}(\mathbb{R})} \qquad (\rho > 0),$$

(3.1.18) $$\lim_{\rho \to \infty} \|\sigma(f; \circ; \rho) - f(\circ)\|_{\mathsf{X}(\mathbb{R})} = 0.$$

There is a close connection between the integrals of Fejér introduced in (1.2.25) and (3.1.14). In fact, a well-known result of the theory of meromorphic functions states that the series

(3.1.19) $$\frac{1}{\sin^2 z} = \sum_{k=-\infty}^{\infty} \frac{1}{(z+k\pi)^2}$$

converges absolutely and uniformly on every compact set of the complex plane which does not contain any of the points $k\pi$, $k \in \mathbb{Z}$ (see Problem 3.1.13). It follows that

(3.1.20) $$\frac{\sin^2(n+1)x}{\sin^2 x} = \sum_{k=-\infty}^{\infty} \frac{\sin^2(n+1)x}{(x+k\pi)^2} \qquad (n \in \mathbb{P})$$

converges absolutely and uniformly on the real line. Therefore, beginning with the integral (1.2.25) for some $f \in C_{2\pi}$, we obtain by the periodicity of f

$$\begin{aligned}\sigma_n(f; x) &= \frac{1}{2\pi(n+1)} \int_{-\pi}^{\pi} f(x-u) \sum_{k=-\infty}^{\infty} \frac{\sin^2((n+1)u/2)}{((u/2)+k\pi)^2}\, du \\ &= \frac{1}{\pi(n+1)} \sum_{k=-\infty}^{\infty} \int_{(2k-1)\pi/2}^{(2k+1)\pi/2} f(x-2u) \frac{\sin^2(n+1)u}{u^2}\, du \\ &= \frac{2}{\pi(n+1)} \int_{-\infty}^{\infty} f(x-u) \frac{\sin^2((n+1)u/2)}{u^2}\, du,\end{aligned}$$

the last integral converging since f is bounded. Thus in terms of (3.1.14)

$$\sigma_n(f; x) = \sigma(f; x; n+1) \qquad (n \in \mathbb{P}). \tag{3.1.21}$$

This relation shows that the periodic integral (1.2.25) of Fejér can be classified under the integral $\sigma(f; x; \rho)$ of (3.1.14), the continuous parameter $\rho > 0$ being replaced by the discrete $(n+1)$. But in the form (3.1.14), the periodicity of f does not enter explicitly, and so, in studying the convergence of $\sigma(f; x; \rho)$ towards f as $\rho \to \infty$, the version (3.1.14) will also give a result for nonperiodic functions as well. Furthermore, (3.1.14) has the technical advantage that the analytical dependence on the parameter, in contrast to (1.2.25), is a very simple one. Indeed, the kernel of the integral (3.1.14) is generated by the function $F(x)$ of one variable through the simple scale change $\rho F(\rho x)$.

In the following we shall see that there are many situations in which a possible 'real-line-form' of a given periodic singular integral is more convenient. It is therefore very useful to know that with every kernel on the real line one may associate a periodic kernel.

To show this, let $g \in L^1$. Setting

$$g^*(x) = \sqrt{2\pi} \sum_{k=-\infty}^{\infty} g(x + 2k\pi), \tag{3.1.22}$$

it follows that g^* belongs to $L^1_{2\pi}$. Indeed, by Prop. 0.3.3

$$\int_{-\pi}^{\pi} \sum_{k=-\infty}^{\infty} |g(u+2k\pi)|\, du = \sum_{k=-\infty}^{\infty} \int_{-\pi}^{\pi} |g(u+2k\pi)|\, du = \int_{-\infty}^{\infty} |g(u)|\, du, \tag{3.1.23}$$

and since the last term is finite, the series (3.1.22) converges absolutely for almost all $x \in (-\pi, \pi)$. Thus the resulting sum $g^*(x)$ exists almost everywhere, and since it is independent of the order of the terms of the series, it is periodic. Moreover, it follows by (3.1.23) that

$$\text{(i)}\ \|g^*\|_{L^1_{2\pi}} \le \|g\|_{L^1}, \qquad \text{(ii)}\ \frac{1}{2\pi} \int_{-\pi}^{\pi} g^*(u)\, du = \frac{1}{\sqrt{2\pi}} \int_{-\infty}^{\infty} g(u)\, du. \tag{3.1.24}$$

Furthermore, if $f \in X_{2\pi}$, then

$$\frac{1}{2\pi} \int_{-\pi}^{\pi} f(x-u) g^*(u)\, du = \frac{1}{\sqrt{2\pi}} \int_{-\infty}^{\infty} f(x-u) g(u)\, du \quad \text{(a.e.)}. \tag{3.1.25}$$

Apart from these facts we shall need the following

Proposition 3.1.11. *Let $g \in \mathsf{L}^1 \cap \mathsf{BV}$. Then g^* is a 2π-periodic function of BV_{loc}, the series (3.1.22) being absolutely and uniformly convergent. Furthermore, (3.1.25) holds for all x, and thus in particular for every $f \in \mathsf{X}_{2\pi}$*

$$\frac{1}{2\pi}\int_{-\pi}^{\pi} f(u)g^*(u)\,du = \frac{1}{\sqrt{2\pi}}\int_{-\infty}^{\infty} f(u)g(u)\,du. \tag{3.1.26}$$

Proof. Setting

$$g_k(x) = \frac{1}{2\pi}\int_{x+2k\pi}^{x+2(k+1)\pi} g(u)\,du,$$

we have for $n < m$

$$\sum_{k=n}^{m} |g_k(x)| \le \frac{1}{2\pi}\int_{x+2n\pi}^{x+2(m+1)\pi} |g(u)|\,du.$$

Since $g \in \mathsf{L}^1$, the series $\sum_{k=-\infty}^{\infty} |g_k(x)|$ converges uniformly on every finite interval. Furthermore

$$g(x+2k\pi) - g_k(x) = \frac{1}{2\pi}\int_{x+2k\pi}^{x+2(k+1)\pi} \{g(x+2k\pi) - g(u)\}\,du,$$

and thus

$$|g(x+2k\pi)| \le |g_k(x)| + \varepsilon_k(x),$$

where $\varepsilon_k(x)$ is the total variation of g over $[x+2k\pi, x+2(k+1)\pi]$. Since $g \in \mathsf{BV}$, the series $\sum_{k=-\infty}^{\infty} \varepsilon_k(x)$ converges uniformly on every finite interval. Therefore $\sqrt{2\pi}\sum_{k=-\infty}^{\infty} g(x+2k\pi)$ converges absolutely and uniformly on every finite interval. Since its sum $g^*(x)$ is already known to be 2π-periodic, it follows that the series (3.1.22) is absolutely and uniformly convergent on the whole real line. Moreover, for any finite partition $-\pi = x_0 < x_1 < x_2 < \ldots < x_n = \pi$ of the interval $[-\pi, \pi]$

$$\sum_{j=0}^{n-1} |g^*(x_{j+1}) - g^*(x_j)| \le \sqrt{2\pi}\sum_{k=-\infty}^{\infty}\Big(\sum_{j=0}^{n-1} |g(x_{j+1}+2k\pi) - g(x_j+2k\pi)|\Big)$$

$$\le \sqrt{2\pi}\sum_{k=-\infty}^{\infty} [\operatorname{Var} g]_{(2k-1)\pi}^{(2k+1)\pi} = 2\pi\,\|g\|_{\mathsf{BV}}.$$

Thus g^* is of bounded variation on $[-\pi, \pi]$, in fact $g^* \in \mathsf{BV}_{\text{loc}}$ and

$$\|g^*\|_{\mathsf{BV}_{2\pi}} \le \|g\|_{\mathsf{BV}}. \tag{3.1.27}$$

In particular, g^* is a bounded function. Therefore the convolution $(f * g^*)(x)$ exists everywhere by Prop. 0.4.1, and (3.1.25) holds for all x in view of the uniform convergence of the series (3.1.22).

We proceed with the case that g is a kernel of Fejér's type (for general kernels see Problem 3.1.11). Thus, let $\chi \in \mathsf{NL}^1$ and set

$$\chi_\rho^*(x) = \sqrt{2\pi}\sum_{k=-\infty}^{\infty} \rho\chi(\rho(x+2k\pi)) \qquad (\rho > 0). \tag{3.1.28}$$

Then χ_ρ^* belongs to $\mathsf{L}_{2\pi}^1$ and

$$\text{(i)}\ \|\chi_\rho^*\|_{\mathsf{L}_{2\pi}^1} \le \|\chi\|_{\mathsf{L}^1}, \qquad \text{(ii)}\ \frac{1}{2\pi}\int_{-\pi}^{\pi} \chi_\rho^*(u)\,du = 1 \tag{3.1.29}$$

for every $\rho > 0$. As an immediate consequence we note that for every $\rho > 0$ and $h \in \mathbb{R}$

$$\|\chi_\rho^*(\circ + h) - \chi_\rho^*(\circ)\|_{\mathsf{L}_{2\pi}^1} \le \|\chi(\circ + h) - \chi(\circ)\|_{\mathsf{L}^1}.$$

This implies that the functions χ_ρ^* are continuous in $L^1_{2\pi}$-norm, uniformly with respect to $\rho > 0$. Moreover, for any (fixed) δ with $0 < \delta < \pi$

$$\frac{1}{2\pi}\int_{\delta\le|u|\le\pi}|\chi_\rho^*(u)|\,du \le \frac{1}{\sqrt{2\pi}}\int_{\delta\le|u|}|\rho\chi(\rho u)|\,du = \frac{1}{\sqrt{2\pi}}\int_{\rho\delta\le|u|}|\chi(u)|\,du,$$

which tends to zero as $\rho\to\infty$ since $\chi\in L^1$. Therefore

Proposition 3.1.12. *If $\chi\in NL^1$, then $\{\chi_\rho^*(x)\}$ as defined by (3.1.28) is a periodic approximate identity with parameter set $\mathbb{A} = (0,\infty)$ and $\rho_0 = \infty$. The functions χ_ρ^* are continuous in $L^1_{2\pi}$-norm, uniformly with respect to $\rho > 0$.*

In this case, we say that the periodic approximate identity $\{\chi_\rho^*(x)\}$ is *generated* via (3.1.28) by χ. Concerning (3.1.25), if we set

$$I_\rho^*(f;x) = \frac{1}{2\pi}\int_{-\pi}^{\pi} f(x-u)\chi_\rho^*(u)\,du, \tag{3.1.30}$$

and if $J(f;x;\rho)$ is given by (3.1.8), then for every $f\in X_{2\pi}$ and $\rho > 0$

$$I_\rho^*(f;x) = J(f;x;\rho) \quad \text{(a.e.)}. \tag{3.1.31}$$

The foregoing considerations concerning the two Fejér kernels, in particular (3.1.21), may serve as a first illustration to the above. They show that the dependence of the periodic kernel $\{\chi_\rho^*(x)\}$ upon the parameter ρ may be badly arranged though the corresponding nonperiodic kernel is of Fejér's type. On the other hand, it is actually this aspect which makes the change from a periodic kernel to its nonperiodic version so useful (see (5.1.58)). In the following, the periodic kernel $\{\chi_\rho^*(x)\}$ as given through (3.1.28) and the corresponding singular integral (3.1.30) are also said to be of *Fejér's type.*

3.1.3 Singular Integral of Gauss–Weierstrass

We continue with the *singular integral of Gauss–Weierstrass* associated with the function $f\in X(\mathbb{R})$ and defined by

$$W(f;x;t) = \frac{1}{\sqrt{4\pi t}}\int_{-\infty}^{\infty} f(x-u)\,e^{-u^2/4t}\,du \tag{3.1.32}$$

with parameter $t\to 0+$. (3.1.32) is of Fejér's type with

$$\chi(x) = 2^{-1/2}\,e^{-x^2/4} \equiv w(x); \qquad \rho = t^{-1/2}. \tag{3.1.33}$$

Indeed, $w(x)$ is an even, positive function belonging to $C_0\cap L^1$, and in fact $w\in NL^1$ since $\int_{-\infty}^{\infty}\exp\{-u^2\}\,du = \sqrt{\pi}$. Therefore we have

Corollary 3.1.13. *For $f\in X(\mathbb{R})$ the singular integral of Gauss–Weierstrass exists for all $x\in\mathbb{R}$ and $t > 0$ and defines a function in $X(\mathbb{R})\cap C$ satisfying*

$$\|W(f;\circ;t)\|_{X(\mathbb{R})} \le \|f\|_{X(\mathbb{R})} \qquad (t > 0), \tag{3.1.34}$$

$$\lim_{t\to 0+}\|W(f;\circ;t) - f(\circ)\|_{X(\mathbb{R})} = 0. \tag{3.1.35}$$

The periodic kernel of Weierstrass (see Problem 1.3.10) and the nonperiodic kernel of Gauss–Weierstrass are also related via (3.1.28). Indeed

$$\sum_{k=-\infty}^{\infty} e^{-|k|^2 t}\cos kx = \sqrt{\frac{\pi}{t}} \sum_{k=-\infty}^{\infty} e^{-(x+2k\pi)^2/4t} \qquad (t>0), \tag{3.1.36}$$

which will be established in Sec. 5.1.5.

Let us mention another important aspect of the singular integral of Gauss–Weierstrass. According to Problem 3.1.5(iv), $W(f; x; t) \in \mathsf{C}^\infty$ for every $f \in \mathsf{X}(\mathbb{R})$ and $t > 0$. Thus, in view of (3.1.35), the set of functions $\{W(f; x; t)\}$ forms a well-behaved, dense subset of $\mathsf{X}(\mathbb{R})$. Moreover, as is easily checked, the integral (3.1.32) is a solution of the initial value problem of the heat equation for an infinite rod, namely of

$$\begin{aligned} &\frac{\partial u(x,t)}{\partial t} = \frac{\partial^2 u(x,t)}{\partial x^2} \qquad (x \in \mathbb{R},\ t>0), \\ &\lim_{t\to 0+} \|u(\circ, t) - f(\circ)\|_{\mathsf{X}(\mathbb{R})} = 0, \end{aligned} \tag{3.1.37}$$

$f \in \mathsf{X}(\mathbb{R})$ being the prescribed initial temperature distribution. For a more detailed discussion we refer to Sec. 7.2.1.

3.1.4 Singular Integral of Cauchy–Poisson

As a final example in this section we investigate the *singular integral of Cauchy–Poisson* corresponding to the function $f \in \mathsf{X}(\mathbb{R})$, namely

$$P(f; x; y) = \frac{y}{\pi}\int_{-\infty}^{\infty} \frac{f(x-u)}{y^2+u^2}\,du \tag{3.1.38}$$

with parameter $y \to 0+$. This is again a particular case of a singular integral of Fejér's type with

$$\chi(x) = \sqrt{2/\pi}(1+x^2)^{-1} \equiv p(x), \quad \rho = y^{-1}. \tag{3.1.39}$$

Certainly, $\{\rho p(\rho x)\}$ is an even, positive, and continuous kernel, and we have

Corollary 3.1.14. *For $f \in \mathsf{X}(\mathbb{R})$ the singular integral of Cauchy–Poisson exists for all $x \in \mathbb{R}$ and $y > 0$ and defines a function in $\mathsf{X}(\mathbb{R}) \cap \mathsf{C}^\infty$ satisfying*

$$\|P(f; \circ; y)\|_{\mathsf{X}(\mathbb{R})} \le \|f\|_{\mathsf{X}(\mathbb{R})} \qquad (y>0), \tag{3.1.40}$$

$$\lim_{y\to 0+} \|P(f; \circ; y) - f(\circ)\|_{\mathsf{X}(\mathbb{R})} = 0. \tag{3.1.41}$$

Similarly as for the singular integral of Gauss–Weierstrass the integral (3.1.38) also plays an important rôle in the theory of partial differential equations. As may easily be verified, it is a solution of Dirichlet's problem for the upper half-plane with prescribed boundary value $f \in \mathsf{X}(\mathbb{R})$, thus of

$$\begin{aligned} &\frac{\partial^2 u(x,y)}{\partial x^2} + \frac{\partial^2 u(x,y)}{\partial y^2} = 0 \qquad (x \in \mathbb{R},\ y>0), \\ &\lim_{y\to 0+} \|u(\circ, y) - f(\circ)\|_{\mathsf{X}(\mathbb{R})} = 0. \end{aligned} \tag{3.1.42}$$

For further details see Sec. 7.2.2.

There is a close connection between the singular integral (1.2.36) of Abel–Poisson, which is a solution of Dirichlet's problem for the unit disc, and the singular integral (3.1.38) of Cauchy–Poisson. Indeed, setting $r = e^{-y}$, the kernels $\{p_r(x)\}$ of (1.2.37) and $\{\rho p(\rho x)\}$ of (3.1.39) are again related via (3.1.28) according to (see Sec. 5.1.5)

$$\sum_{k=-\infty}^{\infty} e^{-|k|y} \cos kx = 2y \sum_{k=-\infty}^{\infty} \frac{1}{y^2 + (x + 2k\pi)^2} \qquad (y > 0). \tag{3.1.43}$$

As we have seen in Sec. 1.2, the singular integral (1.2.36) of Abel–Poisson for real-valued f may be regarded as the real part of a function $F(z)$ holomorphic in the unit disc. The analog holds for the singular integral $P(f; x; y)$ of Cauchy–Poisson, too. Indeed, for any real-valued $f \in \mathsf{X}(\mathbb{R})$ the function

$$H(z) = \frac{i}{\pi} \int_{-\infty}^{\infty} \frac{f(u)}{z - u} \, du \qquad (z = x + iy) \tag{3.1.44}$$

is holomorphic in the upper half-plane $\mathbb{R}^{2,+} \equiv \{(x, y) \mid x \in \mathbb{R}, y > 0\}$ and

$$H(z) = P(f; x; y) + iQ(f; x; y) \qquad (x \in \mathbb{R}, y > 0), \tag{3.1.45}$$

where $P(f; x; y)$ is the integral (3.1.38) and $Q(f; x; y)$ is given by

$$Q(f; x; y) = \frac{1}{\pi} \int_{-\infty}^{\infty} f(x - u) \frac{u}{y^2 + u^2} \, du \qquad (x \in \mathbb{R}, y > 0). \tag{3.1.46}$$

As in the periodic situation of (1.2.52) we call $Q(f; x; y)$, also denoted by $P^{\sim}(f; x; y)$, the *conjugate* of the singular integral of Cauchy–Poisson. If we set

$$p^{\sim}(x) = \sqrt{2/\pi}\, x(1 + x^2)^{-1}, \tag{3.1.47}$$

then $p^{\sim}(x)$ is called the *conjugate function* (or Hilbert transform) of $p(x)$. $\{\rho p^{\sim}(\rho x)\}$ is not a kernel in the sense of Def. 3.1.1 since $p^{\sim} \notin \mathsf{L}^1$. For further information about $Q(f; x; y)$ we refer to Sec. 8.1, 8.2. There the definition (cf. (8.2.10)) of the conjugate $I^{\sim}(f; x; \rho)$ of a general singular integral $I(f; x; \rho)$ will be given, followed by a detailed discussion of its properties.

Problems

1. (i) Prove Theorem 3.1.6 for approximate identities satisfying the weaker condition (3.1.2) instead of (3.1.1).
   (ii) If the kernel is of Fejér's type, prove Theorem 3.1.6 with the aid of Lebesgue's dominated convergence theorem. (Hint: Use the estimate
   $$\|J(f; \circ; \rho) - f(\circ)\|_{\mathsf{X}(\mathbb{R})} \le (2\pi)^{-1/2} \int_{-\infty}^{\infty} \|f(\circ - \rho^{-1}u) - f(\circ)\|_{\mathsf{X}(\mathbb{R})} |\chi(u)| \, du,$$
   compare with the proof of Prop. 3.2.1; see also Rudin [4, p. 186])
   (iii) Let $\{f_n\}_{n=1}^{\infty}$ be a sequence of functions in $\mathsf{X}(\mathbb{R})$ which are bounded and continuous in $\mathsf{X}(\mathbb{R})$-norm, uniformly with respect to $n \in \mathbb{N}$. If $\{\chi(x; \rho)\}$ is an approximate identity, show that $\lim_{\rho\to\infty} \|I(f_n; \circ; \rho) - f_n(\circ)\|_{\mathsf{X}(\mathbb{R})} = 0$ uniformly for $n \in \mathbb{N}$.
   (iv) Let f be bounded and continuous, but not necessarily uniformly continuous. If $\{\chi(x; \rho)\}$ is an approximate identity, show that $\lim_{\rho\to\infty} I(f; x; \rho) = f(x)$ uniformly on each bounded interval.

2. (i) Let $\zeta(x)$ be defined through $\zeta(x) = \exp\{1/(x^2 - 1)\}$ for $|x| < 1$, $= 0$ for $|x| \geq 1$. Show that ζ belongs to C_{00}^{∞}.
   (ii) Show that C_{00}^{∞} is dense in C_{00}, that is, every function in C_{00} can be approximated uniformly by functions in C_{00}^{∞}. (Hint: For any $f \in \mathsf{C}_{00}$ consider the singular integral $J(f; x; \rho)$ of Fejér's type with (positive) kernel $\chi(x) = \zeta(x)/\|\zeta\|_1$, ζ being given as in (i). Show that $J(f; x; \rho) \in \mathsf{C}_{00}^{\infty}$ for each $\rho > 0$ and $\lim_{\rho\to\infty} J(f; x; \rho) = f(x)$ uniformly for all x; see also ZEMANIAN [1, p. 3])
   (iii) Show that C_{00}^{∞} is dense in C_0 and L^p, $1 \leq p < \infty$. (Hint: Use (ii) and Prop. 0.3.4)
   (iv) Let $f \in \mathsf{X}(\mathbb{R})$ and suppose $\int_{-\infty}^{\infty} f(u)\phi(u)\,du = 0$ for every $\phi \in \mathsf{C}_{00}^{\infty}$. Show that $\|f\|_{\mathsf{X}(\mathbb{R})} = 0$, i.e., $f(x) = 0$ (a.e.). (Hint: By (ii) one has $\int_{-\infty}^{\infty} f(u)\phi(u)\,du = 0$ for every $\phi \in \mathsf{C}_{00}$. By Prop. 0.3.4, $\int_{-\infty}^{\infty} f(u)\phi(u)\,du = 0$ for every $\phi \in \mathsf{L}^1$ if $\mathsf{X}(\mathbb{R}) = \mathsf{C}$, for every $\phi \in \mathsf{C}_0$ if $\mathsf{X}(\mathbb{R}) = \mathsf{L}^1$, and for every $\phi \in \mathsf{L}^{p'}$ if $\mathsf{X}(\mathbb{R}) = \mathsf{L}^p$, $1 < p < \infty$. Now use (3.1.35), for example)
3. Let $h_n(x)$ be the nth Hermite polynomial (3.1.12) and $H_n(x)$ the nth Hermite function (3.1.13).
   (i) Show that $h_n(x) = 2xh_{n-1}(x) - h'_{n-1}(x)$ for each $n \in \mathbb{N}$. As obviously $h_0(x) = 1$, it therefore follows that h_n is an algebraic polynomial of degree n of the form $h_n(x) = 2^n x^n + \sum_{k=0}^{n-1} a_k x^k$. In particular, $h_1(x) = 2x$, $h_2(x) = 4x^2 - 2, \dots$
   (ii) Show that for each $n \in \mathbb{N}$
   $$h'_n(x) = 2nh_{n-1}(x), \qquad h_{n+1}(x) - 2xh_n(x) + 2nh_{n-1}(x) = 0,$$
   $$h''_n(x) - 2xh'_n(x) + 2nh_n(x) = 0.$$
   (iii) Show that $H''_n(x) = (x^2 - 2n - 1)H_n(x)$ for each $n \in \mathbb{P}$.
   (iv) Show that the Hermite functions form an *orthogonal* sequence on $\mathbb{R}$, i.e., $\int_{-\infty}^{\infty} H_n(x)H_m(x)\,dx = 0$ for $n \neq m$. (Hint: Use (iii) and partial integration) The Hermite functions and thus the Hermite polynomials are linearly independent.
   (v) Show that $\int_{-\infty}^{\infty} H_n^2(x)\,dx = 2n\int_{-\infty}^{\infty} H_{n-1}^2(x)\,dx$. (Hint: Partial integration and the first relation of (ii).) Therefore $\int_{-\infty}^{\infty} H_n^2(x)\,dx = \sqrt{\pi}2^n n!$.
   (vi) Show that $H_{n+1}(x) = xH_n(x) - H'_n(x)$ for each $n \in \mathbb{P}$. This recursion formula, together with $H_0(x) = e^{-x^2/2}$, determines the sequence of Hermite functions uniquely.

   (Hint: For these elementary properties of Hermite polynomials and functions one may consult SZ.–NAGY [5, p. 334 ff], HEWITT–STROMBERG [1, p. 243 ff])

The following three Problems deal with counterparts to Sec. 1.1.2. Thus, if r is a natural number, we introduce the classes

$$\text{(3.1.48)} \qquad \mathsf{W}_{\mathsf{X}(\mathbb{R})}^r = \begin{cases} \{f \in \mathsf{C} \mid f \in \mathsf{C}^r\} \\ \{f \in \mathsf{L}^p \mid f = \phi \text{ a.e., } \phi \in \mathsf{AC}_{\text{loc}}^{r-1}, \phi^{(k)} \in \mathsf{L}^p, 0 \leq k \leq r\} \quad (1 \leq p < \infty). \end{cases}$$

For $p = 1$ this is equivalently expressed by $\mathsf{W}_{\mathsf{L}^1}^r = \{f \in \mathsf{L}^1 \mid f = \phi \text{ a.e., } \phi \in \mathsf{AC}^{r-1}\}$. Analogously, the definitions of an rth *strong derivative* $D_s^{(r)}f$ and an rth *weak derivative* $D_w^{(r)}f$ of $f \in \mathsf{X}(\mathbb{R})$ are quite obvious: In Def. 1.1.9 and 1.1.11 one only has to replace $\mathsf{X}_{2\pi}$ by $\mathsf{X}(\mathbb{R})$, $\mathsf{BV}_{2\pi}$ by BV, and $\mathsf{L}_{2\pi}^{p'}$ by $\mathsf{L}^{p'}$. Sometimes, we consider functions χ belonging to $\mathsf{W}_{\mathsf{C}}^r \cap \mathsf{W}_{\mathsf{L}^1}^r$; in this case, we abbreviate the notation and write $\chi \in \mathsf{W}_{\mathsf{C}\cap\mathsf{L}^1}^r$.

4. (i) Show that $f \in \mathsf{W}_{\mathsf{X}(\mathbb{R})}^r$ implies the existence of the rth strong derivative $D_s^{(r)}f$ of f, which in turn implies the existence of the rth weak derivative $D_w^{(r)}f$ of f. If $\mathsf{X}(\mathbb{R}) = \mathsf{C}$, then $f^{(r)}(x) = (D_s^{(r)}f)(x) = (D_w^{(r)}f)(x)$ for all $x \in \mathbb{R}$; if $\mathsf{X}(\mathbb{R}) = \mathsf{L}^p$, $1 \leq p < \infty$, then $\phi^{(r)}(x) = (D_s^{(r)}f)(x) = (D_w^{(r)}f)(x)$ a.e., $\phi \in \mathsf{AC}_{\text{loc}}^{r-1}$ with $\phi^{(k)} \in \mathsf{L}^p$, $0 \leq k \leq r$, being such that $f(x) = \phi(x)$ a.e. (for the converse of the above statements the reader is referred to Prop. 10.5.3, Problem 10.6.16).
   (ii) Show that $\chi \in \mathsf{W}_{\mathsf{C}\cap\mathsf{L}^1}^r$ implies $\chi \in \mathsf{W}_{\mathsf{L}^p}^r$ for every $p > 1$.

5. (i) Let $f \in X(\mathbb{R})$ and $\chi \in NL^1$ be the kernel of the singular integral $J(f; x; \rho)$. Show that if $\chi \in W^r_{C \cap L^1}$, then $J(f; x; \rho)$ is an r-times continuously differentiable function of x. In particular, for every $x \in \mathbb{R}$ and $\rho > 0$

$$(3.1.49) \qquad [J(f; \circ; \rho)]^{(r)}(x) = \frac{\rho^{r+1}}{\sqrt{2\pi}} \int_{-\infty}^{\infty} f(x-u)\chi^{(r)}(\rho u)\, du,$$

thus $J(f; x; \rho) \in W^r_C$ for every $f \in X(\mathbb{R})$ and $\rho > 0$.

(ii) Let the kernel $\chi \in NL^1$ of the singular integral $J(f; x; \rho)$ be bounded. If $f \in W^r_{X(\mathbb{R})}$, show that $J(f; x; \rho)$ is an r-times continuously differentiable function of x, and for every $x \in \mathbb{R}$ and $\rho > 0$

$$(3.1.50) \qquad [J(f; \circ; \rho)]^{(r)}(x) = J(D_s^{(r)}f; x; \rho) = J(\phi^{(r)}; x; \rho),$$

where $\phi \in AC^{r-1}_{loc}$ with $\phi^{(k)} \in X(\mathbb{R})$, $0 \le k \le r$, is such that $\phi(x) = f(x)$ (a.e.). Moreover,

$$(3.1.51) \qquad \lim_{\rho \to \infty} \|[J(f; \circ; \rho)]^{(r)}(\circ) - (D_s^{(r)}f)(\circ)\|_{X(\mathbb{R})} = 0.$$

(iii) State and prove counterparts for the (general) singular integral $I(f; x; \rho)$ with kernel $\{\chi(x; \rho)\}$.

(iv) As an application of (i) show that the Weierstrass integral $W(f; x; t)$ belongs to C^∞ for every $f \in X(\mathbb{R})$ and $t > 0$. Give applications to further examples.

6. Let $f \in W^r_{X(\mathbb{R})}$ and $g \in L^1$. Show that the rth strong derivative (in $X(\mathbb{R})$) of $f * g$ exists and that $D_s^{(r)}[f * g] = (D_s^{(r)}f * g)$. (Prop. 10.5.3, Problem 10.6.16 indeed give $f * g \in W^r_{X(\mathbb{R})}$, in other words, if $f \in W^r_{X(\mathbb{R})}$ and $g \in L^1$, then also $f * g \in W^r_{X(\mathbb{R})}$.) State and prove further counterparts to Problems 1.1.8, 1.1.9.

7. The *first integral means* (or moving averages) of $f \in X(\mathbb{R})$ are defined as for periodic functions in (1.1.15). However, in accordance with our different notation for periodic and nonperiodic singular integrals we now employ the notation $A(f; x; h)$, thus

$$A(f; x; h) = h^{-1} \int_{x-(h/2)}^{x+(h/2)} f(u)\, du = h^{-1} \int_{-h/2}^{h/2} f(x-u)\, du.$$

Show that $A(f; x; h)$ is a singular integral of Fejér's type with even and positive kernel $\chi(x) = \sqrt{2\pi}\kappa_{[-1/2,1/2]}(x)$ and parameter $\rho = h^{-1}$, $h \to 0+$. Hence for every $f \in X(\mathbb{R})$

$$\lim_{h \to 0+} \|A(f; \circ; h) - f(\circ)\|_{X(\mathbb{R})} = 0.$$

For further properties see the comments connected with (1.1.15).

8. (i) Let $\{\chi(x; \rho)\}$ be a kernel and the *iterated singular integral* $I^2(f; x; \rho)$ of $f \in X(\mathbb{R})$ be defined by

$$I^2(f; x; \rho) = \frac{1}{\sqrt{2\pi}} \int_{-\infty}^{\infty} f(x-u)[\chi(\circ; \rho) * \chi(\circ; \rho)](u)\, du.$$

Prove counterparts of Problem 1.1.4 for $I^2(f; x; \rho)$.

(ii) The *rth integral means* $A^r(f; x; h)$ of $f \in X(\mathbb{R})$ are defined as for periodic functions by $A^r(f; x; h) = A(A^{r-1}(f; \circ; h); x; h)$. State and prove counterparts to Problems 1.1.5, 1.1.7.

(iii) Since the kernel of the rth integral means has compact support, $A^r(f; x; h)$ is well-defined for every $f \in L^1_{loc}$. For those functions f show that $A^r(f; x; h) \in C^{r-1}_{loc}$ for each $h > 0$ and $\lim_{h \to 0+} A^r(f; x; h) = f(x)$ a.e. (see also Prop. 0.3.1). If $f \in C_{loc}$, then $\lim_{h \to 0+} A^r(f; x; h) = f(x)$ for every $x \in \mathbb{R}$.

9. (i) Analogously to (3.1.3) we may assign to each function $\mu \in BV$ a singular integral by

$$(3.1.52) \qquad I(d\mu; x; \rho) = \frac{1}{\sqrt{2\pi}} \int_{-\infty}^{\infty} \chi(x-u; \rho)\, d\mu(u).$$

Show that if $\{\chi(x; \rho)\}$ is a kernel, then $I(d\mu; x; \rho) \in \mathsf{L}^1$ for each $\rho > 0$ and $\|I(d\mu; \circ; \rho)\|_1 \le \|\chi(\circ; \rho)\|_1 \|\mu\|_{\mathsf{BV}}$. Moreover, if $\{\chi(x; \rho)\}$ is an approximate identity, then for every $h \in \mathsf{C}$

$$\lim_{\rho\to\infty} \int_{-\infty}^{\infty} h(x) I(d\mu; x; \rho)\, dx = \int_{-\infty}^{\infty} h(x)\, d\mu(x). \tag{3.1.53}$$

(Hint: Compare with Prop. 1.1.7)

(ii) If the kernel is of Fejér's type, the notation will be

$$J(d\mu; x; \rho) = \frac{\rho}{\sqrt{2\pi}} \int_{-\infty}^{\infty} \chi(\rho(x-u))\, d\mu(u). \tag{3.1.54}$$

Show that if $\chi \in \mathsf{NL}^1$, then $\|J(d\mu; \circ; \rho)\|_1 \le \|\chi\|_1 \|\mu\|_{\mathsf{BV}}$ for all $\rho > 0$, and (3.1.53) is valid as well.

10. (i) Show that if $g \in \mathsf{L}^1$ is even (odd), then also g^* as defined by (3.1.22) is even (odd).

(ii) Let $g \in \mathsf{L}^1$, and suppose that the αth absolute moment $m(g; \alpha)$ of g exists for some $\alpha > 0$ (cf. (3.3.6)). Show that $m(g^*; \alpha) < \infty$ (cf. (1.6.9)) and

$$(2\pi)^{-1} \int_{-\pi}^{\pi} |x|^{\alpha} |g^*(x)|\, dx \le (2\pi)^{-1/2} \int_{-\infty}^{\infty} |x|^{\alpha} |g(x)|\, dx.$$

In particular, if $\chi \in \mathsf{NL}^1$ satisfies $m(\chi; \alpha) < \infty$, then $m(\chi_\rho^*; \alpha) \le \rho^{-\alpha} m(\chi; \alpha)$ for all $\rho > 0$, χ_ρ^* being given by (3.1.28). On this basis, examine the estimates of Lemma 1.6.4 once more.

(iii) Let $\chi \in \mathsf{L}^1$ and χ_ρ^* be defined through (3.1.28). Show that $\lim_{\rho\to\infty} \|\chi_\rho^*\|_{\mathsf{L}^1_{2\pi}} = \|\chi\|_{\mathsf{L}^1}$.

(iv) If $\chi \in \mathsf{NL}^1$ is positive, show that $\{\chi_\rho^*(x)\}$ as given through (3.1.28) is a positive (periodic) approximate identity.

11. Let $\{\chi(x; \rho)\}$ be a kernel (on the real line) and set

$$\chi_\rho(x) = \sqrt{2\pi} \sum_{k=-\infty}^{\infty} \chi(x + 2k\pi; \rho). \tag{3.1.55}$$

Show that $\{\chi_\rho(x)\}$ defines a periodic kernel with parameter set $\mathbb{A} = (0, \infty)$ and $\rho_0 = \infty$. In particular, $\|\chi_\rho\|_{\mathsf{L}^1_{2\pi}} \le \|\chi(\circ; \rho)\|_{\mathsf{L}^1}$ for each $\rho > 0$. Furthermore, if $\{\chi(x; \rho)\}$ is an approximate identity on the real line, then $\{\chi_\rho(x)\}$ is a periodic approximate identity.

12. Let $\mu \in \mathsf{BV}$ and define a function μ^* through

$$\mu^*(x) = \sqrt{2\pi} \sum_{k=-\infty}^{\infty} [\mu(x + 2k\pi) - \mu(2k\pi)]. \tag{3.1.56}$$

Show that $\mu^* \in \mathsf{BV}_{2\pi}$. If $f \in \mathsf{C}_{2\pi}$, then

$$\frac{1}{2\pi} \int_{-\pi}^{\pi} f(u)\, d\mu^*(u) = \frac{1}{\sqrt{2\pi}} \int_{-\infty}^{\infty} f(u)\, d\mu(u). \tag{3.1.57}$$

(Hint: See also KATZNELSON [1, p. 134], BERENS–GÖRLICH [1])

13. Prove (3.1.19). (Hint: HILLE [4I, p. 261])

14. The *singular integral of Jackson–de La Vallée Poussin* is defined through

$$N(f; x; \rho) = \frac{12}{\pi\rho^3} \int_{-\infty}^{\infty} f(x-u) \frac{\sin^4(\rho u/2)}{u^4}\, du \tag{3.1.58}$$

with kernel $N(x) = (3/\sqrt{8\pi})(x/2)^{-4} \sin^4(x/2)$. Show that $N \in \mathsf{NL}^1$ (compare with Problem 5.2.8). Thus $\{\rho N(\rho x)\}$ is an even, positive, continuous kernel of Fejér's type. Show that $\|N(f; \circ; \rho)\|_{\mathsf{X}(\mathbb{R})} \le \|f\|_{\mathsf{X}(\mathbb{R})}$ and $\lim_{\rho\to\infty} \|N(f; \circ; \rho) - f(\circ)\|_{\mathsf{X}(\mathbb{R})} = 0$ for every $f \in \mathsf{X}(\mathbb{R})$ (see also DE LA VALLÉE POUSSIN [3, p. 45], ACHIESER [2, p. 138, p. 375], BUTZER–GÖRLICH [2, p. 378]).

15. (i) Show that $3 \sum_{k=-\infty}^{\infty} (z + k\pi)^{-4} = (2 + \cos 2z) \sin^{-4} z$. (Hint: Differentiate (3.1.19))

(ii) Replacing the continuous parameter ρ in (3.1.58) by the discrete $n \in \mathbb{N}$, show that the *periodic* singular integral $N_n^*(f; x)$ of Jackson–de La Vallée Poussin as associated with $N(f; x; n)$ through (3.1.28), (3.1.30) is given by

$$(3.1.59)\quad N_n^*(f; x) = \frac{1}{4\pi n^3} \int_{-\pi}^{\pi} f(x-u)(2+\cos u) \left[\frac{\sin(nu/2)}{\sin(u/2)}\right]^4 du$$

with kernel $\{N_n^*(x)\}$, $N_n^*(x) = (2n^3)^{-1}(2+\cos x)[\sin(nx/2)/\sin(x/2)]^4$. Show that $N_n^*(x)$ is an even, positive trigonometric polynomial of degree $2n-1$. Compare with the (original) singular integral of Jackson as defined in Problem 1.3.9. For the coefficients of the polynomial $N_n^*(x)$ see Problem 5.1.2(v). Show that $\lim_{n\to\infty} \|N_n^*(f; \circ) - f(\circ)\|_{X_{2\pi}} = 0$ for every $f \in X_{2\pi}$.

16. Let $\nu \in \mathsf{NBV}$. Show that for every $f \in X(\mathbb{R})$

$$\lim_{\rho\to\infty} \left\| \frac{1}{\sqrt{2\pi}} \int_{-\infty}^{\infty} f(\circ - u)\, d\nu(\rho u) - f(\circ) \right\|_{X(\mathbb{R})} = 0.$$

(Hint: Since $\nu \in \mathsf{NBV}$, it follows by the Hölder–Minkowski inequality that

$$\left\| \frac{1}{\sqrt{2\pi}} \int_{-\infty}^{\infty} f(\circ - u)\, d\nu(\rho u) - f(\circ) \right\|_{X(\mathbb{R})} \le \frac{1}{\sqrt{2\pi}} \int_{-\infty}^{\infty} \|f(\circ - \rho^{-1}u) - f(\circ)\|_{X(\mathbb{R})} |d\nu(u)|.$$

Now use the continuity of f in $X(\mathbb{R})$-norm and Lebesgue's dominated convergence theorem; see also the proof of Prop. 3.2.1)

17. Let the kernel $\{\chi(x; \rho)\}$ of the singular integral $I(f; x; \rho)$ be positive. Suppose that at some point $x_0 \in \mathbb{R}$

$$\lim_{\rho\to\infty} I(u; x_0; \rho) = x_0, \qquad \lim_{\rho\to\infty} I(u^2; x_0; \rho) = x_0^2.$$

Show that $\lim_{\rho\to\infty} \|I(f; \circ; \rho) - f(\circ)\|_{X(\mathbb{R})} = 0$ for every $f \in X(\mathbb{R})$. Apply to the singular integral of Gauss–Weierstrass. (Hint: Although x and x^2 do not belong to $X(\mathbb{R})$, the hypothesis assumes that these two functions belong to the domain of definition of the singular integral under consideration, in other words, that the second moment $m(\chi(\circ; \rho); 2)$ of $\chi(x; \rho)$ (cf. (3.3.6)) exists for each $\rho > 0$. Hence proceed as in the proofs of Theorem 1.3.7, Prop. 1.3.9 and use the identity

$$x_0^2 - 2x_0 I(u; x_0; \rho) + I(u^2; x_0; \rho) = \frac{1}{\sqrt{2\pi}} \int_{-\infty}^{\infty} u^2 \chi(u; \rho)\, du,$$

which tends to zero as $\rho \to \infty$. Thus $\{\chi(x; \rho)\}$ satisfies (3.1.6), and the assertion follows by Theorem 3.1.6; see also Prop. 3.3.1)

18. (i) Show that the function $e^{-n\alpha x}$, $n \in \mathbb{N}$, $\alpha > 0$, can be uniformly approximated on $0 \le x < \infty$ by functions of the form $p(x)\, e^{-\alpha x}$, where $p(x)$ is an algebraic polynomial. (Hint: Suppose $\alpha = 1$ and proceed via mathematical induction; see STONE [1, p. 72], also Natanson [8II, p. 155])

(ii) For each $\alpha > 0$, the functions of the form $p(x)\, e^{-\alpha x}$, where $p(x)$ is any algebraic polynomial, are dense in the spaces $C_0[0, \infty)$ and $L^p(0, \infty)$, $1 \le p < \infty$. (Hint: Consider first the case $C_0[0, \infty)$, the space of all continuous functions f on $[0, \infty)$ with $\lim_{x\to\infty} f(x) = 0$. Introducing a new variable $t = e^{-\alpha x}$, the function $\eta(t) = f((-1/\alpha)\log t)$, $\eta(0) = 0$ is continuous on $0 \le t \le 1$, so that the (ordinary) Weierstrass approximation theorem (cf. Problem 1.2.12) applies. Now use (i). In $L^p(0, \infty)$ the discussion may be reduced to functions $f \in C_{00}[0, \infty)$ (compare Prop. 0.3.4), and then use the result for $C_0[0, \infty)$; see also STONE [1, p. 74] where the results are also derived from the general Stone–Weierstrass approximation theorem)

19. The *Laguerre polynomials* are defined by ($n \in \mathbb{P}$)

$$(3.1.60)\qquad l_n(x) = (1/n!)\, e^x\, (d/dx)^n [x^n e^{-x}]$$

and the *Laguerre functions* by

(3.1.61) $$L_n(x) = l_n(x)\, e^{-x/2}.$$

(i) Show that $l_n(x) = \sum_{k=0}^{n} (-1)^k \binom{n}{k} x^k/k!$. Thus $l_n(x)$ is an algebraic polynomial of degree n. In particular, $l_0(x) = 1$, $l_1(x) = -x + 1$.

(ii) Show that $\int_0^\infty e^{-x} l_n(x) l_m(x)\, dx = 0$ for $n \neq m$, $=1$ for $n = m$. Thus the Laguerre functions form an *orthonormal* system on $[0, \infty)$. The Laguerre functions and polynomials are linearly independent.

(iii) As an application of the previous Problem show that the Laguerre functions form a fundamental set in the spaces $C_0[0, \infty)$ and $L^p(0, \infty)$, $1 \leq p < \infty$.

(Hint: Among others, one may consult Tricomi [1, p. 212 ff], Helmberg [1, p. 55 ff])

20. Prove Prop. 3.1.8. (Hint: For a proof by an application of the general Stone–Weierstrass approximation theorem see Stone [1, p. 79]; this also contains a variant which makes appropriate use of the (ordinary) Weierstrass approximation theorem for functions continuous on a compact set of the plane. For reductions to (generalized) Laguerre functions one may also consult Natanson [8II, p. 158], Tricomi [1, p. 239], compare also Helmberg [1, p. 55 ff])

3.2 Pointwise Convergence

In this section we establish results for singular integrals on the real line which we have already shown for periodic singular integrals in Sec. 1.4. In what follows we shall restrict our discussion to singular integrals of Fejér's type. Analogous results for the general singular integral (3.1.3) are stated in Problem 3.2.1.

Proposition 3.2.1. *Let* $f \in L^\infty$ *and* $\chi \in NL^1$.

(i) *At every point* x_0 *of continuity of* f, $\lim_{\rho\to\infty} J(f; x_0; \rho) = f(x_0)$.

(ii) *If* f *is continuous on* $(a - \eta, b + \eta)$ *for some* $\eta > 0$, $a < b$, $a, b \in \mathbb{R}$, *then* $\lim_{\rho\to\infty} J(f; x; \rho) = f(x)$ *uniformly on* $[a, b]$.

(iii) *If* χ *is even and* x_0 *is such that* $\lim_{h\to 0+} [f(x_0 + h) + f(x_0 - h)] = 2f(x_0)$, *then* $\lim_{\rho\to\infty} J(f; x_0; \rho) = f(x_0)$.

Proof. We shall only prove (iii). We have

$$J(f; x_0; \rho) - f(x_0) = \frac{1}{\sqrt{2\pi}} \int_0^\infty [f(x_0 + \rho^{-1}u) + f(x_0 - \rho^{-1}u) - 2f(x_0)]\chi(u)\, du.$$

According to the hypotheses, the integrand is dominated by $4\,\|f\|_\infty |\chi(u)|$ and converges to zero pointwise for almost all u as $\rho \to \infty$. Hence the conclusion follows by Lebesgue's dominated convergence theorem.

We now proceed with convergence almost everywhere. The first result is concerned with convergence at D-points (compare (0.3.3)).

Proposition 3.2.2. *Let* $f \in L^p$, $1 \leq p \leq \infty$, *and* $\chi \in NL^1$ *be an even, positive function, monotonely decreasing on* $[0, \infty)$. *Then at each point* x *for which*

(3.2.1) $$\int_0^h [f(x + u) + f(x - u) - 2f(x)]\, du = o(h) \qquad (h \to 0+),$$

thus for almost all x, one has

(3.2.2) $$\lim_{\rho\to\infty} J(f; x; \rho) = f(x).$$

Proof. Since for each $x > 0$

$$\int_{x/2}^{x} \chi(u)\, du \geq \chi(x) \int_{x/2}^{x} du = (x/2)\chi(x),$$

and since the left-hand side tends to zero as $x \to \infty$, one first of all obtains that

(3.2.3) $$\lim_{x\to\infty} x\chi(x) = 0.$$

Setting

(3.2.4) $$G(u) = \int_0^u [f(x + t) + f(x - t) - 2f(x)]\, dt,$$

then, according to (3.2.1), given any $\varepsilon > 0$ there is a $\delta > 0$ such that $|G(u)| \leq \varepsilon u$ for all $0 < u \leq \delta$. Therefore, in view of

(3.2.5) $$J(f; x; \rho) - f(x)$$
$$= \frac{\rho}{\sqrt{2\pi}}\left(\int_0^{\delta} + \int_{\delta}^{\infty}\right)[f(x + u) + f(x - u) - 2f(x)]\chi(\rho u)\, du \equiv I_1 + I_2,$$

we have by partial integration

$$|I_1| = \frac{\rho}{\sqrt{2\pi}}\left|\int_0^{\delta} \chi(\rho u)\, dG(u)\right| = \frac{1}{\sqrt{2\pi}}\left|\rho\chi(\rho\delta)G(\delta) + \rho \int_0^{\delta} G(u)\, d[-\chi(\rho u)]\right|$$
$$\leq \frac{\varepsilon}{\sqrt{2\pi}}\left\{(\rho\delta)\chi(\rho\delta) + \rho \int_0^{\delta} u\, d[-\chi(\rho u)]\right\} = \varepsilon \frac{\rho}{\sqrt{2\pi}} \int_0^{\delta} \chi(\rho u)\, du \leq \varepsilon\, \|\chi\|_1.$$

Regarding I_2 we have

$$|I_2| \leq \frac{\rho}{\sqrt{2\pi}} \int_{\delta}^{\infty} |f(x + u) + f(x - u)|\, \chi(\rho u)\, du$$
$$+ 2\,|f(x)|\, \frac{\rho}{\sqrt{2\pi}} \int_{\delta}^{\infty} \chi(\rho u)\, du \equiv I_2^1 + I_2^2,$$

say. Then $I_2^2 \leq 2\,|f(x)| \int_{\rho\delta}^{\infty} \chi(u)\, du$. Furthermore, for $p = 1$

$$I_2^1 \leq \rho\chi(\rho\delta) \frac{1}{\sqrt{2\pi}} \int_{\delta}^{\infty} |f(x + u) + f(x - u)|\, du \leq \frac{2\,\|f\|_1}{\delta}(\rho\delta)\chi(\rho\delta),$$

whereas for $1 < p < \infty$

$$I_2^1 \leq \left\{\frac{\rho}{\sqrt{2\pi}} \int_{\delta}^{\infty} |f(x + u) + f(x - u)|^p \chi(\rho u)\, du\right\}^{1/p} \left\{\frac{\rho}{\sqrt{2\pi}} \int_{\delta}^{\infty} \chi(\rho u)\, du\right\}^{1/p'}$$
$$\leq \frac{2\,\|f\|_p}{\delta^{1/p}}\{(\rho\delta)\chi(\rho\delta)\}^{1/p} \left\{\frac{1}{\sqrt{2\pi}} \int_{\rho\delta}^{\infty} \chi(u)\, du\right\}^{1/p'},$$

and for $p = \infty$, $I_2^1 \le 2 \|f\|_\infty \int_{\rho\delta}^\infty \chi(u)\,du$. Combining these results, since δ is now fixed, we obtain $\lim_{\rho\to\infty} I_2 = 0$, proving (3.2.2).

If we consider the singular integrals of Gauss–Weierstrass and Cauchy–Poisson, then one immediately sees that all the assumptions of Prop. 3.2.2 are satisfied. Thus we have

Corollary 3.2.3. *Let $f \in \mathsf{L}^p$, $1 \le p \le \infty$, and $W(f; x; t)$ be the singular integral of Gauss–Weierstrass, $P(f; x; y)$ that of Cauchy–Poisson. Then for almost all $x \in \mathbb{R}$*

$$\lim_{t\to 0+} W(f; x; t) = f(x), \qquad \lim_{y\to 0+} P(f; x; y) = f(x), \tag{3.2.6}$$

in particular at each x for which (3.2.1) *holds.*

If χ does not satisfy all of the assumptions of Prop. 3.2.2, yet possesses a majorant χ^* satisfying them, then the conclusion of Prop. 3.2.2 remains valid at all L-points (compare (0.3.4)). Indeed, as a counterpart to Prop. 1.4.6

Proposition 3.2.4. *Let $f \in \mathsf{L}^p$, $1 \le p \le \infty$, and $\chi \in \mathsf{NL}^1$ be an even function for which there exists $\chi^* \in \mathsf{L}^1$ monotonely decreasing on* $[0, \infty)$ *such that $|\chi(x)| \le \chi^*(x)$* a.e. *Then at each point x for which*

$$\int_0^h |f(x+u) + f(x-u) - 2f(x)|\,du = o(h) \qquad (h \to 0+), \tag{3.2.7}$$

thus for almost all $x \in \mathbb{R}$, we have (3.2.2).

The proof is left to the reader. As an example we consider the singular integral of Fejér. Since for $F(x)$ as given by (3.1.15) there exists a constant C such that $F(x) \le C(1 + x^2)^{-1}$, one has

Corollary 3.2.5. *Let $f \in \mathsf{L}^p$, $1 \le p \le \infty$, and $\sigma(f; x; \rho)$ its singular integral of Fejér. Then for almost all $x \in \mathbb{R}$*

$$\lim_{\rho\to\infty} \sigma(f; x; \rho) = f(x), \tag{3.2.8}$$

in particular at each x for which (3.2.7) *holds.*

Let us conclude the section with the counterpart of Prop. 1.4.8 concerning pointwise convergence of the integral $J(d\mu; x; \rho)$ introduced in Problem 3.1.9.

Proposition 3.2.6. *Let $\mu \in \mathsf{BV}$ and $\chi \in \mathsf{NL}^1$ be an even and positive function, monotonely decreasing on* $[0, \infty)$. *Then at each point x for which $\mu'(x)$ exists*

$$\lim_{\rho\to\infty} J(d\mu; x; \rho) = \mu'(x). \tag{3.2.9}$$

In particular, relation (3.2.9) *holds for almost all x.*

Proof. Since χ is even, it follows that

$$J(d\mu; x; \rho) = \frac{\rho}{\sqrt{2\pi}} \int_0^\infty \chi(\rho u)\, d[\mu(x+u) - \mu(x-u)],$$

and therefore by partial integration

$$J(d\mu;x;\rho)=\frac{\rho}{\sqrt{2\pi}}\Big([\chi(\rho u)(\mu(x+u)-\mu(x-u))]_{u=0}^{\infty}-\int_0^{\infty}[\mu(x+u)-\mu(x-u)]\,d\chi(\rho u)\Big).$$

If x is a point at which $\mu'(x)$ exists, then, in particular, x is a point of continuity of μ. This implies the first relation of

$$\lim_{u\to 0+}\chi(\rho u)[\mu(x+u)-\mu(x-u)]=0,\qquad \lim_{u\to\infty}\chi(\rho u)[\mu(x+u)-\mu(x-u)]=0,$$

since $\chi(0+)$ exists as a finite number by hypothesis. The second relation follows since $\lim_{u\to\infty}\chi(u)=0$ (see (3.2.3)) and since $\lim_{u\to\infty}[\mu(x+u)-\mu(x-u)]=\mu(\infty)$ exists as a finite number. Therefore

$$J(d\mu;x;\rho)=-\frac{\rho}{\sqrt{2\pi}}\int_0^{\infty}[\mu(x+u)-\mu(x-u)]\,d\chi(\rho u).$$

Since μ is differentiable at x, we have that

$$\mu(x+u)-\mu(x-u)=2u\big(\mu'(x)+\eta(u)\big),$$

η being a bounded function satisfying $\lim_{u\to 0+}\eta(u)=0$. Hence

$$J(d\mu;x;\rho)=\mu'(x)\frac{\rho}{\sqrt{2\pi}}\int_0^{\infty}2u\,d[-\chi(\rho u)]+\frac{\rho}{\sqrt{2\pi}}\int_0^{\infty}2u\eta(u)\,d[-\chi(\rho u)].$$

A further integration by parts delivers by (3.2.3)

$$\frac{\rho}{\sqrt{2\pi}}\int_0^{\infty}2u\,d[-\chi(\rho u)]=\frac{2}{\sqrt{2\pi}}\Big([-(\rho u)\chi(\rho u)]_{u=0}^{\infty}+\rho\int_0^{\infty}\chi(\rho u)\,du\Big)=1,$$

which implies that

$$|J(d\mu;x;\rho)-\mu'(x)|\le\frac{2\rho}{\sqrt{2\pi}}\Big(\int_0^{\delta}+\int_{\delta}^{\infty}\Big)u\,|\eta(u)|\,d[-\chi(\rho u)]\equiv I_1+I_2,$$

say. Given $\varepsilon>0$, let $\delta>0$ be such that $|\eta(u)|<\varepsilon$ for all $0\le u\le\delta$. Then

$$I_1\le\varepsilon\frac{2\rho}{\sqrt{2\pi}}\int_0^{\delta}u\,d[-\chi(\rho u)]\le\varepsilon.$$

Concerning I_2, since η is bounded, it follows by partial integration that

$$I_2\le\|\eta\|_{\infty}\frac{2}{\sqrt{2\pi}}\Big([-(\rho u)\chi(\rho u)]_{u=\delta}^{\infty}+\rho\int_{\delta}^{\infty}\chi(\rho u)\,du\Big)$$

$$\le\|\eta\|_{\infty}\sqrt{2/\pi}\Big((\rho\delta)\chi(\rho\delta)+\int_{\rho\delta}^{\infty}\chi(u)\,du\Big)$$

which tends to zero as $\rho\to\infty$. This proves the assertion, since μ is differentiable almost everywhere by Prop. 0.5.3.

For applications and extensions we refer to Problems 3.2.3, 3.2.4.

Problems

1. (i) Let $f\in X(\mathbb{R})$, and let the kernel $\{\chi(x;\rho)\}$ of the singular integral $I(f;x;\rho)$ be an approximate identity satisfying (3.1.7). State and prove counterparts to Prop. 3.2.1; compare also with Prop. 1.4.1.
   (ii) Let $f\in L^p$, $1\le p\le\infty$, and let the kernel $\{\chi(x;\rho)\}$ of the singular integral $I(f;x;\rho)$ be an even, positive approximate identity such that $\chi(x;\rho)$ is monotonely decreasing on $[0,\infty)$ for each $\rho>0$. Show that $\lim_{\rho\to\infty}I(f;x;\rho)=f(x)$ at each point x for which (3.2.1) holds. (Hint: Compare the proofs of Prop. 1.4.4 and 3.2.2)

2. (i) Let $f \in L^p$, $1 \le p \le \infty$, and χ satisfy the assumptions of Prop. 3.2.2. Suppose that the αth moment $m(\chi; \alpha)$ of χ exists as a finite number for some $\alpha > 0$ (cf. (3.3.6)). Show that $|J(f; x; \rho) - f(x)| = o(\rho^{-\alpha})$, $\rho \to \infty$, at each point x for which $\int_0^h [f(x+u) + f(x-u) - 2f(x)]\, du = o(h^{1+\alpha})$, $h \to 0+$ (compare with Problem 1.4.3). (Hint: Show that $\lim_{x\to\infty} x^{1+\alpha}\chi(x) = 0$ and proceed as in the proof of Prop. 3.2.2; see also BUTZER [6])

   (ii) If f is continuous throughout an interval and if at every point x of that interval
$$\int_0^h [f(x+u) + f(x-u) - 2f(x)]\, du = o(h^3) \qquad (h \to 0+),$$
show that f is a linear function throughout that interval. Thus (i) of this Problem is of interest only in case $0 < \alpha \le 2$. (Hint: If $\phi(x) = \int_a^x f(u)\, du$, then the hypothesis is equivalent to $\phi(x+h) - \phi(x-h) - 2h\phi'(x) = o(h^3)$. According to a known result (see CHAUNDY [1, p. 138]) this implies that ϕ is a quadratic function; see also BUTZER [6], BARLAZ [1], and Problem 10.1.3)

3. (i) Let $\mu \in \mathsf{BV}$. Show that for the singular integrals of Gauss–Weierstrass and Cauchy–Poisson
$$\lim_{t\to 0+} W(d\mu; x; t) = \mu'(x), \qquad \lim_{y\to 0+} P(d\mu; x; y) = \mu'(x)$$
at each point x for which $\mu'(x)$ exists, thus almost everywhere.

   (ii) Let χ be absolutely continuous and satisfy the assumptions of Prop. 3.2.6. Show that $-x\chi'(x) \in \mathsf{NL}^1$, thus $\{-\rho^2 x\chi'(\rho x)\}$ is an even, positive kernel of Fejér's type. Now compare the proofs of Prop. 1.4.8 and 3.2.6.

   (iii) Let χ satisfy the assumptions of (ii) and suppose that $f \in \mathsf{L}^\infty$ is differentiable at x_0. Show that $\lim_{\rho\to\infty} J'(f; x_0; \rho) = f'(x_0)$. As an application show that $\lim_{y\to 0+} P'(f; x_0; y) = f'(x_0)$. (Hint: Apply (3.1.49), also valid under the present assumptions, and use (ii); compare with Problem 1.4.2)

4. Let $\mu \in \mathsf{BV}$ and χ satisfy the assumptions of Prop. 3.2.4. State and prove a counterpart to Prop. 3.2.6 so as to obtain as an application to the Fejér integral that $\lim_{\rho\to\infty} \sigma(d\mu; x; \rho) = \mu'(x)$ a.e. (Hint: Compare Prop. 1.4.10)

5. The (special) *singular integral of Picard* is defined through

(3.2.10)
$$C_2(f; x; \rho) = \frac{\rho}{2} \int_{-\infty}^{\infty} f(x-u)\, e^{-\rho|u|}\, du$$
with kernel $c_2(x) = \sqrt{\pi/2}\, e^{-|x|}$. Show that $c_2 \in \mathsf{NL}^1$. Thus $\{\rho c_2(\rho x)\}$ is an even, positive, continuous kernel of Fejér's type. Show that $\|C_2(f; \circ; \rho)\|_{\mathsf{X}(\mathbb{R})} \le \|f\|_{\mathsf{X}(\mathbb{R})}$ and $\lim_{\rho\to\infty} \|C_2(f; \circ; \rho) - f(\circ)\|_{\mathsf{X}(\mathbb{R})} = 0$ for every $f \in \mathsf{X}(\mathbb{R})$. Furthermore, $\lim_{\rho\to\infty} C_2(f; x; \rho) = f(x)$ at each point x of type (3.2.1).

3.3 Order of Approximation

In this section certain results of Sec. 1.5 for periodic functions are carried over to the real line. Apart from replacing $\mathsf{X}_{2\pi}$ by $\mathsf{X}(\mathbb{R})$-norms, practically no changes occur in the definition of the moduli of continuity and Lipschitz classes so that we leave their formal definition together with the elementary properties to the reader (compare Problem 3.3.1). In Sec. 1.5 the rate of approximation was essentially expressed through the behaviour of the Fourier coefficients of the kernel. In the discussion concerning approximation on the real line, the moments of the kernel, more or less, take the place of the Fourier coefficients.

First of all we give an approximation theorem for the general singular integral (3.1.3) which corresponds to Prop. 1.5.10 concerning the method of test functions.

Proposition 3.3.1. *Let the kernel* $\{\chi(x; \rho)\}$ *of the singular integral* $I(f; x; \rho)$ *be positive. If at some point* $x_0 \in \mathbb{R}$

$$(3.3.1)\qquad \begin{aligned} I(u; x_0; \rho) &= x_0 + \beta(x_0; \rho), \qquad \lim_{\rho\to\infty} \beta(x_0; \rho) = 0,\\ I(u^2; x_0; \rho) &= x_0^2 + \gamma(x_0; \rho), \qquad \lim_{\rho\to\infty} \gamma(x_0; \rho) = 0, \end{aligned}$$

then for each $f \in X(\mathbb{R})$

$$(3.3.2)\qquad \|I(f; \circ; \rho) - f(\circ)\|_{X(\mathbb{R})} = O(\omega(X(\mathbb{R}); f; \sqrt{\gamma(x_0; \rho) - 2x_0\beta(x_0; \rho)})) \qquad (\rho \to \infty).$$

Proof. First we note that the hypotheses (3.3.1) include the existence of the integrals $I(u; x; \rho)$ and $I(u^2; x; \rho)$ for each $\rho > 0$. Similarly as in Sec. 1.5 we conclude for any $\lambda > 0$

$$\|I(f; \circ; \rho) - f(\circ)\|_{X(\mathbb{R})} \le \omega(X(\mathbb{R}); f; \lambda^{-1})\left[1 + \lambda\left\{\frac{1}{\sqrt{2\pi}}\int_{-\infty}^{\infty} u^2\chi(u; \rho)\, du\right\}^{1/2}\right].$$

But for the second moment of the kernel we have (see Problem 3.1.17)

$$\frac{1}{\sqrt{2\pi}}\int_{-\infty}^{\infty} u^2\chi(u; \rho)\, du = \gamma(x_0; \rho) - 2x_0\beta(x_0; \rho),$$

which implies (3.3.2) by taking $\lambda^{-1} = \sqrt{\gamma(x_0; \rho) - 2x_0\beta(x_0; \rho)}$.

According to Prop. 0.1.9 the assumption (3.3.1) implies that the singular integral $I(f; x; \rho)$ converges to f in the metric of the space $X(\mathbb{R})$ for every $f \in X(\mathbb{R})$. In Sec. 1.3 and 1.5 we already remarked that the test functions which 'test' the approximation for the full class $X(\mathbb{R})$ need not necessarily belong to $X(\mathbb{R})$. Prop. 3.3.1 serves as an illustration to this fact. Of course, we then only obtain a sufficient condition for approximation in contrast to Theorem 1.3.7 which also gives necessary conditions.

Before applying Prop. 3.3.1 to the singular integral of Gauss–Weierstrass we observe that if the kernel $\{\chi(x; \rho)\}$ is also even, then $I(u; x; \rho) \equiv x$, i.e., $\beta(x; \rho) \equiv 0$. As is easily seen, we then have an estimate in terms of the generalized modulus of continuity given by

$$(3.3.3)\qquad \|I(f; \circ; \rho) - f(\circ)\|_{X(\mathbb{R})} = O(\omega^*(X(\mathbb{R}); f; \sqrt{\gamma(x_0; \rho)})) \qquad (\rho \to \infty).$$

Since the kernel of the integral $W(f; x; t)$ is even and since $W(u^2; x; t) = x^2 + 2t$, we therefore obtain

Corollary 3.3.2. *For the singular integral* $W(f; x; t)$ *of Gauss–Weierstrass one has for every* $f \in X(\mathbb{R})$

$$(3.3.4)\qquad \|W(f; \circ; t) - f(\circ)\|_{X(\mathbb{R})} = O(\omega^*(X(\mathbb{R}); f; t^{1/2})) \qquad (t \to 0+).$$

If furthermore $f \in \mathrm{Lip}^*(X(\mathbb{R}); \alpha)$ *for some* $0 < \alpha \le 2$, *then*

$$(3.3.5)\qquad \|W(f; \circ; t) - f(\circ)\|_{X(\mathbb{R})} = O(t^{\alpha/2}) \qquad (t \to 0+).$$

We conclude with a short discussion concerning asymptotic expansions of singular integrals on the real line. The methods of proof of Sec. 1.5.4 may be carried over to obtain analogous theorems for singular integrals $I(f; x; \rho)$ of type (3.1.3), keeping in mind that Fourier coefficients are replaced by moments (compare Problem 3.3.4). Here we shall give a simple result concerning expansions for singular integrals of Fejér's type. If α is a positive number, the αth *absolute moment* of χ is defined by

$$m(\chi; \alpha) = \frac{1}{\sqrt{2\pi}} \int_{-\infty}^{\infty} |u|^{\alpha} |\chi(u)|\, du. \tag{3.3.6}$$

Proposition 3.3.3. *Let $\chi \in \mathsf{NL}^1$ be an even and positive function for which $m(\chi; 2)$ exists as a finite number. If the second derivative f'' of $f \in \mathsf{L}^\infty$ exists at some point $x_0 \in \mathbb{R}$, then*

$$\lim_{\rho\to\infty} \rho^2[J(f; x_0; \rho) - f(x_0)] = (m(\chi; 2)/2) f''(x_0). \tag{3.3.7}$$

Proof. Since the kernel is even, we have

$$J(f; x; \rho) - f(x) = \frac{\rho}{\sqrt{2\pi}} \int_0^\infty [f(x+u) + f(x-u) - 2f(x)]\chi(\rho u)\, du. \tag{3.3.8}$$

Obviously, the hypothesis implies that f is continuous at x_0. Therefore the quotient $\overline{\Delta}_u^2 f(x_0)/u^2$ takes the indeterminate form 0/0 as u approaches zero. To eliminate this indeterminacy, the existence of f'' at x_0 includes the existence of f' in a neighbourhood of x_0. Therefore L'Hospital's rule may be applied, and it follows that

$$\lim_{u\to 0} \frac{f(x_0+u) + f(x_0-u) - 2f(x_0)}{u^2} = \lim_{u\to 0} \frac{f'(x_0+u) - f'(x_0-u)}{2u} = f''(x_0),$$

the latter relation being valid by definition of $f''(x_0)$. Hence

$$f(x_0+u) + f(x_0-u) - 2f(x_0) = u^2 f''(x_0) + u^2\eta(u),$$

η being an essentially bounded function satisfying $\lim_{u\to 0}\eta(u) = 0$. Therefore by (3.3.8)

$$\begin{aligned} J(f; x_0; \rho) - f(x_0) &= f''(x_0)\frac{\rho}{\sqrt{2\pi}}\int_0^\infty u^2\chi(\rho u)\, du + \frac{\rho}{\sqrt{2\pi}}\int_0^\infty \eta(u) u^2 \chi(\rho u)\, du \\ &= \rho^{-2}(m(\chi; 2)/2) f''(x_0) + \rho^{-2} R_\rho, \end{aligned}$$

say. Given $\varepsilon > 0$, let $\delta > 0$ be such that $|\eta(u)| \le \varepsilon$ for $0 \le u \le \delta$. Then

$$|R_\rho| \le \frac{\rho^3}{\sqrt{2\pi}}\left(\int_0^\delta + \int_\delta^\infty\right)|\eta(u)|\, u^2\chi(\rho u)\, du \le \varepsilon m(\chi; 2) + \frac{\|\eta\|_\infty}{\sqrt{2\pi}}\int_{\rho\delta}^\infty u^2\chi(u)\, du,$$

the latter integral tending to zero as $\rho \to \infty$. Thereby relation (3.3.7) is established.

Applying this result to the singular integral of Gauss–Weierstrass we obtain

Corollary 3.3.4. *If the second derivative of $f \in \mathsf{L}^\infty$ exists at some point x_0, then for the singular integral of Gauss–Weierstrass there holds the asymptotic expansion*

$$\lim_{t\to 0+} \frac{W(f; x_0; t) - f(x_0)}{t} = f''(x_0). \tag{3.3.9}$$

Let us finally return to Nikolskiĭ constants for periodic singular integrals as introduced in Sec. 1.6.3. It is the aim here to determine these in the fractional case for the rather general class of singular integrals of Fejér's type.

Proposition 3.3.5. *Let the periodic approximate identity* $\{\chi_\rho^*(x)\}$ *be generated via* (3.1.28) *by* $\chi \in \mathsf{NL}^1$, *and let* $I_\rho^*(f; x)$ *be the corresponding (periodic) singular integral* (3.1.30). *Suppose that* χ *is even, positive and that the* α*th absolute moment* $m(\chi; \alpha)$ *exists as a finite number for some* $0 < \alpha \leq 2$. *Then*

$$\lim_{\rho\to\infty} \rho^\alpha \Delta^*(\chi_\rho^*; \alpha) = m(\chi; \alpha). \tag{3.3.10}$$

Thus the Nikolskiĭ constant of the integral $I_\rho^*(f; x)$ *for the class* $\mathsf{Lip}^*(\mathsf{C}_{2\pi}; \alpha)_2$ *is equal to the moment* $m(\chi; \alpha)$ *of* χ.

Proof. Let $f \in \mathsf{Lip}^*(\mathsf{C}_{2\pi}; \alpha)_2$. With $J(f; x; \rho)$ of (3.1.8) it follows by (3.1.31) that for each $\rho > 0$

$$I_\rho^*(f; x) \equiv \frac{1}{2\pi}\int_{-\pi}^{\pi} f(x-u)\chi_\rho^*(u)\,du = \frac{\rho}{\sqrt{2\pi}}\int_{-\infty}^{\infty} f(x-u)\chi(\rho u)\,du \equiv J(f; x; \rho)$$

for all $x \in \mathbb{R}$. Therefore by (3.3.8)

$$\begin{aligned}\|I_\rho^*(f; \circ) - f(\circ)\|_{\mathsf{C}_{2\pi}} &= \|J(f; \circ; \rho) - f(\circ)\|_{\mathsf{C}}\\ &\leq \frac{\rho}{\sqrt{2\pi}}\int_0^\infty \|f(\circ+u) + f(\circ-u) - 2f(\circ)\|_{\mathsf{C}}\,\chi(\rho u)\,du\\ &\leq \frac{2\rho}{\sqrt{2\pi}}\int_0^\infty u^\alpha\chi(\rho u)\,du = m(\chi; \alpha)\rho^{-\alpha},\end{aligned}$$

giving $\Delta^*(\chi_\rho^*; \alpha) \leq m(\chi; \alpha)\rho^{-\alpha}$. On the other hand, using the function $g_\alpha^*(x) = |2\sin(x/2)|^\alpha$ which belongs to $\mathsf{Lip}^*(\mathsf{C}_{2\pi}; \alpha)_2$ for $0 < \alpha \leq 2$ by Problem 1.6.9, we obtain by (1.6.17) and (3.1.31) that

$$\Delta^*(\chi_\rho^*; \alpha) \geq \frac{1}{2\pi}\int_{-\pi}^{\pi}\left|2\sin\frac{u}{2}\right|^\alpha \chi_\rho^*(u)\,du = \frac{\rho}{\sqrt{2\pi}}\int_{-\infty}^{\infty}\left|2\sin\frac{u}{2}\right|^\alpha\chi(\rho u)\,du.$$

Therefore

$$m(\chi; \alpha)\rho^{-\alpha} - R(\rho) \leq \Delta^*(\chi_\rho^*; \alpha) \leq m(\chi; \alpha)\rho^{-\alpha}, \tag{3.3.11}$$

where we have set

$$R(\rho) = \frac{\rho}{\sqrt{2\pi}}\int_{-\infty}^{\infty}\left[|u|^\alpha - \left|2\sin\frac{u}{2}\right|^\alpha\right]\chi(\rho u)\,du.$$

Hence the proof of (3.3.10) would be complete if one could show that $R(\rho) = o(\rho^{-\alpha})$, $\rho \to \infty$. To this end, by an elementary substitution

$$\begin{aligned}\rho^\alpha R(\rho) &= \frac{1}{\sqrt{2\pi}}\int_{-\infty}^{\infty}\left[|u|^\alpha - \left|2\rho\sin\frac{u}{2\rho}\right|^\alpha\right]\chi(u)\,du\\ &= \frac{1}{\sqrt{2\pi}}\int_{-\infty}^{\infty}\left[1 - \left|\frac{\sin(u/2\rho)}{u/2\rho}\right|^\alpha\right]|u|^\alpha\chi(u)\,du.\end{aligned}$$

The term in brackets is bounded by 1 and tends to zero as $\rho \to \infty$ for each $u \in \mathbb{R}$. Since $m(\chi; \alpha)$ is finite, Lebesgue's dominated convergence theorem yields $\rho^\alpha R(\rho) = o(1)$ as $\rho \to \infty$. This establishes (3.3.10) completely.

To give an application, let us consider the singular integral of Fejér. In view of (3.1.20), the (periodic) kernel $\{F_n(x)\}$ of (1.2.24) is generated via (3.1.28) by $F(x)$ of (3.1.15) with $\rho = n + 1$. Obviously, the αth absolute moment of F exists for $0 < \alpha < 1$. Moreover, by partial integration and Problem 11.1.1

$$\text{(3.3.12)} \qquad m(F; \alpha) = \frac{2^{1+\alpha}}{\pi} \int_0^\infty u^{\alpha-2} \sin^2 u \, du$$

$$= \frac{2}{\pi(1-\alpha)} \int_0^\infty u^{\alpha-1} \sin u \, du = \frac{2}{\pi(1-\alpha)} \Gamma(\alpha) \sin \frac{\pi\alpha}{2}.$$

Thus, since $\Delta(F_n; \alpha) = \Delta^*(F_n; \alpha)$ for $0 < \alpha < 1$ by Problem 1.6.10(i), Prop. 3.3.5 and (3.3.12) establish the original result of Nikolskiĭ as formulated in Problem 1.6.11(ii). For the case $\alpha = 1$ one may consult Problem 3.3.7.

Problems

1. This Problem deals with counterparts of Sec. 1.5.1 and Problems 1.5.1–1.5.4.
   (i) Give the definition of the *modulus of continuity* $\omega(\mathsf{X}(\mathbb{R}); f; \delta)$, the generalized modulus $\omega^*(\mathsf{X}(\mathbb{R}); f; \delta)$, and the rth modulus of continuity (smoothness) $\omega_r(\mathsf{X}(\mathbb{R}); f; \delta)$ of a function $f \in \mathsf{X}(\mathbb{R})$. Prove counterparts of parts (i)–(iii) of Lemmata 1.5.2, 1.5.4, and of parts (i)–(iv) of Problem 1.5.3. As to (iv) of Lemma 1.5.2, if $\omega(\mathsf{C}; f; \delta) = o(\delta)$, $\delta \to 0+$, show that f is constant. In case $\mathsf{X}(\mathbb{R}) = \mathsf{L}^p$, $1 \le p < \infty$, see Cor. 10.5.2, Problem 10.6.18 which assert that $\omega_r(\mathsf{L}^p; f; \delta) = o(\delta^r)$, $\delta \to 0+$, implies $f(x) = 0$ a.e.
   (ii) Give the definition of the classes $\mathsf{Lip}(\mathsf{X}(\mathbb{R}); \alpha)$, $\mathsf{Lip}^*(\mathsf{X}(\mathbb{R}); \alpha)$, $\mathsf{Lip}_r(\mathsf{X}(\mathbb{R}); \alpha)$, $\mathsf{W}^{r,\alpha}_{\mathsf{X}(\mathbb{R})}$, $^*\mathsf{W}^{r,\alpha}_{\mathsf{X}(\mathbb{R})}$, $\mathsf{Lip}(\mathsf{X}(\mathbb{R}); \alpha)_M$, $\mathsf{Lip}^*(\mathsf{X}(\mathbb{R}); \alpha)_{M^*}$, etc. Prove counterparts of Problem 1.5.4.
2. (i) Let Z be the set of those functions $f \in \mathsf{C}_{\text{loc}}$ for which, given $\varepsilon > 0$, there exists $\delta > 0$ such that $|\overline{\Delta}_h^2 f(x)| < \varepsilon$ for all $|h| < \delta$ and uniformly for all $x \in \mathbb{R}$. Obviously, $\mathsf{C} \subset \mathsf{Z}$ but not conversely as the example $f(x) = x^2$ shows. For $f \in \mathsf{Z}$ the second modulus of continuity is defined by
$$\omega^*(\mathsf{Z}; f; \delta) = \sup_{0 \le |h| \le \delta} \|f(\circ + h) + f(\circ - h) - 2f(\circ)\|_{\mathsf{C}}.$$
   $\omega^*(\mathsf{Z}; f; \delta)$ is well-defined and $\lim_{\delta \to 0+} \omega^*(\mathsf{Z}; f; \delta) = 0$ for every $f \in \mathsf{Z}$. Show that for each $\lambda > 0$
$$\omega^*(\mathsf{Z}; f; \lambda\delta) \le (1 + \lambda)^2 \omega^*(\mathsf{Z}; f; \delta).$$
   Use this estimate to show that $|f(x)| = O(x^2)$, $|x| \to \infty$, for each $f \in \mathsf{Z}$. The class of functions $f \in \mathsf{Z}$ for which $\omega^*(\mathsf{Z}; f; \delta) \le M^* \delta^\alpha$ is denoted by $\mathsf{Lip}^*(\mathsf{Z}; \alpha)_{M^*}$ (see also Taberski [1]).
   (ii) Let the kernel $\chi \in \mathsf{NL}^1$ of the singular integral $J(f; x; \rho)$ be even and positive, and suppose that $m(\chi; 2)$ exists as a finite number so that $J(f; x; \rho)$ is well-defined for every $f \in \mathsf{Z}$. Show that for $0 < \alpha \le 2$
$$\sup_{f \in \mathsf{Lip}^*(\mathsf{Z}; \alpha)_2} \| J(f; \circ; \rho) - f(\circ) \|_{\mathsf{C}} = m(\chi; \alpha)\rho^{-\alpha}.$$
   Thus the Nikolskiĭ constant of the integral $J(f; x; \rho)$ for the class $\mathsf{Lip}^*(\mathsf{Z}; \alpha)_2$ is equal to the moment $m(\chi; \alpha)$ of χ. (Hint: Proceed as in the proof of (1.6.13) and use that $|x|^\alpha$ belongs to $\mathsf{Lip}^*(\mathsf{Z}; \alpha)_2$; see also Mamedov [3])

(iii) As an application of (ii) to the singular integral of Gauss–Weierstrass show that (compare Problem 1.6.12)

$$\sup_{f\in \mathsf{Lip}^*(\mathsf{Z};\alpha)_2} \|W(f;\circ;t)-f(\circ)\|_{\mathsf{C}} = 2^{\alpha}\pi^{-1/2}\Gamma((1+\alpha)/2)t^{\alpha/2} \qquad (0<\alpha\le 2).$$

(Hint: To evaluate the moment of the kernel use the definition of the gamma function, given in Problem 11.1.3; see also TABERSKI [1])

3. Let $\chi \in \mathsf{NL}^1$ be an even and positive function and $f \in \mathsf{L}^\infty$.
(i) If $m(\chi; 1)$ exists as a finite number, show that $\lim_{\rho\to\infty} \rho[J(f; x_0; \rho) - f(x_0)] = 0$ at each point x_0 for which the first derivative $f'(x_0)$ exists.
(ii) If $m(\chi; 2r)$ exists as a finite number for some $r \in \mathbb{N}$, show that

$$\lim_{\rho\to\infty} \rho^{2r}\Big[J(f;x_0;\rho) - f(x_0) - \sum_{k=1}^{r-1}\frac{m(\chi;2k)}{(2k)!\rho^{2k}}f^{(2k)}(x_0)\Big] = \frac{m(\chi;2r)}{(2r)!}f^{(2r)}(x_0)$$

at each point x_0 for which $f^{(2r)}(x_0)$ exists. (Hint: Use the expansion

$$(3.3.13)\qquad f(x_0+u) - f(x_0) = \sum_{k=1}^{2r}(u^k/k!)f^{(k)}(x_0) + (u^{2r}/(2r)!)\eta(u),$$

where η belongs to L^∞ and satisfies $\lim_{u\to 0}\eta(u) = 0$; see also BUTZER [6])

4. (i) Let $\mathsf{Z}^{2r}(x_0)$ be the set of those f on $\mathbb{R}$ which are $2r$-times differentiable at x_0 and satisfy relation (3.3.13) with η as specified there. Show that $f(x) = x^{2r}$ belongs to $\mathsf{Z}^{2r}(x_0)$ for each $x_0 \in \mathbb{R}$.
(ii) Let the kernel $\{\chi(x;\rho)\}$ of the singular integral $I(f; x; \rho)$ be positive, and suppose that $m(\chi(\circ;\rho); 2(r+1))$ exists as a finite number. Show that for every $f, g \in \mathsf{Z}^{2r}(x_0)$

$$\lim_{\rho\to\infty}\frac{I(f;x_0;\rho) - f(x_0) - \sum_{k=1}^{2r-1}(m(\chi(\circ;\rho);k)/k!)f^{(k)}(x_0)}{I(g;x_0;\rho) - g(x_0) - \sum_{k=1}^{2r-1}(m(\chi(\circ;\rho);k)/k!)g^{(k)}(x_0)} = \frac{f^{(2r)}(x_0)}{g^{(2r)}(x_0)},$$

provided that $\lim_{\rho\to\infty} m(\chi(\circ;\rho); 2(r+1))/m(\chi(\circ;\rho); 2r) = 0$ and the denumerators are different from zero. Specialize to even kernels of Fejér's type and compare with Problem 3.3.3. (Hint: Compare with Theorems 1.5.13, 1.5.15; see also MAMEDOV [6], MÜLLER [1, p. 51 f])

5. (i) Let $f \in \mathsf{L}^\infty$ and the kernel $\chi \in \mathsf{NL}^1$ of the singular integral $J(f; x; \rho)$ be even and positive. Suppose that $m(\chi; 1)$ exists as a finite number. If the one-sided derivatives $f'_+(x_0)$, $f'_-(x_0)$ exist at x_0, show that

$$\lim_{\rho\to\infty}\rho[J(f;x_0;\rho) - f(x_0)] = (m(\chi;1)/2)[f'_+(x_0) - f'_-(x_0)].$$

(Hint: Use (3.3.8) and the comment to Problem 1.5.16(i); see also MAMEDOV [3])
(ii) Show that if f satisfies the assumptions of (i), then for the integral of Gauss–Weierstrass

$$\lim_{t\to 0+}[W(f;x_0;t) - f(x_0)]/\sqrt{t} = [f'_+(x_0) - f'_-(x_0)]/\sqrt{\pi}.$$

6. Establish Problem 1.6.12 as an application of Prop. 3.3.5; likewise for Problem 1.6.14 in case $0 < \alpha < 1$. (Hint: Use (3.1.36), (3.1.43))

7. (i) Let the periodic approximate identity $\{\chi^*_\rho(x)\}$ be generated via (3.1.28) by $\chi \in \mathsf{NL}^1$, and let $I^*_\rho(f; x)$ be the corresponding (periodic) singular integral (3.1.30). Suppose that χ is positive and that there exist constants β, c with $0 < \beta \le 1$, $c > 0$ such that $\chi(x) = \mathsf{O}(|x|^{-1-\beta})$ for $|x| \to \infty$ and

$$\lim_{\rho\to\infty}\frac{1}{\sqrt{2\pi}\log\rho}\int_{-\rho}^{\rho}|u|^\beta\chi(u)\,du = c.$$

Show that $\Delta(\chi^*_\rho; \beta) = c\rho^{-\beta}\log\rho + \mathsf{O}(\rho^{-\beta})$, $\rho\to\infty$. Thus the Nikolskiĭ constant of the integrals $I^*_\rho(f; x)$ for the class $\mathsf{Lip}(\mathsf{C}_{2\pi}; \beta)_1$ is given by the constant c.

(Hint: By the periodicity of f, $\omega(\mathsf{C}_{2\pi}; f; \delta) \le \pi^{\beta}$ for every $f \in \mathsf{Lip}(\mathsf{C}_{2\pi}; \beta)_1$ and $\delta > 0$, and thus by (3.1.31) (compare also Problem 3.4.2)

$$\|I_\rho^*(f; \circ) - f(\circ)\|_{\mathsf{C}_{2\pi}} \le (\rho^{-\beta}/\sqrt{2\pi}) \int_{-\rho}^{\rho} |u|^\beta \chi(u)\, du + (\pi^\beta/\sqrt{2\pi}) \int_{|u| \ge \rho} \chi(u)\, du.$$

On the other hand, $\Delta(\chi_\rho^*; \beta) \ge (1/\sqrt{2\pi}) \int_{-\rho}^{\rho} |2 \sin(u/2\rho)|^\beta \chi(u)\, du$, and thus

$$\frac{\rho^{-\beta}}{\sqrt{2\pi}} \int_{-\rho}^{\rho} |u|^\beta \chi(u)\, du - R(\rho) \le \Delta(\chi_\rho^*; \beta) \le \frac{\rho^{-\beta}}{\sqrt{2\pi}} \int_{-\rho}^{\rho} |u|^\beta \chi(u)\, du + O(\rho^{-\beta}),$$

where $R(\rho) = (1/\sqrt{2\pi}) \int_{-\rho}^{\rho} [|u/\rho|^\beta - |2 \sin(u/2\rho)|^\beta] \chi(u)\, du$; see also NESSEL [4])

(ii) Show that the Fejér and Abel–Poisson kernels satisfy the assumptions of (i) with $\beta = 1$, $c = 2/\pi$. Therefore an application of (i) again delivers Problems 1.6.11 (i), 1.6.14 (case $\alpha = 1$).

3.4 Further Direct Approximation Theorems

In this and the following section only singular integrals of Fejér's type are considered. Results for the integral (3.1.3) are left to the reader. However, in contrast to the preceding section the kernel need not be positive.

Proposition 3.4.1. *Let $f \in \mathsf{X}(\mathbb{R})$ and $\chi \in \mathsf{NL}^1$ be an even function for which the second absolute moment exists as a finite number. Then for the corresponding singular integral $J(f; x; \rho)$ one has*

$$\text{(3.4.1)} \qquad \|J(f; \circ; \rho) - f(\circ)\|_{\mathsf{X}(\mathbb{R})} = O(\omega^*(\mathsf{X}(\mathbb{R}); f; \rho^{-1})) \qquad (\rho \to \infty).$$

Proof. It readily follows by (3.3.8), the Hölder–Minkowski inequality and Problem 3.3.1 that

$$\begin{aligned}\|J(f; \circ; \rho) - f(\circ)\|_{\mathsf{X}(\mathbb{R})} &\le \frac{\rho}{\sqrt{2\pi}} \int_0^\infty \|f(\circ + u) + f(\circ - u) - 2f(\circ)\|_{\mathsf{X}(\mathbb{R})} |\chi(\rho u)|\, du \\ &\le \omega^*(\mathsf{X}(\mathbb{R}); f; \rho^{-1}) \frac{\rho}{\sqrt{2\pi}} \int_0^\infty (1 + \rho u)^2 |\chi(\rho u)|\, du \\ &= \left\{\frac{1}{\sqrt{2\pi}} \int_0^\infty (1 + u)^2 |\chi(u)|\, du\right\} \omega^*(\mathsf{X}(\mathbb{R}); f; \rho^{-1}),\end{aligned}$$

which proves the assertion.

As an example, we treat the singular integral of Picard (cf. Problem 3.2.5).

Corollary 3.4.2. *For the singular integral $C_2(f; x; \rho)$ of Picard one has for every $f \in \mathsf{X}(\mathbb{R})$*

$$\|C_2(f; \circ; \rho) - f(\circ)\|_{\mathsf{X}(\mathbb{R})} = O(\omega^*(\mathsf{X}(\mathbb{R}); f; \rho^{-1})) \qquad (\rho \to \infty).$$

If furthermore $f \in \mathsf{Lip}^* (\mathsf{X}(\mathbb{R}); \alpha)$ for some $0 < \alpha \le 2$, then

$$\text{(3.4.2)} \qquad \|C_2(f; \circ; \rho) - f(\circ)\|_{\mathsf{X}(\mathbb{R})} = O(\rho^{-\alpha}) \qquad (\rho \to \infty).$$

If the second moment of the kernel does not exist, but the first, a corresponding direct approximation theorem holds. But even this weaker hypothesis is not satisfied by many particular singular integrals. For instance, the Fejér kernel (3.1.15) does not possess a first moment. Here the counterpart of Prop. 1.6.3 may be of interest.

Proposition 3.4.3. *Let $\chi \in \mathsf{NL}^1$ be an even function for which the αth absolute moment $m(\chi; \alpha)$ exists as a finite number for some $0 < \alpha \le 2$. Then for every $f \in \mathsf{Lip}^*(\mathsf{X}(\mathbb{R}); \alpha)$*

$$\|J(f; \circ; \rho) - f(\circ)\|_{\mathsf{X}(\mathbb{R})} = O(\rho^{-\alpha}) \qquad (\rho \to \infty). \tag{3.4.3}$$

Indeed, if $f \in \mathsf{Lip}^*(\mathsf{X}(\mathbb{R}); \alpha)_{M^*}$, then

$$\begin{aligned}\|J(f; \circ; \rho) - f(\circ)\|_{\mathsf{X}(\mathbb{R})} &\le (\rho/\sqrt{2\pi}) \int_0^\infty \|f(\circ + u) + f(\circ - u) - 2f(\circ)\|_{\mathsf{X}(\mathbb{R})} |\chi(\rho u)|\, du \\ &\le (\rho/\sqrt{2\pi}) M^* \int_0^\infty u^\alpha |\chi(\rho u)|\, du = (M^*/2) m(\chi; \alpha)\rho^{-\alpha},\end{aligned}$$

establishing (3.4.3).

As an immediate consequence we state

Corollary 3.4.4. *Let $\sigma(f; x; \rho)$ be the singular integral of Fejér. For every $f \in \mathsf{Lip}^*(\mathsf{X}(\mathbb{R}); \alpha)$, $0 < \alpha < 1$,*

$$\|\sigma(f; \circ; \rho) - f(\circ)\|_{\mathsf{X}(\mathbb{R})} = O(\rho^{-\alpha}) \qquad (\rho \to \infty). \tag{3.4.4}$$

For the case $\alpha = 1$ compare Problem 3.4.2.

Concerning higher order approximation we here only discuss those singular integrals $J(f; x; \rho)$ for which the kernel χ belongs to the class M^r, defined for $r \in \mathbb{N}$ by

$$\mathsf{M}^r = \left\{\chi \in \mathsf{NL}^1 \mid m(\chi; r) < \infty, \quad \int_{-\infty}^\infty u^k \chi(u)\, du = 0 \quad \text{for} \quad k = 1, \dots, r\right\}. \tag{3.4.5}$$

Lemma 3.4.5. *If $\chi \in \mathsf{M}^r$, then $f \in \mathsf{W}_{\mathsf{X}(\mathbb{R})}$ implies*

$$J(f; x; \rho) - f(x) = \frac{\rho}{\sqrt{2\pi}} \int_{-\infty}^\infty R(x, u)\chi(\rho u)\, du. \tag{3.4.6}$$

Here $R(x, u)$ is given by

$$R(x, u) = \frac{1}{(r-1)!} \int_x^{x-u} [\phi^{(r)}(t) - \phi^{(r)}(x)](x - u - t)^{r-1}\, dt, \tag{3.4.7}$$

where $\phi \in \mathsf{AC}_{\mathrm{loc}}^{r-1}$ with $\phi^{(k)} \in \mathsf{X}(\mathbb{R})$, $k = 0, 1, \dots, r$, is such that $\phi(x) = f(x)$ (a.e.). In particular,

$$\|J(f; \circ; \rho) - f(\circ)\|_{\mathsf{X}(\mathbb{R})} \le (2/r!) m(\chi; r) \|\phi^{(r)}\|_{\mathsf{X}(\mathbb{R})} \rho^{-r}. \tag{3.4.8}$$

Proof. We have

$$J(f; x; \rho) - f(x) = \frac{\rho}{\sqrt{2\pi}} \int_{-\infty}^\infty [f(x - u) - f(x)]\chi(\rho u)\, du. \tag{3.4.9}$$

With ϕ and $R(x, u)$ as specified in this Lemma, we obtain by partial integration

$$f(x - u) - f(x) = \sum_{k=1}^r (\phi^{(k)}(x)/k!)(-u)^k + R(x, u) \quad \text{(a.e.)}. \tag{3.4.10}$$

Thus (3.4.6) follows. Moreover, by an elementary substitution and the Hölder–Minkowski inequality

$$(3.4.11)\quad \|R(\circ, u)\|_{X(\mathbb{R})} = \left\| (u^r/(r-1)!) \int_0^1 [\phi^{(r)}(\circ - ut) - \phi^{(r)}(\circ)](t-1)^{r-1}\,dt \right\|_{X(\mathbb{R})}$$
$$\leq (|u|^r/(r-1)!) \int_0^1 \|\phi^{(r)}(\circ - ut) - \phi^{(r)}(\circ)\|_{X(\mathbb{R})}(1-t)^{r-1}\,dt,$$

which implies that

$$(3.4.12)\qquad \|R(\circ, u)\|_{X(\mathbb{R})} \leq 2\,\|\phi^{(r)}\|_{X(\mathbb{R})}|u|^r/r!.$$

Substitution into the estimate

$$(3.4.13)\qquad \|J(f; \circ; \rho) - f(\circ)\|_{X(\mathbb{R})} \leq (\rho/\sqrt{2\pi}) \int_{-\infty}^{\infty} \|R(\circ; u)\|_{X(\mathbb{R})}|\chi(\rho u)|\,du$$

delivers (3.4.8).

Proposition 3.4.6. (i) *Let* $r \in \mathbb{N}$ *and* $0 \leq \alpha \leq 1$. *If the* $(r+\alpha)$*th absolute moment of* $\chi \in \mathsf{M}^r$ *exists, then* $f \in \mathsf{W}^{r;\alpha}_{X(\mathbb{R})}$ *implies*

$$(3.4.14)\qquad \|J(f; \circ; \rho) - f(\circ)\|_{X(\mathbb{R})} = O(\rho^{-(r+\alpha)}) \qquad (\rho \to \infty).$$

(ii) *Let* $r = 2s$, $s \in \mathbb{N}$, *and* $0 < \alpha \leq 2$. *If the* $(2s+\alpha)$*th absolute moment of the even kernel* $\chi \in \mathsf{M}^r$ *exists, then* $f \in {}^*\mathsf{W}^{r;\alpha}_{X(\mathbb{R})}$ *implies* (3.4.14).

(iii) *Let* $r \in \mathbb{N}$ *and* $0 < \alpha \leq 2$. *If* $\chi \in \mathsf{M}^{r+2}$, *then for* $f \in \mathsf{W}^r_{X(\mathbb{R})}$

$$\|J(f; \circ; \rho) - f(\circ)\|_{X(\mathbb{R})} \leq (2/r!)[m(\chi; r+2) + m(\chi; r)]\rho^{-r}\omega^*(\mathsf{X}(\mathbb{R}); \phi^{(r)}; \rho^{-1}),$$

ϕ *being defined as in Lemma* 3.4.5. *In particular,* $f \in {}^*\mathsf{W}^{r;\alpha}_{X(\mathbb{R})}$ *implies* (3.4.14).

Proof. (i) Since $f \in \mathsf{W}^{r;\alpha}_{X(\mathbb{R})}$, there exists a constant B such that $\|\phi^{(r)}(\circ - u) - \phi^{(r)}(\circ)\|_{X(\mathbb{R})} \leq B\,|u|^\alpha$ for all $u \in \mathbb{R}$. Therefore by (3.4.11)

$$\|R(\circ, u)\|_{X(\mathbb{R})} \leq (|u|^r/(r-1)!)B \int_0^1 |ut|^\alpha(1-t)^{r-1}\,dt \leq (B/r!)\,|u|^{r+\alpha}.$$

This implies by (3.4.13) that

$$\|J(f; \circ; \rho) - f(\circ)\|_{X(\mathbb{R})} \leq \frac{B\rho}{r!\sqrt{2\pi}} \int_{-\infty}^{\infty} |u|^{r+\alpha}|\chi(\rho u)|\,du = \frac{B}{r!}\,m(\chi; r+\alpha)\rho^{-(r+\alpha)}.$$

(ii) Since χ is an even function, it follows by (3.4.6) that

$$J(f; x; \rho) - f(x) = (\rho/\sqrt{2\pi}) \int_0^{\infty} [R(x, u) + R(x, -u)]\chi(\rho u)\,du.$$

An elementary calculation yields

$$R(x, u) + R(x, -u) = \frac{u^r}{(r-1)!} \int_0^1 [\phi^{(r)}(x + ut) - \phi^{(r)}(x)$$
$$+ (-1)^r\phi^{(r)}(x - ut) - (-1)^r\phi^{(r)}(x)](1-t)^{r-1}\,dt.$$

Thus, if $r = 2s$, $s \in \mathbb{N}$, then

$$\|R(\circ, u) + R(\circ, -u)\|_{X(\mathbb{R})}$$
$$\leq \frac{|u|^r}{(r-1)!} \int_0^1 \|\phi^{(r)}(\circ + ut) + \phi^{(r)}(\circ - ut) - 2\phi^{(r)}(\circ)\|_{X(\mathbb{R})}(1-t)^{r-1}\,dt.$$

Now the hypothesis $f \in {}^*\mathsf{W}^{r;\alpha}_{X(\mathbb{R})}$ yields the existence of a constant B^* such that

$$\|R(\circ, u) + R(\circ, -u)\|_{X(\mathbb{R})} \leq (|u|^r/(r-1)!)B^* \int_0^1 |ut|^\alpha(1-t)^{r-1}\,dt \leq (B^*/r!)\,|u|^{r+\alpha},$$

which implies (3.4.14) as in (i).

(iii) The proof follows by (3.4.8) as for (2.2.2). Indeed, if $A^2(f; x; h)$ denote the second integral means of f (see Problem 3.1.8), we write $f(x) = A^2(f; x; h) + g(x)$. Again $f \in \mathsf{W}^r_{\mathsf{X}(\mathbb{R})}$ implies $A^2(f; x; h) \in \mathsf{W}^{r+2}_{\mathsf{X}(\mathbb{R})}$ and

$$\|[A^2(f; \circ; h)]^{(r+2)}(\circ)\|_{\mathsf{X}(\mathbb{R})} \le h^{-2}\omega^*(\mathsf{X}(\mathbb{R}); \phi^{(r)}; h).$$

On the other hand, $g \in \mathsf{W}^r_{\mathsf{X}(\mathbb{R})}$, and since by Problem 3.1.5(ii) and the counterpart of Problem 1.5.5

$$\|[A^2(\phi; \circ; h) - \phi(\circ)]^{(r)}(\circ)\|_{\mathsf{X}(\mathbb{R})} = \|A^2(\phi^{(r)}; \circ; h) - \phi^{(r)}(\circ)\|_{\mathsf{X}(\mathbb{R})} \le \omega^*(\mathsf{X}(\mathbb{R}); \phi^{(r)}; h),$$

Minkowski's inequality and (3.4.8) give

$$\|J(f; \circ; \rho) - f(\circ)\|_{\mathsf{X}(\mathbb{R})} \le \|J(A^2(f; \circ; h); \circ; \rho) - A^2(f; \circ; h)\|_{\mathsf{X}(\mathbb{R})} + \|J(g; \circ; \rho) - g(\circ)\|_{\mathsf{X}(\mathbb{R})}$$

$$\le \frac{2}{(r+2)!}\, m(\chi; r+2)\, \|[A^2(\phi; \circ; h)]^{(r+2)}(\circ)\|_{\mathsf{X}(\mathbb{R})}\, \rho^{-(r+2)} + \frac{2}{r!}\, m(\chi; r)\omega^*(\mathsf{X}(\mathbb{R}); \phi^{(r)}; h)\rho^{-r}.$$

This implies the assertion by setting $h = \rho^{-1}$.

For applications of the preceding results we refer to Sec. 6.4.

Problems

1. Formulate and prove the results of this section so as to obtain estimates in terms of the first modulus of continuity $\omega(\mathsf{X}(\mathbb{R}); f; \delta)$ (instead of ω^*), for $\mathsf{Lip}\,(\mathsf{X}(\mathbb{R}); \alpha)$-classes (instead of Lip^*), etc.
2. (i) Show that for every $\chi \in \mathsf{NL}^1$ and $f \in \mathsf{Lip}\,(\mathsf{X}(\mathbb{R}); \alpha)_M$

$$\|J(f; \circ; \rho) - f(\circ)\|_{\mathsf{X}(\mathbb{R})} \le M\rho^{-\alpha}\int_{-\rho a}^{\rho a} |u|^\alpha |\chi(u)|\, du + 2\,\|f\|_{\mathsf{X}(\mathbb{R})} \int_{|u| \ge \rho a} |\chi(u)|\, du,$$

where a is a positive number which one is free to choose suitably (compare also H. S. Shapiro [1, p. 26]).
   (ii) As an application prove that for the singular integral of Fejér $f \in \mathsf{Lip}\,(\mathsf{X}(\mathbb{R}); 1)$ implies $\|\sigma(f; \circ; \rho) - f(\circ)\|_{\mathsf{X}(\mathbb{R})} = O(\rho^{-1} \log \rho)$, $\rho \to \infty$. Show by an example that this estimate cannot in general be improved for the class $\mathsf{Lip}\,(\mathsf{X}(\mathbb{R}); 1)$.
3. Let $f \in \mathsf{W}^{2r}_{\mathsf{X}(\mathbb{R})}$ and $\chi \in \mathsf{NL}^1$ be an even, positive function for which $m(\chi; 2r+1) < \infty$. Show that

$$\Big\|J(f; \circ; \rho) - f(\circ) - \sum_{k=1}^{r} (m(\chi; 2k)/\rho^{2k}(2k)!)\phi^{(2k)}(\circ)\Big\|_{\mathsf{X}(\mathbb{R})}$$
$$\le [(m(\chi; 2r) + m(\chi; 2r+1))/(2r)!]\rho^{-2r}\omega(\mathsf{X}(\mathbb{R}); \phi^{(2r)}; \rho^{-1}),$$

where ϕ is given as in Lemma 3.4.5. Extend to kernels which are not necessarily even and positive. (Hint: See also Butzer [6])
4. Let $\chi \in \mathsf{NL}^1$ be such that $m(\chi; 1) < \infty$. If $f \in \mathsf{L}^\infty$ is differentiable at x_0, show that

$$\lim_{\rho \to \infty} \rho[J(f; x_0; \rho) - f(x_0)] = -\Big[(1/\sqrt{2\pi})\int_{-\infty}^{\infty} u\chi(u)\, du\Big] f'(x_0).$$

(Hint: Use Lebesgue's dominated convergence theorem and (see also H. S. Shapiro [1, p. 39])

$$\rho[J(f; x_0; \rho) - f(x_0)] = \frac{1}{\sqrt{2\pi}}\int_{-\infty}^{\infty} \frac{f(x_0 - \rho^{-1}u) - f(x_0)}{\rho^{-1}u}\, u\chi(u)\, du)$$

5. (i) Let $\chi \in \mathsf{NL}^1$ be such that $m(\chi; r) < \infty$ for some $r \in \mathbb{N}$. Show that for $f \in \mathsf{W}^r_{\mathsf{X}(\mathbb{R})}$ and $\rho \to \infty$

$$\Big\|\rho^r\Big[J(f; \circ; \rho) - f(\circ) - \sum_{k=1}^{r-1} \frac{(-1)^k \gamma(\chi; k)}{\rho^k k!}\, \phi^{(k)}(\circ)\Big] - \frac{(-1)^r\gamma(\chi; r)}{r!}\, \phi^{(r)}(\circ)\Big\|_{\mathsf{X}(\mathbb{R})} = o(1),$$

where ϕ is given as in Lemma 3.4.5 and $\gamma(\chi; k) \equiv (1/\sqrt{2\pi}) \int_{-\infty}^{\infty} u^k \chi(u)\, du$. (Hint: Apply (3.4.10) and show that the left-hand side of the assertion may be estimated by

$$(\rho^{r+1}/\sqrt{2\pi})\Big(\int_{|u|\le\delta} + \int_{|u|\ge\delta}\Big) \|R(\circ, u)\|_{X(\mathbb{R})} |\chi(\rho u)|\, du.$$

Using (3.4.11) and the continuity of $\phi^{(r)}$ in the mean for the first, (3.4.12) and the existence of $m(\chi; r)$ for the second integral, the assertion follows)

(ii) Let $\chi \in \mathsf{NL}^1$ be an even function for which $m(\chi; 2) < \infty$. Show that for $f \in \mathsf{W}^2_{X(\mathbb{R})}$

$$\lim_{\rho\to\infty} \|\rho^2[J(f; \circ; \rho) - f(\circ)] - (m(\chi; 2)/2)\phi^{(2)}(\circ)\|_{X(\mathbb{R})} = 0.$$

Compare with Prop. 3.3.3.

(iii) Show that for the singular integral of Gauss–Weierstrass of $f \in \mathsf{W}^2_{X(\mathbb{R})}$

$$\lim_{t\to 0+} \|t^{-1}[W(f; \circ; t) - f(\circ)] - \phi^{(2)}(\circ)\|_{X(\mathbb{R})} = 0.$$

6. (i) Let $f \in \mathsf{L}^\infty$ and suppose that $\int_0^\infty u^{-2}\overline{\Delta}_u^2 f(x_0)\, du$ exists as a Lebesgue integral at some point x_0. Show that for the singular integral of Cauchy–Poisson

$$\lim_{y\to 0+} \frac{P(f; x_0; y) - f(x_0)}{y} = \frac{1}{\pi}\int_0^\infty \frac{f(x_0 + u) + f(x_0 - u) - 2f(x_0)}{u^2}\, du.$$

In case of the Fejér integral show that $\rho[\sigma(f; x_0; \rho) - f(x_0)]$ tends to the same limit as $\rho \to \infty$. (Hint: Use (3.3.8) and Lebesgue's dominated convergence theorem for $P(f; x; y)$, the Riemann–Lebesgue lemma (Prop. 5.1.2) for $\sigma(f; x; \rho)$; see also H. S. Shapiro [1, p. 40 f])

(ii) Let $f \in \mathsf{X}(\mathbb{R})$. Show that for the Fejér integral

$$\Big\| \sigma(f; \circ; \rho) - f(\circ) - \frac{1}{\pi\rho}\int_{1/\rho}^\infty \frac{\overline{\Delta}_u^2 f(\circ)}{u^2}\, du \Big\|_{X(\mathbb{R})} = O(\omega^*(\mathsf{X}(\mathbb{R}); f; \rho^{-1})).$$

(Hint: Efimov [1], see also Zamansky [1], Golinskiĭ [2])

3.5 Inverse Approximation Theorems

Bernstein's Theorem 2.3.3, particularly its method of proof, was the starting point for a series of investigations on the inference of structural properties of a function from a given order of approximation. In the present section we discuss one of these generalizations by considering the approximation of $f \in \mathsf{X}(\mathbb{R})$ by singular integrals $J(f; x; \rho)$ of Fejér's type.

Theorem 3.5.1. *Let the kernel* $\chi \in \mathsf{NL}^1$ *of the singular integral* $J(f; x; \rho)$ *belong to* $\mathsf{W}^2_{C\cap L^1}$. *If, for some* $\alpha > 0$, $f \in \mathsf{X}(\mathbb{R})$ *can be approximated by* $J(f; x; \rho)$ *such that*

(3.5.1) $$\|J(f; \circ; \rho) - f(\circ)\|_{X(\mathbb{R})} = O(\rho^{-\alpha}) \qquad (\rho \to \infty),$$

then

(3.5.2) $$\omega^*(\mathsf{X}(\mathbb{R}); f; \delta) = \begin{cases} O(\delta^\alpha), & 0 < \alpha < 2 \\ O(\delta^2\, |\log \delta|), & \alpha = 2 \\ O(\delta^2), & \alpha > 2 \end{cases} \qquad (\delta \to 0+).$$

Proof. The technique to be used is a direct adaptation of Bernstein's idea used in the proofs of Theorem 2.3.3, 2.3.5 and 2.5.4. Indeed, there exists a constant $B > 0$ such that $\|J(f; \circ; \rho) - f(\circ)\|_{X(\mathbb{R})} \le B\rho^{-\alpha}$ for all $\rho > 0$. Setting

$$U_2(x) = J(f; x; 2^2), \qquad U_n(x) = J(f; x; 2^n) - J(f; x; 2^{n-1}) \quad (n = 3, 4, \dots),$$

we obtain

$$\begin{aligned}(3.5.3) \qquad \|U_n\|_{X(\mathbb{R})} &\le \|J(f; \circ; 2^n) - f(\circ)\|_{X(\mathbb{R})} + \|J(f; \circ; 2^{n-1}) - f(\circ)\|_{X(\mathbb{R})} \\ &\le (1 + 2^\alpha)B2^{-n\alpha}\end{aligned}$$

for each $n = 3, 4, \dots$. Since $\sum_{k=2}^{n} U_k(x) = J(f; x; 2^n)$, by Theorem 3.1.6

$$(3.5.4) \qquad \lim_{n\to\infty} \Big\| f - \sum_{k=2}^{n} U_k \Big\|_{X(\mathbb{R})} = 0.$$

In particular, $\|f\|_{X(\mathbb{R})} \le \sum_{k=2}^{\infty} \|U_k\|_{X(\mathbb{R})}$, and for every $h \in \mathbb{R}$ and integer $m \ge 2$

$$\begin{aligned}(3.5.5) \quad &\|f(\circ + h) + f(\circ - h) - 2f(\circ)\|_{X(\mathbb{R})} \\ &\le \sum_{k=2}^{m} \|U_k(\circ + h) + U_k(\circ - h) - 2U_k(\circ)\|_{X(\mathbb{R})} + 4 \sum_{k=m+1}^{\infty} \|U_k\|_{X(\mathbb{R})}.\end{aligned}$$

Since $\chi \in \mathsf{W}^2_{\mathsf{C}\cap\mathsf{L}^1}$, it follows by Problem 3.1.5(i) that $J(f; x; \rho) \in \mathsf{W}^2_{\mathsf{C}}$ for every $f \in \mathsf{X}(\mathbb{R})$ and $\rho > 0$, and thus

$$U_k(x + h) + U_k(x - h) - 2U_k(x) = \int_{-h/2}^{h/2} \int_{-h/2}^{h/2} U_k''(x + u_1 + u_2)\, du_1\, du_2.$$

Therefore by the Hölder–Minkowski inequality

$$\|U_k(\circ + h) + U_k(\circ - h) - 2U_k(\circ)\|_{X(\mathbb{R})} \le h^2 \|U_k''\|_{X(\mathbb{R})}.$$

By the commutative law for convolutions the identity

$$(3.5.6) \quad U_k(x) = J(f(\circ) - J(f; \circ; 2^{k-1}); x; 2^k) - J(f(\circ) - J(f; \circ; 2^k); x; 2^{k-1})$$

follows for every $k = 3, 4, \dots$. Since $\|J''(f; \circ; \rho)\|_{X(\mathbb{R})} \le \rho^2 \|\chi''\|_1 \|f\|_{X(\mathbb{R})}$ by (3.1.49), this implies

$$\begin{aligned}\|U_k''\|_{X(\mathbb{R})} &\le \|J''(f(\circ) - J(f; \circ; 2^{k-1}); \circ; 2^k)\|_{X(\mathbb{R})} + \|J''(f(\circ) - J(f; \circ; 2^k); \circ; 2^{k-1})\|_{X(\mathbb{R})} \\ &\le 2^{2k}\|\chi''\|_1\{\|f(\circ) - J(f; \circ; 2^{k-1})\|_{X(\mathbb{R})} + 2^{-2}\|f(\circ) - J(f; \circ; 2^k)\|_{X(\mathbb{R})}\} \\ &\le 2^{2k}\|\chi''\|_1\{B2^{-(k-1)\alpha} + 2^{-2}B2^{-k\alpha}\} \le 2^{1+\alpha}B\,\|\chi''\|_1 2^{k(2-\alpha)}.\end{aligned}$$

Together with (3.5.3) and (3.5.5) it therefore follows that

$$\omega^*(\mathsf{X}(\mathbb{R}); f; \delta) \le C\Big\{\delta^2 \sum_{k=2}^{m} 2^{k(2-\alpha)} + \sum_{k=m+1}^{\infty} 2^{-k\alpha}\Big\},$$

where C is a suitable constant. Thus the second modulus of continuity of f admits the estimate (2.3.10), and the proof is completed by the routine arguments.

We recall Problem 3.1.5(i) which states that if $\chi \in \mathsf{W}^r_{\mathsf{C}\cap\mathsf{L}^1}$ for some positive integer r, then $J(f; x; \rho) \in \mathsf{W}^r_{\mathsf{C}}$ for every $f \in \mathsf{X}(\mathbb{R})$ and $\rho > 0$ and

$$\|[J(f; \circ; \rho)]^{(r)}(\circ)\|_{\mathsf{X}(\mathbb{R})} \le \rho^r \|\chi^{(r)}\|_1 \|f\|_{\mathsf{X}(\mathbb{R})}. \tag{3.5.7}$$

From this Bernstein-type inequality it is possible to infer an estimate for the rth (rather than the second) modulus of continuity of f, given the approximation (3.5.1) (see Problem 3.5.1).

Here we are interested in the analog of Theorems 2.3.4, 2.3.6, furnishing the existence of derivatives in case the approximation (3.5.1) is nice enough.

Theorem 3.5.2. *Let the kernel $\chi \in \mathsf{NL}^1$ of the singular integral $J(f; x; \rho)$ belong to $\mathsf{W}^{r+2}_{\mathsf{C}\cap\mathsf{L}^1}$ for some positive integer r. If, for some $\alpha > 0$, $f \in \mathsf{X}(\mathbb{R})$ can be approximated by $J(f; x; \rho)$ such that*

$$\|J(f; \circ; \rho) - f(\circ)\|_{\mathsf{X}(\mathbb{R})} = O(\rho^{-(r+\alpha)}) \qquad (\rho \to \infty), \tag{3.5.8}$$

then $f \in \mathsf{W}^r_{\mathsf{X}(\mathbb{R})}$. Moreover, if $\phi \in \mathsf{AC}^{r-1}_{\mathrm{loc}}$ with $\phi^{(k)} \in \mathsf{X}(\mathbb{R})$, $k = 0, 1, \ldots, r$, is such that $\phi(x) = f(x)$ (a.e.), then

$$\omega^*(\mathsf{X}(\mathbb{R}); \phi^{(r)}; \delta) = \begin{cases} O(\delta^\alpha), & 0 < \alpha < 2 \\ O(\delta^2\,|\log \delta|), & \alpha = 2 \\ O(\delta^2), & \alpha > 2 \end{cases} \qquad (\delta \to 0+). \tag{3.5.9}$$

Proof. By hypothesis $\|J(f; \circ; \rho) - f(\circ)\|_{\mathsf{X}(\mathbb{R})} \le B\rho^{-r-\alpha}$ for some constant B. Defining $U_n(x)$ as in the previous proof, then $\chi \in \mathsf{W}^r_{\mathsf{C}\cap\mathsf{L}^1}$ implies by Problem 3.1.5(i) that $U_n \in \mathsf{W}^j_{\mathsf{C}}$ for every $j = 1, 2, \ldots, r$ and $n = 2, 3, \ldots$. Furthermore, in view of (3.5.6) and (3.5.7) for every $n = 3, 4, \ldots$

$$\begin{aligned} \|U_n^{(j)}\|_{\mathsf{X}(\mathbb{R})} &\le \|J^{(j)}(f(\circ) - J(f; \circ; 2^{n-1}); \circ; 2^n)\|_{\mathsf{X}(\mathbb{R})} \\ &\quad + \|J^{(j)}(f(\circ) - J(f; \circ; 2^n); \circ; 2^{n-1})\|_{\mathsf{X}(\mathbb{R})} \\ &\le 2^{nj}\|\chi^{(j)}\|_1\{\|f(\circ) - J(f; \circ; 2^{n-1})\|_{\mathsf{X}(\mathbb{R})} + 2^{-j}\|f(\circ) - J(f; \circ; 2^n)\|_{\mathsf{X}(\mathbb{R})}\} \\ &\le 2^{nj}\|\chi^{(j)}\|_1\{B2^{-(n-1)(r+\alpha)} + 2^{-j}B2^{-n(r+\alpha)}\} \le 2^{r+\alpha+1}B\,\|\chi^{(j)}\|_1 2^{-n(r-j)-n\alpha}. \end{aligned} \tag{3.5.10}$$

Thus the series $\sum_{k=2}^\infty \|U_k^{(j)}\|_{\mathsf{X}(\mathbb{R})}$ converge for each $j = 0, 1, \ldots, r$.

In case $\mathsf{X} = \mathsf{C}$, this implies that each series $\sum_{k=2}^\infty U_k^{(j)}(x)$ converges uniformly, and since $f(x) = \sum_{k=2}^\infty U_k(x)$ by (3.5.4), $f^{(r)} \in \mathsf{C}$ and $f^{(r)}(x) = \sum_{k=2}^\infty U^{(r)}(x)$.

If $\mathsf{X}(\mathbb{R}) = \mathsf{L}^p$, $1 \le p < \infty$, then $\sum_{k=2}^\infty \|U_k^{(j)}\|_{\mathsf{X}(\mathbb{R})} < \infty$ implies that for each $j = 0, 1, \ldots, r$ the sequence of partial sums $\sum_{k=2}^n U_k^{(j)}(x)$ converges in L^p-norm to a function $g_j \in \mathsf{L}^p$. Hence there exists a subsequence $\sum_{k=2}^{n_l} U_k^{(j)}(x)$ which converges almost everywhere to $g_j(x)$ for each $j = 0, 1, \ldots, r$. For a common point of convergence x_0 we set

$$R_{j-1}(x) = \int_{x_0}^x g_j(u)\,du + g_{j-1}(x_0).$$

Since by Hölder's inequality for $x > x_0$, for example,

$$\begin{aligned} \left|\int_{x_0}^x \left[\sum_{k=2}^n U_k^{(j)}(u) - g_j(u)\right] du\right| &\le \int_{x_0}^x \left|\sum_{k=2}^n U_k^{(j)}(u) - g_j(u)\right| du \\ &\le (x - x_0)^{1/p'}\left\{\int_{x_0}^x \left|\sum_{k=2}^n U_k^{(j)}(u) - g_j(u)\right|^p du\right\}^{1/p} \le (x - x_0)^{1/p'}\left\|\sum_{k=}^n U_k^{(j)} - g_j\right\|_p, \end{aligned}$$

it follows that

$$\lim_{n\to\infty} \int_{x_0}^x \sum_{k=2}^n U_k^{(j)}(u)\,du = \int_{x_0}^x g_j(u)\,du = R_{j-1}(x) - g_{j-1}(x_0). \tag{3.5.11}$$

On the other hand, $U_k^{(j)}$ is an absolutely continuous function for each $j = 0, 1, \ldots, r-1$. Therefore

$$\int_{x_0}^{x} \sum_{k=2}^{n_l} U_k^{(j)}(u)\,du = \sum_{k=2}^{n_l} U_k^{(j-1)}(x) - \sum_{k=2}^{n_l} U_k^{(j-1)}(x_0),$$

and letting $l \to \infty$, the right-hand side converges to $[g_{j-1}(x) - g_{j-1}(x_0)]$ for almost all $x \in \mathbb{R}$. Together with (3.5.11) this shows that $R_{j-1}(x) = g_{j-1}(x)$ a.e., and since $g_0(x) = f(x)$ a.e. by (3.5.4), the assertion $f \in \mathsf{W}^r_{\mathsf{X}(\mathbb{R})}$ is established (compare (2.3.8)).

If $\phi \in \mathsf{AC}^{r-1}_{\mathrm{loc}}$ with $\phi^{(j)} \in \mathsf{X}(\mathbb{R})$ for each $j = 0, 1, \ldots, r$ is such that $\phi(x) = f(x)$ (a.e.), then $\lim_{n\to\infty} \|\phi^{(r)} - \sum_{k=2}^{n} U_k^{(r)}\|_{\mathsf{X}(\mathbb{R})} = 0$ by the preceding arguments. Thus $\|\phi^{(r)}\|_{\mathsf{X}(\mathbb{R})} \le \sum_{k=2}^{\infty} \|U_k^{(r)}\|_{\mathsf{X}(\mathbb{R})}$ and

$$\|\phi^{(r)}(\circ + h) + \phi^{(r)}(\circ - h) - 2\phi^{(r)}(\circ)\|_{\mathsf{X}(\mathbb{R})}$$
$$\le \sum_{k=2}^{m} \|U_k^{(r)}(\circ + h) + U_k^{(r)}(\circ - h) - 2U_k^{(r)}(\circ)\|_{\mathsf{X}(\mathbb{R})} + 4 \sum_{k=m+1}^{\infty} \|U_k^{(r)}\|_{\mathsf{X}(\mathbb{R})}.$$

Since $\chi \in \mathsf{W}^{r+2}_{\mathsf{C}\cap\mathsf{L}^1}$, $U_k^{(r)} \in \mathsf{W}^2_{\mathsf{C}}$ and $\|U_k^{(r)}(\circ + h) + U_k^{(r)}(\circ - h) - 2U_k^{(r)}(\circ)\|_{\mathsf{X}(\mathbb{R})} \le h^2 \|U_k^{(r+2)}\|_{\mathsf{X}(\mathbb{R})}$ for every $h \in \mathbb{R}$ and $k = 2, 3, \ldots$. Moreover, as in (3.5.10)

$$\|U_k^{(r+2)}\|_{\mathsf{X}(\mathbb{R})} \le 2^{r+\alpha+1} B\, \|\chi^{(r+2)}\|_1 2^{k(2-\alpha)}.$$

Combining the results, one has for every integer $m \ge 2$

$$\omega^*(\mathsf{X}(\mathbb{R}); \phi^{(r)}; \delta) \le C\Big\{\delta^2 \sum_{k=2}^{m} 2^{k(2-\alpha)} + \sum_{k=m+1}^{\infty} 2^{-k\alpha}\Big\},$$

where C is a constant. By considering different cases as in the proofs of Sec. 2.3, (3.5.9) follows.

For many of the particular singular integrals so far considered it is easily seen that the kernels are of class C_0^∞, all derivatives being absolutely integrable. Therefore the previous results may be applied for any positive integer r.

Corollary 3.5.3. *Let $W(f; x; t)$ be the singular integral of Gauss–Weierstrass of $f \in \mathsf{X}(\mathbb{R})$.*

(i) *For* $0 < \alpha < 2$, $f \in \mathsf{Lip}^*(\mathsf{X}(\mathbb{R}); \alpha) \Leftrightarrow \|W(f; \circ; t) - f(\circ)\|_{\mathsf{X}(\mathbb{R})} = O(t^{\alpha/2}),\quad t \to 0+,$

(ii) $f \in \mathsf{Lip}^*(\mathsf{X}(\mathbb{R}); 2) \Rightarrow \|W(f; \circ; t) - f(\circ)\|_{\mathsf{X}(\mathbb{R})} = O(t),\quad t \to 0+.$

The proof follows by Cor. 3.3.2 and Theorem 3.5.1. The inverse implication of (ii) is valid as well. But for this purpose other methods of proof must be employed which will be presented in Chapters 12, 13. There it is shown that an order of approximation beyond the critical value $O(t)$ is impossible for nontrivial functions f. Indeed, Prop. 12.3.2, 12.4.3, 13.2.1, and Problem 13.2.1 assert that $\|W(f; \circ; t) - f(\circ)\|_{\mathsf{X}(\mathbb{R})} = o(t)$ as $t \to 0+$ necessarily implies f to be constant (which must be zero in L^p-spaces).

Corollary 3.5.4. *Let $\sigma(f; x; \rho)$ be the singular integral of Fejér of $f \in \mathsf{X}(\mathbb{R})$.*

(i) *For* $0 < \alpha < 1$, $\|\sigma(f; \circ; \rho) - f(\circ)\|_{\mathsf{X}(\mathbb{R})} = O(\rho^{-\alpha}) \Leftrightarrow f \in \mathsf{Lip}(\mathsf{X}(\mathbb{R}); \alpha)$,

(ii) $\|\sigma(f; \circ; \rho) - f(\circ)\|_{\mathsf{X}(\mathbb{R})} = O(\rho^{-1}) \Rightarrow f \in \mathsf{Lip}^*(\mathsf{X}(\mathbb{R}); 1)$.

The proof is given by Cor. 3.4.4 and Problem 3.5.1. The result in (ii) is not best possible as is to be shown in Sec. 12.4.1, 13.2.5, where the exact characterization of the class of functions $\{f \in \mathsf{X}(\mathbb{R}) \mid \|\sigma(f; \circ; \rho) - f(\circ)\|_{\mathsf{X}(\mathbb{R})} = O(\rho^{-1})\}$ will be given. For approximation by Fejér's integral the value $\alpha = 1$ is the critical one. For L^p, $1 \le p \le 2$, this will be a result of Sec. 12.3.1, where it is shown that $\|\sigma(f; \circ; \rho) - f(\circ)\|_p = o(\rho^{-1})$ implies $\|f\|_p = 0$.

Problems

1. Let the kernel $\chi \in \mathsf{NL}^1$ of the singular integral $J(f; x; \rho)$ belong to $\mathsf{W}^r_{\mathsf{C} \cap \mathsf{L}^1}$ for some $r \in \mathbb{N}$. If, for some $\alpha > 0$, $f \in \mathsf{X}(\mathbb{R})$ can be approximated by $J(f; x; \rho)$ such that (3.5.1) holds, show that
$$\omega_r(\mathsf{X}(\mathbb{R}); f; \delta) = \begin{cases} O(\delta^\alpha), & 0 < \alpha < r \\ O(\delta^r\,|\log \delta|), & \alpha = r \\ O(\delta^r), & \alpha > r \end{cases} \qquad (\delta \to 0+).$$
2. Let the kernel $\chi \in \mathsf{NL}^1$ of the singular integral $J(f; x; \rho)$ belong to $\mathsf{W}^{r+1}_{\mathsf{C} \cap \mathsf{L}^1}$ for some $r \in \mathbb{N}$. If, for some $\alpha > 0$, $f \in \mathsf{X}(\mathbb{R})$ can be approximated by $J(f; x; \rho)$ such that (3.5.8) holds, then $f \in \mathsf{W}^r_{\mathsf{X}(\mathbb{R})}$. If ϕ is as in Theorem 3.5.2, show that $\omega(\mathsf{X}(\mathbb{R}); \phi^{(r)}; \delta)$ admits estimates as in the previous Problem for $r = 1$.
3. Let $f \in \mathsf{X}(\mathbb{R})$. Show that
   (i) for $0 < \alpha < 1$, $f \in \mathsf{Lip}^* (\mathsf{X}(\mathbb{R}); \alpha) \Leftrightarrow f \in \mathsf{Lip}(\mathsf{X}(\mathbb{R}); \alpha)$,
   (ii) $f \in \mathsf{Lip}^* (\mathsf{X}(\mathbb{R}); 1) \Rightarrow \omega(\mathsf{X}(\mathbb{R}); f; \delta) = O(\delta\,|\log \delta|)$.
4. Let $N(f; x; \rho)$ be the singular integral of Jackson–de La Vallée Poussin of $f \in \mathsf{X}(\mathbb{R})$. Show that
   (i) for $0 < \alpha < 2$, $f \in \mathsf{Lip}^* (\mathsf{X}(\mathbb{R}); \alpha) \Leftrightarrow \|N(f; \circ; \rho) - f(\circ)\|_{\mathsf{X}(\mathbb{R})} = O(\rho^{-\alpha})$, $\rho \to \infty$,
   (ii) $f \in \mathsf{Lip}^* (\mathsf{X}(\mathbb{R}); 2) \Rightarrow \|N(f; \circ; \rho) - f(\circ)\|_{\mathsf{X}(\mathbb{R})} = O(\rho^{-2})$, $\rho \to \infty$.
   (Hint: Apply Prop. 3.4.1, Theorem 3.5.1. For the inverse implication of (ii) see Problems 12.3.4, 12.3.5, and Sec. 10.6, 13.2)
5. Let $P(f; x; y)$ be the singular integral of Cauchy–Poisson of $f \in \mathsf{X}(\mathbb{R})$. Show that
   (i) for $0 < \alpha < 1$, $\|P(f; \circ; y) - f(\circ)\|_{\mathsf{X}(\mathbb{R})} = O(y^\alpha) \Leftrightarrow f \in \mathsf{Lip}(\mathsf{X}(\mathbb{R}); \alpha)$,
   (ii) $\|P(f; \circ; y) - f(\circ)\|_{\mathsf{X}(\mathbb{R})} = O(y) \Rightarrow f \in \mathsf{Lip}^* (\mathsf{X}(\mathbb{R}); 1)$.

3.6 Shape Preserving Properties

In this section we will give a brief account concerned with the following general problem: Assume that the graph of a function f has a certain shape; for example, assume f to be monotone or convex. What can be said about the approximation by a certain process? One result is that the approximation of a convex function f by the Gauss–Weierstrass integral is monotone. Furthermore, one asks for those approximation processes which assume the shape of the graph of f as well. This problem will be discussed in terms of the variation diminishing property. Roughly speaking, an approximation process is called variation diminishing if the approximators do not oscillate more often about any straight line than the function to be approximated. The result is that a singular integral is variation diminishing if and only if the kernel is totally positive.

3.6.1 Singular Integral of Gauss–Weierstrass

The discussion of the problems of this subsection will be restricted to the representative example of the singular integral $W(f; x; t)$ of Gauss–Weierstrass as defined in

Sec. 3.1.3. Let W denote the class of all functions $f \in C_{loc}$ which grow to infinity in such a way that for each $\alpha > 0$

$$\lim_{|x| \to \infty} e^{-\alpha x^2} |f(x)| = 0. \tag{3.6.1}$$

Obviously, every algebraic polynomial belongs to W. For every $f \in W$ and $t > 0$ the integral $W(f; x; t)$ is well-defined as a function of class C_{loc} (see also Problem 3.6.1). In particular,

$$W(1; x; t) = 1, \qquad W(u; x; t) = x. \tag{3.6.2}$$

A real-valued function f is said to be *convex* on $\mathbb{R}$ if $\Delta_h^2 f(x) \geq 0$ for all $x \in \mathbb{R}$, $h > 0$ (compare Sec. 6.3.1). Note that a function convex on $\mathbb{R}$ cannot be bounded unless it is a constant.

Proposition 3.6.1. *If the function $f \in W$ is convex on $\mathbb{R}$, then $W(f; x; t)$ is convex on $\mathbb{R}$ for each $t > 0$.*

Indeed, one has

$$\Delta_h^2 W(f; x; t) = \frac{1}{\sqrt{4\pi t}} \int_{-\infty}^{\infty} \Delta_h^2 f(x - u)\, e^{-u^2/4t}\, du,$$

so that the assertion follows since the kernal is positive.

Proposition 3.6.2. *If $f \in W$ is convex, then $W(f; x; t) \geq f(x)$ for all $x \in \mathbb{R}$, $t > 0$.*

Proof. Since f is continuous and convex on $\mathbb{R}$, there exists a right-hand derivative $f'_+(x)$ and a left-hand derivative $f'_-(x)$ (cf. Problem 1.5.12 for the definition) for all $x \in \mathbb{R}$; these derivatives are monotonely increasing, and $f'_+(x) = f'_-(x)$ for almost all x (compare (6.3.3), Problem 6.3.5). Let x_0 be fixed and choose c such that

$$c = \begin{cases} f'(x_0) & \text{if } f'(x_0) \text{ exist,} \\ f'_-(x_0) < c < f'_+(x_0) & \text{otherwise.} \end{cases} \tag{3.6.3}$$

In view of the monotonicity of the derivatives one has

$$\begin{aligned} c &\leq f'_+(x) \quad \text{for } x > x_0, \\ c &\geq f'_-(x) \quad \text{for } x < x_0. \end{aligned} \tag{3.6.4}$$

For the first degree polynomial $p(x) = c(x - x_0) + f(x_0)$, (3.6.2) implies $W(p; x; t) = p(x)$ and thus

$$W(p; x_0; t) = f(x_0). \tag{3.6.5}$$

By (3.6.4) it follows that (cf. (6.3.3))

$$f(x) - f(x_0) = \int_{x_0}^{x} f'_+(u)\, du \geq c(x - x_0) \qquad \text{for } x > x_0,$$

$$f(x) - f(x_0) = -\int_{x}^{x_0} f'_-(u)\, du \geq c(x - x_0) \quad \text{for } x < x_0,$$

and therefore $f(x) \geq p(x)$ for all $x \in \mathbb{R}$. Since the Gauss–Weierstrass kernel is positive, this implies $W(f; x; t) \geq W(p; x; t)$, and thus $W(f; x_0; t) \geq f(x_0)$ by (3.6.5). Since x_0 is arbitrary, the assertion follows.

Lemma 3.6.3. *Let $f \in \mathsf{W}$. Then for each $x \in \mathbb{R}$ and every fixed pair $0 < \delta < \eta$*

$$\lim_{t\to 0+} \frac{\int_\eta^\infty |f(x \pm u)|\, e^{-u^2/4t}\, du}{\int_\delta^\eta e^{-u^2/4t}\, du} = 0. \tag{3.6.6}$$

Proof. Condition (3.6.1) implies that for each $\alpha > 0$ there exists a constant c such that $|f(x + u)| \leq c \exp\{2\alpha u^2\}$ for all $u \in \mathbb{R}$. Let α be fixed and choose $t < 1/8\alpha$. Then

$$\begin{aligned}\int_\eta^\infty |f(x + u)|\, e^{-u^2/4t}\, du &\leq c \int_\eta^\infty e^{-u^2[(1/4t) - 2\alpha]}\, du\\ &\leq \frac{c}{\eta}\int_\eta^\infty u\, e^{-u^2[(1/4t) - 2\alpha]}\, du\\ &= \frac{c}{2\eta[(1/4t) - 2\alpha]}\, e^{2\alpha\eta^2}\, e^{-\eta^2/4t}.\end{aligned}$$

On the other hand,

$$\int_\delta^\eta e^{-u^2/4t}\, du \geq (1/\eta)\int_\delta^\eta u\, e^{-u^2/4t}\, du = (2t/\eta)\{e^{-\delta^2/4t} - e^{-\eta^2/4t}\}.$$

For $t \leq (\eta^2 - \delta^2)/4 \log 2$ one has $\exp\{-\eta^2/4t\} \leq \frac{1}{2}\exp\{-\delta^2/4t\}$, and therefore

$$\int_\delta^\eta e^{-u^2/4t}\, du \geq (t/\eta)\, e^{-\delta^2/4t}.$$

Upon collecting the results, it follows that

$$\begin{aligned}\frac{\int_\eta^\infty |f(x + u)|\, e^{-u^2/4t}\, du}{\int_\delta^\eta e^{-u^2/4t}\, du} &\leq \frac{c}{2t[(1/4t) - 2\alpha]}\, e^{2\alpha\eta^2}\, e^{-(\eta^2 - \delta^2)/4t}\\ &\leq c_1\, e^{-(\eta^2 - \delta^2)/4t},\end{aligned}$$

where c_1 is a suitable constant. Since the right-hand side tends to zero as $t \to 0+$, this proves one of the assertions (3.6.6), the proof for the other being similar.

With the aid of the previous lemma one proves by the usual technique (compare also Problem 3.6.1).

Lemma 3.6.4. *If $f \in \mathsf{W}$, then $\lim_{t\to 0+} W(f; x; t) = f(x)$ for every $x \in \mathbb{R}$.*

We may now turn to the converse of Prop. 3.6.2.

Proposition 3.6.5. *Let $f \in \mathsf{W}$. If $W(f; x; t) \geq f(x)$ for all $x \in \mathbb{R}$, $t > 0$, then f is convex.*

Proof. Assume that f is not convex. Then there exist points $x_1 < x_2 < x_3$ and a first degree (algebraic) polynomial $p(x)$ which intersects $f(x)$ in x_1 and x_3 but for which $f(x_2) > p(x_2)$ (see Problem 6.3.5(v)). Defining the function g by $g(x) = f(x) - p(x)$, then $g(x_1) = g(x_3) = 0$ and $g(x_2) > 0$. Let $M = \max_{x_1 \leq x \leq x_3} g(x)$ and

$$y = \max\{x \in [x_1, x_3] \mid g(x) = M\}, \qquad z = \min\{x \in [x_1, x_3] \mid g(x) = M\}.$$

If the first degree polynomial $q(x)$ is defined by $q(x) = p(x) + M$, then

$$q(y) = f(y), \qquad q(x) \geq f(x) \text{ for } x \in [x_1, x_3], \quad q(x) > f(x) \text{ for } x \in [x_1, z) \cup (y, x_3].$$

Furthermore, let

$$m_0 = \min\,(q(x) - f(x)) \quad \text{for } x \in [x_1, (x_1 + z)/2] \cup [(y + x_3)/2, x_3],$$

$$\eta_1 = y - x_1, \quad \delta_1 = y - (x_1 + z)/2, \qquad \eta_2 = x_3 - y, \quad \delta_2 = (x_3 + y)/2 - y.$$

Then $\eta_1 > \delta_1 > 0$, $\eta_2 > \delta_2 > 0$ and

$$q(x) - f(x) \ge 0 \qquad \text{for } x \in [y - \delta_1, y + \delta_2],$$
$$q(x) - f(x) \ge m_0 > 0 \quad \text{for } x \in [y - \eta_1, y - \delta_1] \cup [y + \delta_2, y + \eta_2].$$

This implies that

$$\int_{-\eta_2}^{\eta_1} [q(y-u) - f(y-u)]\, e^{-u^2/4t}\, du$$
$$\ge \left(\int_{-\eta_2}^{-\delta_2} + \int_{\delta_1}^{\eta_1}\right)[q(y-u) - f(y-u)]\, e^{-u^2/4t}\, du$$
$$\ge m_0\left\{\int_{-\eta_2}^{-\delta_2} e^{-u^2/4t}\, du + \int_{\delta_1}^{\eta_1} e^{-u^2/4t}\right\}.$$

Since $q(x) - f(x) \in \mathsf{W}$, it follows by Lemma 3.6.3 that there exists t_0 such that for all $0 < t < t_0$

$$\left(\int_{-\infty}^{-\eta_2} + \int_{\eta_1}^{\infty}\right) |q(y-u) - f(y-u)|\, e^{-u^2/4t}\, du \le \frac{m_0}{2}\left(\int_{-\eta_2}^{-\delta_2} + \int_{\delta_1}^{\eta_1}\right) e^{-u^2/4t}\, du.$$

Upon collecting the results, one has for $0 < t < t_0$

$$W(q - f; y; t) \ge \int_{-\eta_2}^{\eta_1} [q(y-u) - f(y-u)]\, e^{-u^2/4t}$$
$$- \left(\int_{-\infty}^{-\eta_2} + \int_{\eta_1}^{\infty}\right) |q(y-u) - f(y-u)|\, e^{-u^2/4t}\, du$$
$$\ge \frac{m_0}{2}\left(\int_{-\eta_2}^{-\delta_2} + \int_{\delta_1}^{\eta_1}\right) e^{-u^2/4t}\, du > 0.$$

Therefore $f(y) = q(y) = W(q; y; t) > W(f; y; t)$ for $0 < t < t_0$, a contradiction to the assumption. Thus f is convex on $\mathbb{R}$.

Next we turn to monotonicity properties of the approximation of f by the Gauss-Weierstrass integral.

Proposition 3.6.6. *If $f \in \mathsf{W}$ is convex, then $W(f; x; t_1) \ge W(f; x; t_2)$ for all $x \in \mathbb{R}$ and $0 < t_2 < t_1$.*

Proof. For fixed $0 < t_2 < t_1$ consider the function

$$g(u) = (1/\sqrt{t_1})\, e^{-u^2/4t_1} - (1/\sqrt{t_2})\, e^{-u^2/4t_2}.$$

g has exactly two zeros given by $\pm u_0$ where

$$u_0 = \sqrt{\frac{2t_1t_2}{t_1 - t_2} \log \frac{t_1}{t_2}}.$$

Since $g(0) < 0$, this implies $g(u) \le 0$ for $|u| \le u_0$ and $g(u) > 0$ for $|u| > u_0$. Let $x \in \mathbb{R}$ be fixed. Then

$$W(f; x; t_1) - W(f; x; t_2) = (4\pi)^{-1/2} \int_{-\infty}^{\infty} f(u) g(x - u)\, du.$$

Setting $u_1 = x - u_0$, $u_2 = x + u_0$, one has $g(x - u) \le 0$ for $u_1 \le u \le u_2$ and $g(x - u) > 0$ for $u < u_1$, $u > u_2$. On the other hand, let $p(u)$ be the (algebraic) polynomial of first degree which intersects $f(u)$ in u_1 and u_2. Then for

$$h(u) = \begin{vmatrix} 1 & u & f(u) \\ 1 & u_1 & f(u_1) \\ 1 & u_2 & f(u_2) \end{vmatrix} = (u_2 - u_1)f(u) + u(f(u_1) - f(u_2)) + u_1 f(u_2) - u_2 f(u_1)$$

one has

$$h(u) = (u_2 - u_1)[f(u) - p(u)]. \tag{3.6.7}$$

Since f is convex, it follows (cf. Problem 6.3.5(v)) that $h(u) \le 0$ for $u_1 \le u \le u_2$ and $h(u) \ge 0$ for $u < u_1$, $u > u_2$. Therefore $h(u)g(x - u) \ge 0$ for all $u \in \mathbb{R}$, and thus $(h * g)(x) \ge 0$. But

$$(4\pi)^{-1/2} \int_{-\infty}^{\infty} p(u)g(x - u)\, du = W(p; x; t_1) - W(p; x; t_2) = p(x) - p(x) = 0.$$

This implies by (3.6.7) that

$$\int_{-\infty}^{\infty} h(u)g(x - u)\, du = (u_2 - u_1) \int_{-\infty}^{\infty} f(u)g(x - u)\, du.$$

Hence $\int_{-\infty}^{\infty} f(u)g(x - u)\, du \ge 0$ which proves the assertion.

Proposition 3.6.7. *Let $f \in \mathsf{W}$. If $W(f; x; t_1) \ge W(f; x; t_2)$ for all $x \in \mathbb{R}$ and $0 < t_2 < t_1$, then f is convex.*

Proof. This is an immediate consequence of Prop. 3.6.5; for letting t_2 tend to zero, the assumption implies by Lemma 3.6.4 that $W(f; x; t_1) \ge f(x)$ for all $x \in \mathbb{R}$ and $t_1 > 0$.

The results of this subsection may be summarized in the following

Theorem 3.6.8. *Let $f \in \mathsf{W}$, and $W(f; x; t)$ be the singular integral of Gauss–Weierstrass.*

(i) *A necessary and sufficient condition for f to be convex on $\mathbb{R}$ is that $W(f; x; t) \ge f(x)$ for all $x \in \mathbb{R}$, $t > 0$.*

(ii) *A necessary and sufficient condition for f to be convex on $\mathbb{R}$ is that $W(f; x; t_1) \ge W(f; x; t_2)$ for all $x \in \mathbb{R}$ and $0 < t_2 < t_1$.*

3.6.2 Variation Diminishing Kernels

In this subsection it is always assumed that f belongs to the space C and that the kernel $\chi \in \mathsf{NL}^1$ of the singular integral $J(f; x; \rho)$ is continuous. The number $v(f)$ of sign changes of $f(x)$ on $\mathbb{R}$ is defined in the following manner: If $x_1 < x_2 < \cdots < x_n$ is any finite increasing sequence of reals, let $v(f(x_j))$ denote the number of sign changes in the finite sequence $\{f(x_j)\}$. Then $v(f)$ is defined by $v(f) = \sup v(f(x_j))$, the supremum being formed for all ordered finite sets $\{x_j\}$. The singular integral $J(f; x; \rho)$ is said to be *variation diminishing* if for every $f \in \mathsf{C}$ and $\rho > 0$

$$v(J(f; \circ; \rho)) \le v(f). \tag{3.6.8}$$

In this case one also says that the kernel χ is variation diminishing. The problem now is to characterize such kernels χ.

At first the elementary

Proposition 3.6.9. *Let the continuous kernel $\chi \in \mathsf{NL}^1$ be variation diminishing. If $f \in \mathsf{C}$ is monotone on $\mathbb{R}$, then $J(f; x; \rho)$ is a monotone function of x for each $\rho > 0$.*

Proof. For any real α consider the relation

$$J(f; x; \rho) - \alpha = \frac{\rho}{\sqrt{2\pi}} \int_{-\infty}^{\infty} [f(x - u) - \alpha]\chi(\rho u)\, du.$$

Since f is monotone, $f(x) - \alpha$ changes sign at most once. By the variation diminishing property of the kernel the same is true for $J(f; x; \rho) - \alpha$. Since this holds for every real α, $J(f; x; \rho)$ is a monotone function of x for each $\rho > 0$.

A real-valued, continuous kernel $\chi \in \mathsf{NL}^1$ is said to be *totally positive* if for any $n \in \mathbb{N}$ and sets of reals $x_1 < x_2 < \cdots < x_n$, $u_1 < u_2 < \cdots < u_n$ one has

$$\det (\chi(x_j - u_k)) \geq 0. \tag{3.6.9}$$

Obviously, if $\chi(x)$ is totally positive, so is $\rho\chi(\rho x)$ for each $\rho > 0$. Every totally positive function is positive (take $n = 1$ in (3.6.9)). Sometimes a totally positive kernel is called *Pólya frequency*; a positive kernel $\chi \in \mathsf{NL}^1$ is then called *frequency*.

To prove the main result, we need some facts concerning variation diminishing properties of finite matrix transformations.

Let $A = (a_{jk})$ be a real (m, n)-matrix. The linear transformation

$$y_j = \sum_{k=1}^{n} a_{jk}x_k \qquad (j = 1, 2, \ldots, m), \tag{3.6.10}$$

or the matrix A, is said to be variation diminishing if $v(y_j) \leq v(x_k)$ for any finite sequence of reals $x_1, \ldots, x_n$. The matrix A is called totally positive if all minors of A of any order are positive†. If rank A denotes the rank of A which is supposed throughout to be different from zero, then

Lemma 3.6.10. *If the matrix A is totally positive, then $v(y_j) \leq \operatorname{rank} A - 1$ for all real values of $x_1, \ldots, x_n$.*

Proof. Consider first the particular cases that $\operatorname{rank} A = 1$ and $\operatorname{rank} A = m - 1$. Let $\operatorname{rank} A = 1$. Since A is totally positive, all elements of A are positive. Since $\operatorname{rank} A = 1$, one of the linear forms (3.6.10) does not vanish identically and all the others differ from that one only by a positive factor. Thus $v(y_j) = 0 = \operatorname{rank} A - 1$.

Let $\operatorname{rank} A = m - 1$. Certainly, since y is an m-dimensional vector, one has $v(y_j) \leq m - 1$, and it is to be shown that equality is impossible. To this end, assume that there are $x_1, \ldots, x_n$ such that $v(y_j) = m - 1 = \operatorname{rank} A$. The matrix A has at least one nonzero minor B of order $m - 1$. Let $A_1, A_2, \ldots, A_m$ be the m minors of A of order $m - 1$ which may be constructed from the $m - 1$ columns of B. Obviously, B is one of these minors. With a property of rank one has $\sum_{j=1}^{m} (-1)^{j+1}A_j y_j = 0$. As A is totally positive, all minors A_j are positive; as $v(y_j) = m - 1$, the sequence $\{y_j\}$ alternates in sign. Therefore $\sum_{j=1}^{m} |y_j|\, A_j = 0$. But this is impossible since no y_j vanishes and since at least one A_j is different from zero. Thus $v(y_j) \leq m - 2 = \operatorname{rank} A - 1$.

† Recall that a quantity c is called positive if $c \geqslant 0$.

The proof of Lemma 3.6.10 now follows by induction. Suppose that the assertion is valid in case rank $A = 1, 2, \ldots, l - 1$, and let rank $A = l$. One may assume that $l < m$; for, if $l = m$, then trivially $v(y_j) \le m - 1 = \text{rank } A - 1$. Suppose that the assertion is not valid for rank $A = l$, i.e., there exist $x_1, \ldots, x_n$ such that $v(y_j) \ge l$. Then one may select $l + 1$ linear forms of (3.6.10) such that $v(y_{j_1}, \ldots, y_{j_{l+1}}) = l$. Let us denote this transformation by

$$y_j^* = \sum_{k=1}^{n} a_{jk}^* x_k \qquad (j = 1, 2, \ldots, l + 1).$$

Setting $A^* = (a_{jk}^*)$, then certainly A^* is again totally positive and rank $A^* \le \text{rank } A = l$. However, if rank $A^* < l$, then $v(y_j^*) = l > \text{rank } A^*$, which is impossible by the induction hypothesis. On the other hand, if rank $A^* = l$, then the second particular case applies (with A replaced by A^* and m by $l + 1$) which would imply

$$l = v(y_j^*) \le \text{rank } A^* - 1 = l - 1,$$

again a contradiction. Therefore $v(y_j) < l$ so that $v(y_j) \le l - 1 = \text{rank } A - 1$. This completes the proof.

Lemma 3.6.11. *If the matrix A is totally positive, then A is variation diminishing.*

Proof. It is sufficient to prove the assertion in case $v(y_j) = m - 1$. For, if $v(y_j) < m - 1$, one may select $m^* = v(y_j) + 1$ elements of the sequence $y_1, \ldots, y_m$ and denote them by $y_1^*, \ldots, y_{m^*}^*$ such that $v(y_j^*) = v(y_j) = m^* - 1$. Then $v(y_j^*) \le v(x_k)$ would imply $v(y_j) \le v(x_k)$, and thus the assertion. Therefore one may assume that the $x_1, \ldots, x_n$ are such that $v(y_j) = m - 1$, and one has to show that necessarily $v(x_k) \ge m - 1$.

By Lemma 3.6.10 and the assumption it follows that $m - 1 = v(y_j) \le \text{rank } A - 1$. Therefore rank $A = m$, and in particular $n \ge m$. The proof now follows by induction on n. Suppose $n = m$. Then no x_k vanishes; for, if $x_k = 0$ for some k, one would have a new system with $m - 1$ columns, in contradiction to rank $A = m$. Since A is totally positive and the y_j alternate in sign, it follows from $\det(a_{jk}) \ne 0$ and Cramer's rule that the x_k alternate in sign as well. Therefore $v(x_k) = n - 1 = m - 1$, and the assertion is shown for $n = m$.

Now suppose that the assertion has been established for $n = m, m + 1, \ldots, l - 1$, and let $n = l$. Then there are two cases: (i) The x_k alternate in sign; then $v(x_k) = n - 1 > m - 1$, and the assertion is shown. (ii) The x_k do not alternate in sign. But then either $x_s = 0$ for some s of $1 \le s \le n$ or $x_s x_{s+1} > 0$ for some s of $1 \le s < n$. If $x_s = 0$, one may cancel the sth column of A and obtain a new system which satisfies all assumptions but has less than l columns so that the induction hypothesis may be applied. If $x_s x_{s+1} > 0$, then there exists $\lambda > 0$ such that $x_{s+1} = \lambda x_s$. Setting

$$x_k^* = \begin{cases} x_k & \text{for } k \le s \\ x_{k+1} & \text{for } k > s, \end{cases} \qquad a_{jk}^* = \begin{cases} a_{jk} & \text{for } k < s \\ a_{jk} + \lambda a_{j\,k+1} & \text{for } k = s \\ a_{j\,k+1} & \text{for } k > s, \end{cases}$$

one has $v(x_k^*) = v(x_k)$ and

$$y_j = \sum_{k=1}^{n} a_{jk} x_k = \sum_{k=1}^{n-1} a_{jk}^* x_k^* \qquad (j = 1, 2, \ldots, m).$$

Since the matrix $A^* = (a_{jk}^*)$ is again totally positive but has only $n - 1 = l - 1$ columns, the induction hypothesis applies, giving the assertion for $n = l$. This completes the proof.

On the basis of the preceding lemma it is possible to prove the following

Proposition 3.6.12. *Let the continuous kernel $\chi \in \mathsf{NL}^1$ be totally positive. Then the singular integral $J(f; x; \rho)$ is variation diminishing.*

Proof. One may assume $v(f) < \infty$ and $v(J(f; \circ; \rho)) > 0$, for otherwise the assertion (3.6.8) holds trivially. Let $x_0 < x_1 < \cdots < x_m$ be such that the numbers

$$J(f; x_0; \rho), J(f; x_1; \rho), \ldots, J(f; x_m; \rho)$$

alternate in sign, i.e., $v(J(f; x_j; \rho)) = m$. In order to show that $m \leq v(f)$, one has for every $x \in \mathbb{R}$

$$J(f; x; \rho) = \lim_{a \to -\infty,\, b \to \infty} (\rho/\sqrt{2\pi}) \int_a^b f(u)\, \chi(\rho(x - u))\, du.$$

Obviously, the convergence is uniform for the $m + 1$ values $x_0, \ldots, x_m$. Therefore one may choose a, b such that the function $g(x)$, defined by

$$g(x) = \int_a^b f(u)\, \chi(\rho(x - u))\, du, \tag{3.6.11}$$

will also alternate in sign over the $m + 1$ points $x_0, \cdots, x_m$ (note that $J(f; x; \rho) \in \mathsf{C}$ for each $f \in \mathsf{C}$, $\rho > 0$ by Prop. 3.1.3). The interval $[a, b]$ is then subdivided into n equal parts by $a = u_0 < u_1 < \cdots < u_n = b$ so that the length of each subinterval is given by $u_k - u_{k-1} = (b - a)/n$. Considering the corresponding Riemann sum of the integral (3.6.11), the numbers $g(x_j)$, $j = 0, 1, \ldots, m$, are approximable by

$$\kappa_j = [(b - a)/n] \sum_{k=1}^{n} \chi(\rho x_j - \rho u_k) f(u_k) \tag{3.6.12}$$

for $n \to \infty$. Hence one may choose n so large that also the numbers $\kappa_0, \kappa_1, \cdots, \kappa_m$ alternate in sign, i.e., $v(\kappa_j) = m$. Since χ is totally positive, all minors of the matrix $A = (\chi(\rho x_j - \rho u_k))$ of the transformation (3.6.12) are positive. Therefore Lemma 3.6.11 applies, giving $v(\kappa_j) \leq v(f(u_k))$. Since $v(f(u_k)) \leq v(f)$ by definition and $v(\kappa_j) = v(J(f; x_j; \rho))$ by construction, it follows that $v(J(f; x_j; \rho)) \leq v(f)$. Noting that the finite sequence of reals x_j was arbitrary, this indeed implies $v(J(f; \circ; \rho)) \leq v(f)$ for each $f \in \mathsf{C}$ and $\rho > 0$, and the proof is complete.

Theorem 3.6.13. *Let $\chi \in \mathsf{NL}^1$ be continuous. The corresponding singular integral $J(f; x; \rho)$ is variation diminishing if and only if the kernel χ is (up to the sign) totally positive.*

That total positivity is sufficient for the variation diminishing property is given by the previous proposition; for the necessity the reader may consult the literature cited in Sec. 3.7.

Let us again consider the integral $W(f; x; t)$ of Gauss–Weierstrass. By Problem 3.6.5 the kernel $w(x) = (1/\sqrt{2}) \exp\{-x^2/4\}$ is totally positive. Prop. 3.6.12 then gives that the approximation by the Gauss–Weierstrass integral is variation diminishing, i.e., $v(W(f; \circ; t)) \leq v(f)$ for every $f \in \mathsf{C}$, $t > 0$.

Problems

1. Let $f \in \mathsf{W}$. Show that $W(f; x; t) \in \mathsf{C}_{100}$ for each $t > 0$ and $\lim_{t \to 0+} W(f; x; t) = f(x)$ for each $x \in \mathbb{R}$.
2. Let $f \in \mathsf{C}$ and suppose that f does not reduce to a constant. Show that for each $x \in \mathbb{R}$

there exists a monotone null-sequence $\{t_j\}$ such that $W(f; x; t_j) < f(x)$ and $W(f; x; t_j) < W(f; x; t_{j+1})$ for every $j \in \mathbb{N}$. (Hint: Use Theorem 3.6.8 and the fact that a function convex on $\mathbb{R}$ cannot be bounded unless it is a constant; see also ZIEGLER [1])

3. A kernel $\{\chi(x; \rho)\}$ is said to be *strongly centered* at the origin if for each fixed pair of values $0 < \delta < \eta$

$$\lim_{\rho\to\infty} \frac{\int_\eta^\infty |\chi(u; \rho)|\, du}{\int_\delta^\eta |\chi(u; \rho)|\, du} = 0, \qquad \lim_{\rho\to\infty} \frac{\int_{-\infty}^{-\eta} |\chi(u; \rho)|\, du}{\int_{-\eta}^{-\delta} |\chi(u; \rho)|\, du} = 0.$$

Show that if $\{\chi(x; \rho)\}$ is strongly centered at the origin and satisfies (3.1.5), then it is an approximate identity. Show that the Gauss-Weierstrass kernel is strongly centered at the origin.

4. Concerning the results of Sec. 3.6.1, state and prove a similar analysis for the singular integral of Picard, defined in Problem 3.2.5.

5. Show that the Gauss-Weierstrass kernel $w(x)$ of (3.1.33) is a totally positive kernel. (Hint: Obviously,

$$\begin{aligned}\det\,(w(x_j - u_k)) &= 2^{-n/2} \det\,(e^{-x_j^2/4}\, e^{-t_k^2/4}\, e^{x_j t_k/2}) \\ &= 2^{-n/2} \exp\Big\{-\sum_{j=1}^n x_j^2/4\Big\} \exp\Big\{-\sum_{k=1}^n t_k^2/4\Big\} \det\,(e^{x_j t_k/2}).\end{aligned}$$

Now use PÓLYA–SZEGÖ [1 II, p. 49], see also SCHOENBERG [4])

6. (i) Let $\chi_1, \chi_2 \in \mathsf{NL}^1$ be two totally positive functions. Show that the convolution $\chi_1 * \chi_2$ is again totally positive. (Hint: SCHOENBERG [4])
   (ii) Show that the Picard kernel $c_2(x)$ of Problem 3.2.5 is totally positive.

3.7 Notes and Remarks

Sec. 3.1. The notion of an approximate identity on the line and Theorem 3.1.6 are standard; see, e.g., BOCHNER [6, p. 1 ff; 7, p. 57 ff], TITCHMARSH [6, p. 34], DUNFORD–SCHWARTZ [1I, p. 218 ff], HEWITT [1, p. 186 ff], and H. S. SHAPIRO [1, p. 10 f]. ACHIESER [2, p. 133 ff] apparently coined the term 'kernel of Fejér's type' and in his treatment only assumed $f(x)/(1 + x^2) \in \mathsf{L}^1$; in the latter respect see also BOCHNER [7, p. 138 ff]. The concept of an approximate identity is connected with Friedrich's mollifiers; see FRIEDMAN [2, p. 274 f].

The results on singular integrals on the line group play an important rôle in the summability of Fourier inversion integrals to be treated in Chapter 5. In this respect the reader is also referred to TITCHMARSH [6, p. 26 ff]. The corresponding problems for periodic singular integrals were already touched upon in connection with summability theory of Fourier series.

For Prop. 3.1.8, Cor. 3.1.9 see STONE [1, p. 78 ff]. Prop. 3.1.11 is due to G. H. HARDY, see also ACHIESER [2, p. 126 ff]. For this material see especially BOCHNER [6, p. 19 f] and ZYGMUND [7I, p. 68]. The fundamental conversion relation (3.1.21) seems first to have been observed by DE LA VALLÉE POUSSIN [3, p. 30 ff] (see also Problem 3.1.15); he was also the first to recognize the importance of the property of (periodic) kernels to be of Fejér's type, particularly in solving periodic problems by unrolling them onto the real axis via (3.1.31) (compare the comments to Prop. 3.3.5). For further results leading to the Poisson summation formula see Sec. 5.1.5.

Concerning Problem 3.1.4, there is a close connection between weak derivatives and *distributional* derivatives. A distribution (or generalized function) is an element of the set D' of all continuous linear functionals over D ($= \mathsf{C}_{00}^\infty$), the set of Schwartz' test functions. The usual notation for the operation of $f \in \mathsf{D}'$ on $\phi \in \mathsf{D}$ is $\langle f, \phi\rangle$, more specifically $\langle f(x), \phi(x)\rangle$, thus emphasizing that the distribution f acts on the function ϕ, considered as a function of the independent variable x. In this terminology, the translate $f(x + h)$, $h \in \mathbb{R}$, of $f \in \mathsf{D}'$ is then defined by $\langle f(x+h), \phi(x)\rangle = \langle f(x), \phi(x-h)\rangle$. A distribution f is called regular if it is generated by some (ordinary) function $f \in \mathsf{L}^1_{\text{loc}}$ via $\langle f, \phi\rangle = \int_{-\infty}^{\infty} f(x)\phi(x)\,dx$, $\phi \in \mathsf{D}$. The δ-distribution (also called δ-functional or even δ-function) is not regular; it is defined through $\langle \delta, \phi\rangle = \phi(0)$, $\phi \in \mathsf{D}$. The distributional derivative of $f \in \mathsf{D}'$ is defined as the distribution f' given through $\langle f', \phi\rangle = -\langle f, \phi'\rangle$, $\phi \in \mathsf{D}$; equivalently, f' is the limit in the D'-topology of $h^{-1}[f(x+h) - f(x)]$, i.e.,

$$\langle f', \phi\rangle = \lim_{h\to 0}\left\langle \frac{f(x+h) - f(x)}{h}, \phi(x)\right\rangle = \lim_{h\to 0}\left\langle f(x), \frac{\phi(x-h) - \phi(x)}{h}\right\rangle.$$

It follows that every $f \in \mathsf{D}'$ is arbitrarily often differentiable; see SCHWARTZ [1], FRIEDMAN [1, 2], ZEMANIAN [1], BREMERMANN [1]. Let f be a regular distribution and suppose that the first distributional derivative of f is also regular, generated by $g \in \mathsf{L}^1_{\text{loc}}$, say. The definition of a distributional derivative then takes on the form $\int_{-\infty}^{\infty} f(x)\phi'(x)\,dx = -\int_{-\infty}^{\infty} g(x)\phi(x)\,dx$, thus

$$\lim_{h\to 0}\int_{-\infty}^{\infty} \frac{f(x+h) - f(x)}{h}\phi(x)\,dx = \int_{-\infty}^{\infty} g(x)\phi(x)\,dx.$$

This is the definition of the weak derivative of $f \in \mathsf{X}(\mathbb{R})$ if one furthermore assumes that g belongs to $\mathsf{X}(\mathbb{R})$ and that the above limit does not hold for all $\phi \in \mathsf{D}$ but for all $\phi \in \mathsf{L}^{p'}$ in case $\mathsf{X}(\mathbb{R}) = \mathsf{L}^p$, for example. For details see Vol. II of this treatise (see also the account in LOOMIS [3, pp. 164–181]).

Sec. 3.2. For the material of this section the reader may compare the comments to Sec. 1.4. The references here are TITCHMARSH [6, pp. 28–32], ACHIESER [2, p. 133], BOCHNER–CHANDRASEKHARAN [1, p. 13 ff], BUTZER [4, 6], and MAMEDOV [5, 7, 9]; compare also BRUCKNER–WEISS [1]. For Prop. 3.2.6 see also H. S. SHAPIRO [1, p. 14 f].

Sec. 3.3. For Prop. 3.3.1 compare MAMEDOV [1], MÜLLER [1, p. 65], and the comments to Sec. 1.5. For Prop. 3.3.3 see BUTZER [4], H. S. SHAPIRO [1, p. 32 f]. Prop. 3.3.5 is given in NESSEL [4]; see also the literature cited there. This result may serve as an illustration to our earlier remarks concerning the property of periodic kernels to be of Fejér's type. To this end, in Sec. 1.6.3 it was shown that $\Delta^*(\chi_\rho^*; \alpha) = m(\chi_\rho^*; \alpha)$ for $0 < \alpha \le 1$, $m(\chi_\rho^*; \alpha)$ being the αth moment (1.6.9) of χ_ρ^*. But this does not solve the problem since $m(\chi_\rho^*; \alpha)$ still depends on ρ. Indeed, in almost all cases of interest the dependence of the periodic kernel $\{\chi_\rho^*\}$ upon the parameter ρ is so difficult to deal with that the expansion of $m(\chi_\rho^*; \alpha)$ in terms of, e.g., powers of ρ is the actual problem (compare the literature given around Sec. 1.6.3). On the other hand, Prop. 3.3.5 reveals that the measure of approximation $\Delta^*(\chi_\rho^*; \alpha)$ is asymptotically equal to $m(\rho\chi(\rho\,\circ); \alpha)$, the αth moment (3.3.6) of $\rho\chi(\rho x)$. But here no further expansion is needed, since the parameter may now be separated by an elementary substitution, giving the constant $m(\chi; \alpha)$. Later on, the property of periodic kernels to be of Fejér's type will often be used to deduce important results by elementary methods; compare Cor. 6.3.12, Problems 6.4.2, 6.4.3, Theorem 12.2.5. In Prop. 3.3.5, the assumption that the αth moment $m(\chi; \alpha)$ exists as a finite number is not essential; an analogous assertion concerning $\Delta^*(\chi_\rho^*; \alpha)$ is possible in case one only knows how $m(\chi; \alpha)$ is asymptotically divergent (compare Problem 3.3.7). Let us emphasize that in this text Nikolskiĭ constants are only determined with regard to classical Lipschitz classes. On the other hand, one may consider measures of approximation for any other classes of functions. A case, often discussed in the literature, is that of the classes $\mathsf{W}^\alpha(\beta)$ (cf. Sec. 11.6). For this one may consult the fundamental papers of SZ.-NAGY [7; 3] as well as TELJAKOVSKIĬ [2], STEČKIN–TELJAKOVSKIĬ [1]; see also the detailed discussion in NESSEL [4, 5].

Sec. 3.4. For the results on higher order approximation see BUTZER [4, 6], H. S. SHAPIRO [1, p. 55 f]; also BUTZER–TILLMANN [1, 2]. The proof of Prop. 3.4.6(iii) follows along the lines of that of (2.2.2), thus of ZYGMUND [3]; see also ALJANČIĆ [2]. For higher order direct approximation theorems for periodic singular integrals the kernels of which are generated via (3.1.28) by some $\chi \in \mathsf{NL}^1$, the reader is referred to Problem 6.4.2.

Sec. 3.5. The methods of proof of this section are intimately connected with those of Sec. 2.3, 2.5; see in particular the comments to Sec. 2.5. There we were concerned with inverse approximation theorems for particular periodic singular integrals. However, on the basis of the present results one may establish inverse approximation theorems for the general class of those periodic singular integrals for which the kernel is generated via (3.1.28) by some $\chi \in \mathsf{NL}^1$. See Problem 6.4.3 for the formulation of the result. At this stage the assumption $\chi \in \mathsf{W}^2_{\mathsf{C}\cap\mathsf{L}^1}$ in Theorem 3.5.1 seems to be essential. Thus inverse approximation theorems for the integral means $A(f; x; h)$ cannot be deduced here. This gap will be filled in Sec. 13.3.3 (see in particular Problem 13.3.3) after the study of saturation theory. There it will be shown (Prop. 13.3.16) that the hypothesis $\chi \in \mathsf{W}^2_{\mathsf{C}\cap\mathsf{L}^1}$ is quite dispensable.

Sec. 3.6. Some first results concerning shape preserving properties of approximation processes may be found in FEJÉR [2] and SZÁSZ [3, 4]. For the present results of Sec. 3.6.1 concerning the Gauss–Weierstrass integral see ZIEGLER [1]; there they are mostly derived as consequences of general theorems. The proof of Prop. 3.6.6 is the concrete version of that of a general result given in KARLIN–NOVIKOFF [1]. For an analog of Theorem 3.6.8 concerning Bernstein polynomials one may consult PÓLYA–SCHOENBERG [1], SCHOENBERG [7], ZIEGLER [1], and the literature cited there. Along the lines of the proofs of Sec. 3.6.1, one may deduce similar results concerning the approximation of periodic functions by singular integrals (1.1.3). Thus let $f \in \mathsf{C}_{2\pi}$ and $\{\chi_\rho(x)\}$ be an even, positive approximate identity for $\rho \in (0, \infty)$, $\rho \to \infty$, strongly centered at the origin (cf. Problem 3.6.3). Then f is convex on $[-\pi, \pi]$ if and only if $I_\rho(f; x) \geq f(x)$ for all $x \in [-\pi, \pi]$ and $\rho > 0$. This, together with a corresponding result concerning monotone convergence, may be applied to the de La Vallée Poussin means $V_n(f; x)$ of (2.5.11), giving the equivalence of the following assertions for $f \in \mathsf{C}_{2\pi}$: (i) f is convex on $[-\pi, \pi]$; (ii) $V_n(f; x) \geq f(x)$ for all $x \in [-\pi, \pi]$, $n \in \mathbb{N}$; (iii) $V_n(f; x) \geq V_{n+1}(f; x)$ for all $x \in [-\pi. \pi]$, $n \in \mathbb{N}$. See also PÓLYA–SCHOENBERG [1], SCHOENBERG [7].

Concerning Sec. 3.6.2, the theory of variation diminishing transformations is especially connected with the name of I. J. SCHOENBERG; important precursory material may be found in the work of G. PÓLYA. For Lemmata 3.6.10, 3.6.11 see SCHOENBERG [1] where also the converse to Lemma 3.6.11 is given to a certain extent. The necessary and sufficient conditions for the matrix A to be variation diminishing were found by MOTZKIN [1]. For further developments concerning variation diminishing properties of finite matrix transformations see GANTMACHER–KREIN [1] and SCHOENBERG–WHITNEY [1].

For Prop. 3.6.12 see SCHOENBERG [3, 4]. The proof of the converse uses the results of MOTZKIN [1], see SCHOENBERG [3]. I. J. SCHOENBERG gave a further remarkable characterization for the class of variation diminishing kernels. If one asks for the class of entire functions $\mu(s)$ which are limits, uniform in every finite domain, of real polynomials with only real zeros, then it follows from the classical work of E. LAGUERRE and G. PÓLYA that this class consists precisely of those functions μ which are of the form

$$\mu(s) = C\, e^{-\gamma s^2 + \delta s} s^m \prod_{k=1}^{\infty} (1 + \delta_k s)\, e^{-\delta_k s} \qquad \Big(\gamma \geq 0;\ C, \delta, \delta_k \in \mathbb{R};\ \sum_{k=1}^{\infty} \delta_k^2 < \infty\Big).$$

Concerning the subclass B of all μ with $\mu(0) > 0$ which do not reduce to the form $C\, e^{\delta s}$, i.e., for which $m = 0$, $C > 0$, $\gamma \geq 0$, $\delta, \delta_k \in \mathbb{R}$, $0 < \gamma + \sum_{k=1}^{\infty} \delta_k^2 < \infty$, SCHOENBERG [4] showed the following: If $\mu \in \mathsf{B}$, then its reciprocal may be represented as a bilateral Laplace transform: $1/\mu(s) = \int_{-\infty}^{\infty} e^{-su} \chi(u)\, du$ within the vertical strip of regularity of $1/\mu(s)$ which contains the origin, and where χ is a Pólya frequency function. Conversely, given a Pólya frequency function χ, then its bilateral Laplace transform converges within such a vertical strip and represents the reciprocal of a function of class B.

For various extensions, including the study of a wide class of integral transformations (not necessarily of convolution type) which are variation diminishing and of which the Bernstein construction is but a very special example, one may refer to the important work of S. KARLIN (e.g. [1]) and his collaborators and students. For results concerning the wider class of (generalized) convexity preserving transformations (cf. Prop. 3.6.1) one may consult SCHOENBERG [7], KARLIN [2], ZIEGLER [1], and the literature cited there; see also the monograph of KARLIN–STUDDEN [1].

For totally positive $\chi \in \mathsf{NL}^1$, I. I. HIRSCHMAN and D. V. WIDDER investigated the convolution transform $f^\wedge(x) = \int_{-\infty}^{\infty} \chi(x-u)f(u)\,du$ which assigns to each f of a definite class its $(\chi - t)$ transform $f^\wedge$. The unilateral Laplace transform, the Stieltjes transform, and the Gauss–Weierstrass transform, i.e.,

$$\int_0^\infty e^{-su} f(u)\,du, \qquad \int_0^\infty \frac{f(u)}{x+u}\,du, \qquad \int_{-\infty}^\infty e^{-(x-u)^2} f(u)\,du,$$

respectively, may be considered as particular examples of this general convolution transform (corresponding to the kernels $\exp\{x - e^x\}$, $[\cosh x]^{-1}$, and $\exp\{-x^2\}$, respectively). Extending classical results for the unilateral Laplace transform (compare WIDDER [1]), HIRSCHMAN–WIDDER [1–4] and WIDDER [2–4] obtained an inversion and representation theory for the general convolution transform; see in particular their monograph [5] and the literature cited there.

Corresponding to Sec. 3.6.2, there is a similar theory for approximation processes of periodic functions. The relevant notion now is the number $v_c(f)$ of cyclic variations of sign of the (real-valued) $f \in \mathsf{C}_{2\pi}$, defined as $v_c(f) = \sup v_c(f(x_k))$. Here the points of the finite set $\{x_k\}$ are in cyclic order along the unit circumference (or mod 2π), and $v_c(f(x_k))$ is the (always even) number of sign changes encountered on describing the circumference just once. The singular integral $I_\rho(f; x)$ of (1.1.3) or the kernel $\{\chi_\rho(x)\}$, which is assumed to be positive and to belong to BV_{loc}, is called cyclic variation diminishing provided that $v_c(I_\rho(f; \circ)) \le v_c(f)$ for each $f \in \mathsf{C}_{2\pi}$ and $\rho \in \mathbb{A}$. The first paper mainly concerned with variation diminishing transformations on the circle is that of PÓLYA–SCHOENBERG [1]; they considered the de La Vallée Poussin means $V_n(f; x)$ of (2.5.11). Using the variation diminishing property of the $V_n(f; x)$, MAIRHUBER–SCHOENBERG–WILLIAMSON [1] treated the general case: The convolution integral $I_\rho(f; x)$ is cyclic variation diminishing if and only if χ_ρ is cyclic totally positive for each $\rho \in \mathbb{A}$. Here the (periodic) function $\chi_\rho(x)$ is said to be cyclic totally positive provided that it satisfies $D_{2n+1}(x, u) \equiv \det(\chi_\rho(x_j - u_k)) \ge 0$ for all $n \in \mathbb{P}$, where $x_1 < \cdots < x_{2n+1} < x_1 + 2\pi$, $u_1 < \cdots < u_{2n+1} < u_1 + 2\pi$ are arbitrary sets of points on the circle in the counter clockwise order. As in this text, particular attention is paid to the case that the periodic kernel $\{\chi_\rho(x)\}$ is of Fejér's type, thus generated via (3.1.28) by a Pólya frequency function $\chi \in \mathsf{NL}^1$; see MAIRHUBER–SCHOENBERG–WILLIAMSON [1] for the details.

Concerning Pólya frequency sequence transformations $y_j = \sum_{k=-\infty}^{\infty} a_{j-k} x_k$ one may consult AISSEN–SCHOENBERG–WHITNEY [1], EDREI [1, 2], SCHOENBERG [5, 6].

These few comments may serve as a first guide for the reader interested in the topics of this section; all have been dealt with in book-form, and so our treatment is rather concise. Apart from the monographs of GANTMACHER–KREIN [1], HIRSCHMAN–WIDDER [5], KARLIN–STUDDEN [1] as well as the survey articles of SCHOENBERG [5, 7] already mentioned above, particular reference can be made to KARLIN [3]. He investigates variation diminishing transformations extensively from a general point of view; compare in particular the notes and references given in that text.

Part II

Fourier Transforms

For any (complex-valued) function $f \in \mathsf{L}^1_{2\pi}$, the kth complex Fourier coefficient of f was defined (cf. Sec. 1.2) by

$$f^\wedge(k) = \frac{1}{2\pi}\int_{-\pi}^{\pi} f(u)\, e^{-iku}\, du \qquad (k \in \mathbb{Z}). \tag{II.1}$$

This† associates with each $f \in \mathsf{L}^1_{2\pi}$ a sequence $\{f^\wedge(k)\}_{k=-\infty}^{\infty}$ of complex numbers; in other words, (II.1) actually defines a complex-valued function $f^\wedge$ on the group $\mathbb{Z}$ of all integers. It is called the *finite Fourier transform* of f. Since $|f^\wedge(k)| \le \|f\|_1$, $k \in \mathbb{Z}$, this transform defines a bounded mapping of $\mathsf{L}^1_{2\pi}$ into l^∞; since $[c_1 f_1 + c_2 f_2]^\wedge(k) = c_1 f_1^\wedge(k) + c_2 f_2^\wedge(k)$ ($c_1, c_2 \in \mathbb{C}$, $k \in \mathbb{Z}$), the transform is also linear.

It will be of importance to study the operational rules of this transform in detail. Certain operations on functions will correspond to (more suitable) operations on their transforms. In particular, the finite Fourier transform converts convolution of two functions into the (pointwise) product of their transforms, differentiation into multiplication of the transform by ik.

This correspondence will be of particular value if it is possible to recapture f from its transform $f^\wedge$, that is to say, if there is an inversion formula. The most natural way of returning from $f^\wedge$ to f is by means of the Fourier series of f, the resulting inversion formula being

$$f(x) = \sum_{k=-\infty}^{\infty} f^\wedge(k)\, e^{ikx}. \tag{II.2}$$

However, the question arises whether the Fourier series of f converges to $f(x)$, that is to say, whether equality holds in (II.2). As we have seen in Chapter 1, this problem is rather delicate. Thus the Fourier series of a function $f \in \mathsf{C}_{2\pi}$ is either divergent at a given point x_0 or converges to $f(x_0)$ (see Problem 1.2.16). If f is sufficiently smooth (of bounded variation, for example), then the Fourier series of $f \in \mathsf{C}_{2\pi}$ converges uniformly to f. If $f \in \mathsf{L}^1_{2\pi}$ and $f^\wedge \in \mathsf{l}^1$, then (II.2) holds almost everywhere (see Problem 1.2.13 and

† The definition of $f^\wedge$ on $\mathbb{Z}$ may be extended to all of $\mathbb{R}$. Moreover, the integral is well-defined provided only $f \in \mathsf{L}^1(-\pi, \pi)$.

Prop. 4.1.5(i)). On the other hand, there are functions $f \in \mathsf{L}^1_{2\pi}$ such that $f^\wedge \notin \mathsf{l}^1$ (see Problem 4.1.5). Moreover, (II.2) does not necessarily hold in the sense of pointwise convergence if f is just continuous (see Problem 1.3.5).

A second way to recapture f from $f^\wedge$ is to take the arithmetic means of the (symmetric) partial sums $\sum_{k=-n}^{n} f^\wedge(k)e^{ikx}$. Indeed, Fejér's theorem (see Cor. 1.2.4) states that these means, defined by $\sum_{k=-n}^{n} (1 - |k|/(n+1))f^\wedge(k)e^{ikx}$, tend uniformly to f in case f just belongs to $\mathsf{C}_{2\pi}$.

The finite Fourier transform, which is a particular instance of an *integral transform*, is of significance in its use as an example for the *integral transform method.* The method is as follows: To solve problems of mathematical analysis by transforming them into another function space, solve (which should be simpler) the problem in the transformed state, and then apply a suitable inversion formula to obtain the solution in the original function space. The operational rules of a transform as well as inversion formulae assume particular importance in connection with the integral transform method.

Whereas the finite Fourier transform is appropriate for problems involving periodic functions, a further example of an integral transform is the *Fourier transform* which is a suitable tool for functions f defined on the real line $\mathbb{R}$. For f in L^1 it is the function $f^\wedge$ on $\mathbb{R}$ defined by

$$f^\wedge(v) = \frac{1}{\sqrt{2\pi}} \int_{-\infty}^{\infty} e^{-ivu} f(u)\, du. \tag{II.3}$$

Let us first consider a heuristic connection between the finite Fourier transform and the Fourier transform associated with the real line. If f is a $2\pi\lambda$-periodic function integrable over $[-\pi\lambda, \pi\lambda]$ with $\lambda > 0$, then the corresponding finite Fourier transform on $\mathbb{Z}$ is defined by

$$f^\wedge(k) = \frac{1}{2\pi\lambda} \int_{-\pi\lambda}^{\pi\lambda} e^{-iku/\lambda} f(u)\, du \qquad (k \in \mathbb{Z}). \tag{II.4}$$

If we assume that the Fourier series of f is convergent to f, thus

$$f(x) = \sum_{k=-\infty}^{\infty} f^\wedge(k)\, e^{ikx/\lambda},$$

inserting for each coefficient $f^\wedge(k)$ its formula (II.4), we obtain

$$f(x) = \sum_{k=-\infty}^{\infty} \frac{e^{ikx/\lambda}}{2\pi\lambda} \int_{-\pi\lambda}^{\pi\lambda} e^{-iku/\lambda} f(u)\, du.$$

Setting $v_k = k/\lambda$, $\Delta v_k = v_{k+1} - v_k = 1/\lambda$, then

$$f(x) = \frac{1}{2\pi} \sum_{k=-\infty}^{\infty} e^{ixv_k} \left\{ \int_{-\pi\lambda}^{\pi\lambda} e^{-iv_k u} f(u)\, du \right\} \Delta v_k,$$

and letting $\lambda \to \infty$, the sum passes formally into an integral, giving

$$f(x) = \frac{1}{\sqrt{2\pi}} \int_{-\infty}^{\infty} e^{ixv} \left\{ \frac{1}{\sqrt{2\pi}} \int_{-\infty}^{\infty} e^{-ivu} f(u)\, du \right\} dv.$$

The formal inversion of (II.3) is then given by the formula

$$f(x) = \frac{1}{\sqrt{2\pi}} \int_{-\infty}^{\infty} e^{ixv} f^{\wedge}(v)\, dv. \tag{II.5}$$

Thus there is a formal analogy between the finite Fourier transform (II.1) and the Fourier transform (II.3), between the inversion formula given by the Fourier series (II.2) and (II.5).

Indeed, if $f \in \mathsf{L}^1$, the integral (II.3) is well defined for every real v and $\|f^{\wedge}\|_{\infty} \le \|f\|_1$. As a matter of fact, the Fourier transform defines a bounded linear transformation of L^1 into C_0, and most of the operational rules for finite Fourier transforms are valid as well. Moreover, it will be shown that for functions $f \in \mathsf{L}^1$ for which $f^{\wedge} \in \mathsf{L}^1$ the inversion formula (II.5) holds almost everywhere; however, $f^{\wedge}$ need not be in L^1. On the other hand, it is again possible to recapture f from $f^{\wedge}$ by the formula

$$f(x) = \lim_{\rho \to \infty} \frac{1}{\sqrt{2\pi}} \int_{-\rho}^{\rho} \left(1 - \frac{|v|}{\rho}\right) f^{\wedge}(v)\, e^{ixv}\, dv,$$

valid for almost all x if $f \in \mathsf{L}^1$. Thus for Fourier transforms of functions f in L^1 one meets essentially the same problems as for the finite $\mathsf{L}^1_{2\pi}$-theory.

On the other hand, the difference between the two theories becomes more apparent when one tries to see what happens for L^p, $p > 1$. The finite Fourier transform is defined over the compact interval $[-\pi, \pi]$, the Fourier transform over the noncompact real axis $\mathbb{R}$. This entails that $\mathsf{X}_{2\pi} \subset \mathsf{L}^1_{2\pi}$, in particular $\mathsf{L}^p_{2\pi} \subset \mathsf{L}^1_{2\pi}$ for $p > 1$, while $\mathsf{L}^p \not\subset \mathsf{L}^1$ for $p > 1$. Whereas a large part of finite Fourier transform theory may be subsumed under the $\mathsf{L}^1_{2\pi}$-theory, not even the definition of the Fourier transform by formula (II.3) is directly applicable to every $f \in \mathsf{L}^p$. A new approach is needed; it even turns out to be quite different for $1 < p \le 2$ and $p > 2$. However, the resulting L^p-theory has rather more symmetry than in the case $p = 1$. In particular, f and $f^{\wedge}$ play exactly the same rôle in L^2.

Nevertheless, a large part of the theory of Fourier series and finite Fourier transforms is not only intimately connected with but very similar to the theory of Fourier transforms, and it is the purpose of this presentation to develop the common features.

To this end, it is often useful to think of the periodic functions as defined on the additive group of real numbers modulo 2π or on the perimeter $\mathbb{T}$ of the unit disc of the complex plane. The theory of the finite Fourier transform is then often referred to as the harmonic analysis associated with this circle, or the reals modulo 2π. In the same way, the Fourier transform is associated with the additive group of real numbers.

Harmonic analysis can be associated with a variety of domains. In particular, let us briefly consider the transform for functions defined on the group of integers. This leads to a rather simple but illuminating theory and provides additional motivation for the preceding.

Consider the integers as a measure space in which each point has measure 1, and an integrable function f defined on this measure space; that is, f is a sequence $\{f(k)\}_{k=-\infty}^{\infty}$ of complex numbers such that $\sum_{k=-\infty}^{\infty} |f(k)| < \infty$, or $f \in \mathsf{l}^1$ in the notation of Sec. 0.3. The Fourier transform of $f \in \mathsf{l}^1$ is then the function $f^{\wedge}$ whose value at x is

$$f^{\wedge}(x) = \sum_{k=-\infty}^{\infty} f(k)\, e^{-ikx}. \tag{II.6}$$

Since this series is absolutely and uniformly convergent, the transform $f^\wedge$ belongs to $C_{2\pi}$. Furthermore, term-by-term integration is possible, and thus the inversion formula

$$f(k) = \frac{1}{2\pi}\int_{-\pi}^{\pi} e^{iku} f^\wedge(u)\,du \qquad (k \in \mathbb{Z}) \tag{II.7}$$

is immediate; here we do not encounter any of the difficulties met above when trying to express f in terms of $f^\wedge$.

At this point we call attention to the duality that exists between the interval $[-\pi, \pi]$ and $\mathbb{Z}$ in case of formulae (II.1), (II.2) for the Fourier transform associated with the circle group, between $\mathbb{Z}$ and $[-\pi, \pi]$ in case of formulae (II.6), (II.7) for the transform on l^1, and between $\mathbb{R}$ and $\mathbb{R}$ in formulae (II.3), (II.5). This observation is important in the study of Fourier transforms on more general groups. The analogy between Fourier transforms associated with the different groups becomes especially apparent by comparing the theory on the Hilbert spaces of square-integrable functions.

Whereas Chapter 4 is concerned with the finite Fourier transform, Chapter 5 is reserved to the Fourier transform on the real line $\mathbb{R}$. The material is presented in such a fashion that the parallel results of the two theories come to light. These results are analogous in statement and often in proof. To avoid repetition, however, emphasis is often laid upon different methods of proof.

Chapter 6 is concerned with a detailed treatment of representation theorems. Necessary and sufficient conditions such that a function $f(k)$ on $\mathbb{Z}$ is representable as a finite Fourier or Fourier–Stieltjes transform are given; this problem is also considered for functions $f(v)$ on the line group. Moreover, rather convenient sufficient conditions for representation are supplied as well as a short account on (classical) multiplier theory.

Chapter 7 is devoted to the first and best-known application of Fourier transform methods, namely to the solution of partial differential equations. The technique is to be further developed and refined in Parts IV and V so that profounder and also more theoretical problems can be handled.

4

Finite Fourier Transforms

4.0 Introduction

The plan of this chapter can be outlined as follows: Sec. 4.1 is concerned with a detailed treatment of the fundamental operational properties of the finite Fourier transform, including the Riemann–Lebesgue lemma and convolution theorem. Discussion of the inversion problem is kept to a minimum since it turns out to be the convergence problem for Fourier series. Indeed, the theory of periodic singular integrals of Chapter 1 enters in when considering summation processes. Of particular importance in later applications is Theorem 4.1.10 on the transform of derivatives of f. Sec. 4.2 is devoted to the special features of the finite Fourier transform in $L^p_{2\pi}$, $p > 1$. Although the definition of the transform for $L^1_{2\pi}$-functions also applies to $X_{2\pi}$-functions, nevertheless several of its important properties are only valid under the additional assumption $f \in L^p_{2\pi}$, $p > 1$. The Parseval equation (4.2.6) and Riesz–Fischer theorem for $p = 2$ and the Hausdorff–Young inequality (4.2.15) for $1 < p < 2$ can, for example, be mentioned in this connection. This section also contains a few words on the harmonic analysis associated with functions in l^p for $p > 1$. It may be regarded as precursory to Sec. 5.2, reserved to the definition and properties of Fourier transforms on L^p, $p > 1$. In particular, the clear and elegant results of Sec. 4.2 may serve as models of those to be expected in Sec. 5.2. Sec. 4.3 deals with the definition and properties of the finite Fourier–Stieltjes transform, including a detailed inversion theory. The classes $V[X_{2\pi}; \psi(k)]$ are introduced, and the fundamental Theorem 4.3.13 is derived in case $\psi(k) = (ik)^r$.

4.1 $L^1_{2\pi}$-Theory

4.1.1 Fundamental Properties

We recall that the finite Fourier transform of a function $f \in L^1_{2\pi}$ is the function $f^\wedge$, defined on $\mathbb{Z}$, whose value at k is the kth Fourier coefficient

(4.1.1) $$f^\wedge(k) = \frac{1}{2\pi}\int_{-\pi}^{\pi} f(u)\, e^{-iku}\, du \equiv [f(\circ)]^\wedge(k).$$

Some of the elementary operational properties of this transform are collected in the following

Proposition 4.1.1. *For $f \in \mathsf{L}^1_{2\pi}$ we have*

(i) $[f(\circ + h)]^\wedge(k) = e^{ihk}f^\wedge(k)$ $\qquad (h \in \mathbb{R}, k \in \mathbb{Z})$,

(ii) $[e^{-ij\circ}f(\circ)]^\wedge(k) = f^\wedge(k+j)$ $\qquad (j, k \in \mathbb{Z})$,

(iii) $[\overline{f(-\circ)}]^\wedge(k) = \overline{f^\wedge(k)}$ $\qquad (k \in \mathbb{Z})$.

The proof follows immediately by direct substitution into (4.1.1). Regarding the finite Fourier transform as a transformation from one Banach space into another we have

Proposition 4.1.2. *If $f, g \in \mathsf{L}^1_{2\pi}$ and $c \in \mathbb{C}$, then*

(i) $[f+g]^\wedge(k) = f^\wedge(k) + g^\wedge(k), \quad [cf]^\wedge(k) = cf^\wedge(k)$ $\qquad (k \in \mathbb{Z})$,

(ii) $\lim_{|k|\to\infty} f^\wedge(k) = 0$,

(iii) $f^\wedge(k) = 0$ *on* $\mathbb{Z}$ *implies* $f(x) = 0$ a.e.,

(iv) *there are sequences of class l^∞_0 which are not the finite Fourier transform of a function $f \in \mathsf{L}^1_{2\pi}$, yet the set $[\mathsf{L}^1_{2\pi}]^\wedge \equiv \{\alpha \in \mathsf{l}^\infty_0 \mid \alpha = f^\wedge, f \in \mathsf{L}^1_{2\pi}\}$ is dense in l^∞_0.*

In other words, the finite Fourier transform defines a one-to-one bounded linear transformation of $\mathsf{L}^1_{2\pi}$ into (but not onto) l^∞_0. In the following the set of all those elements of l^∞_0 which are the finite Fourier transform of some $f \in \mathsf{X}_{2\pi}$ will be denoted by $[\mathsf{X}_{2\pi}]^\wedge$.

Concerning the proof, part (i) being obvious, it follows from

$$|f^\wedge(k)| \le \|f\|_1 \qquad (k \in \mathbb{Z}) \tag{4.1.2}$$

that the finite Fourier transform defines a bounded linear transformation of $\mathsf{L}^1_{2\pi}$ into l^∞. Moreover, we have for any $k \in \mathbb{Z}$, $k \neq 0$,

$$2\pi f^\wedge(k) = -\int_{-\pi}^{\pi} f\Big(u + \frac{\pi}{k}\Big)\, e^{-iku}\, du = \frac{1}{2}\int_{-\pi}^{\pi}\Big[f(u) - f\Big(u + \frac{\pi}{k}\Big)\Big]\, e^{-iku}\, du.$$

Hence

$$|f^\wedge(k)| \le \frac{1}{2}\,\omega\Big(\mathsf{L}^1_{2\pi}; f; \frac{\pi}{|k|}\Big) \qquad (k \in \mathbb{Z}, k \neq 0), \tag{4.1.3}$$

and thus Lemma 1.5.2(iii) implies (ii) which is known as the *Riemann–Lebesgue lemma*. Property (iii), often referred to as the *uniqueness theorem* of the finite Fourier transform, is already given by Cor. 1.2.7, whereas the proof of (iv) is left to Problem 4.1.3.

Theorem 4.1.3. *If $f, g \in \mathsf{L}^1_{2\pi}$, then*

$$[f * g]^\wedge(k) = f^\wedge(k)g^\wedge(k) \qquad (k \in \mathbb{Z}). \tag{4.1.4}$$

Proof. By Prop. 0.4.1 we have $f * g \in \mathsf{L}^1_{2\pi}$, and thus by Fubini's theorem for any $k \in \mathbb{Z}$

$$[f * g]^\wedge(k) = \frac{1}{2\pi}\int_{-\pi}^{\pi} g(u)\, e^{-iku}\, du\Big\{\frac{1}{2\pi}\int_{-\pi}^{\pi} f(x-u)\, e^{-ik(x-u)}\, dx\Big\} = f^\wedge(k)g^\wedge(k).$$

This proves the *convolution theorem* which in particular shows that the finite Fourier transform converts convolutions into pointwise products and actually defines an isomorphism of the commutative Banach algebra $L^1_{2\pi}$ (with convolution as multiplication) into the commutative Banach algebra l^∞_0 (with pointwise multiplication). Moreover, in Sec. 0.7 we already mentioned that the Banach algebra $L^1_{2\pi}$ has no unit element. The convolution and uniqueness theorem of the finite Fourier transform enables us to give a simple proof of this fact. We have

Proposition 4.1.4. $L^1_{2\pi}$ *is a commutative Banach algebra without unit element. But there are* ***approximate identities****, thus sets of functions* $\chi_\rho \in L^1_{2\pi}$, $\rho \in \mathbb{A}$, *such that for every* $f \in L^1_{2\pi}$, $\lim_{\rho\to\rho_0}\|f * \chi_\rho - f\|_1 = 0$.

Proof. Suppose there exists $e \in L^1_{2\pi}$ such that $e * f = f$ for every $f \in L^1_{2\pi}$. Then, taking $f(x) = \exp\{ikx\}$, $k \in \mathbb{Z}$, it would follow that $e^\wedge(k) = 1$ for all $k \in \mathbb{Z}$. But this would be a contradiction to the Riemann–Lebesgue lemma which asserts $\lim_{|k|\to\infty} e^\wedge(k) = 0$. Theorem 1.1.5 and the various examples of Chapter 1 then complete the proof.

In Problem 1.2.13 we saw that every uniformly convergent Fourier series of a function $f \in C_{2\pi}$ represents f at each x. Let $f \in L^1_{2\pi}$ be such that $f^\wedge \in l^1$. Then $\sum_{k=-\infty}^{\infty} f^\wedge(k)\, e^{ikx}$ converges uniformly and thus represents a function $f_0 \in C_{2\pi}$, i.e., $f_0(x) = \sum_{k=-\infty}^{\infty} f^\wedge(k)\, e^{ikx}$ for all x. Moreover, for the Fourier coefficients of f_0 we have $f_0^\wedge(k) = f^\wedge(k)$ for every $k \in \mathbb{Z}$, and hence $f_0(x) = f(x)$ a.e. by the uniqueness theorem. Therefore (for (ii) below see the Jordan criterion of Problem 1.2.10)

Proposition 4.1.5. (i) *Let* $f \in L^1_{2\pi}$ *be such that* $f^\wedge \in l^1$. *Then*

$$f(x) = \sum_{k=-\infty}^{\infty} f^\wedge(k)\, e^{ikx} \quad \text{a.e.} \tag{4.1.5}$$

Hence f is equal a.e. *to a function in* $C_{2\pi}$. *If* $f \in C_{2\pi}$, *then* (4.1.5) *holds everywhere.*

(ii) *If* $f \in BV_{loc}$ *is* 2π*-periodic, then for every x*

$$f(x) = \lim_{n\to\infty} \sum_{k=-n}^{n} f^\wedge(k)\, e^{ikx}. \tag{4.1.6}$$

It follows by direct substitution that (cf. Problem 1.2.14(ii))

$$(f * t_n)(x) = \sum_{k=-n}^{n} f^\wedge(k) c_k\, e^{ikx} \tag{4.1.7}$$

for every $f \in L^1_{2\pi}$ and any complex trigonometric polynomial t_n with coefficients $c_k \in \mathbb{C}$. We shall now give several extensions of (4.1.7).

Proposition 4.1.6. *If* $f, g \in L^1_{2\pi}$ *and g is such that* $g^\wedge \in l^1$, *then for every x*

$$(f * g)(x) = \sum_{k=-\infty}^{\infty} f^\wedge(k) g^\wedge(k)\, e^{ikx}. \tag{4.1.8}$$

Proof. Since $g^\wedge \in l^1$, Prop. 4.1.5(i) implies $g(x) = g_0(x)$ a.e. with $g_0 \in C_{2\pi}$. Concerning the convolution $f * g$, we may replace g by g_0 and, since $f * g_0 \in C_{2\pi}$ by Prop. 0.4.1, we first of all obtain that $(f * g)(x)$ exists for all x and $f * g \in C_{2\pi}$.

According to Theorem 4.1.3 the right-hand side of (4.1.8) is the Fourier series of $f * g$. Since it converges absolutely and uniformly and since $f * g$ is continuous, the equality (4.1.8) again follows for all x by Prop. 4.1.5.

Let us mention that under the hypotheses of Prop. 4.1.6

$$\frac{1}{2\pi}\int_{-\pi}^{\pi} f(u)\overline{g(u)}\,du = \sum_{k=-\infty}^{\infty} f^\wedge(k)\overline{g^\wedge(k)}. \tag{4.1.9}$$

In particular, if $f = g$, then (4.1.9) shows that the *Parseval equation*

$$\|f\|_{\mathsf{L}^2_{2\pi}} = \|f^\wedge\|_{\mathsf{l}^2} \tag{4.1.10}$$

is valid for all functions $f \in \mathsf{L}^1_{2\pi}$ for which $f^\wedge \in \mathsf{l}^1$.

However, the latter relations are also valid under weaker hypotheses. Here we shall show

Proposition 4.1.7. *Let $f \in \mathsf{L}^1_{2\pi}$ and $g \in \mathsf{BV}_{\text{loc}}$ be 2π-periodic. Then*

$$(f * g)(x) = \lim_{n\to\infty} \sum_{k=-n}^{n} f^\wedge(k)g^\wedge(k)\, e^{ikx} \tag{4.1.11}$$

for all x. In particular,

$$\frac{1}{2\pi}\int_{-\pi}^{\pi} f(u)\overline{g(u)}\,du = \lim_{n\to\infty} \sum_{k=-n}^{n} f^\wedge(k)\overline{g^\wedge(k)}. \tag{4.1.12}$$

Proof. First of all we observe that, since g is bounded, $(f * g)(x)$ exists everywhere by Prop. 0.4.1. For the nth partial sum of the Fourier series of g, namely $S_n(g; x) = \sum_{k=-n}^{n} g^\wedge(k)\, e^{ikx}$, we have by Problem 1.2.11 that, for all x and $n \in \mathbb{N}$, $|S_n(g; x)| \le \|g\|_{\mathsf{L}^\infty_{2\pi}} + 2\|g\|_{\mathsf{BV}_{2\pi}} \equiv M$, say. Hence $|f(x-u)S_n(g; u)| \le M\,|f(x-u)|$ uniformly for all $n \in \mathbb{N}$. Moreover, $\lim_{n\to\infty} S_n(g; u) = g(u)$ for all u by Jordan's criterion (cf. Prop. 4.1.5(ii)). Therefore by Lebesgue's dominated convergence theorem

$$\frac{1}{2\pi}\int_{-\pi}^{\pi} f(x-u)g(u)\,du = \lim_{n\to\infty}\frac{1}{2\pi}\int_{-\pi}^{\pi} f(x-u)\sum_{k=-n}^{n} g^\wedge(k)\, e^{iku}\,du,$$

which proves (4.1.11). On replacing $g(u)$ by $\overline{g(-u)}$, (4.1.11) implies (4.1.12) by setting $x = 0$.

For a further set of conditions which ensure (4.1.8)–(4.1.10) we refer to Prop. 4.2.2.

If we begin with some sequence $\alpha \in \mathsf{l}^1$ and define its Fourier transform $\alpha^\wedge(x)$ by (II.6), then $\alpha^\wedge \in \mathsf{C}_{2\pi}$ as we saw, and according to (II.7) we obtain for the finite Fourier transform of $\alpha^\wedge$ that $[\alpha^\wedge]^\wedge(k) = \alpha(-k)$ for every $k \in \mathbb{Z}$. Thus, if we apply Prop. 4.1.6, we have every $f \in \mathsf{L}^1_{2\pi}$

$$(f * \alpha^\wedge)(x) = \sum_{k=-\infty}^{\infty} \alpha(-k)f^\wedge(k)\, e^{ikx}, \tag{4.1.13}$$

and in particular

$$\frac{1}{2\pi}\int_{-\pi}^{\pi} f(u)\alpha^\wedge(u)\,du = \sum_{k=-\infty}^{\infty} \alpha(k)f^\wedge(k). \tag{4.1.14}$$

We shall refer to relations such as (4.1.9), (4.1.14) as *Parseval formulae.*

4.1.2 Inversion Theory

So far, given a function $f \in \mathsf{L}^1_{2\pi}$, we have defined $f^\wedge$ as a function on $\mathbb{Z}$. We shall now study the inversion problem of the finite Fourier transform, in other words, if we know that a function in l^∞_0 is the finite Fourier transform of some $f \in \mathsf{L}^1_{2\pi}$, we wish to determine the original function f from the values of $f^\wedge$ on $\mathbb{Z}$. According to (4.1.1),

$$f^\wedge(k) = \frac{1}{2\pi} \int_{-\pi}^{\pi} f(u)\, e^{-iku}\, du,$$

and the formal inversion would be given (see (4.1.5)) by

$$f(x) = \sum_{k=-\infty}^{\infty} f^\wedge(k)\, e^{ikx}, \tag{4.1.15}$$

i.e., the inversion problem of the finite Fourier transform is nothing but the convergence problem for Fourier series. It is clear from the results of Chapter 1 that (4.1.15) does not hold in general but must be interpreted in some generalized sense as has already been specified in Sec. 1.2 and 1.4. Although we only need to refer to the relevant sections of Chapter 1, we shall, for the reader's convenience, recall some of the results in the new terminology.

In Sec. 1.2 we introduced θ-factors (1.2.28) and summed the series (4.1.15) in the form

$$U_\rho(f; x) = \sum_{k=-\infty}^{\infty} \theta_\rho(k) f^\wedge(k)\, e^{ikx}. \tag{4.1.16}$$

According to (4.1.13) we have

$$U_\rho(f; x) = \frac{1}{2\pi} \int_{-\pi}^{\pi} f(x-u)\theta_\rho^\wedge(u)\, du \tag{4.1.17}$$

which is (1.2.31), since in view of (1.2.32) and (II.6)

$$C_\rho(x) \equiv 1 + 2 \sum_{k=1}^{\infty} \theta_\rho(k) \cos kx = \theta_\rho^\wedge(x). \tag{4.1.18}$$

Hence, if the assumptions of Prop. 1.2.8 are satisfied, we have for every $f \in \mathsf{X}_{2\pi}$

$$\lim_{\rho\to\rho_0} \left\| \sum_{k=-\infty}^{\infty} \theta_\rho(k) f^\wedge(k)\, e^{ik\circ} - f(\circ) \right\|_{\mathsf{X}_{2\pi}} = 0, \tag{4.1.19}$$

which may be regarded as a certain type of inversion formula for the finite Fourier transform. If we are more interested in recapturing the original function f by a pointwise limit, the results of Sec. 1.4 assure that for each $f \in \mathsf{L}^1_{2\pi}$

$$\lim_{\rho\to\rho_0} \sum_{k=-\infty}^{\infty} \theta_\rho(k) f^\wedge(k)\, e^{ikx} = f(x) \quad \text{a.e.} \tag{4.1.20}$$

in case the hypotheses of Prop. 1.4.2 or 1.4.6, for example, are satisfied by the kernel $\{\theta_\rho^\wedge(x)\}$. For explicit formulae of the most important examples of θ-factors we refer to Problem 4.1.4.

Let us finally observe that if the restrictive hypotheses of Prop. 4.1.5 are satisfied, then of course we need not introduce convergence factors. For a further result on inversion see Problem 4.1.5.

4.1.3 Fourier Transforms of Derivatives

In what follows r always denotes a natural number.

Proposition 4.1.8. *If* $f \in \mathsf{AC}_{2\pi}^{r-1}$, *then*

$$[f^{(r)}]^\wedge(k) = (ik)^r f^\wedge(k) \qquad (k \in \mathbb{Z}).$$

Proof. If $r = 1$, then f is absolutely continuous and an integration by parts gives

$$2\pi[f']^\wedge(k) = f(u)\, e^{-iku}|_{-\pi}^{\pi} + ik \int_{-\pi}^{\pi} f(u)\, e^{-iku}\, du = 2\pi ik f^\wedge(k).$$

The result for general r follows by induction.

Next we consider the converse of the latter assertion.

Proposition 4.1.9. *If for* $f \in \mathsf{X}_{2\pi}$ *there exists* $g \in \mathsf{X}_{2\pi}$ *such that*

$$(4.1.21) \qquad (ik)^r f^\wedge(k) = g^\wedge(k) \qquad (k \in \mathbb{Z}, k \neq 0),$$

then $f \in \mathsf{W}_{\mathsf{X}_{2\pi}}^r$ *(for the definition see* (1.1.16)).

Proof. We set $G_0(x) = g(x)$,

$$G_1(x) = \int_{-\pi}^{x} [g(x_r) - g^\wedge(0)]\, dx_r, \qquad G_2(x) = \int_{-\pi}^{x} [G_1(x_{r-1}) - G_1^\wedge(0)]\, dx_{r-1}, \ldots,$$

$$(4.1.22) \quad G_r(x) = \int_{-\pi}^{x} [G_{r-1}(x_1) - G_{r-1}^\wedge(0)]\, dx_1$$
$$= \int_{-\pi}^{x} dx_1 \Big[-G_{r-1}^\wedge(0) + \int_{-\pi}^{x_1} dx_2 \Big[-G_{r-2}^\wedge(0) + \cdots$$
$$+ \int_{-\pi}^{x_{r-1}} dx_r[-g^\wedge(0) + g(x_r)] \ldots\Big]\Big].$$

Then $G_r \in \mathsf{AC}_{2\pi}^{r-1}$ if we can show that G_r is 2π-periodic. Since g is 2π-periodic, we have $G_1(x + 2\pi) - G_1(x) = \int_x^{x+2\pi} g(x_r)\, dx_r - 2\pi g^\wedge(0) = 0$, and thus G_1 is 2π-periodic. If we now assume G_{k-1} for some k with $1 \leq k \leq r$ to be 2π-periodic, then again

$$G_k(x + 2\pi) = G_k(x) + \int_{x}^{x+2\pi} [G_{k-1}(u) - G_{k-1}^\wedge(0)]\, du$$
$$= G_k(x) + \int_{-\pi}^{\pi} G_{k-1}(u)\, du - 2\pi G_{k-1}^\wedge(0) = G_k(x).$$

Thus $G_r \in \mathsf{AC}_{2\pi}^{r-1}$ and $G_r^{(r)}(x) = g(x)$ (a.e.).

By (4.1.21) and Prop. 4.1.8 it follows that

$$(ik)^r f^\wedge(k) = g^\wedge(k) = [G_r^{(r)}]^\wedge(k) = (ik)^r G_r^\wedge(k) \qquad (k \in \mathbb{Z}, k \neq 0).$$

This implies $[f - G_r]^\wedge(k) = 0$ for all $k = \pm 1, \pm 2, \ldots$, and therefore by the uniqueness theorem

(4.1.23) $$f(x) = \text{const} + G_r(x) \quad \text{(a.e.)},$$

giving $f \in \mathsf{W}^r_{X_{2\pi}}$.

It is now convenient to introduce the following notation: Let $\psi(k)$ be an arbitrary complex-valued function on $\mathbb{Z}$. Then $\mathsf{W}[X_{2\pi}; \psi(k)]$ is the set of all functions $f \in X_{2\pi}$ for which there exists $g \in X_{2\pi}$ such that $\psi(k)f^\wedge(k) = g^\wedge(k)$ for every $k \in \mathbb{Z}$, i.e.,

(4.1.24) $$\mathsf{W}[X_{2\pi}; \psi(k)] = \{f \in X_{2\pi} \mid \psi(k)f^\wedge(k) = g^\wedge(k), g \in X_{2\pi}\}.$$

If we now combine the results of the last two propositions, we arrive at the following characterizations of the class $\mathsf{W}^r_{X_{2\pi}}$.

Theorem 4.1.10. *Let $f \in X_{2\pi}$. The following statements are equivalent:*

(i) $f \in \mathsf{W}^r_{X_{2\pi}}$, (ii) $f \in \mathsf{W}[X_{2\pi}; (ik)^r]$,
(iii) *there exist constants* $\alpha_k \in \mathbb{C}$, $0 \le k \le r - 1$, *and* $g \in X_{2\pi}$ *such that* (a.e.)

$$f(x) = \alpha_0 + \int_{-\pi}^{x} du_1\Big[\alpha_1 + \int_{-\pi}^{u_1} du_2\Big[\alpha_2 + \cdots + \int_{-\pi}^{u_{r-2}} du_{r-1}\Big[\alpha_{r-1} + \int_{-\pi}^{u_{r-1}} g(u_r)\, du_r\Big]\ldots\Big]\Big].$$

The classes $\mathsf{W}^r_{X_{2\pi}}$ and their various representations will play a significant rôle in our later considerations. In particular, the fact that the finite Fourier transform converts differentiation to multiplication by (ik) makes the finite Fourier transform a useful tool in the study of differential equations as we shall see in Chapter 7.

Problems

1. (i) Let $f \in L^1_{2\pi}$. Show that $f^\wedge(k)$ is an even (odd) function on $\mathbb{Z}$ if and only if $f(x)$ is an even (odd) function on $[-\pi, \pi]$.
   (ii) Let $f, f_n \in L^1_{2\pi}$ be such that $\lim_{n\to\infty} \|f - f_n\|_1 = 0$. Show that $\lim_{n\to\infty} f_n^\wedge(k) = f^\wedge(k)$ uniformly for $k \in \mathbb{Z}$.
   (iii) Let $f \in X_{2\pi}$. Show that $|f^\wedge(k)| \le \|f\|_{X_{2\pi}}$ for all $k \in \mathbb{Z}$.
2. Let $f \in X_{2\pi}$ and $f^\wedge(k) = 0$ for $|k| > n$. Show that $f(x) = t_n(x)$ (a.e.) with $t_n \in \mathsf{T}_n$.
3. Prove Prop. 4.1.2(iv). (Hint: Hewitt [1, p. 16] examines the example $\alpha(k) = [\log k]^{-1}$ for $k = 2, 3, \ldots, = 0$ otherwise; see also Rudin [4, p. 104]).
4. (i) Give examples of functions $f \in L^1_{2\pi}$ for which $f^\wedge \notin l^1$ (cf. Problem 4.1.5(i)).
   (ii) Let $f \in X_{2\pi}$. Show that the Fourier series of f is Cesàro, Abel, and Gauss summable to f (almost everywhere), i.e.

$$\lim_{n\to\infty} \sum_{k=-n}^{n} \Big(1 - \frac{|k|}{n+1}\Big) f^\wedge(k)\, e^{ikx} = f(x) \quad \text{(a.e.)},$$

$$\lim_{r\to 1-} \sum_{k=-\infty}^{\infty} r^{|k|} f^\wedge(k)\, e^{ikx} = f(x) \quad \text{(a.e.)},$$

$$\lim_{t\to 0+} \sum_{k=-\infty}^{\infty} e^{-t|k|^2} f^\wedge(k)\, e^{ikx} = f(x) \quad \text{(a.e.)}.$$

In particular, all these relations are valid at each point of continuity of f.

5. (i) Let $m^*(x)$ be the 2π-periodic function which is defined on $[-\pi, \pi]$ by $2\pi\kappa_{[-1,0]}$, $\kappa_{[-1,0]}$ being the characteristic function of the interval $[-1, 0]$. Show that $[m^*]^\wedge(k) = (ik)^{-1}(e^{ik} - 1)$ for $k \neq 0$, $=1$ for $k = 0$.
   (ii) Let $0 < h < \pi$. Show that for $f \in X_{2\pi}$
$$\int_x^{x+h} f(u)\, du = \frac{1}{2\pi}\int_{-\pi}^{\pi} f(x-u)m^*(u/h)\, du = (f * m^*(\circ/h))(x)$$
and $[\int_0^h f(\circ + u)\, du]^\wedge(k) = (ik)^{-1}(e^{ihk} - 1)f^\wedge(k)$ for $k \neq 0$, $= hf^\wedge(0)$ for $k = 0$.
   (iii) Let $f \in L^1_{2\pi}$. Show that $\int_0^h f(x+u)\, du$ (as a function of x) is locally of bounded variation and
$$\int_0^h f(x+u)\, du = hf^\wedge(0) + \lim_{n\to\infty} \sum_{k=-n}^{n}{}' \frac{e^{ihk} - 1}{ik} f^\wedge(k)\, e^{ikx}$$
for all x and h. In other words, the Fourier series of a function $f \in L^1_{2\pi}$ may be integrated term by term (whether the Fourier series itself is convergent or not), i.e.
$$\int_0^x [f(u) - f^\wedge(0)]\, du = \lim_{n\to\infty} \sum_{k=-n}^{n}{}' f^\wedge(k) \frac{e^{ikx} - 1}{ik},$$
and hence for almost all x
$$f(x) = f^\wedge(0) + \frac{d}{dx}\left(\lim_{n\to\infty} \sum_{k=-n}^{n}{}' f^\wedge(k) \frac{e^{ikx} - 1}{ik}\right).$$
(Hint: Apply Prop. 4.1.7 with g replaced by $m^*(\circ/h)$ or, to be independent, use Jordan's criterion; see also HARDY–ROGOSINSKI [1, p. 30], ASPLUND–BUNGART [1, p. 436], EDWARDS [1I, p. 92])
6. Show that $W[X_{2\pi}; (ik)^r] \subset W[X_{2\pi}; (ik)^j]$ for every $j = 1, 2, \ldots, r-1$.
7. (i) Let $f \in \mathrm{Lip}(X_{2\pi}; \alpha)$, $0 < \alpha \leq 1$. Show that $f^\wedge(k) = O(|k|^{-\alpha})$, $k \to \infty$. (Hint: Use (4.1.3) and Problem 4.1.1(iii))
   (ii) Show that $4|f^\wedge(k)| \leq \omega^*(X_{2\pi}; f; \pi/|k|)$ for every $f \in X_{2\pi}$ and $k \neq 0$. Thus, $f^\wedge(k) = O(|k|^{-\alpha})$, $k \to \infty$, for every $f \in \mathrm{Lip}^*(X_{2\pi}; \alpha)$, $0 < \alpha \leq 2$.
   (iii) Let $f \in BV_{loc}$ be 2π-periodic. Show that $f^\wedge(k) = O(|k|^{-1})$, $k \to \infty$. (Hint: Compare with Problem 1.2.11)
   (iv) Let $f \in AC_{2\pi}$. Show that $f^\wedge(k) = o(|k|^{-1})$, $k \to \infty$. (Hint: Use Prop. 4.1.2(ii), 4.1.8; see also HARDY–ROGOSINSKI [1, p. 26])

4.2 $L^p_{2\pi}$-Theory, $p > 1$

4.2.1 The Case $p = 2$

Having considered the finite Fourier transform mainly as a transform on $L^1_{2\pi}$, we shall here establish further results in case the functions in question are square-integrable, for example. It will be seen in particular that the inversion problem is completely solvable in $L^2_{2\pi}$-space and, as a matter of fact, solvable without the introduction of convergence factors.

We recall that the definition of the Fourier transform for $L^1_{2\pi}$-functions as given in (4.1.1) also applies to $X_{2\pi}$-functions. The following proposition deals with an interesting minimal property of the partial sums $S_n(f; x)$ of the Fourier series of f.

Proposition 4.2.1. *Let $f \in L^2_{2\pi}$. Then for any $t_n \in T_n$ (with coefficients $c_k \in \mathbb{C}$)*

(4.2.1) $$\|S_n(f; \circ) - f(\circ)\|_2 \leq \|t_n(\circ) - f(\circ)\|_2,$$

equality holding if and only if $c_k = f^\wedge(k)$, $|k| \le n$. *Furthermore*

$$\|f^\wedge\|_{l^2} \le \|f\|_{L^2_{2\pi}}. \tag{4.2.2}$$

The latter inequality is known as *Bessel's inequality.*

Proof. Using the Hilbert space notations of Sec. 0.7 we have

$$\|t_n - f\|_2^2 = (t_n - f, t_n - f) = (t_n, t_n) - \overline{(f, t_n)} - (f, t_n) + \|f\|_2^2$$
$$= \sum_{k=-n}^{n} |c_k|^2 - \sum_{k=-n}^{n} c_k\overline{f^\wedge(k)} - \sum_{k=-n}^{n} \overline{c_k}f^\wedge(k) + \|f\|_2^2.$$

Substituting $S_n(f; x)$ for $t_n(x)$ yields

$$\|S_n(f; \circ) - f(\circ)\|_2^2 = \|f\|_2^2 - \sum_{k=-n}^{n} |f^\wedge(k)|^2. \tag{4.2.3}$$

Moreover

$$\|t_n - f\|_2^2 - \|S_n(f; \circ) - f(\circ)\|_2^2 = \sum_{k=-n}^{n} |c_k - f^\wedge(k)|^2,$$

and thus (4.2.1) is established. (4.2.2) is an immediate consequence of (4.2.3).

Next we establish assertions (4.1.8)–(4.1.10) under different hypotheses.

Proposition 4.2.2. *If* $f, g \in L^2_{2\pi}$, *then*

$$(f * g)(x) = \sum_{k=-\infty}^{\infty} f^\wedge(k)g^\wedge(k)\, e^{ikx} \tag{4.2.4}$$

for all x, the series being absolutely and uniformly convergent. In particular,

$$\frac{1}{2\pi}\int_{-\pi}^{\pi} f(u)\overline{g(u)}\, du = \sum_{k=-\infty}^{\infty} f^\wedge(k)\overline{g^\wedge(k)}, \tag{4.2.5}$$

$$\|f\|_{L^2_{2\pi}} = \|f^\wedge\|_{l^2}. \tag{4.2.6}$$

Proof. By Prop. 0.4.1 we obtain that $f * g \in C_{2\pi}$, and by Hölder's inequality that the series in (4.2.4) converges absolutely and uniformly. Since it is the Fourier series of $f * g$ by Theorem 4.1.3, (4.2.4) follows by Prop. 4.1.5. Relations (4.2.5) and (4.2.6), known as the *generalized Parseval equation* and the *Parseval equation*, respectively, are now easy consequences of (4.2.4).

As an immediate application of (4.2.3) and (4.2.6) we have

Proposition 4.2.3. *If* $f \in L^2_{2\pi}$, *then*

$$\lim_{n\to\infty} \|S_n(f; \circ) - f(\circ)\|_2 = 0. \tag{4.2.7}$$

According to Prop. 1.2.3 and Theorem 1.3.5, the analog of the latter proposition in $L^1_{2\pi}$-space is not valid. Therefore, in order to produce convergence of the Fourier series, thus to have inversion of the finite Fourier transform on $L^1_{2\pi}$, we either supposed the functions to be smooth enough (see e.g. Prop. 4.1.5) or we introduced a summation process. But for $p = 2$, the nth partial sum of the Fourier series always converges to the original function in the mean of order 2, no convergence factor being needed. Although the Dirichlet kernel $\{D_n(x)\}$ is not an approximate identity for $\mathsf{L}^1_{2\pi}$-space, it behaves like one for $\mathsf{L}^2_{2\pi}$-space, and even for $\mathsf{L}^p_{2\pi}$-space, $p > 1$, as we shall see in Sec. 9.3.3.

Up to the present we began with a function $f \in \mathsf{L}^2_{2\pi}$ and formed its finite Fourier transform obtaining a function $f^\wedge \in \mathsf{l}^2$. The question arises whether every element of l^2 is representable as the finite Fourier transform of a function $f \in \mathsf{L}^2_{2\pi}$, that is to say, whether the finite Fourier transform defines a bounded linear transformation of $\mathsf{L}^2_{2\pi}$ *onto* l^2. The answer is affirmative and given by the following *theorem of Riesz–Fischer.*

Theorem 4.2.4. *Let* $\alpha \in \mathsf{l}^2$. *Then there exists* $f \in \mathsf{L}^2_{2\pi}$ *such that*

$$\alpha(k) = f^\wedge(k) \qquad (k \in \mathbb{Z}) \tag{4.2.8}$$

and

$$\|\alpha\|_{\mathsf{l}^2} = \|f\|_{\mathsf{L}^2_{2\pi}} = \|f^\wedge\|_{\mathsf{l}^2}. \tag{4.2.9}$$

Proof. If we set

$$s_n(x) = \sum_{k=-n}^{n} \alpha(k)\, e^{ikx}, \tag{4.2.10}$$

then we obtain for $m > n$

$$\|s_m - s_n\|^2_{\mathsf{L}^2_{2\pi}} = \sum_{k=-m}^{-(n+1)} |\alpha(k)|^2 + \sum_{k=n+1}^{m} |\alpha(k)|^2.$$

Thus the functions $s_n(x)$ form a Cauchy sequence in $\mathsf{L}^2_{2\pi}$, and the completeness of $\mathsf{L}^2_{2\pi}$ implies that there is $f \in \mathsf{L}^2_{2\pi}$ such that

$$\lim_{n\to\infty} \|s_n - f\|_{\mathsf{L}^2_{2\pi}} = 0. \tag{4.2.11}$$

Let $k \in \mathbb{Z}$ be arbitrary and choose $n > |k|$. Then by Hölder's inequality

$$|f^\wedge(k) - \alpha(k)| = \left|\frac{1}{2\pi}\int_{-\pi}^{\pi} [f(u) - s_n(u)]\, e^{-iku}\, du\right| \le \|f - s_n\|_{\mathsf{L}^2_{2\pi}},$$

and (4.2.8) follows by (4.2.11). Moreover, this establishes the theorem by (4.2.6).

Combining the results so far obtained for square-summable functions we may state

Corollary 4.2.5. *The finite Fourier transform defines a bounded linear transformation of the Hilbert space* $\mathsf{L}^2_{2\pi}$ *onto the Hilbert space* l^2 *which preserves inner products, i.e.*

$$(f, g) = (f^\wedge, g^\wedge) \tag{4.2.12}$$

for any $f, g \in \mathsf{L}^2_{2\pi}$.

Thus the map $f \to f^\wedge$ defines an isomorphism of the Hilbert space $\mathsf{L}^2_{2\pi}$ onto the Hilbert space l^2.

4.2.2 The Case $p \neq 2$

Until now we discussed the finite Fourier transform of functions $f \in \mathsf{L}^p_{2\pi}$ for $p = 1, 2$. The transform is a function on $\mathbb{Z}$ belonging to $\mathsf{l}^{p'}$. In particular, by (4.1.2) and (4.2.6),

$$\|f^\wedge\|_{\mathsf{l}^\infty} \leq \|f\|_{\mathsf{L}^1_{2\pi}}, \tag{4.2.13}$$

$$\|f^\wedge\|_{\mathsf{l}^2} = \|f\|_{\mathsf{L}^2_{2\pi}}, \tag{4.2.14}$$

respectively. It is natural to inquire whether these results can be extended to exponents other than 1 or 2. This is partially possible by the M. Riesz–Thorin convexity theorem.

Proposition 4.2.6. *Let* $1 < p < 2$ *and* $f \in \mathsf{L}^p_{2\pi}$. *Then* $f^\wedge \in \mathsf{l}^{p'}$ *and*

$$\|f^\wedge\|_{\mathsf{l}^{p'}} \leq \|f\|_{\mathsf{L}^p_{2\pi}}. \tag{4.2.15}$$

The assertion of this proposition is referred to as the *Hausdorff–Young inequality*; in fact, this phrase will also be used to cover the cases $p = 1$ and $p = 2$ of (4.2.13) and (4.2.14).

Proof. To apply the convexity theorem (cf. Sec. 4.4), let $\mathbb{R}_1 = (-\pi, \pi)$ with ordinary Lebesgue measure and $\mathbb{R}_2 = \mathbb{Z}$ where, in the usual way, $\mathbb{Z}$ is considered as a measure space in which each point has measure 1. For $f \in \mathsf{L}^p_{2\pi}$ let T be defined as the finite Fourier transform: $Tf \equiv f^\wedge$. Then, since $\mathsf{S}_{2\pi} \subset \mathsf{L}^p_{2\pi}$, it follows by (4.2.13) and (4.2.14) that $\|Th\|_{\mathsf{l}^\infty} \leq \|h\|_{\mathsf{L}^1_{2\pi}}$ and $\|Th\|_{\mathsf{l}^2} = \|h\|_{\mathsf{L}^2_{2\pi}}$ for every $h \in \mathsf{S}_{2\pi}$. Thus T is of strong type $(1; \infty)$, $(2; 2)$ on $\mathsf{S}_{2\pi}$ with constants $M_1 = M_2 = 1$, and (4.2.15) follows for every $h \in \mathsf{S}_{2\pi}$ by the M. Riesz–Thorin convexity theorem.

Let $f \in \mathsf{L}^p_{2\pi}$ be arbitrary. Then (see Sec. 0.4) there exists a sequence $\{h_j\}$ of functions in $\mathsf{S}_{2\pi}$ such that $\lim_{j\to\infty} \|f - h_j\|_{\mathsf{L}^p_{2\pi}} = 0$. By Hölder's inequality $|f^\wedge(k) - h_j^\wedge(k)| \leq \|f - h_j\|_{\mathsf{L}^p_{2\pi}}$, $k \in \mathbb{Z}$, and therefore $\lim_{j\to\infty} h_j^\wedge(k) = f^\wedge(k)$ for each $k \in \mathbb{Z}$. Hence by Fatou's lemma

$$\|f^\wedge\|^{p'}_{\mathsf{l}^{p'}} \leq \liminf_{j\to\infty} \|h_j^\wedge\|^{p'}_{\mathsf{l}^{p'}} \leq \liminf_{j\to\infty} \|h_j\|^{p'}_{\mathsf{L}^p_{2\pi}} = \|f\|^{p'}_{\mathsf{L}^p_{2\pi}},$$

where we have used the fact that (4.2.15) is already valid for functions in $\mathsf{S}_{2\pi}$. Thus (4.2.15) is completely established.

The restriction to $1 \leq p \leq 2$ for (4.2.15) to be valid is essential. There is $f \in \mathsf{C}_{2\pi}$ such that $\|f^\wedge\|_{\mathsf{l}^q} = \infty$ for all $q < 2$ (see Problem 4.2.2).

The Hausdorff–Young inequality (4.2.15) may be regarded as an extension of the original Parseval equation (4.2.6) to exponents $1 < p < 2$. But, in view of the Riesz–Fischer theorem, the Parseval equation contains a further assertion, namely, for any $\alpha \in \mathsf{l}^2$ there is $f \in \mathsf{L}^2_{2\pi}$ such that (4.2.8) and thus (4.2.9) holds. Regarding this aspect we have the following

Proposition 4.2.7. *Let* $\alpha \in \mathsf{l}^p$, $1 < p < 2$. *Then there exists* $f \in \mathsf{L}^{p'}_{2\pi}$ *such that*

$$\alpha(k) = f^\wedge(k) \qquad (k \in \mathbb{Z}) \tag{4.2.16}$$

and

$$\|f\|_{\mathsf{L}^{p'}_{2\pi}} \leq \|\alpha\|_{\mathsf{l}^p}. \tag{4.2.17}$$

Proof. If $s_n(x)$ is defined by (4.2.10), then we have by the Hölder and Hausdorff–Young inequalities for every $h \in \mathsf{L}^p_{2\pi}$

$$\left|\frac{1}{2\pi}\int_{-\pi}^{\pi} h(u)\overline{s_n(u)}\,du\right| = \left|\sum_{k=-n}^{n} \overline{\alpha(k)}h^\wedge(k)\right|$$
$$\leq \left\{\sum_{k=-n}^{n} |\alpha(k)|^p\right\}^{1/p}\left\{\sum_{k=-n}^{n} |h^\wedge(k)|^{p'}\right\}^{1/p'}$$
$$\leq \|h^\wedge\|_{\mathsf{l}^{p'}}\|\alpha\|_{\mathsf{l}^p} \leq \|h\|_{\mathsf{L}^p_{2\pi}}\|\alpha\|_{\mathsf{l}^p}.$$

This implies by (0.8.5) that

$$\|s_n\|_{\mathsf{L}^{p'}_{2\pi}} \leq \|\alpha\|_{\mathsf{l}^p}. \tag{4.2.18}$$

Since (4.2.18) is valid for each $\alpha \in \mathsf{l}^p$ and each $n \in \mathbb{N}$, it follows that for $m > n$

$$\|s_m - s_n\|^{p}_{\mathsf{L}^{p'}_{2\pi}} \leq \sum_{k=-m}^{-(n+1)} |\alpha(k)|^p + \sum_{k=n+1}^{m} |\alpha(k)|^p.$$

Thus the functions $s_n(x)$ form a Cauchy sequence in $\mathsf{L}^{p'}_{2\pi}$, and hence the completeness of $\mathsf{L}^{p'}_{2\pi}$ assures the existence of $f \in \mathsf{L}^{p'}_{2\pi}$ such that

$$\lim_{n\to\infty} \|s_n - f\|_{\mathsf{L}^{p'}_{2\pi}} = 0, \tag{4.2.19}$$

which together with (4.2.18) implies (4.2.17). Moreover, if $k \in \mathbb{Z}$ is arbitrary and $n \in \mathbb{N}$ such that $n > |k|$, then by Hölder's inequality

$$|f^\wedge(k) - \alpha(k)| = \left|\frac{1}{2\pi}\int_{-\pi}^{\pi} [f(u) - s_n(u)]\,e^{-iku}\,du\right| \leq \|f - s_n\|_{\mathsf{L}^{p'}_{2\pi}},$$

and (4.2.16) follows by (4.2.19).

Concerning the case $p = 1$, we recall the introduction to this Part. Indeed, to every $\alpha \in \mathsf{l}^1$ one may assign the continuous function $f(x) = \sum_{k=-\infty}^{\infty} \alpha(k)\,e^{ikx}$. Since this series converges uniformly, one easily deduces that $\alpha(k) = f^\wedge(k)$ and $\|f\|_{\mathsf{C}_{2\pi}} \leq \|\alpha\|_{\mathsf{l}^1}$.

Again the restriction of Prop. 4.2.7 to $1 \leq p \leq 2$ is essential. For there is a sequence $\{\alpha(k)\}$ that belongs to l^q for all $q > 2$ and yet is not the finite Fourier transform of any function in $\mathsf{L}^1_{2\pi}$ (cf. Problem 4.2.3).

In the introduction to this Part we defined the Fourier transform on l^1, and the question arises whether it is possible to define a Fourier transform on l^p for other values of p. Since $\mathsf{l}^p \subset \mathsf{l}^q$ for $q > p$, but not conversely, definition (II.6) does not apply. But Theorem 4.2.4 and Prop. 4.2.7 give us the feasibility of the following

Definition 4.2.8. *The **Fourier transform** $\alpha^\wedge$ of $\alpha \in \mathsf{l}^p$, $1 < p \leq 2$, is defined by*

$$\alpha^\wedge(x) = \underset{n\to\infty}{\overset{(p')}{\text{l.i.m.}}} \sum_{k=-n}^{n} \alpha(k)\,e^{-ikx}.$$

In other words, the Fourier transform of $\alpha \in \mathsf{l}^p$, $1 < p \leq 2$, is defined as the uniquely determined function $\alpha^\wedge \in \mathsf{L}^{p'}_{2\pi}$ given by

$$\lim_{n\to\infty} \left\|\alpha^\wedge(\circ) - \sum_{k=-n}^{n} \alpha(k)\,e^{-ik\circ}\right\|_{p'} = 0 \tag{4.2.20}$$

(see (4.2.11) and (4.2.19)). It is an easy consequence (cf. Prop. 0.1.10) that the definitions (II.6) and (4.2.20) are consistent for $\alpha \in l^1 \cap l^p$.

The last few remarks are important in so far as they shall lead us (cf. Sec. 5.2) to the solution of the corresponding problems for the Fourier transform associated with the real line.

Problems

1. Let $f, g \in L^2_{2\pi}$. Show that $fg \in L^1_{2\pi}$ and $[fg]^\wedge(k) = \sum_{j=-\infty}^{\infty} f^\wedge(j) g^\wedge(k-j)$. (Hint: HARDY–ROGOSINSKI [1, p. 23])
2. (i) Show that there are functions $f \in C_{2\pi}$ such that $\|f^\wedge\|_{l^q} = \infty$ for all $q < 2$. (Hint: ZYGMUND [7II, p. 101])
   (ii) Let $f \in L^p_{2\pi}$, $1 < p < 2$. Show that $\|f^\wedge\|_{l^{p'}} = \|f\|_{L^p_{2\pi}}$ if and only if $f(x) = c \exp\{imx\}$ for some $c \in \mathbb{C}$ and $m \in \mathbb{Z}$. (Hint: HARDY–LITTLEWOOD [1], see also HEWITT [1, p. 110])
3. (i) Show that there are sequences $\{\alpha(k)\}_{k=-\infty}^{\infty}$ that belong to l^q for all $q > 2$, yet are not the finite Fourier transform of any function in $L^1_{2\pi}$. (Hint: ZYGMUND [5, p. 190, 7II, p. 102], HEWITT [1, p. 110])
   (ii) Show that equality occurs in (4.2.17) if and only if $\alpha(k) = c\delta_{m,k}$ for some $c \in \mathbb{C}$, $m \in \mathbb{Z}$, where $\delta_{m,k}$ is Kronecker's symbol. (Hint: HEWITT–HIRSCHMAN [1], see also HEWITT [1, p. 110])
4. Let f be a $2\pi\lambda$-periodic function belonging to $L^2(-\pi\lambda, \pi\lambda)$, $\lambda > 0$. Show that $\sum_{k=-\infty}^{\infty} |f^\wedge(k)|^2 = (1/2\pi\lambda)\int_{-\pi\lambda}^{\pi\lambda} |f(u)|^2\,du$, where $f^\wedge$ is defined by (II.4).

4.3 Finite Fourier–Stieltjes Transforms

4.3.1 Fundamental Properties

In Sec. 1.2.7 we introduced the kth complex Fourier–Stieltjes coefficient of a (complex-valued) function $\mu \in BV_{2\pi}$ by

$$\mu^\vee(k) = \frac{1}{2\pi}\int_{-\pi}^{\pi} e^{-iku}\,d\mu(u) \equiv [\mu(\circ)]^\vee(k) \qquad (k \in \mathbb{Z}), \tag{4.3.1}$$

the integral being a Riemann-Stieltjes integral. For every† $\mu \in BV_{2\pi}$, (4.3.1) defines a function $\mu^\vee$ on $\mathbb{Z}$, called the *finite Fourier–Stieltjes transform* of μ. We are going to derive some of its operational properties.

Proposition 4.3.1. *For $\mu \in BV_{2\pi}$ we have*

(i) $[\mu(\circ + h)]^\vee(k) = e^{ihk}\mu^\vee(k)$ $\qquad (h \in \mathbb{R}, k \in \mathbb{Z})$,

(ii) $[\bar{\mu}]^\vee(-k) = \overline{\mu^\vee(k)}$ $\qquad (k \in \mathbb{Z})$,

(iii) *if μ is absolutely continuous, i.e. $\mu(x) = \int_{-\pi}^{x} f(u)\,du + \mu(-\pi)$, $f \in L^1_{2\pi}$, then $\mu^\vee(k) = f^\wedge(k)$ for all $k \in \mathbb{Z}$,*

(iv) *if μ is 2π-periodic, then $\mu \in L^1_{2\pi}$ and $(ik)\mu^\wedge(k) = \mu^\vee(k)$ for all $k \in \mathbb{Z}$.*

† Obviously, the integral defining $\mu^\vee(k)$ is also meaningful for every $\mu \in BV[-\pi, \pi]$. This will sometimes be used though some of the very elementary properties may then fail to hold (e.g. Prop. 4.3.1(i)).

The proof is left to Problem 4.3.1. Further information about the finite Fourier–Stieltjes transform is given by

Proposition 4.3.2. *If* $\mu, \nu \in \mathsf{BV}_{2\pi}$ *and* $c \in \mathbb{C}$, *then*

(i) $[\mu + \nu]^\vee(k) = \mu^\vee(k) + \nu^\vee(k), \qquad [c\mu]^\vee(k) = c\mu^\vee(k) \qquad (k \in \mathbb{Z})$,
(ii) $|\mu^\vee(k)| \le \|\mu\|_{\mathsf{BV}_{2\pi}} \qquad (k \in \mathbb{Z})$,
(iii) $\mu^\vee(k) = 0$ *on* $\mathbb{Z}$ *implies* $\mu(x) \equiv$ const,
(iv) *there are sequences of class* l^∞ *which are not the finite Fourier–Stieltjes transform of a function* $\mu \in \mathsf{BV}_{2\pi}$.

Concerning the proof, parts (i) and (ii) are immediate consequences of the definition (4.3.1), whereas (iii) follows by Prop. 1.2.12. Indeed, for the Fejér means of the corresponding Fourier–Stieltjes series the hypothesis implies $\sigma_n(d\mu; x) \equiv 0$ for every $n \in \mathbb{N}$. Therefore (1.2.59) yields $\int_{-\pi}^{\pi} h(x)\, d\mu(x) = 0$ for every $h \in \mathsf{C}_{2\pi}$ from which (iii) follows by Prop. 0.6.1. For the proof of (iv) we refer to Problem 4.3.2.

Part (iii) of the foregoing proposition only reveals a partial *uniqueness* property for the finite Fourier–Stieltjes transform, since we did not normalize the function $\mu \in \mathsf{BV}_{2\pi}$ by $\mu(-\pi) = 0$. But if we regard μ as an element of the Banach space $\mathsf{BV}_{2\pi}^0$, then $\mu^\vee(k) = 0$ on $\mathbb{Z}$ implies that μ is the zero-element of $\mathsf{BV}_{2\pi}^0$, and thus we would have uniqueness. Hence the finite Fourier–Stieltjes transform defines a one-to-one bounded linear transformation of the Banach space $\mathsf{BV}_{2\pi}^0$ into (but not onto) l^∞.

The Riemann–Lebesgue lemma does not hold for the finite Fourier–Stieltjes transform. In fact, the Fourier–Stieltjes transform of $\delta^* \in \mathsf{BV}_{2\pi}$ as defined on $[-\pi, \pi]$ by

$$\delta^*(x) = \begin{cases} 0, & -\pi \le x < 0 \\ \pi, & x = 0 \\ 2\pi, & 0 < x \le \pi \end{cases} \tag{4.3.2}$$

is given by

$$[\delta^*]^\vee(k) = 1 \qquad (k \in \mathbb{Z}). \tag{4.3.3}$$

There are also continuous functions $\mu \in \mathsf{BV}_{2\pi}$ for which $\mu^\vee(k)$ does not tend to zero as $|k| \to \infty$.

Proposition 4.3.3. *Let* $f \in \mathsf{L}_{2\pi}^1$, $\mu \in \mathsf{BV}_{2\pi}$. *Then for the convolution* $(f * d\mu)(x)$ (*defined by* (0.6.8)) *we have*

$$[f * d\mu]^\wedge(k) = f^\wedge(k)\mu^\vee(k) \qquad (k \in \mathbb{Z}). \tag{4.3.4}$$

Proof. In view of Prop. 0.6.3, $f * d\mu \in \mathsf{L}_{2\pi}^1$ and thus both sides of (4.3.4) are meaningful. Furthermore, by Fubini's theorem

$$[f * d\mu]^\wedge(k) = \frac{1}{2\pi}\int_{-\pi}^{\pi} e^{-iku}\, d\mu(u) \left\{\frac{1}{2\pi}\int_{-\pi}^{\pi} f(x - u)\, e^{-ik(x-u)}\, dx\right\} = f^\wedge(k)\mu^\vee(k),$$

establishing (4.3.4).

For the proof of the general *convolution theorem* for the finite Fourier–Stieltjes transform, which is to follow, we need

Lemma 4.3.4. *Let $\mu \in \mathsf{BV}_{2\pi}$, $h \in \mathbb{R}$, and set*

$$q(x) = \mu(x+h) - \mu(x). \tag{4.3.5}$$

Then q is 2π-periodic, belongs to $\mathsf{L}^1_{2\pi} \cap \mathsf{BV}_{2\pi}$, and

$$\|\mu(\circ + h) - \mu(\circ)\|_{\mathsf{L}^1_{2\pi}} \le |h| \, \|\mu\|_{\mathsf{BV}_{2\pi}}, \tag{4.3.6}$$

$$q^\wedge(k) = \begin{cases} \dfrac{e^{ihk}-1}{ik}\, \mu^\vee(k), & k \neq 0 \\ h\mu^\vee(0), & k = 0. \end{cases} \tag{4.3.7}$$

Proof. By the definition of the class $\mathsf{BV}_{2\pi}$ it easily follows that q is a 2π-periodic function, and thus $q \in \mathsf{L}^1_{2\pi} \cap \mathsf{BV}_{2\pi}$. Suppose that μ is monotonely increasing and $h > 0$. Then

$$\|\mu(\circ + h) - \mu(\circ)\|_{\mathsf{L}^1_{2\pi}} = \frac{1}{2\pi} \int_\pi^{\pi+h} [\mu(x) - \mu(x - 2\pi)]\, dx$$
$$= \frac{h}{2\pi} [\mu(\pi) - \mu(-\pi)] = h \, \|\mu\|_{\mathsf{BV}_{2\pi}}.$$

In the general case, we use the Jordan decomposition of μ, and (4.3.6) follows. Finally, by Prop. 4.3.1

$$ikq^\wedge(k) = q^\vee(k) = [\mu(\circ + h) - \mu(\circ)]^\vee(k) = (e^{ihk} - 1)\mu^\vee(k),$$

and the proof is complete.

Theorem 4.3.5. *For the convolution $(\mu * d\nu)(x)$ of μ, $\nu \in \mathsf{BV}_{2\pi}$ we have*

$$[\mu * d\nu]^\vee(k) = \mu^\vee(k)\nu^\vee(k) \qquad (k \in \mathbb{Z}). \tag{4.3.8}$$

Proof. By Prop. 0.6.2, $\mu * d\nu \in \mathsf{BV}_{2\pi}$, and thus both terms of (4.3.8) are well-defined. To prove the actual equality, let q be given by (4.3.5). Since $q \in \mathsf{L}^1_{2\pi} \cap \mathsf{BV}_{2\pi}$, it follows that $q * d\nu \in \mathsf{L}^1_{2\pi} \cap \mathsf{BV}_{2\pi}$. Moreover,

$$[q * d\nu]^\wedge(k) = q^\wedge(k)\nu^\vee(k) = \begin{cases} \dfrac{e^{ihk}-1}{ik}\, \mu^\vee(k)\nu^\vee(k), & k \neq 0 \\ h\mu^\vee(0)\nu^\vee(0), & k = 0 \end{cases} \tag{4.3.9}$$

by Prop. 4.3.3 and (4.3.7). On the other hand, $(q * d\nu)(x) = (\mu * d\nu)(x + h) - (\mu * d\nu)(x)$. Since the right-hand side is of type (4.3.5), μ being replaced by $\mu * d\nu$, we may apply (4.3.7) to obtain

$$[q * d\nu]^\wedge(k) = \begin{cases} \dfrac{e^{ihk}-1}{ik}\, [\mu * d\nu]^\vee(k), & k \neq 0 \\ h[\mu * d\nu]^\vee(0), & k = 0, \end{cases}$$

which, together with (4.3.9), proves the theorem.

It follows from (4.3.8) and Prop. 4.3.2(iii) that $(\mu * d\nu)(x) - (\nu * d\mu)(x) \equiv \text{const}$, a result which we already obtained directly from the definition of $\mu * d\nu$ by partial integration (compare (0.6.7)). Thus $\mathsf{BV}^0_{2\pi}$ becomes a Banach algebra under convolution as multiplication which is commutative. It even has a unit element which is given by

(4.3.2). Furthermore, the finite Fourier–Stieltjes transform is a non norm-increasing isomorphism of the Banach algebra $\mathsf{BV}_{2\pi}^0$ into l^∞, the algebra of bounded functions on $\mathbb{Z}$ with pointwise operations and supremum-norm.

Proposition 4.3.6. *If $\mu \in \mathsf{BV}_{2\pi}$ is such that $\mu^\vee \in \mathsf{l}^1$, then μ is absolutely continuous and*

$$\mu'(x) = \sum_{k=-\infty}^{\infty} \mu^\vee(k)\, e^{ikx}. \tag{4.3.10}$$

Proof. Since $\mu^\vee \in \mathsf{l}^1$, the series on the right of (4.3.10) converges uniformly and thus defines a function $f \in \mathsf{C}_{2\pi}$ for which $f^\wedge(k) = \mu^\vee(k)$, $k \in \mathbb{Z}$. Therefore $\mu(x) = \int_{-\pi}^{x} f(u)\, du + \mu(-\pi)$ by Prop. 4.3.2(iii).

Proposition 4.3.7. *If $\mu \in \mathsf{BV}_{2\pi}$, and $f \in \mathsf{L}_{2\pi}^1$ is such that $f^\wedge \in \mathsf{l}^1$, then for all x*

$$(f * d\mu)(x) = \sum_{k=-\infty}^{\infty} f^\wedge(k)\mu^\vee(k)\, e^{ikx}. \tag{4.3.11}$$

Proof. By Prop. 4.1.5 we have $f(x) = \sum_{k=-\infty}^{\infty} f^\wedge(k)\, e^{ikx}$ a.e. Since the series converges uniformly, we obtain by term-by-term integration

$$(f * d\mu)(x) = \frac{1}{2\pi}\int_{-\pi}^{\pi} \left\{ \sum_{k=-\infty}^{\infty} f^\wedge(k)\, e^{ikx}\, e^{-iku} \right\} d\mu(u) = \sum_{k=-\infty}^{\infty} f^\wedge(k)\mu^\vee(k)\, e^{ikx},$$

which holds everywhere since both sides represent continuous functions.

4.3.2 Inversion Theory

In view of the definition (4.3.1) of the finite Fourier–Stieltjes transform of $\mu \in \mathsf{BV}_{2\pi}$, we might expect by (4.3.10) that an inversion is given by

$$\mu'(x) = \sum_{k=-\infty}^{\infty} \mu^\vee(k)\, e^{ikx}.$$

But even if this series is understood as the limit for $n \to \infty$ of the symmetrical partial sums

$$\sum_{k=-n}^{n} \mu^\vee(k)\, e^{ikx} = \frac{1}{2\pi}\int_{-\pi}^{\pi} D_n(x-u)\, d\mu(u) \equiv S_n(d\mu; x)$$

(see Sec. 1.2.7), the limit does by no means exist for every x and $\mu \in \mathsf{BV}_{2\pi}$; it must be interpreted in some generalized sense.

Proposition 4.3.8. *Let $\{\theta_\rho(k)\}$ satisfy (1.2.28) such that the corresponding kernel $\{\theta_\rho^\wedge(x)\}$ of (4.1.18) is an approximate identity. Then for $\mu \in \mathsf{BV}_{2\pi}$*

$$\lim_{\rho\to\rho_0} \int_{-\pi}^{\pi} h(x)\left[\sum_{k=-\infty}^{\infty} \theta_\rho(k)\mu^\vee(k)\, e^{ikx}\right] dx = \int_{-\pi}^{\pi} h(x)\, d\mu(x)$$

for every $h \in \mathsf{C}_{2\pi}$.

The proof follows by Prop. 1.2.12 by making use of

$$U_\rho(d\mu; x) \equiv \sum_{k=-\infty}^{\infty} \theta_\rho(k)\mu^\vee(k)\, e^{ikx} = \frac{1}{2\pi}\int_{-\pi}^{\pi} \theta_\rho^\wedge(x-u)\, d\mu(u), \tag{4.3.12}$$

which is valid in virtue of Prop. 4.3.7.

The next proposition is an application of Prop. 1.4.8.

Proposition 4.3.9. *Let $\mu \in \mathsf{BV}_{2\pi}$ and $\{\theta_\rho(k)\}$ be a θ-factor such that the corresponding kernel $\{\theta_\rho^\wedge(x)\}$ of (4.1.18) is an absolutely continuous, even, positive approximate identity, monotonely decreasing on $[0, \pi]$. Then*

$$\lim_{\rho\to\rho_0} \sum_{k=-\infty}^{\infty} \theta_\rho(k)\mu^\vee(k)\, e^{ikx} = \mu'(x) \quad \text{a.e.} \tag{4.3.13}$$

For the application of further results of Sec. 1.4 as well as for explicit formulae for some important examples of θ-factors we refer to Problem 4.3.3.

If one wishes to obtain μ instead of μ', formula (4.3.13) suggests that an inversion of the type

$$\mu(x+h) - \mu(x) = h\mu^\vee(0) + \lim_{\rho\to\rho_0} \sum_{k=-\infty}^{\infty}{}' \theta_\rho(k)\frac{e^{ihk}-1}{ik}\mu^\vee(k)\, e^{ikx} \tag{4.3.14}$$

may be valid. But we already know (cf. Problem 4.1.5) that term-by-term integration converts the Fourier series of any $f \in \mathsf{L}^1_{2\pi}$ into a (uniformly) convergent series. Thus we may expect that (4.3.14) holds without the convergence-factor $\{\theta_\rho(k)\}$. In fact, we have

Proposition 4.3.10. *If $\mu \in \mathsf{BV}_{2\pi}$, then for all x, h*

$$\mu(x+h) - \mu(x) = h\mu^\vee(0) + \lim_{n\to\infty} \sum_{k=-n}^{n}{}' \frac{e^{ihk}-1}{ik}\mu^\vee(k)\, e^{ikx}. \tag{4.3.15}$$

Proof. Let $h \in \mathbb{R}$ be fixed. If q is defined by (4.3.5), then q is a 2π-periodic function belonging to $\mathsf{L}^1_{2\pi} \cap \mathsf{BV}_{2\pi}$, and thus by Prop. 4.1.5(ii) $q(x) = \lim_{n\to\infty} \sum_{k=-n}^{n} q^\wedge(k)\, e^{ikx}$ for all x. (4.3.15) now follows by (4.3.7).

4.3.3 Fourier–Stieltjes Transforms of Derivatives

In this subsection r always denotes a natural number.

Proposition 4.3.11. *If $f \in \mathsf{AC}^{r-2}_{2\pi}$ and $f^{(r-1)} \in \mathsf{BV}_{2\pi}$, then*

$$(ik)^r f^\wedge(k) = [f^{(r-1)}]^\vee(k) \qquad (k \in \mathbb{Z}). \tag{4.3.16}$$

For the proof we observe that for $r = 1$ the assumptions are that $f \in \mathsf{BV}_{2\pi}$ is 2π-periodic. In this case, (4.3.16) is Prop. 4.3.1(iv). If $r \ge 2$, then Prop. 4.1.8 shows that $(ik)^{r-1}f^\wedge(k) = [f^{(r-1)}]^\wedge(k)$, $k \in \mathbb{Z}$, and (4.3.16) follows by an application of Prop. 4.3.1(iv) to $f^{(r-1)}$.

In order to establish the converse we introduce the following classes of functions:

$$\text{(4.3.17)}\qquad \mathsf{V}^r_{\mathsf{X}_{2\pi}} = \begin{cases} \{f\in \mathsf{C}_{2\pi} \mid f\in \mathsf{AC}^{r-1}_{2\pi}, f^{(r)}\in \mathsf{L}^\infty_{2\pi}\} \\ \{f\in \mathsf{L}^1_{2\pi} \mid f=\phi \quad \text{a.e.}, \phi\in \mathsf{AC}^{r-2}_{2\pi}, \phi^{(r-1)}\in \mathsf{BV}_{2\pi}\} \\ \{f\in \mathsf{L}^p_{2\pi} \mid f=\phi \quad \text{a.e.}, \phi\in \mathsf{AC}^{r-1}_{2\pi}, \phi^{(r)}\in \mathsf{L}^p_{2\pi}\} \qquad (1<p<\infty). \end{cases}$$

In case of the reflexive spaces $\mathsf{X}_{2\pi} = \mathsf{L}^p_{2\pi}$, $1<p<\infty$, the classes $\mathsf{V}^r_{\mathsf{L}^p_{2\pi}}$ and $\mathsf{W}^r_{\mathsf{L}^p_{2\pi}}$ are equal by definition.

Proposition 4.3.12. *If for $f\in \mathsf{L}^1_{2\pi}$ there is $\mu\in \mathsf{BV}_{2\pi}$ such that*

$$\text{(4.3.18)}\qquad (ik)^r f^\wedge(k) = \mu^\vee(k) \qquad (k\in\mathbb{Z}),$$

then $f\in \mathsf{V}^r_{\mathsf{L}^1_{2\pi}}$.

Proof. Since (4.3.18) implies $\mu^\vee(0)=0$, μ is a 2π-periodic function and thus belongs to $\mathsf{L}^1_{2\pi}\cap \mathsf{BV}_{2\pi}$. Hence, by Prop. 4.3.1(iv), $(ik)^{r-1}f^\wedge(k) = \mu^\wedge(k)$ for $k\in\mathbb{Z}$, $k\neq 0$. Now an application of Prop. 4.1.9 gives $f\in \mathsf{W}^{r-1}_{\mathsf{L}^1_{2\pi}}$. Furthermore, if $G_{r-1}(x)$ is defined via (4.1.22), g being replaced by μ, then (4.1.23) implies $f(x) - G_{r-1}(x) = \text{const a.e.}$, $G_{r-1}\in \mathsf{AC}^{r-2}_{2\pi}$ and $G^{(r-1)}_{r-1}(x) = \mu(x)$. This already establishes the present assertion.

The treatment of the classes $\mathsf{W}^r_{\mathsf{X}_{2\pi}}$ in Sec. 4.1.3 now suggests that one introduces, apart from $\mathsf{V}^r_{\mathsf{X}_{2\pi}}$, the classes

$$\text{(4.3.19)}\qquad \mathsf{V}[\mathsf{X}_{2\pi};\psi(k)] = \begin{cases} \{f\in \mathsf{C}_{2\pi} \mid \psi(k)f^\wedge(k) = g^\wedge(k), g\in \mathsf{L}^\infty_{2\pi}\} \\ \{f\in \mathsf{L}^1_{2\pi} \mid \psi(k)f^\wedge(k) = \mu^\vee(k), \mu\in \mathsf{BV}_{2\pi}\} \\ \{f\in \mathsf{L}^p_{2\pi} \mid \psi(k)f^\wedge(k) = g^\wedge(k), g\in \mathsf{L}^p_{2\pi}\} \qquad (1<p<\infty) \end{cases}$$

for an arbitrary complex-valued function $\psi(k)$ on $\mathbb{Z}$. Then we may summarize the results of Prop. 4.3.11, 4.3.12 as well as of Problem 4.3.4 (and of Theorem 4.1.10 for $\mathsf{L}^p_{2\pi}$, $1<p<\infty$) to

Theorem 4.3.13. *Let $f\in \mathsf{X}_{2\pi}$. The following statements are equivalent:*

(i) $f\in \mathsf{V}^r_{\mathsf{X}_{2\pi}}$, (ii) $f\in \mathsf{V}[\mathsf{X}_{2\pi};(ik)^r]$,

(iii) *there exists $g\in \mathsf{L}^\infty_{2\pi}$ if $\mathsf{X}_{2\pi} = \mathsf{C}_{2\pi}$, $\mu\in \mathsf{BV}_{2\pi}$ if $\mathsf{X}_{2\pi} = \mathsf{L}^1_{2\pi}$, $g\in \mathsf{L}^p_{2\pi}$ if $\mathsf{X}_{2\pi} = \mathsf{L}^p_{2\pi}$, $1<p<\infty$, and constants $\alpha_k\in\mathbb{C}$, $0\le k\le r-1$, such that* (a.e.)

$$f(x) = \alpha_0 + \int_{-\pi}^{x} du_1\Bigg[\alpha_1 + \int_{-\pi}^{u_1} du_2\Bigg[\alpha_2 + \cdots + \int_{-\pi}^{u_{r-2}} du_{r-1}\Bigg[\alpha_{r-1} + \int_{-\pi}^{u_{r-1}} \begin{Bmatrix} g(u_r)\,du_r \\ d\mu(u_r) \\ g(u_r)\,du_r \end{Bmatrix}\Bigg]\cdots\Bigg]\Bigg].$$

The classes $\mathsf{V}[\mathsf{X}_{2\pi};\psi(k)]$ and their various representations for a particular choice of $\psi(k)$ will play a fundamental rôle in our later investigations on saturation theory.

Problems

1. Prove Prop. 4.3.1 (compare the comments given to the definition of the class $\mathsf{BV}_{2\pi}$ in Sec. 0.6.).
2. Complete the proof of Prop. 4.3.2. (Hint: For (iv) use the hint of Problem 4.1.3; see also HEWITT [1, pp. 23–27], where the problems of identifying the Fourier–Stieltjes transforms in l^∞ as well as those elements $\alpha \in \mathsf{l}^\infty$ which admit uniform approximation on $\mathbb{Z}$ by Fourier–Stieltjes transforms are discussed, cf. also ZYGMUND [7I, p. 194 ff])
3. (i) Let $\mu \in \mathsf{BV}_{2\pi}$. Show that

$$\lim_{r\to 1-} \sum_{k=-\infty}^{\infty} r^{|k|}\mu^\vee(k)\, e^{ikx} = \mu'(x) \quad \text{a.e.}$$

   (ii) State and prove the counterpart of Prop. 4.3.9 for θ-factors for which the corresponding kernel $\{\theta_\rho^\wedge(x)\}$ satisfies the assumptions of Prop. 1.4.10.

   (iii) Let $\mu \in \mathsf{BV}_{2\pi}$. Show that

$$\lim_{n\to\infty} \sum_{k=-n}^{n} \left(1 - \frac{|k|}{n+1}\right)\mu^\vee(k)\, e^{ikx} = \mu'(x) \quad \text{a.e.}$$

   (iv) Prove (4.3.14).
4. (i) Let $f \in \mathsf{AC}_{2\pi}^{r-1}$ be such that $f^{(r)} \in \mathsf{L}_{2\pi}^\infty$. Show that $(ik)^r f^\wedge(k) = [f^{(r)}]^\wedge(k)$.

   (ii) Show that if for $f \in \mathsf{C}_{2\pi}$ there exists $g \in \mathsf{L}_{2\pi}^\infty$ such that $(ik)^r f^\wedge(k) = g^\wedge(k)$, then $f \in \mathsf{V}_{\mathsf{C}_{2\pi}}^r$.
5. Show that if $f \in \mathsf{V}[\mathsf{X}_{2\pi}; (ik)^r]$ for some $r \in \mathbb{N}$, then $f \in \mathsf{W}[\mathsf{X}_{2\pi}; (ik)^j]$ for every $j = 1, 2, \ldots, r-1$.

4.4 Notes and Remarks

The classical standard references to this chapter are the treatises of ZYGMUND [5, 7] and BARI [1] on trigonometric series as well as the brief but excellent treatment by HARDY–ROGOSINSKI [1]. For a modern introduction to the theory of Fourier series we must mention EDWARDS [1]. These books include comprehensive bibliographies on the subject so that our discussion here may be confined to a few, selected comments (see also BURKHARDT [1], HILB–RIESZ [1] for the older literature). Since Fourier analysis occupies a central position in analysis, almost all texts on advanced calculus, theory of functions, approximation theory, etc. include accounts on Fourier series and integrals. Particular mention can be made of ACHIESER [2], ASPLUND–BUNGART [1], HEWITT–STROMBERG [1], RUDIN [4], SEELEY [1], and SZ.-NAGY [5]. On the other hand, KATZNELSON [1] and HEWITT [1] have extensively covered the theory of Fourier series as well as integrals. The approach of these books is primarily oriented to harmonic analysis on groups as treated in HEWITT–ROSS [1], LOOMIS [2], REITER [1], and RUDIN [3].

In contrast to Sec. 1.2, Fourier or Fourier–Stieltjes coefficients have here been regarded as a transform of one function space into another. This point of view is useful in order to give a unified approach to the theory of Fourier series, Fourier integrals, and their many analogs and extensions. For this emphasis particular reference can be made to the excellent survey article of WEISS [2]. Fourier series in the frame of distribution theory are considered in SCHWARTZ [1, 3] and EDWARDS [1], for example.

Sec. 4.1. Concerning Prop. 4.1.2, it is even true that the Fourier transform is a one-to-one bounded linear transformation of $\mathsf{X}_{2\pi}$ onto $[\mathsf{X}_{2\pi}]^\wedge$, the latter being a proper subset of l_0^∞; compare RUDIN [4, p. 103]. There are other proofs of the uniqueness theorem, see, for example, HARDY–ROGOSINSKI [1, p. 18]. Many results on the order of magnitude of Fourier coefficients (cf. Problem 4.1.7) are known. Thus, the estimate (4.1.3) may be strengthened as follows: if $f^{(r)} \in \mathsf{L}^p_{2\pi}$, then $2\,|f^\wedge(k)| \le |k|^{-r}\omega(\mathsf{L}^p_{2\pi}; f^{(r)}; \pi/|k|)$, $k \in \mathbb{Z}$; see ACHIESER [2, p. 207 ff]. Converse-type results are also possible: If $f^\wedge(k) = O(|k|^{-l})$ with $f \in \mathsf{L}^1_{2\pi}$, then f is r-times differentiable with $f^{(r)} \in \mathsf{L}^2_{2\pi}$ provided $2(l - r) > 1$; compare KATZNELSON [1, p. 30] and Prop. 4.1.9.

In connection with Prop. 4.1.4, although the Banach algebra $\mathsf{L}^1_{2\pi}$ contains no unit element, every $f \in \mathsf{L}^1_{2\pi}$ can be factorized into a convolution product $f_1 * f_2$ with $f_1, f_2 \in \mathsf{L}^1_{2\pi}$. More generally, $\mathsf{X}_{2\pi} = \mathsf{L}^1_{2\pi} * \mathsf{X}_{2\pi}$; see EDWARDS [1I, p. 53, p. 117 ff] and the literature cited there. For Prop. 4.1.7, due to W. H. YOUNG, see ACHIESER [2, p. 102] or EDWARDS [1I, p. 91].

In case f is not absolutely continuous in Prop. 4.1.8 but has jump discontinuities, see ZYGMUND [7I, p. 41]; thus there exists $f \in \mathsf{L}^1_{2\pi}$ with $f' \in \mathsf{L}^1_{2\pi}$, f not absolutely continuous such that $[f']^\wedge(k) \ne (ik)f^\wedge(k)$.

Sec. 4.2. The treatment of Sec. 4.2.1 intrinsically depends upon the fact that the underlying spaces are Hilbert spaces. In fact, these methods may be employed for expansions by orthonormal systems in arbitrary Hilbert spaces, a standard topic in texts on functional analysis, see e.g. TAYLOR [1, p. 106 ff], DUNFORD–SCHWARTZ [1I, p. 247].

The results of Sec. 4.2.2 (cf. e.g. ZYGMUND [7II, p. 101 ff], HEWITT [1, p. 106 ff]) rest upon the *M. Riesz–Thorin convexity theorem: Let* $\mathbb{R}_1$ *and* $\mathbb{R}_2$ *be two measure spaces with measures* μ_1 *and* μ_2, *respectively. Let T be a linear operator defined for all simple functions f on* $\mathbb{R}_1$ *and taking values in the set of (complex-valued)* μ_2*-measurable functions on* $\mathbb{R}_2$. *Suppose that T is simultaneously of strong type* $(1/p_1; 1/q_1)$ *and* $(1/p_2; 1/q_2)$, *i.e., that*

$$\|Tf\|_{1/q_1} \le M_1\|f\|_{1/p_1}, \qquad \|Tf\|_{1/q_2} \le M_2\|f\|_{1/p_2},$$

the points (p_1, q_1) *and* (p_2, q_2) *belonging to the square* $0 \le p \le 1$, $0 \le q \le 1$. *Then T is also of strong type* $(1/p; 1/q)$ *for all*

$$p = (1 - t)p_1 + tp_2, \qquad q = (1 - t)q_1 + tq_2 \qquad (0 < t < 1),$$

and $\|Tf\|_{1/q} \le M_1^{1-t}M_2^t\|f\|_{1/p}$. *In particular, if* $p > 0$, *the operator T can be uniquely extended to the whole space* $\mathsf{L}^{1/p}_{\mu_1}$ *preserving norm.* Here we call any function *simple* if it takes on only a finite number of values (on μ_1-measurable sets) and (if $\mu_1(\mathbb{R}_1)$ is infinite) vanishes outside a subset of $\mathbb{R}_1$ of finite measure. We recall that the set of all simple functions was denoted by S_{00} if $\mathbb{R}_1 = \mathbb{R}$ and $\mu_1 =$ Lebesgue measure, and by $\mathsf{S}_{2\pi}$ if $\mathbb{R}_1 = (-\pi, \pi)$ and $\mu_1 =$ Lebesgue measure (in the latter case we extended the functions by periodicity). The operator T is said to be of *strong type* $(r; s)$ if $\|Tf\|_{\mathsf{L}^s_{\mu_2}} \le M\|f\|_{\mathsf{L}^r_{\mu}}$ for all simple functions on $\mathbb{R}_1$. T is said to be of *weak type* $(r; s)$ if $\mu_2(\{x \in \mathbb{R}_2 \mid |(Tf)(x)| > y > 0\}) \le (M\,\|f\|_{\mathsf{L}^r_{\mu_1}}/y)^s$ for all simple functions on $\mathbb{R}_1$ with M independent of f and y. The least value of M may be called the weak $(r; s)$ norm of T. Every linear operator T of strong type is also of weak type, but not conversely. For the latter operators there holds the following *convexity theorem of Marcinkiewicz: Let* (p_1, q_1) *and* (p_2, q_2) *be any two points of the triangle* $0 \le q \le p \le 1$ *such that* $q_1 \ne q_2$. *Suppose that a linear operator T is simultaneously of weak type* $(1/p_1; 1/q_1)$ *and* $(1/p_2; 1/q_2)$ *with norms* M_1 *and* M_2, *respectively. Then for any point* (p, q) *with*

$$p = (1 - t)p_1 + tp_2, \qquad q = (1 - t)q_1 + tq_2 \qquad (0 < t < 1)$$

the operator T is of strong type $(1/p; 1/q)$, *and* $\|Tf\|_{1/q} \le KM_1^{1-t}M_2^t\|f\|_{1/p}$, *where* $K = K_{t,p_1,q_1,p_2,q_2}$ *is independent of f and is bounded if* p_1, q_1, p_2, q_2 *are fixed and t stays away from* 0 *and* 1. This theorem remains valid if T is only quasi-linear. For this and further comments as well as for the proofs of the two convexity theorems we refer to ZYGMUND [7II, Chapter 12], EDWARDS [1II, Chapter 13], BUTZER–BERENS [1, p. 187 ff]. In the last book the above classical convexity theorems are discussed in the general setting of intermediate spaces and interpolation.

Most of the results of this section generalize to general orthogonal series of functions; compare KACZMARZ–STEINHAUS [1]. For further and more recent results on such series see the important work by the Hungarian school, in particular ALEXITS [3], FREUD [3] and the literature cited there.

Contributions to harmonic analysis on the group $\mathbb{Z}$ are found widely scattered in the literature. As is the case in this text, they mainly serve as illustrations, at particular places, for the results to be expected on other groups.

Sec. 4.3. The material of this section, though standard, is somewhat scattered in the literature. As in the preceding sections the emphasis lies upon the fact that the finite Fourier–Stieltjes transform defines a mapping from one function space into another. This has influenced the selection of the material given here, which is symmetrical with that of Sec. 4.1 and 5.3, so that the general references given there are applicable.

In connection with (4.3.2), for an example of a continuous function $\mu \in \mathsf{BV}_{2\pi}$ such that $\mu^\vee(k)$ does not tend to zero; see ZYGMUND [7I, p. 194 ff]. Another proof of the general convolution theorem is to be found in ZYGMUND [7I, p. 39]. For the results of Sec. 4.3.3 reference may also be made to BUTZER–GÖRLICH [1].

5

Fourier Transforms Associated with the Line Group

5.0 Introduction

In the preceding chapter we have regarded the finite Fourier transform as a transform of one function space into another. This emphasis is a useful one in order to give a unified approach to Fourier analysis on different groups. This chapter is devoted to the study of the line group. Parallel to Sec. 4.1, Sec. 5.1 is concerned with the operational rules of the Fourier transform in L^1. The inversion theory will follow by the theory of singular integrals presented in Chapter 3. Included are results on generalized derivatives (Peano and Riemann) and connections with Fourier transforms and moments of positive functions. The relation between Fourier transforms and Fourier coefficients given by the Poisson summation formula is developed in Sec. 5.1.5. Sec. 5.2 is devoted to the definition of the Fourier transform for functions in L^p, $1 < p \leq 2$, including the Titchmarsh inequality (Theorem 5.2.9), Parseval's formula (Prop. 5.2.13), and Plancherel's theorem (Theorem 5.2.23). The operational rules are developed, together with the central Theorem 5.2.21. Sec. 5.3 is concerned with a thorough investigation of the Fourier–Stieltjes transform with its basic properties. We specifically mention the Lévy inversion formula (Theorem 5.3.9) and the uniqueness theorem (Prop. 5.3.11).

5.1 L^1-Theory

5.1.1 Fundamental Properties

With every $f \in \mathsf{L}^1$ we have associated (cf. (II.3)) its *Fourier transform* $f^\wedge$ defined by

$$f^\wedge(v) = \frac{1}{\sqrt{2\pi}} \int_{-\infty}^{\infty} f(u)\, e^{-ivu}\, du \equiv [f(\circ)]^\wedge(v) \equiv F^1[f](v) \qquad (v \in \mathbb{R}). \tag{5.1.1}$$

In comparison with (4.1.1), there is a slight discrepancy regarding the constant factor in the definition of $f^\wedge$. Again this factor is chosen such that it will lead to a symmetric inversion formula and L^2-theory.

The first proposition which may be shown by direct substitution in (5.1.1) deals with some elementary operational properties of the transform.

Proposition 5.1.1. *For $f \in L^1$ we have*

(i) $[f(\circ + h)]^\wedge(v) = e^{ihv} f^\wedge(v)$ $(h, v \in \mathbb{R})$,

(ii) $[e^{-ih\circ} f(\circ)]^\wedge(v) = f^\wedge(v + h)$ $(h, v \in \mathbb{R})$,

(iii) $[\rho f(\rho\, \circ)]^\wedge(v) = f^\wedge(v/\rho)$ $(\rho > 0, v \in \mathbb{R})$,

(iv) $[\overline{f(-\circ)}]^\wedge(v) = \overline{f^\wedge(v)}$ $(v \in \mathbb{R})$.

It follows immediately by definition that the Fourier transform $f^\wedge$ of $f \in L^1$ exists for all $v \in \mathbb{R}$ as a bounded function satisfying

$$|f^\wedge(v)| \le \|f\|_1 \qquad (v \in \mathbb{R}), \tag{5.1.2}$$

thus $\|f^\wedge\|_\infty \le \|f\|_1$. Moreover,

Proposition 5.1.2. *The Fourier transform defines a bounded linear transformation of L^1 into C_0.*

Indeed, we have for all $h, v \in \mathbb{R}$

$$\sqrt{2\pi}\, |f^\wedge(v + h) - f^\wedge(v)| \le \int_{-\infty}^{\infty} |e^{-ihu} - 1|\, |f(u)|\, du.$$

Since the integrand is bounded by $2\,|f(u)|$, tends to zero as $h \to 0$ for every u, and is independent of v, uniform continuity of $f^\wedge$ follows by Lebesgue's dominated convergence theorem, and thus $f^\wedge \in C$. Furthermore

$$f^\wedge(v) = \frac{1}{2\sqrt{2\pi}} \int_{-\infty}^{\infty} \left[f(u) - f\left(u + \frac{\pi}{v}\right)\right] e^{-ivu}\, du,$$

and therefore

$$|f^\wedge(v)| \le \frac{1}{2}\, \omega\left(L^1; f; \frac{\pi}{|v|}\right) \qquad (v \ne 0). \tag{5.1.3}$$

By Prop. 0.1.9 this implies $\lim_{|v| \to \infty} f^\wedge(v) = 0$, a result which is known as the *Riemann–Lebesgue lemma*. Hence $f^\wedge \in C_0$.

Theorem 5.1.3. *If $f, g \in L^1$, then*

$$[f * g]^\wedge(v) = f^\wedge(v) g^\wedge(v) \qquad (v \in \mathbb{R}). \tag{5.1.4}$$

Proof. Since by Prop. 0.2.2 the convolution $f * g$ exists a.e. as a function of L^1, we have for every $v \in \mathbb{R}$

$$[f * g]^\wedge(v) = \frac{1}{\sqrt{2\pi}} \int_{-\infty}^{\infty} \left\{\frac{1}{\sqrt{2\pi}} \int_{-\infty}^{\infty} f(x - u) g(u)\, du\right\} e^{-ivx}\, dx$$

$$= \frac{1}{\sqrt{2\pi}} \int_{-\infty}^{\infty} g(u)\, e^{-ivu}\, du \left\{\frac{1}{\sqrt{2\pi}} \int_{-\infty}^{\infty} f(x - u)\, e^{-iv(x-u)}\, dx\right\} = f^\wedge(v) g^\wedge(v),$$

the inversion of the order of integration being justified by Fubini's theorem.

The latter theorem is known as the *convolution theorem* for Fourier transforms in L^1. It in particular shows that the Fourier transform converts convolutions to pointwise products.

Proposition 5.1.4. *If $f, g \in L^1$, then*

$$\int_{-\infty}^{\infty} f^\wedge(v)g(v)\,dv = \int_{-\infty}^{\infty} f(v)g^\wedge(v)\,dv. \tag{5.1.5}$$

Proof. It follows by Fubini's theorem that

$$\int_{-\infty}^{\infty} f^\wedge(v)g(v)\,dv = \frac{1}{\sqrt{2\pi}}\int_{-\infty}^{\infty} g(v)\,dv \int_{-\infty}^{\infty} f(u)\,e^{-ivu}\,du = \int_{-\infty}^{\infty} f(u)g^\wedge(u)\,du.$$

In future we shall refer to a formula of type (5.1.5) as a *Parseval formula.*

5.1.2 Inversion Theory

So far, given a function $f \in L^1$, we defined its Fourier transform $f^\wedge$ and considered some of its fundamental properties. We now take up the inversion, problem, i.e., the problem of reconstructing the original function f from the values $f^\wedge(v)$ of $f^\wedge$. In correspondence with the finite Fourier transform and with (II.5), if $f \in L^1$ and

$$f^\wedge(v) = \frac{1}{\sqrt{2\pi}}\int_{-\infty}^{\infty} f(u)\,e^{-ivu}\,du,$$

we might expect a formula like

$$f(x) = \frac{1}{\sqrt{2\pi}}\int_{-\infty}^{\infty} f^\wedge(v)\,e^{ixv}\,dv \tag{5.1.6}$$

to be valid. But just as for the finite Fourier transform we immediately encounter the problem of giving the *Fourier inversion integral* (5.1.6) a suitable interpretation since the Fourier transform $f^\wedge$ of $f \in L^1$ need not belong to L^1 (see Problem 5.1.4). Although we do not intend to give a general treatment of summability of integrals on the real line, we shall discuss the particular case concerned with (5.1.6) in some detail.

Definition 5.1.5. *An even function $\theta \in L^1$ is called a **θ-factor** (on the real line) if $\theta^\wedge \in L^1$ such that*

$$\int_{-\infty}^{\infty} \theta^\wedge(v)\,dv = \sqrt{2\pi}. \tag{5.1.7}$$

*If θ is continuous, we call it a **continuous** θ-factor.*

Important examples of θ-factors are given by the *Cesàro*, *Abel*, and *Gauss factor*, i.e., by

$$\text{(i) } \theta_1(x) = \begin{cases} 1-|x|, & |x| \le 1 \\ 0, & |x| > 1 \end{cases}, \quad \text{(ii) } \theta_2(x) = e^{-|x|}, \quad \text{(iii) } \theta_3(x) = e^{-x^2}, \tag{5.1.8}$$

respectively (see Problem 5.1.2(i)).

Theorem 5.1.6. *Let $f \in \mathsf{L}^1$. Then for a θ-factor the **θ-means** of the integral* (5.1.6) *defined for each $\rho > 0$ by*

$$U(f; x; \rho) = \frac{1}{\sqrt{2\pi}} \int_{-\infty}^{\infty} \theta\left(\frac{v}{\rho}\right) f^\wedge(v)\, e^{ixv}\, dv \tag{5.1.9}$$

exist for all $x \in \mathbb{R}$, belong to L^1, and satisfy

$$\|U(f; \circ; \rho)\|_1 \le \|\theta^\wedge\|_1 \|f\|_1 \qquad (\rho > 0), \tag{5.1.10}$$

$$\lim_{\rho \to \infty} \|U(f; \circ; \rho) - f(\circ)\|_1 = 0. \tag{5.1.11}$$

Thus the Fourier inversion integral is *θ-summable* to f in L^1-norm.

Proof. Since θ is even, we have by Prop. 5.1.1, 5.1.4, and Problem 5.1.1 for each $\rho > 0$

$$U(f; x; \rho) = \frac{\rho}{\sqrt{2\pi}} \int_{-\infty}^{\infty} f(x - u)\theta^\wedge(\rho u)\, du, \tag{5.1.12}$$

and hence the assertions follow immediately by Prop. 0.2.1, 0.2.2, Lemma 3.1.5, and Theorem 3.1.6 if we take $\chi(x; \rho) = \rho\theta^\wedge(\rho x)$.

In contrast to the periodic case of Prop. 1.2.8 the definition of a θ-factor on the real line implies that the kernel $\{\rho\theta^\wedge(\rho x)\}$ of the singular integral (5.1.12) is an approximate identity. Thus, via (5.1.9), a θ-factor on the real line always defines a *summation process* of the integral (5.1.6) with respect to convergence in L^1-space.

As applications of the results of Sec. 3.2 we obtain

Theorem 5.1.7. *Let $f \in \mathsf{L}^1$. If for a θ-factor $\theta^\wedge$ is moreover positive and monotonely decreasing on* $[0, \infty)$, *then the Fourier inversion integral is θ-summable* a.e. *to $f(x)$, i.e.*

$$\lim_{\rho \to \infty} \frac{1}{\sqrt{2\pi}} \int_{-\infty}^{\infty} \theta\left(\frac{v}{\rho}\right) f^\wedge(v)\, e^{ixv}\, dv = f(x) \quad \text{a.e.} \tag{5.1.13}$$

Theorem 5.1.8. *Let $f \in \mathsf{L}^1$. If for a θ-factor $\theta^\wedge$ has a majorant $(\theta^\wedge)_M$: $|\theta^\wedge(v)| \le (\theta^\wedge)_M(v)$ such that $(\theta^\wedge)_M$ belongs to L^1 and is monotonely decreasing on* $[0, \infty)$, *then the Fourier inversion integral is θ-summable* a.e. *to $f(x)$.*

For the proof of the last two theorems we only mention that in view of (5.1.12)

$$U(f; x; \rho) - f(x) = \frac{\rho}{\sqrt{2\pi}} \int_0^{\infty} [f(x + u) + f(x - u) - 2f(x)]\theta^\wedge(\rho u)\, du. \tag{5.1.14}$$

Therefore (5.1.13) follows by Prop. 3.2.2 and 3.2.4, respectively.

Concerning the examples (5.1.8) of θ-factors we have at the same time shown

Corollary 5.1.9. *For $f \in L^1$ the Fourier inversion integral is Cesàro, Abel, and Gauss summable to $f(x)$ for almost all $x \in \mathbb{R}$, i.e.*

$$\lim_{\rho\to\infty} \frac{1}{\sqrt{2\pi}} \int_{-\rho}^{\rho} \left(1 - \frac{|v|}{\rho}\right) f^\wedge(v)\, e^{ixv}\, dv = f(x) \quad \text{a.e.,} \tag{5.1.15}$$

$$\lim_{\rho\to\infty} \frac{1}{\sqrt{2\pi}} \int_{-\infty}^{\infty} e^{-|v|/\rho} f^\wedge(v)\, e^{ixv}\, dv = f(x) \quad \text{a.e.,} \tag{5.1.16}$$

$$\lim_{\rho\to\infty} \frac{1}{\sqrt{2\pi}} \int_{-\infty}^{\infty} e^{-(v/\rho)^2} f^\wedge(v)\, e^{ixv}\, dv = f(x) \quad \text{a.e.,} \tag{5.1.17}$$

respectively. In particular, all these relations are valid at each point of continuity of f.

In fact, in view of Problem 5.1.2(i) we have $\theta_1^\wedge(x) = F(x)$, $\theta_2^\wedge(x) = p(x)$, and $\theta_3^\wedge(x) = w(x)$, where the functions F, p and w are defined by (3.1.15), (3.1.39), and (3.1.33), respectively. Thus, if we denote the θ_k-means of (5.1.6) by $U_k(f; x; \rho)$, $k = 1, 2, 3$, we obtain via (5.1.12) the following correspondence with the singular integrals of Fejér, Cauchy–Poisson and Gauss–Weierstrass as introduced in (3.1.14), (3.1.38) and (3.1.32), respectively:

$$U_1(f; x; \rho) = \sigma(f; x; \rho), \tag{5.1.18}$$

$$U_2(f; x; \rho) = P(f; x; y) \qquad (\rho = y^{-1}), \tag{5.1.19}$$

$$U_3(f; x; \rho) = W(f; x; t) \qquad (\rho = t^{-1/2}). \tag{5.1.20}$$

Therefore Cor. 5.1.9 is nothing but a collection of the results of Cor. 3.2.3 and 3.2.5. One observes that formulae (5.1.18)–(5.1.20) indicate the important rôle that the singular integrals considered in Chapter 3 play in the theory of summation of the Fourier inversion integral.

The preceding corollary on the pointwise summability of (5.1.6) gives three possible interpretations for the inversion formula (5.1.6). A further interpretation is given by

Proposition 5.1.10. *If f and $f^\wedge$ belong to L^1, then*

$$\frac{1}{\sqrt{2\pi}} \int_{-\infty}^{\infty} f^\wedge(v)\, e^{ixv}\, dv = f(x) \tag{5.1.21}$$

for almost all $x \in \mathbb{R}$. Therefore f is equal a.e. *to a function in $L^1 \cap C_0$. If f is continuous on $\mathbb{R}$, then the inversion formula* (5.1.21) *holds everywhere.*

Proof. By Cor. 5.1.9 the left-hand side of (5.1.21) is Abel summable to f a.e. On the other hand, as

$$|e^{-|v|/\rho} f^\wedge(v)\, e^{ixv}| \le |f^\wedge(v)| \in L^1$$

for all $\rho > 0$ and

$$\lim_{\rho\to\infty} e^{-|v|/\rho} f^\wedge(v)\, e^{ixv} = f^\wedge(v)\, e^{ixv},$$

it follows by Lebesgue's dominated convergence theorem that

$$\lim_{\rho\to\infty} \frac{1}{\sqrt{2\pi}} \int_{-\infty}^{\infty} e^{-|v|/\rho} f^\wedge(v)\, e^{ixv}\, dv = \frac{1}{\sqrt{2\pi}} \int_{-\infty}^{\infty} f^\wedge(v)\, e^{ixv}\, dv$$

for all $x \in \mathbb{R}$. This proves the result.

Another interpretation of (5.1.6) follows from Jordan's criterion (see Problem 5.1.5(iv)).

A further consequence of Cor. 5.1.9 is the *uniqueness theorem* of the Fourier transform in L^1. It states that two distinct functions f_1 and f_2 in L^1 have distinct Fourier transforms $f_1^\wedge$ and $f_2^\wedge$. Indeed,

Proposition 5.1.11. *If $f \in \mathsf{L}^1$ and $f^\wedge(v) \equiv 0$ on $\mathbb{R}$, then $f(x) = 0$ a.e.*

Together with Problem 5.1.3 we may therefore strengthen Prop. 5.1.2 to: The Fourier transform defines a one-to-one bounded linear transformation of L^1 into (but not onto) C_0. However, the set of functions $[\mathsf{L}^1]^\wedge \equiv \{g \in \mathsf{C}_0 \mid g = f^\wedge, f \in \mathsf{L}^1\}$ is dense in C_0. Moreover, according to Theorem 5.1.3 the Fourier transform defines an isomorphism of the commutative Banach algebra L^1 (with convolution as multiplication) into the Banach algebra C_0 (with pointwise multiplication).

Concerning the Banach algebra L^1 itself, the properties of the Fourier transform so far obtained enable us to show the following proposition the proof of which may be regarded as a first example of the integral transform method.

Proposition 5.1.12. *The Banach algebra L^1 has no unit element. However, there exist approximate identities, i.e., there are sets of functions $\{\chi(x;\rho)\} \subset \mathsf{L}^1$, $\rho > 0$, such that $\lim_{\rho\to\infty} \|\chi(\circ;\rho) * f - f\|_1 = 0$ for every $f \in \mathsf{L}^1$.*

Proof. Indeed, if $e \in \mathsf{L}^1$ existed such that $e * f = f$ for every $f \in \mathsf{L}^1$, then, in particular, $e * e = e$. By Theorem 5.1.3 it would follow that $(e^\wedge(v))^2 = e^\wedge(v)$, and thus, for each $v \in \mathbb{R}$, $e^\wedge(v)$ would equal zero or one. But $e^\wedge(v)$ is continuous, and therefore $e^\wedge(v)$ would be identically zero or one, and in fact $e^\wedge(v) \equiv 0$ since $e^\wedge$ must vanish at $\pm\infty$. Then the uniqueness theorem implies $e(x) = 0$ a.e. which contradicts $e * f = f$ for every $f \in \mathsf{L}^1$.

The existence of approximate identities is given by Lemma 3.1.5 and Theorem 3.1.6.

As a consequence of Prop. 5.1.10 we now prove the following analogs of (4.1.8)–(4.1.10).

Proposition 5.1.13. *If $f, g \in \mathsf{L}^1$ and g is such that $g^\wedge \in \mathsf{L}^1$, then*

$$(f * g)(x) = \frac{1}{\sqrt{2\pi}} \int_{-\infty}^{\infty} f^\wedge(v) g^\wedge(v)\, e^{ixv}\, dv \qquad (x \in \mathbb{R}), \tag{5.1.22}$$

$$\int_{-\infty}^{\infty} f(u)\overline{g(u)}\, du = \int_{-\infty}^{\infty} f^\wedge(u)\overline{g^\wedge(u)}\, du. \tag{5.1.23}$$

Proof. According to Prop. 5.1.10, g is equal a.e. to a function of class C_0. Therefore by Prop. 0.2.1 the convolution $f * g$ exists for every $x \in \mathbb{R}$ as a continuous function. By the convolution theorem $[f * g]^\wedge(v) = f^\wedge(v) g^\wedge(v)$, and since the hypotheses assure

$\overline{[f*g]^\wedge \in \mathsf{L}^1}$, a further application of Prop. 5.1.10 gives (5.1.22). Replacing g by $\overline{g(-x)}$ and using Prop. 5.1.1(iv), we obtain (5.1.23) by setting $x = 0$.

The particular case $f = g$ is of some interest; formula (5.1.23) then states that the *Parseval equation*

$$\|f\|_{\mathsf{L}^2} = \|f^\wedge\|_{\mathsf{L}^2} \tag{5.1.24}$$

is valid for all $f \in \mathsf{L}^1$ with $f^\wedge \in \mathsf{L}^1$.

5.1.3 Fourier Transforms of Derivatives

We have seen in Sec. 5.1.1 that certain operations on functions correspond to more suitable operations on their Fourier transforms. We continue these investigations on the operational properties of the Fourier transform by fitting in derivatives.

Proposition 5.1.14. *Let* $f \in \mathsf{L}^1 \cap \mathsf{AC}_{\text{loc}}^{r-1}$ *and* $f^{(r)} \in \mathsf{L}^1$. *Then*

$$[f^{(r)}]^\wedge(v) = (iv)^r f^\wedge(v) \qquad (v \in \mathbb{R}). \tag{5.1.25}$$

Thus, if the hypotheses of the above proposition are satisfied, then $f^\wedge(v) = o(|v|^{-r})$ as $|v| \to \infty$ (compare also Problems 5.1.9(ii) and 5.3.5).

Proof. Since $f \in \mathsf{AC}_{\text{loc}}^{r-1}$, integration by parts ($r$-times) yields

$$\sum_{k=0}^{r} (-1)^{r-k}\binom{r}{k} f(x+k) = \int_0^1 \cdots \int_0^1 f^{(r)}(x + u_1 + \cdots + u_r)\, du_1 \ldots du_r \tag{5.1.26}$$

for all $x \in \mathbb{R}$. If we introduce the function $m(x) \equiv \sqrt{2\pi}\kappa_{[-1,0]}(x)$, thus

$$m(x) = \begin{cases} \sqrt{2\pi}, & x \in [-1,0] \\ 0, & x \notin [-1,0], \end{cases} \tag{5.1.27}$$

we may rewrite the right-hand side of (5.1.26) as a convolution product. Indeed (cf. Problem 5.1.6(i)),

$$\sum_{k=0}^{r} (-1)^{r-k}\binom{r}{k} f(x+k) = ([m*]^r * f^{(r)})(x), \tag{5.1.28}$$

where the right side, being the convolution of the product (r-times) $m * \cdots * m \in \mathsf{L}^1$ with $f^{(r)} \in \mathsf{L}^1$, exists for all $x \in \mathbb{R}$ by Prop. 0.2.1. Since

$$m^\wedge(v) = \begin{cases} \dfrac{e^{iv}-1}{iv}, & v \neq 0 \\ 1, & v = 0, \end{cases} \tag{5.1.29}$$

the Fourier transform of equation (5.1.28) is given in view of Prop. 5.1.1 and Theorem 5.1.3 by

$$(e^{iv} - 1)^r f^\wedge(v) = \left(\frac{e^{iv}-1}{iv}\right)^r [f^{(r)}]^\wedge(v)$$

for all $v \neq 0$. Therefore (5.1.25) holds at all points $v \neq 2k\pi$, $k \in \mathbb{Z}$. Since this set is denumerable and since, in virtue of Prop. 5.1.2, both sides of equation (5.1.25) represent continuous functions, this equation is actually established for all $v \in \mathbb{R}$.

Proposition 5.1.15. *If for $f \in \mathsf{L}^1$ there exists $g \in \mathsf{L}^1$ such that*

$$(iv)^r f^\wedge(v) = g^\wedge(v) \qquad (v \in \mathbb{R}) \tag{5.1.30}$$

for some natural number r, then $f \in \mathsf{W}^r_{\mathsf{L}^1}$ (for the definition see (3.1.48)).

Proof. Let $r = 1$. Then we may invert the single steps of the proof of the latter proposition to obtain the result. In fact, the assumption (5.1.30) together with (5.1.29) and the convolution theorem implies for each $h > 0$ and $v \neq 0$

$$[f(x) - f(x-h)]^\wedge(v) = \frac{1 - e^{-ihv}}{iv} g^\wedge(v) = \left[\int_{-h}^{0} g(x+u)\,du\right]^\wedge(v). \tag{5.1.31}$$

Therefore it follows by the uniqueness theorem of Fourier transforms that for each fixed $h > 0$

$$f(x) - f(x-h) = \int_{x-h}^{x} g(u)\,du \tag{5.1.32}$$

for almost all $x \in \mathbb{R}$. We then have for every $y > 0$, $h > 0$

$$\int_0^y [f(x) - f(x-h)]\,dx = \int_0^y \left[\int_{x-h}^{x} g(u)\,du\right] dx.$$

Since $f \in \mathsf{L}^1$, on the one hand we obtain for each fixed $y > 0$

$$\lim_{h\to\infty} \int_0^y [f(x) - f(x-h)]\,dx = \int_0^y f(x)\,dx - \lim_{h\to\infty} \int_{-h}^{y-h} f(x)\,dx = \int_0^y f(x)\,dx.$$

On the other hand, by Lebesgue's dominated convergence theorem

$$\lim_{h\to\infty} \int_0^y \left[\int_{x-h}^{x} g(u)\,du\right] dx = \int_0^y \left[\int_{-\infty}^{x} g(u)\,du\right] dx.$$

Therefore it follows that

$$f(x) = \int_{-\infty}^{x} g(u)\,du \quad \text{a.e.,} \tag{5.1.33}$$

and if we set $\phi(x) \equiv \int_{-\infty}^{x} g(u)\,du$, Prop. 5.1.15 is established for $r = 1$.

Regarding the general case $r \geq 2$, we first show that the assumption (5.1.30) implies that $ivf^\wedge(v)$ is the Fourier transform of some L^1-function. Let

$$\eta(x) = \begin{cases} \sqrt{2\pi}\, e^{-x}, & 0 \leq x < \infty \\ 0, & -\infty < x < 0. \end{cases} \tag{5.1.34}$$

Then $\eta^\wedge(v) = (1 + iv)^{-1}$, and therefore by the convolution theorem

$$(1 + iv)^{1-r} \in [\mathsf{L}^1]^\wedge, \qquad \frac{(iv)^r}{(1+iv)^{r-1}} f^\wedge(v) \in [\mathsf{L}^1]^\wedge.$$

But for certain constant coefficients a_k, $1 \le k \le r - 1$,

$$\frac{(iv)^r}{(1 + iv)^{r-1}} f^\wedge(v) = ivf^\wedge(v) + \left\{-(r-1)f^\wedge(v) + \sum_{k=1}^{r-1} \frac{a_k}{(1+iv)^k} f^\wedge(v)\right\}.$$

Now the left-hand side of the latter equation belongs to $[\mathsf{L}^1]^\wedge$ as well as the terms in curly brackets. This implies $ivf^\wedge(v) \in [\mathsf{L}^1]^\wedge$, i.e., there exists $g_1 \in \mathsf{L}^1$ such that $ivf^\wedge(v) = g_1^\wedge(v)$ for all $v \in \mathbb{R}$. Thus the case $r = 1$ applies, and (5.1.33) in particular shows that $f(x) = \int_{-\infty}^x g_1(u)\, du$ a.e.

The proof for $r \ge 2$ now proceeds iteratively as follows. If (5.1.30) holds, then there exists $g_1 \in \mathsf{L}^1$ such that $(iv)^{r-1} g_1^\wedge(v) = g^\wedge(v)$ and $f(x) = \int_{-\infty}^x g_1(u_1)\, du_1$. But this in turn implies $ivg_1^\wedge(v) \in [\mathsf{L}^1]^\wedge$, i.e. $ivg_1^\wedge(v) = g_2^\wedge(v)$, say, with $g_2 \in \mathsf{L}^1$ and $g_1(x) = \int_{-\infty}^x g_2(u)\, du$ a.e. Hence

$$f(x) = \int_{-\infty}^x du_1 \int_{-\infty}^{u_1} g_2(u_2)\, du_2.$$

Applying this method successively we obtain a sequence of functions $g_k \in \mathsf{L}^1$ for $1 \le k \le r - 1$ such that for almost all $x \in \mathbb{R}$

$$f(x) = \int_{-\infty}^x g_1(u)\, du, \quad g_k(x) = \int_{-\infty}^x g_{k+1}(u)\, du, \quad 1 \le k \le r-2, \quad g_{r-1}(x) = \int_{-\infty}^x g(u)\, du,$$

and therefore

$$f(x) = \int_{-\infty}^x du_1 \int_{-\infty}^{u_1} du_2 \ldots \int_{-\infty}^{u_{r-1}} g(u_r)\, du_r \quad \text{a.e.} \tag{5.1.35}$$

If we abbreviate the right side as $\phi(x)$, then $\phi \in \mathsf{AC}_{\text{loc}}^{r-1}$, $\phi^{(k)} = g_k \in \mathsf{L}^1$, $1 \le k \le r - 1$, $\phi^{(r)} = g$ a.e., and the proof is complete.

As in the periodic case it is now convenient to introduce the class

$$\mathsf{W}[\mathsf{L}^1; \psi(v)] = \{f \in \mathsf{L}^1 \mid \psi(v) f^\wedge(v) = g^\wedge(v),\, g \in \mathsf{L}^1\}, \tag{5.1.36}$$

$\psi(v)$ being an arbitrary complex-valued (continuous) function on $\mathbb{R}$. Then

Theorem 5.1.16. *For* $f \in \mathsf{L}^1$ *the following assertions are equivalent:*

(i) $f \in \mathsf{W}_{\mathsf{L}^1}^r$, (ii) $f \in \mathsf{W}[\mathsf{L}^1; (iv)^r]$,

(iii) *there exists* $g \in \mathsf{L}^1$ *such that the representation* (5.1.35) *holds, each of the iterated integrals existing as a function in* L^1.

These results in particular show that if $f \in \mathsf{L}^1$ is r-times continuously differentiable with $f^{(r)} \in \mathsf{L}^1$, then all derivatives $f^{(k)}$, $1 \le k \le r - 1$, belong to L^1.

5.1.4 Derivatives of Fourier Transforms, Moments of Positive Functions, Peano and Riemann Derivatives

Whereas the last subsection dealt with the Fourier transform of derivatives of the original function f, we are now interested in results concerning smoothness properties of the transform $f^\wedge$. Prop. 5.1.2 may be regarded as a first contribution in this direction.

Proposition 5.1.17. *If $f \in \mathsf{L}^1$ is such that $x^r f(x) \in \mathsf{L}^1$ for some $r \in \mathbb{P}$, then the Fourier transform $f^\wedge$ of f has an rth derivative of class C_0, and*

$$(f^\wedge)^{(k)}(v) = \frac{(-i)^k}{\sqrt{2\pi}} \int_{-\infty}^{\infty} u^k f(u)\, e^{-ivu}\, du \qquad (k = 0, 1, \ldots, r), \tag{5.1.37}$$

$$(f^\wedge)^{(k)}(0) = (-i)^k \gamma_k \qquad (k = 0, 1, \ldots, r), \tag{5.1.38}$$

where γ_k is the kth **moment** *of f, namely*

$$\gamma_k = \frac{1}{\sqrt{2\pi}} \int_{-\infty}^{\infty} u^k f(u)\, du \qquad (k \in \mathbb{P}). \tag{5.1.39}$$

Proof. Suppose $r = 1$. Then

$$\frac{f^\wedge(v+h) - f^\wedge(v)}{h} = \frac{-i}{\sqrt{2\pi}} \int_{-\infty}^{\infty} e^{-ihu/2} \frac{\sin(hu/2)}{hu/2} uf(u)\, e^{-ivu}\, du.$$

Since the integrand converges to $uf(u)\exp\{-ivu\}$ a.e. as $h \to 0$, and since it has an integrable majorant independent of h, given by $|u|\,|f(u)|$, we obtain by Lebesgue's dominated convergence theorem that $(f^\wedge)'$ exists at each $v \in \mathbb{R}$ and

$$(f^\wedge)'(v) = \frac{-i}{\sqrt{2\pi}} \int_{-\infty}^{\infty} uf(u)\, e^{-ivu}\, du.$$

The proof for general r follows by mathematical induction.

We observe that the moment γ_r exists (as a Lebesgue integral) if and only if $x^r f(x) \in \mathsf{L}^1$, i.e., if and only if the absolute moment (cf. (3.3.6))

$$m(f; r) = \frac{1}{\sqrt{2\pi}} \int_{-\infty}^{\infty} |u|^r\, |f(u)|\, du \tag{5.1.40}$$

exists as a finite number. Furthermore, if γ_r exists, so do all the moments γ_k of order $k = 0, 1, \ldots, r$.

It is important to note that for positive functions a certain converse of Prop. 5.1.17 is valid, and it is then possible to weaken the assumptions upon the differentiability of the transform $f^\wedge$ considerably. For this purpose we need the following notations. For an arbitrary function f on the real line we define its first *central difference* at x with respect to the increment $h \in \mathbb{R}$ by

$$\bar{\Delta}_h^1 f(x) = f\left(x + \frac{h}{2}\right) - f\left(x - \frac{h}{2}\right)$$

and the higher differences by

$$\bar{\Delta}_h^{r+1} f(x) = \bar{\Delta}_h^1 \bar{\Delta}_h^r f(x) \qquad (r \in \mathbb{N}).$$

It can be shown (cf. Problem 1.1.6) that

$$\bar{\Delta}_h^r f(x) = \sum_{k=0}^{r} (-1)^k \binom{r}{k} f\left(x + \left(\frac{r}{2} - k\right)h\right) \qquad (h \in \mathbb{R}, r \in \mathbb{N}). \tag{5.1.41}$$

As a generalization of the concept of ordinary derivatives we extend Def. 1.5.14 to

Definition 5.1.18. *Let $f(x)$ be defined in a neighbourhood of the point x_0. The rth* **Riemann derivative** *of f at x_0 (with respect to the central difference* (5.1.41)) *is defined by*

$$f^{[r]}(x_0) = \lim_{h\to 0} \frac{1}{h^r} \overline{\Delta}_h^r f(x_0), \tag{5.1.42}$$

in case the limit exists and is finite.

In other words, if the rth central difference quotient $h^{-r}\overline{\Delta}_h^r f(x_0)$ of f at x_0 has a limit as $h \to 0$, then we call this limit the rth Riemann derivative of f at x_0. The notation $f^{[r]}$ is used to distinguish the rth Riemann derivative from the rth ordinary derivative $f^{(r)}$.

The concept of a Riemann derivative is more general than that of the ordinary derivative. Indeed, it follows by Problem 5.1.11 that if the rth ordinary derivative $f^{(r)}(x_0)$ exists, so does the rth Riemann derivative $f^{[r]}(x_0)$, and we have $f^{[r]}(x_0) = f^{(r)}(x_0)$, but not conversely. We emphasize that, in order to define $f^{[r]}$, we need not suppose the existence of Riemann derivatives of lower order whereas the ordinary derivative $f^{(r)}$ is defined successively by means of the derivatives $f^{(k)}$, $1 \le k \le r - 1$.

After these preliminaries we may now state the following

Proposition 5.1.19. *Let $f \in \mathsf{L}^1$ be positive. If the 2rth central difference quotient of $f^\wedge$ at the origin satisfies*

$$\liminf_{h\to 0} \left| \frac{1}{h^{2r}} \overline{\Delta}_h^{2r} f^\wedge(0) \right| \equiv M < \infty,$$

then the 2rth moment γ_{2r} of f exists as do all moments γ_k of order $k < 2r$. Moreover, the derivatives $(f^\wedge)^{(k)}$, $1 \le k \le 2r$, exist as functions in C_0 and are given by (5.1.37). (5.1.38) *holds as well.*

Proof. In view of the definition we have by Prop. 5.1.1(ii)

$$\begin{aligned}\overline{\Delta}_h^{2r} f^\wedge(v) &= \frac{1}{\sqrt{2\pi}} \int_{-\infty}^{\infty} \{e^{-ihu/2} - e^{ihu/2}\}^{2r} f(u)\, e^{-ivu}\, du \\ &= \frac{1}{\sqrt{2\pi}} \int_{-\infty}^{\infty} \left\{-2i \sin \frac{hu}{2}\right\}^{2r} f(u)\, e^{-ivu}\, du.\end{aligned}$$

Hence we obtain for the $2r$th central difference quotient of $f^\wedge$ at the origin

$$\left| \frac{\overline{\Delta}_h^{2r} f^\wedge(0)}{h^{2r}} \right| = \frac{1}{\sqrt{2\pi}} \int_{-\infty}^{\infty} \left[\frac{\sin (hu/2)}{h/2}\right]^{2r} f(u)\, du.$$

By Fatou's lemma it follows that

$$\frac{1}{\sqrt{2\pi}} \int_{-\infty}^{\infty} u^{2r} f(u)\, du \le \liminf_{h\to\infty} \frac{1}{\sqrt{2\pi}} \int_{-\infty}^{\infty} \left[\frac{\sin (hu/2)}{h/2}\right]^{2r} f(u)\, du$$

which is bounded by hypothesis. Thus γ_{2r} exists which, in view of Prop. 5.1.17, completes the proof.

Upon combining the assertions of the last two propositions we obtain

Corollary 5.1.20. *The 2rth moment γ_{2r} of a positive function $f \in \mathsf{L}^1$ exists if and only if the Fourier transform $f^\wedge$ has a 2rth ordinary derivative at the origin.*

Moreover, if we pose the problem of finding conditions upon the Riemann derivative of a function which assure the existence of its ordinary derivative, we have as a first answer

Corollary 5.1.21. *If a function is representable as the Fourier transform of a positive function of class L^1, then its 2rth ordinary derivative exists if and only if it has a 2rth Riemann derivative.*

Let us assume that the rth moment of $f \in \mathsf{L}^1$ exists. Then by Prop. 5.1.17 and Taylor's formula the Fourier transform $f^\wedge$ admits the following expansion at the origin:

$$f^\wedge(v) = \sum_{k=0}^{r} \frac{v^k}{k!} (f^\wedge)^{(k)}(0) + \frac{1}{r!} [(f^\wedge)^{(r)}(\eta v) - (f^\wedge)^{(r)}(0)]v^r \quad (0 \le \eta < 1).$$

Therefore

Proposition 5.1.22. *If the rth moment γ_r of a function $f \in \mathsf{L}^1$ exists absolutely as a finite number, then*

$$f^\wedge(v) = \sum_{k=0}^{r} \frac{(-iv)^k}{k!} \gamma_k + o(v^r) \qquad (v \to 0). \tag{5.1.43}$$

Again a converse is valid for positive functions f. In order to derive this result we introduce Peano derivatives.

For an arbitrary function f on the real line an expression of the form

$$\Diamond_h^r f(x) = r!\left\{f(x+h) - f(x) - \sum_{k=1}^{r-1} \frac{h^k}{k!} l_k\right\} \qquad (h \in \mathbb{R}, r \in \mathbb{N}) \tag{5.1.44}$$

with arbitrary (constant) coefficients $l_k \in \mathbb{C}$, $1 \le k \le r-1$, is called an rth *Peano difference* of f at x with respect to the increment h (and the constants l_k).

Definition 5.1.23. *Let $f(x)$ be defined in a neighbourhood of the point x_0. If there are constants l_k, $1 \le k \le r-1$, such that the corresponding rth Peano difference quotient $h^{-r}\Diamond_h^r f(x_0)$ of f at x_0 tends to a limit as $h \to 0$, then this limit is called the rth* ***Peano derivative*** *of f at x_0 and denoted by $f^{\langle r\rangle}(x_0)$, i.e.*

$$f^{\langle r\rangle}(x_0) = \lim_{h\to 0} \frac{r!}{h^r}\left\{f(x_0+h) - f(x_0) - \sum_{k=1}^{r-1} \frac{h^k}{k!} l_k\right\}. \tag{5.1.45}$$

Again, in order to define a Peano derivative of order r we do not need any explicit assumptions on Peano derivatives of lower order. Only the existence of certain constants l_k, $1 \le k \le r-1$, is assumed for which the limit (5.1.45) exists. But we still have to show that the definition of a Peano derivative of *a function at a point* is meaningful, that is to say, that $f^{\langle r\rangle}(x_0)$ only depends on f and x_0, but is independent of the particular choice of constants l_k, $1 \le k \le r-1$, occurring in the definition. This is a consequence of the fact that for $r = 1$ the definitions of a Peano and of an ordinary derivative coincide, and of the following

Lemma 5.1.24. *If the rth Peano derivative $f^{\langle r\rangle}(x_0)$ exists, then f has Peano derivatives of order k, $1 \le k \le r-1$, and if $f^{\langle r\rangle}(x_0)$ is given by* (5.1.45), *then*

$$f^{\langle k\rangle}(x_0) = l_k \qquad (k = 1, 2, \dots, r-1), \tag{5.1.46}$$

$$f(x_0 + h) = f(x_0) + \sum_{k=1}^{r} \frac{h^k}{k!} f^{\langle k\rangle}(x_0) + o(h^r) \qquad (h \to 0). \tag{5.1.47}$$

Proof. If there are constants l_k, $1 \le k \le r-1$, such that (5.1.45) holds, then for $h \to 0$

$$\frac{(r-1)!}{h^{r-1}}\left\{f(x_0 + h) - f(x_0) - \sum_{k=1}^{r-2} \frac{h^k}{k!} l_k\right\} - l_{r-1} = \frac{1}{rh^{r-1}} \diamondsuit_h^r f(x_0) = O(h),$$

which shows that the $(r-1)$th Peano derivative exists and that $f^{\langle r-1\rangle}(x_0) = l_{r-1}$. The rest of the proof of (5.1.46) is obvious. In particular we have

$$f'(x_0) \equiv f^{\langle 1\rangle}(x_0) = l_1. \tag{5.1.48}$$

The expansion (5.1.47) now follows from the definition of an rth Peano derivative.

Thus the Peano derivative of a function f at a point x_0 is well-defined, in particular, we are justified in speaking of *the* Peano derivative of f at x_0.

If we compare the Peano derivative with the Riemann and the ordinary derivative, we find that it is more general than the ordinary but less general than the Riemann derivative. The existence of an ordinary derivative $f^{(r)}$ at a point x_0 implies that of the Peano derivative $f^{\langle r\rangle}(x_0)$ and $f^{\langle r\rangle}(x_0) = f^{(r)}(x_0)$. But the converse is not necessarily valid (see Problem 5.1.13). On the other hand, the Riemann derivative is a true generalization of the Peano derivative. In fact, the example of Problem 5.1.14(ii) shows that Riemann differentiability does not imply Peano differentiability whereas the converse is true in view of

Lemma 5.1.25. *If f has an rth Peano derivative $f^{\langle r\rangle}(x_0)$, then it also possesses an rth Riemann derivative $f^{[r]}(x_0)$, and $f^{[r]}(x_0) = f^{\langle r\rangle}(x_0)$.*

Proof. Suppose (5.1.45) and thus (5.1.47) is valid. Then

$$f\left(x_0 + \left(\frac{r}{2} - k\right)h\right) = f(x_0) + \sum_{j=1}^{r} \frac{l_j}{j!}\left[\left(\frac{r}{2} - k\right)h\right]^j + o\left(\left[\left(\frac{r}{2} - k\right)h\right]^r\right),$$

and we obtain according to (5.1.41)

$$\begin{aligned}\overline{\Delta}_h^r f(x_0) &= \sum_{k=0}^{r} (-1)^k \binom{r}{k} \sum_{j=1}^{r} \frac{l_j}{j!}\left[\left(\frac{r}{2} - k\right)h\right]^j + o(h^r)\\ &= \sum_{j=1}^{r} \frac{l_j}{j!} h^j \sum_{k=0}^{r} (-1)^k \binom{r}{k}\left(\frac{r}{2} - k\right)^j + o(h^r) \qquad (h \to 0).\end{aligned}$$

But in view of

$$\sum_{k=0}^{r} (-1)^k \binom{r}{k} k^j = \begin{cases} 0, & j = 0, 1, \dots, r-1\\ (-1)^r r!, & j = r\end{cases} \tag{5.1.49}$$

(see Problem 5.1.14(i)), we have

$$\sum_{k=0}^{r}(-1)^k\binom{r}{k}\left(\frac{r}{2}-k\right)^j=\begin{cases}0, & j=0,1,\ldots,r-1\\ r!, & j=r,\end{cases}$$

and therefore

$$\overline{\Delta}_h^r f(x_0)=h^r f^{\langle r\rangle}(x_0)+o(h^r) \qquad (h\to 0). \tag{5.1.50}$$

Thus the rth Riemann derivative of f at x_0 exists and $f^{[r]}(x_0)=f^{\langle r\rangle}(x_0)$.

Now we may establish the following converse to Prop. 5.1.22.

Proposition 5.1.26. *Let $f\in \mathsf{L}^1$ be positive. If the Fourier transform $f^\wedge$ has a $2r$th Peano derivative at the origin, then the $2r$th moment γ_{2r} of f exists as do all moments γ_k of order $k<2r$. The derivatives $(f^\wedge)^k$, $1\le k\le 2r$, exist as functions in C_0 and are given by* (5.1.37). *Moreover,*

$$(f^\wedge)^{\langle k\rangle}(x_0)=(f^\wedge)^{[k]}(x_0)=(f^\wedge)^{(k)}(x_0) \qquad (k=1,2,\ldots,2r).$$

Proof. As the $2r$th Peano derivative of $f^\wedge$ exists at the origin, also the $2r$th Riemann derivative of $f^\wedge(v)$ exists at $v=0$ by Lemma 5.1.25. Therefore the assumptions of Prop. 5.1.19 are satisfied with $M=|(f^\wedge)^{\langle 2r\rangle}(0)|$, and the proof is complete.

We observe that the results of Prop. 5.1.22, 5.1.26 again establish Cor. 5.1.20. Moreover,

Corollary 5.1.27. *If a function is representable as the Fourier transform of a positive function of class L^1, then its $2r$th ordinary derivative exists if and only if it has a $2r$th Peano derivative.*

For a further discussion concerning generalized derivatives of a function we refer to Chapter 10.

5.1.5 Poisson Summation Formula

In Sec. 3.1.2 we associated a periodic function g^* with every $g\in\mathsf{L}^1$ by means of formula (3.1.22) and derived some first results, for instance (3.1.25). This in particular states that

$$\frac{1}{2\pi}\int_{-\pi}^{\pi} f(u)g^*(u)\,du=\frac{1}{\sqrt{2\pi}}\int_{-\infty}^{\infty} f(u)g(u)\,du \tag{5.1.51}$$

for every $f\in\mathsf{C}_{2\pi}$. We continue these investigations by introducing the respective Fourier transforms.

Proposition 5.1.28. *If $g\in\mathsf{L}^1$ and g^* is given by* (3.1.22), *then the finite Fourier transform $[g^*]^\wedge$ is the restriction to the integers of the Fourier transform $g^\wedge$, i.e.*

$$[g^*]^\wedge(k)=g^\wedge(k) \qquad (k\in\mathbb{Z}). \tag{5.1.52}$$

The proof follows immediately from (5.1.51) by taking $f(x)=\exp\{-ikx\}$, $k\in\mathbb{Z}$.

Prop. 5.1.28 may be expressed in the following way: If $g^\wedge$ denotes the L^1-Fourier transform of $g \in \mathsf{L}^1$, then

$$\sqrt{2\pi} \sum_{k=-\infty}^{\infty} g(x + 2k\pi) \sim \sum_{k=-\infty}^{\infty} g^\wedge(k)\, e^{ikx} \tag{5.1.53}$$

in the sense, that the series on the right is the Fourier series of g^*, thus of the function on the left. We are now interested in establishing an actual equality in (5.1.53) in which case it is known as the *Poisson summation formula.* To this end, we may employ summation by a θ-factor and use the results of Sec. 1.2.4 and 4.1.2. But here we only give the following result which is an easy consequence of Prop. 4.1.5 and 3.1.11.

Proposition 5.1.29. (i) *Let $g \in \mathsf{L}^1$ be continuous on $\mathbb{R}$. If the series* (3.1.22) *defining g^* is uniformly convergent and if $[g^*]^\wedge \in \mathsf{l}^1$, then for all $x \in \mathbb{R}$*

$$\sqrt{2\pi} \sum_{k=-\infty}^{\infty} g(x + 2k\pi) = \sum_{k=-\infty}^{\infty} g^\wedge(k)\, e^{ikx}. \tag{5.1.54}$$

(ii) *If $g \in \mathsf{L}^1 \cap \mathsf{BV}$, then for all $x \in \mathbb{R}$*

$$\sqrt{2\pi} \sum_{k=-\infty}^{\infty} g(x + 2k\pi) = \lim_{n\to\infty} \sum_{k=-n}^{n} g^\wedge(k)\, e^{ikx}.$$

The following proposition may be seen in connection with Prop. 4.1.7 and 5.1.13.

Proposition 5.1.30. *Let $f \in \mathsf{L}^1_{2\pi}$ and $g \in \mathsf{L}^1 \cap \mathsf{BV}$. Then for every $x \in \mathbb{R}$*

$$\frac{1}{\sqrt{2\pi}} \int_{-\infty}^{\infty} f(x-u) g(u)\, du = \lim_{n\to\infty} \sum_{k=-n}^{n} f^\wedge(k) g^\wedge(k)\, e^{ikx}. \tag{5.1.55}$$

In particular,

$$\frac{1}{\sqrt{2\pi}} \int_{-\infty}^{\infty} f(u)\overline{g(u)}\, du = \lim_{n\to\infty} \sum_{k=-n}^{n} f^\wedge(k)\overline{g^\wedge(k)}. \tag{5.1.56}$$

Proof. According to Prop. 3.1.11, g^* as defined by (3.1.22) is a 2π-periodic function of BV_{loc}. Therefore it follows by Prop. 4.1.7 that

$$\frac{1}{2\pi} \int_{-\pi}^{\pi} f(x-u) g^*(u)\, du = \lim_{n\to\infty} \sum_{k=-n}^{n} f^\wedge(k)[g^*]^\wedge(k)\, e^{ikx},$$

which already implies (5.1.55) by (3.1.25) and (5.1.52).

Let us conclude with the particular situation that g is a kernel of Fejér's type. Thus let $\chi \in \mathsf{NL}^1$ and $\{\chi_\rho^*(x)\}$ be the periodic approximate identity which is generated by χ via (3.1.28), i.e., for $\rho > 0$

$$\chi_\rho^*(x) = \sqrt{2\pi} \sum_{k=-\infty}^{\infty} \rho\chi(\rho(x + 2k\pi)) \quad \text{a.e.} \tag{5.1.57}$$

Then by Prop. 5.1.1(iii) and (5.1.52)

$$[\chi_\rho^*]^\wedge(k) = \chi^\wedge(k/\rho) \qquad (k \in \mathbb{Z},\ \rho > 0). \tag{5.1.58}$$

It is actually this property which makes periodic kernels $\{\chi_\rho^*(x)\}$ generated by some $\chi \in \mathsf{NL}^1$ so useful in applications. Though the dependence of the kernel itself upon the parameter ρ may be badly arranged, the Fourier transforms of χ_ρ^*, $\rho > 0$, are obtained by one and the same function $\chi^\wedge$ by a simple scale change.

If furthermore $\chi \in \mathsf{BV}$, we may apply Prop. 5.1.30 to deduce

$$\text{(5.1.59)} \qquad \frac{\rho}{\sqrt{2\pi}} \int_{-\infty}^{\infty} f(x-u)\chi(\rho u)\,du = \lim_{n\to\infty} \sum_{k=-n}^{n} \chi^\wedge\left(\frac{k}{\rho}\right) f^\wedge(k)\, e^{ikx},$$

which holds for every $f \in \mathsf{X}_{2\pi}$ and all $x \in \mathbb{R}$, $\rho > 0$. Thus the singular integral $J(f; x; \rho)$ for $f \in \mathsf{X}_{2\pi}$ may be obtained by introducing the *factor* $\chi^\wedge(k/\rho)$ into the Fourier series of f in which case we may consider $J(f; x; \rho)$ as a method of summation of the Fourier series of f with summation function $\chi^\wedge(k/\rho)$. A comparison with our definition (1.2.28) of a θ-factor shows that this would correspond to the particular situation that one has the representation $\theta_\rho(k) = \theta(k/\rho)$ with $\mathbb{A} = (0, \infty)$, $\rho_0 = \infty$. But note that (5.1.59) does not assume $\chi^\wedge(k/\rho) \in \mathsf{l}^1$.

Let χ satisfy the assumptions of Prop. 5.1.29(i). Then the Poisson summation formula gives

$$\text{(5.1.60)} \qquad \sqrt{2\pi} \sum_{k=-\infty}^{\infty} \rho\chi(\rho(x+2k\pi)) = \sum_{k=-\infty}^{\infty} \chi^\wedge(k/\rho)\, e^{ikx} \qquad (x \in \mathbb{R},\ \rho > 0).$$

To illustrate the importance of this formula, we give proofs of (3.1.20), (3.1.36), and (3.1.43) which connect the periodic and nonperiodic kernels of Fejér, Weierstrass, and Poisson, respectively.

(i) Fejér: Let $\chi(x) = F(x)$ as given by (3.1.15) with $\rho = n+1$, $n \in \mathbb{P}$. Then by Problem 5.1.2(i) and Prop. 5.1.10

$$F^\wedge(v) = \begin{cases} 1 - |v|, & |v| \le 1 \\ 0, & |v| > 1, \end{cases}$$

and since all the assumptions of Prop. 5.1.29 are satisfied, (5.1.60) is valid. For the right-hand side we have by (1.2.24)

$$\sum_{k=-\infty}^{\infty} F^\wedge\left(\frac{k}{n+1}\right) e^{ikx} = \sum_{k=-n}^{n} \left(1 - \frac{|k|}{n+1}\right) e^{ikx} = \frac{1}{n+1} \left[\frac{\sin((n+1)x/2)}{\sin(x/2)}\right]^2$$

for all $x \ne 2j\pi$, $j \in \mathbb{Z}$, and for the left-hand side

$$\sqrt{2\pi} \sum_{k=-\infty}^{\infty} (n+1)F((n+1)(x+2k\pi)) = \frac{1}{n+1} \sum_{k=-\infty}^{\infty} \frac{\sin^2((n+1)x/2)}{((x/2)+k\pi)^2}.$$

Therefore for all x

$$\sum_{k=-\infty}^{\infty} \frac{\sin^2(n+1)x}{(x+k\pi)^2} = \frac{\sin^2(n+1)x}{\sin^2 x},$$

which proves (3.1.20) (without using the theory of meromorphic functions).

(ii) Weierstrass: Let $\chi(x) = w(x)$ as given by (3.1.33) with $\rho = t^{-1/2}$, $t > 0$. Then $w^\wedge(v) = \exp\{-v^2\}$, and (5.1.60) turns out to be

$$\sqrt{\frac{\pi}{t}} \sum_{k=-\infty}^{\infty} e^{-(x+2k\pi)^2/4t} = \sum_{k=-\infty}^{\infty} e^{-k^2 t} e^{ikx} \qquad (x \in \mathbb{R}, t > 0), \tag{5.1.61}$$

which is (3.1.36).

(iii) Poisson: Here we set $\chi(x) = p(x)$, where p is given by (3.1.39), and $\rho = y^{-1}$, $y > 0$. Then $p^\wedge(v) = \exp\{-|v|\}$ by Problem 5.1.2 and Prop. 5.1.10, and (5.1.60) gives

$$2 \sum_{k=-\infty}^{\infty} \frac{y}{y^2 + (x + 2k\pi)^2} = \sum_{k=-\infty}^{\infty} e^{-|k|y} e^{ikx} \qquad (x \in \mathbb{R}, y > 0),$$

which proves (3.1.43).

Apart from these connections between periodic and nonperiodic kernels, the results of this subsection (and of Sec. 3.1.2), in particular formula (5.1.58), may also be used to link the theory of Fourier integrals to that of Fourier series, thus to obtain many facts about the Fourier transform from the corresponding facts about the finite Fourier transform. For examples see Problem 5.1.16.

Problems

1. (i) Let $f \in \mathsf{L}^1$. Show that $f^\wedge(v)$ is an even (odd) function on $\mathbb{R}$ if and only if $f(x)$ is an even (odd) function on $\mathbb{R}$. Show that $[f(-\circ)]^\wedge(v) = f^\wedge(-v)$.
   (ii) Let $f \in \mathsf{L}^1$ and $h \in \mathbb{R}$, $h \neq 0$. Show that $[f(h\circ)]^\wedge(v) = |h|^{-1} f^\wedge(v/h)$.
   (iii) Let $f, f_n \in \mathsf{L}^1$ be such that $\lim_{n\to\infty} \|f - f_n\|_1 = 0$. Show that $\lim_{n\to\infty} f_n^\wedge(v) = f^\wedge(v)$ uniformly for $v \in \mathbb{R}$. In particular, show that if $\{\chi(x; \rho)\}$ is an approximate identity, then $\lim_{\rho\to\infty} \chi^\wedge(v; \rho) = 1$ for all $v \in \mathbb{R}$.
   (iv) Let $f, g \in \mathsf{L}^1$ and suppose that $g^\wedge(v) = 0$ for $|v| \geq 1$. Show that $[f * g]^\wedge(v) = 0$ for $|v| \geq 1$.
2. (i) In the notation of (5.1.8) show that

$$\theta_1^\wedge(v) = \frac{1}{\sqrt{2\pi}} \left[\frac{\sin (v/2)}{v/2}\right]^2, \qquad \theta_2^\wedge(v) = \sqrt{\frac{2}{\pi}} \frac{1}{1 + v^2}, \qquad \theta_3^\wedge(v) = \frac{1}{\sqrt{2}} e^{-v^2/4}.$$

   (Hint: See e.g. Bochner [7, p. 58], Hewitt–Stromberg [1, p. 407])
   (ii) Given real numbers $a < b$ and $\varepsilon > 0$, show that there exists $g \in \mathsf{L}^1$ such that $g^\wedge(v) = 1$ for $a \leq v \leq b$, $= 0$ for $v \leq a - \varepsilon$, $v \geq b + \varepsilon$, and $g^\wedge$ is linear on $[a - \varepsilon, a]$, $[b, b + \varepsilon]$. Evaluate the Fourier transform of $(1 - ix)^{-2}$, and show that there exist functions $g \in \mathsf{L}^1$ such that $g^\wedge(v) > 0$ for $v > 0$, $= 0$ for $v \leq 0$. (Hint: For the first question use θ_1, see also R. R. Goldberg [1, pp. 23–26])
   (iii) Show that the Hermite functions $H_n(x)$ of (3.1.13) satisfy $H_n^\wedge(v) = (-i)^n H_n(v)$. Thus the Hermite functions $H_n(x)$ are eigenfunctions of the Fourier transform with the corresponding eigenvalues $(-i)^n$. (Hint: Use Problem 3.1.3(iv), Prop. 5.1.14, 5.1.17, and show that $i^n H_n^\wedge(v)$ satisfies the same recursion formula as $H_n(v)$; see also Hewitt [1, p. 160], Sz.–Nagy [5, p. 350], Titchmarsh [6, p. 81])
   (iv) Show that

$$\frac{1}{\pi} \int_{-\infty}^{\infty} \frac{\sin au}{u}\, du = \begin{cases} 1, & a > 0 \\ 0, & a = 0 \\ -1, & a < 0. \end{cases}$$

   (Hint: See e.g. Bochner [7, p. 16], Hewitt [1, p. 163])

(v) If $N(x)$ denotes the kernel of the singular integral (3.1.58) of Jackson–de La Vallée Poussin, show that

$$N^\wedge(v) = \begin{cases} 1 - (3/2)v^2 + (3/4)\,|v|^3, & |v| \le 1 \\ (2 - |v|)^3/4, & 1 < |v| \le 2 \\ 0, & |v| > 2 \end{cases}$$

(see ACHIESER [2, p. 138]). Thus, in view of (5.1.58), the coefficients of the polynomial $N_n^*(x)$, constituting the kernel of the periodic singular integral (3.1.59) of Jackson–de La Vallée Poussin, are given through $[N_n^*]^\wedge(k) = N^\wedge(k/n)$.

3. Show that there exist functions $g \in \mathsf{C}_0$ which are not the Fourier transform of some function in L^1, yet the set $[\mathsf{L}^1]^\wedge \equiv \{g \in \mathsf{C}_0 \mid g(v) = f^\wedge(v), f \in \mathsf{L}^1\}$ is dense in C_0. (Hint: As to the first part, R. R. GOLDBERG [1, p. 8] examines the example $g(x) = 1/\log x$ for $x > e$, $= x/e$ for $0 \le x \le e$, $= -g(-x)$ for $x < 0$; see also RUDIN [4, p. 195]. For the second part, one may use Problem 5.1.2(iii) and Cor. 3.1.9; see also LEVITAN [1])

4. (i) Give examples of functions $f \in \mathsf{L}^1$ for which $f^\wedge \notin \mathsf{L}^1$ (cf. (5.1.27), (5.1.34)).

(ii) Suppose that $f \in \mathsf{L}^1 \cap \mathsf{L}^\infty$ and that $f^\wedge$ is real-valued and non-negative. Show that $f^\wedge$ belongs to L^1. (Hint: Use Lebesgue's monotone convergence theorem and Cor. 5.1.9; compare BOCHNER–CHANDRASEKHARAN [1, p. 20], HEWITT–STROMBERG [1, p. 409])

(iii) Suppose that $f \in \mathsf{L}^1 \cap \mathsf{L}^\infty$. Show that $\|f\|_2 = \|f^\wedge\|_2$. (Hint: Use (ii); compare BOCHNER–CHANDRASEKHARAN [1, p. 23])

(iv) Let $\rho > 0$ and suppose that f is bounded and integrable in $[-\rho, \rho]$, and that

$$\phi(v) \equiv \frac{1}{\sqrt{2\pi}} \int_{-\rho}^{\rho} e^{-ivu} f(u)\, du \ge 0$$

for all $v \in \mathbb{R}$. Show that ϕ belongs to L^1. (Hint: Compare also LINNIK [1, p. 13])

5. Let $f \in \mathsf{L}^1$. (i) Show that for every $\rho > 0$

$$S(f; x; \rho) \equiv \frac{1}{\sqrt{2\pi}} \int_{-\rho}^{\rho} f^\wedge(v)\, e^{ixv}\, dv = \frac{\rho}{\pi} \int_{-\infty}^{\infty} f(x-u) \frac{\sin \rho u}{\rho u}\, du.$$

(ii) If there is a point x_0 and $\delta > 0$ such that

$$\int_0^\delta \left| \frac{f(x_0+u) + f(x_0-u) - 2f(x_0)}{u} \right| du < \infty, \tag{5.1.62}$$

show that $\lim_{\rho\to\infty} S(f; x_0; \rho) = f(x_0)$. (5.1.62) is *Dini's condition* for the convergence of the Fourier inversion integral. This result states that if $f \in \mathsf{L}^1$, then the convergence or divergence of $S(f; x; \rho)$ at a particular point is governed entirely by the behaviour of f in a neighbourhood of that point. This is *Riemann's localization theorem.* (Hint: See also BOCHNER–CHANDRASEKHARAN [1, p. 10])

(iii) If f is of bounded variation in an interval including the point x_0, show that $\lim_{\rho\to\infty} S(f; x_0; \rho) = 2^{-1}[f(x_0+) + f(x_0-)]$. This is *Jordan's theorem* for the convergence of the Fourier inversion integral. (Hint: By Problem 5.1.2(iv)

$$S(f; x_0; \rho) - f(x_0) = \frac{1}{\pi} \int_0^\infty [f(x_0+u) + f(x_0-u) - 2f(x_0)] \frac{\sin \rho u}{u}\, du.$$

Splitting into the two parts $\int_0^\delta$, $\int_\delta^\infty$ for some $\delta > 0$, the first tends to $2^{-1}[f(x_0+) + f(x_0-)]$ as $\rho \to \infty$ (compare Problem 1.2.10), whereas the second tends to zero by the Riemann–Lebesgue lemma; see e.g. R. R. GOLDBERG [1, p. 10 ff])

(iv) If $f \in \mathsf{L}^1 \cap \mathsf{BV}$, then for every x

$$f(x) = \lim_{\rho\to\infty} \frac{1}{\sqrt{2\pi}} \int_{-\rho}^{\rho} f^\wedge(v)\, e^{ixv}\, dv,$$

which is referred to as *Jordan's criterion* (for Fourier integrals).

(v) Examine the particular case where f is the characteristic function of $[-h, h]$, $h > 0$, thus $f(x) = \kappa_{[-h,h]}(x)$. (Hint: Compare with Problem 5.1.2(iv))

6. (i) Let $f \in X(\mathbb{R})$ and $h > 0$. Show that (cf. (5.1.27))

$$\int_x^{x+h} f(u)\,du = \frac{1}{\sqrt{2\pi}} \int_{-\infty}^{\infty} f(x-u)m(u/h)\,du = (f * m(\circ/h))(x),$$

thus $\int_0^h f(\circ + u)\,du \in X(\mathbb{R})$.

(ii) If $f \in L^1$ and $h \in \mathbb{R}$, show that for all $v \in \mathbb{R}$

$$\left[\int_0^h f(\circ + u)\,du\right]^\wedge(v) = \begin{cases} \dfrac{e^{ihv}-1}{iv} f^\wedge(v), & v \neq 0 \\ hf^\wedge(0), & v = 0. \end{cases}$$

(iii) Let $f \in L^1$. Show that $\int_0^h f(x+u)\,du$ (as a function of x) is of bounded variation and

$$\int_0^h f(x+u)\,du = \lim_{\rho\to\infty} \frac{1}{\sqrt{2\pi}} \int_{-\rho}^{\rho} \frac{e^{ihv}-1}{iv} f^\wedge(v)\,e^{ixv}\,dv$$

for all $x, h \in \mathbb{R}$. In other words, one has the inversion formula

$$f(x) = \frac{d}{dx}\left[\lim_{\rho\to\infty} \frac{1}{\sqrt{2\pi}} \int_{-\rho}^{\rho} \frac{e^{ixv}-1}{iv} f^\wedge(v)\,dv\right] \quad \text{a.e.}$$

(Hint: Use Jordan's criterion; see also ACHIESER [2, p. 144])

7. (i) Find two functions $f, g \in L^1$, neither of which vanishes anywhere, such that $(f * g)(x) = 0$. (Hint: Consider θ_1 and use Prop. 5.1.1(ii), the convolution and uniqueness theorem; see also HEWITT–STROMBERG [1, p. 414])

(ii) Suppose that $f \in L^1$ and $(f * f)(x) = f(x)$ a.e. Prove that $f(x) = 0$ a.e.

(iii) Suppose that $f \in L^1$ and $(f * f)(x) = 0$ a.e. Prove that $f(x) = 0$ a.e.

8. Prove the *Bernstein inequality* for the derivative of a low frequency function, i.e.: Let $\chi \in W^1_{L^1}$ and suppose that $\chi^\wedge(v) = 0$ for $|v| \geq n$. For $f \in X(\mathbb{R})$ show that $\|(f * \chi)'\|_{X(\mathbb{R})} \leq An\,\|f * \chi\|_{X(\mathbb{R})}$, where A is a certain constant. (Hint: Let $l \in W^1_{L^1}$ be such that $l^\wedge(v) = 1$ for $|v| \leq 1$. Show that $\chi = \chi * (nl(n\,\circ))$ and apply Problem 3.1.5 to prove the assertion with $A = \|l'\|_1$; see H. S. SHAPIRO [1, p. 95])

9. (i) Let $f(x), f(x)/x \in L^1$. Show that $[f(\circ)/\circ]^\wedge(v) = i\int_v^\infty f^\wedge(u)\,du$.

(ii) Let S be the space of all *rapidly decreasing functions*, i.e., the set of all those C^∞-functions f for which $\lim_{|x|\to\infty} |x|^k f^{(r)}(x) = 0$ for all $k, r \in \mathbb{P}$. Show that the Fourier transform maps S onto S. (Hint: See e.g. ZEMANIAN [1, p. 182])

10. For $f \in L^1(0, \infty)$ the *Fourier-cosine* [*-sine*] *transform* is defined for $v \in \mathbb{R}$ by

$$f_c^\wedge(v) = \sqrt{\frac{2}{\pi}} \int_0^\infty f(u)\cos vu\,du \left[f_s^\wedge(v) = \sqrt{\frac{2}{\pi}} \int_0^\infty f(u)\sin vu\,du\right].$$

Show that if $f \in L^1$ is even [odd], then $f^\wedge(v) = f_c^\wedge(v)$ [$f^\wedge(v) = -if_s^\wedge(v)$].

11. (i) Show that if the rth ordinary derivative $f^{(r)}(x_0)$ exists, so does the rth Riemann derivative $f^{[r]}(x_0)$, and we have $f^{[r]}(x_0) = f^{(r)}(x_0)$.

(ii) For the function $f(x) = x\sin(x^{-2})$ for $x \neq 0$, $=0$ for $x = 0$ show that the second Riemann derivative exists at $x = 0$ but not $f''(0)$. (Hint: CHAUNDY [1, p. 126])

12. If the rth one-sided difference $\Delta_h^r f(x)$ of f at x with increment $h \in \mathbb{R}$ is defined successively by $\Delta_h^1 f(x) = f(x+h) - f(x)$, $\Delta_h^r f(x) = \Delta_h^1 \Delta_h^{r-1} f(x)$, it may be shown by induction that

$$\Delta_h^r f(x) = \sum_{k=0}^{r} (-1)^{r-k} \binom{r}{k} f(x+kh).$$

The rth *Riemann derivative* of f at x_0 with respect to the one-sided difference is then defined by

$$f^{[r]}(x_0) = \lim_{h\to 0} \frac{1}{h^r} \Delta_h^r f(x_0),$$

in case the limit exists. We use the same notation since no confusion may occur. The latter definition is particularly useful in Chapter 10.

(i) Show that if the rth ordinary derivative $f^{(r)}(x_0)$ exists, so does the rth Riemann derivative with respect to the one-sided difference, and both are equal.

(ii) Show that

$$f(x) = \begin{cases} x \sin[(2\pi/\log 2)\log x], & x \neq 0 \\ 0, & x = 0 \end{cases}$$

has a second Riemann derivative (with respect to the one-sided difference) at $x = 0$, but there exists no ordinary derivative at the origin. (Hint: CHAUNDY [1, p. 126])

(iii) Show that for the function f of Problem 5.1.11(ii) the second Riemann derivative with respect to the one-sided difference does not exist at $x = 0$.

13. (i) Show that if the rth ordinary derivative $f^{(r)}(x_0)$ exists, so does the rth Peano derivative $f^{\langle r\rangle}(x_0)$, and we have $f^{\langle r\rangle}(x_0) = f^{(r)}(x_0)$.

(ii) For the function $f(x) = \exp\{-x^{-2}\}\sin(\exp\{x^{-2}\})$ for $x \neq 0$, $=0$ for $x = 0$ show that Peano derivatives of every order exist at $x = 0$ (and are zero), whereas the second ordinary derivative does not exist at $x = 0$. (Hint: CHAUNDY [1, p. 119])

14. (i) Prove (5.1.49). (Hint: Use the formula

$$(e^{ix} - 1)^r = (-1)^r \sum_{k=0}^{r} \binom{r}{k}(-1)^k e^{ikx},$$

evaluate the jth derivative and set $x = 0$)

(ii) For $f(x) = |x|\, x^{r-1}$ show that f has only Peano derivatives up to the order $r - 1$ at $x = 0$ (and these are zero), but Riemann derivatives (with respect to the central difference) of higher order exist (and are zero). (Hint: See also BUTZER [10])

15. If f is defined in some neighbourhood of the point x_0 and if the $(r - 1)$th ordinary derivative $f^{(r-1)}(x_0)$ exists, then we call

$$f^{\circledcirc}(x_0) = \lim_{h\to 0} \frac{r!}{h^r}\left[f(x_0 + h) - \sum_{k=0}^{r-1} \frac{h^k}{k!} f^{(k)}(x_0)\right]$$

the *rth Taylor derivative* of f at x_0 if the limit exists. The following shows that the Taylor derivative is more general than the ordinary but less general than the Peano derivative.

(i) Show that if the rth ordinary derivative $f^{(r)}(x_0)$ exists, so does the rth Taylor derivative $f^{\circledcirc}(x_0)$, and we have $f^{\circledcirc}(x_0) = f^{(r)}(x_0)$. On the other hand, the existence of the rth Taylor derivative at x_0 implies the existence of the Peano derivative at x_0, and both are equal.

(ii) For the function ($r, n \in \mathbb{N}$, $r, n \geq 2$)

$$f(x) = \begin{cases} x^{(r-1)(n+1)} \sin x^{-n}, & x \neq 0 \\ 0, & x = 0 \end{cases}$$

show that there exist only the first $(r - 1)$ ordinary derivatives at $x = 0$ (which are zero), but the rth Taylor derivative $f^{\circledcirc}(0)$ exists (and is zero). On the other hand, f has Peano derivatives up to the order $[(r - 1)(n + 1) - 1]$ at $x = 0$, whereas the $(r + 1)$th Taylor derivative does not exist at $x = 0$. (Hint: DENJOY [1], CHAUNDY [1, p. 137])

16. Use (5.1.58) to deduce the uniqueness theorem [and Riemann–Lebesgue lemma] for L^1-Fourier transforms from that for the finite Fourier transform. (Hint: Apply Problem 3.1.10(iii) and the uniform continuity of L^1-Fourier transforms; see KATZNELSON [1, p. 129])

5.2 L^p-Theory, $1 < p \le 2$

The definition of the Fourier transform for a function $f \in \mathsf{L}^p$, $1 < p \le 2$, is more complicated than for $f \in \mathsf{L}^1$. In fact, recalling the definition (5.1.1) of the Fourier transform for $f \in \mathsf{L}^1$,

$$f^{\wedge}(v) = \frac{1}{\sqrt{2\pi}} \int_{-\infty}^{\infty} f(u)\, e^{-ivu}\, du \qquad (v \in \mathbb{R}), \tag{5.2.1}$$

we observe that for arbitrary functions $f \in \mathsf{L}^p$, $p > 1$, the integral defining the transform $f^{\wedge}$ need not exist as an ordinary Lebesgue integral (see Problem 5.2.1), and thus the problem is to give a suitable interpretation of (5.2.1).

The question of defining the Fourier transform on L^p, $1 < p \le 2$, is divided into the cases $p = 2$ and $1 < p < 2$. We first study $p = 2$ and then that for $1 < p < 2$ as an application of the case $p = 2$ and the M. Riesz–Thorin convexity theorem. The treatment is quite analogous to Def. 4.2.8 of the Fourier transform on l^p, $1 < p \le 2$.

5.2.1 The Case $p = 2$

We begin by discussing the case that f belongs to both L^1 and L^2 and reduce the general case hereto. The following proposition will turn out to be an extension of (5.1.24).

Proposition 5.2.1. *If $f \in \mathsf{L}^1 \cap \mathsf{L}^2$, then $f^{\wedge} \in \mathsf{L}^2$ and*

$$\|f^{\wedge}\|_2 = \|f\|_2. \tag{5.2.2}$$

Proof. If we set $f^*(x) = \overline{f(-x)}$ and $h(x) = (f * f^*)(x)$, then it follows by Prop. 0.2.1, 0.2.2 that $h \in \mathsf{C}_0 \cap \mathsf{L}^1$, and an application of (5.1.16) at $x = 0$ gives

$$\lim_{\rho \to \infty} \frac{1}{\sqrt{2\pi}} \int_{-\infty}^{\infty} e^{-|v|/\rho} h^{\wedge}(v)\, dv = h(0).$$

But according to Prop. 5.1.1(iv) and Theorem 5.1.3 we have $h^{\wedge}(v) = f^{\wedge}(v)\overline{f^{\wedge}(v)} = |f^{\wedge}(v)|^2$, and thus the integrand is a positive function which increases monotonely to $h^{\wedge}$ as $\rho \to \infty$. Hence we may apply Lebesgue's monotone convergence theorem to deduce $h^{\wedge} \in \mathsf{L}^1$ and $\int_{-\infty}^{\infty} h^{\wedge}(v)\, dv = \sqrt{2\pi} h(0)$. Since $h(0) = \|f\|_2^2$ and $h^{\wedge}(v) = |f^{\wedge}(v)|^2$, the result follows.

Proposition 5.2.2. *Let $f \in \mathsf{L}^2$. Defining f_ρ by*

$$f_\rho(x) = \begin{cases} f(x), & |x| \le \rho \\ 0, & |x| > \rho \end{cases} \qquad (\rho > 0), \tag{5.2.3}$$

then $f_\rho \in \mathsf{L}^1 \cap \mathsf{L}^2$ and $f_\rho^{\wedge} \in \mathsf{L}^2$ for each $\rho > 0$. Moreover, $f_\rho^{\wedge}$ converges in L^2-norm to a function in L^2 as $\rho \to \infty$.

Proof. It follows immediately by Hölder's inequality that $f_\rho \in \mathsf{L}^1 \cap \mathsf{L}^2$ for all $\rho > 0$, and thus $f_\rho^\wedge \in L^2$ by Prop. 5.2.1. Moreover, since $f_{\rho_1}^\wedge - f_{\rho_2}^\wedge$ is the Fourier transform of $f_{\rho_1} - f_{\rho_2}$, (5.2.2) gives for $\rho_1 < \rho_2$

$$\|f_{\rho_1}^\wedge - f_{\rho_2}^\wedge\|_2^2 = \|f_{\rho_1} - f_{\rho_2}\|_2^2 = \frac{1}{\sqrt{2\pi}}\int_{-\rho_2}^{-\rho_1} |f(u)|^2\, du + \frac{1}{\sqrt{2\pi}}\int_{\rho_1}^{\rho_2} |f(u)|^2\, du.$$

Since $f \in L^2$, the latter sum tends to zero as $\rho_1, \rho_2 \to \infty$, giving $\lim_{\rho_1,\rho_2\to\infty}\|f_{\rho_1}^\wedge - f_{\rho_2}^\wedge\|_2 = 0$. Thus by completeness of L^2 there exists a unique function in L^2 such that the functions $f_\rho^\wedge$ tend to it in the L^2-metric as $\rho \to \infty$.

We may now define the *Fourier transform* of any L^2-function.

Definition 5.2.3. *For $f \in L^2$ we define the Fourier transform $F^2[f]$ of f as*

$$F^2[f](v) = \underset{\rho\to\infty}{\overset{(2)}{\text{l.i.m.}}}\, f_\rho^\wedge(v) \equiv \underset{\rho\to\infty}{\overset{(2)}{\text{l.i.m.}}}\, \frac{1}{\sqrt{2\pi}}\int_{-\rho}^{\rho} f(u)\, e^{-ivu}\, du.$$

Thus the Fourier transform of $f \in L^2$ is the uniquely determined function $F^2[f] \in L^2$ given by

$$\lim_{\rho\to\infty} \|F^2[f](\circ) - \frac{1}{\sqrt{2\pi}}\int_{-\rho}^{\rho} f(u)\, e^{-i\circ u}\, du\|_2 = 0. \tag{5.2.4}$$

The latter definition is meaningful by Prop. 5.2.2. Note that the Fourier transform $F^2[f]$ as an element of L^2 needs to be defined only a.e. Using the same arguments as above, one might give the definition of the L^2-Fourier transform in the seemingly more general form

$$\lim_{\rho_1,\rho_2\to\infty} \left\| F^2[f](\circ) - \frac{1}{\sqrt{2\pi}}\int_{-\rho_1}^{\rho_2} f(u)\, e^{-i\circ u}\, du \right\|_2 = 0,$$

where ρ_1, ρ_2 tend to ∞ independently. But it is easy to see that both definitions would be equivalent.

For the Fourier transform on L^2 we obtain as an extension of Prop. 5.2.1

Theorem 5.2.4 *The Fourier transform* (5.2.4) *defines a bounded linear transformation of L^2 into L^2 which preserves norms, i.e., for $f \in L^2$*

$$\|F^2[f]\|_2 = \|f\|_2. \tag{5.2.5}$$

Proof. Since $\lim_{\rho\to\infty} \|F^2[f] - f_\rho^\wedge\|_2 = 0$, it follows that $\lim_{\rho\to\infty} \|f_\rho^\wedge\|_2 = \|F^2[f]\|_2$. On the other hand, $\lim_{\rho\to\infty} \|f_\rho\|_2 = \|f\|_2$, and since $f_\rho \in L^1 \cap L^2$, Prop. 5.2.1 implies $\|f_\rho^\wedge\|_2 = \|f_\rho\|_2$. Upon collecting results the proof is complete.

We remark that we shall later on show that the Fourier transform (5.2.4) actually defines an isomorphism of the Hilbert space L^2 *onto* itself. Equation (5.2.5) is again referred to as *Parseval's equation* (cf. (5.1.24)).

5.2.2 The Case $1 < p < 2$

We have defined the Fourier transform in L^p for the values $p = 1, 2$. It is natural to inquire whether this is also possible for other values of p. The answer given in the following will be affirmative for $1 < p < 2$. As a significant application of the M. Riesz–Thorin convexity theorem we obtain

Proposition 5.2.5. *Let* $1 \le p \le 2$. *If* $h \in \mathsf{S}_{00}$, *then* $h^\wedge \in \mathsf{L}^{p'}$ *and*

$$\|h^\wedge\|_{p'} \le \|h\|_p. \tag{5.2.6}$$

Proof. In order to apply the convexity theorem (cf. Sec. 4.4), let $\mathbb{R}_1 = \mathbb{R}_2 = \mathbb{R}$, $\mu_1 = \mu_2 =$ Lebesgue measure on $\mathbb{R}$, and let T be the transformation which assigns to each simple function h its Fourier transform $h^\wedge$ of (5.2.1):

$$(Th)(v) \equiv h^\wedge(v) \equiv \frac{1}{\sqrt{2\pi}} \int_{-\infty}^{\infty} h(u)\, e^{-ivu}\, du \qquad (h \in \mathsf{S}_{00}). \tag{5.2.7}$$

Obviously, since $\mathsf{S}_{00} \subset \mathsf{L}^1$, T is well-defined as a linear transformation on S_{00} which by (5.1.2) satisfies

$$\|Th\|_\infty \le \|h\|_1, \tag{5.2.8}$$

and by (5.2.2), since $\mathsf{S}_{00} \subset \mathsf{L}^2$,

$$\|Th\|_2 = \|h\|_2. \tag{5.2.9}$$

Thus T is of strong type $(1; \infty)$ and $(2; 2)$ on S_{00} with constants $M_1 = M_2 = 1$. The M. Riesz–Thorin convexity theorem then implies that T is also of strong type $(p; p')$ satisfying (5.2.6).

As an immediate consequence of Prop. 0.3.4, 0.7.1 we have

Proposition 5.2.6. *The operator T as defined in* (5.2.7) *on* S_{00} *can be uniquely extended to a bounded linear transformation of* L^p, $1 \le p \le 2$, *into* $\mathsf{L}^{p'}$ *such that*

$$\|Tf\|_{p'} \le \|f\|_p \tag{5.2.10}$$

holds for all $f \in \mathsf{L}^p$.

Thus we are able to assign to each $f \in \mathsf{L}^p$, $1 < p < 2$, a uniquely determined function $Tf \in \mathsf{L}^{p'}$ for which (5.2.10) holds and which is of the form (5.2.7) on S_{00}. We may furthermore prove

Proposition 5.2.7. *Let* $f \in \mathsf{L}^p$, $1 < p < 2$, *and* f_ρ *be defined as in* (5.2.3). *Then*

$$\lim_{\rho\to\infty} \|Tf - f_\rho^\wedge\|_{p'} = 0. \tag{5.2.11}$$

Proof. By Hölder's inequality it follows that $f_\rho \in \mathsf{L}^1 \cap \mathsf{L}^p$, and therefore $f_\rho^\wedge$ is well-defined by (5.2.1). Moreover,

$$\lim_{\rho\to\infty} \|f - f_\rho\|_p = 0. \tag{5.2.12}$$

Let $\rho > 0$ be fixed. Then by Prop. 0.3.4 we choose a sequence $\{h_k\}$ of functions in S_{00} which vanish outside of $[-\rho, \rho]$ and which approximate f_ρ in the L^p-metric, thus

$$\lim_{k\to\infty} \|f_\rho - h_k\|_p = 0. \tag{5.2.13}$$

By Hölder's inequality this implies $\lim_{k\to\infty} \|f_\rho - h_k\|_1 = 0$. Therefore we obtain $\lim_{k\to\infty} \|f_\rho^\wedge - h_k^\wedge\|_{\mathsf{C}} = 0$ by (5.1.2), i.e. $\lim_{k\to\infty} h_k^\wedge(v) = f_\rho^\wedge(v)$ for all $v \in \mathbb{R}$.

Since $\|Tf_\rho - h_k^\wedge\|_{p'} \le \|f_\rho - h_k\|_p$ by equations (5.2.7) and (5.2.10), we have $\lim_{k\to\infty} \|Tf_\rho - h_k^\wedge\|_{p'} = 0$ by (5.2.13). Hence by Prop. 0.1.10 there exists a subsequence $\{h_{k_j}\}$ such that $\lim_{j\to\infty} h_{k_j}^\wedge(v) = (Tf_\rho)(v)$ for almost all $v \in \mathbb{R}$. Therefore it follows for $f \in \mathsf{L}^p$, $1 < p < 2$, and for each $\rho > 0$ that

$$(Tf_\rho)(v) = f_\rho^\wedge(v) \quad \text{a.e.,} \tag{5.2.14}$$

extending definition (5.2.7) on S_{00} to all functions f_ρ. But by (5.2.10) this implies $\|Tf - f_\rho^\wedge\|_{p'} \le \|f - f_\rho\|_p$ from which the assertion (5.2.11) follows by (5.2.12).

We are now justified in defining Tf as the *Fourier transform* of $f \in \mathsf{L}^p$, $1 < p < 2$.

Definition 5.2.8. *For $f \in \mathsf{L}^p$, $1 < p < 2$, we define the Fourier transform $F^p[f]$ of f by*

$$F^p[f](v) = \underset{\rho\to\infty}{\overset{(p')}{\text{l.i.m.}}}\, f_\rho^\wedge(v) \equiv \underset{\rho\to\infty}{\overset{(p')}{\text{l.i.m.}}}\, \frac{1}{\sqrt{2\pi}} \int_{-\rho}^{\rho} f(u)\, e^{-ivu}\, du. \tag{5.2.15}$$

In other words, the Fourier transform of $f \in \mathsf{L}^p$, $1 < p < 2$, is defined as the uniquely determined function $F^p[f] \in \mathsf{L}^{p'}$ given by (5.2.15). We have

Theorem 5.2.9. *The Fourier transform* (5.2.15) *defines a bounded linear transformation of L^p, $1 < p < 2$, into $\mathsf{L}^{p'}$ which contracts norms, i.e., for $f \in \mathsf{L}^p$*

$$\|F^p[f]\|_{p'} \le \|f\|_p. \tag{5.2.16}$$

The inequality (5.2.16) is often called the *Titchmarsh inequality*. In the following we shall sometimes call (5.1.2) and (5.2.5) by the same name so as to cover inequalities of this type for all $1 \le p \le 2$.

Proposition 5.2.10. *If $f \in \mathsf{L}^1 \cap \mathsf{L}^p$, $1 < p \le 2$, then $F^1[f](v) = F^p[f](v)$* a.e.

Proof. According to (5.2.4), (5.2.15), and Prop. 0.1.10 there exists a subsequence such that

$$F^p[f](v) = \lim_{k\to\infty} \frac{1}{\sqrt{2\pi}} \int_{-\rho_k}^{\rho_k} f(u)\, e^{-ivu}\, du \quad \text{a.e.}$$

Since $f \in \mathsf{L}^1$, the limit exists for all $v \in \mathbb{R}$ and

$$F^p[f](v) = \frac{1}{\sqrt{2\pi}} \int_{-\infty}^{\infty} f(u)\, e^{-ivu}\, du \equiv F^1[f](v),$$

giving the result.

Let us mention that the latter proposition extends (5.2.7) and (5.2.14) to all $f \in \mathsf{L}^1 \cap \mathsf{L}^p$, $1 < p < 2$. Prop. 5.2.10 is important as it shows that the definition (5.2.1) for $p = 1$ on the one hand and Def. 5.2.3, 5.2.8 for $1 < p \le 2$ on the other hand are consistent. We are therefore justified in employing the more simple notation $f^\wedge$ to denote the Fourier transform for all spaces L^p, $1 \le p \le 2$. It will be clear from the context in which sense $f^\wedge$ will be taken. If any confusion should occur we shall return to $F^p[f]$.

As in case of the definition of Fourier transforms in l^p-spaces (cf. Problem 4.2.3) it is not generally possible to define a Fourier transform for functions $f \in \mathsf{L}^p$, $p > 2$, by the methods so far employed (see Problem 5.2.2). See also Sec. 5.4.

5.2.3 Fundamental Properties

Up to this stage we have defined the Fourier transform for $f \in \mathsf{L}^p$, $1 < p \leq 2$, and showed that it obeys the Parseval equation (5.2.5) in case $p = 2$ and the Titchmarsh inequality (5.2.16) in case $1 < p < 2$. We shall now study further properties of these transforms.

Proposition 5.2.11. *If $f \in \mathsf{L}^p$, $1 < p \leq 2$, then for each fixed $h \in \mathbb{R}$, $\rho > 0$:*

(i) $$[f(\circ + h)]^\wedge(v) = e^{ihv} f^\wedge(v) \quad \text{a.e.,}$$

(ii) $$[e^{-ih\circ} f(\circ)]^\wedge(v) = f^\wedge(v + h) \quad \text{a.e.,}$$

(iii) $$[f(\circ/\rho)]^\wedge(v) = \rho f^\wedge(\rho v) \quad \text{a.e.,}$$

(iv) $$\overline{[f(-\circ)]^\wedge(v)} = \overline{f^\wedge(v)} \quad \text{a.e.}$$

The proof is left to Problem 5.2.3.

The following result establishes the *convolution theorem* for Fourier transforms in L^p, $1 < p \leq 2$.

Theorem 5.2.12. *If $f \in \mathsf{L}^p$, $1 < p \leq 2$, and $g \in \mathsf{L}^1$, then*

$$F^p[f * g](v) = F^p[f](v) F^1[g](v) \quad \text{a.e.} \tag{5.2.17}$$

Proof. In Prop. 0.2.2 we have already shown that the convolution $f * g$ of $f \in \mathsf{L}^p$, $1 < p \leq 2$, and $g \in \mathsf{L}^1$ exists a.e. and again belongs to L^p. Therefore both sides of (5.2.17) are well-defined. To prove equality we first suppose $f \in \mathsf{L}^1 \cap \mathsf{L}^p$. Then $f * g \in \mathsf{L}^1 \cap \mathsf{L}^p$ by Prop. 0.2.2, and $F^1[f * g](v) = F^1[f](v) F^1[g](v)$ for all $v \in \mathbb{R}$ by Theorem 5.1.3. (5.2.17) now follows in virtue of Prop. 5.2.10.

In general, if $f \in \mathsf{L}^p$, $1 < p \leq 2$, then the functions f_ρ of (5.2.3) belong to $\mathsf{L}^1 \cap \mathsf{L}^p$ for each $\rho > 0$ and thus $F^p[f_\rho * g](v) = F^p[f_\rho](v) F^1[g](v)$ for almost all $v \in \mathbb{R}$. This implies by the Minkowski and Titchmarsh inequality that for each $\rho > 0$

$$\begin{aligned} &\|F^p[f * g] - F^p[f] F^1[g]\|_{p'} \\ &\quad \leq \|F^p[f * g] - F^p[f_\rho * g]\|_{p'} + \|F^p[f_\rho] F^1[g] - F^p[f] F^1[g]\|_{p'} \\ &\quad \leq \|(f - f_\rho) * g\|_p + \|F^1[g]\|_\infty \|F^p[f_\rho - f]\|_{p'} \leq 2\, \|g\|_1 \|f - f_\rho\|_p, \end{aligned}$$

which by (5.2.12) tends to zero as $\rho \to \infty$, proving (5.2.17).

We next prove the *Parseval formula* for L^p-functions.

Proposition 5.2.13. *If $f, g \in \mathsf{L}^p$, $1 < p \leq 2$, then*

$$\int_{-\infty}^{\infty} f^\wedge(v) g(v)\, dv = \int_{-\infty}^{\infty} f(v) g^\wedge(v)\, dv. \tag{5.2.18}$$

Proof. For positive ρ_1 and ρ_2 let f_{ρ_1} and g_{ρ_2} be defined by (5.2.3). Since $f_{\rho_1}, g_{\rho_2} \in \mathsf{L}^1$ we may apply Prop. 5.2.10, 5.1.4 to deduce

$$\int_{-\infty}^{\infty} f_{\rho_1}^\wedge(v) g_{\rho_2}(v)\, dv = \int_{-\infty}^{\infty} f_{\rho_1}(v) g_{\rho_2}^\wedge(v)\, dv.$$

Moreover, $\lim_{\rho_1\to\infty} \|f^\wedge - f_{\rho_1}^\wedge\|_{p'} = 0$, $\lim_{\rho_1\to\infty} \|f - f_{\rho_1}\|_p = 0$ by Titchmarsh's inequality and (5.2.12). Therefore, taking the limit as $\rho_1 \to \infty$ for each side separately and using Hölder's inequality, we obtain for each $\rho_2 > 0$

$$\int_{-\infty}^{\infty} f^\wedge(v) g_{\rho_2}(v)\, dv = \int_{-\infty}^{\infty} f(v) g_{\rho_2}^\wedge(v)\, dv$$

from which the result follows by letting $\rho_2 \to \infty$.

We remark that in case $p = 2$ formula (5.2.18) reads $(f^\wedge, \overline{g}) = (f, \overline{g^\wedge})$.

5.2.4 Summation of the Fourier Inversion Integral

Next, we turn to the inversion problem of the Fourier transform in L^p, $1 < p \le 2$. Whereas we have so far assigned to each $f \in \mathsf{L}^p$, $1 < p \le 2$, its Fourier transform $f^\wedge$ and studied some of its fundamental properties, we are now interested in determining f explicitly, given $f^\wedge$. As in the case $p = 1$ (see (5.1.6)), the formal inversion will be given by

$$f(x) = \frac{1}{\sqrt{2\pi}} \int_{-\infty}^{\infty} f^\wedge(v)\, e^{ixv}\, dv. \tag{5.2.19}$$

But since $f^\wedge \in \mathsf{L}^{p'}$, the *Fourier inversion integral* (5.2.19) does not exist in general as an ordinary Lebesgue integral, and the problem again is to interpret (5.2.19) suitably.

Theorem 5.2.14. *Let $f \in \mathsf{L}^p$, $1 < p \le 2$. Then for a θ-factor the **θ-means** of the integral (5.2.19), defined for each $\rho > 0$ by*

$$U(f; x; \rho) = \frac{1}{\sqrt{2\pi}} \int_{-\infty}^{\infty} \theta\Big(\frac{v}{\rho}\Big) f^\wedge(v)\, e^{ixv}\, dv, \tag{5.2.20}$$

exist for all $x \in \mathbb{R}$, belong to L^p and satisfy

$$\|U(f; \circ; \rho)\|_p \le \|\theta^\wedge\|_1 \|f\|_p \qquad (\rho > 0), \tag{5.2.21}$$

$$\lim_{\rho\to\infty} \|U(f; \circ; \rho) - f(\circ)\|_p = 0. \tag{5.2.22}$$

Thus the inversion integral (5.2.19) is *θ-summable* to f in L^p-norm.

Proof. Since $\|\theta\|_\infty \le \|\theta^\wedge\|_1$ by (5.1.21), and as $\theta \in \mathsf{L}^1 \cap \mathsf{L}^\infty$ implies $\theta \in \mathsf{L}^q$, $1 < q < \infty$, it follows by Prop. 5.2.10, 5.2.11, 5.2.13 that

$$U(f; x; \rho) = \frac{\rho}{\sqrt{2\pi}} \int_{-\infty}^{\infty} f(x - u)\theta^\wedge(\rho u)\, du. \tag{5.2.23}$$

Since $\theta^\wedge \in \mathsf{L}^1 \cap \mathsf{L}^\infty$, too, the proof is now an immediate consequence of Prop. 0.2.1, 0.2.2, Lemma 3.1.5, and Theorem 3.1.6.

Theorem 5.2.15. *Let $f \in \mathsf{L}^p$, $1 < p \le 2$. If for a θ-factor $\theta^\wedge$ is moreover positive and monotonely decreasing on $[0, \infty)$, then the Fourier inversion integral is θ-summable* a.e. *to $f(x)$, i.e.*

$$\lim_{\rho\to\infty} \frac{1}{\sqrt{2\pi}} \int_{-\infty}^{\infty} \theta\Big(\frac{v}{\rho}\Big) f^\wedge(v)\, e^{ixv}\, dv = f(x) \quad \text{a.e.} \tag{5.2.24}$$

For the proof as well as for the explicit formulae of some important examples of θ-factors we refer to Problem 5.2.4.

Proposition 5.2.16. *Let* $f \in \mathsf{L}^p$, $1 < p \leq 2$. *If* $f^\wedge \in \mathsf{L}^1$, *then*

$$\frac{1}{\sqrt{2\pi}} \int_{-\infty}^{\infty} f^\wedge(v)\, e^{ixv}\, dv = f(x) \quad \text{a.e.} \tag{5.2.25}$$

The proof is that of Prop. 5.1.10. One consequence of (5.2.25) is the *uniqueness theorem* for Fourier transforms in L^p, $1 < p \leq 2$.

Proposition 5.2.17. *Let* $f, g \in \mathsf{L}^p$, $1 < p \leq 2$. *If* $f^\wedge(v) = g^\wedge(v)$ *for almost all* $v \in \mathbb{R}$, *then* $f(x) = g(x)$ a.e.

This shows that the Fourier transform defines a one-to-one bounded linear transformation of the Banach space L^p, $1 < p \leq 2$, into the Banach space $\mathsf{L}^{p'}$. If $1 < p < 2$, the corresponding mapping is into, but not onto (see Problem 5.2.6), yet the set $[\mathsf{L}^p]^\wedge \equiv \{g \in \mathsf{L}^{p'} \mid g = f^\wedge \text{ a.e.}, f \in \mathsf{L}^p\}$ is dense in $\mathsf{L}^{p'}$. In case $p = 2$ the Fourier transform is a transformation of L^2 *onto* L^2 as we shall see in Sec. 5.2.6.

Proposition 5.2.18. *Let* $f \in \mathsf{L}^p$, $1 < p \leq 2$. *If* $g \in \mathsf{L}^1$ *is such that* $g^\wedge \in \mathsf{L}^1$, *then*

$$(f * g)(x) = \frac{1}{\sqrt{2\pi}} \int_{-\infty}^{\infty} f^\wedge(v) g^\wedge(v)\, e^{ixv}\, dv \qquad (x \in \mathbb{R}), \tag{5.2.26}$$

$$\int_{-\infty}^{\infty} f(u)\overline{g(u)}\, du = \int_{-\infty}^{\infty} f^\wedge(u)\overline{g^\wedge(u)}\, du. \tag{5.2.27}$$

In case $f^\wedge \in \mathsf{L}^1$ *we have furthermore*

$$\|f\|_2 = \|f^\wedge\|_2. \tag{5.2.28}$$

The proof is left to Problem 5.2.7.

5.2.5 Fourier Transforms of Derivatives

In this subsection r always denotes a natural number.

Proposition 5.2.19. *For* $1 < p \leq 2$ *let* $f \in \mathsf{L}^p \cap \mathsf{AC}_{\text{loc}}^{r-1}$ *and* $f^{(r)} \in \mathsf{L}^p$. *Then*

$$[f^{(r)}]^\wedge(v) = (iv)^r f^\wedge(v) \quad \text{a.e.} \tag{5.2.29}$$

Proof. As in the case $p = 1$, $f \in \mathsf{AC}_{\text{loc}}^{r-1}$ implies (5.1.26) for all $x \in \mathbb{R}$, and if we introduce the function m by (5.1.27), we have (cf. (5.1.28))

$$\sum_{k=0}^{r} (-1)^{r-k} \binom{r}{k} f(x+k) = ([m*]^r * f^{(r)})(x). \tag{5.2.30}$$

Since $m \in \mathsf{L}^1 \cap \mathsf{L}^\infty$ and therefore $m \in \mathsf{L}^q$, $1 \leq q \leq \infty$, the right-hand side exists by Prop. 0.2.1 for all $x \in \mathbb{R}$ as a function in C_0 and by Prop. 0.2.2 as a function in L^p.

Taking the Fourier transform of equation (5.2.30), we obtain by (5.1.29), Prop. 5.2.11, and Theorem 5.2.12

$$(e^{iv} - 1)^r f^\wedge(v) = \left(\frac{e^{iv} - 1}{iv}\right)^r [f^{(r)}]^\wedge(v)$$

for almost all $v \in \mathbb{R}$, which implies (5.2.29).

Proposition 5.2.20. *If for* $f \in \mathsf{L}^p$, $1 < p \leq 2$, *there exists* $g \in \mathsf{L}^p$ *such that*

$$(iv)^r f^\wedge(v) = g^\wedge(v) \quad \text{a.e.,} \tag{5.2.31}$$

then $f \in \mathsf{W}^r_{\mathsf{L}^p}$ *(for the definition see* (3.1.48)).

Proof. Let $r = 1$. In conjunction with the proof for $p = 1$ (cf. (5.1.31)) the assumption (5.2.31) gives for each $h > 0$

$$[f(x) - f(x - h)]^\wedge(v) = \left[\int_{-h}^{0} g(x + u)\, du\right]^\wedge_1(v) \quad \text{a.e.,} \tag{5.2.32}$$

the integral being the convolution of $m(-\circ/h) \in \mathsf{L}^1$ and $g \in \mathsf{L}^p$. As an application of the uniqueness theorem for Fourier transforms in L^p we obtain for each fixed $h > 0$

$$f(x) - f(x - h) = \int_{x-h}^{x} g(u)\, du \quad \text{a.e.} \tag{5.2.33}$$

Since $f, g \in \mathsf{L}^p$, $1 < p \leq 2$, implies $f^\wedge, g^\wedge \in \mathsf{L}^{p'}$, we have $(1 + |v|)f^\wedge(v) \in \mathsf{L}^{p'}$ by (5.2.31), and since $(1 + |v|)^{-1} \in \mathsf{L}^q$, $q > 1$, it follows by Hölder's inequality that

$$f^\wedge(v) = \frac{1}{1 + |v|}(1 + |v|)f^\wedge(v) \in \mathsf{L}^1.$$

Hence we obtain by Prop. 5.2.16 for almost all $x \in \mathbb{R}$

$$f(x) = \frac{1}{\sqrt{2\pi}} \int_{-\infty}^{\infty} f^\wedge(v)\, e^{ixv}\, dv \equiv \phi(x),$$

the latter integral defining a continuous function ϕ which belongs to C_0 by Prop. 5.1.2. Now (5.2.33) delivers

$$\phi(x) - \phi(x - h) = \int_{x-h}^{x} g(u)\, du$$

which holds for all $x \in \mathbb{R}$, $h > 0$, since both sides are continuous functions. But $\phi \in \mathsf{C}_0$ implies $\lim_{h\to\infty} \phi(x - h) = 0$, i.e., the limit $\lim_{h\to\infty} \int_{x-h}^{x} g(u)\, du$ exists and

$$\phi(x) = \int_{-\infty}^{x} g(u)\, du, \tag{5.2.34}$$

the integral being in general only conditionally convergent (note that $g \in \mathsf{L}^p$, $p > 1$, only). Since $f = \phi$ a.e., Prop. 5.2.20 is established for $r = 1$.

The proof for arbitrary integers $r \geq 2$ follows along the same lines as for the case $p = 1$. Again, $(1 + iv)^{1-r} \in [\mathsf{L}^1]^\wedge$ and $(1 + iv)^{1-r}(iv)^r f^\wedge(v) \in [\mathsf{L}^p]^\wedge$ by Theorem 5.2.12 and (5.2.31). This implies $ivf^\wedge(v) \in [\mathsf{L}^p]^\wedge$, i.e., there exists $g_1 \in \mathsf{L}^p$ such that

$ivf^\wedge(v) = g_1^\wedge(v)$. If we now apply the result for $r = 1$ and repeat this method successively, we obtain a sequence of functions $g_k \in \mathsf{L}^p$, $1 \le k \le r-1$, such that $f(x) = \int_{-\infty}^x g_1(u)\,du$, $g_k(x) = \int_{-\infty}^x g_{k+1}(u)\,du$, $1 \le k \le r-2$, $g_{r-1}(x) = \int_{-\infty}^x g(u)\,du$, and hence

$$f(x) = \int_{-\infty}^{x} du_1 \int_{-\infty}^{u_1} du_2 \ldots \int_{-\infty}^{u_{r-1}} g(u_r)\,du_r \quad \text{a.e.,} \tag{5.2.35}$$

the integrals being in general only conditionally convergent. If we define the right-hand side as the function ϕ, then $f = \phi$ a.e. and $\phi \in \mathsf{AC}_{\mathrm{loc}}^{r-1}$, $\phi^{(k)} = g_k \in \mathsf{L}^p$, $1 \le k \le r-1$, and $\phi^{(r)} = g$ a.e. This proves the proposition. Note that $\phi^{(k)} \in \mathsf{C}_0$, $0 \le k \le r-1$.

The preceding results actually give equivalent characterizations of the function class $\mathsf{W}_{\mathsf{L}^p}^r$ for $1 < p \le 2$. Indeed, extending the definition (5.1.36), let

$$\mathsf{W}[\mathsf{L}^p; \psi(v)] = \{f \in \mathsf{L}^p \mid \psi(v) f^\wedge(v) = g^\wedge(v),\, g \in \mathsf{L}^p\} \qquad (1 \le p \le 2) \tag{5.2.36}$$

for an arbitrary complex-valued (continuous) function $\psi(v)$ on $\mathbb{R}$. Then together with Theorem 5.1.16,

Theorem 5.2.21. *Let $f \in \mathsf{L}^p$, $1 \le p \le 2$. The following assertions are equivalent:*

(i) $f \in \mathsf{W}_{\mathsf{L}^p}^r$, (ii) $f \in \mathsf{W}[\mathsf{L}^p; (iv)^r]$,

(iii) *there exists $g \in \mathsf{L}^p$ such that the representation* (5.2.35) *holds, each of the iterated integrals existing (for $1 < p \le 2$ only conditionally) as a function in L^p.*

We again have the following consequence: If $f \in \mathsf{L}^p$, $1 \le p \le 2$, is r-times continuously differentiable with $f^{(r)} \in \mathsf{L}^p$, then all derivatives $f^{(k)}$, $1 \le k \le r-1$, belong to L^p.

5.2.6 Theorem of Plancherel

We return to the inversion problem for Fourier transforms. We obtained the original function $f \in \mathsf{L}^p$, $1 < p \le 2$, from the Fourier transform $f^\wedge$ by summation of the Fourier inversion integral. This led to the inversion formulae (5.2.22) and (5.2.24) which are also valid in L^1. It is a very important feature of the theory of Fourier transforms in L^p, $1 < p \le 2$, that we may interpret the integral (5.2.19) in the following, more direct fashion which is completely symmetrical to Def. 5.2.3, 5.2.8:

$$\lim_{\rho \to \infty} \left\| f(\circ) - \frac{1}{\sqrt{2\pi}} \int_{-\rho}^{\rho} f^\wedge(v)\, e^{i\circ v}\,dv \right\|_p = 0. \tag{5.2.37}$$

In this section we shall prove (5.2.37) only in case $p = 2$, whereas the proof for the cases $1 < p < 2$ is essentially dependent upon the theory of Hilbert transforms and thus left to Sec. 8.3.4.

Theorem 5.2.22. *If $f \in \mathsf{L}^2$, then*

$$\lim_{\rho \to \infty} \left\| f(\circ) - \frac{1}{\sqrt{2\pi}} \int_{-\rho}^{\rho} f^\wedge(v)\, e^{i\circ v}\,dv \right\|_2 = 0. \tag{5.2.38}$$

Proof. In view of Theorem 5.2.4 we have $f^\wedge \in \mathsf{L}^2$ and thus $[\overline{f^\wedge}]^\wedge$ exists as an element of L^2. According to Parseval's formula (5.2.18) and Parseval's equation (5.2.5) the (Hilbert space) product $\|\bar{f} - [\overline{f^\wedge}]^\wedge\|_2^2$ vanishes. Therefore $\overline{f(x)} = [\overline{f^\wedge}]^\wedge(x)$ a.e., and it follows by the definition of the Fourier transform in L^2 that

$$\lim_{\rho\to\infty} \left\| \overline{f(\circ)} - \frac{1}{\sqrt{2\pi}} \int_{-\rho}^{\rho} \overline{f^\wedge(u)}\, e^{-i\circ u}\, du \right\|_2 = 0,$$

which implies (5.2.38) by taking complex conjugates.

In the following we summarize the results of Theorem 5.2.4, 5.2.22 to obtain the *theorem of Plancherel.*

Theorem 5.2.23. *For each $f \in \mathsf{L}^2$ there exists a function $f^\wedge \in \mathsf{L}^2$, called the Fourier transform of f, such that $\|f^\wedge - f_\rho^\wedge\|_2 \to 0$ as $\rho \to \infty$, where f_ρ is defined by* (5.2.3). *Similarly, the relation* (5.2.38) *holds. The functions f and $f^\wedge$ satisfy the Parseval equation $\|f\|_2 = \|f^\wedge\|_2$. Every $f \in \mathsf{L}^2$ is the Fourier transform of a unique element of L^2.*

In other words, the Fourier transform on L^2 is a one-to-one, norm-preserving linear transformation of L^2 onto L^2. Moreover, it preserves inner products, thus is *unitary*, as is shown by

Proposition 5.2.24. *If $f, g \in \mathsf{L}^2$, then $(f, g) = (f^\wedge, g^\wedge)$, i.e.*

$$\int_{-\infty}^{\infty} f(u)\overline{g(u)}\, du = \int_{-\infty}^{\infty} f^\wedge(u)\overline{g^\wedge(u)}\, du. \tag{5.2.39}$$

The proof follows immediately by Parseval's formula (5.2.18). Indeed, since $\overline{g(x)} = [\overline{g^\wedge}]^\wedge(x)$ a.e., we have

$$\int_{-\infty}^{\infty} f(u)\overline{g(u)}\, du = \int_{-\infty}^{\infty} f(u)[\overline{g^\wedge}]^\wedge(u)\, du = \int_{-\infty}^{\infty} f^\wedge(u)\overline{g^\wedge(u)}\, du.$$

Therefore the Fourier transform defines an isomorphism of the Hilbert space L^2 onto itself.

Problems

1. Show that the function $[\sqrt{|x|}\,(1 + |\log|x||)]^{-1}$ belongs to L^2 but not to L^p for any other value of p.
2. Construct examples of functions in order to show that the definition of a Fourier transform as given in this section cannot be extended to L^p-space, $p > 2$. (Hint: Titchmarsh [6, p. 111], Zygmund [7II, p. 258])
3. Prove Prop. 5.2.11.
4. (i) Prove Theorem 5.2.15 and show that

$$\lim_{y\to 0+} \frac{1}{\sqrt{2\pi}} \int_{-\infty}^{\infty} e^{-y|v|} f^\wedge(v)\, e^{ixv}\, dv = f(x) \quad \text{a.e.}$$

(ii) Formulate and prove the counterpart of Theorem 5.2.15 in case $\theta^\wedge$ satisfies the assumptions of Prop. 3.2.4. As a consequence, show that the Fourier inversion integral of $f \in \mathsf{L}^p$, $1 < p \le 2$, is Cesàro summable to $f(x)$ almost everywhere, i.e.

$$\lim_{\rho\to\infty} \frac{1}{\sqrt{2\pi}} \int_{-\rho}^{\rho} \left(1 - \frac{|v|}{\rho}\right) f^\wedge(v)\, e^{ixv}\, dv = f(x) \quad \text{a.e.}$$

5. Show that there are functions $f \in \mathsf{L}^p$, $1 < p \le 2$, such that $f \notin \mathsf{L}^1$ but $f^\wedge \in \mathsf{L}^1$. (Hint: Consider the function $f(x) = (ix)^{-1}(e^{ihx} - 1)$, $h > 0$. Show that $f \in \mathsf{L}^p$, $1 < p \le 2$, and $\sqrt{2\pi} f^\wedge(v) = 1$ for $0 \le v \le h$, $= 0$ for $v < 0$, $v > h$)
6. Let $1 < p < 2$. Show that there exist functions $g \in \mathsf{L}^{p'}$ which are not the Fourier transform of some function in L^p, yet the set $[\mathsf{L}^p]^\wedge \equiv \{g \in \mathsf{L}^{p'} \mid g(v) = f^\wedge(v) \text{ a.e.}, f \in \mathsf{L}^p\}$ is dense in $\mathsf{L}^{p'}$. (Hint: cf. Problem 5.1.3, compare also with Hewitt [1, p. 177])
7. (i) Prove Prop. 5.2.18. Show in particular that if $\chi \in \mathsf{NL}^1$ is such that $\chi^\wedge \in \mathsf{L}^1$, then for the singular integral $J(f; x; \rho)$

$$\frac{\rho}{\sqrt{2\pi}} \int_{-\infty}^{\infty} f(x-u)\chi(\rho u)\, du = \frac{1}{\sqrt{2\pi}} \int_{-\infty}^{\infty} \chi^\wedge\Big(\frac{v}{\rho}\Big) f^\wedge(v)\, e^{ixv}\, dv$$

for every $f \in \mathsf{L}^p$, $1 \le p \le 2$. Apply this to the standard examples of kernels χ.

(ii) Let $f, g \in \mathsf{L}^2$. Show that for all $x \in \mathbb{R}$

$$(f * g)(x) = \frac{1}{\sqrt{2\pi}} \int_{-\infty}^{\infty} f^\wedge(v) g^\wedge(v)\, e^{ixv}\, dv.$$

(Hint: See e.g. Achieser [2, p. 116], Zygmund [7II, p. 253])
8. Use the Parseval equation to evaluate

$$\int_{-\infty}^{\infty} \frac{\sin^2 u}{u^2}\, du = \pi, \qquad \int_{-\infty}^{\infty} \frac{\sin^4 u}{u^4}\, du = \frac{2\pi}{3}.$$

(Hint: Apply (5.2.5) to $f(x) = \kappa_{[-1,1]}(x)$, $f(x) = (1 - |x|)\kappa_{[-1,1]}(x)$, respectively. Compare with Hewitt–Stromberg [1, p. 413 ff], Zygmund [7II, p. 251])
9. Let $f \in \mathsf{L}^p$, $1 < p \le 2$.

(i) Show that for every $x, h \in \mathbb{R}$

$$\int_x^{x+h} f^\wedge(u)\, du = \frac{1}{\sqrt{2\pi}} \int_{-\infty}^{\infty} \frac{e^{-ihu} - 1}{-iu} f(u)\, e^{-ixu}\, du.$$

In particular,

$$f^\wedge(v) = \frac{d}{dv}\Big[\frac{1}{\sqrt{2\pi}} \int_{-\infty}^{\infty} \frac{e^{-ivu} - 1}{-iu} f(u)\, du\Big] \quad \text{a.e.},$$

which may be used to provide a further interpretation of (5.2.1) for functions $f \in \mathsf{L}^p$, $1 < p \le 2$. (Hint: Use Problem 5.1.6(i) and apply (5.2.18) with $g(u) = m((x - u)/h)$, m being given by (5.1.27))

(ii) Show that for every $x, h \in \mathbb{R}$

$$\int_x^{x+h} f(u)\, du = \frac{1}{\sqrt{2\pi}} \int_{-\infty}^{\infty} \frac{e^{ihv} - 1}{iv} f^\wedge(v)\, e^{ixv}\, dv.$$

In particular,

$$f(x) = \frac{d}{dx}\Big[\frac{1}{\sqrt{2\pi}} \int_{-\infty}^{\infty} \frac{e^{ixv} - 1}{iv} f^\wedge(v)\, dv\Big] \quad \text{a.e.},$$

which may be regarded as a further interpretation of the Fourier inversion integral for functions $f \in \mathsf{L}^p$, $1 < p \le 2$. (Hint: Again apply (5.2.18) but with $g(u) = (iu)^{-1}(e^{ihu} - 1)$, cf. Problem 5.2.5. The formulae of this Problem are also derived in e.g. Achieser [2, p. 120], Zygmund [7II, p. 254], Hewitt [1, p. 173 ff], Weinberger [1, p. 314])
10. Let $f \in \mathsf{L}^p$, $1 < p \le 2$, be such that $(ix)f(x) \in \mathsf{L}^p$. Show that $[f^\wedge]'(v)$ exists almost everywhere as a function in $\mathsf{L}^{p'}$ and $[f^\wedge]'(v) = -[(i\circ)f(\circ)]^\wedge(v)$ a.e. (Hint: Show that $f \in \mathsf{L}^1$ and thus $f^\wedge \in \mathsf{C}_0$. Then use the preceding Problem to derive $f^\wedge(v) = -\int_{-\infty}^{v} [(i\circ)f(\circ)]^\wedge(u)\, du$, the integral being only conditionally convergent; see also Bochner–Chandrasekharan [1, p. 126])

11. If $f \in C_{00}$ has a piecewise continuous derivative, show that $f^\wedge \in L^1$ and thus $f(x) = [f^\wedge]^\vee(-x)$. (Hint: Use the proof for $r = 1$ of Prop. 5.2.19, 5.2.20)
12. If $f \in W[L^p; (iv)^r]$ for some $r \in \mathbb{N}$, show that $f \in W[L^p; (iv)^j]$ for $j = 1, 2, \ldots, r - 1$.

5.3 Fourier-Stieltjes Transforms

5.3.1 Fundamental Properties

For $\mu \in BV$ (cf. Sec. 0.5.) the *Fourier–Stieltjes transform* $\mu^\vee(v)$ is defined by

$$\mu^\vee(v) = \frac{1}{\sqrt{2\pi}} \int_{-\infty}^{\infty} e^{-ivu}\, d\mu(u) \equiv [\mu(\circ)]^\vee(v) \qquad (v \in \mathbb{R}). \tag{5.3.1}$$

This integral is absolutely and uniformly convergent for all $v \in \mathbb{R}$ and thus defines a function $\mu^\vee$ which is defined at each point of $\mathbb{R}$. We shall first give some of its operational properties.

Proposition 5.3.1. *For* $\mu \in BV$

(i) $\left[\mu\left(\frac{\circ + h}{\rho}\right)\right]^\vee(v) = e^{ihv}\, \mu^\vee(\rho v) \qquad (\rho > 0,\ h, v \in \mathbb{R}),$

(ii) $[\bar{\mu}]^\vee(-v) = \overline{\mu^\vee(v)} \qquad (v \in \mathbb{R}),$

(iii) *if* μ *is absolutely continuous, i.e.* $\mu(x) = \int_{-\infty}^{x} f(u)\, du$, $f \in L^1$, *then* $\mu^\vee(v) = f^\wedge(v)$ *for all* $v \in \mathbb{R}$.

The proof is left to Problem 5.3.1.

Proposition 5.3.2. *The Fourier–Stieltjes transform defines a bounded linear transformation of* BV *into* C.

Proof. The fact that $\mu^\vee(v)$ is a bounded function on $\mathbb{R}$ follows by

$$|\mu^\vee(v)| \le \|\mu\|_{BV} \qquad (v \in \mathbb{R}). \tag{5.3.2}$$

Furthermore, we have for all $v, h \in \mathbb{R}$

$$\begin{aligned}
|\mu^\vee(v + h) - \mu^\vee(v)| &\le \frac{1}{\sqrt{2\pi}} \int_{-\infty}^{\infty} |e^{-ihu} - 1|\, |d\mu(u)| \\
&= \frac{2}{\sqrt{2\pi}} \left(\int_{|u| \le \rho} + \int_{|u| > \rho} \right) \left| \sin \frac{hu}{2} \right| |d\mu(u)| \\
&\le |h|\, \rho \cdot \frac{1}{\sqrt{2\pi}} \int_{-\rho}^{\rho} |d\mu(u)| + 2 \cdot \frac{1}{\sqrt{2\pi}} \int_{|u| > \rho} |d\mu(u)| \equiv I_1 + I_2,
\end{aligned}$$

say. Given $\varepsilon > 0$, we choose ρ so large that $I_2 < \varepsilon/2$ and then choose h so small that $I_1 < \varepsilon/2$. Thus $\mu^\vee$ is uniformly continuous, and the proof is complete.

We observe that the Riemann–Lebesgue lemma does not hold for Fourier–Stieltjes transforms. Indeed, the Fourier–Stieltjes transform of

$$\delta(x) = \begin{cases} 0, & -\infty < x < 0 \\ \sqrt{\pi/2}, & x = 0 \\ \sqrt{2\pi}, & 0 < x < \infty \end{cases} \tag{5.3.3}$$

is given by

$$\delta^{\vee}(v) = 1 \qquad (v \in \mathbb{R}). \tag{5.3.4}$$

The function δ, often referred to as the *Dirac measure* (or *Heaviside function*), is discontinuous, but there are also continuous functions $\mu \in \mathsf{BV}$ for which $\mu^{\vee}(v)$ does not tend to zero as $|v| \to \infty$.

Proposition 5.3.3. *If* $f \in \mathsf{L}^p$, $1 \le p \le 2$, *and* $\mu \in \mathsf{BV}$, *then*

$$[f * d\mu]^{\wedge}(v) = f^{\wedge}(v)\mu^{\vee}(v) \quad \text{(a.e.)}. \tag{5.3.5}$$

Proof. According to Prop. 0.5.5, $f * d\mu \in \mathsf{L}^p$ and thus both sides of (5.3.5) are meaningful. Let $p = 1$. By Fubini's theorem

$$\begin{aligned}[f * d\mu]^{\wedge}(v) &= \frac{1}{\sqrt{2\pi}} \int_{-\infty}^{\infty} \left\{\frac{1}{\sqrt{2\pi}} \int_{-\infty}^{\infty} f(x-u)\, d\mu(u)\right\} e^{-ivx}\, dx \\ &= \frac{1}{\sqrt{2\pi}} \int_{-\infty}^{\infty} e^{-ivu}\, d\mu(u) \left\{\frac{1}{\sqrt{2\pi}} \int_{-\infty}^{\infty} f(x-u)\, e^{-iv(x-u)}\, dx\right\} \\ &= f^{\wedge}(v)\mu^{\vee}(v)\end{aligned}$$

for all $v \in \mathbb{R}$. If $1 < p \le 2$, let $\{f_n\}$ be a sequence of functions in $\mathsf{L}^1 \cap \mathsf{L}^p$ such that $\lim_{n\to\infty} \|f_n - f\|_p = 0$ (cf. (5.2.12)). Then by the previous case $[f_n * d\mu]^{\wedge}(v) = f_n^{\wedge}(v)\mu^{\vee}(v)$ for each n, and thus by the Minkowski and Titchmarsh inequality

$$\begin{aligned}\|[f * d\mu]^{\wedge} - f^{\wedge}\mu^{\vee}\|_{p'} &\le \|[f * d\mu]^{\wedge} - [f_n * d\mu]^{\wedge}\|_{p'} + \|(f_n^{\wedge} - f^{\wedge})\mu^{\vee}\|_{p'} \\ &\le \|\{f - f_n\} * d\mu\|_p + \|\mu^{\vee}\|_{\infty}\|f_n^{\wedge} - f^{\wedge}\|_{p'} \le 2\,\|\mu\|_{\mathsf{BV}}\|f_n - f\|_p,\end{aligned}$$

which tends to zero as $n \to \infty$. This completes the proof.

Our next aim is to prove the *convolution theorem* for Fourier–Stieltjes transforms. This is the extension of Prop. 5.3.3 (case $p = 1$) with the convolution $f * d\mu$, $f \in \mathsf{L}^1$, $\mu \in \mathsf{BV}$, replaced by the general Stieltjes convolution $\mu * d\nu$ of μ, $\nu \in \mathsf{BV}$. But first the following counterpart of Lemma 4.3.4.

Lemma 5.3.4. *Let* $\mu \in \mathsf{BV}$, $h \in \mathbb{R}$ *and set*

$$q(x) = \mu(x+h) - \mu(x). \tag{5.3.6}$$

Then q *belongs to* $\mathsf{BV} \cap \mathsf{L}^1$ *with* $\lim_{|x|\to\infty} q(x) = 0$ *and*

$$\|\mu(\circ + h) - \mu(\circ)\|_1 \le |h|\, \|\mu\|_{\mathsf{BV}}, \tag{5.3.7}$$

$$q^{\wedge}(v) = \begin{cases} \dfrac{e^{ihv} - 1}{iv}\, \mu^{\vee}(v), & v \ne 0 \\ h\mu^{\vee}(0), & v = 0. \end{cases} \tag{5.3.8}$$

Proof. Since $\mu \in \mathsf{BV}$, q belongs to BV with $\lim_{|x|\to\infty} q(x) = 0$. If m is defined by (5.1.27), we may write for $h > 0$, for example,

$$q(x) = \int_x^{x+h} d\mu(u) = ([m(\circ/h)] * d\mu)(x).$$

Thus q is the convolution of $m(\circ/h) \in \mathsf{L}^1$ with $\mu \in \mathsf{BV}$ and hence belongs to L^1 by Prop. 0.5.5. Moreover, $\|q\|_1 \le \|m(\circ/h)\|_1 \|\mu\|_{\mathsf{BV}}$ by (0.5.9), proving (5.3.7). Finally (5.3.8) follows by (5.1.29), Prop. 5.3.3, and Problem 5.1.1(ii).

Theorem 5.3.5. *For* $\mu, \nu \in \mathsf{BV}$ *one has*

$$[\mu * d\nu]^\vee(v) = \mu^\vee(v)\nu^\vee(v) \qquad (v \in \mathbb{R}). \tag{5.3.9}$$

Proof. By Prop. 0.5.2, $\mu * d\nu \in \mathsf{BV}$ and thus both sides of (5.3.9) are well-defined. Let q be given by (5.3.6). Since $q \in \mathsf{L}^1$, it follows that $q * d\nu \in \mathsf{L}^1$, and by Prop. 5.3.3 and (5.3.8)

$$[q * d\nu]^\wedge(v) = q^\wedge(v)\nu^\vee(v) = \begin{cases} \dfrac{e^{ihv} - 1}{iv}\, \mu^\vee(v)\nu^\vee(v), & v \ne 0 \\ h\mu^\vee(0)\nu^\vee(0), & v = 0. \end{cases} \tag{5.3.10}$$

On the other hand, $(q * d\nu)(x) = (\mu * d\nu)(x + h) - (\mu * d\nu)(x)$. Since the right-hand side is of type (5.3.6), μ being replaced by $\mu * d\nu$, we may apply (5.3.8) to deduce

$$[q * d\nu]^\wedge(v) = \begin{cases} \dfrac{e^{ihv} - 1}{iv}\, [\mu * d\nu]^\vee(v), & v \ne 0 \\ h[\mu * d\nu]^\vee(0), & v = 0, \end{cases}$$

which, together with (5.3.10), proves the theorem.

Next we derive certain types of *Parseval formulae* for the Fourier–Stieltjes transform.

Proposition 5.3.6. (i) *For* $f \in \mathsf{L}^1$, $\mu \in \mathsf{BV}$ *we have*

$$\int_{-\infty}^{\infty} f^\wedge(v)\, d\mu(v) = \int_{-\infty}^{\infty} \mu^\vee(v) f(v)\, dv. \tag{5.3.11}$$

(ii) *If* $\mu, \nu \in \mathsf{BV}$, *then*

$$\int_{-\infty}^{\infty} \nu^\vee(v)\, d\mu(v) = \int_{-\infty}^{\infty} \mu^\vee(v)\, d\nu(v). \tag{5.3.12}$$

Proof. Regarding (i), in virtue of Fubini's theorem

$$\int_{-\infty}^{\infty} f^\wedge(v)\, d\mu(v) = \int_{-\infty}^{\infty} f(u)\, du \left\{\frac{1}{\sqrt{2\pi}} \int_{-\infty}^{\infty} e^{-ivu}\, d\mu(v)\right\} = \int_{-\infty}^{\infty} f(u)\mu^\vee(u)\, du,$$

which establishes (5.3.11). The proof of (5.3.12), which in fact includes (5.3.11), follows similarly.

5.3.2 Inversion Theory

Recalling the definition of the Fourier–Stieltjes transform of $\mu \in \mathsf{BV}$ we might, in view of the results for periodic functions, expect that the formal inversion would be given by

$$\mu'(x) = \frac{1}{\sqrt{2\pi}} \int_{-\infty}^{\infty} e^{ixv} \mu^{\vee}(v)\, dv. \tag{5.3.13}$$

Since the Fourier–Stieltjes transform $\mu^{\vee}$ of $\mu \in \mathsf{BV}$ need not belong to L^1 (cf. (5.3.4)), the Fourier–Stieltjes *inversion integral* (5.3.13) has to be interpreted in some generalized sense.

Proposition 5.3.7. *Let* $\mu \in \mathsf{BV}$. *Then for a* θ*-factor the* θ***-means** *of the integral* (5.3.13), *defined for each* $\rho > 0$ *by*

$$U(d\mu; x; \rho) = \frac{1}{\sqrt{2\pi}} \int_{-\infty}^{\infty} \theta\left(\frac{v}{\rho}\right) e^{ixv} \mu^{\vee}(v)\, dv, \tag{5.3.14}$$

exist for all $x \in \mathbb{R}$, *belong to* L^1 *and satisfy*

$$\| U(d\mu; \circ; \rho)\|_1 \leq \|\theta^{\wedge}\|_1 \|\mu\|_{\mathsf{BV}} \qquad (\rho > 0), \tag{5.3.15}$$

$$\lim_{\rho\to\infty} \int_{-\infty}^{\infty} h(x) U(d\mu; x; \rho)\, dx = \int_{-\infty}^{\infty} h(x)\, d\mu(x) \tag{5.3.16}$$

for every $h \in \mathsf{C}$.

According to (5.3.11) we have for each $\rho > 0$

$$U(d\mu; x; \rho) = \frac{\rho}{\sqrt{2\pi}} \int_{-\infty}^{\infty} \theta^{\wedge}(\rho(x-u))\, d\mu(u), \tag{5.3.17}$$

and hence the assertions follow immediately by Prop. 0.5.5, Lemma 3.1.5, and Problem 3.1.9 upon taking $\chi(x; \rho) = \rho\theta^{\wedge}(\rho x)$.

Thus the Fourier–Stieltjes inversion integral is θ*-summable* to μ in the weak* topology of $(\mathsf{C}_0)^*$. Concerning generalized pointwise convergence of (5.3.13) we have in view of (5.3.17) and Prop. 3.2.6

Proposition 5.3.8. *Let* $\mu \in \mathsf{BV}$. *If for a* θ*-factor* $\theta^{\wedge}$ *is moreover positive and monotonely decreasing on* $[0, \infty)$, *then the Fourier–Stieltjes inversion integral is* θ*-summable* a.e. *to* $\mu'(x)$, *i.e.*

$$\lim_{\rho\to\infty} \frac{1}{\sqrt{2\pi}} \int_{-\infty}^{\infty} \theta\left(\frac{v}{\rho}\right) e^{ixv} \mu^{\vee}(v)\, dv = \mu'(x) \quad \text{a.e.} \tag{5.3.18}$$

Thus, for example, for the Abel-means of (5.3.13) we have (cf. (5.1.8))

$$\lim_{y\to 0+} \frac{1}{\sqrt{2\pi}} \int_{-\infty}^{\infty} e^{-y|v|} e^{ixv} \mu^{\vee}(v)\, dv = \mu'(x) \quad \text{a.e.} \tag{5.3.19}$$

For further results see Problem 5.3.3.

We continue our investigations on the convergence of the Fourier–Stieltjes inversion integral by establishing the *theorem of Lévy*. In connection with (5.3.18) it is natural to look for inversion formulae delivering μ instead of μ'. For this purpose we may integrate (5.3.18), but Problem 5.1.6 then suggests that we may avoid the use of convergence-factors. These heuristic considerations are confirmed by

Theorem 5.3.9. *If* $\mu \in \mathsf{BV}$, *then for any* $h \in \mathbb{R}$

$$\mu(x+h) - \mu(x) = \lim_{\rho\to\infty} \frac{1}{\sqrt{2\pi}} \int_{-\rho}^{\rho} \frac{e^{ihv}-1}{iv} e^{ixv} \mu^{\vee}(v)\, dv \qquad (x \in \mathbb{R}). \tag{5.3.20}$$

Furthermore,

$$\int_0^h [\mu(x+u) - \mu(x-u)]\, du = \frac{2}{\sqrt{2\pi}} \int_{-\infty}^{\infty} \frac{1-\cos hv}{v^2} e^{ixv} \mu^{\vee}(v)\, dv, \tag{5.3.21}$$

the integral being absolutely convergent.

Proof. If we define $q(x)$ by (5.3.6), then $q \in \mathsf{BV} \cap \mathsf{L}^1$ and it follows by Jordan's criterion (cf. Problem 5.1.5) that

$$q(x) = \lim_{\rho\to\infty} \frac{1}{\sqrt{2\pi}} \int_{-\rho}^{\rho} q^{\wedge}(v)\, e^{ixv}\, dv \qquad (x \in \mathbb{R}). \tag{5.3.22}$$

This, in virtue of (5.3.8), implies (5.3.20).

Intuitively, (5.3.21) follows from (5.3.20) by a further integration. For a rigorous proof we introduce for $a > 0$, say, the function

$$q_1(x) = \int_x^{x+a} q(u)\, du = \left(\left\{m\left(\frac{\circ}{a}\right)\right\} * q\right)(x) \tag{5.3.23}$$

which again belongs to $\mathsf{BV} \cap \mathsf{L}^1$. By the convolution theorem for L^1-functions we obtain by (5.1.29) and (5.3.8)

$$q_1^{\wedge}(v) = \begin{cases} \dfrac{e^{iav}-1}{iv}\dfrac{e^{ihv}-1}{iv}\mu^{\vee}(v), & v \neq 0 \\ ah\mu^{\vee}(0), & v = 0. \end{cases} \tag{5.3.24}$$

If we apply (5.3.22) to q_1 and set $a = h$, then

$$\begin{aligned} \int_0^h [\mu(x+h+u) - \mu(x+u)]\, du &= \lim_{\rho\to\infty} \frac{1}{\sqrt{2\pi}} \int_{-\rho}^{\rho} \left[\frac{e^{ihv}-1}{iv}\right]^2 e^{ixv} \mu^{\vee}(v)\, dv \\ &= \lim_{\rho\to\infty} \frac{2}{\sqrt{2\pi}} \int_{-\rho}^{\rho} \frac{1-\cos hv}{v^2} e^{i(x+h)v} \mu^{\vee}(v)\, dv. \end{aligned}$$

Since $(1-\cos hv)/v^2 \in \mathsf{L}^1$, the integral on the right converges absolutely. Passing to the limit and replacing $x+h$ by x, we obtain

$$\int_0^h [\mu(x+u) - \mu(x-h+u)]\, du = \frac{2}{\sqrt{2\pi}} \int_{-\infty}^{\infty} \frac{1-\cos hv}{v^2} e^{ixv} \mu^{\vee}(v)\, dv,$$

which, after an obvious change of variables, establishes (5.3.21).

Proposition 5.3.10. *Let $\mu \in \mathsf{BV}$ be such that $\mu^\vee \in \mathsf{L}^1$. Then μ is uniformly continuous with derivative $\mu' \in \mathsf{C}_0$ given by* (5.3.13).

Proof. Since $\mu^\vee \in \mathsf{L}^1$ and $(\exp\{ihv\} - 1)/ihv$ is bounded, it follows from (5.3.20) that

$$\frac{\mu(x+h) - \mu(x)}{h} = \frac{1}{\sqrt{2\pi}} \int_{-\infty}^{\infty} \frac{e^{ihv} - 1}{ihv} e^{ixv} \mu^\vee(v)\, dv. \tag{5.3.25}$$

Hence $|\mu(x+h) - \mu(x)| \le |h|\, \|\mu^\vee\|_1$ and μ is uniformly continuous. Moreover, since the integrand is dominated by $|\mu^\vee|$ and converges to $\exp\{ixv\}\mu^\vee(v)$ as $h \to 0$, we may apply Lebesgue's dominated convergence theorem to deduce that $\mu'(x)$ exists for all $x \in \mathbb{R}$ and is given by (5.3.13). By Prop. 5.1.2 this implies $\mu' \in \mathsf{C}_0$.

Now to the *uniqueness theorem* for the Fourier–Stieltjes transform.

Proposition 5.3.11. *If $\mu \in \mathsf{BV}$ and $\mu^\vee(v) \equiv 0$ on $\mathbb{R}$, then $\mu(x) \equiv 0$ on $\mathbb{R}$.*

In other words, if μ_1 and μ_2 are two functions of class BV such that $\mu_1^\vee(v) \equiv \mu_2^\vee(v)$, then $\mu_1(x) \equiv \mu_2(x)$. For the proof we may use Prop. 5.3.10. Indeed, (5.3.13) implies $\mu'(x) \equiv 0$, and thus $\mu(x)$ is identically equal to a constant which must be zero since μ, as an element of BV, is normalized by $\mu(-\infty) = 0$.

Together with Prop. 5.3.2, Theorem 5.3.5, and Problem 5.3.2 we may therefore state

Corollary 5.3.12. *The Fourier–Stieltjes transform defines a one-to-one bounded linear transformation of* BV *into (but not onto)* C. *Moreover, it is an isomorphism of the Banach algebra* BV *(with convolution as multiplication) into the Banach algebra* C *(with pointwise multiplication).*

Note that the Banach algebra BV is not only commutative, but also has a unit element given by (5.3.3), in distinction to the Banach algebra L^1 which has no unit (cf. Prop. 5.1.12).

5.3.3 Fourier–Stieltjes Transforms of Derivatives

Again, in this subsection r always denotes a natural number.

Proposition 5.3.13. *Let $f \in \mathsf{L}^1 \cap \mathsf{AC}_{\text{loc}}^{r-2}$ and $f^{(r-1)} \in \mathsf{BV}$. Then*

$$(iv)^r f^\wedge(v) = [f^{(r-1)}]^\vee(v) \qquad (v \in \mathbb{R}). \tag{5.3.26}$$

Proof. As in the proof of Prop. 5.1.14, integration by parts ($(r-1)$-times) yields

$$\Delta_1^{r-1} f(x) = \int_0^1 \cdots \int_0^1 f^{(r-1)}(x + u_1 + \cdots + u_{r-1})\, du_1 \ldots du_{r-1}. \tag{5.3.27}$$

This implies by (5.1.27)

$$\begin{aligned}\Delta_1^r f(x) &= \int_0^1 du_1 \ldots \int_0^1 du_{r-1} \int_0^1 d_{u_r} f^{(r-1)}(x + u_1 + \cdots + u_r) \\ &= ([m\, *]^r * df^{(r-1)})(x).\end{aligned} \tag{5.3.28}$$

The last term is the convolution of the product (r-times) $m * \cdots * m \in \mathsf{L}^1$ with $f^{(r-1)} \in \mathsf{BV}$. Now it follows by Prop. 5.1.1(i), (5.3.5), and (5.1.29) that

$$(e^{iv} - 1)^r f^\wedge(v) = \left(\frac{e^{iv} - 1}{iv}\right)^r [f^{(r-1)}]^\vee(v)$$

for all $v \neq 0$. This proves (5.3.26) in view of the continuity of all expressions involved.

In order to prove the converse, let us introduce the following classes of functions:

$$\text{(5.3.29)} \quad \mathsf{V}^r_{\mathsf{X}(\mathbb{R})} = \begin{cases} \{f \in \mathsf{C} \mid f \in \mathsf{AC}^{r-1}_{\text{loc}} \cap \mathsf{C}^{r-1}, f^{(r)} \in \mathsf{L}^\infty\} \\ \{f \in \mathsf{L}^1 \mid f = \phi \text{ a.e.}, \phi \in \mathsf{AC}^{r-2}_{\text{loc}}, \phi^{(k)} \in \mathsf{L}^1, 1 \le k \le r-1, \phi^{(r-1)} \in \mathsf{BV}\} \\ \{f \in \mathsf{L}^p \mid f = \phi \text{ a.e.}, \phi \in \mathsf{AC}^{r-1}_{\text{loc}}, \phi^{(k)} \in \mathsf{L}^p, 1 \le k \le r\} \quad (1 < p < \infty). \end{cases}$$

Note that the classes $\mathsf{W}^r_{\mathsf{X}(\mathbb{R})}$ and $\mathsf{V}^r_{\mathsf{X}(\mathbb{R})}$ are equal by definition for the reflexive spaces $\mathsf{X}(\mathbb{R}) = \mathsf{L}^p$, $1 < p < \infty$.

Proposition 5.3.14. *If for $f \in \mathsf{L}^1$ there exists $\mu \in \mathsf{BV}$ such that*

$$\text{(5.3.30)} \qquad (iv)^r f^\wedge(v) = \mu^\vee(v) \qquad (v \in \mathbb{R}),$$

then $f \in \mathsf{V}^r_{\mathsf{L}^1}$.

Proof. Let $r = 1$. With $q(x)$ as defined by (5.3.6) we have for $v \neq 0$

$$q^\wedge(v) = \frac{e^{ihv} - 1}{iv} \mu^\vee(v) = (e^{ihv} - 1) f^\wedge(v) = [f(\circ + h) - f(\circ)]^\wedge(v).$$

Therefore we obtain by the uniqueness theorem of the Fourier transform in L^1 for each $h \in \mathbb{R}$

$$\mu(x + h) - \mu(x) = f(x + h) - f(x) \quad \text{a.e.}$$

On replacing h by $-h$ this implies

$$\int_0^x [\mu(u) - \mu(u - h)]\, du = \int_0^x f(u)\, du - \int_{-h}^{x-h} f(u)\, du$$

for every $x \in \mathbb{R}$. Taking the limit for $h \to \infty$, the right-hand side tends to $\int_0^x f(u)\, du$ since $f \in \mathsf{L}^1$. On the other hand, by Lebesgue's dominated convergence theorem

$$\lim_{h \to \infty} \int_0^x du \left[\int_{u-h}^u d\mu(y)\right] = \int_0^x du \left[\int_{-\infty}^u d\mu(y)\right],$$

and thus $f(x) = \int_{-\infty}^x d\mu(u)$ a.e. or $f(x) = \mu(x)$ a.e.

Let $r \ge 2$. As in the proof of Prop. 5.1.15, $(1 + iv)^{1-r} \in [\mathsf{L}^1]^\wedge$, and hence $(1 + iv)^{1-r}(iv)^r f^\wedge(v) \in [\mathsf{L}^1]^\wedge$ by Prop. 5.3.3 and (5.3.30). This implies $ivf^\wedge(v) \in [\mathsf{L}^1]^\wedge$, i.e., there exists $g_1 \in \mathsf{L}^1$ such that $ivf^\wedge(v) = g_1^\wedge(v)$. Thus the case $r = 1$ of Prop. 5.1.15 applies, giving $f(x) = \int_{-\infty}^x g_1(u)\, du$ a.e. Furthermore, $(iv)^{r-1} g_1^\wedge(v) = \mu^\vee(v)$. If $r > 2$, this in turn implies $ivg_1^\wedge(v) \in [\mathsf{L}^1]^\wedge$, i.e., $ivg_1^\wedge(v) = g_2^\wedge(v)$ for some $g_2 \in \mathsf{L}^1$. Hence

$$f(x) = \int_{-\infty}^x du_1 \int_{-\infty}^{u_1} g_2(u_2)\, du_2 \quad \text{a.e.}$$

Applying this method successively, we obtain a sequence of functions $g_k \in \mathsf{L}^1$, $1 \le k \le r-1$, such that $(iv)^{r-k} g_k^{\wedge}(v) = \mu^{\vee}(v)$ and

$$f(x) = \int_{-\infty}^{x} g_1(u)\,du, \qquad g_k(x) = \int_{-\infty}^{x} g_{k+1}(u)\,du, \quad 1 \le k \le r-2,$$
$$g_{r-1}(x) = \mu(x) \quad \text{a.e.}$$

Therefore $\mu \in \mathsf{L}^1$ and

$$f(x) = \int_{-\infty}^{x} du_1 \int_{-\infty}^{u_1} du_2 \ldots \int_{-\infty}^{u_{r-2}} du_{r-1} \int_{-\infty}^{u_{r-1}} d\mu(u_r) \quad \text{a.e.} \tag{5.3.31}$$

If we abbreviate the right-hand side by $\phi(x)$, then $\phi \in \mathsf{AC}_{\text{loc}}^{r-2}$, $\phi^{(k)} = g_k \in \mathsf{L}^1$, $1 \le k \le r-1$, $\phi^{(r-1)} = \mu \in \mathsf{BV}$, and the proof is complete.

If we introduce the classes of functions

$$\mathsf{V}[\mathsf{L}^p; \psi(v)] = \begin{cases} \{f \in \mathsf{L}^1 \mid \psi(v) f^{\wedge}(v) = \mu^{\vee}(v), \mu \in \mathsf{BV}\} \\ \{f \in \mathsf{L}^p \mid \psi(v) f^{\wedge}(v) = g^{\wedge}(v), g \in \mathsf{L}^p\} \qquad (1 < p \le 2) \end{cases} \tag{5.3.32}$$

for an arbitrary complex-valued (continuous) function $\psi(v)$ on $\mathbb{R}$, the preceding results together with Theorem 5.2.21 may be summarized to

Theorem 5.3.15. *Let $f \in \mathsf{L}^p$, $1 \le p \le 2$. The following assertions are equivalent:*

(i) $f \in \mathsf{V}_{\mathsf{L}^p}^r$, (ii) $f \in \mathsf{V}[\mathsf{L}^p; (iv)^r]$,

(iii) *there exists $\mu \in \mathsf{BV}$ if $p = 1$ and $g \in \mathsf{L}^p$ if $1 < p \le 2$ such that the representations* (5.3.31) *and* (5.2.35) *hold, respectively, each of the iterated integrals existing (for $1 < p \le 2$ only conditionally) as a function in L^p.*

Problems

1. (i) Prove Prop. 5.3.1.
   (ii) The *dipole measure* is the function $\delta_1 \in \mathsf{BV}$ defined by $\delta_1(x) = 0$ for $x < -1$, $x > 0$, $= \sqrt{2\pi}$ for $-1 < x < 0$, $= \sqrt{\pi/2}$ for $x = -1$, $x = 0$. Show that $\delta_1^{\vee}(v) = (\exp\{iv\} - 1)$.
   (iii) The *binomial measure* is the function $\delta_r \in \mathsf{BV}$, $r \in \mathbb{N}$, defined by $\delta_r(x) = 0$ for $x < -r$, $x > 0$, $= \sqrt{2\pi}(-1)^{r-k}\binom{r}{k}$ for $-k < x < -k+1$, $k = 1, \ldots, r$, $= \sqrt{\pi/2}(-1)^{r-k}\binom{r}{k}$ for $x = -k$, $k = 0, 1, \ldots, r$. Show that $\delta_r^{\vee}(v) = (\exp\{iv\} - 1)^r$.
2. Show that there exist functions $g \in \mathsf{C}$ which are not the Fourier–Stieltjes transform of some function in BV. (Hint: Use the example of Problem 5.1.3, see also HEWITT [1, p. 157] for a discussion of uniform approximation on $\mathbb{R}$ of elements $g \in \mathsf{C}$ by Fourier–Stieltjes transforms)
3. (i) Let $\mu \in \mathsf{BV}$. Show that
$$\lim_{t \to 0+} \frac{1}{\sqrt{2\pi}} \int_{-\infty}^{\infty} e^{-tv^2} \mu^{\vee}(v)\, e^{ixv}\, dv = \mu'(x) \quad \text{a.e.}$$
   (ii) State and prove the counterpart of Prop. 5.3.8 for θ-factors for which $\theta^{\wedge}$ satisfies the assumptions of Problem 3.2.4.
   (iii) Let $\mu \in \mathsf{BV}$. Show that
$$\lim_{\rho \to \infty} \frac{1}{\sqrt{2\pi}} \int_{-\rho}^{\rho} \left(1 - \frac{|v|}{\rho}\right) \mu^{\vee}(v)\, e^{ixv}\, dv = \mu'(x) \quad \text{a.e.}$$

4. Let $\mu \in \mathsf{BV}$ and $f \in \mathsf{L}^1$ be such that $f^\wedge \in \mathsf{L}^1$. Show that

$$(f * d\mu)(x) = \frac{1}{\sqrt{2\pi}} \int_{-\infty}^{\infty} f^\wedge(v)\mu^\vee(v)\, e^{ixv}\, dv$$

for all $x \in \mathbb{R}$ and that (see also KATZNELSON [1, p. 132])

$$\int_{-\infty}^{\infty} \overline{f(u)}\, d\mu(u) = \int_{-\infty}^{\infty} \overline{f^\wedge(v)}\mu^\vee(v)\, dv.$$

5. Let $\mu \in \mathsf{BV}$ and suppose that $\mu^\vee(v) = O(|v|^r)$ as $v \to 0$ for some integer $r \geq 2$. Show that

$$\frac{1}{\sqrt{2\pi}} \int_{-\infty}^{\infty} \frac{\mu^\vee(v)}{(iv)^r} e^{ixv}\, dv = \int_{-\infty}^{x} du_1 \int_{-\infty}^{u_1} du_2 \ldots \int_{-\infty}^{u_{r-2}} du_{r-1} \int_{-\infty}^{u_{r-1}} d\mu(u_r).$$

(Hint: COOPER [2])

6. Show that $\mu \in \mathsf{BV}$ is continuous if and only if $\lim_{\rho\to\infty} (1/2\rho) \int_{-\rho}^{\rho} |\mu^\vee(v)|^2\, dv = 0$. (Hint: LUKACS [1, p. 47], KATZNELSON [1, p. 138])
7. If $f \in \mathsf{V}[\mathsf{L}^p; (iv)^r]$ for some $r \in \mathbb{N}$, show that $f \in \mathsf{W}[\mathsf{L}^p; (iv)^j]$ for every $j = 1, 2, \ldots, r-1$.
8. Let $\mu \in \mathsf{BV}$ and $\mu^* \in \mathsf{BV}_{2\pi}$ be given by (3.1.56). Show that the finite Fourier–Stieltjes transform $[\mu^*]^\vee$ is the restriction to the integers of the Fourier–Stieltjes transform $\mu^\vee$, i.e. $[\mu^*]^\vee(k) = \mu^\vee(k)$, $k \in \mathbb{Z}$. (Hint: Use (3.1.57); see also KATZNELSON [1, p. 134])
9. Let $f \in \mathsf{X}(\mathbb{R})$ and $\xi = \mu * dv$, $\mu, \nu \in \mathsf{BV}$. Show that

$$\frac{1}{\sqrt{2\pi}} \int_{-\infty}^{\infty} f(x-u)\, d\xi(u) = \frac{1}{2\pi} \int_{-\infty}^{\infty} \int_{-\infty}^{\infty} f(x-u-y)\, d\mu(u)\, d\nu(y) \quad \text{(a.e.)}.$$

In particular, (0.8.8) holds. (Hint: The result follows for $\mathsf{X}(\mathbb{R}) = \mathsf{L}^1$ by the convolution and uniqueness theorem. For $\mathsf{X}(\mathbb{R}) = \mathsf{L}^p$, $1 < p < \infty$, approximate f in L^p-norm by a sequence $\{f_n\} \subset \mathsf{L}^1 \cap \mathsf{L}^p$ (cf. (5.2.3)). To cover the case $\mathsf{X}(\mathbb{R}) = \mathsf{C}$, let $f \in \mathsf{C}_0$ and take a sequence $\{f_n\} \subset \mathsf{C}_{00}$ which approximates f (cf. Prop. 0.3.4); then it follows in particular for the Weierstrass integral that for every $x \in \mathbb{R}$, $t > 0$

$$W(d\xi; x; t) = \frac{1}{\sqrt{2\pi}} \int_{-\infty}^{\infty} W(d\mu; x-y; t)\, d\nu(y);$$

this together with (3.1.53) gives the assertion for every $f \in \mathsf{C}$)

5.4 Notes and Remarks

The main references to this chapter are the books by BOCHNER [7], WIENER [2], TITCHMARSH [6], CARLEMAN [1], BOCHNER–CHANDRASEKHARAN [1] (see also BURKHARDT [1] and the literature cited there), ZYGMUND [7II, Chapter 16], HEWITT [1], R. R. GOLDBERG [1], WEISS [1], KATZNELSON [1]. Shorter accounts may be found in most of the texts on analysis, see for example the relevant chapters in ACHIESER [2], HEWITT–STROMBERG [1], RUDIN [4], SZ.–NAGY [4], or the survey articles of WEISS [2] and DOETSCH [5].

Sec. 5.1. For the Riemann–Lebesgue lemma see also ACHIESER [2, p. 114] and BOCHNER–CHANDRASEKHARAN [1, p. 3]. In connection with the inversion theory of Sec. 5.1.2, summability of divergent integrals can of course be considered under a more general setting. If the integral $\int_{-\infty}^{\infty} g(v)\, dv$ does not exist in the ordinary (Lebesgue or principal value) sense, various conventional definitions can be employed which assign to it a definite meaning. Such a definition should satisfy the condition of permanency, that is, if the integral exists in the ordinary sense with value l, then the value of the integral in the generalized sense should also equal l. Then the present integral will be said to be θ-summable to the value l if $\lim_{\rho\to\infty}\int_{-\infty}^{\infty}\theta(u/\rho)g(u)\,du = l$, the integral existing for each finite $\rho > 0$. Particular examples of convergence factors are θ_1 (Cesàro) and θ_2 (Abel) of (5.1.8). It follows readily that the Cesàro summability of $\int_{-\infty}^{\infty} g(v)\,dv$ to l implies its Abel summability to l. In this respect see HOBSON [2II, p. 384 ff], HARDY [2, p. 110 ff]. The present results on the Fourier inversion integral follow as easy consequences of the theory on singular integrals of Chapter 3. They are of course standard, see, for example, BOCHNER–CHANDRASEKHARAN [1], HEWITT–STROMBERG [1, p. 400 ff], WEISS [1, p. 15 ff].

There is a theory of trigonometric integrals which corresponds to the theory of trigonometric series considered in Sec. 1.2. If $g \in \mathsf{L}^1(-\rho, \rho)$ for all $\rho > 0$, then $\int_{-\infty}^{\infty} g(u)\, e^{ixu}\,du$ is called a *trigonometric integral*, with value at $x \in \mathbb{R}$ equal to $\lim_{\rho\to\infty}\int_{-\rho}^{\rho} g(u)\, e^{ixu}\,du$ wherever this limit exists. The integral $\int_{-\infty}^{\infty}\{-i\operatorname{sgn} u\}g(u)\,e^{ixu}\,du$ is called the *conjugate integral*. (Note that $g(u)$ is not necessarily a Fourier transform.) Correspondingly, one may consider the integrals

$$\sqrt{2/\pi}\int_0^{\infty} g(u)\cos xu\,du, \qquad \sqrt{2/\pi}\int_0^{\infty} g(u)\sin xu\,du.$$

Compare HOBSON [2II, p. 720 ff], TITCHMARSH [6, p. 152 ff], ACHIESER [2, p. 111 ff], ZYGMUND [7II, p. 244 ff], HEWITT [1, p. 152 ff]. A class of functions for which a theory for the above integrals may be developed is that of locally integrable functions $g(u)$ which tend monotonely to zero as $|u| \to \infty$, see BOCHNER [7, p. 1 ff].

There is an interesting connection between Fourier series and Fourier inversion integrals which often permits derivation of a result for one from the other. Indeed, let $f \in \mathsf{L}^1$, $a \in \mathbb{R}$, and let $f_a(x)$ be the 2π-periodic function equal to $f(x)$ in $[a, a + 2\pi)$. Then

$$\lim_{\rho\to\infty}\left\{\frac{1}{\sqrt{2\pi}}\int_{-\rho}^{\rho} f^\wedge(v)\, e^{ixv}\,dv - \sum_{k=-[\rho]}^{[\rho]} f_a^\wedge(k)\, e^{ikx}\right\} = 0$$

uniformly in x for $x \in [a + \varepsilon, a + 2\pi - \varepsilon]$, $\varepsilon > 0$. The assertion remains valid if the hypothesis $f \in \mathsf{L}^1$ is replaced by $f(x)/(1 + |x|) \in \mathsf{L}^1$, and a similar result also holds for conjugate Fourier series and integrals. See ZYGMUND [7II, p. 242], HEWITT [1, p. 180]. A further connection which links the theory of Fourier integrals to that of Fourier series is given by the results of Sec. 3.1.2 and 5.1.5 (cf. Problem 5.1.16). For a partial converse of the latter see Prop. 6.1.10.

Prop. 5.1.11 on the uniqueness of Fourier transforms can also be deduced from the parallel result for the finite transform, see SZ.–NAGY [5, p. 316] (for further methods compare Problem 5.1.16 or BOCHNER–CHANDRASEKHARAN [1, p. 11]). It can moreover be refined considerably. OFFORD [2, 3] showed that if $f \in \mathsf{L}^1_{\text{loc}}$ and $\lim_{\rho\to\infty}\int_{-\rho}^{\rho} f(u)\, e^{-ivu}\,du = 0$ for all $v \in \mathbb{R}$, then $f(x) = 0$ a.e. He was even able to weaken the hypothesis to

$$\lim_{\rho\to\infty}\int_{-\rho}^{\rho}\left(1 - \frac{|u|}{\rho}\right) f(u)\, e^{-ivu}\,du = 0$$

for all $v \in \mathbb{R}$; compare TITCHMARSH [6, p. 164 f].

In connection with Prop. 5.1.12, although the Banach algebra L^1 has no unit element, every $f \in \mathsf{L}^1$ can be factorized into a convolution product $f_1 * f_2$ with $f_1, f_2 \in \mathsf{L}^1$. See RUDIN [1, 2], and also [4, p. 192 ff] for the representation of the complex homomorphism on L^1.

An important topic not considered here is Wiener's theorem on the closure of translates in L^1 (WIENER [2, p. 97]). It has significant applications to Wiener's Tauberian theorem and enables a ready proof for the prime number theorem; it is treated in almost every book on Fourier analysis. The reader is referred to HARDY [2], WIDDER [1], BOCHNER–CHANDRASEKHARAN [1] for the applications; we here follow R. R. GOLDBERG [1, p. 32 ff] for a brief formulation of the problems and results (see also ACHIESER [2, p. 150 ff], REITER [1, p. 8 ff], EDWARDS [1II, p. 6 ff]):

Suppose $f \in \mathsf{L}^1$. Let T_f denote the set of all $g \in \mathsf{L}^1$ such that g is a finite linear combination of translates of f. That is, g belongs to T_f if $g(x) = \sum \alpha_k f(x + c_k)$ for some finite set of real c_k and complex α_k. The theorem of Wiener asserts: For $f \in \mathsf{L}^1$ the closure of T_f in the L^1-topology, namely $\overline{\mathsf{T}_f}$, is equal to all of L^1 if and only if $f^\wedge(v) \neq 0$ for $v \in \mathbb{R}$. This is connected with the theory of *translation invariant subspaces*, initiated by BEURLING [2]. A (closed linear) subspace M of L^1 is said to be translation invariant if $f \in \mathsf{M}$ implies that every translate of f is also in M. Then M is a translation invariant subspace if and only if M is a closed ideal of L^1 (ideal means: M is an algebra with respect to the operations of L^1 and $g * h \in \mathsf{M}$ whenever $g \in \mathsf{M}$, $h \in \mathsf{L}^1$). Setting out to determine what are the closed maximal ideals in L^1, let M_λ, $\lambda \in \mathbb{R}$, be the set of all $f \in \mathsf{L}^1$ such that $f^\wedge(\lambda) = 0$. If M is any closed maximal ideal of L^1, then $\mathsf{M} = \mathsf{M}_\lambda$ for some λ. A further interesting problem is posed if we assume that $\overline{\mathsf{T}_f}$ is a proper ideal for some $f \in \mathsf{L}^1$. Then the question is: which $f \in \mathsf{L}^1$ have the property that $\overline{\mathsf{T}_f}$ is precisely the intersection of the maximal ideals containing it? This problem, in a reformulation involving bounded functions, is often referred to as the problem of spectral synthesis. For the bounded function formulation see H. POLLARD [1] and the literature cited there (see also KATZNELSON [1, p. 159 ff]). For results in L^2 compare the notes and remarks to Sec. 5.2.

A further classical result of Wiener not treated here is concerned with analytic functions of Fourier transforms (cf. R. R. GOLDBERG [1, p. 26 ff]). As a particular result of this theory we cite: Let $\mathbb{E}$ be a compact set of $\mathbb{R}$. If $f \in \mathsf{L}^1$ and $f^\wedge(v) \neq 0$ for $v \in \mathbb{E}$, then there exists $g \in \mathsf{L}^1$ such that $1/f^\wedge(v) = g^\wedge(v)$ for $v \in \mathbb{E}$.

For Fourier transforms in the complex domain we refer to PALAY–WIENER [1], TITCHMARSH [6], CARLEMAN [1].

For Sec. 5.1.3 see BOCHNER–CHANDRASEKHARAN [1, p. 7 ff]. Prop. 5.1.14 is commonly treated (cf. RUDIN [4, p. 182]). The proof of case $r = 1$ of Prop. 5.1.15 makes use of ideas of COOPER [2] as modified by BERENS (unpublished, see also NESSEL [2, p. 102 ff]). The proof for $r \geq 2$ is taken from BOCHNER–CHANDRASEKHARAN [1, p. 28].

For the material of Sec. 5.1.4 we refer to books on probability theory, for example, CRAMÉR [2, p. 89 ff], LINNIK [1, p. 49], and LUKACS [1, p. 27 ff]; it need not be emphasized that such coverage is often given from a somewhat different point of view. For Prop. 5.1.17–5.1.19 compare also BOCHNER [6, p. 70 ff]. For the connection between the existence of symmetric moments $\lim_{\rho\to\infty} \int_{-\rho}^{\rho} u^k f(u)\, du$ and Riemann derivatives of $f^\wedge$ we refer to ZYGMUND [4], see also PITMAN [1]. Concerning the literature to generalized derivatives see BUTZER–BERENS [1, Sec. 2.2]. The (second) Riemann derivative plays a fundamental rôle in RIEMANN's Habilitationsschrift [1] on the theory of trigonometric series (see the account in ZYGMUND [7I, Chapter 9] or BARI [1I, p. 192 ff]). The notion of a Peano derivative was introduced by PEANO [1, pp. 204–209]. For Lemma 5.1.25 see also OLIVER [1].

For Sec. 5.1.5 on the Poisson summation formula see BOCHNER [7, p. 39 ff], [6, p. 19, 30 ff]. ZYGMUND [7I, p. 68, 160], ACHIESER [2, p. 126 ff], FELLER [2II, p. 592 f], KATZNELSON [1, p. 128] (also GOOD [1]). For applications of Prop. 5.1.30, due to W. H. YOUNG and G. H. HARDY, to singular integrals see ACHIESER [2, p. 137], BUTZER [6]. Formula (5.1.61) is the well-known transformation formula for the theta funtion $\theta_3(x, t)$; see BELLMAN [1], DOETSCH [4].

Sec. 5.2. The L^2-theory was first given by M. PLANCHEREL in 1910–15; the approach of Sec. 5.2.1 is due to F. RIESZ [2]. There are other approaches due to TITCHMARSH, BOCHNER, and WIENER; see the literature cited in TITCHMARSH [6, p. 69 ff]. The method of Wiener proceeds via Hermite functions, see WIENER [2, Chapter 1] or SZ.-NAGY [5, p. 349 ff]. For a generalization of the Plancherel theory due to WATSON compare ACHIESER [2, p. 117 ff], BOCHNER–CHANDRASEKHARAN [1, Chapter 5].

Concerning the L^p-theory for $1 < p < 2$ which depends upon the M. Riesz–Thorin interpolation theorem we mention especially WEISS [2, p. 168 ff], ZYGMUND [7II, p. 254 ff], and KATZNELSON [1, p. 141 ff]. The inequality (5.2.16), due to TITCHMARSH [1], is also often referred to as the Hausdorff–Young inequality.

The present method cannot be used to define a Fourier transform on L^p, $p > 2$ (see TITCHMARSH [6, p. 111]). However, there are several different ways to define the Fourier transform on L^p for $p > 2$, namely by a method of OFFORD [1], Bochner–Wiener's *generalized harmonic analysis*, and via distribution theory. The latter is the most reasonable way; one uses Parseval's formulae of type (5.1.5), that is, duality, to define the Fourier transform of tempered distributions. Details are to be found in any text on distribution theory, for example, SCHWARTZ [1, 3], ZEMANIAN [1], BREMERMAN [1], DONOGHUE [1]. See also KATZNELSON [1, p. 146 ff] or Vol. II of this treatise. Whereas the Offord approach is classical, having the Cesàro means of the Fourier inversion integral as its starting point, that of Bochner–Wiener may be regarded as a precursor to the distributional approach (see the review by BOCHNER [5] of L. SCHWARTZ's treatise on distributions). Let the set F_r be defined by $F_r = \{f \in L^1_{loc} | f(x)(1 + |x|^r)^{-1} \in L^1\}$, $r \in \mathbb{P}$. Obviously, $F_r \subset F_{r+1}$ and $X_{2\pi}, X(\mathbb{R}) \subset F_2$, for example. For $f \in F_r$ the Fourier transform is then defined by

$$\frac{1}{\sqrt{2\pi}} \int_{-1}^{1} f(u) \frac{e^{-ivu} - \sum_{k=0}^{r-1} (-ivu)^k/k!}{(-iu)^r}\, du + \frac{1}{\sqrt{2\pi}} \left(\int_{-\infty}^{-1} + \int_{1}^{\infty} \right) \frac{f(u)}{(-iu)^r} e^{-ivu}\, du.$$

This transform is uniquely determined up to an algebraic polynomial of degree $(r - 1)$ at most. After suitable modifications, many of the operational rules of the ordinary Fourier transform are transferable to this generalized one. Thus a uniqueness theorem holds, a convolution is introduced and transforms of derivatives are considered. However, for all details we refer to BOCHNER [7, Chapter 6], WIENER [2, Chapter 4]; see also MASANI [1] (and the literature cited there), particularly for the connection with filter and prediction theory.

For the operational properties of the Fourier transform in L^p-space as given in Sec. 5.2.3 we cite TITCHMARSH [6, Chapter 4]. For Theorem 5.2.12 see also WEISS [1, p. 32]. The inversion theory of Sec. 5.2.4, despite of its independent interest, is not as necessary as in L^1-space. This is due to the standard theorems of Plancherel ($p = 2$) of Sec. 5.2.6 and of Hille–Tamarkin ($1 < p < 2$) of Sec. 8.3.4. For Sec. 5.2.5 see the references given to Sec. 5.1.3.

For translation invariant subspaces in L^2 see WIENER [2, p. 100 ff], BOCHNER–CHANDRASEKHARAN [1, p. 148 ff], RUDIN [4, p. 190 ff], for example. One result states that, if $f \in L^2$ and $f^\wedge(v) \neq 0$ almost everywhere, then every $g \in L^2$ belongs to the L^2-closure of T_f. For more on invariant subspaces see HELSON [1].

Sec. 5.3. Standard references to the material of this section are texts on probability theory such as LÉVY [2], CRAMÉR [2], FELLER [2], LOÈVE [1], BOCHNER [6], LUKACS [1], LINNIK [1], EISEN [1]; nevertheless, see also the accounts in BOCHNER [7, Chapter 4], ZYGMUND [7II, p. 258 ff], KATZNELSON [1, p. 131 ff].

There exist continuous functions $\mu \in BV$ such that $\lim_{|v| \to \infty} \mu^\vee(v) \neq 0$, see e.g. ZYGMUND [7II, p. 259]. The notation 'Dirac measure' for the function δ of (5.3.3) will be cleared up in Vol. II while introducing the 'δ-distribution'. For Theorem 5.3.5 see also WIDDER [1, p. 252]. The summation of the integral (5.3.13) by a θ-factor is suggested by L^1-theory (compare with BOCHNER [6, p. 23]). The results follow again as easy consequences of those of Chapter 3 on singular integrals. Concerning the theorem of LÉVY (Theorem 5.3.9) we mention, among others, ZYGMUND [7II, p. 260], LUKACS [1, p. 38]. For Prop. 5.3.10 see LUKACS [1, p. 40]. There are other proofs of the uniqueness theorem; for example, LUKACS [1, p. 35] uses the Weierstrass approximation theorem. For the results of Sec. 5.3.3 we refer to COOPER [2], BUTZER–TREBELS [2, p. 36 ff]. As to Sec. 5.1.3, 5.2.5, the treatment in BOCHNER–CHANDRASEKHARAN [1] supplied motivation. Finally, for global and local divisibility in the Wiener ring of Fourier–Stieltjes transforms we refer to Chapter 13 of this volume.

6

Representation Theorems

6.0 Introduction

Suppose we are given a trigonometric series

$$\sum_{k=-\infty}^{\infty} f(k)e^{ikx} \tag{6.0.1}$$

with arbitrary complex coefficients $f(k)$. How can one tell whether this series is the Fourier series of an $\mathsf{X}_{2\pi}$-function, in other words, whether the numbers $f(k)$ are the Fourier coefficients $g^{\wedge}(k)$ of some function $g \in \mathsf{X}_{2\pi}$? The problem may be restated as follows: Given an arbitrary function f on $\mathbb{Z}$, to determine conditions under which f admits a representation as the Fourier transform $g^{\wedge}$ of some function $g \in \mathsf{X}_{2\pi}$ or as the finite Fourier–Stieltjes transform of some $\mu \in \mathsf{BV}_{2\pi}$. We have seen that f on $\mathbb{Z}$ has to satisfy certain necessary conditions in order to be the finite Fourier or Fourier–Stieltjes transform. Thus f on $\mathbb{Z}$ must be bounded in view of (4.1.2) and Prop. 4.3.2(ii). On the other hand, for $\mathsf{L}^2_{2\pi}$ we already know that a necessary and sufficient condition for a function f on $\mathbb{Z}$ to be the finite Fourier transform of some $g \in \mathsf{L}^2_{2\pi}$ is that $f \in \mathsf{l}^2$. This is a consequence of the Parseval equation and the theorem of Riesz–Fischer. But nothing as simple seems possible for other classes such as $\mathsf{L}^p_{2\pi}$, $p \neq 2$.

However, if we consider the inversion formulae involving θ-factors (see Sec. 4.1.2, 4.3.2), it is possible to give a reasonably satisfactory answer to the representation problem that includes necessary and sufficient conditions. Suitable sufficient conditions are also important. As a first contribution we recall Prop. 4.2.7.

From the parallel point of view, it is frequently of interest to decide whether a given, complex-valued function $f(v)$ on the line group $\mathbb{R}$ is, or is not, the Fourier–Stieltjes transform of some $\mu \in \mathsf{BV}$ or the Fourier transform of some $g \in \mathsf{L}^p$, $1 \leq p \leq 2$. For $p = 2$ this problem is again completely solved. Indeed, the theorem of Plancherel states that f is the Fourier transform of some $g \in \mathsf{L}^2$ if and only if $f \in \mathsf{L}^2$. In the other cases we have only established certain necessary conditions that f has to satisfy. Thus Prop. 5.3.2 asserts that every Fourier–Stieltjes transform is bounded, while Prop.

5.1.2, Theorems 5.2.4, 5.2.9 state that the Fourier transform of some $g \in \mathsf{L}^p$, $1 \le p \le 2$, is necessarily of class $\mathsf{L}^{p'}$. But it will be seen that the results of Sec. 5.1.2, 5.2.4, and 5.3.2 on inversion theory again provide a method to solve the representation problem.

Chapter 6 is composed of five sections. Each section treats a definite problem on the circle group and the corresponding one on the line group. Since the material on the circle group is rather standard, it will mainly serve as motivation for the counterparts on the line group. The latter are treated in detail, the results on the circle group being sometimes relegated to the Problems.

Sec. 6.1.1 is concerned with classical necessary and sufficient conditions for representation of sequences as finite Fourier transforms, mainly due to W. H. and G. C. YOUNG, W. GROSS, and H. STEINHAUS. Sec. 6.1.2 deals with the analogous problem on the line group, including results of H. CRAMÉR; of further interest is a certain converse to the Poisson summation formula (Prop. 6.1.10). Sec. 6.2 is reserved to another type of characterization of Fourier–Stieltjes transforms among all bounded and uniformly continuous functions on $\mathbb{R}$, due to S. BOCHNER; Theorem 6.2.3 is concerned with one aspect of the so-called continuity theorem for Fourier–Stieltjes transforms. As a matter of fact, the necessary and sufficient conditions for sequences or functions to be Fourier transforms are somewhat involved (except for the case $p = 2$). Therefore, apart from the principal importance of these results in theoretical problems, we shall treat further conditions for representation in Sec. 6.3. Although these are only sufficient, they are more readily applicable to special sequences or functions. They depend upon results on convex and quasi-convex functions defined on $\mathbb{Z}$ and $\mathbb{R}$ which are established in Sec. 6.3.1. The Pólya-type characterization of Theorem 6.3.11 deserves particular mention. Sec. 6.3.4 discusses an important reduction theorem, resting upon a lemma due to E. M. STEIN. It states that if for some $\alpha > 0$ and $f \in \mathsf{L}^1$ there exists $\mu \in \mathsf{BV}$ such that $|v|^\alpha f^\wedge(v) = \mu^\vee(v)$, then for every $0 < \beta < \alpha$ there exists $g_\beta \in \mathsf{L}^1$ such that $|v|^\beta f^\wedge(v) = g_\beta^\wedge(v)$. The first applications of Theorem 6.3.11 are to be found in Sec. 6.4. There it is shown that a general class of means of Fourier inversion integrals and Fourier series are representable as singular integrals. Sec. 6.5 gives a brief account on the representation of multipliers, including a theorem of K. DE LEEUW on multipliers of type $(\mathsf{L}^p, \mathsf{L}^p)$, $1 < p < 2$.

6.1 Necessary and Sufficient Conditions

6.1.1 Representation of Sequences as Finite Fourier or Fourier–Stieltjes Transforms

Given the finite Fourier transform $g^\wedge$ of some $g \in \mathsf{X}_{2\pi}$, the inversion formulae (4.1.19) and (4.1.20) show how to recapture the original function g. Thus starting off with an arbitrary $f \in \mathsf{l}^\infty$, the question arises whether there exist necessary and sufficient conditions upon the *means*

$$u_\rho(x) = \sum_{k=-\infty}^{\infty} \theta_\rho(k) f(k)\, e^{ikx} \tag{6.1.1}$$

which guarantee the existence of a function g, belonging to some definite class, for which the finite Fourier transform $g^\wedge$ is equal to f.

Theorem 6.1.1. *Let $f \in \mathsf{l}^\infty$, and let $\{\theta_\rho(k)\}$ be a θ-factor such that $\{\theta_\rho^\wedge(x)\}$ is an approximate identity. Then the condition*

$$\Big\| \sum_{k=-\infty}^{\infty} \theta_\rho(k) f(k)\, e^{ik\circ} \Big\|_p = O(1) \qquad (\rho \to \rho_0) \tag{6.1.2}$$

is necessary and sufficient such that f is the finite Fourier–Stieltjes transform of a function $\mu \in \mathsf{BV}_{2\pi}$ if $p = 1$, and f is the finite Fourier transform of a function $g \in \mathsf{L}_{2\pi}^p$ if $1 < p \le \infty$.

Proof. *Necessity.* Let $1 < p \le \infty$. If f admits a representation as the finite Fourier transform of some $g \in \mathsf{L}_{2\pi}^p$, thus

$$f(k) = g^\wedge(k) \equiv \frac{1}{2\pi} \int_{-\pi}^{\pi} g(u)\, e^{-iku}\, du \qquad (k \in \mathbb{Z}),$$

then, according to (4.1.16) and (4.1.17),

$$u_\rho(x) = \sum_{k=-\infty}^{\infty} \theta_\rho(k) g^\wedge(k)\, e^{ikx} \equiv U_\rho(g; x) = \frac{1}{2\pi} \int_{-\pi}^{\pi} g(x-u) \theta_\rho^\wedge(u)\, du, \tag{6.1.3}$$

where $\theta_\rho^\wedge$ is defined by (4.1.18). Hence by (1.1.4)

$$\Big\| \sum_{k=-\infty}^{\infty} \theta_\rho(k) f(k)\, e^{ik\circ} \Big\|_p \le \|\theta_\rho^\wedge\|_1 \|g\|_p,$$

which proves (6.1.2) since $\{\theta_\rho^\wedge(x)\}$ is an approximate identity (cf. (1.1.5)).

If $p = 1$ and thus $f(k) = \mu^\vee(k)$, $k \in \mathbb{Z}$, for some $\mu \in \mathsf{BV}_{2\pi}$, then by (4.3.12)

$$u_\rho(x) = \sum_{k=-\infty}^{\infty} \theta_\rho(k) \mu^\vee(k)\, e^{ikx} \equiv U_\rho(d\mu; x) = \frac{1}{2\pi} \int_{-\pi}^{\pi} \theta_\rho^\wedge(x-u)\, d\mu(u), \tag{6.1.4}$$

and hence by (1.1.13) and (1.1.5)

$$\Big\| \sum_{k=-\infty}^{\infty} \theta_\rho(k) f(k)\, e^{ik\circ} \Big\|_1 \le \|\theta_\rho^\wedge\|_1 \|\mu\|_{\mathsf{BV}_{2\pi}} = O(1) \qquad (\rho \to \rho_0).$$

Sufficiency. Since $f \in \mathsf{l}^\infty$ and (cf. (1.2.28)) $\theta_\rho \in \mathsf{l}^1$, $\rho \in \mathbb{A}$, the series (6.1.1) is absolutely and uniformly convergent, and thus defines, for each $\rho \in \mathbb{A}$, a function $u_\rho \in \mathsf{C}_{2\pi}$ for which $u_\rho^\wedge(k) = \theta_\rho(k) f(k)$, $k \in \mathbb{Z}$. Furthermore, since $\{\theta_\rho^\wedge(x)\}$ is an approximate identity, $\lim_{\rho \to \rho_0} \theta_\rho(k) = 1$, $k \in \mathbb{Z}$, by Problem 1.2.14(iv). Therefore in any case

$$\lim_{\rho \to \rho_0} \frac{1}{2\pi} \int_{-\pi}^{\pi} u_\rho(x)\, e^{-ikx}\, dx = f(k) \qquad (k \in \mathbb{Z}). \tag{6.1.5}$$

On the other hand, if $p = 1$ and if we set $\mu_\rho(x) = \int_{-\pi}^{x} u_\rho(v)\, dv$, this defines a family of absolutely continuous functions $\mu_\rho(x)$ the total variation (over $[-\pi, \pi]$) of which is uniformly bounded as $\rho \to \rho_0$ according to (6.1.2). Moreover, $\mu_\rho(x + 2\pi) - \mu_\rho(x) = 2\pi f(0)$ for all x and $\rho \in \mathbb{A}$ by (1.2.28). Therefore by Prop. 0.8.13 there exists $\{\rho_j\} \subset \mathbb{A}$ with $\lim_{j \to \infty} \rho_j = \rho_0$ and $\mu \in \mathsf{BV}_{2\pi}$ such that

$$\lim_{j \to \infty} \int_{-\pi}^{\pi} h(x) u_{\rho_j}(x)\, dx = \int_{-\pi}^{\pi} h(x)\, d\mu(x)$$

for every $h \in C_{2\pi}$. In particular, for $h(x) = (1/2\pi)\exp\{-ikx\}$, $k \in \mathbb{Z}$, we obtain $\lim_{j\to\infty} \theta_{\rho_j}(k)f(k) = \mu^\vee(k)$, $k \in \mathbb{Z}$, which, together with (6.1.5), proves the theorem for $p = 1$.

If $1 < p \le \infty$, we apply Prop. 0.8.12 and obtain $\{\rho_j\} \subset \mathbb{A}$ with $\lim_{j\to\infty} \rho_j = \rho_0$ and $g \in L^p_{2\pi}$ such that

$$\lim_{j\to\infty} \int_{-\pi}^{\pi} h(x)u_{\rho_j}(x)\,dx = \int_{-\pi}^{\pi} h(x)g(x)\,dx$$

for every $h \in L^{p'}_{2\pi}$. Again the specialization for $h(x) = (1/2\pi)\exp\{-ikx\}k \in \mathbb{Z}$, leads to $\lim_{j\to\infty} \theta_{\rho_j}(k)f(k) = g^\wedge(k)$, $k \in \mathbb{Z}$, which, according to (6.1.5), proves the required representation for $1 < p \le \infty$.

In the preceding theorem it has been *a priori* assumed that the function f to be represented should be bounded, i.e. $f \in l^\infty$. But this assumption has only been used to ensure the absolute and uniform convergence of the sum (6.1.1) which defines $u_\rho(x)$. Thus, if we consider row-finite θ-factors (cf. Sec. 1.2.5), no boundedness hypothesis is needed. In particular, for the Fejér means

$$\sigma_n(x) = \sum_{k=-n}^{n} \left(1 - \frac{|k|}{n+1}\right) f(k)\, e^{ikx} \tag{6.1.6}$$

Theorem 6.1.2. *An arbitrary (complex-valued) function f on $\mathbb{Z}$ is the finite Fourier transform of a function g belonging to*

(i) $C_{2\pi} \Leftrightarrow \|\sigma_n(\circ) - \sigma_m(\circ)\|_{C_{2\pi}} = o(1)$ $(m, n \to \infty)$,

(ii) $L^1_{2\pi} \Leftrightarrow \|\sigma_n(\circ) - \sigma_m(\circ)\|_1 = o(1)$ $(m, n \to \infty)$,

(iii) $L^p_{2\pi}$, $1 < p \le \infty \Leftrightarrow \|\sigma_n(\circ)\|_p = O(1)$ $(n \to \infty)$.

A function f on $\mathbb{Z}$ is the finite Fourier–Stieltjes transform of a function μ belonging to

(iv) $BV_{2\pi} \Leftrightarrow \|\sigma_n(\circ)\|_1 = O(1)$ $(n \to \infty)$.

We only sketch the proof. If f is the finite Fourier transform of some $g \in X_{2\pi}$, then it follows by (6.1.3) and (4.1.19) that

$$\lim_{n\to\infty} \left\| \sum_{k=-n}^{n} \left(1 - \frac{|k|}{n+1}\right) f(k)\, e^{ik\circ} - g(\circ) \right\|_{X_{2\pi}} = 0.$$

On the other hand, if the Fejér means $\sigma_n(x)$ as defined by (6.1.6) form a Cauchy sequence in $X_{2\pi}$, they converge to a function $g \in X_{2\pi}$ in view of the completeness of $X_{2\pi}$. Together with Problem 4.1.1(ii) and (6.1.5) this shows $f(k) = g^\wedge(k)$, $k \in \mathbb{Z}$.

Of course, if the given function f on $\mathbb{Z}$ satisfies one of the conditions of Theorem 6.1.2, then it in particular follows that f is bounded. But sometimes it is of advantage to decide whether a given f admits a representation without knowing explicitly that f is bounded.

Indeed, to decide whether a given $f \in X_{2\pi}$ belongs to the class $W[X_{2\pi}; \psi(k)]$ (cf. (4.1.24) for the definition) we have to examine whether for a fixed but arbitrary complex-valued function ψ (on $\mathbb{Z}$) $\psi(k)f^\wedge(k)$ is the finite Fourier transform of some $g \in X_{2\pi}$. An example of interest (cf. Theorem 4.1.10) is $\psi(k) = (ik)^r$, $r \in \mathbb{N}$, which is in fact unbounded on $\mathbb{Z}$. Nevertheless, if f belongs to $W[X_{2\pi}; (ik)^r]$, then $(ik)^r f^\wedge(k)$ is bounded on $\mathbb{Z}$.

We conclude with a characterization of the class $V[X_{2\pi}; \psi(k)]$, introduced by (4.3.19).

Corollary 6.1.3. *Let* $f \in X_{2\pi}$. *The following assertions are equivalent:*

(i) $f \in V[X_{2\pi}; \psi(k)]$,

(ii) $\left\| \sum_{k=-n}^{n} \left(1 - \frac{|k|}{n+1}\right)\psi(k) f^\wedge(k)\, e^{ik\circ} \right\|_{X_{2\pi}} = O(1) \qquad (n \to \infty).$

In particular, for $\psi(k) = (ik)^r$, $r \in \mathbb{N}$, we have by Theorem 4.3.13

Corollary 6.1.4. *Let* $f \in X_{2\pi}$. *The following assertions are equivalent:*

(i) $f \in V^r_{X_{2\pi}}$, (ii) $f \in V[X_{2\pi}; (ik)^r]$, (iii) $\|\sigma_n^{(r)}(f; \circ)\|_{X_{2\pi}} = O(1)$,

(iv) $\left\| \sum_{k=-n}^{n} \left(1 - \frac{|k|}{n+1}\right)(ik)^r f^\wedge(k)\, e^{ik\circ} \right\|_{X_{2\pi}} = O(1) \qquad (n \to \infty).$

Concerning the corresponding results for the classes $W[X_{2\pi}; \psi(k)]$ we refer to Problem 6.1.3, 6.1.4.

6.1.2 Representation of Functions as Fourier or Fourier–Stieltjes Transforms

Concerning the representation of a given function $f(v)$ on $\mathbb{R}$ as the Fourier–Stieltjes transform of a function $\mu \in BV$ or as the Fourier transform of some $g \in L^p$, $1 \le p \le 2$, we may proceed as in the periodic case. Then the results of Sec. 5.1.2, 5.2.4, 5.3.2 on the inversion theory again provide a method in solving the representation problem; necessary and sufficient conditions are stated in terms of the *means*

$$u(x; \rho) = \frac{1}{\sqrt{2\pi}} \int_{-\infty}^{\infty} \theta\left(\frac{v}{\rho}\right) f(v)\, e^{ixv}\, dv. \tag{6.1.7}$$

Here θ is an arbitrary continuous θ-factor (Def. 5.1.5).

Theorem 6.1.5. *A necessary and sufficient condition for* $f \in L^\infty$ *to be representable almost everywhere on* $\mathbb{R}$ *as a Fourier–Stieltjes transform, i.e.*

$$f(v) = \frac{1}{\sqrt{2\pi}} \int_{-\infty}^{\infty} e^{-ivu}\, d\mu(u) \quad \text{a.e.} \tag{6.1.8}$$

with $\mu \in BV$, *is that for any continuous* θ*-factor*

$$\left\| \frac{1}{\sqrt{2\pi}} \int_{-\infty}^{\infty} \theta\left(\frac{v}{\rho}\right) f(v)\, e^{i\circ v}\, dv \right\|_1 = O(1) \qquad (\rho > 0). \tag{6.1.9}$$

If f is continuous, (6.1.8) *holds for all* $v \in \mathbb{R}$.

Proof. *Necessity.* If f admits the representation (6.1.8), then

$$\int_{-\infty}^{\infty} \theta\left(\frac{v}{\rho}\right) \mu^\vee(v)\, e^{ixv}\, dv = \rho \int_{-\infty}^{\infty} \theta^\wedge(\rho(x-u))\, d\mu(u)$$

by (5.3.17), and therefore by (5.3.15)

$$\left\| \frac{1}{\sqrt{2\pi}} \int_{-\infty}^{\infty} \theta\left(\frac{v}{\rho}\right) f(v)\, e^{i\circ v}\, dv \right\|_1 \le \|\theta^\wedge\|_1 \|\mu\|_{BV} \qquad (\rho > 0).$$

Sufficiency. Let (6.1.9) be valid for some continuous θ-factor. Since $\theta \in L^1$, we have $\theta(\circ/\rho) f(\circ) \in L^1$ so that Prop. 5.1.2 and the hypothesis (6.1.9) imply that $u(x; \rho)$ as defined by (6.1.7) belongs to $\mathsf{C}_0 \cap \mathsf{L}^1$ for each $\rho > 0$. Moreover, $[u(\circ; \rho)]^\wedge(v) = \theta(v/\rho)f(v)$ a.e. by Prop. 5.1.10. Now if we set

(6.1.10) $$\mathsf{F} = \{\phi \in \mathsf{L}^1 \mid \phi \in \mathsf{C}_0, \phi^\wedge \in \mathsf{L}^1\},$$

we obtain by the Parseval formula (5.1.23) and by (6.1.9) for every $\phi \in \mathsf{F}$

$$\left|\int_{-\infty}^{\infty} \theta\left(\frac{v}{\rho}\right) f(v)\overline{\phi^\wedge(v)}\, dv\right| = \left|\int_{-\infty}^{\infty} u(v; \rho)\overline{\phi(v)}\, dv\right|$$
$$\leq \sqrt{2\pi}\cdot \|u(\circ; \rho)\|_1 \|\phi\|_\infty \leq M\|\phi\|_\infty.$$

Furthermore, since θ is continuous, we have $\lim_{\rho\to\infty} \theta(v/\rho) = \theta(0) = 1$, and since $|\theta(v/\rho)f(v)\overline{\phi^\wedge(v)}| \leq \|\theta^\wedge\|_1\|f\|_\infty|\phi^\wedge(\circ)| \in \mathsf{L}^1$, Lebesgue's dominated convergence theorem gives

(6.1.11) $$\left|\int_{-\infty}^{\infty} f(v)\overline{\phi^\wedge(v)}\, dv\right| \leq M\|\phi\|_\infty.$$

Thus the integral on the left defines a bounded linear functional on F, considered as a subspace of C_0. Since F is a dense subspace of C_0 (cf. Problem 6.1.5(i)), we may, according to Prop. 0.7.1, extend this functional such that it is bounded on C_0. Now the Riesz representation theorem for bounded linear functionals on C_0 applies and gives the existence of $\mu \in \mathsf{BV}$ such that for all $\phi \in \mathsf{F}$

$$\int_{-\infty}^{\infty} f(v)\overline{\phi^\wedge(v)}\, dv = \int_{-\infty}^{\infty} \overline{\phi(v)}\, d\mu(v).$$

In particular, we obtain for $\phi(v) = \exp\{-tv^2 + ixv\}$, $t > 0$, $x \in \mathbb{R}$, that

(6.1.12) $$\frac{1}{\sqrt{4\pi t}}\int_{-\infty}^{\infty} f(x-u)\, e^{-u^2/4t}\, du = \frac{1}{\sqrt{2\pi}}\int_{-\infty}^{\infty} e^{-tv^2}\, e^{-ixv}\, d\mu(v).$$

Taking $t \to 0+$, the left-hand side converges by Cor. 3.2.3 to $f(x)$ a.e. and the right-hand side by Lebesgue's dominated convergence theorem to $\mu^\vee(x)$ for all $x \in \mathbb{R}$, establishing the representation (6.1.8).

Theorem 6.1.6. *Let* $1 < p \leq 2$ *and* $f \in \mathsf{L}^{p'}$. *A necessary and sufficient condition that*

(6.1.13) $$f(v) = \underset{\rho\to\infty}{\overset{(p')}{\text{l.i.m.}}}\ \frac{1}{\sqrt{2\pi}}\int_{-\rho}^{\rho} g(u)\, e^{-ivu}\, du$$

with $g \in \mathsf{L}^p$, *is that for any continuous* θ-*factor*

(6.1.14) $$\left\|\frac{1}{\sqrt{2\pi}}\int_{-\infty}^{\infty} \theta\left(\frac{v}{\rho}\right) f(v)\, e^{i\circ v}\, dv\right\|_p = \mathsf{O}(1) \qquad (\rho > 0).$$

Let us recall that for $p = 2$ condition (6.1.14) is indeed superfluous since $f \in \mathsf{L}^2$ implies $f \in [\mathsf{L}^2]^\wedge$ by Plancherel's theorem.

Proof. For the necessity we observe that, if f is the Fourier transform of $g \in \mathsf{L}^p$, we have in view of Theorem 5.2.14

$$\left\| \frac{1}{\sqrt{2\pi}} \int_{-\infty}^{\infty} \theta\left(\frac{v}{\rho}\right) f(v)\, e^{i\circ v}\, dv \right\|_p = \left\| \frac{\rho}{\sqrt{2\pi}} \int_{-\infty}^{\infty} g(\circ - u)\theta^\wedge(\rho u)\, du \right\|_p \leq \|\theta^\wedge\|_1 \|g\|_p.$$

To prove the sufficiency we may proceed as in the proof of Theorem 6.1.5 using the Riesz representation theorem for bounded linear functionals on $\mathsf{L}^{p'}$, $2 \leq p' < \infty$. But we may equally well use the weak* compactness theorem for L^p, $1 < p \leq 2$. Thus, let (6.1.14) be valid for some continuous θ-factor. Since $\theta \in \mathsf{L}^1 \cap \mathsf{C}_0$, we have $\theta(\circ/\rho)f(\circ) \in \mathsf{L}^1$ by Hölder's inequality, and therefore $u(x; \rho)$ as defined by (6.1.7) belongs to $\mathsf{C}_0 \cap \mathsf{L}^p$ for each $\rho > 0$. In view of the stated theorems, assumption (6.1.14) implies that there exist a sequence $\{\rho_j\}$ with $\lim_{j\to\infty} \rho_j = \infty$ and $g \in \mathsf{L}^p$ such that

$$\lim_{j\to\infty} \int_{-\infty}^{\infty} h(v)u(v; \rho_j)\, dv = \int_{-\infty}^{\infty} h(v)g(v)\, dv$$

for every $h \in \mathsf{L}^{p'}$. Therefore, if we take $h = \bar{\phi}$, $\phi \in \mathsf{F}$, we have by Parseval formulae (5.2.27), (5.1.5), and Prop. 5.1.1(iv)

$$\int_{-\infty}^{\infty} \overline{\phi^\wedge(v)}g^\wedge(v)\, dv = \int_{-\infty}^{\infty} \overline{\phi(v)}g(v)\, dv = \lim_{j\to\infty} \int_{-\infty}^{\infty} \overline{\phi(v)}u(v; \rho_j)\, dv$$
$$= \lim_{j\to\infty} \int_{-\infty}^{\infty} \overline{\phi^\wedge(v)}\theta\left(\frac{v}{\rho_j}\right) f(v)\, dv = \int_{-\infty}^{\infty} \overline{\phi^\wedge(v)}f(v)\, dv,$$

the last equality being valid by Lebesgue's dominated convergence theorem. If we again take $\phi(v) = \exp\{-tv^2 + ixv\}$, then for every $t > 0$ and $x \in \mathbb{R}$

$$\frac{1}{\sqrt{4\pi t}} \int_{-\infty}^{\infty} e^{-(x-u)^2/4t}\{f(u) - g^\wedge(u)\}\, du = 0,$$

and since $f - g^\wedge \in \mathsf{L}^{p'}$, Cor. 3.2.3 finally implies that $f(x) = g^\wedge(x)$ a.e.

In the preceding theorems it has been *a priori* assumed that $f \in \mathsf{L}^{p'}$ to ensure the absolute convergence of the integral (6.1.7) defining $u(x; \rho)$. But this condition, though a necessary one, may be explicitly avoided. Thus, if we consider θ-factors with compact support, we need only assume $f \in \mathsf{L}^1_{\mathrm{loc}}$. In particular, for the Cesàro factor we have (for a further extension see Problem 6.1.10)

Theorem 6.1.7. *Let f be measurable on $\mathbb{R}$ and integrable over every finite interval. Then for $1 \leq p \leq 2$ the condition*

$$\left\| \frac{1}{\sqrt{2\pi}} \int_{-\rho}^{\rho} \left(1 - \frac{|v|}{\rho}\right) f(v)\, e^{i\circ v}\, dv \right\|_p = O(1) \qquad (\rho > 0) \tag{6.1.15}$$

is necessary and sufficient such that f is almost everywhere equal to the Fourier–Stieltjes transform of a function $\mu \in \mathsf{BV}$ if $p = 1$, and f is the Fourier transform of a function $g \in \mathsf{L}^p$ if $1 < p \leq 2$.

Proof. We only have to show that the assumptions $f \in \mathsf{L}^1_{\mathrm{loc}}$ and (6.1.15) imply $f \in \mathsf{L}^{p'}$. Let us set

$$F_\rho(v) = \begin{cases} \left(1 - \dfrac{|v|}{\rho}\right) f(v), & |v| \le \rho \\ 0, & |v| > \rho \end{cases}$$

and

$$\sigma(x; \rho) = \frac{1}{\sqrt{2\pi}} \int_{-\rho}^{\rho} \left(1 - \frac{|v|}{\rho}\right) f(v)\, e^{ixv}\, dv. \tag{6.1.16}$$

It follows from $f \in \mathsf{L}^1_{\mathrm{loc}}$ that $F_\rho \in \mathsf{L}^1$ for every $\rho > 0$.

If $p = 1$, (6.1.15) states that the Fourier transform $\sigma(-x; \rho)$ of $F_\rho(v)$ belongs to L^1. By Prop. 5.1.10 we then have

$$F_\rho(v) = \frac{1}{\sqrt{2\pi}} \int_{-\infty}^{\infty} \sigma(x; \rho)\, e^{-ivx}\, dx,$$

and hence $|F_\rho(v)| \le \|\sigma(\circ; \rho)\|_1$ for every $\rho > 0$ and almost all $v \in \mathbb{R}$. Since there is a constant M such that $\|\sigma(\circ; \rho)\|_1 \le M$ uniformly with respect to $\rho > 0$ and since $\lim_{\rho \to \infty} F_\rho(v) = f(v)$ a.e., we conclude $f \in \mathsf{L}^\infty$.

If $1 < p \le 2$, then $\sigma(\circ; \rho) \in \mathsf{L}^p$ by (6.1.15). Therefore we can form the Fourier transform $[\sigma(\circ; \rho)]^\wedge(v)$ and obtain by the Parseval formulae (5.2.27) and (5.1.5)

$$\frac{1}{\sqrt{4\pi t}} \int_{-\infty}^{\infty} e^{-(x-u)^2/4t} [\sigma(\circ; \rho)]^\wedge(u)\, du = \frac{1}{\sqrt{2\pi}} \int_{-\infty}^{\infty} e^{-tu^2}\, e^{-ixu} \sigma(u; \rho)\, du$$

$$= \frac{1}{\sqrt{4\pi t}} \int_{-\infty}^{\infty} e^{-(x-u)^2/4t} F_\rho(u)\, du.$$

By Cor. 3.2.3 the left-hand side tends to $[\sigma(\circ; \rho)]^\wedge(x)$ a.e. as $t \to 0+$, because $[\sigma(\circ; \rho)]^\wedge \in \mathsf{L}^{p'}$, whereas the right-hand side tends to $F_\rho(x)$ a.e. as $t \to 0+$, because $F_\rho \in \mathsf{L}^1$. Therefore $F_\rho(x) = [\sigma(\circ; \rho)]^\wedge(x)$ a.e., and by Titchmarsh's inequality $\|F_\rho\|_{p'} \le \|\sigma(\circ; \rho)\|_p$. Hence $\|F_\rho\|_{p'}$ is uniformly bounded with respect to $\rho > 0$. Since $\lim_{\rho \to \infty} F_\rho(x) = f(x)$ a.e., Fatou's lemma gives $f \in \mathsf{L}^{p'}$.

Now we may complete the proof by an application of Theorems 6.1.5 and 6.1.6.

We observe that for $p = 2$ Theorem 6.1.7 is a true improvement of the corresponding assertion of Plancherel's theorem since we do not assume $f \in \mathsf{L}^2$ explicitly.

The fact that we may avoid the explicit assumption $f \in \mathsf{L}^{p'}$ is of interest if we look at the classes $\mathsf{W}[\mathsf{L}^p; \psi(v)]$, $\mathsf{V}[\mathsf{L}^p; \psi(v)]$ of (5.2.36), (5.3.32). Indeed, if f belongs to $\mathsf{W}[\mathsf{L}^p; \psi(v)]$, then the representation $\psi(v) f^\wedge(v) = g^\wedge(v)$, $g \in \mathsf{L}^p$, of course implies $\psi(v) f^\wedge(v) \in \mathsf{L}^{p'}$. But starting off with some $f \in \mathsf{L}^p$, to decide whether f belongs to $\mathsf{W}[\mathsf{L}^p; \psi(v)]$, it is sometimes not convenient to commence with $\psi(v) f^\wedge(v) \in \mathsf{L}^{p'}$, since $\psi(v)$ may be unbounded, e.g. if we take $\psi(v) = (iv)^r$.

Regarding the classes $\mathsf{V}[\mathsf{L}^p; \psi(v)]$ we have

Corollary 6.1.8. *Let $f \in \mathsf{L}^p$, $1 \le p \le 2$. The following assertions are equivalent:*

(i) $f \in \mathsf{V}[\mathsf{L}^p; \psi(v)]$,

(ii) $$\left\| \frac{1}{\sqrt{2\pi}} \int_{-\rho}^{\rho} \left(1 - \frac{|v|}{\rho}\right) \psi(v) f^{\wedge}(v)\, e^{i\circ v}\, dv \right\|_p = O(1) \qquad (\rho \to \infty).$$

For $\psi(v) = (iv)^r$, $r \in \mathbb{N}$, it follows by Theorem 5.3.15

Corollary 6.1.9. *Let $f \in \mathsf{L}^p$, $1 \le p \le 2$. The following assertions are equivalent:*

(i) $f \in \mathsf{V}_{\mathsf{L}^p}^r$, (ii) $f \in \mathsf{V}[\mathsf{L}^p; (iv)^r]$, (iii) $\|\sigma^{(r)}(f; \circ; \rho)\|_p = O(1)$,

(iv) $$\left\| \frac{1}{\sqrt{2\pi}} \int_{-\rho}^{\rho} \left(1 - \frac{|v|}{\rho}\right) (iv)^r f^{\wedge}(v)\, e^{i\circ v}\, dv \right\|_p = O(1) \qquad (\rho \to \infty).$$

For corresponding results concerning the classes $\mathsf{W}[\mathsf{L}^p; \psi(v)]$ we refer to Problems 6.1.11, 6.1.12.

Here we conclude with a certain converse to the Poisson summation formula. In Sec. 3.1.2 we have associated a periodic approximate identity $\{\chi_\rho^*(x)\}$ via (3.1.28) with every $\chi \in \mathsf{NL}^1$. In particular, it follows by Prop. 3.1.12 that the functions χ_ρ^* are bounded and continuous in $\mathsf{L}_{2\pi}^1$-norm, uniformly with respect to $\rho > 0$. Moreover, $[\chi_\rho^*]^{\wedge}(k) = \chi^{\wedge}(k/\rho)$ for every $k \in \mathbb{Z}$, $\rho > 0$ by (5.1.58). After obvious changes of notation, this proves one direction of the following

Proposition 6.1.10. *A function f, defined and continuous on $\mathbb{R}$, is an L^1-Fourier transform if and only if, for each $r \in \mathbb{N}$, the functions $f(k/r)$ on $\mathbb{Z}$ are the finite Fourier transforms of $\mathsf{L}_{2\pi}^1$-functions which are bounded and continuous in $\mathsf{L}_{2\pi}^1$-norm, uniformly with respect to $r \in \mathbb{N}$.*

To complete the proof, let g_r, $r \in \mathbb{N}$, be functions in $\mathsf{L}_{2\pi}^1$ such that (i) $g_r^{\wedge}(k) = f(k/r)$ for all $k \in \mathbb{Z}$, $r \in \mathbb{N}$, (ii) there exists a constant $M > 0$ such that $\|g_r\|_{\mathsf{L}_{2\pi}^1} \le M$ for all $r \in \mathbb{N}$, (iii) given $\varepsilon > 0$, there exists $\delta > 0$ such that $\|g_r(\circ + h) - g_r(\circ)\|_{\mathsf{L}_{2\pi}^1} < \varepsilon$ for all $|h| < \delta$ and $r \in \mathbb{N}$. Then it follows by Problem 1.1.3 that for the integral of Fejér $\lim_{n\to\infty} \|\sigma_{nr}(g_r; \circ) - g_r(\circ)\|_{\mathsf{L}_{2\pi}^1} = 0$ uniformly with respect to $r \in \mathbb{N}$. Therefore $\{\sigma_{nr}(g_r; x)\}_{n=1}^{\infty}$ is a Cauchy sequence in $\mathsf{L}_{2\pi}^1$, uniformly for $r \in \mathbb{N}$. This implies that, given $\varepsilon > 0$, there is $l \in \mathbb{N}$ such that for all $n, m > l$ and $r \in \mathbb{N}$ (cf. Theorem 6.1.2(ii))

$$\int_{-\pi}^{\pi} \left| \sum_{|k| \le nr} \left(1 - \frac{|k|}{nr}\right) f\left(\frac{k}{r}\right) e^{ikx} - \sum_{|k| \le mr} \left(1 - \frac{|k|}{mr}\right) f\left(\frac{k}{r}\right) e^{ikx} \right| dx < \varepsilon.$$

By a change of variables

$$\int_{-\pi r}^{\pi r} \left| \sum_{|k/r| \le n} \left(1 - \frac{|k/r|}{n}\right) f\left(\frac{k}{r}\right) e^{ix(k/r)} \frac{1}{r} - \sum_{|k/r| \le m} \left(1 - \frac{|k/r|}{m}\right) f\left(\frac{k}{r}\right) e^{ix(k/r)} \frac{1}{r} \right| dx < \varepsilon.$$

Since each term of the difference is an approximate Riemann sum of an integral, letting $r \to \infty$, it follows by Fatou's lemma that

$$\int_{-\infty}^{\infty} \left| \int_{-n}^{n} \left(1 - \frac{|v|}{n}\right) f(v)\, e^{ixv}\, dv - \int_{-m}^{m} \left(1 - \frac{|v|}{m}\right) f(v)\, e^{ixv}\, dv \right| dx < \varepsilon$$

for all $n, m > l$. By Problem 6.1.8 this implies the representation $f(v) = g^{\wedge}(v)$ with some $g \in \mathsf{L}^1$, and Prop. 6.1.10 is completely established.

There is an analog of Prop. 6.1.10 if Fourier transforms are replaced by Fourier–Stieltjes transforms. For this we refer to Problem 6.1.13.

Problems

1. Let $f \in \mathsf{l}^\infty$, $\{\theta_\rho(k)\}$ be a θ-factor such that $\{\theta_\rho^\frown(x)\}$ is an approximate identity, and let the means $u_\rho(x)$ be defined by (6.1.1).
   (i) Show that f is the finite Fourier transform of a function in $\mathsf{X}_{2\pi}$ if and only if the means u_ρ form a Cauchy sequence in $\mathsf{X}_{2\pi}$, i.e., $\lim_{\rho_1,\rho_2\to\rho_0} \|u_{\rho_1}(\circ) - u_{\rho_2}(\circ)\|_{\mathsf{X}_{2\pi}} = 0$.
   (ii) For $1 < p < \infty$ show that $\|u_\rho(\circ)\|_p = O(1)$, $\rho \to \rho_0$, if and only if $\|u_{\rho_1}(\circ) - u_{\rho_2}(\circ)\|_p = o(1)$, $\rho_1, \rho_2 \to \rho_0$.
   (iii) Show that f is the finite Fourier–Stieltjes transform of a monotonely increasing function if and only if $u_\rho(x) \ge 0$ for all x and $\rho \in \mathbb{A}$, provided the approximate identity $\{\theta_\rho^\frown(x)\}$ is positive.
2. (i) Complete the proof of Theorem 6.1.2 and show that $\|\sigma_n(\circ)\|_p = O(1)$, $n \to \infty$, for $1 < p < \infty$ if and only if $\|\sigma_n(\circ) - \sigma_m(\circ)\|_p = o(1)$, $n, m \to \infty$.
   (ii) Show that a function f on $\mathbb{Z}$ is the finite Fourier–Stieltjes transform of a monotonely increasing function if and only if $\sigma_n(x) \ge 0$ for all x and n.
3. Let $f \in \mathsf{X}_{2\pi}$. Show that f belongs to $\mathsf{W}[\mathsf{X}_{2\pi}; \psi(k)]$ if and only if

$$\lim_{m,n\to\infty} \left\| \sum_{k=-n}^{n} \left(1 - \frac{|k|}{n+1}\right)\psi(k) f^\frown(k)\, e^{ik\circ} - \sum_{k=-m}^{m} \left(1 - \frac{|k|}{m+1}\right)\psi(k) f^\frown(k)\, e^{ik\circ} \right\|_{\mathsf{X}_{2\pi}} = 0.$$

4. Show that for $f \in \mathsf{X}_{2\pi}$ the following assertions are equivalent:
   (i) $f \in \mathsf{W}^r_{\mathsf{X}_{2\pi}}$, (ii) $f \in \mathsf{W}[\mathsf{X}_{2\pi}; (ik)^r]$, (iii) $\lim_{n,m\to\infty} \|\sigma_n^{(r)}(f; \circ) - \sigma_m^{(r)}(f; \circ)\|_{\mathsf{X}_{2\pi}} = 0$.
5. (i) Show that the set F as defined by (6.1.10) is dense in C_0 and L^p, $1 \le p < \infty$. (Hint: Use Problem 5.1.2(iii) and Cor. 3.1.9)
   (ii) Let $f \in \mathsf{L}^{p'}$. Show that if (6.1.9) [or (6.1.14)] is satisfied for some continuous θ-factor, then the condition is satisfied for every continuous θ-factor. (Hint: CRAMÉR [1, p. 192])
6. Let $f \in \mathsf{L}^{p'}$ and the means $u(x; \rho)$ be defined by (6.1.7) with some continuous θ-factor.
   (i) Show that f is the Fourier transform of a function in L^p, $1 \le p \le 2$, if and only if $\lim_{\rho_1,\rho_2\to\infty} \|u(\circ; \rho_1) - u(\circ; \rho_2)\|_p = 0$.
   (ii) For $1 < p \le 2$ show that $\|u(\circ; \rho)\|_p = O(1)$ for $\rho \to \infty$, if and only if one has $\|u(\circ; \rho_1) - u(\circ; \rho_2)\|_p = o(1)$, $\rho_1, \rho_2 \to \infty$.
   (iii) Show that $\|u(\circ; \rho)\|_p = o(1)$, $\rho \to \infty$, for some $1 \le p \le 2$ implies $f(x) = 0$ a.e.
7. Let $f \in \mathsf{L}^\infty$ and the means $u(x; \rho)$ be defined by (6.1.7) with some continuous θ-factor for which $\theta^\frown$ is positive. Show that f is the Fourier–Stieltjes transform of a monotonely increasing function of BV if and only if $u(x; \rho) \ge 0$ for all x and $\rho > 0$. (Hint: CRAMÉR [1, p. 192])
8. Let f be continuous on $\mathbb{R}$ and consider (for $\rho = n \in \mathbb{N}$) the means $\sigma(x; n)$ of (6.1.16). Show that f is an L^1-Fourier transform if and only if $\lim_{n,m\to\infty} \|\sigma(\circ; n) - \sigma(\circ; m)\|_1 = 0$.
9. Let $f \in \mathsf{L}^1_{\text{loc}}$ satisfy (6.1.15) for some $1 < p \le 2$. Show that the functions $\sigma(x; \rho)$ of (6.1.16) converge in L^p-norm and almost everywhere to a function $g \in \mathsf{L}^p$ as $\rho \to \infty$ and $\|g\|_p = \lim_{\rho\to\infty} \|\sigma(\circ; \rho)\|_p$. State and prove the analogous result for $p = 1$.
10. Let f be measurable on $\mathbb{R}$ such that $\theta f \in \mathsf{L}^1$ for some continuous θ-factor. Show that (6.1.9) [(6.1.14)] is necessary and sufficient for the representation (6.1.8) [(6.1.13)]. (Hint: Proceed as in the proof of Theorem 6.1.7; see also BUTZER–TREBELS [2, p. 14 ff]) State and prove counterparts for periodic functions.
11. Let $f \in \mathsf{L}^p$, $1 \le p \le 2$. Show that f belongs to $\mathsf{W}[\mathsf{L}^p; \psi(v)]$ if and only if

$$\lim_{\rho_1,\rho_2\to\infty} \left\| \frac{1}{\sqrt{2\pi}} \int_{-\rho_1}^{\rho_1} \left(1 - \frac{|v|}{\rho_1}\right)\psi(v) f^\frown(v)\, e^{i\circ v}\, dv - \frac{1}{\sqrt{2\pi}} \int_{-\rho_2}^{\rho_2} \left(1 - \frac{|v|}{\rho_2}\right)\psi(v) f^\frown(v)\, e^{i\circ v}\, dv \right\|_p = 0.$$

12. Show that for $f \in \mathsf{L}^p$, $1 \le p \le 2$, the following assertions are equivalent:
   (i) $f \in \mathsf{W}^r_{\mathsf{L}^p}$, (ii) $f \in \mathsf{W}[\mathsf{L}^p; (iv)^r]$, (iii) $\lim_{\rho_1,\rho_2\to\infty} \|\sigma^{(r)}(f; \circ; \rho_1) - \sigma^{(r)}(f; \circ; \rho_2)\|_p = 0$.

13. Show that a function f, defined and continuous on $\mathbb{R}$, is the Fourier–Stieltjes transform of a function in BV if and only if there exists a constant $M > 0$ such that, for each $r \in \mathbb{N}$, the functions $f(k/r)$ on $\mathbb{Z}$ are finite Fourier–Stieltjes transforms of $\mathsf{BV}_{2\pi}$-functions with total variation on $[-\pi, \pi]$ bounded by M, uniformly for $r \in \mathbb{N}$. (Hint: Use Problems 3.1.12, 5.3.8; see KATZNELSON [1, p. 135], SUNOUCHI [8])

6.2 Theorems of Bochner

We now turn to another type of necessary and sufficient condition for representation in terms of Fourier–Stieltjes transforms, due to S. BOCHNER.

Theorem 6.2.1. *Let $f(v)$ be a complex-valued function, defined and continuous for all real v. Then f is representable as the Fourier–Stieltjes transform of a function $\mu \in \mathsf{BV}$ if and only if there exists a constant $M > 0$ such that*

$$\left| \sum_{k=1}^{n} c_k f(v_k) \right| \leq M \left\| \sum_{k=1}^{n} c_k \, e^{-iv_k \circ} \right\|_{\infty} \tag{6.2.1}$$

for all finite sets of real numbers $\{v_k\}$ and complex numbers $\{c_k\}$.

The necessity of condition (6.2.1) is quite obvious. Indeed, if there is $\mu \in \mathsf{BV}$ such that $f(v) = \mu^\vee(v)$ for all $v \in \mathbb{R}$, then for any $n \in \mathbb{N}$ and any real $v_1, \ldots, v_n$ and complex $c_1, \ldots, c_n$

$$\sum_{k=1}^{n} c_k f(v_k) = \frac{1}{\sqrt{2\pi}} \int_{-\infty}^{\infty} \left(\sum_{k=1}^{n} c_k \, e^{-iv_k u} \right) d\mu(u),$$

which implies (6.2.1) with $M = \|\mu\|_{\mathsf{BV}}$.

In the opposite direction we prove the following generalization, due to R. S. PHILLIPS.

Theorem 6.2.2. *If $f(v)$ is measurable on $\mathbb{R}$ and satisfies* (6.2.1), *then there exists a unique $\mu \in \mathsf{BV}$ such that*

$$f(v) = \frac{1}{\sqrt{2\pi}} \int_{-\infty}^{\infty} e^{-ivu} \, d\mu(u) \quad \text{a.e.} \tag{6.2.2}$$

with $\|\mu\|_{\mathsf{BV}} \leq M^*$, M^* being the smallest value of M satisfying (6.2.1). *If f is continuous on $\mathbb{R}$, then* (6.2.2) *holds for all $v \in \mathbb{R}$ and $\|\mu\|_{\mathsf{BV}} = M^*$.*

Consequently, if $f(v)$ is measurable and satisfies (6.2.1), it can differ from the continuous function $\mu^\vee(v)$ only on a set of measure zero. Thus $\mu^\vee$ and hence, by Prop. 5.3.11, μ is uniquely determined.

Proof. For arbitrary $n \in \mathbb{N}$, $\{v_k\} \subset \mathbb{R}$, $\{c_k\} \subset \mathbb{C}$ we set

$$\mathsf{T} = \left\{ t \in \mathsf{C} \mid t(x) = \sum_{k=1}^{n} c_k \, e^{iv_k x} \right\}. \tag{6.2.3}$$

Then any function f satisfying (6.2.1) can be used to define a bounded linear functional L on T by

$$L\Big[\sum_{k=1}^{N} c_k\, e^{iv_k \circ}\Big] = \sum_{k=1}^{n} \overline{c_k} f(v_k). \tag{6.2.4}$$

By definition $\|L\| = M^*$.

As in the proof of Theorem 6.1.5 we shall show that f can be used to define a bounded linear functional L_0 on C_0. To this end, let $\phi \in \mathsf{B}$ where

$$\mathsf{B} = \{\phi \in \mathsf{C}_{00} | \phi' \text{ is piecewise continuous}\}. \tag{6.2.5}$$

By Problem 3.1.2, B is dense in C_0. Furthermore, the L^1–Fourier transform $\phi^\wedge$ of $\phi \in \mathsf{B}$ is well-defined and $\phi^\wedge \in \mathsf{C}_0 \cap \mathsf{L}^1$ (cf. Problem 5.2.11).

We now define the functional L_0 on B by (cf. (6.1.11))

$$L_0(\phi) = \frac{1}{\sqrt{2\pi}} \int_{-\infty}^{\infty} f(v)\overline{\phi^\wedge(v)}\, dv. \tag{6.2.6}$$

Since f is measurable and uniformly bounded by M, it follows that L_0 is linear on B. It remains to show that L_0 is bounded on B, considered as a subset of C_0, and in fact that $\|L_0\| \le M^*$. Then, by Prop. 0.7.1, L_0 will be uniquely extended to a bounded linear functional on C_0 having the same norm.

Let $\phi \in \mathsf{B}$. If the support of ϕ is contained in $(-l, l)$, $l > 0$, ϕ coincides on $(-l, l)$ with a $2l$-periodic, piecewise continuously differentiable function ψ which may be expanded in its Fourier series (cf. (II.4))

$$\psi(x) = \sum_{k=-\infty}^{\infty} \psi^\wedge(k)\, e^{ik\pi x/l}, \tag{6.2.7}$$

where $\psi(x) = \phi(x)$ for $x \in (-l, l)$ and

$$\psi^\wedge(k) = \frac{1}{2l} \int_{-l}^{l} \phi(u)\, e^{-ik\pi u/l}\, du = \frac{\sqrt{2\pi}}{2l} \phi^\wedge\Big(\frac{k\pi}{l}\Big). \tag{6.2.8}$$

The series (6.2.7) converges uniformly on $\mathbb{R}$. In fact, since ϕ' is piecewise continuous and has compact support, it follows by Prop. 5.1.14 that $iv\phi^\wedge(v) = [\phi']^\wedge(v)$ for all $v \in \mathbb{R}$, and therefore $i\pi k\psi^\wedge(k) = l[\psi']^\wedge(k)$ for all $k \in \mathbb{Z}$. By the Parseval equation for $2l$-periodic functions (cf. Problem 4.2.4) we have

$$\sum_{k=-\infty}^{\infty} |[\psi']^\wedge(k)|^2 = \frac{1}{2l} \int_{-l}^{l} |\psi'(u)|^2\, du = \frac{1}{2l} \int_{-\infty}^{\infty} |\phi'(u)|^2\, du,$$

and hence by Hölder's inequality

$$\begin{aligned} \sum_{k=-\infty}^{\infty}{}' |\psi^\wedge(k)| &\le \Big\{\sum_{k=-\infty}^{\infty}{}' k^{-2}\Big\}^{1/2} \Big\{\sum_{k=-\infty}^{\infty} |k\psi^\wedge(k)|^2\Big\}^{1/2} \\ &= \frac{l}{\sqrt{3}} \Big\{\sum_{k=-\infty}^{\infty} |[\psi']^\wedge(k)|^2\Big\}^{1/2} \le l^{1/2}\|\phi'\|_2. \end{aligned} \tag{6.2.9}$$

Now we consider the auxiliary function

$$\Omega(l; \phi) = \frac{1}{\sqrt{2\pi}} \sum_{k=-\infty}^{\infty} f\Big(\frac{k\pi}{l}\Big)\overline{\phi^\wedge\Big(\frac{k\pi}{l}\Big)} \cdot \frac{\pi}{l}, \tag{6.2.10}$$

which, roughly speaking, approximates the integral (6.2.6). By Prop. 0.7.1 we may extend the bounded linear functional L as defined on T by (6.2.4) to a bounded linear functional on the set of all $2l$-periodic and continuous functions, with norms remaining unchanged. This is possible since T contains all $2l$-periodic trigonometric polynomials. By the definition of L and (6.2.8), (6.2.9) we obtain

$$\Omega(l; \phi) = L\Big[\frac{1}{\sqrt{2\pi}} \sum_{k=-\infty}^{\infty} \phi^\wedge\Big(\frac{k\pi}{l}\Big)\, e^{ik\pi \circ/l} \cdot \frac{\pi}{l}\Big]. \tag{6.2.11}$$

Hence by (6.2.7) and (6.2.8)

(6.2.12) $$|\Omega(l;\phi)| \le \|L\|\;\|\phi\|_\infty.$$

We next consider

$$\Omega\left(\frac{1}{u};\phi\right) = \frac{1}{\sqrt{2\pi}}\sum_{k=-\infty}^{\infty} f(k\pi u)\overline{\phi^\wedge(k\pi u)}\cdot\pi u \qquad (u > 0).$$

As a function of u, $\Omega(u^{-1};\phi)$ being the sum of measurable functions is measurable. Furthermore, by (6.2.8) and (6.2.9)

$$\sum_{k=-\infty}^{\infty} |f(k\pi u)\overline{\phi^\wedge(k\pi u)}\cdot\pi u| \le M^*\left(|\phi^\wedge(0)|\pi u + \sqrt{2\pi}\sum_{k=-\infty}^{\infty}{}' |\psi^\wedge(k)|\right)$$
$$\le M^*\left(|\phi^\wedge(0)|\pi u + \sqrt{2\pi}\|\phi'\|_2 u^{-1/2}\right)$$

Hence the following integral exists and we can interchange the order of summation and integration

$$\frac{1}{\varepsilon}\int_0^\varepsilon \Omega\left(\frac{1}{u};\phi\right)du = \sum_{k=-\infty}^{\infty}\frac{1}{\varepsilon\sqrt{2\pi}}\int_0^\varepsilon f(k\pi u)\overline{\phi^\wedge(k\pi u)}\cdot\pi u\,du$$
$$= f(0)\overline{\phi^\wedge(0)}\sqrt{\frac{\pi}{8}}\,\varepsilon + \frac{1}{\sqrt{2\pi}}\sum_{k=1}^{\infty}\frac{1}{\pi\varepsilon k^2}\int_{-k\pi\varepsilon}^{k\pi\varepsilon} f(v)\overline{\phi^\wedge(v)}\,|v|\,dv.$$

If we now define for $(k-1)\pi\varepsilon \le |v| < k\pi\varepsilon$

(6.2.13) $$g_\varepsilon(v) = \frac{|v|}{\pi\varepsilon}\sum_{j=k}^{\infty}\frac{1}{j^2} \qquad (k\in\mathbb{N}),$$

then $g_\varepsilon(v) = \pi|v|/6\varepsilon \le 2$ for $|v| < \pi\varepsilon$, and since

(6.2.14) $$\frac{1}{k} \le \sum_{j=k}^{\infty}\frac{1}{j^2} \le \frac{1}{k-1},$$

we have for $(k-1)\pi\varepsilon \le |v| < k\pi\varepsilon$, $k \ge 2$,

$$\frac{k-1}{k} = \frac{(k-1)\pi\varepsilon}{\pi\varepsilon}\cdot\frac{1}{k} \le g_\varepsilon(v) \le \frac{k\pi\varepsilon}{\pi\varepsilon}\cdot\frac{1}{k-1} = \frac{k}{k-1}.$$

Thus $0 \le g_\varepsilon(v) \le 2$ for all $v\in\mathbb{R}$ and $\lim_{\varepsilon\to 0+} g_\varepsilon(v) = 1$ for any fixed $v \ne 0$. According to $\sum_{k=1}^{\infty} a_k \sum_{j=1}^{k} b_j = \sum_{j=1}^{\infty} b_j \sum_{k=j}^{\infty} a_k$ it follows that

$$\sum_{k=1}^{\infty}\frac{1}{\pi\varepsilon k^2}\int_{-k\pi\varepsilon}^{k\pi\varepsilon} f(v)\overline{\phi^\wedge(v)}\,|v|\,dv$$
$$= \sum_{j=1}^{\infty}\left\{\left(\int_{(j-1)\pi\varepsilon}^{j\pi\varepsilon} + \int_{-j\pi\varepsilon}^{-(j-1)\pi\varepsilon}\right) f(v)\overline{\phi^\wedge(v)}\left[\frac{|v|}{\pi\varepsilon}\sum_{k=j}^{\infty}\frac{1}{k^2}\right]dv\right\} = \int_{-\infty}^{\infty} f(v)\overline{\phi^\wedge(v)}g_\varepsilon(v)\,dv.$$

Hence

$$\frac{1}{\varepsilon}\int_0^\varepsilon \Omega\left(\frac{1}{u};\phi\right)du = f(0)\overline{\phi^\wedge(0)}\sqrt{\frac{\pi}{8}}\,\varepsilon + \frac{1}{\sqrt{2\pi}}\int_{-\infty}^{\infty} f(v)\overline{\phi^\wedge(v)}g_\varepsilon(v)\,dv,$$

and since we have dominated convergence,

(6.2.15) $$\lim_{\varepsilon\to 0+}\frac{1}{\varepsilon}\int_0^\varepsilon \Omega\left(\frac{1}{u};\phi\right)du = \frac{1}{\sqrt{2\pi}}\int_{-\infty}^{\infty} f(v)\overline{\phi^\wedge(v)}\,dv.$$

Finally, applying (6.2.12) to the left side, we obtain

$$|L_0(\phi)| = \left|\frac{1}{\sqrt{2\pi}}\int_{-\infty}^{\infty} f(v)\overline{\phi^\wedge(v)}\,dv\right| \le \|L\|\;\|\phi\|_\infty.$$

Hence $\|L_0\| \le \|L\| = M^*$.

Thus L_0 as defined by (6.2.6) is a bounded linear functional on B (considered as a subset of C_0) and can be extended in a unique fashion to a bounded linear functional on C_0 satisfying $\|L_0\| \leq M^*$. Moreover, according to Prop. 0.8.10, it can be uniquely represented by a function $\mu \in \mathsf{BV}$ as $L_0(\phi) = (2\pi)^{-1/2} \int_{-\infty}^{\infty} \overline{\phi(u)}\, d\mu(u)$ for all $\phi \in \mathsf{C}_0$ with $\|L_0\| = \|\mu\|_{\mathsf{BV}} \leq M^*$. In particular, for every $\phi \in \mathsf{B}$

$$\int_{-\infty}^{\infty} f(v)\overline{\phi^\wedge(v)}\, dv = \int_{-\infty}^{\infty} \overline{\phi(v)}\, d\mu(v). \tag{6.2.16}$$

To prove the actual representation (6.2.2), let $\phi(v) = \theta_1(v/\rho) \exp\{ixv\}$, $x \in \mathbb{R}$, $\rho > 0$ where θ_1 is defined by (5.1.8). Since $\theta_1 \in \mathsf{B}$, we have by (6.2.16) and in view of Problem 5.1.2(i)

$$\frac{2}{\pi\rho}\int_{-\infty}^{\infty} f(x-u)\,\frac{\sin^2(\rho u/2)}{u^2}\, du = \frac{1}{\sqrt{2\pi}}\int_{-\rho}^{\rho}\left(1 - \frac{|v|}{\rho}\right) e^{-ixv}\, d\mu(v)$$

for every $x \in \mathbb{R}$, $\rho > 0$. For $\rho \to \infty$ the left side converges by Cor. 3.2.5 to $f(x)$ a.e., since $f \in \mathsf{L}^\infty$, whereas the right side tends to $\mu^\vee(x)$ for all $x \in \mathbb{R}$ by Lebesgue's dominated convergence theorem. This proves (6.2.2) which holds everywhere if f is continuous. In this case, let $t \in \mathsf{T}$ be arbitrary. Then

$$|L[t]| = \left|\sum_{k=1}^{n} \overline{c_k} f(v_k)\right| = \left|\frac{1}{\sqrt{2\pi}}\int_{-\infty}^{\infty} \overline{t(x)}\, d\mu(x)\right| \leq \|\mu\|_{\mathsf{BV}}\|t\|_\infty,$$

and hence $\|\mu\|_{\mathsf{BV}} \geq M^*$ which implies $\|\mu\|_{\mathsf{BV}} = M^*$. This proves Theorem 6.2.2, which in turn completes the proof of the Bochner result of Theorem 6.2.1.

There is another famous and important representation theorem, due to BOCHNER, which connects positive-definite functions and Fourier–Stieltjes transforms of monotonely increasing functions $\mu \in \mathsf{BV}$. For this one we refer to Problem 6.2.3.

Here we conclude with the following

Theorem 6.2.3. *Let $\{\mu_n(x)\}_{n=1}^{\infty}$ be a sequence of functions in* BV *such that $\|\mu_n\|_{\mathsf{BV}} \leq M$ uniformly for all n. If the Fourier–Stieltjes transforms $\mu_n^\vee(v)$ converge to a function $f(v)$ almost everywhere as $n \to \infty$, then $f(v)$ is almost everywhere equal to the Fourier–Stieltjes transform of a function $\mu \in \mathsf{BV}$. If the sequence $\{\mu_n^\vee(v)\}$ converges uniformly on every finite interval to $f(v)$, then $f(v) = \mu^\vee(v)$ for all $v \in \mathbb{R}$.*

Proof. Let $\phi \in \mathsf{F}$ be arbitrary (cf. (6.1.10)). Since $\phi \in \mathsf{C}_0$, the assumption $\|\mu_n\|_{\mathsf{BV}} \leq M$ implies by Prop. 0.8.15 that there exists a subsequence $\{\mu_{n_k}(x)\}$, $\lim_{k\to\infty} n_k = \infty$, and $\mu \in \mathsf{BV}$ such that

$$\lim_{k\to\infty} (\phi * d\mu_{n_k})(x) = (\phi * d\mu)(x) \qquad (x \in \mathbb{R}). \tag{6.2.17}$$

On the other hand, since $\phi \in \mathsf{L}^1$, we have $\phi * d\mu_n \in \mathsf{L}^1$ and thus $[\phi * d\mu_n]^\wedge(v) = \phi^\wedge(v)\mu_n^\vee(v)$ by Prop. 5.3.3. Also, by hypothesis, $|[\phi * d\mu_n]^\wedge(v)| \leq M|\phi^\wedge(v)|$ and $\lim_{n\to\infty} [\phi * d\mu_n]^\wedge(v) = f(v)\phi^\wedge(v)$ a.e. Therefore, since $\phi^\wedge \in \mathsf{L}^1$, Lebesgue's dominated convergence theorem gives $f\phi^\wedge \in \mathsf{L}^1$ and

$$\lim_{n\to\infty} \int_{-\infty}^{\infty} |[\phi * d\mu_n]^\wedge(v) - f(v)\phi^\wedge(v)|\, dv = 0. \tag{6.2.18}$$

In view of $[\phi * d\mu_n]^\wedge \in \mathsf{L}^1$ we have by Prop. 5.1.10

$$(\phi * d\mu_n)(x) = \frac{1}{\sqrt{2\pi}} \int_{-\infty}^{\infty} [\phi * d\mu_n]^\wedge(v)\, e^{ixv}\, dv,$$

and hence by (6.2.17), (6.2.18)

$$(\phi * d\mu)(x) = \lim_{k\to\infty} \frac{1}{\sqrt{2\pi}} \int_{-\infty}^{\infty} [\phi * d\mu_{n_k}]^\wedge(v)\, e^{ixv}\, dv = \frac{1}{\sqrt{2\pi}} \int_{-\infty}^{\infty} f(v)\phi^\wedge(v)\, e^{ixv}\, dv.$$

By Prop. 5.1.10, 5.3.3, and the uniqueness theorem this implies $f(v)\phi^\wedge(v) = [\phi * d\mu]^\wedge(v) = \mu^\vee(v)\phi^\wedge(v)$ a.e., and thus $f(v) = \mu^\vee(v)$ a.e. if we choose ϕ such that $\phi^\wedge(v) \neq 0$ (e.g. $\phi(x) = \exp\{-x^2\}$). This completes the proof.

For the counterpart for finite Fourier–Stieltjes transforms we refer to Problem 6.2.8. Theorem 6.2.3 shall often be applied in the following particular form:

Corollary 6.2.4. *Let $\{g_n(x)\}_{n=1}^{\infty}$ be a sequence of functions in L^1 such that $\|g_n\|_1 \leq M$ uniformly for all n. If the Fourier transforms $g_n^\wedge(v)$ converge to a function $f(v)$ uniformly on every finite interval, then $f(v)$ is the Fourier–Stieltjes transform of some $\mu \in \mathsf{BV}$.*

Problems

1. Let $f(k)$ be a given function on $\mathbb{Z}$.
   (i) Show that f is the finite Fourier–Stieltjes transform of a function $\mu \in \mathsf{BV}_{2\pi}$ if and only if there exists a constant $M > 0$ such that
   $$\left| \sum_{k=1}^{n} c_k f(k) \right| \leq M \left\| \sum_{k=1}^{n} c_k\, e^{-ik\circ} \right\|_\infty$$
   for every finite set of complex numbers $\{c_k\}$.
   (ii) Show that f is the finite Fourier transform of a function $g \in \mathsf{L}^p$, $1 < p \leq \infty$, if and only if there exists a constant $M > 0$ such that
   $$\left| \sum_{k=1}^{n} c_k f(k) \right| \leq M \left\| \sum_{k=1}^{n} c_k\, e^{-ik\circ} \right\|_{p'}$$
   for every finite set of complex $\{c_k\}$. (Hint: KACZMARZ–STEINHAUS [1, p. 31 ff])
2. (i) A function $f(k)$ on $\mathbb{Z}$ is said to be *positive-definite* if for every finite subset $c_1, \ldots, c_n$ of complex numbers and all $n \in \mathbb{N}$ the inequality $\sum_{j=1}^{n} \sum_{k=1}^{n} c_j \bar{c}_k f(j-k) \geq 0$ holds. Show that f on $\mathbb{Z}$ is positive-definite if and only if f is the finite Fourier–Stieltjes transform of a monotonely increasing function $\mu \in \mathsf{BV}_{2\pi}$. (Hint: See e.g. ZYGMUND [7I, p. 138], HEWITT [1, p. 29 ff])
   (ii) Show that a function $f \in \mathbb{Z}$ is the finite Fourier–Stieltjes transform of some $\mu \in \mathsf{BV}_{2\pi}$ if and only if there are positive-definite functions $f_1, \ldots, f_4$ on $\mathbb{Z}$ such that $f(k) = f_1(k) - f_2(k) + i[f_3(k) - f_4(k)]$. (Hint: Use the Jordan decomposition)
3. A function $f(v)$ on $\mathbb{R}$ is said to be *positive-definite* if (i) $f \in \mathsf{C}$, (ii) $f(v) = \overline{f(-v)}$, (iii) $\sum_{j=1}^{n} \sum_{k=1}^{n} c_j \bar{c}_k f(v_j - v_k) \geq 0$ for every finite collection $v_1, \ldots, v_n \in \mathbb{R}$, $c_1, \ldots, c_n \in \mathbb{C}$, and all $n \in \mathbb{N}$. Show that $f(v)$ on $\mathbb{R}$ is positive-definite if and only if f is the Fourier–Stieltjes transform of a monotonely increasing function $\mu \in \mathsf{BV}$. (Hint: BOCHNER [7, p. 92 ff], R. R. GOLDBERG [1, p. 59 ff])
4. (i) Let $f \in \mathsf{L}^\infty$. Show that there exists $\mu \in \mathsf{BV}$ such that $f(v) = \mu^\vee(v)$ a.e. if and only if there exists a constant $M > 0$ such that $|\int_{-\infty}^{\infty} f(v)\overline{\phi^\wedge(v)}\, dv| \leq M\|\phi\|_\infty$ for every

$\phi \in \mathsf{F}$. Modify the class F so as to obtain a necessary and sufficient condition for the representation in case $f \in \mathsf{L}^1_{\text{loc}}$ only. (Hint: Use Problem 5.3.4 and argue as for (6.1.11), (6.1.12); compare KATZNELSON [1, p. 134])

(ii) Let $f \in \mathsf{L}^\infty$. Show that there exists $\mu \in \mathsf{BV}$ such that $f(v) = \mu^\vee(v)$ a.e. if and only if there exists a constant $M > 0$ such that $|\int_{-\infty}^{\infty} f(v)\phi(v)\, dv| \leq M\|\phi^\wedge\|_\infty$ for every $\phi \in \mathsf{L}^1$. (Hint: Use (5.3.11); compare BERRY [1], SCHOENBERG [2])

5. (i) Let $1 < p \leq 2$ and $f \in \mathsf{L}^{p'}$. Show that there exists $g \in \mathsf{L}^p$ such that $f(v) = g^\wedge(v)$ a.e. if and only if there exists a constant $M > 0$ such that $|\int_{-\infty}^{\infty} f(v)\overline{\phi^\wedge(v)}\, dv| \leq M\|\phi\|_{p'}$ for every $\phi \in \mathsf{F}$. Extend to functions $f \in \mathsf{L}^1_{\text{loc}}$. (Hint: Use (5.2.27) and proceed as for the preceding Problem, compare also with Theorem 6.5.7)

(ii) Let $1 < p \leq 2$ and $f \in \mathsf{L}^{p'}$. Show that there exists $g \in \mathsf{L}^p$ such that $f(v) = g^\wedge(v)$ a.e. if and only if there exists a constant $M > 0$ such that $|\int_{-\infty}^{\infty} f(v)\phi(v)\, dv| \leq M\|\phi^\wedge\|_{p'}$ for every $\phi \in \mathsf{L}^p$.

6. Let $f \in \mathsf{L}^\infty$. Show that there exists $g \in \mathsf{L}^1$ such that $f(v) = g^\wedge(v)$ a.e. if and only if (i) there exists a constant $M > 0$ such that $|\int_{-\infty}^{\infty} f(v)\phi(v)\, dv| \leq M\|\phi^\wedge\|_\infty$ for every $\phi \in \mathsf{L}^1$, (ii) to each $\varepsilon > 0$ there exists $\delta > 0$ such that $|\int_{-\infty}^{\infty} f(v)\phi(v)\, dv| \leq \varepsilon\|\phi^\wedge\|_\infty$ for every $\phi \in \mathsf{F}$ with $\|\phi^\wedge\|_1 \leq \delta\|\phi^\wedge\|_\infty$. (Hint: BERRY [1])

7. Prove the following representation theorem: A necessary and sufficient condition that $f(v)$ be the Fourier transform of a function in L^p, $1 < p \leq 2$, is that $f \in \mathsf{L}^{p'}$ and

$$\sum_{k=1}^{n-1} \frac{|(1/\sqrt{2\pi}) \int_{-\infty}^{\infty} f(v)[(e^{ix_{k+1}v} - e^{ix_k v})/iv]\, dv|^p}{(x_{k+1} - x_k)^{p-1}} \leq M$$

where M is independent of the subdivision $-\infty < x_1 < x_2 < \cdots < x_n < \infty$. (Hint: Use a criterion of F. RIESZ (RIESZ–SZ.-NAGY [1, p. 75]): f is an indefinite integral of a function in L^q, $1 < q < \infty$, if and only if $\sum_{k=1}^{n-1} |f(x_{k+1}) - f(x_k)|^q/(x_{k+1} - x_k)^{q-1} \leq M$, see BERRY [1])

8. Let $\{\mu_n(x)\}_{n=1}^{\infty}$ be a sequence of functions in $\mathsf{BV}_{2\pi}$ such that $\|\mu_n\|_{\mathsf{BV}_{2\pi}} \leq M$ uniformly for all n. If the finite Fourier–Stieltjes transforms $\mu_n^\vee(k)$ converge to a function $f(k)$ on $\mathbb{Z}$ for every $k \in \mathbb{Z}$, show that $f(k)$ is the finite Fourier–Stieltjes transform of a function $\mu \in \mathsf{BV}_{2\pi}$.

6.3 Sufficient Conditions

In the preceding sections we have been concerned with necessary and sufficient conditions for representation. But sometimes one is more interested in sufficient conditions which are readily applicable to decide whether a given function f is the Fourier transform of an L^1-function. To this end we start off by gathering some facts on convex and quasi-convex functions defined on $\mathbb{Z}$ or $\mathbb{R}$.

6.3.1 Quasi-Convexity

First we deal with (even) functions $f(k)$ on $\mathbb{Z}$. If, as usual, $\Delta f(k) = f(k+1) - f(k)$ and $\Delta^2 f(k) = f(k+2) - 2f(k+1) + f(k)$, then f is said to be *convex* on $\mathbb{P}$ if $f(k)$ is real and $\Delta^2 f(k) \geq 0$ for all $k \in \mathbb{P}$.

Lemma 6.3.1. *Let f be bounded and convex on $\mathbb{P}$. Then f is monotonely decreasing on $\mathbb{P}$, i.e. $\Delta f(k) \leq 0$ for $k \in \mathbb{P}$, $\lim_{k\to\infty} k\Delta f(k) = 0$, and*

$$\sum_{k=0}^{\infty} (k+1)\Delta^2 f(k) < \infty. \tag{6.3.1}$$

Proof. Suppose that $\Delta f(k_0) \equiv a > 0$ for some $k_0 \in \mathbb{P}$. Then $\Delta f(k) \geq a$ for all $k \geq k_0$ since $\Delta f(k+1) \geq \Delta f(k)$ for all $k \in \mathbb{P}$ in view of the convexity of f on $\mathbb{P}$. Hence $f(k) \geq (k - k_0)a + f(k_0)$ for $k \geq k_0$, and thus $\lim_{k\to\infty} f(k) = \infty$, which is a contradiction to the boundedness of f on $\mathbb{P}$. Therefore $\Delta f(k) \leq 0$ for $k \in \mathbb{P}$.

Since f is bounded and monotonely decreasing on $\mathbb{P}$, $\lim_{k\to\infty} f(k) = f(\infty)$ exists as a finite number. Now $\sum_{k=0}^{n} \Delta f(k) = f(n+1) - f(0)$ so that $\sum_{k=0}^{\infty} \Delta f(k) = f(\infty) - f(0)$. Thus the series is convergent, and since the terms of the series are negative and monotonely increasing, it follows by Cauchy's criterion that $\lim_{k\to\infty} k\Delta f(k) = 0$. Using Abel's transformation (cf. Problem 1.2.7) we have

$$f(n+1) - f(0) = \sum_{k=0}^{n} \Delta f(k) = -\sum_{k=0}^{n-1} (k+1)\Delta^2 f(k) + (n+1)\Delta f(n),$$

and the convergence of (6.3.1) follows. In fact,

$$\sum_{k=0}^{\infty} (k+1)\Delta^2 f(k) = f(0) - f(\infty).$$

For examples of functions which are convex on $\mathbb{P}$ we refer to Problem 6.3.1.

Definition 6.3.2. *A function f, defined on $\mathbb{P}$ and satisfying*

$$\sum_{k=0}^{\infty} (k+1)\,|\Delta^2 f(k)| < \infty, \tag{6.3.2}$$

is called ***quasi-convex***.

Thus every bounded and convex function on $\mathbb{P}$ is quasi-convex, but not conversely.

Proposition 6.3.3. *Let $f \in \mathsf{l}_0^\infty(\mathbb{P})$ be quasi-convex. Then* $\lim_{k\to\infty} k\,|\Delta f(k)| = 0$.

Proof. We have $\sum_{k=n}^{\infty} \Delta^2 f(k) = -\Delta f(n)$, and therefore

$$n\,|\Delta f(n)| \leq \sum_{k=n}^{\infty} \frac{n}{k+1}(k+1)\,|\Delta^2 f(k)| \leq \sum_{k=n}^{\infty} (k+1)\,|\Delta^2 f(k)|$$

which tends to zero as $n \to \infty$.

If we now turn to functions defined on $\mathbb{R}$, any real-valued function $f(x)$ for which $2f((x_1 + x_2)/2) \leq f(x_1) + f(x_2)$ for all x_1, x_2 with $a \leq x_1 < x_2 \leq b$ is said to be *convex* on the interval $[a, b]$. Equivalently, f is convex on $[a, b)$, if $\Delta_h^2 f(x) \geq 0$ for every $x \in [a, b)$, $h > 0$, provided h is so small that $x + 2h < b$ (see also Sec. 3.6.1).

We shall only consider convex functions which are continuous. This condition is not very restrictive, since a convex function is either very regular or very irregular, and in particular, since a convex function which is not entirely irregular is necessarily continuous (cf. Problem 6.3.5). Any continuous convex function is absolutely continuous. In fact,

$$f(x) = \int_a^x \psi(u)\,du + f(a), \tag{6.3.3}$$

where ψ is monotonely increasing (cf. Problem 6.3.5).

Lemma 6.3.4. *Let $f(x) \in \mathsf{C}_0[0, \infty)$ be convex for $x > 0$, and let ψ be given through (6.3.3). Then $\psi(x) \leq 0$ for $x > 0$ and $\lim_{x\to\infty} x\psi(x) = 0$. Moreover, $f(x)$ is monotonely decreasing for $x > 0$ and*

$$\int_0^\infty u\,d\psi(u) < \infty. \tag{6.3.4}$$

Proof. Since $\lim_{x\to\infty} f(x) = 0$, we have $-f(a) = \lim_{x\to\infty} \int_a^x \psi(u)\,du$, and since ψ is monotonely increasing, $\lim_{x\to\infty} \psi(x) = 0$. Therefore $\psi(x) \le 0$ for all $x > 0$, and $f(x)$ is monotonely decreasing for $x > 0$. Furthermore,

$$f(a) - f(x) = \int_a^x [-\psi(u)]\,du \ge (x - a)[-\psi(x)],$$

and hence $f(a) \ge \limsup_{x\to\infty} [-x\psi(x)]$. In other words, $\limsup_{x\to\infty} x\,|\psi(x)| \le f(a)$, and since this is true for every $a > 0$, we conclude $\lim_{x\to\infty} x\,|\psi(x)| = 0$ since $\lim_{a\to\infty} f(a) = 0$.

In view of the continuity of f at $x = 0+$, the integral $\int_{0+}^{\varepsilon} \psi(u)\,du$ exists and

$$f(0) - f(\varepsilon) = \int_0^{\varepsilon} [-\psi(u)]\,du \ge -\varepsilon\psi(\varepsilon).$$

Therefore $0 \ge \varepsilon\psi(\varepsilon) \ge f(\varepsilon) - f(0)$, and thus $\lim_{\varepsilon\to 0+} \varepsilon\psi(\varepsilon) = 0$. By a repeated integration by parts it follows that

$$\int_{\varepsilon}^{\rho} u\,d\psi(u) = \rho\psi(\rho) - \varepsilon\psi(\varepsilon) - f(\rho) + f(\varepsilon).$$

Hence $\lim_{\rho\to\infty,\varepsilon\to 0+} \int_{\varepsilon}^{\rho} u\,d\psi(u)$ exists, and since the integrand is positive, (6.3.4) is established.

Under the assumptions of Lemma 6.3.4 we have by (6.3.3) that $f(x + h) - f(x) = \int_0^h \psi(x + u)\,du$. Therefore $f'(x)$ exists almost everywhere. In fact, $f'(x) = \psi(x)$ for all $x > 0$ except perhaps for the denumerable set of discontinuities of the first kind (finite jumps) of ψ. But, after a suitable convention concerning the value of f' at these points, we may assume without loss of generality that $f'(x)$ is monotonely increasing for $x > 0$ and $\int_0^\infty u\,df'(u) < \infty$. This motivates the following

Definition 6.3.5. *A function f, defined on* $(0, \infty)$, *is called* ***quasi-convex*** *if* $f \in \mathsf{AC}_{\mathrm{loc}}(0, \infty)$ *with derivative* $f' \in \mathsf{BV}_{\mathrm{loc}}(0, \infty)$ *such that*

$$\int_0^\infty u\,|df'(u)| < \infty. \tag{6.3.5}$$

In many cases of interest f' is furthermore continuous on $(0, \infty)$, except perhaps for a finite set of discontinuities of the first kind, and absolutely continuous in every bounded subinterval of $(0, \infty)$ which does not contain any of these points. In this case, (6.3.5) is satisfied if $\int_0^\infty u\,|f''(u)|\,du < \infty$.

Proposition 6.3.6. *Let* $f \in \mathsf{C}_0[0, \infty)$ *be quasi-convex. Then* $\lim_{\rho\to\infty} \rho\,|f'(\rho)| = 0$, $\lim_{\varepsilon\to 0+} \varepsilon^2|f'(\varepsilon)| = 0$.

Proof. If $0 < \varepsilon < \rho < \infty$, we have as above

$$\int_{\varepsilon}^{\rho} u\,df'(u) = \rho f'(\rho) - \varepsilon f'(\varepsilon) - f(\rho) + f(\varepsilon). \tag{6.3.6}$$

Since the left-hand side is bounded as $\rho \to \infty$ and since $\lim_{\rho\to\infty} f(\rho) = 0$, it follows that $|\rho f'(\rho)| = \mathsf{O}(1)$ as $\rho \to \infty$, and thus $\lim_{\rho\to\infty} f'(\rho) = 0$. This in turn implies

$$|\rho f'(\rho)| = \left|\rho \int_{\rho}^{\infty} df'(u)\right| \le \int_{\rho}^{\infty} u\,|df'(u)|.$$

Therefore $\lim_{\rho\to\infty} \rho\,|f'(\rho)| = 0$. Now it follows by (6.3.6) that

$$\int_{\varepsilon}^{\infty} u\,df'(u) = f(\varepsilon) - \varepsilon f'(\varepsilon),$$

which in particular shows that $\lim_{\varepsilon\to 0+} \varepsilon^2 f'(\varepsilon) = 0$.

6.3.2 Representation as $L^1_{2\pi}$-Transform

We now turn to the actual problem of this section.

Theorem 6.3.7. *If $f \in l_0^\infty$ is even and quasi-convex, then there exists an even function $g \in L^1_{2\pi}$ such that $f(k) = g^\wedge(k)$, $k \in \mathbb{Z}$. If f is convex on $\mathbb{P}$, g is positive.*

Proof. Let $x \in (0, \pi]$ be fixed and set

$$s_n(x) = \sum_{k=-n}^{n} f(k)\, e^{ikx} = f(0) + 2 \sum_{k=1}^{n} f(k) \cos kx. \tag{6.3.7}$$

Applying Abel's transformation twice we obtain by (1.2.20) and (1.2.24)

$$\begin{aligned} s_n(x) &= -\sum_{k=0}^{n-1} D_k(x)\Delta f(k) + D_n(x) f(n) \\ &= \sum_{k=0}^{n-2} (k+1) F_k(x)\Delta^2 f(k) - nF_{n-1}(x)\Delta f(n-1) + D_n(x) f(n). \end{aligned}$$

By Problems 1.2.5, 1.2.6 the functions $D_k(x)$ and $F_k(x)$ are both bounded by $\sin^{-1}(x/2)$ for all $k \in \mathbb{P}$. Furthermore, the assumptions imply that $\lim_{n\to\infty} f(n) = 0$, and by Prop. 6.3.3 $\lim_{n\to\infty} n\,|\Delta f(n)| = 0$. Therefore $\lim_{n\to\infty} [F_{n-1}(x) n\Delta f(n-1) - D_n(x) f(n)] = 0$. On the other hand, the quasi-convexity of f and the boundedness of $F_k(x)$ by $\sin^{-1}(x/2)$ show that the series $\sum_{k=0}^{\infty} (k+1)\Delta^2 f(k) F_k(x)$ converges absolutely for each $x \in (0, \pi]$ and even uniformly on each interval $[\delta, \pi]$, $0 < \delta < \pi$. Hence it defines an even function g on $[-\pi, \pi]$, $x \neq 0$, which is 2π-periodic and continuous on $(0, \pi]$. Moreover, $\lim_{n\to\infty} s_n(x) = g(x)$, i.e., the series $f(0) + 2\sum_{k=1}^{\infty} f(k) \cos kx$ converges on $[-\pi, \pi]$ save for $x = 0$ and

$$f(0) + 2 \sum_{k=1}^{\infty} f(k) \cos kx = g(x) \equiv \sum_{k=0}^{\infty} (k+1)\Delta^2 f(k) F_k(x). \tag{6.3.8}$$

We note that if f is convex on $\mathbb{P}$, then g is positive since $\Delta^2 f(k) \geq 0$ and $F_k(x) \geq 0$ for all $k \in \mathbb{P}$.

To prove that g belongs to $L^1_{2\pi}$, we have

$$|g(x)| \leq \sum_{k=0}^{\infty} (k+1)\, |\Delta^2 f(k)|\, F_k(x).$$

Therefore by Lebesgue's theorem on integration of series of positive functions (compare Prop. 0.3.3)

$$\frac{1}{2\pi}\int_{-\pi}^{\pi} |g(x)|\, dx \leq \sum_{k=0}^{\infty} (k+1)\, |\Delta^2 f(k)| \frac{1}{2\pi}\int_{-\pi}^{\pi} F_k(x)\, dx = \sum_{k=0}^{\infty} (k+1) |\Delta^2 f(k)|. \tag{6.3.9}$$

Thus $g \in L^1_{2\pi}$ and $\int_{-\pi}^{\pi} g(x)\, dx = 2\pi \sum_{k=0}^{\infty} (k+1)\Delta^2 f(k)$.

To prove the actual representation $f(k) = g^\wedge(k)$, $k \in \mathbb{Z}$, we consider for any $j \in \mathbb{N}$

$$f(0)(1 - \cos jx) + 2 \sum_{k=1}^{\infty} f(k) \cos kx (1 - \cos jx). \tag{6.3.10}$$

We shall show that the latter series is uniformly convergent on $[-\pi, \pi]$ to $g(x)(1 - \cos jx)$. Since it is even and takes on the value 0 at $x = 0$, it suffices to prove uniform convergence on $(0, \pi]$. Thus, let $0 < x \leq \pi$ and $m < n$. Then by Problem 1.2.4

$$\begin{aligned} \left| 2 \sum_{k=m+1}^{n} f(k) \cos kx (1 - \cos jx) \right| &\leq \frac{(jx)^2}{2} \left| 2 \sum_{k=m+1}^{n} f(k) \cos kx \right| \\ &= \frac{(jx)^2}{2} \Bigg| \sum_{k=m-1}^{n-2} (k+1)\Delta^2 f(k) F_k(x) - n\Delta f(n-1) F_{n-1}(x) + D_n(x) f(n) \\ &\qquad + m\Delta f(m-1) F_{m-1}(x) - D_m(x) f(m) \Bigg| \\ &\leq \frac{(j\pi)^2}{2} \left[\sum_{k=m-1}^{n-2} (k+1)\, |\Delta^2 f(k)| + n\, |\Delta f(n-1)| + m\, |\Delta f(m-1)| + |f(n)| + |f(m)| \right], \end{aligned}$$

which tends to 0 as $m, n \to \infty$. The uniform convergence of (6.3.10) to $g(x)(1 - \cos jx)$ now follows. Therefore

$$\frac{1}{2\pi}\int_{-\pi}^{\pi} g(x)(1-\cos jx)\,dx = f(0) + \sum_{k=1}^{\infty} f(k)\frac{1}{\pi}\int_{-\pi}^{\pi}\cos kx(1-\cos jx)\,dx = f(0) - f(j).$$

If we let $j \to \infty$ and apply the Riemann–Lebesgue lemma, we obtain $\int_{-\pi}^{\pi} g(x)\,dx = 2\pi f(0)$, and hence

$$g^{\wedge}(j) = \frac{1}{2\pi}\int_{-\pi}^{\pi} g(x)\cos jx\,dx = f(j) \qquad (j \in \mathbb{Z}).$$

We note that if f is convex on $\mathbb{P}$, the last part of the proof is rather easy and follows from Lemma 6.3.1 and Problem 1.2.7. Moreover, the last proof gives

Corollary 6.3.8. *If the assumptions of Theorem 6.3.7 are satisfied, then*

$$g(x) = f(0) + 2\sum_{k=1}^{\infty} f(k)\cos kx,$$

where the series converges save perhaps for $x = 2\pi j, j \in \mathbb{Z}$. It converges even uniformly on each interval $[\delta, \pi]$, $0 < \delta < \pi$, and is the Fourier series of g.

Let us finally consider a family of even functions f_ρ on $\mathbb{Z}$ depending on a parameter $\rho \in \mathbb{A}$. It is sometimes useful to decide whether the series $\sum f_\rho(k)\exp\{ikx\}$ are the Fourier series of functions $g_\rho \in \mathsf{L}^1_{2\pi}$ for which $\|g_\rho\|_1$ is uniformly bounded. It turns out that this will happen if the functions f_ρ are *uniformly quasi-convex*, i.e., if

$$\sum_{k=0}^{\infty}(k+1)\,|\Delta^2 f_\rho(k)| \le M \qquad (\rho \in \mathbb{A}), \tag{6.3.11}$$

where M is a constant independent of $\rho \in \mathbb{A}$. Indeed, the following assertion is an immediate consequence of (6.3.9):

Corollary 6.3.9. *Let $f_\rho \in \mathsf{l}_0^\infty$ be even functions on $\mathbb{Z}$ which are uniformly quasi-convex relative to $\rho \in \mathbb{A}$. Then there exist even functions $g_\rho \in \mathsf{L}^1_{2\pi}$ such that f_ρ is the finite Fourier transform of g_ρ and $\|g_\rho\|_1$ is uniformly bounded for $\rho \in \mathbb{A}$.*

6.3.3 Representation as L^1-Transform

Next we turn to the representation of functions f defined on $\mathbb{R}$. We commence with a very elementary sufficient condition.

Proposition 6.3.10. *Let $f \in \mathsf{L}^2$ be locally absolutely continuous on $\mathbb{R}$ such that $f' \in \mathsf{L}^2$. Then there exists a function $g \in \mathsf{L}^1$ such that for all $v \in \mathbb{R}$*

$$f(v) = \frac{1}{\sqrt{2\pi}}\int_{-\infty}^{\infty} g(u)\,e^{-ivu}\,du. \tag{6.3.12}$$

Proof. Since $ivf^{\wedge}(v) = [f']^{\wedge}(v)$ a.e. by Prop. 5.2.19, we may proceed as in the proof of Prop. 5.2.20 (case $r = 1$) to show that $f^{\wedge} \in \mathsf{L}^1 \cap \mathsf{L}^2$. Therefore Prop. 5.2.16 applies, and since f is continuous, we have $f(v) = (1/\sqrt{2\pi})\int_{-\infty}^{\infty} f^{\wedge}(u)\,e^{ivu}\,du$ for all $v \in \mathbb{R}$. If we set $g(u) = f^{\wedge}(-u)$, the representation (6.3.12) follows.

The counterpart to Theorem 6.3.7 is given by

Theorem 6.3.11. *If $f \in \mathsf{C}_0$ is even and quasi-convex, then f is representable as the Fourier transform of an even function $g \in \mathsf{L}^1$, i.e., there exists an even function $g \in \mathsf{L}^1$ such that (6.3.12) holds for all $v \in \mathbb{R}$. In fact, for $x \neq 0$*

$$g(x) = \frac{1}{\sqrt{2\pi}} \int_0^\infty \left(\frac{\sin(xu/2)}{x/2}\right)^2 df'(u). \tag{6.3.13}$$

If f is convex, g is positive.

Proof. Let $x \neq 0$ be arbitrary, fixed. Then by partial integration for $0 < \varepsilon < \rho < \infty$

$$\int_\varepsilon^\rho f(u) \cos xu \, du = \frac{\sin x\rho}{x} f(\rho) - \frac{\sin x\varepsilon}{x} f(\varepsilon) + \frac{\cos x\rho - 1}{x^2} f'(\rho) - \frac{\cos x\varepsilon - 1}{x^2} f'(\varepsilon) + \int_\varepsilon^\rho \frac{1 - \cos xu}{x^2} df'(u).$$

According to Prop. 6.3.6 we have $\lim_{\rho\to\infty} \rho f'(\rho) = 0$ and $\lim_{\varepsilon\to 0+} \varepsilon^2 f'(\varepsilon) = 0$. Therefore

$$\lim_{\substack{\rho\to\infty \\ \varepsilon\to 0+}} \left[\frac{\sin x\rho}{x} f(\rho) - \frac{\sin x\varepsilon}{x} f(\varepsilon) + \frac{\cos x\rho - 1}{x^2} f'(\rho) - \frac{\cos x\varepsilon - 1}{x^2} f'(\varepsilon)\right] = 0. \tag{6.3.14}$$

On the other hand, the quasi-convexity of f implies that

$$\int_0^1 u^2 |df'(u)| \leq \int_0^1 u\, |df'(u)| < \infty, \qquad \int_1^\infty |df'(u)| \leq \int_1^\infty u\, |df'(u)| < \infty.$$

Hence for $x \neq 0$ the integral in (6.3.13) defines an even and continuous function g on $\mathbb{R}$. We have by (6.3.14) that

$$\frac{2}{\sqrt{2\pi}} \int_0^\infty f(u) \cos xu \, du = g(x) \equiv \frac{1}{\sqrt{2\pi}} \int_0^\infty \left(\frac{\sin(xu/2)}{x/2}\right)^2 df'(u), \tag{6.3.15}$$

where the integral on the left exists as an improper Riemann integral and the integral on the right absolutely (thus as a Lebesgue integral) for every $x \neq 0$. (We note that (6.3.15) is to be compared with (6.3.8).) Moreover, if f is convex, then f' is monotonely increasing (cf. (6.3.3)), and thus $g(x) \geq 0$.

To show that g possesses the required properties, we have by Fubini's theorem

$$\int_{-\infty}^\infty |g(x)| \, dx \leq \frac{1}{\sqrt{2\pi}} \int_0^\infty u\, |df'(u)| \int_{-\infty}^\infty \frac{1}{u} \left(\frac{\sin(xu/2)}{x/2}\right)^2 dx,$$

and hence $g \in \mathsf{L}^1$. In fact,

$$\|g\|_1 \leq \int_0^\infty u\, |df'(u)|. \tag{6.3.16}$$

Furthermore, by Problem 5.1.2(i) we have for $v \neq 0$

$$g^\wedge(v) = \int_0^\infty u\, df'(u)\left\{\frac{1}{2\pi}\int_{-\infty}^\infty \left(\frac{\sin(t/2)}{t/2}\right)^2 e^{-i(v/u)t}\,dt\right\}$$

$$= \int_{|v|}^\infty u\left(1 - \frac{|v|}{u}\right) df'(u) = f(|v|),$$

the latter equality following by partial integration. Therefore $g^\wedge(v) = f(v)$ for all $v \neq 0$ since f is an even function. Moreover, equality holds for all $v \in \mathbb{R}$ since both functions are continuous. This completes the proof.

The last result may just as well be applied to the problem of representing a function on $\mathbb{Z}$ as the finite Fourier transform of a function in $L^1_{2\pi}$. Indeed, let $f(k)$ be given on $\mathbb{Z}$, and let $f(v)$ be an extension on $\mathbb{R}$ such that f satisfies the assumptions of Theorem 6.3.11. Then there exists $g \in L^1$ such that $f(v) = g^\wedge(v)$ for all $v \in \mathbb{R}$, and if $g^*(x)$ is defined by (3.1.22), we have by (3.1.24) and (5.1.52) that $g^* \in L^1_{2\pi}$ and $[g^*]^\wedge(k) = g^\wedge(k) = f(k)$, $k \in \mathbb{Z}$. Moreover,

Corollary 6.3.12. *Let $f_\rho \in l_0^\infty$, $\rho > 0$, be even functions on $\mathbb{Z}$, and let there exist a function f on $\mathbb{R}$ which satisfies the assumptions of Theorem* 6.3.11 *such that $f_\rho(k) = f(k/\rho)$, $k \in \mathbb{Z}$, $\rho > 0$. Then the f_ρ are the finite Fourier transforms of even functions of class $L^1_{2\pi}$, the norms of which are uniformly bounded with respect to $\rho > 0$.*

In fact, if $g \in L^1$ is such that $f(v) = g^\wedge(v)$, $v \in \mathbb{R}$, and if $g_\rho^*(x)$ is given by

$$g_\rho^*(x) = \sqrt{2\pi}\sum_{k=-\infty}^{\infty} \rho g(\rho(x + 2k\pi)), \tag{6.3.17}$$

then, in view of (3.1.29) and (5.1.58), $\|g_\rho^*\|_{L^1_{2\pi}} \leq \|g\|_{L^1}$ and $[g_\rho^*]^\wedge(k) = g^\wedge(k/\rho) = f(k/\rho) = f_\rho(k)$, $k \in \mathbb{Z}$, $\rho > 0$.

We shall see that Cor. 6.3.12 is very convenient and useful for applications, more so than Cor. 6.3.9.

6.3.4 A Reduction Theorem

In the following chapters it turns out that the classes $W[L^p; \psi(v)]$, $V[L^p; \psi(v)]$ of (5.2.36), (5.3.32) for $\psi(v) = |v|^\alpha$, $\alpha > 0$, play a fundamental rôle in our investigations on approximation theory. Preparatorily, we are here concerned with the following question which we elucidate for the class $W[L^1; |v|^\alpha]$: If for $f \in L^1$ there is some $g \in L^1$ such that $|v|^\alpha f^\wedge(v) = g^\wedge(v)$, does there exist $g_\beta \in L^1$ for each $0 < \beta < \alpha$ such that $|v|^\beta f^\wedge(v) = g_\beta^\wedge(v)$? In other words, does $f \in W[L^1; |v|^\alpha]$ imply $f \in W[L^1; |v|^\beta]$ for every $0 < \beta < \alpha$? In Sec. 11.2 we shall see that g may be interpreted as an αth derivative of f in some generalized (fractional) sense. In the light of this, an affirmative answer to the above question would therefore assure that the existence of an αth derivative implies the existence of the βth derivative for every $0 < \beta < \alpha$.

This corresponds quite naturally to the procedure by which we proved Prop. 5.1.15. There it was shown that if for $f \in L^1$ there exists $g \in L^1$ such that $(iv)^r f^\wedge(v) = g^\wedge(v)$ for some $r \in \mathbb{N}$, then there exist $g_k \in L^1$ such that $(iv)^k f^\wedge(v) = g_k^\wedge(v)$ for every

$k = 1, 2, \ldots, r-1$; that is to say, $\mathsf{W}[\mathsf{L}^1; (iv)^r] \subset \mathsf{W}[\mathsf{L}^1; (iv)^k]$ for $k = 1, 2, \ldots, r-1$. There, the key was the function η of (5.1.34), here for the reduction of the power $|v|^\alpha$ we use the functions G_α of Problem 6.3.8 in connection with the following lemma of E. M. Stein.

Lemma 6.3.13. *Let* $\alpha > 0$. *There exist* $\mu_{j,\alpha} \in \mathsf{BV}$, $j = 1, 2, 3$, *such that for all* $v \in \mathbb{R}$

$$|v|^\alpha = (1 + v^2)^{\alpha/2}\mu_{1,\alpha}^{\vee}(v), \tag{6.3.18}$$

$$(1 + v^2)^{\alpha/2} = \mu_{2,\alpha}^{\vee}(v) + |v|^\alpha \mu_{3,\alpha}^{\vee}(v). \tag{6.3.19}$$

Proof. In view of Problem 6.3.8 there exists $G_2 \in \mathsf{L}^1$ such that $G_2^\wedge(v) = (1 + v^2)^{-1}$ for every $v \in \mathbb{R}$. In fact, $G_2(x) = \sqrt{\pi/2}\exp\{-|x|\}$ by Problem 5.1.2(i), and thus $\|G_2\|_1 = 1$. We now consider the binomial expansion

$$(1 - t)^{\alpha/2} = \sum_{k=0}^{\infty} (-1)^k \binom{\alpha/2}{k} t^k, \tag{6.3.20}$$

the series being absolutely and uniformly convergent for $|t| \le 1$ since (cf. Problem 6.3.12)

$$\sum_{k=0}^{\infty} \left|\binom{\alpha/2}{k}\right| < \infty \qquad (\alpha > 0). \tag{6.3.21}$$

Setting $t = (1 + v^2)^{-1}$, (6.3.20) gives

$$\frac{|v|^\alpha}{(1 + v^2)^{\alpha/2}} = \sum_{k=0}^{\infty} (-1)^k \binom{\alpha/2}{k} [G_2^\wedge(v)]^k, \tag{6.3.22}$$

the series being absolutely and uniformly convergent for all $v \in \mathbb{R}$. Considering the partial sums

$$\sum_{k=0}^{n} (-1)^k \binom{\alpha/2}{k} [G_2^\wedge(v)]^k,$$

these are, by the convolution theorem, Fourier(-Stieltjes) transforms of (cf. (5.3.3))

$$\delta(x) + s_{n,\alpha}(x) \equiv \delta(x) + \sum_{k=1}^{n} (-1)^k \binom{\alpha/2}{k} [G_2 *]^k(x).$$

It follows that $\lim_{n\to\infty} [\delta^v(v) + s_{n,\alpha}^\wedge(v)] = |v|^\alpha(1 + v^2)^{-\alpha/2}$ for all $v \in \mathbb{R}$ and

$$\|s_{n,\alpha}\|_1 \le \sum_{k=1}^{n} \left|\binom{\alpha/2}{k}\right| \|G_2\|_1^k \le \sum_{k=0}^{\infty} \left|\binom{\alpha/2}{k}\right| < \infty$$

uniformly for all $n \in \mathbb{N}$ since $\|G_2\|_1 = 1$. Therefore Theorem 6.2.3 applies and guarantees the existence of some $\mu_{1,\alpha} \in \mathsf{BV}$ such that $\lim_{n\to\infty} [\delta^v(v) + s_{n,\alpha}^\wedge(v)] = \mu_{1,\alpha}^{\vee}(v)$. This proves (6.3.18).

To establish (6.3.19), let $\alpha > 0$ be a given fixed number, and choose $n \in \mathbb{N}$ and $0 < \gamma < 2$ such that $\alpha = n\gamma$. We have

$$\begin{aligned}(1 + v^2)^{\gamma/2} - |v|^\gamma \left(\frac{v^2}{1 + v^2}\right)^{n-(\gamma/2)} &= \left(\frac{1}{1 + v^2}\right)^{n-(\gamma/2)} \sum_{k=0}^{n-1} \binom{n}{k} v^{2k} \\ &= \sum_{k=0}^{n-1} \binom{n}{k} \left(\frac{v^2}{1 + v^2}\right)^k \left(\frac{1}{1 + v^2}\right)^{n-k-(\gamma/2)}.\end{aligned}$$

Now $v^2/(1+v^2) \in [\mathsf{BV}]^\vee$ by (6.3.18) and $(1+v^2)^{-(n-k-(\gamma/2))} \in [\mathsf{L}^1]^\wedge$ by Problem 6.3.8 since $n-k-(\gamma/2) > 0$ for $k = 0, 1, \ldots, n-1$. Therefore a repeated application of the convolution theorems (Theorems 5.1.3, 5.3.5, Prop. 5.3.3) gives the existence of some $\mu_{2,\alpha} \in \mathsf{BV}$ (in fact, $\mu_{2,\alpha}$ may be assumed to be absolutely continuous) such that

$$\mu_{2,\alpha}^\vee(v) = \left[(1+v^2)^{\gamma/2} - |v|^\gamma\left(\frac{v^2}{1+v^2}\right)^{n-(\gamma/2)}\right]^n$$

$$= (1+v^2)^{n\gamma/2} + \sum_{k=1}^{n} (-1)^k \binom{n}{k} |v|^{\gamma k}\left(\frac{v^2}{1+v^2}\right)^{(n-(\gamma/2))k} (1+v^2)^{(n-k)\gamma/2}$$

$$= (1+v^2)^{\alpha/2} - |v|^\alpha \sum_{k=1}^{n} (-1)^{k-1}\binom{n}{k}\left(\frac{v^2}{1+v^2}\right)^{n(k-(\gamma/2))}.$$

Again, $[v^2/(1+v^2)]^{n(k-(\gamma/2))} \in [\mathsf{BV}]^\vee$ by (6.3.18) since $n(k-(\gamma/2)) > 0$ for $k = 1, 2, \ldots, n$. Therefore the latter sum is the Fourier–Stieltjes transform of some $\mu_{3,\alpha} \in \mathsf{BV}$, which proves (6.3.19).

Now to the problem posed at the beginning of this subsection.

Theorem 6.3.14. *Let $1 \le p \le 2$ and $\alpha > 0$. If $f \in \mathsf{W}[\mathsf{L}^p; |v|^\alpha]$, then $f \in \mathsf{W}[\mathsf{L}^p; |v|^\beta]$ for every $0 < \beta < \alpha$. Moreover, if $f \in \mathsf{V}[\mathsf{L}^p; |v|^\alpha]$, then $f \in \mathsf{W}[\mathsf{L}^p; |v|^\beta]$ for every $0 < \beta < \alpha$.*

Proof. Let $f \in \mathsf{V}[\mathsf{L}^1; |v|^\alpha]$, i.e., for $f \in \mathsf{L}^1$ there exists $\mu \in \mathsf{BV}$ such that $|v|^\alpha f^\wedge(v) = \mu^\vee(v)$. By (6.3.19) there are $\mu_{2,\alpha}, \mu_{3,\alpha} \in \mathsf{BV}$ such that

$$(1+v^2)^{\alpha/2} f^\wedge(v) = f^\wedge(v)\mu_{2,\alpha}^\vee(v) + \mu^\vee(v)\mu_{3,\alpha}^\vee(v).$$

Therefore $(1+v^2)^{\alpha/2} f^\wedge(v) \in [\mathsf{BV}]^\vee$ by the convolution theorems. Since by Problem 6.3.8 $(1+v^2)^{-(\alpha-\beta)/2} \in [\mathsf{L}^1]^\wedge$ for every $0 < \beta < \alpha$, it follows that $(1+v^2)^{\beta/2} f^\wedge(v) \in [\mathsf{L}^1]^\wedge$ by Prop. 5.3.3. Now (6.3.18) implies the existence of $\mu_{1,\beta} \in \mathsf{BV}$ such that

$$|v|^\beta f^\wedge(v) = (1+v^2)^{\beta/2} f^\wedge(v)\mu_{1,\beta}^\vee(v).$$

Thus $|v|^\beta f^\wedge(v) \in [\mathsf{L}^1]^\wedge$ by Prop. 5.3.3, i.e., there exists some $g_\beta \in \mathsf{L}^1$ such that $|v|^\beta f^\wedge(v) = g_\beta^\wedge(v)$. This proves the second assertion for $p = 1$. The proof of the other cases follows along the same lines.

Concerning the classes $\mathsf{W}[\mathsf{X}_{2\pi}; |k|^\alpha]$, $\mathsf{V}[\mathsf{X}_{2\pi}; |k|^\alpha]$ of periodic functions, the corresponding problem may be treated in the same manner using the periodic version of Lemma 6.3.13 (see Problem 6.3.10). But the reduction will now be more elementary since for every $\beta > 0$ there exists $\varphi_\beta \in \mathsf{L}_{2\pi}^1$ such that $\varphi_\beta^\wedge(k) = |k|^{-\beta}$ for $k \ne 0$, 0 for $k = 0$ (see Problem 6.3.1). Compare Problem 6.3.11 for the formulation of the result.

Problems

1. (i) Let $f(v)$ be a real-valued function defined on $[0, \infty)$ and suppose that f', f'' exist and $f''(v) \ge 0$ for $v \in [0, \infty)$. Show that the sequence $\{f(k)\}_{k=0}^\infty$ is convex on $\mathbb{P}$.
   (ii) Let $\alpha > 0$. Show that the sequence $\{k^{-\alpha}\}_{k=1}^\infty$ is convex on $\mathbb{N}$.
   (iii) Let $\alpha > 0$. Show that there exists an even function $\varphi_\alpha \in \mathsf{L}_{2\pi}^1$ such that $\varphi_\alpha^\wedge(k) = |k|^{-\alpha}$ for $k \ne 0$, 0 for $k = 0$. In fact,

$$\varphi_\alpha(x) = \sideset{}{'}\sum_{k=-\infty}^{\infty} |k|^{-\alpha} e^{ikx} = 2\sum_{k=1}^{\infty} k^{-\alpha}\cos kx.$$

2. (i) Let $\{a_k\}_{k=0}^{\infty}$ be a convex sequence on $\mathbb{P}$ such that $\lim_{k\to\infty} a_k = 0$. Show that the series $(a_0/2) + \sum_{k=1}^{\infty} a_k \cos kx$ converges on $[-\pi, \pi]$, save possibly at $x = 0$, to a positive and integrable sum $f(x)$, and is the Fourier series of f. (Hint: Theorem 6.3.7, compare with Problem 1.2.7; see also ZYGMUND [7I, p. 183], EDWARDS [1I, p. 112 ff])
   (ii) Show that $\sum_{k=2}^{\infty} (\log k)^{-1} \cos kx$ is the Fourier series of an $\mathsf{L}_{2\pi}^1$-function.
   (iii) Show that there are series $(a_0/2) + \sum_{k=1}^{\infty} a_k \cos kx$ with coefficients monotonely decreasing to zero such that the sum $f(x)$, which exists save for $x = 0$ by Problem 1.2.7, does not belong to $\mathsf{L}_{2\pi}^1$. (Hint: ZYGMUND [7I, p. 184])
3. Let $\{a_k\}_{k=1}^{\infty}$ be a sequence of real numbers monotonely decreasing to zero as $k \to \infty$.
   (i) The sum $f(x)$ of $\sum_{k=1}^{\infty} a_k \sin kx$, which exists save for $x = 0$ by Problem 1.2.7, belongs to $\mathsf{L}_{2\pi}^1$ if and only if $\sum_{k=1}^{\infty} (\Delta a_k) \log k < \infty$. If this condition is satisfied, then $\sum_{k=1}^{\infty} a_k \sin kx$ is the Fourier series of f.
   (ii) Show that $\sum_{k=1}^{\infty} (\Delta a_k) \log k < \infty$ if and only if $\sum_{k=1}^{\infty} k^{-1} a_k < \infty$.
   (iii) Let $\alpha > 0$. Show that there exists an odd function $\psi_\alpha \in \mathsf{L}_{2\pi}^1$ such that $\psi_\alpha^\wedge(k) = \{-i \operatorname{sgn} k\} |k|^{-\alpha}$ for $k \neq 0$, 0 for $k = 0$. In fact,

$$\psi_\alpha(x) = \sum_{k=-\infty}^{\infty}{}' \{-i \operatorname{sgn} k\} |k|^{-\alpha} e^{ikx} = 2 \sum_{k=1}^{\infty} k^{-\alpha} \sin kx.$$

   (iv) Show that $\sum_{k=2}^{\infty} (\log k)^{-1} \sin kx$ is not the Fourier series of an $\mathsf{L}_{2\pi}^1$-function. (Hint: HARDY–ROGOSINSKI [1, p. 33], BARI [1II, p. 201])
4. Let $f \in \mathsf{l}_0^\infty$ be even and quasi-convex. Show that f is of *bounded variation* on $\mathbb{Z}$, i.e., $\sum_{k=-\infty}^{\infty} |\Delta f(k)| < \infty$. (Hint: BARI [1II, p. 202])
5. (i) Show that if $f(x)$ is twice differentiable on (a, b) such that $f''(x) \geq 0$ on (a, b), then $f(x)$ is convex on (a, b).
   (ii) Suppose that $f(x)$ is convex on (a, b) and bounded from above in some subinterval of (a, b). Show that $f(x)$ is continuous and has left-hand and right-hand derivatives on (a, b). The right-hand derivative is not less than the left-hand derivative and both derivatives increase with x. (Hint: HARDY–LITTLEWOOD–PÓLYA [1, p. 91], ZYGMUND [7I, p. 22])
   (iii) Prove (6.3.3). (Hint: HARDY–LITTLEWOOD–PÓLYA [1, p. 94], ZYGMUND [7I, p. 24])
   (iv) Let $f(x)$ be a real-valued function on (a, b), $-\infty \leq a < b \leq \infty$, such that $f((1 - \lambda)x_1 + \lambda x_2) \leq (1 - \lambda)f(x_1) + \lambda f(x_2)$ for every $x_1, x_2 \in (a, b)$, $0 \leq \lambda \leq 1$. Show that f is convex and continuous on (a, b). (Hint: RUDIN [4, p. 60])
   (v) Let $f(x)$ be convex and continuous on (a, b). Show that $f((1 - \lambda)x_1 + \lambda x_2) \leq (1 - \lambda)f(x_1) + \lambda f(x_2)$ for all $x_1, x_2 \in (a, b)$, $0 \leq \lambda \leq 1$. (Hint: KRASNOSEL'SKIĬ–RUTICKIĬ [1, p. 1])
   (vi) Let $f, g \in \mathsf{L}^1$ and suppose that g is positive. If f is convex on $\mathbb{R}$, so is the convolution $f * g$. Thus, if f is convex, so is the integral of Fejér.
6. Let $f(v) \in \mathsf{C}_0$ be an even function such that $f \in \mathsf{AC}_{\text{loc}}(0, \infty)$. Furthermore, suppose that f' is piecewise continuous on $(0, \infty)$ with n discontinuities of the first kind (finite jumps) at $0 < v_1 < v_2 < \cdots < v_n < \infty$ and absolutely continuous on every compact subinterval of $(0, \infty)$ which does not contain any of these points. If $\int_0^\infty v\, |f''(v)|\, dv < \infty$, show that there exists an even function $g \in \mathsf{L}^1$ such that $f(v) = g^\wedge(v)$. In fact, for $x \neq 0$

$$2 \int_0^\infty f(u) \cos xu\, du = \sqrt{2\pi}\, g(x) \equiv \sum_{k=1}^{n} [f'(v_k+) - f'(v_k-)]\left(\frac{\sin (v_k x/2)}{x/2}\right)^2 + \int_0^\infty \left(\frac{\sin (xu/2)}{x/2}\right)^2 f''(u)\, du,$$

which is to be compared with (6.3.15). (Hint: BERENS–GÖRLICH [1])

7. (i) Prove the *theorem of Pólya:* Let $f(x)$ be an even function of C_0 which is convex for $x > 0$. Then f is the Fourier transform of a positive function of L^1. (Hint: Theorem 6.3.11; see also TITCHMARSH [6, p. 170], LUKACS [1, p. 70], LINNIK [1, p. 37])
   (ii) Show that there is an even function $f(x)$ of C_0 monotonely decreasing on $(0, \infty)$ which is not the Fourier transform of an L^1-function. (Hint: See TITCHMARSH [6, p. 171])
8. (i) Show that, for every $\gamma > 0$, $(1 + v^2)^{-\gamma}$ is the Fourier transform of an L^1-function.
   (ii) Let $\alpha > 0$. Show that there exists $G_\alpha \in \mathsf{NL}^1$ such that $G_\alpha^\wedge(v) = (1 + v^2)^{-\alpha/2}$ for every $v \in \mathbb{R}$. Furthermore, $\|G_\alpha\|_1 \le M$ with constant M, independent of α. (Hint: Use (6.3.16))
   (iii) Let $\alpha > 0$. Show that there exists $G_\alpha^* \in \mathsf{L}^1_{2\pi}$ such that $[G_\alpha^*]^\wedge(k) = (1 + k^2)^{-\alpha/2}$ for every $k \in \mathbb{Z}$.
9. Let $\alpha > 0$. Show that $(1 + |v|^\alpha)^{-1}$ is the Fourier transform of an L^1-function.
10. Let $\alpha > 0$. Show that there exists $\mu_{j,\alpha}^* \in \mathsf{BV}_{2\pi}$, $j = 1, 2, 3$, such that for all $k \in \mathbb{Z}$

$$|k|^\alpha = (1 + k^2)^{\alpha/2}[\mu_{1,\alpha}^*]^\vee(k), \qquad (1 + k^2)^{\alpha/2} = [\mu_{2,\alpha}^*]^\vee(k) + |k|^\alpha[\mu_{3,\alpha}^*]^\vee(k).$$

   (Hint: Either copy the proof of Lemma 6.3.13 by using Problems 6.2.8, 6.3.8(iii) or apply Problem 5.3.8)
11. (i) Show that if $f \in \mathsf{W}[\mathsf{X}_{2\pi}; |k|^\alpha]$ for some $\alpha > 0$, then $f \in \mathsf{W}[\mathsf{X}_{2\pi}; |k|^\beta]$ for every $0 < \beta < \alpha$.
   (ii) Show that if $f \in \mathsf{V}[\mathsf{X}_{2\pi}; |k|^\alpha]$ for some $\alpha > 0$, then $f \in \mathsf{W}[\mathsf{X}_{2\pi}; |k|^\beta]$ for every $0 < \beta < \alpha$.
12. Prove (6.3.21). (Hint: If $(c_{k+1}/c_k) = 1 + (\gamma/k) + (\beta_k/k^2)$, $|\beta_k| \le M$, for all large k, then the series $\sum_{k=0}^\infty c_k$ converges for $\gamma < -1$ but diverges for $\gamma \ge -1$; see HILLE [4I, p. 105], also KNOPP [1, p. 440])
13. Show that every $f \in \mathsf{W}^2_{\mathsf{L}^1}$ is representable as the Fourier transform of an L^1-function. (Hint: Use Theorem 5.1.16 to deduce $f^\wedge \in \mathsf{L}^1$, and then Prop. 5.1.10)
14. Let $f \in \mathsf{C}_0$ be an even function. Suppose that $f' \in \mathsf{L}^1$, f' is of bounded variation on every compact subinterval of $(0, \infty)$ which does not contain any of the points $0 < v_1 < v_2 < \cdots < v_n < \infty$, and assume that $\int_{0+} u\,|df'(u)| < \infty$,

$$\left(\int^{v_j-} + \int_{v_j+}\right) |u - v_j| \log(1/|u - v_j|)\, |df'(u)| < \infty$$

   for $j = 1, 2, \ldots, n$, $\int^\infty u\,|df'(u)| < \infty$. Show that f is the Fourier transform of an even function in L^1. (Hint: SZ.-NAGY [3])

6.4 Applications to Singular Integrals

The aim of this section is to introduce further important singular integrals and to supply some results on the order of approximation of these integrals which are available at this stage.

In Chapter 3, a number of examples of singular integrals of Fejér's type (cf. (3.1.8))

$$J(f; x; \rho) = \frac{\rho}{\sqrt{2\pi}} \int_{-\infty}^{\infty} f(x - u)\chi(\rho u)\, du \tag{6.4.1}$$

has been introduced via the explicit representation of the kernel $\chi \in \mathsf{NL}^1$. On the other hand, if $f \in \mathsf{L}^p$, $1 \le p \le 2$, and if χ is such that $\chi^\wedge \in \mathsf{L}^1$, then $J(f; x; \rho)$ may be rewritten (see (5.1.22), (5.2.26)) as

$$J(f; x; \rho) = \frac{1}{\sqrt{2\pi}} \int_{-\infty}^{\infty} \chi^\wedge(v/\rho) f^\wedge(v)\, e^{ixv}\, dv, \tag{6.4.2}$$

which is a summation process of the Fourier inversion integral. More generally, we may commence with the means

(6.4.3) $$\frac{1}{\sqrt{2\pi}}\int_{-\infty}^{\infty} \eta(v/\rho)f^\wedge(v)\, e^{ixv}\, dv,$$

η being any even continuous function in L^1 with $\eta(0) = 1$. If, for some given η, we want to represent (6.4.3) as a singular integral of type (6.4.1), we have to decide whether η is the Fourier transform of an L^1-function χ, that is to say, whether η is a θ-factor (cf. Def. 5.1.5). This will be achieved by the representation theorems, particularly Theorem 6.3.11, even if there is no closed representation of χ. This is the case for the following two representative examples. (For further ones we refer to the Problems.)

6.4.1 General Singular Integral of Weierstrass

As a first example we consider the means

(6.4.4) $$\frac{1}{\sqrt{2\pi}}\int_{-\infty}^{\infty} e^{-t|v|^\kappa} f^\wedge(v)\, e^{ixv}\, dv$$

of the Fourier inversion integral. Here $\kappa > 0$ is an arbitrary but fixed number. These are of type (6.4.3) with $\eta_\kappa(v) = \exp\{-|v|^\kappa\}$ and $\rho = t^{-1/\kappa}$, $t \to 0+$. Obviously, η_κ is an even function belonging to C_0 for each $\kappa > 0$. Moreover, η_κ is quasi-convex. Indeed, $\eta_\kappa \in AC^2_{loc}(0, \infty)$ and $\int_0^\infty v\, |\eta_\kappa''(v)|\, dv < \infty$ for each $\kappa > 0$. Therefore by Theorem 6.3.11, Prop. 5.1.10, and (5.1.22), (5.2.26)

Proposition 6.4.1. *For each $\kappa > 0$ there exists an even $w_\kappa \in NL^1$ such that $w_\kappa^\wedge(v) = \exp\{-|v|^\kappa\}$. In fact,*

(6.4.5) $$w_\kappa(x) = \frac{1}{\sqrt{2\pi}}\int_{-\infty}^{\infty} e^{-|v|^\kappa}\, e^{ixv}\, dv.$$

For every $f \in L^p$, $1 \le p \le 2$, and $t > 0$,

(6.4.6) $$\frac{1}{\sqrt{2\pi}}\int_{-\infty}^{\infty} e^{-t|v|^\kappa} f^\wedge(v)\, e^{ixv}\, dv = \frac{t^{-1/\kappa}}{\sqrt{2\pi}}\int_{-\infty}^{\infty} f(x-u)\, w_\kappa(t^{-1/\kappa}u)\, du.$$

Since $w_\kappa \in NL^1$, the right-hand side of (6.4.6) is now defined for any $f \in X(\mathbb{R})$. It is an integral of type (6.4.1) with $\rho = t^{-1/\kappa}$, denoted by $W_\kappa(f; x; t)$, thus

(6.4.7) $$W_\kappa(f; x; t) = \frac{t^{-1/\kappa}}{\sqrt{2\pi}}\int_{-\infty}^{\infty} f(x-u)\, w_\kappa(t^{-1/\kappa}u)\, du.$$

It is called the *general singular integral of Weierstrass*.

Theorem 6.3.11 assures $w_\kappa \in NL^1$ for each $\kappa > 0$; but in view of (6.4.5) we may moreover assume $w_\kappa \in C_0$. It is furthermore a positive function for $0 < \kappa \le 2$. For $0 < \kappa \le 1$ this follows immediately by Theorem 6.3.11 and the convexity of $w_\kappa^\wedge(v)$ on $(0, \infty)$. However, the extension to $1 < \kappa \le 2$ is by no means trivial (cf. Problem 6.4.1(i)). On the other hand, w_κ fails to be positive for $2 < \kappa < \infty$ (see Problem

6.4.1(ii)). The integral (6.4.7) subsumes those of Cauchy–Poisson and Gauss–Weierstrass as the particular cases $\kappa = 1$ and $\kappa = 2$. Indeed, $w_1(x) = \sqrt{2/\pi}(1 + v^2)^{-1}$ and $w_2(x) = 2^{-1/2} \exp\{-x^2/4\}$ by Problem 5.1.2(i). For the representation of w_κ in terms of an infinite series we refer to Problem 6.4.1(iii).

Since $w_\kappa \in \mathsf{NL}^1$ for each $\kappa > 0$, it follows by Lemma 3.1.5 and Theorem 3.1.6 that

$$\lim_{t\to 0+} \|W_\kappa(f; \circ; t) - f(\circ)\|_{\mathsf{X}(\mathbb{R})} = 0 \tag{6.4.8}$$

for every $f \in \mathsf{X}(\mathbb{R})$. It is now natural to ask for an order of approximation in (6.4.8) and to study its dependence upon structural properties of f. Here the results of Sec. 3.4 and 3.5 will be of use. Indeed, we have the following inverse approximation theorem:

Proposition 6.4.2. *Let $f \in \mathsf{X}(\mathbb{R})$ and $\kappa > 0$. Then $\|W_\kappa(f; \circ; t) - f(\circ)\|_{\mathsf{X}(\mathbb{R})} = O(t^{\alpha/\kappa})$, $t \to 0+$, for some $0 < \alpha < \kappa$ implies that $f \in {}^*\mathsf{W}^{r,\beta}_{\mathsf{X}(\mathbb{R})}$, where $r \in \mathbb{P}$ and $0 < \beta < 2$ are such that $\alpha = r + \beta$.*

Proof. Since $\rho = t^{-1/\kappa}$, the assertion will be a consequence of Theorems 3.5.1, 3.5.2. Indeed, we show that w_κ belongs to $\mathsf{W}^{r+2}_{\mathsf{L}^1\cap\mathsf{C}}$. To this end, it follows by Theorem 6.3.11 that $v^{2s}w_\kappa^\wedge(v)$ is the Fourier transform of an L^1-function for each $s \in \mathbb{P}$. Thus $w_\kappa \in \mathsf{W}[\mathsf{L}^1; (iv)^{2s}]$ implying that $w_\kappa \in \mathsf{W}^{2s}_{\mathsf{L}^1}$ for each $s \in \mathbb{P}$ by Theorem 5.1.16. In view of Problem 5.2.12 this gives $w_\kappa \in \mathsf{W}^{r+2}_{\mathsf{L}^1}$ for every $r \in \mathbb{P}$. Moreover, $w_\kappa \in \mathsf{C}_0$ by (6.4.5), and since the absolute moments of $\exp\{-|v|^\kappa\}$ exist for any positive order, it follows by Prop. 5.1.17 that w_κ has derivatives of every order belonging to C_0. Thus $w_\kappa \in \mathsf{W}^{r+2}_{\mathsf{L}^1\cap\mathsf{C}}$ for every $r \in \mathbb{P}$, and Prop. 6.4.2 is completely established.

In applying the direct approximation theorems of Sec. 3.4 we immediately encounter the problem of the existence of the αth absolute moment of w_κ for $0 < \alpha < \kappa$. To this end, the next three lemmata supply sufficient conditions upon quasi-convex functions f such that the αth absolute moment of g exists, $g \in \mathsf{L}^1$ being such that $f(v) = g^\wedge(v)$. We distinguish two cases: For $0 < \alpha < 1$ we obtain a result for arbitrary quasi-convex functions, whereas for $\alpha \geq 1$ we have to assume the absolute continuity of f for a sufficiently high order.

Lemma 6.4.3. *Let $f \in \mathsf{C}_0$ be even and quasi-convex, and let g be such that $f(v) = g^\wedge(v)$. Then for $0 < \alpha < 1$ condition $\int_0^\infty u^{1-\alpha}|df'(u)| < \infty$ implies the existence of the αth absolute moment $m(g; \alpha)$ of g.*

Proof. Since $f \in \mathsf{C}_0$ is even and quasi-convex, Theorem 6.3.11 assures the existence of some $g \in \mathsf{L}^1$ with $f(v) = g^\wedge(v)$. In fact, the estimate (6.3.16) holds. Moreover, since g is even,

$$\frac{1}{\sqrt{2\pi}} \int_{-\infty}^{\infty} |x|^\alpha |g(x)|\, dx \leq \frac{1}{\pi} \int_0^\infty u^{1-\alpha}|df'(u)| \int_0^\infty u^{\alpha-1}\left(\frac{\sin(xu/2)}{x/2}\right)^2 x^\alpha\, dx$$

$$= \left\{\frac{2^{1+\alpha}}{\pi} \int_0^\infty t^\alpha \left(\frac{\sin t}{t}\right)^2 dt\right\} \int_0^\infty u^{1-\alpha}|df'(u)|,$$

which is bounded by hypothesis. This proves the assertion.

Since the αth moment of the Fejér kernel exists only for $0 < \alpha < 1$, one cannot immediately use the estimate (6.3.16) in case $\alpha \geq 1$. Indeed, the assumptions and proof of Theorem 6.3.11 will be modified; the proof is given for $\alpha < 3$, the case α arbitrary being left to the reader.

Lemma 6.4.4. *Let $f \in \mathsf{C}_0$ be even and quasi-convex, and let g be such that $f(v) = g^\wedge(v)$. Suppose that $f \in \mathsf{AC}^3_{\mathrm{loc}}(0, \infty)$ with $\lim_{\rho\to\infty} f^{(j)}(\rho) = 0$, $\lim_{\varepsilon\to 0+} \varepsilon^{j-1}f^{(j)}(\varepsilon) = 0$ for $j = 1, 2, 3$. Then for $1 < \alpha < 3$ condition $\int_0^\infty u^{3-\alpha}|f^{(4)}(u)|\, du < \infty$ implies the existence of the αth absolute moment of g.*

Proof. The proof follows by a straightforward modification of that of Theorem 6.3.11. Indeed, let $x \neq 0$ be arbitrary, fixed. Since $f \in \mathsf{AC}^3_{loc}(0, \infty)$, a repeated integration by parts gives for $0 < \varepsilon < \rho < \infty$

$$\int_\varepsilon^\rho f(u) \cos xu \, du = \left[\frac{\sin xu}{x} f(u) + \frac{\cos xu}{x^2} f'(u) - \frac{\sin xu}{x^3} f''(u) + \frac{1 - \cos xu}{x^4} f'''(u)\right]_{u=\varepsilon}^{\rho} - \int_\varepsilon^\rho \frac{1 - \cos xu}{x^4} f^{(4)}(u) \, du.$$

Letting $\rho \to \infty$, $\varepsilon \to 0+$, the left-hand side tends to $\sqrt{\pi/2}\, g(x)$ as given by (6.3.15). On the other hand, the present hypotheses assure that the terms in brackets tend to zero. Therefore

$$g(x) = -\frac{4}{\sqrt{2\pi}} \int_0^\infty \frac{\sin^2 (xu/2)}{x^4} f^{(4)}(u) \, du, \tag{6.4.9}$$

the integral being absolutely convergent since for $1 < \alpha < 3$

$$\int_0^1 u^2 |f^{(4)}(u)| \, du \le \int_0^1 u^{3-\alpha} |f^{(4)}(u)| \, du, \qquad \int_1^\infty |f^{(4)}(u)| \, du \le \int_1^\infty u^{3-\alpha} |f^{(4)}(u)| \, du.$$

The representation (6.4.9) implies (see (3.3.6))

$$m(g; \alpha) \le \frac{2^{\alpha-1}}{\pi} \int_0^\infty u^{3-\alpha} |f^{(4)}(u)| \, du \int_0^\infty \left(\frac{xu}{2}\right)^\alpha \frac{\sin^2 (xu/2)}{(xu/2)^4} \frac{u}{2} \, dx$$
$$\le \left\{\frac{2^{\alpha-1}}{\pi} \int_0^\infty t^{\alpha-4} \sin^2 t \, dt\right\} \int_0^\infty u^{3-\alpha} |f^{(4)}(u)| \, du,$$

the constant in brackets being finite for $1 < \alpha < 3$. This proves the lemma.

Lemma 6.4.5. *Let $f \in \mathsf{C}_0$ be even and quasi-convex, and let g be such that $f(v) = g^\wedge(v)$. Suppose for some $r \in \mathbb{N}$ that $f \in \mathsf{AC}^{2r+1}_{loc}(0, \infty)$ with $\lim_{\rho \to \infty} f^{(j)}(\rho) = 0$ for $j = 1, 2, \ldots, 2r+1$, $\lim_{\varepsilon \to 0+} f^{(j)}(\varepsilon) = 0$ for $j = 1, 2, \ldots, 2r-1$, $\lim_{\varepsilon \to 0+} \varepsilon^j f^{(2r-1+j)}(\varepsilon) = 0$ for $j = 1, 2$. Then for $2r - 1 < \alpha < 2r + 1$ condition $\int_0^\infty u^{2r+1-\alpha} |f^{(2r+2)}(u)| \, du < \infty$ implies the existence of the αth absolute moment of g. In this case, g admits the representation*

$$g(x) = \frac{(-1)^r 4}{\sqrt{2\pi}} \int_0^\infty \frac{\sin^2 (xu/2)}{x^{2r+2}} f^{(2r+2)}(u) \, du. \tag{6.4.10}$$

Obviously, if the assumptions of Lemma 6.4.4 or 6.4.5 are satisfied for some α, then the βth absolute moment of g exists for every $0 < \beta < \alpha$.

Let us now return to the direct approximation problem for the general singular integral of Weierstrass. By an elementary calculation it follows from Lemma 6.4.3 if $\kappa \le 1$, and Lemma 6.4.5 if $\kappa > 1$ (choosing $r \in \mathbb{N}$ such that $2r-1 < \kappa$):

Proposition 6.4.6. *If $\kappa > 0$, the αth absolute moment of w_κ exists for every $0 < \alpha < \kappa$.*

This result and Prop. 5.1.17 imply that the kernel w_κ for $\kappa > 1$ belongs to M^s (cf. (3.4.5)) for every $s \in \mathbb{N}$ with $s < \kappa$. Now we may state as immediate consequences of Prop. 3.4.3 and 3.4.6

Proposition 6.4.7. *Let $f \in \mathsf{X}(\mathbb{R})$ and $\kappa > 0$, $\alpha > 0$ being such that $\alpha < \kappa$.*

(i) *If $0 < \alpha \le 2$ and $f \in \mathsf{Lip}^* (\mathsf{X}(\mathbb{R}); \alpha)$ then*

$$\| W_\kappa(f; \circ; t) - f(\circ)\|_{\mathsf{X}(\mathbb{R})} = O(t^{\alpha/\kappa}) \qquad (t \to 0+). \tag{6.4.11}$$

(ii) *If $\alpha = r + \beta$ with $r \in \mathbb{N}$, $0 \le \beta \le 1$, then $f \in \mathsf{W}^{r,\beta}_{\mathsf{X}(\mathbb{R})}$ implies* (6.4.11).

(iii) *If $\alpha = r + \beta$ with $r \in \mathbb{N}$, $0 \le \beta \le 2$, and if $r + 2 < \kappa$, then $f \in {}^*\mathsf{W}^{r,\beta}_{\mathsf{X}(\mathbb{R})}$ implies* (6.4.11).

(iv) *If $\alpha = 2s + \beta$ with $s \in \mathbb{N}$, $0 \le \beta \le 2$, then $f \in {}^*\mathsf{W}^{r,\beta}_{\mathsf{X}(\mathbb{R})}$ implies* (6.4.11).

It is easy to combine the results of Prop. 6.4.2 and 6.4.7 so as to obtain equivalence theorems for the approximation of f by the integral $\mathsf{W}_\kappa(f; x; t)$. As an example we state

Corollary 6.4.8. *Let* $f \in \mathsf{X}(\mathbb{R})$ *and* $0 < \alpha < \kappa \le 2$. *Then*

$$\|W_\kappa(f; \circ; t) - f(\circ)\|_{\mathsf{X}(\mathbb{R})} = O(t^{\alpha/\kappa}) \Leftrightarrow f \in \mathsf{Lip}^*\,(\mathsf{X}(\mathbb{R}); \alpha).$$

Certainly it is possible to obtain equivalence theorems for $\kappa > 2$. But we have to emphasize that Prop. 6.4.2, 6.4.7 do not cover all cases simultaneously. Thus for $\kappa = 3$, $\alpha = 2$ the methods so far employed do not enable us to establish an equivalence theorem. Indeed, Prop. 6.4.2 states that the approximation $\|W_3(f; \circ; t) - f(\circ)\|_{\mathsf{X}(\mathbb{R})} = O(t^{2/3})$ implies $f \in {}^*\mathsf{W}^{1,1}_{\mathsf{X}(\mathbb{R})}$, but conversely, Prop. 6.4.7(ii) shows that only the stronger hypothesis $f \in \mathsf{W}^{1,1}_{\mathsf{X}(\mathbb{R})}$ implies the latter approximation. This gap will be removed in Sec. 13.3.3 as a consequence of results on saturation (the case $\alpha = \kappa$) for the integral $W_\kappa(f; x; t)$.

We conclude with a brief introduction of the corresponding periodic integral. As above, $\kappa > 0$ is a fixed parameter. Via (3.1.28) we may associate with $w_\kappa \in \mathsf{NL}^1$ a periodic kernel which, with a slight abuse of notation (omitting the star), is given for each $t > 0$ by

$$w_{t,\kappa}(x) = \sqrt{2\pi} \sum_{k=-\infty}^{\infty} t^{-1/\kappa} w_\kappa(t^{-1/\kappa}(x + 2k\pi)) \quad \text{a.e.} \tag{6.4.12}$$

In view of (5.1.58), $w^\wedge_{t,\kappa}(k) = \exp\{-t\,|k|^\kappa\}$. Thus the Fourier series of $w_{t,\kappa} \in \mathsf{L}^1_{2\pi}$ converges absolutely and uniformly in x for each $t > 0$, and (see Prop. 4.1.5(i), 5.1.29)

$$w_{t,\kappa}(x) = \sum_{k=-\infty}^{\infty} e^{-t|k|^\kappa} e^{ikx}. \tag{6.4.13}$$

In the following we shall regard $w_{t,\kappa}$ as given for all x by (6.4.13). In view of Prop. 3.1.12 and Problem 3.1.10, $\{w_{t,\kappa}(x)\}$ is then an even continuous periodic approximate identity with parameter t tending to $0+$. The corresponding singular integral

$$W_{t,\kappa}(f; x) = \frac{1}{2\pi} \int_{-\pi}^{\pi} f(x - u) w_{t,\kappa}(u)\,du \tag{6.4.14}$$

is called the *general* (periodic) *singular integral of Weierstrass*. The particular case $\kappa = 2$ has already been introduced in Problem 1.3.10. Since $w^\wedge_{t,\kappa} \in \mathsf{l}^1$ for each $t > 0$, it follows by Prop. 4.1.6 that for $f \in \mathsf{X}_{2\pi}$

$$W_{t,\kappa}(f; x) = \sum_{k=-\infty}^{\infty} e^{-t|k|^\kappa} f^\wedge(k)\, e^{ikx}. \tag{6.4.15}$$

Thus $W_{t,\kappa}(f; x)$ is a summation process of type (4.1.16) for Fourier series with θ-factor (cf. (1.2.28)) given by $\{\exp(-t\,|k|^\kappa)\}$. In this form, $W_{t,\kappa}(f; x)$ is also known as the *Abel–Cartwright means* of the Fourier series of f. Sometimes one sets $e^{-t} = r$ which turns (6.4.15) into

$$\sum_{k=-\infty}^{\infty} r^{|k|^\kappa} f^\wedge(k)\, e^{ikx}, \tag{6.4.16}$$

where $r \in (0, 1)$, $r \to 1-$. For $\kappa = 1$ we recognize the Abel means as introduced in Sec. 1.2.4. Indeed, $w_{\log(1/r),1}(x)$ corresponds to $p_r(x)$ and $W_{\log(1/r),1}(f; x)$ to $P_r(f; x)$, the integral of Abel–Poisson; $w_{\log(1/r),1}(x)$ admits the elementary representation (1.2.37).

Concerning the approximation of f by the integral $W_{t,\kappa}(f; x)$ for $t \to 0+$, it follows by Theorem 1.1.5 that for every $f \in X_{2\pi}$

$$\lim_{t\to 0+} \| W_{t,\kappa}(f; \circ) - f(\circ)\|_{X_{2\pi}} = 0. \tag{6.4.17}$$

Regarding direct approximation theorems, by Problem 3.1.10 and (6.4.12)

$$\frac{1}{2\pi}\int_{-\pi}^{\pi} |u|^{\alpha}|w_{t,\kappa}(u)|\,du \le \frac{1}{\sqrt{2\pi}}\int_{-\infty}^{\infty} |u|^{\alpha}|t^{-1/\kappa}w_{\kappa}(t^{-1/\kappa}u)|\,du = t^{\alpha/\kappa}\frac{1}{\sqrt{2\pi}}\int_{-\infty}^{\infty} |u|^{\alpha}|w_{\kappa}(u)|\,du,$$

which is finite for $0 < \alpha < \kappa$ by Prop. 6.4.6. Therefore, if $0 < \alpha < \kappa$ and, in addition, $0 < \alpha \le 2$, $f \in \mathsf{Lip}^*(X_{2\pi}; \alpha)$ implies $\| W_{t,\kappa}(f; \circ) - f(\circ)\|_{X_{2\pi}} = O(t^{\alpha/\kappa})$ by Prop. 1.6.3 (see also Problem 6.4.2(i)). Concerning higher order direct approximation theorems it is not hard to establish a periodic version of Prop. 3.4.6, applicable to the integral $W_{t,\kappa}(f; x)$ (compare Problems 6.4.2, 6.4.4). To prove inverse approximation theorems one may proceed as in Sec. 2.5.1 and 3.5 (see Problem 6.4.3 for the formulation of the result). Thus, keeping in mind that the periodic kernel $\{w_{t,\kappa}(x)\}$ is generated via (6.4.12) by $w_\kappa \in \mathsf{NL}^1$, one may develop results for the periodic integral $W_{t,\kappa}(f; x)$ to the extent given for the non-periodic integral $W_\kappa(f; x; t)$. These concern order of approximation with α subject to the restriction $0 < \alpha < \kappa$. In Sec. 13.3 we shall return to this nonoptimal approximation case, the saturation case $\alpha = \kappa$ being treated in Sec. 12.2.3.

6.4.2 Typical Means

The *typical means* of the Fourier inversion integral are given by

$$\frac{1}{\sqrt{2\pi}}\int_{-\rho}^{\rho}\left(1 - \left|\frac{v}{\rho}\right|^{\kappa}\right) f^{\wedge}(v)\, e^{ixv}\,dv, \tag{6.4.18}$$

$\kappa > 0$ again being an arbitrary but fixed number. The integral (6.4.18) is of type (6.4.3) with $\eta_\kappa(v) = (1 - |v|^\kappa)$ for $|v| \le 1$, 0 for $|v| > 1$, which is an even and quasi-convex function belonging to C_0. Therefore by Theorem 6.3.11, Prop. 5.1.10, and (5.1.22), (5.2.26)

Proposition 6.4.9. *For each $\kappa > 0$ there exists an even $r_\kappa \in \mathsf{NL}^1$ such that $r_\kappa^{\wedge}(v) = (1 - |v|^\kappa)$ for $|v| \le 1$, 0 for $|v| > 1$. In fact,*

$$r_\kappa(x) = \frac{1}{\sqrt{2\pi}}\int_{-1}^{1} (1 - |v|^\kappa)\, e^{ixv}\,dv. \tag{6.4.19}$$

For every $f \in \mathsf{L}^p$, $1 \le p \le 2$, and $\rho > 0$,

$$\frac{1}{\sqrt{2\pi}}\int_{-\rho}^{\rho}\left(1 - \left|\frac{v}{\rho}\right|^{\kappa}\right) f^{\wedge}(v)\, e^{ixv}\,dv = \frac{\rho}{\sqrt{2\pi}}\int_{-\infty}^{\infty} f(x-u)\, r_\kappa(\rho u)\,du. \tag{6.4.20}$$

The right-hand side is meaningful for any $f \in X(\mathbb{R})$ and defines a singular integral of Fejér's type:

$$R_\kappa(f; x; \rho) = \frac{\rho}{\sqrt{2\pi}}\int_{-\infty}^{\infty} f(x-u)\, r_\kappa(\rho u)\,du. \tag{6.4.21}$$

We also call $R_\kappa(f; x; \rho)$ *typical means*.

In view of (6.4.19) the kernel r_κ may be assumed to belong to C_0. It is positive for $0 < \kappa \le 1$ since $r_\kappa^\wedge$ is convex on $(0, \infty)$. The particular case $\kappa = 1$ gives Fejér's kernel $F(x)$ in which case $r_1(x)$ admits the explicit representation (3.1.15) (cf. also Problem 5.1.2(i)).

Since $r_\kappa \in \mathsf{NL}^1$ for each $\kappa > 0$, by Lemma 3.1.5 and Theorem 3.1.6

$$\lim_{\rho\to\infty} \|R_\kappa(f; \circ; \rho) - f(\circ)\|_{\mathsf{X}(\mathbb{R})} = 0 \tag{6.4.22}$$

for every $f \in \mathsf{X}(\mathbb{R})$. As an inverse approximation theorem we have as for Prop. 6.4.2

Proposition 6.4.10. *Let $f \in \mathsf{X}(\mathbb{R})$ and $\kappa > 0$. Then $\|R_\kappa(f; \circ; \rho) - f(\circ)\|_{\mathsf{X}(\mathbb{R})} = O(\rho^{-\alpha})$, $\rho \to \infty$, for some $0 < \alpha < \kappa$ implies $f \in {}^*\mathsf{W}^{r,\beta}_{\mathsf{X}(\mathbb{R})}$, where $r \in \mathbb{P}$ and $0 < \beta < 2$ are such that $\alpha = r + \beta$.*

Concerning direct approximation theorems, Lemma 6.4.3 gives that the αth absolute moment of r_κ exists for $0 < \alpha < 1$, $\alpha < \kappa$. Therefore Prop. 3.4.3 is applicable. This leads to the following equivalence theorem:

Corollary 6.4.11. *Let $f \in \mathsf{X}(\mathbb{R})$ and $0 < \alpha < \kappa \le 1$. Then*

$$\|R_\kappa(f; \circ; \rho) - f(\circ)\|_{\mathsf{X}(\mathbb{R})} = O(\rho^{-\alpha}) \Leftrightarrow f \in \mathsf{Lip}^*\,(\mathsf{X}(\mathbb{R}); \alpha).$$

However, Lemmata 6.4.4, 6.4.5 are not applicable since $[r_\kappa^\wedge]'$ does not belong to $\mathsf{AC}_{\text{loc}}(0, \infty)$ (cf. also Problem 6.3.6). This implies that the results of Sec. 3.4 do not apply at this stage in case $\alpha \ge 1$. Therefore for the typical means higher order ($1 \le \alpha < \kappa$) direct approximation theorems are entirely postponed to Sec. 13.3.3. The saturation case $\alpha = \kappa$ is discussed in Sec. 12.4.5.

We turn to the periodic counterpart. As for the Weierstrass integral we may associate with r_κ via (3.1.28) an even periodic approximate identity. It being customary to replace the continuous parameter $\rho > 0$ by the discrete $(n + 1)$ with $n \in \mathbb{N}$, $n \to \infty$, with slight changes of notation we are led to

$$r_{n,\kappa}(x) = \sqrt{2\pi} \sum_{k=-\infty}^{\infty} (n + 1) r_\kappa((n + 1)(x + 2k\pi)) \quad \text{a.e.} \tag{6.4.23}$$

Since $r_{n,\kappa}^\wedge(k) = (1 - |k/(n + 1)|^\kappa)$ for $|k| \le n$, 0 for $|k| > n$ by (5.1.58), the Fourier series of $r_{n,\kappa}(x)$ reduces to a trigonometric polynomial of degree n; we may regard

$$r_{n,\kappa}(x) = \sum_{k=-n}^{n} \left(1 - \left|\frac{k}{n+1}\right|^\kappa\right) e^{ikx} \tag{6.4.24}$$

as the definition of $r_{n,\kappa}(x)$ for all x. This implies that $\{r_{n,\kappa}(x)\}$ is a continuous kernel (for each fixed $\kappa > 0$). The corresponding singular integral of $f \in \mathsf{X}_{2\pi}$ is denoted by

$$R_{n,\kappa}(f; x) = \frac{1}{2\pi} \int_{-\pi}^{\pi} f(x - u) r_{n,\kappa}(u)\, du; \tag{6.4.25}$$

in view of (4.1.7) and (6.4.24) it may be rewritten as

$$R_{n,\kappa}(f; x) = \sum_{k=-n}^{n} \left(1 - \left|\frac{k}{n+1}\right|^\kappa\right) f^\wedge(k)\, e^{ikx}, \tag{6.4.26}$$

known as the *typical means* of the Fourier series of $f \in \mathsf{X}_{2\pi}$. These are of type (1.2.30) with row-finite θ-factor given by $r_{n,\kappa}^\wedge$. Obviously, $\kappa = 1$ corresponds to the Fejér means of Sec. 1.2.2.

By Theorem 1.1.5, for every $f \in X_{2\pi}$,

(6.4.27) $$\lim_{n\to\infty} \|R_{n,\kappa}(f; \circ) - f(\circ)\|_{X_{2\pi}} = 0.$$

Since the αth absolute moment of r_κ exists for $0 < \alpha < 1$,

$$\frac{1}{2\pi}\int_{-\pi}^{\pi} |u|^\alpha |r_{n,\kappa}(u)|\, du \le \frac{1}{\sqrt{2\pi}}\int_{-\infty}^{\infty} |u|^\alpha |(n+1) r_\kappa((n+1)u)|\, du = O(n^{-\alpha})$$

by Problem 3.1.10 and (6.4.23). Therefore, if $0 < \alpha < \kappa$ and, in addition, $0 < \alpha < 1$, then $f \in \mathrm{Lip}^*(X_{2\pi}; \alpha)$ implies $\|R_{n,\kappa}(f; \circ) - f(\circ)\|_{X_{2\pi}} = O(n^{-\alpha})$, $n \to \infty$, by Prop. 1.6.3. As for the nonperiodic integral, higher order direct approximation theorems for $R_{n,\kappa}(f; x)$ are postponed to Sec. 13.3. Concerning inverse theorems, the situation for the typical means is quite different to that for the Abel–Cartwright means. Since for every $f \in X_{2\pi}$ $R_{n,\kappa}(f; x)$ is a trigonometric polynomial of degree n at most, the theorems of Bernstein on best approximation by trigonometric polynomials are applicable. Indeed, if $\|R_{n,\kappa}(f; \circ) - f(\circ)\|_{X_{2\pi}} = O(n^{-\alpha})$ for some $0 < \alpha < \kappa$, then $E_n(X_{2\pi}; f) = O(n^{-\alpha})$. Theorem 2.3.6 now asserts $f \in {}^*W^{r,\beta}_{X_{2\pi}}$, where $r \in \mathbb{P}$ and $0 < \beta < 2$ are such that $\alpha = r + \beta$. Thus there only remains the saturation case $\alpha = \kappa$ which is investigated in Sec. 12.2.2.

Problems

1. Let $\kappa > 0$ and $w_\kappa \in \mathsf{NL}^1$ be given by (6.4.5).
   (i) Show that $w_\kappa(x)$ is a positive function of x for each $1 < \kappa \le 2$. (Hint: Lévy [2, pp. 272–274], Bochner [3, 4])
   (ii) Show that $w_\kappa(x)$ is not a positive function of x in case $2 < \kappa < \infty$. (Hint: Bochner [7, p. 96], Cramér [2, p. 91])
   (iii) Show that for $x \ne 0$

$$w_\kappa(x) = \begin{cases} -\dfrac{2}{|x|}\displaystyle\sum_{k=0}^{\infty}\left(\frac{-1}{|x|^\kappa}\right)^k \frac{\Gamma(1+k\kappa)}{k!}\sin\frac{k\kappa\pi}{2}, & 0 < \kappa < 1. \\ -2\displaystyle\sum_{k=0}^{\infty}\frac{|x|^k \sin(3(k+1)\pi/2)}{(k+1)!}\,\Gamma\!\left(\frac{k+1}{\kappa}+1\right), & 1 \le \kappa. \end{cases}$$

   (Hint: Hardy [2, p. 385], Bergström [1], Feller [1], Lukacs [1, p. 105])
2. Let the periodic approximate identity $\{\chi^*_\rho(x)\}$ be generated via (3.1.28) by an even $\chi \in \mathsf{NL}^1$ and let $I^*_\rho(f; x)$, $J(f; x; \rho)$ be the singular integrals corresponding to the kernels $\{\chi^*_\rho(x)\}$, $\{\rho\chi(\rho x)\}$, i.e.

$$I^*_\rho(f; x) = \frac{1}{2\pi}\int_{-\pi}^{\pi} f(x-u)\chi^*_\rho(u)\, du, \qquad J(f; x; \rho) = \frac{\rho}{\sqrt{2\pi}}\int_{-\infty}^{\infty} f(x-u)\chi(\rho u)\, du$$

   (then $I^*_\rho(f; x) = J(f; x; \rho)$ (a.e.) by (3.1.31) for each $f \in X_{2\pi}$ and $\rho > 0$). Establish the following periodic counterparts of Prop. 3.4.3, Lemma 3.4.5, Prop. 3.4.6:
   (i) Let $f \in X_{2\pi}$. If the αth absolute moment of χ exists for some $0 < \alpha \le 2$, then $f \in \mathrm{Lip}^*(X_{2\pi}; \alpha)$ implies $\|I^*_\rho(f; \circ) - f(\circ)\|_{X_{2\pi}} = O(\rho^{-\alpha})$, $\rho \to \infty$.
   (ii) If $\chi \in \mathsf{M}^r$ (cf. (3.4.5)), then $f \in W^r_{X_{2\pi}}$ implies

$$I^*_\rho(f; x) - f(x) = \frac{\rho}{\sqrt{2\pi}}\int_{-\infty}^{\infty} R(x, u)\chi(\rho u)\, du.$$

   Here $R(x, u)$ is given by

$$R(x, u) = \frac{1}{(r-1)!}\int_x^{x+u} [\phi^{(r)}(t) - \phi^{(r)}(x)](x + u - t)^{r-1}\, dt,$$

   where $\phi \in \mathsf{AC}^{r-1}_{2\pi}$ with $\phi^{(r)} \in X_{2\pi}$ is such that $f(x) = \phi(x)$ (a.e.). In particular,

$$\|I^*_\rho(f; \circ) - f(\circ)\|_{X_{2\pi}} \le \frac{2}{r!}\, m(\chi; r)\|\phi^{(r)}\|_{X_{2\pi}}\rho^{-r}.$$

(iii) Let $f \in X_{2\pi}$, $r \in \mathbb{N}$, and $0 \le \alpha \le 1$. If the $(r+\alpha)$th absolute moment of $\chi \in M^r$ exists, then $f \in W_{X_{2\pi}}^{r,\alpha}$ implies

$$\|I_\rho^*(f; \circ) - f(\circ)\|_{X_{2\pi}} = O(\rho^{-(r+\alpha)}) \qquad (\rho \to \infty). \tag{6.4.28}$$

(iv) Let $f \in X_{2\pi}$, $r = 2s$, $s \in \mathbb{N}$, and $0 < \alpha \le 2$. If the $(2s+\alpha)$th absolute moment of $\chi \in M^r$ exists, then $f \in {}^*W_{X_{2\pi}}^{r,\alpha}$ implies (6.4.28).

(v) Let $f \in X_{2\pi}$, $r \in \mathbb{N}$, and $0 < \alpha \le 2$. If $\chi \in M^{r+2}$, then for $f \in W_{X_{2\pi}}^r$

$$\|I_\rho^*(f; \circ) - f(\circ)\|_{X_{2\pi}} \le \frac{2}{r!}\left[\frac{m(\chi; r+2)}{(r+1)(r+2)} + m(\chi; r)\right]\rho^{-r}\omega^*(X_{2\pi}; \phi^{(r)}; \rho^{-1}).$$

In particular, $f \in {}^*W_{X_{2\pi}}^{r,\alpha}$ implies (6.4.28).

3. Let $\chi, \chi_\rho^*, J(f; x; \rho), I_\rho^*(f; x)$ be given as in Problem 6.4.2. Prove the following periodic counterpart of Theorems 3.5.1, 3.5.2: Let χ satisfy the assumptions of these theorems. If, for some $r \in \mathbb{P}$, $0 < \alpha < 2$, $f \in X_{2\pi}$ can be approximated by $I_\rho^*(f; x)$ such that $\|I_\rho^*(f; \circ) - f(\circ)\|_{X_{2\pi}} = O(\rho^{-(r+\alpha)})$, $\rho \to \infty$, then $f \in {}^*W_{X_{2\pi}}^{r,\alpha}$.
4. Apply the preceding Problems so as to obtain equivalence theorems for the approximation of $f \in X_{2\pi}$ by the general (periodic) integral $W_{t,\kappa}(f; x)$ of Weierstrass. Prove, for example, that if $\alpha = r + \beta$ with $r \in \mathbb{N}$, $0 < \beta < 2$, and if $r + 2 < \kappa$, then $\|W_{t,\kappa}(f; \circ) - f(\circ)\|_{X_{2\pi}} = O(t^{\alpha/\kappa})$, $t \to 0+$, if and only if $f \in {}^*W_{X_{2\pi}}^{r,\beta}$.
5. Let $\kappa > 0$ and $c_\kappa \in NL^1$ be such that $c_\kappa^\wedge(v) = (1 + |v|^\kappa)^{-1}$ (cf. Problem 6.3.9). The integral

$$C_\kappa(f; x; \rho) = \frac{\rho}{\sqrt{2\pi}}\int_{-\infty}^{\infty} f(x-u)c_\kappa(\rho u)\,du \qquad (f \in X(\mathbb{R}), \rho > 0)$$

is called the *generalized singular integral of Picard* (see BERENS–GÖRLICH [1]).

(i) Let $\kappa > 1$. Show that $c_\kappa(x) = (1/\sqrt{2\pi})\int_{-\infty}^{\infty}(1 + |v|^\kappa)^{-1} e^{ixv}\,dv$ and that for $f \in L^p$, $1 \le p \le 2$,

$$C_\kappa(f; x; \rho) = \frac{1}{\sqrt{2\pi}}\int_{-\infty}^{\infty}\left(1 + \left|\frac{v}{\rho}\right|^\kappa\right)^{-1} f^\wedge(v)\, e^{ixv}\,dv \qquad (\rho > 0).$$

(ii) Show that $c_\kappa(x)$ is a positive function of x for each $0 < \kappa \le 2$. (Hint: For $0 < \kappa \le 1$ use the convexity of $c_\kappa^\wedge(v)$ on $(0, \infty)$, for $1 < \kappa \le 2$ see LINNIK [1, p. 40])

(iii) Show that $c_2(x) = \sqrt{\pi/2}\exp\{-|x|\}$. Thus $C_2(f; x; \rho)$ is the special singular integral of Picard as introduced in Problem 3.2.5.

(iv) Try to apply the results of Sec. 3.4, 3.5 so as to obtain equivalence theorems for the approximation of $f \in X(\mathbb{R})$ by the integral $C_\kappa(f; x; \rho)$ as $\rho \to \infty$.

(v) Discuss the periodic version of the generalized singular integral of Picard as obtained via (3.1.28), (3.1.31) from the nonperiodic $C_\kappa(f; x; \rho)$.

6. Let $\kappa > 0$ and $G_\kappa \in NL^1$ be such that $G_\kappa^\wedge(v) = (1 + v^2)^{-\kappa/2}$ (cf. Problem 6.3.8(ii)). The integral

$$L_\kappa(f; x; \rho) = \frac{\rho}{\sqrt{2\pi}}\int_{-\infty}^{\infty} f(x-u)G_\kappa(\rho u)\,du \qquad (f \in X(\mathbb{R}), \rho > 0)$$

is called the *Bessel potential* of f.

(i) Let $\kappa > 1$. Show that $G_\kappa(x) = (1/\sqrt{2\pi})\int_{-\infty}^{\infty}(1 + v^2)^{-\kappa/2} e^{ixv}\,dv$ and that for $f \in L^p$, $1 \le p \le 2$,

$$L_\kappa(f; x; \rho) = \frac{1}{\sqrt{2\pi}}\int_{-\infty}^{\infty}\left(1 + \frac{v^2}{\rho^2}\right)^{-\kappa/2} f^\wedge(v)\, e^{ixv}\,dv \qquad (\rho > 0).$$

(ii) Show that G_κ admits the representation

$$G_\kappa(x) = [2^{(\kappa-2)/2}\Gamma(\kappa/2)]^{-1}|x|^{(\kappa-1)/2}K_{(1-\kappa)/2}(|x|) \qquad (\kappa > 0),$$

where K_γ is the modified Bessel function of the third kind of order γ:

$$K_\gamma(x) = \frac{\pi}{2}\frac{J_{-\gamma}(x) - J_\gamma(x)}{\sin \gamma\pi}, \qquad J_\gamma(x) = \sum_{k=0}^{\infty} \frac{(x/2)^{\gamma+2k}}{k!\Gamma(\gamma + k + 1)}.$$

(Hint: ARONSZAJN–SMITH [1, p. 414 ff], CALDERÓN [1], CALDERÓN–ZYGMUND [3])

(iii) Show that $G_\kappa(x)$ is a positive function of x for each $\kappa > 0$. (Hint: For this and further properties see the above literature as well as ARONSZAJN–MULLA–SZEPTYCKI [1])

(iv) Formulate and prove counterparts of Problem 6.4.5(iv) [(v)] for [periodic] Bessel potentials.

7. Let $\kappa > 0$, $\lambda > 0$. Show that there exists $r_{\kappa,\lambda} \in \mathsf{NL}^1$ such that $r^\wedge_{\kappa,\lambda}(v) = (1 - |v|^\kappa)^\lambda$ for $|v| \le 1$, 0 for $|v| > 1$. (Hint: For $\kappa > 0$, $\lambda \ge 1$ use Theorem 6.3.11, for $\kappa > 0$, $0 < \lambda < 1$ Problem 6.3.14)

8. For $f \in \mathsf{X}(\mathbb{R})$ the *Riesz means* are defined by

$$R_{\kappa,\lambda}(f; x; \rho) = \frac{\rho}{\sqrt{2\pi}} \int_{-\infty}^{\infty} f(x - u) r_{\kappa,\lambda}(\rho u)\, du \qquad (\rho > 0),$$

where the kernel $r_{\kappa,\lambda}$ is given for $\kappa > 0$, $\lambda > 0$ by Problem 6.4.7.

(i) Show that $r_{\kappa,\lambda}(x) = (1/\sqrt{2\pi}) \int_{-1}^{1} (1 - |v|^\kappa)^\lambda e^{ixv}\, dv$ and that for $f \in \mathsf{L}^p$, $1 \le p \le 2$,

$$R_{\kappa,\lambda}(f; x; \rho) = \frac{1}{\sqrt{2\pi}} \int_{-\rho}^{\rho} \left(1 - \left|\frac{v}{\rho}\right|^\kappa\right)^\lambda f^\wedge(v)\, e^{ixv}\, dv \qquad (\rho > 0).$$

(ii) Show that $r_{\kappa,\lambda}(x)$ is a positive function of x for each $0 < \kappa \le 1$, $\lambda \ge 1$.

(iii) Try to apply the results of Sec. 3.4, 3.5 so as to obtain equivalence theorems for the approximation of $f \in \mathsf{X}(\mathbb{R})$ by the Riesz means $R_{\kappa,\lambda}(f; x; \rho)$ as $\rho \to \infty$. (Hint: Compare the discussion on the typical means in Sec. 6.4.2; these are indeed the particular case $\lambda = 1$ of the Riesz means)

(iv) Discuss the periodic version of the Riesz means. In this case, we employ the notation

$$R_{n,\kappa,\lambda}(f; x) = \frac{1}{2\pi} \int_{-\pi}^{\pi} f(x - u) r_{n,\kappa,\lambda}(u)\, du \qquad (f \in \mathsf{X}_{2\pi};\ \kappa, \lambda > 0;\ n \in \mathbb{N}),$$

where the periodic kernel $\{r_{n,\kappa,\lambda}(x)\}$ is generated by $r_{\kappa,\lambda}$ via (3.1.28) by setting $\rho = n + 1$, $n \in \mathbb{N}$, $n \to \infty$, thus

$$r_{n,\kappa,\lambda}(x) = \sqrt{2\pi} \sum_{k=-\infty}^{\infty} (n + 1) r_{\kappa,\lambda}((n + 1)(x + 2k\pi)).$$

Show that $r_{n,\kappa,\lambda}(x) = \sum_{k=-n}^{n} (1 - |k/(n + 1)|^\kappa)^\lambda e^{ikx}$ and

$$R_{n,\kappa,\lambda}(f; x) = \sum_{k=-n}^{n} \left(1 - \left|\frac{k}{n + 1}\right|^\kappa\right)^\lambda f^\wedge(k)\, e^{ikx},$$

which are known as the *Riesz means* of the Fourier series of f. Moreover, $\lim_{n\to\infty} \|R_{n,\kappa,\lambda}(f; \circ) - f(\circ)\|_{\mathsf{X}_{2\pi}} = 0$ for every $f \in \mathsf{X}_{2\pi}$.

9. The particular Riesz means for $\kappa = 2$ are known as the *singular integral of Bochner–Riesz*. In this case we use the notation

$$B_\lambda(f; x; \rho) = \frac{\rho}{\sqrt{2\pi}} \int_{-\infty}^{\infty} f(x - u) b_\lambda(\rho u)\, du,$$

where the kernel b_λ is defined through $b_\lambda(x) = r_{2,\lambda}(x)$. Thus $b_\lambda \in \mathsf{NL}^1$ for each $\lambda > 0$. Show that $b_\lambda(x) = 2^\lambda \Gamma(1 + \lambda)|x|^{-((1/2)+\lambda)} J_{(1/2)+\lambda}(|x|)$, where $J_\gamma(x)$ is the Bessel function of order γ. (Hint: BOCHNER [2], V. L. SHAPIRO [1])

6.5 Multipliers

The representation problem of the preceding sections is connected with the problem of the representation of multipliers. Indeed, it will be shown that a function λ on $\mathbb{Z}$ is the finite Fourier–Stieltjes transform of some $\mu \in \mathsf{BV}_{2\pi}$ if and only if for each $f \in \mathsf{C}_{2\pi}$ $\lambda(k)f^\wedge(k)$ is the finite Fourier transform of some $g \in \mathsf{C}_{2\pi}$. A function λ on $\mathbb{Z}$ satisfying this condition will be called a multiplier for Fourier coefficients of type $(\mathsf{C}_{2\pi}, \mathsf{C}_{2\pi})$. Correspondingly, it will be seen that a function λ on $\mathbb{R}$ is a Fourier–Stieltjes transform if and only if, for each $f \in \mathsf{L}^1$, $\lambda(v)f^\wedge(v)$ is the Fourier transform of some $g \in \mathsf{L}^1$, in other words, if and only if $\lambda(v)$ is a multiplier for Fourier transforms of type $(\mathsf{L}^1, \mathsf{L}^1)$. So one is led to investigate multipliers of various classes of functions and their representation.

6.5.1 Multipliers of Classes of Periodic Functions

Let Y_1 and Y_2 be one of the spaces $\mathsf{C}_{2\pi}$, $\mathsf{L}^p_{2\pi}$, $1 \le p \le \infty$. A function λ on $\mathbb{Z}$ is said to be a *multiplier* of type $(\mathsf{Y}_1, \mathsf{Y}_2)$ if for each $f \in \mathsf{Y}_1$ there exists a function $g \in \mathsf{Y}_2$ such that $\lambda(k)f^\wedge(k) = g^\wedge(k)$, $k \in \mathbb{Z}$. In other words, a sequence $\{\lambda(k)\}_{k\in\mathbb{Z}}$ of complex numbers is a multiplier of type $(\mathsf{Y}_1, \mathsf{Y}_2)$ if for every $f \in \mathsf{Y}_1$ the (formal) trigonometric series $\sum_{k=-\infty}^{\infty} \lambda(k)f^\wedge(k)\, e^{ikx}$ is the Fourier series of a function $g \in \mathsf{Y}_2$.

Proposition 6.5.1. *Let λ be a multiplier of type $(\mathsf{Y}_1, \mathsf{Y}_2)$, and let U be defined as the transformation of Y_1 into Y_2 which assigns to each $f \in \mathsf{Y}_1$ the function $Uf = g$ where $g \in \mathsf{Y}_2$ is given by $\lambda(k)f^\wedge(k) = g^\wedge(k)$, $k \in \mathbb{Z}$. Then U is linear, bounded, and translation-invariant.*

An operator U is called *translation-invariant* on Y_1 if $U[f(\circ + h)](x) = (Uf)(x + h)$ for every $f \in \mathsf{Y}_1$ and $h \in \mathbb{R}$.

Proof. It follows immediately by Prop. 4.1.1(i), 4.1.2(i), and the uniqueness theorem that U is well-defined as a transformation of Y_1 into Y_2 which is linear and commutes with the group of real translations. According to the closed graph theorem U will be bounded if it can be shown that U is closed. For this purpose suppose that

$$\lim_{n\to\infty} \|f_n - f\|_{\mathsf{Y}_1} = 0, \qquad \lim_{n\to\infty} \|Uf_n - g\|_{\mathsf{Y}_2} = 0.$$

Then by Problem 4.1.1

$$\lim_{n\to\infty} f_n^\wedge(k) = f^\wedge(k), \qquad \lim_{n\to\infty} [Uf_n]^\wedge(k) = g^\wedge(k) \qquad (k \in \mathbb{Z}).$$

But $\lambda(k)f_n^\wedge(k) = [Uf_n]^\wedge(k)$, $k \in \mathbb{Z}$, by definition. Therefore $\lambda(k)f^\wedge(k) = g^\wedge(k)$, $k \in \mathbb{Z}$, i.e. $Uf = g$, and thus U is closed.

Having associated with each multiplier λ the corresponding *multiplier operator* U $(= U_\lambda)$, the problem is to characterize the functions λ of a definite multiplier class $(\mathsf{Y}_1, \mathsf{Y}_2)$ and to develop a representation of the operators U_λ. There are complete solutions of numerous important particular cases, but here we shall confine ourselves to two representative examples. For further results we refer to the Problems.

Proposition 6.5.2. *A function λ on $\mathbb{Z}$ is a multiplier of type $(\mathsf{L}^2_{2\pi}, \mathsf{L}^2_{2\pi})$ if and only if $\lambda \in \mathsf{l}^\infty$. In this case, we have for the multiplier operator*

$$(6.5.1) \qquad \|U\|_{[\mathsf{L}^2_{2\pi}, \mathsf{L}^2_{2\pi}]} = \|\lambda\|_{\mathsf{l}^\infty}.$$

Proof. Let λ be a multiplier of type $(\mathsf{L}^2_{2\pi}, \mathsf{L}^2_{2\pi})$. Then by Prop. 6.5.1 there exists a constant M such that $\|Uf\|_{\mathsf{L}^2_{2\pi}} \le M\,\|f\|_{\mathsf{L}^2_{2\pi}}$ for all $f \in \mathsf{L}^2_{2\pi}$. Therefore by the Parseval equation

$$\sum_{k=-\infty}^{\infty} |\lambda(k)f^\wedge(k)|^2 \le M^2 \sum_{k=-\infty}^{\infty} |f^\wedge(k)|^2.$$

If we take $f(x) = \exp\{imx\}$, $m \in \mathbb{Z}$, this implies that $|\lambda(m)| \le M$ for all $m \in \mathbb{Z}$. Conversely, if $\lambda \in \mathsf{l}^\infty$, then $\lambda f^\wedge \in \mathsf{l}^2$ for every $f \in \mathsf{L}^2_{2\pi}$, and thus it follows by the theorem of Riesz–Fischer that λ is a multiplier of type $(\mathsf{L}^2_{2\pi}, \mathsf{L}^2_{2\pi})$. Moreover, $\|Uf\|_{\mathsf{L}^2_{2\pi}} \le \|\lambda\|_{\mathsf{l}^\infty}\|f\|_{\mathsf{L}^2_{2\pi}}$ by the Parseval equation. On the other hand, for $f(x) = \exp\{imx\}$, $m \in \mathbb{Z}$, we obtain $\|Uf\|_{\mathsf{L}^2_{2\pi}} = |\lambda(m)|$, which proves (6.5.1).

Proposition 6.5.3. *A necessary and sufficient condition for a function λ on $\mathbb{Z}$ to be a multiplier of type $(\mathsf{C}_{2\pi}, \mathsf{C}_{2\pi})$ is that λ is the finite Fourier–Stieltjes transform of a function $\mu \in \mathsf{BV}_{2\pi}$. In this case,*

$$(6.5.2) \qquad (Uf)(x) = \frac{1}{2\pi}\int_{-\pi}^{\pi} f(x-u)\,d\mu(u)$$

for every $f \in \mathsf{C}_{2\pi}$ and

$$(6.5.3) \qquad \|U\|_{[\mathsf{C}_{2\pi}, \mathsf{C}_{2\pi}]} = \|\mu\|_{\mathsf{BV}_{2\pi}}.$$

Proof. *Necessity.* Let $\sigma_n(f; x)$ be the Fejér means of the Fourier series of f (cf. (1.2.25)), and let us denote the Fejér means of the series $\sum_{k=-\infty}^{\infty} \lambda(k)\exp\{ikx\}$ and $\sum_{k=-\infty}^{\infty}\lambda(k)f^\wedge(k)\exp\{ikx\}$ by $\tau_n(x)$ and $\pi_n(x)$, respectively, i.e.

$$(6.5.4) \qquad \begin{aligned} &\text{(i)} \qquad \tau_n(x) = \sum_{k=-n}^{n}\left(1 - \frac{|k|}{n+1}\right)\lambda(k)\,e^{ikx} \\ &\text{(ii)} \qquad \pi_n(x) = \sum_{k=-n}^{n}\left(1 - \frac{|k|}{n+1}\right)\lambda(k)f^\wedge(k)\,e^{ikx}. \end{aligned}$$

Then by (4.1.7)

$$(6.5.5) \qquad \pi_n(x) = \frac{1}{2\pi}\int_{-\pi}^{\pi} f(x-u)\tau_n(u)\,du.$$

If λ is a multiplier of type $(\mathsf{C}_{2\pi}, \mathsf{C}_{2\pi})$, then for every $f \in \mathsf{C}_{2\pi}$ there exists $g \in \mathsf{C}_{2\pi}$ such that $\pi_n(x) = \sigma_n(g; x)$. Therefore $\|\pi_n\|_{\mathsf{C}_{2\pi}} \le \|g\|_{\mathsf{C}_{2\pi}}$ which in particular implies by (6.5.5) that for each $f \in \mathsf{C}_{2\pi}$

$$(6.5.6) \qquad \left|\frac{1}{2\pi}\int_{-\pi}^{\pi} f(u)\tau_n(-u)\,du\right| \le M,$$

where the constant M depends on f but not on n. For each $n \in \mathbb{N}$ the integral on the left defines a bounded linear functional on $\mathsf{C}_{2\pi}$ the norm of which is given by $\|\tau_n\|_1$ (see Prop. 0.8.8). Hence (6.5.6) and the uniform boundedness principle imply that the sequence $\{\|\tau_n\|_1\}$ is bounded, that is to say that

$$\left\|\sum_{k=-n}^{n}\left(1 - \frac{|k|}{n+1}\right)\lambda(k)\,e^{ik\circ}\right\|_1 = O(1) \qquad (n \to \infty).$$

Theorem 6.1.2 then shows that λ is representable as the finite Fourier–Stieltjes transform of a function $\mu \in \mathsf{BV}_{2\pi}$.

Sufficiency. If there exists $\mu \in \mathsf{BV}_{2\pi}$ such that $\lambda(k) = \mu^{\vee}(k)$, $k \in \mathbb{Z}$, then for every $f \in \mathsf{C}_{2\pi}$

$$\pi_n(x) = \frac{1}{2\pi}\int_{-\pi}^{\pi} \sigma_n(f; x-u)\, d\mu(u). \tag{6.5.7}$$

Since $[\sigma_m(f; x) - \sigma_n(f; x)]$ tends uniformly to zero as $m, n \to \infty$ by Theorem 6.1.2, so does $[\pi_m(x) - \pi_n(x)]$. Hence Theorem 6.1.2 applies again and shows that $\lambda(k)f^{\wedge}(k)$ are the Fourier coefficients of a $\mathsf{C}_{2\pi}$-function.

Now the representation (6.5.2) is valid since both sides have the same finite Fourier transform, whereas the relation (6.5.3) is a consequence of Problem 1.3.1. We finally note that the function μ may be expressed in terms of the multiplier λ by

$$\mu(x) - \mu(0) = \lambda(0)x + \lim_{n\to\infty} \sum_{k=-n}^{n}{}' \lambda(k)\frac{e^{ikx}-1}{ik}. \tag{6.5.8}$$

This is given by Prop. 4.3.10.

6.5.2 Multipliers on L^p

A function $\lambda(v)$ on $\mathbb{R}$ is said to be a *multiplier* of type $(\mathsf{L}^p, \mathsf{L}^q)$, $1 \le p, q \le 2$, if for every $f \in \mathsf{L}^p$ there exists a function $g \in \mathsf{L}^q$ such that $\lambda(v)f^{\wedge}(v) = g^{\wedge}(v)$. As in the periodic case one may prove

Proposition 6.5.4. *Let λ be a multiplier of type $(\mathsf{L}^p, \mathsf{L}^q)$, $1 \le p, q \le 2$, and let U be defined as the operator of L^p into L^q which assigns to each $f \in \mathsf{L}^p$ the function $Uf = g$ where $g \in \mathsf{L}^q$ is given by $\lambda(v)f^{\wedge}(v) = g^{\wedge}(v)$. Then the* ***multiplier operator*** *is linear, bounded, and translation-invariant.*

Parallel to the treatment in the preceding subsection we now discuss the important case of multiplier classes $(\mathsf{L}^p, \mathsf{L}^p)$, $1 \le p \le 2$.

Theorem 6.5.5. *A function λ is a multiplier of type $(\mathsf{L}^2, \mathsf{L}^2)$ if and only if $\lambda \in \mathsf{L}^\infty$. In this case,*

$$\|U\|_{[\mathsf{L}^2, \mathsf{L}^2]} = \|\lambda\|_\infty. \tag{6.5.9}$$

Proof. Let λ be a multiplier of type $(\mathsf{L}^2, \mathsf{L}^2)$. Then by Prop. 6.5.4 there exists a constant $M > 0$ such that $\|Uf\|_2 \le M\,\|f\|_2$ for all $f \in \mathsf{L}^2$. Hence by the Parseval equation

$$\int_{-\infty}^{\infty} |\lambda(v)f^{\wedge}(v)|^2\, dv \le M^2 \int_{-\infty}^{\infty} |f^{\wedge}(v)|^2\, dv. \tag{6.5.10}$$

Moreover, by Plancherel's theorem we may choose any convenient L^2-function for $f^{\wedge}$. Thus, if we put $f^{\wedge} = \kappa_{[-\rho,\rho]}$, the characteristic function of the interval $[-\rho, \rho]$, then it follows from (6.5.10) that $\int_{-\rho}^{\rho} |\lambda(v)|^2\, dv \le 2M^2\rho$, i.e. λ is measurable and integrable over every finite interval. On the other hand, if we substitute

$$\kappa_{[-\rho,\rho]}(v)(4\pi t)^{-1/4}\exp\{-(x-v)^2/8t\} \qquad (x \in \mathbb{R}, t > 0)$$

for $f^{\wedge}$, we obtain for $\lambda_\rho(v) \equiv \lambda(v)\kappa_{[-\rho,\rho]}(v)$ that

$$\frac{1}{\sqrt{4\pi t}}\int_{-\infty}^{\infty} |\lambda_\rho(v)|^2 e^{-(x-v)^2/4t}\, dv \le \frac{M^2}{\sqrt{4\pi t}}\int_{-\rho}^{\rho} e^{-(x-v)^2/4t}\, dv \le M^2.$$

By Cor. 3.2.3 the left side tends to $|\lambda_\rho(x)|^2$ a.e. as $t \to 0+$, since $\lambda_\rho \in \mathsf{L}^1$ for each $\rho > 0$. Therefore $\|\lambda_\rho\|_\infty \le M$, and since this is true for all $\rho > 0$, we have $\|\lambda\|_\infty \le M$.

Conversely, if $\lambda \in \mathsf{L}^\infty$, then $\lambda f^\wedge \in \mathsf{L}^2$ for every $f \in \mathsf{L}^2$, and it follows by Plancherel's theorem that there exists $g \in \mathsf{L}^2$ such that $\lambda(v)f^\wedge(v) = g^\wedge(v)$, i.e. λ is a multiplier of type $(\mathsf{L}^2, \mathsf{L}^2)$. Moreover, $\|Uf\|_2 = \|\lambda f^\wedge\|_2 \le \|\lambda\|_\infty \|f\|_2$ by the Parseval equation, and therefore $\|U\| \le \|\lambda\|_\infty$. Together with the opposite inequality $\|\lambda\|_\infty \le \|U\|$ obtained above, this proves (6.5.9).

Theorem 6.5.6. *A necessary and sufficient condition for a function* λ *to be a multiplier of type* $(\mathsf{L}^1, \mathsf{L}^1)$ *is that* λ *is the Fourier–Stieltjes transform of a function* $\mu \in \mathsf{BV}$. *In this case,*

$$(6.5.11)\qquad (Uf)(x) = \frac{1}{\sqrt{2\pi}} \int_{-\infty}^{\infty} f(x-u)\, d\mu(u) \quad \text{a.e.}$$

for every $f \in \mathsf{L}^1$ *and*

$$(6.5.12)\qquad \|U\|_{[\mathsf{L}^1, \mathsf{L}^1]} = \|\mu\|_{\mathsf{BV}}.$$

Proof. *Necessity.* If λ is a multiplier of type $(\mathsf{L}^1, \mathsf{L}^1)$, then the corresponding multiplier operator U is a linear, bounded, translation-invariant operator of L^1 into L^1 by Prop. 6.5.4. Let $f, g \in \mathsf{L}^1$. Then $f * g \in \mathsf{L}^1$ by Prop. 0.2.2, and thus $(U[f*g])(x)$ is well-defined as a function in L^1. By the definition of U and the convolution theorem we have $[U[f*g]]^\wedge(v) = \lambda(v)f^\wedge(v)g^\wedge(v) = [f*Ug]^\wedge(v)$, and hence by the uniqueness theorem

$$(6.5.13)\qquad (U[f*g])(x) = (f*Ug)(x) \quad \text{a.e.}$$

Let $\{\chi(x;\rho)\}$ be a positive approximate identity. Then $\lim_{\rho\to\infty} \|I(f;\circ;\rho) - f(\circ)\|_1 = 0$ for every $f \in \mathsf{L}^1$ by Theorem 3.1.6, and since $(U[I(f;\circ;\rho)])(x) = (f * U[\chi(\circ;\rho)])(x)$ a.e. by (6.5.13), we have for every $f \in \mathsf{L}^1$

$$(6.5.14)\qquad \lim_{\rho\to\infty} \|f * U[\chi(\circ;\rho)] - Uf\|_1 = 0.$$

If we set $\mu_\rho(x) = \int_{-\infty}^x (U[\chi(\circ;\rho)])(u)\, du$, this defines a set of absolutely continuous functions μ_ρ which are of uniformly bounded variation since $\|\chi(\circ;\rho)\|_1 = 1$ and

$$(6.5.15)\qquad \|\mu_\rho\|_{\mathsf{BV}} = \|U[\chi(\circ;\rho)](\circ)\|_1 \le \|U\| \qquad (\rho > 0).$$

Therefore by the weak* compactness theorem for C_0 there exists $\{\rho_j\}$ with $\lim_{j\to\infty} \rho_j = \infty$ and $\mu \in \mathsf{BV}$ with $\|\mu\|_{\mathsf{BV}} \le \|U\|$ such that for every $f \in \mathsf{C}_0$

$$\lim_{j\to\infty} \int_{-\infty}^{\infty} f(x-u)\, d\mu_{\rho_j}(u) = \int_{-\infty}^{\infty} f(x-u)\, d\mu(u).$$

This together with (6.5.14) and Prop. 0.1.10 shows that (6.5.11) holds for every $f \in \mathsf{C}_0 \cap \mathsf{L}^1$. However, both members in (6.5.11) define bounded linear operators of L^1 into L^1 (cf. Prop. 0.5.6), and thus, since $\mathsf{C}_0 \cap \mathsf{L}^1$ is dense in L^1, it follows that (6.5.11) holds for all $f \in \mathsf{L}^1$. In particular, we have $\|U\| \le \|\mu\|_{\mathsf{BV}}$, which together with the opposite inequality obtained above shows that (6.5.12) is valid. Furthermore, by the definition of U and by Prop. 5.3.3, the representation (6.5.11) implies that $\lambda f^\wedge = [Uf]^\wedge = f^\wedge \mu^\vee$ for every $f \in \mathsf{L}^1$. Therefore λ is the Fourier–Stieltjes transform of $\mu \in \mathsf{BV}$.

Sufficiency. Suppose there exists $\mu \in \mathsf{BV}$ such that $\lambda = \mu^\vee$. Then with each $f \in \mathsf{L}^1$ we may associate the function $g(x) \equiv (f * d\mu)(x)$ which belongs to L^1 by Prop. 0.5.6. In view of Prop. 5.3.3 we then have $\mu^\vee f^\wedge = g^\wedge$. Thus λ is a multiplier of type $(\mathsf{L}^1, \mathsf{L}^1)$, and the proof is complete.

As a consequence of Lévy's theorem we mention that μ is uniquely determined by the multiplier λ and given by

$$\mu(x) - \mu(0) = \lim_{\rho\to\infty} \frac{1}{\sqrt{2\pi}} \int_{-\rho}^{\rho} \frac{e^{ixv} - 1}{iv} \lambda(v)\, dv.$$

We now turn to the problem of deriving necessary and sufficient conditions for a bounded measurable function to be a multiplier of type $(\mathsf{L}^p, \mathsf{L}^p)$, $1 < p < 2$.

Theorem 6.5.7. *Let λ be a bounded measurable function on R. Then λ is a multiplier of type $(\mathsf{L}^p, \mathsf{L}^p)$, $1 < p < 2$, if and only if there is a constant M such that*

$$\left|\frac{1}{\sqrt{2\pi}}\int_{-\infty}^{\infty} \lambda(v)\phi_1^\wedge(v)\phi_2^\wedge(v)\, dv\right| \leq M\, \|\phi_1\|_p\|\phi_2\|_{p'} \tag{6.5.16}$$

for all $\phi_1, \phi_2 \in \mathsf{F}$. In this case, $\|U\|_{[\mathsf{L}^p,\mathsf{L}^p]} = M^*$, where M^* is the smallest constant satisfying (6.5.16).

Proof. Let λ be a multiplier of type $(\mathsf{L}^p, \mathsf{L}^p)$, $1 < p < 2$. Then, in particular, for each $\phi_1 \in \mathsf{F}$ (cf. (6.1.10)) there exists $g_1 \in \mathsf{L}^p$ such that $\lambda\phi_1^\wedge = g_1^\wedge$. Since $\lambda \in \mathsf{L}^\infty$ and $\phi_1^\wedge \in \mathsf{L}^1$, we have $g_1^\wedge \in \mathsf{L}^1$, and hence by Prop. 5.2.16

$$g_1(x) = \frac{1}{\sqrt{2\pi}}\int_{-\infty}^{\infty} \lambda(v)\phi_1^\wedge(v)\, e^{ixv}\, dv \quad \text{a.e.},$$

i.e. $g_1(x) = [\lambda\phi_1^\wedge]^\wedge(-x)$ a.e. Therefore, if U is the multiplier operator corresponding to λ, then $U\phi_1 = g_1$, and we have by Parseval's formula (5.2.18) and Hölder's inequality for every $\phi_1, \phi_2 \in \mathsf{F}$

$$\left|\frac{1}{\sqrt{2\pi}}\int_{-\infty}^{\infty} \lambda(v)\phi_1^\wedge(v)\phi_2^\wedge(v)\, dv\right| = \left|\frac{1}{\sqrt{2\pi}}\int_{-\infty}^{\infty} [\lambda\phi_1^\wedge]^\wedge(v)\phi_2(v)\, dv\right|$$
$$\leq \|[\lambda\phi_1^\wedge]^\wedge\|_p\|\phi_2\|_{p'} = \|U\phi_1\|_p\|\phi_2\|_{p'} \leq \|U\|\ \|\phi_1\|_p\|\phi_2\|_{p'}.$$

Hence (6.5.16) is valid with $M^* \leq \|U\|$.

Conversely, if (6.5.16) holds, then for every fixed $\phi_1 \in \mathsf{F}$ the integral there defines a bounded linear functional over F considered as a subset of $\mathsf{L}^{p'}$. Since F is dense in $\mathsf{L}^{p'}$, it may be extended to all of $\mathsf{L}^{p'}$. Therefore by Prop. 0.8.11 there exists $g_1 \in \mathsf{L}^p$ with $\|g_1\|_p \leq M^*\|\phi_1\|_p$ such that for every $\phi_2 \in \mathsf{F}$

$$\int_{-\infty}^{\infty} \lambda(v)\phi_1^\wedge(v)\phi_2^\wedge(v)\, dv = \int_{-\infty}^{\infty} \phi_2(v)g_1(-v)\, dv = \int_{-\infty}^{\infty} g_1^\wedge(v)\phi_2^\wedge(v)\, dv$$

upon using Parseval's formula (5.2.27). If we take $\phi_2(u) = \exp\{-tu^2 + ixu\}$, $x \in \mathbb{R}$, $t > 0$, as in the proof of Theorem 6.1.6, it hence follows that to every $\phi_1 \in \mathsf{F}$ there exists $g_1 \in \mathsf{L}^p$ such that $\lambda\phi_1^\wedge = g_1^\wedge$. Let T be the operator on F (considered as a subspace of L^p) into L^p given by $T\phi_1 = g_1$. Upon taking the supremum over all $\phi_1 \in \mathsf{F}$ with $\|\phi_1\|_p \leq 1$, we have by (0.7.5) $\|T\|_{[\mathsf{F},\mathsf{L}^p]} = \sup \|T\phi_1\|_p = \sup \|g_1\|_p \leq M^*$. By Prop. 0.7.1, T may be extended to all of L^p with the same norm, thus $\|T\|_{[\mathsf{L}^p,\mathsf{L}^p]} \leq M^*$.

Let $f \in \mathsf{L}^p$ and $Tf = g$. Then there exists a sequence $\{\phi_n\} \subset \mathsf{F}$ such that $\lim_{n\to\infty} \|f - \phi_n\|_p = 0$. Furthermore, $\lim_{n\to\infty} \|Tf - T\phi_n\|_p = 0$ by the continuity of T, and $\lim_{n\to\infty} \|f^\wedge - \phi_n^\wedge\|_{p'} = 0$, $\lim_{n\to\infty} \|g^\wedge - [T\phi_n]^\wedge\|_{p'} = 0$ by Titchmarsh's inequality. Since by definition $\lambda\phi_n^\wedge = [T\phi_n]^\wedge$, we finally obtain $\|\lambda f^\wedge - g^\wedge\|_{p'} = 0$. Hence λ is a multiplier of type $(\mathsf{L}^p, \mathsf{L}^p)$. Moreover, T is nothing but U, and $\|U\|_{[\mathsf{L}^p,\mathsf{L}^p]} = M^*$.

Theorem 6.5.8. *Let λ be a bounded continuous function on $\mathbb{R}$. Then λ is a multiplier of type $(\mathsf{L}^p, \mathsf{L}^p)$, $1 < p < 2$, if and only if there is a constant M_1 such that*

$$\left|\sum_{k=1}^{n} c_k\, d_k\, \lambda(v_k)\right| \tag{6.5.17}$$
$$\leq M_1\left\{\limsup_{\rho\to\infty}\frac{1}{2\rho}\int_{-\rho}^{\rho}\left|\sum_{k=1}^{n} c_k\, e^{iv_k u}\right|^p du\right\}^{1/p}\left\{\limsup_{\rho\to\infty}\frac{1}{2\rho}\int_{-\rho}^{\rho}\left|\sum_{k=1}^{n} d_k\, e^{iv_k u}\right|^{p'} du\right\}^{1/p'}$$

for all finite sets of real numbers $\{v_k\}$ and complex numbers $\{c_k\}$, $\{d_k\}$. In this case,

$$\|U\|_{[L^p, L^p]} = M_1^*, \tag{6.5.18}$$

where M_1^* is the smallest constant satisfying (6.5.17).

Proof. *Necessity.* Let the bounded continuous function λ be a multiplier of type $(\mathsf{L}^p, \mathsf{L}^p)$. Then there is a constant M such that (6.5.16) holds. If for any $n \in \mathbb{N}$, arbitrary sets of real numbers $v_1, \dots, v_n$ and complex numbers $c_1, \dots, c_n, d_1, \dots, d_n$, and any $t > 0$ we take

$$\phi_1(x) = e^{-tp'x^2} \sum_{k=1}^n c_k e^{iv_k x}, \qquad \phi_2(x) = e^{-tpx^2} \sum_{k=1}^n d_k e^{iv_k x},$$

then by Problem 5.1.2(i)

$$\phi_1^{\wedge}(u) = \frac{1}{\sqrt{2tp'}} \sum_{k=1}^n c_k e^{-(v_k - u)^2/4tp'}, \qquad \phi_2^{\wedge}(u) = \frac{1}{\sqrt{2tp}} \sum_{k=1}^n d_k e^{-(v_k - u)^2/4tp}.$$

Therefore by (6.5.16)

$$\begin{aligned} &\left| \frac{1}{\sqrt{4\pi t}} \int_{-\infty}^{\infty} \lambda(u) \left\{ \sum_{j=1}^n c_j e^{-(v_j - u)^2/4tp'} \right\} \left\{ \sum_{k=1}^n d_k e^{-(v_k - u)^2/4tp} \right\} du \right| \\ &\quad \le M \left\{ \sqrt{\frac{tpp'}{\pi}} \int_{-\infty}^{\infty} \left| \sum_{k=1}^n c_k e^{iv_k u} \right|^p e^{-tpp'u^2} du \right\}^{1/p} \\ &\qquad \left\{ \sqrt{\frac{tpp'}{\pi}} \int_{-\infty}^{\infty} \left| \sum_{k=1}^n d_k e^{iv_k u} \right|^{p'} e^{-tpp'u^2} du \right\}^{1/p'}. \end{aligned} \tag{6.5.19}$$

Let $x_1 \ne x_2$, $x_1, x_2 \in \mathbb{R}$. Then

$$\lim_{t \to 0+} \frac{1}{\sqrt{4\pi t}} \int_{-\infty}^{\infty} e^{-(x_1 - u)^2/4tp'} e^{-(x_2 - u)^2/4tp} du = 0, \tag{6.5.20}$$

and therefore for every $\lambda \in \mathsf{L}^\infty$

$$\lim_{t \to 0+} \frac{1}{\sqrt{4\pi t}} \int_{-\infty}^{\infty} \lambda(u) e^{-(x_1 - u)^2/4tp'} e^{-(x_2 - u)^2/4tp} du = 0. \tag{6.5.21}$$

Since we may assume without loss of generality that $v_j \ne v_k$ for $j \ne k$, $1 \le j, k \le n$, we obtain by Prop. 3.2.1

$$\begin{aligned} &\lim_{t \to 0+} \frac{1}{\sqrt{4\pi t}} \int_{-\infty}^{\infty} \lambda(u) \left\{ \sum_{j=1}^n c_j e^{-(v_j - u)^2/4tp'} \right\} \left\{ \sum_{k=1}^n d_k e^{-(v_k - u)^2/4tp} \right\} du \\ &\quad = \sum_{k=1}^n c_k d_k \lim_{t \to 0+} \frac{1}{\sqrt{4\pi t}} \int_{-\infty}^{\infty} \lambda(u) e^{-(v_k - u)^2/4t} du = \sum_{k=1}^n c_k d_k \lambda(v_k). \end{aligned} \tag{6.5.22}$$

On the other hand,

$$\frac{1}{2\rho} \int_{-\rho}^{\rho} \left| \sum_{k=1}^n c_k e^{iv_k u} \right|^p du = \frac{1}{2} \int_{-1}^{1} \left| \sum_{k=1}^n e^{i\rho v_k u} \right|^p du,$$

and upon setting $1/t = \rho^2 pp'$

$$\sqrt{\frac{tpp'}{\pi}} \int_{-\infty}^{\infty} \left| \sum_{k=1}^n c_k e^{iv_k u} \right|^p e^{-tpp'u^2} du = \frac{1}{\sqrt{\pi}} \int_{-\infty}^{\infty} \left| \sum_{k=1}^n c_k e^{i\rho v_k u} \right|^p e^{-u^2} du.$$

Thus taking $\limsup_{\rho \to \infty}$, an application of Lebesgue's dominated convergence theorem shows that

$$\limsup_{t \to 0+} \sqrt{\frac{tpp'}{\pi}} \int_{-\infty}^{\infty} \left| \sum_{k=1}^n c_k e^{iv_k u} \right|^p e^{-tpp'u^2} du = \limsup_{\rho \to \infty} \frac{1}{2\rho} \int_{-\rho}^{\rho} \left| \sum_{k=1}^n c_k e^{iv_k u} \right|^p du.$$

Since the expressions containing the d_k satisfy a similar relation, inequality (6.5.17) is a consequence of (6.5.19) and (6.5.22) with $M_1^* \le \|U\|$.

Sufficiency. Suppose (6.5.17) holds. In order to show that λ is a multiplier of type $(\mathsf{L}^p, \mathsf{L}^p)$, $1 < p < 2$, it suffices to establish (6.5.16) for any two functions $\phi_1, \phi_2 \in \mathsf{F}$ or even for any $\phi_1, \phi_2 \in \mathsf{B}$ (cf. (6.2.5)).

To this end we proceed as in the proof of Theorem 6.2.2. Let $\phi_1, \phi_2 \in \mathsf{B}$ be arbitrary and $l > 0$ be such that the supports of ϕ_1 and ϕ_2 are both contained in $(-l, l)$. If ψ_1, ψ_2 are the $2l$-periodic extensions of ϕ_1, ϕ_2, then (6.2.7)–(6.2.9) hold for ϕ_i, ψ_i, $i = 1, 2$. Furthermore, for

$$\Omega(l; \overline{(\phi_1 * \phi_2)(-\circ)}) = \frac{1}{\sqrt{2\pi}} \sum_{k=-\infty}^{\infty} \lambda\Big(\frac{k\pi}{l}\Big)\phi_1^\wedge\Big(\frac{k\pi}{l}\Big)\phi_2^\wedge\Big(\frac{k\pi}{l}\Big)\cdot\frac{\pi}{l} \tag{6.5.23}$$

we have in view of (6.2.15)

$$\lim_{\varepsilon\to 0+} \frac{1}{\varepsilon}\int_0^\varepsilon \Omega\Big(\frac{1}{u}; \overline{(\phi_1 * \phi_2)(-\circ)}\Big)\,du = \frac{1}{\sqrt{2\pi}}\int_{-\infty}^{\infty} \lambda(v)\phi_1^\wedge(v)\phi_2^\wedge(v)\,dv. \tag{6.5.24}$$

On the other hand, if (6.5.17) holds for arbitrary finite sets $v_1, \ldots, v_n \in \mathbb{R}$, $c_1, \ldots, c_n$, $d_1, \ldots, d_n \in \mathbb{C}$, it moreover holds for all countable sets in case the series occurring are absolutely convergent. Therefore, in virtue of (6.2.7)–(6.2.9)

$$\begin{aligned}|\Omega(l; \overline{(\phi_1 * \phi_2)(-\circ)})| &= \Big|\frac{l}{\pi\sqrt{2\pi}} \sum_{k=-\infty}^{\infty} \phi_1^\wedge\Big(\frac{k\pi}{l}\Big)\frac{\pi}{l}\cdot\phi_2^\wedge\Big(\frac{k\pi}{l}\Big)\frac{\pi}{l}\cdot\lambda\Big(\frac{k\pi}{l}\Big)\Big|\\ &\le 2lM_1\Big\{\limsup_{\rho\to\infty}\frac{1}{2\rho\sqrt{2\pi}}\int_{-\rho}^{\rho}|\psi_1(u)|^p\,du\Big\}^{1/p}\\ &\qquad\Big\{\limsup_{\rho\to\infty}\frac{1}{2\rho\sqrt{2\pi}}\int_{-\rho}^{\rho}|\psi_2(u)|^{p'}\,du\Big\}^{1/p'}.\end{aligned}$$

If $\rho = ln$, $n \in \mathbb{N}$, then in view of the $2l$-periodicity of ψ_1

$$\begin{aligned}\lim_{\rho\to\infty}\frac{1}{2\rho}\int_{-\rho}^{\rho}|\psi_1(u)|^p\,du &= \lim_{n\to\infty}\frac{1}{2ln}\sum_{k=0}^{n-1}\int_{-nl+2kl}^{-nl+2(k+1)l}|\psi_1(u)|^p\,du\\ &= \lim_{n\to\infty}\frac{1}{2ln}\sum_{k=0}^{n-1}\int_{-l}^{l}|\psi_1(u)|^p\,du = \frac{1}{2l}\int_{-l}^{l}|\psi_1(u)|^p\,du,\end{aligned}$$

which now is easily seen to be a bound also for $\limsup_{\rho\to\infty}$ in case $\rho > 0$ is arbitrary. Since a similar relation holds for ψ_2 and since $\psi_i(u) = \phi_i(u)$, $i = 1, 2$, for $|u| \le l$, it therefore follows by (6.5.24) that

$$\Big|\frac{1}{\sqrt{2\pi}}\int_{-\infty}^{\infty}\lambda(v)\phi_1^\wedge(v)\phi_2^\wedge(v)\,dv\Big| \le M_1\|\phi_1\|_p\|\phi_2\|_{p'}$$

for every $\phi_1, \phi_2 \in \mathsf{B}$. Thus λ is a multiplier of type $(\mathsf{L}^p, \mathsf{L}^p)$. Moreover, $\|U\| \le M_1^*$, and since we have already seen that the opposite inequality is valid as well, (6.5.18) holds.

We observe that in case $p = 1$ Theorems 6.2.1 and 6.5.6 state that a bounded continuous function λ is a multiplier of type $(\mathsf{L}^1, \mathsf{L}^1)$ if and only if λ satisfies (6.2.1). In this respect the result of Theorem 6.5.8 may be regarded as a certain extension of the representation theorem of S. Bochner to $1 < p < 2$.

Problems

1. (i) Give the definition of a multiplier of type $(\mathsf{BV}_{2\pi}, \mathsf{BV}_{2\pi})$, $(\mathsf{BV}_{2\pi}, \mathsf{Y}_1)$, $(\mathsf{Y}_1, \mathsf{BV}_{2\pi})$.
   (ii) Show that $\lambda \in (\mathsf{BV}_{2\pi}, \mathsf{BV}_{2\pi})$ if and only if λ is the finite Fourier–Stieltjes transform of some $\mu \in \mathsf{BV}_{2\pi}$, i.e. $(\mathsf{BV}_{2\pi}, \mathsf{BV}_{2\pi}) = [\mathsf{BV}_{2\pi}]^\vee$. (Hint: For the necessity use (4.3.2), (4.3.3); see also Zygmund [7I, p. 176])

(iii) Show that $(\mathsf{L}^\infty_{2\pi}, \mathsf{L}^\infty_{2\pi}) = (\mathsf{L}^1_{2\pi}, \mathsf{L}^1_{2\pi}) = (\mathsf{L}^1_{2\pi}, \mathsf{BV}_{2\pi}) = (\mathsf{C}_{2\pi}, \mathsf{L}^\infty_{2\pi}) = [\mathsf{BV}_{2\pi}]^\vee$ and that for the corresponding multiplier operator U one has the representation (6.5.2) (a.e.). (Hint: See e.g. EDWARDS [1II, p. 254 ff])

2. (i) Show that $(\mathsf{L}^\infty_{2\pi}, \mathsf{L}^\infty_{2\pi}) \subset (\mathsf{L}^p_{2\pi}, \mathsf{L}^p_{2\pi})$ for every $1 < p < \infty$.
(ii) Show that $\mathsf{l}^1 \subset (\mathsf{BV}_{2\pi}, \mathsf{C}_{2\pi})$, $\mathsf{l}^2 \subset (\mathsf{BV}_{2\pi}, \mathsf{L}^2_{2\pi})$, and $\mathsf{l}^p \subset (\mathsf{BV}_{2\pi}, \mathsf{L}^{p'}_{2\pi})$, $1 < p < 2$. (Hint: Use Prop. 4.3.2(ii), the theorem of Riesz–Fischer, and Prop. 4.2.7)
(iii) Let $\lambda \in \mathsf{l}^\infty_0$ be even and quasi-convex. Show that λ is a multiplier of type $(\mathsf{L}^p_{2\pi}, \mathsf{L}^p_{2\pi})$ and that the corresponding multiplier operator U is expressible as the convolution with an integrable function. (Hint: Theorem 6.3.7)

3. Let λ be a multiplier of type $(\mathsf{Y}_1, \mathsf{Y}_2)$ and U the corresponding multiplier operator.
(i) Show that $U(f*t_n) = Uf*t_n = f*Ut_n$ for every $f \in \mathsf{Y}_1$, $t_n \in \mathsf{T}_n$.
(ii) If $\mathsf{Y}_1 = \mathsf{Y}_2 = \mathsf{L}^1_{2\pi}$, show that $U(f_1*f_2) = Uf_1*f_2 = f_1*Uf_2$ for every $f_1, f_2 \in \mathsf{L}^1_{2\pi}$.

4. (i) Let $1 < p \le \infty$. Show that $\lambda \in (\mathsf{L}^1_{2\pi}, \mathsf{L}^p_{2\pi})$ if and only if $\lambda \in [\mathsf{L}^p]^\wedge$. In this case there exists $g \in \mathsf{L}^p_{2\pi}$ such that $Uf = f*g$ for every $f \in \mathsf{L}^1_{2\pi}$.
(ii) Let $1 \le p \le \infty$. Show that $\lambda \in (\mathsf{L}^p_{2\pi}, \mathsf{C}_{2\pi})$ if and only if $\lambda \in [\mathsf{L}^{p'}_{2\pi}]^\wedge$. In this case there exists $g \in \mathsf{L}^{p'}_{2\pi}$ such that $Uf = f*g$ for every $f \in \mathsf{L}^p_{2\pi}$.
(iii) Show that $\lambda \in (\mathsf{BV}_{2\pi}, \mathsf{C}_{2\pi})$ if and only if $\lambda \in [\mathsf{C}_{2\pi}]^\wedge$. In this case there exists $g \in \mathsf{C}_{2\pi}$ such that $Uf = g*\,df$ for every $f \in \mathsf{BV}_{2\pi}$.
(iv) Show that $\lambda \in (\mathsf{BV}_{2\pi}, \mathsf{L}^1_{2\pi})$ if and only if $\lambda \in [\mathsf{L}^1_{2\pi}]^\wedge$. In this case there exists $g \in \mathsf{L}^1_{2\pi}$ such that $Uf = g*\,df$ for every $f \in \mathsf{BV}_{2\pi}$.
(Hint: ZYGMUND [7I, pp. 175–177], EDWARDS [1II, pp. 255–257])

5. (i) Give the definition of a multiplier of type $(\mathsf{BV}, \mathsf{BV})$.
(ii) Show that $\lambda \in (\mathsf{BV}, \mathsf{BV})$ if and only if $\lambda \in [\mathsf{BV}]^\vee$. In this case there exists $\mu \in \mathsf{BV}$ such that $Uf = f*\,d\mu$ for every $f \in \mathsf{BV}$. (Hint: For the necessity use (5.3.3), (5.3.4))

6. (i) Show that $(\mathsf{L}^1, \mathsf{L}^1) \subset (\mathsf{L}^p, \mathsf{L}^p)$ for every $1 < p \le 2$.
(ii) Let $\lambda \in \mathsf{C}_0$ be even and quasi-convex. Show that λ is a multiplier of type $(\mathsf{L}^p, \mathsf{L}^p)$, $1 \le p \le 2$, and that the corresponding multiplier operator U is expressible as the convolution with a function $g \in \mathsf{L}^1$. (Hint: Theorem 6.3.11)

7. Let λ be a multiplier of type $(\mathsf{L}^p, \mathsf{L}^p)$, $1 \le p \le 2$, and U the corresponding multiplier operator. For $g \in \mathsf{L}^1$ show that $U(f*g) = Uf*g$ for every $f \in \mathsf{L}^p$.

8. Prove (6.5.20), (6.5.21).

9. State and prove (if necessary at all) counterparts of Theorems 6.5.7, 6.5.8 for the limiting cases $p = 1$, $p = 2$.

10. Let $\mathsf{C}_{2\pi,N}$ be the set of those $f \in \mathsf{C}_{2\pi}$ for which the Fourier series converges uniformly. If λ is an even function on $\mathbb{Z}$, show that $\lambda \in (\mathsf{C}_{2\pi}, \mathsf{C}_{2\pi,N})$ if and only if
$$\int_0^\pi \Big|\lambda(0) + \sum_{k=1}^n \lambda(k)\cos ku\Big|\,du = O(1) \qquad (n \to \infty).$$
(Hint: Use the theorem of Banach–Steinhaus and Prop. 1.3.1; see also KARAMATA [1], GOES [1])

11. Let $1 < p, q < \infty$. Show that $(\mathsf{L}^p_{2\pi}, \mathsf{L}^q_{2\pi}) = (\mathsf{L}^{q'}_{2\pi}, \mathsf{L}^{p'}_{2\pi})$. In particular, $(\mathsf{L}^p_{2\pi}, \mathsf{L}^p_{2\pi}) = (\mathsf{L}^{p'}_{2\pi}, \mathsf{L}^p_{2\pi})$. (Hint: See ZYGMUND [7I, p. 178], EDWARDS [1II, p. 265])

6.6 Notes and Remarks

Sec. 6.1. The material of Sec. 6.1.1 is standard and may be found in ZYGMUND [7I, p. 136 ff], HEWITT [1, p. 89 ff], HOFFMAN [1, p. 22 ff] (see also the references cited there). In many of the books the conditions for representation are expressed in terms of the particular Cesàro or Abel means. In the latter case see also FICHTENHOLZ [1].

The representation problem is connected with the moment problem for a finite interval. Given a sequence of numbers $\{\alpha_k\}_{k=0}^{\infty}$, it is required to find necessary and sufficient conditions which determine $f \in \mathsf{L}^p(0, 1)$ such that $\alpha_k = \int_0^1 u^k f(u)\, du$ for all $k \in \mathbb{P}$. More generally, the powers $\{x^k\}_{k=0}^{\infty}$ are replaced by an arbitrary sequence of functions $\{\varphi_k(x)\}_{k=0}^{\infty} \subset \mathsf{L}^{p'}(a, b)$. Correspondingly, the problem is to find conditions for the existence of $\mu \in \mathsf{BV}[a, b]$ such that $\alpha_k = \int_a^b \varphi_k(u)\, d\mu(u)$ for all $k \in \mathbb{P}$, $\{\varphi_k\}_{k=0}^{\infty}$ being a sequence in $\mathsf{C}[a, b]$. In the particular case that $\varphi_k(x) = (1/2\pi)\, e^{-ikx}$ and $[a, b] = [-\pi, \pi]$ one speaks of the trigonometric moment problem (see ACHIESER [1, p. 178 ff]). Our conditions are connected with those given by HAUSDORFF [1]. Thus $\alpha_k = (1/2\pi) \int_{-\pi}^{\pi} e^{-iku}\, d\mu(u)$, $k \in \mathbb{Z}$, with $\mu \in \mathsf{BV}_{2\pi}$ if and only if for $x_l = x_0 + (l\pi/(n + 1))$, $x_0 \in \mathbb{R}$ arbitrary,

$$\frac{1}{n+1} \sum_{l=0}^{2n+1} \left| \sum_{k=-n}^{n} \left(1 - \frac{|k|}{n+1}\right) \alpha_k\, e^{ix_l k} \right| = O(1) \qquad (n \to \infty).$$

For further details see KACZMARZ–STEINHAUS [1, p. 31 ff], WIDDER [1, p. 100 ff], LORENTZ [1, p. 57 f, p. 77 f], ACHIESER [1], ACHIESER–KREIN [1]. The term moment problem is first found in the work of T. STIELTJES of 1894–1895. He considers the case that (a, b) is the infinite interval $(0, \infty)$, while H. HAMBURGER was concerned with the interval $(-\infty, \infty)$.

The problem of Sec. 6.1.1 has also been considered in case the system $\{e^{ikx}\}_{k=-\infty}^{\infty}$ is replaced by a general orthogonal system of functions, thus determining necessary and sufficient conditions such that an orthogonal series is the Fourier series (with respect to the underlying orthogonal system) of a function belonging to a definite class. See KACZMARZ–STEINHAUS [1, p. 214 ff].

Concerning Sec. 6.1.2, Theorem 6.1.5 is due to CRAMÉR [1] (1939). His proof made use of Theorem 6.2.3; the present proof (see PFLUGER [1], BUTZER–TREBELS [2, p. 15 f]) rests upon the Riesz representation theorem for bounded linear functionals on C_0 and certain Parseval formulae. It is indeed a straightforward modification of the method used originally by F. RIESZ [3] (1933) for the representation of positive-definite functions. DOSS [1] also treats the criterion of Cramér but he proves the sufficiency by reduction to a theorem of BERRY [1] of 1931. The Cramér result is a standard topic in most books on probability theory, see e.g. LUKACS [1, p. 65] and LINNIK [1, p. 45]; see also KATZNELSON [1, p. 132]. The proof of Theorem 6.1.6, the L^p-version of the Cramér criterion, was suggested by work of OFFORD [1] devoted to the problem of defining a Fourier transform for L^p, $p > 2$. It is given in COOPER [3] and makes use of a selection principle, namely the weak* compactness theorem for L^p plus the Parseval formulae. Although it is related to the Riesz method, we regard it as a third method of proof of the representation theorems. It is, of course, also possible to prove Theorem 6.1.5 using the theorems of Helly and Helly–Bray. For functions of several variables the problem is considered in CRAMÉR [1], BUTZER–NESSEL [2], and Vol. II. For related results concerning the representation as a Laplace transform see WIDDER [1, p. 276 ff], BERENS–BUTZER [2], BERENS–WESTPHAL [1]; for the representation as a Weierstrass transform see HIRSCHMAN–WIDDER [5, p. 170 ff] and NESSEL [1].

Prop. 6.1.10 in the formulation of Problem 6.1.13 is given in KATZNELSON [1, p. 135] (who used Problem 6.2.4(i) in his proof). The present proof which rests upon the Cramér criterion is due to SUNOUCHI [8].

Sec. 6.2. The proof of the sufficiency part of Theorem 6.2.1 was originally given by BOCHNER [1] (1934) via Theorem 6.2.3. Our proof of the sufficiency is that of R. S. PHILLIPS [1] and is valid if f is only bounded and measurable on $\mathbb{R}$. It proceeds via the Riesz method; it has also been generalized by PHILLIPS so as to give a representation theorem for Banach space-valued functions.

The results of S. BOCHNER in 1932–1934 actually initiated the many papers on representation theory and its widespread application in various branches of harmonic analysis. Thus SCHOENBERG [2] using the Riesz method gave an integral analog of Theorem 6.2.1 which may be compared with a paper of BERRY [1] on the representation as L^p-Fourier transforms (cf. Problems 6.2.4(ii), 6.2.6, 6.2.7). The results of BERRY, seldom cited, are some of the earliest on the subject. Theorem 6.2.1 plays an important rôle in abstract harmonic analysis; it may be restated in terms of the Bohr compactification of the real line: a bounded function f on $\mathbb{R}$ is the Fourier–Stieltjes transform of a bounded measure on $\mathbb{R}$ if and only if it is continuous and the Fourier–Stieltjes transform of a measure on the Bohr compactification of $\mathbb{R}$. See EBERLEIN [1] and RUDIN [3, p. 32].

The periodic version of Theorem 6.2.1, due to F. RIESZ and S. BANACH, is given by Problem 6.2.1(i). Since $\exp(-ikx) \in \mathsf{L}^q_{2\pi}$ for all $1 \le q < \infty$, an analog for the representation as an $\mathsf{L}^p_{2\pi}$–Fourier transform is possible (see Problem 6.2.1(ii)). This result may also be regarded as a contribution to the trigonometric moment problem.

A further fundamental representation theorem is the one connected with positive–definite functions. For functions on the circle group this is to be found in Problem 6.2.2, on the line group in Problem 6.2.3. In the former case it is due to CARATHEODORY [1], HERGLOTZ [1], and TOEPLITZ [1] (all papers appearing in 1911), in the latter to BOCHNER [7, p. 92 ff, 325 ff]. It is commonly treated in books on the subject, e.g. KATZNELSON [1, p. 38, p. 137], R. R. GOLDBERG [1, p. 59 ff], LUKACS [1, p. 62 ff], LINNIK [1, p, 42 f]. BOCHNER was the first to recognize the key rôle of Problem 6.2.3 in harmonic analysis; he originally assumed f to be continuous. F. RIESZ [3] first showed that measurability was sufficient. For results concerning an integral analog of the positive–definite condition see also COOPER [1]. Problem 6.2.3 has furthermore been carried over (WEIL [1]) to locally compact abelian groups, see RUDIN [3, p. 17], REITER [1, p. 106] and the literature cited there. Theorem 6.2.3 is connected with the fundamental *continuity theorem* of probability theory due to P. LÉVY [1] (1922). A detailed discussion is given in BOCHNER [7, p. 85 ff, p. 321 ff; 6, p. 16 ff] and CRAMÉR [2, p. 96 ff]. The following formulation is to be found in LUKACS [1, p. 54]: *A sequence* $\{\mu_n(x)\}_{n=1}^{\infty}$ *of distribution functions converges weakly to a distribution function* μ *if and only if the sequence* $\{\mu_n^{\vee}(v)\}$ *of characteristic functions converges for every* v *to a function* $f(v)$ *which is continuous at* $v = 0$. *The limiting function* f *is then the characteristic function* (*i.e. Fourier–Stieltjes transform*) *of* μ. Here μ is called a distribution function if $\mu \in \mathsf{BV}$ increases monotonely with $\|\mu\|_{\mathsf{BV}} = 1$. A sequence $\{\mu_n(x)\}$ of distribution functions is said to converge weakly to a distribution function μ if $\lim_{n\to\infty} \mu_n(x) = \mu(x)$ for all continuity points of μ. Thus the continuity theorem indicates that the one-to-one correspondence between distribution functions and characteristic functions is continuous. A further criterion for characteristic functions is the MATHIAS–KHINTCHINE theorem found in all books on probability.

ROONEY [1, 2] is concerned with representation conditions of a somewhat different type.

Sec. 6.3. The parallel remarks in Sec. 6.3.1 on convex and quasi-convex functions on $\mathbb{Z}$ and $\mathbb{R}$ and especially their applications in the subsections to follow once again exhibit the common structure of harmonic analysis on different groups. For functions defined on $\mathbb{Z}$ these notions are to be found in BARI [1I, p. 3 f, II, p. 202], HEWITT [1, p. 61 ff], EDWARDS [1I, p. 110 ff]. Quasi-convexity (on $\mathbb{P}$) together with $f(k) \to 0$ as $k \to \infty$ does not imply that $f(k)$ is monotonely decreasing on $\mathbb{P}$; but it follows that f is of bounded variation on $\mathbb{P}$ (see Problem 6.3.4). Consequently Problem 1.2.7 cannot be applied in the proof of Theorem 6.3.7 as would be the case if f were convex on $\mathbb{P}$. On the other hand, if $f \in \mathsf{l}_0^{\infty}$ is even and monotonely decreasing on $\mathbb{P}$, f is not necessarily the Fourier transform of an $\mathsf{L}^1_{2\pi}$-function (see Problem 6.3.2).

For convex functions on $\mathbb{R}$ see especially HARDY–LITTLEWOOD–PÓLYA [1, p. 70 ff], ZYGMUND [7I, p. 21 f]. Frequently, f on $[a, b]$ is said to be convex if for any two points p_1, p_2 of the curve of f all the points of the arc p_1p_2 lie either below the chord p_1p_2 or on it. This (stronger) definition per se implies that f is continuous on (a, b) (see Problem 6.3.5(iv)). For Lemma 6.3.4 and Prop. 6.3.6 see also BERENS–GÖRLICH [1].

The material of Sec. 6.3.2 is standard and much of it due to A. N. KOLMOGOROV and W. H. YOUNG. See BARI [1II, p. 202 ff], EDWARDS [1I, p. 113 ff], ZYGMUND [7I, p. 182 ff]. For applications of Theorem 6.3.7 to factorization problems see EDWARDS [1I, p. 117 ff], to saturation theory TABERSKI [2], SUNOUCHI [3]. Although the material will not be made use of in this text, it may motivate the results of Sec. 6.3.3. The latter are also well-known but are treated less commonly. For Prop. 6.3.10 and Theorem 6.3.11 compare BEURLING [1]. The present proof of Theorem 6.3.11 is taken from BERENS–GÖRLICH [1], a paper largely concerned with applications, in particular to saturation theory (see Sec. 12.2.2, Sec. 12.4.5). For a more general version (Problem 6.3.14) of the theorem see SZ.-NAGY [3], TELJAKOVSKIĬ [2]. The related result on convex (double monotonic) functions (see the final assertion of Theorem 6.3.11 and Problem 6.3.7) is to be found in ACHIESER [2, p. 154 ff], TITCHMARSH [6, p. 170]. In probability theory it is known as Pólya's condition; for extensions and applications to probability see LUKACS [1, p. 70 ff], LINNIK [1, p. 37 ff]. For further sufficient conditions concerning the representation as L^1-Fourier transform the reader is referred to Sec. 11.3.2.

Lemma 6.3.13 of Sec. 6.3.4 is due to STEIN [1I]. A proof was communicated to the authors in April 1967. The present proof of (6.3.19) is more elementary and due to H. JOHNEN (unpublished). For the periodic version in Problem 6.3.10 see also TAIBLESON [1].

Sec. 6.4. For background material on the general singular integral of Weierstrass as well as the other examples of this section, the reader is referred to BERENS–GÖRLICH [1] and the literature cited there. The w_κ of (6.4.5) are stable density functions in the sense of P. LÉVY. Lemmata 6.4.3–6.4.5 are elementary versions of general results which were obtained by B. SZ.-NAGY in his fundamental paper [3]. Concerning the periodic version of (6.4.14) of the Weierstrass integral, since the particular case $\kappa = 1$ gives Abel's method of summation of the Fourier series of f, the means (6.4.15) are often referred to as the Abel–Cartwright or (A, κ) means of the Fourier series (see CARTWRIGHT [1], HARDY [2, p. 71 ff, p. 380 ff]).

The typical means (6.4.18) of the Fourier inversion integral are the particular case $\lambda = 1$ of the Riesz means defined in Problem 6.4.8. For background material in the periodic case see HARDY–RIESZ [1] and CHANDRASEKHARAN–MINAKSHISUNDARAM [1]. First detailed investigations on approximation by the typical means are to be found in ZYGMUND [3]. In the case $\kappa \in \mathbb{N}$ he showed, for example, that if $f^{(r)} \in \mathsf{Lip}\,(\mathsf{C}_{2\pi}; \alpha)$ for $r \leq \kappa - 1$, $0 < \alpha \leq 1$, then $\|R_{n,\kappa}(f; \circ) - f(\circ)\|_{\mathsf{C}_{2\pi}} = O(n^{-r-\alpha})$, except if κ is odd and $r = \kappa - 1$, $\alpha = 1$, when the (multiplicative) factor $\log n$ must be added to the O-term. For fractional κ see ALJANČIĆ [2]. Further important contributions to the approximation by typical means may be found in SZ.-NAGY [3]; see also EFIMOV [3], TELJAKOVSKIĬ [2], GUO ZHU-RUI [1], CHEN TIAN-PING [1].

Concerning the generalized integral of Picard defined in Problem 6.4.5 see BERENS–GÖRLICH [1] and LINNIK [1, p. 39]. Bessel potentials (see Problem 6.4.6) play an important rôle in modern analysis. For a recent discussion see DONOGHUE [1, p. 297 ff]. For the original literature compare Problem 6.4.6 as well as GÖRLICH [2, 3], WESTPHAL [2]. Periodic Bessel potentials—these are generated via (3.1.28), (3.1.31) by G_κ—do not seem to receive as much attention; nevertheless see TAIBLESON [1], WAINGER [1] as well as NESSEL–PAWELKE [1]. The results of this section on nonoptimal approximation ($\alpha < \kappa$) of the different processes may also be obtained, without any restrictions, as consequences of general theorems on interpolation spaces given in BUTZER–BERENS [1], BUTZER–SCHERER [1], and BERENS [3].

Sec. 6.5. For the classical results on multipliers (or factor sequences) for finite Fourier transforms the reader is referred to ZYGMUND [7I, p. 177 ff], for a detailed modern treatment to EDWARDS [1II, pp. 243–299]. For these reasons our account is brief. In Prop. 6.5.3 the functions λ of the multiplier class $(\mathsf{C}_{2\pi}, \mathsf{C}_{2\pi})$ are characterized, the norm of the corresponding multiplier operator U is evaluated and even Uf is represented as the convolution $f * d\mu$ of $f \in \mathsf{C}_{2\pi}$ with some $\mu \in \mathsf{BV}_{2\pi}$, determining U. On the other hand, in Prop. 6.5.2 no representation of Uf as convolution is given. Yet it can be shown that there is a unique distribution (actually a pseudomeasure) A such that $Uf = f * A$ for all $f \in \mathsf{L}^2_{2\pi}$. For specific multiplier classes the real task is indeed to represent U as convolution with some distribution A from which λ is easily recaptured as the Fourier transform of A, and to characterize

A as closely as possible. However, in this section we have confined the discussion to some of those elementary situations where A belongs to one of our traditional function classes (compare also Problems 6.5.1, 6.5.4). For multipliers of type $(\mathsf{L}^p_{2\pi}, \mathsf{L}^q_{2\pi})$ there is as yet no complete and effective solution to the problem. Inclusion relations are known between classes $(\mathsf{L}^p_{2\pi}, \mathsf{L}^q_{2\pi})$ for different values of p and q some of the proofs depending upon interpolation theorems such as the Riesz–Thorin convexity theorem; see EDWARDS [1II, p. 263 ff], HIRSCHMAN [1], and STEIN [2]. A related multiplier problem is the following: Which function λ on $\mathbb{Z}$ has the property that, for each $f \in \mathsf{C}_{2\pi}$, the series $\sum_{k=-\infty}^{\infty} \lambda(k) f^\wedge(k)\, e^{ikx}$ has uniformly bounded partial sums? For an answer the reader is referred to EDWARDS [1II, p. 257 ff]. A further problem is to characterize those functions λ such that, for each $f \in \mathsf{C}_{2\pi}$, $\lambda(k) f^\wedge(k)$ is the finite Fourier transform of some $g \in \mathsf{Lip}\,(\mathsf{C}_{2\pi}; \alpha)$. For this so-called Lipschitz multiplier problem see EDWARDS [1II, p. 260] as well as ALJANČIĆ–TOMIĆ [1] and the literature cited there. For a recent extended report on related problems see especially BOAS [1].

Concerning Sec. 6.5.2 the first treatment in book-form of multipliers (or factor functions) for Fourier transforms of functions in L^p, $1 \le p \le 2$, seems to be that of HILLE [2, p. 361 ff], see also HILLE–PHILLIPS [1, p. 566]. Theorem 6.5.6 is to be found there as well as in RUDIN [3, p. 73]. Theorems 6.5.7, 6.5.8 are due to DE LEEUW [3]. The extension of Theorem 6.5.8 to bounded measurable functions λ, analogous to the PHILLIPS result of Theorem 6.2.2, is equally valid and given there. The paper by DE LEEUW is written in the setting of abstract harmonic analysis; he considers multipliers on subgroups and quotient groups. The assertions of Theorem 6.5.8 (in particular also (6.2.3)) are connected with the theory of almost periodic functions on the line. To each such function f there corresponds a unique number $M(f)$, called the mean value of f, such that $\lim_{\rho\to\infty} (1/2\rho) \int_{-\rho}^{\rho} f(u)\, du = M(f)$. See WIENER [2, pp. 185–199], KATZNELSON [1, pp. 155–169] (compare also Problem 5.3.6).

Multipliers of type $(\mathsf{L}^p, \mathsf{L}^q)$ have been the subject of many recent investigations. It may be mentioned here that most of these speak of *translation-invariant operators*. Thus, in his fundamental and detailed paper [1], L. HÖRMANDER notes that to each bounded linear translation-invariant operator U from L^p to L^q there may be associated a unique tempered distribution A such that $Uf = A * f$ for all $f \in \mathsf{S}$ (see also LOOMIS [3, p. 253]). The space of tempered distributions A such that $\|A * f\|_q \le C \|f\|_p$ for all $f \in \mathsf{S}$ is denoted by L^q_p. The set of Fourier transforms $A^\wedge$ of $A \in \mathsf{L}^q_p$ is denoted by M^q_p. The elements of M^q_p are called multipliers of type (p, q). HÖRMANDER's studies begin on this basis. Obviously, this definition of multipliers of type (p, q) is not restricted to $1 \le p, q \le 2$ as was the case in our classical setting. For further literature on multipliers a very incomplete list is added. Those preceding the milestone paper by L. HÖRMANDER are MARCINKIEWICZ [1], SCHWARTZ [2], BEURLING–HELSON [1], MIKHLIN [1, 2], KRABBE [1]; for those following see FIGÀ-TALAMANCA [1, 2], KRÉE [1], CORDES [1], BRAINERD–EDWARDS [1], CALDERÓN [1], GAUDRY [1] PEETRE [3], LITTMAN–MCCARTHY–RIVIÈRE [1, 2], RIEFFEL [1], EDWARDS [2], FIGÀ-TALAMANCA–GAUDRY [1], LARSEN [1]. Compare also the literature cited in EDWARDS [1].

7

Fourier Transform Methods and Second-Order Partial Differential Equations

7.0 Introduction

Integral transform methods have proven of great utility in the solution of initial and boundary value problems in the theory of partial differential equations. The general method is to transform a given partial differential equation, involving an unknown function, into an equation involving the transform of this function. This transformed equation is then solved, and an appropriate inverse formula is applied to obtain the solution in terms of the original function space.

More specifically, we have seen that the Fourier transform converts differentiation of order r into multiplication by $(iv)^r$ (cf. Prop. 5.1.14), so transforming an ordinary differential equation into an algebraic equation (if the range of the independent variable is $\mathbb{R}$). In the case of partial differential equations the method reduces by one the number of variables with respect to which differentiation occurs. Since such equations have several independent variables, which may also range over intervals of different type, the first question is concerned with respect to what variable one transforms the original equations and what transform can be used at all.

Since the interval of integration of the transform must coincide with the range of a variable of the differential equation in question, one may generally apply: (i) the Fourier transform with respect to a variable x if it varies in the interval $-\infty < x < \infty$, (ii) the finite Fourier transform (the Fourier transform in the form (II.4)) in the case that one or more of the independent variables, say x, ranges over a finite interval $-L \leq x \leq L$, say, (iii) half-range transforms (such as the cosine transform) for intervals of type $0 \leq x \leq L$, (iv) the Laplace transform with respect to t if it varies in $t \geq 0$, this transform being particularly suitable for solving initial value problems. Of further importance are Mellin, Hankel, and Legendre transforms.

In order to describe the method, let us consider a nonhomogeneous linear second-order partial differential equation in two independent variables x, y

$$\text{(7.0.1)}\quad A\frac{\partial^2 u(x, y)}{\partial x^2} + B\frac{\partial^2 u(x, y)}{\partial x\,\partial y} + C\frac{\partial^2 u(x, y)}{\partial y^2} + D\frac{\partial u(x, y)}{\partial x} + E\frac{\partial u(x, y)}{\partial y} + Fu(x, y) = G(x, y)$$

with constant coefficients $A, B, \ldots, F$ and domain $-\infty < x < \infty$, $y_0 < y < y_1$, say (y_0, y_1 may be $-\infty$, $+\infty$, respectively). In general, the unknown function $u(x, y)$ is not uniquely determined by the differential equation alone. Furthermore, 'one-point' (or initial) conditions and 'two-point' (or boundary) conditions must be prescribed.

As the variable x ranges over $\mathbb{R}$, the first step of the procedure is to apply the Fourier transform to both sides of equation (7.0.1) with respect to x. Setting $[u(\circ, y)]^\wedge(v) = u^\wedge(v, y)$, $[G(\circ, y)]^\wedge(v) = G^\wedge(v, y)$, this yields

$$\text{(7.0.2)}\quad A\,(iv)^2 u^\wedge(v, y) + B\,(iv)\frac{\partial u^\wedge(v, y)}{\partial y} + C\frac{\partial^2 u^\wedge(v, y)}{\partial y^2} + D\,(iv)u^\wedge(v, y) + E\frac{\partial u^\wedge(v, y)}{\partial y} + Fu^\wedge(v, y) = G^\wedge(v, y),$$

provided one assumes $u(\circ, y), G(\circ, y) \in \mathsf{L}^1$ for every $y_0 < y < y_1$ and imposes further assumptions such that for $r = 1, 2$

$$\text{(7.0.3)}\quad \begin{aligned} &\left[\frac{\partial^r u(x, y)}{\partial x^r}\right]^\wedge(v) = (iv)^r u^\wedge(v, y), \qquad \left[\frac{\partial^r u(\circ, y)}{\partial y^r}\right]^\wedge(v) = \frac{\partial^r u^\wedge(v, y)}{\partial y^r},\\ &\left[\frac{\partial^2 u(x, y)}{\partial x\,\partial y}\right]^\wedge(v) = (iv)\frac{\partial u^\wedge(v, y)}{\partial y}. \end{aligned}$$

Thus the second-order partial differential equation converts into the equation (7.0.2), which is really a continuous system with one (ordinary differential) equation for each value of the parameter $v \in \mathbb{R}$. In order to apply (7.0.3), the coefficients $A, B, \ldots, F$ in equation (7.0.1) may be functions of y. They may even depend on x provided that rules corresponding to (7.0.3) can be established, converting differentiation with respect to x into more suitable operations.

The second step is to solve this second-order ordinary differential equation for $u^\wedge(v, y)$, v being a parameter. According to classical theory, this equation has a unique solution provided that either two initial conditions or two-point (boundary) conditions are prescribed. These must be deductible from corresponding conditions upon a solution of the original equation (7.0.1). Thus we further assume, for example, that (7.0.1) satisfies the one-point conditions

$$\lim_{y\to y_0+} u(x, y) = a_0(x), \qquad \lim_{y\to y_0+} \frac{\partial u(x, y)}{\partial y} = a_1(x)$$

which must be interpreted in such a way that they convert into the following one-point conditions for (7.0.2)

$$\lim_{y\to y_0+} u^\wedge(v, y) = a_0^\wedge(v), \qquad \lim_{y\to y_0+} \frac{\partial u^\wedge(v, y)}{\partial y} = a_1^\wedge(v).$$

The third step is to reconstruct the solution $u(x, y)$ of the original equation (7.0.1) from the values of $u^\wedge(v, y)$. For this purpose the inverse Fourier transform of $u^\wedge(v, y)$ is evaluated via a suitable inversion theorem (such as Prop. 5.1.10).

The method as a whole may be indicated by the following diagram:

Partial Differential Equation + initial and boundary conditions	$\dashrightarrow$	Solution $u(x, y)$
↓ Transform		↑ Inverse Transform
Ordinary Differential Equation + initial and/or boundary conditions	$\longrightarrow$	Solution $u^\wedge(v, y)$

Instead of solving the equation directly in the original function space, we take the indirect way indicated by the arrows. The essential point of the method is that the original equation is converted into the transformed equation and that the latter should be more readily solvable. The advantage of the transform method is that it greatly facilitates the solution of such problems to a standard procedure. Finally, the method is not only concerned with the construction of the solutions in question but it delivers the uniqueness at the same time. The method to be presented not only provides a formal procedure but also a completely rigorous approach to the subject.

By analogy with conic sections, in the theory of differential equations the general form of the equation (7.0.1) is said to be hyperbolic, parabolic, or elliptic according as the discriminant $B^2 - 4AC$ is positive, zero, or negative. One can show that the standard forms of the terms of second-order of a hyperbolic, parabolic, or elliptic equation are given by

$$\frac{\partial^2 u}{\partial \xi^2} - \frac{\partial^2 u}{\partial \eta^2}, \qquad \frac{\partial^2 u}{\partial \eta^2}, \qquad \frac{\partial^2 u}{\partial \xi^2} + \frac{\partial^2 u}{\partial \eta^2},$$

respectively (by affine transformation $\xi = \alpha x + \beta y$, $\eta = \gamma x + \delta y$, $\alpha\delta - \beta\gamma \neq 0$). For this purpose it suffices to study three standard forms of second-order equations. This fits in with the preceding discussion since the method of solution depends not only upon the auxiliary conditions but also upon the nature of the coefficients and the domain of the independent variables.

Sec. 7.1 is concerned with the finite Fourier transform method in the solution of three standard equations which are of parabolic, elliptic, and hyperbolic type. These are, in order, the equation of heat flow, Laplace's equation, and the equation of motion. Further, the nonhomogeneous problem is considered in connection with the first equation, the Neumann problem with Laplace's equation. Sec. 7.2 is devoted to the Fourier transform method as applied to the problem of heat flow on an infinite bar, to Dirichlet's problem for the upper half-plane as well as to the equation of motion for an infinite string. Fourier transform methods are also applicable to certain difference and integral equations and to mixed problems that are of Fourier convolution type. Several such examples are given in the Problems.

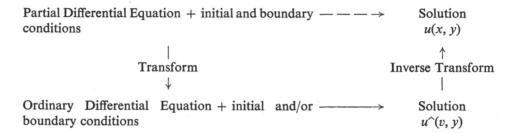

7.1 Finite Fourier Transform Method

7.1.1 Solution of Heat Conduction Problems

Given a homogeneous isotropic ring of unit radius whose diameter is small in comparison with its length, let $f(x)$, $-\pi \le x \le \pi$, be any prescribed temperature distribution at time $t = 0$ with $f(-\pi) = f(\pi)$. *Fourier's problem of the ring* is to determine the temperature at any point of the ring at time $t > 0$ if there is no radiation at the surface. This temperature function $u(x, t)$ satisfies the differential equation of heat conduction†

(7.1.1) $$\frac{\partial u(x, t)}{\partial t} = \frac{\partial^2 u(x, t)}{\partial x^2} \qquad (-\pi \le x \le \pi,\ t > 0)$$

with the boundary conditions

(7.1.2) $$u(-\pi, t) = u(\pi, t), \qquad \frac{\partial u(-\pi, t)}{\partial x} = \frac{\partial u(\pi, t)}{\partial x} \qquad (t > 0)$$

and the initial condition

(7.1.3) $$u(x, 0) = f(x) \qquad (-\pi \le x \le \pi).$$

We interpret this problem in the following sense:

7.1.4) *Given a function* $f \in X_{2\pi}$, *we call for a function* $u(x, t)$, *defined and periodic in* x *for all* x *and* $t > 0$, *such that*

(i) $\partial u(x, t)/\partial x$, $\partial^2 u(x, t)/\partial x^2$, $\partial u(x, t)/\partial t$ *exist for all* x *and* $t > 0$,

(ii) $u(x, t)$ *satisfies equation* (7.1.1) *for all* x *and* $t > 0$,

(iii) $\lim_{t\to 0+} \|u(\circ, t) - f(\circ)\|_{X_{2\pi}} = 0$,

(iv) *for each* $t > 0$, $u(x, t) \in W^2_{X_{2\pi}}$ *as a function of* x,

(v) *for each* $t > 0$, $\partial u(x, t)/\partial t \in X_{2\pi}$ *as a function of* x *and*

$$\lim_{\tau\to 0} \left\| \frac{u(\circ, t + \tau) - u(\circ, t)}{\tau} - \frac{\partial u(\circ, t)}{\partial t} \right\|_{X_{2\pi}} = 0.$$

We remark that the boundary conditions (7.1.2) are contained in (7.1.4) in view of the periodicity of the functions involved. In particular, $u(x, t)$ belongs to $C_{2\pi}$ as a function of x for each $t > 0$.

To solve this problem we assume that there exists a function $u(x, t)$ with these properties and apply the finite Fourier transform to each side of the equation (7.1.1). Setting $[u(\circ, t)]^\wedge(k) = u^\wedge(k, t)$, we have for the right-hand side by (7.1.4)(iv) and Prop. 4.1.8

$$\left[\frac{\partial^2 u(\circ, t)}{\partial x^2}\right]^\wedge(k) = (ik)^2 u^\wedge(k, t) \qquad (t > 0,\ k \in \mathbb{Z}),$$

whereas for the left-hand side

$$\left[\frac{\partial u(\circ, t)}{\partial t}\right]^\wedge(k) = \frac{\partial u^\wedge(k, t)}{\partial t} \qquad (t > 0,\ k \in \mathbb{Z}),$$

† The physical constants of density, specific heat, and conductivity have been omitted since they do not affect the nature of the solution.

since by (7.1.4)(v) and Problem 4.1.1

$$\left|\frac{u^\wedge(k, t+\tau) - u^\wedge(k, t)}{\tau} - \left[\frac{\partial u(\circ, t)}{\partial t}\right]^\wedge(k)\right|$$
$$\leq \left\|\frac{u(\circ, t+\tau) - u(\circ, t)}{\tau} - \frac{\partial u(\circ, t)}{\partial t}\right\|_{X_{2\pi}} = o(1) \qquad (\tau \to 0).$$

Hence $\partial u^\wedge(k, t)/\partial t$ exists for each $k \in \mathbb{Z}$ and $t > 0$ and is equal to the finite Fourier transform of $\partial u(x, t)/\partial t$†. In particular, for every $k \in \mathbb{Z}$, $u^\wedge(k, t)$ is a continuous function of t on $(0, \infty)$.

Thus the finite Fourier transform of a solution $u(x, t)$ of (7.1.4) satisfies for each $k \in \mathbb{Z}$ the ordinary differential equation

(7.1.5) $$\frac{\partial u^\wedge(k, t)}{\partial t} + k^2 u^\wedge(k, t) = 0 \qquad (t > 0)$$

having the classical solution

$$u^\wedge(k, t) = A(k)\, e^{-k^2 t}$$

with constant $A(k)$ independent of t. The initial condition (7.1.4)(iii) yields

$$|u^\wedge(k, t) - f^\wedge(k)| \leq \|u(\circ, t) - f(\circ)\|_{X_{2\pi}} = o(1) \qquad (t \to 0+);$$

therefore $A(k) = f^\wedge(k)$ and

(7.1.6) $$u^\wedge(k, t) = f^\wedge(k)\, e^{-k^2 t} \qquad (k \in \mathbb{Z},\ t > 0).$$

Hence, if our problem has a solution, then its finite Fourier transform is given by (7.1.6). In order to represent it in terms of the original functions we have to apply a suitable inversion theorem. According to Prop. 4.1.5(i) we have for all x and $t > 0$

(7.1.7) $$u(x, t) = \sum_{k=-\infty}^{\infty} e^{-k^2 t} f^\wedge(k)\, e^{ikx},$$

since the series converges absolutely and uniformly in x for each $t > 0$. Thus, if a solution of (7.1.4) exists, it is *unique* by Cor. 1.2.7.

To prove that (7.1.7) actually *is* a solution of our problem, it remains to show that the function $u(x, t)$ given by (7.1.7) satisfies (7.1.4). This fact is left to Problem 7.1.1.

Finally we may represent the solution of (7.1.4) in form of a singular integral. Indeed

(7.1.8) $$u(x, t) = \sum_{k=-\infty}^{\infty} e^{-k^2 t} \frac{1}{2\pi} \int_{-\pi}^{\pi} f(s)\, e^{-ik(s-x)}\, ds$$
$$= \frac{1}{2\pi} \int_{-\pi}^{\pi} f(x - s) \left\{1 + 2 \sum_{k=1}^{\infty} e^{-k^2 t} \cos ks\right\} ds,$$

† We note that instead of (7.1.4)(v) it would have been sufficient to suppose that the limit on the left side of the last inequality is zero, for we here only need the fact that the finite Fourier transform commutes with differentiation with respect to the time-parameter t. In other words, whereas (7.1.4)(v) assumes that the vector-valued function $u(x, t)$, defined on $t > 0$ with values in the Banach space $X_{2\pi}$, is strongly differentiable for $t > 0$, it is sufficient to suppose that the difference quotient converges in the weak topology towards $\partial u(\circ, t)/\partial t$.

the interchange of integration and summation being justified by the uniform convergence of the series. Thus in view of Problem 1.3.10 we have

Theorem 7.1.1. *Fourier's problem* (7.1.4) *of the ring has a unique solution. It is given by the singular integral of Weierstrass, i.e.*

$$u(x, t) = W_t(f; x) \equiv \frac{1}{2\pi} \int_{-\pi}^{\pi} f(x - s)\theta_3(s, t)\, ds, \tag{7.1.9}$$

$\theta_3(x, t)$ being the Jacobi theta-function.

In Sec. 12.2.1 we shall study the dependence of the solution of (7.1.4) upon its initial value f and examine the rate at which it approximates f for small values of t.

Now let us briefly discuss a problem with sources distributed along the ring. This situation is covered by the nonhomogeneous equation

$$\frac{\partial u(x, t)}{\partial t} = \frac{\partial^2 u(x, t)}{\partial x^2} + F(x, t) \qquad (-\pi \le x \le \pi, t > 0). \tag{7.1.10}$$

Suppose that, for each $t > 0$, $F(x, t)$ belongs to $X_{2\pi}$ as a function of x and satisfies

$$\text{(7.1.11)} \quad \begin{aligned} &\text{(i)} && \|F(\circ, t)\|_{X_{2\pi}} \le M && (0 < t \le 1), \\ &\text{(ii)} && \lim_{\tau \to 0} \|F(\circ, t + \tau) - F(\circ, t)\|_{X_{2\pi}} = 0 && (t > 0). \end{aligned}$$

Then we seek for a solution $u(x, t)$ of (7.1.10) which satisfies (7.1.4)(i), (iii)–(v).

Let us assume that this problem has a solution $u(x, t)$. As in the homogeneous case we obtain that its finite Fourier transform $u^\wedge(k, t)$ satisfies for each $k \in \mathbb{Z}$ the ordinary differential equation

$$\frac{\partial u^\wedge(k, t)}{\partial t} + k^2 u^\wedge(k, t) = F^\wedge(k, t) \qquad (t > 0). \tag{7.1.12}$$

To solve these equations we first of all note that in view of (7.1.11) and Problem 4.1.1 $F^\wedge(k, t)$ is a continuous function of t on $(0, \infty)$ for each $k \in \mathbb{Z}$. In consideration of the initial condition the homogeneous part of (7.1.12) has the solution $u_1^\wedge(k, t) = f^\wedge(k) \exp\{-k^2 t\}$ (cf. (7.1.6)).

Furthermore, by classical methods a particular solution of the nonhomogeneous equation (7.1.12) is given for each $k \in \mathbb{Z}$ by

$$u_2^\wedge(k, t) = \int_0^t F^\wedge(k, t - \tau)\, e^{-k^2\tau}\, d\tau, \tag{7.1.13}$$

the integral existing since there is a constant M^* for each $t > 0$ such that for $k \neq 0$

$$\left| \int_0^t F^\wedge(k, t - \tau)\, e^{-k^2\tau}\, d\tau \right| \le \sup_{0 < \tau \le t} \|F(\circ, \tau)\|_{X_{2\pi}} \int_0^t e^{-k^2\tau}\, d\tau \le \frac{M^*}{k^2}.$$

Hence, if the equation (7.1.10) has a solution $u(x, t)$ in the above sense, then its finite

Fourier transform is given by $u_1^{\wedge}(k, t) + u_2^{\wedge}(k, t)$. In order to represent this solution in terms of the original functions, by Prop. 4.1.5(i)

$$u(x, t) = \sum_{k=-\infty}^{\infty} \left\{ f^{\wedge}(k)\, e^{-k^2 t} + \int_0^t F^{\wedge}(k, t - \tau)\, e^{-k^2 \tau}\, d\tau \right\} e^{ikx},$$

since the series converges absolutely and uniformly in x for each $t > 0$. Moreover, by Fubini's theorem

$$\begin{aligned} u_2(x, t) &= \sum_{k=-\infty}^{\infty} \left\{ \int_0^t \left[\frac{1}{2\pi} \int_{-\pi}^{\pi} F(s, t - \tau)\, e^{-ik(s-x)}\, e^{-k^2 \tau}\, ds \right] d\tau \right\} \\ &= \frac{1}{2\pi} \int_{-\pi}^{\pi} \int_0^t F(x - s, t - \tau) \left\{ 1 + 2 \sum_{k=1}^{\infty} e^{-k^2 \tau} \cos ks \right\} d\tau\, ds, \end{aligned}$$

giving (cf. Problem 7.1.1)

Theorem 7.1.2. *If the source distribution* $F(x, t)$ *satisfies* (7.1.11), *then equation* (7.1.10) *has a unique solution satisfying* (7.1.4)(i), (iii)–(v). *It is given for all* x *and* $t > 0$ *by*

$$(7.1.14)\quad u(x, t) = \frac{1}{2\pi} \int_{-\pi}^{\pi} f(x - s)\theta_3(s, t)\, ds + \frac{1}{2\pi} \int_{-\pi}^{\pi} \int_0^t F(x - s, t - \tau)\theta_3(s, \tau)\, d\tau\, ds\dagger.$$

7.1.2 Dirichlet's and Neumann's Problem for the Unit Disc

Given a thin plate of a homogeneous and isotropic material in form of a disc of unit radius, let $f(x)$, $-\pi \le x \le \pi$, with $f(-\pi) = f(\pi)$ be a constant temperature distribution on the boundary of the disc. What is the temperature distribution in the interior of the disc in the stationary, i.e., time-independent case?

The solution $w(\xi, \eta)$ must be independent of the time t and satisfy the two-dimensional heat equation which now reduces to the elliptic equation

$$(7.1.15)\qquad \frac{\partial^2 w(\xi, \eta)}{\partial \xi^2} + \frac{\partial^2 w(\xi, \eta)}{\partial \eta^2} = 0 \qquad (\xi^2 + \eta^2 < 1).$$

Here it is convenient to use polar coordinates. If we write

$$w(\xi, \eta) = w(r \cos x, r \sin x) \equiv u(x, r) \quad (-\pi \le x \le \pi, 0 \le r < 1),$$

then the differential equation for $w(\xi, \eta)$ passes into the form

$$(7.1.16)\qquad \frac{\partial^2 u(x, r)}{\partial r^2} + \frac{1}{r} \frac{\partial u(x, r)}{\partial r} + \frac{1}{r^2} \frac{\partial^2 u(x, r)}{\partial x^2} = 0 \quad (-\pi \le x \le \pi, 0 < r < 1),$$

† The second term is indeed of Fourier convolution-type in the first variable and of Laplace convolution-type in the second variable, in accordance with the domain of definition of the functions involved.

and we are interested in a solution of the following (interior) *Dirichlet problem* for the unit disc†:

(7.1.17) *Given a function* $f \in X_{2\pi}$, *we seek for a function* $u(x, r)$, *defined and periodic in* x *for all* x *and* $r \in (0, 1)$, *such that*

(i) $\partial^2 u(x, r)/\partial x^2$, $\partial^2 u(x, r)/\partial r^2$ *exist for all* x *and* $r \in (0, 1)$,

(ii) $u(x, r)$ *satisfies equation* (7.1.16) *for all* x *and* $r \in (0, 1)$,

(iii) $\|u(\circ, r)\|_{X_{2\pi}} \le M$ *uniformly for* $r \in (0, 1)$ *and*

(7.1.18) $$\lim_{r \to 1-} \|u(\circ, r) - f(\circ)\|_{X_{2\pi}} = 0,$$

(iv) *for each* $r \in (0, 1)$, $u(x, r) \in W^2_{X_{2\pi}}$ *as a function of* x,

(v) *for each* $r \in (0, 1)$, $\partial u(x, r)/\partial r$ *and* $\partial^2 u(x, r)/\partial r^2$ *belong to* $X_{2\pi}$ *as functions of* x *and*

$$\lim_{\rho \to 0} \left\| \frac{u(\circ, r + \rho) - u(\circ, r)}{\rho} - \frac{\partial u(\circ, r)}{\partial r} \right\|_{X_{2\pi}} = 0 \qquad (0 < r < 1),$$

$$\lim_{\rho \to 0} \left\| \frac{\dfrac{\partial u(\circ, r + \rho)}{\partial r} - \dfrac{\partial u(\circ, r)}{\partial r}}{\rho} - \frac{\partial^2 u(\circ, r)}{\partial r^2} \right\|_{X_{2\pi}} = 0 \qquad (0 < r < 1).$$

We again assume that there exists a solution $u(x, r)$ of the above problem. In order to transform the equation (7.1.16) we denote the kth (complex) Fourier coefficient of $u(x, r)$ with respect to x by $u^\wedge(k, r)$ and infer from (7.1.17)(v) and Problem 4.1.1 that

$$\left| \frac{u^\wedge(k, r + \rho) - u^\wedge(k, r)}{\rho} - \left[\frac{\partial u(\circ, r)}{\partial r}\right]^\wedge (k) \right| = o(1) \qquad (\rho \to 0).$$

Thus

(7.1.19) $$\left[\frac{\partial u(\circ, r)}{\partial r}\right]^\wedge (k) = \frac{\partial u^\wedge(k, r)}{\partial r} \qquad (k \in \mathbb{Z},\ r \in (0, 1)),$$

and similarly

$$\left[\frac{\partial^2 u(\circ, r)}{\partial r^2}\right]^\wedge (k) = \frac{\partial^2 u^\wedge(k, r)}{\partial r^2} \qquad (k \in \mathbb{Z},\ r \in (0, 1)).$$

Hence, for each $k \in \mathbb{Z}$, $\partial u^\wedge(k, r)/\partial r$ and $\partial^2 u^\wedge(k, r)/\partial r^2$ exist for $r \in (0, 1)$; in particular, for each $k \in \mathbb{Z}$, $u^\wedge(k, r)$ and $\partial u^\wedge(k, r)/\partial r$ are continuous functions of r on $(0, 1)$. By (7.1.17)(iv) and Prop. 4.1.8 we have

$$\left[\frac{\partial^2 u(\circ, r)}{\partial x^2}\right]^\wedge (k) = (ik)^2 u^\wedge(k, r) \qquad (k \in \mathbb{Z},\ r \in (0, 1)).$$

Therefore the kth Fourier coefficient of a solution $u(x, r)$ of (7.1.17) necessarily satisfies the ordinary differential equation

(7.1.20) $$r^2 \frac{\partial^2 u^\wedge(k, r)}{\partial r^2} + r \frac{\partial u^\wedge(k, r)}{\partial r} - k^2 u^\wedge(k, r) = 0 \qquad (r \in (0, 1)).$$

† We note that the problem is raised in such a way that certain peculiarities, produced by introducing polar coordinates, are already removed. In particular, a solution of Laplace's equation (7.1.15) is harmonic at the origin, whereas the differential equation (7.1.16) does not apply at $r = 0$. Therefore we pose the condition $\|u(\circ, r)\|_{X_{2\pi}} \le M$ for each $r \in (0, 1)$, obviously a necessary condition from the physical point of view, too.

By classical methods one obtains as its solution

$$u^\wedge(k, r) = \begin{cases} A(k)r^k + B(k)r^{-k}, & k \neq 0 \\ A(0) + B(0)\log r, & k = 0. \end{cases}$$

To specify the constants $A(k)$, $B(k)$, we have by (7.1.17)(iii) that for each $k \in \mathbb{Z}$

$$\text{(7.1.21)} \qquad \begin{aligned} &\text{(i)} \qquad |u^\wedge(k, r)| \le M \qquad (r \in (0, 1)), \\ &\text{(ii)} \qquad \lim_{r \to 1-} u^\wedge(k, r) = f^\wedge(k). \end{aligned}$$

Let $k = 0$. Then (7.1.21) implies $B(0) = 0$ and $A(0) = f^\wedge(0)$, respectively. If $k \neq 0$, then it follows by (7.1.21)(ii) that $A(k) + B(k) = f^\wedge(k)$ and by (7.1.21)(i), upon letting $r \to 0+$, that $A(k) = 0$ for $k < 0$ and $B(k) = 0$ for $k > 0$. Summarizing, we have that the finite Fourier transform of any solution of (7.1.17) is given by

$$\text{(7.1.22)} \qquad u^\wedge(k, r) = f^\wedge(k) r^{|k|} \qquad (k \in \mathbb{Z},\ r \in (0, 1)).$$

To reconstruct the full solution we deduce by Prop. 4.1.5(i) and the uniform convergence of the series that

$$\text{(7.1.23)} \qquad u(x, r) = \sum_{k=-\infty}^{\infty} r^{|k|} f^\wedge(k)\, e^{ikx} \qquad (x \in \mathbb{R},\ r \in (0, 1)).$$

This may be rewritten in the form

$$u(x, r) = \frac{1}{2\pi} \int_{-\pi}^{\pi} f(x - s) \left\{1 + 2 \sum_{k=1}^{\infty} r^k \cos ks\right\} ds,$$

which by (1.2.36) is the singular integral of Abel–Poisson. By direct computation (cf. Problem 7.1.2) (7.1.23) actually is a solution† of (7.1.17), and since we have shown that any solution of (7.1.17) is of the form (7.1.23), we may strengthen Prop. 1.2.10 to

Theorem 7.1.3. *Dirichlet's problem* (7.1.17) *for the unit disc has a unique solution. It is given by the singular integral* (1.2.36) *of Abel–Poisson, thus for* $x \in \mathbb{R}$, $r \in [0, 1)$ *by*

$$\text{(7.1.24)} \qquad u(x, r) = P_r(f; x) \equiv \frac{1}{2\pi} \int_{-\pi}^{\pi} f(x - s) \frac{1 - r^2}{1 - 2r\cos s + r^2}\, ds.$$

The problem of finding a solution of (7.1.17) has been called a Dirichlet problem (or first boundary value problem of potential theory). This phrase covers a large variety of problems with certain features in common with the one discussed by us, but specifically in this text it means the problem of finding a solution u of Laplace's equation $\Delta u = 0$ in some region, when the values of u are prescribed on the boundary of the region.

Let us now consider the *Neumann problem* or second boundary value problem of potential theory. This arises when given amounts of heat are supplied to various parts

† The solution obtained is defined at $r = 0$, too. Actually, it is harmonic in the whole interior of the unit disc. Moreover, since $u(x, 0) = (2\pi)^{-1} \int_{-\pi}^{\pi} f(s)\, ds$, the temperature at the center of the disc with a steady state distribution is the average of the temperatures on the boundary. This illustrates the well-known mean value property of harmonic functions.

of the boundary of some region, and we are interested in a stationary temperature distribution in the region. The heat supplied to the boundary will be conducted throughout the medium in such a way as to maintain the stationary (and in general nonconstant) temperature profile. However, if the situation is stationary, the total heat content of the region must be stationary, i.e., constant in time. Therefore the net supply of heat at the boundary, thus $\int_{-\pi}^{\pi} f(s)\,ds$, must be zero. Our mathematical formulation of the (interior) Neumann problem of the unit disc is then:

(7.1.25) *Given a function $f \in X_{2\pi}$ such that $\int_{-\pi}^{\pi} f(s)\,ds = 0$, find a solution of* (7.1.17) *with condition* (7.1.18) *replaced by*

$$\lim_{r\to 1-} \left\| \frac{\partial u(\circ, r)}{\partial r} - f(\circ) \right\|_{X_{2\pi}} = 0. \tag{7.1.26}$$

Again we suppose that there exists a solution $u(x, r)$. Then the finite Fourier transform $u^\wedge(k, r)$ is given by

$$u^\wedge(k, r) = A(k) r^{|k|},$$

where, in view of (7.1.21)(i), $A(k)$ is a bounded function on $\mathbb{Z}$. According to (7.1.19) and (7.1.26) we have for each $k \in \mathbb{Z}$

$$\lim_{r\to 1-} \frac{\partial u^\wedge(k, r)}{\partial r} = f^\wedge(k),$$

giving

$$|k|\, A(k) = f^\wedge(k) \qquad (k \in \mathbb{Z}). \tag{7.1.27}$$

Note that we necessarily have to suppose that $f^\wedge(0) = 0$. Furthermore, for all x

$$\begin{aligned} u(x, r) &= A(0) + \sum_{k\neq 0} \frac{r^{|k|}}{|k|} f^\wedge(k)\, e^{ikx} \\ &= A(0) + \frac{1}{\pi}\int_{-\pi}^{\pi} f(x-s) \left\{ \sum_{k=1}^{\infty} \frac{r^k}{k} \cos ks \right\} ds \qquad (r \in [0, 1)). \end{aligned}$$

In virtue of Problem 1.2.18(iv) we may therefore state

Theorem 7.1.4. *Neumann's problem* (7.1.25) *for the unit disc has, apart from an additive constant, a unique solution. For arbitrary $\alpha \in \mathbb{C}$ it is given for all x by*

$$u(x, r) = \alpha - \frac{1}{2\pi}\int_{-\pi}^{\pi} f(x-s) \log\,[1 - 2r\cos s + r^2]\, ds \qquad (r \in [0, 1)). \tag{7.1.28}$$

For the question of simultaneous solutions of Dirichlet's and Neumann's problem we refer to Problem 12.2.13.

7.1.3 Vibrating String Problems

Consider a string which is stretched between the points $(0, 0)$ and $(\pi, 0)$ and vibrates in the (x, y)-plane. Let the displacement from the x-axis at position x and time t be

$u(x, t)$, that is, at time t the string passes through the point $(x, u(x, t))$. Then, for small displacements and constant line density, $u(x, t)$ satisfies the differential equation of hyperbolic type

$$\frac{\partial^2 u(x,t)}{\partial t^2} = c^2 \frac{\partial^2 u(x,t)}{\partial x^2} \qquad (0 < x < \pi,\ t > 0), \tag{7.1.29}$$

where $c^2 = \tau/\rho$, τ and ρ being the tension and density of the string, respectively. Let us suppose that both ends of the string are fixed. Then we have the boundary conditions

$$u(0, t) = 0, \qquad u(\pi, t) = 0 \qquad (t > 0). \tag{7.1.30}$$

Finally, to describe the motion completely, we need to know the initial position and velocity of the string, i.e.

$$u(x, 0) = f(x), \qquad \frac{\partial u(x, 0)}{\partial t} = g(x) \qquad (0 \le x \le \pi), \tag{7.1.31}$$

where f and g are given functions. In view of (7.1.30) we assume

$$f(0) = f(\pi) = 0, \qquad g(0) = g(\pi) = 0. \tag{7.1.32}$$

As in the preceding cases this intuitive introduction of the problem is now completed by the following exact interpretation:

(7.1.33) *For given functions $f, g \in \mathsf{X}[0, \pi]$ satisfying (7.1.32), find a function $u(x, t)$ defined on $0 \le x \le \pi$, $t > 0$ such that*

(i) *$\partial^2 u(x, t)/\partial x^2$, $\partial^2 u(x, t)/\partial t^2$ exist for all $x \in (0, \pi)$, $t > 0$,*

(ii) *$u(x, t)$ satisfies equation (7.1.29) for all $x \in (0, \pi)$, $t > 0$,*

(iii)
$$\lim_{x\to 0+} u(x,t) = 0, \qquad \lim_{x\to \pi-} u(x,t) = 0 \qquad (t > 0),$$
$$\lim_{x\to 0+} \frac{\partial u(x,t)}{\partial t} = 0, \qquad \lim_{x\to \pi-} \frac{\partial u(x,t)}{\partial t} = 0 \qquad (t > 0),$$
$$\lim_{x\to 0+} \frac{\partial^2 u(x,t)}{\partial t^2} = 0 \qquad (t > 0),$$

it moreover being assumed that the limits $\lim_{x\to 0+} \partial u(x, t)/\partial x$, $\lim_{x\to\pi-} \partial u(x, t)/\partial x$ and $\lim_{x\to 0+} \partial^2 u(x, t)/\partial x^2$ exist for each $t > 0$,

(iv) *for each $t > 0$, $u(x, t)$ and $\partial u(x, t)/\partial t$ belong to $\mathsf{X}[0, \pi]$ as functions of x and*

$$\lim_{t\to 0+} \|u(\circ, t) - f(\circ)\|_{\mathsf{X}[0,\pi]} = 0, \qquad \lim_{t\to 0+} \left\| \frac{\partial u(\circ, t)}{\partial t} - g(\circ) \right\|_{\mathsf{X}[0,\pi]} = 0,$$

(v) *for each $t > 0$, $\partial u(x, t)/\partial x$ is absolutely continuous on $0 < x < \pi$ as a function of x and $\partial^2 u(x, t)/\partial x^2$ belongs to $\mathsf{L}^1(0, \pi)$,*

(vi) *for each $t > 0$, $\partial^2 u(x, t)/\partial t^2 \in \mathsf{L}^1(0, \pi)$ as a function of x and*

$$\lim_{\tau\to 0} \left\| \frac{u(\circ, t+\tau) - u(\circ, t)}{\tau} - \frac{\partial u(\circ, t)}{\partial t} \right\|_{\mathsf{L}^1(0,\pi)} = 0 \qquad (t > 0),$$

$$\lim_{\tau\to 0} \left\| \frac{\dfrac{\partial u(\circ, t+\tau)}{\partial t} - \dfrac{\partial u(\circ, t)}{\partial t}}{\tau} - \frac{\partial^2 u(\circ, t)}{\partial t^2} \right\|_{\mathsf{L}^1(0,\pi)} = 0 \qquad (t > 0).$$

Let $u(x, t)$ be a solution of (7.1.33). Since the problem deals with the interval $0 \leq x \leq \pi$, one is led to apply a finite half-range transform, i.e., the finite Fourier-sine or -cosine transform. By Problem 7.1.8, the transform of $\partial^2 u(x, t)/\partial x^2$ would then not only include the boundary values of $u(x, t)$ but also those of $\partial u(x, t)/\partial x$ which we actually do not know.

Therefore we proceed by extending the functions $u(x, t)$, $f(x)$, $g(x)$ to the interval $[-\pi, 0]$, for then we are able to express the nonperiodic problem (7.1.33) as a periodic one. Obviously, the extension is determined by the local behaviour of $u(x, t)$, $f(x)$, $g(x)$ at $x = 0$.

As immediate consequences of the boundary conditions (7.1.33)(iii) it follows that, for each $t > 0$, $u(x, t)$, $\partial u(x, t)/\partial x$, $\partial u(x, t)/\partial t$ are continuous functions on $0 \leq x \leq \pi$. Here the derivatives at the end points are of course defined in the one-sided sense. Moreover, $\partial^2 u(x, t)/\partial x^2$ and $\partial^2 u(x, t)/\partial t^2$ exist and are continuous at $x = 0$.

The essential point is that the continuity guarantees that the differential equation (7.1.29) holds at $x = 0$, too. Therefore by (7.1.33)(iii)

$$c^2 \frac{\partial^2 u(0, t)}{\partial x^2} = c^2 \lim_{x \to 0+} \frac{\partial^2 u(x, t)}{\partial x^2} = \lim_{x \to 0+} \frac{\partial^2 u(x, t)}{\partial t^2} = 0.$$

Since $u(0, t) = 0$ and $\partial^2 u(0, t)/\partial x^2 = 0$ for each $t > 0$ (in general $\partial u(0, t)/\partial x \neq 0$), only an odd extension is possible if we desire that $\partial u(x, t)/\partial x$ be absolutely continuous on $[-\pi, \pi]$. Thus we set

$$(7.1.34) \qquad u(-x, t) \equiv -u(x, t), \qquad f(-x) \equiv -f(x), \qquad g(-x) \equiv -g(x) \qquad (0 \leq x \leq \pi,\ t > 0).$$

Then (7.1.30) and (7.1.32) imply $u(-\pi, t) = u(\pi, t)$, $f(-\pi) = f(\pi)$ and $g(-\pi) = g(\pi)$ for each $t > 0$. Since $\partial u(-x, t)/\partial x = \partial u(x, t)/\partial x$, $\partial^2 u(-x, t)/\partial x^2 = -\partial^2 u(x, t)/\partial x^2$ and $\partial u(-x, t)/\partial t = -\partial u(x, t)/\partial t$, we obtain that $u(x, t)$, $\partial u(x, t)/\partial x$ and $\partial u(x, t)/\partial t$ are continuous for each $t > 0$ on $-\pi \leq x \leq \pi$, moreover $\partial u(-\pi, t)/\partial x = \partial u(\pi, t)/\partial x$ and $\partial u(-\pi, t)/\partial t = \partial u(\pi, t)/\partial t = 0$, and the differential equation (7.1.29) is satisfied for all $x \in (-\pi, \pi)$, $t > 0$.

Now our original problem (7.1.33) is solved if we are able to determine a solution of the following problem:

(7.1.35) *For given functions $f, g \in X_{2\pi}$, find a function $u(x, t)$, defined and periodic in x for all x and $t > 0$, such that*

(i) *$\partial u(x, t)/\partial x$ and $\partial u(x, t)/\partial t$ exist for all x and $t > 0$, $\partial^2 u(x, t)/\partial x^2$ and $\partial^2 u(x, t)/\partial t^2$ exist for all $x \in (-\pi, \pi)$, $t > 0$,*

(ii) *$u(x, t)$ satisfies equation (7.1.29) for all $x \in (-\pi, \pi)$, $t > 0$,*

(iii) *for each $t > 0$, $\partial u(x, t)/\partial t \in X_{2\pi}$ as a function of x and*

$$\lim_{t \to 0+} \|u(\circ, t) - f(\circ)\|_{X_{2\pi}} = 0, \qquad \lim_{t \to 0+} \left\| \frac{\partial u(\circ, t)}{\partial t} - g(\circ) \right\|_{X_{2\pi}} = 0,$$

(iv) *for each $t > 0$, $u(x, t) \in W^2_{L^1_{2\pi}}$ as a function of x,*

(v) *for each* $t > 0$, $\partial^2 u(x, t)/\partial t^2 \in \mathsf{L}^1_{2\pi}$ *and*

$$\lim_{\tau \to 0} \left\| \frac{u(\circ, t + \tau) - u(\circ, t)}{\tau} - \frac{\partial u(\circ, t)}{\partial t} \right\|_{\mathsf{L}^1_{2\pi}} = 0,$$

$$\lim_{\tau \to 0} \left\| \frac{\dfrac{\partial u(\circ, t + \tau)}{\partial t} - \dfrac{\partial u(\circ, t)}{\partial t}}{\tau} - \frac{\partial^2 u(\circ, t)}{\partial t^2} \right\|_{\mathsf{L}^1_{2\pi}} = 0.$$

To solve this problem we take the finite Fourier transform of equation (7.1.29). Then we obtain for $k \in \mathbb{Z}$, $t > 0$

$$\left[\frac{\partial^2 u(\circ, t)}{\partial x^2}\right]^\wedge(k) = (ik)^2 u^\wedge(k, t)$$

as well as

$$\left[\frac{\partial u(\circ, t)}{\partial t}\right]^\wedge(k) = \frac{\partial u^\wedge(k, t)}{\partial t}, \qquad \left[\frac{\partial^2 u(\circ, t)}{\partial t^2}\right]^\wedge(k) = \frac{\partial^2 u^\wedge(k, t)}{\partial t^2}. \tag{7.1.36}$$

Thus the finite Fourier transform $u^\wedge(k, t)$ of any solution of (7.1.35) satisfies for each $k \in \mathbb{Z}$ the ordinary differential equation

$$\frac{\partial^2 u^\wedge(k, t)}{\partial t^2} + (ck)^2 u^\wedge(k, t) = 0 \qquad (t > 0), \tag{7.1.37}$$

which has the general solution

$$u^\wedge(k, t) = \begin{cases} A(k)\, e^{ickt} + B(k)\, e^{-ickt}, & k \neq 0 \\ A(0) + B(0)t, & k = 0. \end{cases}$$

To determine $A(k)$ and $B(k)$ we note that $\lim_{t \to 0+} u^\wedge(k, t) = f^\wedge(k)$ which implies

$$f^\wedge(k) = \begin{cases} A(k) + B(k), & k \neq 0 \\ A(0), & k = 0. \end{cases}$$

Furthermore, since $\lim_{t \to 0+} [\partial u(\circ, t)/\partial t]^\wedge(k) = g^\wedge(k)$, it follows by (7.1.36) that

$$\lim_{t \to 0+} \frac{\partial u^\wedge(k, t)}{\partial t} = g^\wedge(k) \qquad (k \in \mathbb{Z}).$$

Therefore,

$$g^\wedge(k) = \begin{cases} ickA(k) - ickB(k), & k \neq 0 \\ B(0), & k = 0, \end{cases}$$

and the finite Fourier transform of any solution $u(x, t)$ of (7.1.35) is given by

$$u^\wedge(k, t) = \begin{cases} f^\wedge(k)\left[\dfrac{e^{ickt} + e^{-ickt}}{2}\right] + g^\wedge(k)\left[\dfrac{e^{ickt} - e^{-ickt}}{2ick}\right], & k \neq 0 \\ f^\wedge(0) + g^\wedge(0)t, & k = 0. \end{cases} \tag{7.1.38}$$

By Prop. 4.1.1(i) and Problem 4.1.5

$$u^\wedge(k, t) = \left[\frac{f(\circ + ct) + f(\circ - ct)}{2}\right]^\wedge(k) + \left[\frac{1}{2c}\int_{-ct}^{ct} g(\circ + s)\, ds\right]^\wedge(k)$$

for all $k \in \mathbb{Z}$, $t > 0$. The uniqueness theorem of the finite Fourier transform then gives for each $t > 0$ the solution

$$(7.1.39) \qquad u(x, t) = \frac{1}{2}[f(x + ct) + f(x - ct)] + \frac{1}{2c}\int_{x-ct}^{x+ct} g(s)\, ds \quad \text{(a.e.)}.$$

Thus, if (7.1.35) has a solution $u(x, t)$, then it is of the form (7.1.39). In order to show that the right-hand side of (7.1.39) actually is a solution we have to pose certain conditions upon the initial values f and g, in contrast to the parabolic and elliptic situation of Sec. 7.1.1 and 7.1.2, respectively. There, the corresponding series (7.1.7) and (7.1.23) always define solutions since they are absolutely and uniformly convergent for any $f \in \mathsf{X}_{2\pi}$. But here the Fourier series of (7.1.39) does not necessarily converge.

Theorem 7.1.5. *Let $f \in \mathsf{AC}^1_{2\pi}$ and $g \in \mathsf{AC}_{2\pi}$ be such that f'' and g' exist for all x and belong to $\mathsf{X}_{2\pi}$. Then problem* (7.1.35) *has a unique solution. It is given for all x and $t > 0$ by* (7.1.39). *Moreover, we have*

$$(7.1.40) \quad u(x, t) = f^{\wedge}(0) + g^{\wedge}(0)t + \sum_{k \neq 0}\left[f^{\wedge}(k)\cos ckt + \frac{1}{ck}g^{\wedge}(k)\sin ckt\right]e^{ikx},$$

the series being absolutely and uniformly convergent.

The proof is left to Problem 7.1.3. As an application we obtain the following solution to problem (7.1.33).

Theorem 7.1.6. *Let $f, g \in \mathsf{X}[0, \pi]$ satisfy* (7.1.32) *and let f and g be extended to the interval $-\pi \le x \le 0$ as odd functions and then to the whole real axis as periodic functions. Suppose that for these extensions $f \in \mathsf{AC}^1_{2\pi}$ and $g \in \mathsf{AC}_{2\pi}$ such that f'' and g' exists for all x and belong to $\mathsf{X}_{2\pi}$. Then the* **vibrating string problem** (7.1.33) *has a unique solution $u(x, t)$ given by* (7.1.39). *The representation* (7.1.39) *holds for all x and $t > 0$. Moreover,*

$$(7.1.41) \qquad u(x, t) = 2\sum_{k=1}^{\infty}\left[f_s^{\wedge}(k)\cos ckt + \frac{1}{ck}g_s^{\wedge}(k)\sin ckt\right]\sin kx,$$

the series being absolutely and uniformly convergent.

The proof is left to Problem 7.1.3. $f_s^{\wedge}(k)$ and $g_s^{\wedge}(k)$ denote the kth Fourier sine-coefficients (cf. (1.2.10) and Problem 7.1.8) of f and g, respectively.

Finally, we shall briefly discuss the problem connected with a vibrating string with both ends at $x = 0$ and $x = \pi$ free to slide along fixed parallel rails. Again the displacement $u(x, t)$ satisfies equation (7.1.29) with initial conditions (7.1.31). But the boundary conditions (7.1.30) are now replaced by

$$(7.1.42) \qquad \frac{\partial u(0, t)}{\partial x} = 0, \quad \frac{\partial u(\pi, t)}{\partial x} = 0 \qquad (t > 0).$$

The exact formulation of the problem is very similar to (7.1.33). Indeed, conditions (i), (ii), (iv), (v), and (vi) are unchanged but the initial conditions $f, g \in \mathsf{X}[0, \pi]$ need not satisfy (7.1.32), and (iii) must be replaced by

$$\text{(7.1.43)} \qquad \lim_{x\to 0+} \frac{\partial u(x, t)}{\partial x} = 0, \qquad \lim_{x\to\pi-} \frac{\partial u(x, t)}{\partial x} = 0 \qquad (t > 0),$$

it being moreover assumed that the limits $\lim_{x\to 0+} u(x, t)$, $\lim_{x\to\pi-} u(x, t)$, $\lim_{x\to 0+} \partial u(x, t)/\partial t$, $\lim_{x\to\pi-} \partial u(x, t)/\partial t$, $\lim_{x\to 0+} \partial^2 u(x, t)/\partial x^2$, *and* $\lim_{x\to 0+} \partial^2 u(x, t)/\partial t^2$ *exist for each* $t > 0$.

As in the case of fixed ends it follows by (7.1.43) that, for each $t > 0$, $u(x, t)$, $\partial u(x, t)/\partial x$ and $\partial u(x, t)/\partial t$ are continuous functions on $0 \le x \le \pi$ and that the differential equation (7.1.29) is satisfied at $x = 0$. Therefore we may again extend the problem to the interval $-\pi \le x \le 0$. Now this extension has to be even. Then, in particular, $\partial u(x, t)/\partial x$ is an odd function in correspondence with (7.1.42). Thus the present problem is again a special case of (7.1.35), namely the case of even initial conditions f and g.

Theorem 7.1.7. *Let* $f, g \in \mathsf{X}[0, \pi]$ *be extended to the interval* $-\pi \le x \le 0$ *as even functions and then to the whole real axis as* 2π*-periodic functions. Suppose that for these extensions* $f \in \mathsf{AC}^1_{2\pi}$ *and* $g \in \mathsf{AC}_{2\pi}$ *such that* f'' *and* g' *exist for all* x *and belong to* $\mathsf{X}_{2\pi}$. *Then the* ***sliding string problem*** *has a unique solution* $u(x, t)$ *given by* (7.1.39). *The representation* (7.1.39) *holds for all* x *and* $t > 0$. *Moreover,*

$$\text{(7.1.44)} \quad u(x, t) = f_c^\wedge(0) + g_c^\wedge(0)t + 2\sum_{k=1}^{\infty}\left[f_c^\wedge(k)\cos ckt + \frac{1}{ck}g_c^\wedge(k)\sin ckt\right]\cos kx,$$

the series being absolutely and uniformly convergent.

The proof is left to Problem 7.1.3. $f_c^\wedge(k)$ and $g_c^\wedge(k)$ denote the kth Fourier cosine-coefficients (cf. (1.2.10) and Problem 7.1.8) of f and g, respectively.

Problems

1. (i) Show by direct evaluation that $u(x, t)$ as given by (7.1.7) is a solution of (7.1.4).
   (ii) If the source distribution $F(x, t)$ satisfies (7.1.11), show that $u(x, t)$ as given by (7.1.14) is a solution of (7.1.10) satisfying (7.1.4)(i), (iii)–(v).
2. Show that $u(x, r)$ as given by (7.1.23) is a solution of (7.1.17).
3. Complete the proof of Theorems 7.1.5–7.1.7.

In the following we shall give some further examples of differential, difference and integral equations (or of mixed type) which may be solved by integral transform methods. In any case, the reader is asked to supplement the formal statement by a rigorous interpretation. The references given contain at least the formal construction of solutions.

4. Let $f \in \mathsf{X}_{2\pi}$. Determine the solution of the first-order equation
$$\frac{\partial u(x, t)}{\partial t} - \frac{\partial u(x, t)}{\partial x} - u(x, t) = 0 \qquad (-\pi \le x \le \pi,\ t \in \mathbb{R})$$
with initial condition $u(x, 0) = f(x)$ and boundary condition $u(-\pi, t) = u(\pi, t)$. (Hint: Weinberger [1, p. 299])
5. By means of the finite Fourier transform evaluate the solution $u(x) \in \mathsf{C}_{2\pi}$ of the differential-difference equations
(i) $u'(x) + u(x) + u(x + \pi) = \sin^3 x$, (ii) $u'(x + \pi) - u(x) = \sin x$.

6. Find the solution $u \in X_{2\pi}$ of the integral equation $\cos x = \int_{-\pi}^{\pi} s\, u(x - s)\, ds.$

7. Determine the solution $u \in X_{2\pi}$ of

$$u(x) - \frac{1}{\pi}\int_{-\pi/2}^{\pi/2} u(x + s)\, ds = |\sin x| - \frac{2}{\pi}.$$

8. The *finite Fourier-cosine* [*-sine*] *transform* of $f \in L^1(0, \pi)$ is the function $f_c^\wedge(k)$ [$f_s^\wedge(k)$], defined on $\mathbb{P}$ [$\mathbb{N}$], whose value at $k \in \mathbb{P}$ [$k \in \mathbb{N}$] is

$$f_c^\wedge(k) = \frac{2}{\pi}\int_0^\pi f(u)\cos ku\, du \qquad \left[f_s^\wedge(k) = \frac{2}{\pi}\int_0^\pi f(u)\sin ku\, du\right],$$

the notation being consistent with (1.2.10) as applied to even [odd] 2π-periodic functions.

(i) Let $f \in L^1(0, \pi)$. Show that if $f_c^\wedge(k) = 0$ [$f_s^\wedge(k) = 0$] for all $k \in \mathbb{P}$ [$k \in \mathbb{N}$], then $f(x) = 0$ for almost all $x \in (0, \pi)$. (Hint: HARDY–ROGOSINSKI [1, p. 20])

(ii) Let $f \in L^1(0, \pi)$ be such that $f_c^\wedge \in l^1(\mathbb{P})$ [$f_s^\wedge \in l^1(\mathbb{N})$]. Show that for almost all $x \in (0, \pi)$

$$f(x) = \tfrac{1}{2}f_c^\wedge(0) + \sum_{k=1}^\infty f_c^\wedge(k)\cos kx \qquad \left[f(x) = \sum_{k=1}^\infty f_s^\wedge(k)\sin kx\right].$$

(iii) Let f, f' be absolutely continuous on $[0, \pi]$ and $f'' \in L^1(0, \pi)$. Show that

$$[f'']_c^\wedge(k) = -k^2 f_c^\wedge(k) - \frac{2}{\pi}\{f'(0) - (-1)^k f'(\pi)\} \qquad (k \in \mathbb{P}),$$

$$[f'']_s^\wedge(k) = -k^2 f_s^\wedge(k) + \frac{2}{\pi}k\{f(0) - (-1)^k f(\pi)\} \qquad (k \in \mathbb{N}).$$

(Hint: CHURCHILL [2, p. 290 ff])

(iv) Let $f, g \in L^1(0, \pi)$ and f_1, g_1 [f_2, g_2] be their 2π-periodic extensions which are odd [even] on $(-\pi, \pi)$. If $*$ denotes the usual convolution of 2π-periodic functions (cf. (0.4.5)), show that

$$-2[f_1 * g_1]_c^\wedge(k) = f_s^\wedge(k)g_s^\wedge(k) \qquad (k \in \mathbb{P}),$$
$$2[f_1 * g_2]_s^\wedge(k) = f_s^\wedge(k)g_c^\wedge(k) \qquad (k \in \mathbb{P}),$$
$$2[f_2 * g_2]_c^\wedge(k) = f_c^\wedge(k)g_c^\wedge(k) \qquad (k \in \mathbb{N}),$$

(Hint: CHURCHILL [2, p. 298])

9. Apply the finite Fourier-sine transform in order to solve the following problem: If α is a fixed complex number, determine the solution $u(x, y)$ of

$$\frac{\partial^2 u(x, y)}{\partial x^2} + \frac{\partial^2 u(x, y)}{\partial y^2} = -\alpha \qquad (0 < x < \pi,\ y > 0),$$

for which $u(0, y) = 0$, $u(\pi, y) = 1$, $u(x, 0) = 0$, and $\|u(\circ, y)\|_{X[0,\pi]}$ is bounded for $y > 0$. (Hint: CHURCHILL [2, p. 301])

10. The following three-dimensional problem can be solved by a repeated use of the finite Fourier-sine transform: given $\alpha \in \mathbb{C}$, find $u(x, y, z)$ such that

$$\frac{\partial^2 u(x, y, z)}{\partial x^2} + \frac{\partial^2 u(x, y, z)}{\partial y^2} + \frac{\partial^2 u(x, y, z)}{\partial z^2} = 0 \qquad (0 < x, y, z < \pi)$$

with boundary conditions $u(x, \pi, z) = \alpha$, $u(0, y, z) = u(\pi, y, z) = u(x, 0, z) = u(x, y, 0) = u(x, y, \pi) = 0$. (Hint: TRANTER [1, p. 83])

11. Apply the finite Fourier-cosine transform in order to determine the solution $u(x, t)$ of

$$\frac{\partial u(x, t)}{\partial t} = \frac{\partial^2 u(x, t)}{\partial x^2} \qquad (0 < x < \pi,\ t > 0)$$

with boundary conditions $\partial u(0, t)/\partial x = \partial u(\pi, t)/\partial x = 0$ and initial condition $u(x, 0) = f(x)$. Here f is a given function of $\mathsf{X}[0, \pi]$. (Hint: TRANTER [1, p. 84])

12. Apply the finite Fourier-sine transform in order to solve the following problem: Let $F \in \mathsf{X}[0, \pi]$. Find $u(x, t)$ such that

$$\frac{\partial^2 u(x, t)}{\partial t^2} = -\frac{\partial^4 u(x, t)}{\partial x^4} + F(x) \qquad (0 < x < \pi, t > 0)$$

with initial conditions $u(x, 0) = \partial u(x, 0)/\partial t = 0$ and boundary conditions $u(0, t) = u(\pi, t) = \partial^2 u(0, t)/\partial x^2 = \partial^2 u(\pi, t)/\partial x^2 = 0$. (Hint: CHURCHILL [2, p. 308])

13. Use the finite Fourier-sine transform in order to determine the even solution u of the integral equation

$$\int_{-\pi}^{\pi} \{\operatorname{sgn} s\} \frac{\sinh(\pi - |s|)}{\sinh \pi} u(x - s)\, ds = \{\operatorname{sgn} x\} \frac{|x|(\pi^2 - x^2)}{6\pi}.$$

(Hint: Problem 7.1.8(iv))

7.2 Fourier Transform Method in L^1

7.2.1 Diffusion on an Infinite Rod

Our first application of the Fourier transform method will be to the classical problem of the *flow of heat in an infinite rod* with a given initial temperature distribution. Denoting the temperature at the point $x \in \mathbb{R}$ at time $t > 0$ by $u(x, t)$ and the initial temperature distribution by $f(x)$, the problem is to obtain a solution of the normalized parabolic differential equation

$$\text{(7.2.1)} \qquad \frac{\partial u(x, t)}{\partial t} = \frac{\partial^2 u(x, t)}{\partial x^2} \qquad (x \in \mathbb{R}, t > 0)$$

satisfying the initial condition

$$\text{(7.2.2)} \qquad u(x, 0) = f(x).$$

We shall solve this problem subject to the following conditions:

(7.2.3) *Given a function $f \in \mathsf{L}^1$, we call for a function $u(x, t)$, defined for all $x \in \mathbb{R}$ and $t > 0$, such that*

(i) $\partial u(x, t)/\partial x$, $\partial^2 u(x, t)/\partial x^2$, $\partial u(x, t)/\partial t$ *exist for all* $x \in \mathbb{R}$, $t > 0$,

(ii) $u(x, t)$ *satisfies equation* (7.2.1) *for all* $x \in \mathbb{R}$, $t > 0$,

(iii) *for each* $t > 0$, $u(x, t)$ *belongs to* L^1 *as a function of* x *and*

$$\lim_{t \to 0+} \|u(\circ, t) - f(\circ)\|_1 = 0,$$

(iv) *for each* $t > 0$, $u(x, t) \in \mathsf{W}^2_{\mathsf{L}^1}$ *as a function of* x,

(v) *for each* $t > 0$, $\partial u(x, t)/\partial t \in \mathsf{L}^1$ *as a function of* x *and*

$$\lim_{\tau \to 0} \left\| \frac{u(\circ, t + \tau) - u(\circ, t)}{\tau} - \frac{\partial u(\circ, t)}{\partial t} \right\|_1 = 0.$$

First of all, the fact that all functions involved belong to L^1 implicitly implies that certain boundary conditions are already satisfied; for instance, $\lim_{|x|\to\infty} u(x, t) = 0$ for each $t > 0$. Indeed, (i) and (iv) give

$$u(x + h, t) - u(x, t) = \int_0^h \frac{\partial u(x + s, t)}{\partial x}\, ds.$$

Since $\partial u(\circ, t)/\partial x \in \mathsf{L}^1$ for each $t > 0$, $\lim_{|h|\to\infty} u(x + h, t)$ exists as a finite number which must be zero because $u(\circ, t) \in \mathsf{L}^1$. Similarly one shows $\lim_{|x|\to\infty} \partial u(x, t)/\partial x = 0$.

Let us suppose that there exists a solution $u(x, t)$ of (7.2.3). Denoting, for each $t > 0$, the Fourier transform of $u(x, t)$ by $u^\wedge(v, t)$, we may apply the Fourier transform to the differential equation (7.2.1). Then by (7.2.3)(iv) and Prop. 5.1.14

$$\left[\frac{\partial^2 u(\circ, t)}{\partial x^2}\right]^\wedge(v) = (iv)^2 u^\wedge(v, t) \qquad (v \in \mathbb{R},\ t > 0),$$

and by (7.2.3)(v) and (5.1.2)

$$\left[\frac{\partial u(\circ, t)}{\partial t}\right]^\wedge(v) = \frac{\partial u^\wedge(v, t)}{\partial t} \qquad (v \in \mathbb{R},\ t > 0).$$

In particular, $u^\wedge(v, t)$ is a continuous function of t on $(0, \infty)$ for each $v \in \mathbb{R}$. Thus the transform of equation (7.2.1) passes into

$$\frac{\partial u^\wedge(v, t)}{\partial t} = -v^2 u^\wedge(v, t) \qquad (v \in \mathbb{R},\ t > 0), \tag{7.2.4}$$

which is really a continuous system with one equation for each value of the parameter $v \in \mathbb{R}$ and is to be compared with the discrete system of (7.1.5). The latter (ordinary) differential equation may be solved by classical methods giving for each $v \in \mathbb{R}$ the solution

$$u^\wedge(v, t) = A(v)\, e^{-v^2 t}$$

with constant $A(v)$ independent of t. To determine $A(v)$ we have by (7.2.3)(iii) and (5.1.2) that $\lim_{t\to 0+} u^\wedge(v, t) = f^\wedge(v)$ uniformly with respect to $v \in \mathbb{R}$. Hence, if our problem (7.2.3) has a solution, then its Fourier transform is given by

$$u^\wedge(v, t) = f^\wedge(v)\, e^{-v^2 t} \qquad (v \in \mathbb{R},\ t > 0). \tag{7.2.5}$$

In order to represent it in terms of the original functions, since $f^\wedge(v) \exp\{-v^2 t\} \in \mathsf{L}^1$ for each $t > 0$, we may use Prop. 5.1.10 to deduce

$$u(x, t) = \frac{1}{\sqrt{2\pi}} \int_{-\infty}^{\infty} e^{-tv^2} f^\wedge(v)\, e^{ixv}\, dv \tag{7.2.6}$$

for all x and $t > 0$. Thus if a solution of (7.2.3) exists, it is unique by Prop. 5.1.11. The fact that $u(x, t)$ as given by (7.2.6) is actually a solution of (7.2.3) is left to Problem 7.2.1. Using the Parseval formula (5.1.5) and Problem 5.1.2 we may rewrite this solution as

$$u(x, t) = \frac{1}{\sqrt{4\pi t}} \int_{-\infty}^{\infty} f(x - s)\, e^{-s^2/4t}\, ds. \tag{7.2.7}$$

Therefore

Theorem 7.2.1. *Problem* (7.2.3) *on the flow of heat in an infinite rod has a unique solution. It is given by the singular integral* (3.1.32) *of Gauss–Weierstrass associated with the initial value f.*

Let us also examine the related problem with heat sources $F(x, t)$ distributed along the rod. This phenomenon is described by the nonhomogeneous equation

$$\frac{\partial u(x,t)}{\partial t} = \frac{\partial^2 u(x,t)}{\partial x^2} + F(x,t) \qquad (x \in \mathbb{R},\ t > 0). \tag{7.2.8}$$

Suppose that, for each $t > 0$, $F(x, t)$ belongs to L^1 as a function of x and satisfies

(7.2.9)

(i) $$\|F(\circ, t)\|_1 \le M \qquad (0 < t \le 1),$$

(ii) $$\lim_{\tau \to 0} \|F(\circ, t+\tau) - F(\circ, t)\|_1 = 0 \qquad (t > 0).$$

Then we seek for a function $u(x, t)$ which solves the problem (7.2.3), equation (7.2.1) being replaced by (7.2.8).

As in the homogeneous case the assumption that there exists a solution $u(x, t)$ yields that its Fourier transform $u^\wedge(v, t)$ satisfies for each $v \in \mathbb{R}$ the ordinary differential equation

$$\frac{\partial u^\wedge(v,t)}{\partial t} + v^2 u^\wedge(v,t) = F^\wedge(v,t) \qquad (t > 0). \tag{7.2.10}$$

In view of the initial condition the homogeneous part of (7.2.10) has the solution $u_1^\wedge(v, t) = f^\wedge(v) \exp\{-v^2 t\}$. Furthermore, as $F^\wedge(v, t)$ is a continuous function of t on $(0, \infty)$ uniformly with respect to $v \in \mathbb{R}$ in view of (7.2.9), a particular solution of the nonhomogeneous equation (7.2.10) is given for each $v \in \mathbb{R}$ by

$$u_2^\wedge(v,t) = \int_0^t F^\wedge(v, t-\tau)\, e^{-v^2\tau}\, d\tau. \tag{7.2.11}$$

Moreover, by (7.2.9) there exists for each $t > 0$ a constant M^* such that for $v \neq 0$

$$\left| \int_0^t F^\wedge(v, t-\tau)\, e^{-v^2\tau}\, d\tau \right| \le \sup_{0<\tau\le t} \|F(\circ, \tau)\|_1 \cdot \int_0^t e^{-v^2\tau}\, d\tau \le \frac{M^*}{v^2}.$$

Therefore $u_2^\wedge(v, t)$ belongs for each $t > 0$ to L^1 as a function of v. Now Prop. 5.1.10 and Fubini's theorem yield

$$\begin{aligned} u_2(x,t) &= \frac{1}{\sqrt{2\pi}} \int_{-\infty}^{\infty} \left\{ \int_0^t F^\wedge(v, t-\tau)\, e^{-v^2\tau}\, d\tau \right\} e^{ixv}\, dv \\ &= \int_0^t \left\{ \frac{1}{\sqrt{4\pi\tau}} \int_{-\infty}^{\infty} F(x-s, t-\tau)\, e^{-s^2/4\tau}\, ds \right\} d\tau. \end{aligned}$$

Thus, introducing the Heaviside function $\delta(x)$ of (5.3.3), we may state

Theorem 7.2.2. *If the source distribution $F(x, t)$ is defined for all $x \in \mathbb{R}$, $t > 0$ and satisfies* (7.2.9), *then equation* (7.2.8) *has a unique solution satisfying* (7.2.3)(i), (iii)–(v). *It is given for all x and $t > 0$ by*

(7.2.12)
$$u(x, t) = W(f; x; t) + \frac{1}{2\pi} \int_{-\infty}^{\infty} \int_{-\infty}^{\infty} F(x - s, t - \tau)\delta(t - \tau) \left\{ \frac{1}{\sqrt{4\pi\tau}} e^{-s^2/4\tau} \delta(\tau) \right\} ds\, d\tau.$$

7.2.2 Dirichlet's Problem for the Half-Plane

Suppose there is a steady state temperature distribution in an infinitely large plate, and the temperature distribution along one edge is known. Do these conditions determine the temperature in the plate? The mathematical formulation of this question is *Dirichlet's problem* for the upper half-plane. We raise it in the following way:

(7.2.13) *Given a function $f \in \mathsf{L}^1$, we call for a function $u(x, y)$, defined for all $x \in \mathbb{R}$ and $y > 0$, such that*

(i) *$\partial^2 u(x, y)/\partial x^2$, $\partial^2 u(x, y)/\partial y^2$ exist for all $x \in \mathbb{R}$, $y > 0$,*

(ii) *$u(x, y)$ satisfies for all $x \in \mathbb{R}$, $y > 0$ the equation*

(7.2.14)
$$\frac{\partial^2 u(x, y)}{\partial x^2} + \frac{\partial^2 u(x, y)}{\partial y^2} = 0,$$

(iii) *for each $y > 0$, $u(x, y) \in \mathsf{L}^1$ as a function of x, $\|u(\circ, y)\|_1 \leq M$ uniformly for $y > 0$ and*

(7.2.15)
$$\lim_{y \to 0+} \|u(\circ, y) - f(\circ)\|_1 = 0,$$

(iv) *for each $y > 0$, $u(x, y) \in \mathsf{W}^2_{\mathsf{L}^1}$ as a function of x,*

(v) *for each $y > 0$, $\partial u(x, y)/\partial y$ and $\partial^2 u(x, y)/\partial y^2$ belong to L^1 as functions of x and*

$$\lim_{\eta \to 0} \left\| \frac{u(\circ, y + \eta) - u(\circ, y)}{\eta} - \frac{\partial u(\circ, y)}{\partial y} \right\|_1 = 0,$$

$$\lim_{\eta \to 0} \left\| \frac{\dfrac{\partial u(\circ, y + \eta)}{\partial y} - \dfrac{\partial u(\circ, y)}{\partial y}}{\eta} - \frac{\partial^2 u(\circ, y)}{\partial y^2} \right\|_1 = 0.$$

By arguments similar to those used in Sec. 7.1.2 and 7.2.1 we obtain for the Fourier transform of any solution $u(x, t)$ of (7.2.13)

(7.2.16)
$$\frac{\partial^2 u^\wedge(v, y)}{\partial y^2} - v^2 u^\wedge(v, y) = 0 \qquad (v \in \mathbb{R}, y > 0),$$

which has as its solution

$$u^\wedge(v, y) = \begin{cases} A(v)\, e^{vy} + B(v)\, e^{-vy}, & v \neq 0 \\ A(0) + B(0) y, & v = 0. \end{cases}$$

To determine the constants $A(v)$, $B(v)$ we note that $|u^\wedge(v, y)| \leq \|u(\circ, y)\|_1 \leq M$ for all $v \in \mathbb{R}$, $y > 0$, and hence by letting $y \to \infty$ we obtain that $A(v) = 0$ for $v > 0$ and $B(v) = 0$ for $v \leq 0$. On the other hand, the relation $\lim_{y\to 0+} u^\wedge(v, y) = f^\wedge(v)$ implies $A(v) + B(v) = f^\wedge(v)$ for all $v \in \mathbb{R}$. Summarizing, we have that the Fourier transform of any solution $u(x, y)$ of (7.2.13) is given by

$$u^\wedge(v, y) = f^\wedge(v)\, e^{-|v|y} \qquad (v \in \mathbb{R},\ y > 0). \tag{7.2.17}$$

Since $f^\wedge(v) \exp\{-|v|\, y\} \in \mathsf{L}^1$ as a function of v for each $y > 0$, we have by Prop. 5.1.10 for every $x \in \mathbb{R}$ and $y > 0$

$$u(x, y) = \frac{1}{\sqrt{2\pi}} \int_{-\infty}^{\infty} e^{-y|v|} f^\wedge(v)\, e^{ixv}\, dv \tag{7.2.18}$$

or, in view of (5.1.19),

$$u(x, y) = \frac{y}{\pi} \int_{-\infty}^{\infty} \frac{f(x - s)}{y^2 + s^2}\, ds \qquad (x \in \mathbb{R},\ y > 0), \tag{7.2.19}$$

which is the singular integral of Cauchy–Poisson as defined by (3.1.38). Thus, any solution of (7.2.13) is given as the Cauchy–Poisson integral of the boundary value f. Since, by virtue of Problem 7.2.2, the function $u(x, y)$ of (7.2.19) actually is a solution of (7.2.13), we have shown (cf. Sec. 3.1.4)

Theorem 7.2.3. *The Dirichlet problem* (7.2.13) *for the upper half-plane has a unique solution. It is given by the singular integral of Cauchy–Poisson of the boundary value f.*

In Sec. 12.4.3 we shall discuss the behaviour of the solution $u(x, y)$ of Dirichlet's problem (7.2.13) on the boundary $y = 0$. In particular, from assumptions upon the degree of approximation of the boundary value f conclusions upon the boundary value itself will be drawn.

7.2.3 Motion of an Infinite String

The final application of the Fourier transform method is concerned with the problem of the motion of an infinite string with a given initial displacement f and velocity g, $u(x, t)$ being the displacement at distance x along the string at time t. The precise formulation of the problem is given by:

(7.2.20) *For given functions $f, g \in \mathsf{L}^1$, find a function $u(x, t)$, defined for all $x \in \mathbb{R}$ and $t > 0$, such that*

(i) *$\partial^2 u(x, t)/\partial x^2$, $\partial u^2(x, t)/\partial t^2$, exist for all $x \in \mathbb{R}$, $t > 0$,*

(ii) *$u(x, t)$ satisfies for all $x \in \mathbb{R}$, $t > 0$ the equation* ($c > 0$)

$$\frac{\partial^2 u(x, t)}{\partial t^2} = c^2 \frac{\partial^2 u(x, t)}{\partial x^2}, \tag{7.2.21}$$

(iii) *for each $t > 0$, $u(x, t)$ and $\partial u(x, t)/\partial t$ belong to L^1 as functions of x and*

$$\lim_{t\to 0+} \|u(\circ, t) - f(\circ)\|_1 = 0, \qquad \lim_{t\to 0+} \left\| \frac{\partial u(\circ, t)}{\partial t} - g(\circ) \right\|_1 = 0,$$

(iv) *for each* $t > 0$, $u(x, t) \in W^2_{L^1}$ *as a function of* x,
(v) *for each* $t > 0$, $\partial^2 u(x, t)/\partial t^2 \in L^1$ *as a function of* x *and*

$$\lim_{\tau \to 0} \left\| \frac{u(\circ, t + \tau) - u(\circ, t)}{\tau} - \frac{\partial u(\circ, t)}{\partial t} \right\|_1 = 0,$$

$$\lim_{\tau \to 0} \left\| \frac{\dfrac{\partial u(\circ, t + \tau)}{\partial t} - \dfrac{\partial u(\circ, t)}{\partial t}}{\tau} - \frac{\partial^2 u(\circ, t)}{\partial t^2} \right\|_1 = 0.$$

We proceed as in the foregoing sections and assume that there exists a solution $u(x, t)$ of (7.2.20). Then the Fourier transform $u^\wedge(v, t)$ of this solution satisfies for each $v \in \mathbb{R}$ the ordinary differential equation

$$\frac{\partial^2 u^\wedge(v, t)}{\partial t^2} + (cv)^2 u^\wedge(v, t) = 0 \qquad (t > 0), \tag{7.2.22}$$

which has the general solution

$$u^\wedge(v, t) = \begin{cases} A(v)\, e^{icvt} + B(v)\, e^{-icvt}, & v \neq 0 \\ A(0) + B(0)t, & v = 0, \end{cases}$$

$A(v)$, $B(v)$ being arbitrary complex constants independent of t. To determine these constants we note that by (7.2.20), (iii) and (v),

$$\lim_{t \to 0+} u^\wedge(v, t) = f^\wedge(v), \qquad \lim_{t \to 0+} \frac{\partial u^\wedge(v, t)}{\partial t} = g^\wedge(v)$$

uniformly with respect to $v \in \mathbb{R}$. This implies as in Sec. 7.1.3 that the Fourier transform of any solution $u(x, t)$ of (7.2.20) is given by

$$u^\wedge(v, t) = \begin{cases} f^\wedge(v) \cos cvt + \dfrac{1}{cv} g^\wedge(v) \sin cvt, & v \neq 0 \\ f^\wedge(0) + g^\wedge(0)t, & v = 0 \end{cases} \qquad (t > 0). \tag{7.2.23}$$

To reconstruct the full solution we may, in view of (5.1.29), rewrite (7.2.23) as

$$u^\wedge(v, t) = \left[\frac{f(\circ + ct) + f(\circ - ct)}{2}\right]^\wedge(v) + \left[\frac{1}{2c}\int_{-ct}^{ct} g(\circ + s)\, ds\right]^\wedge(v),$$

which holds for all $v \in \mathbb{R}$, $t > 0$. Now the uniqueness theorem of the Fourier transform in L^1 yields that for each $t > 0$

$$u(x, t) = \frac{1}{2}[f(x + ct) + f(x - ct)] + \frac{1}{2c}\int_{x-ct}^{x+ct} g(s)\, ds \tag{7.2.24}$$

for almost all $x \in \mathbb{R}$.

In order to show that the right-hand side of (7.2.24) actually is a solution, just as in Sec. 7.1.3, we have to pose certain conditions upon the initial values f and g.

Theorem 7.2.4. *Let* $f \in W^2_{L^1}$ *and* $g \in W^1_{L^1}$ *such that* $f''(x)$ *and* $g'(x)$ *exist for every* $x \in \mathbb{R}$. *Then problem* (7.2.20) *has a unique solution; it is given by* (7.2.24) *which holds for all* $x \in \mathbb{R}$, $t > 0$.

The proof is left to Problem 7.2.3.

Problems

1. (i) Show by direct evaluation that $u(x, t)$ as given by (7.2.6) is a solution of (7.2.3).
   (ii) If the source distribution $F(x, t)$ satisfies (7.2.9), show that $u(x, t)$ as given by (7.2.12) is a solution of (7.2.8) satisfying (7.2.3)(i), (iii)–(v).
2. Show that $u(x, y)$ as given by (7.2.19) is a solution of (7.2.13).
3. Complete the proof of Theorem 7.2.4.

Concerning the following Problems the reader is again asked to supplement the formal statement by a rigorous interpretation.

4. Let $f \in \mathsf{L}^1$. Find the solution $u(\circ, t) \in \mathsf{L}^1$ of the difference-differential equation
$$u(x+1, t) - 2u(x, t) + u(x-1, t) = \frac{\partial u(x, t)}{\partial t} \qquad (x, t \in \mathbb{R}),$$
which satisfies $u(x, 0) = f(x)$. (Hint: WEINBERGER [1, pp. 326, 435])
5. Find the solution $u \in \mathsf{L}^1$ of the ordinary differential equation
$$xu''(x) + u'(x) - xu(x) = 0 \qquad (x \in \mathbb{R}).$$
(Hint: WEINBERGER [1, pp. 326, 340, 435])
6. Find the solution $u \in \mathsf{L}^2$ of the integral equation
$$u(x) = e^{-|x|} + \tfrac{1}{2} e^{x} \int_x^\infty e^{-s} u(s)\, ds \qquad (x \in \mathbb{R}).$$
(Hint: TITCHMARSH [6, p. 303 ff])
7. Determine the solution $u \in \mathsf{L}^2$ of the integral equation
$$e^{-x^2/4} = \int_{-\infty}^{\infty} \frac{1}{1 + (x - s)^2}\, u(s)\, ds \qquad (x \in \mathbb{R}).$$
(Hint: TITCHMARSH [6, p. 314])
8. Determine the solution $u(\circ, y) \in \mathsf{L}^1$ of the partial differential equation
$$\frac{\partial^2 u(x, y)}{\partial x^2} + \frac{\partial^2 u(x, y)}{\partial y^2} - u(x, y) = 0 \qquad (x \in \mathbb{R}, 0 < y < 1),$$
satisfying $u(x, 1) = \exp\{-x^2\}$, $\partial u(x, 0)/\partial y = 0$. (Hint: WEINBERGER [1, pp. 320, 434])
9. Determine the solution $u(\circ, t) \in \mathsf{L}^1$ of the fourth-order equation
$$\frac{\partial^4 u(x, t)}{\partial x^4} + \frac{\partial^2 u(x, t)}{\partial t^2} = 0 \qquad (x \in \mathbb{R}, t > 0),$$
which satisfies $u(x, 0) = \exp\{-x^2\}$, $\partial u(x, 0)/\partial t = 0$. (Hint: SNEDDON [3, p. 129])

7.3 Notes and Remarks

For basic results on partial differential equations of second-order, in particular on the classification of the three types, see e.g. COURANT–HILBERT [1II], HELLWIG [1], PETROVSKY [1, p. 45], WEINBERGER [1, p. 41 ff]. For a useful account of transform methods, particularly the Laplace transform, see DOETSCH [3, p. 339 ff], [4III, p. 13 ff].

Sec. 7.1. The finite Fourier transform method for solving partial differential equations was first considered by DOETSCH [2] in 1935. But the method does not seem to have found its way into the many text-books on the subject, exceptions being CHURCHILL [2], SNEDDON [1, 2] (see also WEINBERGER [1, p. 126] and SEELEY [1]). Our approach differs from that of the above authors in that convergence to the initial and boundary values is not considered in the pointwise sense but in the norm of the space in question. This allows a rigorous treatment (including uniqueness), inspired by the study of BOCHNER–CHANDRASEKHARAN [1, p. 40 ff] on Fourier transforms.

Our definition of a solution of a given differential equation is really raised as an *abstract Cauchy problem*. This calls for a solution of the following (see BUTZER–BERENS [1, p. 3 ff] and the literature cited there): Given a Banach space X and a linear operator U whose domain $\mathsf{D}(U)$ and range $\mathsf{R}(U)$ belong to X and given an element $f \in \mathsf{X}$, to find a (vector-valued) function $w(t) \equiv w(t; f)$ on $(0, \infty)$ into X such that

(i) $w(t)$ is continuously differentiable on $(0, \infty)$ in the norm,
(ii) $w(t) \in \mathsf{D}(U)$ and $w'(t) = Uw(t)$ for each $t > 0$,
(iii) $\lim_{t\to 0+} \|w(t) - f\|_{\mathsf{X}} = 0$.

For instance, for the Fourier problem of the ring in $\mathsf{C}_{2\pi}$, we have $w(t) = u(\circ, t)$ and $[Uw(t)](x) = \partial^2 u(x, t)/\partial x^2$ with $\mathsf{D}(U) = \mathsf{W}^2_{\mathsf{C}_{2\pi}}$. Thus condition (i) of the above problem corresponds to (v) of (7.1.4), (ii) to (iv) and (ii), and (iii) to (iii). However, in the light of an abstract Cauchy problem, Fourier's problem in $\mathsf{L}^p_{2\pi}$ already shows that it would be more appropriate also to interpret differentiation with respect to the space-variable x in the norm-sense. Here we prefer the mixed formulation (7.1.4) which assumes in (i) and (ii) that the differential equation is satisfied in the classical pointwise sense. Similar remarks apply to the other examples treated, e.g. problem (7.1.17) corresponds to an abstract Cauchy problem of second order.

Finite Hankel and Legendre transforms have been used in the solution of boundary value problems involving cylinders and spheres of finite radius by SNEDDON [1, 2] and TRANTER [1], respectively. Legendre transforms have also been applied to problems of potential theory by CHURCHILL [1]. For finite Laplace transform methods see e.g. PINNEY [1], DOETSCH [4III, p. 225 ff]. One disadvantage of transform methods as a whole is that they do not deliver singular or 'explosive' solutions that may be important from a physical point of view (see DOETSCH [4III, pp. 31–38]).

The classical method for solving problems of this section is D. Bernoulli's method of *separation of variables* which is treated in every text-book on the subject, e.g. SAGAN [1], SEELEY [1], H. F. DAVIS [1]. It leads to an eigenvalue-problem for which we tacitly assume that the solution is known, since we did not actually discuss the question how to find the suitable transform for a given problem (cf. CHURCHILL [2, 3]).

For Fourier's problem of the ring see the classical accounts in CARSLAW [1, p. 211 ff], DOETSCH [2], MINAKSHISUNDARAM [1], HILLE [2, p. 402]. For nonhomogeneous problems compare e.g. WEINBERGER [1], TYCHONOV–SAMARSKII [1]. For the (interior) Dirichlet and Neumann problem see e.g. DUFF–NAYLOR [1, p. 133 ff], SNEDDON [3], SEELEY [1]. In an analogous fashion the third boundary value problem of potential theory as well as the exterior problems may be considered. The treatment of Sec. 7.1.3 was given by DOETSCH [2] for the differential equation of heat conduction describing the heat transfer in a bar of finite length. The task was, roughly speaking, to describe a typical situation where it is possible to unroll a 'finite' problem on the circle group, thus to deal with it as a periodic one. One advantage here is that the vibrating string problem as well as the sliding string problem are particular cases of the periodic problem (7.1.35) (for $\mathsf{L}^2_{2\pi}$-space see also EDWARDS [1I, p. 136]). Normally, the conditions are posed such that one has to apply the finite Fourier-sine or -cosine transform separately. A further method would be to apply the Laplace transform with respect to the time-variable t (cf. DOETSCH [4]).

Sec. 7.2. The classical standard reference is TITCHMARSH [6]. The material of the first two subsections may be found in BOCHNER–CHANDRASEKHARAN [1, pp. 40–47], SEELEY [1, pp. 65–96] (see also IVANOV [1]). The former authors were probably the first to give a really rigorous proof (at least in book form) of these results, using the Fourier transform method. It is to be noted that we have here restricted discussion to the space L^1, since a treatment in L^p-space $(1 < p \le 2)$ would involve a study of vector-valued differential equations (cf.

HILLE–PHILLIPS [1, p. 67 ff]). In this respect the authors cannot follow the reasoning given in BOCHNER–CHANDRASEKHARAN [1, p. 135] in the L^2-case.

One-dimensional transform methods may also be applied to partial differential equations in three and more independent variables. The procedure reduces an equation in n independent variables $x_1, x_2, \ldots, x_n$ to one in $n-1$ independent variables $x_2, \ldots, x_n$ and a parameter v. By a successive use of integral transforms (perhaps of a different type) the given partial differential equation is eventually reduced to an ordinary differential or even an algebraic equation, which may be solved easily.

The fact that one might solve linear partial differential equations more effectively by, for example, applying the Fourier transform method depends upon Prop. 5.1.14 which states that the Fourier transform converts differentiation (r-times) into multiplication with $(iv)^r$. On the other hand, for the solution of integral equations of Fourier convolution type (cf. Problems 7.2.6, 7.2.7) Theorem 5.1.3 is essential; it asserts that the Fourier transform converts convolution into pointwise multiplication. But when considered from a distributional point of view, convolution turns out to be the more general process. Various types of differential equations, difference equations, and integral equations are all special cases of *convolution equations*. For example, if δ is the 'delta function' (defined through $(\delta, \phi) = \phi(0)$, $\phi \in \mathsf{D}$) and f any distribution (of D'), then $f = f * \delta$, $Lf = f * L\delta$, L being any linear differential operator. Thus shifting and differentiation may be written as convolutions, emphasizing the importance of convolution theorems. These aspects are now standard topics of modern treatments on partial differential equations. Fourier transform machinery, in particular its extension to generalized functions, plays an important rôle in these theoretical investigations and not only in the treatment of particular examples as given in this chapter. As a (partial) list of references we can cite SCHWARTZ [1, 3], GELFAND–SHILOV [1], LIONS [1], HÖRMANDER [2], FRIEDMAN [1, 2], BREMERMANN [1], ZEMANIAN [1, 2], TRÈVES [1], SHILOV [1], STAKGOLD [1]. (See also Vol. II of this treatise.)

Part III

Hilbert Transforms

It was shown in Sec. 1.2, 1.4, 4.1 that the Fourier series of $f \in X_{2\pi}$ is θ-summable to f, thus

$$U_\rho(f; x) = \sum_{k=-\infty}^{\infty} \theta_\rho(k) f^\wedge(k)\, e^{ikx}$$

tends in $X_{2\pi}$-norm or pointwise to $f(x)$ for $\rho \to \rho_0$, in case $\{\theta_\rho(k)\}$ is a convergence-factor satisfying suitable conditions. On the other hand, an application of the θ-factor to the conjugate Fourier series $S^\sim[f]$ of f gave the expression

$$\begin{aligned} U_\rho^\sim(f; x) &= \sum_{k=-\infty}^{\infty} \theta_\rho(k)\{-i \operatorname{sgn} k\} f^\wedge(k)\, e^{ikx} \\ &= \frac{1}{2\pi}\int_{-\pi}^{\pi} f(x-u)\Big[2\sum_{k=1}^{\infty} \theta_\rho(k) \sin ku\Big]\, du, \end{aligned}$$

called the conjugate of $U_\rho(f; x)$. A particular instance of such an expression is the conjugate of the Abel–Poisson integral of (1.2.52). This suggests strongly an examination of the limit of $U_\rho^\sim(f; x)$ for $\rho \to \rho_0$. Indeed, this limit is called the *conjugate function* of f and denoted by $f^\sim$.

For $\lim_{\rho\to\rho_0} \|U_\rho(f; \circ) - f(\circ)\|_{X_{2\pi}} = 0$ to be valid for every $f \in X_{2\pi}$ it was necessary that $\lim_{\rho\to\rho_0} \theta_\rho(k) = 1$, $k \in \mathbb{Z}$. Therefore, proceeding formally, we have by (1.2.48)

$$\lim_{\rho\to\rho_0} 2\sum_{k=1}^{\infty} \theta_\rho(k) \sin ku = 2\sum_{k=1}^{\infty} \sin ku = \lim_{n\to\infty}\Big[\cot\frac{u}{2} - \frac{\cos(2n+1)(u/2)}{\sin(u/2)}\Big],$$

and thus

$$\lim_{\rho\to\rho_0} U_\rho^\sim(f; x) = \frac{1}{2\pi}\int_{-\pi}^{\pi} f(x-u)\cot\frac{u}{2}\, du$$

$$- \lim_{n\to\infty} \frac{1}{2\pi}\int_{-\pi}^{\pi} f(x-u)\frac{\cos(2n+1)(u/2)}{\sin(u/2)}\, du.$$

In view of the singularity at $u = 0$, in order to evaluate the latter limit, one is led to study the Cauchy principal value. Then by the Riemann–Lebesgue lemma

$$\lim_{n\to\infty} \int_0^\pi \frac{f(x-u) - f(x+u)}{\sin (u/2)} \cos \frac{(2n+1)u}{2}\, du = 0$$

if f is sufficiently smooth. Thus we formally come to

$$f^\sim(x) = \frac{1}{2\pi} \int_{-\pi}^{\pi} f(x-u) \cot \frac{u}{2}\, du,$$

which is to be understood in the Cauchy principal value sense. As a matter of fact, $f^\sim$ exists finitely for almost all x for any $f \in \mathsf{X}_{2\pi}$ (Theorem 9.1.1).

It will be shown that the mapping $f \to f^\sim$ is a bounded linear transformation of $\mathsf{L}^2_{2\pi}$ into itself with $\|f^\sim\|_2 \leq \|f\|_2$. However, there are functions in $\mathsf{L}^1_{2\pi}$ for which $f^\sim$ is not integrable, and thus the mapping is not bounded as an operator from $\mathsf{L}^1_{2\pi}$ into itself. Moreover, there exist (absolutely) continuous f such that $f^\sim \notin \mathsf{L}^\infty_{2\pi}$. On the other hand, the theorem of Marcel Riesz (1927) asserts that for $1 < p < \infty$ there exists a constant $M_p \geq 1$ such that $f^\sim \in \mathsf{L}^p_{2\pi}$ and $\|f^\sim\|_p \leq M_p\|f\|_p$ for each $f \in \mathsf{L}^p_{2\pi}$ (Theorem 9.1.3).

Chapter 9 is devoted to a study of the conjugate function, or *Hilbert transform* as it is often called, on the circle group; Chapter 8 is concerned with the parallel theory of Hilbert transforms on the line group. At this stage we must emphasize that although we speak of a Hilbert transform, it will be useful for us not to regard the entity as a transform but as a function. Indeed, if it will not be possible to characterize certain function classes in terms of f, it will often be so in terms of the conjugate function $f^\sim$. Many authors work with the Hilbert transform on the line as if it were an operational transform, obtain an inversion formula, etc.; but this is not our point of view.

8

Hilbert Transforms on the Real Line

8.0 Introduction

Recalling some results of Sec. 3.1 we have noted that for any real $f \in \mathsf{L}^p$, $1 \le p < \infty$, the integral

$$H(z) = \frac{i}{\pi} \int_{-\infty}^{\infty} \frac{f(u)}{z-u}\, du \qquad (z = x + iy)$$

defines a function $H(z) = P(f; x; y) + iQ(f; x; y)$ which is holomorphic in the upper-half plane $\mathbb{R}^{2,+}$. Thus it follows that the real part

$$P(f; x; y) = \frac{1}{\pi} \int_{-\infty}^{\infty} f(x-u) \frac{y}{u^2 + y^2}\, du \qquad (x \in \mathbb{R}, y > 0)$$

and imaginary part

$$Q(f; x; y) = \frac{1}{\pi} \int_{-\infty}^{\infty} f(x-u) \frac{u}{u^2 + y^2}\, du$$

of $H(z)$ are conjugate harmonic functions (in the usual sense of complex function theory). We have seen that $P(f; x; y)$ tends to the limit f as $y \to 0+$; the limit of $Q(f; x; y)$ for $y \to 0+$ yields a function which is called the *conjugate function* of f, denoted by $f^{\sim}$. If we let $y \to 0+$ we obtain, formally,

$$f^{\sim}(x) = \frac{1}{\pi} \int_{-\infty}^{\infty} f(x-u) \frac{1}{u}\, du = \frac{1}{\pi} \int_{-\infty}^{\infty} \frac{f(u)}{x-u}\, du. \tag{8.0.1}$$

This convolution integral, however, is not defined even when f is an extremely well-behaved function. In fact, if $f \in \mathsf{L}^p$, $1 \le p < \infty$, we have

$$f^{\sim}(x) = \frac{1}{\pi} \int_{|u| \le 1} \frac{f(x-u)}{u}\, du + \frac{1}{\pi} \int_{|u| \ge 1} \frac{f(x-u)}{u}\, du,$$

and the second integral exists everywhere and represents a bounded function (by

Hölder's inequality). The main difficulty lies with the first integral, and it clearly need not exist as a Lebesgue integral at any point. For example, if f is constant and nonzero in a neighbourhood of a point x_0, then this integral fails to exist in this neighbourhood of x_0. Thus in order to give (8.0.1) a meaning one has to interpret the integral in a suitable way.

Definition 8.0.1. *The **Hilbert transform** of a function* $f \in \mathsf{L}^p$, $1 \le p < \infty$, *is defined as the Cauchy principal value of the integral* (8.0.1) *whenever it exists, i.e.* $f^{\sim}(x) \equiv (Hf)(x)$ *is defined as*

$$f^{\sim}(x) = \mathrm{PV}\left[\frac{1}{\pi}\int_{-\infty}^{\infty}\frac{f(x-u)}{u}\,du\right] \equiv \lim_{\delta\to 0+} f_\delta^{\sim}(x) \tag{8.0.2}$$

at every point x *for which this limit exists. Hereby,*

$$f_\delta^{\sim}(x) = \frac{1}{\pi}\int_{|u|\ge\delta}\frac{f(x-u)}{u}\,du = \frac{1}{\pi}\int_{\delta}^{\infty}\frac{f(x-u)-f(x+u)}{u}\,du. \tag{8.0.3}$$

For $f \in \mathsf{L}^p$ and any $\delta > 0$, $f_\delta^{\sim}(x)$ is well-defined everywhere as a bounded function. The limit in (8.0.2) will not only be interpreted in the pointwise sense, but also in the norm.

In Sec. 8.1 we show that the Hilbert–Stieltjes transform of $\mu \in \mathsf{BV}$ exists a.e., treat the existence problem for $f \in \mathsf{L}^p$, $1 \le p < \infty$, and derive fundamental properties of $f^{\sim}$, including the Marcel Riesz inequality, the selection partly depending upon their use in our later investigations in approximation theory. It will be important to evaluate the Fourier transform of $f^{\sim}$. Noting that

$$\mathrm{PV}\left[\frac{1}{\sqrt{2\pi}}\int_{-\infty}^{\infty}\frac{e^{-ivx}}{x}\,dx\right] = \lim_{\varepsilon\to 0}\frac{-2i}{\sqrt{2\pi}}\int_{\varepsilon}^{\infty}\frac{\sin vx}{x}\,dx$$
$$= \{-i\,\mathrm{sgn}\,v\}\sqrt{\frac{2}{\pi}}\int_0^{\infty}\frac{\sin y}{y}\,dy = \{-i\,\mathrm{sgn}\,v\}\sqrt{\pi/2},$$

we have formally

$$[f^{\sim}]^{\wedge}(v) = \sqrt{\frac{2}{\pi}}\left[f * \frac{1}{x}\right]^{\wedge}(v) = \{-i\,\mathrm{sgn}\,v\}f^{\wedge}(v). \tag{8.0.4}$$

It will be shown that this formula is valid for almost all v if $f \in \mathsf{L}^p$, $1 < p \le 2$ (Theorem 8.1.7, Prop. 8.3.1). If $p = 1$, it is valid for all v provided f as well as $f^{\sim}$ belong to L^1 (Prop. 8.3.1). These facts lead the way to characterizations of the classes $\mathsf{W}[\mathsf{L}^p; |v|^r]$, $\mathsf{V}[\mathsf{L}^p; |v|^r]$, $1 \le p \le 2$, $r \in \mathbb{N}$. Indeed (Theorem 8.3.6),

$$f \in \mathsf{W}[\mathsf{L}^p; |v|] \iff f^{\sim} \in \mathsf{AC}_{\mathrm{loc}} \quad\text{and}\quad (f^{\sim})' \in \mathsf{L}^p.$$

Sec. 8.2 is devoted to Hilbert formulae (preparatory to Prop. 8.3.1) and to a detailed study of convergence (pointwise and norm sense) of conjugates of general singular integrals. Furthermore, theorems of Privalov-type are considered. Sec. 8.3 also deals with summation of allied integrals as well as with a theorem of Hille and Tamarkin (Theorem 8.3.8) on the norm-convergence of the Fourier inversion integral.

8.1 Existence of the Transform

8.1.1 Existence Almost Everywhere

We are first concerned with the existence almost everywhere of the Hilbert transform of $f \in \mathsf{L}^p$, $1 \le p < \infty$. In fact, we establish the existence of the *Hilbert–Stieltjes transform*.

Definition 8.1.1. *The Hilbert–Stieltjes transform $\mu^\sim$ of a function $\mu \in \mathsf{BV}$ is defined by the Cauchy principal value*

$$\mu^\sim(x) = \lim_{\delta\to 0+} \frac{1}{\pi} \int_{|x-u|\ge\delta} \frac{d\mu(u)}{x-u} \equiv \mathrm{PV}\frac{1}{\pi}\int_{-\infty}^{\infty} \frac{d\mu(u)}{x-u} \tag{8.1.1}$$

at each $x \in \mathbb{R}$ for which the limit exists.

To show that the Hilbert–Stieltjes transform exists almost everywhere we need the following lemmata.

Lemma 8.1.2. *Let $d_k > 0$, $a_k \in \mathbb{R}$, $1 \le k \le n$, and*

$$g(x) = \sum_{k=1}^{n} \frac{d_k}{x-a_k}. \tag{8.1.2}$$

Then

$$\operatorname{meas}\{x \mid |g(x)| > M > 0\} = \frac{2}{M}\sum_{k=1}^{n} d_k, \tag{8.1.3}$$

and this set consists precisely of $2n$ intervals.

Proof. Since $g(a_k-) = -\infty$, $g(a_k+) = +\infty$ and $g'(x) < 0$ for all $x \ne a_k$, $1 \le k \le n$, there are precisely n points m_k such that $g(m_k) = M$ with $a_k < m_k < a_{k+1}$, $1 \le k \le n-1$, and $a_n < m_n$. The set where $g(x) > M$ therefore consists of the intervals (a_k, m_k) and has total length

$$\sum_{k=1}^{n} (m_k - a_k) = \sum_{k=1}^{n} m_k - \sum_{k=1}^{n} a_k. \tag{8.1.4}$$

But the numbers m_k are the roots of the equation

$$\sum_{k=1}^{n} \frac{d_k}{x-a_k} = M,$$

and hence by a multiplication with $\prod_{k=1}^{n}(x-a_k)$ we have

$$\sum_{k=1}^{n} d_k\Big[\prod_{j\ne k}(x-a_j)\Big] = M\prod_{k=1}^{n}(x-a_k)$$

or

$$Mx^n - \Big[M\sum_{k=1}^{n} a_k + \sum_{k=1}^{n} d_k\Big]x^{n-1} + \cdots = 0.$$

If we compare the coefficients of the last equation with those of $\prod_{k=1}^{n}(x-m_k) = 0$, then

$$\sum_{k=1}^{n} m_k = \sum_{k=1}^{n} a_k + \frac{1}{M}\sum_{k=1}^{n} d_k.$$

If we substitute into (8.1.4) and combine this with the corresponding result for the set where $g(x) < -M$, (8.1.3) follows.

Lemma 8.1.3. *Let $\mu \in \mathsf{BV}$ be real-valued. If $(x_k - \delta_k, x_k + \delta_k)$, $1 \le k \le n$, are disjoint intervals such that*

$$\left|\left(\int_{-\infty}^{x_k-\delta_k} + \int_{x_k+\delta_k}^{\infty}\right)\frac{d\mu(u)}{x_k - u}\right| > M > 0, \tag{8.1.5}$$

then $\sum_{k=1}^{n} \delta_k \le \sqrt{2\pi}(16/M)\|\mu\|_{\mathsf{BV}}$.

Proof. First we observe that the integrals (8.1.5) are well-defined and bounded by $\delta_k^{-1}\|\mu\|_{\mathsf{BV}}$, $1 \le k \le n$. Hence they may be approximated arbitrarily closely by finite Riemann–Stieltjes sums, if the norm of the subdivision is sufficiently small. Thus let u_j, $1 \le j \le m$, be a finite subdivision including the points $x_k - \delta_k$, x_k, $x_k + \delta_k$, $1 \le k \le n$, such that, according to (8.1.5), for each k, $1 \le k \le n$,

$$\left|\sum_{j \notin I_k} \frac{\mu(u_{j+1}) - \mu(u_j)}{y - u_j}\right| > M > 0 \tag{8.1.6}$$

holds for $y = x_k$, where the set I_k of indices omitted is defined by

$$\bigcup_{j \in I_k} (u_j, u_{j+1}) = (x_k - \delta_k, x_k + \delta_k).$$

Now suppose that μ is monotonely increasing. Then the left member of (8.1.6) is a decreasing function of y for $x_k - \delta_k < y < x_k + \delta_k$. Hence, if the sum is positive (negative), the inequality (8.1.6) holds for $x_k - \delta_k < y \le x_k$ ($x_k \le y < x_k + \delta_k$). For every such y one of the following inequalities is therefore satisfied:

$$\left|\sum_{j=1}^{m-1} \frac{\mu(u_{j+1}) - \mu(u_j)}{y - u_j}\right| > \frac{M}{2}, \qquad \left|\sum_{j \in I_k} \frac{\mu(u_{j+1}) - \mu(u_j)}{y - u_j}\right| > \frac{M}{2}. \tag{8.1.7}$$

Since $\mu(u_{j+1}) - \mu(u_j) \ge 0$, we may apply Lemma 8.1.2, and summing over k it follows that

$$\sum_{k=1}^{n} \delta_k \le \frac{4}{M}\sum_{j=1}^{m-1}\{\mu(u_{j+1}) - \mu(u_j)\} + \frac{4}{M}\sum_{k=1}^{n}\sum_{j \in I_k}\{\mu(u_{j+1}) - \mu(u_j)\}$$

$$\le \frac{8}{M}\sum_{j=1}^{m-1}\{\mu(u_{j+1}) - \mu(u_j)\} \le \frac{8}{M}\sqrt{2\pi}\,\|\mu\|_{\mathsf{BV}}.$$

If μ is of bounded variation, we may by Jordan's theorem decompose μ into the difference of two increasing functions, $\mu = \mu_1 - \mu_2$, such that $\|\mu\|_{\mathsf{BV}} = \|\mu_1\|_{\mathsf{BV}} + \|\mu_2\|_{\mathsf{BV}}$. Then the hypothesis (8.1.5) means that

$$\left|\left\{\left(\int_{-\infty}^{x_k-\delta_k} + \int_{x_k+\delta_k}^{\infty}\right)\frac{d\mu_1(u)}{x_k - u}\right\} - \left\{\left(\int_{-\infty}^{x_k-\delta_k} + \int_{x_k+\delta_k}^{\infty}\right)\frac{d\mu_2(u)}{x_k - u}\right\}\right| > M > 0.$$

Thus one of the two terms in the curly brackets must in the absolute value be greater than $M/2$, and since μ_1 and μ_2 are now increasing, we may apply our foregoing considerations in order to obtain the result.

Lemma 8.1.4. *The Hilbert transform of any step function vanishing outside of compact sets exists except at a finite number of points. In particular, the Hilbert transform of the characteristic function $\kappa_{[a,b]}$ exists except at the two points a and b and is given by*

$$[\kappa_{[a,b]}]^{\sim}(x) = \frac{1}{\pi}\log\left|\frac{x-b}{x-a}\right|. \tag{8.1.8}$$

Indeed, according to Def. 8.0.1, (8.1.8) follows by an elementary calculation.

Theorem 8.1.5. *The Hilbert–Stieltjes transform* (8.1.1) *of* $\mu \in \mathsf{BV}$ *exists almost everywhere. Moreover, for every* $M > 0$,

$$\text{meas}\,\{x \mid |\mu^{\sim}(x)| > M\} \le \frac{128}{M}\,\|\mu\|_{\mathsf{BV}}. \tag{8.1.9}$$

Proof. Let μ be real-valued. We first prove the existence a.e. of $\mu^{\sim}(x)$. By Cauchy's criterion, it is sufficient to show that, given $\varepsilon > 0$, for every x except in a set of measure less than ε

$$\left|\left(\int_{x-\delta}^{x-\delta'} + \int_{x+\delta'}^{x+\delta}\right)\frac{d\mu(u)}{x-u}\right| \le \varepsilon \tag{8.1.10}$$

for all sufficiently small δ and δ', $0 < \delta' < \delta$.

Let $\mu(x) = \mu_{\mathsf{AC}}(x) + \mu_{\text{sing}}(x)$ be the Lebesgue decomposition of μ into its absolutely continuous part μ_{AC} and its singular part μ_{sing}, and let $\mu_{\mathsf{AC}}(x) = \int_{-\infty}^{x} g(u)\,du$ with $g \in \mathsf{L}^1$. Since the step functions which vanish outside of compact sets are dense in L^1, we may, to given $\varepsilon' > 0$, choose a function h such that $\|g - h\|_1 < \varepsilon'$. Thus, if $\mu_1(x) = \int_{-\infty}^{x} h(u)\,du$, then $\|\mu_{\mathsf{AC}} - \mu_1\|_{\mathsf{BV}} < \varepsilon'$. Furthermore, μ_{sing} can be approximated to within ε' by a singular function μ_2 with variation confined to a closed set of measure zero, that is, which is constant on the intervals of an open set $\mathbb{B}$ whose complement has measure zero: $\|\mu_{\text{sing}} - \mu_2\|_{\mathsf{BV}} < \varepsilon'$. Thus, taking $\varepsilon' = \varepsilon^2/384\sqrt{2\pi}$, we set $\mu = \mu_1 + \mu_2 + \mu_3$ with $\|\mu_3\|_{\mathsf{BV}} < \varepsilon^2/192\sqrt{2\pi}$.

Let $\mathbb{E}_\varepsilon$ be the set of x for which the inequality

$$\left|\left(\int_{x-\delta}^{x-\delta'} + \int_{x+\delta'}^{x+\delta}\right)\frac{d\mu_3(u)}{x-u}\right| \le \frac{\varepsilon}{3} \tag{8.1.11}$$

fails to hold for all sufficiently small δ and δ', $0 < \delta' < \delta$. Then, for every $x \in \mathbb{E}_\varepsilon$,

$$\left|\left(\int_{-\infty}^{x-\eta} + \int_{x+\eta}^{\infty}\right)\frac{d\mu_3(u)}{x-u}\right| > \frac{\varepsilon}{6} \tag{8.1.12}$$

for sufficiently small η. By Vitali's theorem a sequence of disjoint intervals $(x_k - \eta_k, x_k + \eta_k)$ satisfying (8.1.12) can be chosen so as to cover $\mathbb{E}_\varepsilon$ except for a set of measure zero. Then by Lemma 8.1.3

$$\text{meas}\,\{\mathbb{E}_\varepsilon\} \le 2\sum_k \eta_k \le 2\,\frac{16}{\varepsilon/6}\,\sqrt{2\pi}\,\|\mu_3\|_{\mathsf{BV}} < \varepsilon.$$

Since μ_1 is the integral of a step function h which vanishes outside of a compact set, its Hilbert–Stieltjes transform, i.e., the Hilbert transform of h, exists by Lemma 8.1.4 except for a finite number of points which we add to $\mathbb{E}_\varepsilon$. Since μ_2 is constant on the intervals of $\mathbb{B}$, its Hilbert–Stieltjes transform obviously exists except on the complement of $\mathbb{B}$ which we add to $\mathbb{E}_\varepsilon$. Thus, if x is not in the enlarged $\mathbb{E}_\varepsilon$, there is a $\delta_0 > 0$ such that for all δ and δ' with $0 < \delta' < \delta < \delta_0$ (8.1.11) holds for μ_1, μ_2, and μ_3, and hence (8.1.10) holds as was to be shown.

The relation (8.1.9) now follows immediately from Lemma 8.1.3. In fact, if x is such that $\mu^{\sim}(x)$ exists and $|\mu^{\sim}(x)| > M$, then

$$\left|\left(\int_{-\infty}^{x-\delta} + \int_{x+\delta}^{\infty}\right)\frac{d\mu(u)}{x-u}\right| > \pi M \tag{8.1.13}$$

for sufficiently small $\delta > 0$. Again Vitali's theorem gives a sequence of disjoint intervals $(x_k - \delta_k, x_k + \delta_k)$ satisfying (8.1.13) which covers the set of x with $|\mu^{\sim}(x)| > M$ up to a set of measure zero. Therefore

$$\text{meas}\,\{x \mid |\mu^{\sim}(x)| > M > 0\} \le 2\sum_k \delta_k \le 2\cdot\frac{16}{M}\|\mu\|_{\mathsf{BV}}.$$

If μ is complex-valued, then the above considerations apply to the real and imaginary part of μ, respectively. Thus Theorem 8.1.5 is completely established.

Theorem 8.1.6. *The Hilbert transform $f^{\sim}$ of a function $f \in \mathsf{L}^p$, $1 \le p < \infty$, exists almost everywhere.*

Proof. The case $p = 1$ follows by Theorem 8.1.5 since the Hilbert transform $f^{\sim}$ may be regarded as the Hilbert–Stieltjes transform $\mu^{\sim}$ of the absolutely continuous function $\mu(x) = \int_{-\infty}^{x} f(u)\,du$. If $1 < p < \infty$, suppose $x \in [-a, a]$, $a > 0$, and set $f = f_1 + f_2$ with $f_1(x) = \kappa_{[-2a,2a]}(x)f(x)$. Then

$$f_\delta^{\sim}(x) = \frac{1}{\pi}\int_{|x-u|\ge\delta}\frac{f_1(u)}{x-u}\,du + \frac{1}{\pi}\int_{|x-u|\ge\delta}\frac{f_2(u)}{x-u}\,du \equiv I_1 + I_2.$$

But

$$\lim_{\delta\to 0+} I_2 = \frac{1}{\pi}\int_{|u|\ge 2a}\frac{f(u)}{x-u}\,du$$

for all $x \in [-a, a]$, and since $f_1 \in \mathsf{L}^1$, $\lim_{\delta\to 0+} I_1$ exists almost everywhere. Thus the Hilbert transform $f^{\sim}$ exists for almost all $x \in [-a, a]$, and since a was arbitrary, the theorem is established.

8.1.2 Existence in L^2-Norm

So far we have discussed the pointwise limit in (8.0.2) and thus the pointwise existence of the Hilbert transform $f^{\sim}$ of functions $f \in \mathsf{L}^p$, $1 \le p < \infty$. We shall now turn to the question of mean convergence in (8.0.2). This will be part of the general problem of carrying over certain properties of the original function f to its Hilbert transform $f^{\sim}$. In particular, we shall show that $f \in \mathsf{L}^p$, $1 < p < \infty$, implies $f^{\sim} \in \mathsf{L}^p$. We proceed with the case $p = 2$.

Let $f \in \mathsf{L}^2$ and set

$$k_{\delta,\eta}(x) = \begin{cases}\dfrac{1}{x}, & 0 < \delta \le |x| \le \eta < \infty\\ 0, & \text{elsewhere,}\end{cases} \tag{8.1.14}$$

$$f_{\delta,\eta}^{\sim}(x) = \frac{1}{\pi}\int_{-\infty}^{\infty} f(x-u)k_{\delta,\eta}(u)\,du = \frac{1}{\pi}\int_{\delta\le|u|\le\eta}\frac{f(x-u)}{u}\,du. \tag{8.1.15}$$

Then, since $k_{\delta,\eta} \in L^1 \cap L^2$ for all $0 < \delta < \eta < \infty$, it follows by Prop. 0.2.1 and 0.2.2 that $f_{\delta,\eta}^{\sim}(x)$ exists everywhere as a function in $C_0 \cap L^2$. Furthermore,

(8.1.16) $$\lim_{\eta\to\infty} f_{\delta,\eta}^{\sim}(x) = f_\delta^{\sim}(x).$$

By Theorem 5.2.12 we conclude

$$[f_{\delta,\eta}^{\sim}]^\wedge(v) = \sqrt{\frac{2}{\pi}}\, f^\wedge(v)[k_{\delta,\eta}]^\wedge(v) \quad \text{a.e.}$$

Since

$$[k_{\delta,\eta}]^\wedge(v) = \frac{1}{\sqrt{2\pi}} \int_{\delta\le|x|\le\eta} \frac{e^{-ivx}}{x}\,dx = \{-i \operatorname{sgn} v\}\sqrt{\frac{2}{\pi}} \int_{\delta|v|}^{\eta|v|} \frac{\sin u}{u}\,du,$$

we have

(8.1.17) $$\lim_{\delta\to0+}\left\{\lim_{\eta\to\infty} k_{\delta,\eta}^\wedge(v)\right\} = \{-i \operatorname{sgn} v\}\sqrt{\frac{\pi}{2}}$$

and $|k_{\delta,\eta}^\wedge(v)| \le M$ with constant M independent of v, δ, η (see also Problem 1.2.10). Therefore

$$\lim_{\delta\to0+}\left\{\lim_{\eta\to\infty} [f_{\delta,\eta}^{\sim}]^\wedge(v)\right\} = \{-i \operatorname{sgn} v\} f^\wedge(v) \quad \text{a.e.,}$$

and since $|[f_{\delta,\eta}^{\sim}]^\wedge(v)| \le M\,|f^\wedge(v)|$ is bounded by an L^2-function, we may apply Lebesgue's dominated convergence theorem to deduce

(8.1.18) $$\lim_{\delta\to0+}\left\{\lim_{\eta\to\infty} \|[f_{\delta,\eta}^{\sim}]^\wedge(\circ) - \{-i \operatorname{sgn} \circ\} f^\wedge(\circ)\|_2\right\} = 0.$$

Since $\{-i \operatorname{sgn} v\} f^\wedge(v) \in L^2$, by Plancherel's theorem there exists $g \in L^2$ such that $g^\wedge(v) = \{-i \operatorname{sgn} v\} f^\wedge(v)$ a.e. Therefore by the Parseval equation

(8.1.19) $$\|[f_{\delta,\eta}^{\sim}]^\wedge(\circ) - \{-i \operatorname{sgn} \circ\} f^\wedge(\circ)\|_2 = \|f_{\delta,\eta}^{\sim} - g\|_2.$$

Thus (8.1.18) implies $\text{l.i.m.}^{(2)}_{\delta\to0+,\,\eta\to\infty} f_{\delta,\eta}^{\sim}(x) = g(x)$. Moreover, by (8.1.16) and Theorem 8.1.6 it follows that $\lim_{\delta\to0+,\,\eta\to\infty} f_{\delta,\eta}^{\sim}(x) = f^{\sim}(x)$ a.e. Therefore $g(x) = f^{\sim}(x)$ a.e. in view of Prop. 0.1.10. Since $\|f^{\sim}\|_2 = \|g^\wedge\|_2 = \|\{-i \operatorname{sgn} \circ\} f^\wedge(\circ)\|_2 = \|f\|_2$ by (5.2.5), we may summarize our results in the following

Theorem 8.1.7. *For $f \in L^2$ the Hilbert transform $f^{\sim}$ exists almost everywhere, belongs to* L^2, and satisfies

(8.1.20) $$\|f^{\sim}\|_2 = \|f\|_2.$$

Moreover,

(8.1.21) $$\lim_{\delta\to0+} \|f^{\sim} - f_\delta^{\sim}\|_2 = 0,$$

and the Fourier transform of $f^{\sim}$ is given by

(8.1.22) $$[f^{\sim}]^\wedge(v) = \{-i \operatorname{sgn} v\} f^\wedge(v) \quad \text{a.e.}$$

It only remains to show (8.1.21). In view of Fatou's lemma we have by (8.1.16), (8.1.18), and (8.1.19) that

$$\|f_\delta^\sim - f^\sim\|_2^2 \le \liminf_{\eta\to\infty} \|f_{\delta,\eta}^\sim - f^\sim\|_2^2 = o(1) \qquad (\delta \to 0+),$$

proving (8.1.21). In particular, $f_\delta^\sim \in \mathsf{L}^2$ with $\|f_\delta^\sim\|_2 \le M\,\|f\|_2$ for every $\delta > 0$, $f \in \mathsf{L}^2$.

Thus the mapping $f \to f^\sim$ is a bounded linear transformation of L^2 into L^2 which preserves norms. It is actually *onto* as follows by (8.1.22).

8.1.3 Existence in L^p-Norm, $1 < p < \infty$

To this end we apply the theorem of Marcinkiewicz (cf. Sec. 4.4) with $\mathbb{R}_1 = \mathbb{R}_2 = \mathbb{R}$ and $\mu_1 = \mu_2 =$ Lebesgue measure on $\mathbb{R}$. The operator T which assigns to each simple function h its Hilbert transform $h^\sim$,

$$(8.1.23) \qquad (Th)(x) = h^\sim(x) = \lim_{\delta\to 0+} \frac{1}{\pi} \int_{|u|\ge\delta} \frac{h(x-u)}{u}\,du \qquad (h \in \mathsf{S}_{00}),$$

is well-defined by Theorem 8.1.6 as a linear operator on S_{00} which satisfies

$$(8.1.24) \qquad \text{meas}\,\{x \mid |(Th)(x)| > M\} \le \frac{128}{M}\,\|h\|_1$$

by (8.1.9), and furthermore

$$(8.1.25) \qquad \|Th\|_2 = \|h\|_2$$

by (8.1.20), since $h \in \mathsf{S}_{00}$ implies $h \in \mathsf{L}^q$, $1 \le q \le \infty$. Thus T is of weak type (1; 1) and of strong type (2; 2) on S_{00} (with constants $M_1 = 128$ and $M_2 = 1$). The theorem of Marcinkiewicz then implies that T is of strong type $(p;p)$, $1 < p < 2$, on S_{00}, and that $\|Th\|_p \le M_p\|h\|_p$ for all $h \in \mathsf{S}_{00}$ with constant M_p depending only upon p. Hence

Proposition 8.1.8. *Let* $1 < p < 2$. *If* $h \in \mathsf{S}_{00}$, *then* $h^\sim \in \mathsf{L}^p$ *and*

$$(8.1.26) \qquad \|h^\sim\|_p \le M_p\|h\|_p,$$

the constant M_p *being independent of* h.

Moreover,

Proposition 8.1.9. *The operator* T *as defined on* S_{00} *by* (8.1.23) *has a unique extension as a bounded linear transformation of* L^p *into* L^p, $1 < p < \infty$, *such that*

$$(8.1.27) \qquad \|Tf\|_p \le M_p\|f\|_p$$

holds for all $f \in \mathsf{L}^p$ *with constant* M_p *independent of* f.

Proof. By Theorem 8.1.7 and Prop. 8.1.8 the operator T as defined on S_{00} by (8.1.23) is of strong type $(p;p)$, $1 < p \le 2$, satisfying (8.1.27). To show that T is of strong type

$(p;p)$ on S_{00} for $2 < p < \infty$, let $h, g \in \mathsf{S}_{00}$ and $\delta > 0$. Then we have by Fubini's theorem

$$(8.1.28) \qquad \int_{-\infty}^{\infty} h_\delta^\sim(x)g(x)\,dx = -\int_{-\infty}^{\infty} h(x)g_\delta^\sim(x)\,dx.$$

Since $h, g \in \mathsf{L}^2$, we obtain by (8.1.21) and Hölder's inequality

$$(8.1.29) \qquad \int_{-\infty}^{\infty} (Th)(x)g(x)\,dx = -\int_{-\infty}^{\infty} h(x)(Tg)(x)\,dx$$

for all $h, g \in \mathsf{S}_{00}$. Now, let $2 < p < \infty$. Then $Tg \in \mathsf{L}^{p'}$ with $\|Tg\|_{p'} \le M_{p'}\|g\|_{p'}$ by Prop. 8.1.8. Therefore, by (8.1.29) and Hölder's inequality,

$$\left|\int_{-\infty}^{\infty} (Th)(x)g(x)\,dx\right| \le \|h\|_p\|Tg\|_{p'} \le M_{p'}\|h\|_p\|g\|_{p'}.$$

Hence Th defines a bounded linear functional on the dense subset S_{00} of $\mathsf{L}^{p'}$ which may therefore be extended to a bounded linear functional on all of $\mathsf{L}^{p'}$ (having the same bound). The F. Riesz representation theorem then implies $Th \in \mathsf{L}^p$ and

$$(8.1.30) \qquad \|Th\|_p \le M_{p/(p-1)}\|h\|_p$$

for all $h \in \mathsf{S}_{00}$.

Thus T defines a bounded linear transformation of S_{00} into L^p for all $1 < p < \infty$. Therefore it can be extended to a bounded linear transformation on all of L^p into L^p satisfying (8.1.25), (8.1.26), and (8.1.30), respectively.

So far we have defined the Hilbert transform $f^\sim$ of $f \in \mathsf{L}^p$, $1 < p < \infty$, as a function which by Theorem 8.1.6 exists almost everywhere. On the other hand, by Prop. 8.1.9 we have assigned to each $f \in \mathsf{L}^p$, $1 < p < \infty$, a uniquely determined function $Tf \in \mathsf{L}^p$ which satisfies (8.1.27) and coincides with $f^\sim$ for all f of the subset S_{00} by (8.1.23). In the following we shall show that $Tf = f^\sim$ for all $f \in \mathsf{L}^p$. To this end we need two elementary lemmata.

Lemma 8.1.10. *For $f \in \mathsf{L}^p$, $1 \le p < \infty$,*

$$(8.1.31) \qquad \lim_{y \to 0+}\left\|\frac{1}{\pi}\int_{-\infty}^{\infty} f(\circ - u)\frac{u}{u^2 + y^2}\,du - f_y^\sim(\circ)\right\|_p = 0.$$

In particular, for all $y > 0$

$$(8.1.32) \qquad \left\|\frac{1}{\pi}\int_{-\infty}^{\infty} f(\circ - u)\frac{u}{u^2 + y^2}\,du - f_y^\sim(\circ)\right\|_p \le \|f\|_p.$$

Proof. Firstly, all terms exist everywhere as continuous functions by Prop. 0.2.1. We have for $y > 0$

$$\begin{aligned}(8.1.33) \quad \frac{1}{\pi}\int_{-\infty}^{\infty} f(x - u)\frac{u}{u^2 + y^2}\,du - f_y^\sim(x)\\
&= \frac{1}{\pi}\int_0^y [f(x - u) - f(x + u)]\frac{u}{u^2 + y^2}\,du\\
&\quad - \frac{y^2}{\pi}\left(\int_y^\delta + \int_\delta^\infty\right)[f(x - u) - f(x + u)]\frac{1}{u(u^2 + y^2)}\,du\\
&\equiv I_1 + I_2 + I_3,\end{aligned}$$

say. Given $\varepsilon > 0$, choose $\delta > 0$ such that $\|f(\circ + u) - f(\circ)\|_p < \varepsilon$ for all $|u| \le \delta$ (see Prop. 0.1.9). If $y < \delta$, then by the Hölder–Minkowski inequality

$$\|I_1\|_p \le \frac{1}{\pi}\int_0^y \frac{\|f(\circ - u) - f(\circ + u)\|_p |u|}{u^2 + y^2}\,du \le \frac{2\varepsilon}{\pi y}\int_0^y du = \frac{2}{\pi}\varepsilon,$$

$$\|I_2\|_p \le \frac{2\varepsilon y^2}{\pi}\int_y^\delta \frac{1}{|u|\,(u^2 + y^2)}\,du \le \frac{2\varepsilon y^2}{\pi}\int_y^\delta \frac{du}{u^3} \le \frac{\varepsilon}{\pi}.$$

If we now fix δ, we may take y so small that

$$\|I_3\|_p \le \frac{y^2}{\pi}\int_\delta^\infty \frac{\|f(\circ - u) - f(\circ + u)\|_p}{u^3}\,du \le \frac{\|f\|_p}{\pi\delta^2}\,y^2 < \varepsilon,$$

proving (8.1.31). For (8.1.32) we note that $\|I_1\|_p \le (2/\pi)\|f\|_p$ and $\|I_2 + I_3\|_p \le (1/\pi)\|f\|_p$.

Let us recall (see (3.1.46)) the notation

$$Q(f; x; y) = \frac{1}{\pi}\int_{-\infty}^{\infty} f(x - u)\frac{u}{u^2 + y^2}\,du \qquad (y > 0) \tag{8.1.34}$$

of the *conjugate of the Cauchy–Poisson integral of f*. If $f \in \mathsf{L}^p$, $1 \le p < \infty$, then $Q(f; x; y) \in \mathsf{C}$ by Prop. 0.2.1, since $\sqrt{\pi/2}\,q(x) \equiv x/(1 + x^2)$ (cf. (3.1.47)) belongs to L^r, $1 < r \le \infty$. But note that $q \notin \mathsf{L}^1$. If, as usual, $P(f; x; y)$ denotes the Cauchy–Poisson integral (3.1.38) of f, then

Lemma 8.1.11. *For $f \in \mathsf{L}^p$, $1 < p < \infty$, we have for all $x \in \mathbb{R}$, $y > 0$*

$$P(Tf; x; y) = Q(f; x; y). \tag{8.1.35}$$

In particular, $Q(f; \circ; y), f_\delta^\sim \in \mathsf{L}^p$ and

$$\|Q(f; \circ; y)\|_p \le M_p\|f\|_p, \qquad \|f_\delta^\sim\|_p \le (1 + M_p)\|f\|_p \tag{8.1.36}$$

uniformly for $y > 0$, $\delta > 0$, respectively, the constant M_p being independent of f.

Proof. Since $1/(1 + x^2) \in \mathsf{L}^1 \cap \mathsf{L}^\infty$ and

$$\left[\frac{1}{1 + (\circ)^2}\right]^\sim(x) = \frac{x}{1 + x^2} \qquad (x \in \mathbb{R}) \tag{8.1.37}$$

by Problem 8.1.2, we may argue as in the proof of (8.1.29) to obtain

$$\int_{-\infty}^{\infty} h^\sim(x - u)\frac{y}{u^2 + y^2}\,du = \int_{-\infty}^{\infty} h(x - u)\frac{u}{u^2 + y^2}\,du \tag{8.1.38}$$

for every $h \in \mathsf{S}_{00}$. Thus $P(h^\sim; x; y) = Q(h; x; y)$ for all $x \in \mathbb{R}$, $y > 0$ and $h \in \mathsf{S}_{00}$.

To establish (8.1.35) for all $f \in \mathsf{L}^p$, $1 < p < \infty$, let $\{f_n\} \subset \mathsf{S}_{00}$ be a sequence such that $\lim_{n\to\infty}\|f - f_n\|_p = 0$. Then $P(Tf_n; x; y) = Q(f_n; x; y)$ by (8.1.38). Since it follows by (8.1.27) that $\lim_{n\to\infty}\|T(f - f_n)\|_p = 0$, we have by Hölder's inequality and (8.1.23) for each $x \in \mathbb{R}$ and $y > 0$ that

$$|P(Tf; x; y) - P(f_n^\sim; x; y)| \le y^{-1/p}\,\|p(\circ)\|_{p'}\|Tf - f_n^\sim\|_p = o(1) \qquad (n \to \infty),$$

$$|Q(f; x; y) - Q(f_n; x; y)| \le y^{-1/p}\,\|p^\sim(\circ)\|_{p'}\|f - f_n\|_p = o(1) \qquad (n \to \infty).$$

Combining the results we obtain (8.1.35). Since for $f \in \mathsf{L}^p$, $1 < p < \infty$, $Tf \in \mathsf{L}^p$ and therefore $P(Tf; x; y) \in \mathsf{L}^p$ for all $y > 0$, we have $Q(f; x; y) \in \mathsf{L}^p$ for all $y > 0$, and by (8.1.32) $f_\delta^\sim \in \mathsf{L}^p$ for all $\delta > 0$. Finally (8.1.27), (8.1.32), and (8.1.35) imply (8.1.36).

Let us mention that we will soon generalize the results of the last two lemmata considerably.

Theorem 8.1.12. *For $f \in \mathsf{L}^p$, $1 < p < \infty$, the Hilbert transform as defined by* (8.0.2) *exists almost everywhere, belongs to* L^p *and satisfies*

$$\|f^{\sim}\|_p \le M_p \|f\|_p \tag{8.1.39}$$

with some constant M_p independent of f. Moreover,

$$\lim_{\delta \to 0+} \|f^{\sim} - f_\delta^{\sim}\|_p = 0, \tag{8.1.40}$$

i.e., the limit (8.0.2) *exists not only pointwise* a.e. *but also in the mean of order p.*

Thus for $1 < p < \infty$ the Hilbert transform defines a bounded linear mapping of L^p into L^p. (8.1.39) is often referred to as the *Marcel Riesz inequality* for Hilbert transforms.

Proof. Let $1 < p < \infty$. By Theorem 8.1.6 the Hilbert transform $f^{\sim}$ of $f \in \mathsf{L}^p$ exists almost everywhere, in fact $\lim_{\delta \to 0+} f_\delta^{\sim}(x) = f^{\sim}(x)$ a.e. According to Prop. 8.1.9, to each $f \in \mathsf{L}^p$ we may assign a uniquely determined function $Tf \in \mathsf{L}^p$ such that (8.1.27) holds. We now show that

$$\lim_{\delta \to 0+} \|Tf - f_\delta^{\sim}\|_p = 0. \tag{8.1.41}$$

Indeed, $f_\delta^{\sim} \in \mathsf{L}^p$ for all $\delta > 0$ by Lemma 8.1.11, and by (8.1.35)

$$\|Tf - f_\delta^{\sim}\|_p \le \|(Tf)(\circ) - P(Tf; \circ; \delta)\|_p + \|Q(f; \circ; \delta) - f_\delta^{\sim}(\circ)\|_p,$$

which tends to zero as $\delta \to 0+$ by (3.1.41), (8.1.31). Now it follows as an application of Prop. 0.1.10 that

$$(Tf)(x) = f^{\sim}(x) \quad \text{a.e.} \tag{8.1.42}$$

for all $f \in \mathsf{L}^p$, $1 < p < \infty$. In virtue of Prop. 8.1.9 and of (8.1.41) the proof is complete.

Problems

1. Let $f, g \in \mathsf{L}^p$, $1 \le p < \infty$. Show that
   (i) $[\alpha f + \beta g]^{\sim}(x) = \alpha f^{\sim}(x) + \beta g^{\sim}(x)$ a.e. for every $\alpha, \beta \in \mathbb{C}$,
   (ii) if f is an even function, then $f^{\sim}$ is odd,
   (iii) $[\overline{f}]^{\sim}(x) = \overline{f^{\sim}(x)}$ a.e.,
   (iv) $[f(\circ + h)]^{\sim}(x) = f^{\sim}(x + h)$ a.e. for each $h \in \mathbb{R}$,
   (v) $[\rho f(\rho \circ)]^{\sim}(x) = \rho f^{\sim}(\rho x)$ a.e. for each $\rho > 0$.
2. By means of (8.0.2) evaluate the following Hilbert transforms:
   (i) $[\kappa_{[a,b]}]^{\sim}(x) = (1/\pi) \log(|x - b|/|x - a|)$ for all $x \ne a, b$,
   (ii) if $p(x) = \sqrt{2/\pi}(1/(1 + x^2))$, then $p^{\sim}(x) = \sqrt{2/\pi}(x/(1 + x^2))$ $(\equiv q(x))$ for all $x \in \mathbb{R}$. (For a further method of evaluating Hilbert transforms see Problem 8.3.2.)
3. Give examples of functions $f \in \mathsf{L}^1$ for which $f^{\sim}$ does not belong to L^1. (Hint: Use e.g. $f(x) = 1/(1 + x^2)$ or TITCHMARSH [6, p. 143])
4. Let $f \in \mathsf{L}^p$, $1 \le p < \infty$. If f belongs to $\mathsf{Lip}(\mathsf{C}; \alpha)$ for some $\alpha > 0$, show that $f^{\sim}(x)$ exists for all $x \in \mathbb{R}$.

5. (i) Show that $g(u) = (e^{-ihu} - 1)/|u|$ belongs to L^2 and $g^\wedge(v) = \sqrt{2/\pi}\log|v/(h+v)|$.
(ii) Let $f \in \mathsf{L}^2$. Show that

$$f^\sim(x) = \frac{d}{dx}\left(\frac{1}{\pi}\int_{-\infty}^{\infty} f(u)\log\left|1 - \frac{x}{u}\right| du\right) \quad \text{a.e.}$$

(Hint: Use Theorem 8.1.7, Problem 5.2.9, Prop. 5.2.13; see TITCHMARSH [6, p. 121])

8.2 Hilbert Formulae, Conjugates of Singular Integrals, Iterated Hilbert Transforms

8.2.1 Hilbert Formulae

During the course of the preceding proof we have shown some useful relations which we now generalize.

Proposition 8.2.1. *Let $f \in \mathsf{L}^p$, $1 < p < \infty$, and $g \in \mathsf{L}^{p'}$. Then*

$$\int_{-\infty}^{\infty} f^\sim(x)g(x)\,dx = -\int_{-\infty}^{\infty} f(x)g^\sim(x)\,dx, \tag{8.2.1}$$

$$\int_{-\infty}^{\infty} f(x)\overline{g(x)}\,dx = \int_{-\infty}^{\infty} f^\sim(x)\overline{g^\sim(x)}\,dx. \tag{8.2.2}$$

Relations of this type are referred to as *Hilbert formulae*.

Proof. Since $f^\sim \in \mathsf{L}^p$ and $g^\sim \in \mathsf{L}^{p'}$ by Theorem 8.1.12, all integrals are well-defined. Next we observe that by (8.1.29) the relation (8.2.1) is already known to hold for functions $f, g \in \mathsf{S}_{00}$ (see also (8.1.35)). Let $\{f_n\} \subset \mathsf{S}_{00}$ and $\{g_m\} \subset \mathsf{S}_{00}$ be such that $\lim_{n\to\infty} \|f - f_n\|_p = 0$ and $\lim_{m\to\infty} \|g - g_m\|_{p'} = 0$. Then $\lim_{n\to\infty} \|f^\sim - f_n^\sim\|_p = 0$ and $\lim_{m\to\infty} \|g^\sim - g_m^\sim\|_{p'} = 0$ by (8.1.39), and since for all $n, m \in \mathbb{N}$

$$\int_{-\infty}^{\infty} f_n^\sim(x)g_m(x)\,dx = -\int_{-\infty}^{\infty} f_n(x)g_m^\sim(x)\,dx,$$

equation (8.2.1) follows by Hölder's inequality.

In order to prove (8.2.2), let $p = 2$. Then $\|f + g\|_2 = \|f^\sim + g^\sim\|_2$ and $\|f + ig\|_2 = \|f^\sim + ig^\sim\|_2$ by (8.1.20), and therefore $(f, g) + (g, f) = (f^\sim, g^\sim) + (g^\sim, f^\sim)$ and $(g, f) - (f, g) = (g^\sim, f^\sim) - (f^\sim, g^\sim)$, which implies (8.2.2) for $p = 2$. For general p we choose sequences $\{f_n\}$ and $\{g_m\}$ as above. Since these functions also belong to L^2, we have

$$\int_{-\infty}^{\infty} f_n(x)\overline{g_m(x)}\,dx = \int_{-\infty}^{\infty} f_n^\sim(x)\overline{g_m^\sim(x)}\,dx$$

for all $n, m \in \mathbb{N}$ so that the proof again follows by an application of Hölder's inequality.

Proposition 8.2.2. *Let $f \in \mathsf{L}^p$ and $g \in \mathsf{L}^1$. Then for $1 < p < \infty$*

$$[f * g]^\sim(x) = (f^\sim * g)(x) \quad \text{a.e.} \tag{8.2.3}$$

For $p = 1$ this remains true if one furthermore supposes $f^\sim \in \mathsf{L}^1$.

Proof. Let $1 < p < \infty$. According to Prop. 0.2.2 and Theorem 8.1.12 we have that $f * g, f^{\sim} * g$, and $[f * g]^{\sim}$ belong to L^p so that both sides of (8.2.3) are well-defined. In view of (8.2.1) we obtain by Fubini's theorem for all $h \in \mathsf{L}^{p'}$

$$\begin{aligned}\int_{-\infty}^{\infty} [f * g]^{\sim}(x)h(x)\,dx &= -\int_{-\infty}^{\infty} h^{\sim}(x)(f * g)(x)\,dx\\ &= -\frac{1}{\sqrt{2\pi}}\int_{-\infty}^{\infty} g(u)\,du \int_{-\infty}^{\infty} f(x-u)h^{\sim}(x)\,dx\\ &= \frac{1}{\sqrt{2\pi}}\int_{-\infty}^{\infty} g(u)\,du \int_{-\infty}^{\infty} f^{\sim}(x-u)h(x)\,dx\\ &= \int_{-\infty}^{\infty} (f^{\sim} * g)(x)h(x)\,dx,\end{aligned}$$

which implies (8.2.3) by Prop. 0.8.11.

We first treat the case $p = 1$ under the additional assumption that $g \in \mathsf{C}_{00}$. Let $Z\colon \cdots < t_k < t_{k+1} < \cdots$ be a partition of $(-\infty, \infty)$ and

$$\sigma_Z(x) = \frac{1}{\sqrt{2\pi}} \sum_{k=-\infty}^{\infty} f(x - t_k) \int_{t_k}^{t_{k+1}} g(u)\,du.$$

Since $g \in \mathsf{C}_{00}$, the sum is actually finite, and therefore $\sigma_Z(x)$ is well-defined and belongs to L^1. Furthermore, to each $\varepsilon > 0$ there exists $\delta > 0$ such that according to the continuity of f in L^1-norm

$$\|(f * g)(\circ) - \sigma_Z(\circ)\|_1 \le \frac{1}{\sqrt{2\pi}} \sum_{k=-\infty}^{\infty} \int_{t_k}^{t_{k+1}} |g(u)|\, \|f(\circ - u) - f(\circ - t_k)\|_1\,du \le \varepsilon\,\|g\|_1$$

for all Z with $\|Z\| \equiv \sup_k |t_{k+1} - t_k| < \delta$. Thus $\lim_{\|Z\|\to 0} \|(f * g)(\circ) - \sigma_Z(\circ)\|_1 = 0$. Since $\sigma_Z(x)$ is a finite sum and the Hilbert transform is linear, we have

$$\sigma_Z^{\sim}(x) = \frac{1}{\sqrt{2\pi}} \sum_{k=-\infty}^{\infty} f^{\sim}(x - t_k) \int_{t_k}^{t_{k+1}} g(u)\,du.$$

Just as for $\sigma_Z(x)$ it follows by $f^{\sim} \in \mathsf{L}^1$ that $\lim_{\|Z\|\to 0} \|(f^{\sim} * g)(\circ) - \sigma_Z^{\sim}(\circ)\|_1 = 0$. Furthermore, in view of (8.1.9)

$$\operatorname{meas}\{x \mid |[f * g]^{\sim}(x) - \sigma_Z^{\sim}(x)| > M > 0\} \le \frac{128}{M}\,\|(f * g)(\circ) - \sigma_Z(\circ)\|_1,$$

which tends to zero as $\|Z\| \to 0$ for each $M > 0$. Therefore, by Prop. 0.1.10, there exists a sequence $\{Z_j\}$, $\lim_{j\to\infty} \|Z_j\| = 0$, such that

$$[f * g]^{\sim}(x) = \lim_{j\to\infty} \sigma_{Z_j}^{\sim}(x) = (f^{\sim} * g)(x) \quad \text{a.e.,}$$

establishing (8.2.3) for $p = 1$ and $g \in \mathsf{C}_{00}$.

If $g \in \mathsf{L}^1$ only, let $\{g_n\} \subset \mathsf{C}_{00}$ such that $\lim_{n\to\infty} \|g - g_n\|_1 = 0$ (Problem 3.1.2 (iii)). Then $[f * g_n]^{\sim}(x) = (f^{\sim} * g_n)(x)$ a.e. for each $n \in \mathbb{N}$. Since $f^{\sim} \in \mathsf{L}^1$, we have $f^{\sim} * g \in \mathsf{L}^1$ and

$$\|(f^{\sim} * g) - (f^{\sim} * g_n)\|_1 \le \|f^{\sim}\|_1 \|g - g_n\|_1 = o(1) \qquad (n \to \infty).$$

On the other hand, by (8.1.9)

$$\text{meas}\,\{x \mid |[f*g]^\sim(x) - [f*g_n]^\sim(x)| > M > 0\} \le \frac{128}{M}\,\|f\|_1\|g - g_n\|_1 = o(1)$$

as $n \to \infty$. Therefore there exists a subsequence $\{g_{n_k}\}$ such that

$$[f*g]^\sim(x) = \lim_{k\to\infty} [f*g_{n_k}]^\sim(x) = (f^\sim * g)(x) \quad \text{a.e.}$$

For examples of functions which together with their Hilbert transforms belong to L^1 we refer to Lemma 11.2.2.

Proposition 8.2.3. *Let* $f \in \mathsf{L}^p$ *and* $g \in \mathsf{L}^1 \cap \mathsf{L}^{p'}$.

(i) *If* $1 \le p < \infty$, *then*

$$[f*g]^\sim(x) = (f*g^\sim)(x) \quad \text{a.e.} \tag{8.2.4}$$

(ii) If $1 < p < \infty$, *then furthermore*

$$(f*g^\sim)(x) = (f^\sim * g)(x) \quad \text{a.e.} \tag{8.2.5}$$

(iii) *If* $p = 1$, *then* (8.2.5) *remains valid under the additional assumption* $f^\sim \in \mathsf{L}^1$.

Proof. The assertion (i) follows for $p = 1$ by (8.2.3), the rôles of f and g being interchanged. If $1 < p < \infty$, then $[f*g]^\sim = f^\sim * g = f * g^\sim$ by (8.2.3) and (8.2.1), which at the same time implies (8.2.5). Finally, part (iii) is an immediate consequence of (8.2.3) and part (i).

As a first application of formulae (8.2.4) and (8.2.5) we have (see Problem 8.1.2) for the singular integral (3.1.38) of Cauchy–Poisson and its conjugate (8.1.34) that for each $y > 0$

$$[P(f; \circ; y)]^\sim(x) = Q(f; x; y) = P(f^\sim; x; y) \quad \text{a.e.,} \tag{8.2.6}$$

which holds for all $f \in \mathsf{L}^p$, $1 < p < \infty$, extending (8.1.35). In case $p = 1$ only the first equation of (8.2.6) holds in general.

Proposition 8.2.4. *Let* $f \in \mathsf{L}^1$ *be such that* $f^\sim \in \mathsf{L}^1$. *Then* $f_\delta^\sim \in \mathsf{L}^1$ *for all* $\delta > 0$ *and*

$$\|f_\delta^\sim\|_1 \le \|f\|_1 + \|f^\sim\|_1, \tag{8.2.7}$$

$$\lim_{\delta\to 0+} \|f^\sim - f_\delta^\sim\|_1 = 0. \tag{8.2.8}$$

Proof. Since $P(f^\sim; x; y) = Q(f; x; y)$ by Prop. 8.2.3(iii), we have by (8.1.32) and (3.1.40) that $\|f_\delta^\sim\|_1 \le \|f\|_1 + \|Q(f; \circ; \delta)\|_1 \le \|f\|_1 + \|f^\sim\|_1$. In virtue of the inequality $\|f^\sim - f_\delta^\sim\|_1 \le \|f^\sim(\circ) - P(f^\sim; \circ; \delta)\|_1 + \|Q(f; \circ; \delta) - f_\delta^\sim(\circ)\|_1$, (8.2.8) is an easy consequence of (3.1.41) and (8.1.31).

8.2.2 Conjugates of Singular Integrals: $1 < p < \infty$

Relation (8.2.6) on the singular integral of Cauchy–Poisson again calls our attention to the convergence problem for general singular integrals which we have already studied in some detail in Chapter 3. In Sec. 3.1 and 3.2 we in particular considered the convergence of the integral

$$J(f; x; \rho) = \frac{\rho}{\sqrt{2\pi}} \int_{-\infty}^{\infty} f(x-u)\chi(\rho u)\, du \tag{8.2.9}$$

towards $f \in \mathsf{L}^p$, $1 \le p < \infty$, as $\rho \to \infty$. We saw that if $\chi \in \mathsf{NL}^1$, then

$$\lim_{\rho\to\infty} \|J(f; \circ; \rho) - f(\circ)\|_p = 0$$

for all $f \in \mathsf{L}^p$. If χ furthermore is even, positive, and monotonely decreasing on $[0, \infty)$, then $\lim_{\rho\to\infty} J(f; x; \rho) = f(x)$ a.e. The problem now is to establish similar results for the Hilbert transform $f^\sim$.

In particular, since the explicit evaluation of the Hilbert transform $f^\sim$ of a given $f \in \mathsf{L}^p$, $1 \le p < \infty$, by means of definition (8.0.2) is often troublesome, we are interested in singular integrals of type (8.2.9) involving the original function f which nevertheless approximate $f^\sim$ in the mean and pointwise a.e. The Hilbert formulae so far established suggest that for χ in (8.2.9) one takes the Hilbert transform $\chi^\sim$, and thus that one considers the integral

$$J^\sim(f; x; \rho) = \frac{\rho}{\sqrt{2\pi}} \int_{-\infty}^{\infty} f(x-u)\chi^\sim(\rho u)\, du. \tag{8.2.10}$$

As an example, for the integral $P(f; x; y)$ of Cauchy–Poisson $\chi(x) = p(x)$, $\chi^\sim(x) = p^\sim(x) = q(x)$, and $P^\sim(f; x; y) \equiv Q(f; x; y)$, the conjugate of the Cauchy–Poisson integral. Generally, we call (8.2.10) the *conjugate of the singular integral* $J(f; x; \rho)$, which by definition is not to be mistaken for the Hilbert transform $[J(f; \circ; \rho)]^\sim(x)$ of $J(f; x; \rho)$. But in our future results the hypotheses upon χ are so strong that $J^\sim(f; x; \rho) = [J(f; \circ; \rho)]^\sim(x)$ a.e. for every $f \in \mathsf{L}^p$, $1 \le p < \infty$.

Proposition 8.2.5. *Let* $1 < p < \infty$ *and* $f \in \mathsf{L}^p$. *If* $\chi \in \mathsf{NL}^1 \cap \mathsf{L}^\infty$, *then*

$$\lim_{\rho\to\infty} \|J^\sim(f; \circ; \rho) - f^\sim(\circ)\|_p = 0. \tag{8.2.11}$$

If χ *furthermore is even, positive, and monotonely decreasing on* $[0, \infty)$, *then*

$$\lim_{\rho\to\infty} J^\sim(f; x; \rho) = f^\sim(x) \quad \text{a.e.} \tag{8.2.12}$$

Thus under the additional assumption $\chi \in \mathsf{L}^\infty$ not only does the singular integral $J(f; x; \rho)$ of (8.2.9) converge to f but also $J^\sim(f; x; \rho)$ converges to the Hilbert transform $f^\sim$.

Proof. Since for each $\rho > 0$, $[\rho\chi(\rho\circ)]^\sim(x) = \rho\chi^\sim(\rho x)$ a.e. by Problem 8.1.1, an application of (8.2.5) gives $J^\sim(f; x; \rho) = J(f^\sim; x; \rho)$. Therefore (8.2.11) follows by Theorem 3.1.6 as applied to $f^\sim$. If χ is even, positive, and monotonely decreasing on $[0, \infty)$, then, by Prop. 3.2.2, $\lim_{\rho\to\infty} J(f^\sim; x; \rho) = f^\sim(x)$ at all D-points of $f^\sim$ and hence a.e., which establishes (8.2.12).

8.2.3 Conjugates of Singular Integrals: $p = 1$

Since the first assumption of the last proposition upon the kernel is usually satisfied in the applications, the problem posed above for the Hilbert transform $f^\sim$ may be considered as solved for $1 < p < \infty$. It remains to treat the case $p = 1$. Since $f \in \mathsf{L}^1$ does not necessarily imply $f^\sim \in \mathsf{L}^1$ and since, in particular, the assumption $\chi \in \mathsf{L}^1 \cap \mathsf{L}^\infty$ does not ensure $\chi^\sim \in \mathsf{L}^1$, we cannot use the results for the original function f as in the case $1 < p < \infty$. Therefore we can hardly expect to prove the counterpart of Prop. 8.2.5 for $p = 1$. But we shall carry over the weaker results of Problem 8.2.2.

Proposition 8.2.6. *Let $f \in \mathsf{L}^1$ and $\chi \in \mathsf{L}^1 \cap \mathsf{L}^\infty$ be an even function. If Θ as defined by*

$$\Theta(x) = \frac{1}{x} - \sqrt{\frac{\pi}{2}}\,\chi^\sim(x) \tag{8.2.13}$$

belongs to $\mathsf{L}^1(1, \infty)$, then

$$\lim_{\rho\to\infty} \|J^\sim(f; \circ; \rho) - f^\sim_{1/\rho}(\circ)\|_1 = 0. \tag{8.2.14}$$

Proof. First of all, since $\chi \in \mathsf{L}^q$ for every $1 < q < \infty$, we have $\chi^\sim \in \mathsf{L}^q$, and therefore the expressions occurring in (8.2.14) are meaningful. Since χ is even, it follows by Problem 8.1.1 that $\chi^\sim$ is odd, and thus as in (8.1.33)

$$\begin{aligned} J^\sim(f; x; \rho) - f^\sim_{1/\rho}(x) &= \frac{\rho}{\sqrt{2\pi}} \int_0^{1/\rho} [f(x-u) - f(x+u)]\chi^\sim(\rho u)\,du \\ &\quad - \frac{\rho}{\pi}\left(\int_{1/\rho}^{\delta} + \int_{\delta}^{\infty}\right)[f(x-u) - f(x+u)]\Theta(\rho u)\,du \equiv I_1 + I_2 + I_3. \end{aligned} \tag{8.2.15}$$

With ε, δ being defined as in the proof of Lemma 8.1.10 and ρ being so large that $1 < \rho\delta$, it follows by Fubini's theorem and Hölder's inequality that

$$\|I_1\|_1 \le \frac{2\varepsilon}{\sqrt{2\pi}} \int_0^1 |\chi^\sim(u)|\,du \le 2\varepsilon\,\|\chi^\sim\|_q,$$

$$\|I_2\|_1 \le \frac{2\varepsilon}{\pi} \int_1^\infty |\Theta(u)|\,du, \qquad \|I_3\|_1 \le \frac{2\,\|f\|_1}{\pi} \int_{\rho\delta}^\infty |\Theta(u)|\,du,$$

the last integral tending to zero as $\rho \to \infty$ for each fixed δ. Hereby the proof is complete.

Regarding pointwise convergence we have

Proposition 8.2.7. *Let $f \in \mathsf{L}^1$, $\chi \in \mathsf{L}^1 \cap \mathsf{L}^\infty$ be an even function, and let $\chi^\sim$ be positive on $(0, \infty)$, monotonely increasing on $[0, 1]$. If Θ, defined by (8.2.13), belongs to $\mathsf{L}^1(1, \infty)$ and is a positive, monotonely decreasing function on $[1, \infty)$, then, at each point x for which*

$$\int_0^t [f(x+u) - f(x-u)]\,du = o(t) \qquad (t \to 0+), \tag{8.2.16}$$

we have

$$\lim_{\rho\to\infty} [J^\sim(f; x; \rho) - f^\sim_{1/\rho}(x)] = 0. \tag{8.2.17}$$

In particular,

$$\lim_{\rho\to\infty} J^\sim(f; x; \rho) = f^\sim(x) \quad \text{a.e.} \tag{8.2.18}$$

Proof. If we set

$$\Psi(t) = \int_0^t [f(x+u) - f(x-u)]\, du, \tag{8.2.19}$$

then assumption (8.2.16) shows that to every $\varepsilon > 0$ there exists $\delta > 0$ such that $|\Psi(t)| < \varepsilon t$ for all $0 < t \le \delta$. We now fix this δ and choose ρ so large that $\rho\delta > 1$. Then in the notation of (8.2.15) and by partial integration

$$I_1 = \frac{\rho}{\sqrt{2\pi}} \int_0^{1/\rho} [-\chi^\sim(\rho u)]\, d\Psi(u)$$

$$= \frac{\rho}{\sqrt{2\pi}} [-\chi^\sim(1)\Psi(1/\rho)] + \frac{\rho}{\sqrt{2\pi}} \int_0^{1/\rho} \Psi(u)\, d\chi^\sim(\rho u) \equiv I_1^1 + I_1^2.$$

Now $\sqrt{2\pi}\, |I_1^1| \le \varepsilon\chi^\sim(1)$ and

$$|I_1^2| \le \frac{\varepsilon\rho}{\sqrt{2\pi}} \int_0^{1/\rho} u\, d\chi^\sim(\rho u) \le \frac{\varepsilon}{\sqrt{2\pi}} \left[\chi^\sim(1) + \int_0^1 \chi^\sim(u)\, du\right] \le \frac{2\varepsilon}{\sqrt{2\pi}} \chi^\sim(1),$$

since $\chi^\sim$ is monotonely increasing on $[0, 1]$. Regarding I_2, we have

$$I_2 = \frac{\rho}{\pi} \int_{1/\rho}^{\delta} \Theta(\rho u)\, d\Psi(u)$$

$$= \frac{\rho}{\pi} [\Theta(\rho\delta)\Psi(\delta) - \Theta(1)\Psi(1/\rho)] + \frac{\rho}{\pi} \int_{1/\rho}^{\delta} \Psi(u)\, d[-\Theta(\rho u)] \equiv I_2^1 + I_2^2.$$

Since Θ is monotonely decreasing on $[1, \infty)$, it follows for $x > 2$ that

$$\int_{x/2}^{x} \Theta(u)\, du \ge \Theta(x) \frac{x}{2},$$

and since the left-hand side tends to zero as $x \to \infty$, we obtain

$$\lim_{x\to\infty} x\Theta(x) = 0. \tag{8.2.20}$$

Therefore

$$|I_2^1| \le \frac{\varepsilon}{\pi} [(\rho\delta)\Theta(\rho\delta) + \Theta(1)] \le 2\Theta(1)\, \varepsilon$$

for all ρ sufficiently large. Furthermore,

$$|I_2^2| \le \frac{\varepsilon\rho}{\pi} \int_{1/\rho}^{\delta} u\, d[-\Theta(\rho u)] \le \varepsilon\left[(\rho\delta)\Theta(\rho\delta) + \Theta(1) + \int_1^{\rho\delta} \Theta(u)\, du\right]$$

$$\le \varepsilon\left[2\Theta(1) + \int_1^{\infty} \Theta(u)\, du\right].$$

Finally we deduce

$$|I_3| \le \frac{\rho}{\pi} \int_\delta^\infty \{|f(x+u)| + |f(x-u)|\}\Theta(\rho u)\, du \le (\rho\delta)\Theta(\rho\delta)\cdot\frac{2\,\|f\|_1}{\pi\delta},$$

which by (8.2.20) tends to zero as $\rho \to \infty$, establishing (8.2.17). Since $\lim_{\delta\to 0+} f_\delta^\sim(x) = f^\sim(x)$ a.e. by Theorem 8.1.6, relation (8.2.18) follows.

We observe that the proofs of the last two propositions may of course be carried over to the case $1 < p < \infty$ as well.

As an example we again treat the singular integral $P(f; x; y)$ of Cauchy–Poisson. In view of Problem 8.2.3 we have as a generalization of Lemma 8.1.10

Corollary 8.2.8. *The conjugate of the singular integral of Cauchy–Poisson of $f \in \mathsf{L}^p$, $1 \le p < \infty$, satisfies:*

$$\lim_{y\to 0+} Q(f; x; y) = f^\sim(x) \quad \text{a.e.,} \tag{8.2.21}$$

$$\lim_{y\to 0+} \|Q(f; \circ; y) - f_y^\sim(\circ)\|_p = 0. \tag{8.2.22}$$

For $1 < p < \infty$ we furthermore have

$$\lim_{y\to 0+} \|Q(f; \circ; y) - f^\sim(\circ)\|_p = 0, \tag{8.2.23}$$

which remains valid for $p = 1$ under the additional assumption $f^\sim \in \mathsf{L}^1$.

Let us conclude with the following result of Privalov-type which indicates that, in general, the operation of taking Hilbert transforms leaves invariant (norm-) smoothness properties of the original function f (cf. also Problems 8.2.5–8.2.7 and Prop. 8.3.7).

Proposition 8.2.9. *Let $f \in \mathsf{L}^p$, $1 \le p < \infty$, and $0 < \alpha < 1$. Then $f \in \mathsf{Lip}(\mathsf{L}^p; \alpha)$ implies $f^\sim \in \mathsf{Lip}(\mathsf{L}^p; \alpha)$.*†

Proof. For $1 < p < \infty$ the proof follows immediately by the fact that the operation of taking Hilbert transform is a bounded linear transformation of L^p into L^p. Indeed, $\|f^\sim(\circ + h) - f^\sim(\circ)\|_p \le M_p\|f(\circ + h) - f(\circ)\|_p$.

Let $p = 1$. Since $f^\sim$ exists almost everywhere by Theorem 8.1.6, we have

$$\begin{aligned} Q(f; x; y) - f^\sim(x) &= \frac{1}{\pi}\int_{-\infty}^{\infty} f(x-u)\frac{u}{u^2+y^2}\,du - \mathrm{PV}\Big[\frac{1}{\pi}\int_{-\infty}^{\infty}\frac{f(x-u)}{u}\,du\Big] \\ &= \frac{y^2}{\pi}\int_0^{\infty}\frac{f(x+u) - f(x-u)}{u(u^2+y^2)}\,du \quad \text{a.e.} \end{aligned}$$

Now the hypothesis implies that $[Q(f; x; y) - f^\sim(x)]$ belongs to L^1 and‡

$$\begin{aligned} \|Q(f; \circ; y) - f^\sim(\circ)\|_1 &\le \frac{y^2}{\pi}\int_0^{\infty}\frac{\|f(\circ + u) - f(\circ - u)\|_1}{u(u^2+y^2)}\,du \\ &\le \frac{2M}{\pi}y^2\int_0^{\infty}\frac{u^{\alpha-1}}{u^2+y^2}\,du = y^\alpha\frac{2M}{\pi}\int_0^{\infty}\frac{u^{\alpha-1}}{1+u^2}\,du \equiv M_1 y^\alpha, \end{aligned}$$

where M is the Lipschitz constant. Furthermore, since $q(x) \in \mathsf{W}^1_{\mathsf{C}}$, it follows by Problem 3.1.6 that

$$\frac{dQ(f; x; y)}{dx} = \frac{1}{\pi}\int_{-\infty}^{\infty} f(x-u)\frac{y^2-u^2}{(y^2+u^2)^2}\,du = \frac{1}{\pi}\int_{-\infty}^{\infty}[f(x-u) - f(x)]\frac{y^2-u^2}{(y^2+u^2)^2}\,du,$$

and therefore

$$\left\|\frac{dQ(f; \circ; y)}{dx}\right\|_1 \le \frac{M}{\pi}\int_{-\infty}^{\infty}\frac{|u|^\alpha|y^2-u^2|}{(y^2+u^2)^2}\,du \equiv M_2 y^{\alpha-1}.$$

† For $p = 1$ we indeed use the following slight generalization of the concept of Lipschitz classes: g belongs to $\mathsf{Lip}(\mathsf{L}^p; \alpha)$ if $[g(\circ + h) - g(\circ)] \in \mathsf{L}^p$ for every $h \in \mathbb{R}$ and there exists a constant M with $\|g(\circ + h) - g(\circ)\|_p \le M|h|^\alpha$. Thus not each term of the difference $g(x + h) - g(x)$ need belong to L^p.

‡ In fact, the following inequality gives a direct approximation theorem for the integral $Q(f; x; y)$, stating that $f \in \mathsf{Lip}(\mathsf{L}^p; \alpha)$ implies $\|Q(f; \circ; y) - f^\sim(\circ)\|_p = O(y^\alpha)$.

Hence we obtain by the Minkowski inequality for $h > 0$

$$\begin{aligned}\|f^\sim(\circ + h) - f^\sim(\circ)\|_1 &\le \|f^\sim(\circ + h) - Q(f; \circ + h; h)\|_1 \\ &\quad + \|Q(f; \circ + h; h) - Q(f; \circ; h)\|_1 + \|Q(f; \circ; h) - f^\sim(\circ)\|_1 \\ &\le 2M_1 h^\alpha + \left\| \int_0^h \frac{dQ(f; \circ + u; h)}{du}\, du \right\|_1 \le (2M_1 + M_2)h^\alpha,\end{aligned}$$

proving the assertion.

8.2.4 Iterated Hilbert Transforms

Thus far we have assigned to a function f its Hilbert transform. If $f \in \mathsf{L}^p$, $1 < p < \infty$, then one of our results implies that $f^\sim \in \mathsf{L}^p$. We may therefore take the Hilbert transform of $f^\sim$ to obtain $[f^\sim]^\sim \in \mathsf{L}^p$. Our next proposition states that the iterated Hilbert transform $[f^\sim]^\sim$ is equal to the original function f apart from a minus sign.

Proposition 8.2.10. *Let* $1 < p < \infty$ *and* $f \in \mathsf{L}^p$. *Then* $[f^\sim]^\sim(x) = -f(x)$ a.e. *This remains valid for* $p = 1$ *if one furthermore assumes that* $f^\sim \in \mathsf{L}^1$.

Proof. If $1 < p < \infty$, then in virtue of the Hilbert formulae (8.2.1), (8.2.2) we obtain for every $g \in \mathsf{L}^{p'}$

$$\int_{-\infty}^{\infty} [f^\sim]^\sim(x)\overline{g(x)}\, dx = -\int_{-\infty}^{\infty} f^\sim(x)\overline{g^\sim(x)}\, dx = -\int_{-\infty}^{\infty} f(x)\overline{g(x)}\, dx.$$

This implies $[f^\sim]^\sim(x) = -f(x)$, i.e.

$$\lim_{\delta\to 0+} \frac{1}{\pi} \int_{|u|\ge\delta} \frac{f^\sim(x-u)}{u}\, du = -f(x) \quad \text{a.e.,} \tag{8.2.24}$$

which may be regarded as an *inversion formula* for the Hilbert transform.

Concerning $p = 1$, the Cauchy–Poisson kernel $p(x)$ satisfies $[p^\sim]^\sim(x) = -p(x)$. This implies by Cor. 8.2.8, Problem 8.2.1, Cor. 3.2.3 that

$$\begin{aligned}[f^\sim]^\sim(x) &= \lim_{y\to 0+} Q(f^\sim; x; y) = \lim_{y\to 0+} \left(f * \left[\frac{1}{y} p^\sim\left(\frac{\circ}{y}\right)\right]^\sim\right)(x) \\ &= -\lim_{y\to 0+} P(f; x; y) = -f(x)\end{aligned}$$

for almost all $x \in \mathbb{R}$, and Prop. 8.2.10 is completely established.

Thus we have in particular shown that the mapping $f \to f^\sim$ is a bounded linear transformation of L^p *onto* L^p for $1 < p < \infty$. Moreover, if we employ the notation $Hf = f^\sim$, $H^r f = H(H^{r-1}f)$, then

$$H^r f = (-1)^{[r/2]} \cdot \begin{cases} f, & r \text{ even} \\ f^\sim, & r \text{ odd} \end{cases} \tag{8.2.25}$$

for every $f \in \mathsf{L}^p$, $1 < p < \infty$, and $r \in \mathbb{N}$.

Problems

1. Let $f, f^{\sim} \in \mathsf{L}^1$ and $g \in \mathsf{L}^p$ for some $1 < p < \infty$. Show that $[f * g]^{\sim} = f^{\sim} * g = f * g^{\sim}$.
2. Let $f \in \mathsf{L}^p$, $1 < p < \infty$, and $\chi \in \mathsf{NL}^1 \cap \mathsf{L}^\infty$. Show $\lim_{\rho\to\infty} \|J^{\sim}(f; \circ; \rho) - f^{\sim}_{1/\rho}(\circ)\|_p = 0$, and if χ is furthermore even, positive, and monotonely decreasing on $[0, \infty)$, then $\lim_{\rho\to\infty} [J^{\sim}(f; x; \rho) - f^{\sim}_{1/\rho}(x)] = 0$ a.e.
3. If $p(x)$ denotes the Cauchy–Poisson kernel, then
$$p(x) = \sqrt{\frac{2}{\pi}}\frac{1}{1 + x^2}, \qquad p^{\sim}(x) = \sqrt{\frac{2}{\pi}}\frac{x}{1 + x^2}, \qquad \Theta(x) = \frac{1}{x(1 + x^2)}.$$
Show that $p(x)$ satisfies the assumptions of Prop. 8.2.5–8.2.7.
4. (i) Let $f \in \mathsf{L}^1$, $\chi \in \mathsf{L}^1 \cap \mathsf{L}^\infty$ be an even function and $\chi^{\sim}$ be positive on $(0, \infty)$, monotonely increasing on $[0, 1]$. If Θ is defined by (8.2.13), assume that there is a majorant Θ^*: $|\Theta(x)| \le \Theta^*(x)$ on $[1, \infty)$ such that Θ^* is monotonely decreasing and integrable on $[1, \infty)$. Show that at each point x for which $\int_0^t |f(x + u) - f(x - u)|\, du = o(t)$, $t \to 0+$, one has $\lim_{\rho\to\infty} [J^{\sim}(f; x; \rho) - f^{\sim}_{1/\rho}(x)] = 0$. In particular, $\lim_{\rho\to\infty} J^{\sim}(f; x; \rho) = f^{\sim}(x)$ a.e.
   (ii) State and prove the counterpart of Cor. 8.2.8 for the singular integral of Fejér. (Hint: It follows by Problem 8.3.2 that
$$F(x) = \frac{1}{\sqrt{2\pi}}\left[\frac{\sin(x/2)}{(x/2)}\right]^2, \qquad F^{\sim}(x) = \sqrt{\frac{2}{\pi}}\frac{x - \sin x}{x^2}, \qquad \Theta(x) = \frac{\sin x}{x^2}\ )$$
5. Let $1 < p < \infty$ and $0 < \alpha \le 1$. Show that $f \in \mathsf{Lip}(\mathsf{L}^p; \alpha)$ if and only if $f^{\sim} \in \mathsf{Lip}(\mathsf{L}^p; \alpha)$.
6. (i) Show that $f \in \mathsf{Lip}(\mathsf{L}^1; 1)$ implies $f^{\sim} \in \mathsf{Lip}^*(\mathsf{L}^1; 1)$. (Hint: Use the method of proof of Prop. 8.2.9; cf. Ogieveckiĭ–Boĭcun [1], Butzer–Trebels [2, p. 21])
   (ii) Show furthermore that $f \in \mathsf{Lip}^*(\mathsf{L}^1; 1)$ implies $f^{\sim} \in \mathsf{Lip}^*(\mathsf{L}^1; 1)$ (see also Butzer–Berens [1, p. 252], Butzer–Scherer [1, p. 146]).
7. Let $1 \le p < \infty$, $0 < \alpha < 1$, and $f \in \mathsf{L}^p \cap \mathsf{Lip}(\mathsf{C}; \alpha)$. Show that $f^{\sim}$ exists everywhere and belongs to $\mathsf{Lip}(\mathsf{C}; \alpha)$. (Hint: Titchmarsh [6, p. 145])
8. Let $f \in \mathsf{L}^p$, $1 \le p < \infty$. Show that
$$\left\| \sigma^{\sim}(f; \circ; \rho) - f^{\sim}(\circ) + \frac{1}{\pi}\int_0^{1/\rho} \frac{f(\circ - u) - f(\circ + u)}{u}\, du \right\|_p = O(\omega(\mathsf{L}^p; f; \rho^{-1})).$$
(Hint: Compare the proof of Prop. 8.2.6; see also the literature given in Problem 3.4.6 (ii))

8.3 Fourier Transforms of Hilbert Transforms

8.3.1 Signum Rule

We begin with the proof of relation (8.0.4), the importance of which we already emphasized in the introduction to this chapter.

Proposition 8.3.1. *Let $f \in \mathsf{L}^p$, $1 < p \le 2$. Then the Fourier transform of the Hilbert transform $f^{\sim}$ is given by*

$$[f^{\sim}]^{\wedge}(v) = \{-i \operatorname{sgn} v\} f^{\wedge}(v) \quad \text{a.e.} \tag{8.3.1}$$

For $p = 1$ this remains valid and then holds everywhere under the additional assumption $f^{\sim} \in \mathsf{L}^1$.

Proof. The case $p = 2$ is already given by (8.1.22). Let $f \in \mathsf{L}^p$, $1 \le p < 2$. The assumptions imply that (8.2.3) holds for every $g \in \mathsf{L}^1 \cap \mathsf{L}^\infty$. Since $f * g \in \mathsf{L}^p \cap \mathsf{C}$ by Prop. 0.2.1, 0.2.2 and hence $f * g \in \mathsf{L}^2$, $[f * g]^\sim \in \mathsf{L}^2$, we have by the convolution theorem and (8.1.22)

$$\begin{aligned}[][f^\sim]^\wedge(v)g^\wedge(v) = [f^\sim * g]^\wedge(v) = [[f * g]^\sim]^\wedge(v) = \{-i \operatorname{sgn} v\}[f * g]^\wedge(v)\\ = \{-i \operatorname{sgn} v\}f^\wedge(v)g^\wedge(v).\end{aligned}$$

Thus, if g is such that $g^\wedge(v) \neq 0$ (e.g. $g(x) = \exp\{-x^2\}$), (8.3.1) follows.

Although the next proposition is an easy consequence of the Hilbert formulae, it will be of some use in the actual evaluation of Hilbert transforms.

Proposition 8.3.2. *Let $f \in \mathsf{L}^1 \cap \mathsf{L}^p$, $1 < p \le 2$, be an even function. Then*

$$f^\wedge(v) = f_c^\wedge(v), \tag{8.3.2}$$

$$[f^\wedge]^\sim(v) = f_s^\wedge(v) \quad \text{a.e.} \tag{8.3.3}$$

Proof. In view of Problem 5.1.10 we only need to prove (8.3.3). Let $h \in \mathsf{L}^p$. On the one hand, by (8.3.1), (5.2.18), and (8.2.1)

$$\int_{-\infty}^{\infty} f(u)\{-i \operatorname{sgn} u\}h^\wedge(u)\,du = \int_{-\infty}^{\infty} f(u)[h^\sim]^\wedge(u)\,du = -\int_{-\infty}^{\infty} [f^\wedge]^\sim(u)h(u)\,du.$$

On the other hand, since

$$[\{-i \operatorname{sgn} \circ\}f(\circ)]^\wedge(v) = -f_s^\wedge(v), \tag{8.3.4}$$

we have

$$\int_{-\infty}^{\infty} f(u)\{-i \operatorname{sgn} u\}h^\wedge(u)\,du = -\int_{-\infty}^{\infty} f_s^\wedge(u)h(u)\,du,$$

which implies (8.3.3) by Prop. 0.8.11.

8.3.2 Summation of Allied Integrals

The preceding results may be regarded as preparatory to the summability of the *allied integral*

$$\frac{1}{\sqrt{2\pi}}\int_{-\infty}^{\infty} \{-i \operatorname{sgn} v\}f^\wedge(v)\, e^{ixv}\,dv. \tag{8.3.5}$$

In Chapter 5 we assigned to each $f \in \mathsf{L}^p$, $1 \le p \le 2$, the Fourier transform $f^\wedge$, and the problem then was to recapture the original function f from the values of its Fourier transform $f^\wedge$. The formal inversion was given by (5.1.6). But since the Fourier inversion integral does in general not exist as an ordinary Lebesgue integral, it was summed up via (5.1.9). Concerning the allied integral (8.3.5) we expect in view of (8.3.1) that it converges to the Hilbert transform $f^\sim$.

In fact, if $1 < p \le 2$, then Theorems 5.2.14 and 5.2.15 apply here, too, and introducing

$$U^{\sim}(f; x; \rho) = \frac{1}{\sqrt{2\pi}} \int_{-\infty}^{\infty} \theta\Big(\frac{v}{\rho}\Big)\{-i \operatorname{sgn} v\} f^{\wedge}(v)\, e^{ixv}\, dv \qquad (\rho > 0) \tag{8.3.6}$$

(cf. (5.2.20)), we have

Proposition 8.3.3. *Let $f \in \mathsf{L}^p$, $1 < p \le 2$. Then for a θ-factor the θ-means* (8.3.6) *of the allied integral exist for all $x \in \mathbb{R}$, belong to L^p, and satisfy*

$$\|U^{\sim}(f; \circ; \rho)\|_p \le M_p \|\theta^{\wedge}\|_1 \|f\|_p \qquad (\rho > 0), \tag{8.3.7}$$

$$\lim_{\rho\to\infty} \|U^{\sim}(f; \circ; \rho) - f^{\sim}(\circ)\|_p = 0. \tag{8.3.8}$$

If θ furthermore satisfies the conditions of Theorem 5.2.15 (*or Problem* 5.2.4), *then*

$$\lim_{\rho\to\infty} U^{\sim}(f; x; \rho) = f^{\sim}(x) \quad \text{a.e.} \tag{8.3.9}$$

According to (8.3.1) and the Parseval formula (5.2.18) we have

$$U^{\sim}(f; x; \rho) = \frac{1}{\sqrt{2\pi}} \int_{-\infty}^{\infty} \theta\Big(\frac{v}{\rho}\Big)[f^{\sim}]^{\wedge}(v)\, e^{ixv}\, dv \tag{8.3.10}$$
$$= \frac{\rho}{\sqrt{2\pi}} \int_{-\infty}^{\infty} f^{\sim}(x - v)\theta^{\wedge}(\rho v)\, dv,$$

and the proof follows since $f^{\sim} \in \mathsf{L}^p$.

The last argument does not apply to the case $p = 1$. Here we have

Proposition 8.3.4. *Let $f \in \mathsf{L}^1$ and assume for a θ-factor that $\Theta(x) \equiv (1/x) - \sqrt{\pi/2}\, [\theta^{\wedge}]^{\sim}(x)$ belongs to $\mathsf{L}^1(1, \infty)$. Then*

$$\lim_{\rho\to\infty} \|U^{\sim}(f; \circ; \rho) - f^{\sim}_{1/\rho}(\circ)\|_1 = 0. \tag{8.3.11}$$

If all further hypotheses of Prop. 8.2.7 (*or Problem* 8.2.4) *are satisfied* (*with $\chi \equiv \theta^{\wedge}$*), *then*

$$\lim_{\rho\to\infty} U^{\sim}(f; x; \rho) = f^{\sim}(x) \quad \text{a.e.} \tag{8.3.12}$$

Proof. Since $\theta \in \mathsf{L}^1 \cap \mathsf{L}^\infty$, we obtain by (5.1.5), (8.3.3), and (8.3.4)

$$U^{\sim}(f; x; \rho) = \frac{\rho}{\sqrt{2\pi}} \int_{-\infty}^{\infty} f(x - u)[\theta^{\wedge}]^{\sim}(\rho u)\, du, \tag{8.3.13}$$

and the proof follows by Prop. 8.2.6, 8.2.7, and Problem 8.2.4, respectively.

In particular, in view of (5.2.23) and (8.2.10), the notation (8.3.6) is justified by (8.3.13). To evaluate $[\theta^{\wedge}]^{\sim}$ for important examples of θ-factors (such as those of Cesàro, Abel or Gauss, cf. (5.1.8)) one may use Prop. 8.3.2 or Problem 8.3.2. Then we have

Corollary 8.3.5. *Let $f \in \mathsf{L}^p$, $1 \le p \le 2$. The allied integral is Cesàro and Abel summable to $f^\sim(x)$ for almost all $x \in \mathbb{R}$, i.e.*

$$\text{(8.3.14)} \qquad \lim_{\rho\to\infty} \frac{1}{\sqrt{2\pi}} \int_{-\rho}^{\rho} \left(1 - \frac{|v|}{\rho}\right)\{-i \operatorname{sgn} v\} f^\wedge(v)\, e^{ixv}\, dv = f^\sim(x) \quad \text{a.e.,}$$

$$\text{(8.3.15)} \qquad \lim_{y\to 0+} \frac{1}{\sqrt{2\pi}} \int_{-\infty}^{\infty} e^{-y|v|}\{-i \operatorname{sgn} v\} f^\wedge(v)\, e^{ixv}\, dv = f^\sim(x) \quad \text{a.e.}$$

The proof is left to Problem 8.3.1.

8.3.3 Fourier Transforms of Derivatives of Hilbert Transforms, the Classes $(\mathsf{W}^\sim)^r_{\mathsf{L}^p}$, $(\mathsf{V}^\sim)^r_{\mathsf{L}^p}$

We now investigate the Fourier transform of derivatives of the Hilbert transform $f^\sim$ (as considered for the original function $f \in \mathsf{L}^p$, $1 \le p \le 2$, in Sec. 5.1.3, 5.2.5, and 5.3.3). The definitions of the classes $\mathsf{W}^r_{\mathsf{X}(\mathbb{R})}$, $\mathsf{V}^r_{\mathsf{X}(\mathbb{R})}$ in Sec. 3.1 and 5.3.3 suggest the introduction of ($r \in \mathbb{N}$, $1 \le p < \infty$)

$$\text{(8.3.16)} \quad (\mathsf{W}^\sim)^r_{\mathsf{L}^p} = \begin{cases} \{f \in \mathsf{L}^1 \mid f^\sim = \phi \text{ a.e., } \phi \in \mathsf{AC}^{r-1}_{\text{loc}},\ \phi^{(k)} \in \mathsf{L}^1,\ 1 \le k \le r\} \\ \{f \in \mathsf{L}^p \mid f^\sim \in \mathsf{W}^r_{\mathsf{L}^p}\} & (1 < p < \infty), \end{cases}$$

$$\text{(8.3.17)} \quad (\mathsf{V}^\sim)^r_{\mathsf{L}^p} = \begin{cases} \{f \in \mathsf{L}^1 \mid f^\sim = \phi \text{ a.e., } \phi \in \mathsf{AC}^{r-2}_{\text{loc}},\ \phi^{(k)} \in \mathsf{L}^1,\ 1 \le k \le r-1,\ \phi^{(r-1)} \in \mathsf{BV}\} \\ \{f \in \mathsf{L}^p \mid f^\sim \in \mathsf{V}^r_{\mathsf{L}^p}\} & (1 < p < \infty). \end{cases}$$

By Theorem 8.1.6, $f \in \mathsf{L}^p$, $1 \le p < \infty$, implies that $f^\sim$ exists almost everywhere. Therefore the above definitions are meaningful. They are particularly easy for $1 < p < \infty$, since then $f^\sim$ belongs to L^p by Theorem 8.1.12. But for $p = 1$, $f^\sim$ does not necessarily belong to L^1 so that the relevant function ϕ is not assumed to be an element of L^1 (cf. footnote to Prop. 8.2.9).

The following theorem gives equivalent characterizations of $(\mathsf{V}^\sim)^r_{\mathsf{L}^p}$ for $1 \le p \le 2$, whereas those of $(\mathsf{W}^\sim)^r_{\mathsf{L}^p}$ are left to Problem 8.3.3.

Theorem 8.3.6. *For $f \in \mathsf{L}^p$, $1 \le p \le 2$, the following assertions are equivalent:*

(i) $f \in (\mathsf{V}^\sim)^r_{\mathsf{L}^p}$, (ii) $f \in \mathsf{V}[\mathsf{L}^p; \{-i \operatorname{sgn} v\}(iv)^r]$,
(iii) *there exists $\mu \in \mathsf{BV}$ if $p = 1$ and $g \in \mathsf{L}^p$ if $1 < p \le 2$ such that*

$$f^\sim(x) = \int_{-\infty}^{x} du_1 \int_{-\infty}^{u_1} du_2 \ldots du_{r-1} \int_{-\infty}^{u_{r-1}} \begin{Bmatrix} d\mu(u_r) \\ g(u_r)\, du_r \end{Bmatrix} \quad \text{a.e.,}$$

the integrals existing in the sense of Theorem 5.3.15 (*apart possibly from the first one in case* $p = 1$).

Proof. Let $1 < p \le 2$. If f satisfies (i), then $f^\sim \in \mathsf{V}^r_{\mathsf{L}^p}$, and therefore by Theorem 5.2.21† there exists $g \in \mathsf{L}^p$ such that $(iv)^r[f^\sim]^\wedge(v) = g^\wedge(v)$ a.e. Since $p > 1$, we have $[f^\sim]^\wedge(v) = \{-i \operatorname{sgn} v\} f^\wedge(v)$ by Prop. 8.3.1, and thus $f \in \mathsf{V}[\mathsf{L}^p; \{-i \operatorname{sgn} v\}(iv)^r]$. On the

† Note that $\mathsf{V}^r_{\mathsf{L}^p}$ and $\mathsf{W}^r_{\mathsf{L}^p}$ are equal for $1 < p \le 2$ by definition.

other hand, if f belongs to the latter class, then $\{-i \operatorname{sgn} v\}(iv)^r f^\wedge(v) = g^\wedge(v)$, $g \in L^p$, implies $(iv)^r[f^\sim]^\wedge(v) = g^\wedge(v)$ by Theorem 8.1.12 and Prop. 8.3.1. Thus (iii) and (i) follow by Theorem 5.2.21 as applied to $f^\sim$.

Let $p = 1$. If $f \in (V^\sim)^r_{L^1}$, then as in the proof of Prop. 5.3.13

$$\text{(8.3.18)} \quad \Delta_1^r f^\sim(x) = \int_x^{x+1} du_1 \int_{u_1}^{u_1+1} du_2 \ldots du_{r-1} \int_{u_{r-1}}^{u_{r-1}+1} d\phi^{(r-1)}(u_r) \quad \text{a.e.}$$

Since the right-hand side belongs to L^1 by the convolution theorem (cf. Prop. 5.3.3, Lemma 5.3.4), $\Delta_1^r f^\sim \in L^1$, and since obviously $\Delta_1^r f \in L^1$, it follows by Prop. 8.3.1 that

$$[\Delta_1^r f^\sim]^\wedge(v) = \{-i \operatorname{sgn} v\}(e^{iv} - 1)^r f^\wedge(v).$$

Therefore by (5.3.8) and (8.3.18)

$$\{-i \operatorname{sgn} v\}(e^{iv} - 1)^r f^\wedge(v) = \left(\frac{e^{iv} - 1}{iv}\right)^r [\phi^{(r-1)}]^\vee(v),$$

i.e. $f \in V[L^1; \{-i \operatorname{sgn} v\}(iv)^r]$.

Conversely, let there exist $\mu \in BV$ such that $\{-i \operatorname{sgn} v\}(iv)^r f^\wedge(v) = \mu^\vee(v)$ for all $v \in \mathbb{R}$. To prove that (ii) implies (iii) (and thus (i)) we consider three cases. Let $r = 1$. Then by (8.3.14), (5.1.15), and Lemma 5.3.4 for each $h \in \mathbb{R}$

$$\begin{aligned} f^\sim(x + h) - f^\sim(x) &= \lim_{\rho \to \infty} \frac{1}{\sqrt{2\pi}} \int_{-\rho}^{\rho} \left(1 - \frac{|v|}{\rho}\right) \{-i \operatorname{sgn} v\}(e^{ihv} - 1) f^\wedge(v)\, e^{ixv}\, dv \\ &= \lim_{\rho \to \infty} \frac{1}{\sqrt{2\pi}} \int_{-\rho}^{\rho} \left(1 - \frac{|v|}{\rho}\right) \frac{e^{ihv} - 1}{iv}\, \mu^\vee(v)\, e^{ixv}\, dv \\ &= \mu(x + h) - \mu(x) \quad \text{a.e.} \end{aligned}$$

To show that $f^\sim(x) = \mu(x)$ a.e., we integrate the latter equation with respect to x over $[0, y]$, thus consider

$$\int_0^y [\mu(x) - f^\sim(x)]\, dx = \int_0^y \mu(x + h)\, dx - \int_h^{y+h} f^\sim(x)\, dx$$

for any $y > 0$. Now $f^\wedge \in L^2$ since

$$|f^\wedge(v)| \le \begin{cases} \|f\|_1, & |v| \le 1 \\ \|\mu\|_{BV}/|v|, & |v| > 1. \end{cases}$$

Hence $f \in L^2$ by Plancherel's theorem and $f^\sim \in L^2$ by Theorem 8.1.7. Therefore $\lim_{h \to -\infty} \int_h^{y+h} f^\sim(x)\, dx = 0$. Moreover, $\lim_{h \to -\infty} \mu(x + h) = 0$, since μ is normalized. This implies $\lim_{h \to -\infty} \int_0^y \mu(x + h)\, dx = 0$ by Lebesgue's dominated convergence theorem. Thus $f^\sim(x) = \mu(x)$ a.e., in other words $f^\sim(x) = \int_{-\infty}^x d\mu(u_1)$ a.e.

Secondly, let r be odd, i.e. $r = 2n + 1$, $n \in \mathbb{N}$. Then $|v|^r f^\wedge(v) = (-1)^n \mu^\vee(v)$, and it follows by Theorem 6.3.14 that there exists $g \in L^1$ such that $|v|\, f^\wedge(v) = g^\wedge(v)$. Therefore $(iv)^{2n} g^\wedge(v) = \mu^\vee(v)$, and $g \in V^{2n}_{L^1}$ by Theorem 5.3.15. In particular,

$$g(u_1) = \int_{-\infty}^{u_1} du_2 \int_{-\infty}^{u_2} du_3 \ldots du_{r-1} \int_{-\infty}^{u_{r-1}} d\mu(u_r) \quad \text{a.e.}$$

Moreover, $|v| f^\wedge(v) = g^\wedge(v)$ implies $f^\sim(x) = \int_{-\infty}^{x} g(u_1)\, du_1$ a.e. by the first case, which proves (iii).

Finally, let r be even, i.e. $r = 2n$, $n \in \mathbb{N}$. Then $\{-i \operatorname{sgn} v\}|v|^r f^\wedge(v) = (-1)^n \mu^\vee(v)$. Since $(1 + iv)^{-1} \in [\mathsf{L}^1]^\wedge$ (cf. (5.1.34)), it follows that

$$\{-i \operatorname{sgn} v\} \frac{(iv)^{2n} + (iv)^{2n+1}}{1 + iv} f^\wedge(v) \in [\mathsf{BV}]^\vee, \qquad \{-i \operatorname{sgn} v\} \frac{(iv)^{2n} f^\wedge(v)}{1 + iv} \in [\mathsf{L}^1]^\wedge.$$

Therefore $|v|^{2n+1}(1 + iv)^{-1} f^\wedge(v) \in [\mathsf{BV}]^\vee$, and since $(1 + iv)^{-1} f^\wedge(v) \in [\mathsf{L}^1]^\wedge$, Theorem 6.3.14 implies $|v|^{2n-1}(1 + iv)^{-1} f^\wedge(v) \in [\mathsf{L}^1]^\wedge$. Furthermore, since

$$|v|^{2n-1} f^\wedge(v) = |v|^{2n-1}(1 + iv)^{-1} f^\wedge(v) - \{-i \operatorname{sgn} v\}|v|^{2n}(1 + iv)^{-1} f^\wedge(v),$$

we have $|v|^{2n-1} f^\wedge(v) \in [\mathsf{L}^1]^\wedge$. Thus, let $g \in \mathsf{L}^1$ be such that $|v|^{r-1} f^\wedge(v) = (-1)^{n-1} g^\wedge(v)$. Then by the previous (odd) case there exists $\phi \in \mathsf{AC}_{\text{loc}}^{r-2}$ with $\phi^{(k)} \in \mathsf{L}^1$ for $1 \le k \le r - 1$ such that $f^\sim(x) = \phi(x)$ and $\phi^{(r-1)}(x) = g(x)$ a.e. Moreover,

$$f^\sim(x) = \int_{-\infty}^{x} du_1 \int_{-\infty}^{u_1} du_2 \ldots du_{r-2} \int_{-\infty}^{u_{r-2}} g(u_{r-1})\, du_{r-1}.$$

However, $-\{-i \operatorname{sgn} v\}|v|\, g^\wedge(v) = \mu^\vee(v)$, that is, $(iv) g^\wedge(v) = \mu^\vee(v)$ which implies $g(u_{r-1}) = \int_{-\infty}^{u_{r-1}} d\mu(u_r)$ a.e. by Theorem 5.3.15. Thus, (iii) is completely established, proving the theorem.

Let us conclude with a brief account of the problem whether the operations of taking Hilbert transform and differentiation may be interchanged, i.e., under what conditions the relation $(f^\sim)' = (f')^\sim$ is valid.

Proposition 8.3.7. *Let* $f \in \mathsf{L}^p$, $1 < p < \infty$. *Then* $f \in \mathsf{W}^1_{\mathsf{L}^p}$ *if and only if* $f^\sim \in \mathsf{W}^1_{\mathsf{L}^p}$; *and if* $f(x) = \phi(x)$ a.e. *with* $\phi \in \mathsf{AC}_{\text{loc}}$, $\phi' \in \mathsf{L}^p$ *and* $f^\sim(x) = \psi(x)$ a.e. *with* $\psi \in \mathsf{AC}_{\text{loc}}$, $\psi' \in \mathsf{L}^p$, *then* $(\phi')^\sim(x) = \psi'(x)$ a.e.

Proof. Let $1 < p \le 2$. If $f \in \mathsf{W}^1_{\mathsf{L}^p}$ and $f(x) = \phi(x)$ a.e. with $\phi \in \mathsf{AC}_{\text{loc}}$, $\phi' \in \mathsf{L}^p$, then $iv f^\wedge(v) = [\phi']^\wedge(v)$ by Prop. 5.2.19. Since $f^\sim$, $(\phi')^\sim \in \mathsf{L}^p$ by Theorem 8.1.12, it follows by Prop. 8.3.1 that $iv[f^\sim]^\wedge(v) = [(\phi')^\sim]^\wedge(v)$. Thus $f^\sim \in \mathsf{W}^1_{\mathsf{L}^p}$ by Theorem 5.2.21, and if $f^\sim(x) = \psi(x)$ a.e. with $\psi \in \mathsf{AC}_{\text{loc}}$, $\psi' \in \mathsf{L}^p$, then $\psi'(x) = (\phi')^\sim(x)$ a.e. The converse assertion follows by the same reasons.

If $2 < p < \infty$, then the proof is obtained by a duality argument which however is left to Problem 8.3.5.

8.3.4 Norm-Convergence of the Fourier Inversion Integral

We shall finally give an application of the theory of Hilbert transforms to the theory of Fourier transforms of functions $f \in \mathsf{L}^p$, $1 < p \le 2$. We shall show that the inversion formula (5.2.37) already established for $p = 2$ also holds for $1 < p < 2$. This is the famous result of E. Hille and J. D. Tamarkin given by

Theorem 8.3.8. *If* $f \in \mathsf{L}^p$, $1 < p \le 2$, *then*

$$\lim_{\rho \to \infty} \left\| f(\circ) - \frac{1}{\sqrt{2\pi}} \int_{-\rho}^{\rho} f^\wedge(v)\, e^{i \circ v}\, dv \right\|_p = 0. \tag{8.3.19}$$

Proof. By (5.2.18) we have (cf. Problem 5.1.5)

$$S(f;x;\rho) \equiv \frac{1}{\sqrt{2\pi}} \int_{-\rho}^{\rho} f^{\wedge}(v)\, e^{ixv}\, dv = \frac{1}{\pi} \int_{-\infty}^{\infty} f(u) \frac{\sin \rho(x-u)}{x-u}\, du,$$

where the right-hand side exists by Hölder's inequality as an ordinary Lebesgue integral for every $x \in \mathbb{R}$. Hence by Theorem 8.1.6

$$S(f;x;\rho) = \sin \rho x\, [f(\circ) \cos \rho\circ]^{\sim}(x) - \cos \rho x\, [f(\circ) \sin \rho\circ]^{\sim}(x) \quad \text{a.e.}$$

By (8.1.39) this implies

$$\|S(f;\circ;\rho)\|_p \le 2M_p \|f\|_p \qquad (\rho > 0).$$

Given $\varepsilon > 0$, there exists a function $f_1 \in \mathsf{S}_{00}$ such that $f = f_1 + f_2$ with $\|f_2\|_p < \varepsilon$, and since

$$\|S(f;\circ;\rho) - f(\circ)\|_p \le \|S(f_1;\circ;\rho) - f_1(\circ)\|_p + (2M_p + 1)\,\varepsilon,$$

it suffices to prove (8.3.19) for functions $f_1 \in \mathsf{S}_{00}$. Moreover, it suffices to consider the case $f_1 = \kappa_{[a,b]}$. For such f_1 we have

$$S(f_1;x;\rho) = \frac{1}{\pi} \int_{(x-b)\rho}^{(x-a)\rho} \frac{\sin v}{v}\, dv. \tag{8.3.20}$$

Let us choose A such that $A > \max\{|a|, |b|\}$. Then, if $x > A$, by the second mean value theorem (cf. Problem 1.2.10), we obtain for some ξ with $(x-b)\rho \le \xi \le (x-a)\rho$

$$S(f_1;x;\rho) = \frac{1}{\pi(x-b)\rho} \int_{(x-b)\rho}^{\xi} \sin v\, dv,$$

which implies $|S(f_1;x;\rho)| \le 2/\pi(x-b)\rho$. Therefore

$$\int_A^{\infty} |S(f_1;x;\rho) - f_1(x)|^p\, dx \le \left(\frac{2}{\pi\rho}\right)^p \int_A^{\infty} \frac{dx}{(x-b)^p},$$

which tends to zero as $\rho \to \infty$. Similarly $\int_{-\infty}^{-A} |S(f_1;x;\rho) - f_1(x)|^p\, dx = o(1)$ for $\rho \to \infty$. Thus it remains to show

$$\int_{-A}^{A} |S(f_1;x;\rho) - f_1(x)|^p\, dx = o(1) \qquad (\rho \to \infty). \tag{8.3.21}$$

But it follows from (8.3.20) (cf. Problems 1.2.10(i), 5.1.2(iv)) that $S(f_1;x;\rho)$ is uniformly bounded in x and ρ and tends to $f_1(x)$ for $x \ne a, b$ as $\rho \to \infty$. Therefore (8.3.21) follows by Lebesgue's dominated convergence theorem, and (8.3.19) is established.

Problems

1. Prove Cor. 8.3.5.
2. (i) Show that for $f, f^{\wedge} \in \mathsf{L}^1$

$$f^{\sim}(x) = \frac{1}{\sqrt{2\pi}} \int_{-\infty}^{\infty} \{-i \operatorname{sgn} v\} f^{\wedge}(v)\, e^{ixv}\, dv \quad \text{a.e.} \tag{8.3.22}$$

   (ii) As an application of (8.3.22) (or Prop. 8.3.2) evaluate the Hilbert transform of the kernel of Cauchy–Poisson (cf. Problem 8.1.2), Fejér (cf. Problem 8.2.4), Weierstrass, Jackson etc. (see also GOLINSKIĬ [2] for further examples).

   (iii) Show that if $f \in \mathsf{L}^p$, $1 \le p \le 2$, then

$$\frac{1}{\pi\rho} \int_{-\infty}^{\infty} f(x-u) \frac{\rho u - \sin \rho u}{u^2}\, du = \frac{1}{\sqrt{2\pi}} \int_{-\rho}^{\rho} \left(1 - \frac{|v|}{\rho}\right) \{i \operatorname{sgn} v\} f^{\wedge}(v) e^{ixv}\, dv.$$

3. Show that for $f \in L^p$, $1 \le p \le 2$, the following assertions are equivalent:
   (i) $f \in (W^\sim)_{L^p}$, (ii) $f \in W[L^p; \{-i \operatorname{sgn} v\}(iv)^r]$,
   (iii) there exists $g \in L^p$ such that
$$f^\sim(x) = \int_{-\infty}^{x} du_1 \int_{-\infty}^{u_1} du_2 \dots du_{r-1} \int_{-\infty}^{u_{r-1}} g(u_r)\, du_r \quad \text{a.e.},$$
   the integrals existing in the sense of Theorem 5.2.21 (apart possibly from the first one in case $p = 1$).
4. State and prove corresponding results for the classes $W[L^p; |v|^r]$, $V[L^p; |v|^r]$, $r \in \mathbb{N}$. (Hint: Distinguish the cases r even, odd)
5. (i) Complete the proof of Prop. 8.3.7. (Hint: HILLE [3]; see also Sec. 10.6, 13.2)
   (ii) State and prove the counterpart of Prop. 8.3.7 for the rth derivative (see also BUTZER–TREBELS [2, p. 19]).
   (iii) Under suitable stronger hypotheses state and prove corresponding results in L^1-space. (Hint: BUTZER–TREBELS [2, p. 44 ff])
6. (i) Let $f \in L^1$. Show that $f^\sim \in L^1$ necessarily implies $\int_{-\infty}^{\infty} f(u)\, du = 0$.
   (ii) Show that the set $\{f \in L^1 \mid f^\sim \in L^1\}$ is not dense in L^1. (Hint: KOBER [1])

8.4 Notes and Remarks

Although many of the results on the Hilbert transform presented in this chapter are more or less standard, they are somewhat scattered in the literature. The main references are TITCHMARSH [6], ZYGMUND [7], HEWITT [1], WEISS [1], and the literature cited there. We again emphasize that TITCHMARSH, for example, regarded the Hilbert transform as a transform rather than a function, in contrast to our point of view. The Hilbert transform will be required here to solve equations of type $|v|\, f^\wedge(v) = g^\wedge(v)$, our final aim in this chapter; in subsequent chapters this will be important for our investigations on saturation theory.

Sec. 8.1. The proof of Theorem 8.1.5, which not only gives the existence a.e. of the Hilbert–Stieltjes transform but also that it is of weak type, is due to LOOMIS [1]. Whereas the existence a.e. of $\mu^\sim(x)$ of $\mu \in BV$ is connected with the name of S. POLLARD [1], the weak type property (8.1.9) is often referred to as a KOLMOGOROV bound for $\mu^\sim$. The present treatment studies the Hilbert transform along strictly real variable lines, an aspect for which we also refer to HARDY [1], COSSAR [1], and the fundamental (higher dimensional) investigations of CALDERÓN–ZYGMUND [1,2], CALDERÓN [2], ZYGMUND [6]. (See also DUNFORD–SCHWARTZ [1II, pp. 1044–1073].) There are other proofs for the PLESSNER result on the existence of $f^\sim$ for $f \in L^1$ which depend on complex function theory (see e.g. TITCHMARSH [6, p. 132]) or on a reduction to the (periodic) Hilbert transform of $f \in L^1_{2\pi}$ (cf. e.g. ZYGMUND [7II, p. 242 ff]). There is a further proof by STEIN–WEISS [1] to the effect that the Hilbert transform of $f \in L^p$, $1 \le p < \infty$, is of restricted weak type (p, p), $1 \le p < \infty$ (see also BUTZER–BERENS [1, p. 240 ff]).

Theorem 8.1.7 is standard, e.g. TITCHMARSH [6, p. 121 ff], WEISS [1, p. 49 ff]. The L^p-theory for $1 < p < \infty$ was developed by M. RIESZ [1] (cf. e.g. TITCHMARSH [6, pp. 132–143], HEWITT [1, p. 195 ff]). The present proof rests upon the theorem of MARCINKIEWICZ (cf. also WEISS [1, p. 71]).

In our approach, in which the Hilbert transform $f^\sim$ of $f \in L^p$, $1 < p < \infty$, was first defined as a pointwise limit through (8.0.2), it must of course be shown that the extension of T in (8.1.23) to all of L^p as given by the theorem of Marcinkiewicz coincides with $f^\sim$. Our proof leading to Theorem 8.1.12 was suggested by work of WEISS [1, pp. 46–56, 72]. For Lemma 8.1.10 compare also TITCHMARSH [6, p. 124].

Theorem 8.1.12 breaks down for $p = 1$. Indeed, as we have seen, if $f \in \mathsf{L}^1$, the Hilbert transform $f^\sim$ need not be integrable, even locally. However, A. KOLMOGOROV has shown that if $f \in \mathsf{L}^1$, then $|f^\sim(x)|^\alpha$ is locally integrable for every $0 < \alpha < 1$ (for extensions see KOBER [1], KOIZUMI [1, 2], and the literature cited there). There are many investigations concerning the class of functions f for which $|f(x)|(1 + \log^+ |f(x)|)$ belongs to L^1. For this class one can prove the local integrability of $f^\sim$ (compare ZYGMUND [5, p. 150], KOBER [2], CALDERÓN–ZYGMUND [1]). In view of the fact that we wish to have the property (8.0.4) at our disposal, we commence with $f^\sim \in \mathsf{L}^1$.

The Hilbert transform can also be defined for more general classes of functions. Following ACHIESER [2, p. 159], let W^2 be the class† of all measurable functions f for which $f(x)/(1 + |x|) \in \mathsf{L}^2$. (For instance, every function belonging to $\mathsf{L}^2_{2\pi}$, L^2 or C is contained in W^2.) Let η be the L^2-Fourier transform of $f(x)/(x - i)$. Then by (5.2.38)

$$f(x) = (x - i) \underset{\rho\to\infty}{\overset{(2)}{\mathrm{l.i.m.}}} \frac{1}{\sqrt{2\pi}} \int_{-\rho}^{\rho} \eta(v)\, e^{ixv}\, dv.$$

In view of (8.3.22) and the theorem of Plancherel a generalized Hilbert transform may be introduced through

$$f^\sim(x) = (x - i) \underset{\rho\to\infty}{\overset{(2)}{\mathrm{l.i.m.}}} \frac{1}{\sqrt{2\pi}} \int_{-\rho}^{\rho} \{-i \operatorname{sgn} v\}\eta(v)\, e^{ixv}\, dv.$$

It then follows by the Parseval equation that

$$\int_{-\infty}^{\infty} \frac{|f^\sim(x)|^2}{1 + x^2}\, dx = \int_{-\infty}^{\infty} \frac{|f(x)|^2}{1 + x^2}\, dx.$$

If f belongs to $\mathsf{L}^2_{2\pi}$ or L^2, ACHIESER [2, p. 160] shows that the above definition is essentially equal to the usual one. On the other hand, since $\mathsf{C} \subset \mathsf{W}^2$, the definition may be used to associate a Hilbert transform with every continuous bounded function. Then an analysis quite similar to that in L^p, $1 \le p < \infty$, may be developed. Further generalizations are due to KOIZUMI [1, 2]. He introduces a Hilbert transform for $f \in \mathsf{W}^2$ through

$$f^\sim(x) = \frac{x + i}{\pi} \int_{-\infty}^{\infty} \frac{f(u)}{u + i} \frac{du}{x - u}.$$

Moreover, he considers Hilbert transforms on classes of functions f for which $|f(x)|^p/(1 + |x|^\alpha) \in \mathsf{L}^1$ for some $p \ge 1$, $\alpha > 0$.

There is also a theory of discrete Hilbert transforms. Let $\{\alpha_k\}_{k=-\infty}^{\infty}$ be a two-way infinite sequence of numbers. Then the Hilbert transform of $\{\alpha_k\}$ is defined by

$$\alpha_n^\sim = \sum_{k=-\infty}^{\infty}{}' \frac{\alpha_k}{n - k} \qquad (n \in \mathbb{Z}),$$

where the prime indicates that the index $k = n$ is omitted. A classical result asserts that $\|\alpha^\sim\|_2 \le \pi \|\alpha\|_2$ for every $\alpha \in \mathsf{l}^2$. The result was extended by M. RIESZ who showed that $\|\alpha^\sim\|_p \le M_p \|\alpha\|_p$ for every $\alpha \in \mathsf{l}^p$, $1 < p < \infty$. See ZYGMUND [6] and, for further generalizations, KOIZUMI [1, 2].

Hilbert transforms in connection with distribution theory may be found in BELTRAMI–WOHLERS [1, 2], GÜTTINGER [1], HORVÁTH [1], LAUWERIER [1], NEWCOMB [1], TILLMANN [1, 2]. See also Vol. II of this treatise.

Sec. 8.2. For the Hilbert formulae of Prop. 8.2.1 see e.g. Titchmarsh [6, p. 138]. The proofs are similar to the arguments used for the Parseval formulae for L^p-Fourier transforms (Prop. 5.2.13). Prop. 8.2.2 is preliminary to one of our main results, namely Prop. 8.3.1. The proof presented is due to A. P. CALDERÓN and was communicated to the authors by G. WEISS in Nov. 1965. For a paper largely concerned with the set $\{f \in \mathsf{L}^1 \mid f^\sim \in \mathsf{L}^1\}$ we

† We recall that the class W^2 plays an important rôle in the generalized harmonic analysis of BOCHNER–WIENER, see e.g. WIENER [2, p. 138 ff], BOCHNER [7, p. 138 ff].

refer to KOBER [1]. For conditions sufficient to ensure $f^{\sim} \in L^1$ see also KÖNIG [1]. Prop. 8.2.5–8.2.7 in this general form do not appear to be stated elsewhere; they are reminiscent of work on singular integrals (compare Sec. 1.4 and 3.2) and are suggested by some remarks of HARDY–ROGOSINSKI [1, p. 98] in the periodic case. Of course, the applications given to the integrals of Cauchy–Poisson (Cor. 8.2.8) and Fejér (Problem 8.2.4) are well-known and contained in all texts cited here. For the Privalov-type result of Prop. 8.2.9 we refer to BUTZER–TREBELS [2, p. 20] and the literature cited in Problems 8.2.5–8.2.7. For the second footnote to Prop. 8.2.9 compare also BUTZER [4]. For Prop. 8.2.10 see e.g. TITCHMARSH [6, p. 132] or WEISS [1, p. 49].

Sec. 8.3. Concerning Prop. 8.3.1 for $1 < p < 2$ see HILLE [3]. The case $p = 1$ is the difficult one; see the notes to Sec. 8.2 above. This material is also covered in BUTZER–TREBELS [2, p. 21 ff] and BUTZER–BERENS [1, Sec. 4.2]. There is considerable literature on allied integrals. TITCHMARSH [6, p. 147] e.g. shows that the allied integral is (C, α) summable to $f^{\sim}(x)$ a.e. for every $\alpha > 0$. Thus in particular Cor. 8.3.5 is well-known. GOLINSKIĬ [1, 2] defines a further transform through the allied integral (8.3.5) and compares it with the traditional definition (8.0.2). In particular, he discusses applications to Prop. 8.3.2. For Theorem 8.3.8, due to HILLE–TAMARKIN [1], see also TITCHMARSH [6, p. 148] and HEWITT [1, p. 199].

Concerning Theorem 8.3.6 in the more difficult case $p = 1$, see BUTZER–TREBELS [2, p. 41]; first results are due to HILLE [3], COOPER [2]. This calls to mind the results of BOCHNER–CHANDRASEKHARAN of Sec. 5.1.3. For extensions of Theorem 8.3.6 to L^p-spaces, $2 < p < \infty$, we refer to the treatment of Sec. 10.6 and 13.2.

In our future investigations the classes $W[L^p; |v|^\alpha]$, $V[L^p; |v|^\alpha]$, $1 \le p \le 2$, $\alpha > 0$, shall play the key rôle. Since these are defined in terms of the Fourier transform of f, the problem is to find equivalent characterizations in terms of the original function f. If r is an even positive integer, the results of Theorems 5.1.16, 5.2.21, 5.3.15 will do. If r is odd, we may apply Theorem 8.3.6. For extensions to fractional $\alpha > 0$ we refer to Chapter 11.

Prop. 8.3.7 is due to HILLE [3]; see also TITCHMARSH [4] and BUTZER–TREBELS [2].

9

Hilbert Transforms of Periodic Functions

9.0 Introduction

This chapter is concerned with the periodic version of the theory presented in the preceding chapter. We commence with our definition of the *conjugate function*, thus with the correct interpretation of

(9.0.1)
$$f^{\sim}(x) = \frac{1}{2\pi}\int_{-\pi}^{\pi} f(x-u)\cot\frac{u}{2}\,du$$

considered in the informal approach of the introduction to this Part.

Definition 9.0.1. *The **Hilbert transform**† (or conjugate function) of a function $f \in X_{2\pi}$ is defined as the Cauchy principal value of the integral* (9.0.1) *whenever it exists, i.e. $f^{\sim}(x) \equiv (Hf)(x)$ is defined as*

(9.0.2)
$$f^{\sim}(x) = \mathrm{PV}\Big[\frac{1}{2\pi}\int_{-\pi}^{\pi} f(x-u)\cot\frac{u}{2}\,du\Big] \equiv \lim_{\delta\to 0+} f_{\delta}^{\sim}(x)$$

at every point x for which this limit exists. Hereby,

(9.0.3)
$$f_{\delta}^{\sim}(x) = \frac{1}{2\pi}\int_{\delta\le|u|\le\pi} f(x-u)\cot\frac{u}{2}\,du$$
$$= \frac{1}{2\pi}\int_{\delta}^{\pi}[f(x-u)-f(x+u)]\cot\frac{u}{2}\,du.$$

Again, $f_{\delta}^{\sim}(x)$ is well-defined everywhere as a bounded periodic function for every $f \in X_{2\pi}$ and $0 < \delta < \pi$. The limit in (9.0.2) will not only be interpreted in the pointwise sense, but also in the norm.

If $f \in L^p$, $1 \le p < \infty$, the Hilbert transform $f^{\sim}$ was defined as the convolution of f with the 'kernel' $\{\sqrt{2/\pi}\,x^{-1}\}$, the integral being taken in the Cauchy principal value

† The Hilbert transform (8.0.1) of $f \in L^p$, $1 \le p < \infty$, as well as the Hilbert transform (9.0.1) of $f \in X_{2\pi}$ are denoted by the same symbol $f^{\sim}$ and, generally, by the same name. But there is no danger of confusion since the class of functions under consideration specifies the meaning of $f^{\sim}$.

sense (Def. 8.0.1). Recalling Sec. 3.1.2, we used results on meromorphic functions in order to convert the periodic singular integral of Fejér into its nonperiodic version, thus to unroll results on the circle group onto the line (and conversely). The expansion†

$$\frac{1}{2}\cot\frac{z}{2} = \frac{1}{z} + \sum_{k=-\infty}^{\infty}{}' \left\{\frac{1}{z-2\pi k} + \frac{1}{2\pi k}\right\} \qquad (z = x + iy) \tag{9.0.4}$$

of the periodic 'Hilbert-kernel' $\{\cot(x/2)\}$ strongly suggests that many of the fundamental properties concerning the Hilbert transform on the circle group may be obtained by reduction to those on the line group.

This program will be outlined in Sec. 9.1.1, concerned with the existence of the transform. But to evaluate the Fourier transform of the conjugate function, we already make use of different methods, in particular using the fact that the exponentials $\{\exp(ikx)\}$, $k \in \mathbb{Z}$, form a fundamental set in $\mathsf{X}_{2\pi}$. This has the disadvantage of being somewhat unsystematic, but it has the advantage of avoiding duplication of proofs and of exhibiting the simple structure of harmonic analysis on the circle group. While Sec. 9.1.2 is reserved to Hilbert formulae, Sec. 9.2 deals with the convergence problem of conjugates of singular integrals, the treatment being parallel to Sec. 8.2.2, 8.2.3. Finally, Sec. 9.3 is concerned with summation of allied series and norm-convergence of Fourier series as well as with characterizations of the classes $\mathsf{W}[\mathsf{X}_{2\pi}; \{-i \operatorname{sgn} k\}(ik)^r]$, $\mathsf{V}[\mathsf{X}_{2\pi}; \{-i \operatorname{sgn} k\}(ik)^r]$, $r \in \mathbb{N}$. The latter result enables one to characterize the classes $\mathsf{W}[\mathsf{X}_{2\pi}; |k|^r]$, $\mathsf{V}[\mathsf{X}_{2\pi}; |k|^r]$ which will be of interest in our later investigations in saturation theory.

9.1 Existence and Basic Properties

9.1.1 Existence

We first prove a fundamental result due originally to N. LUSIN and I. I. PRIVALOV.

Theorem 9.1.1. *The Hilbert transform $f^\sim$ of $f \in \mathsf{X}_{2\pi}$ exists almost everywhere.*

Proof. According to (9.0.4) we have for $|x| \le \pi$

$$\left|\frac{1}{2}\cot\frac{x}{2} - \frac{1}{x}\right| = \left|\sum_{k=-\infty}^{\infty}{}' \frac{x}{2k\pi(x-2k\pi)}\right| \le \frac{\pi}{4\pi^2}\sum_{k=-\infty}^{\infty}{}' \left|\frac{1}{1-(x/2k\pi)}\right|\frac{1}{k^2} \le \frac{\pi}{6}. \tag{9.1.1}$$

To each $f \in \mathsf{X}_{2\pi}$ we define f_1 on $\mathbb{R}$ by $f_1(x) = \kappa_{[-2\pi,2\pi]}(x)f(x)$. Then $f_1 \in \mathsf{L}^1$ and $f(x-u) \equiv f_1(x-u)$ for all $|x|$, $|u| \le \pi$. By the definition of $f_\delta^\sim$ (cf. (9.0.3)) we have that the limit of

† This well-known Laurent series expansion may be found in any text on complex function theory, cf. e.g. HILLE [4, p. 261]. The prime indicates that the value $k = 0$ is omitted. Note that the series converges absolutely and uniformly in every compact set of the complex plane not containing any of the points $2\pi k$, $k \in \mathbb{Z}$, $k \neq 0$.

$$(9.1.2)\quad f_\delta^{\sim}(x) = \frac{1}{\pi}\int_{\delta\le|u|\le\pi} \frac{f_1(x-u)}{u}\,du + \frac{1}{\pi}\int_{\delta\le|u|\le\pi} f(x-u)\sum_{k=-\infty}^{\infty}{}' \left\{\frac{1}{u-2k\pi}+\frac{1}{2k\pi}\right\}du \equiv I_1(x)+I_2(x)$$

for $\delta\to 0+$ exists for some $|x|\le\pi$ if and only if $I_1(x)$ converges, since $I_2(x)$ is everywhere convergent by (9.1.1). Moreover

$$(9.1.3)\qquad |f_\delta^{\sim}(x) - I_1(x)| \le (\pi/3)\|f\|_1.$$

On the other hand, if we replace the intervals $\delta\le|u|\le\pi$ in $I_1(x)$ by $\delta\le|u|<\infty$, we obtain $[f_1]_\delta^{\sim}(x)$ (cf. (8.0.3)). It follows for $|x|\le\pi$ that $I_1(x)$ converges as $\delta\to 0+$ if and only if $f_1^{\sim}(x)$ exists. Moreover,

$$(9.1.4)\qquad |[f_1]_\delta^{\sim}(x) - I_1(x)| = \frac{1}{\pi}\left|\int_{\pi\le|u|<\infty}\frac{f_1(x-u)}{u}\,du\right| \le (2/\pi)\|f\|_1.$$

Collecting the results we see that for $|x|\le\pi$ the Hilbert transform $f^{\sim}(x)$ of $f\in X_{2\pi}$ exists if and only if the Hilbert transform $f_1^{\sim}(x)$ of $f_1\in L^1$ exists, which, according to Theorem 8.1.6, is the case for almost all $x\in\mathbb{R}$. Thus, since $f_\delta^{\sim}$ is periodic, $f^{\sim}(x)$ exists as a periodic function for almost all x.

Proposition 9.1.2. *If $f\in L^1_{2\pi}$, then there exists a constant M such that*

$$(9.1.5)\qquad \operatorname{meas}\{x\in[-\pi,\pi] \mid |f^{\sim}(x)|>y>0\} \le \frac{M}{y}\|f\|_1.$$

Proof. We may suppose that $\|f\|_1 = 1$. If again $f_1 = \kappa_{[-2\pi,2\pi]}f$, then by (8.1.9) there exists a constant A such that

$$(9.1.6)\qquad \operatorname{meas}\{x \mid |f_1^{\sim}(x)|>y>0\} \le \frac{A}{y}\|f_1\|_1.$$

On the other hand, by (9.1.3) and (9.1.4)

$$(9.1.7)\qquad |f^{\sim}(x) - f_1^{\sim}(x)| \le 2\,\|f\|_1 \qquad (x\in[-\pi,\pi]).$$

Therefore we see that for all $y>4$ the set of points $x\in[-\pi,\pi]$ for which $|f^{\sim}(x)|>y$ is contained in the set of points $x\in\mathbb{R}$ for which $|f_1^{\sim}(x)|>y/2$, and so by (9.1.6) has measure not exceeding $4A/y$. This gives (9.1.5) with constant $4A$ provided $y>4$.

If we choose $M=\max\{4A, 8\pi\}$, then $My^{-1}\ge 2\pi$ for all $0<y\le 4$, and (9.1.5) is trivially satisfied.

Theorem 9.1.3. *For $f\in L^p_{2\pi}$, $1<p<\infty$, the Hilbert transform $f^{\sim}$ belongs to $L^p_{2\pi}$, and there exists a constant M_p such that*

$$(9.1.8)\qquad \|f_\delta^{\sim}\|_p \le M_p\|f\|_p \qquad (0<\delta<\pi),$$

$$(9.1.9)\qquad \|f^{\sim}\|_p \le M_p\|f\|_p.$$

Proof. If f_1 is defined as above, we obtain by (9.1.3), (9.1.4), and (8.1.36) that for any $0 < \delta < \pi$

$$\|f_\delta^\sim\|_p \le 2\,\|f\|_1 + \left(\frac{1}{2\pi}\int_{-\pi}^{\pi} |[f_1]_\delta^\sim(x)|^p\,dx\right)^{1/p}$$
$$\le 2\,\|f\|_p + (1 + M_p)\|f_1\|_p.\dagger$$

Hence (9.1.8) is established, giving (9.1.9) by Fatou's lemma.

Thus the mapping $f \to f^\sim$ is a bounded linear transformation of $\mathsf{L}_{2\pi}^p$ into $\mathsf{L}_{2\pi}^p$ for $1 < p < \infty$.

The next propositions, which are also of independent interest, may be considered as preparatory material for the proofs of the periodic counterparts of (8.1.40) and (8.2.8) as well as of (8.3.1).

Proposition 9.1.4. *For each $m \in \mathbb{Z}$,*

$$\lim_{\delta\to 0+} \frac{1}{2\pi} \int_{\delta \le |u| \le \pi} \cot\frac{u}{2}\, e^{-imu}\,du = -i \operatorname{sgn} m. \tag{9.1.10}$$

Proof. First of all, (9.1.10) is trivially true for $m = 0$. For arbitrary integers m, we have in view of (9.0.4) and the uniform convergence of the series that the left-hand side of (9.1.10) is equal to

$$\lim_{\delta\to 0+} \frac{1}{\pi} \int_{\delta \le |u| \le \pi} \frac{e^{-imu}}{u}\,du + \frac{1}{\pi}\int_{-\pi}^{\pi} \left[\sum_{k=-\infty}^{\infty}{}' \left\{\frac{1}{u - 2k\pi} + \frac{1}{2k\pi}\right\}\right] e^{-imu}\,du \tag{9.1.11}$$
$$= \lim_{\delta\to 0+} \frac{1}{\pi} \int_{\delta \le |u| \le \pi} \frac{e^{-imu}}{u}\,du + \sum_{k=-\infty}^{\infty}{}' \frac{1}{\pi}\int_{-\pi}^{\pi} \frac{e^{-imu}}{u - 2k\pi}\,du$$
$$= \lim_{\delta\to 0+} \frac{1}{\pi} \int_{\delta \le |u|} \frac{e^{-imu}}{u}\,du.$$

Then (9.1.10) follows by (8.1.17).

Corollary 9.1.5. *If $f \in \mathsf{L}_{2\pi}^1$, then*

$$\lim_{\delta\to 0+} [f_\delta^\sim]^\wedge(k) = \{-i \operatorname{sgn} k\} f^\wedge(k) \qquad (k \in \mathbb{Z}). \tag{9.1.12}$$

Proof. By the convolution theorem we have for any $0 < \delta < \pi$

$$[f_\delta^\sim]^\wedge(k) = \left\{\frac{1}{2\pi} \int_{\delta \le |u| \le \pi} \cot\frac{u}{2}\, e^{-iku}\,du\right\} \cdot f^\wedge(k) \qquad (k \in \mathbb{Z}).$$

Thus (9.1.10) implies (9.1.12). Note that for any $0 < \delta < \pi$

$$\int_{-\pi}^{\pi} f_\delta^\sim(x)\,dx = 0. \tag{9.1.13}$$

† It is obvious that the constant M_p in the *Marcel Riesz inequality* (9.1.9) need not be the same as the constant in (8.1.39).

Proposition 9.1.6. *If t_n belongs to T_n, then its Hilbert transform $t_n^\sim$ belongs to T_n, in fact,*

$$t_n(x) = \sum_{k=-n}^{n} c_k e^{ikx}, \qquad t_n^\sim(x) = \sum_{k=-n}^{n} \{-i \operatorname{sgn} k\} c_k e^{ikx}.$$

Moreover, the functions $[t_n]_\delta^\sim$ of (9.0.3) *converge uniformly towards $t_n^\sim$ as $\delta \to 0+$.*

The proof follows immediately since for any $0 < \delta < \pi$ and $k \in \mathbb{Z}$

$$\left| \frac{1}{2\pi} \int\limits_{\delta \le |u| \le \pi} \cot \frac{u}{2} e^{ik(x-u)} \, du + \{i \operatorname{sgn} k\} e^{ikx} \right| = \left| \frac{1}{2\pi} \int\limits_{\delta \le |u| \le \pi} \cot \frac{u}{2} e^{-iku} \, du + \{i \operatorname{sgn} k\} \right|,$$

which by (9.1.10) tends to zero for $\delta \to 0+$, and indeed uniformly in x. In particular, we have for any $t_n \in \mathsf{T}_n$ and $1 \le p < \infty$ that

$$\lim_{\delta \to 0+} \|t_n^\sim - [t_n]_\delta^\sim\|_p = 0, \tag{9.1.14}$$

which we are now going to extend to arbitrary functions $f \in \mathsf{L}_{2\pi}^p$.

Proposition 9.1.7. *Let $f \in \mathsf{L}_{2\pi}^p$, $1 < p < \infty$. Then*

$$\lim_{\delta \to 0+} \|f^\sim - f_\delta^\sim\|_p = 0, \tag{9.1.15}$$

i.e., the limit (9.0.2) *does not only exist pointwise* a.e. *but also in the mean of order p.*

Proof. By Theorem 1.2.5, given $\varepsilon > 0$, there exists $t_n \in \mathsf{T}_n$ such that $\|f - t_n\|_p < \varepsilon$. Hence it follows by Theorem 9.1.3 and (9.1.14) that

$$\|f^\sim - f_\delta^\sim\|_p \le \|[f - t_n]^\sim\|_p + \|t_n^\sim - [t_n]_\delta^\sim\|_p + \|[t_n - f]_\delta^\sim\|_p = o(1) \quad (\delta \to 0+),$$

and (9.1.15) is established.

Proposition 9.1.8. *Let $f \in \mathsf{L}_{2\pi}^p$, $1 < p < \infty$. Then*

$$[f^\sim]^\wedge(k) = \{-i \operatorname{sgn} k\} f^\wedge(k) \qquad (k \in \mathbb{Z}). \tag{9.1.16}$$

The proof of (9.1.16) follows immediately by (9.1.12) and (9.1.15) since the latter relation in particular implies

$$\lim_{\delta \to 0+} \|f^\sim - f_\delta^\sim\|_1 = 0 \tag{9.1.17}$$

for all $f \in \mathsf{L}_{2\pi}^p$, $1 < p < \infty$. By the way, we note that in view of (9.1.13), relation (9.1.17) implies that

$$\int_{-\pi}^{\pi} f^\sim(x) \, dx = 0 \tag{9.1.18}$$

for all $f \in \mathsf{L}_{2\pi}^p$, $1 < p < \infty$, which in fact is the case $k = 0$ of (9.1.16).

9.1.2 Hilbert Formulae

Proposition 9.1.9. *If $f \in \mathsf{L}_{2\pi}^2$, then*

$$\frac{1}{2\pi} \int_{-\pi}^{\pi} |f(x)|^2 \, dx = \left| \frac{1}{2\pi} \int_{-\pi}^{\pi} f(x) \, dx \right|^2 + \frac{1}{2\pi} \int_{-\pi}^{\pi} |f^\sim(x)|^2 \, dx. \tag{9.1.19}$$

For the proof we use the Parseval equation (4.2.6), Theorem 9.1.3, and (9.1.16) to deduce

$$\|f\|_2^2 = \sum_{k=-\infty}^{\infty} |f^\wedge(k)|^2 = |f^\wedge(0)|^2 + \sum_{k=-\infty}^{\infty} |\{-i \operatorname{sgn} k\} f^\wedge(k)|^2$$
$$= |f^\wedge(0)|^2 + \|f^\sim\|_2^2,$$

which is (9.1.19).

Proposition 9.1.10. *Let* $f \in \mathsf{L}_{2\pi}^p$, $1 < p < \infty$, *and* $g \in \mathsf{L}_{2\pi}^{p'}$. *Then*

$$\int_{-\pi}^{\pi} f^\sim(x) g(x)\, dx = -\int_{-\pi}^{\pi} f(x) g^\sim(x)\, dx, \tag{9.1.20}$$

$$\int_{-\pi}^{\pi} f(x)\overline{g(x)}\, dx = \frac{1}{2\pi}\left\{\int_{-\pi}^{\pi} f(x)\, dx\right\}\left\{\int_{-\pi}^{\pi} \overline{g(x)}\, dx\right\} + \int_{-\pi}^{\pi} f^\sim(x)\overline{g^\sim(x)}\, dx. \tag{9.1.21}$$

Again, relations of this type are referred to as *Hilbert formulae.*

Proof. Since by (9.1.16) and Prop. 9.1.6

$$\int_{-\pi}^{\pi} f^\sim(x)\, e^{-ikx}\, dx = -\int_{-\pi}^{\pi} f(x)[e^{-ik\circ}]^\sim(x)\, dx,$$

(9.1.20) follows for any $g \in \mathsf{T}_n$. Since T_n is dense in $\mathsf{L}_{2\pi}^{p'}$ by Theorem 1.2.5, (9.1.20) is valid for general g by (9.1.9). Concerning (9.1.21), we have for any

$$t_n(x) = \sum_{k=-n}^{n} c_k\, e^{ikx}, \qquad t_m^*(x) = \sum_{k=-m}^{m} c_k^*\, e^{ikx}$$

with $n < m$, say, that

$$\frac{1}{2\pi}\int_{-\pi}^{\pi} t_n(x)\overline{t_m^*(x)}\, dx = \sum_{k=-n}^{n} c_k\overline{c_k^*} = c_0\overline{c_0^*} + \sum_{k=-n}^{n} [\{-i \operatorname{sgn} k\}c_k]\overline{[\{i \operatorname{sgn} k\}c_k^*]}$$
$$= \left\{\frac{1}{2\pi}\int_{-\pi}^{\pi} t_n(x)\, dx\right\}\left\{\frac{1}{2\pi}\int_{-\pi}^{\pi} \overline{t_m^*(x)}\, dx\right\} + \frac{1}{2\pi}\int_{-\pi}^{\pi} t_n^\sim(x)\overline{[t_m^*]^\sim(x)}\, dx.$$

Therefore (9.1.21) follows by Theorem 1.2.5 and (9.1.9).

Proposition 9.1.11. *Let* $f \in \mathsf{L}_{2\pi}^p$, $1 < p < \infty$. *Then*

$$[f^\sim]^\sim(x) = -f(x) + \frac{1}{2\pi}\int_{-\pi}^{\pi} f(u)\, du \quad \text{a.e.} \tag{9.1.22}$$

Proof. By Prop. 9.1.6 we have for any $t_n \in \mathsf{T}_n$

$$[t_n^\sim]^\sim(x) = -\sum_{k=-n}^{n}{}' c_k\, e^{ikx} = -t_n(x) + \frac{1}{2\pi}\int_{-\pi}^{\pi} t_n(u)\, du,$$

and (9.1.22) again follows by Theorem 1.2.5 and (9.1.9).

Relation (9.1.22) may be regarded as an *inversion formula* for the Hilbert transform. It shows that the mapping $f \to f^\sim$ is a bounded linear transformation of $\mathsf{L}_{2\pi}^p$ *onto*

$L_{2\pi}^{p;0}$ for $1 < p < \infty$, where $L_{2\pi}^{p;0}$ is the set $\{f \in L_{2\pi}^p \mid \int_{-\pi}^{\pi} f(u)\, du = 0\}$. Moreover, using the notation $Hf = f^\sim$, $H^r f = H(H^{r-1}f)$, we have just shown that

$$H^r f = (-1)^{[r/2]} \begin{cases} f, & r \text{ even} \\ f^\sim, & r \text{ odd} \end{cases} \tag{9.1.23}$$

for every $f \in L_{2\pi}^{p;0}$, $1 < p < \infty$, and $r \in \mathbb{N}$.

Proposition 9.1.12. *Let $f \in L_{2\pi}^p$ and $g \in L_{2\pi}^1$. Then for $1 < p < \infty$*

$$[f * g]^\sim(x) = (f^\sim * g)(x) \quad \text{a.e.} \tag{9.1.24}$$

The conclusion remains valid for $p = 1$ under the further assumption $f^\sim \in L_{2\pi}^1$.

Proof. Let $1 < p < \infty$. By (4.1.7), Prop. 9.1.6, and (9.1.16) we have for any $t_n \in T_n$

$$[f * t_n]^\sim(x) = \Big[\sum_{k=-n}^{n} f^\wedge(k) c_k\, e^{ik\circ}\Big]^\sim(x) = \sum_{k=-n}^{n} \{-i \operatorname{sgn} k\} f^\wedge(k) c_k\, e^{ikx}$$
$$= (f^\sim * t_n)(x),$$

and thus (9.1.24) holds if $g \in T_n$. By the routine density argument, (9.1.24) follows for general g.

To prove (9.1.24) for $p = 1$, let $Z\colon -\pi = t_0 < t_1 < \cdots < t_l = \pi$ be any partition of the interval $[-\pi, \pi]$ and set

$$\sigma_Z(x) = \frac{1}{2\pi} \sum_{k=0}^{l-1} f(x - t_k) \int_{t_k}^{t_{k+1}} g(u)\, du.$$

Certainly, $\sigma_Z(x) \in L_{2\pi}^1$ and

$$\lim_{\|Z\| \to 0} \|(f * g)(\circ) - \sigma_Z(\circ)\|_1 = 0,$$

which, as well as the rest of the proof, follows as in the proof of Prop. 8.2.2.

Proposition 9.1.13. (i) *If $f \in L_{2\pi}^1$ and $g \in L_{2\pi}^r$, $1 < r < \infty$, then*

$$[f * g]^\sim(x) = (f * g^\sim)(x) \quad \text{a.e.} \tag{9.1.25}$$

(ii) *If $f \in L_{2\pi}^p$, $1 < p < \infty$, and $g \in L_{2\pi}^{p'}$, then furthermore*

$$[f * g]^\sim(x) = (f * g^\sim)(x) = (f^\sim * g)(x) \quad \text{a.e.} \tag{9.1.26}$$

(iii) *If $f \in L_{2\pi}^1$ and $g \in L_{2\pi}^r$, $1 < r < \infty$, then* (9.1.26) *remains valid under the additional assumption $f^\sim \in L_{2\pi}^1$.*

Proof. Part (i) is (9.1.24) with the rôles of f and g interchanged. (9.1.26) follows immediately by (9.1.20), and finally part (iii) is an easy consequence of part (i) and (9.1.24).

Problems

1. State and prove counterparts of Problem 8.1.1 for (periodic) Hilbert transforms.
2. Let $\{p_r(x)\}$ be the Abel–Poisson kernel of (1.2.37). Evaluate $p_r^\sim(x)$ and show that for each $0 < r < 1$ and all $x \in \mathbb{R}$

$$p_r^\sim(x) = \sum_{k=-\infty}^{\infty} \{-i \operatorname{sgn} k\} r^{|k|} e^{ikx} = \frac{2r \sin x}{1 - 2r \cos x + r^2}.$$

(Hint: Problem 1.2.18)

3. Show that $\sum_{k=2}^{\infty} \{\cos kx/\log k\}$ is the Fourier series of a function $f \in \mathsf{L}_{2\pi}^1$ for which $f^\sim$ does not belong to $\mathsf{L}_{2\pi}^1$. Show that $f(x) = \sum_{k=2}^{\infty} \{\sin kx/k \log k\}$ is an example of an absolutely continuous function for which $f^\sim$ does not belong to $\mathsf{L}_{2\pi}^{\infty}$. (Hint: Problems 6.3.2, 6.3.3; see also EDWARDS [1II, p. 90])
4. Give a proof of (9.1.15) following upon Theorem 9.1.3 by reduction from (8.1.40).
5. Show that the set $\{f \in \mathsf{L}_{2\pi}^1 \mid f^\sim \in \mathsf{L}_{2\pi}^1\}$ is dense in $\mathsf{L}_{2\pi}^1$ (cf. Problem 8.3.6).

9.2 Conjugates of Singular Integrals

9.2.1 The Case $1 < p < \infty$

We recall that a general singular integral $I_\rho(f; x)$ of $f \in \mathsf{X}_{2\pi}$ (see Sec. 1.1) converges towards f as $\rho \to \rho_0$ on the one hand in the norm of the space $\mathsf{X}_{2\pi}$ if the kernel is a (periodic) approximate identity (cf. Theorem 1.1.5), and on the other hand pointwise almost everywhere if the kernel furthermore satisfies the assumptions of, for example, Prop. 1.4.4 or 1.4.6. In order to establish analogous results for the Hilbert transform $f^\sim$, the same reasons as in Sec. 8.2.2 call for the introduction of the convolution integral

$$I_\rho^\sim(f; x) = \frac{1}{2\pi}\int_{-\pi}^{\pi} f(x-u)\chi_\rho^\sim(u)\,du. \tag{9.2.1}$$

$I_\rho^\sim(f; x)$ will be referred to as the *conjugate of the singular integral* $I_\rho(f; x)$ of $f \in \mathsf{X}_{2\pi}$. In view of (9.1.18), $\{\chi_\rho^\sim(x)\}$ is *not* a kernel.

Proposition 9.2.1. *Let $1 < p < \infty$ and $f \in \mathsf{L}_{2\pi}^p$. If the kernel $\{\chi_\rho(x)\}$ is a bounded approximate identity, then*

$$\lim_{\rho\to\rho_0} \|I_\rho^\sim(f; \circ) - f^\sim(\circ)\|_p = 0. \tag{9.2.2}$$

If $\{\chi_\rho(x)\}$ furthermore satisfies the assumptions of Prop. 1.4.4 *or* 1.4.6, *then*

$$\lim_{\rho\to\rho_0} I_\rho^\sim(f; x) = f^\sim(x) \quad \text{a.e.} \tag{9.2.3}$$

Since $I_\rho^\sim(f; x) = I_\rho(f^\sim; x)$ a.e. by (9.1.26), the proof follows immediately by Theorem 1.1.6, Prop. 1.4.4, 1.4.6.

9.2.2 Convergence in $\mathsf{C}_{2\pi}$ and $\mathsf{L}_{2\pi}^1$

Although $\chi_\rho \in \mathsf{L}_{2\pi}^1 \cap \mathsf{L}_{2\pi}^{\infty}$ now implies $\chi_\rho^\sim \in \mathsf{L}_{2\pi}^1$, the situation for $p = 1$ is nevertheless quite similar to that of Sec. 8.2.3. The counterpart of Prop. 8.2.6 now reads

Proposition 9.2.2. *Let $f \in \mathsf{X}_{2\pi}$ and the kernel $\{\chi_\rho(x)\}$ of the singular integral* (1.1.3) *be even and bounded. Let $\Theta_\rho(x)$ be defined by*

$$\Theta_\rho(x) = \cot\frac{x}{2} - \chi_\rho^\sim(x), \tag{9.2.4}$$

and assume that there exists a positive function $\varphi(\rho)$ *tending monotonely to zero as* $\rho \to \rho_0$ *such that*

$$\int_0^{\varphi(\rho)} |\chi_\rho^\sim(u)|\, du \le M, \qquad \int_{\varphi(\rho)}^{\pi} |\Theta_\rho(u)|\, du \le M' \qquad (\rho \in \mathbb{A}), \tag{9.2.5}$$

$$\lim_{\rho \to \rho_0} \int_\delta^\pi |\Theta_\rho(u)|\, du = 0 \tag{9.2.6}$$

for every $0 < \delta < \pi$. *Then*

$$\lim_{\rho \to \rho_0} \|I_\rho^\sim(f; \circ) - f_{\varphi(\rho)}^\sim(\circ)\|_{X_{2\pi}} = 0, \tag{9.2.7}$$

and for all $\rho \in \mathbb{A}$

$$\|I_\rho^\sim(f; \circ) - f_{\varphi(\rho)}^\sim(\circ)\|_{X_{2\pi}} \le \frac{M + M'}{\pi} \|f\|_{X_{2\pi}}. \tag{9.2.8}$$

Proof. First of all, $\chi_\rho^\sim \in L_{2\pi}^1$ and therefore $I_\rho^\sim(f; x) \in X_{2\pi}$ for each $\rho \in \mathbb{A}$. Since $\chi_\rho^\sim$ is an odd function, we have (cf. (8.2.15))

$$I_\rho^\sim(f; x) - f_{\varphi(\rho)}^\sim(x) = \frac{1}{2\pi} \int_0^{\varphi(\rho)} [f(x - u) - f(x + u)]\chi_\rho^\sim(u)\, du$$
$$- \frac{1}{2\pi}\left(\int_{\varphi(\rho)}^{\delta} + \int_\delta^\pi\right)[f(x - u) - f(x + u)]\Theta_\rho(u)\, du \equiv I_1 + I_2 + I_3, \tag{9.2.9}$$

where $0 < \delta < \pi$ is chosen so that for given $\varepsilon > 0$ $\|f(\circ + u) - f(\circ)\|_{X_{2\pi}} < \varepsilon$ for all $|u| < \delta$, and ρ is so close to ρ_0 that $0 < \varphi(\rho) < \delta$. Then, in virtue of the hypotheses and the Hölder–Minkowski inequality,

$$\|I_1\|_{X_{2\pi}} \le \frac{\varepsilon}{\pi} \int_0^{\varphi(\rho)} |\chi_\rho^\sim(u)|\, du \le \frac{M}{\pi}\varepsilon, \qquad \|I_2\|_{X_{2\pi}} \le \frac{\varepsilon}{\pi} \int_{\varphi(\rho)}^{\delta} |\Theta_\rho(u)|\, du \le \frac{M'}{\pi}\varepsilon,$$
$$\|I_3\|_{X_{2\pi}} \le \frac{\|f\|_{X_{2\pi}}}{\pi} \int_\delta^\pi |\Theta_\rho(u)|\, du = o(1) \qquad (\rho \to \rho_0),$$

establishing (9.2.7) as well as (9.2.8).

Note that in view of (9.2.2) the last proposition is only interesting in case $X_{2\pi} = C_{2\pi}$ and $X_{2\pi} = L_{2\pi}^1$. (9.2.8) remains true for $p = \infty$.

Concerning pointwise convergence we have

Proposition 9.2.3. *Let* $f \in L_{2\pi}^1$ *and the kernel* $\{\chi_\rho(x)\}$ *of the singular integral* (1.1.3) *be even and bounded such that* $\chi_\rho^\sim$ *is positive on* $[0, \pi]$. *Let* $\Theta_\rho(x)$ *be defined by* (9.2.4), *and a function* $\varphi(\rho)$ *be given as in Prop.* 9.2.2 *such that* $\chi_\rho^\sim(x)$ *is monotonely increasing on* $[0, \varphi(\rho)]$ *with*

$$\varphi(\rho)\chi_\rho^\sim(\varphi(\rho)) \le M_1 \qquad (\rho \in \mathbb{A}), \tag{9.2.10}$$

and that $\Theta_\rho(x)$ *is positive and monotonely decreasing on* $[\varphi(\rho), \pi]$ *with*

$$\varphi(\rho)\Theta_\rho(\varphi(\rho)) \le M_2, \qquad \int_{\varphi(\rho)}^{\pi} \Theta_\rho(u)\, du \le M_3 \qquad (\rho \in \mathbb{A}),$$
$$\lim_{\rho \to \rho_0} \Theta_\rho(x) = 0 \qquad (x \in (0, \pi]). \tag{9.2.11}$$

Then at each point x for which

$$\int_0^t [f(x + u) - f(x - u)]\, du = o(t) \qquad (t \to 0+) \tag{9.2.12}$$

one has

$$\lim_{\rho \to \rho_0} [I_\rho^\sim(f; x) - f_{\varphi(\rho)}^\sim(x)] = 0. \tag{9.2.13}$$

In particular,

(9.2.14)
$$\lim_{\rho\to\rho_0} I_\rho^\sim(f; x) = f^\sim(x) \quad \text{a.e.}$$

The proof is left to Problem 9.2.1.

As an example we consider the singular integral $P_r(f; x)$ of Abel–Poisson with kernel $\{p_r(x)\}$ given by (1.2.37). Since $p_r(x) \in C_{2\pi}$, $0 < r < 1$, it follows by Theorem 9.1.3 that the Hilbert transform $p_r^\sim(x)$ exists as a function in $L^p_{2\pi}$, $1 < p < \infty$. Hence (9.1.16) and (1.2.34) imply

$$[p_r^\sim]^\wedge(k) = \{-i \operatorname{sgn} k\}p_r^\wedge(k) = \{-i \operatorname{sgn} k\}r^{|k|}.$$

Since the Fourier series of $p_r^\sim(x)$ converges absolutely and uniformly for each $0 < r < 1$, its sum must be equal to $p_r^\sim(x)$ a.e. by Prop. 4.1.5(i). Moreover, by Problem 9.1.2

(9.2.15)
$$p_r^\sim(x) = 2\sum_{k=1}^\infty r^k \sin kx = \frac{2r \sin x}{1 - 2r\cos x + r^2},$$

which is *a posterior* justification of the terminology of Sec. 1.2.6 in speaking of $p_r^\sim(x)(\equiv q_r(x))$ as the conjugate function of $p_r(x)$. Thus we have

(9.2.16)
$$P_r^\sim(f; x) \equiv Q_r(f; x) = \frac{1}{2\pi}\int_{-\pi}^{\pi} f(x-u)\,\frac{2r\sin u}{1 - 2r\cos u + r^2}\,du,$$

in accordance with the notation used in Sec. 1.2.6 and (9.2.1).

Since all assumptions of Prop. 9.2.1–9.2.3 are satisfied by the Abel–Poisson kernel (cf. Problem 9.2.1), we have

Corollary 9.2.4. *For the conjugate* (9.2.16) *of the Abel–Poisson integral of $f \in X_{2\pi}$ we have (for the definition of $\varphi(r)$ see Problem* 9.2.1*):*

(9.2.17)
$$\lim_{r\to 1-} Q_r(f; x) = f^\sim(x) \quad \text{a.e.},$$

(9.2.18)
$$\lim_{r\to 1-} \|Q_r(f; \circ) - f^\sim_{\varphi(r)}(\circ)\|_{X_{2\pi}} = 0,$$

(9.2.19)
$$\|Q_r(f; \circ) - f^\sim_{\varphi(r)}(\circ)\|_{X_{2\pi}} \le M_{r_0}\|f\|_{X_{2\pi}} \qquad (0 < r_0 \le r < 1).$$

If $X_{2\pi} = L^p_{2\pi}$, $1 < p < \infty$, then furthermore

(9.2.20)
$$\lim_{r\to 1-} \|Q_r(f; \circ) - f^\sim(\circ)\|_p = 0,$$

which remains true for $p = 1$ under the additional assumption $f^\sim \in L^1_{2\pi}$.

We may now extend some results already established for $L^p_{2\pi}$-space, $1 < p < \infty$, to functions belonging to $L^1_{2\pi}$ or $C_{2\pi}$.

Proposition 9.2.5. *Let $f \in L^1_{2\pi}$ be such that $f^\sim \in L^1_{2\pi}$. Then $f^\sim_\delta \in L^1_{2\pi}$ for all $0 < \delta < \pi$ and*

(9.2.21)
$$\|f^\sim_\delta\|_1 \le M_{\delta_0}\|f\|_1 + \|f^\sim\|_1 \qquad \left(0 < \delta \le \delta_0 < \frac{\pi}{2}\right),$$

(9.2.22)
$$\lim_{\delta\to 0+} \|f^\sim - f^\sim_\delta\|_1 = 0,$$

(9.2.23) $$[f^{\sim}]^{\sim}(x) = -f(x) + \frac{1}{2\pi}\int_{-\pi}^{\pi} f(u)\,du \quad \text{a.e.},$$

(9.2.24) $$[f^{\sim}]^{\wedge}(k) = \{-i \operatorname{sgn} k\} f^{\wedge}(k) \qquad (k \in \mathbb{Z}).$$

If $f \in C_{2\pi}$ is such that $f^{\sim} \in C_{2\pi}$, then the assertions remain valid with $L^1_{2\pi}$-norm replaced by $C_{2\pi}$-norm.

Proof. In view of Prop. 9.1.13(iii) we have for the Abel–Poisson integral

(9.2.25) $$[P_r(f; \circ)]^{\sim}(x) = Q_r(f; x) = P_r(f^{\sim}; x).$$

Since $\|f_\delta^{\sim}\|_1 \le \cot(\delta/2)\|f\|_1$, $f_\delta^{\sim} \in L^1_{2\pi}$ for all $\delta > 0$. Moreover, if r is such that $\delta = \varphi(r)$ (see Problem 9.2.1), then

$$\|f_\delta^{\sim}\|_1 \le M_{\delta_0}\|f\|_1 + \|Q_r(f; \circ)\|_1 \le M_{\delta_0}\|f\|_1 + \|f^{\sim}\|_1$$

by (9.2.19), and

$$\|f^{\sim} - f_\delta^{\sim}\|_1 \le \|f^{\sim}(\circ) - P_r(f^{\sim}; \circ)\|_1 + \|Q_r(f; \circ) - f^{\sim}_{\varphi(r)}(\circ)\|_1,$$

which, in virtue of Cor. 1.2.9 and (9.2.18), immediately implies (9.2.22). The proof of (9.2.23) follows as for Prop. 8.2.10, whereas (9.2.24) is an easy consequence of (9.1.12) and (9.2.22). Likewise, the proof in $C_{2\pi}$-space is given.

Proposition 9.2.6. *If for $f \in C_{2\pi}$ there exists $g \in C_{2\pi}$ such that $\{-i \operatorname{sgn} k\} f^{\wedge}(k) = g^{\wedge}(k)$, then $f^{\sim} \in C_{2\pi}$.*

Proof. Since $f \in C_{2\pi}$ implies by Theorem 9.1.3 that $f^{\sim} \in L^s_{2\pi}$ for any $1 \le s < \infty$, we have $f^{\sim}(x) = g(x)$ a.e. by (9.1.16) and the uniqueness theorem. Since $g \in C_{2\pi}$, $\lim_{r\to 1-} \|P_r(g; \circ) - g(\circ)\|_{C_{2\pi}} = 0$. Consequently, $\lim_{r\to 1-} \|P_r(f^{\sim}; \circ) - g(\circ)\|_{C_{2\pi}} = 0$, and it follows by (9.2.18), (9.2.25) that

$$\|g - f^{\sim}_{\varphi(r)}\|_{C_{2\pi}} \le \|g(\circ) - P_r(f^{\sim}; \circ)\|_{C_{2\pi}} + \|Q_r(f; \circ) - f^{\sim}_{\varphi(r)}(\circ)\|_{C_{2\pi}} = o(1)$$

as $r \to 1-$. Thus $f_\delta^{\sim}$ converges uniformly towards $g \in C_{2\pi}$ as $\delta \to 0+$. By Def. 9.0.1 this limit is the Hilbert transform $f^{\sim}$.

In discussing the convergence problem for conjugates of singular integrals we have to look at the 'positive' case. Thus far we have seen that the results on norm- as well as on pointwise convergence of $I_\rho^{\sim}(f; x)$ correspond quite naturally to those given for the original integrals $I_\rho(f; x)$ in Sec. 1.1 and 1.4. Now it is interesting to note that also the theory of Sec. 1.5 on singular integrals with positive kernels $\{\chi_\rho(x)\}$ is, *mutatis mutandis*, transferable to the conjugate $I_\rho^{\sim}(f; x)$. For instance, we have as a counterpart to Theorem 1.5.15

Proposition 9.2.7. *Let the kernel $\{\chi_\rho(x)\}$ of the singular integral $I_\rho(f; x)$ be even and bounded, and suppose that $\Theta_\rho(x)$ as given by (9.2.4) is positive on $(0, \pi]$ for each $\rho \in \mathbb{A}$. If $f \in C_{2\pi}$ has a continuous derivative such that $f'(x + \pi) = -f'(x)$, then for $\rho \to \rho_0$*

(9.2.26) $$I_\rho^{\sim}(f; x) - f^{\sim}(x) = \tfrac{1}{2}(2 - [\chi_\rho^{\sim}]_s^{\wedge}(1)) f'(x) + o(2 - [\chi_\rho^{\sim}]_s^{\wedge}(1))$$

if and only if

(9.2.27) $$\lim_{\rho\to\rho_0} \frac{[\chi_\rho^{\sim}]_s^{\wedge}(3) - 2}{[\chi_\rho^{\sim}]_s^{\wedge}(1) - 2} = 3.$$

Proof. First we observe that the assumptions upon f ensure that $f^\sim$ exists everywhere as a continuous function (Problem 9.2.3). Furthermore, we shall frequently make use of (compare (1.2.10))

$$\text{(9.2.28)}\quad \frac{1}{\pi}\int_0^\pi \Theta_\rho(u)\sin u\,du = \frac{2}{\pi}\int_0^\pi \left[\frac{\cos(u/2)}{\sin(u/2)} - \chi_\rho^\sim(u)\right]\sin\frac{u}{2}\cos\frac{u}{2}\,du$$
$$= \frac{2}{\pi}\int_0^\pi \cos^2\frac{u}{2}\,du - \frac{1}{\pi}\int_0^\pi \chi_\rho^\sim(u)\sin u\,du = 1 - \tfrac{1}{2}[\chi_\rho^\sim]_s^\wedge(1),$$

as well as of the relation

$$\text{(9.2.29)}\quad \frac{1}{\pi}\int_0^\pi \Theta_\rho(u)\sin^3 u\,du = \frac{1}{\pi}\int_0^\pi (1+\cos u)\sin^2 u\,du - \frac{1}{\pi}\int_0^\pi \chi_\rho^\sim(u)\sin^3 u\,du$$
$$= \frac{1}{2} - \frac{1}{4\pi}\int_0^\pi \chi_\rho^\sim(u)[3\sin u - \sin 3u]\,du = \frac{3(2-[\chi_\rho^\sim]_s^\wedge(1)) - (2-[\chi_\rho^\sim]_s^\wedge(3))}{8}.$$

Necessity. If we take $x = 0$ and $g(x) = \sin^3 x$, then $g'(0) = 0$, and the hypothesis gives

$$I_\rho^\sim(g;0) - g^\sim(0) = o(2-[\chi_\rho^\sim]_s^\wedge(1)) \qquad (\rho\to\rho_0).$$

On the other hand, we have

$$I_\rho^\sim(g;0) - g^\sim(0) = \frac{1}{2\pi}\int_0^\pi [g(u)-g(-u)]\Theta_\rho(u)\,du = \frac{1}{\pi}\int_0^\pi \Theta_\rho(u)\sin^3 u\,du$$

so that (9.2.27) necessarily follows by (9.2.29).

Sufficiency. In view of the assumptions upon f, the function $\psi_x(u)$ defined by

$$\text{(9.2.30)}\qquad \psi_x(u) = \frac{f(x+u) - f(x-u) - 2f'(x)\sin u}{\sin u}$$

is bounded on $[0,\pi]$ with $\lim_{u\to 0+}\psi_x(u) = \lim_{u\to\pi-}\psi_x(u) = 0$. We have by (9.2.28)

$$I_\rho^\sim(f;x) - f^\sim(x) = \frac{1}{2\pi}\int_0^\pi \psi_x(u)\sin u\,\Theta_\rho(u)\,du + f'(x)\frac{1}{\pi}\int_0^\pi \Theta_\rho(u)\sin u\,du$$
$$= \tfrac{1}{2}(2-[\chi_\rho^\sim]_s^\wedge(1))f'(x) + \frac{1}{2\pi}\int_0^\pi \psi_x(u)\sin u\,\Theta_\rho(u)\,du.$$

Now, given $\varepsilon > 0$, there is δ with $0 < \delta < \pi/2$ such that $|\psi_x(u)| < \varepsilon$ for every u with $0 \le u \le \delta$, $\pi - \delta \le u \le \pi$. Therefore, if M is a constant such that $|\psi_x(u)| \le M$ for all $0 \le u \le \pi$, then

$$\left|\frac{1}{2\pi}\int_0^\pi \psi_x(u)\sin u\,\Theta_\rho(u)\,du\right| \le \frac{\varepsilon}{2\pi}\left(\int_0^\delta + \int_{\pi-\delta}^\pi\right)\Theta_\rho(u)\sin u\,du$$
$$+ \frac{M}{2\pi\sin^2\delta}\int_\delta^{\pi-\delta}\Theta_\rho(u)\sin^3 u\,du$$
$$\le \frac{\varepsilon}{4}(2-[\chi_\rho^\sim]_s^\wedge(1)) + \frac{M}{16\sin^2\delta}[3(2-[\chi_\rho^\sim]_s^\wedge(1)) - (2-[\chi_\rho^\sim]_s^\wedge(3))],$$

and (9.2.26) follows by (9.2.27).

Obviously, one may weaken the differentiability hypothesis on f in the sense of Theorem 1.5.15. Thus it suffices to assume that f has a first Riemann derivative at the point x (cf. Def. 5.1.18). Furthermore, if f' does not satisfy the symmetry condition $f'(x+\pi) = -f'(x)$, then analogous results may nevertheless be obtained (with $\sin u$ replaced by $\sin(u/2)$ in (9.2.30)). For further comments we refer to Sec. 9.4. Here we end this brief outline of the 'positive' theory by an application to summation processes of Fourier series.

Thus, let the kernel $\{\chi_\rho(x)\}$ be generated via (1.2.32) by a θ-factor, i.e.

$$\text{(9.2.31)}\qquad \chi_\rho(x) = 1 + 2\sum_{k=1}^\infty \theta_\rho(k)\cos kx.$$

It follows by Prop. 9.2.6 (cf. (1.2.47), (9.3.8)) that

$$\chi_\rho^\sim(x) = 2 \sum_{k=1}^{\infty} \theta_\rho(k) \sin kx.$$

Hence $[\chi_\rho^\sim]_s^\wedge(k) = 2\theta_\rho(k)$, and therefore

Proposition 9.2.8. *Let* $\{\chi_\rho(x)\}$ *of* $I_\rho(f; x)$ *be generated via* (9.2.31) *by the* θ*-factor* $\{\theta_\rho(k)\}$, *and suppose that*

$$\cot\frac{x}{2} - 2 \sum_{k=1}^{\infty} \theta_\rho(k) \sin kx \geq 0$$

for all $0 < x \leq \pi$ *and* $\rho \in \mathbb{A}$. *If* f *satisfies the assumptions of Prop.* 9.2.7, *then*

$$I_\rho^\sim(f; x) - f^\sim(x) = (1 - \theta_\rho(1))f'(x) + o(1 - \theta_\rho(1)) \qquad (\rho \to \rho_0)$$

if and only if

(9.2.32) $$\lim_{\rho\to\rho_0} \frac{\theta_\rho(3) - 1}{\theta_\rho(1) - 1} = 3.$$

Condition (9.2.32) is to be compared with (1.5.27), i.e., with

$$\lim_{\rho\to\rho_0} \frac{\theta_\rho(2) - 1}{\theta_\rho(1) - 1} = 4,$$

which covers the asymptotic behaviour of the original integral $I_\rho(f; x)$.

Corollary 9.2.9. *Let* f *satisfy the assumptions of Prop.* 9.2.7. *Then*

$$n[\sigma_n^\sim(f; x) - f^\sim(x)] = f'(x) + o(1) \qquad (n \to \infty),$$
$$P_r^\sim(f; x) - f^\sim(x) = (1 - r)f'(x) + o(1 - r) \qquad (r \to 1-).$$

These asymptotic relations on the conjugates of the Fejér and Abel means follow immediately by Prop. 9.2.8.

Problems

1. (i) Prove Prop. 9.2.3.
   (ii) It follows for the Abel–Poisson kernel that

$$p_r(x) = \frac{1 - r^2}{1 - 2r\cos x + r^2}, \qquad p_r^\sim(x) = \frac{2r \sin x}{1 - 2r\cos x + r^2},$$
$$\Theta_r(x) = \frac{(1 - r)^2}{1 - 2r\cos x + r^2} \cot\frac{x}{2}.$$

   Show that $\{p_r(x)\}$ satisfies the assumptions of Prop. 9.2.1–9.2.3. (Hint: Set $\varphi(r) = \arccos(2r/(1 + r^2))$, $0 \leq \varphi(r) < (\pi/2)$; cf. HEWITT [1, p. 124], see also HARDY–ROGOSINSKI [1, p. 66], BARI [1II, p. 60], ZYGMUND [7I, p. 103])
   (iii) Prove Cor. 9.2.4.
2. (i) Let $f \in L_{2\pi}^1$ and the kernel $\{\chi_\rho(x)\}$ of the singular integral (1.1.3) be even and bounded. Let $\Theta_\rho(x)$ be defined by (9.2.4), and a function $\varphi(\rho)$ be given as in Prop. 9.2.2 such that there exists a majorant of $\chi_\rho^\sim$: $|\chi_\rho^\sim(x)| \leq (\chi_\rho^\sim)^*(x)$, monotonely increasing on $[0, \varphi(\rho)]$ satisfying (9.2.10), and of Θ_ρ: $|\Theta_\rho(x)| \leq \Theta_\rho^*(x)$, monotonely decreasing on $[\varphi(\rho), \pi]$ satisfying (9.2.11). Show that at every point x for which $\int_0^t |f(x + u) - f(x - u)|\, du = o(t)$, one has $\lim_{\rho\to\rho_0}[I_\rho^\sim(f; x) - f_{\varphi(\rho)}^\sim(x)] = 0$. In particular, (9.2.14) holds.

(ii) State and prove the counterpart of Cor. 9.2.4 for the singular integral of Fejér. (Hint: It follows by Prop. 9.1.6 that (cf. (1.2.49), (9.3.8))

$$F_n(x) = \frac{\sin^2 (n+1)(x/2)}{(n+1)\sin^2 (x/2)}, \qquad F_n^{\sim}(x) = \cot\frac{x}{2} - \frac{\sin (n+1)x}{2(n+1)\sin^2 (x/2)},$$
$$\Theta_n(x) = \frac{\sin (n+1)x}{2(n+1)\sin^2 (x/2)}.$$

Thus take $\varphi(n) = n^{-1}$, $(F_n^{\sim})^*(x) = n$, $\Theta_n^*(x) = (\pi^2/(2n+2))x^{-2}$; see also HARDY–ROGOSINSKI [1, p. 63], HEWITT [1, p. 83], BARI [1II, p. 58], ZYGMUND [7I, p. 91])

3. (i) Prove the following theorem of I. I. Privalov: Let $f \in \mathsf{Lip}(\mathsf{C}_{2\pi}; \alpha)$. If $0 < \alpha < 1$, then also $f^{\sim} \in \mathsf{Lip}(\mathsf{C}_{2\pi}; \alpha)$. If $\alpha = 1$, then $\omega(\mathsf{C}_{2\pi}; f^{\sim}; \delta) = O(\delta\,|\log \delta|)$, $\delta \to 0+$. (Hint: Use the method of proof of Prop. 8.2.9; see also ACHIESER [2, p. 210], BARI [1II, p. 99], ZYGMUND [5, p. 156])
   (ii) State and prove counterparts of Problems 8.2.5, 8.2.6 for periodic functions (see also ZYGMUND [2]).
4. (i) Let $f \in \mathsf{C}_{2\pi}$. Show that $f^{\sim}$ belongs to $\mathsf{C}_{2\pi}$ if and only if $f_\delta^{\sim}(x)$ converges uniformly in x as $\delta \to 0+$. (Hint: ZAMANSKY [3])
   (ii) Let $f \in \mathsf{L}_{2\pi}^1$. Show that $f^{\sim}$ belongs to $\mathsf{L}_{2\pi}^1$ if and only if $f_\delta^{\sim}$ tends to a limit in $\mathsf{L}_{2\pi}^1$-norm as $\delta \to 0+$. (Hint: cf. ZYGMUND [7I, p. 180])
5. State conditions upon f such that the asymptotic relations

$$n[f(x) - \sigma_n(f; x)] = (f^{\sim})'(x) + o(1) \qquad (n \to \infty),$$
$$f(x) - P_r(f; x) = (1-r)(f^{\sim})'(x) + o(1-r) \qquad (r \to 1-)$$

for the singular integral of Fejér and Abel–Poisson hold at a point x. (Compare also with Cor. 9.2.9 and Problems 1.5.12, 12.2.12)
6. Show that $\|(t_n^{\sim})'\|_{\mathsf{X}_{2\pi}} \le 2n\,\|t_n\|_{\mathsf{X}_{2\pi}}$ for every $t_n \in \mathsf{T}_n$. (Hint: ZYGMUND [7I, p. 118])
7. Let $f \in \mathsf{X}_{2\pi}$. Show that

$$\|\sigma_n^{\sim}(f; \circ) - f^{\sim}(\circ) + \frac{1}{\pi}\int_0^{1/n} \frac{f(\circ - u) - f(\circ + u)}{u}\,du\|_{\mathsf{X}_{2\pi}} = O(\omega(\mathsf{X}_{2\pi}; f; n^{-1})).$$

(Hint: Compare Problems 3.4.6(ii), 8.2.8, and the references cited there)

9.3 Fourier Transforms of Hilbert Transforms

9.3.1 Conjugate Fourier Series

Prop. 9.1.6 on Hilbert transforms of trigonometric polynomials suggests that one associates to each trigonometric series (cf. Sec. 1.2.1)

$$\sum_{k=-\infty}^{\infty} c_k\, e^{ikx} \tag{9.3.1}$$

its *conjugate* (or *allied*) *series*

$$\sum_{k=-\infty}^{\infty} \{-i \operatorname{sgn} k\}\, c_k\, e^{ikx}, \tag{9.3.2}$$

and that one considers the questions treated in connection with (9.3.1) also for (9.3.2).

In particular, if (9.3.1) is the Fourier series of some $f \in \mathsf{L}^1_{2\pi}$, i.e.

$$f(x) \sim \sum_{k=-\infty}^{\infty} f^\wedge(k)\, e^{ikx}, \tag{9.3.3}$$

then the question arises whether the *conjugate Fourier series*† of f, namely (cf. Def. 1.2.2)

$$\sum_{k=-\infty}^{\infty} \{-i \operatorname{sgn} k\} f^\wedge(k)\, e^{ikx}, \tag{9.3.4}$$

is again the Fourier series of some function in $\mathsf{L}^1_{2\pi}$. As may be expected by Prop. 9.1.6, and in fact shown by Prop. 9.2.5, we have

Proposition 9.3.1. *Let $f \in \mathsf{L}^1_{2\pi}$ be such that $f^\sim \in \mathsf{L}^1_{2\pi}$. Then the conjugate Fourier series (9.3.4) of f is the Fourier series of the Hilbert transform (conjugate function) $f^\sim$, i.e.*

$$f^\sim(x) \sim \sum_{k=-\infty}^{\infty} \{-i \operatorname{sgn} k\} f^\wedge(k)\, e^{ikx}. \tag{9.3.5}$$

We observe that (9.3.5) holds if f belongs to $\mathsf{L}^p_{2\pi}$ for some $1 < p < \infty$, for example.

Although the sign $\sim$ in (9.3.3) only denotes a definite correspondence and does not imply in the least the convergence of the series, we saw in Sec. 1.2, 1.4, and 4.1 that, if this series is summed up suitably, then it converges to the original function f. In view of (9.3.5), the informal approach of Sec. 1.2.6 and the results on the integral (9.2.1) suggest that one applies—whether $f^\sim$ belongs to $\mathsf{L}^1_{2\pi}$ or not—the same summation process to the conjugate Fourier series of f expecting that it produces convergence to the Hilbert transform $f^\sim$.

Hence (cf. (1.2.45)) for a θ-factor $\{\theta_\rho(k)\}$ and $f \in \mathsf{X}_{2\pi}$ we introduce the *θ-means*

$$U_\rho^\sim(f; x) = \sum_{k=-\infty}^{\infty} \theta_\rho(k)\, \{-i \operatorname{sgn} k\} f^\wedge(k)\, e^{ikx} \tag{9.3.6}$$

of the conjugate Fourier series of f. Since $\theta_\rho \in \mathsf{l}^1$ for each $\rho \in \mathbb{A}$, the series (9.3.6) converges absolutely and uniformly, and thus $U_\rho^\sim(f; x) \in \mathsf{C}_{2\pi}$ for each $\rho \in \mathbb{A}$ (cf. Sec. 1.2.6 and 8.3.2).

Moreover, the Fourier transform of the sequence $\{\theta_\rho(k)\}$, given by the uniformly convergent series (cf. (4.1.18))

$$\theta_\rho^\wedge(x) = 1 + 2 \sum_{k=1}^{\infty} \theta_\rho(k) \cos kx, \tag{9.3.7}$$

belongs to $\mathsf{C}_{2\pi}$ for each $\rho \in \mathbb{A}$, and thus by Theorem 9.1.3 possesses a Hilbert transform $[\theta_\rho^\wedge]^\sim \in \mathsf{L}^1_{2\pi}$ with (complex) Fourier coefficients $\{-i \operatorname{sgn} k\}\theta_\rho(k)$. Since the Fourier series of $[\theta_\rho^\wedge]^\sim$ again converges absolutely and uniformly, we have by Prop. 9.2.6

$$[\theta_\rho^\wedge]^\sim(x) = 2 \sum_{k=1}^{\infty} \theta_\rho(k) \sin kx \qquad (\rho \in \mathbb{A}), \tag{9.3.8}$$

† It would be more correct to speak of the conjugate *of the* Fourier series of f, since for $f \in \mathsf{L}^1_{2\pi}$ (9.3.4) need not be a Fourier series.

which justifies the terminology of (1.2.47) in speaking of $C_\rho^\sim$ as the conjugate function of C_ρ. (Relations (9.3.7) and (9.3.8) are the counterparts of (8.3.2) and (8.3.3).) In particular, (9.3.6) may be rewritten in the form

$$U_\rho^\sim(f; x) = \frac{1}{2\pi}\int_{-\pi}^{\pi} f(x-u)[\theta_\rho^\wedge]^\sim(u)\,du \tag{9.3.9}$$

which, in view of (1.1.3) and (9.2.1) on the one hand and of (4.1.17) on the other hand, also justifies the notation $U_\rho^\sim(f; x)$ of (9.3.6) and (1.2.46).

Proposition 9.3.2. *Let $f \in \mathsf{L}_{2\pi}^p$, $1 < p < \infty$, and let $\{\theta_\rho(k)\}$ be a θ-factor such that the corresponding kernel $\{\theta_\rho^\wedge(x)\}$ of (9.3.7) forms an approximate identity. Then*

$$\lim_{\rho\to\rho_0} \|U_\rho^\sim(f; \circ) - f^\sim(\circ)\|_p = 0. \tag{9.3.10}$$

If the kernel $\{\theta_\rho^\wedge(x)\}$ furthermore satisfies the assumptions of Prop. 1.4.4 *or* 1.4.6, *then*

$$\lim_{\rho\to\rho_0} U_\rho^\sim(f; x) = f^\sim(x) \quad \text{a.e.} \tag{9.3.11}$$

For the proof we only mention that in view of (9.3.9) and (9.1.26)

$$U_\rho^\sim(f; x) = \frac{1}{2\pi}\int_{-\pi}^{\pi} f^\sim(x-u)\theta_\rho^\wedge(u)\,du. \tag{9.3.12}$$

Proposition 9.3.3. *Let $f \in \mathsf{L}_{2\pi}^1$ and let $\{\theta_\rho(k)\}$ be a θ-factor such that the corresponding kernel $\{\theta_\rho^\wedge(x)\}$ satisfies the assumptions of Prop.* 9.2.2 *and* 9.2.3. *Then*

$$\lim_{\rho\to\rho_0} \|U_\rho^\sim(f; \circ) - f_{\varphi(\rho)}^\sim(\circ)\|_1 = 0, \tag{9.3.13}$$

$$\lim_{\rho\to\rho_0} U_\rho^\sim(f; x) = f^\sim(x) \quad \text{a.e.} \tag{9.3.14}$$

As an application of Problem 9.2.2 and (9.2.17) we have

Corollary 9.3.4. *Let $f \in \mathsf{X}_{2\pi}$. The conjugate Fourier series of f is Cesàro and Abel summable to $f^\sim(x)$ for almost all x, i.e.*

$$\lim_{n\to\infty} \sum_{k=-n}^{n} \left(1 - \frac{|k|}{n+1}\right)\{-i \operatorname{sgn} k\} f^\wedge(k)\, e^{ikx} = f^\sim(x) \quad \text{a.e.,} \tag{9.3.15}$$

$$\lim_{r\to 1-} \sum_{k=-\infty}^{\infty} r^{|k|}\{-i \operatorname{sgn} k\} f^\wedge(k)\, e^{ikx} = f^\sim(x) \quad \text{a.e.} \tag{9.3.16}$$

9.3.2 Fourier Transforms of Derivatives of Conjugate Functions, the Classes $(\mathsf{W}^\sim)_{\mathsf{X}_{2\pi}}^r$, $(\mathsf{V}^\sim)_{\mathsf{X}_{2\pi}}^r$

We now consider Fourier transforms of derivatives of Hilbert transforms (as considered for the original function $f \in \mathsf{X}_{2\pi}$ in Sec. 4.1.3 and 4.3.3). The definitions of the classes $\mathsf{W}_{\mathsf{X}_{2\pi}}^r$ and $\mathsf{V}_{\mathsf{X}_{2\pi}}^r$ in Sec. 1.1.2 and 4.3.3 suggest that one introduces for $r \in \mathbb{N}$

$$(\mathsf{W}^\sim)_{\mathsf{X}_{2\pi}}^r = \{f \in \mathsf{X}_{2\pi} \mid f^\sim \in \mathsf{W}_{\mathsf{X}_{2\pi}}^r\}, \tag{9.3.17}$$

$$(\mathsf{V}^\sim)_{\mathsf{X}_{2\pi}}^r = \{f \in \mathsf{X}_{2\pi} \mid f^\sim \in \mathsf{V}_{\mathsf{X}_{2\pi}}^r\}. \tag{9.3.18}$$

By Theorem 9.1.1, $f \in X_{2\pi}$ implies that $f^{\sim}$ exists almost everywhere. Therefore the above definitions are meaningful. They are particularly evident for $X_{2\pi} = L^p_{2\pi}$, $1 < p < \infty$, since then $f^{\sim}$ belongs to $L^p_{2\pi}$ by Theorem 9.1.3. But $f \in C_{2\pi}$ ($f \in L^1_{2\pi}$) does not necessarily imply that $f^{\sim}$ belongs to $C_{2\pi}$ ($L^1_{2\pi}$).

Theorem 9.3.5. *For $f \in X_{2\pi}$ the following assertions are equivalent:*

(i) $f \in (V^{\sim})^r_{X_{2\pi}}$, (ii) $f \in V[X_{2\pi}; \{-i \operatorname{sgn} k\}(ik)^r]$,

(iii) *there exists $g \in L^\infty_{2\pi}$ if $X_{2\pi} = C_{2\pi}$, $\mu \in BV_{2\pi}$ if $X_{2\pi} = L^1_{2\pi}$, $g \in L^p_{2\pi}$ if $X_{2\pi} = L^p_{2\pi}$, $1 < p < \infty$, and constants $\alpha_j \in \mathbb{C}$, $0 \le j \le r-1$, such that*

$$f^{\sim}(x) = \alpha_0 + \int_{-\pi}^{x} du_1\Big[\alpha_1 + \int_{-\pi}^{u_1} du_2\Big[\alpha_2 + \cdots$$

$$+ \int_{-\pi}^{u_{r-2}} du_{r-1}\Big[\alpha_{r-1} + \int_{-\pi}^{u_{r-1}} \begin{Bmatrix} g(u_r)\, du_r \\ d\mu(u_r) \\ g(u_r)\, du_r \end{Bmatrix}\Big]\cdots\Big]\Big] \quad \text{(a.e.)}.$$

Proof. Let $f \in X_{2\pi}$ be such that $f^{\sim} \in V^r_{X_{2\pi}}$. Then the assumption in particular implies $f^{\sim} \in X_{2\pi}$, and thus $f^{\sim} \in V[X_{2\pi}; (ik)^r]$ by Theorem 4.3.13. Since $f^{\sim} \in X_{2\pi}$, we have $[f^{\sim}]^{\wedge}(k) = \{-i \operatorname{sgn} k\} f^{\wedge}(k)$ by Prop. 9.1.8 and 9.2.5, and (ii) follows.

On the other hand, let $f \in V[X_{2\pi}; \{-i \operatorname{sgn} k\}(ik)^r]$. In case $X_{2\pi} = L^p_{2\pi}$, $1 < p < \infty$, it follows by Theorem 9.1.3 that $f^{\sim} \in L^p_{2\pi}$, and by Prop. 9.1.8 that $\{-i \operatorname{sgn} k\} f^{\wedge}(k) = [f^{\sim}]^{\wedge}(k)$. Hence the hypothesis means that for $f^{\sim} \in L^p_{2\pi}$ there exists $g \in L^p_{2\pi}$ such that $(ik)^r [f^{\sim}]^{\wedge}(k) = g^{\wedge}(k)$. This establishes assertions (iii) and (i) by Theorem 4.3.13.

For $X_{2\pi} = C_{2\pi}$, $L^1_{2\pi}$ the main trouble is that we have to show $f^{\sim} \in X_{2\pi}$.

Let $X_{2\pi} = C_{2\pi}$. Using the function $G_r(x)$ of (4.1.22), assumption (ii) implies $\{-i \operatorname{sgn} k\} f^{\wedge}(k) = G_r^{\wedge}(k)$ for every $k \neq 0$. Thus the conjugate Fourier series of $f \in C_{2\pi}$ is the Fourier series of a continuous function, namely of $G_r(x) - G_r^{\wedge}(0)$. In view of Prop. 9.2.6 this implies $f^{\sim} \in C_{2\pi}$. Now, $\{-i \operatorname{sgn} k\} f^{\wedge}(k) = [f^{\sim}]^{\wedge}(k)$ by e.g. Prop. 9.1.8. Hence the hypothesis (ii) states that there exists $g \in L^\infty_{2\pi}$ such that $(ik)^r [f^{\sim}]^{\wedge}(k) = g^{\wedge}(k)$, and since $f^{\sim} \in C_{2\pi}$, assertions (iii) and (i) follow for $C_{2\pi}$-space by Theorem 4.3.13.

Let $f \in L^1_{2\pi}$ and $\mu \in BV_{2\pi}$ be such that $\{-i \operatorname{sgn} k\}(ik)^r f^{\wedge}(k) = \mu^{\vee}(k)$. Then $\sum_{k=-\infty}^{\infty} |f^{\wedge}(k)|^2 < \infty$. Therefore, $f \in L^2_{2\pi}$ by Theorem 4.2.4 and $f^{\sim} \in L^2_{2\pi}$ by Theorem 9.1.3, which in particular shows $f^{\sim} \in L^1_{2\pi}$. Moreover, $[f^{\sim}]^{\wedge}(k) = \{-i \operatorname{sgn} k\} f^{\wedge}(k)$ by (9.1.16). Now the assumption states that for $f^{\sim} \in L^1_{2\pi}$ there exists $\mu \in BV_{2\pi}$ such that $(ik)^r [f^{\sim}]^{\wedge}(k) = \mu^{\vee}(k)$ which gives (iii) and (i) by Theorem 4.3.13. This completes the proof.

9.3.3 Norm-Convergence of Fourier Series

Let us now give applications of the theory of Hilbert transforms to the convergence of Fourier series. As a generalization of (4.2.7) we have

Theorem 9.3.6. *Let $f \in L^p_{2\pi}$, $1 < p < \infty$. Then*

(9.3.19)
$$\lim_{n \to \infty} \Big\| f(\circ) - \sum_{k=-n}^{n} f^{\wedge}(k)\, e^{ik\circ} \Big\|_p = 0.$$

Proof. Let us recall the modified partial sums of a Fourier series

$$S_n^*(f; x) = \frac{1}{2\pi} \int_{-\pi}^{\pi} f(x-u) \cot \frac{u}{2} \sin nu \, du. \tag{9.3.20}$$

Since $\lim_{n\to\infty} \|S_n(f; \circ) - S_n^*(f; \circ)\|_{C_{2\pi}} = 0$ by Problem 1.2.17, we only need to establish (9.3.19) with $S_n(f; x)$ replaced by $S_n^*(f; x)$. Since $\{\cot(u/2) \sin nu\}$ is a continuous function on $[-\pi, \pi]$, the integral in (9.3.20) exists everywhere as an ordinary Lebesgue integral. Hence by Theorem 9.1.1

$$S_n^*(f; x) = \sin nx \, [f(\circ) \cos n\circ]^\sim(x) - \cos nx \, [f(\circ) \sin n\circ]^\sim(x) \quad \text{a.e.,}$$

and therefore by (9.1.9)

$$\|S_n^*(f; \circ)\|_p \leq 2M_p \|f\|_p \qquad (n \in \mathbb{N}). \tag{9.3.21}$$

Let $t_m \in \mathsf{T}_m$ be such that $\|f - t_m\|_p \leq \varepsilon$. Then

$$\|S_n^*(f; \circ) - f(\circ)\|_p \leq \|S_n^*(t_m; \circ) - t_m(\circ)\|_p + (2M_p + 1)\,\varepsilon.$$

But since

$$S_n^*(f; x) = S_n(f; x) - (1/2)[f^\wedge(n)\, e^{inx} + f^\wedge(-n)\, e^{-inx}],$$

we have $S_n^*(t_m; x) = t_m(x)$ for $n > m$, and the theorem is established.

Proposition 9.3.7. *Let $f \in \mathsf{L}_{2\pi}^p$, $1 < p < \infty$, and $g \in \mathsf{L}_{2\pi}^{p'}$. Then*

$$\frac{1}{2\pi} \int_{-\pi}^{\pi} f(u)\overline{g(u)}\, du = \lim_{n\to\infty} \sum_{k=-n}^{n} f^\wedge(k)\overline{g^\wedge(k)}. \tag{9.3.22}$$

Since

$$\frac{1}{2\pi} \int_{-\pi}^{\pi} [f(u) - S_n(f; u)][\overline{g(u) - S_n(g; u)}]\, du \tag{9.3.23}$$
$$= \frac{1}{2\pi} \int_{-\pi}^{\pi} f(u)\overline{g(u)}\, du - \sum_{k=-n}^{n} f^\wedge(k)\overline{g^\wedge(k)},$$

the proof of (9.3.22) immediately follows from Hölder's inequality and (9.3.19). For $p = 2$ see (4.2.5).

At this stage, let us recall the result of Theorem 2.4.3 which states that $\|S_n(f; \circ) - f(\circ)\|_{X_{2\pi}} = O(\log n \, E_n(X_{2\pi}; f))$ for every $f \in X_{2\pi}$. But if we restrict $X_{2\pi}$ to $\mathsf{L}_{2\pi}^p$, $1 < p < \infty$, we furthermore have

Proposition 9.3.8. *If $f \in \mathsf{L}_{2\pi}^p$, $1 < p < \infty$, then there exists a constant A such that*

$$\|S_n(f; \circ) - f(\circ)\|_p \leq A\, \|f - t_n\|_p$$

for every $t_n \in \mathsf{T}_n$. In particular,

$$\|S_n(f; \circ) - f(\circ)\|_p \leq A E_n(\mathsf{L}_{2\pi}^p; f). \tag{9.3.24}$$

The proof follows by Problems 1.2.14(iii), 9.3.6 since

$$\|S_n(f; \circ) - f(\circ)\|_p \leq \|S_n(f - t_n; \circ)\|_p + \|f - t_n\|_p.$$

Thus in $\mathsf{L}_{2\pi}^p$-space, $1 < p < \infty$, the nth partial sum of the Fourier series of f furnishes an example of a trigonometrical polynomial of degree n at most which actually approximates f with the order of best approximation. In particular,

Corollary 9.3.9. *Let $f \in \mathsf{L}_{2\pi}^p$, $1 < p < \infty$. Then for $0 < \alpha < 2$*

$$\|S_n(f; \circ) - f(\circ)\|_p = O(n^{-\alpha}) \Leftrightarrow f \in \mathrm{Lip}^*(\mathsf{L}_{2\pi}^p; \alpha).$$

Proof. If $\|S_n(f;\circ) - f(\circ)\|_p = O(n^{-\alpha})$, $n \to \infty$, then *a fortiori* $E_n(\mathsf{L}^p_{2\pi}; f) = O(n^{-\alpha})$, and thus $f \in \mathsf{Lip}^*(\mathsf{L}^p_{2\pi}; \alpha)$ by Theorem 2.3.5. Conversely, if $f \in \mathsf{Lip}^*(\mathsf{L}^p_{2\pi}; \alpha)$, then $E_n(\mathsf{L}^p_{2\pi}; f) = O(n^{-\alpha})$ by Cor. 2.2.2, and the assertion follows by Prop. 9.3.8.

Cor. 9.3.9 is a far-reaching extension of Cor. 2.4.6 in case $\mathsf{X}_{2\pi} = \mathsf{L}^p_{2\pi}$, $1 < p < \infty$. However, we recall that for $\mathsf{X}_{2\pi} = \mathsf{L}^1_{2\pi}$ (or $\mathsf{C}_{2\pi}$) there are functions in $\mathsf{Lip}(\mathsf{X}_{2\pi}; 1)$ for which $\|S_n(f;\circ) - f(\circ)\|_{\mathsf{X}_{2\pi}} \neq o(n^{-1}\log n)$ (see Problem 2.4.12(iii)).

As an application to summation processes of Fourier series we note the following result on the Fejér means $\sigma_n(f; x)$.

Corollary 9.3.10. *Let* $f \in \mathsf{L}^p_{2\pi}$, $1 < p < \infty$. *If* $f \in \mathsf{Lip}(\mathsf{L}^p_{2\pi}; 1)$, *then* $\|\sigma_n(f;\circ) - f(\circ)\|_p = O(n^{-1})$, $n \to \infty$.

Proof. We need the fact that if $f \in \mathsf{Lip}(\mathsf{L}^p_{2\pi}; 1)$, $1 < p < \infty$, then f is equivalent to the integral of a function $g \in \mathsf{L}^p_{2\pi}$ (for a proof we refer to Sec. 10.2). This implies $ikf^\wedge(k) = g^\wedge(k)$ by Prop. 4.1.8 and $|k|\, f^\wedge(k) = [g^\sim]^\wedge(k)$ by Prop. 9.1.8. Therefore by Problem 9.3.6

$$\|\sigma_n(f;\circ) - S_n(f;\circ)\|_p = \Big\| \frac{1}{n+1} \sum_{k=-n}^{n} |k|\, f^\wedge(k)\, e^{ik\circ} \Big\|_p$$
$$= \frac{1}{n+1} \|S_n(g^\sim;\circ)\|_p \le \frac{A_p \|g^\sim\|_p}{n+1}.$$

The assertion now follows by Cor. 9.3.9 and the inequality

$$\|\sigma_n(f;\circ) - f(\circ)\|_p \le \|\sigma_n(f;\circ) - S_n(f;\circ)\|_p + \|S_n(f;\circ) - f(\circ)\|_p.$$

A combination of the result of Cor. 9.3.10 with that of Theorem 2.4.7(ii) gives for $1 < p < \infty$

$$f \in \mathsf{Lip}(\mathsf{L}^p_{2\pi}; 1) \Rightarrow \|\sigma_n(f;\circ) - f(\circ)\|_p = O(n^{-1}) \Rightarrow f \in \mathsf{Lip}^*(\mathsf{L}^p_{2\pi}; 1).$$

Thus the problem of characterizing the subclass of functions $f \in \mathsf{X}_{2\pi}$ for which $\|\sigma_n(f;\circ) - f(\circ)\|_{\mathsf{X}_{2\pi}} = O(n^{-1})$ is still unsolved, even in $\mathsf{L}^p_{2\pi}$, $1 < p < \infty$. For the solution the reader is referred to Chapter 12.

Problems

1. Prove Cor. 9.3.4.
2. State and prove counterparts of Problem 8.3.2 for periodic functions.
3. Show that for $f \in \mathsf{X}_{2\pi}$ the following assertions are equivalent:
   (i) $f \in (\mathsf{W}^\sim)^r_{\mathsf{X}_{2\pi}}$, (ii) $f \in \mathsf{W}[\mathsf{X}_{2\pi}; \{-i \operatorname{sgn} k\}(ik)^r]$,
   (iii) there exist $g \in \mathsf{X}_{2\pi}$ and constants $\alpha_j \in \mathbb{C}$, $0 \le j \le r-1$, such that
   $$f^\sim(x) = \alpha_0 + \int_{-\pi}^{x} du_1 \Big[\alpha_1 + \int_{-\pi}^{u_1} du_2 \Big[\alpha_2 + \dots$$
   $$+ \int_{-\pi}^{u_{r-2}} du_{r-1} \Big[\alpha_{r-1} + \int_{-\pi}^{u_{r-1}} g(u_r)\, du_r\Big] \dots\Big]\Big] \quad \text{(a.e.)}.$$
4. State and prove corresponding results for the classes $\mathsf{W}[\mathsf{X}_{2\pi}; |k|^r]$, $\mathsf{V}[\mathsf{X}_{2\pi}; |k|^r]$, $r \in \mathbb{N}$. (Hint: Distinguish the cases r even, odd)
5. State and prove counterparts of Prop. 8.3.7 and Problem 8.3.5 for periodic functions.
6. Let $f \in \mathsf{L}^p_{2\pi}$, $1 < p < \infty$. Show that there is a constant A_p such that for the nth partial sum of the Fourier series of f: $\|S_n(f;\circ)\|_p \le A_p \|f\|_p$.
7. Let $F(x)$ be 2π-periodic and the indefinite integral of a function $f \in \mathsf{L}^p$, $1 \le p < \infty$. Show that $\|S_n(F;\circ) - F(\circ)\|_p \le An^{-1} \|S_n(f;\circ) - f(\circ)\|_p$ if $p > 1$, and in case $p = 1$ $\|S_n(F;\circ) - F(\circ)\|_p \le A(1 + \log n)n^{-1} \|S_n(f;\circ) - f(\circ)\|_p$ (Hint: QUADE [1, p. 542])

9.4 Notes and Remarks

Most of the notes and remarks to Chapter 8 also apply *mutatis mutandis* to Hilbert transforms of periodic functions. The main general references to the subject are ZYGMUND [5, 7], HARDY–ROGOSINSKI [1], BARI [1], HEWITT [1], WEISS [2], EDWARDS [1], KATZNELSON [1]. The present choice of material has essentially been dictated by the condition that the topics in question once again exhibit the common structure of harmonic analysis on the line and circle group, and that they lead readily to our main result, namely Theorem 9.3.5.

Sec. 9.1. For the reduction of the basic properties of Hilbert transforms of periodic functions from those of Chapter 8 we have followed CALDERÓN–ZYGMUND [2], a paper mainly interested in extensions to the several variable case. For a given 'kernel' $k(x)$ on the line group one associates a periodic 'kernel' $k^*(x)$ defined through

$$k^*(x) = k(x) + \sum_{j=-\infty}^{\infty}{}' \{k(x + 2\pi j) - k(2\pi j)\}. \tag{9.4.1}$$

The procedure assumes that k possesses certain properties that are transferable to k^*. Thus, for example, (9.0.4) is the case $k(x) = (1/x)$ of (9.4.1). In Sec. 11.5 we shall discuss the situation when $k(x)$ is equal to $|x|^{\alpha-1}$, $\{\text{sgn } x\}|x|^{\alpha-1}$, and $\kappa_{[0,\infty)}(x)x^{\alpha-1}$, respectively. In case $k \in \mathsf{L}^1$ we refer to the parallel treatment in Sec. 3.1.2.

Other proofs for the existence of the conjugate function may be given by complex methods (see e.g. ZYGMUND [7, Chapter VII], KATZNELSON [1, p. 62 ff]) or by a separate application of the theorem of Marcinkiewicz (see e.g. EDWARDS [1II, p. 168 ff]). According to Problem 9.1.3, the Hilbert transform is *not* a bounded linear transformation of $\mathsf{X}_{2\pi}$ into $\mathsf{X}_{2\pi}$ if $\mathsf{X}_{2\pi}$ is $\mathsf{C}_{2\pi}$ or $\mathsf{L}^1_{2\pi}$. However, one can show that $f \in \mathsf{L}^1_{2\pi}$ implies $|f^\sim(x)|^\alpha \in \mathsf{L}^1_{2\pi}$ for every $0 < \alpha < 1$. Moreover, if $|f(x)| \log^+ |f(x)| \in \mathsf{L}^1_{2\pi}$, then $f^\sim \in \mathsf{L}^1_{2\pi}$; see ZYGMUND [5, p. 150] and the literature cited there (for further extensions cf. also O'NEIL [1]).

Hilbert formulae in this or another form are contained in all texts cited.

Sec. 9.2. The same comments as to Sec. 8.2.2, 8.2.3 apply here. See also the references given in the Problems. Prop. 9.2.7 is due to TABERSKI [3] who considers asymptotic relations (of higher order) for summation processes of Fourier series. He also discusses the asymptotic behaviour of the quantity sup $\|I_\rho^\sim(f; \circ) - f^\sim(\circ)\|_{\mathsf{C}_{2\pi}}$, where the supremum is to be taken over all $f \in \mathsf{Lip}(\mathsf{C}_{2\pi}; \alpha)$, $0 < \alpha \le 1$, with Lipschitz constant $M \le 1$. For the application to the Abel–Poisson integral we refer to the earlier result of SZ.-NAGY [4] (see also NATANSON [3], TIMAN [1]). The asymptotic behaviour of the singular integral of Fejér is considered by ZAMANSKY [1] (cf. also SUNOUCHI [1II]).

Problem 9.2.6 establishes the *Bernstein inequality* for the conjugate of a trigonometric polynomial and plays an important rôle in the theory of best approximation. Indeed, given a certain behaviour of the best approximation $E_n(\mathsf{X}_{2\pi}; f)$ of f as $n \to \infty$, we have deduced smoothness properties of f in Sec. 2.3. It is also possible to deduce those for the conjugate $f^\sim$. The interested reader is referred to BARI–STEČKIN [1], STEČKIN [2], BUTZER–NESSEL [1], and in particular to TIMAN [2, p. 389 ff] and the literature cited there. For corresponding Jackson-type results see TIMAN [2, p. 315 ff].

Sec. 9.3. Conjugate Fourier series are standard topics of all texts on trigonometric series. In particular, Cor. 9.3.4 is well-known (cf. the references given to Problems 9.2.1, 9.2.2). Concerning (C, α) summation of conjugate Fourier series, see ZYGMUND [1]. The informal approach as given in the introduction to this Part may find a rigorous interpretation in the setting of distribution theory as, for example, in the modern treatment of EDWARDS [1II, p. 86 ff]. For the characterizations of the classes $(\mathsf{W}^\sim)^r_{\mathsf{X}_{2\pi}}$, $(\mathsf{V}^\sim)^r_{\mathsf{X}_{2\pi}}$ see also GÖRLICH [1], BUTZER–GÖRLICH [1]. These results pave the way for equivalent descriptions of the classes $\mathsf{W}[\mathsf{X}_{2\pi}; |k|^r]$, $\mathsf{V}[\mathsf{X}_{2\pi}; |k|^r]$, $r \in \mathbb{N}$, in terms of the original functions or their conjugates. These classes play an intrinsic rôle in our later investigations on saturation theory.

In this section, as throughout the volume, the order of performing the conjugate operation and differentiation is always fixed. For conditions which ensure interchange (and its correct interpretation) compare Prop. 8.3.7, Problem 8.3.5 (see also Butzer–Görlich [1]).

Theorem 9.3.6 is the periodic version of the famous result of Hille–Tamarkin [1], compare also Zygmund [7I, p. 266], Hewitt [1, p. 140]. For the Parseval formula (9.3.22) see e.g. Zygmund [7I, p. 267], Hewitt [1, p. 142], for Prop. 9.3.8 and Cor. 9.3.9, 9.3.10 Quade [1] and the literature cited there.

Part IV

Characterization of Certain Function Classes

The function classes that are to be characterized in this Part are those that arise in connection with saturation theory—a central theme of the later chapters. These classes are actually *Favard* (or saturation) classes associated with convolution integrals, thus precisely those functions which furnish the best possible (optimal) order of approximation by a given convolution integral.

The function classes are considered both for functions defined on the circle group (the 2π-periodic case) as well as on the line group. But the classes are also of independent interest. Indeed, our study (in the integral case) may be motivated as follows. In Chapter 2 we characterized the class $W^{r;\alpha}_{C_{2\pi}}$, $r \in \mathbb{N}$, $0 < \alpha < 1$, in terms of the best approximation $E_n(C_{2\pi}; f)$. In particular, $f \in W^{r;\alpha}_{C_{2\pi}}$ if and only if $E_n(C_{2\pi}; f) = O(n^{-r-\alpha})$ for $0 < \alpha < 1$, whereas for $\alpha = 1$ no simple characterization in terms of $E_n(C_{2\pi}; f)$ was possible. On the other hand, $f \in W^{r;\alpha}_{C_{2\pi}}$ implies $\|f^{(r)}(\circ + h) - f^{(r)}(\circ)\|_{C_{2\pi}} = o(h)$ in case $\alpha > 1$, and thus f is a constant. From this point of view $\alpha = 1$ may be regarded as an extremal index for the $W^{r;\alpha}_{C_{2\pi}}$-class. We recall that to bypass the difficulties in the characterization of the function class $\{f \in C_{2\pi} \mid E_n(C_{2\pi}; f) = O(n^{-r-1})\}$ we introduced the Zygmund class $^*W^{r;\alpha}_{C_{2\pi}}$ (but then $\alpha = 2$ is an extremal index). Here, on the other hand, we shall characterize the original class $W^{r;\alpha}_{C_{2\pi}}$ for $\alpha = 1$ not in terms of the order of best approximation but in terms of other properties upon f. In particular, $f \in W^{r;1}_{C_{2\pi}}$ if and only if the Riemann derivative (cf. Sec. 5.1.4) of order $(r + 1)$ of f exists in $C_{2\pi}$-norm.

In Sec. 2.4–2.5 we considered equivalence theorems for particular convolution integrals in the case of 'nonoptimal' approximation. Thus for the integral of Fejér $\|\sigma_n(f; \circ) - f(\circ)\|_{X_{2\pi}} = O(n^{-\alpha})$ for $0 < \alpha < 1$ if and only if $f \in \mathrm{Lip}\,(X_{2\pi}; \alpha)$. In this equivalence theorem the order of approximation is 'nonoptimal'. In case it is 'optimal' —we recall that $\|\sigma_n(f; \circ) - f(\circ)\|_{X_{2\pi}} = o(n^{-1})$ implies f is constant—we speak of a *saturation theorem*, and theorems of this type are one object of Part V. For the given example, the optimal order is $O(n^{-1})$ and this order is reached precisely by those functions which make up the *saturation class*. Such classes are to be studied here. Saturation theorems may also be viewed as extremal cases of the corresponding results on nonoptimal approximation in the sense that $\alpha = 1$ is the extremal index for Fejér's integral.

The classes are also to be distinguished according to whether they are associated with small-o or large-O approximation theorems. In the case of periodic functions this gives rise to the classes

$$\mathsf{W}[\mathsf{X}_{2\pi}; \psi(k)] = \{f \in \mathsf{X}_{2\pi} \mid \psi(k) f^\wedge(k) = g^\wedge(k), g \in \mathsf{X}_{2\pi}\}$$

and

$$\mathsf{V}[\mathsf{X}_{2\pi}; \psi(k)] = \begin{cases} \{f \in \mathsf{C}_{2\pi} \mid \psi(k) f^\wedge(k) = g^\wedge(k), g \in \mathsf{L}^\infty_{2\pi}\} \\ \{f \in \mathsf{L}^1_{2\pi} \mid \psi(k) f^\wedge(k) = \mu^\vee(k), \mu \in \mathsf{BV}_{2\pi}\} \\ \{f \in \mathsf{L}^p_{2\pi} \mid \psi(k) f^\wedge(k) = g^\wedge(k), g \in \mathsf{L}^p_{2\pi}\} \quad (1 < p < \infty), \end{cases}$$

respectively, where ψ is a given function on $\mathbb{Z}$. (Actually, we would like these classes to be defined in terms of the functions f directly, instead of through their Fourier transforms.) The instance that ψ is some power of k is the most interesting one. Chapter 10 is concerned with the case that the power α is integral, Chapter 11 with fractional α.

10

Characterization in the Integral Case

10.0 Introduction

The classes $W[X_{2\pi}; (ik)^r]$ or $W^r_{X_{2\pi}}$ (cf. Theorem 4.1.10) are connected with various generalizations of the classical rth derivative. In Sec. 5.1.4 we have shown that the concept of a Riemann derivative of order r of a function f at some point x_0 is more general than that of the Peano derivative (of f at x_0), which in turn is more general than that of the Taylor derivative and this again being more general than the (ordinary) derivative of order r of f at x_0. We shall first prove the (apparently surprising) fact that all four definitions of a derivative of order r of f are equivalent to another provided limits are taken in the strong topology and not in the pointwise sense. In particular, a function $f \in X_{2\pi}$ will belong to $W^r_{X_{2\pi}}$ (we recall its pointwise definition in Sec. 1.1.2) if and only if f has an rth Peano derivative in $X_{2\pi}$-norm. We shall also consider ordinary and generalized derivatives in the weak sense, thus in the weak topology. We shall, for example, prove the (seemingly perplexing) fact that the weak Riemann derivative of order r of f exists if and only if its strong (ordinary) derivative of order r exists. Then both are equal.

The classes $V[X_{2\pi}; (ik)^r]$ or $V^r_{X_{2\pi}}$ (cf. Theorem 4.3.13) are connected with generalizations of some significant theorems of G. H. Hardy and J. E. Littlewood. Thus $f \in X_{2\pi}$ belongs to $V^r_{X_{2\pi}}$ if and only if for the rth Riemann difference $\|\Delta^r_h f\|_{X_{2\pi}} = O(|h|^r)$ $(h \to 0)$ and in turn if and only if the quotient $h^{-r}\Delta^r_h f$ is weak* convergent for $h \to 0$.

Results of the type just discussed are also considered for the conjugate function $f^\sim$ of $f \in X_{2\pi}$. This enables one to give characterizations of the classes $V[X_{2\pi}; |k|^r]$, $r \in \mathbb{N}$.

The W- and V-classes are themselves connected through the important concept of relative completion. Indeed, the completion of $W[X_{2\pi}; \psi(k)]$ relative to $X_{2\pi}$ is equivalent to $V[X_{2\pi}; \psi(k)]$ for an arbitrary function ψ on $\mathbb{Z}$.

The counterparts of these results are also discussed for functions defined on the real line $\mathbb{R}$. At this stage let us just mention that we modify the corresponding proofs at a number of points. In the case of the circle group we use Fourier transform methods together with weak* compactness arguments for $C_{2\pi}$ as well as $L^p_{2\pi}$ for all $1 \le p < \infty$.

For the line group we employ Fourier transform methods plus the representation theorems of Sec. 6.1.2 for functions in L^p, p restricted to $1 \le p \le 2$. However, for $p > 2$ we must turn to dual arguments.

Sec. 10.1 is devoted to a discussion on generalized derivatives as well as to characterizations of the classes $W^r_{X_{2\pi}}$. Sec. 10.2 and 10.3 are reserved for characterizations of the classes $V^r_{X_{2\pi}}$ and $(V^\sim)^r_{X_{2\pi}}$, respectively. In Sec. 10.4 fundamental properties associated with the concept of relative completion are given. Sec. 10.5 is concerned with the case of functions defined in L^p, $1 \le p \le 2$, and Sec. 10.6 with the case $p > 2$.

10.1 Generalized Derivatives, Characterization of the Classes $W^r_{X_{2\pi}}$

In this section we introduce the (finite) Fourier transform method in its simplest form as applied to the Riemann derivative, for example. We recall Sec. 5.1.4 for the pointwise definition of this derivative as well as for its basic properties. From Problems 5.1.11, 5.1.12 it follows that the existence of the ordinary derivative of order r of some function at some point x_0 implies the existence of the Riemann derivative of order r at x_0. But the converse assertion is not necessarily true.

But in replacing the pointwise limit in (5.1.42) by the limit in norm of the spaces $X_{2\pi}$ we obtain a type of generalized derivative which in some sense has lost some of those special properties which distinguishes it from the ordinary derivative. We shall prove for example that the existence of the rth Riemann derivative in the norm of $C_{2\pi}$ is equivalent to the existence of the continuous rth ordinary derivative.

10.1.1 Riemann Derivatives in $X_{2\pi}$-Norm

For a function f defined in a neighbourhood of a point x_0 the rth Riemann derivative at x_0 with respect to the rth one-sided difference

(10.1.1)
$$\Delta_h^r f(x) = \sum_{k=0}^{r} (-1)^{r-k}\binom{r}{k} f(x+kh) \qquad (h \in \mathbb{R})$$

is defined by

(10.1.2)
$$f^{[r]}(x_0) = \lim_{h\to 0} \frac{1}{h^r}\,\Delta_h^r f(x_0),$$

in case this limit exists.

Throughout this chapter Riemann derivatives are defined via the difference (10.1.1) unless otherwise stated. We shall also simply refer to $\Delta_h^r f$ as the rth difference unless we explicitly distinguish between central, forward ($h > 0$) or backward ($h < 0$) differences (cf. (5.1.41)). We prefer the definition via the difference (10.1.1) since for $r = 1$ it reduces to the first ordinary derivative and may therefore be considered as a more genuine generalization of the concept of an rth derivative than (5.1.42).

Definition 10.1.1. *If for $f \in C_{2\pi}$ there exists $D_s^{[r]} f \in C_{2\pi}$ such that*

(10.1.3)
$$\lim_{h\to 0} \left\| \frac{1}{h^r}\,\Delta_h^r f - D_s^{[r]} f \right\|_{C_{2\pi}} = 0,$$

then $D_s^{[r]}f$ is called the **rth uniform Riemann derivative** *of f. Similarly, if f and $D_s^{[r]}f$ are functions belonging to $L_{2\pi}^p$, $1 \le p < \infty$, and if (10.1.3) holds with the $C_{2\pi}$-norm replaced by the $L_{2\pi}^p$-norm, then we speak of $D_s^{[r]}f$ as the* **rth Riemann derivative of f in the mean of order p**. *Both derivatives will also be called* **rth Riemann derivatives in $X_{2\pi}$-norm** *or* **strong Riemann derivatives**† *of order* **r**.

Obviously, the existence of the pointwise Riemann derivative at every x_0 does not necessarily imply that of the strong Riemann derivative. But the existence of the rth uniform Riemann derivative $D_s^{[r]}f$ of a function $f \in C_{2\pi}$ also implies that of the pointwise limit (10.1.2) at every x_0, whereas this is not so for the Riemann derivative in the mean of order p. Although this reveals that there is a difference between the uniform Riemann derivative and the Riemann derivative in the mean of order p, both do have properties in common which will be illustrated by the following results.

As an introductory example of the problems which may be treated by the (finite) Fourier transform method we consider the relationship between the continuous second ordinary derivative of $f \in C_{2\pi}$ and the second uniform Riemann derivative of f. Our first result states that these two concepts are equivalent.

Proposition 10.1.2. *The function $f \in C_{2\pi}$ has a continuous second ordinary derivative if and only if it has a second uniform Riemann derivative. In this event, $(D_s^{[2]}f)(x) = f^{(2)}(x)$ for all x.*

Proof. (i) Suppose $f^{(2)}$ exists and belongs to $C_{2\pi}$. Then by the mean value theorem

$$f(x+2h) - 2f(x+h) + f(x) = h^2 f^{(2)}(x+2\vartheta h) \qquad (0 < \vartheta < 1),$$

and therefore by the uniform continuity of $f^{(2)}$

$$\left\| \frac{1}{h^2} \Delta_h^2 f - f^{(2)} \right\|_{C_{2\pi}} = \|f^{(2)}(\circ + 2\vartheta h) - f^{(2)}(\circ)\|_{C_{2\pi}} = o(1) \qquad (h \to 0).$$

Thus $D_s^{[2]}f$ exists in $C_{2\pi}$-norm and $(D_s^{[2]}f)(x) = f^{(2)}(x)$ for all x.

(ii) Supposing $D_s^{[2]}f$ exists in $C_{2\pi}$-norm, we have

$$\left[\frac{f(\circ + 2h) - 2f(\circ + h) + f(\circ)}{h^2} - (D_s^{[2]}f)(\circ)\right]^{\wedge}(k) = \left(\frac{e^{ihk} - 1}{h}\right)^2 f^{\wedge}(k) - [D_s^{[2]}f]^{\wedge}(k)$$

in view of Prop. 4.1.1(i), and by (4.1.2)

$$\left|\left(\frac{e^{ihk} - 1}{h}\right)^2 f^{\wedge}(k) - [D_s^{[2]}f]^{\wedge}(k)\right| \le \left\| \frac{1}{h^2} \Delta_h^2 f - D_s^{[2]}f \right\|_{C_{2\pi}} \qquad (k \in \mathbb{Z}).$$

Since the second uniform Riemann derivative exists it follows that

$$\lim_{h \to 0} \left(\frac{e^{ihk} - 1}{h}\right)^2 f^{\wedge}(k) = [D_s^{[2]}f]^{\wedge}(k),$$

and so $(ik)^2 f^{\wedge}(k) = [D_s^{[2]}f]^{\wedge}(k)$, $k \in \mathbb{Z}$. Now an application of Prop. 4.1.9 for $r = 2$ shows that $f^{(2)}(x)$ exists, is continuous, and $f^{(2)}(x) = (D_s^{[2]}f)(x)$ for all x.

† The subscript s in $D_s^{[r]}f$ refers to convergence in the *strong* topology and is to be distinguished from w in $D_w^{[r]}f$ indicating convergence in the *weak* topology (cf. Def. 10.1.13). On the contrary, $f^{[r]}$ is always defined through pointwise convergence (cf. Sec. 1.1.2).

The proof of part (ii) of the above proposition exhibits the essential technique of the Fourier transform method. The problem given by $\lim_{h\to 0} h^{-2}\Delta_h^2 f(x) = (D_s^{[2]}f)(x)$ is transformed by the finite Fourier transform into the relation $\lim_{h\to 0} h^{-2}(e^{ihk}-1)^2 f^\wedge(k) = [D_s^{[2]}f]^\wedge(k)$, involving Fourier transforms. The solution of the latter relation is $(ik)^2 f^\wedge(k) = [D_s^{[2]}f]^\wedge(k)$, and Prop. 4.1.9 is applied as an inversion formula to deduce the solution of the original problem.

Considering the case for integers $r > 2$ we have

Theorem 10.1.3. *The function $f \in C_{2\pi}$ has a continuous rth ordinary derivative if and only if it has an rth uniform Riemann derivative. In this case, all Riemann derivatives of lower order exist in $C_{2\pi}$-norm and $(D_s^{[j]}f)(x) = f^{(j)}(x)$ for every $x, j = 1, 2, \ldots, r$.*

The proof of this theorem which is essentially that of Prop. 10.1.2 is left to Problem 10.1.1. As a consequence, Theorem 10.1.3 contains the following sufficient condition for $f \in C_{2\pi}$ to have a continuous rth ordinary derivative.

Corollary 10.1.4. *If f and g both belong to $C_{2\pi}$, and if for some integer $r > 0$*

$$\lim_{h\to 0} \frac{1}{h^r}\Delta_h^r f(x) = g(x)$$

uniformly for all x, then the rth (ordinary) derivative $f^{(r)}$ exists, is continuous, and $f^{(r)}(x) = g(x)$ for all x. In particular, $g(x) \equiv 0$ implies $f(x) \equiv$ const.

The latter assertion as formulated with respect to the central difference (cf. Problem 10.1.2) is related to the well-known *lemma of H. A. Schwarz* (see Problem 10.1.3). For periodic functions it states that if $f \in C_{2\pi}$ and

$$\lim_{h\to 0} \frac{f(x+h) - 2f(x) + f(x-h)}{h^2} = 0 \tag{10.1.4}$$

for each fixed $x \in \mathbb{R}$, then f is a constant on $\mathbb{R}$. Thus Problem 10.1.2 in particular asserts that the lemma of Schwarz is valid under the stronger assumption that (10.1.4) holds uniformly. But Problem 10.1.2 holds for general $r \in \mathbb{N}$ whereas the lemma of Schwarz fails to hold for $r > 2$. For further comments we refer to Sec. 10.7.

On the other hand, Theorem 10.1.3 may be regarded as a further characterization of the class $W_{C_{2\pi}}^r$ (cf. (1.1.16)). Indeed

Corollary 10.1.5. *Necessary and sufficient for $f \in C_{2\pi}$ to belong to $W_{C_{2\pi}}^r$ is that f has an rth uniform Riemann derivative.*

It is this latter aspect which may be carried over to Riemann derivatives in the mean of order p. In fact

Theorem 10.1.6. *A function $f \in L_{2\pi}^p$, $1 \le p < \infty$, belongs to $W_{L_{2\pi}^p}^r$ if and only if it has an rth Riemann derivative in the mean of order p. In this case, all strong Riemann derivatives of lower order exist as well.*

Proof. (i) Let $f \in W_{L_{2\pi}^p}^r$, $1 \le p < \infty$, i.e., there exists $\phi \in AC_{2\pi}^{r-1}$ such that $f = \phi$ a.e. a nd $\phi^{(r)} \in L_{2\pi}^p$. Then integration by parts (r-times) yields

$$\Delta_h^r f(x) = \int_0^h \cdots \int_0^h \phi^{(r)}(x + u_1 + \cdots + u_r)\, du_1 \ldots du_r \quad \text{a.e.,} \tag{10.1.5}$$

and hence by the Hölder–Minkowski inequality

$$\left\| \frac{1}{h^r} \Delta_h^r f - \phi^{(r)} \right\|_p = \left\| \int_0^1 \cdots \int_0^1 [\phi^{(r)}(\circ + hu_1 + \cdots + hu_r) - \phi^{(r)}(\circ)]\, du_1 \ldots du_r \right\|_p$$
$$\leq \sup_{|u| \leq r|h|} \|\phi^{(r)}(\circ + u) - \phi^{(r)}(\circ)\|_p,$$

which tends to zero as $h \to 0$ in view of the continuity in the mean of $\phi^{(r)}$.

(ii) Conversely, we have by Prop. 4.1.1(i) that

$$[\Delta_h^r f]^\wedge(k) = (e^{ihk} - 1)^r f^\wedge(k),$$

and hence

$$\lim_{h \to 0} \left[\frac{1}{h^r} \Delta_h^r f\right]^\wedge(k) = (ik)^r f^\wedge(k) \qquad (k \in \mathbb{Z}).$$

The existence of $D_s^{[r]} f$ implies that

$$\left| \left[\frac{1}{h^r} \Delta_h^r f - D_s^{[r]} f\right]^\wedge(k) \right| \leq \left\| \frac{1}{h^r} \Delta_h^r f - D_s^{[r]} f \right\|_p = o(1) \qquad (h \to 0)$$

for each $k \in \mathbb{Z}$ and so $(ik)^r f^\wedge(k) = [D_s^{[r]} f]^\wedge(k)$, $k \in \mathbb{Z}$, which according to Prop. 4.1.9 shows that $f \in \mathsf{W}^r_{\mathsf{L}^p_{2\pi}}$.

As an immediate consequence of the particular instance that $D_s^{[r]} f$ exists and vanishes a.e. we note

Corollary 10.1.7. *If for* $f \in \mathsf{L}^p_{2\pi}$, $1 \leq p < \infty$,

$$\|\Delta_h^r f\|_p = o(|h|^r) \qquad (h \to 0),$$

then $f = \text{const}$ a.e.

Let us mention that in Definition 10.1.1 there was no restriction upon how h approaches zero. It may easily be seen that the preceding results also remain valid for the (seemingly more general) concept of a Riemann derivative obtained by restricting h to positive values and considering the lim inf as $h \to 0+$ (cf. Problem 10.1.7).

10.1.2 Strong Peano Derivatives

Another concept of a generalized derivative for which similar results may be established is the Peano derivative (cf. (5.1.45) for its pointwise definition). Extending the pointwise definition (5.1.44), an *rth Peano difference of* $f \in \mathsf{X}_{2\pi}$ is an expression of the form

$$(10.1.6) \qquad \Diamond_h^r f(x) = r! \left\{ f(x+h) - f(x) - \sum_{k=1}^{r-1} \frac{h^k}{k!} l_k(x) \right\},$$

where l_k, $1 \leq k \leq r - 1$, are arbitrary functions in $\mathsf{X}_{2\pi}$.

Definition 10.1.8. *Let* $f \in \mathsf{X}_{2\pi}$. *If there exist* $l_k \in \mathsf{X}_{2\pi}$, $1 \leq k \leq r$, *such that*

$$(10.1.7) \qquad \lim_{h \to 0} \left\| \frac{r!}{h^r} \left\{ f(\circ + h) - f(\circ) - \sum_{k=1}^{r-1} \frac{h^k}{k!} l_k(\circ) \right\} - l_r(\circ) \right\|_{\mathsf{X}_{2\pi}} = 0,$$

then l_r is called the ***rth uniform Peano derivative of f*** *if $X_{2\pi} = C_{2\pi}$, and the* ***rth Peano derivative of f in the mean of order p*** *if $X_{2\pi} = L^p_{2\pi}$, $1 \le p < \infty$. In short, we shall speak of these derivatives as of* ***rth Peano derivatives in $X_{2\pi}$-norm*** *or* ***strong Peano derivatives of order r*** *and denote l_r by $D_s^{\langle r\rangle}f$.*

If in the sequel we shall say that there exists an rth Peano difference $\Diamond_h^r f$ of $f \in X_{2\pi}$ or that f has an rth Peano derivative in $X_{2\pi}$-norm, we always tacitly assume that there exist functions $l_k \in X_{2\pi}$ such that (10.1.6) and (10.1.7) have a meaning.

Just as for the Riemann derivatives we need not suppose explicitly that the Peano derivatives of lower order exist. Indeed, as for the pointwise definition (cf. Lemma 5.1.24) we have that the existence of $D_s^{\langle r\rangle}f$ in $X_{2\pi}$-norm directly implies the existence of derivatives $D_s^{\langle k\rangle}f$, $1 \le k \le r-1$, in $X_{2\pi}$-norm as well, and that if $D_s^{\langle r\rangle}f$ is given by (10.1.7) then $D_s^{\langle k\rangle}f = l_k$ in $X_{2\pi}$. In particular, this shows that the rth Peano derivative of f in $X_{2\pi}$ is defined uniquely, i.e., Peano differentiability of f in $X_{2\pi}$ is a genuine property of f and is independent of the particular choice of functions l_k. Thus we may actually speak of *the* rth strong Peano derivative.

In Sec. 5.1.4 we saw that the definition of the pointwise Peano derivative is a true generalization of the pointwise ordinary derivative. But if we sharpen pointwise convergence to uniform convergence, both concepts of a derivative turn out to be equivalent.

Theorem 10.1.9. *The function $f \in C_{2\pi}$ has a continuous rth ordinary derivative if and only if f has an rth uniform Peano derivative. In this event, $(D_s^{\langle r\rangle}f)(x) = f^{(r)}(x)$ for all $x \in \mathbb{R}$.*

Proof. If f is r-times continuously differentiable, it follows by partial integration that

$$f(x+h) - f(x) - \sum_{k=1}^{r-1} \frac{h^k}{k!} f^{(k)}(x) = \int_0^h \frac{(h-u)^{r-1}}{(r-1)!} f^{(r)}(x+u)\,du, \tag{10.1.8}$$

and therefore

$$\left\| \frac{r!}{h^r}\left\{ f(\circ + h) - f(\circ) - \sum_{k=1}^{r-1} \frac{h^k}{k!} f^{(k)}(\circ)\right\} - f^{(r)}(\circ) \right\|_{C_{2\pi}}$$

$$= \left\| r\int_0^1 (1-u)^{r-1}[f^{(r)}(\circ + hu) - f^{(r)}(\circ)]\,du \right\|_{C_{2\pi}} \le \sup_{|u|\le|h|} \|f^{(r)}(\circ + u) - f^{(r)}(\circ)\|_{C_{2\pi}},$$

which tends to zero as $h \to 0$. Thus the rth uniform Peano derivative $D_s^{\langle r\rangle}f$ exists and $(D_s^{\langle r\rangle}f)(x) = f^{(r)}(x)$ for all x.

Conversely, if $D_s^{\langle r\rangle}f$ exists, it follows (cf. Problem 10.1.4) that

$$\Delta_h^r f(x) = h^r (D_s^{\langle r\rangle}f)(x) + o(h^r) \qquad (h \to 0) \tag{10.1.9}$$

uniformly for all x. Hence f has an rth uniform Riemann derivative and $(D_s^{[r]}f)(x) = (D_s^{\langle r\rangle}f)(x)$ for all x. Now we only need to apply Theorem 10.1.3 to complete the proof.

The preceding result may be interpreted as a further characterization of the class $W^r_{C_{2\pi}}$. By the theorem to follow this is established for general $X_{2\pi}$.

Theorem 10.1.10. *A function $f \in X_{2\pi}$ belongs to $W^r_{X_{2\pi}}$ if and only if it has an rth strong Peano derivative.*

Proof. We only need to establish the case $X_{2\pi} = L^p_{2\pi}$, $1 \le p < \infty$. Thus let $f \in W^r_{L^p_{2\pi}}$, i.e. $f = \phi$ a.e. with $\phi \in AC^{r-1}_{2\pi}$, $\phi^{(r)} \in L^p_{2\pi}$. Since (10.1.8) now holds with f replaced by ϕ, the Hölder–Minkowski inequality gives

$$\left\| \frac{r!}{h^r}\left\{f(\circ + h) - f(\circ) - \sum_{k=1}^{r-1} \frac{h^k}{k!}\phi^{(k)}(\circ)\right\} - \phi^{(r)}(\circ)\right\|_p \le \sup_{|u| \le |h|} \|\phi^{(r)}(\circ + u) - \phi^{(r)}(\circ)\|_p,$$

which tends to zero as $h \to 0$. Therefore the rth Peano derivative $D_s^{\langle r\rangle} f$ exists in the mean of order p and $D_s^{\langle r\rangle} f = \phi^{(r)}$ a.e. On the other hand, if $D_s^{\langle r\rangle} f$ exists in the mean of order p, formula (cf. Problem 10.1.4)

$$\Delta_h^r f(x) = \frac{1}{r!}\sum_{k=0}^{r} (-1)^{r-k}\binom{r}{k} \Diamond_{kh}^r f(x) \quad \text{a.e.} \tag{10.1.10}$$

and (5.1.49) imply that

$$\left\| \frac{1}{h^r}\Delta_h^r f - D_s^{\langle r\rangle} f\right\|_p \le \frac{1}{r!}\sum_{k=1}^{r}\binom{r}{k}k^r \left\|\frac{1}{(kh)^r}\Diamond_{kh}^r f - D_s^{\langle r\rangle} f\right\|_p = o(1) \quad (h \to 0).$$

Thus f has an rth Riemann derivative in the mean of order p, and $f \in W^r_{L^p_{2\pi}}$ in view of Theorem 10.1.6.

For later references we formulate the particular case that the strong Peano derivative of order r exists and vanishes.

Corollary 10.1.11. *If for $f \in X_{2\pi}$ there exists an rth Peano difference such that*

$$\|\Diamond_h^r f\|_{X_{2\pi}} = o(|h|^r) \qquad (h \to 0),$$

then $f =$ const in $X_{2\pi}$.

The results on the Peano derivative so far obtained were established by a reduction to the respective statements concerning Riemann derivatives. But it is clear that a separate approach by the Fourier transform method is possible (cf. Problem 10.1.5). We observe further that analogous results may be shown for the Taylor derivative (cf. Problem 10.1.6).

10.1.3 Strong and Weak Derivatives, Weak Generalized Derivatives

In Sec. 1.1.2 we defined strong and weak (ordinary) derivatives of $f \in X_{2\pi}$ and showed that if $f \in W^r_{X_{2\pi}}$, then the rth strong derivative $D_s^{(r)} f$ and thus the rth weak derivative $D_w^{(r)} f$ of f exists. It is interesting to note that also the converse result is true. This may be shown via the Fourier transform method already employed above. Indeed

Theorem 10.1.12. *Let $f \in X_{2\pi}$. The following statements are equivalent:*

(i) *$f \in W^r_{X_{2\pi}}$, i.e. $f = \phi$ in $X_{2\pi}$ where $\phi \in AC^{r-1}_{2\pi}$ and $\phi^{(r)} \in X_{2\pi}$,*
(ii) *f has an rth strong derivative $D_s^{(r)} f$,*
(iii) *f has an rth weak derivative $D_w^{(r)} f$.*

In this event, $\phi^{(r)} = D_s^{(r)} f = D_w^{(r)} f$.

Proof. Since the implications (i) $\Rightarrow$ (ii) $\Rightarrow$ (iii) are already established in Sec. 1.1.2, we only have to show that (iii) implies (i). By hypothesis, the first r weak (ordinary) derivatives $D_w^{(j)}f$, $j = 1, \ldots, r$, exist as elements of $X_{2\pi}$. In other words, if $X_{2\pi} = L_{2\pi}^p$, $1 \le p < \infty$, there exist functions $D_w^{(j)}f \in L_{2\pi}^p$, $j = 0, 1, \ldots, r$, with $D_w^{(0)}f \equiv f$ such that

$$\lim_{h\to 0} \int_{-\pi}^{\pi} \frac{(D_w^{(j-1)}f)(x+h) - (D_w^{(j-1)}f)(x)}{h} s(x)\, dx = \int_{-\pi}^{\pi} (D_w^{(j)}f)(x)s(x)\, dx$$

for each $s \in L_{2\pi}^{p'}$. If we take $s(x) = (2\pi)^{-1} \exp\{-ikx\}$, $k \in \mathbb{Z}$, then

$$ik[D_w^{(j-1)}f]^\wedge(k) = [D_w^{(j)}]^\wedge(k),$$

and since this holds for each $j = 1, 2, \ldots, r$, we have $(ik)^r f^\wedge(k) = [D_w^{(r)}f]^\wedge(k)$ for every $k \in \mathbb{Z}$. Thus (i) follows by Prop. 4.1.9.

If $X_{2\pi} = C_{2\pi}$, then the hypothesis ensures the existence of functions $D_w^{(j)}f \in C_{2\pi}$, $j = 0, 1, \ldots, r$, such that

$$\lim_{h\to 0} \int_{-\pi}^{\pi} \frac{(D_w^{(j-1)}f)(x+h) - (D_w^{(j-1)}f)(x)}{h} d\eta(x) = \int_{-\pi}^{\pi} (D_w^{(j)}f)(x)\, d\eta(x)$$

for each $\eta \in BV_{2\pi}$. Taking

(10.1.11) $$\eta(x) = \frac{1}{2\pi} \int_{-\pi}^{x} e^{-iku}\, du \qquad (k \in \mathbb{Z}),$$

then as above $(ik)^r f^\wedge(k) = [D_w^{(r)}f]^\wedge(k)$ for every $k \in \mathbb{Z}$, and the proof is complete by Prop. 4.1.9.

Having introduced strong and weak (ordinary) derivatives of $f \in X_{2\pi}$ we may equally well define an rth weak Riemann derivative $D_w^{[r]}f$ and an rth weak Peano derivative $D_w^{\langle r\rangle}f$ of f. Let us consider the Riemann case in some detail.

Definition 10.1.13. *If for $f \in X_{2\pi}$ there exists $D_w^{[r]}f \in X_{2\pi}$ such that*

(10.1.12) $$\lim_{h\to 0} \int_{-\pi}^{\pi} \frac{1}{h^r} \Delta_h^r f(x)\, d\eta(x) = \int_{-\pi}^{\pi} (D_w^{[r]}f)(x)\, d\eta(x)$$

for every $\eta \in BV_{2\pi}$ if $X_{2\pi} = C_{2\pi}$, and

(10.1.13) $$\lim_{h\to 0} \int_{-\pi}^{\pi} \frac{1}{h^r} \Delta_h^r f(x)s(x)\, dx = \int_{-\pi}^{\pi} (D_w^{[r]}f)(x)s(x)\, dx$$

for every $s \in L_{2\pi}^{p'}$ if $X_{2\pi} = L_{2\pi}^p$, $1 \le p < \infty$, then $D_w^{[r]}f$ is called the ***rth weak Riemann derivative of f.***

Proposition 10.1.14. *For $f \in X_{2\pi}$ the rth strong Riemann derivative $D_s^{[r]}f$ exists if and only if the rth weak Riemann derivative $D_w^{[r]}f$ exists, and both are equal.*

Proof. Certainly, the existence of $D_s^{[r]}f$ implies that of $D_w^{[r]}f$. Conversely, if $D_w^{[r]}f$ exists, then in view of the definitions (10.1.12) (as applied to (10.1.11)) and (10.1.13) (as applied to $s(x) = (2\pi)^{-1} \exp\{-ikx\}$, $k \in \mathbb{Z}$)

$$\lim_{h\to 0} \left[\frac{e^{ihk} - 1}{h}\right]^r f^\wedge(k) = [D_w^{[r]}f]^\wedge(k)$$

by Prop. 4.1.1(i). Hence $(ik)^r f^\wedge(k) = [D_w^{[r]}f]^\wedge(k)$ for every $k \in \mathbb{Z}$ and $D_s^{[r]}f$ exists in view of Prop. 4.1.9, Theorems 10.1.3, and 10.1.6.

By the same technique it may be shown that the concepts of weak and strong Peano (Taylor) derivatives are equivalent as well. Summarizing, we therefore have the following characterizations of the classes $W_{X_{2\pi}}^r$.

Corollary 10.1.15. *Let $f \in X_{2\pi}$. The following statements are equivalent:*

(i) $f \in W_{X_{2\pi}}^r$, (ii) $f \in W[X_{2\pi}; (ik)^r]$,

(iii) *there exist constants $\alpha_k \in \mathbb{C}$, $0 \le k \le r-1$, and $g \in X_{2\pi}$ such that*

$$f(x) = \alpha_0 + \int_{-\pi}^{x} du_1\Big[\alpha_1 + \int_{-\pi}^{u_1} du_2\Big[\alpha_2 + \cdots + \int_{-\pi}^{u_{r-2}} du_{r-1}\Big[\alpha_{r-1} + \int_{-\pi}^{u_{r-1}} g(u_r)\, du_r\Big]\ldots\Big]\Big],$$

(iv) *f has an rth strong (weak) derivative $D_s^{(r)}f$ ($D_w^{(r)}f$),*

(v) *f has an rth strong (weak) Taylor derivative $D_s^{\textcircled{r}}f$ ($D_w^{\textcircled{r}}f$), provided $f^{(r-1)}$ exists and belongs to $X_{2\pi}$,*

(vi) *f has an rth strong (weak) Peano derivative $D_s^{\langle r\rangle}f$ ($D_w^{\langle r\rangle}f$),*

(vii) *f has an rth strong (weak) Riemann derivative $D_s^{[r]}f$ ($D_w^{[r]}f$).*

Problems

1. Prove Theorem 10.1.3.
2. If $\bar{\Delta}_h^r f(x)$ denotes the central difference of f at x (cf. (5.1.41)) show that f belongs to $W_{X_{2\pi}}^r$ if and only if there exists $g \in X_{2\pi}$ such that $\lim_{h\to 0} \|h^{-r}\bar{\Delta}_h^r f - g\|_{X_{2\pi}} = 0$.
3. Prove the *lemma of H. A. Schwarz*: If f is continuous on (a, b) and for each fixed $x \in (a, b)$ $\lim_{h\to 0} h^{-2}[f(x+h) + f(x-h) - 2f(x)] = 0$, then f is a linear function. (Hint: TITCHMARSH [5, p. 431])
4. Prove (10.1.9) and (10.1.10) (cf. (5.1.50)). (Hint: OLIVER [1])
5. Establish the results on strong and weak Peano derivatives of Sec. 10.1.2 by the Fourier transform method directly, i.e., without reduction to the results on Riemann derivatives. In particular, show that if $D_s^{\langle r\rangle}f$ exists in $X_{2\pi}$-norm, then $(ik)^r f^\wedge(k) = [D_s^{\langle r\rangle}f]^\wedge(k)$, $k \in \mathbb{Z}$.
6. Let $f \in X_{2\pi}$ be such that $f^{(r-1)}$ exists and belongs to $X_{2\pi}$. The expression

(10.1.14)
$$\nabla_h^r f(x) = r!\Big\{f(x+h) - f(x) - \sum_{k=1}^{r-1} \frac{h^k}{k!} f^{(k)}(x)\Big\} \qquad (h \in \mathbb{R})$$

is called the *rth Taylor difference of $f \in X_{2\pi}$*. If there exists $D_s^{\textcircled{r}}f \in X_{2\pi}$ such that

(10.1.15)
$$\lim_{h\to 0} \Big\|\frac{1}{h^r}\nabla_h^r f - D_s^{\textcircled{r}}f\Big\|_{X_{2\pi}} = 0,$$

then $D_s^{\textcircled{r}}f$ is called the *rth strong Taylor derivative of f* (cf. Problem 5.1.15 for its pointwise definition).

Prove that f belongs to $W_{X_{2\pi}}^r$ if and only if f has an rth strong Taylor derivative. Define an rth weak Taylor derivative $D_w^{\textcircled{r}}f$ (cf. Def. 10.1.13) and show that $D_w^{\textcircled{r}}f$ exists if and only if $f \in W_{X_{2\pi}}^r$.

7. Let $f \in X_{2\pi}$. The following statements are equivalent:

   (i) f belongs to $W_{X_{2\pi}}^r$,

   (ii) there exists $g \in X_{2\pi}$ such that $\liminf_{h\to 0+} \|h^{-r}\Delta_h^r f - g\|_{X_{2\pi}} = 0$ (or, such that $\lim_{j\to\infty} \|h_j^{-r}\Delta_{h_j}^r f - g\|_{X_{2\pi}} = 0$ for at least one null-sequence $\{h_j\}$),

   (iii) there exists an rth Peano difference of f such that $\liminf_{h\to 0}\|h^{-r}\diamondsuit_h^r f - g\|_{X_{2\pi}} = 0$, for some $g \in X_{2\pi}$.

   (iv) there exists $g \in X_{2\pi}$ such that $\liminf_{h\to 0+} \|h^{-r}\nabla_h^r f - g\|_{X_{2\pi}} = 0$ under the additional assumption that $f^{(r-1)}$ exists and belongs to $X_{2\pi}$.

8. Let $f \in X_{2\pi}$. Then

$$f \in W^r_{X_{2\pi}} \Leftrightarrow \begin{cases} \|\sigma_m^{(r)}(f;\circ) - \sigma_n^{(r)}(f;\circ)\|_{C_{2\pi}} = o(1) & (m, n \to \infty, X_{2\pi} = C_{2\pi}), \\ \|\sigma_m^{(r)}(f;\circ) - \sigma_n^{(r)}(f;\circ)\|_{L^1_{2\pi}} = o(1) & (m, n \to \infty, X_{2\pi} = L^1_{2\pi}), \\ \|\sigma_n^{(r)}(f;\circ)\|_p = O(1) & (n \to \infty, X_{2\pi} = L^p_{2\pi}, 1 < p < \infty). \end{cases}$$

(Hint: Theorem 6.1.2, Prop. 4.1.9)

10.2 Characterization of the Classes $V^r_{X_{2\pi}}$

In this section we examine the Fourier transform method in its wider and more general form as applied to give unified proofs of some classical results of G. H. Hardy and L. E. Littlewood on Lipschitz classes. Together with theorems on weak* compactness this method moreover yields generalizations of these results in several directions.

The following results have their source in Cor. 10.1.7 which states that for $f \in L^p_{2\pi}$, $1 \le p < \infty$,

$$\|f(\circ + h) - f(\circ)\|_p = o(|h|) \qquad (h \to 0)$$

implies $f = \text{const}$ a.e. The particular case $p = 1$ is the periodic version of a result of E. C. Titchmarsh. The question then arises as to what can be said about the function f if the 'small' $o(|h|)$ is replaced by 'large' $O(|h|)$. In this direction, Hardy and Littlewood have proved:

(i) *Let $f \in L^1_{2\pi}$. A necessary and sufficient condition for f to satisfy*

$$\|f(\circ + h) - f(\circ)\|_1 = O(|h|) \qquad (h \to 0)$$

is that f should coincide almost everywhere with a function of bounded variation.

(ii) *Let $f \in L^p_{2\pi}$, $1 < p < \infty$. A necessary and sufficient condition for f to satisfy*

$$\|f(\circ + h) - f(\circ)\|_p = O(|h|) \qquad (h \to 0)$$

is that f should be equivalent to the integral of a function in $L^p_{2\pi}$.

If we recall the definition (4.3.17) of the classes $V^r_{X_{2\pi}}$, the above results may be formulated as

Proposition 10.2.1. *Let $f \in X_{2\pi}$. A necessary and sufficient condition for f to satisfy*

$$\|f(\circ + h) - f(\circ)\|_{X_{2\pi}} = O(|h|) \qquad (h \to 0) \tag{10.2.1}$$

is that f belongs to $V^1_{X_{2\pi}}$.

We now generalize this proposition by replacing the difference $[f(x + h) - f(x)]$ in (10.2.1) by the rth differences $\Delta^r_h f(x)$, $\Diamond^r_h f(x)$ and $\nabla^r_h f(x)$ of Riemann, Peano and Taylor, respectively. We begin with a theorem on the Riemann differences.

Theorem 10.2.2. *Let $f \in X_{2\pi}$. The following assertions are equivalent:*

(i) $$f \in V^r_{X_{2\pi}},$$

(ii) $$\|\Delta^r_h f\|_{X_{2\pi}} = O(|h|^r) \qquad (h \to 0).$$

Proof. (i) $\Rightarrow$ (ii). If $X_{2\pi} = C_{2\pi}$, integration by parts yields

$$(10.2.2)\qquad \Delta_h^r f(x) = \int_0^h \cdots \int_0^h f^{(r)}(x + u_1 + \cdots + u_r)\, du_1 \ldots du_r$$

and thus $\|\Delta_h^r f\|_{C_{2\pi}} \leq \|f^{(r)}\|_\infty |h|^r$.

If $X_{2\pi} = L_{2\pi}^p$, $1 < p < \infty$, then (10.1.5) holds, and by the Hölder-Minkowski inequality $\|\Delta_h^r f\|_p \leq \|\phi^{(r)}\|_p\, |h|^r$.

If $X_{2\pi} = L_{2\pi}^1$, then again by partial integration

$$(10.2.3)\qquad \Delta_h^{r-1} f(x) = h^{r-1} \int_0^1 \cdots \int_0^1 \phi^{(r-1)}(x + hu_1 + \cdots + hu_{r-1})\, du_1 \ldots du_{r-1}$$

almost everywhere, and therefore by Fubini's theorem

$$\|\Delta_h^r f\|_1 \leq |h|^{r-1} \int_0^1 \cdots \int_0^1 \|\phi^{(r-1)}(\circ + h + hu_1 + \cdots + hu_{r-1}) - \phi^{(r-1)}(\circ + hu_1 + \cdots + hu_{r-1}\|_1\, du_1 \ldots du_{r-1}$$
$$= \|\phi^{(r-1)}(\circ + h) - \phi^{(r-1)}(\circ)\|_1\, |h|^{r-1}.$$

But in view of (4.3.6)

$$(10.2.4)\qquad \|\phi^{(r-1)}(\circ + h) - \phi^{(r-1)}(\circ)\|_1 \leq \|\phi^{(r-1)}\|_{BV_{2\pi}}\, |h|,$$

and thus $\|\Delta_h^r f\|_1 \leq \|\phi^{(r-1)}\|_{BV_{2\pi}}\, |h|^r$.

(ii) $\Rightarrow$ (i). Let $X_{2\pi} = C_{2\pi}$. Since $\|h^{-r}\Delta_h^r f\|_\infty = O(1)$ as $h \to 0$, the weak* compactness theorem for $L_{2\pi}^\infty$ furnishes a sequence $\{h_j\}$ with $\lim_{j\to\infty} h_j = 0$ and $g \in L_{2\pi}^\infty$ such that

$$(10.2.5)\qquad \lim_{j\to\infty} \int_{-\pi}^{\pi} [h_j^{-r}\Delta_{h_j}^r f(u)] s(u)\, du = \int_{-\pi}^{\pi} g(u) s(u)\, du$$

for every $s \in L_{2\pi}^1$. Choosing $s(u) = (2\pi)^{-1} \exp\{-iku\}$, $k \in \mathbb{Z}$, we obtain

$$\lim_{j\to\infty} h_j^{-r}(e^{ih_jk} - 1)^r f^\wedge(k) = g^\wedge(k),$$

and thus $(ik)^r f^\wedge(k) = g^\wedge(k)$, $k \in \mathbb{Z}$. Now Theorem 4.3.13 ensures $f \in V_{C_{2\pi}}^r$.

If $X_{2\pi} = L_{2\pi}^p$, $1 < p < \infty$, the weak* compactness theorem for $L_{2\pi}^p$ implies that there exists $\{h_j\}$ with $\lim_{j\to\infty} h_j = 0$ and $g \in L_{2\pi}^p$ such that (10.2.5) holds for every $s \in L_{2\pi}^{p'}$. Again it follows that $(ik)^r f^\wedge(k) = g^\wedge(k)$, $k \in \mathbb{Z}$, and hence $f \in V_{L_{2\pi}^p}^r$ by Theorem 4.3.13.

Let $X_{2\pi} = L_{2\pi}^1$. If we set

$$(10.2.6)\qquad \mu_h(x) = \int_{-\pi}^{x} h^{-r}\Delta_h^r f(u)\, du,$$

this defines a family $\{\mu_h(x)\}$ of 2π-periodic, absolutely continuous functions. Since $\|\mu_h\|_{BV_{2\pi}} = \|h^{-r}\Delta_h^r f\|_1$, the assumption implies that $\|\mu_h\|_{BV_{2\pi}} = O(1)$ as $h \to 0$. By the theorem of Helly-Bray there exists a sequence $\{h_j\}$ with $\lim_{h\to 0} h_j = 0$ and $\mu \in BV_{2\pi}$ such that for every $s \in C_{2\pi}$

$$(10.2.7)\qquad \lim_{j\to\infty} \int_{-\pi}^{\pi} s(u)[h_j^{-r}\Delta_{h_j}^r f(u)]\, du = \int_{-\pi}^{\pi} s(u)\, d\mu(u).$$

Taking s as above we have $\lim_{j\to\infty} h_j^{-r}(e^{ih_jk}-1)^r f^\wedge(k) = \mu^\vee(k)$, which yields $(ik)^r f^\wedge(k) = \mu^\vee(k)$, $k \in \mathbb{Z}$. Hence $f \in \mathsf{V}^r_{\mathsf{L}^1_{2\pi}}$ by Prop. 4.3.12.

As a simple consequence of the fact that $\mathsf{V}^r_{\mathsf{X}_{2\pi}} = \mathsf{W}^r_{\mathsf{X}_{2\pi}}$ for $\mathsf{X}_{2\pi} = \mathsf{L}^p_{2\pi}$, $1 < p < \infty$, and of Theorem 10.1.6 we have

Corollary 10.2.3. *Let $f \in \mathsf{L}^p_{2\pi}$, $1 < p < \infty$. The following statements are equivalent:*

(i)
$$\left\| \frac{1}{h^r} \Delta^r_h f \right\|_p = \mathrm{O}(1) \qquad (h \to 0),$$

(ii) *there exists $g \in \mathsf{L}^p_{2\pi}$ such that*
$$\lim_{h\to 0} \left\| \frac{1}{h^r} \Delta^r_h f - g \right\|_p = 0.$$

In other words, when $f \in \mathsf{L}^p_{2\pi}$, $1 < p < \infty$, we have $\|h^{-r}\Delta^r_h f\|_p = \mathrm{O}(1)$ if and only if f has a strong Riemann derivative of order r, thus, by Prop. 10.1.14, if and only if f has a weak Riemann derivative of order r. This is a consequence of the reflexivity and weak* compactness theorem for $\mathsf{L}^p_{2\pi}$, $1 < p < \infty$. For the space $\mathsf{L}^1_{2\pi}$ we have a characterization via the following type of convergence (actually of weak*-type) of the quotient $h^{-r}\Delta^r_h f$.

Proposition 10.2.4. *Let $f \in \mathsf{L}^1_{2\pi}$. The following statements are equivalent:*

(i)
$$\left\| \frac{1}{h^r} \Delta^r_h f \right\|_1 = \mathrm{O}(1) \qquad (h \to 0),$$

(ii) *there exists $\mu \in \mathsf{BV}_{2\pi}$ such that*

$$\lim_{h\to 0} \int_{-\pi}^{\pi} s(x) \frac{1}{h^r} \Delta^r_h f(x)\, dx = \int_{-\pi}^{\pi} s(x)\, d\mu(x) \tag{10.2.8}$$

for every $s \in \mathsf{C}_{2\pi}$.

Proof. If (i) holds, then in view of Theorem 10.2.2 and Prop. 4.3.11 there exists $\mu \in \mathsf{BV}_{2\pi}$ such that $(ik)^r f^\wedge(k) = \mu^\vee(k)$, $k \in \mathbb{Z}$. By Prop. 4.3.2(iii), μ is uniquely determined by f (up to const). On the other hand, the assumption (i) implies by the theorem of Helly-Bray that the family of functions $\{\mu_h(x)\}$ of (10.2.6) has a weak* accumulation point (cf. (10.2.7)) given by μ. But a set having exactly one weak* accumulation point is weak* convergent. Conversely, if (ii) is valid, then we apply (10.2.8) to $s(x) = (2\pi)^{-1}\exp\{-ikx\}$, $k \in \mathbb{Z}$, and obtain $(ik)^r f^\wedge(k) = \mu^\vee(k)$, which implies (i) by Prop. 4.3.12 and Theorem 10.2.2.

For the counterpart in $\mathsf{C}_{2\pi}$-space we refer to Problem 10.2.8.

Next we turn to Peano differences, the situation being completely analogous.

Theorem 10.2.5. *Let $f \in \mathsf{X}_{2\pi}$. The following assertions are equivalent:*

(i) *f belongs to $\mathsf{V}^r_{\mathsf{X}_{2\pi}}$,*

(ii) *there exists an rth Peano difference of f such that*
$$\| \Diamond^r_h f \|_{\mathsf{X}_{2\pi}} = \mathrm{O}(|h|^r) \qquad (h \to 0).$$

Proof. (i) $\Rightarrow$ (ii). When $\mathsf{X}_{2\pi} = \mathsf{C}_{2\pi}$ formula (10.1.8) holds again, and hence
$$\Big\| r!\Big\{f(\circ + h) - f(\circ) - \sum_{k=1}^{r-1} \frac{h^k}{k!} f^{(k)}(\circ)\Big\}\Big\|_{\mathsf{C}_{2\pi}} \le \|f^{(r)}\|_\infty\, |h|^r,$$
whereas for $\mathsf{X}_{2\pi} = \mathsf{L}^p_{2\pi}$, $1 < p < \infty$, (10.1.8) holds with f replaced by ϕ and
$$\left\| r!\Big\{f(\circ + h) - f(\circ) - \sum_{k=1}^{r-1} \frac{h^k}{k!} \phi^{(k)}(\circ)\Big\} \right\|_p \le \|\phi^{(r)}\|_p\, |h|^r.$$

Thus there is an rth Peano difference of f such that (ii) is satisfied.

If $X_{2\pi} = L^1_{2\pi}$, then $f \in V^r_{L^1_{2\pi}}$ implies

$$(10.2.9)\qquad f(x+h) - f(x) - \sum_{k=1}^{r-1} \frac{h^k}{k!}\phi^{(k)}(x) = \int_0^h \frac{(h-u)^{r-2}}{(r-2)!}[\phi^{(r-1)}(x+u) - \phi^{(r-1)}(x)]\,du \quad \text{a.e.}$$

Therefore we have by Fubini's theorem and (10.2.4)

$$\left\| r!\{f(\circ+h) - f(\circ) - \sum_{k=1}^{r-1}\frac{h^k}{k!}\phi^{(k)}(\circ)\}\right\|_1 \le \sup_{|u|\le|h|} \|\phi^{(r-1)}(\circ+u) - \phi^{(r-1)}(\circ)\|_1\, r\,|h|^{r-1} \le r\|\phi^{(r-1)}\|_{BV_{2\pi}}|h|^r.$$

(ii) $\Rightarrow$ (i). As an immediate consequence of (10.1.10), (ii) implies condition (ii) of Theorem 10.2.2 so that the proof follows by an application of this theorem.

Corollary 10.2.6. *Let $f \in L^p_{2\pi}$, $1 < p < \infty$. The following statements are equivalent:*

(i) *there exists an rth Peano difference of f such that*

$$\left\|\frac{1}{h^r}\diamondsuit^r_h f\right\|_p = O(1) \qquad (h \to 0),$$

(ii) *there exists an rth Peano difference of f and a function $g \in L^p_{2\pi}$ such that*

$$\lim_{h\to 0}\left\|\frac{1}{h^r}\diamondsuit^r_h f - g\right\|_p = 0.$$

This is an easy consequence of Theorems 10.1.10 and 10.2.5. We have thus shown that $\|h^{-r}\diamondsuit^r_h f\|_p = O(1)$ if and only if the rth weak Peano derivative of f exists provided $f \in L^p_{2\pi}$, $1 < p < \infty$. But this equivalence is no longer true for arbitrary $X_{2\pi}$. However, we still have

Proposition 10.2.7. *Let $f \in C_{2\pi}$. The following statements are equivalent:*

(i) *there exists an rth Peano difference of f such that*

$$\left\|\frac{1}{h^r}\diamondsuit^r_h f\right\|_{C_{2\pi}} = O(1) \qquad (h \to 0),$$

(ii) *there exists an rth Peano difference of f and a function $g \in L^\infty_{2\pi}$ such that*

$$(10.2.10)\qquad \lim_{h\to 0}\int_{-\pi}^{\pi} s(x)\frac{1}{h^r}\diamondsuit^r_h f(x)\,dx = \int_{-\pi}^{\pi} s(x)g(x)\,dx$$

for every $s \in L^1_{2\pi}$.

Proof. If (i) holds, then in view of Theorems 10.2.5 and 4.3.13 there exists $g \in L^\infty_{2\pi}$ so that $(ik)^r f^\wedge(k) = g^\wedge(k)$, $k \in \mathbb{Z}$. By Prop. 4.1.2, g is uniquely determined by f. On the other hand, the assumption (i) implies by the weak* compactness for $L^\infty_{2\pi}$ that the set of functions $\{h^{-r}\diamondsuit^r_h f(x)\}$ has a weak* accumulation point (cf. (10.2.5)) given by g; since this is uniquely determined, (ii) follows. The implication (ii) $\Rightarrow$ (i) is settled as usual by taking the functions $s(x) = (2\pi)^{-1}\exp\{-ikx\}$.

For the case $X_{2\pi} = L^1_{2\pi}$ we refer to Problem 10.2.9.

Let us once more emphasize that strong and weak convergence of the respective quotients are characteristic for the classes $W^r_{X_{2\pi}}$, whilst the type of convergence of (10.2.8), (10.2.10), for example, corresponds to the classes $V^r_{X_{2\pi}}$. In reflexive spaces there is no distinction. For further characterizations, including the nonreflexive case, we refer to Sec. 10.4.

Finally, let us return to the open problem of Sec. 2.4 on the characterization of the class $\mathsf{Lip}(\mathsf{X}_{2\pi}; 1)$. We noted that there is actually no simple characterization in terms of the best approximation $E_n(\mathsf{X}_{2\pi}; f)$. But the above arguments and results reveal $f \in \mathsf{Lip}(\mathsf{X}_{2\pi}; 1)$ if and only if $f \in \mathsf{V}^1_{\mathsf{X}_{2\pi}}$ or if and only if $f \in \mathsf{V}[\mathsf{X}_{2\pi}; ik]$. Thus, in particular, f belongs to $\mathsf{Lip}(\mathsf{C}_{2\pi}; 1)$ if and only if it is the integral of a function in $\mathsf{L}^\infty_{2\pi}$. More generally

Corollary 10.2.8. *Let $f \in \mathsf{X}_{2\pi}$. The following statements are equivalent:*

(i) $f \in \mathsf{Lip}_r(\mathsf{X}_{2\pi}; r)$, (ii) $f \in \mathsf{V}^r_{\mathsf{X}_{2\pi}}$, (iii) $f \in \mathsf{V}[\mathsf{X}_{2\pi}; (ik)^r]$.

The proof follows by Theorems 10.2.2 and 4.1.13. For the definition of $\mathsf{Lip}_r(\mathsf{X}_{2\pi}; \alpha)$ we recall Problem 1.5.4.

Problems

1. Prove the analog of Prop. 10.2.1 for nonperiodic functions in $\mathsf{L}^p(a, b)$, $1 \le p < \infty$. (Hint: For $p = 1$ see TITCHMARSH [5, p. 372]. For $p > 1$ the methods of proof of HARDY and LITTLEWOOD [2I, p. 599; 3, p. 619] may be applied)
2. Show that if $f \in \mathsf{V}^r_{\mathsf{X}_{2\pi}}$ then for the central Riemann difference $\|\bar{\Delta}^r_h f\|_{\mathsf{X}_{2\pi}} = O(|h|^r)$, $h \to 0$. Conversely, if there is a null-sequence $\{h_j\}$ such that $\liminf_{j\to\infty} \|\bar{\Delta}^r_{h_j} f\|_{\mathsf{X}_{2\pi}} < \infty$, then $f \in \mathsf{V}^r_{\mathsf{X}_{2\pi}}$.
3. Show that $f \in \mathsf{X}_{2\pi}$ belongs to $\mathsf{V}^r_{\mathsf{X}_{2\pi}}$ if and only if for the Taylor difference $\|\nabla^r_h f\|_{\mathsf{X}_{2\pi}} = O(|h|^r)$, $h \to 0$, provided $f^{(r-1)}$ exists and belongs to $\mathsf{X}_{2\pi}$.
4. Show that $f \in \mathsf{X}_{2\pi}$ belongs to $\mathsf{V}^r_{\mathsf{X}_{2\pi}}$ if and only if $\|\sigma^{(r)}_n(f; \circ)\|_{\mathsf{X}_{2\pi}} = O(1)$, $n \to \infty$.
5. The fact that $f \in \mathsf{V}[\mathsf{X}_{2\pi}; (ik)^r]$ implies $\|\Delta^r_h f\|_{\mathsf{X}_{2\pi}} = O(|h|^r)$, $h \to 0$, may be shown directly by using the uniqueness theorem and the fact that, e.g., for $\mathsf{X}_{2\pi} = \mathsf{C}_{2\pi}$ and $k \neq 0$
$$[\Delta^r_h f]^\wedge(k) = \left(\frac{e^{ihk} - 1}{ik}\right)^r g^\wedge(k) = \left[\int_0^h \cdots \int_0^h g(\circ + u_1 + \cdots + u_r)\, du_1 \ldots du_r\right]^\wedge(k).$$
6. Give another proof of the implication (ii) $\Rightarrow$ (i) of Theorem 10.2.2 by considering the limit $h \to 0$ of the quantity $\sigma_n(\Delta^r_h f; x)$ and using Problem 10.2.4. (Hint: See BUTZER [10, p. 294])
7. Show that if there is an rth Peano difference of $f \in \mathsf{X}_{2\pi}$ such that $\|\Diamond^r_h f\|_{\mathsf{X}_{2\pi}} = O(|h|^r)$, $h \to 0$, then $D^{\langle r-1\rangle}_s f$ exists. (Hint: cf. Lemma 5.1.24)
8. Let $f \in \mathsf{C}_{2\pi}$. Show that the following statements are equivalent:

   (i) $$\|h^{-r}\Delta^r_h f\|_{\mathsf{C}_{2\pi}} = O(1) \qquad (h \to 0),$$

   (ii) there exists $g \in \mathsf{L}^\infty_{2\pi}$ such that for every $s \in \mathsf{L}^1_{2\pi}$
   $$\lim_{h\to 0} \int_{-\pi}^{\pi} s(x)[h^{-r}\Delta^r_h f(x)]\, dx = \int_{-\pi}^{\pi} s(x)g(x)\, dx.$$
9. Let $f \in \mathsf{L}^1_{2\pi}$. Show that the following statements are equivalent:

   (i) there exists an rth Peano difference of f such that
   $$\|h^{-r}\Diamond^r_h f\|_{\mathsf{L}^1_{2\pi}} = O(1) \qquad (h \to 0),$$

   (ii) there exists an rth Peano difference of f and a function $\mu \in \mathsf{BV}_{2\pi}$ such that
   $$\lim_{h\to 0} \int_{-\pi}^{\pi} s(x)[h^{-r}\Diamond^r_h f(x)]\, dx = \int_{-\pi}^{\pi} s(x)\, d\mu(x)$$
   for every $s \in \mathsf{C}_{2\pi}$.

10. Let $f \in X_{2\pi}$. Show that the following statements are equivalent:

(i) $f \in V^r_{X_{2\pi}}$, (ii) $f \in V[X_{2\pi}; (ik)^r]$, (iii) $\|\sigma_n^{(r)}(f; \circ)\|_{X_{2\pi}} = O(1)$, $n \to \infty$,
(iv) $\|\bar{\Delta}_h^r f\|_{X_{2\pi}} = O(|h|^r)$, $h \to 0$, (v) $\|\Delta_h^r f\|_{X_{2\pi}} = O(|h|^r)$,
(vi) there exists an rth Peano difference such that $\|\Diamond_h^r f\|_{X_{2\pi}} = O(|h|^r)$,
(vii) $\|\nabla_h^r f\|_{X_{2\pi}} = O(|h|^r)$, provided $f^{(r-1)}$ exists and belongs to $X_{2\pi}$.

10.3 Characterization of the Classes $(V^\sim)^r_{X_{2\pi}}$

We establish results similar to those of the preceding section for the Hilbert transform $f^\sim$ of $f \in X_{2\pi}$. Thus we are interested in some further characterizations of the classes $(V^\sim)^r_{X_{2\pi}}$ as introduced in (9.3.18).

Proposition 10.3.1. *Let $f \in X_{2\pi}$. The following assertions are equivalent:*

(i) $$f \in (V^\sim)^r_{X_{2\pi}},$$

(ii) $$\|\Delta_h^r f^\sim\|_{X_{2\pi}} = O(|h|^r) \qquad (h \to 0).$$

Proof. (i) $\Rightarrow$ (ii). By definition, the assumption $f \in (V^\sim)^r_{X_{2\pi}}$ means that $f^\sim \in V^r_{X_{2\pi}}$, and thus (ii) follows by the implication (i) $\Rightarrow$ (ii) of Theorem 10.2.2 with f replaced by $f^\sim$.

(ii) $\Rightarrow$ (i). For $X_{2\pi} = L^p_{2\pi}$, $1 < p < \infty$, Theorem 9.1.3 implies $f^\sim \in L^p_{2\pi}$. Hence we may apply Theorem 10.2.2 to the Hilbert transform $f^\sim$ in order to show that $f^\sim \in V^r_{X_{2\pi}}$. By definition this is (i).

For the cases $X_{2\pi} = C_{2\pi}, L^1_{2\pi}$ we first observe that we do not explicitly assume $f^\sim \in X_{2\pi}$. This entails that we cannot apply Theorem 10.2.2 directly but have to proceed via Theorem 9.3.5.

Thus let $X_{2\pi} = C_{2\pi}$. Then by Theorem 9.1.3, $f \in C_{2\pi}$ yields $f^\sim \in L^q_{2\pi}$ for each $1 \le q < \infty$. By the assumption $\|h^{-r}\Delta_h^r f^\sim\|_\infty = O(1)$ and weak* compactness for $L^\infty_{2\pi}$ we deduce (cf. proof of Theorem 10.2.2) that there exists $g \in L^\infty_{2\pi}$ such that $(ik)^r[f^\sim]^\wedge(k) = g^\wedge(k)$, $k \in \mathbb{Z}$. Using (9.1.16) it follows that $f \in (V^\sim)^r_{C_{2\pi}}$ by Theorem 9.3.5.

When $X_{2\pi} = L^1_{2\pi}$ we first recall that, by Theorem 9.1.1, $f \in L^1_{2\pi}$ only says that the Hilbert transform $f^\sim$ exists almost everywhere. But condition (ii) assumes that at least the difference $\Delta_h^r f^\sim$ belongs to $L^1_{2\pi}$, and since $\Delta_h^r f^\sim = (\Delta_h^r f)^\sim$ a.e. and $\Delta_h^r f \in L^1_{2\pi}$, we have by (9.2.24)

$$[\Delta_h^r f^\sim]^\wedge(k) = \{-i \operatorname{sgn} k\}[\Delta_h^r f]^\wedge(k) \qquad (k \in \mathbb{Z}).$$

Again as in the proof of Theorem 10.2.2 we may conclude the existence of a null-sequence $\{h_j\}$ and $\mu \in BV_{2\pi}$ so that

$$\begin{aligned}\mu^\vee(k) &= \lim_{j\to\infty}\Big[\frac{1}{h_j^r}\Delta_{h_j}^r f^\sim\Big]^\wedge(k) = \{-i \operatorname{sgn} k\}\lim_{j\to\infty}\Big[\frac{1}{h_j^r}\Delta_{h_j}^r f\Big]^\wedge(k)\\ &= \{-i \operatorname{sgn} k\}\lim_{j\to\infty}\Big(\frac{e^{ih_jk}-1}{h_j}\Big)^r f^\wedge(k) = \{-i \operatorname{sgn} k\}(ik)^r f^\wedge(k) \qquad (k \in \mathbb{Z}),\end{aligned}$$

which by Theorem 9.3.5 ensures $f \in (V^\sim)^r_{L^1_{2\pi}}$.

We recall that in this section the order in forming the operation of Hilbert transforms and differentiation is always fixed since we do not investigate here the problem of commutativity of these operations, i.e., under what conditions the relation $(f^{\sim})^{(r)} = (f^{(r)})^{\sim}$ is valid (cf. Problem 9.3.5). Furthermore, let us mention that results similar to those of Prop. 10.3.1 also hold for central Riemann (5.1.41), Peano (10.1.6) and Taylor differences (10.1.14), respectively. We leave them to Problems 10.3.1–10.3.3 as they may easily be inferred from the results of this and the preceding section. Problem 10.3.5 contains a collection of characterizations of the classes $(\mathsf{V}^{\sim})^r_{X_{2\pi}}$. We compile the analogous results concerning the classes $(\mathsf{W}^{\sim})^r_{X_{2\pi}}$ in Problem 10.3.6.

Problems

1. Show that $f \in X_{2\pi}$ belongs to $(\mathsf{V}^{\sim})^r_{X_{2\pi}}$ if and only if $\|\bar{\Delta}^r_h f^{\sim}\|_{X_{2\pi}} = O(|h|^r)$, $h \to 0$.
2. Show that $f \in X_{2\pi}$ belongs to $(\mathsf{V}^{\sim})^r_{X_{2\pi}}$ if and only if there exists an rth Peano difference† of $f^{\sim}$ such that $\|\Diamond^r_h f^{\sim}\|_{X_{2\pi}} = O(|h|^r)$, $h \to 0$.
3. Show that $f \in X_{2\pi}$ belongs to $(\mathsf{V}^{\sim})^r_{X_{2\pi}}$ if and only if $\|\nabla^r_h f^{\sim}\|_{X_{2\pi}} = O(|h|^r)$, $h \to 0$, provided $(f^{\sim})^{(r-1)}$ exists and belongs to $X_{2\pi}$.
4. Show that $f \in X_{2\pi}$ belongs to $(\mathsf{V}^{\sim})^r_{X_{2\pi}}$ if and only if $\|(\sigma^{\sim}_n(f; \circ))^{(r)}\|_{X_{2\pi}} = O(1)$, $n \to \infty$.
5. Let $f \in X_{2\pi}$. Show that the following statements are equivalent:

   (i) $f \in (\mathsf{V}^{\sim})_{X_{2\pi}}$, (ii) $f \in \mathsf{V}[X_{2\pi}; \{-i \operatorname{sgn} k\}(ik)^r]$, (iii) $\|(\sigma^{\sim}_n(f; \circ))^{(r)}\|_{X_{2\pi}} = O(1)$, $n \to \infty$,
   (iv) $\|\bar{\Delta}^r_h f^{\sim}\|_{X_{2\pi}} = O(|h|^r)$, $h \to 0$, (v) $\|\Delta^r_h f^{\sim}\|_{X_{2\pi}} = O(|h|^r)$,
   (vi) there exists an rth Peano difference of $f^{\sim}$ such that $\|\Diamond^r_h f^{\sim}\|_{X_{2\pi}} = O(|h|^r)$,
   (vii) $\|\nabla^r_h f^{\sim}\|_{X_{2\pi}} = O(|h|^r)$, $h \to 0$, provided $(f^{\sim})^{(r-1)}$ exists and belongs to $X_{2\pi}$.
6. Let $f \in X_{2\pi}$. Show that the following statements are equivalent:

   (i) $f \in (\mathsf{W}^{\sim})^r_{X_{2\pi}}$, (ii) $f \in \mathsf{W}[X_{2\pi}; \{-i \operatorname{sgn} k\}(ik)^r]$,
   (iii) there exists $g \in X_{2\pi}$ such that $\lim_{h \to 0} \|h^{-r}\Delta^r_h f^{\sim} - g\|_{X_{2\pi}} = 0$‡,
   (iv) there exists an rth Peano difference of $f^{\sim}$ and a function $g \in X_{2\pi}$ such that $\lim_{h \to 0} \|h^{-r}\Diamond^r_h f^{\sim} - g\|_{X_{2\pi}} = 0$,
   (v) $\|(\sigma^{\sim}_m(f; \circ))^{(r)} - (\sigma^{\sim}_n(f; \circ))^{(r)}\|_{X_{2\pi}} = o(1)$, $m, n \to \infty$.
7. Let $f \in L^p_{2\pi}$, $1 < p < \infty$. Show that $\|\Delta^r_h f\|_p = O(|h|^r)$ if and only if $\|\Delta^r_h f^{\sim}\|_p = O(|h|^r)$.
8. Let $f \in L^p_{2\pi}$, $1 < p < \infty$. Show that there is an rth Peano difference of f such that $\|\Diamond^r_h f\|_p = O(|h|^r)$ if and only if there is an rth Peano difference of $f^{\sim}$ such that $\|\Diamond^r_h f^{\sim}\|_p = O(|h|^r)$.
9. State and prove characterizations of the classes $\mathsf{W}[X_{2\pi}; |k|^r]$ and $\mathsf{V}[X_{2\pi}; |k|^r]$, $r \in \mathbb{N}$. (Hint: Use the characterizations of, e.g., $\mathsf{V}^r_{X_{2\pi}}$ and $(\mathsf{V}^{\sim})^r_{X_{2\pi}}$; see also Sec. 11.5)

† The definition of Peano differences of the Hilbert transform $f^{\sim}$ of functions f belonging to $C_{2\pi}$ or $L^1_{2\pi}$ tacitly assumes a slight generalization of our definition (10.1.6). Indeed, we assigned Peano differences to functions $f \in L^1_{2\pi}$, say, and defined $\Diamond^r_h f$ as any expression of the form (10.1.6), where $l_k \in L^1_{2\pi}$, $1 \le k \le r - 1$. Replacing f by its Hilbert transform $f^{\sim}$ it is not necessarily true that $f^{\sim} \in L^1_{2\pi}$. Nevertheless, we call

$$\Diamond^r_h f^{\sim}(x) = r!\{f^{\sim}(x+h) - f^{\sim}(x) - \sum_{k=1}^{r-1} \frac{h^k}{k!} l_k(x)\}$$

an rth Peano difference of $f^{\sim}$, if $l_k \in L^1_{2\pi}$, $1 \le k \le r - 1$, and at least $[f^{\sim}(x+h) - f^{\sim}(x)] \in L^1_{2\pi}$ for every $h \in \mathbb{R}$. Corresponding modifications have to be made for the Peano (and Riemann) derivative of the Hilbert transform $f^{\sim}$ of $f \in L^1_{2\pi}$. The same remarks are necessary in $C_{2\pi}$-space, whereas in $L^p_{2\pi}$, $1 < p < \infty$, there is no difference with the definitions involving f, since then $f^{\sim} \in L^p_{2\pi}$ by Theorem 9.1.3.

‡ g may be referred to as the rth strong Riemann derivative of $f^{\sim}$. Here we again tacitly assume that at least $\Delta^r_h f^{\sim}$ belongs to $X_{2\pi}$.

10.4 Relative Completion

So far we have established characterizations of the classes $\mathsf{W}[\mathsf{X}_{2\pi};\psi(k)]$ and $\mathsf{V}[\mathsf{X}_{2\pi};\psi(k)]$ (for their definition cf. (4.1.24), (4.3.19)) separately for particular functions ψ. Now we shall briefly discuss how these types of classes are connected, the relevant notion being that of *relative completion.*

Let X be an arbitrary Banach space endowed with norm $\|\circ\|_{\mathsf{X}}$. A linear manifold Y of X is called a *Banach subspace* of X if it is a Banach space (with norm $\|\circ\|_{\mathsf{Y}}$) continuously embedded in X, in other words, if there exists a constant c such that $\|f\|_{\mathsf{X}} \le c\,\|f\|_{\mathsf{Y}}$ for all $f \in \mathsf{Y}$, or the identity considered as a mapping of Y into X is a bounded linear operator of Y into X. Without loss of generality we may assume in the sequel that Y is normalized (relative to X), i.e.

$$\|f\|_{\mathsf{X}} \le \|f\|_{\mathsf{Y}} \qquad (f \in \mathsf{Y}). \tag{10.4.1}$$

Definition 10.4.1. *Let* X *be a Banach space and* Y *a normalized Banach subspace of* X. *The* ***completion of*** Y ***relative to*** X, *denoted by* $\tilde{\mathsf{Y}}^{\mathsf{X}}$, *is the set of those elements* $f \in \mathsf{X}$ *for which there is a sequence* $\{f_n\}_{n=1}^{\infty} \subset \mathsf{Y}$ *and a constant* $M > 0$ *such that* $\|f_n\|_{\mathsf{Y}} \le M$ *for all* n *and* $\lim_{n\to\infty} \|f - f_n\|_{\mathsf{X}} = 0$.

In other words, $\tilde{\mathsf{Y}}^{\mathsf{X}}$ is the set of all elements $f \in \mathsf{X}$ which are contained in the closure in X of some bounded sphere in Y, i.e.

$$\tilde{\mathsf{Y}}^{\mathsf{X}} = \bigcup_{\rho>0} \overline{S_{\mathsf{Y}}(\rho)}^{\mathsf{X}}, \quad S_{\mathsf{Y}}(\rho) = \{f \in \mathsf{Y} \mid \|f\|_{\mathsf{Y}} \le \rho\}. \tag{10.4.2}$$

Here $\overline{S_{\mathsf{Y}}(\rho)}^{\mathsf{X}}$ denotes the closure of $S_{\mathsf{Y}}(\rho)$ in X-norm (cf. Sec. 0.7).

Proposition 10.4.2. *If* X *is a Banach space and* Y *a normalized Banach subspace of* X, *then* $\tilde{\mathsf{Y}}^{\mathsf{X}}$ *is a normalized Banach subspace of* X *under the norm*

$$\begin{aligned}\|f\|_{\tilde{\mathsf{Y}}^{\mathsf{X}}} &= \inf\{\rho > 0 \mid f \in \overline{S_{\mathsf{Y}}(\rho)}^{\mathsf{X}}\}\\ &\equiv \inf\{\sup_{n\in\mathbb{N}}\|f_n\|_{\mathsf{Y}} \mid \{f_n\} \subset \mathsf{Y},\ \lim_{n\to\infty} \|f - f_n\|_{\mathsf{X}} = 0\}.\end{aligned} \tag{10.4.3}$$

Proof. The fact that $\tilde{\mathsf{Y}}^{\mathsf{X}}$ is a linear manifold and $\|f\|_{\tilde{\mathsf{Y}}^{\mathsf{X}}}$ is homogeneous and satisfies the triangular inequality follows immediately from the observation that if $f_1 \in \overline{S_{\mathsf{Y}}(\rho_1)}^{\mathsf{X}}$ and $f_2 \in \overline{S_{\mathsf{Y}}(\rho_2)}^{\mathsf{X}}$, then $\alpha_1 f_1 + \alpha_2 f_2 \in \overline{S_{\mathsf{Y}}(|\alpha_1|\,\rho_1 + |\alpha_2|\,\rho_2)}^{\mathsf{X}}$. Furthermore, in view of (10.4.1) we have $\|f\|_{\mathsf{X}} \le \|f\|_{\tilde{\mathsf{Y}}^{\mathsf{X}}}$. Hence (10.4.3) defines a norm, and $\tilde{\mathsf{Y}}^{\mathsf{X}}$ is a normalized (normed linear) subspace of X. It remains to show that $\tilde{\mathsf{Y}}^{\mathsf{X}}$ is complete. Let $\{f_n\}$ be a Cauchy sequence in $\tilde{\mathsf{Y}}^{\mathsf{X}}$. It is, *a fortiori*, a Cauchy sequence in X with limit f, say. Certainly, $f \in \tilde{\mathsf{Y}}^{\mathsf{X}}$. For any $\varepsilon > 0$ we can choose an $n(\varepsilon)$ such that $f_m - f_n \in \overline{S_{\mathsf{Y}}(\varepsilon)}^{\mathsf{X}}$ for all $m, n > n(\varepsilon)$. Hence also $f_m - f \in \overline{S_{\mathsf{Y}}(\varepsilon)}^{\mathsf{X}}$, and thus f is the limit of $\{f_n\}$ in the $\tilde{\mathsf{Y}}^{\mathsf{X}}$-metric.

Proposition 10.4.3. *If* Y *is reflexive, then the Banach spaces* Y *and* $\tilde{\mathsf{Y}}^{\mathsf{X}}$ *are equal with equal norms.*

For a proof as well as for further general properties of $\tilde{\mathsf{Y}}^{\mathsf{X}}$ we refer to the references given in Sec. 10.7.

We now turn to a discussion of the relation between the classes $\mathsf{W}[\mathsf{X}_{2\pi}; \psi(k)]$ and $\mathsf{V}[\mathsf{X}_{2\pi}; \psi(k)]$, illustrating the usefulness of the above concepts. In what follows, ψ is an arbitrary complex-valued function on $\mathbb{Z}$.

Proposition 10.4.4. (a) *The class*

$$\mathsf{W}[\mathsf{X}_{2\pi}; \psi(k)] = \{f \in \mathsf{X}_{2\pi} \mid \psi(k)f^\wedge(k) = g^\wedge(k), g \in \mathsf{X}_{2\pi}\} \tag{10.4.4}$$

becomes a normalized Banach subspace of $\mathsf{X}_{2\pi}$ *under the norm*

$$\|f\|_{\mathsf{W}[\mathsf{X}_{2\pi};\psi(k)]} = \|f\|_{\mathsf{X}_{2\pi}} + \|g\|_{\mathsf{X}_{2\pi}}. \tag{10.4.5}$$

(b) *The set* T_n *of (trigonometric) polynomials is dense in* $\mathsf{W}[\mathsf{X}_{2\pi}; \psi(k)]$ (*with respect to norm* (10.4.5)).

Proof. (a) Obviously, $\mathsf{W}[\mathsf{X}_{2\pi}; \psi(k)]$ is a linear manifold of $\mathsf{X}_{2\pi}$ on which a normalized norm is defined by (10.4.5). It remains to show that $\mathsf{W}[\mathsf{X}_{2\pi}; \psi(k)]$ is complete. Let $\{f_n\}$ be a Cauchy sequence in $\mathsf{W}[\mathsf{X}_{2\pi}; \psi(k)]$, i.e. $\|f_m - f_n\|_{\mathsf{W}[\mathsf{X}_{2\pi};\psi(k)]} \to 0$ as $m, n \to \infty$. If $g_n \in \mathsf{X}_{2\pi}$ is such that $\psi(k)f_n^\wedge(k) = g_n^\wedge(k)$, $k \in \mathbb{Z}$, the former limit means $\|f_m - f_n\|_{\mathsf{X}_{2\pi}} + \|g_m - g_n\|_{\mathsf{X}_{2\pi}} \to 0$ as $m, n \to \infty$. Thus $\{f_n\}$ and $\{g_n\}$ are Cauchy sequences in $\mathsf{X}_{2\pi}$. The completeness of $\mathsf{X}_{2\pi}$ now ensures the existence of two functions $f, g \in \mathsf{X}_{2\pi}$ such that f_n and g_n tend to f and g in $\mathsf{X}_{2\pi}$-norm, respectively. By (4.1.2) this implies that $\lim_{n\to\infty} f_n^\wedge(k) = f^\wedge(k)$ and $\lim_{n\to\infty} g_n^\wedge(k) = g^\wedge(k)$ for every $k \in \mathbb{Z}$. Therefore $\psi(k)f^\wedge(k) = g^\wedge(k)$, i.e. $f \in \mathsf{W}[\mathsf{X}_{2\pi}; \psi(k)]$.

(b) First of all, $\mathsf{T}_n \subset \mathsf{W}[\mathsf{X}_{2\pi}; \psi(k)]$ for every n. For, if $f(x) = \sum_{k=-n}^{n} c_k \exp\{ikx\}$, then $\psi(k)f^\wedge(k) = g^\wedge(k)$ with g given by $g(x) = \sum_{k=-n}^{n} \psi(k)c_k \exp\{ikx\}$. Therefore, let $f \in \mathsf{W}[\mathsf{X}_{2\pi}; \psi(k)]$ and $g \in \mathsf{X}_{2\pi}$ be such that $\psi(k)f^\wedge(k) = g^\wedge(k)$, $k \in \mathbb{Z}$. For the Fejér means of f we have $\psi(k)[\sigma_n(f; \circ)]^\wedge(k) = [\sigma_n(g; \circ)]^\wedge(k)$, and therefore

$$\|\sigma_n(f; \circ) - f(\circ)\|_{\mathsf{W}[\mathsf{X}_{2\pi};\psi(k)]} = \|\sigma_n(f; \circ) - f(\circ)\|_{\mathsf{X}_{2\pi}} + \|\sigma_n(g; \circ) - g(\circ)\|_{\mathsf{X}_{2\pi}},$$

which tends to zero as $n \to \infty$ by Cor. 1.2.4. This completes the proof.

Proposition 10.4.5. *The class*

$$\mathsf{V}[\mathsf{X}_{2\pi}; \psi(k)] = \begin{cases} \{f \in \mathsf{C}_{2\pi} \mid \psi(k)f^\wedge(k) = g^\wedge(k), g \in \mathsf{L}_{2\pi}^\infty\} \\ \{f \in \mathsf{L}_{2\pi}^1 \mid \psi(k)f^\wedge(k) = \mu^\vee(k), \mu \in \mathsf{BV}_{2\pi}\} \\ \{f \in \mathsf{L}_{2\pi}^p \mid \psi(k)f^\wedge(k) = g^\wedge(k), g \in \mathsf{L}_{2\pi}^p\} \quad (1 < p < \infty). \end{cases} \tag{10.4.6}$$

becomes a normalized Banach subspace of $\mathsf{X}_{2\pi}$ *under the norm*

$$\|f\|_{\mathsf{X}[\mathsf{V}_{2\pi};\psi(k)]} = \begin{cases} \|f\|_{\mathsf{C}_{2\pi}} + \|g\|_{\mathsf{L}_{2\pi}^\infty}, & \mathsf{X}_{2\pi} = \mathsf{C}_{2\pi} \\ \|f\|_{\mathsf{L}_{2\pi}^1} + \|\mu\|_{\mathsf{BV}_{2\pi}}, & \mathsf{X}_{2\pi} = \mathsf{L}_{2\pi}^1 \\ \|f\|_{\mathsf{L}_{2\pi}^p} + \|g\|_{\mathsf{L}_{2\pi}^p}, & \mathsf{X}_{2\pi} = \mathsf{L}_{2\pi}^p, \quad (1 < p < \infty). \end{cases} \tag{10.4.7}$$

The proof follows as for Prop. 10.4.4 by exploiting the completeness of the underlying spaces $\mathsf{X}_{2\pi}$, $\mathsf{BV}_{2\pi}$, $\mathsf{L}_{2\pi}^\infty$. But part (b) of Prop. 10.4.4 is not necessarily true for the classes $\mathsf{V}[\mathsf{C}_{2\pi}; \psi(k)]$ and $\mathsf{V}[\mathsf{L}_{2\pi}^1; \psi(k)]$ (cf. Prop. 1.1.6).

Theorem 10.4.6. *A function* $f \in \mathsf{X}_{2\pi}$ *belongs to* $\mathsf{V}[\mathsf{X}_{2\pi}; \psi(k)]$ *if and only if* f *belongs to the completion of* $\mathsf{W}[\mathsf{X}_{2\pi}; \psi(k)]$ *relative to* $\mathsf{X}_{2\pi}$.

Proof. For $X_{2\pi} = L^p_{2\pi}$, $1 < p < \infty$, the reflexivity of the spaces settles the assertion.

Let $X_{2\pi} = L^1_{2\pi}$ and suppose that $f \in V[L^1_{2\pi}; \psi(k)]$ with $\psi(k)f^\wedge(k) = \mu^\vee(k)$, $\mu \in BV_{2\pi}$. Since T_n is a subset of $W[L^1_{2\pi}; \psi(k)]$ it follows that $\sigma_n(f; x) \in W[L^1_{2\pi}; \psi(k)]$; also $\lim_{n\to\infty} \|\sigma_n(f; \circ) - f(\circ)\|_1 = 0$ by Cor. 1.2.4. Moreover we have, $\psi(k)[\sigma_n(f; \circ)]^\wedge(k) = [\sigma_n(d\mu; \circ)]^\wedge(k)$, and thus by (1.1.4) and (1.1.13)

$$\|\sigma_n(f; \circ)\|_{W[L^1_{2\pi};\psi(k)]} = \|\sigma_n(f; \circ)\|_1 + \|\sigma_n(d\mu; \circ)\|_1 \le \|f\|_1 + \|\mu\|_{BV_{2\pi}}$$

uniformly for all $n \in \mathbb{N}$. This implies that $f \in \widetilde{W[L^1_{2\pi}; \psi(k)]}^{L^1_{2\pi}}$. Conversely, let $f \in \widetilde{W[L^1_{2\pi}; \psi(k)]}^{L^1_{2\pi}}$. By definition, there exists a sequence $\{f_n\} \subset W[L^1_{2\pi}; \psi(k)]$ such that $\lim_{n\to\infty} \|f_n - f\|_1 = 0$ and $\|f_n\|_1 + \|g_n\|_1 \le M$ uniformly for all n where $g_n \in L^1_{2\pi}$ is such that $\psi(k)f_n^\wedge(k) = g_n^\wedge(k)$. We have

$$\left\| \sum_{k=-m}^{m} \left(1 - \frac{|k|}{m+1}\right)\psi(k)f_n^\wedge(k)\, e^{ik\circ} \right\|_1 = \|\sigma_m(g_n; \circ)\|_1 \le M$$

uniformly for all $m, n \in \mathbb{N}$. Moreover, since $\lim_{n\to\infty} \psi(k)f_n^\wedge(k) = \psi(k)f^\wedge(k)$ for each $k \in \mathbb{Z}$, we have by Fatou's lemma

$$\left\| \sum_{k=-m}^{m} \left(1 - \frac{|k|}{m+1}\right)\psi(k)f^\wedge(k)\, e^{ik\circ} \right\|_1 \le M \qquad (m \in \mathbb{N}).$$

Now Theorem 6.1.2(iv) ensures the existence of $\mu \in BV_{2\pi}$ such that $\psi(k)f^\wedge(k) = \mu^\vee(k)$ for every $k \in \mathbb{Z}$.

When $X_{2\pi} = C_{2\pi}$ we may proceed in similar fashion, establishing the assertions.

By the way, the proof of Theorem 10.4.6 in particular shows that

$$\|f\|_{V[X_{2\pi};\psi(k)]} = \|f\|_{\widetilde{W[X_{2\pi};\psi(k)]}^{X_{2\pi}}}.$$

Thus the Banach subspaces $V[X_{2\pi}; \psi(k)]$ and $\widetilde{W[X_{2\pi}; \psi(k)]}^{X_{2\pi}}$ are equal with equal norms.

The latter remark leads to an interesting interpretation of many of our equivalent assertions in connection with continuous embeddings of Banach spaces. Let X be a linear system, Y a linear subset of X, and suppose that two norms $\|\circ\|_1$, $\|\circ\|_2$ are defined on Y. These norms are said to be equivalent (or define the same topology) on Y if there exist two positive constants c_1, c_2 such that $c_1\|f\|_1 \le \|f\|_2 \le c_2\|f\|_1$ for every $f \in Y$; in other words, if the identity mapping $f \to f$ of Y, endowed with norm $\|\circ\|_1$, onto Y, endowed with norm $\|\circ\|_2$, is continuous together with its inverse. If $c_1 = c_2$, thus $\|f\|_1 = \|f\|_2$ for every $f \in Y$, then the two norms are said to be equal. From this point of view, let us once more consider the results of Theorem 4.3.13 in case $X_{2\pi} = C_{2\pi}$, for example.

By Prop. 10.4.5, $V[C_{2\pi}; (ik)^r]$ is a normalized Banach subspace of $C_{2\pi}$ under the norm (10.4.7). On the other hand, $V^r_{C_{2\pi}}$ is also a linear subset of $C_{2\pi}$ on which one may introduce the norm

$$\|f\|_{V^r_{C_{2\pi}}} = \|f\|_{C_{2\pi}} + \|f^{(r)}\|_{L^\infty_{2\pi}} \qquad (f \in V^r_{C_{2\pi}}). \tag{10.4.8}$$

Obviously, $V^r_{C_{2\pi}}$ becomes a normalized normed linear subspace of $C_{2\pi}$ under (10.4.8).

$\mathsf{V}^r_{\mathsf{C}_{2\pi}}$ even becomes a Banach subspace which may, for example, be established via the proof of Theorem 4.3.13. Indeed, let $\{f_n\}$ be a Cauchy sequence in $\mathsf{V}^r_{\mathsf{C}_{2\pi}}$, thus

$$\lim_{m,n\to\infty} [\|f_m - f_n\|_{\mathsf{C}_{2\pi}} + \|f_m^{(r)} - f_n^{(r)}\|_{\mathsf{L}^\infty_{2\pi}}] = 0.$$

By the completeness of $\mathsf{C}_{2\pi}$, $\mathsf{L}^\infty_{2\pi}$ there exist functions $f \in \mathsf{C}_{2\pi}$, $g \in \mathsf{L}^\infty_{2\pi}$ such that $\lim_{n\to\infty} \|f_n - f\|_{\mathsf{C}_{2\pi}} = 0$, $\lim_{n\to\infty} \|f_n^{(r)} - g\|_{\mathsf{L}^\infty_{2\pi}} = 0$. On the other hand, passing to Fourier transforms, $(ik)^r f_n^\wedge(k) = [f_n^{(r)}]^\wedge(k)$ by Problem 4.3.4(i), and therefore $(ik)^r f^\wedge(k) = g^\wedge(k)$ by (4.1.2). Now Problem 4.3.4(ii) applies and gives $f \in \mathsf{V}^r_{\mathsf{C}_{2\pi}}$ with $f^{(r)}(x) = g(x)$ a.e.

In Theorem 4.3.13 it has been stated that $f \in \mathsf{V}^r_{\mathsf{C}_{2\pi}}$ if and only if $f \in \mathsf{V}[\mathsf{C}_{2\pi}; (ik)^r]$. In view of the above considerations this result may be rephrased and even strengthened by saying that the Banach spaces $\mathsf{V}^r_{\mathsf{C}_{2\pi}}$ (with norm (10.4.8)) and $\mathsf{V}[\mathsf{C}_{2\pi}; (ik)^r]$ (with norm (10.4.7)) are equal with equal norms.

Likewise, one may rephrase the result of Theorem 10.2.2. Indeed, the set $\mathsf{Lip}_r(\mathsf{C}_{2\pi}; \alpha)$, $0 < \alpha \le r$, is a linear subset of $\mathsf{C}_{2\pi}$ on which one may define a norm by

$$\|f\|_{\mathsf{Lip}_r(\mathsf{C}_{2\pi};\alpha)} = \|f\|_{\mathsf{C}_{2\pi}} + \sup_{h\neq 0} \|h^{-\alpha}\Delta_h^r f\|_{\mathsf{C}_{2\pi}} \qquad (f \in \mathsf{Lip}_r(\mathsf{C}_{2\pi}; \alpha)). \tag{10.4.9}$$

By elementary direct arguments it follows that $\mathsf{Lip}_r(\mathsf{C}_{2\pi}; \alpha)$ is a normalized Banach subspace of $\mathsf{C}_{2\pi}$ under (10.4.9) (cf. Problem 10.4.5). Theorem 10.2.2 then states that $f \in \mathsf{V}^r_{\mathsf{C}_{2\pi}}$ if and only if $\|h^{-r}\Delta_h^r f\|_{\mathsf{C}_{2\pi}} = O(1)$ as $h \to 0$. However, following the lines of the proof of this theorem, it was actually shown (cf. Prop. 0.8.5(iii), 0.8.9) that the Banach spaces $\mathsf{V}^r_{\mathsf{C}_{2\pi}}$ (with norm (10.4.8)) and $\mathsf{Lip}_r(\mathsf{C}_{2\pi}; r)$ (with norm (10.4.9)) are equal with equal norms.

Let us finally emphasize that almost all of the equivalent assertions so far obtained as well as many yet to be established may be reformulated in the above way in terms of different norms on linear sets; compare the corresponding Problems.

Problems

1. Prove Prop. 10.4.5.
2. Prove Theorem 10.4.6 in $\mathsf{C}_{2\pi}$-space.
3. Show that $\mathsf{V}^r_{\mathsf{X}_{2\pi}} = \widetilde{\mathsf{W}^r_{\mathsf{X}_{2\pi}}}^{\mathsf{X}_{2\pi}}$ and $(\mathsf{V}^\sim)^r_{\mathsf{X}_{2\pi}} = \widetilde{(\mathsf{W}^\sim)^r_{\mathsf{X}_{2\pi}}}^{\mathsf{X}_{2\pi}}$.
4. Define a norm on $\mathsf{V}^r_{\mathsf{X}_{2\pi}}$ analogously to (10.4.8) and show that $\mathsf{V}^r_{\mathsf{X}_{2\pi}}$ is a normalized Banach subspace of $\mathsf{X}_{2\pi}$. Show that the Banach spaces $\mathsf{V}^r_{\mathsf{X}_{2\pi}}$ and $\mathsf{V}[\mathsf{X}_{2\pi}; (ik)^r]$ are equal with equal norms. (Hint: Theorem 4.3.13)
5. (i) Show that $\mathsf{Lip}_r(\mathsf{X}_{2\pi}; \alpha)$ is a normalized Banach subspace of $\mathsf{X}_{2\pi}$ under the norm (10.4.9) (with $\mathsf{C}_{2\pi}$ replaced by $\mathsf{X}_{2\pi}$). (Hint: See also LORENTZ [3, p. 50])
   (ii) Show that the Banach spaces $\mathsf{V}^r_{\mathsf{X}_{2\pi}}$ and $\mathsf{Lip}_r(\mathsf{X}_{2\pi}; r)$ are equal with equal norms. (Hint: Theorem 10.2.2, Prop. 0.8.5(iii), 0.8.8, (0.8.9))
   (iii) State and prove a corresponding reformulation of Theorem 10.2.5.

10.5 Generalized Derivatives in L^p-Norm and Characterizations for $1 \le p \le 2$

The aim of the next two sections is to establish analogs of the results of the preceding sections for functions belonging to $\mathsf{X}(\mathbb{R})$. Here we may employ the Fourier transform method on $\mathsf{X}(\mathbb{R})$. But, since the Fourier transform has only been introduced on the spaces L^p, $1 \le p \le 2$, the application of this method is limited to these spaces, which are to be considered in the present section. The next section is devoted to a study of the corresponding problems in the spaces C and L^p, $2 < p < \infty$, using elementary duality arguments.

We start off with characterizations of the classes $\mathsf{W}^r_{\mathsf{L}^p}$ (cf. 3.1.48). To this end the definitions of an *rth Riemann derivative*, an *rth Peano difference* and an *rth Peano derivative in* $\mathsf{X}(\mathbb{R})$*-norm* are quite obvious; in Def. 10.1.1 and 10.1.8 we only have to replace $\mathsf{X}_{2\pi}$ by $\mathsf{X}(\mathbb{R})$.

Theorem 10.5.1. *Let $f \in \mathsf{L}^p$, $1 \le p \le 2$. The following statements are equivalent:*

(i) *f belongs to $\mathsf{W}^r_{\mathsf{L}^p}$,*
(ii) *f has an rth Peano derivative $D_s^{\langle r\rangle}f$ in L^p-norm,*
(iii) *f has an rth Riemann derivative $D_s^{[r]}f$ in L^p-norm.*

Proof. (i) $\Rightarrow$ (ii). Since $f \in \mathsf{W}^r_{\mathsf{L}^p}$, there exists $\phi \in \mathsf{AC}^{r-1}_{\text{loc}}$ such that $f = \phi$ a.e. and $\phi^{(k)} \in \mathsf{L}^p$, $1 \le k \le r$. Hence integration by parts gives

$$(10.5.1) \qquad f(x+h) - f(x) - \sum_{k=1}^{r-1} \frac{h^k}{k!}\phi^{(k)}(x) = \int_0^h \frac{(h-u)^{r-1}}{(r-1)!}\phi^{(r)}(x+u)\,du \quad \text{a.e.}$$

By the Hölder–Minkowski inequality and continuity in the mean this implies that

$$\left\| \frac{r!}{h^r}\left\{ f(\circ+h) - f(\circ) - \sum_{k=1}^{r-1}\frac{h^k}{k!}\phi^{(k)}(\circ)\right\} - \phi^{(r)}(\circ)\right\|_p$$
$$\le r\int_0^1 (1-u)^{r-1}\|\phi^{(r)}(\circ+hu) - \phi^{(r)}(\circ)\|_p\,du = o(1) \qquad (h \to 0).$$

Thus the rth Peano derivative of f in L^p-norm exists and $D_s^{\langle r\rangle}f = \phi^{(r)}$ a.e.

(ii) $\Rightarrow$ (iii). This part follows immediately by (10.1.10).

(iii) $\Rightarrow$ (i). Let $p = 1$. Then by Prop. 5.1.1(i)

$$(10.5.2) \qquad [\Delta_h^r f]^\wedge(v) = (e^{ihv} - 1)^r f^\wedge(v),$$

and thus for each $v \in \mathbb{R}$

$$(10.5.3) \qquad \lim_{h\to 0}\left[\frac{1}{h^r}\Delta_h^r f\right]^\wedge(v) = (iv)^r f^\wedge(v).$$

On the other hand, by (5.1.2) and the hypothesis

$$\left|\left[\frac{1}{h^r}\Delta_h^r f\right]^\wedge(v) - [D_s^{[r]}f]^\wedge(v)\right| \le \left\|\frac{1}{h^r}\Delta_h^r f - D_s^{[r]}f\right\|_1 = o(1)$$

as $h \to 0$, and therefore

$$(10.5.4) \qquad (iv)^r f^\wedge(v) = [D_s^{[r]}f]^\wedge(v) \qquad (v \in \mathbb{R}),$$

which implies $f \in \mathsf{W}^r_{\mathsf{L}^1}$ by Prop. 5.1.15.

If $1 < p \le 2$, it follows by Prop. 5.2.11(i) that (10.5.2) and therefore (10.5.3) hold almost everywhere. Since by Titchmarsh' inequality and the assumption

$$\left\|\left[\frac{1}{h^r}\Delta_h^r f\right]^\wedge - [D_s^{[r]}f]^\wedge\right\|_{p'} \le \left\|\frac{1}{h^r}\Delta_h^r f - D_s^{[r]}f\right\|_p = o(1)$$

as $h \to 0$, Fatou's lemma implies that

$$\left\| (i\circ)^r f^\wedge(\circ) - [D_s^{[r]}f]^\wedge(\circ) \right\|_{p'}^{p'} \le \liminf_{h\to 0} \left\| \left[\frac{1}{h^r}\Delta_h^r f\right]^\wedge - [D_s^{[r]}f]^\wedge \right\|_{p'}^{p'} = 0,$$

which shows that (10.5.4) is valid almost everywhere. In virtue of Prop. 5.2.20 the proof is complete.

It is worthwhile to note the particular case that the rth Peano or Riemann derivative in the mean of order p vanishes.

Corollary 10.5.2. *Let $f \in \mathsf{L}^p$, $1 \le p \le 2$. The following statements are equivalent:*

(i) *f equals zero almost everywhere,*
(ii) *there exists an rth Peano difference of f such that*

$$\| \Diamond_h^r f \|_p = o(|h|^r) \qquad (h \to 0),$$

(iii)
$$\| \Delta_h^r f \|_p = o(|h|^r) \qquad (h \to 0).$$

For the sake of completeness let us mention

Proposition 10.5.3. *Let $f \in \mathsf{L}^p$, $1 \le p \le 2$. The following statements are equivalent:*
(i) *f belongs to $\mathsf{W}_{\mathsf{L}^p}^r$,*
(ii) *f has an rth strong (weak) derivative $D_s^{(r)}f$ $(D_w^{(r)}f)$,*
(iii) *f has an rth strong (weak) Riemann derivative $D_s^{[r]}f$ $(D_w^{[r]}f)$.*

Proof. We shall only show that the existence of the rth weak Riemann derivative implies (i); the other implications follow by routine arguments (cf. Cor. 10.1.15). The existence of $D_w^{[r]}f$ for $f \in \mathsf{L}^p$ states that $D_w^{[r]}f \in \mathsf{L}^p$ and

$$\lim_{h\to 0} \int_{-\infty}^{\infty} \frac{1}{h^r}\Delta_h^r f(u)s(u)\,du = \int_{-\infty}^{\infty} (D_w^{[r]}f)(u)s(u)\,du \tag{10.5.5}$$

for every $s \in \mathsf{L}^{p'}$. If $p = 1$ we take $s(u) = (2\pi)^{-1/2}\exp\{-ivu\}$, $v \in \mathbb{R}$, and have

$$\lim_{h\to 0} \left(\frac{e^{ihv} - 1}{h}\right)^r f^\wedge(v) = [D_w^{[r]}f]^\wedge(v).$$

Thus $(iv)^r f^\wedge(v) = [D_w^{[r]}]^\wedge(v)$, and (i) follows by Prop. 5.1.15. If $1 < p \le 2$ we take

$$s(u) = \frac{2}{\pi\rho}\frac{\sin^2[\rho(x-u)/2]}{(x-u)^2} \qquad (x \in \mathbb{R}, \rho > 0) \tag{10.5.6}$$

and obtain by Parseval's formula (5.2.18), Problem 5.1.2(i), and Lebesgue's dominated convergence theorem that

$$\begin{aligned}
\frac{2}{\pi\rho}\int_{-\infty}^{\infty} (D_w^{[r]}f)(u)\frac{\sin^2[\rho(x-u)/2]}{(x-u)^2}\,du &= \lim_{h\to 0}\frac{2}{\pi\rho}\int_{-\infty}^{\infty}\frac{1}{h^r}\Delta_h^r f(u)\frac{\sin^2[\rho(x-u)/2]}{(x-u)^2}\,du \\
&= \lim_{h\to 0}\frac{1}{\sqrt{2\pi}}\int_{-\rho}^{\rho}\left(1 - \frac{|v|}{\rho}\right)\left(\frac{e^{ihv}-1}{h}\right)^r f^\wedge(v)e^{ixv}\,dv \\
&= \frac{1}{\sqrt{2\pi}}\int_{-\rho}^{\rho}\left(1 - \frac{|v|}{\rho}\right)(iv)^r f^\wedge(v)\,e^{ixv}\,dv.
\end{aligned}$$

Therefore we have by (3.1.17) uniformly for all $\rho > 0$

$$\left\| \frac{1}{\sqrt{2\pi}}\int_{-\rho}^{\rho}\left(1 - \frac{|v|}{\rho}\right)(iv)^r f^\wedge(v)\,e^{i\circ v}\,dv \right\|_p \le \| D_w^{[r]}f \|_p,$$

which implies the existence of $g \in \mathsf{L}^p$ such that $(iv)^r f^\wedge(v) = g^\wedge(v)$ a.e. by Theorem 6.1.7. According to Prop. 5.2.20 assertion (i) follows.

Next we proceed to characterizations of the classes $\mathsf{V}^r_{\mathsf{L}^p}$, $1 \le p \le 2$, introduced in (5.3.29).

Theorem 10.5.4. *Let $f \in \mathsf{L}^p$, $1 \le p \le 2$. The following statements are equivalent:*

(i) *f belongs to $\mathsf{V}^r_{\mathsf{L}^p}$,*

(ii) *there exists an rth Peano difference of f such that*

$$\|\Diamond^r_h f\|_p = O(|h|^r) \qquad (h \to 0),$$

(iii)
$$\|\Delta^r_h f\|_p = O(|h|^r) \qquad (h \to 0).$$

Proof. Since the proofs of the implications (i) ⇒ (ii) ⇒ (iii) follow by routine arguments we only prove (iii) ⇒ (i). For this purpose, we consider the Fejér means of $h^{-r}\Delta^r_h f$, obtaining by (3.1.17)

$$\left\|\sigma\left(\frac{1}{h^r}\Delta^r_h f;\circ;\rho\right)\right\|_p \le \left\|\frac{1}{h^r}\Delta^r_h f\right\|_p.$$

Thus the hypothesis and the Parseval formulae (5.1.5) and (5.2.18) yield

$$\left\|\frac{1}{\sqrt{2\pi}}\int_{-\rho}^{\rho}\left(1-\frac{|v|}{\rho}\right)\left(\frac{e^{ihv}-1}{h}\right)^r f^\wedge(v)\, e^{i\circ v}\, dv\right\|_p = O(1) \qquad (\rho > 0, h \to 0).$$

On the other hand, by Lebesgue's dominated convergence theorem we have

$$\lim_{h\to 0}\int_{-\rho}^{\rho}\left(1-\frac{|v|}{\rho}\right)\left(\frac{e^{ihv}-1}{h}\right)^r f^\wedge(v)\, e^{ixv}\, dv = \int_{-\rho}^{\rho}\left(1-\frac{|v|}{\rho}\right)(iv)^r f^\wedge(v)\, e^{ixv}\, dv$$

for each $x \in \mathbb{R}$ and $\rho > 0$. Hence by Fatou's lemma

$$\left\|\frac{1}{\sqrt{2\pi}}\int_{-\rho}^{\rho}\left(1-\frac{|v|}{\rho}\right)(iv)^r f^\wedge(v)\, e^{i\circ v}\, dv\right\|_p^p$$
$$\le \liminf_{h\to 0}\left\|\frac{1}{\sqrt{2\pi}}\int_{-\rho}^{\rho}\left(1-\frac{|v|}{\rho}\right)\left(\frac{e^{ihv}-1}{h}\right)^r f^\wedge(v)\, e^{i\circ v}\, dv\right\|_p^p = O(1) \qquad (\rho > 0),$$

so that Theorems 6.1.7 and 5.3.15 give $f \in \mathsf{V}^r_{\mathsf{L}^p}$.

In conjunction with Cor. 10.2.3 and Prop. 10.2.4 let us state

Proposition 10.5.5. *Let $f \in \mathsf{L}^p$, $1 \le p \le 2$. The following assertions are equivalent:*

(i)
$$\left\|\frac{1}{h^r}\Delta^r_h f\right\|_p = O(1) \qquad (h \to 0),$$

(ii) *for $p = 1$: there exists $\mu \in \mathsf{BV}$ such that for each $s \in \mathsf{C}_0$*

$$\lim_{h\to 0}\int_{-\infty}^{\infty} s(u)\,\frac{1}{h^r}\Delta^r_h f(u)\, du = \int_{-\infty}^{\infty} s(u)\, d\mu(u),$$

for $1 < p \le 2$: there exists $g \in \mathsf{L}^p$ such that

$$\lim_{h\to 0}\left\|\frac{1}{h^r}\Delta^r_h f - g\right\|_p = 0.$$

We only indicate the proof of (ii) $\Rightarrow$ (i) for $p = 1$. Taking s as in (10.5.6) we have as in the proof of Prop. 10.5.3

$$\frac{2}{\pi\rho}\int_{-\infty}^{\infty}\frac{\sin^2[\rho(x-u)/2]}{(x-u)^2}\,d\mu(u) = \frac{1}{\sqrt{2\pi}}\int_{-\rho}^{\rho}\left(1-\frac{|v|}{\rho}\right)(iv)^r f^\wedge(v)\,e^{ixv}\,dv.$$

Therefore by Problem 3.1.9 uniformly for all $\rho > 0$

$$\left\|\frac{1}{\sqrt{2\pi}}\int_{-\rho}^{\rho}\left(1-\frac{|v|}{\rho}\right)(iv)^r f^\wedge(v)\,e^{i\circ v}\,dv\right\|_1 \le \|\mu\|_{\mathsf{BV}}.$$

By Theorem 6.1.7 this implies the existence of $\nu \in \mathsf{BV}$ such that $(iv)^r f^\wedge(v) = \nu^\vee(v)$ for all $v \in \mathbb{R}$ (in fact, μ and ν are equal). Assertion (i) follows by Prop. 5.3.14 and Theorem 10.5.4.

Theorem 10.5.6. *Let $f \in \mathsf{L}^p$, $1 \le p \le 2$. The following statements are equivalent:*

(i) *f belongs to $(\mathsf{V}^\sim)^r_{\mathsf{L}^p}$,*

(ii) *there exists an rth Peano difference† of the Hilbert transform $f^\sim$ such that*

$$\|\Diamond^r_h f^\sim\|_p = O(|h|^r) \qquad (h \to 0),$$

(iii)
$$\|\Delta^r_h f^\sim\|_p = O(|h|^r) \qquad (h \to 0).$$

Proof. If $1 < p \le 2$, then $f \in \mathsf{L}^p$ implies $f^\sim \in \mathsf{L}^p$ by Theorem 8.1.12, and therefore the proof follows by Theorem 10.5.4 with f replaced by $f^\sim$.

Thus it remains to prove the case $p = 1$ for which $f \in \mathsf{L}^1$ *a priori* only implies the existence of $f^\sim$ almost everywhere (cf. Theorem 8.1.6). Let $f \in (\mathsf{V}^\sim)^r_{\mathsf{L}^1}$, i.e., there exists $\phi \in \mathsf{AC}^{r-2}_{\text{loc}}$ with $\phi^{(k)} \in \mathsf{L}^1$, $1 \le k \le r-1$, $\phi^{(r-1)} \in \mathsf{BV}$ such that $f^\sim = \phi$ a.e. By partial integration we have (cf. (10.2.9))

$$\text{(10.5.7)} \qquad f^\sim(x+h) - f^\sim(x) = \sum_{k=1}^{r-1}\frac{h^k}{k!}\phi^{(k)}(x)$$
$$+ \int_0^h \frac{(h-u)^{r-2}}{(r-2)!}\,[\phi^{(r-1)}(x+u) - \phi^{(r-1)}(x)]\,du \quad \text{a.e.}$$

Since the right-hand side of (10.5.7) belongs to L^1, $[f^\sim(x+h) - f^\sim(x)]$ represents a function of L^1 and therefore by (5.3.7)

$$\left\|r!\left\{f^\sim(\circ+h) - f^\sim(\circ) - \sum_{k=1}^{r-1}\frac{h^k}{k!}\phi^{(k)}(\circ)\right\}\right\|_1$$
$$\le \sup_{|u|\le|h|}\|\phi^{(r-1)}(\circ+u) - \phi^{(r-1)}(\circ)\|_1 r\,|h|^{r-1} \le r\|\phi^{(r-1)}\|_{\mathsf{BV}}\,|h|^r.$$

Thus there exists an rth Peano difference of the Hilbert transform $f^\sim$ such that (ii) is satisfied.

The proof of (ii) $\Rightarrow$ (iii) again follows by (10.1.10), and so in particular $\Delta^r_h f^\sim \in \mathsf{L}^1$.

Finally, let f be such that (iii) is satisfied. Since $\Delta^r_h f^\sim \in \mathsf{L}^1$ by hypothesis, and since obviously $\Delta^r_h f \in \mathsf{L}^1$, we have by Prop. 8.3.1 that

$$[\Delta^r_h f^\sim]^\wedge(v) = \{-i\,\mathrm{sgn}\,v\}[\Delta^r_h f]^\wedge(v) = \{-i\,\mathrm{sgn}\,v\}(e^{ihv}-1)^r f^\wedge(v)$$

for all $v \in \mathbb{R}$. Since

$$\left\|\sigma\left(\frac{1}{h^r}\Delta^r_h f^\sim;\circ;\rho\right)\right\|_1 = O(1) \qquad (\rho > 0,\ h \to 0),$$

† Cf. footnote to Prop. 8.2.9 and Problem 10.5.6.

it follows as in the proof of Theorem 10.5.4 that

$$\left\| \frac{1}{\sqrt{2\pi}} \int_{-\rho}^{\rho} \left(1 - \frac{|v|}{\rho}\right)\{-i \operatorname{sgn} v\}\left(\frac{e^{ihv}-1}{h}\right)^r f^\wedge(v)\, e^{i\circ v}\, dv \right\|_1 = O(1) \quad (\rho > 0, h \to 0)$$

and

$$\left\| \frac{1}{\sqrt{2\pi}} \int_{-\rho}^{\rho} \left(1 - \frac{|v|}{\rho}\right)\{-i \operatorname{sgn} v\}(iv)^r f^\wedge(v)\, e^{i\circ v}\, dv \right\|_1 = O(1) \qquad (\rho > 0).$$

Theorem 6.1.7 shows that there exists $\mu \in \mathsf{BV}$ such that $\{-i \operatorname{sgn} v\}(iv)^r f^\wedge(v) = \mu^\vee(v)$, which in view of Theorem 8.3.6 implies $f \in (\mathsf{V}^\sim)^r_{\mathsf{L}^1}$.

As in the periodic case, we may generally interpret the V-classes as relative completions of the relevant W-classes. We state

Proposition 10.5.7. *Let $f \in \mathsf{L}^p$, $1 \le p \le 2$, and ψ be an arbitrary complex-valued function defined on $\mathbb{R}$.*

(a) *The class*

$$\mathsf{W}[\mathsf{L}^p; \psi(v)] = \{f \in \mathsf{L}^p \mid \psi(v) f^\wedge(v) = g^\wedge(v), g \in \mathsf{L}^p\} \tag{10.5.8}$$

becomes a normalized Banach subspace of L^p under the norm

$$\|f\|_{\mathsf{W}[\mathsf{L}^p;\psi(v)]} = \|f\|_p + \|g\|_p. \tag{10.5.9}$$

(b) *The class*

$$\mathsf{V}[\mathsf{L}^p; \psi(v)] = \begin{cases} \{f \in \mathsf{L}^1 \mid \psi(v) f^\wedge(v) = \mu^\vee(v), & \mu \in \mathsf{BV}\} \\ \{f \in \mathsf{L}^p \mid \psi(v) f^\wedge(v) = g^\wedge(v), & g \in \mathsf{L}^p\} \end{cases} \quad (1 < p \le 2) \tag{10.5.10}$$

becomes a normalized Banach subspace of L^p under the norm

$$\|f\|_{\mathsf{V}[\mathsf{L}^p;\psi(v)]} = \begin{cases} \|f\|_1 + \|\mu\|_{\mathsf{BV}}, & p = 1 \\ \|f\|_p + \|g\|_p, & 1 < p \le 2. \end{cases} \tag{10.5.11}$$

(c) *A function f belongs to $\mathsf{V}[\mathsf{L}^p; \psi(v)]$ if and only if f belongs to the completion of $\mathsf{W}[\mathsf{L}^p; \psi(v)]$ relative to L^p.*

We finally formulate some characterizations of the classes $\mathsf{V}[\mathsf{L}^p; |v|^r]$.

Corollary 10.5.8. *Let $f \in \mathsf{L}^p$, $1 \le p \le 2$. The following assertions are equivalent for r even (odd):*

(i) $f \in \mathsf{V}[\mathsf{L}^p; |v|^r]$, (ii) $f \in \widetilde{\mathsf{W}[\mathsf{L}^p; |v|^r]}^{\mathsf{L}^p}$, (iii) $f \in \mathsf{V}^r_{\mathsf{L}^p}((\mathsf{V}^\sim)^r_{\mathsf{L}^p})$,
(iv) $\|\nabla^r_h f\|_p\, (\|\nabla^r_h f^\sim\|_p) = O(|h|^r)$, $h \to 0$, *under the additional assumption that $f^{(r-1)}((f^\sim)^{(r-1)})$ exists and $f^{(k)}((f^\sim)^{(k)})$, $1 \le k \le r-1$, belong to L^p in case r is even (odd),*
(v) *there exists an rth Peano difference such that* $\|\Diamond^r_h f\|_p\, (\|\Diamond^r_h f^\sim\|_p) = O(|h|^r)$,
(vi) $\|\bar{\Delta}^r_h f\|_p\, (\|\bar{\Delta}^r_h f^\sim\|_p) = O(|h|^r)$, (vii) $\|\Delta^r_h f\|_p\, (\|\Delta^r_h f^\sim\|_p) = O(|h|^r)$,
(viii)
$$\left\| \frac{1}{\sqrt{2\pi}} \int_{-\rho}^{\rho} \left(1 - \frac{|v|}{\rho}\right)|v|^r f^\wedge(v)\, e^{i\circ v}\, dv \right\|_p = O(1) \qquad (\rho \to \infty).$$

The proof follows by a more or less immediate application of the preceding results. For a further characterization see also Theorem 11.5.2, Problems 11.5.1, 11.5.2.

Problems

1. Prove Prop. 10.5.7.
2. Show that f belongs to $\mathsf{V}^1_{\mathsf{L}^1}$ if and only if there exists a sequence $\{f_n\} \subset \mathsf{C}_0^\infty \cap \mathsf{L}^1$ such that $\lim_{n\to\infty} \|f - f_n\|_1 = 0$ and $\int_{-\infty}^{\infty} |f_n'(x)|\, dx < \infty$ uniformly for all $n \in \mathbb{N}$. (Hint: H. S. SHAPIRO [4, Sec. 3.5])
3. Let $1 \le p \le 2$ and $r, k \in \mathbb{N}$ be such that $0 \le k \le r-1$. Show that $\|\Delta_h^r f\|_p = O(|h|^r)$ if and only if $f \in \mathsf{W}^k_{\mathsf{L}^p}$ and $\|\Delta_h^{r-k}\phi^{(k)}\|_p = O(|h|^{r-k})$, where $\phi \in \mathsf{AC}^{k-1}_{\mathrm{loc}}$ is such that $\phi = f$ a.e. and $\phi^{(j)} \in \mathsf{L}^p$, $0 \le j \le k$.
4. Let $f \in \mathsf{L}^p$, $1 \le p \le 2$. The following statements are equivalent: (i) $f \in \mathsf{V}^r_{\mathsf{L}^p}$, (ii) $\|\nabla_h^r f\|_p = O(|h|^r)$ provided $f \in \mathsf{C}_{\mathrm{loc}} \cap \mathsf{W}^{r-1}_{\mathsf{L}^p}$, (iii) $\|\overline{\Delta}_h^r f\|_p = O(|h|^r)$.
5. Let $f \in \mathsf{L}^p$, $1 \le p \le 2$. The following statements are equivalent:
   (i) there exists a Peano difference of f such that $\|h^{-r}\Diamond_h^r f\|_p = O(1)$,
   (ii) $p = 1$: there exists a Peano difference of f and $\mu \in \mathsf{BV}$ such that for each $s \in \mathsf{C}_0$
$$\lim_{h\to 0} \int_{-\infty}^{\infty} s(u)[h^{-r}\Diamond_h^r f(u)]\, du = \int_{-\infty}^{\infty} s(u)\, d\mu(u),$$
   $1 < p \le 2$: there exists a Peano difference of f and $g \in \mathsf{L}^p$ such that
$$\lim_{h\to 0} \|h^{-r}\Diamond_h^r f - g\|_p = 0.$$
6. Let $f \in \mathsf{L}^1$. Any expression of the form
$$r!\left\{f^\sim(x+h) - f^\sim(x) - \sum_{k=1}^{r-1} \frac{h^k}{k!} l_k(x)\right\}$$
   is called an rth Peano difference of the Hilbert transform $f^\sim$ if $[f^\sim(\circ + h) - f^\sim(\circ)] \in \mathsf{L}^1$ for each $h \in \mathbb{R}$ and $l_k \in \mathsf{L}^1$, $1 \le k \le r-1$. If there also exists $l_r \in \mathsf{L}^1$ such that
$$\lim_{h\to 0} \left\| \frac{r!}{h^r}\left\{f^\sim(\circ + h) - f^\sim(\circ) - \sum_{k=1}^{r-1} \frac{h^k}{k!} l_k(\circ)\right\} - l_r(\circ)\right\|_1 = 0,$$
   then l_r is called the rth strong Peano derivative of $f^\sim$ and denoted by $D_s^{\langle r\rangle} f^\sim$. Show that if there is an rth Peano difference of $f^\sim$ such that $\|\Diamond_h^r f^\sim\|_1 = O(|h|^r)$, then $D_s^{\langle r-1\rangle} f^\sim$ exists.
7. Let $f \in \mathsf{L}^1$. If $\Delta_h^r f^\sim \in \mathsf{L}^1$ for each $h \in \mathbb{R}$ and $\lim_{h\to 0} \|h^{-r}\Delta_h^r f^\sim - g\|_1 = 0$ for some function $g \in \mathsf{L}^1$, then g is called the rth strong Riemann derivative of $f^\sim$ and denoted by $D_s^{[r]} f^\sim$. Show that f belongs to $(\mathsf{W}^\sim)^r_{\mathsf{L}^1}$ if and only if the rth strong Riemann derivative of $f^\sim$ exists. In this case $D_s^{[k]} f^\sim$ exists for every $1 \le k \le r-1$.
8. Let $1 \le p \le 2$. Show that f belongs to $(\mathsf{W}^\sim)^r_{\mathsf{L}^p}$ if and only if the rth strong Peano derivative of $f^\sim$ exists or if and only if the rth strong Riemann derivative of $f^\sim$ exists.
9. State and prove characterizations of $\mathsf{W}[\mathsf{L}^p; |v|^r]$, $r \in \mathbb{N}$ (cf. Cor. 10.5.8).
10. Define norms on $\mathsf{V}^r_{\mathsf{X}(\mathbb{R})}$, $\mathsf{Lip}_r(\mathsf{X}(\mathbb{R}); \alpha)$ analogously to (10.4.8), (10.4.9) and show that these sets become normalized Banach subspaces of $\mathsf{X}(\mathbb{R})$. Show that the Banach spaces $\mathsf{V}^r_{\mathsf{L}^p}$ and $\mathsf{Lip}_r(\mathsf{X}(\mathbb{R}); r)$ are equal with equal norms for $1 \le p \le 2$. (Hint: Sec. 10.4, Theorem 10.5.4)

10.6 Generalized Derivatives in $\mathsf{X}(\mathbb{R})$-Norm and Characterizations of the Classes $\mathsf{W}^r_{\mathsf{X}(\mathbb{R})}$ and $\mathsf{V}^r_{\mathsf{X}(\mathbb{R})}$

The results of the last section are meaningful not only for the space L^p, $1 \le p \le 2$, but also for C and L^p, $2 < p < \infty$, so that the restriction to $1 \le p \le 2$ seems to depend upon the method of proof. To extend the results to the latter spaces, we may proceed in two different ways. One method, namely the extension of the definition of the Fourier transform from L^p, $1 \le p \le 2$, to all spaces $\mathsf{X}(\mathbb{R})$, is left to Vol. II; here we prefer a direct approach.

For $f \in \mathsf{L}^1$ the value $f^\wedge(v)$ of the Fourier transform at the point $v \in \mathbb{R}$ may be interpreted as a bounded linear functional on L^1. Thus the Fourier transform is an operation which assigns to each $f \in \mathsf{L}^1$ a certain set of bounded linear functionals, namely $f^\wedge(v)$ for each $v \in \mathbb{R}$. For our purposes it is again sufficient to use a particular set of bounded linear functionals on $\mathsf{X}(\mathbb{R})$; these are not characterized by the exponentials $\exp\{-ivx\}$ but by the functions $\phi \in \mathsf{C}_{00}^r$ (cf. (10.6.13)).

We first establish some auxiliary results which may be regarded as substitutes for the characterization of the classes $\mathsf{W}_{\mathsf{L}^p}^r$ and $\mathsf{V}_{\mathsf{L}^p}^r$ given by $\mathsf{W}[\mathsf{L}^p; (iv)^r]$ and $\mathsf{V}[\mathsf{L}^p; (iv)^r]$, respectively, which is only valid for $1 \le p \le 2$. Subsequently, P_n denotes the set of all algebraic polynomials of degree n at most.

Lemma 10.6.1. *If for* $f \in \mathsf{L}_{\text{loc}}^1$

$$\Delta_h^r f(x) = 0 \quad \text{a.e.} \tag{10.6.1}$$

for each $h > 0$, *then there exists* $p_{r-1} \in \mathsf{P}_{r-1}$ *such that* $f(x) = p_{r-1}(x)$ a.e. *If* $f \in \mathsf{C}_{\text{loc}}$, *then* (10.6.1) *implies that* $f(x) = p_{r-1}(x)$ *for all* $x \in \mathbb{R}$.

Proof. Let us first assume $f \in \mathsf{C}_{\text{loc}}^r$. Then by the mean value theorem

$$\lim_{h \to 0+} \frac{1}{h^r} \Delta_h^r f(x) = f^{(r)}(x) \qquad (x \in \mathbb{R}).$$

Thus (10.6.1) implies that $f^{(r)}(x) = 0$ on $\mathbb{R}$, and consequently $f(x) = p_{r-1}(x)$ for all $x \in \mathbb{R}$.

If now $f \in \mathsf{L}_{\text{loc}}^1$, we consider the $(r + 1)$th integral mean

$$A^{r+1}(f; x; t) = \frac{1}{t^{r+1}} \int_{-t/2}^{t/2} \cdots \int_{-t/2}^{t/2} f(x + u_1 + \cdots + u_{r+1})\, du_1 \ldots du_{r+1} \quad (t > 0) \tag{10.6.2}$$

of f (cf. Problem 3.1.8(iii)). Since obviously $\Delta_h^r A^{r+1}(f; x; t) = A^{r+1}(\Delta_h^r f; x; t)$, condition (10.6.1) implies that $\Delta_h^r A^{r+1}(f; x; t) = 0$ for all $x \in \mathbb{R}$ and $h, t > 0$. Since in view of the fact that $A^{r+1}(f; x; t) \in \mathsf{C}_{\text{loc}}^r$ for each $t > 0$, it follows from the initial step that there exists $p_{r-1}(x; t) \in \mathsf{P}_{r-1}$ for each $t > 0$ such that $A^{r+1}(f; x; t) = p_{r-1}(x; t)$ for all $x \in \mathbb{R}$. Furthermore, $\lim_{t \to 0+} A^{r+1}(f; x; t) = f(x)$ a.e. by Problem 3.1.8(iii). Therefore $\lim_{t \to 0+} p_{r-1}(x; t)$ exists a.e., and this latter limit must be an element of P_{r-1} (cf. Problem 10.6.5). Consequently, $f(x) = p_{r-1}(x)$ a.e.

If $f \in \mathsf{C}_{\text{loc}}$, then $\lim_{t \to 0+} A^{r+1}(f; x; t) = f(x)$ for all $x \in \mathbb{R}$, and therefore $\lim_{t \to 0+} p_{r-1}(x; t)$ exists for all $x \in \mathbb{R}$ as an element of P_{r-1}. Thus $f(x) = p_{r-1}(x)$ for all $x \in \mathbb{R}$.

Lemma 10.6.2. *If for* $f \in \mathsf{L}_{\text{loc}}^1$ *there is* $g \in \mathsf{L}_{\text{loc}}^1$ *such that*

$$\Delta_h^r f(x) = \int_0^h \cdots \int_0^h g(x + u_1 + \ldots + u_r)\, du_1 \ldots du_r \quad \text{a.e.} \tag{10.6.3}$$

for each $h > 0$, *then there exists* $p_{r-1} \in \mathsf{P}_{r-1}$ *such that*

$$f(x) = p_{r-1}(x) + \int_0^x du_1 \int_0^{u_1} du_2 \ldots du_{r-1} \int_0^{u_{r-1}} g(u_r)\, du_r \quad \text{a.e.} \tag{10.6.4}$$

If $f \in \mathsf{C}_{\text{loc}}$, *then the representation* (10.6.4) *holds everywhere.*

Proof. If we set

$$G_r(x) = \int_0^x du_1 \int_0^{u_1} du_2 \ldots du_{r-1} \int_0^{u_{r-1}} g(u_r)\, du_r, \tag{10.6.5}$$

then (cf. Problem 10.6.6)

$$G_r(x) = \int_0^x \frac{(x-u)^{r-1}}{(r-1)!} g(u)\, du = \frac{x^r}{(r-1)!} \int_0^1 (1-u)^{r-1} g(xu)\, du \tag{10.6.6}$$

and

$$\Delta_h^r G_r(x) = \int_0^h \cdots \int_0^h g(x + u_1 + \cdots + u_r)\, du_1 \ldots du_r \tag{10.6.7}$$

for every $x \in \mathbb{R}$ and $h > 0$. Therefore, the function $[f(x) - G_r(x)]$ is locally integrable and satisfies $\Delta_h^r[f - G_r](x) = 0$ a.e. Thus Lemma 10.6.1 implies the representation (10.6.4) which may equally well be expressed by saying that there exist constants $\alpha_k \in \mathbb{C}$, $0 \le k \le r - 1$, such that for almost all $x \in \mathbb{R}$

$$(10.6.8) \qquad f(x) = \alpha_0 + \int_0^x du_1\Big[\alpha_1 + \int_0^{u_1} du_2\Big[\alpha_2 + \cdots + \int_0^{u_{r-2}} du_{r-1}\Big[\alpha_{r-1} + \int_0^{u_{r-1}} g(u_r)\,du_r\Big]\ldots\Big]\Big].$$

If, in particular, $f \in \mathsf{C}_{\mathrm{loc}}$, then $[f(x) - G_r(x)] \in \mathsf{C}_{\mathrm{loc}}$, and (10.6.4) or (10.6.8) holds everywhere by Lemma 10.6.1.

Lemma 10.6.3. *If for $f \in \mathsf{X}(\mathbb{R})$ there exists $g \in \mathsf{X}(\mathbb{R})$ such that* (10.6.3) *holds for each $h > 0$, then $f \in \mathsf{W}^r_{\mathsf{X}(\mathbb{R})}$.*

Proof. Let $\mathsf{X}(\mathbb{R}) = \mathsf{L}^p$, $1 \le p < \infty$. Then by Lemma 10.6.2 and in particular by the representation (10.6.8) there is a function ϕ such that $f = \phi$ a.e. with $\phi \in \mathsf{AC}^{r-1}_{\mathrm{loc}}$ and $\phi^{(r)} = g$ a.e. In view of definition (3.1.43) it only remains to show that $\phi^{(k)} \in \mathsf{L}^p$, $1 \le k < r$.

Integration by parts gives

$$(10.6.9) \qquad \Delta^1_{h_1}\ldots\Delta^1_{h_r}\phi(x) = \int_0^{h_1}\cdots\int_0^{h_r} g(x + u_1 + \cdots + u_r)\,du_1\ldots du_r$$

for each $h_k \ge 0$, $1 \le k \le r$, and all $x \in \mathbb{R}$. We now show via mathematical induction that all derivatives $\phi^{(k)}$, $1 \le k \le r$, belong to L^p, if (10.6.9) holds for $\phi \in \mathsf{AC}^{r-1}_{\mathrm{loc}} \cap \mathsf{L}^p$ and $g \in \mathsf{L}^p$.

For $r = 1$ the assertion is trivial. Assuming now that it is valid for $r - 1$, then, putting

$$(10.6.10) \qquad \phi_{h_r}(x) = \phi(x + h_r) - \phi(x), \qquad g_{h_r}(x) = \int_0^{h_r} g(x + u_r)\,du_r,$$

the hypothesis (10.6.9) and the assumption as applied to $\phi_{h_r} \in \mathsf{AC}^{r-2}_{\mathrm{loc}} \cap \mathsf{L}^p$ and $g_{h_r} \in \mathsf{L}^p$ imply that $\phi^{(k)}_{h_r}(x) = [\phi^{(k)}(x + h_r) - \phi^{(k)}(x)] \in \mathsf{L}^p$, $1 \le k \le r - 1$, and $\phi^{(r-1)}_{h_r}(x) = g_{h_r}(x)$ for each fixed $h_r \ge 0$ and all $x \in \mathbb{R}$. We shall show that $\|\phi'_{h_r}\|_p$ is a continuous function of h_r. Since ϕ and g belong to L^p, we obtain by Prop. 0.1.9 and 0.1.7 that $\|\phi_{h_r}\|_p$ and $\|\phi^{(r-1)}_{h_r}\|_p = \|g_{h_r}\|_p$ are continuous functions of h_r on $[0, t]$, $t > 0$, respectively. Since

$$\phi^{(r-3)}_{h_r}(x + t) - \phi^{(r-3)}_{h_r}(x) - t\phi^{(r-2)}_{h_r}(x) = \int_0^t (t - u)\phi^{(r-1)}_{h_r}(x + u)\,du$$

and $\phi_{h_r}(x + t) - \phi_{h_r}(x) = \phi_t(x + h_r) - \phi_t(x)$, we have (cf. (10.6.5), (10.6.6))

$$t\phi^{(r-2)}_{h_r}(x) = \phi^{(r-3)}_t(x + h_r) - \phi^{(r-3)}_t(x) - \int_0^t du_1\int_0^{u_1} du\int_0^{h_r} g(x + u + u_r)\,du_r$$

and, setting $t = 1$,

$$\phi^{(r-2)}_{h_r}(x) = \phi^{(r-3)}_1(x + h_r) - \phi^{(r-3)}_1(x) - \int_0^{h_r}\Big[\int_0^1 g_{u_1}(x + u_r)\,du_1\Big]\,du_r.$$

Therefore it follows that

$$\|\phi^{(r-2)}_{h_r}\|_p \le \|\phi^{(r-3)}_1(\circ + h_r) - \phi^{(r-3)}_1(\circ)\|_p + h_r \sup_{0 \le u_1 \le 1}\|g_{u_1}\|_p.$$

But since $\phi^{(r-3)}_1 \in \mathsf{L}^p$, the right-hand side is a continuous function of h_r on $[0, b]$ for each $b > 0$, and therefore $\|\phi^{(r-2)}_{h_r}\|_p$ is one, too. Applying this argument successively, it follows that $\|\phi'_{h_r}\|_p$ is a continuous function of h_r on $[0, b]$. Now

$$\phi(x + 1) - \phi(x) - \phi'(x) = \int_0^1 \phi'_{h_r}(x)\,dh_r,$$

and therefore

$$\|\phi'\|_p \le 2\|\phi\|_p + \sup_{0 \le h_r \le 1}\|\phi'_{h_r}\|_p$$

which is bounded since $\|\phi'_{h_r}\|_p$ is continuous on $0 \le h_r \le 1$. Thus we obtain that $\phi' \in \mathsf{L}^p$. Now we may divide (10.6.9) by h_r and take the limit $h_r \to 0+$. Then

$$\Delta^1_{h_1}\Delta^1_{h_2}\dots\Delta^1_{h_{r-1}}\phi'(x) = \int_0^{h_1}\dots\int_0^{h_{r-1}} g(x+u_1+\cdots+u_{r-1})\,du_1\dots du_{r-1}.$$

As $\phi', g \in \mathsf{L}^p$, the induction hypothesis implies that all derivatives $\phi^{(k)}$, $1 \le k \le r$, belong to L^p.

If $\mathsf{X}(\mathbb{R}) = \mathsf{C}$, then an application of Lemma 10.6.2 shows that $f \in \mathsf{AC}^{r-1}_{\text{loc}}$ and $f^{(r)}(x) = g(x)$ for all $x \in \mathbb{R}$. Moreover, it follows similarly that $f^{(k)} \in \mathsf{C}$, $1 \le k \le r$, and thus $f \in \mathsf{W}^r_{\mathsf{C}}$

Proposition 10.6.4. *Let $f \in \mathsf{X}(\mathbb{R})$. The following statements are equivalent:*

(i) *f belongs to $\mathsf{W}^r_{\mathsf{X}(\mathbb{R})}$,*

(ii) *there exists $g \in \mathsf{X}(\mathbb{R})$ such that*

$$\int_{-\infty}^{\infty} f(u)\phi^{(r)}(u)\,du = (-1)^r\int_{-\infty}^{\infty} g(u)\phi(u)\,du \tag{10.6.11}$$

for every $\phi \in \mathsf{C}^r_{00}$.

Proof. Let $\mathsf{X}(\mathbb{R}) = \mathsf{C}$. If $f \in \mathsf{W}^r_{\mathsf{C}}$ and $\phi \in \mathsf{C}^r_{00}$, then successive integration by parts yields

$$\int_{-\infty}^{\infty} f(u)\phi^{(r)}(u)\,du = (-1)^r\int_{-\infty}^{\infty} f^{(r)}(u)\phi(u)\,du,$$

since ϕ has compact support, i.e., (10.6.11) is satisfied with $g \equiv f^{(r)}$. Conversely, if (10.6.11) holds for all elements of C^r_{00}, then

$$(-1)^r\int_{-\infty}^{\infty} f(x)\phi^{(r)}(x-u_1-\cdots-u_r)\,dx = \int_{-\infty}^{\infty} g(x+u_1+\dots+u_r)\phi(x)\,dx, \tag{10.6.12}$$

since with $\phi(\circ) \in \mathsf{C}^r_{00}$ also $\phi(\circ - u_1 - \dots - u_r) \in \mathsf{C}^r_{00}$ for all $u_k \in \mathbb{R}$, $1 \le k \le r$. By Fubini's theorem, integration of (10.6.12) over $u_k \in [0, h]$, $1 \le k \le r$, yields after an obvious change of variables

$$\int_{-\infty}^{\infty} \Delta^r_h f(x)\phi(x)\,dx = \int_{-\infty}^{\infty}\Big[\int_0^h\dots\int_0^h g(x+u_1+\cdots+u_r)\,du_1\dots du_r\Big]\phi(x)\,dx.$$

But Problem 3.1.2(iv) implies that

$$\Delta^r_h f(x) = \int_0^h\dots\int_0^h g(x+u_1+\cdots+u_r)\,du_1\dots du_r$$

for each $h > 0$ and all $x \in \mathbb{R}$. Hence it follows by Lemma 10.6.3 that $f \in \mathsf{W}^r_{\mathsf{C}}$.

After evident modifications the above arguments also give Prop. 10.6.4 for the spaces $\mathsf{X}(\mathbb{R}) = \mathsf{L}^p$, $1 \le p < \infty$.

We are now in a position to restate Theorem 10.5.1 for all spaces $\mathsf{X}(\mathbb{R})$.

Theorem 10.6.5. *Let $f \in \mathsf{X}(\mathbb{R})$. The following statements are equivalent:*

(i) *f belongs to $\mathsf{W}^r_{\mathsf{X}(\mathbb{R})}$,*

(ii) *f has an rth Peano derivative $D^{\langle r\rangle}_s f$ in $\mathsf{X}(\mathbb{R})$-norm,*

(iii) *f has an rth Riemann derivative $D^{[r]}_s f$ in $\mathsf{X}(\mathbb{R})$-norm.*

Proof. The proof of the implications (i) ⇒ (ii) ⇒ (iii) follows by the standard argument (see proof of Theorem 10.5.1). Thus let (iii) be satisfied. For any $\phi \in \mathsf{C}^r_{00}$ and $h \in \mathbb{R}$ we define

$$\Lambda_h(f) = \int_{-\infty}^{\infty} f(u)\Big[\frac{1}{h^r}\Delta^r_{-h}\phi(u)\Big]\,du \qquad (f \in \mathsf{X}(\mathbb{R})). \tag{10.6.13}$$

Obviously, $\Lambda_h(f)$ is a bounded linear functional on $\mathsf{X}(\mathbb{R})$ for each $\phi \in \mathsf{C}_{00}^r$ and $h \in \mathbb{R}$. We have

$$(10.6.14) \qquad \Lambda_h(f) = \int_{-\infty}^{\infty} \left[\frac{1}{h^r} \Delta_h^r f(u)\right] \phi(u)\, du.$$

Since by the mean value theorem

$$\lim_{h \to 0} \left\| \frac{1}{(-h)^r} \Delta_{-h}^r \phi - \phi^{(r)} \right\|_{\mathsf{C}} = 0,$$

it follows by (10.6.13) that for every $f \in \mathsf{X}(\mathbb{R})$

$$(10.6.15) \qquad \lim_{h \to 0} \Lambda_h(f) = (-1)^r \int_{-\infty}^{\infty} f(u) \phi^{(r)}(u)\, du.$$

On the other hand, the existence of the rth Riemann derivative $D_s^{[r]} f$ in $\mathsf{X}(\mathbb{R})$-norm implies by (10.6.14) that

$$(10.6.16) \qquad \lim_{h \to 0} \Lambda_h(f) = \int_{-\infty}^{\infty} (D_s^{[r]} f)(u) \phi(u)\, du.$$

Therefore

$$(10.6.17) \qquad \int_{-\infty}^{\infty} f(u) \phi^{(r)}(u)\, du = (-1)^r \int_{-\infty}^{\infty} (D_s^{[r]} f)(u) \phi(u)\, du$$

for every $\phi \in \mathsf{C}_{00}^r$, and (i) follows by Prop. 10.6.4.

Concerning the extension of Theorem 10.5.4 to all $\mathsf{X}(\mathbb{R})$-spaces we need the following analog of Prop. 10.6.4 for the class $\mathsf{V}_{\mathsf{C}}^r$.

Proposition 10.6.6. *Let $f \in \mathsf{C}$. The following statements are equivalent:*

(i) *f belongs to $\mathsf{V}_{\mathsf{C}}^r$,*

(ii) *there exists $g \in \mathsf{L}^\infty$ such that*

$$(10.6.18) \qquad \int_{-\infty}^{\infty} f(u) \phi^{(r)}(u)\, du = (-1)^r \int_{-\infty}^{\infty} g(u) \phi(u)\, du$$

for every $\phi \in \mathsf{C}_{00}^r$.

We leave the proof to Problem 10.6.9.

Theorem 10.6.7. *Let $f \in \mathsf{X}(\mathbb{R})$. The following statements are equivalent:*

(i) *f belongs to $\mathsf{V}_{\mathsf{X}(\mathbb{R})}^r$,*

(ii) *there exists an rth Peano difference of f such that*

$$\| \diamondsuit_h^r f \|_{\mathsf{X}(\mathbb{R})} = O(|h|^r) \qquad (h \to 0),$$

(iii)
$$\| \Delta_h^r f \|_{\mathsf{X}(\mathbb{R})} = O(|h|^r) \qquad (h \to 0).$$

Proof. The proof of the implications (i) $\Rightarrow$ (ii) $\Rightarrow$ (iii) is again standard. In view of Theorem 10.5.4 only the spaces C and L^p, $2 < p < \infty$, need to be considered.

Let (iii) be satisfied and $\mathsf{X}(\mathbb{R}) = \mathsf{L}^p$, $2 < p < \infty$. Again we consider the bounded linear functionals $\Lambda_h(f)$ of (10.6.13) to deduce (10.6.14) and (10.6.15). The weak* compactness for L^p implies that there exists a null-sequence $\{h_j\}$ and $g \in \mathsf{L}^p$ such that

$$\lim_{j\to\infty} \int_{-\infty}^{\infty} [h_j^{-r}\Delta_{h_j}^r f(u)]s(u)\, du = \int_{-\infty}^{\infty} g(u)s(u)\, du \tag{10.6.19}$$

for every $s \in \mathsf{L}^{p'}$, in particular for every $\phi \in \mathsf{C}_{00}^r$. Therefore

$$\lim_{j\to\infty} \Lambda_{h_j}(f) = \int_{-\infty}^{\infty} g(u)\phi(u)\, du, \tag{10.6.20}$$

and we have that for every $\phi \in \mathsf{C}_{00}^r$

$$\int_{-\infty}^{\infty} f(u)\phi^{(r)}(u)\, du = (-1)^r \int_{-\infty}^{\infty} g(u)\phi(u)\, du. \tag{10.6.21}$$

Then (i) follows by Prop. 10.6.4 since $\mathsf{V}_{\mathsf{L}^p}^r = \mathsf{W}_{\mathsf{L}^p}^r$ for $1 < p < \infty$ by definition.

If $\mathsf{X} = \mathsf{C}$, then (iii) states that $\|h^{-r}\Delta_h^r f\|_\infty = O(1)$ as $h \to 0$. Therefore the weak* compactness for L^∞ may be used to deduce the existence of a null-sequence $\{h_j\}$ and of $g \in \mathsf{L}^\infty$ such that (10.6.19) holds for all $s \in \mathsf{L}^1$. Thus (10.6.20) and (10.6.21) are again valid, and (i) follows by Prop. 10.6.6.

Thus we have extended the results of Theorems 10.5.1 and 10.5.4 to all spaces $\mathsf{X}(\mathbb{R})$. By the way, we have found a further characterization of $\mathsf{W}_{\mathsf{X}(\mathbb{R})}^r$ by the condition (10.6.11) and of $\mathsf{V}_{\mathsf{C}}^r$ and $\mathsf{V}_{\mathsf{L}^p}^r$, $1 < p < \infty$, by (10.6.18) and (10.6.11), respectively. The still missing characterization of this kind for $\mathsf{V}_{\mathsf{L}^1}^r$ is given by

Proposition 10.6.8. *Let $f \in \mathsf{L}^1$. The following statements are equivalent:*

(i) *f belongs to $\mathsf{V}_{\mathsf{L}^1}^r$,*
(ii) *there exists $\mu \in \mathsf{BV}$ such that*

$$\int_{-\infty}^{\infty} f(u)\phi^{(r)}(u)\, du = (-1)^r \int_{-\infty}^{\infty} \phi(u)\, d\mu(u) \tag{10.6.22}$$

for every $\phi \in \mathsf{C}_{00}^r$.

Proof. The proof of (i) $\Rightarrow$ (ii) follows as usual by partial integration. Conversely, if (10.6.22) is satisfied, then by partial integration

$$\int_{-\infty}^{\infty} f(x)\phi^{(r)}(x)\, dx = (-1)^{r-1} \int_{-\infty}^{\infty} \mu(x)\phi'(x)\, dx.$$

Replacing $\phi(x)$ by $\phi(x - u_1 - \cdots - u_r)$, $u_k \in \mathbb{R}$, $1 \le k \le r$, we obtain as in the proof of Prop. 10.6.4 that

$$(-1)^r \int_{-\infty}^{\infty} f(x)\phi^{(r)}(x - u_1 - \cdots - u_r)\, dx = -\int_{-\infty}^{\infty} \mu(x + u_1 + \cdots + u_{r-1})\phi'(x - u_r)\, dx$$

and

$$\int_{-\infty}^{\infty} \Delta_h^r f(x)\phi(x)\, dx = -\int_{-\infty}^{\infty} \Big[\int_0^h \cdots \int_0^h \mu(x + u_1 + \cdots + u_{r-1})\, du_1 \ldots du_{r-1}\Big] \Big[\int_0^h \phi'(x - u_r) du_r\Big] dx$$

$$= \int_{-\infty}^{\infty} \Big[\int_0^h \cdots \int_0^h \{\mu(x + h + u_1 + \cdots + u_{r-1}) - \mu(x + u_1 + \cdots + u_{r-1})\}\, du_1 \ldots du_{r-1}\Big]\phi(x)\, dx.$$

Again Problem 3.1.2(iv) implies that for each $h \in \mathbb{R}$

$$\text{(10.6.23)} \qquad \Delta_h^r f(x) = \int_0^h \cdots \int_0^h \{\mu(x + h + u_1 + \cdots + u_{r-1}) - \mu(x + u_1 + \cdots + u_{r-1})\}\, du_1 \ldots du_{r-1} \quad \text{a.e.}$$

But since $\|\mu(\circ + h) - \mu(\circ)\|_1 \le |h|\, \|\mu\|_{\mathsf{BV}}$ for each $h \in \mathbb{R}$ and $\mu \in \mathsf{BV}$ by (5.3.7), it follows by (10.6.23) that

$$\|\Delta_h^r f\|_1 \le |h|^{r-1} \int_0^1 \cdots \int_0^1 \|\mu(\circ + h) - \mu(\circ)\|_1\, du_1 \ldots du_{r-1} \le \|\mu\|_{\mathsf{BV}}\, |h|^r.$$

An application of Theorem 10.6.7 now yields $f \in \mathsf{V}_{L^1}^r$.

For another proof of Prop. 10.6.8 which is independent of Theorem 10.6.7 and thus of 10.5.4, we refer to Problem 10.6.12.

Problems

1. Show that if $f \in \mathsf{L}^p$, $1 \le p < \infty$, and $\Delta_h^r f(x) = 0$ a.e. for each $h \in \mathbb{R}$, then $f(x) = 0$ a.e. State and prove the analogous assertion in C-space.
2. Show that if $f \in \mathsf{L}_{\text{loc}}^1$ and $\int_{-\infty}^{\infty} f(x)\phi^{(r)}(x)\, dx = 0$ for every $\phi \in \mathsf{C}_{00}^r$, then there exists $p_{r-1} \in \mathsf{P}_{r-1}$ such that $f(x) = p_{r-1}(x)$ a.e.
3. Let $\phi_0 \in \mathsf{C}_{00}^1$ be such that $\int_{-\infty}^{\infty} \phi_0(u)\, du = 1$. Show that every $\phi \in \mathsf{C}_{00}^1$ may be decomposed uniquely as $\phi(x) = M\phi_0(x) + \eta(x)$, where η is the derivative of a function belonging to C_{00}^1 and the constant M is given by $M = \int_{-\infty}^{\infty} \phi(u)\, du$.
4. Show that for $\phi \in \mathsf{C}_{00}^\infty$ there exists $\psi \in \mathsf{C}_{00}^\infty$ such that $\phi(x) = \psi'(x)$ if and only if $\int_{-\infty}^{\infty} \phi(u)\, du = 0$.
5. For each $t > 0$ let $p_{r-1}(x; t)$ be an algebraic polynomial of degree $r - 1$ at most. Suppose that $\lim_{t\to 0+} p_{r-1}(x; t)$ exists almost everywhere. Then this limit is again an algebraic polynomial $p_{r-1}(x)$ of degree $r - 1$ at most and $\lim_{t\to 0+} p_{r-1}(x; t) = p_{r-1}(x)$ for all $x \in \mathbb{R}$.
6. Show that G_r as defined by (10.6.5) satisfies (10.6.6), (10.6.7), and (10.6.9) with G_r in place of ϕ.
7. Show that the converses of Lemmata 10.6.1–10.6.3 are valid as well.
8. Let f, g belong to $\mathsf{X}(\mathbb{R})$ and satisfy (10.6.9) with f in place of ϕ. Show that the kth strong derivative $D_s^{(k)} f$ of f exists for each $1 \le k \le r$ and $D_s^{(r)} f = g$. (Hint: BERENS [2])
9. Prove Prop. 10.6.6.
10. If for $f \in \mathsf{L}_{\text{loc}}^1$ there is $\mu \in \mathsf{BV}_{\text{loc}}$ such that

    $$\text{(10.6.24)} \qquad \Delta_h^r f(x) = \int_0^h du_1 \int_0^h du_2 \ldots \int_0^h du_{r-1} \int_0^h d_{u_r}\mu(x + u_1 + \cdots + u_r) \quad \text{a.e.}$$

    for each $h > 0$, show that there exists $p_{r-1} \in \mathsf{P}_{r-1}$ such that

    $$f(x) = p_{r-1}(x) + \int_0^x du_1 \int_0^{u_1} du_2 \ldots du_{r-1} \int_0^{u_{r-1}} d\mu(u_r) \quad \text{a.e.}$$

11. If for $f \in \mathsf{L}^1$ there exists $\mu \in \mathsf{BV}$ such that (10.6.24) holds for each $h > 0$, show that $f \in \mathsf{V}_{L^1}^r$.
12. Give a proof of Prop. 10.6.8 which is independent of Theorem 10.5.4. (Hint: Use the preceding two Problems)
13. Let $f \in \mathsf{L}^p$, $2 < p < \infty$. Show that the following statements are equivalent: (i) $f \in (\mathsf{V}^\sim)_{L^p}^r$, (ii) there exists an rth Peano difference of the Hilbert transform $f^\sim$ such that $\|\diamondsuit_h^r f^\sim\|_p = O(|h|^r)$, (iii) $\|\Delta_h^r f^\sim\|_p = O(|h|^r)$.

14. Show that the following statements are equivalent for $f \in \mathsf{C}$: (i) $\|h^{-r}\Delta_h^r f\|_{\mathsf{C}} = O(1)$, (ii) there exists $g \in \mathsf{L}^\infty$ such that for every $s \in \mathsf{L}^1$
$$\lim_{h\to 0} \int_{-\infty}^{\infty} s(u)[h^{-r}\Delta_h^r f(u)]\,du = \int_{-\infty}^{\infty} s(u)g(u)\,du.$$
15. Show that the following statements are equivalent for $f \in \mathsf{L}^p$, $1 < p < \infty$: (i) $\|h^{-r}\Delta_h^r f\|_p = O(1)$, (ii) there exists $g \in \mathsf{L}^p$ such that $\lim_{h\to 0} \|h^{-r}\Delta_h^r f - g\|_p = 0$.
16. Let $f \in \mathsf{X}(\mathbb{R})$. Show that the following statements are equivalent:
   (i) f belongs to $\mathsf{W}^r_{\mathsf{X}(\mathbb{R})}$,
   (ii) there exists $g \in \mathsf{X}(\mathbb{R})$ such that for every $\phi \in \mathsf{C}^r_{00}$
$$\int_{-\infty}^{\infty} f(u)\phi^{(r)}(u)\,du = (-1)^r \int_{-\infty}^{\infty} g(u)\phi(u)\,du,$$
   (iii) f has an rth strong (weak) derivative $D_s^{(r)}f$ $(D_w^{(r)}f)$,
   (iv) f has an rth strong (weak) Taylor derivative $D_s^{(r)}f$ $(D_w^{(r)}f)$ provided that $f^{(r-1)}$ exists and $f^{(k)} \in \mathsf{X}(\mathbb{R})$ for each $1 \le k \le r-1$,
   (v) f has an rth strong (weak) Peano derivative $D_s^{\langle r\rangle}f$ $(D_w^{\langle r\rangle}f)$,
   (vi) f has an rth strong (weak) Riemann derivative $D_s^{[r]}f$ $(D_w^{[r]}f)$.
17. Formulate and prove results corresponding to Problem 16 for the class $\mathsf{V}^r_{\mathsf{X}(\mathbb{R})}$.
18. Show that the following assertions are equivalent for $f \in \mathsf{L}^p$, $1 \le p < \infty$: (i) $f(x) = 0$ a.e., (ii) there exists an rth Peano difference of f such that $\|\Diamond_h^r f\|_p = o(|h|^r)$, (iii) $\|\Delta_h^r f\|_p = o(|h|^r)$.
   State and prove corresponding results in C-space.
19. Let $f \in \mathsf{L}^p$, $1 \le p < \infty$. Show that if $D_s^{(1)}f$ exists and vanishes a.e., then $f(x) = 0$ a.e. State and prove corresponding results in C-space.
20. (i) Show that the set of all $f \in \mathsf{L}^1$ which satisfy (10.6.22) becomes a normalized Banach subspace of L^1 under the norm $\|f\|_{\mathsf{L}^1} + \|\mu\|_{\mathsf{BV}}$. Show that this Banach space and the Banach space $\mathsf{V}^r_{\mathsf{L}^1}$ are equal with equal norms. State and prove corresponding results for arbitrary $\mathsf{X}(\mathbb{R})$-spaces. (Hint: Sec. 10.4, Problem 10.5.10, Prop. 10.6.4, 10.6.6, 10.6.8)
   (ii) Show that the Banach spaces $\mathsf{V}^r_{\mathsf{X}(\mathbb{R})}$ and $\mathsf{Lip}_r\,(\mathsf{X}(\mathbb{R}); r)$ are equal with equal norms. (Hint: Sec. 10.4, Problem 10.5.10, Theorem 10.6.7)

10.7 Notes and Remarks

The material of this chapter is largely based upon BUTZER [8, 10]. In these papers, a first systematic use of Fourier transform methods was given in connection with Riemann and Taylor derivatives and with theorems of Hardy-Littlewood-type. For a parallel treatment of some of these topics using a very different approach, namely semi-group theory, see BUTZER–BERENS [1, pp. 92–94, 106–111]. Both the Fourier transform as well as the semi-group approach to the material in question allow a general and unified presentation of the proofs as well as of the results.

Sec. 10.1. For general comments concerning Riemann and Peano derivatives we first recall Sec. 5.1.4. For Theorems 10.1.3 and 10.1.6 see BUTZER [10]. These are related to a number of classical results including those of A. DENJOY, A. MARCHAUD, C. DE LA VALLÉE POUSSIN, W. T. REID, S. SAKS, S. VERBLUNSKY. For these as well as Schwarz' lemma we refer to BUTZER–BERENS [1, p. 107 f] and the references cited there (cf. also VERBLUNSKY [1], KEMPERMAN [1]). For extensions of Schwarz' lemma of Problem 10.1.3 compare also WHITNEY [1, 2]. Peano derivatives from the point of view of Theorems 10.1.9, 10.1.10 were first considered in GÖRLICH–NESSEL [1]. For elementary properties concerning strong derivatives see ASPLUND–BUNGART [1, p. 456]. For an $\mathsf{L}^2_{2\pi}$-version of Cor. 10.1.15 for $r = 1$ see EDWARDS [1I, pp. 129–130].

Sec. 10.2. Concerning the classical results of TITCHMARSH [2] and HARDY–LITTLEWOOD [2I, 3] we refer to the comments given in BUTZER–BERENS [1, p. 148] as well as the papers listed there. The weak* compactness theorem which plays an important rôle in the proofs was suggested in a written communication to the first-named author by KAREL DE LEEUW in 1959 in connection with the integral of de La Vallée Poussin; it has previously been used by SUNOUCHI [1II] in solving related questions (cf. Sec. 12.1). But it may be replaced by a different argument, see BUTZER [8] and Problem 10.2.6. Theorem 10.2.5 is due to GÖRLICH–NESSEL [1].

Sec. 10.3. Most of this material is to be found in BUTZER–GÖRLICH [1]. The proof of Prop. 10.3.1, (ii) ⇒ (i), is due to H. BERENS (unpublished).

Sec. 10.4. The concept of relative completion in the form of an explicit definition was introduced by GAGLIARDO [1] in his study on interpolation spaces. For Prop. 10.4.2, a proof of Prop. 10.4.3 and further properties connected with this concept see ARONSZAJN–GAGLIARDO [1]. This concept is stronger than that of closure in X yet weaker than compactness in Y, and fits in precisely in our study on the connection between $\mathsf{W}[\mathsf{X}_{2\pi}; \psi(k)]$ and $\mathsf{V}[\mathsf{X}_{2\pi}; \psi(k)]$. In Part V (Sec. 13.4) it will be seen that the notion of relative completion is especially adaptable to the phenomenon of saturation. The idea of using the concept of relative completion to study such problems goes back to BERENS (see his Habilitationsschrift [3], also the remarks in BUTZER–BERENS [1, p. 223]). For a separate interpretation (employing a language of the theory of partial differential equations) see H. S. SHAPIRO [4, Sec. 3.5], where Y is the set of strong solutions of a certain problem and $\tilde{\mathsf{Y}}^{\mathsf{X}}$ that of weak solutions. The Friedrich's mollifier technique used there enables one to prove theorems about these weak solutions by operating with the strong solutions and passing to a limit.
The proofs of Prop. 10.4.4, 10.4.5 are straightforward, see also BUTZER–GÖRLICH [2, p. 381 f], BERENS [3, p. 75]. For the proof of Theorem 10.4.6 see also BERENS [3, p. 76].

Sec. 10.5. The results are largely taken from BUTZER [8, 10], GÖRLICH–NESSEL [1], BUTZER–NESSEL [2]. The proof of Theorem 10.5.6, (iii) ⇒ (i), is due to H. BERENS; see BUTZER–GÖRLICH [1, p. 41]. For Prop. 10.5.7 see also BERENS [3, p. 75 f]. We emphasize that the results of BOCHNER–CHANDRASEKHARAN (cf. Sec. 5.1., 5.2) influenced us.

Sec. 10.6. For the proofs of Lemmata 10.6.1, 10.6.2 see BUTZER–KOZAKIEWICZ [1], also LOOMIS [3, p. 176], WIENER [1]. These are connected with the problem of defining a primitive of a distribution (cf. SCHWARTZ [1I, p. 51 ff], ZEMANIAN [1, p. 68]) and may be compared with a fundamental lemma in the calculus of variation (cf. COURANT–HILBERT [1I, p. 201]). Problem 10.6.8, thus in particular Lemma 10.6.3, is due to BERENS [2] who, more generally, treated the problem of defining the rth infinitesimal generator of a semi-group by rth Peano or rth Riemann differences (cf. Sec. 13.4.2). For Prop. 10.6.4 see S. GOLDBERG [1, p. 180]. Concerning the connection of (10.6.11) (and of (10.6.18), (10.6.22)) with the definition of distributional derivatives we mention the Notes and Remarks to Sec. 1.1 and 3.1. Furthermore, the class of functions f satisfying (10.6.11) may be regarded as a genuine extension of $\mathsf{V}[\mathsf{L}^p; (iv)^r]$, $\mathsf{W}[\mathsf{L}^p; (iv)^r]$ to L^p, $2 < p < \infty$ (see also (13.1.4) and Sec. 13.5)).
For the remaining results see GÖRLICH–NESSEL [1] and the literature cited there (compare also with H. S. SHAPIRO [4, Sec. 3.5]). The results there are also treated in the setting of SCHWARTZ' distribution theory; for a thorough discussion of various definitions of a (ordinary) derivative in connection with distribution theory see also LOOMIS [3, p. 164 ff]. For pointwise and norm-convergence of distributions we also refer to BELTRAMI [1]. For related results concerning functions defined on an open interval (a, b) we mention BUTZER–KOZAKIEWICZ [1] and GÖRLICH–NESSEL [1].

11

Characterization in the Fractional Case

11.0 Introduction

In Chapter 10 the class $\mathsf{W}[\mathsf{L}^p; |v|^r]$, $1 \le p \le 2$, was characterized in terms of differentiability properties upon f (r even) or $f^\sim$ (r odd). The question arises whether for fractional $\alpha > 0$ the class $\mathsf{W}[\mathsf{L}^p; |v|^\alpha]$ is connected with a derivative of f of fractional order α. This will be shown to be the case. It is opportune to define fractional differentiation through integration of fractional order. There are at least two such definitions. If $[L_1 f](x)$ is the integral of f over (a, x), and $[L_\alpha f](x)$ the integral of $[L_{\alpha-1} f](x)$ over (a, x), $\alpha = 2, 3, \ldots$, then

$$[L_\alpha f](x) = \frac{1}{\Gamma(\alpha)} \int_a^x (x-u)^{\alpha-1} f(u)\, du \qquad (x > a). \tag{11.0.1}$$

Here a is finite or $-\infty$, in which case one assumes that the growth of f at infinity will not cause the integrals to diverge. J. Liouville and B. Riemann took (11.0.1) as a definition of $[L_\alpha f](x)$ for every $\alpha > 0$. This integral of fractional order possesses the property $[L_\alpha(L_\beta f)](x) = [L_{\alpha+\beta} f](x)$ $(\alpha, \beta > 0)$. If $\alpha > 0$ is arbitrary and $n > \alpha$ $(n \in \mathbb{N})$ one could define a (pointwise) derivative of f of fractional order α by $(d/dx)^n [L_{n-\alpha} f](x)$. However, we shall not follow this approach to the definition any further since it is connected (cf. Problems 11.1.2, 11.2.7) with the class $\mathsf{W}[\mathsf{L}^p; (iv)^\alpha]$ instead of $\mathsf{W}[\mathsf{L}^p; |v|^\alpha]$ which is our aim.

For this reason we shall consider another definition, introduced by Marcel Riesz and more convenient for our purpose. For $0 < \alpha < 1$ set

$$[R_\alpha f](x) = \frac{1}{2\Gamma(\alpha)\cos(\pi\alpha/2)} \int_{-\infty}^{\infty} \frac{f(u)}{|x-u|^{1-\alpha}}\, du,$$

where f is such that no difficulties arise in the calculations involved. This integral of fractional order also has the property $[R_\alpha(R_\beta f)](x) = [R_{\alpha+\beta} f](x)$ $(\alpha > 0, \beta > 0, \alpha + \beta < 1)$, and we could define a pointwise derivative of fractional order α, $0 < \alpha < 1$, by $(d/dx)[R_{1-\alpha} f](x)$. But this definition would be associated with the class $\mathsf{W}[\mathsf{L}^p; \{i \operatorname{sgn} v\}|v|^\alpha]$ (cf. Prop. 11.2.3). However, if we replace f by its Hilbert transform in the latter definition, then it would be associated with the class $\mathsf{W}[\mathsf{L}^p; |v|^\alpha]$. There is one important drawback here: one must suppose that f belongs to $L^1 \cap L^2$ instead of only to L^p for some $1 \le p \le 2$.

Since $[R_{1-\alpha}f^{\sim}](x) = [R_{1-\alpha}^{\sim}f](x)$ if $f \in L^1 \cap L^2$, and the conjugate of the Riesz integral is also meaningful for $f \in L^1$, one would be led to define the fractional derivative by $(d/dx)[R_{1-\alpha}^{\sim}f](x)$. However, for $0 < \alpha \le 1/p'$, $[R_{1-\alpha}^{\sim}f](x)$ does not necessarily exist (cf. Problem 11.2.3). Noting that

$$[R_{1-\alpha}^{\sim}f](x+h) - [R_{1-\alpha}^{\sim}f](x) = f * m_{h,1-\alpha}^{\sim}(x) \tag{11.0.2}$$

where

$$m_{h,\alpha}(x) = \frac{\sqrt{2\pi}}{2\Gamma(\alpha)\cos(\pi\alpha/2)}\left\{\frac{1}{|x+h|^{1-\alpha}} - \frac{1}{|x|^{1-\alpha}}\right\},$$

then the expression on the right in (11.0.2) exists as a function in L^p for each $1 \le p < \infty$ and $0 < \alpha < 1$ (cf. Lemma 11.2.2). At last we come to the desired definition of a (strong) derivative of order α which for want of a suitable name we call the *strong Riesz derivative* $D_s^{\{\alpha\}}f$ *of order* α, $0 < \alpha < 1$. It is the limit in L^p-norm of $h^{-1}(f * m_{h,1-\alpha}^{\sim})$ as $h \to 0$ if it exists.

If $\alpha > 0$ is arbitrary, then the αth strong Riesz derivative of $f \in L^p$ is defined successively (cf. Def. 11.2.8), noting that $D_s^{\{1\}}f = (f^{\sim})'$ (at least formally). This fractional derivative has the desired properties. Indeed, if $f \in L^p$, $1 \le p \le 2$, and $\alpha > 0$, then $f \in W[L^p; |v|^\alpha]$ if and only if $D_s^{\{\alpha\}}f$ exists. Moreover $[D_s^{\{\alpha\}}f]^\wedge(v) = |v|^\alpha f^\wedge(v)$.

In distinction to Chapter 10, if $\alpha > 0$ is arbitrary we shall characterize the class $W[L^p; |v|^\alpha]$ in terms of

$$\int_\varepsilon^\infty \frac{\overline{\Delta}_u^{2j}f(x)}{u^{1+\alpha}}\,du \qquad (0 < \alpha < 2j;\ \varepsilon > 0), \tag{11.0.3}$$

which may be interpreted as a fractional (Riemann) difference quotient of f of order α. Our fundamental result is the following (cf. Theorems 11.2.6–11.2.9, 11.3.7, 11.5.1–11.5.3, Problem 11.5.2).

Theorem 11.0.1. *Let $f \in L^p$, $1 \le p \le 2$, and $\alpha > 0$ arbitrary. The following assertions are equivalent:*

(i) $f \in W[L^p; |v|^\alpha]$,

(ii) *there exists $g \in L^p$ such that*

$$\lim_{\varepsilon\to 0}\left\|\frac{1}{C_{\alpha,2j}}\int_\varepsilon^\infty \frac{\overline{\Delta}_u^{2j}f(\circ)}{u^{1+\alpha}}\,du - g(\circ)\right\|_p = 0,$$

where $j \in \mathbb{N}$ is chosen so that $0 < \alpha < 2j$, and $C_{\alpha,2j}$ is a constant defined by (11.3.14),

(iii) *$D_s^{\{\alpha\}}f$ exists and equals g,*

(iv) *there exists $g \in L^p$ so that*

for $0 < \alpha < 1, p = 1$: $\quad f(x) = [R_\alpha g](x)$ a.e.

$1 < p \le 2$: $\quad R_{\alpha/p'}g \in L^p$ *and* $f(x) = R_{\alpha/p}[(R_{\alpha/p'}g)](x)$ a.e.

for $\alpha \ge 1, p = 1$: $\quad$ *f belongs to L^2, $R_\beta g$ belongs to L^1, and*

$$[H^{[\alpha]}f](x) = \int_{-\infty}^x du_1 \int_{-\infty}^{u_1} du_2 \ldots \int_{-\infty}^{u_{[\alpha]-1}} [R_\beta g](u_{[\alpha]})\,du_{[\alpha]} \quad \text{a.e.}$$

$1 < p \le 2$: *$R_{\beta/p'}g$, $R_{\beta/p}(R_{\beta/p'}g)$ belong to L^p and*

$$[H^{[\alpha]}f](x) = \int_{-\infty}^{x} du_1 \int_{-\infty}^{u_1} du_2 \ldots \int_{-\infty}^{u_{[\alpha]-1}} [R_{\beta/p}(R_{\beta/p'}g)](u_{[\alpha]})\, du_{[\alpha]} \quad \text{a.e.},$$

where $\alpha = [\![\alpha]\!] + \beta$, $0 \le \beta < 1$, $R_0 g = g$, and H denotes the operation of taking Hilbert transforms.

We note that the integral representation of f given in assertion (iv) above may be compared with the fact that $f \in \mathsf{W}[\mathsf{L}^p; (iv)^r]$, $r \in \mathbb{N}$, if and only if there is $g \in \mathsf{L}^p$ so that (Theorem 5.2.21) $f(x) = \int_{-\infty}^{x} du_1 \int_{-\infty}^{u_1} du_2 \ldots \int_{-\infty}^{u_{r-1}} g(u_r)\, du_r$ a.e.

Concerning characterizations of $\mathsf{V}[\mathsf{L}^p; |v|^\alpha]$ we refer to Theorems 11.2.10–11.2.12, 11.3.6, and Problem 11.5.1. The extension to L^p-space, $2 < p < \infty$ is given in Sec. 13.2.

Sec. 11.1 is concerned with Riemann–Liouville and Riesz fractional integrals and their basic properties. Sec. 11.2 is devoted to the discussion on derivatives of fractional order. This leads to characterizations of the fundamental classes $\mathsf{W}[\mathsf{L}^p; |v|^\alpha]$ and $\mathsf{V}[\mathsf{L}^p; |v|^\alpha]$ in terms of strong Riesz derivatives (Theorems 11.2.6–11.2.12). Sec. 11.3.1 is reserved to characterizations in terms of the integrals (11.0.3). In Sec. 11.3.2 these results are applied to give an extension (Cor. 11.3.9) of a theorem of E. C. Titchmarsh, the periodic version of which is the famous theorem of S. N. Bernstein on absolute convergence of Fourier series, as well as to establish the L^1-version of a theorem of H. Weyl on Lipschitz conditions of 2π-periodic functions (Theorem 11.3.10). Sec. 11.4 is concerned with the counterparts of the results of Sec. 11.3 to the case of periodic functions. The fundamental results are given by Theorems 11.4.6 and 11.4.7. Sec. 11.5 deals with extensions of the results of Sec. 11.2 to periodic functions, and these are deduced by reduction from the nonperiodic case. In particular, fractional derivatives in the sense of M. Riesz and H. Weyl are treated in terms of the classes $\mathsf{W}[\mathsf{X}_{2\pi}; |k|^\alpha]$, $\mathsf{V}[\mathsf{X}_{2\pi}; |k|^\alpha]$, $\alpha > 0$. Particular stress is laid upon characterizations of these classes in terms of integral representations of f (cf. Theorems 11.5.4, 11.5.5). The latter representations are first considered for L^p-functions (Theorems 11.5.1–11.5.3, Problems 11.5.1–11.5.2).

11.1 Integrals of Fractional Order

11.1.1 Integral of Riemann–Liouville

The starting-point for the definition of integration of fractional order was for many authors, in particular for Riemann, the following property of the rth iterated indefinite integral (cf. (10.6.6))

$$[L_r f](x) \equiv \int_a^x du_1 \int_a^{u_1} du_2 \ldots \int_a^{u_{r-1}} f(u_r)\, du_r \tag{11.1.1}$$

$$= \frac{1}{(r-1)!} \int_a^x f(u)(x-u)^{r-1}\, du \qquad (x > a).$$

Since $(r-1)! = \Gamma(r)$, one may conjecture that the expression

(11.1.2)
$$[L_\alpha f](x) = \frac{1}{\Gamma(\alpha)} \int_a^x \frac{f(u)}{(x-u)^{1-\alpha}}\,du$$

will in some sense be an extension of the integral (11.1.1) to fractional α. We call $L_\alpha f$ the *Riemann–Liouville integral of order* α *of* f.

To fix matters we suppose that f is a continuous function on the interval of integration and that a is finite or negatively infinite; in the latter case it is assumed that the growth of f at infinity will not cause the integrals to diverge. We wish to discuss the integral (11.1.2) and its relations to (11.1.1).

Obviously, the integral (11.1.2) exists for $\alpha > 0$ and coincides with (11.1.1) for positive integers α. It also possesses the fundamental property

(11.1.3)
$$[L_\alpha(L_\beta f)](x) = [L_{\alpha+\beta} f](x) \qquad (\alpha, \beta > 0).$$

Indeed, by Dirichlet's formula on the change of the order of integration we have

$$[L_\alpha(L_\beta f)](x) = \frac{1}{\Gamma(\alpha)} \int_a^x (x-u)^{\alpha-1}\left\{\frac{1}{\Gamma(\beta)}\int_a^u f(v)(u-v)^{\beta-1}\,dv\right\} du$$
$$= \frac{1}{\Gamma(\alpha)\Gamma(\beta)} \int_a^x f(v)\left\{\int_v^x (x-u)^{\alpha-1}(u-v)^{\beta-1}\,du\right\} dv.$$

Substituting $u = (x-v)t + v$, the inner integral is equal to (cf. Problem 11.1.3)

$$(x-v)^{\alpha+\beta-1}\int_0^1 (1-t)^{\alpha-1}t^{\beta-1}\,dt = (x-v)^{\alpha+\beta-1}\frac{\Gamma(\alpha)\Gamma(\beta)}{\Gamma(\alpha+\beta)},$$

and (11.1.3) is established.

Next we wish to describe the connections between fractional integration and differentiation. Let $r \in \mathbb{N}$ and $\alpha > 0$. Then by the usual rules of differentiation we have

$$\frac{d}{dx}[L_{r+\alpha} f](x) = \frac{1}{\Gamma(r+\alpha-1)}\int_a^x f(u)(x-u)^{r+\alpha-2}\,du = [L_{r+\alpha-1} f](x),$$

and more generally

(11.1.4)
$$\frac{d^r}{dx^r}[L_{r+\alpha} f](x) = [L_\alpha f](x).$$

If we would set

(11.1.5)
$$[L_0 f](x) = f(x),$$

then (11.1.4) would formally convert into $[L_r f]^{(r)}(x) = f(x)$, a relation which is obvious by (11.1.1). Indeed, we shall show that for each fixed $x > a$

(11.1.6)
$$\lim_{\alpha\to 0+}[L_\alpha f](x) = f(x).$$

For this purpose, we first let a be finite. Then

$$[L_\alpha f](x) = \frac{1}{\Gamma(\alpha)}\int_a^x [f(u) - f(x)](x-u)^{\alpha-1}\,du + \frac{f(x)}{\Gamma(\alpha+1)}(x-a)^\alpha \equiv I_1 + I_2.$$

Here, the last term tends to $f(x)$ as $\alpha \to 0+$ whereas the first one tends to zero. For, given $\varepsilon > 0$, let $\delta > 0$ be such that $|f(x) - f(u)| < \varepsilon$ for all u with $a < x - \delta \leq u \leq x$. Then

$$\begin{aligned} |I_1| &\leq \frac{1}{\Gamma(\alpha)} \int_a^{x-\delta} |f(u) - f(x)|(x - u)^{\alpha-1}\, du + \frac{1}{\Gamma(\alpha)} \int_{x-\delta}^{x} |f(u) - f(x)|(x - u)^{\alpha-1}\, du \\ &\leq \frac{\delta^{\alpha-1}}{\Gamma(\alpha)} \int_a^{x-\delta} |f(u) - f(x)|\, du + \frac{\varepsilon}{\Gamma(\alpha)} \int_{x-\delta}^{x} (x - u)^{\alpha-1}\, du \\ &\leq \frac{2(x - a)}{\delta^{1-\alpha}} \sup_{a \leq u \leq x} |f(u)| \frac{1}{\Gamma(\alpha)} + \frac{\varepsilon\delta^{\alpha}}{\Gamma(\alpha + 1)}, \end{aligned}$$

the last expression being arbitrarily small since $\lim_{\alpha\to 0+} \Gamma(\alpha) = \infty$. It is also due to the factor $1/\Gamma(\alpha)$ that the infinite interval case immediately reduces to the preceding one. For, if $a = -\infty$, then the part corresponding to the interval $(-\infty, b)$, where b is any finite real number with $b < x$, again tends to zero with α.

We should remark that the above results also hold if the real parameter α is allowed to be complex. The condition $\alpha > 0$ is then replaced by $\mathrm{Re}\,(\alpha) > 0$, and for any fixed $x > a$ we may consider the integral in (11.1.2) as a holomorphic function of α for $\mathrm{Re}\,(\alpha) > 0$. Then for sufficiently smooth functions a process of analytic continuation permits us to attach a meaning to the operation $[L_\alpha f](x)$ also for those values of α, namely $\mathrm{Re}\,(\alpha) \leq 0$, for which the integral in (11.1.2) diverges.

To this end, let us suppose that the derivatives $f^{(k)}(x)$ exist for $k \leq r$ and possess properties analogous to those of f. Then for $\alpha > 0$ we obtain by successive integration by parts for $n \leq r$

$$[L_\alpha f](x) = \sum_{k=0}^{n-1} \frac{f^{(k)}(a)}{\Gamma(\alpha + k + 1)} (x - a)^{\alpha+k} + [L_{\alpha+n} f^{(n)}](x). \tag{11.1.7}$$

In case $a = -\infty$ the last formula reduces to

$$[L_\alpha f](x) = [L_{\alpha+n} f^{(n)}](x) \qquad (0 \leq n \leq r). \tag{11.1.8}$$

Since all the expressions on the right in (11.1.7) and (11.1.8) exist as holomorphic functions of α for $\mathrm{Re}\,(\alpha) > -n$, $0 \leq n \leq r$, we may interpret $[L_\alpha f](x)$ for all values α with $\mathrm{Re}\,(\alpha) > -r$ as the analytic continuation of the integral in (11.1.2), considered as a holomorphic function of α.

By complex function theory it is easily shown that the fundamental property (11.1.3) remains valid for all α, β with $\mathrm{Re}\,(\beta)$, $\mathrm{Re}\,(\alpha + \beta) > -r$ with the possible exception (at least at this stage) of those values of α, β for which α, β or $\alpha + \beta$ are zero or negative integers. It is this exceptional case that we shall examine.

We note that for $\alpha = 0$ (11.1.7) reduces to the classical formula of Cauchy

$$f(x) = \sum_{k=0}^{n-1} \frac{f^{(k)}(a)}{k!} (x - a)^k + \frac{1}{(n - 1)!} \int_a^x f^{(n)}(u)(x - u)^{n-1}\, du.$$

If we wish to interpret $[L_{-n} f](x)$ for $n \in \mathbb{N}$ we suppose that $f^{(n)}(x)$ satisfies the same hypotheses as $f(x)$ in case $\alpha = 0$. Then (11.1.7) suggests that one sets

$$[L_{-n} f](x) = \lim_{\substack{\alpha\to -n \\ \mathrm{Re}\,(\alpha) > -n}} [L_\alpha f](x) = f^{(n)}(x),$$

i.e. L_{-n} effects an n-times differentiation. Furthermore, we see that relation (11.1.4) may be expressed by

$$[L_{-r}(L_\alpha f)](x) = [L_{-r+\alpha} f](x).$$

Thus for $a = -\infty$ the fundamental formula (11.1.3) is valid for all values of α and β for which $\operatorname{Re}(\beta)$, $\operatorname{Re}(\alpha + \beta) > -r$. If a is finite, this formula holds for all these values of α and β, except for those values of β which are negative integers. In the obvious case that α is zero or a negative integer, the formula still remains valid.

Let us finally give a further formula which is useful for the process of analytic continuation to be discussed below and which is close to the Hadamard approach to the subject. If we set

$$p(u) = \sum_{k=0}^{r-1} \frac{f^{(k)}(x)}{k!}(u - x)^k,$$

then for $\alpha > 0$

$$\begin{aligned}[L_\alpha f](x) &= \frac{1}{\Gamma(\alpha)} \int_a^x f(u)(x - u)^{\alpha-1}\, du \\ &= \frac{1}{\Gamma(\alpha)} \int_a^x [f(u) - p(u)](x - u)^{\alpha-1}\, du + \frac{1}{\Gamma(\alpha)} \int_a^x p(u)(x - u)^{\alpha-1}\, du.\end{aligned}$$

Thus

(11.1.9)
$$\begin{aligned}[L_\alpha f](x) = \frac{1}{\Gamma(\alpha)} \int_a^x &[f(u) - p(u)](x - u)^{\alpha-1}\, du \\ &+ \frac{1}{\Gamma(\alpha)} \sum_{k=0}^{r-1} \frac{(-1)^k f^{(k)}(x)}{(k + \alpha)k!}(x - a)^{k+\alpha}.\end{aligned}$$

If $f^{(r)}(x)$ exists and is continuous, the latter integral obviously converges for $\alpha > -r$, and the right-hand side of (11.1.9) defines the analytic continuation for all these values. $[L_\alpha f](x)$ is a holomorphic function of α since $(k + \alpha)\Gamma(\alpha) \neq 0$ throughout. In particular for $\alpha = -k$ we have $(k + \alpha)k!\Gamma(\alpha) = (-1)^k$ (cf. Problem 11.1.3) from which we again derive

(11.1.10)
$$[L_0 f](x) = f(x), \qquad [L_{-r} f](x) = f^{(r)}(x).$$

For the latter arguments to be valid it is not necessary that the derivative $f^{(r)}$ exists at x in the classical sense. It suffices that there exists an algebraic polynomial $p(u)$ such that $[f(u) - p(u)](x - u)^{-r}$ remains bounded as $u \to x-$. For example, it is sufficient that the rth Peano derivative $f^{\langle r\rangle}(x)$ exists.

11.1.2 Integral of M. Riesz

In his monumental treatise [2], M. Riesz introduced a modified form of an *integral of fractional order*. For $0 < \alpha < 1$ he considers

(11.1.11)
$$[R_\alpha f](x) = \frac{1}{M(\alpha)} \int_{-\infty}^{\infty} \frac{f(u)}{|x - u|^{1-\alpha}}\, du,$$

the function f being assumed to be so well-behaved that no difficulties arise in the calculations that are based upon the definition. We shall show that the constant $M(\alpha)$, which only depends on α and plays the rôle of $\Gamma(\alpha)$ in the case of the Riemann–Liouville integral, may be determined in such a way that the fundamental formula

(11.1.12)
$$[R_\alpha(R_\beta f)](x) = [R_{\alpha+\beta} f](x) \qquad (\alpha > 0, \beta > 0, \alpha + \beta < 1)$$

is satisfied.

First we show that we may evaluate $M(\alpha)$ such that (11.1.12) is valid for the particular function $f(x) = \exp\{ix\}$. Indeed, for $0 < \alpha < 1$

$$[R_\alpha e^{i\circ}](x) = \frac{e^{ix}}{M(\alpha)} \int_{-\infty}^{\infty} \frac{e^{iu}}{|u|^{1-\alpha}}\, du.$$

Thus if we set

$$M(\alpha) = 2\int_0^\infty \frac{\cos u}{u^{1-\alpha}}\, du,$$

then

$$[R_\alpha e^{i\circ}](x) = e^{ix}, \tag{11.1.13}$$

and the result follows. Note that, since

$$[R_\alpha e^{iv\circ}](x) = |v|^{-\alpha} e^{ivx} \qquad (0 < \alpha < 1) \tag{11.1.14}$$

for all $v \neq 0$, we may connect the integral of M. Riesz with the theory of Fourier transforms (cf. Lemma 11.2.2).

To evaluate $M(\alpha)$ defined above we use Problem 11.1.1 and obtain $M(\alpha) = 2\Gamma(\alpha)\cos(\pi\alpha/2)$ which holds for $0 < \alpha < 1$ to ensure the convergence of the integral. Thus, if we set

$$\Lambda_c(\alpha) = 2\Gamma(\alpha)\cos\frac{\pi\alpha}{2}, \qquad \Lambda_s(\alpha) = 2\Gamma(\alpha)\sin\frac{\pi\alpha}{2}, \tag{11.1.15}$$

the *Riesz integral of order* α *of* f takes on the form

$$[R_\alpha f](x) = \frac{1}{\Lambda_c(\alpha)} \int_{-\infty}^{\infty} \frac{f(u)}{|x-u|^{1-\alpha}}\, du \qquad (0 < \alpha < 1). \tag{11.1.16}$$

To examine the fundamental formula (11.1.12) let us moreover suppose that f is such that all integrals which occur in the following are absolutely convergent. We have

$$[R_\alpha(R_\beta f)](x) = \frac{1}{\Lambda_c(\alpha)} \int_{-\infty}^{\infty} |x-u|^{\alpha-1} \left[\frac{1}{\Lambda_c(\beta)} \int_{-\infty}^{\infty} f(t)|u-t|^{\beta-1}\, dt\right] du$$

$$= \frac{1}{\Lambda_c(\alpha)\Lambda_c(\beta)} \int_{-\infty}^{\infty} f(t)\left[\int_{-\infty}^{\infty} |x-u|^{\alpha-1}|u-t|^{\beta-1}\, du\right] dt.$$

Again by the substitution $u = (x-t)v + t$ we obtain

$$\int_{-\infty}^{\infty} |x-u|^{\alpha-1}|u-t|^{\beta-1}\, du = |x-t|^{\alpha+\beta-1} \int_{-\infty}^{\infty} |1-v|^{\alpha-1}|v|^{\beta-1}\, dv,$$

and if we set

$$A(\alpha,\beta) = \int_{-\infty}^{\infty} |1-u|^{\alpha-1}|u|^{\beta-1}\, du,$$

then

$$[R_\alpha(R_\beta f)](x) = \frac{A(\alpha,\beta)}{\Lambda_c(\alpha)\Lambda_c(\beta)} \int_{-\infty}^{\infty} \frac{f(t)}{|x-t|^{1-\alpha-\beta}}\, dt. \tag{11.1.17}$$

Obviously, $A(\alpha, \beta)$ only depends on α and β and exists for $\alpha > 0, \beta > 0, \alpha + \beta < 1$. If we can show that

$$A(\alpha, \beta) = \frac{\Lambda_c(\alpha)\Lambda_c(\beta)}{\Lambda_c(\alpha+\beta)} \qquad (\alpha > 0, \beta > 0, \alpha + \beta < 1), \tag{11.1.18}$$

then (11.1.17) reduces to the fundamental formula (11.1.12) which would thereby be established.

To verify (11.1.18) we have

$$\begin{aligned} A(\alpha, \beta) &= \int_{-\infty}^{0} (1-u)^{\alpha-1}(-u)^{\beta-1}\,du + \int_0^1 (1-u)^{\alpha-1}u^{\beta-1}\,du + \int_1^{\infty} (u-1)^{\alpha-1}u^{\beta-1}\,du \\ &= \int_0^{\infty} (1+t)^{\alpha-1}t^{\beta-1}\,dt + B(\beta, \alpha) + \int_0^{\infty} t^{\alpha-1}(1+t)^{\beta-1}\,dt, \end{aligned}$$

where $B(\alpha, \beta)$ is the Beta function (cf. Problem 11.1.3). Substituting $t = u/(1-u)$, it follows that

$$\begin{aligned} A(\alpha, \beta) &= \int_0^1 (1-u)^{-\alpha-\beta}u^{\beta-1}\,du + B(\alpha, \beta) + \int_0^1 (1-u)^{-\alpha-\beta}u^{\alpha-1}\,du \\ &= B(\beta, 1-\alpha-\beta) + B(\alpha, \beta) + B(\alpha, 1-\alpha-\beta), \end{aligned}$$

and thus in view of Problem 11.1.3

$$\begin{aligned} \frac{\Gamma(\alpha+\beta)}{\Gamma(\alpha)\Gamma(\beta)} A(\alpha, \beta) &= \frac{\Gamma(\alpha+\beta)\Gamma(1-\alpha-\beta)}{\Gamma(\alpha)\Gamma(1-\alpha)} + 1 + \frac{\Gamma(\alpha+\beta)\Gamma(1-\alpha-\beta)}{\Gamma(\beta)\Gamma(1-\beta)} \\ &= \frac{\sin \pi\alpha}{\sin \pi(\alpha+\beta)} + 1 + \frac{\sin \pi\beta}{\sin \pi(\alpha+\beta)} = \frac{2\cos(\pi\alpha/2)\cos(\pi\beta/2)}{\cos(\pi(\alpha+\beta)/2)}, \end{aligned}$$

which in view of (11.1.15) proves (11.1.18).

As for the Riemann–Liouville integral (11.1.2) we recall that the function f has to satisfy certain regularity requirements, in particular at infinity, in order that the integral (11.1.11) converges. If e.g. $f \in \mathsf{C}_{00}$, then (11.1.16) exists and represents a holomorphic function of α for all complex α with $\operatorname{Re}(\alpha) > 0$ with the exception of $\alpha = 1 + 2k$, $k \in \mathbb{P}$ (cf. (11.1.15)). At these points $[R_\alpha f](x)$ has poles of order one.

We now proceed by discussing the case $\alpha = 1$. Then the integral in (11.1.16) is independent of x, but the factor $1/\Lambda_c(\alpha)$ is infinite. In the particular case that $\int_{-\infty}^{\infty} f(u)\,du = 0$, we may then define

$$\begin{aligned} [R_1 f](x) = \lim_{\varepsilon\to 0} [R_{1-\varepsilon} f](x) &= \lim_{\varepsilon\to 0} \frac{1}{2\Gamma(1-\varepsilon)} \int_{-\infty}^{\infty} f(u) \frac{|x-u|^{-\varepsilon} - 1}{\sin(\pi\varepsilon/2)}\,du \\ &= \frac{1}{\pi}\int_{-\infty}^{\infty} f(u) \log\frac{1}{|x-u|}\,du. \end{aligned} \tag{11.1.19}$$

Thus $[R_1 f](x)$ is defined as a logarithmic potential.

In the general case that $\int_{-\infty}^{\infty} f(u)\,du \neq 0$ we may always write

$$[R_\alpha f](x) = \frac{c_{-1}}{\alpha - 1} + c_0 + \sum_{k=1}^{\infty} c_k(\alpha-1)^k$$

in the neighbourhood of $\alpha = 1$ for any fixed value of x. Here,

$$c_{-1} = -\frac{1}{\pi}\int_{-\infty}^{\infty} f(u)\,du$$

and, except for an additive constant, c_0 is equal to the logarithmic potential

$$\frac{1}{\pi}\int_{-\infty}^{\infty} f(u)\log\frac{1}{|x-u|}\,du = \lim_{\varepsilon\to 0}\tfrac{1}{2}\{[R_{1-\varepsilon}f](x) + [R_{1+\varepsilon}f](x)\}.$$

We note that for $\alpha = 1 + 2k$, $k \in \mathbb{N}$, we obtain kernels of the form $|x-u|^{2k}\log(1/|x-u|)$ by an analogous procedure.

Concerning the connection with differentiation, we have for $\alpha > 0$, $\alpha \neq 1 + 2k$, $k \in \mathbb{P}$, that

$$\begin{aligned}\frac{1}{\Lambda_c(\alpha+2)}\frac{d^2}{dx^2}|x-u|^{\alpha+1} &= \frac{(\alpha+1)\alpha}{-2(\alpha+1)\alpha\Gamma(\alpha)\cos(\pi\alpha/2)}|x-u|^{\alpha-1}\\ &= \frac{-1}{\Lambda_c(\alpha)}|x-u|^{\alpha-1},\end{aligned}$$

and thus (see also (11.1.13))

$$\frac{d^2}{dx^2}[R_{\alpha+2}f](x) = -[R_\alpha f](x) \tag{11.1.20}$$

holds under suitable† regularity conditions on f.

We finally turn to the problem of extending the definition of $[R_\alpha f](x)$ to $\alpha \leq 0$. As for the Riemann–Liouville integral, under certain conditions upon f this extension is constructed by analytic continuation. For $\alpha = 0$ we may consider a limiting process.

Supposing that f is continuous at x we obtain by the same arguments as applied to (11.1.6)

$$[R_0 f](x) = \lim_{\alpha\to 0+}[R_\alpha f](x) = f(x). \tag{11.1.21}$$

Concerning the case $\alpha < 0$, let us assume that the function f is $2r$-times continuously differentiable and that f and its derivatives behave at infinity in such a way that the integrals occurring in the following are absolutely convergent and that the integrations by parts are justified. Then we obtain, first of all for $\alpha > 0$, $\alpha \neq 1 + 2k$, $k \in \mathbb{P}$, that

$$\begin{aligned}[R_\alpha f](x) &= \frac{-1}{\Lambda_c(\alpha+2)}\int_{-\infty}^{\infty} f(u)\frac{d^2}{du^2}|x-u|^{\alpha+1}\,du\\ &= \frac{-1}{\Lambda_c(\alpha+2)}\int_{-\infty}^{\infty} f''(u)|x-u|^{\alpha+1}\,du,\end{aligned}$$

and generally,

$$[R_\alpha f](x) = (-1)^r[R_{\alpha+2r}f^{(2r)}](x). \tag{11.1.22}$$

Thus $[R_\alpha f](x)$ possesses an analytic continuation to all values of α with $\operatorname{Re}(\alpha) > -2r$. Furthermore, by the continuity of $f^{(2r)}$ at x (which is not necessary for the validity of (11.1.22)) the extension to $\alpha = -2r$ follows as above from the limit

$$[R_{-2r}f](x) = \lim_{\alpha\to -2r+}[R_\alpha f](x) = (-1)^r f^{(2r)}(x).$$

An analogous formula of course holds for all values $k < r$, $k \in \mathbb{N}$. The fundamental

† We observe that the property $d^2[R_{\alpha+2}f](x)/dx^2 = +[R_\alpha f](x)$ may also be deduced if we replace $\Lambda_c(\alpha)$ by $\exp\{i\pi\alpha/2\}\Lambda_c(\alpha)$. In a certain sense this value is more suitable than that used by us since it gives $[R_\alpha e^{i\circ}](x) = i^{-\alpha}e^{ix}$. The present choice is justified by the fact that $\Lambda_c(\alpha)$ is real for α real.

formula (11.1.12) may also be readily extended to the domain $\alpha > -2r$, $\beta > -2r$, $\alpha + \beta > -2r$.

Problems

1. Show that for $0 < \alpha < 1$

$$\int_0^\infty \frac{\cos u}{u^{1-\alpha}}\,du = \Gamma(\alpha)\cos\frac{\pi\alpha}{2}, \qquad \int_0^\infty \frac{\sin u}{u^{1-\alpha}}\,du = \Gamma(\alpha)\sin\frac{\pi\alpha}{2},$$

$$\int_0^\infty \frac{e^{ixu}}{u^{1-\alpha}}\,du = \Gamma(\alpha)|x|^{-\alpha}\exp\left\{i\frac{\pi\alpha}{2}\operatorname{sgn} x\right\}.$$

(Hint: Compare with the definition of the gamma function given in Problem 11.1.3; see also Bochner [7, p. 51], Achieser [2, p. 112])

2. Let $a = -\infty$ and $0 < \alpha < 1$. Show that†

(i) $[L_\alpha e^{iv\circ}](x) = (iv)^{-\alpha}e^{ivx}$ $(v \neq 0)$,

(ii) $[L_\alpha e^{k\circ}](x) = k^{-\alpha}e^{kx}$ $(k > 0)$.

3. The gamma and beta functions are defined for Re (z), Re (α), Re $(\beta) > 0$ by

$$\Gamma(z) = \int_0^\infty e^{-t}t^{z-1}\,dt, \qquad B(\alpha,\beta) = \int_0^1 t^{\alpha-1}(1-t)^{\beta-1}\,dt,$$

respectively. Show that

(i) $\Gamma(1+z) = z\Gamma(z)$, (ii) $\Gamma(z)\Gamma(1-z) = \pi\sin^{-1}(\pi z)$,

(iii) $B(\alpha,\beta) = \dfrac{\Gamma(\alpha)\Gamma(\beta)}{\Gamma(\alpha+\beta)}$.

Furthermore, the gamma function is a meromorphic function with poles of order one at $z = 0, -1, -2, \ldots$. Show that $(j + x)\Gamma(x) \neq 0$ for all $x \in \mathbb{R}$, $j \in \mathbb{N}$ and $\lim_{x\to -j}(j + x)\Gamma(x) = (-1)^j/j!$.
(Hint: Erdélyi [1 I, p. 1 ff])

11.2 Characterizations of the Classes $\mathsf{W}[\mathsf{L}^p; |v|^\alpha]$, $\mathsf{V}[\mathsf{L}^p; |v|^\alpha]$, $1 \le p \le 2$

11.2.1 Derivatives of Fractional Order

The aim of this subsection is to prepare the way for characterizations of the classes $\mathsf{W}[\mathsf{L}^p; |v|^\alpha]$ and $\mathsf{V}[\mathsf{L}^p; |v|^\alpha]$ for fractional α, in particular in terms of certain smoothness conditions upon the original function f. To this end we consider the Riesz integral of order $1 - \alpha$, $0 < \alpha < 1$,

$$[R_{1-\alpha}f](x) = \frac{1}{\Lambda_c(1-\alpha)}\int_{-\infty}^\infty \frac{f(u)}{|x-u|^\alpha}\,du. \tag{11.2.1}$$

In comparison with the preceding section we now try to derive some properties of (11.2.1) for functions $f \in \mathsf{L}^p$, $1 \le p \le 2$. We first restrict the discussion to functions f in $\mathsf{L}^1 \cap \mathsf{L}^2$ and show that it is possible to define a *pointwise* derivative of fractional

† $(iv)^{-\alpha}$ is defined by $(iv)^{-\alpha} = |v|^{-\alpha}\exp\{-i(\pi\alpha/2)\operatorname{sgn} v\}$, the principal branch.

order. Then we remove this restriction to obtain suitable definitions of fractional differentiation in L^p-norm.

Lemma 11.2.1. *Let* $f \in \mathsf{L}^1 \cap \mathsf{L}^2$ *and* $0 < \alpha < 1$. *Then* $[R_{1-\alpha}f](x)$, $[R_{1-\alpha}f^\sim](x)$ *exist almost everywhere and are locally absolutely integrable. Furthermore*

$$\lim_{|h|\to\infty} \int_0^y [R_{1-\alpha}f](x+h)\,dx = 0 \qquad (y \in \mathbb{R}), \tag{11.2.2}$$

$$\lim_{|h|\to\infty} \int_0^y [R_{1-\alpha}f^\sim](x+h)\,dx = 0 \qquad (y \in \mathbb{R}). \tag{11.2.3}$$

Proof. Let $y \in \mathbb{R}$ be fixed. We consider the function

$$\psi_y(x) \equiv \int_0^y |x-u|^{-\alpha}\,du = \frac{1}{1-\alpha}\{\operatorname{sgn}(y-x)\cdot|y-x|^{1-\alpha} + \operatorname{sgn} x\cdot|x|^{1-\alpha}\},$$

which obviously is continuous (as a function of x) for $0 < \alpha < 1$. Since $\psi_y(x) = O(|x|^{-\alpha})$ as $|x| \to \infty$, we have $\psi_y \in \mathsf{C}_0$. Moreover, $\psi_y \in \mathsf{L}^q$ for all q with $\alpha q > 1$.

Since $f \in \mathsf{L}^1 \cap \mathsf{L}^2$ implies $f \in \mathsf{L}^p$ for $1 \le p \le 2$ (Sec. 0.1), by Theorem 8.1.12 $f^\sim \in \mathsf{L}^p$ for all $1 < p \le 2$. Let g be equal to f or $f^\sim$. For given α, $0 < \alpha < 1$, we may choose p, $1 < p < 2$, such that $\alpha p' > 1$. Then $g\psi_y \in \mathsf{L}^1$ by Hölder's inequality. Thus the iterated integral

$$\int_{-\infty}^{\infty} g(u)\left\{\int_0^y |u-x|^{-\alpha}\,dx\right\}du$$

converges absolutely, and by Fubini's theorem the integral

$$\int_0^y \left\{\int_{-\infty}^{\infty} g(u)|x-u|^{-\alpha}\,du\right\}dx$$

as well. Thus $[R_{1-\alpha}g](x)$ exists almost everywhere and $R_{1-\alpha}g \in \mathsf{L}^1_{\text{loc}}$. Moreover

$$\Lambda_c(1-\alpha)\int_0^y [R_{1-\alpha}g](x+h)\,dx = \int_{-\infty}^{\infty} g(u)\psi_y(u-h)\,du,$$

and since $\|\psi_y(\circ - h)\|_{p'} = \|\psi_y\|_{p'} \le M$ uniformly for $h \in \mathbb{R}$ and $\lim_{|h|\to\infty}\psi_y(u-h) = 0$, (11.2.2) and (11.2.3) follow by Prop. 0.1.11.

In terms of the Riesz integral (11.2.1) of order $1-\alpha$ we may define a pointwise derivative of f of order α, $0 < \alpha < 1$, as

$$\lim_{h\to 0} \frac{[R_{1-\alpha}f](x+h) - [R_{1-\alpha}f](x)}{h} \tag{11.2.4}$$

for all $x \in \mathbb{R}$ for which the limit exists. In order to consider the difference quotient the following lemma will be fundamental.

Lemma 11.2.2. *Let* $0 < \alpha < 1$ *and* $h \in \mathbb{R}$ *be fixed.*

(i) *The function* $m_{h,\alpha}(x)$ *defined by*

$$m_{h,\alpha}(x) = \frac{\sqrt{2\pi}}{\Lambda_c(\alpha)}\left\{\frac{1}{|x+h|^{1-\alpha}} - \frac{1}{|x|^{1-\alpha}}\right\} \tag{11.2.5}$$

belongs to L^1 *and*

$$\|m_{h,\alpha}\|_1 = |h|^\alpha \|m_{1,\alpha}\|_1, \qquad m_{h,\alpha}^{\wedge}(v) = (e^{ihv} - 1)|v|^{-\alpha} \qquad (v \neq 0).$$

(ii) *The function* $n_{h,\alpha}(x)$ *defined by*

$$n_{h,\alpha}(x) = \frac{\sqrt{2\pi}}{\Lambda_s(\alpha)} \left\{ \frac{\operatorname{sgn}(x+h)}{|x+h|^{1-\alpha}} - \frac{\operatorname{sgn} x}{|x|^{1-\alpha}} \right\} \tag{11.2.6}$$

belongs to L^1 *and*

$$\|n_{h,\alpha}\|_1 = |h|^\alpha \|n_{1,\alpha}\|_1, \qquad n_{h,\alpha}^{\wedge}(v) = \{-i \operatorname{sgn} v\}(e^{ihv} - 1)|v|^{-\alpha} \qquad (v \neq 0).$$

(iii)
$$m_{h,\alpha}^{\sim}(x) = n_{h,\alpha}(x) \quad \text{a.e.}$$

Proof. (i) Let us first consider the function $m_{1,\alpha}(x)$. Obviously, $m_{1,\alpha} \in L^1_{\text{loc}}$, and since $m_{1,\alpha}(x) = O(|x|^{\alpha-2})$ as $|x| \to \infty$, it follows that $m_{1,\alpha} \in L^1$. The substitution $hx = u$ leads to $m_{h,\alpha} \in L^1$ and $\|m_{h,\alpha}\|_1 = |h|^\alpha \|m_{1,\alpha}\|_1$ for any $h \in \mathbb{R}$. To evaluate the Fourier transform of $m_{h,\alpha}$ we have by Problem 11.1.1 for $v \neq 0$ (cf. (11.1.15))

$$\begin{aligned} m_{h,\alpha}^{\wedge}(v) &= \frac{1}{\Lambda_c(\alpha)} \int_{-\infty}^{\infty} \{|x+h|^{\alpha-1} - |x|^{\alpha-1}\}\, e^{-ivx}\, dx \\ &= (e^{ihv} - 1)|v|^{-\alpha} \frac{2}{\Lambda_c(\alpha)} \int_0^\infty \frac{\cos u}{u^{1-\alpha}}\, du = (e^{ihv} - 1)|v|^{-\alpha}. \end{aligned}$$

Note that besides $m_{h,\alpha} \in L^1$ each term of the difference of the first of the integrals above exists as an improper Riemann integral.

(ii) It follows as above that $n_{h,\alpha} \in L^1$ and $\|n_{h,\alpha}\|_1 = |h|^\alpha \|n_{1,\alpha}\|_1$. For the Fourier transform we obtain for $v \neq 0$

$$\begin{aligned} n_{h,\alpha}^{\wedge} &= \{-i \operatorname{sgn} v\}(e^{ihv} - 1)|v|^{-\alpha} \frac{2}{\Lambda_s(\alpha)} \int_0^\infty \frac{\sin u}{u^{1-\alpha}}\, du \\ &= \{-i \operatorname{sgn} v\}(e^{ihv} - 1)|v|^{-\alpha}. \end{aligned}$$

(iii) An application of (8.3.14) and (5.1.15) gives

$$\begin{aligned} m_{h,\alpha}^{\sim}(x) &= \lim_{\rho \to \infty} \frac{1}{\sqrt{2\pi}} \int_{-\rho}^{\rho} \left(1 - \frac{|v|}{\rho}\right) \{-i \operatorname{sgn} v\} m_{h,\alpha}^{\wedge}(v)\, e^{ixv}\, dv \\ &= \lim_{\rho \to \infty} \frac{1}{\sqrt{2\pi}} \int_{-\rho}^{\rho} \left(1 - \frac{|v|}{\rho}\right) n_{h,\alpha}^{\wedge}(v)\, e^{ixv}\, dv = n_{h,\alpha}(x) \quad \text{a.e.}, \end{aligned}$$

and thus the proof of the lemma is complete.

We observe that Lemma 11.2.2 in particular furnishes examples of functions (e.g. $g * m_{h,\alpha}$, $g \in L^1$, cf. (8.2.3)) which together with their Hilbert transforms belong to L^1.

Recalling the definition of the classes $W^r_{L^p}$ for positive integers r, the introduction (11.2.4) of a derivative of fractional order suggests that for fractional α, $0 < \alpha < 1$, one introduces the class

$$\{f \in L^p \mid R_{1-\alpha} f = \phi \text{ a.e.}, \phi \in AC_{\text{loc}}, \phi' \in L^p\}. \tag{11.2.7}$$

Then we have

Proposition 11.2.3. *Let* $0 < \alpha < 1$ *and* $f \in \mathsf{L}^1 \cap \mathsf{L}^2$. *If there exists* $g \in \mathsf{L}^p$, $1 \le p \le 2$, *such that*

$$\{i \operatorname{sgn} v\}|v|^{\alpha} f^{\wedge}(v) = g^{\wedge}(v), \tag{11.2.8}$$

then f belongs to the class defined by (11.2.7).

Proof. Since

$$[R_{1-\alpha}f](x+h) - [R_{1-\alpha}f](x) = (f * m_{h,1-\alpha})(x),$$

we have by Prop. 0.2.2 and Lemma 11.2.2 that $\{[R_{1-\alpha}f](x+h) - [R_{1-\alpha}f](x)\} \in \mathsf{L}^1 \cap \mathsf{L}^2$ for each $h \in \mathbb{R}$, and therefore by Theorems 5.1.3 and 5.2.12

$$\begin{aligned}[][R_{1-\alpha}f](\circ + h) - [R_{1-\alpha}f](\circ)]^{\wedge}(v) &= (e^{ihv} - 1)|v|^{\alpha-1} f^{\wedge}(v)\\ &= \frac{e^{ihv}-1}{iv} g^{\wedge}(v) = \Big[\int_0^h g(\circ + u)\,du\Big]^{\wedge}(v) \quad \text{a.e.}\end{aligned}$$

(cf. (5.1.29)). By the uniqueness theorem for Fourier transforms in L^p it follows that for each $h \in \mathbb{R}$

$$[R_{1-\alpha}f](x+h) - [R_{1-\alpha}f](x) = \int_0^h g(x+u)\,du \quad \text{a.e.} \tag{11.2.9}$$

Hence by Lemma 11.2.1 for any $y \in \mathbb{R}$

$$\int_0^y \{[R_{1-\alpha}f](x+h) - [R_{1-\alpha}f](x)\}\,dx = \int_0^y \Big\{\int_0^h g(x+u)\,du\Big\}\,dx.$$

In virtue of (11.2.2) the left-hand side tends to $-\int_0^y [R_{1-\alpha}f](x)\,dx$ as $h \to -\infty$, whereas for the right-hand side we shall deduce

$$\lim_{h\to-\infty} \int_0^y \Big\{\int_x^{x+h} g(u)\,du\Big\}\,dx = -\int_0^y \Big\{\int_{-\infty}^x g(u)\,du\Big\}\,dx, \tag{11.2.10}$$

which would imply that $[R_{1-\alpha}f](x) = \int_{-\infty}^x g(u)\,du$ a.e., and thus the assertion.

To prove (11.2.10) we apply Lebesgue's dominated convergence theorem. If $p = 1$, then $\lim_{h\to-\infty} \int_x^{x+h} g(u)\,du = -\int_{-\infty}^x g(u)\,du$ exists and $|\int_0^h g(x+u)\,du| \le \sqrt{2\pi}\,\|g\|_1$. If $1 < p \le 2$, we have $g^{\wedge}(v)/iv \in \mathsf{L}^1$ by (11.2.8). It follows by (5.2.25) and the lemma of Riemann–Lebesgue that

$$\begin{aligned}\lim_{h\to-\infty} \int_x^{x+h} g(u)\,du &= \lim_{h\to-\infty} \frac{1}{\sqrt{2\pi}} \int_{-\infty}^{\infty} \frac{e^{ihv}-1}{iv} g^{\wedge}(v)\, e^{ixv}\,dv\\ &= -\frac{1}{\sqrt{2\pi}} \int_{-\infty}^{\infty} \frac{g^{\wedge}(v)}{iv} e^{ixv}\,dv.\end{aligned}$$

Hence $\int_{-\infty}^x g(u)\,du$ exists and $|\int_0^h g(x+u)\,du| \le 2\|g^{\wedge}(\circ)/(i\circ)\|_1$ is bounded, independently of h.

We remark that $\int_{-\infty}^x g(u)\,du$ converges only conditionally for $1 < p \le 2$, but belongs to C_0. Note that for $\alpha = 1$ conditions (11.2.8) and (5.1.30), (5.2.31) are consistent.

For the characterization of the classes $\mathsf{W}[\mathsf{L}^p; |v|^\alpha]$ which is our final aim, the foregoing result is unsatisfactory in two points. Firstly, we have supposed that $f \in \mathsf{L}^1 \cap \mathsf{L}^2$ instead of $f \in \mathsf{L}^p$; secondly, these classes and (11.2.8) differ by the factor $\{i \operatorname{sgn} v\}$. The latter discrepancy is easily removed by

Proposition 11.2.4. *Let $f \in \mathsf{L}^1 \cap \mathsf{L}^2$ and $0 < \alpha < 1$. Then $f \in \mathsf{W}[\mathsf{L}^p; |v|^\alpha]$ for some $1 \le p \le 2$ implies that $[R_{1-\alpha} f^\sim](x)$ is almost everywhere equal to a function $\phi \in \mathsf{AC}_{\mathrm{loc}}$ such that $\phi' \in \mathsf{L}^p$.*

Indeed, $[R_{1-\alpha} f^\sim] \in \mathsf{L}^1_{\mathrm{loc}}$ by Lemma 11.2.1, and the proof follows immediately by (8.3.1) as in Prop. 11.2.3.

Thus the classes $\mathsf{W}[\mathsf{L}^p; |v|^\alpha]$ are connected with the derivative of order α of the Hilbert transform $f^\sim$. But we intend to weaken the hypothesis $f \in \mathsf{L}^1 \cap \mathsf{L}^2$ to $f \in \mathsf{L}^p$, $1 \le p \le 2$. If we consider the case $p = 1$ first, then $f \in \mathsf{L}^1$ does not imply $f^\sim \in \mathsf{L}^1$, and it seems worthwhile to proceed as in Sec. 8.2.3 and to rewrite the fractional derivative of $f^\sim$. Indeed, for $f \in \mathsf{L}^1 \cap \mathsf{L}^2$ we have by Problem 8.2.1 and Lemma 11.2.2 that for each $h \in \mathbb{R}$

$$[R_{1-\alpha} f^\sim](x+h) - [R_{1-\alpha} f^\sim](x) = (f^\sim * m_{h,1-\alpha})(x) = (f * m^\sim_{h,1-\alpha})(x)$$
$$= \frac{1}{A_s(1-\alpha)} \int_{-\infty}^{\infty} f(u) \left\{ \frac{\operatorname{sgn}(x+h-u)}{|x+h-u|^\alpha} - \frac{\operatorname{sgn}(x-u)}{|x-u|^\alpha} \right\} du \quad \text{a.e.}$$

If we integrate over $[0, y]$, let $h \to \infty$, and set (cf. (8.2.10))

$$[R^\sim_\alpha f](x) = \frac{1}{A_s(\alpha)} \int_{-\infty}^{\infty} f(u) \frac{\operatorname{sgn}(x-u)}{|x-u|^{1-\alpha}}\, du, \tag{11.2.11}$$

then we may use (11.2.3) and the analog of (11.2.2) for $R^\sim_{1-\alpha} f$ in order to deduce

$$[R_{1-\alpha} f^\sim](x) = [R^\sim_{1-\alpha} f](x) \quad \text{a.e.} \tag{11.2.12}$$

Thus for $f \in \mathsf{L}^1 \cap \mathsf{L}^2$ the Riesz integral of order $1 - \alpha$ of $f^\sim$ is almost everywhere equal to $[R^\sim_{1-\alpha} f](x)$, the *conjugate of the Riesz integral of order* $1 - \alpha$ *of* f. We observe that, if $f \in \mathsf{Lip}(\mathsf{C}_0; \beta)$ for some $0 < \beta < 1$, then (11.2.12) holds everywhere by Problem 11.2.4. In particular, in view of (11.1.21)

$$\lim_{\alpha \to 1-} [R^\sim_{1-\alpha} f](x) = f^\sim(x). \tag{11.2.13}$$

The advantage is that the conjugate of the Riesz integral is meaningful for $f \in \mathsf{L}^1$, too, since $[R^\sim_{1-\alpha} f](x)$ exists as a locally integrable function by Problem 11.2.2. Hence for the characterization of $\mathsf{W}[\mathsf{L}^p; |v|^\alpha]$ we are led to discuss (instead of (11.2.4)) the difference-quotient

$$\frac{[R^\sim_{1-\alpha} f](x+h) - [R^\sim_{1-\alpha} f](x)}{h}. \tag{11.2.14}$$

For the case $1 < p \le 2$ the following difficulty enters. As in Lemma 11.2.1 we may show that for given p, $1 \le p < \infty$, $f \in \mathsf{L}^p$ implies that $[R^\sim_{1-\alpha} f](x)$ exists almost everywhere and is locally absolutely integrable for $1/p' < \alpha < 1$ (cf. Problem 11.2.2). But for $0 < \alpha \le 1/p'$ the integral in $[R^\sim_{1-\alpha} f](x)$ does not necessarily exist as is shown by the counterexample of Problem 11.2.3. Therefore for $0 < \alpha \le 1/p'$ we cannot even form the relevant difference-quotient. To free ourselves from this restriction on the exponents we consider the difference (11.2.14) as a whole irrespective of the fact whether each member of the difference exists separately. Thus we rewrite (11.2.14) as

$$[R^{\sim}_{1-\alpha}f](x+h) - [R^{\sim}_{1-\alpha}f](x) = (f * m^{\sim}_{h,1-\alpha})(x),$$

the expression on the right existing as a function in L^p by Lemma 11.2.2 for every $1 \le p < \infty$ and $0 < \alpha < 1$.

Finally we replace pointwise convergence by norm-convergence.

Definition 11.2.5. *Let* $0 < \alpha < 1$ *and* $f \in \mathsf{X}(\mathbb{R})$. *If the limit of* $h^{-1}(f * m^{\sim}_{h,1-\alpha})$ *as* $h \to 0$ *exists in* $\mathsf{X}(\mathbb{R})$*-norm, then we call this limit the* ***αth strong Riesz derivative*** $D_s^{\{\alpha\}}f$ *of* f.

The latter terminology may be justified by the following

Theorem 11.2.6. *Let* $f \in \mathsf{L}^p$, $1 \le p \le 2$, and $0 < \alpha < 1$. *Then* $f \in \mathsf{W}[\mathsf{L}^p; |v|^\alpha]$ *if and only if* f *has an* α*th strong Riesz derivative.*

Proof. Let $f \in \mathsf{W}[\mathsf{L}^p; |v|^\alpha]$, i.e. $|v|^\alpha f^\wedge(v) = g^\wedge(v)$, $g \in \mathsf{L}^p$. Then by Lemma 11.2.2

$$[m^{\sim}_{h,1-\alpha}]^\wedge(v)f^\wedge(v) = \{-i \operatorname{sgn} v\}(e^{ihv} - 1)|v|^{\alpha-1}f^\wedge(v) = \frac{e^{ihv} - 1}{iv}\, g^\wedge(v),$$

and therefore by the uniqueness theorem

$$(f * m^{\sim}_{h,1-\alpha})(x) = \int_0^h g(x+u)\, du \quad \text{a.e.} \tag{11.2.15}$$

This implies by Prop. 0.1.7 and 0.1.9 that

$$\left\| \frac{1}{h}(f * m^{\sim}_{h,1-\alpha}) - g \right\|_p \le \int_0^1 \|g(\circ + hu) - g(\circ)\|_p\, du = o(1) \qquad (h \to 0).$$

Conversely, if f has an αth strong Riesz derivative, then it follows by Fatou's lemma and Titchmarsh's inequality that

$$\begin{aligned} \| \, |\circ|^\alpha f^\wedge(\circ) - [D_s^{\{\alpha\}}f]^\wedge(\circ)\|_{p'} &\le \liminf_{h\to 0} \left\| \frac{e^{ih\circ} - 1}{ih\circ} |\circ|^\alpha f^\wedge(\circ) - [D_s^{\{\alpha\}}f]^\wedge(\circ) \right\|_{p'} \\ &\le \liminf_{h\to 0} \left\| \frac{1}{h}(f * m^{\sim}_{h,1-\alpha}) - D_s^{\{\alpha\}}f \right\|_p = 0. \end{aligned}$$

Thus $f \in \mathsf{W}[\mathsf{L}^p; |v|^\alpha]$, and the proof is complete.

11.2.2 Strong Riesz Derivatives of Higher Order, the Classes $\mathsf{V}[\mathsf{L}^p; |v|^\alpha]$

The results of Sec. 8.3 and 10.5 already furnish equivalent characterizations in case $\alpha = 1$. Indeed, in view of (11.2.13) we are led to denote for $f \in \mathsf{L}^p$, $1 \le p < \infty$, the limit† of $h^{-1}\{f^\sim(x+h) - f^\sim(x)\}$ as $h \to 0$, if it exists in L^p-norm, as the first strong Riesz derivative $D_s^{\{1\}}f$ of f, thus

$$\lim_{h\to 0} \left\| \frac{f^\sim(\circ + h) - f^\sim(\circ)}{h} - [D_s^{\{1\}}f](\circ) \right\|_p = 0. \tag{11.2.16}$$

† Concerning (11.2.16), we note that for $p = 1$ only the difference $h^{-1}\{f^\sim(x+h) - f^\sim(x)\}$ need belong to L^1 (cf. the definition for $0 < \alpha < 1$).

Now an application of the results of Sec. 8.3.3 and 10.5 gives

Theorem 11.2.7. *Let* $f \in \mathsf{L}^p$, $1 \le p \le 2$. *Then* $f \in \mathsf{W}[\mathsf{L}^p; |v|]$ *if and only if f has a first strong Riesz derivative. In this event,* $[D_s^{\{1\}}f](x) = \phi'(x)$ a.e., *where* $\phi(x) = f^{\sim}(x)$ a.e. *and* $\phi \in \mathsf{AC}_{\text{loc}}$, $\phi' \in \mathsf{L}^p$.

The case $\alpha > 1$ reduces to the one for $0 < \alpha \le 1$. Indeed,

Definition 11.2.8. *For* $\alpha > 1$ *the* ***αth strong Riesz*** † ***derivative*** $D_s^{\{\alpha\}}f$ *of* $f \in \mathsf{X}(\mathbb{R})$ *is defined successively by* ‡

$$D_s^{\{\alpha\}}f = \begin{cases} D_s^{\{\alpha - [\![\alpha]\!]\}}(D_s^{\{[\![\alpha]\!]\}}f), & \alpha \ne [\![\alpha]\!], \\ D_s^{\{1\}}(D_s^{\{\alpha - 1\}}f), & \alpha = [\![\alpha]\!]. \end{cases} \tag{11.2.17}$$

Theorem 11.2.9. *Let* $f \in \mathsf{L}^p$, $1 \le p \le 2$, *and* $\alpha > 1$. *Then* $f \in \mathsf{W}[\mathsf{L}^p; |v|^\alpha]$ *if and only if f has an αth strong Riesz derivative.*

Proof. Let $f \in \mathsf{W}[\mathsf{L}^p; |v|^\alpha]$, i.e. $|v|^\alpha f^\wedge(v) = g^\wedge(v)$, $g \in \mathsf{L}^p$. By Theorem 6.3.14 we have $|v|^k f^\wedge(v) \in [\mathsf{L}^p]^\wedge$ for each $1 \le k \le [\![\alpha]\!]$ so that a successive application of Theorem 11.2.7 yields the existence of $D_s^{\{k\}}f$ for each $1 \le k \le [\![\alpha]\!]$. This proves the case for $\alpha = [\![\alpha]\!]$. For fractional α we have $|v|^{\alpha - [\![\alpha]\!]}[D_s^{\{[\![\alpha]\!]\}}f]^\wedge(v) = g^\wedge(v)$. By Theorem 11.2.6 the assertion follows, and $D_s^{\{\alpha\}}f = g$.

Conversely, if f has an αth strong Riesz derivative, then by definition $D_s^{\{k\}}f$ exists for $1 \le k \le [\![\alpha]\!]$ and thus $|v|^k f^\wedge(v) = [D_s^{\{k\}}f]^\wedge(v)$, proving the assertion for $\alpha = [\![\alpha]\!]$. In the fractional case we apply Theorem 11.2.6 and obtain

$$[D_s^{\{\alpha\}}f]^\wedge(v) = |v|^{\alpha - [\![\alpha]\!]}[D_s^{\{[\![\alpha]\!]\}}f]^\wedge(v) = |v|^\alpha f^\wedge(v).$$

Let us observe that $D_s^{\{2k\}}f$, $k \in \mathbb{N}$, is equal to the $2k$th strong derivative of $(-1)^k f$ since the classes $\mathsf{W}[\mathsf{L}^p; (-1)^k|v|^{2k}]$ and $\mathsf{W}_{\mathsf{L}^p}^{2k}$ are equivalent by Theorem 5.2.21.

We turn to characterizations of the classes $\mathsf{V}[\mathsf{L}^p; |v|^\alpha]$.

Theorem 11.2.10. *Let* $0 < \alpha < 1$. *For* $f \in \mathsf{L}^p$, $1 \le p \le 2$, *the following assertions are equivalent:*

(i) $f \in \mathsf{V}[\mathsf{L}^p; |v|^\alpha]$, (ii) $f \in \widetilde{\mathsf{W}[\mathsf{L}^p; |v|^\alpha]}^{\mathsf{L}^p}$,

(iii) $\|f * m_{h,1-\alpha}^{\sim}\|_p = O(|h|)$ $(h \to 0)$.

Proof. First of all we observe that we have to establish Theorem 11.2.10 only for $p = 1$; for in the reflexive case $1 < p \le 2$ the classes $\mathsf{V}[\mathsf{L}^p; |v|^\alpha]$ and $\mathsf{W}[\mathsf{L}^p; |v|^\alpha]$ are equivalent by definition. In particular, Theorem 11.2.6 then shows that (iii) may be strengthened to the strong convergence of the difference quotient $h^{-1}(f * m_{h,1-\alpha}^{\sim})$ as $h \to 0$. Furthermore, the equivalence of (i) and (ii) is given by Prop. 10.5.7.

(ii) ⇒ (iii). Let $f \in \widetilde{\mathsf{W}[\mathsf{L}^1; |v|^\alpha]}^{\mathsf{L}^1}$. Then there exists a sequence $\{f_n\} \in \mathsf{W}[\mathsf{L}^p; |v|^\alpha]$

† The integral case ($D^{\{1\}}f = f^{\sim\prime}$, at least formally) would suggest the notation $(d/dx\ H)^\alpha f$, where H denotes the operation of taking Hilbert transforms. Formally, $[(d/dx\ H)^r]^\wedge(v) = [iv\{-i \operatorname{sgn} v\}]^r = |v|^r$ for $r \in \mathbb{N}$.

‡ Note that the existence of $D_s^{\{\alpha\}}f$ implies that of $D^{\{k\}}f$ for each $1 \le k \le [\![\alpha]\!]$.

such that $\lim_{n\to\infty}\|f_n - f\|_1 = 0$ and $\|f_n\|_1 + \|g_n\|_1 \le M$ uniformly for all n where $g_n \in \mathsf{L}^1$ is such that $|v|^\alpha f_n^\wedge(v) = g_n^\wedge(v)$. We have by (11.2.15) for each n

$$(f_n * m^{\sim}_{h,1-\alpha})(x) = \int_0^h g_n(x+u)\,du \quad \text{a.e.},$$

and therefore $\|f_n * m^{\sim}_{h,1-\alpha}\|_1 \le |h|\, \|g_n\|_1 \le M\,|h|$ uniformly for all n. Since $m^{\sim}_{h,1-\alpha} \in \mathsf{L}^1$ by Lemma 11.2.2, $\lim_{n\to\infty}\|f_n * m^{\sim}_{h,1-\alpha}\|_1 = \|f * m^{\sim}_{h,1-\alpha}\|_1$, and thus (iii) is established.

(iii) $\Rightarrow$ (i). Since by Lemma 11.2.2

$$\left[\frac{1}{h}(f * m^{\sim}_{h,1-\alpha})\right]^\wedge(v) = \frac{e^{ihv}-1}{ihv}\,|v|^\alpha f^\wedge(v)$$

which tends to $|v|^\alpha f^\wedge(v)$ as $h \to 0$, (i) follows as in the proof of (iii) $\Rightarrow$ (i) of Theorem 10.5.4.

For $\alpha = 1$ we refer to Theorem 10.5.6 and have (cf. (11.2.13))

Theorem 11.2.11. *Let $f \in \mathsf{L}^p$, $1 \le p \le 2$. The following assertions are equivalent:*

(i) $f \in \mathsf{V}[\mathsf{L}^p; |v|]$, (ii) $f \in \widetilde{\mathsf{W}[\mathsf{L}^p; |v|]}^{\mathsf{L}^p}$,

(iii) $\|f^\sim(\circ + h) - f^\sim(\circ)\|_p = O(|h|) \qquad (h \to 0).$

The counterpart of Theorem 11.2.9 reads as

Theorem 11.2.12. *Let $\alpha > 1$ and $f \in \mathsf{L}^p$, $1 \le p \le 2$. The following assertions are equivalent:*

(i) $f \in \mathsf{V}[\mathsf{L}^p; |v|^\alpha]$, (ii) $f \in \widetilde{\mathsf{W}[\mathsf{L}^p; |v|^\alpha]}^{\mathsf{L}^p}$,

(iii) *for $\alpha \ne [\![\alpha]\!]$ ($\alpha = [\![\alpha]\!]$) the $[\![\alpha]\!]$th ($[\![\alpha - 1]\!]$th) strong Riesz derivative of f exists and*

$$\alpha \ne [\![\alpha]\!]: \quad \|D_s^{\{[\![\alpha]\!]\}} f * m^{\sim}_{h,[\![\alpha]\!]+1-\alpha}\|_p = O(|h|) \qquad (h \to 0),$$

$$\alpha = [\![\alpha]\!]: \quad \|(D_s^{\{[\![\alpha-1]\!]\}} f)^\sim(\circ + h) - (D_s^{\{[\![\alpha-1]\!]\}} f)^\sim(\circ)\|_p = O(|h|) \qquad (h \to 0).$$

The proof is left to Problem 11.2.6.

For $\alpha = 2k$, $k \in \mathbb{N}$, we again refer to Sec. 10.5, observing that the classes $\mathsf{V}[\mathsf{L}^p; (-1)^k|v|^{2k}]$ and $\mathsf{V}^{2k}_{\mathsf{L}^p}$ are equivalent by Theorem 5.3.15. Note that condition (iii) of the last theorem for e.g. $\alpha = 2$ reduces to $\|(D_s^{\{1\}}f)^\sim(\circ + h) - (D_s^{\{1\}}f)^\sim(\circ)\|_p = O(|h|)$, which converts formally (since $(D_s^{\{1\}}f)^\sim = (f^{\sim\prime})^\sim = -f'$) into $\|f'(\circ + h) - f'(\circ)\|_p = O(|h|)$. The latter assertion is rigorous in view of Problem 10.5.3.

Problems

1. Let $f \in \mathsf{L}^1 \cap \mathsf{L}^2$ and $0 < \alpha < 1$. Show that $[R^{\sim}_{1-\alpha} f](x)$ exists almost everywhere, is locally absolutely integrable, and $\lim_{|h|\to\infty} \int_0^y [R^{\sim}_{1-\alpha} f](x + h)\,dx = 0$ for each $y \in \mathbb{R}$.
2. Let $f \in \mathsf{L}^p$, $1 \le p < \infty$, and $0 < \alpha < 1$. Show that $[R_{1-\alpha} f](x)$, $[R^{\sim}_{1-\alpha} f](x)$ exist almost everywhere and are locally absolutely integrable for $1/p' < \alpha < 1$. Furthermore, show that (11.2.2) is valid as well.
3. Let $p > q > 1$. Show that $f(x) = [|x|(1 + |\log|x||)^q]^{-1/p}$ is an example of a function which belongs to L^p, but for which $[R_\alpha f](x)$ diverges on a set of positive measure for $1/p \le \alpha < 1$.

4. Show that (11.2.12) holds everywhere if $f \in \mathsf{Lip}(\mathsf{C}_0; \beta)$ for some $\beta > 0$.
5. Let $f \in \mathsf{L}^1 \cap \mathsf{L}^2$ and $0 < \alpha < 1$. Show that $[R_{1-\alpha}^{\sim} f]^{\wedge}(x) = [R_{1-\alpha} f]^{\sim}(x)$ a.e.
6. Prove Theorem 11.2.12.
7. (i) Let $0 < \alpha < 1$, $h \in \mathbb{R}$, and η be defined by

$$\eta(x) = \begin{cases} (\sqrt{2\pi}/\Gamma(\alpha))x^{\alpha-1}, & x > 0 \\ 0, & x < 0. \end{cases}$$

Show that $\Delta_h^1 \eta(x) \in \mathsf{L}^1$ and (cf. Problem 11.1.2) $[\Delta_h^1 \eta]^{\wedge}(v) = (e^{ihv} - 1)(iv)^{-\alpha}$.

(ii) Let a in (11.1.2) be $-\infty$ and $0 < \alpha < 1$. If $f \in \mathsf{L}^1 \cap \mathsf{L}^2$ and there exists $g \in \mathsf{L}^p$, $1 \le p \le 2$, such that $(iv)^\alpha f^{\wedge}(v) = g^{\wedge}(v)$, then $[L_{1-\alpha} f](x) = \int_{-\infty}^{x} g(u)\,du$ a.e.

8. Let $[R_\alpha\, d\mu](x)$, $[R_\alpha^{\sim}\, d\mu](x)$ be defined for $\mu \in \mathsf{BV}$ and $0 < \alpha < 1$ by

$$[R_\alpha\, d\mu](x) = \frac{1}{\Lambda_c(\alpha)} \int_{-\infty}^{\infty} \frac{d\mu(u)}{|x-u|^{1-\alpha}}, \qquad [R_\alpha^{\sim}\, d\mu](x) = \frac{1}{\Lambda_s(\alpha)} \int_{-\infty}^{\infty} \frac{\operatorname{sgn}(x-u)}{|x-u|^{1-\alpha}}\, d\mu(u).$$

Show that $[R_\alpha\, d\mu](x)$, $[R_\alpha^{\sim}\, d\mu](x)$ exist almost everywhere, belong to L^1_{loc}, and satisfy (11.2.2).

9. Let $f \in \mathsf{L}^p$, $1 \le p \le 2$.
   (i) If $0 < \alpha < 1$ and $\|f * m_{h,1-\alpha}^{\sim}\|_p = o(|h|)$, show that $f = 0$ a.e.
   (ii) If the αth strong Riesz derivative exists for $\alpha > 0$ and vanishes, show that $f = 0$ almost everywhere.
   (iii) Show that the existence of the αth strong Riesz derivative of f implies that the βth strong Riesz derivative of f exists for all $0 < \beta \le \alpha$ and $[D_s^{\{\beta\}} f]^{\wedge}(v) = |v|^\beta f^{\wedge}(v)$.
   (iv) Show that $f \in \mathsf{V}[\mathsf{L}^p; |v|^\alpha]$ implies the existence of the βth strong Riesz derivative of f for every $0 < \beta < \alpha$.
10. Let $f \in \mathsf{L}^p$, $1 \le p \le 2$, and $0 < \alpha < 1$. Show that $f \in \mathsf{W}[\mathsf{L}^p; |v|^\alpha]$ if and only if there exists $g \in \mathsf{L}^p$ such that for every $s \in \mathsf{L}^{p'}$

$$\lim_{h\to 0} \int_{-\infty}^{\infty} \frac{(f * m_{h,1-\alpha}^{\sim})(x)}{h}\, s(x)\,dx = \int_{-\infty}^{\infty} g(x)s(x)\,dx.$$

11. Let $f \in \mathsf{L}^1$ and $0 < \alpha < 1$. Show that $f \in \mathsf{V}[\mathsf{L}^1; |v|^\alpha]$ if and only if there exists $\mu \in \mathsf{BV}$ such that for every $s \in \mathsf{C}_0$

$$\lim_{h\to 0} \int_{-\infty}^{\infty} s(x)\, \frac{(f * m_{h,1-\alpha}^{\sim})(x)}{h}\,dx = \int_{-\infty}^{\infty} s(x)\,d\mu(x).$$

12. Let $f \in \mathsf{L}^p$, $1 \le p \le 2$, and $\alpha > 0$. Show that $f \in \mathsf{V}[\mathsf{L}^p; |v|^\alpha]$ if and only if $\|\sigma^{\{\alpha\}}(f; \circ; \rho)\|_p = O(1)$, $\rho \to \infty$. (Hint: For $0 < \alpha < 1$ the pointwise derivative $f^{\{\alpha\}}$ of f is defined by the limit $h^{-1}\{[R_{1-\alpha}^{\sim} f](x+h) - [R_{1-\alpha}^{\sim} f](x)\}$ as $h \to 0$ if it exists. For $\alpha = 1$ we set $f^{\{1\}}(x) = \lim_{h\to 0} h^{-1}\{f^{\sim}(x+h) - f^{\sim}(x)\}$ if the limit exists. The definition may be extended to $\alpha > 1$ successively as usual. Show that

$$\sigma^{\{\alpha\}}(f; x; \rho) = \frac{1}{\sqrt{2\pi}} \int_{-\rho}^{\rho} \left(1 - \frac{|v|}{\rho}\right) |v|^\alpha f^{\wedge}(v) e^{ixv}\,dv.)$$

13. (i) Let $\phi \in \mathsf{C}_{00}^{\infty}$. Show that ϕ has strong Riesz derivatives in L^p, $1 \le p \le 2$, of every order $\alpha > 0$ and $[D_s^{\{\alpha\}}\phi]^{\wedge} \in \mathsf{L}^1$.
    (ii) Let p_α, q_α be defined for $\alpha > 0$ by $p_\alpha(x) = \sqrt{2/\pi}\,\Gamma(1+\alpha)\operatorname{Re}\{(1-ix)^{-1-\alpha}\}$, $q_\alpha(x) = \sqrt{2/\pi}\,\Gamma(1+\alpha)\operatorname{Im}\{(1-ix)^{-1-\alpha}\}$. Show that $p_\alpha^{\wedge}(v) = |v|^\alpha \exp\{-|v|\}$, $q_\alpha^{\wedge}(v) = \{-i \operatorname{sgn} v\}|v|^\alpha \exp\{-|v|\}$.

11.3 The Operators $R_\varepsilon^{\{\alpha\}}$ on L^p, $1 \le p \le 2$

11.3.1 Characterizations

We now turn to characterizations of the classes $\mathsf{W}[\mathsf{L}^p; |v|^\alpha]$ and $\mathsf{V}[\mathsf{L}^p; |v|^\alpha]$ in terms of the integrals

$$\int_\varepsilon^\infty \frac{\overline{\Delta}_u^{2j} f(x)}{u^{1+\alpha}}\,du \qquad (0 < \alpha < 2j). \tag{11.3.1}$$

We first consider the case $0 < \alpha < 2$, thus†

$$[R_\varepsilon^{\{\alpha\}} f](x) = \frac{1}{\Lambda_c(-\alpha)} \int_{|u| \ge \varepsilon} \frac{f(x-u) - f(x)}{|u|^{1+\alpha}}\,du \qquad (\varepsilon > 0). \tag{11.3.2}$$

By the Hölder and Hölder–Minkowski inequality it follows immediately that $R_\varepsilon^{\{\alpha\}} f \in \mathsf{C} \cap \mathsf{X}(\mathbb{R})$ for every $f \in \mathsf{X}(\mathbb{R})$. In particular, for each $\varepsilon > 0$, $R_\varepsilon^{\{\alpha\}}$ is a bounded linear operator of $\mathsf{X}(\mathbb{R})$ into $\mathsf{X}(\mathbb{R})$ satisfying

$$\|R_\varepsilon^{\{\alpha\}} f\|_{\mathsf{X}(\mathbb{R})} \le \frac{4}{\alpha\varepsilon^\alpha |\Lambda_c(-\alpha)|} \|f\|_{\mathsf{X}(\mathbb{R})}. \tag{11.3.3}$$

In this section we suppose throughout that χ satisfies the following conditions:

(11.3.4) (i) *χ belongs to $\mathsf{C} \cap \mathsf{NL}^1$, is even, and $\chi \in \mathsf{W}_{\mathsf{C}}^2 \cap \mathsf{W}_{\mathsf{L}^1}^2$,*
(ii) *$m(\chi; \alpha)$ and $m(\chi''; \alpha)$ are finite, where*
$m(\chi; \alpha) = (1/\sqrt{2\pi}) \int_{-\infty}^{\infty} |u|^\alpha |\chi(u)|\,du,$
(iii) $[\chi'']^\wedge \in \mathsf{L}^1$.

The kernels of Gauss–Weierstrass and Jackson–de La Vallée Poussin are examples satisfying (11.3.4) (cf. Problem 11.3.1). In the following, $J(f; x; \rho)$ always denotes the singular integral (3.1.8) having kernel $\{\rho\chi(\rho x)\}$.

Lemma 11.3.1. *If χ satisfies* (11.3.4), *then $f \in \mathsf{Lip}^*(\mathsf{X}(\mathbb{R}); \alpha)$, $0 < \alpha < 2$, implies that*

$$\|J(\overline{\Delta}_h^2 f; \circ; \rho)\|_{\mathsf{X}(\mathbb{R})} = O(h^2 \rho^{2-\alpha}) \qquad (h, \rho > 0). \tag{11.3.5}$$

Proof. First we observe that χ belongs to $\mathsf{W}_{\mathsf{L}^p}^2$ for all $1 < p < \infty$ and thus has a second strong derivative in $\mathsf{X}(\mathbb{R})$-norm (cf. Problem 3.1.4). It follows that all interchanges of integration and differentiation with respect to a parameter occurring below are permissible by Hölder's inequality.

To prove (11.3.5) we apply the Taylor formula

$$g(h) = g(0) + hg'(0) + \int_0^h (h - t) g''(t)\,dt \tag{11.3.6}$$

to $g(h) = J(\overline{\Delta}_h^2 f; x; \rho)$, namely to

$$g(h) = \frac{\rho}{\sqrt{2\pi}} \int_{-\infty}^{\infty} [f(x+h-u) + f(x-h-u) - 2f(x-u)]\chi(\rho u)\,du.$$

† Note that $\Lambda_c(-\alpha)$ has a removable singularity at $\alpha = 1$ since $\Lambda_c(-\alpha)\Lambda_c(1+\alpha) = 2\pi$.

By Prop. 0.2.1, $J(\overline{\Delta}_h^2 f; x; \rho) \in \mathsf{C}$ as a function of x. By the preceding remark we have

$$g'(h) = \frac{\rho^2}{\sqrt{2\pi}} \int_{-\infty}^{\infty} [f(x+h-u) - f(x-h-u)]\chi'(\rho u)\, du,$$

$$g''(h) = \frac{\rho^3}{\sqrt{2\pi}} \int_{-\infty}^{\infty} [f(x+h-u) + f(x-h-u)]\chi''(\rho u)\, du.$$

Obviously, $g(0) = g'(0) = 0$ and thus

(11.3.7) $J(\overline{\Delta}_h^2 f; x; \rho)$

$$= \int_0^h (h-t)\left\{\frac{\rho^3}{\sqrt{2\pi}} \int_{-\infty}^{\infty} [f(x+t-u) + f(x-t-u)]\chi''(\rho u)\, du\right\} dt.$$

Moreover, since χ'' is an even function and $\int_0^\infty \chi''(u)\, du = 0$ (which follows by considering $g(h)$ and $g''(h)$ for $f(x) \equiv \text{const}$), we have

$$J(\overline{\Delta}_h^2 f; x; \rho) = \frac{\rho^3}{\sqrt{2\pi}} \int_0^h (h-t)\left\{\int_0^\infty [\overline{\Delta}_u^2 f(x+t) + \overline{\Delta}_u^2 f(x-t)]\chi''(\rho u)\, du\right\} dt.$$

If M is such that $\|\overline{\Delta}_u^2 f\|_{\mathsf{X}(\mathbb{R})} \le M u^\alpha$ for all $u \ge 0$, we obtain by the Hölder–Minkowski inequality

(11.3.8)
$$\|J(\overline{\Delta}_h^2 f; \circ; \rho)\|_{\mathsf{X}(\mathbb{R})} \le \frac{\rho^3}{\sqrt{2\pi}} \int_0^h (h-t)\left\{\int_0^\infty [\|\overline{\Delta}_u^2 f(\circ + t)\|_{\mathsf{X}(\mathbb{R})} + \|\overline{\Delta}_u^2 f(\circ - t)\|_{\mathsf{X}(\mathbb{R})}]|\chi''(\rho u)|\, du\right\} dt$$

$$\le \frac{2M\rho^3}{\sqrt{2\pi}} \int_0^h (h-t)\left\{\int_0^\infty u^\alpha |\chi''(\rho u)|\, du\right\} dt = M m(\chi''; \alpha) h^2 \rho^{2-\alpha},$$

establishing (11.3.5).

Lemma 11.3.2. *If* χ *satisfies* (11.3.4), *then* $f \in \mathsf{Lip}^*(\mathsf{L}^p; \alpha)$, $1 \le p \le 2$, $0 < \alpha < 2$, *implies that*

$$\left\| [R_{1/\rho}^{\{\alpha\}} f](\circ) - \frac{1}{\sqrt{2\pi}} \int_{-\infty}^{\infty} \chi^\wedge\!\left(\frac{v}{\rho}\right) |v|^\alpha f^\wedge(v)\, e^{i\circ v}\, dv \right\|_p = O(1) \qquad (\rho > 0).$$

Proof. Let $\rho > 0$ be fixed. Since $\chi \in \mathsf{W}_{\mathsf{L}^1}^2$, $(iv)^2 \chi^\wedge(v) = [\chi'']^\wedge(v)$, $\chi'' \in \mathsf{L}^1$, and thus $\chi^\wedge \in \mathsf{L}^1$. Therefore we may apply (5.1.22) and (5.2.26) to deduce

$$J(\overline{\Delta}_h^2 f; x; \rho) = \frac{1}{\sqrt{2\pi}} \int_{-\infty}^{\infty} \chi^\wedge\!\left(\frac{v}{\rho}\right) [\overline{\Delta}_h^2 f]^\wedge(v)\, e^{ixv}\, dv$$

$$= \frac{-4}{\sqrt{2\pi}} \int_{-\infty}^{\infty} \chi^\wedge\!\left(\frac{v}{\rho}\right) \sin^2 \frac{hv}{2} f^\wedge(v)\, e^{ixv}\, dv.$$

This implies that $J(\overline{\Delta}_h^2 f; x; \rho) = O(h^2)$, $h \to 0$, for each $f \in \mathsf{L}^p$, $\rho > 0$, $x \in \mathbb{R}$ since $|v^2|\, |\chi^\wedge(v)| = |[\chi'']^\wedge(v)| \in \mathsf{L}^1$. Therefore the integral

$$\int_0^\infty u^{-(1+\alpha)} \left\{\int_{-\infty}^{\infty} \chi^\wedge\!\left(\frac{v}{\rho}\right) \sin^2 \frac{uv}{2} f^\wedge(v)\, e^{ixv}\, dv\right\} du$$

converges absolutely, and Fubini's theorem yields

$$(11.3.9)\quad \int_0^\infty u^{-(1+\alpha)} J(\overline{\Delta}_u^2 f; x; \rho)\, du = \Lambda_c(-\alpha)\, \frac{1}{\sqrt{2\pi}} \int_{-\infty}^{\infty} \chi^\wedge\Big(\frac{v}{\rho}\Big) |v|^\alpha f^\wedge(v)\, e^{ixv}\, dv.$$

Here we may use (cf. Problem 11.1.1)

$$(11.3.10)\quad \int_0^\infty u^{-(1+\alpha)} \sin^2 \frac{vu}{2}\, du = \Big(\frac{|v|}{2}\Big)^\alpha \int_0^\infty u^{-(1+\alpha)} \sin^2 u\, du$$

$$= \frac{|v|^\alpha}{2\alpha} \int_0^\infty \frac{\sin u}{u^\alpha}\, du = -\tfrac{1}{4}\Lambda_c(-\alpha)|v|^\alpha.$$

Thus, as

$$\int_{|u|\ge\varepsilon} \frac{f(x-u)-f(x)}{|u|^{1+\alpha}}\, du = \int_\varepsilon^\infty \frac{f(x+u)+f(x-u)-2f(x)}{u^{1+\alpha}}\, du,$$

we have

$$|\Lambda_c(-\alpha)| \Big\| [R_{1/\rho}^{\{\alpha\}} f](\circ) - \frac{1}{\sqrt{2\pi}} \int_{-\infty}^{\infty} \chi^\wedge\Big(\frac{v}{\rho}\Big) |v|^\alpha f^\wedge(v)\, e^{i\circ v}\, dv \Big\|_p$$

$$\le \Big\| \int_0^{1/\rho} u^{-(1+\alpha)} J(\overline{\Delta}_u^2 f; \circ; \rho)\, du \Big\|_p + \Big\| \int_{1/\rho}^\infty u^{-(1+\alpha)} [\overline{\Delta}_u^2 f(\circ) - J(\overline{\Delta}_u^2 f; \circ; \rho)]\, du \Big\|_p \equiv I.$$

Since $m(\chi; \alpha) < \infty$ and $f \in \mathsf{Lip}^*(\mathsf{L}^p; \alpha)$, it follows by Prop. 3.4.3 that

$$\|J(\overline{\Delta}_h^2 f; \circ; \rho) - \overline{\Delta}_h^2 f(\circ)\|_p = O(\rho^{-\alpha}) \qquad (\rho \to \infty)$$

uniformly for $h \in \mathbb{R}$. Therefore we obtain by the Hölder–Minkowski inequality and Lemma 11.3.1

$$I \le \int_0^{1/\rho} u^{-(1+\alpha)} \|J(\overline{\Delta}_u^2 f; \circ; \rho)\|_p\, du + \int_{1/\rho}^\infty u^{-(1+\alpha)} \|J(\overline{\Delta}_u^2 f; \circ; \rho) - \overline{\Delta}_u^2 f(\circ)\|_p\, du$$

$$\le M_1 \rho^{2-\alpha} \int_0^{1/\rho} u^{1-\alpha}\, du + M_2 \rho^{-\alpha} \int_{1/\rho}^\infty u^{-(1+\alpha)}\, du = O(1) \qquad (\rho \to \infty),$$

which proves the lemma.

Theorem 11.3.3. *Let* $0 < \alpha < 2$. *For* $f \in \mathsf{L}^p$, $1 \le p \le 2$, *the following assertions are equivalent:*

(i) $$f \in \mathsf{V}[\mathsf{L}^p; |v|^\alpha],$$

(ii) $$\Big\| \frac{1}{\Lambda_c(-\alpha)} \int_{|u|\ge\varepsilon} \frac{f(\circ-u)-f(\circ)}{|u|^{1+\alpha}}\, du \Big\|_p = O(1) \qquad (\varepsilon \to 0+).$$

Proof. (ii) ⇒ (i). By (11.3.9), Fatou's lemma, Fubini's theorem, and (3.1.4) it follows that

$$|\Lambda_c(-\alpha)| \left\| \frac{1}{\sqrt{2\pi}} \int_{-\infty}^{\infty} \chi^\wedge\left(\frac{v}{\rho}\right) |v|^\alpha f^\wedge(v)\, e^{i \circ v}\, dv \right\|_p$$

$$\leq \liminf_{\varepsilon\to 0+} \left\| \int_\varepsilon^\infty u^{-(1+\alpha)} J(\overline{\Delta}_u^2 f; \circ; \rho)\, du \right\|_p$$

$$\leq \liminf_{\varepsilon\to 0+} \left\| \frac{\rho}{\sqrt{2\pi}} \int_{-\infty}^{\infty} \chi(\rho t) \left\{ \int_\varepsilon^\infty \frac{\overline{\Delta}_u^2 f(\circ - t)}{u^{1+\alpha}}\, du \right\} dt \right\|_p$$

$$\leq \liminf_{\varepsilon\to 0+} \left\| \int_{|u|\geq\varepsilon} \frac{f(\circ - u) - f(\circ)}{|u|^{1+\alpha}}\, du \right\|_p = O(1)$$

uniformly for $\rho > 0$. Since $|v|^2 |\chi^\wedge(v)| \in \mathsf{L}^1$, Problem 6.1.10 implies that there exists $\mu \in \mathsf{BV}$ for $p = 1$ and $g \in \mathsf{L}^p$ for $1 < p \leq 2$ such that $|v|^\alpha f^\wedge(v) = \mu^\vee(v)$ and $|v|^\alpha f^\wedge(v) = g^\wedge(v)$, respectively. Thus $f \in \mathsf{V}[\mathsf{L}^p; |v|^\alpha]$.

(i) ⇒ (ii). By Problem 6.1.10, $f \in \mathsf{V}[\mathsf{L}^p; |v|^\alpha]$ implies

$$\left\| \frac{1}{\sqrt{2\pi}} \int_{-\infty}^{\infty} \chi^\wedge\left(\frac{v}{\rho}\right) |v|^\alpha f^\wedge(v)\, e^{i \circ v}\, dv \right\|_p = O(1) \qquad (\rho > 0).$$

Under the assumption that $f \in \mathsf{Lip}^*(\mathsf{L}^p; \alpha)$ we obtain by Lemma 11.3.2

$$\|R_{1/\rho}^{\{\alpha\}} f\|_p \leq O(1) + \left\| \frac{1}{\sqrt{2\pi}} \int_{-\infty}^{\infty} \chi^\wedge\left(\frac{v}{\rho}\right) |v|^\alpha f^\wedge(v)\, e^{i \circ v}\, dv \right\|_p = O(1) \qquad (\rho > 0)$$

which establishes (ii) by setting $\rho = 1/\varepsilon$.

Thus it remains to show

Lemma 11.3.4. *If* $f \in \mathsf{V}[\mathsf{L}^p; |v|^\alpha]$, *then* $f \in \mathsf{Lip}^*(\mathsf{L}^p; \alpha)$ *for* $0 < \alpha < 2$.

Proof. Let $p = 1$ and $\mu \in \mathsf{BV}$ be such that $|v|^\alpha f^\wedge(v) = \mu^\vee(v)$. Then by Lemma 11.2.2

$$(e^{ihv} - 1)^2 f^\wedge(v) = \left(\frac{e^{ihv} - 1}{|v|^{\alpha/2}}\right)^2 \mu^\vee(v) = [m_{h,(\alpha/2)} * m_{h,(\alpha/2)} * d\mu]^\wedge(v),$$

and by the uniqueness theorem

$$\|\Delta_h^2 f\|_1 \leq \|m_{h,(\alpha/2)}\|_1^2 \|\mu\|_{\mathsf{BV}} = O(|h|^\alpha) \qquad (h \to 0).$$

The case $1 < p \leq 2$ is treated similarly.

If we turn to characterizations of the classes $\mathsf{W}[\mathsf{L}^p; |v|^\alpha]$, we observe that the informal limit $\rho \to \infty$ in (11.3.9) yields

$$\frac{1}{\Lambda_c(-\alpha)} \int_{-\infty}^{\infty} \frac{f(x - u) - f(x)}{|u|^{1+\alpha}}\, du = \frac{1}{\sqrt{2\pi}} \int_{-\infty}^{\infty} |v|^\alpha f^\wedge(v)\, e^{ixv}\, dv = [D_s^{\{\alpha\}} f](x)$$

(cf. Theorems 11.2.6–11.2.9). Rigorously, we have

Theorem 11.3.5. *Let* $0 < \alpha < 2$. *For* $f \in \mathsf{L}^p$, $1 \leq p \leq 2$, *the following assertions are equivalent:*

(i) $f \in \mathsf{W}[\mathsf{L}^p; |v|^\alpha]$,
(ii) *there exists* $g \in \mathsf{L}^p$ *such that* $\lim_{\varepsilon \to 0+} \|R_\varepsilon^{\{\alpha\}} f - g\|_p = 0$.

In this event, f has an αth strong Riesz derivative and $[D_s^{\{\alpha\}} f](x) = g(x)$ a.e.

Proof. (i) ⇒ (ii). In Prop. 10.5.7 it has been stated that $\mathsf{W}[\mathsf{L}^p; |v|^\alpha]$ becomes a normalized Banach subspace under the norm (10.5.9). By (11.3.3) the operator $R_\varepsilon^{\{\alpha\}}$ is a bounded linear mapping of L^p into L^p for each $\varepsilon > 0$. Furthermore, $\{R_\varepsilon^{\{\alpha\}}\}$ forms a set of uniformly bounded (with respect to $\varepsilon > 0$) linear operators of $\mathsf{W}[\mathsf{L}^p; |v|^\alpha]$ into L^p by Theorem 11.3.3; indeed there exists a constant M_f for each $f \in \mathsf{W}[\mathsf{L}^p; |v|^\alpha]$ such that $\|R_\varepsilon^{\{\alpha\}} f\|_p \leq M_f$, and the uniform boundedness principle then implies the existence of a constant M such that

$$\|R_\varepsilon^{\{\alpha\}} f\|_p \leq M \|f\|_{\mathsf{W}[\mathsf{L}^p; |v|^\alpha]} \qquad (f \in \mathsf{W}[\mathsf{L}^p; |v|^\alpha])$$

uniformly for all $\varepsilon > 0$.

Thus if we are able to show that

$$\lim_{\varepsilon_1, \varepsilon_2 \to 0+} \|R_{\varepsilon_2}^{\{\alpha\}} d - R_{\varepsilon_1}^{\{\alpha\}} d\|_p = 0 \qquad (11.3.11)$$

for all functions d of a dense subset of $\mathsf{W}[\mathsf{L}^p; |v|^\alpha]$, we may apply the theorem of Banach–Steinhaus to deduce (ii).

To establish (11.3.11), consider the set

$$\mathsf{D} = \{d \in \mathsf{C} \,|\, d(x) = J(f; x; \rho), f \in \mathsf{W}[\mathsf{L}^p; |v|^\alpha], \rho > 0\}.$$

D is a dense subset of the Banach space $\mathsf{W}[\mathsf{L}^p; |v|^\alpha]$. Indeed, $f \in \mathsf{W}[\mathsf{L}^p; |v|^\alpha]$ implies by the convolution theorem that

$$|v|^\alpha [J(f; \circ; \rho)]^\wedge(v) = \chi^\wedge\left(\frac{v}{\rho}\right) |v|^\alpha f^\wedge(v) = [J(g; \circ; \rho)]^\wedge(v),$$

i.e. $J(f; x; \rho) \in \mathsf{W}[\mathsf{L}^p; |v|^\alpha]$ for every $\rho > 0$. Moreover,

$$\|J(f; \circ; \rho) - f(\circ)\|_{\mathsf{W}[\mathsf{L}^p; |v|^\alpha]} = \|J(f; \circ; \rho) - f(\circ)\|_p + \|J(g; \circ; \rho) - g(\circ)\|_p,$$

and thus

$$\lim_{\rho \to 0+} \|J(f; \circ; \rho) - f(\circ)\|_{\mathsf{W}[\mathsf{L}^p; |v|^\alpha]} = 0 \qquad (f \in \mathsf{W}[\mathsf{L}^p; |v|^\alpha]).$$

In order to establish (11.3.11) for $d(x) = J(f; x; \rho)$ we use Lemma 11.3.1 and obtain by (11.3.8) for $\varepsilon_1 < \varepsilon_2$

$$|A_c(-\alpha)| \, \|[R_{\varepsilon_2}^{\{\alpha\}} J(f; \circ; \rho)](\circ) - [R_{\varepsilon_1}^{\{\alpha\}} J(f; \circ; \rho)](\circ)\|_p$$
$$= \left\| \int_{\varepsilon_1}^{\varepsilon_2} h^{-(1+\alpha)} J(\overline{\Delta}_h^2 f; \circ; \rho) \, dh \right\|_p \leq M m(\chi''; \alpha) \rho^{2-\alpha} \int_{\varepsilon_1}^{\varepsilon_2} h^{1-\alpha} \, dh,$$

which tends to zero as $\varepsilon_1, \varepsilon_2 \to 0+$ for each fixed $\rho > 0$.

(ii) ⇒ (i). If $1 < p \leq 2$, then (ii) in particular states that $\|R_\varepsilon^{\{\alpha\}} f\|_p = O(1)$ as $\varepsilon \to 0+$, and $f \in \mathsf{W}[\mathsf{L}^p; |v|^\alpha]$ by Theorem 11.3.3.

Let $p = 1$. If we set

$$\eta_\varepsilon(v) = \frac{-4}{A_c(-\alpha)} \int_\varepsilon^\infty u^{-(1+\alpha)} \sin^2 \frac{vu}{2}\, du, \tag{11.3.12}$$

it follows by Fubini's theorem that

$$\begin{aligned}[R_\varepsilon^{\{\alpha\}}f]^\wedge(v) &= \frac{1}{A_c(-\alpha)} \int_\varepsilon^\infty u^{-(1+\alpha)} \left\{\frac{1}{\sqrt{2\pi}} \int_{-\infty}^\infty \overline{\Delta}_u^2 f(x)\, e^{-ivx}\, dx\right\} du \\ &= \eta_\varepsilon(v) f^\wedge(v).\end{aligned} \tag{11.3.13}$$

Therefore, in view of the hypothesis and (5.1.2) we conclude that

$$|\eta_\varepsilon(v) f^\wedge(v) - g^\wedge(v)| \le \|R_\varepsilon^{\{\alpha\}}f - g\|_1 = o(1) \qquad (\varepsilon \to 0+),$$

and since $\lim_{\varepsilon\to 0+} \eta_\varepsilon(v) = |v|^\alpha$ for all $v \in \mathbb{R}$ by (11.3.10), this implies $|v|^\alpha f^\wedge(v) = g^\wedge(v)$, i.e. $f \in \mathsf{W}[\mathsf{L}^p; |v|^\alpha]$. In particular, Theorems 11.2.6–11.2.9 show that f has an αth strong Riesz derivative and $D_s^{\{\alpha\}}f = g$.

Having discussed the case $0 < \alpha < 2$, it is clear how the extension to arbitrary $\alpha > 0$ would proceed. We state

Theorem 11.3.6. *Let $f \in \mathsf{L}^p$, $1 \le p \le 2$, and $\alpha > 0$. The following assertions are equivalent:*

(i) $f \in \mathsf{V}[\mathsf{L}^p; |v|^\alpha]$,

(ii) $$\left\| \frac{1}{C_{\alpha,2j}} \int_\varepsilon^\infty \frac{\overline{\Delta}_u^{2j} f(\circ)}{u^{1+\alpha}}\, du \right\|_p = O(1) \qquad (\varepsilon \to 0+),$$

where $j \in \mathbb{N}$ is chosen such that $0 < \alpha < 2j$ and

$$C_{\alpha,2j} = (-1)^j 2^{2j-\alpha} \int_0^\infty \frac{\sin^{2j} u}{u^{1+\alpha}}\, du. \tag{11.3.14}$$

Theorem 11.3.7. *Let $f \in \mathsf{L}^p$, $1 \le p \le 2$, and $\alpha > 0$. The following assertions are equivalent:*

(i) $f \in \mathsf{W}[\mathsf{L}^p; |v|^\alpha]$,

(ii) *there exists $g \in \mathsf{L}^p$ such that* $(0 < \alpha < 2j)$

$$\lim_{\varepsilon\to 0+} \left\| \frac{1}{C_{\alpha,2j}} \int_\varepsilon^\infty \frac{\overline{\Delta}_u^{2j} f(\circ)}{u^{1+\alpha}}\, du - g(\circ) \right\|_p = 0.$$

In this event, f has an αth strong Riesz derivative and $D_s^{\{\alpha\}}f = g$.

For the proof of the last two theorems we refer to Problem 11.3.3.

11.3.2 Theorems of Bernstein–Titchmarsh and H. Weyl

We shall now give two applications of the preceding results on the operators $R_\varepsilon^{\{\alpha\}}$. The first one is concerned with a theorem of E. C. Titchmarsh which extends the famous theorem of S. N. Bernstein on the absolute convergence of Fourier series (cf. Problem 11.4.4) to Fourier integrals.

Theorem 11.3.8. *Let* $f \in \mathsf{L}^p$, $1 \le p \le 2$. *Then* $f \in \mathsf{Lip}^* (\mathsf{L}^p; \alpha)$, $0 < \alpha \le 2$, *implies* $f^\wedge \in \mathsf{L}^q$ *for all* q *with* $p/(p - 1 + \alpha p) < q \le p/(p - 1)$.

Proof. Let β, $0 < \beta < \alpha$, be arbitrary. By the inequality of Hölder–Minkowski

$$\left\| \int_\varepsilon^\infty u^{-(1+\beta)} \overline{\Delta}_u^2 f(\circ)\, du \right\|_p \le \int_\varepsilon^1 u^{-(1+\beta)} \|\overline{\Delta}_u^2 f(\circ)\|_p \, du + 4\|f\|_p \int_1^\infty u^{-(1+\beta)}\, du$$

$$\le M \int_\varepsilon^1 u^{-1-\beta+\alpha}\, du + \frac{4\|f\|_p}{\beta} = O(1) \qquad (\varepsilon > 0).$$

Thus, if $1 < p \le 2$, Theorem 11.3.3 guarantees the existence of $g \in \mathsf{L}^p$ such that $|v|^\beta f^\wedge(v) = g^\wedge(v)$ a.e. Since $f^\wedge \in \mathsf{L}^q_{loc}$ for $1 \le q \le p'$ and

$$\int_N^\infty |f^\wedge(v)|^q\, dv \le \left\{ \int_N^\infty |g^\wedge(v)|^{p'} \right\}^{q/p'} \left\{ \int_N^\infty |v|^{-\beta q p'/(p'-q)}\, dv \right\}^{(p'-q)/p'}$$

for any $N > 1$ by Hölder's inequality, we have $f^\wedge \in \mathsf{L}^q$ for $-\beta q p'/(p' - q) < -1$, i.e., for all q with $p/(p - 1 + \beta p) < q \le p/(p - 1)$. Since this is valid for all β with $0 < \beta < \alpha$, the assertion follows. The proof for $p = 1$ proceeds similarly, and Theorem 11.3.8 is established.

Corollary 11.3.9. *Let* $f \in \mathsf{L}^p$, $1 \le p \le 2$. *Then* $f \in \mathsf{Lip}^* (\mathsf{L}^p; \alpha)$, $\alpha > 1/p$, *implies* $f^\wedge \in \mathsf{L}^1$.

Next we consider the L^1-version of a theorem of H. Weyl concerning the connection between Lipschitz conditions and representations of continuous 2π-periodic functions as Riemann–Liouville integrals (cf. Theorem 11.5.4, Problem 11.5.3).

Theorem 11.3.10. *Let* $f \in \mathsf{L}^1$. *If there exists* $\mu \in \mathsf{BV}$ *such that* $f(x) = [R_\alpha\, d\mu](x)$ a.e., $0 < \alpha < 1$, *then* $f \in \mathsf{Lip}(\mathsf{L}^1; \alpha)$. *Conversely, if* $f \in \mathsf{Lip}^*(\mathsf{L}^1; \alpha)$, $0 < \alpha \le 1$, *then for each* β *with* $0 < \beta < \alpha$ *there exists* $g \in \mathsf{L}^1$ *such that* $f(x) = [R_\beta g](x)$ a.e.

Proof. If $f(x) = [R_\alpha\, d\mu](x)$ for some $\mu \in \mathsf{BV}$ and $0 < \alpha < 1$, then $\Delta_h^1 f(x) = (m_{h,\alpha} * d\mu)(x)$, and thus $\|\Delta_h f\|_1 \le |h|^\alpha \|m_{1,\alpha}\|_1 \|\mu\|_{\mathsf{BV}}$ by Lemma 11.2.2. Conversely, if $f \in \mathsf{Lip}^*(\mathsf{L}^1; \alpha)$, then for any β, $0 < \beta < \alpha$, and $0 < \varepsilon_1 < \varepsilon_2$

$$\|R_{\varepsilon_2}^{\{\beta\}} f - R_{\varepsilon_1}^{\{\beta\}} f\|_1 \le \frac{1}{|A_c(-\alpha)|} \int_{\varepsilon_1}^{\varepsilon_2} u^{-(1+\beta)} \|\overline{\Delta}_u^2 f\|_1\, du,$$

which tends to zero as $\varepsilon_1, \varepsilon_2 \to 0+$. Therefore there exists $g \in \mathsf{L}^1$ such that $\lim_{\varepsilon\to 0+} \|R_\varepsilon^{\{\beta\}} f - g\|_1 = 0$, and Theorem 11.3.5 implies that $|v|^\beta f^\wedge(v) = g^\wedge(v)$ for all $v \in \mathbb{R}$. Hence it follows by Lemma 11.2.2 that $[\Delta_h f]^\wedge(v) = [m_{h,\beta} * g]^\wedge(v)$, and an application of the uniqueness theorem gives $\Delta_h f(x) = (m_{h,\beta} * g)(x)$ a.e. Therefore by an integration over the interval $[0, y]$

$$\int_0^y f(x+h)\, dx - \int_0^y f(x)\, dx = \int_0^y [R_\beta g](x+h)\, dx - \int_0^y [R_\beta g](x)\, dx.$$

Letting $h \to \infty$, Problem 11.2.2 shows that $f(x) = [R_\beta g](x)$ a.e., and the proof is complete.

Problems

1. Show that the kernels (Problem 3.1.14) of Jackson–de La Vallée Poussin and (3.1.33) of Gauss–Weierstrass are of type (11.3.4).
2. Show that $f \in \mathsf{V}[\mathsf{L}^p; |v|^\alpha]$ implies $\|\Delta_h f\|_p = O(|h|^\alpha)$ for $0 < \alpha < r$.
3. Prove Theorems 11.3.6, 11.3.7. (Hint: TREBELS [1])
4. Give a proof of Theorem 6.3.14 via an application of Problem 11.3.2 and Theorem 11.3.7 which is thus independent of the lemma of E. M. Stein and Bessel potentials.
5. Let $f \in \mathsf{L}^p$, $1 \le p \le 2$, and $\alpha > 0$. Show that f belongs to $\mathsf{V}[\mathsf{L}^p; \{-i \operatorname{sgn} v\}|v|^\alpha]$ if and only if

$$\left\| \frac{1}{C_{\alpha,2j+1}} \int_\varepsilon^\infty \frac{\overline{\Delta}_u^{2j+1} f(\circ)}{u^{1+\alpha}}\, du \right\|_p = O(1) \qquad (\varepsilon \to 0+),$$

where $j \in \mathbb{N}$ is chosen such that $0 < \alpha < 2j + 1$ and

$$C_{\alpha,2j+1} = (-1)^{j+1} 2^{2j+1-\alpha} \int_0^\infty \frac{\sin^{2j+1} u}{u^{1+\alpha}}\, du. \tag{11.3.15}$$

6. Let $f \in \mathsf{L}^p$, $1 \le p \le 2$, and $\alpha > 0$. Show that f belongs to $\mathsf{W}[\mathsf{L}^p; \{-i \operatorname{sgn} v\}|v|^\alpha]$ if and only if there exists $g \in \mathsf{L}^p$ such that $(0 < \alpha < 2j + 1)$

$$\lim_{\varepsilon \to 0+} \left\| \frac{1}{C_{\alpha,2j+1}} \int_\varepsilon^\infty \frac{\overline{\Delta}_u^{2j+1} f(\circ)}{u^{1+\alpha}}\, du - g(\circ) \right\|_p = 0.$$

7. Show that if $f \in \mathsf{V}[\mathsf{L}^p; |v|^\alpha]$ for some $\alpha > 1$, then $f \in \mathsf{W}[\mathsf{L}^p; (iv)^k]$ for every integer k, $1 \le k \le [\![\alpha]\!]$.
8. Prove the following theorem of E. C. TITCHMARSH: Let $f \in \mathsf{L}^p$, $1 < p \le 2$. If $\|f(\circ + h) - f(\circ - h)\|_p = O(h^\alpha)$, $0 < \alpha \le 1$, show that $f^\wedge \in \mathsf{L}^q$ for $p/(p - 1 + \alpha p) < q < p/(p - 1)$.
9. Let $f \in \mathsf{L}^p$, $1 \le p \le 2$, and $0 < \alpha < 2$. Show that

$$\left\| \frac{1}{\Lambda_c(-\alpha)} \int_{|u| \ge \varepsilon} \frac{f(\circ - u) - f(\circ)}{|u|^{1+\alpha}}\, du \right\|_p = o(1) \qquad (\varepsilon \to 0+)$$

implies $f(x) = 0$ a.e. (see also H. S. SHAPIRO [1, p. 42]).

10. Show that for $0 < \alpha < 2$ the set of $f \in \mathsf{X}(\mathbb{R})$ with $\|R_\varepsilon^{\{\alpha\}} f\|_{\mathsf{X}(\mathbb{R})} = O(1)$, $\varepsilon \to 0+$, becomes a normalized Banach subspace of $\mathsf{X}(\mathbb{R})$ under the norm

$$\|f\|_{\mathsf{X}(\mathbb{R})} + \sup_{\varepsilon > 0} \|R_\varepsilon^{\{\alpha\}} f\|_{\mathsf{X}(\mathbb{R})} \qquad (0 < \alpha < 2). \tag{11.3.16}$$

Show that in case $\mathsf{X}(\mathbb{R}) = \mathsf{L}^p$, $1 \le p \le 2$, this Banach space and the Banach space $\mathsf{V}[\mathsf{L}^p; |v|^\alpha]$ are equal with equivalent norms. (Hint: Prop. 10.5.7, Theorem 11.3.3)

11.4 The Operators $R_\varepsilon^{\{\alpha\}}$ on $\mathsf{X}_{2\pi}$

Next we turn to counterparts of the preceding results for periodic functions. In this section we are concerned with characterizations of the classes $\mathsf{W}[\mathsf{X}_{2\pi}; |k|^\alpha]$, $\mathsf{V}[\mathsf{X}_{2\pi}; |k|^\alpha]$ in terms of integrals of type (11.3.1). Here we prefer to proceed separately, whilst in the next section we transcribe the results of Sec. 11.2 to periodic functions by reduction.

The treatment of this section will be much the same as for functions belonging to L^p, $1 \le p \le 2$. Indeed, let $f \in \mathsf{X}_{2\pi}$ and $0 < \alpha < 2$. Defining $[R_\varepsilon^{\{\alpha\}} f](x)$ as in (11.3.2) it follows that $R_\varepsilon^{\{\alpha\}}f \in \mathsf{C}_{2\pi}$. For each $\varepsilon > 0$, $R_\varepsilon^{\{\alpha\}}f$ is a bounded linear operator of $\mathsf{X}_{2\pi}$ into itself satisfying

$$(11.4.1) \qquad \|R_\varepsilon^{\{\alpha\}}f\|_{\mathsf{X}_{2\pi}} \le \frac{4}{\alpha\varepsilon^\alpha|\Lambda_c(-\alpha)|}\|f\|_{\mathsf{X}_{2\pi}}.$$

Here we suppose throughout that χ_ρ satisfies the following conditions:

(11.4.2) *For* $\rho \in \mathbb{A}$, $\rho \to \rho_0$, $\{\chi_\rho(x)\}$ *is an even, positive periodic approximate identity such that*

(i) $\chi_\rho \in \mathsf{W}^2_{\mathsf{C}_{2\pi}}$ *for each* $\rho \in \mathbb{A}$,

(ii) $1 - \chi_\rho^\wedge(1) = O(|\rho - \rho_0|^2) \qquad (\rho \to \rho_0),$

$$\int_0^\pi u^\alpha|\chi_\rho''(u)|\,du = O(|\rho - \rho_0|^{\alpha-2}) \qquad (\rho \to \rho_0),$$

(iii) $\{[\chi_\rho'']^\wedge(k)\} \in \mathsf{l}^1$ *for each* $\rho \in \mathbb{A}$.

For examples of kernels satisfying (11.4.2) see Problem 11.4.1. Subsequently, $I_\rho(f; x)$ always denotes the singular integral associated with a kernel of type (11.4.2).

Lemma 11.4.1. *If* $\{\chi_\rho(x)\}$ *satisfies* (11.4.2), *then* $f \in \mathsf{Lip}^*(\mathsf{X}_{2\pi}; \alpha)$, $0 < \alpha < 2$, *implies that*

(i) $\|I_\rho(\bar\Delta_h^2 f; \circ) - \bar\Delta_h^2 f(\circ)\|_{\mathsf{X}_{2\pi}} = O(|\rho - \rho_0|^\alpha) \qquad (h > 0, \rho \to \rho_0),$

(ii) $\|I_\rho(\bar\Delta_h^2 f; \circ)\|_{\mathsf{X}_{2\pi}} = O(h^2|\rho - \rho_0|^{\alpha-2}) \qquad (h > 0, \rho \to \rho_0).$

Proof. Assertion (i) follows by Theorem 1.5.8. To prove (ii) we proceed as in the proof of Lemma 11.3.1 and apply the Taylor formula (11.3.6) to $g(h) = I_\rho(\bar\Delta_h^2 f; x)$. As χ_ρ has a uniform second derivative for each $\rho \in \mathbb{A}$, we may interchange the order of integration and differentiation (with respect to h) to obtain

$$\frac{\partial^2}{\partial h^2} I_\rho(\bar\Delta_h^2 f; x) = \frac{1}{2\pi}\int_0^\pi [\bar\Delta_h^2 f(x + h) + \bar\Delta_h^2 f(x - h)]\chi_\rho''(u)\,du$$

for each $\rho \in \mathbb{A}$, $x \in \mathbb{R}$. Therefore

$$\|I_\rho(\bar\Delta_h^2 f; \circ)\|_{\mathsf{X}_{2\pi}} = \left\| \int_0^h (h - t)\left\{\frac{1}{2\pi}\int_0^\pi [\bar\Delta_u^2 f(\circ + t) + \bar\Delta_u^2 f(\circ - t)]\chi_\rho''(u)\,du\right\} dt \right\|_{\mathsf{X}_{2\pi}}$$

$$= O\left(\int_0^h (h - t)\left\{\int_0^\pi u^\alpha|\chi_\rho''(u)|\,du\right\} dt\right) = O(h^2|\rho - \rho_0|^{\alpha-2}).$$

Lemma 11.4.2. *If* $\{\chi_\rho(x)\}$ *satisfies* (11.4.2), *then* $f \in \mathsf{Lip}^*(\mathsf{X}_{2\pi}; \alpha)$, $0 < \alpha < 2$, *implies that*

$$\left\| [R^{\{\alpha\}}_{|\rho-\rho_0|}f](\circ) - \sum_{k=-\infty}^{\infty} \chi_\rho^\wedge(k)|k|^\alpha f^\wedge(k)\,e^{ik\circ} \right\|_{\mathsf{X}_{2\pi}} = O(1) \qquad (\rho \to \rho_0).$$

Proof. Let $\rho \in \mathbb{A}$ be fixed. Since $\chi_\rho \in \mathsf{W}^2_{\mathsf{C}_{2\pi}}$, we have $(ik)^2\chi_\rho^\wedge(k) = [\chi_\rho'']^\wedge(k)$ by Prop. 4.1.8, and therefore $\{\chi_\rho^\wedge(k)\} \in \mathsf{l}^1$. As in the proof of Lemma 11.3.2 we may apply Prop. 4.1.6 to deduce

$$I_\rho(\bar\Delta_h^2 f; x) = -4\sum_{k=-\infty}^{\infty} \chi_\rho^\wedge(k)\sin^2\frac{kh}{2} f^\wedge(k)\,e^{ikx}.$$

Since $\{(ik)^2\chi_\rho^\wedge(k)\} = \{[\chi_\rho'']^\wedge(k)\} \in \mathsf{l}^1$ by hypothesis, we again have

$$\int_0^\infty u^{-(1+\alpha)} I_\rho(\bar\Delta_u^2 f; x)\,du = -4\sum_{k=-\infty}^{\infty} \chi_\rho^\wedge(k) f^\wedge(k)\,e^{ikx}\int_0^\infty u^{-(1+\alpha)}\sin^2\frac{ku}{2}\,du$$

$$(11.4.3) \qquad = \Lambda_c(-\alpha)\sum_{k=-\infty}^{\infty} \chi_\rho^\wedge(k)|k|^\alpha f^\wedge(k)\,e^{ikx}$$

for each $x \in \mathbb{R}$. Since moreover

$$[R^{\{\alpha\}}_{|\rho-\rho_0|}f](x) = \frac{1}{\Lambda_c(-\alpha)}\int_{|\rho-\rho_0|}^\infty \frac{\bar\Delta_u^2 f(x)}{u^{1+\alpha}}\,du,$$

it follows by Lemma 11.4.1 that

$$|\Lambda_c(-\alpha)| \left\| [R^{\{\alpha\}}_{|\rho-\rho_0|}f](\circ) - \sum_{k=-\infty}^{\infty} \widehat{\chi_\rho}(k)|k|^\alpha f^\wedge(k)\, e^{ik\circ} \right\|_{X_{2\pi}}$$

$$\le \int_0^{|\rho-\rho_0|} u^{-(1+\alpha)} \|I_\rho(\overline{\Delta}^2_u; \circ)\|_{X_{2\pi}}\, du + \int_{|\rho-\rho_0|}^\infty u^{-(1+\alpha)} \|I_\rho(\overline{\Delta}^2_u f; \circ) - \overline{\Delta}^2_u f(\circ)\|_{X_{2\pi}}\, du$$

$$= O\Big(|\rho - \rho_0|^{\alpha-2} \int_0^{|\rho-\rho_0|} u^{1-\alpha}\, du\Big) + O\Big(|\rho - \rho_0|^\alpha \int_{|\rho-\rho_0|}^\infty u^{-(1+\alpha)}\, du\Big) = O(1) \quad (\rho \to \rho_0).$$

Theorem 11.4.3. *Let* $0 < \alpha < 2$, $\{\chi_\rho(x)\}$ *satisfy* (11.4.2), *and* $f \in X_{2\pi}$. *The following assertions are equivalent:*

(i) $f \in V[X_{2\pi}; |k|^\alpha]$,

(ii) $$\left\| \sum_{k=-\infty}^{\infty} \widehat{\chi_\rho}(k)|k|^\alpha f^\wedge(k)\, e^{ik\circ} \right\|_{X_{2\pi}} = O(1) \qquad (\rho \to \rho_0),$$

(iii) $$\left\| \frac{1}{\Lambda_c(-\alpha)} \int_{|u|\ge\varepsilon} \frac{f(\circ - u) - f(\circ)}{|u|^{1+\alpha}}\, du \right\|_{X_{2\pi}} = O(1) \qquad (\varepsilon \to 0+).$$

Proof. In view of (11.4.2), the proof of (i) ⇔ (ii) is given by Problem 6.1.10. To prove the implication (iii) ⇒ (ii) we proceed as in the proof of Theorem 11.3.3. Indeed, by (11.4.3) and the hypothesis

$$|\Lambda_c(-\alpha)| \left\| \sum_{k=-\infty}^{\infty} \widehat{\chi_\rho}(k)|k|^\alpha f^\wedge(k)\, e^{ik\circ} \right\|_{X_{2\pi}} \le \liminf_{\varepsilon\to 0+} \left\| \int_\varepsilon^\infty u^{-(1+\alpha)} I_\rho(\overline{\Delta}^2_u f; \circ)\, du \right\|_{X_{2\pi}}$$

$$\le \liminf_{\varepsilon\to 0+} \left\| \int_\varepsilon^\infty u^{-(1+\alpha)} \overline{\Delta}^2_u f(\circ)\, du \right\|_{X_{2\pi}} = O(1)$$

uniformly for $\rho \in \mathbb{A}$. Conversely, let (ii) be satisfied. Then $\|R^{\{\alpha\}}_{|\rho-\rho_0|} f\|_{X_{2\pi}} = O(1)$ as $\rho \to \rho_0$ by Lemma 11.4.2, which already proves (iii) if we set $|\rho - \rho_0| = \varepsilon$, provided we can show that f belongs to $\mathrm{Lip}^*(X_{2\pi}; \alpha)$. But this is given by the following

Lemma 11.4.4. *Let* $f \in X_{2\pi}$ *and* $0 < \alpha < 2$. *Then* $f \in V[X_{2\pi}; |k|^\alpha]$ *implies* $f \in \mathrm{Lip}^*(X_{2\pi}; \alpha)$.

The proof is left to Problem 11.4.3.

Concerning characterizations of the classes $W[X_{2\pi}; |k|^\alpha]$ we have

Theorem 11.4.5. *Let* $0 < \alpha < 2$ *and* $f \in X_{2\pi}$. *The following assertions are equivalent:*

(i) $f \in W[X_{2\pi}; |k|^\alpha]$,

(ii) *there exists* $g \in X_{2\pi}$ *such that* $\lim_{\varepsilon\to 0+} \|R^{\{\alpha\}}_\varepsilon f - g\|_{X_{2\pi}} = 0$.

The proof may be given as for Theorem 11.3.5, in fact, the argument is simpler.

So far we have seen that in case $0 < \alpha < 2$ the whole program of Sec. 11.3 may be carried over to periodic functions. The same is true for the general case $\alpha > 0$.

Theorem 11.4.6. *Let* $f \in X_{2\pi}$ *and* $\alpha > 0$. *The following assertions are equivalent:*

(i) $f \in V[X_{2\pi}; |k|^\alpha]$,

(ii) $$\left\| \sum_{k=-n}^{n} \Big(1 - \frac{|k|}{n+1}\Big)|k|^\alpha f^\wedge(k)\, e^{ik\circ} \right\|_{X_{2\pi}} = O(1) \qquad (n \to \infty),$$

(iii) $$\left\| \frac{1}{C_{\alpha,2j}} \int_\varepsilon^\infty \frac{\overline{\Delta}^{2j}_u f(\circ)}{u^{1+\alpha}}\, du \right\|_{X_{2\pi}} = O(1) \qquad (\varepsilon \to 0+),$$

where $j \in \mathbb{N}$ *is chosen such that* $0 < \alpha < 2j$, *and the constant* $C_{\alpha,2j}$ *is given by* (11.3.14).

Theorem 11.4.7. *Let* $f \in \mathsf{X}_{2\pi}$ *and* $\alpha > 0$. *The following assertions are equivalent:*

(i) $f \in \mathsf{W}[\mathsf{X}_{2\pi}; |k|^\alpha]$,

(ii) *there exists* $g \in \mathsf{X}_{2\pi}$ *such that* $(0 < \alpha < 2j)$

$$\lim_{\varepsilon \to 0+} \left\| \frac{1}{C_{\alpha, 2j}} \int_\varepsilon^\infty \frac{\overline{\Delta}_u^{2j} f(\circ)}{u^{1+\alpha}}\, du - g(\circ) \right\|_{\mathsf{X}_{2\pi}} = 0.$$

For the proofs of the last two theorems and for related results we refer to the Problems below.

Problems

1. Show that the periodic kernel $\{\theta_3(x, t)\}$ of Weierstrass (cf. Problem 1.3.10) satisfies (11.4.2) with $\rho_0 = 0$ and $\rho = \sqrt{t}$. Rewrite (11.4.2) such that $\rho_0 = \infty$, and show that the kernel $\{j_n(x)\}$ of Jackson (cf. Problem 1.3.9) is an example in this case.
2. Let $0 < \alpha < 1$ and $h \in \mathbb{R}$ be fixed. If $m_{h,\alpha}, n_{h,\alpha}$ are defined by (11.2.5), (11.2.6), respectively, show that $m_{h,\alpha}^*(x) \equiv \sqrt{2\pi} \sum_{j=-\infty}^{\infty} m_{h,\alpha}(x + 2\pi j)$ and $n_{h,\alpha}^*(x) \equiv \sqrt{2\pi} \sum_{j=-\infty}^{\infty} n_{h,\alpha}(x + 2\pi j)$ belong to $\mathsf{L}_{2\pi}^1$, and $\|m_{h,\alpha}^*\|_{\mathsf{L}_{2\pi}^1} \le |h|^\alpha \|m_{1,\alpha}\|_{\mathsf{L}^1}$, $\|n_{h,\alpha}^*\|_{\mathsf{L}_{2\pi}^1} \le |h|^\alpha \|n_{1,\alpha}\|_{\mathsf{L}^1}$. Furthermore, $[m_{h,\alpha}^*]^\wedge(k) = (e^{ihk} - 1)|k|^{-\alpha}$, $[n_{h,\alpha}^*]^\wedge(k) = \{-i \operatorname{sgn} k\}(e^{ihk} - 1)|k|^{-\alpha}$ for $k \neq 0$, and $[m_{h,\alpha}^*]^\wedge(0) = [n_{h,\alpha}^*]^\wedge(0) = 0$. Show that $[m_{h,\alpha}^*]^\sim = n_{h,\alpha}^*$.
3. Prove Lemma 11.4.4.
4. Prove the following theorem of S. N. Bernstein: $f \in \mathsf{Lip}(\mathsf{C}_{2\pi}; \alpha)$ for $\alpha > (1/2)$ implies $f^\wedge \in \mathsf{l}^1$. State and prove extensions analogous to Theorem 11.3.8. (Hint: Hausdorff–Young inequality)
5. Prove Theorems 11.4.6, 11.4.7.
6. Show that $f \in \mathsf{V}[\mathsf{X}_{2\pi}; |k|^\alpha]$ implies $\|\Delta_h^r f\|_{\mathsf{X}_{2\pi}} = O(|h|^\alpha)$ for $0 < \alpha < r$.
7. Let $f \in \mathsf{X}_{2\pi}$ and $\alpha > 0$. Show that f belongs to $\mathsf{V}[\mathsf{X}_{2\pi}; \{-i \operatorname{sgn} k\}|k|^\alpha]$ if and only if
$$\left\| \frac{1}{C_{\alpha, 2j+1}} \int_\varepsilon^\infty \frac{\overline{\Delta}_u^{2j+1} f(\circ)}{u^{1+\alpha}}\, du \right\|_{\mathsf{X}_{2\pi}} = O(1) \qquad (\varepsilon \to 0+),$$
where $j \in \mathbb{N}$ is chosen such that $0 < \alpha < 2j + 1$ and $C_{\alpha, 2j+1}$ is given by (11.3.15).
8. Let $f \in \mathsf{X}_{2\pi}$ and $\alpha > 0$. Show that f belongs to $\mathsf{W}[\mathsf{X}_{2\pi}; \{-i \operatorname{sgn} k\}|k|^\alpha]$ if and only if there exists $g \in \mathsf{X}_{2\pi}$ such that $(0 < \alpha < 2j + 1)$
$$\lim_{\varepsilon \to 0+} \left\| \frac{1}{C_{\alpha, 2j+1}} \int_\varepsilon^\infty \frac{\overline{\Delta}_u^{2j+1} f(\circ)}{u^{1+\alpha}}\, du - g(\circ) \right\|_{\mathsf{X}_{2\pi}} = 0.$$
9. Show that if $f \in \mathsf{V}[\mathsf{X}_{2\pi}; |k|^\alpha]$ for some $\alpha > 1$, then $f \in \mathsf{W}[\mathsf{X}_{2\pi}; (ik)^l]$ for every integer l, $1 \le l \le [\![\alpha]\!]$.
10. Let $f \in \mathsf{X}_{2\pi}$ and $\alpha > 0$. Show that f belongs to $\mathsf{V}[\mathsf{X}_{2\pi}; |k|^\alpha]$ if and only if
$$\left\| \int_{|u| \ge \varepsilon} \frac{f^{\{r\}}(\circ - u) - f^{\{r\}}(\circ)}{|u|^{1+\gamma}}\, du \right\|_{\mathsf{X}_{2\pi}} = O(1) \qquad (\varepsilon \to 0+),$$
where $r \in \mathbb{P}$ is such that $\alpha = r + \gamma$, $0 < \gamma < 2$. (For the definition of $f^{\{r\}}$ see Def. 11.5.10). (Hint: Butzer–Görlich [1])

11.5 Integral Representations, Fractional Derivatives of Periodic Functions

In this section we briefly discuss suitable extensions of the results of Sec. 11.2 to periodic functions. Before we begin with this program, let us return to functions

$f \in \mathsf{L}^p$, $1 \le p \le 2$, for a moment. For instance in the integral case, we have shown that $f \in \mathsf{W}[\mathsf{L}^1; (iv)^r]$ if and only if f belongs to $\mathsf{W}^r_{\mathsf{L}^1}$, the latter class being defined by differentiability conditions upon f. It was this aspect which we tried to extend to the fractional case in Sec. 11.2. But equivalently, we may say (Theorem 5.1.16) that $f \in \mathsf{W}[\mathsf{L}^1; (iv)^r]$ if and only if there exists $g \in \mathsf{L}^1$ such that

$$f(x) = \int_{-\infty}^{x} du_1 \int_{-\infty}^{u_1} du_2 \ldots \int_{-\infty}^{u_{r-1}} g(u_r)\, du_r \equiv (J^r g)(x), \tag{11.5.1}$$

where each of the iterated integrals represents a function of L^1. In other words, we may also arrive at an integral representation of f. In the fractional case we have the following counterpart:

Theorem 11.5.1. *Let $f \in \mathsf{L}^1$ and $0 < \alpha < 1$. Then f belongs to $\mathsf{W}[\mathsf{L}^1; |v|^\alpha]$ if and only if there exists $g \in \mathsf{L}^1$ such that $f(x) = [R_\alpha g](x)$ almost everywhere.*

Proof. Let $f \in \mathsf{W}[\mathsf{L}^1; |v|^\alpha]$ and $g \in \mathsf{L}^1$ be such that $|v|^\alpha f^\wedge(v) = g^\wedge(v)$. Then by Lemma 11.2.2

$$(e^{ihv} - 1) f^\wedge(v) = (e^{ihv} - 1)|v|^{-\alpha} g^\wedge(v) = [m_{h,\alpha} * g]^\wedge(v) \tag{11.5.2}$$

for all $v \ne 0$, and thus by the uniqueness theorem for each $h \in \mathbb{R}$

$$f(x+h) - f(x) = (m_{h,\alpha} * g)(x) = [R_\alpha g](x+h) - [R_\alpha g](x) \quad \text{a.e.} \tag{11.5.3}$$

Integration with respect to x over $[0, y]$ gives

$$\int_0^y \{f(x) - [R_\alpha g](x)\}\, dx = \int_h^{y+h} f(u)\, du - \int_0^y [R_\alpha g](x+h)\, dx. \tag{11.5.4}$$

Letting $h \to -\infty$ we have (cf. Problem 11.2.2) that the right-hand side tends to zero, and therefore $f(x) = [R_\alpha g](x)$ a.e.

Conversely, if there is $g \in \mathsf{L}^1$ such that $f(x) = [R_\alpha g](x)$ a.e., we may invert the single steps of the above arguments to deduce (11.5.3) and (11.5.2), i.e. $f \in \mathsf{W}[\mathsf{L}^1; |v|^\alpha]$.

Regarding the extension to arbitrary values of α we first note

Theorem 11.5.2. *Let $f \in \mathsf{L}^1$ and $r \in \mathbb{N}$. Then f belongs to $\mathsf{W}[\mathsf{L}^1; |v|^r]$ if and only if $f \in \mathsf{L}^2$ and there exists $g \in \mathsf{L}^1$ such that*†

$$[H^r f](x) = \int_{-\infty}^{x} du_1 \int_{-\infty}^{u_1} du_2 \ldots \int_{-\infty}^{u_{r-1}} g(u_r)\, du_r \quad \text{a.e.}, \tag{11.5.5}$$

where each of the iterated integrals exists as a function of L^1.

Proof. Let $f \in \mathsf{W}[\mathsf{L}^1; |v|^r]$ and $g \in \mathsf{L}^1$ be such that $|v|^r f^\wedge(v) = g^\wedge(v)$. Since $r \ge 1$, this implies $f^\wedge \in \mathsf{L}^2$ and thus $f \in \mathsf{L}^2$ by the theorem of Plancherel. Therefore $Hf (\equiv f^\sim)$ and all powers $H^k f$, $k \in \mathbb{N}$, belong to L^2 by Theorem 8.1.7. Moreover, $H^{2k} f = (-1)^k f$ and $H^{2k+1} f = (-1)^k f^\sim$ by Prop. 8.2.10. Let r be even, say $r = 2k$. Then $(iv)^{2k} f^\wedge(v) = [(-1)^k g]^\wedge(v)$ by hypothesis and therefore (cf. (11.5.1)) $f(x) = (-1)^k (J^r g)(x)$ a.e.,

† We recall that H denotes the operation of taking Hilbert transforms (cf. Def. 8.0.1).

which proves (11.5.5) for even values of r. If r is odd, say $r = 2k + 1$, then by Theorem 6.3.14 there exists $g_1 \in \mathsf{L}^1$ such that $|v| f^\wedge(v) = g_1^\wedge(v)$. Therefore $|v|^{2k} g_1^\wedge(v) = g^\wedge(v)$, and thus the result for the even case may be applied to deduce $g_1(x) = (-1)^k (J^{2k} g)(x)$ a.e. But, in view of Problem 8.3.3, $|v|\, f^\wedge(v) = g_1^\wedge(v)$ implies the representation $f^\sim(x) = \int_{-\infty}^x g_1(u)\, du$ a.e. and therefore $f^\sim(x) = (-1)^k (J^r g)(x)$ a.e., which completes the proof of (11.5.5).

Since $f \in \mathsf{L}^1 \cap \mathsf{L}^2$, we may apply Prop. 8.3.1 so that the converse follows in view of (10.6.7).

Theorem 11.5.3. *Let $f \in \mathsf{L}^1$ and $\alpha > 1$ be such that $\alpha = [\![\alpha]\!] + \beta$, $0 < \beta < 1$. Then the following assertions are equivalent:*

(i) $f \in \mathsf{W}[\mathsf{L}^1; |v|^\alpha]$,

(ii) *f has an $[\![\alpha]\!]$th strong Riesz derivative and there exists $g \in \mathsf{L}^1$ such that*

$$(D_s^{\{[\alpha]\}} f)(x) = [R_\beta g](x) \quad \text{a.e.,} \tag{11.5.6}$$

(iii) *$f \in \mathsf{L}^2$ and there exists $g \in \mathsf{L}^1$ such that $R_\beta g \in \mathsf{L}^1$ and*

$$[H^{[\alpha]} f](x) = \int_{-\infty}^{x} du_1 \int_{-\infty}^{u_1} du_2 \ldots \int_{-\infty}^{u_{[\alpha]-1}} [R_\beta g](u_{[\alpha]})\, du_{[\alpha]} \quad \text{a.e.,} \tag{11.5.7}$$

where each of the iterated integrals exists as a function of L^1.

Proof. (i) $\Rightarrow$ (ii). Let $g \in \mathsf{L}^1$ be such that $|v|^\alpha f^\wedge(v) = g^\wedge(v)$. By Theorem 6.3.14 there exists $g_{[\alpha]} \in \mathsf{L}^1$ such that $|v|^{[\alpha]} f^\wedge(v) = g_{[\alpha]}^\wedge(v)$. Hence the $[\![\alpha]\!]$th strong Riesz derivative exists by Theorem 11.2.9 and $D_s^{\{[\alpha]\}} f = g_{[\alpha]}$. Moreover, $|v|^\beta [D_s^{\{[\alpha]\}} f]^\wedge(v) = g^\wedge(v)$, and assertion (ii) follows by Theorem 11.5.1.

(ii) $\Rightarrow$ (iii). In view of Theorem 11.2.9 the existence of the $[\![\alpha]\!]$th strong Riesz derivative of f implies that $|v|^{[\alpha]} f^\wedge(v) = [D_s^{\{[\alpha]\}} f]^\wedge(v)$. Thus we may apply Theorem 11.5.2 to deduce the statement of (iii).

(iii) $\Rightarrow$ (i). According to Theorem 11.5.2 we have $|v|^{[\alpha]} f^\wedge(v) = [R_\beta g]^\wedge(v)$. Moreover,

$$\begin{aligned}(e^{ihv} - 1)|v|^{[\alpha]} f^\wedge(v) &= [[R_\beta g](\circ + h) - [R_\beta g](\circ)]^\wedge(v) = [m_{h,\beta} * g]^\wedge(v) \\ &= (e^{ihv} - 1)|v|^{-\beta} g^\wedge(v)\end{aligned} \tag{11.5.8}$$

by Lemma 11.2.2. Thus $|v|^\alpha f^\wedge(v) = g^\wedge(v)$, and the proof is complete.

Analogous results concerning the classes $\mathsf{W}[\mathsf{L}^p; |v|^\alpha]$, $1 < p \le 2$, and $\mathsf{V}[\mathsf{L}^p; |v|^\alpha]$, $1 \le p \le 2$, may be obtained by similar arguments (cf. Problems 11.5.1 and 11.5.2).

Here we are more interested in deriving related results for periodic functions. To this end, let us consider the series

$$M_\alpha^*(x) = \frac{2\pi}{A_c(\alpha)} \left[\frac{1}{|x|^{1-\alpha}} + \sideset{}{'}\sum_{j=-\infty}^{\infty} \left\{ \frac{1}{|x + 2\pi j|^{1-\alpha}} - \frac{1}{|2\pi j|^{1-\alpha}} \right\} \right] \tag{11.5.9}$$

for $0 < \alpha < 1$, which is to be compared with (3.1.22) and (9.0.4). Since $\big| |x + 2\pi j|^{\alpha-1} - |2\pi j|^{\alpha-1} \big| = O(|j|^{\alpha-2})$ uniformly for $|x| \le \pi$ as $|j| \to \infty$ (cf. the proof of Lemma 11.2.2), it follows that the series

$$\sideset{}{'}\sum_{j=-\infty}^{\infty} \left\{ \frac{1}{|x + 2\pi j|^{1-\alpha}} - \frac{1}{|2\pi j|^{1-\alpha}} \right\} \tag{11.5.10}$$

converges absolutely and uniformly for $|x| \le \pi$. Thus the definition of M_α^* is meaningful for all $|x| \le \pi$, $x \ne 0$. In fact, M_α^* is well-defined for all $x \in \mathbb{R}$, $x \ne 2\pi j$, $j \in \mathbb{Z}$. Moreover, M_α^* is 2π-periodic. Indeed, denoting the Nth partial sum of the series defining M_α^* by $M_{\alpha,N}^*(x)$, then

$$\frac{\Lambda_c(\alpha)}{2\pi} M_{\alpha,N}^*(x + 2\pi) = \frac{1}{|x + 2\pi|^{1-\alpha}} + \sum_{j=-N}^{N}{}' \left\{\frac{1}{|x + 2\pi(1 + j)|^{1-\alpha}} - \frac{1}{|2\pi j|^{1-\alpha}}\right\},$$

and hence

$$\frac{\Lambda_c(\alpha)}{2\pi} (M_{\alpha,N}^*(x + 2\pi) - M_{\alpha,N}^*(x)) = \frac{1}{|x + 2\pi(N + 1)|^{1-\alpha}} - \frac{1}{|x - 2\pi N|^{1-\alpha}}.$$

Letting $N \to \infty$, this implies $M_\alpha^*(x + 2\pi) = M_\alpha^*(x)$, since the sum of an absolutely convergent series $\sum_{k=-\infty}^{\infty} a_k$ is equal to its Cauchy principal value. Thus we have in particular shown that M_α^* belongs to $\mathsf{L}_{2\pi}^1$ for $0 < \alpha < 1$. Moreover, since the series (11.5.10) is bounded for $|x| \le \pi$, it follows that M_α^* belongs to $\mathsf{L}_{2\pi}^q$, $1 \le q < 1/(1 - \alpha)$.

To evaluate the Fourier coefficients of M_α^* we have in view of the uniform convergence of the series (11.5.10) and by Problem 11.1.1 that for $k \ne 0$

$$(11.5.11)\quad [M_\alpha^*]^\wedge(k) = \frac{1}{\Lambda_c(\alpha)} \Big[\int_{-\pi}^{\pi} |x|^{\alpha-1} e^{-ikx}\,dx + \sum_{j=-\infty}^{\infty}{}' \Big\{\int_{-\pi}^{\pi} |x + 2\pi j|^{\alpha-1} e^{-ikx}\,dx - |2\pi j|^{\alpha-1} \int_{-\pi}^{\pi} e^{-ikx}\,dx\Big\}\Big] = \frac{1}{\Lambda_c(\alpha)} \int_{-\infty}^{\infty} \frac{e^{-ikx}}{|x|^{1-\alpha}}\,dx = |k|^{-\alpha}.$$

Setting $\int_{-\pi}^{\pi} M_\alpha^*(x)\,dx = 2\pi \cdot c_{0,\alpha}$ and

$$(11.5.12)\qquad \varphi_\alpha(x) = \sum_{k=-\infty}^{\infty}{}' \frac{e^{ikx}}{|k|^\alpha} = 2 \sum_{k=1}^{\infty} \frac{\cos kx}{k^\alpha},$$

it follows by Problem 6.3.1 that $M_\alpha^*(x) = c_{0,\alpha} + \varphi_\alpha(x)$ for $0 < \alpha < 1$. But, whilst the definition (11.5.9) of M_α^* is not meaningful for $\alpha \ge 1$, the function φ_α of (11.5.12) is well-defined for every $\alpha > 0$. Therefore, as a counterpart of Theorems 11.5.1–11.5.3

Theorem 11.5.4. *Let $f \in \mathsf{X}_{2\pi}$ and $\alpha > 0$. Then f belongs to $\mathsf{W}[\mathsf{X}_{2\pi}; |k|^\alpha]$ if and only if there exists $g \in \mathsf{X}_{2\pi}$ with $g^\wedge(0) = 0$ such that $f(x) = f^\wedge(0) + (\varphi_\alpha * g)(x)$ in $\mathsf{X}_{2\pi}$.*

Since $\varphi_\alpha \in \mathsf{L}_{2\pi}^1$, the proof follows by an application of the convolution theorem. Note that we need not reproduce the successive structure of Theorems 11.5.1–11.5.3 for functions $f \in \mathsf{L}^1$, since φ_α is well-defined for all $\alpha > 0$.

Concerning the classes $\mathsf{V}[\mathsf{X}_{2\pi}; |k|^\alpha]$ we have similarly

Theorem 11.5.5. *Let $f \in \mathsf{X}_{2\pi}$ and $\alpha > 0$. Then f belongs to $\mathsf{V}[\mathsf{X}_{2\pi}; |k|^\alpha]$ if and only if there exists $g \in \mathsf{L}_{2\pi}^\infty$ for $\mathsf{X}_{2\pi} = \mathsf{C}_{2\pi}$, $\mu \in \mathsf{BV}_{2\pi}$ for $\mathsf{X}_{2\pi} = \mathsf{L}_{2\pi}^1$, and $g \in \mathsf{L}_{2\pi}^p$ for $\mathsf{X}_{2\pi} = \mathsf{L}_{2\pi}^p$, $1 < p < \infty$, such that $g^\wedge(0) = 0$, $\mu^\vee(0) = 0$, and f is representable as*

$$f(x) = f^\wedge(0) + \begin{cases} (\varphi_\alpha * g)(x), & \mathsf{X}_{2\pi} = \mathsf{C}_{2\pi} \\ (\varphi_\alpha * d\mu)(x) \quad \text{a.e.}, & \mathsf{X}_{2\pi} = \mathsf{L}_{2\pi}^1 \\ (\varphi_\alpha * g)(x) \quad \text{a.e.}, & \mathsf{X}_{2\pi} = \mathsf{L}_{2\pi}^p,\ 1 < p < \infty. \end{cases}$$

Let us return to the conjugate $R_\alpha^\sim f$ of the Riesz integral of f (cf. (11.2.11)) and set up for $0 < \alpha < 1$

$$(11.5.13)\quad N_\alpha^*(x) = \frac{2\pi}{\Lambda_s(\alpha)}\left[\frac{\operatorname{sgn} x}{|x|^{1-\alpha}} + \sum_{j=-\infty}^{\infty}{}' \left\{\frac{\operatorname{sgn}(x+2\pi j)}{|x+2\pi j|^{1-\alpha}} - \frac{\operatorname{sgn} j}{|2\pi j|^{1-\alpha}}\right\}\right].$$

Arguing as above it follows that

$$\sum_{j=-\infty}^{\infty}{}' \left|\frac{\operatorname{sgn}(x+2\pi j)}{|x+2\pi j|^{1-\alpha}} - \frac{\operatorname{sgn} j}{|2\pi j|^{1-\alpha}}\right|$$

converges uniformly for $|x| \le \pi$ and that N_α^* belongs to $\mathsf{L}_{2\pi}^1$ for $0 < \alpha < 1$. Moreover, for the Fourier coefficients of N_α^* we obtain for $k \neq 0$

$$(11.5.14)\qquad [N_\alpha^*]^\wedge(k) = \{-i \operatorname{sgn} k\}|k|^{-\alpha}$$

and for $k = 0$

$$\frac{\Lambda_s(\alpha)}{2\pi}\int_{-\pi}^{\pi} N_\alpha^*(x)\,dx = \int_{-\pi}^{\pi}\frac{\operatorname{sgn} x}{|x|^{1-\alpha}}\,dx + \lim_{N\to\infty}\sum_{j=-N}^{N}{}'\left\{\int_{-\pi}^{\pi}\frac{\operatorname{sgn}(x+2\pi j)}{|x+2\pi j|^{1-\alpha}}\,dx - \frac{2\pi \operatorname{sgn} j}{|2\pi j|^{1-\alpha}}\right\}$$

$$= |x|^\alpha\Big|_{-\pi}^{\pi} + \lim_{N\to\infty}\sum_{j=-N}^{N}{}' |x+2\pi j|^\alpha\Big|_{-\pi}^{\pi} = \lim_{N\to\infty}\sum_{j=-N}^{N}\{|\pi(2j+1)|^\alpha - |\pi(2j-1)|^\alpha\} = 0.$$

Here we again used the fact that the sum of an absolutely convergent series $\sum_{k=-\infty}^{\infty} a_k$ may be evaluated by its Cauchy principal value.

Therefore, if we define

$$(11.5.15)\qquad \psi_\alpha(x) = \sum_{k=-\infty}^{\infty}{}' \frac{\{-i \operatorname{sgn} k\}}{|k|^\alpha}\, e^{ikx} = 2\sum_{k=1}^{\infty}\frac{\sin kx}{k^\alpha},$$

then it follows by Problem 6.3.3 that $N_\alpha^*(x) = \psi_\alpha(x)$ for $0 < \alpha < 1$. Again, $\psi_\alpha(x)$ is defined for all $\alpha > 0$ and may thus be regarded as the analytic continuation of $N_\alpha^*(x)$ to $\alpha \ge 1$.

Furthermore, we have in view of (9.3.15) and Problem 4.1.4, that

$$\varphi_\alpha^\sim(x) = \lim_{n\to\infty}\sum_{k=-n}^{n}\left(1 - \frac{|k|}{n+1}\right)\{-i \operatorname{sgn} k\}\varphi_\alpha^\wedge(k)\, e^{ikx}$$

$$= \lim_{n\to\infty}\sum_{k=-n}^{n}\left(1 - \frac{|k|}{n+1}\right)\psi_\alpha^\wedge(k)\, e^{ikx} = \psi_\alpha(x).$$

Thus ψ_α is the conjugate function of φ_α. In this respect, we recall that the same relationship holds, at least formally, between the 'kernels' of the Riesz integral $R_\alpha f$ and the conjugate Riesz integral $R_\alpha^\sim f$ of f, namely

$$\frac{1}{\Lambda_c(\alpha)}\frac{1}{|x|^{1-\alpha}}, \qquad \frac{1}{\Lambda_s(\alpha)}\frac{\operatorname{sgn} x}{|x|^{1-\alpha}}.$$

For a rigorous interpretation we refer to Lemma 11.2.2.

We may now turn to the counterparts of Theorems 11.2.6, 11.2.7, 11.2.9. Let $0 < \alpha < 1$. Then, transcribing the procedure by which we defined a strong Riesz derivative of $f \in \mathsf{X}(\mathbb{R})$, we may set up the difference quotient

$$\frac{(f * \varphi_{1-\alpha}^\sim)(x+h) - (f * \varphi_{1-\alpha}^\sim)(x)}{h}$$

for $f \in X_{2\pi}$ and discuss its limit in $X_{2\pi}$-norm as $h \to 0$. But since $\varphi_{1-\alpha}^{\sim} \in L_{2\pi}^1$, each member of the above difference exists as a function in $X_{2\pi}$ for each $0 < \alpha < 1$ and $h \neq 0$, in contrast to the situation for functions f in $X(\mathbb{R})$ for which the difference (11.2.14) only as a whole belongs to $X(\mathbb{R})$. Corresponding to Theorem 11.2.6 we have

Proposition 11.5.6. *Let $f \in X_{2\pi}$ and $0 < \alpha < 1$. Then f belongs to $W[X_{2\pi}; |k|^\alpha]$ if and only if there exists $g \in X_{2\pi}$ such that*

$$(11.5.16) \qquad \lim_{h \to 0} \left\| \frac{(f * \varphi_{1-\alpha}^{\sim})(\circ + h) - (f * \varphi_{1-\alpha}^{\sim})(\circ)}{h} - g(\circ) \right\|_{X_{2\pi}} = 0.$$

Proof. Let $f \in W[X_{2\pi}; |k|^\alpha]$ and $g \in X_{2\pi}$ such that $|k|^\alpha f^\wedge(k) = g^\wedge(k)$. Then by the usual arguments

$$[(f * \varphi_{1-\alpha}^{\sim})(\circ + h) - (f * \varphi_{1-\alpha}^{\sim})(\circ)]^\wedge(k) = (e^{ihk} - 1) f^\wedge(k)\{-i \operatorname{sgn} k\}|k|^{\alpha-1}$$
$$= \frac{e^{ihk} - 1}{ik} g^\wedge(k) = \left[\int_0^h g(\circ + u)\, du\right]^\wedge(k)$$

and

$$(f * \varphi_{1-\alpha}^{\sim})(x + h) - (f * \varphi_{1-\alpha}^{\sim})(x) = \int_0^h g(x + u)\, du \quad \text{a.e.,}$$

which implies (11.5.16). On the other hand, we have for each $k \in \mathbb{Z}$

$$\left| \frac{e^{ihk} - 1}{h} f^\wedge(k)\{-i \operatorname{sgn} k\}|k|^{\alpha-1} - g^\wedge(k) \right|$$
$$\leq \left\| \frac{(f * \varphi_{1-\alpha}^{\sim})(\circ + h) - (f * \varphi_{1-\alpha}^{\sim})(\circ)}{h} - g(\circ) \right\|_{X_{2\pi}}$$

and thus, if (11.5.16) holds, it follows that $|k|^\alpha f^\wedge(k) = g^\wedge(k)$.

Instead of extending the last result to $\alpha \geq 1$ by a successive definition of a strong fractional derivative as for L^p-space, $1 \leq p \leq 2$, we may proceed more directly in the periodic case, since φ_α, $\varphi_\alpha^{\sim}$ belong to $L_{2\pi}^1$ for every $\alpha > 0$.

Thus, let $f \in X_{2\pi}$ and $\alpha > 0$. If there exists $g \in X_{2\pi}$ such that $|k|^\alpha f^\wedge(k) = g^\wedge(k)$, then f belongs to $L_{2\pi}^q$ for some q with $1 < q < \infty$. Indeed, the only case that remains open is $L_{2\pi}^1$. Then $f(x) = f^\wedge(0) + (\varphi_\alpha * g)(x)$ a.e. by Theorem 11.5.4. For $\alpha > 1$ this implies that f is almost everywhere equal to a function of $C_{2\pi}$ since obviously $\varphi_\alpha \in C_{2\pi}$. If $0 < \alpha \leq 1$, then in view of Theorem 11.5.4 we also have that $f(x) = f^\wedge(0) + (\varphi_{\alpha/2} * \varphi_{\alpha/2} * g)(x)$ a.e., and the assertion follows by the fact that $\varphi_{\alpha/2} \in L_{2\pi}^q$ for $1 \leq q < 2/(2 - \alpha)$.

Having shown that $f \in L_{2\pi}^q$ for some $1 < q < \infty$, it now follows by Theorem 9.1.3 that $f^\sim \equiv Hf$ belongs to $L_{2\pi}^q$, and thus $f * \varphi_{1-\alpha}^{\sim} = Hf * \varphi_{1-\alpha}$ by Prop. 9.1.13. Moreover, the powers $H^r f$ are defined for each $r \in \mathbb{N}$ and $H^r f \in L_{2\pi}^q$. Therefore we may set up $H^r f * \varphi_{r-\alpha}$ for $r > \alpha$. It follows that $H^r f * \varphi_{r-\alpha}$ belongs to $X_{2\pi}$ for each $f \in X_{2\pi}$. Indeed, if e.g. $X_{2\pi} = C_{2\pi}$, then f and thus $H^r f$ belong to $L_{2\pi}^s$ for every $1 < s < \infty$. On the other hand, $\varphi_{r-\alpha}$ belongs to $C_{2\pi}$ if $r - \alpha > 1$, and to $L_{2\pi}^q$, $1 \leq q < 2/(2 - r + \alpha)$, if $r - \alpha \leq 1$. Therefore $H^r f * \varphi_{r-\alpha} \in C_{2\pi}$ by Prop. 0.4.1.

Theorem 11.5.7. *Let $f \in X_{2\pi}$ and $\alpha > 0$. Suppose that $r \in \mathbb{N}$ is such that $\alpha < r$. Then $f \in W[X_{2\pi}; |k|^\alpha]$ if and only if $f \in L^q_{2\pi}$ for some q with $1 < q < \infty$ and $H^r f * \varphi_{r-\alpha}$ belongs to $W^r_{X_{2\pi}}$.*

Proof. If $f \in W[X_{2\pi}; |k|^\alpha]$ and $g \in X_{2\pi}$ is such that $|k|^\alpha f^\wedge(k) = g^\wedge(k)$, we have just shown that $H^r f$ exists and belongs to $L^q_{2\pi}$. Therefore by Prop. 9.1.8 and the convolution theorem

$$(ik)^r[H^r f * \varphi_{r-\alpha}]^\wedge(k) = (ik)^r\{-i \operatorname{sgn} k\}^r f^\wedge(k)|k|^{\alpha-r} = g^\wedge(k),$$

and the assertion follows by Prop. 4.1.9, since $H^r f * \varphi_{r-\alpha}$ belongs to $X_{2\pi}$.

Concerning the converse, we have by Prop. 4.1.8 and 9.1.8 that there exists $g \in X_{2\pi}$ such that

$$\begin{aligned} g^\wedge(k) &= (ik)^r[H^r f * \varphi_{r-\alpha}]^\wedge(k) \\ &= (ik)^r\{-i \operatorname{sgn} k\}^r f^\wedge(k)|k|^{\alpha-r} = |k|^\alpha f^\wedge(k), \end{aligned}$$

i.e. f belongs to $W[X_{2\pi}; |k|^\alpha]$.

For the particular case $X_{2\pi} = C_{2\pi}$ we have

Corollary 11.5.8. *Let* $f \in C_{2\pi}$ and $\alpha > 0$, *$r \in \mathbb{N}$ being such that $\alpha < r$. Then $f \in W[C_{2\pi}; |k|^\alpha]$ with $|k|^\alpha f^\wedge(k) = g^\wedge(k)$, $g \in C_{2\pi}$, if and only if for all x*

$$\frac{d^r}{dx^r}(H^r f * \varphi_{r-\alpha})(x) = g(x). \tag{11.5.17}$$

This is to be seen in contrast with the representation $f(x) = f^\wedge(0) + (\varphi_\alpha * g)(x)$ obtained in Theorem 11.5.4. The same methods of proof may be employed to derive the following characterization of the classes $V[X_{2\pi}; |k|^\alpha]$.

Theorem 11.5.9. *Let $f \in X_{2\pi}$ and $\alpha > 0$, $r \in \mathbb{N}$ being such that $\alpha < r$. Then $f \in V[X_{2\pi}; |k|^\alpha]$ if and only if $f \in L^q_{2\pi}$ for some q with $1 < q < \infty$ and $H^r f * \varphi_{r-\alpha}$ belongs to $V^r_{X_{2\pi}}$.*

The following definition is now motivated.

Definition 11.5.10. *Let $f \in X_{2\pi}$ and $\alpha > 0$. If there is $g \in X_{2\pi}$ such that $|k|^\alpha f^\wedge(k) = g^\wedge(k)$, then g is called the **αth Riesz derivative** of f and denoted by $f^{\{\alpha\}}$.*

Thus we have proceeded quite differently to the case of functions belonging to L^p, $1 \le p \le 2$. The present approach to the definition of a Riesz derivative of order α has the advantage that it is easier to deal with, at least in the periodic case. Note that we do not necessarily need the notation $D_s^{\{\alpha\}} f$ since the αth Riesz derivative may be interpreted in the strong sense as well as in the pointwise sense, in the latter case the definition would be given by (11.5.17).

So far we have only discussed fractional derivatives of periodic functions based upon the Riesz integral. Also Weyl's definition of a derivative of fractional order may be treated similarly. Let $\alpha > 0$, $r \in \mathbb{N}$, and $f \in X_{2\pi}$. We recall Prop. 4.1.8 which in particular states that $f \in W^r_{C_{2\pi}}$ implies $(ik)^r f^\wedge(k) = [f^{(r)}]^\wedge(k)$. Therefore, if there exists $g \in X_{2\pi}$ such that $(ik)^\alpha f^\wedge(k) = g^\wedge(k)$†, then g may be regarded as a fractional derivative of f

† $(ik)^\alpha$ is defined by $(ik)^\alpha = |k|^\alpha \exp\{i(\pi\alpha/2)\operatorname{sgn} k\}$, the principal branch.

known as *Weyl's derivative of order* α, in notation $f^{(\alpha)}$. In view of Problem 11.1.2 we may ask for its connection with the Riemann–Liouville integral.

Let us consider

(11.5.18)
$$\eta_\alpha(x) = \sum_{k=-\infty}^{\infty}{}' \frac{e^{ikx}}{(ik)^\alpha}.$$

Since (cf. (11.5.12), (11.5.15))

(11.5.19)
$$\eta_\alpha(x) = \varphi_\alpha(x) \cos\frac{\pi\alpha}{2} + \psi_\alpha(x) \sin\frac{\pi\alpha}{2},$$

it follows that η_α is defined and belongs to $\mathsf{L}^1_{2\pi}$ for every $\alpha > 0$. In fact, the series (11.5.18) is the Fourier series of η_α. Therefore, if the Weyl derivative $f^{(\alpha)}$ of $f \in \mathsf{X}_{2\pi}$ exists for some $\alpha > 0$, then $f(x) = f^\wedge(0) + (\eta_\alpha * f^{(\alpha)})(x)$ since the Fourier coefficients of both sides coincide.

On the other hand, we may proceed for $0 < \alpha < 1$ as in the Riesz case and deduce (cf. Problem 11.1.2) that for $x > 0$

(11.5.20)
$$\eta_\alpha(x) = \frac{2\pi}{\Gamma(\alpha)}\left[\frac{1}{x^{1-\alpha}} + \sum_{j=1}^{\infty}\left\{\frac{1}{(x+2\pi j)^{1-\alpha}} - \frac{1}{(2\pi j)^{1-\alpha}}\right\}\right],$$

where the series is absolutely and uniformly convergent for $0 \le x \le 2\pi$. Moreover,

$$\begin{aligned} f(x) - f^\wedge(0) &= \frac{1}{2\pi}\int_0^{2\pi} \eta_\alpha(u) f^{(\alpha)}(x-u)\,du = \frac{1}{\Gamma(\alpha)} \lim_{N\to\infty}\left[\int_0^{2\pi} u^{\alpha-1} f^{(\alpha)}(x-u)\,du\right. \\ &\quad \left. + \sum_{j=1}^{N}\int_0^{2\pi}(u+2\pi j)^{\alpha-1} f^{(\alpha)}(x-u)\,du - \sum_{j=1}^{N}(2\pi j)^{\alpha-1}\int_0^{2\pi} f^{(\alpha)}(x-u)\,du\right] \\ &= \lim_{N\to\infty}\frac{1}{\Gamma(\alpha)}\sum_{j=0}^{N}\int_{2\pi j}^{2\pi(j+1)} u^{\alpha-1} f^{(\alpha)}(x-u)\,du \\ &= \frac{1}{\Gamma(\alpha)}\int_0^\infty \frac{f^{(\alpha)}(x-u)}{u^{1-\alpha}}\,du = \frac{1}{\Gamma(\alpha)}\int_{-\infty}^{x}\frac{f^{(\alpha)}(u)}{(x-u)^{1-\alpha}}\,du. \end{aligned}$$

Here we have used the fact that $\int_0^{2\pi} f^{(\alpha)}(u)\,du = 0$ which follows since $(ik)^\alpha f^\wedge(k) = [f^{(\alpha)}]^\wedge(k)$ by definition. Thus, if we assume that $f \in \mathsf{X}_{2\pi}$ is such that $f^\wedge(0) = 0$, then the existence of the αth Weyl derivative $f^{(\alpha)}$ of f for $0 < \alpha < 1$ implies that f is almost everywhere equal to the Riemann–Liouville integral ($a = -\infty$) of order α of $f^{(\alpha)}$ (and vice versa).

If α is a natural number, say $\alpha = r$, then we recall Theorem 4.1.10 and (10.6.6) which show that $(ik)^r f^\wedge(k) = g^\wedge(k)$, $g \in \mathsf{X}_{2\pi}$, implies

$$f(x) = p_{r-1}(x) + \frac{1}{\Gamma(r)}\int_{-\pi}^{x}\frac{g(u)}{(x-u)^{1-r}}\,du \quad \text{(a.e.)},$$

where p_{r-1} is an algebraic polynomial of degree $(r-1)$ at most.

For further information concerning the Weyl derivative and Riemann–Liouville integral of order α we refer to Sec. 11.6.

Problems

1. (i) Let $f \in L^1$ and $0 < \alpha < 1$. Then f belongs to $V[L^1; |v|^\alpha]$ if and only if there exists $\mu \in BV$ such that $f(x) = [R_\alpha\, d\mu](x)$ almost everywhere.
   (ii) Let $f \in L^1$ and $\alpha \geq 1$. Then f belongs to $V[L^1; |v|^\alpha]$ if and only if $f \in L^2$ and there exists $\mu \in BV$ such that

$$\alpha = [\![\alpha]\!]: \quad [H^{[\alpha]}f](x) = \int_{-\infty}^{x} du_1 \int_{-\infty}^{u_1} du_2 \ldots \int_{-\infty}^{u_{[\alpha]-1}} d\mu(u_{[\alpha]}) \quad \text{a.e.},$$

$\alpha = [\![\alpha]\!] + \beta,\ 0 < \beta < 1$: $R_\beta\, d\mu$ belongs to L^1 and

$$[H^{[\alpha]}f](x) = \int_{-\infty}^{x} du_1 \int_{-\infty}^{u_1} du_2 \ldots \int_{-\infty}^{u_{[\alpha]-1}} [R_\beta\, d\mu](u_{[\alpha]})\, du_{[\alpha]}.$$

2. (i) Let $f \in L^p$, $1 < p \leq 2$, and $0 < \alpha < (1/p)$. Then f belongs to $W[L^p; |v|^\alpha]$ if and only if there exists $g \in L^p$ such that $f(x) = [R_\alpha g](x)$ almost everywhere.
   (ii) Let $f \in L^p$, $1 < p \leq 2$, and $\alpha > 0$. Then f belongs to $W[L^p; |v|^\alpha]$ if and only if assertion (iv) of Theorem 11.0.1 (for $1 < p \leq 2$) is valid. (Hint: Problem 11.2.2 and Theorem 6.3.14, the integrals exist in the sense of Theorem 5.2.21)
3. Prove the following theorem of H. WEYL: Let $f \in C_{2\pi}$ and $0 < \alpha < 1$. If there exists $g \in L^\infty_{2\pi}$ such that $\int_{-\pi}^{\pi} g(u)\, du = 0$ and

$$f(x) = \frac{1}{\Gamma(\alpha)} \int_{-\infty}^{x} (x - u)^{\alpha-1} g(u)\, du, \tag{11.5.21}$$

   then f belongs to $\mathrm{Lip}\,(C_{2\pi}; \alpha)$. Conversely, if $\int_{-\pi}^{\pi} f(u)\, du = 0$ and $f \in \mathrm{Lip}\,(C_{2\pi}; \alpha)$, then for each β with $0 < \beta < \alpha$ there exists $g \in C_{2\pi}$ such that (11.5.21) holds with α replaced by β.
4. Let $f \in X_{2\pi}$ and $\alpha > 0$, $r \in \mathbb{N}$ being such that $0 < \alpha < r$. If the αth Weyl derivative of f exists (i.e., there exists $g \in X_{2\pi}$ such that $(ik)^\alpha f^\wedge(k) = g^\wedge(k)$), then f belongs to $\mathrm{Lip}_r\,(X_{2\pi}; \alpha)$. Conversely, if $f \in \mathrm{Lip}_r\,(X_{2\pi}; \alpha)$, then the βth Weyl derivative of f exists for every $0 < \beta < \alpha$.
5. (i) Let $\alpha > 0$, $r \in \mathbb{N}$ being such that $0 < \alpha < r$. Show that φ_α of (11.5.12) belongs to $\mathrm{Lip}_r\,(L^1_{2\pi}; \alpha)$. (Hint: For $r = 1$ see Problem 11.4.2, for arbitrary r use the decomposition of $\varphi_\alpha = \varphi_{\alpha/r} * \varphi_{\alpha/r} * \cdots * \varphi_{\alpha/r}$ as an r-times convolution)
   (ii) Discuss the extension of M_α^* of (11.5.9) to $\alpha \geq 1$ by the r-times convolution of $M_{\alpha/r}^*$, where $r \in \mathbb{N}$ is such that $0 < \alpha < r$.
   (iii) Let $f \in V[X_{2\pi}; |k|^\alpha]$ for some $\alpha > 0$. Using φ_α of (11.5.12), show that $f \in W[X_{2\pi}; |k|^\beta]$ for every $0 < \beta < \alpha$.
6. Let $\alpha > 0$ and $\beta > 0$, r_1, r_2 being such that $0 < \alpha < r_1$, $0 < r_1 + \beta < r_2$. Then $f \in \mathrm{Lip}_{r_1}\,(X_{2\pi}; \alpha)$ implies $\varphi_\beta * f \in \mathrm{Lip}_{r_2}\,(X_{2\pi}; \alpha + \beta)$. (Hint: ZYGMUND [7II, p. 136])
7. Let t_n be a trigonometric polynomial of degree n at most, $t_n(x) = \sum_{k=-n}^{n} c_k \exp\{ikx\}$. Show that the αth Weyl and αth Riesz derivative are given by ($\alpha > 0$)

$$t_n^{(\alpha)}(x) = \sum_{k=-n}^{n} (ik)^\alpha c_k\, e^{ikx}, \qquad t_n^{\{\alpha\}}(x) = \sum_{k=-n}^{n} |k|^\alpha c_k\, e^{ikx},$$

   respectively. Prove the following inequalities (of Bernstein type)

$$\|t_n^{(\alpha)}\|_{X_{2\pi}} \leq M_1(\alpha) n^\alpha \|t_n\|_{X_{2\pi}}, \qquad \|t_n^{\{\alpha\}}\|_{X_{2\pi}} \leq M_2(\alpha) n^\alpha \|t_n\|_{X_{2\pi}}.$$

   (Hint: cf. Theorem 2.3.1 and BUTZER–GÖRLICH [2, p. 355], BUTZER–SCHERER [1, p. 107])
8. Let a in (11.1.2) be $-\infty$ and $0 < \alpha < 1$. Then $f \in W[L^1; (iv)^\alpha]$ if and only if there exists $g \in L^1$ such that $f(x) = [L_\alpha g](x)$ a.e. (cf. Problem 11.2.7).

11.6 Notes and Remarks

The notion of fractional differentiation is almost as old as the calculus itself. The origins of the subject are attributed to G. W. LEIBNIZ (see H. T. DAVIS [1], LEIBNIZ [1]) who mentioned it in a letter to G. F. A. DE L'HOSPITAL as early as 1695. N. H. ABEL appears to have made use of the notion in 1823 in solving the tautochrone problem. But the first systematic study is due to H. J. HOLMGREN [1], J. LIOUVILLE [1], and B. RIEMANN [1]. More recent papers of importance are those of J. HADAMARD [1], H. WEYL [1], E. L. POST [1], G. H. HARDY and L. E. LITTLEWOOD [2], and M. RIESZ [2]. (For a rather complete bibliography of 173 items see the doctoral dissertation of M. GAER [1].)

Chapter 11 was written in collaboration with WALTER TREBELS.

Sec. 11.1. The authors have largely followed M. RIESZ [2, p. 10–26] in this section. The material is also treated in books on distribution theory, e.g. GELFAND–SHILOV [1I, p. 115], SCHWARTZ [1]; for the Hadamard approach see also ZEMANIAN [1, p. 15 ff]. For fractional derivatives of functions analytic in a domain of the complex plane see e.g. GAER–RUBEL [1] and OSLER [1], who in particular gives applications to infinite series. For generalizations and applications connected with the theory of ordinary and partial differential equations see WEINSTEIN [1, 2].

Sec. 11.2 and 11.5. The actual goal of Sec. 11.2 is to associate fractional derivatives with the class $\mathsf{W}[\mathsf{L}^p; |v|^\alpha]$ ($\alpha > 0$) by trying to imitate the classical approach to the Riemann–Liouville derivative for periodic functions, whereby for $0 < \alpha < 1$ one first integrates fractionally of order $(1 - \alpha)$ and then takes the first (ordinary) derivative (compare e.g. ZYGMUND [7II, 133 ff]). This is obtained by replacing the Riemann–Liouville by the Riesz integral and the operation d/dx by $(d/dx)H$, keeping in mind that $f * Hm_{h,1-\alpha} = H[f * m_{h,1-\alpha}]$ (H being the Hilbert operator, all limits being understood in the strong sense, see Def. 11.2.5).

The functions $m_{h,\alpha}$, $n_{h,\alpha}$ were first introduced by OKIKIOLU [1]; for their present modified form, particularly the connection between them given by Lemma 11.2.2(iii), see BUTZER–TREBELS [1, 2]. The development leading to Def. 11.2.5 is due to TREBELS, see also NESSEL–TREBELS [1]; for preliminaries in this direction see BUTZER–TREBELS [2].

Whereas we have defined the Riesz derivative on $\mathsf{X}(\mathbb{R})$ as the strong limit of a generalized difference quotient, a more immediate approach would be via the transformed state, i.e., if $f \in \mathsf{L}^p$, $1 \le p \le 2$, it is the (unique) $g \in \mathsf{L}^p$ (if it exists) given by $|v|^\alpha f^\wedge(v) = g^\wedge(v)$. The equivalence of both definitions is then given by Theorems 11.2.6–11.2.9. For Theorems 11.2.6–11.2.12 see also COOPER [2], BUTZER–TREBELS [2].

Whereas the definition via the transformed state has the advantage that a fractional derivative is defined directly for all $\alpha > 0$, Def. 11.2.8 actually reveals the character of $D_s^{\{\alpha\}}f$ as a derivative (in the usual sense). Nevertheless for periodic functions we have chosen Def. 11.5.10 (see also BUTZER–SCHERER [1, p. 101 ff]). The equivalence to a definition via a difference quotient is then given by Prop. 11.5.6 (for $0 < \alpha < 1$). In view of Theorem 11.5.7 the latter definition also works for arbitrary $\alpha > 0$. Indeed, if $f \in \mathsf{W}[\mathsf{X}_{2\pi}; |k|^\alpha]$ and $r \in \mathbb{N}$ is chosen such that $\alpha < r$, then $H^r f * \varphi_{r-\alpha} = f * H^r \varphi_{r-\alpha} = H^r(f * \varphi_{r-\alpha})$. Thus the αth Riesz derivative (in the sense of Def. 11.5.10) is obtained by first integrating f fractionally of order $(r - \alpha)$: $f * \varphi_{r-\alpha}$, followed by the operation $((d/dx)H)^r$.

The integral representations for functions on $\mathbb{R}$ given by Theorems 11.5.1–11.5.3 may be new. In order to obtain their periodic versions we applied the general reduction method of CALDERÓN–ZYGMUND [2] (see Sec. 9.4) to the kernels $|x|^{\alpha-1}$ and $\{\mathrm{sgn}\, x\}|x|^{\alpha-1}$; see also ACHIESER [2, p. 113], ZYGMUND [7I, p. 69]. Though we understood the integral representations mainly as inverse operations to fractional differentiation, we actually defined fractional integration. Thus for functions $f \in \mathsf{X}_{2\pi}$ with $f^\wedge(0) = 0$, the convolution $f * \varphi_\alpha$ ($f * \eta_\alpha$) is the Riesz (Riemann–Liouville–Weyl) fractional integral of order $\alpha > 0$; see also TIMAN [2, p. 118].

The integral representations for periodic functions are well-known and often used to define specific classes of functions. Thus, by $\mathsf{W}^\alpha(\beta)$, $\alpha > 0, \beta$ real, SZ.-NAGY [3, 7],

NIKOLSKIĬ [2], STEČKIN [1] denote the class of $f \in C_{2\pi}$ (with $f^\wedge(0) = 0$) whose representation is $f = \Phi_{\alpha,\beta} * g$, where

$$\Phi_{\alpha,\beta}(x) = 2 \sum_{k=1}^{\infty} \frac{\cos(kx - \beta(\pi/2))}{k^\alpha}$$

and $g^\wedge(0) = 0$, $g \in L_{2\pi}^\infty$. Note that $\Phi_{\alpha,0}(x) = \varphi_\alpha(x)$, $\Phi_{\alpha,\alpha}(x) = \eta_\alpha(x)$, $\Phi_{\alpha,1}(x) = \psi_\alpha(x)$ in our terminology (cf. (11.5.12), (11.5.15), (11.5.18)). The particular classes $W^\alpha(0)$, $W^\alpha(\alpha)$ are of special interest (e.g. the problem of best constants posed by FAVARD [1], see LORENTZ [3, p. 117 ff]). In view of (11.5.19) the Riemann–Liouville–Weyl integral of $f \in X_{2\pi}$, $f^\wedge(0) = 0$, is a linear combination of the Riesz and conjugate Riesz integral of f and vice versa. Correspondingly, for the fractional derivatives in $X_{2\pi}$,

$$f^{(\alpha)} = \cos\frac{\pi\alpha}{2} f^{\{\alpha\}} - \sin\frac{\pi\alpha}{2}(f^\sim)^{\{\alpha\}}, \qquad f^{\{\alpha\}} = \cos\frac{\pi\alpha}{2} f^{(\alpha)} + \sin\frac{\pi\alpha}{2}(f^\sim)^{(\alpha)}.$$

For similar relations for derivatives on $X(\mathbb{R})$ we refer to BUTZER–TREBELS [2] and the literature cited there; see also KOBER [3]; compare furthermore ALJANČIĆ [2], DZJADYK [1], EFIMOV [2, 3], MALOZEMOV [1], OGIEVECKIĬ [1], TELJAKOVSKIĬ [1, 2].

Sec. 11.3–11.4. Integrals of the type (11.3.1) were apparently first considered by MARCHAUD [1] (1927) for continuous functions defined on a finite interval. Theorem 11.4.3 for $C_{2\pi}$-space is due to ZAMANSKY [3] for $\alpha = 1$ and to SUNOUCHI [3] for $0 < \alpha < 1$. For $X_{2\pi}$-space and $0 < \alpha \leq 1$ as well as for a somewhat different generalization for $\alpha > 0$ (cf. Problem 11.4.10) see BUTZER–GÖRLICH [1]; for the present form of Theorem 11.4.6 see SUNOUCHI [8]. The method of proof here is, in all cases, a generalization of one due to SALEM–ZYGMUND [1], modified by ZAMANSKY [3]. In the case of functions defined on the line group these results, particularly Theorem 11.3.3, are given in BUTZER–TREBELS [2]. See also Sec. 13.2.4.

The form of the operation $R_\varepsilon^{\{\alpha\}} f$ for $\varepsilon \to 0+$ suggests that it is connected with the Riesz integral $R_\alpha f$ of (11.1.16). This fractional integral of order α, also known as Riesz potential of order α, was defined for $0 \leq \alpha < 1$ (f sufficiently smooth). If one would ask for a Riesz potential of negative order, one would have to consider

$$[\Lambda_c(-\alpha)]^{-1} \int_{|u| \geq \varepsilon} |u|^{-\alpha-1} f(x-u)\, du \tag{11.6.1}$$

for $\alpha > 0$ and $\varepsilon \to 0+$. Since this expression in general diverges on a set of positive measure, we consider its 'Hadamard finite part' (see e.g. BUREAU [1, 2]) for $0 < \alpha < 2$, namely

$$[R_\varepsilon^{\{\alpha\}} f](x) = [\Lambda_c(-\alpha)]^{-1} \int_{|u| \geq \varepsilon} |u|^{-\alpha-1} [f(x-u) - f(x)]\, du.$$

Thus the limit of $R_\varepsilon^{\{\alpha\}} f$ for $\varepsilon \to 0+$ may be regarded as an extension of the Riesz integral to negative powers, or as a Riesz potential of negative order (compare BALAKRISHNAN [1], DU PLESSIS [1], STEIN [1]); expressed otherwise, the limit of $R_\varepsilon^{\{\alpha\}}$ for $\varepsilon \to 0+$ is the inverse operation to R_α. We have e.g. actually shown for $0 < \alpha < 1$ (see Theorem 11.3.5, 11.5.1): if $g \in L^1$ and $R_\alpha g \in L^1$, then $\text{s-lim}_{\varepsilon \to 0+} R_\varepsilon^{\{\alpha\}}(R_\alpha g) = g$; conversely, if $f \in L^1$ and $\text{s-lim}_{\varepsilon \to 0+} R_\varepsilon^{\{\alpha\}} f$ exists, then $R_\alpha(\text{s-lim}_{\varepsilon \to 0+} R_\varepsilon^{\{\alpha\}} f) = f$. HEYWOOD [1] established an analogous pointwise result for $R_\alpha^\sim f$ in case $\alpha > 0$, $1 < p < 1/\alpha$.

The extension of Riesz potentials to arbitrary negative orders may also be achieved by using Hadamard's finite parts. But we will not regularize the divergent integral (11.6.1) using a Taylor expansion of f (sufficiently smooth; see WHEEDEN [1,2]) but a (central) difference of f of sufficiently high order. The resulting regularization may also be justified by the fact that the operator $R_{\varepsilon,2j}^{\{\alpha\}}$ defined by

$$[R_{\varepsilon,2j}^{\{\alpha\}} f](x) = \frac{1}{C_{\alpha,2j}} \int_\varepsilon^\infty u^{-1-\alpha} \overline{\Delta}_u^{2j} f(x)\, du \qquad (0 < \alpha < 2j;\ \varepsilon > 0)$$

can be considered in the setting of general results due to WESTPHAL [1, 2]; see also BERENS–WESTPHAL [1, 2], BERENS [3]. Here the space L^p, $1 \le p \le 2$, is replaced by a Banach space X, the group of translations by a one-parameter group of uniformly bounded operators $G(\zeta)$ of class (C_0) on X. If the infinitesimal generator A of $G(\zeta)$ is defined by $Af = \text{s-lim}_{\zeta\to 0} \zeta^{-1}[G(\zeta) - I]f$ for all f belonging to the domain $D(A)$ (cf. (13.4.8)), Miss WESTPHAL developed a theory of fractional powers of the generator A. More explicitly, for $\alpha > 0$, $f \in D((-A^2)^{\alpha/2})$ if and only if

$$\text{s-lim}_{\varepsilon\to 0+} \frac{1}{C_{\alpha,2j}} \int_\varepsilon^\infty u^{-1-\alpha}\Big[G\Big(\frac{u}{2}\Big) - G\Big(-\frac{u}{2}\Big)\Big]^{2j} f\,du \tag{11.6.2}$$

exists, $C_{\alpha,2j}$ being defined by (11.3.14). Comparing this general result with ours justifies calling $\text{s-lim}_{\varepsilon\to 0+} R^{\{\alpha\}}_{\varepsilon,2j}f$ a fractional Riesz derivative. In particular, for the translation group on L^p, $1 \le p \le 2$, for which $A = d/dx$, we have shown (Theorem 11.3.7) that the limit (11.6.2) exists if and only if $f \in W[L^p; |v|^\alpha]$, which thus characterizes $D((-d^2/dx^2)^{\alpha/2})$. The operator $(-d^2/dx^2)^{\alpha/2}$ was introduced by BOCHNER [4] (see FELLER [1]), and our results assert (see footnote to Def. 11.2.8) that

$$\Big(-\frac{d^2}{dx^2}\Big)^{\alpha/2} = D_s^{\{\alpha\}} = \Big(\frac{d}{dx}H\Big)^\alpha \qquad (\alpha > 0).$$

Concerning Theorem 11.3.8, the original result of BERNSTEIN [2] (see Problem 11.4.4), also generalized by SZÁSZ [1], may be found in ZYGMUND [7I, p. 240]; for that of TITCHMARSH [3] (Problem 11.3.8) see TITCHMARSH [6, p. 115]; for stronger results, also in the setting of intermediate spaces, see HERZ [1], PEETRE [3], TAIBLESON [1].

Concerning Theorem 11.3.10, the result (1917) of WEYL [1] (see Problem 11.5.3) does not seem to have been considered in the literature apart from BUTZER–TREBELS [2, p. 73], HERZ [1].

For the use of the integrals (11.3.2) in connection with strong asymptotic expansions of singular integrals compare Problems 3.4.6, 8.2.8, 9.2.7, and the literature cited there.

Part V

Saturation Theory

This Part is devoted to a study of saturation theory for convolution integrals. The method to be employed is the Fourier transform method already familiar to us from Chapters 7, 10, 11. Chapter 12 deals with the more classical aspects of the method, thus treating the saturation problem in $X_{2\pi}$ and L^p, $1 \leq p \leq 2$. Chapter 13 gives the extension to all $X(\mathbb{R})$-spaces by duality arguments. Furthermore, the comparison of the error of approximation corresponding to two different processes is studied together with a brief account of saturation theory for strong approximation processes on arbitrary Banach spaces.

12

Saturation for Singular Integrals on $X_{2\pi}$ and L^p, $1 \leq p \leq 2$

12.0 Introduction

The problem of determining the optimal order of approximation of a function f by a sequence of polynomials allows two different interpretations: either one varies the sequence of polynomials that approximates an f satisfying given properties, or, one keeps the approximation process fixed and varies the properties of the function f to obtain the optimal order. In the former case one obtains a result on the best order of approximation $E_n(X_{2\pi}; f)$ for all f satisfying the given properties; but, in general, no information is available concerning the sequence of polynomials for which this optimal approximation is attained. In the latter case the result is that the approximation by the given process will be optimal for all functions belonging to a certain class, the so-called saturation class. The first interpretation is essentially due to P. L. CHEBYCHEFF (1857); the second was posed by J. FAVARD in 1947.

Saturation theory is the study concerned with the critical or optimal order of approximation by a given approximation process. In this respect, let us consider the Fejér means $\sigma_n(f; x)$ of the Fourier series of $f \in C_{2\pi}$. Although these converge uniformly to f as $n \to \infty$ (Cor. 1.2.4), they cannot tend too rapidly to f no matter how smooth f may be. Indeed, $\|\sigma_n(f; \circ) - f(\circ)\|_{C_{2\pi}} = o(n^{-1})$ necessarily implies $f(x) \equiv$ const. Thus for all nonconstant functions $f \in C_{2\pi}$ the $\sigma_n(f; x)$ approximate f with an order of at most $O(n^{-1})$. On the other hand, this critical order is actually attained by particular functions in $C_{2\pi}$. Indeed, for $f_0(x) = \exp\{ix\}$

$$\|\sigma_n(f_0; \circ) - f_0(\circ)\|_{C_{2\pi}} = \left|1 - \frac{1}{n+1} - 1\right| = \frac{1}{n+1}.$$

One says that the Fejér means are saturated in $C_{2\pi}$ with order $O(n^{-1})$.

The problem now is to characterize the class of elements $f \in C_{2\pi}$ for which this optimal order of approximation of f by $\sigma_n(f;x)$, namely $O(n^{-1})$, is in fact attained. For this purpose, we recall Theorem 2.4.7: in the case $0 < \alpha < 1$,

$$\|\sigma_n(f; \circ) - f(\circ)\|_{C_{2\pi}} = O(n^{-\alpha}) \Leftrightarrow f \in \text{Lip}\,(C_{2\pi}; \alpha).$$

But, as we have seen in Sec. 2.4, the methods of proof employed there are not sufficiently powerful to cover the critical case $\alpha = 1$. In fact, we shall show (cf. Problem 12.2.8) that

$$\|\sigma_n(f;\circ) - f(\circ)\|_{C_{2\pi}} = O(n^{-1}) \Leftrightarrow f^{\sim} \in C_{2\pi} \text{ and } (f^{\sim})' \in L^{\infty}_{2\pi}.$$

It may be expected that this particular instance has an analog when $C_{2\pi}$ is replaced by an arbitrary Banach space and $\sigma_n(f; x)$ by a strong approximation process on X satisfying suitable conditions.

Let X be a complex Banach space with norm $\|\circ\|_X$, and let ρ be a positive parameter varying over some set $\mathbb{A} \subset \mathbb{R}$ and tending to ρ_0 (cf. Def. 1.1.1).

Definition 12.0.1. *A family $\{T_\rho\}$, $\rho \in \mathbb{A}$, of bounded linear operators T_ρ of X into X is called a (uniformly bounded)* ***strong approximation process on*** *X if for each $f \in X$*

$$\|T_\rho f\|_X \leq M \|f\|_X, \tag{12.0.1}$$

the constant M being independent of $\rho \in \mathbb{A}$ and $f \in X$, and

$$\lim_{\rho \to \rho_0} \|T_\rho f - f\|_X = 0. \tag{12.0.2}$$

The strong approximation process $\{T_\rho\}$ on X is said to be ***commutative*** *if $T_{\rho_1}(T_{\rho_2} f) = T_{\rho_2}(T_{\rho_1} f)$ for each $\rho_1, \rho_2 \in \mathbb{A}$ and $f \in X$.*

In view of the uniform boundedness principle, condition (12.0.2) alone implies the existence of a constant M such that $\|T_\rho f\|_X \leq M \|f\|_X$ uniformly for all $f \in X$ and all ρ close to ρ_0. Condition (12.0.1) in addition assumes that M is independent of $\rho \in \mathbb{A}$.

Definition 12.0.2. *Let $\{T_\rho\}$, $\rho \in \mathbb{A}$, be a strong approximation process on the Banach space X. We shall say that the process $\{T_\rho\}$ possesses the* ***saturation property*** *if there exists a positive function $\varphi(\rho)$ on $\mathbb{A}$ tending monotonely† to zero as $\rho \to \rho_0$ such that every $f \in X$ for which*

$$\|T_\rho f - f\|_X = o(\varphi(\rho)) \qquad (\rho \to \rho_0) \tag{12.0.3}$$

is an invariant element of $\{T_\rho\}$, i.e. $T_\rho f = f$ for all $\rho \in \mathbb{A}$, and if the set

$$F[X; T_\rho] = \{f \in X \mid \|T_\rho f - f\|_X = O(\varphi(\rho)), \rho \to \rho_0\} \tag{12.0.4}$$

contains at least one noninvariant element. In this event, the approximation process $\{T_\rho\}$ is said to have ***optimal approximation order*** *$O(\varphi(\rho))$ or to be* ***saturated in*** *X* ***with order*** *$O(\varphi(\rho))$, and $F[X; T_\rho]$ is called its* ***Favard*** *or* ***saturation class.***

It should be emphasized that the *saturation problem* actually consists of two different questions: firstly, the question of whether saturation holds, i.e., the establishment of the *existence* of the saturation property of a given strong approximation process $\{T_\rho\}$; secondly, the *characterization* of the Favard class $F[X; T_\rho]$ by structural properties upon f.

† This is not essential; $\varphi(\rho)$ may be any function on $\mathbb{A}$ tending to zero as $\rho \to \rho_0$ (cf. Problems 12.1.4, 12.2.1).

The solution of the latter problem will be referred to as an *equivalence theorem for the Favard class*. (As is to be expected by the example of Fejér means, the methods of proof of an equivalence theorem in the saturation case may be quite different from those employed in the nonoptimal cases.) The complete solution of both problems, thus of the saturation problem, is called a *saturation theorem* for the given approximation process. As in the nonoptimal case of Chapter 2, the *direct part of a saturation theorem* is the direct one of an equivalence theorem for the saturation class. The *inverse part of a saturation theorem* is defined correspondingly.

The first two sections of this chapter are devoted to a study of saturation on the circle group and the next two sections to the line group. Whereas Sec. 12.1 is concerned with the existence and inverse problem of saturation theory for periodic convolution integrals, Sec. 12.2 deals with the direct problem and the associated Favard classes. The submethods presented and the applications thereto can usually be interchanged. Important submethods are those based on positive kernels, uniformly bounded multipliers, and functional equations. Sec. 12.3 treats the theory in L^p, $1 \le p \le 2$, while Sec. 12.4 is reserved to a study of various convolution integrals as illustrations to the theory presented. Although the *general* underlying method of proof of our results in $\mathsf{X}_{2\pi}$ as well as L^p-space is the Fourier transform method, the results for both space types are presented in such a fashion that they are complementary with respect to the set of submethods of proof (each applicable to an important subclass of problems) as well as examples. Sec. 12.5 is concerned with saturation of higher order for both space types.

12.1 Saturation for Periodic Singular Integrals, Inverse Theorems

Let $f \in \mathsf{X}_{2\pi}$ and $I_\rho(f; x)$ be a periodic singular integral as given by (1.1.3) with kernel $\{\chi_\rho(x)\}$. Our first aim is to give a sufficient condition such that the saturation property holds for $I_\rho(f; x)$:

(12.1.1) *Given a kernel* $\{\chi_\rho(x)\}$ *satisfying* (1.1.5), *let there exist a function* $\psi(k)$ *on* $\mathbb{Z}$ *with* $\psi(0) = 0$, $\psi(k) \neq 0$ *for* $k \neq 0$, *and a positive function* $\varphi(\rho)$ *on* $\mathbb{A}$ *tending monotonely to zero as* $\rho \to \rho_0$ *such that*

$$\lim_{\rho \to \rho_0} \frac{\chi_\rho^\wedge(k) - 1}{\varphi(\rho)} = \psi(k) \qquad (k \in \mathbb{Z}).$$

The latter limit-relation holds trivially for $k = 0$ since $\chi_\rho^\wedge(0) = 1$, $\rho \in \mathbb{A}$, by (1.1.1). Furthermore, (12.1.1) implies $\lim_{\rho \to \rho_0} \chi_\rho^\wedge(k) = 1$ for every $k \in \mathbb{Z}$, whereupon (1.1.5) and Problem 1.3.3 guarantee that $\lim_{\rho \to \rho_0} \|I_\rho(f; \circ) - f(\circ)\|_{\mathsf{X}_{2\pi}} = 0$ for every $f \in \mathsf{X}_{2\pi}$.

Proposition 12.1.1. *Let* $f \in \mathsf{X}_{2\pi}$ *and* $\{\chi_\rho(x)\}$ *satisfy* (12.1.1). *If there is* $g \in \mathsf{X}_{2\pi}$ *such that*

$$\text{(12.1.2)} \qquad \lim_{\rho \to \rho_0} \left\| \frac{I_\rho(f; \circ) - f(\circ)}{\varphi(\rho)} - g(\circ) \right\|_{\mathsf{X}_{2\pi}} = 0,$$

then f *belongs to* $\mathsf{W}[\mathsf{X}_{2\pi}; \psi(k)]$.

Proof. We apply the (finite) Fourier transform method as in Sec. 10.1. Indeed, $[I_\rho(f;\circ)]^\frown(k) = \chi_\rho^\frown(k)f^\frown(k)$ by the convolution theorem, and thus by (4.1.2)

$$\left|\frac{\chi_\rho^\frown(k)-1}{\varphi(\rho)}f^\frown(k) - g^\frown(k)\right| \le \left\|\frac{I_\rho(f;\circ)-f(\circ)}{\varphi(\rho)} - g(\circ)\right\|_{X_{2\pi}} \qquad (k \in \mathbb{Z}),$$

which in view of (12.1.1) and (12.1.2) gives $f \in \mathsf{W}[X_{2\pi}; \psi(k)]$.

Proposition 12.1.2. *If $f \in X_{2\pi}$ and $\{\chi_\rho(x)\}$ satisfies* (12.1.1), *then*

$$\|I_\rho(f;\circ) - f(\circ)\|_{X_{2\pi}} = o(\varphi(\rho)) \qquad (\rho \to \rho_0)$$

implies that f is constant. Moreover, the constant functions are the only invariant elements of the singular integral $I_\rho(f; x)$.

Proof. Since (12.1.2) holds with $g(x) = 0$ in $X_{2\pi}$, we conclude $\psi(k)f^\frown(k) = 0$, $k \in \mathbb{Z}$, and since $\psi(k) \ne 0$ for $k \ne 0$, this implies $f^\frown(k) = 0$ for $k \ne 0$. Therefore $f = \text{const}$ by the uniqueness theorem. Since $\chi_\rho^\frown(0) = 1$, $\rho \in \mathbb{A}$, it is obvious that the constant functions are invariant elements of $I_\rho(f; x)$. Conversely, if $f \in X_{2\pi}$ is invariant, then $[\chi_\rho^\frown(k) - 1]f^\frown(k) = 0$, $k \in \mathbb{Z}$, by the convolution theorem. By (12.1.1) it follows that $\psi(k)f^\frown(k) = 0$, $k \in \mathbb{Z}$, and thus $f = \text{const}$ as above.

If the kernel of the singular integral $I_\rho(f; x)$ satisfies (12.1.1), then it is rather obvious that there exist noninvariant elements $f \in X_{2\pi}$ such that actually

(12.1.3) $$\|I_\rho(f;\circ) - f(\circ)\|_{X_{2\pi}} = O(\varphi(\rho)) \qquad (\rho \to \rho_0).$$

For instance, for the function $f_0(x) = \exp\{ix\}$ we have $I_\rho(f_0; x) = \chi_\rho^\frown(1)f_0(x)$, and hence

$$\lim_{\rho\to\rho_0}\left\|\frac{I_\rho(f_0;\circ) - f_0(\circ)}{\varphi(\rho)}\right\|_{X_{2\pi}} = \lim_{\rho\to\rho_0}\left|\frac{\chi_\rho^\frown(1)-1}{\varphi(\rho)}\right| = |\psi(1)|.$$

Therefore we may state

Theorem 12.1.3. *Let $f \in X_{2\pi}$. If the kernel $\{\chi_\rho(x)\}$ of the singular integral $I_\rho(f; x)$ satisfies* (12.1.1), *then $I_\rho(f; x)$ is saturated in $X_{2\pi}$ with order $O(\varphi(\rho))$.*

Hence (12.1.1) is a sufficient condition such that the saturation property holds for the singular integral $I_\rho(f; x)$. On the other hand, there are also certain converse statements.

For the sake of simplicity, let us assume that the kernel $\{\chi_\rho(x)\}$ is even and positive; then the coefficients $\chi_\rho^\frown(k)$ are real numbers, bounded by one (cf. Sec. 1.5.2). Obviously, the assertions of Theorem 12.1.3 remain valid in this case if instead of (12.1.1) we only assume that (with φ and ψ given in (12.1.1), compare also (1.7.2))

(12.1.4) $$\limsup_{\rho\to\rho_0}\frac{1-\chi_\rho^\frown(k)}{\varphi(\rho)} = \psi(k)$$

exists (as a finite number). Concerning the converse, suppose that $I_\rho(f; x)$ is saturated in $X_{2\pi}$ with order $O(\varphi(\rho))$, constants being the only invariant elements. Furthermore, assume that the saturation class contains all trigonometric polynomials. The left side of (12.1.4) defines a certain function $\psi(k)$ on $\mathbb{Z}$ with $\psi(0) = 0$ and $\psi(k) \ge 0$ for $k \ne 0$. In fact, $\psi(k)$ is a finite number for each $k \in \mathbb{Z}$. For, suppose that $\psi(m) = \infty$ for some $m \in \mathbb{Z}$. Since for $f_m(x) = \exp\{imx\}$ (cf. (1.3.14))

$$\left\| \frac{I_\rho(f_m; \circ) - f_m(\circ)}{\varphi(\rho)} \right\|_{X_{2\pi}} = \frac{1 - \chi_\rho^\wedge(m)}{\varphi(\rho)},$$

this would imply that f_m does not belong to the saturation class of $I_\rho(f; x)$, a contradiction to the assumption. Furthermore, $\psi(k) \neq 0$ for $k \neq 0$. For, if there exists $k_0 \neq 0$ such that $\psi(k_0) = 0$, then it follows (as above) that $\|I_\rho(f_0; \circ) - f_0(\circ)\|_{X_{2\pi}} = o(\varphi(\rho))$ for $f_0(x) = \exp\{ik_0x\}$, which is a contradiction to the assumption that the constants are the only invariant elements. Therefore, under the present assumptions, condition (12.1.4) with φ and ψ given in (12.1.1) is not only sufficient but also necessary for the saturation property of $I_\rho(f; x)$ to hold; compare also Sec. 12.5.2, 12.6, and Problems 12.1.4, 12.1.5, 12.2.14.

Having checked that the saturation property holds for every singular integral $I_\rho(f; x)$ with kernel satisfying (12.1.1), we now have to treat the associated problem of characterizing the Favard class. In this respect, the following theorem gives a solution of the inverse part using the Fourier transform method as employed in Sec. 10.2.

Theorem 12.1.4. *If* $f \in X_{2\pi}$ *and* $\{\chi_\rho(x)\}$ *satisfies condition* (12.1.1), *then* (12.1.3) *implies* $f \in V[X_{2\pi}; \psi(k)]$.

Proof. Firstly, let $X_{2\pi}$ be $C_{2\pi}$. By hypothesis there is a constant $M > 0$ such that $\|(\varphi(\rho))^{-1}[I_\rho(f; \circ) - f(\circ)]\|_\infty \leq M$ as $\rho \to \rho_0$. By the weak* compactness theorem for $L_{2\pi}^\infty$ there exists $g \in L_{2\pi}^\infty$ and a sequence $\{\rho_j\}$ with $\lim_{j\to\infty} \rho_j = \rho_0$ such that

$$\lim_{j\to\infty} \frac{1}{2\pi} \int_{-\pi}^{\pi} \frac{I_{\rho_j}(f; x) - f(x)}{\varphi(\rho_j)} e^{-ikx}\, dx = \frac{1}{2\pi} \int_{-\pi}^{\pi} g(x) e^{-ikx}\, dx$$

since $\exp\{-ikx\} \in L_{2\pi}^1$, $k \in \mathbb{Z}$. On the other hand, the left-hand side equals

$$\lim_{j\to\infty} \frac{\chi_{\rho_j}^\wedge(k) - 1}{\varphi(\rho_j)} f^\wedge(k) = \psi(k) f^\wedge(k). \tag{12.1.5}$$

Thus $\psi(k) f^\wedge(k) = g^\wedge(k)$, $k \in \mathbb{Z}$.

For the cases $X_{2\pi} = L_{2\pi}^p$, $1 < p < \infty$, the proof is much the same using the weak* compactness theorem for $L_{2\pi}^p$.

Finally, let $X_{2\pi} = L_{2\pi}^1$. If we set

$$\mu_\rho(x) = \int_{-\pi}^{x} \frac{I_\rho(f; u) - f(u)}{\varphi(\rho)}\, du,$$

then $\{\mu_\rho(x)\}$ is a family of 2π-periodic, absolutely continuous functions, and assumption (12.1.3) means that there is a constant $M > 0$ such that $\|\mu_\rho\|_{BV_{2\pi}} \leq M$ as $\rho \to \rho_0$. We now infer by the theorem of Helly–Bray that there exists $\mu \in BV_{2\pi}$ and a sequence $\{\rho_j\}$ with $\lim_{j\to\infty} \rho_j = \rho_0$ such that

$$\lim_{j\to\infty} \frac{1}{2\pi} \int_{-\pi}^{\pi} e^{-ikx}\, d\mu_{\rho_j}(x) = \frac{1}{2\pi} \int_{-\pi}^{\pi} e^{-ikx}\, d\mu(x)$$

since $\exp\{-ikx\} \in C_{2\pi}$, $k \in \mathbb{Z}$. For the left-hand side we again have (12.1.5) and thus $\psi(k) f^\wedge(k) = \mu^\vee(k)$, $k \in \mathbb{Z}$, which completes the proof.

Thus the Favard class $F[X_{2\pi}; I_\rho]$ of the singular integral $I_\rho(f; x)$ with kernel satisfying (12.1.1) is *contained* in $V[X_{2\pi}; \psi(k)]$. To prove the converse implication (i.e., the direct part of a saturation theorem) which would complete Theorem 12.1.4 to an equivalence theorem, one would need further information upon the kernel. This is postponed to the next section, whilst we here proceed by examining condition (12.1.1) for certain examples.

Let us first consider the singular integral of Fejér as given by (1.2.25). Here the parameter ρ is discrete, namely n, with $\mathbb{A} = \mathbb{N}$ and $\rho_0 = \infty$. Since for the kernel $\{F_n(x)\}$ we have $F_n^\wedge(k) = 1 - (|k|/(n+1))$ for $|k| \le n$ and $F_n^\wedge(k) = 0$ otherwise (cf. (1.2.43)), it follows that

$$\lim_{n\to\infty} \frac{F_n^\wedge(k) - 1}{n^{-1}} = \lim_{n\to\infty} -|k| \frac{n}{n+1} = -|k| \qquad (k \in \mathbb{Z}).$$

Thus $\{F_n(x)\}$ satisfies (12.1.1) with $\varphi(n) = n^{-1}$ and $\psi(k) = -|k|$.

Corollary 12.1.5. *The singular integral of Fejér is saturated in* $\mathsf{X}_{2\pi}$ *with order* $\mathsf{O}(n^{-1})$, *and its Favard class is contained in* $\mathsf{V}[\mathsf{X}_{2\pi}; -|k|]$.

But there are other examples of singular integrals for which (12.1.1) is not so easily evaluated. For instance, for the kernel $\{k_n(x)\}$ of the singular integral of Fejér–Korovkin (cf. Sec. 1.6.1) we have

$$k_n^\wedge(k) = \begin{cases} \left(1 - \dfrac{|k|}{n+2}\right) \cos \dfrac{|k|\,\pi}{n+2} + \dfrac{1}{n+2} \cot \dfrac{\pi}{n+2} \sin \dfrac{|k|\,\pi}{n+2}, & |k| \le n \\ 0, & |k| > n. \end{cases}$$

Certainly, one could show by a Taylor series expansion that this kernel satisfies (12.1.1) with $\varphi(n) = n^{-2}$ and $\psi(k) = -(\pi^2/2)k^2$. On the other hand, if we would exploit the fact that the kernel is positive, we may reduce the verification of (12.1.1) to the particular case $k = 1$ (and $k = 2$). Indeed,

Proposition 12.1.6. *If the kernel* $\{\chi_\rho(x)\}$ *with parameter* $\rho \in \mathbb{A}$ *is even and positive and if* $\varphi(\rho)$ *is a positive function on* $\mathbb{A}$ *tending monotonely to zero as* $\rho \to \rho_0$ *so that*

$$\text{(12.1.6)} \qquad \begin{array}{ll} \text{(i)} & \displaystyle\lim_{\rho\to\rho_0} \frac{\chi_\rho^\wedge(1) - 1}{\varphi(\rho)} = c \ne 0, \\ \text{(ii)} & I_\rho(u^4; 0) = \mathsf{o}(\varphi(\rho)) \qquad (\rho \to \rho_0), \end{array}$$

then the singular integral $I_\rho(f; x)$ *is saturated in* $\mathsf{X}_{2\pi}$ *with order* $\mathsf{O}(\varphi(\rho))$, *and its Favard class is contained in* $\mathsf{V}[\mathsf{X}_{2\pi}; ck^2]$. *In particular,* (12.1.6)(ii) *is satisfied if* (i) *and the condition* (1.5.27), *i.e.*

$$\text{(12.1.7)} \qquad \lim_{\rho\to\rho_0} \frac{\chi_\rho^\wedge(2) - 1}{\chi_\rho^\wedge(1) - 1} = 4,$$

hold.

Proof. We first observe that (12.1.6)(i) implies $\lim_{\rho\to\rho_0} \chi_\rho^\wedge(1) = 1$, which by Prop. 1.3.10 is a necessary and sufficient condition for $\lim_{\rho\to\rho_0} \|I_\rho(f; \circ) - f(\circ)\|_{\mathsf{X}_{2\pi}} = 0$ to be valid for every $f \in \mathsf{X}_{2\pi}$.

Now for $h(x) = k^2(1 - \cos x) - (1 - \cos kx)$, $k \in \mathbb{Z}$, we have (e.g. by the mean value theorem) that $0 \le h(x) \le k^4x^4/4!$. Since the kernel is positive, hypothesis (12.1.6)(ii) implies that $I_\rho(h; 0) = \mathsf{o}(\varphi(\rho))$, $\rho \to \rho_0$. Furthermore, $I_\rho(h; 0) = k^2(1 - \chi_\rho^\wedge(1)) - (1 - \chi_\rho^\wedge(k))$, and thus

$$\frac{\chi_\rho^\wedge(k) - 1}{\varphi(\rho)} = k^2 \frac{\chi_\rho^\wedge(1) - 1}{\varphi(\rho)} + \frac{I_\rho(h; 0)}{\varphi(\rho)}.$$

Hence $\{\chi_\rho(x)\}$ satisfies (12.1.1) with $\psi(k) = c\,k^2$ by (12.1.6), and the conclusion follows by Theorems 12.1.3 and 12.1.4.

Finally, in virtue of (1.5.6) we have

$$\frac{1}{2\pi}\int_{-\pi}^{\pi} u^4\,\chi_\rho(u)\,du \le \frac{\pi^4}{8}(1-\chi_\rho^\wedge(1))\left[4-\frac{\chi_\rho^\wedge(2)-1}{\chi_\rho^\wedge(1)-1}\right],$$

and thus (12.1.6)(ii) follows by (i) and (12.1.7).

Returning to the singular integral of Fejér–Korovkin we have in view of $k_n^\wedge(1) = \cos(\pi/(n+2))$ and (1.6.6) that the kernel $\{k_n(x)\}$ satisfies (12.1.6)(i) and (12.1.7) with $\varphi(n) = n^{-2}$ and $c = -\pi^2/2$. Therefore

Corollary 12.1.7. *The singular integral of Fejér–Korovkin is saturated in* $X_{2\pi}$ *with order* $O(n^{-2})$, *and its Favard class is contained in* $V[X_{2\pi};\,-(\pi^2/2)k^2]$.

Problems

1. Show that Def. 12.0.2 is equivalent to the following: The strong approximation process $\{T_\rho\}$, $\rho\in\mathbb{A}$, on the Banach space X is said to be saturated in X if there exists a positive function $\varphi(\rho)$ on $\mathbb{A}$ tending monotonely to zero as $\rho\to\rho_0$ such that for each $f\in X$ which is not invariant under the given process
$$\|T_\rho f - f\|_X > C_1\varphi(\rho),$$
where $C_1 > 0$ depends only upon f, and if there exists at least one noninvariant element $f_0\in X$ such that
$$\|T_\rho f_0 - f_0\|_X < C_2\varphi(\rho),$$
where $C_2 > 0$ is another constant depending only upon f_0.
2. Prove Theorem 12.1.4 by considering the quantity $\sigma_n((\varphi(\rho))^{-1}[I_\rho(f;\circ) - f(\circ)];x)$. (Hint: Use Cor. 6.1.3; see also the proof of Theorem 12.3.6)
3. Show that Prop. 12.1.1 remains valid if (12.1.2) is replaced by
$$\liminf_{\rho\to\rho_0}\|(\varphi(\rho))^{-1}[I_\rho(f;\circ) - f(\circ)] - g(\circ)\|_{X_{2\pi}} = 0.$$
4. Let the kernel $\{\chi_\rho(x)\}$ of the singular integral $I_\rho(f;x)$ be even and positive.
   (i) Show that $I_\rho(f;x)$ is saturated in $X_{2\pi}$ if and only if (12.1.4) holds, where $\psi(k) > 0$, $k\neq 0$, may be infinite for certain values of k but there exists at least one $m\in\mathbb{N}$ such that $\psi(m) < \infty$. (Hint: As to the sufficiency, compare the proof of Theorem 12.1.3. As to the necessity, let $I_\rho(f;x)$ be saturated with order $O(\varphi(\rho))$, and constants being the only invariant elements. It follows that $\psi(k)\neq 0$ for $k\neq 0$. Furthermore, there exists at least one $m\in\mathbb{N}$ such that $\psi(m) < \infty$; for otherwise, the saturation class does not contain a nonconstant element. Note that in any case $\psi(k)f^\wedge(k)$ is a bounded function on $\mathbb{Z}$ for every $f\in F[X_{2\pi};I_\rho]$)
   (ii) Show that $I_\rho(f;x)$ is saturated in $X_{2\pi}$ if and only if there exists $m\in\mathbb{N}$ such that for all $k\neq 0$
$$\limsup_{\rho\to\rho_0}[1-\chi_\rho^\wedge(k)]/[1-\chi_\rho^\wedge(m)] \ge \psi(k) > 0,$$
   where $\psi(k)$ may be infinite for certain values of k. In this case, the saturation order of $I_\rho(f;x)$ is $O(1-\chi_\rho^\wedge(m))$, $\rho\to\rho_0$.

   Extend to arbitrary kernels $\{\chi_\rho(x)\}$. (Hint: For this Problem see also DE VORE [2])
5. Let $\{\chi_\rho(x)\}$, $\rho\in\mathbb{A}$, be a kernel satisfying (1.1.5). Suppose that there exists $m\in\mathbb{N}$ such that $\chi_\rho^\wedge(k) = 1$ for $|k|\le m$, and that condition (12.1.1) is satisfied for a function ψ with $\psi(k)\neq 0$ for $|k| > m$. Show that the set of invariant elements $f\in X_{2\pi}$ of the singular integral $I_\rho(f;x)$ is given by T_m. Furthermore, show that $I_\rho(f;x)$ is saturated in $X_{2\pi}$ with order $O(\varphi(\rho))$. (Hint: See also TURECKIĬ [3])

6. Let $\{\chi_\rho(x)\}$ satisfy (12.1.1). Show that the kernel $\{(\chi_\rho * \chi_\rho)(x)\}$ satisfies (12.1.1) with $\varphi(\rho)$ and $2\psi(k)$. Thus the strong approximation process $I_\rho^2(f; x)$ given by (1.1.23) is saturated in $X_{2\pi}$ with order $O(\varphi(\rho))$, and has Favard class $F[X_{2\pi}; I_\rho^2]$ contained in $V[X_{2\pi}; 2\psi(k)]$.
7. Give further applications of Prop. 12.1.6 such as to the singular integrals $J_n(f; x)$ of Jackson, $N_n^*(f; x)$ of Jackson–de La Vallée Poussin, etc.
8. Let $\{T_\rho\}$, $\rho \in \mathbb{A}$, be a strong approximation process on the Banach space X and $\varphi(\rho)$ a positive function on $\mathbb{A}$ tending monotonely to zero as $\rho \to \rho_0$. Show that the set
$$X_{\{T_\rho\},\varphi(\rho)} = \{f \in X \mid \|T_\rho f - f\|_X = O(\varphi(\rho)),\ \rho \to \rho_0\}$$
becomes a normalized Banach subspace of X under the norm
$$\|f\|_{\{T_\rho\},\varphi(\rho)} = \|f\|_X + \sup_{\rho \in \mathbb{A}} [\varphi(\rho)]^{-1} \|T_\rho f - f\|_X. \tag{12.1.8}$$
In case $X = X_{2\pi}$, $X = X(\mathbb{R})$ we shall employ the notation $X_{2\pi,\{T_\rho\},\varphi(\rho)}$, $X(\mathbb{R})_{\{T_\rho\},\varphi(\rho)}$, respectively. (Hint: See also Butzer–Berens [1, p. 157 ff], Berens [3, p. 23 ff])

12.2 Favard Classes

In view of Theorem 12.1.3, singular integrals with kernels satisfying (12.1.1) possess the saturation property. By Theorem 12.1.4 the corresponding Favard class is contained in $V[X_{2\pi}; \psi(k)]$. It is the aim of this section to investigate whether $V[X_{2\pi}; \psi(k)]$ precisely makes up the Favard class and to characterize this class by more concrete conditions upon the original functions f. The still missing direct part of a saturation theorem does not permit such a unified treatment as was given by Theorem 12.1.4 for the inverse part. Thus we shall consider the direct problem by different methods according to the diverse properties of the singular integrals in question. Nevertheless, the particular singular integrals to follow need not necessarily be treated by the method given, and the various methods may often be interchanged.

12.2.1 Positive Kernels

Our first method of proving direct theorems is closely connected with the moments of the corresponding kernels. Here we consider the particular case of even and positive kernels for which $\psi(k) = c\, k^2$.

Theorem 12.2.1. *Let $f \in X_{2\pi}$ and $\{\chi_\rho(x)\}$ be an even and positive kernel satisfying (12.1.1) with $\psi(k) = c\, k^2$. Then the singular integral $I_\rho(f; x)$ is saturated in $X_{2\pi}$ with order $O(\varphi(\rho))$, and the Favard class $F[X_{2\pi}; I_\rho]$ is characterized by any one of the following statements:*

(i) $f \in V[X_{2\pi}; c\, k^2]$,

(ii) *there exist* $g \in L_{2\pi}^\infty$ *if* $X_{2\pi} = C_{2\pi}$, $\mu \in BV_{2\pi}$ *if* $X_{2\pi} = L_{2\pi}^1$, $g \in L_{2\pi}^p$ *if* $X_{2\pi} = L_{2\pi}^p$, $1 < p < \infty$, *and constants* a_0, a_1 *such that*
$$f(x) = a_0 + \int_{-\pi}^{x} \left[a_1 + \int_{-\pi}^{u_1} \begin{Bmatrix} g(u_2)\, du_2 \\ d\mu(u_2) \\ g(u_2)\, du_2 \end{Bmatrix} \right] du_1 \quad \text{(a.e.)},$$

(iii) $\|f(\circ + h) + f(\circ - h) - 2f(\circ)\|_{X_{2\pi}} = O(h^2) \qquad (h \to 0).$

The proof is given by a cyclic argument which follows by Theorems 12.1.3, 12.1.4, 10.2.2, 1.5.8, and 4.3.13. Note that there may be added any further characterization of the class $\mathsf{V}[\mathsf{X}_{2\pi}; c\,k^2]$ as given in Sec. 10.2.

Let us mention that in view of Theorem 1.5.8, Prop. 12.1.6, and Problem 12.2.1 only the behaviour of $[\chi_\rho^\wedge(1) - 1]$ and, e.g., of the fourth moment of the kernel determines the saturation property of even, positive singular integrals.

Although from a general point of view the above theorem describes a rather particular situation, it may be applied to such important singular integrals as those of Fejér–Korovkin, de La Vallée Poussin (cf. Sec. 12.2.3), Jackson, and Riemann. But we shall leave all details to the Problems of this section. The singular integral of Rogosinski may also be handled by the same arguments though the corresponding kernel is not positive (cf. Theorem 2.4.9 and Problem 12.2.3).

Here we shall apply Theorem 12.2.1 to the (special) singular integral of Weierstrass as given by (cf. Problem 1.3.10)

$$W_t(f; x) = \frac{1}{2\pi}\int_{-\pi}^{\pi} f(x-u)\theta_3(u, t)\,du \tag{12.2.1}$$

with kernel

$$\theta_3(x, t) = \sum_{k=-\infty}^{\infty} e^{-tk^2} e^{ikx}, \tag{12.2.2}$$

t being a positive parameter tending to zero. Since

$$\lim_{t\to 0+} \frac{\theta_3^\wedge(k, t) - 1}{t} = -k^2 \qquad (k \in \mathbb{Z}), \tag{12.2.3}$$

the kernel $\{\theta_3(x, t)\}$ satisfies (12.1.1) with $\varphi(t) = t$ and $\psi(k) = -k^2$. Thus Theorems 12.1.3 and 12.1.4 apply, and since the kernel is even and positive, so does Theorem 12.2.1, giving a complete saturation theorem for $W_t(f; x)$. Together with Problem 2.5.12 on nonoptimal approximation, this would complete the discussion on the approximation behaviour of the integral $W_t(f; x)$.

As we have seen in Chapter 7, the integral $W_t(f; x)$ of Weierstrass is the unique solution of Fourier's problem of the ring as posed in (7.1.4). At this point we examine the dependence of this solution upon its initial value f and study the rate at which $W_t(f; x)$ approximates f as $t \to 0+$.

Proposition 12.2.2. *Let* $f \in \mathsf{X}_{2\pi}$.

(a) *If there is* $g \in \mathsf{X}_{2\pi}$ *such that* $\liminf_{t\to 0+} \|t^{-1}[W_t(f; \circ) - f(\circ)] - g(\circ)\|_{\mathsf{X}_{2\pi}} = 0$, *then* f *is equal* (a.e.) *to a function in* $\mathsf{AC}_{2\pi}^1$ *with second derivative in* $\mathsf{X}_{2\pi}$. *In particular, if* $\|W_t(f; \circ) - f(\circ)\|_{\mathsf{X}_{2\pi}} = o(t)$, $t \to 0+$, *then* f *is constant* (a.e.).

(b) $\|W_t(f; \circ) - f(\circ)\|_{\mathsf{X}_{2\pi}} = O(t^\alpha) \Leftrightarrow f \in \mathsf{Lip}^*(\mathsf{X}_{2\pi}; 2\alpha)$ $\qquad (0 < \alpha \le 1)$.

This result reveals that if the solution $W_t(f; x)$ tends too rapidly (in $\mathsf{X}_{2\pi}$-norm) towards the initial temperature distribution f, namely with order $o(t)$ for $t \to 0+$, then f must be constant (a.e.). Thus $W_t(f; x)$ approximates f with order of at most $O(t)$; moreover, in case $0 < \alpha \le 1$, the order of approximation is given by $O(t^\alpha)$, $t \to 0+$, if and only if the initial temperature distribution is a $\mathsf{Lip}^*(\mathsf{X}_{2\pi}; 2\alpha)$-function.

12.2.2 Uniformly Bounded Multipliers

Several important singular integrals are not covered by the results of the preceding subsection, so, for example, Fejér's integral. To obtain a general solution of the characterization problem for the Favard class $\mathsf{F}[\mathsf{X}_{2\pi}; I_\rho]$ we shall assume that, apart from (12.1.1), $\{\chi_\rho(x)\}$ satisfies the following condition:

(12.2.4) *Given a kernel* $\{\chi_\rho(x)\}$ *satisfying* (1.1.5), *let there exist* $\psi(k)$ *with* $\psi(0) = 0$, $\psi(k) \neq 0$ *for* $k \neq 0$, *and* $\varphi(\rho) > 0$ *tending monotonely to zero as* $\rho \to \rho_0$ *such that the functions* λ_ρ *defined by*

$$\lambda_\rho(k) = \begin{cases} \dfrac{\chi_\rho^\wedge(k) - 1}{\varphi(\rho)\psi(k)}, & k \neq 0 \\ 1, & k = 0 \end{cases}$$

be multipliers, uniformly bounded with respect to $\rho \in \mathbb{A}$, *of type*

$$\begin{aligned} (\mathsf{L}^\infty_{2\pi}, \mathsf{L}^\infty_{2\pi}) \quad &\text{if} \quad \mathsf{X}_{2\pi} = \mathsf{C}_{2\pi}, \\ (\mathsf{BV}_{2\pi}, \mathsf{BV}_{2\pi}) \quad &\text{if} \quad \mathsf{X}_{2\pi} = \mathsf{L}^1_{2\pi}, \\ (\mathsf{L}^p_{2\pi}, \mathsf{L}^p_{2\pi}) \quad &\text{if} \quad \mathsf{X}_{2\pi} = \mathsf{L}^p_{2\pi}, \quad 1 < p < \infty. \end{aligned}$$

As will be observed, (12.2.4) does not necessarily imply (12.1.1).

Theorem 12.2.3. *Let* $f \in \mathsf{X}_{2\pi}$ *and* $\{\chi_\rho(x)\}$ *satisfy both* (12.1.1) *and* (12.2.4). *Then the integral* $I_\rho(f; x)$ *is saturated in* $\mathsf{X}_{2\pi}$ *with order* $\mathsf{O}(\varphi(\rho))$, *and* $\mathsf{F}[\mathsf{X}_{2\pi}; I_\rho]$ *is characterized by any one of the following statements:*

(i) $f \in \mathsf{V}[\mathsf{X}_{2\pi}; \psi(k)]$,

(ii) $\left\| \sum_{k=-n}^{n} \left(1 - \dfrac{|k|}{n+1}\right)\psi(k)f^\wedge(k)e^{ik\circ} \right\|_{\mathsf{X}_{2\pi}} = \mathsf{O}(1) \qquad (n \to \infty)$,

(iii) $f \in \widetilde{\mathsf{W}[\mathsf{X}_{2\pi}; \psi(k)]}^{\mathsf{X}_{2\pi}}$,

(iv) *there exists* $g \in \mathsf{L}^\infty_{2\pi}$ *if* $\mathsf{X}_{2\pi} = \mathsf{C}_{2\pi}$, $\mu \in \mathsf{BV}_{2\pi}$ *if* $\mathsf{X}_{2\pi} = \mathsf{L}^1_{2\pi}$, *and* $g \in \mathsf{L}^p_{2\pi}$ *if* $\mathsf{X}_{2\pi} = \mathsf{L}^p_{2\pi}$, $1 < p < \infty$, *such that*

$$\lim_{\rho \to \rho_0} \int_{-\pi}^{\pi} s(x) \frac{I_\rho(f; x) - f(x)}{\varphi(\rho)}\, dx = \int_{-\pi}^{\pi} s(x) \begin{Bmatrix} g(x)\, dx \\ d\mu(x) \\ g(x)\, dx \end{Bmatrix}$$

for every $s \in \mathsf{L}^1_{2\pi}$, $s \in \mathsf{C}_{2\pi}$, $s \in \mathsf{L}^{p'}_{2\pi}$, *respectively.*

Proof. In view of Theorems 12.1.3 and 12.1.4 we already know that $I_\rho(f; x)$ is saturated with order $\mathsf{O}(\varphi(\rho))$ and that its Favard class is contained in $\mathsf{V}[\mathsf{X}_{2\pi}; \psi(k)]$. Next we show that conversely $\mathsf{V}[\mathsf{X}_{2\pi}; \psi(k)] \subset \mathsf{F}[\mathsf{X}_{2\pi}; I_\rho]$.

Let $\mathsf{X}_{2\pi} = \mathsf{C}_{2\pi}$ and $\psi(k)f^\wedge(k) = g^\wedge(k)$, $g \in \mathsf{L}^\infty_{2\pi}$. Then by the convolution theorem and (12.2.4) we have for $k \in \mathbb{Z}$

$$\left[\frac{I_\rho(f; \circ) - f(\circ)}{\varphi(\rho)}\right]^\wedge(k) = \lambda_\rho(k)g^\wedge(k) \equiv [U_\rho g]^\wedge(k),$$

thus defining a closed linear operator U_ρ of $\mathsf{L}^\infty_{2\pi}$ into $\mathsf{L}^\infty_{2\pi}$ (cf. Prop. 6.5.1). By the uniqueness theorem it follows that for each $\rho \in \mathbb{A}$

$$\frac{I_\rho(f; x) - f(x)}{\varphi(\rho)} = [U_\rho g](x) \quad \text{a.e.} \tag{12.2.5}$$

Since the left-hand side is continuous and the operators U_ρ are uniformly bounded,

$$\left\| \frac{I_\rho(f;\circ) - f(\circ)}{\varphi(\rho)} \right\|_{C_{2\pi}} = \|U_\rho g\|_{L^\infty_{2\pi}} \le \|U_\rho\|_{[L^\infty_{2\pi}, L^\infty_{2\pi}]} \|g\|_{L^\infty_{2\pi}} = O(1).$$

If $X_{2\pi} = L^p_{2\pi}$, $1 < p < \infty$, it follows as above that (12.2.5) holds, and hence by (12.2.4)

$$\left\| \frac{I_\rho(f;\circ) - f(\circ)}{\varphi(\rho)} \right\|_{p \atop 2\pi} \le \|U_\rho\|_{[L^p_{2\pi}, L^p_{2\pi}]} \|g\|_{L^p_{2\pi}} = O(1).$$

If finally $X_{2\pi} = L^1_{2\pi}$ and $\psi(k)f^\wedge(k) = \mu^\vee(k)$, $\mu \in BV_{2\pi}$, let U_ρ be the closed linear operator of $BV_{2\pi}$ into $BV_{2\pi}$ defined by $[U_\rho\mu]^\vee(k) = \lambda_\rho(k)\mu^\vee(k)$. Then by Prop. 4.3.1(iii)

$$\left[\int_{-\pi}^{\circ} \frac{I_\rho(f;u) - f(u)}{\varphi(\rho)}\,du\right]^\vee(k) = \frac{\chi_\rho^\wedge(k) - 1}{\varphi(\rho)} f^\wedge(k) = [U_\rho\mu]^\vee(k),$$

and thus by the uniqueness theorem for Fourier–Stieltjes transforms

$$\int_{-\pi}^{x} \frac{I_\rho(f;u) - f(u)}{\varphi(\rho)}\,du = [U_\rho\mu](x) + \text{const}$$

for all $x \in \mathbb{R}$, $\rho \in \mathbb{A}$. Therefore

$$\left\| \frac{I_\rho(f;\circ) - f(\circ)}{\varphi(\rho)} \right\|_{L^1_{2\pi}} = \|U_\rho\mu\|_{BV_{2\pi}} \le \|U_\rho\|_{[BV_{2\pi}, BV_{2\pi}]} \|\mu\|_{BV_{2\pi}} = O(1).$$

Thus $f \in F[X_{2\pi}; I_\rho]$ if and only if $f \in V[X_{2\pi}; \psi(k)]$.

For the proof of the equivalences (i) $\Leftrightarrow$ (ii) $\Leftrightarrow$ (iii) we refer to Cor. 6.1.3 and Theorem 10.4.6, that of (i) $\Leftrightarrow$ (iv) is left to Problem 12.2.7.

So far we have shown that conditions (12.1.1) and (12.2.4) are *sufficient* for the solution of the saturation problem of the singular integral $I_\rho(f; x)$. In fact, by Theorem 12.1.3 condition (12.1.1) is sufficient for the existence of the saturation property, whereas it follows by Theorem 12.2.3 that (12.1.1) and (12.2.4) are sufficient for the proof of an equivalence theorem for the Favard class $F[X_{2\pi}; I_\rho]$. Now, the following result shows that (at least for $\psi(k) = |k|^\alpha$) condition (12.2.4) is also *necessary* for the validity of the direct part of the saturation theorem.

Proposition 12.2.4. *Let $f \in X_{2\pi}$, $\{\chi_\rho(x)\}$ be a kernel satisfying* (1.1.5) *and $\varphi(\rho)$ a positive function on $\mathbb{A}$ tending monotonely to zero as $\rho \to \rho_0$. If*

$$V[X_{2\pi}; |k|^\alpha] \subset \{f \in X_{2\pi} \mid \|I_\rho(f;\circ) - f(\circ)\|_{X_{2\pi}} = O(\varphi(\rho))\}$$

for some $\alpha > 0$, then necessarily condition (12.2.4) *holds with $\psi(k) = |k|^\alpha$.*

Proof. Let $X_{2\pi} = C_{2\pi}$. We define an operator S of $V[C_{2\pi}; |k|^\alpha]$ into $L^\infty_{2\pi}$ through $Sf = g$ where $g \in L^\infty_{2\pi}$ is such that $|k|^\alpha f^\wedge(k) = g^\wedge(k)$. Then $g \in L^\infty_{2\pi}$ belongs to the range of S, i.e. $g \in S(V[C_{2\pi}; |k|^\alpha])$ if and only if $g^\wedge(0) = 0$. Indeed, let $g \in L^\infty_{2\pi}$ be such that $g^\wedge(0) = 0$ and φ_α be given by (11.5.12). Then $f_g \equiv \varphi_\alpha * g$ belongs to $C_{2\pi}$ and $f_g^\wedge(k) = |k|^{-\alpha}g^\wedge(k)$ for $k \ne 0$, $f_g^\wedge(0) = 0$. Therefore $|k|^\alpha f_g^\wedge(k) = g^\wedge(k)$ for $k \in \mathbb{Z}$, and thus $g \in S(V[C_{2\pi}; |k|^\alpha])$. The opposite direction follows trivially. Hence, if we set $L^{\infty;0}_{2\pi} = \{g \in L^\infty_{2\pi} \mid g^\wedge(0) = 0\}$, then S maps $V[C_{2\pi}; |k|^\alpha]$ *onto* $L^{\infty;0}_{2\pi}$.

Let $g \in L^\infty_{2\pi}$ be arbitrary. Setting $g(x) = g^\wedge(0) + g_1(x)$, we have $g_1 \in L^{\infty;0}_{2\pi}$. Therefore there exists $f_{g_1} \in C_{2\pi}$ such that $|k|^\alpha f_{g_1}^\wedge(k) = g_1^\wedge(k)$. Consequently, the identity

$$\lambda_\rho(k)g^\wedge(k) = [(\varphi(\rho))^{-1}\{I_\rho(f_{g_1};\circ) - f_{g_1}(\circ)\} + g^\wedge(0)]^\wedge(k)$$

holds for every $k \in \mathbb{Z}$, where $\lambda_\rho(k)$ is defined as in (12.2.4) with $\psi(k) = |k|^\alpha$. In other words, we have for every $g \in \mathsf{L}^\infty_{2\pi}$ that $\lambda_\rho(k)g^\wedge(k)$ is the Fourier transform of a function in $\mathsf{L}^\infty_{2\pi}$, namely of $[(\varphi(\rho))^{-1}\{I_\rho(f_{g_1}; x) - f_{g_1}(x)\} + g^\wedge(0)]$, which is actually an element of $\mathsf{C}_{2\pi}$. Thus $\lambda_\rho(k)$ defines a multiplier U_ρ of type $(\mathsf{L}^\infty_{2\pi}, \mathsf{L}^\infty_{2\pi})$ for each $\rho \in \mathbb{A}$, and since $f_{g_1} \in \mathsf{V}[\mathsf{C}_{2\pi}; |k|^\alpha]$ implies $\|I_\rho(f_{g_1}; \circ) - f_{g_1}(\circ)\|_{\mathsf{C}_{2\pi}} = O(\varphi(\rho))$ by hypothesis, there exists a constant M_g such that $\|U_\rho g\|_{\mathsf{L}^\infty_{2\pi}} \le M_g$. The uniform boundedness principle then shows that the multipliers are uniformly bounded for $\rho \in \mathbb{A}$, giving the proof in $\mathsf{C}_{2\pi}$. Analogously, the cases $\mathsf{X}_{2\pi} = \mathsf{L}^p_{2\pi}$ are treated.

The critical point concerning the application of Theorem 12.2.3 is the verification of condition (12.2.4).† Firstly, we observe that, in view of Problem 6.5.1, λ_ρ is a multiplier of type $(\mathsf{BV}_{2\pi}, \mathsf{BV}_{2\pi})$ if and only if it is of type $(\mathsf{L}^\infty_{2\pi}, \mathsf{L}^\infty_{2\pi})$, and since by Problem 6.5.2 a multiplier of type $(\mathsf{L}^\infty_{2\pi}, \mathsf{L}^\infty_{2\pi})$ is also of type $(\mathsf{L}^p_{2\pi}, \mathsf{L}^p_{2\pi})$, it suffices to prove that λ_ρ is a multiplier of type $(\mathsf{L}^\infty_{2\pi}, \mathsf{L}^\infty_{2\pi})$, bounded uniformly with respect to $\rho \in \mathbb{A}$. Moreover (cf. Prop. 6.5.3), a necessary and sufficient condition upon a function λ_ρ on $\mathbb{Z}$ to be a multiplier of type $(\mathsf{L}^\infty_{2\pi}, \mathsf{L}^\infty_{2\pi})$ is that λ_ρ be the Fourier–Stieltjes transform of a function $\nu_\rho \in \mathsf{BV}_{2\pi}$.

To verify that λ_ρ is the Fourier–Stieltjes transform of some $\nu_\rho \in \mathsf{BV}_{2\pi}$ with $\|\nu_\rho\|_{\mathsf{BV}_{2\pi}} = O(1)$, $\rho \in \mathbb{A}$, we may use any of the criteria given in Chapter 6, thus e.g. Theorem 6.1.2 or Cor. 6.3.9 on uniformly quasi-convex functions on $\mathbb{Z}$.

But for many singular integrals $\lambda_\rho(k)$ is representable as $\lambda(k/\rho)$ and then the following property turns out to be the most convenient sufficient condition for (12.2.4) to be valid:

(12.2.6) *Given a kernel* $\{\chi_\rho(x)\}$, $\rho > 0$, $\rho_0 = \infty$, *satisfying* (1.1.5), *let there exist* $\psi(k)$ *with* $\psi(0) = 0$, $\psi(k) \neq 0$ *for* $k \neq 0$, $\varphi(\rho) > 0$ *tending monotonely to zero as* $\rho \to \infty$, *and* $h \in \mathsf{NL}^1$ *such that the representation*

$$\frac{\chi_\rho^\wedge(k) - 1}{\varphi(\rho)} = \psi(k)h^\wedge\Big(\frac{k}{\rho}\Big)$$

holds for every $k \in \mathbb{Z}$, $\rho > 0$.

Theorem 12.2.5. *Let* $f \in \mathsf{X}_{2\pi}$ *and* $\{\chi_\rho(x)\}$ *satisfy* (12.2.6). *Then* $I_\rho(f; x)$ *is saturated in* $\mathsf{X}_{2\pi}$ *with order* $O(\varphi(\rho))$, *and* $\mathsf{F}[\mathsf{X}_{2\pi}; I_\rho]$ *is characterized by any of the four statements of Theorem* 12.2.3.

Proof. Since h belongs to NL^1 and $h^\wedge$ is continuous, $\lim_{\rho\to\infty} (\chi_\rho^\wedge(k) - 1)/\varphi(\rho) = \psi(k)h^\wedge(0) = \psi(k)$ for each $k \in \mathbb{Z}$, and thus (12.1.1) is satisfied. Moreover, (12.2.6) implies (12.2.4). Indeed, by (3.1.29) and (5.1.58) the function h_ρ^* as defined by (cf. (6.3.17))

$$h_\rho^*(x) = \rho\sqrt{2\pi} \sum_{k=-\infty}^{\infty} h(\rho(x + 2k\pi)) \tag{12.2.7}$$

belongs to $\mathsf{L}^1_{2\pi}$ such that $\|h_\rho^*\|_{\mathsf{L}^1_{2\pi}} \le \|h\|_{\mathsf{L}^1}$ and $[h_\rho^*]^\wedge(k) = h^\wedge(k/\rho) = \lambda_\rho(k)$, $k \in \mathbb{Z}$, $\rho > 0$. Thus (12.2.4) follows by Problems 6.5.1, 6.5.2 (cf. also Prop. 6.5.3). Theorem 12.2.3 then completes the proof.

† In the particular instance of the Hilbert space $\mathsf{L}^2_{2\pi}$ condition (12.2.4) is satisfied (cf. Prop. 6.5.2) if and only if there exists a constant M such that $|\lambda_\rho(k)| \le M$ for all $k \in \mathbb{Z}$, $\rho \in \mathbb{A}$ (cf. Problem 12.2.4). This is easily verified in the applications.

Note that (12.2.6) in particular implies that the operators U_ρ occurring in the proof of Theorem 12.2.3 (cf. (12.2.5)) are convolution operators (cf. (6.5.2)). This enables us to prove

Theorem 12.2.6. *Let $f \in X_{2\pi}$ and $\{\chi_\rho(x)\}$ satisfy (12.2.6). The following assertions are equivalent:*

(i) *There exists $g \in X_{2\pi}$ such that*

$$\lim_{\rho\to\infty} \left\| \frac{I_\rho(f;\circ) - f(\circ)}{\varphi(\rho)} - g(\circ) \right\|_{X_{2\pi}} = 0,$$

(ii) $f \in W[X_{2\pi}; \psi(k)]$,
(iii) *there exists $g \in X_{2\pi}$ such that for each $\rho > 0$*

$$\frac{I_\rho(f;x) - f(x)}{\varphi(\rho)} = \frac{1}{2\pi}\int_{-\pi}^{\pi} g(x-u)h_\rho^*(u)\,du \quad \text{a.e.},$$

where the kernel $\{h_\rho^*(x)\}$ is given by (12.2.7).

Proof. Since (12.2.6) implies (12.1.1), the proof of the implication (i) $\Rightarrow$ (ii) follows by Prop. 12.1.1. Let $g \in X_{2\pi}$ with $\psi(k)f^\wedge(k) = g^\wedge(k)$. If h_ρ^* is defined by (12.2.7), then by the convolution theorem

$$\left[\frac{I_\rho(f;\circ) - f(\circ)}{\varphi(\rho)}\right]^\wedge(k) = h^\wedge\left(\frac{k}{\rho}\right)g^\wedge(k) = [g*h_\rho^*]^\wedge(k),$$

and thus for each $\rho > 0$

(12.2.8)
$$\frac{I_\rho(f;x) - f(x)}{\varphi(\rho)} = \frac{1}{2\pi}\int_{-\pi}^{\pi} g(x-u)h_\rho^*(u)\,du.$$

Moreover, since $h \in NL^1$, $\{\rho h(\rho x)\}$ is an approximate identity on $\mathbb{R}$ in the sense of Def. 3.1.4 which implies that $\{h_\rho^*(x)\}$ is a periodic approximate identity (cf. Sec. 3.1.2). Therefore by Theorem 1.1.5, $\lim_{\rho\to\infty} \|(g*h_\rho^*) - g\|_{X_{2\pi}} = 0$ for every $g \in X_{2\pi}$, which establishes (i).

Regarding the verification of condition (12.2.6), the notion of quasi-convexity of functions *on the real line* (cf. Def. 6.3.5) and in particular Cor. 6.3.12 turn out to be most useful.

As an example, let us consider the typical means of the Fourier series of $f \in X_{2\pi}$ as introduced in Sec. 6.4.2. and given by the singular integral

(12.2.9)
$$R_{n,\kappa}(f;x) = \frac{1}{2\pi}\int_{-\pi}^{\pi} f(x-u)r_{n,\kappa}(u)\,du$$

with kernel

(12.2.10)
$$r_{n,\kappa}(x) = \sum_{k=-n}^{n}\left(1 - \left|\frac{k}{n+1}\right|^\kappa\right)e^{ikx} \qquad (\kappa > 0).$$

This kernel satisfies (12.1.1) with $\varphi(n) = (n+1)^{-\kappa}$ and $\psi(k) = -|k|^\kappa$ since

(12.2.11)
$$\lim_{n\to\infty}\frac{r_{n,\kappa}^\wedge(k) - 1}{(n+1)^{-\kappa}} = -|k|^\kappa \qquad (k \in \mathbb{Z}).$$

Furthermore, (12.2.6) is satisfied, for if we set

$$\eta\Big(\frac{|k|}{n+1}\Big) = \begin{cases} -\dfrac{r^{\wedge}_{n,\kappa}(k) - 1}{(n+1)^{-\kappa}\,|k|^{\kappa}}, & k \neq 0, \\ 1, & k = 0, \end{cases} \qquad \eta(x) = \begin{cases} 1, & 0 \le |x| \le 1 \\ |x|^{-\kappa}, & 1 \le |x|, \end{cases}$$

then η and η' are locally absolutely continuous on $(0, \infty)$ except at $x_1 = 1$ where η' has a jump: $\eta'(1+) - \eta'(1-) = -\kappa$. Furthermore,

$$\int_0^\infty u\,|\eta''(u)|\,du = \int_1^\infty \kappa(\kappa+1)x^{-\kappa-1}\,dx = \kappa + 1,$$

i.e. η is an even and quasi-convex function of C_0. Thus by Theorem 6.3.11 η is representable as the Fourier transform of an even function $h \in \mathsf{L}^1$. Therefore

Proposition 12.2.7. *The typical means $R_{n,\kappa}(f; x)$ of the Fourier series of $f \in \mathsf{X}_{2\pi}$ are saturated in $\mathsf{X}_{2\pi}$ with order $\mathsf{O}(n^{-\kappa})$. The Favard class $\mathsf{F}[\mathsf{X}_{2\pi}; R_{n,\kappa}]$ is characterized by any of the four statements of Theorem* 12.2.3 *with $\psi(k) = -|k|^{\kappa}$ as well as by the following:*

(v) $$\Big\| \int_\varepsilon^\infty \frac{\overline{\Delta}_u^{2j} f(\circ)}{u^{1+\kappa}}\,du \Big\|_{\mathsf{X}_{2\pi}} = \mathsf{O}(1) \qquad (\varepsilon \to 0+),$$

where $j \in \mathbb{N}$ is chosen so that $0 < \kappa < 2j$,

(vi) *there exists $g \in \mathsf{L}^\infty_{2\pi}$ for $\mathsf{X}_{2\pi} = \mathsf{C}_{2\pi}$, $\mu \in \mathsf{BV}_{2\pi}$ for $\mathsf{X}_{2\pi} = \mathsf{L}^1_{2\pi}$, and $g \in \mathsf{L}^p_{2\pi}$ for $\mathsf{X}_{2\pi} = \mathsf{L}^p_{2\pi}$, $1 < p < \infty$, such that $g^\wedge(0) = 0$, $\mu^\vee(0) = 0$, and f is representable as*

$$f(x) = f^\wedge(0) + \begin{cases} (\varphi_\kappa * g)(x) \\ (\varphi_\kappa * d\mu)(x) \quad \text{(a.e.)}, \\ (\varphi_\kappa * g)(x) \end{cases}$$

respectively, where φ_κ is given by (11.5.12),

(vii) *$f \in \mathsf{L}^q_{2\pi}$ for some q with $1 < q < \infty$ and $H^r f * \varphi_{r-\kappa}$ belongs to $\mathsf{V}^r_{\mathsf{X}_{2\pi}}$, where $r \in \mathbb{N}$ is such that $\kappa < r$ and H denotes the operation of taking Hilbert transforms.*

For the proof we cite Theorems 12.2.3, 12.2.5, 11.4.6, 11.5.5, and 11.5.9.

If $\kappa = 1$, $R_{n,\kappa}(f; x)$ reduces to Fejér's singular integral $\sigma_n(f; x)$ which is thus saturated in $\mathsf{X}_{2\pi}$ with order $\mathsf{O}(n^{-1})$. Its Favard class $\mathsf{F}[\mathsf{X}_{2\pi}; \sigma_n]$ is described by $\mathsf{V}[\mathsf{X}_{2\pi}; -|k|]$. Moreover, $\mathsf{F}[\mathsf{X}_{2\pi}; \sigma_n]$ may also be characterized, apart from the seven assertions of Prop. 12.2.7 for $\kappa = 1$, in terms of assertions concerning the class $\mathsf{V}[\mathsf{X}_{2\pi}; -|k|]$ given by Sec. 10.3 (cf. Problem 12.2.8).

The typical means are the particular case $\lambda = 1$ of the Riesz means $R_{n,\kappa,\lambda}(f; x)$ of the Fourier series of $f \in \mathsf{X}_{2\pi}$ for which we refer to Sec. 12.5.

It is left to the reader to give further examples to Theorem 12.2.5. Generally, any of the conditions of Chapter 6 which is sufficient for the representation as the Fourier transform of an L^1-function may be used to examine (12.2.6), and thus to provide exact characterizations of Favard classes. Moreover, it is sufficient that (12.2.6) holds with $h^\wedge$ replaced by the Fourier–Stieltjes transform of a function $\nu \in \mathsf{NBV}$.

Finally, note that of course also those singular integrals for which ρ_0 is finite are covered by Theorem 6.3.11 (cf. Problem 12.2.5).

12.2.3 Functional Equations

Thus far condition (12.2.6) has been an essential tool for characterizing the Favard class of periodic singular integrals. It is actually a condition upon the kernel $\{\chi_\rho(x)\}$ of the singular integral which is expressed by the equation (12.2.8) for the subclass of functions $f \in \mathsf{W}[\mathsf{X}_{2\pi}; \psi(k)]$. On the other hand, the difference $[I_\rho(f; x) - f(x)]$

may also be determined via certain functional equations which are valid for *all* $f \in \mathsf{X}_{2\pi}$ and arise from the specific structure of the process in question.

As an example we consider the general singular integral of Weierstrass of Sec. 6.4.1. It is defined by

$$W_{t,\kappa}(f; x) = \frac{1}{2\pi}\int_{-\pi}^{\pi} f(x-u)w_{t,\kappa}(u)\,du \tag{12.2.12}$$

with kernel

$$w_{t,\kappa}(x) = \sum_{k=-\infty}^{\infty} e^{-t|k|^\kappa} e^{ikx} \qquad (\kappa > 0). \tag{12.2.13}$$

The kernel $\{w_{t,\kappa}(x)\}$ satisfies (12.1.1) with $\varphi(t) = t$ and $\psi(k) = -|k|^\kappa$ since

$$\lim_{t\to 0+} \frac{w_{t,\kappa}^\wedge(k) - 1}{t} = -|k|^\kappa \qquad (k \in \mathbb{Z}). \tag{12.2.14}$$

Thus Theorems 12.1.3 and 12.1.4 apply.

For the direct part of the saturation theorem, let $f \in \mathsf{V}[\mathsf{C}_{2\pi}; -|k|^\kappa]$ and $g \in \mathsf{L}_{2\pi}^\infty$ such that $-|k|^\kappa f^\wedge(k) = g^\wedge(k)$. We have for $k \in \mathbb{Z}$

$$[W_{t,\kappa}(f;\circ) - f(\circ)]^\wedge(k) = (e^{-t|k|^\kappa} - 1)f^\wedge(k) = \int_0^t \{e^{-\tau|k|^\kappa} g^\wedge(k)\}\,d\tau$$
$$= \int_0^t [W_{\tau,\kappa}(g;\circ)]^\wedge(k)\,d\tau = \left[\int_0^t W_{\tau,\kappa}(g;\circ)\,d\tau\right]^\wedge(k).$$

The change of the order of integration in the last equality is justified by Fubini's theorem since $\|W_{\tau,\kappa}(g;\circ)\|_\infty \le M_\kappa \|g\|_\infty$ uniformly for all $\tau > 0$, and thus

$$\int_0^t \left\{\frac{1}{2\pi}\int_{-\pi}^{\pi} |W_{\tau,\kappa}(g;u)|\,du\right\} d\tau \le M_\kappa \|g\|_\infty t.$$

Therefore, by the uniqueness theorem and the continuity of all expressions involved,

$$W_{t,\kappa}(f;x) - f(x) = \int_0^t W_{\tau,\kappa}(g;x)\,d\tau \tag{12.2.15}$$

for all $t > 0$ and $x \in \mathbb{R}$. If $f \in \mathsf{V}[\mathsf{L}_{2\pi}^p; -|k|^\kappa]$, $1 < p < \infty$, then for each $t > 0$ (12.2.15) holds almost everywhere. Finally, if $f \in [\mathsf{L}_{2\pi}^1; -|k|^\kappa]$ and $-|k|^\kappa f^\wedge(k) = \mu^\vee(k)$, $\mu \in \mathsf{BV}_{2\pi}$, then for each $t > 0$

$$W_{t,\kappa}(f;x) - f(x) = \int_0^t W_{\tau,\kappa}(d\mu;x)\,d\tau \quad \text{a.e.} \tag{12.2.16}$$

Thus $f \in \mathsf{V}[\mathsf{X}_{2\pi}; -|k|^\kappa]$ implies $\|W_{t,\kappa}(f;\circ) - f(\circ)\|_{\mathsf{X}_{2\pi}} = \mathsf{O}(t)$, and we may state

Proposition 12.2.8. *The general singular integral $W_{t,\kappa}(f;x)$ of Weierstrass is saturated in $\mathsf{X}_{2\pi}$ with order $\mathsf{O}(t)$, $t \to 0+$, and the Favard class $\mathsf{F}[\mathsf{X}_{2\pi}; W_{t,\kappa}]$ is characterized by any one of the seven statements of Prop. 12.2.7. Moreover, $f \in \mathsf{W}[\mathsf{X}_{2\pi}; -|k|^\kappa]$ if and only if there exists $g \in \mathsf{X}_{2\pi}$ such that*

$$\lim_{t\to 0+} \left\| \frac{W_{t,\kappa}(f;\circ) - f(\circ)}{t} - g(\circ)\right\|_{\mathsf{X}_{2\pi}} = 0. \tag{12.2.17}$$

For the proof of the latter assertion we note that if $f \in \mathsf{W}[\mathsf{X}_{2\pi}; -|k|^{\kappa}]$ then (12.2.15) holds, and therefore

$$\lim_{t\to 0+} \left\| \frac{W_{t,\kappa}(f;\circ) - f(\circ)}{t} - g(\circ) \right\|_{\mathsf{X}_{2\pi}} = \lim_{t\to 0+} \left\| \frac{1}{t} \int_0^t [W_{\tau,\kappa}(g;\circ) - g(\circ)]\, d\tau \right\|_{\mathsf{X}_{2\pi}} = 0$$

in view of the Hölder–Minkowski inequality.

For further characterizations in the particular cases $\kappa = 1$ and $\kappa = 2$ we refer to Sec. 12.2.1 and Problem 12.2.13, respectively.

The above proof of the implication $\mathsf{V}[\mathsf{X}_{2\pi}; -|k|^{\kappa}] \subset \mathsf{F}[\mathsf{X}_{2\pi}; W_{t,\kappa}]$ actually makes use of the so-called semi-group property of the general singular integral of Weierstrass. This means that the operators $W_{t,\kappa}$ satisfy the functional equation (in the parameter t)

$$W_{t_1,\kappa}\{W_{t_2,\kappa}f\} = W_{t_1+t_2,\kappa}f \qquad (t_1, t_2 > 0). \tag{12.2.18}$$

This is easily established since

$$[W_{t,\kappa}f](x) = \sum_{k=-\infty}^{\infty} e^{-t|k|^{\kappa}} f^{\wedge}(k)\, e^{ikx} \tag{12.2.19}$$

for all $f \in \mathsf{X}_{2\pi}$ and $t > 0$, and therefore

$$[W_{t_1,\kappa}\{W_{t_2,\kappa}f\}](x) = \sum_{k=-\infty}^{\infty} e^{-t_1|k|^{\kappa}}\{e^{-t_2|k|^{\kappa}} f^{\wedge}(k)\}\, e^{ikx} = [W_{t_1+t_2,\kappa}f](x).$$

For details we refer to Sec. 13.4.2 where we shall give a short account of approximation- and saturation-theoretical questions for semi-groups of operators on a Banach space.

Apart from the semi-group property, other types of functional equations may be useful to establish characterizations of Favard classes. As an example we consider the singular integral of de La Vallée Poussin defined by (cf. Sec. 2.5.2)

$$V_n(f;x) = \frac{1}{2\pi} \int_{-\pi}^{\pi} f(x-u) v_n(u)\, du \tag{12.2.20}$$

with kernel

$$v_n(x) = \frac{(n!)^2}{(2n)!} \left(2\cos\frac{x}{2}\right)^{2n} = \sum_{k=-n}^{n} \frac{(n!)^2}{(n-|k|)!(n+|k|)!} e^{ikx}. \tag{12.2.21}$$

Condition (12.1.1) is satisfied with $\varphi(n) = n^{-1}$ and $\psi(k) = -k^2$ since

$$\lim_{n\to\infty} \frac{v_n^{\wedge}(k) - 1}{n^{-1}} = -k^2, \tag{12.2.22}$$

and thus Theorems 12.1.3 and 12.1.4 apply.

But for the characterization of the Favard class, Theorem 12.2.5 is not applicable since $v_n^{\wedge}(k)$ does not permit a representation such as $v_n^{\wedge}(k) = \lambda(k/n)$. On the other hand, the kernel is even and positive, and since $\psi(k) = -k^2$, Theorem 12.2.1 may be used. But here we prefer to proceed as follows.

We need the equation (cf. (2.5.19))

$$k^2 v_n^{\wedge}(k) = n^2[v_n^{\wedge}(k) - v_{n-1}^{\wedge}(k)] \qquad (n \in \mathbb{N}, k \in \mathbb{Z}) \tag{12.2.23}$$

which follows by direct computation. Let $f \in \mathsf{V}[\mathsf{C}_{2\pi}; -k^2]$ and $-k^2 f^\wedge(k) = g^\wedge(k)$, $g \in \mathsf{L}^\infty_{2\pi}$. Then for $m > n$ and $k \in \mathbb{Z}$

$$[V_n(f;\circ) - V_m(f;\circ)]^\wedge(k) = -\sum_{j=n+1}^{m} \{v_j^\wedge(k) - v_{j-1}^\wedge(k)\} f^\wedge(k)$$

$$= \sum_{j=n+1}^{m} \frac{1}{j^2} v_j^\wedge(k) g^\wedge(k) = \Big[\sum_{j=n+1}^{m} \frac{1}{j^2} V_j(g;\circ)\Big]^\wedge(k),$$

and therefore by the uniqueness theorem

$$V_n(f;x) - V_m(f;x) = \sum_{j=n+1}^{m} \frac{1}{j^2} V_j(g;x).$$

Letting $m \to \infty$ we have in view of (2.5.15) that for all $n \in \mathbb{N}$ and $x \in \mathbb{R}$

$$V_n(f;x) - f(x) = \sum_{j=n+1}^{\infty} \frac{1}{j^2} V_j(g;x), \tag{12.2.24}$$

the right-hand side being convergent in $\mathsf{C}_{2\pi}$ since

$$\Big\| \sum_{j=n+1}^{\infty} \frac{1}{j^2} V_j(g;\circ) \Big\|_{\mathsf{C}_{2\pi}} \le \|g\|_\infty \sum_{j=n+1}^{\infty} \frac{1}{j^2} \le \frac{\|g\|_\infty}{n}. \tag{12.2.25}$$

If $f \in \mathsf{V}[\mathsf{L}^p_{2\pi}; -k^2]$, $1 < p < \infty$, then for each n (12.2.24) holds almost everywhere, whereas for $f \in \mathsf{V}[\mathsf{L}^1_{2\pi}; -k^2]$ we obtain for each $n \in \mathbb{N}$

$$V_n(f;x) - f(x) = \sum_{j=n+1}^{\infty} \frac{1}{j^2} V_j(d\mu;x) \quad \text{a.e.,} \tag{12.2.26}$$

where $\mu \in \mathsf{BV}_{2\pi}$ is such that $-k^2 f^\wedge(k) = \mu^\vee(k)$. In any case, $f \in \mathsf{V}[\mathsf{X}_{2\pi}; -k^2]$ implies $\|V_n(f;\circ) - f(\circ)\|_{\mathsf{X}_{2\pi}} = O(1/n)$, and therefore (cf. Cor. 2.5.11)

Proposition 12.2.9. *The integral $V_n(f;x)$ of de La Vallée Poussin is saturated in $\mathsf{X}_{2\pi}$ with order $O(n^{-1})$, $n \to \infty$, and the Favard class $\mathsf{F}[\mathsf{X}_{2\pi}; \mathsf{V}_n]$ is characterized by any of the three statements of Theorem* 12.2.1 (*with* $c = -1$). *Moreover, $f \in \mathsf{W}[\mathsf{X}_{2\pi}; -k^2]$ if and only if there exists $g \in \mathsf{X}_{2\pi}$ such that*

$$\lim_{n\to\infty} \|n[V_n(f;\circ) - f(\circ)] - g(\circ)\|_{\mathsf{X}_{2\pi}} = 0. \tag{12.2.27}$$

For the proof of (12.2.27) we observe if $f \in \mathsf{W}[\mathsf{X}_{2\pi}; -k^2]$, then (12.2.24) holds, and if we set $c_n = \sum_{j=n+1}^\infty j^{-2}$, then

$$\Big\| \frac{V_n(f;\circ) - f(\circ)}{c_n} - g(\circ) \Big\|_{\mathsf{X}_{2\pi}} = \Big\| \frac{1}{c_n} \sum_{j=n+1}^{\infty} \frac{1}{j^2} [V_j(g;\circ) - g(\circ)] \Big\|_{\mathsf{X}_{2\pi}}$$

$$\le \sup_{j \ge n+1} \|V_j(g;\circ) - g(\circ)\|_{\mathsf{X}_{2\pi}},$$

which tends to zero as $n \to \infty$ by (2.5.15). Now (12.2.27) follows by (6.2.14).

Whereas for the general singular integral of Weierstrass the functional equation is given by the semi-group property (12.2.18), for the singular integral of de La Vallée Poussin the equation

$$(12.2.28)\qquad [V_n(f;\circ)]''(x) = -n^2\{V_n(f;x) - V_{n-1}(f;x)\}$$

is essential. (12.2.28) holds for every $f \in X_{2\pi}$ and every $x \in \mathbb{R}$, $n \in \mathbb{N}$.

Problems

1. (i) Let $f \in X_{2\pi}$ and the kernel $\{\chi_\rho(x)\}$ of the integral $I_\rho(f;x)$ be even and positive. Complete the proof of the following list of equivalences (cf. Prop. 12.1.6 and Theorem 12.2.1):

   (1) $\lim_{\rho\to\rho_0} \dfrac{\chi_\rho^\wedge(k) - 1}{\chi_\rho^\wedge(1) - 1} = k^2$, (2) $\lim_{\rho\to\rho_0} \dfrac{\chi_\rho^\wedge(2) - 1}{\chi_\rho^\wedge(1) - 1} = 4$,

   (3) $\lim_{\rho\to\rho_0} \dfrac{\int_0^\pi \sin^4(u/2)\chi_\rho(u)\,du}{\int_0^\pi \sin^2(u/2)\chi_\rho(u)\,du} = 0$, (4) $\int_0^\pi u^4\chi_\rho(u)\,du = o(1 - \chi_\rho^\wedge(1))\quad (\rho\to\rho_0)$,

   (5) $\int_0^\pi u^2\chi_\rho(u)\,du = \pi(1 - \chi_\rho^\wedge(1)) + o(1 - \chi_\rho^\wedge(1))\quad (\rho\to\rho_0)$,

   (6) $I_\rho(f;x)$ is saturated in $X_{2\pi}$ with order $O(1 - \chi_\rho^\wedge(1))$, and the Favard class is characterized as in Theorem 12.2.1.

   (Hint: cf. GÖRLICH–STARK [1, 2])

   (ii) Show that the singular integrals $J_n(f;x)$ of Jackson, $N_n^*(f;x)$ of Jackson–de La Vallée Poussin, and $K_n(f;x)$ of Fejér–Korovkin are saturated in $X_{2\pi}$ with order $O(n^{-2})$, $n\to\infty$, the Favard class being characterized as $V[X_{2\pi}; c\,k^2]$ with $c = -3/2$, $-3/2$, $-\pi^2/2$, respectively.
2. Show that the integral means $A_h(f;x)$ of $f \in X_{2\pi}$ are saturated in $X_{2\pi}$ with order $O(h^2)$, $h\to 0+$, with Favard class characterized as $V[X_{2\pi}; c\,k^2]$, $c = -1/6$. Show that the same is true for the rth integral means $A_h^r(f;x)$ with $c = -r/6$. Thus an r-times repeated application of the smoothing operation A_h effects neither the order nor the class of saturation.
3. Show that the singular integral $B_n(f;x)$ of Rogosinski is saturated in $X_{2\pi}$ with order $O(n^{-2})$, with Favard class characterized as $V[X_{2\pi}; (-\pi^2/8)k^2]$. (Hint: Theorem 2.4.9)
4. Given a kernel $\{\chi_\rho(x)\}$, let there exist $\psi(k)$ with $\psi(0) = 0$, $\psi(k) \neq 0$ for $k \neq 0$, and $\varphi(\rho) > 0$ tending monotonely to zero as $\rho\to\rho_0$ such that $|\lambda_\rho(k)| \le M$ for all $k \in \mathbb{Z}$, $\rho \in \mathbb{A}$, $\lambda_\rho(k)$ being defined as in (12.2.4). Show that $f \in V[L^2_{2\pi}; \psi(k)]$ implies $\|I_\rho(f;\circ) - f(\circ)\|_2 = O(\varphi(\rho))$ (cf. also Prop. 6.5.2).
5. Given a kernel $\{\chi_\rho(x)\}$, $\rho \in \mathbb{A}$, $\rho_0 < \infty$, satisfying (1.1.5), let there exist $\psi(k)$ with $\psi(0) = 0$, $\psi(k) \neq 0$ for $k \neq 0$, $\varphi(\rho) > 0$ tending monotonely to zero as $\rho\to\rho_0$, and $h \in NL^1$ such that $\chi_\rho^\wedge(k) - 1 = \varphi(\rho)\,\psi(k)h^\wedge(|\rho - \rho_0|k)$ holds for each $k \in \mathbb{Z}$, $\rho \in \mathbb{A}$. Show that the singular integral $I_\rho(f;x)$ is saturated in $X_{2\pi}$ with order $O(\varphi(\rho))$, $\rho\to\rho_0$, and that the Favard class $F[X_{2\pi}; I_\rho]$ is characterized as $V[X_{2\pi}; \psi(k)]$. (Hint: BERENS–GÖRLICH [1])
6. Let $\{\chi_\rho(x)\}$ satisfy (12.1.1) and (12.2.4). Show that the iterated integral $I_\rho^2(f;x)$ is saturated in $X_{2\pi}$ with order $O(\varphi(\rho))$ and that the Favard class $F[X_{2\pi}; I_\rho^2]$ is characterized as $V[X_{2\pi}; 2\psi(k)]$. Furthermore, if $\{\chi_\rho(x)\}$ satisfies (12.2.6), so does the iterated kernel $\{(\chi_\rho * \chi_\rho)(x)\}$. (Hint: Use Problem 12.1.6 and the estimate (cf. Problem 1.1.4) $\|I_\rho^2(f;\circ) - f(\circ)\|_{X_{2\pi}} \le (\|\chi_\rho\|_1 + 1)\,\|I_\rho(f;\circ) - f(\circ)\|_{X_{2\pi}}$)
7. Establish the equivalence of the statements (i) and (iv) of Theorem 12.2.3.
8. (i) Use the identities ($\sigma_{n,2}(f;x)$ denote the $(C, 2)$ means of the Fourier series of $f \in X_{2\pi}$, cf. Problem 1.4.4)

$$[\sigma_n^\sim(f;\circ)]'(x) = (n+2)(\sigma_n(f;x) - \sigma_{n,2}(f;x)),$$
$$\sigma_{n,2}(f;x) - \sigma_{n-1,2}(f;x) = 2[\sigma_n^\sim(f;\circ)]'(x)/n(n+2)$$

to deduce that $\|\sigma_n(f; \circ) - f(\circ)\|_{X_{2\pi}} = O(n^{-1})$ if and only if $\|\sigma_n^{\{1\}}(f; \circ)\|_{X_{2\pi}} = O(1)$, $n \to \infty$. (Hint: See also ZAMANSKY [2], ZYGMUND [7I, p. 123], LORENTZ [3, p. 100])

(ii) Show that the singular integral $\sigma_n(f; x)$ of Fejér is saturated in $X_{2\pi}$ with order $O(n^{-1})$, the Favard class $F[X_{2\pi}; \sigma_n]$ being characterized by any of the seven statements of Prop. 12.2.7 for $\kappa = 1$. Furthermore, show that the following statements are equivalent for f to belong to $F[X_{2\pi}; \sigma_n]$:

(a) There exist $g \in L_{2\pi}^\infty$ if $X_{2\pi} = C_{2\pi}$, $\mu \in BV_{2\pi}$ if $X_{2\pi} = L_{2\pi}^1$, $g \in L_{2\pi}^p$ if $X_{2\pi} = L_{2\pi}^p$, $1 < p < \infty$, and a constant a such that

$$f^\sim(x) = a + \int_{-\pi}^{x} \begin{Bmatrix} g(u)\,du \\ d\mu(u) \\ g(u)\,du \end{Bmatrix} \quad \text{(a.e.)},$$

(b) $\|f^\sim(\circ + h) - f^\sim(\circ)\|_{X_{2\pi}} = O(|h|)$ $\qquad (h \to 0)$,

(c) $\|f(\circ + h) - f(\circ)\|_{X_{2\pi}} = O(|h|)$, provided $X_{2\pi} = L_{2\pi}^p$, $1 < p < \infty$.

(iii) Show that the (C, α) means $\sigma_{n,\alpha}(f; x)$ of the Fourier series of $f \in X_{2\pi}$ are saturated in $X_{2\pi}$ with order $O(n^{-1})$, the Favard class being characterized as $V[X_{2\pi}; -\alpha\,|k|]$. (Hint: cf. BUTZER–GÖRLICH [2, p. 388])

9. Let $f \in W[X_{2\pi}; k^2]$ and $k^2 f^\wedge(k) = g^\wedge(k)$, $g \in X_{2\pi}$. Show that for the integral means $A_h(f; x)$

$$A_h(f; x) - f(x) = -\frac{1}{h}\int_0^h \left[\int_0^u t\,A_t(g; x)\,dt\right] du,$$

and therefore

$$\lim_{h \to 0+} \left\| \frac{A_h(f; \circ) - f(\circ)}{h^2} - \left(-\frac{1}{6}\right) g(\circ) \right\|_{X_{2\pi}} = 0.$$

10. Let $f \in X_{2\pi}$ and $\{\chi_\rho(x)\}$ satisfy (12.1.1). If $f \in W[X_{2\pi}; \psi(k)]$ implies the approximation $\|I_\rho(f; \circ) - f(\circ)\|_{X_{2\pi}} = O(\varphi(\rho))$, then in fact

$$\lim_{\rho \to \rho_0} \left\| \frac{I_\rho(f; \circ) - f(\circ)}{\varphi(\rho)} - g(\circ) \right\|_{X_{2\pi}} = 0,$$

where $g \in X_{2\pi}$ is such that $\psi(k) f^\wedge(k) = g^\wedge(k)$. (Hint: Use the theorem of Banach–Steinhaus with T_n as the dense subset, cf. BUTZER–GÖRLICH [2])

11. Show that for $f \in W[C_{2\pi}; k^2]$ the following limit relations of Voronovskaja type are valid (cf. (1.5.29), (1.6.7)):

(i) $$\lim_{t \to 0+} \left\| \frac{W_t(f; \circ) - f(\circ)}{t} - f''(\circ) \right\|_{C_{2\pi}} = 0,$$

(ii) $$\lim_{n \to \infty} \left\| n^2[K_n(f; \circ) - f(\circ)] - \frac{\pi^2}{2} f''(\circ) \right\|_{C_{2\pi}} = 0,$$

(iii) $$\lim_{n \to \infty} \left\| n^2[B_n(f; \circ) - f(\circ)] - \frac{\pi^2}{8} f''(\circ) \right\|_{C_{2\pi}} = 0,$$

(iv) $$\lim_{h \to 0+} \left\| \frac{A_h^r(f; \circ) - f(\circ)}{h^2} - \frac{r}{6} f''(\circ) \right\|_{C_{2\pi}} = 0.$$

Show that the converse statement is also valid. State and prove analogous results in $L_{2\pi}^p$-space. State and prove analogous results for the singular integrals $J_n(f; x)$ of Jackson, $N_n^*(f; x)$ of Jackson–de La Vallée Poussin, $V_n(f; x)$ of de La Vallée Poussin, etc. (Hint: BUTZER–GÖRLICH [2])

12. Show that for $f \in \mathsf{W}[\mathsf{C}_{2\pi}; |k|]$ the following limit relations of Voronovskaja type are valid (cf. Problem 9.2.5)

 (i) $\lim_{n\to\infty} \|n[f(\circ) - \sigma_n(f; \circ)] - (f^\sim)'(\circ)\|_{\mathsf{C}_{2\pi}} = 0,$

 (ii) $\lim_{r\to 1-} \left\|\frac{f(\circ) - P_r(f; \circ)}{1-r} - (f^\sim)'(\circ)\right\|_{\mathsf{C}_{2\pi}} = 0.$

 Show that the converse statement is also valid. State and prove analogous results in $\mathsf{L}^p_{2\pi}$-space.

13. (i) Show that the singular integral $P_r(f; x)$ of Abel–Poisson is saturated in $\mathsf{X}_{2\pi}$ with order $O(1 - r)$, $r \to 1-$, with Favard class $\mathsf{F}[\mathsf{X}_{2\pi}; P_r]$ characterized as $\mathsf{V}[\mathsf{X}_{2\pi}; -|k|]$. Give further characterizations of $\mathsf{F}[\mathsf{X}_{2\pi}; P_r]$ (cf. Problem 12.2.8).

 (ii) In the light of (i) and Cor. 2.5.5 examine the dependence of the solution of Dirichlet's problem for the unit disc (cf. Sec. 7.1.2) upon its boundary value f, and study the rate at which $P_r(f; x)$ approximates f as $r \to 1-$.

 (iii) Show that the solution of Dirichlet's problem (7.1.17) for a given boundary value $f_0 \in \mathsf{X}_{2\pi}$ also solves Neumann's problem (7.1.25) for a boundary value $g_0 \in \mathsf{X}_{2\pi}$ if and only if $|k|\, f_0^\wedge(k) = g_0^\wedge(k)$. Conversely, the solution of Neumann's problem for g_0 solves Dirichlet's problem for f_0, where f_0 is determined by $\varphi_1 * g_0$ (except for an additive constant), φ_1 being defined through (11.5.12) (cf. also the proof of Prop. 12.2.4). (Hint: Butzer–Berens [1, p. 119 ff], where also the connection of simultaneous solutions of Dirichlet's and Neumann's problem with the class $\mathsf{V}[\mathsf{X}_{2\pi}; |k|]$ is studied)

14. Let $S_n(f; x)$, $\sigma_n(f; x)$ be the singular integrals of Dirichlet and Fejér of $f \in \mathsf{X}_{2\pi}$, respectively. Show that the kernel of the integral

$$(12.2.29)\qquad I_n(f; x) \equiv \frac{1}{2}[S_n(f; x) + \sigma_n(f; x)] = \sum_{k=-n}^{n}\left(1 - \frac{|k|}{2(n+1)}\right) f^\wedge(k)e^{ikx}$$

 satisfies (12.1.1) and (12.2.6) so that $I_n(f; x)$ is saturated in $\mathsf{X}_{2\pi}$ with order $O(n^{-1})$, the Favard class $\mathsf{F}[\mathsf{X}_{2\pi}; I_n]$ being characterized as $\mathsf{V}[\mathsf{X}_{2\pi}; -\frac{1}{2}|k|]$. Does the kernel form an approximate identity?

15. Treat the saturation problem for the singular integral $\Phi_n(f; x)$ defined in Problem 1.6.16. Apply to the integrals of Fejér, Fejér–Korovkin, and Jackson.

16. Let the kernel $\{\chi_\rho(x)\}$ of the singular integral $I_\rho(f; x)$ satisfy (12.1.1) and (12.2.4). Show that the Banach spaces $\mathsf{V}[\mathsf{X}_{2\pi}; \psi(k)]$, endowed with norm (10.4.7), and $\mathsf{X}_{2\pi,\{I_\rho\},\varphi(\rho)}$, endowed with norm (12.1.8), are equal with equivalent norms. (Hint: Compare with Theorems 12.1.4, 12.2.3, see also Prop. 0.8.5)

12.3 Saturation in L^p, $1 \leq p \leq 2$

12.3.1 Saturation Property

Let $f \in \mathsf{X}(\mathbb{R})$ and $\chi \in \mathsf{NL}^1$. In Sec. 3.1 we have seen that the singular integral (of Fejér's type)

$$(12.3.1)\qquad J(f; x; \rho) = \frac{\rho}{\sqrt{2\pi}} \int_{-\infty}^{\infty} f(x - u)\chi(\rho u)\, du \qquad (\rho > 0)$$

defines a family of bounded linear transformations $J(\rho)$ of $\mathsf{X}(\mathbb{R})$ into itself having uniformly bounded norms, i.e.

$$(12.3.2)\qquad \|J(f; \circ; \rho)\|_{\mathsf{X}(\mathbb{R})} \leq \|\chi\|_1 \|f\|_{\mathsf{X}(\mathbb{R})} \qquad (\rho > 0),$$

and which approximate the identity, i.e.

(12.3.3) $$\lim_{\rho\to\infty} \|J(f;\circ;\rho) - f(\circ)\|_{X(\mathbb{R})} = 0.$$

In the study of the corresponding saturation problem† we shall restrict ourselves in this chapter to the spaces $X(\mathbb{R}) = L^p$, $1 \le p \le 2$. Then saturation theory of periodic singular integrals suggests that the following conditions are of interest:

(12.3.4) *Given* $\chi \in NL^1$, *let there exist constants* $c \neq 0$‡ *and* $\alpha > 0$ *such that*

$$\lim_{v\to 0} \frac{\chi^\wedge(v) - 1}{|v|^\alpha} = c.$$

(12.3.5) *Given* $\chi \in NL^1$, *let there exist constants* $c \neq 0$, $\alpha > 0$, *and a function* $\nu \in NBV$ *such that for all* $v \neq 0$

$$\frac{\chi^\wedge(v) - 1}{c\,|v|^\alpha} = \nu^\vee(v).$$

Since ν is normalized and $\nu^\vee$ is continuous by Prop. 5.3.2, condition (12.3.5) implies (12.3.4). But the converse assertion is not necessarily true (cf. Problem 12.3.1).

Proposition 12.3.1. *Let* χ *satisfy* (12.3.4). *If for* $f \in L^p$, $1 \le p \le 2$, *there exists* $g \in L^p$ *such that*

(12.3.6) $$\lim_{\rho\to\infty} \left\| \frac{J(f;\circ;\rho) - f(\circ)}{\rho^{-\alpha}} - g(\circ) \right\|_p = 0,$$

then f belongs to $W[L^p; c\,|v|^\alpha]$.

Proof. Let $p = 1$. In virtue of Theorem 5.1.3 we have $[J(f;\circ;\rho)]^\wedge(v) = \chi^\wedge(v/\rho)f^\wedge(v)$, and thus

(12.3.7) $$\left[\frac{J(f;\circ;\rho) - f(\circ)}{\rho^{-\alpha}} - g(\circ)\right]^\wedge(v) = \frac{\chi^\wedge(v/\rho) - 1}{\rho^{-\alpha}} f^\wedge(v) - g^\wedge(v)$$

for all $v \in \mathbb{R}$. Hence by (5.1.2)

$$\left|\frac{\chi^\wedge(v/\rho) - 1}{\rho^{-\alpha}} f^\wedge(v) - g^\wedge(v)\right| \le \left\| \frac{J(f;\circ;\rho) - f(\circ)}{\rho^{-\alpha}} - g(\circ) \right\|_1 = o(1) \quad (\rho \to \infty)$$

uniformly for all $v \in \mathbb{R}$. But (12.3.4) in particular implies

(12.3.8) $$\lim_{\rho\to\infty} \frac{\chi^\wedge(v/\rho) - 1}{\rho^{-\alpha}} = c\,|v|^\alpha \qquad (v \in \mathbb{R}),$$

and therefore $c\,|v|^\alpha f^\wedge(v) = g^\wedge(v)$ for all $v \in \mathbb{R}$.

For $1 < p \le 2$ Theorem 5.2.12 yields (12.3.7) for almost all $v \in \mathbb{R}$ which implies

$$\left\| \frac{\chi^\wedge(\circ/\rho) - 1}{\rho^{-\alpha}} f^\wedge(\circ) - g^\wedge(\circ) \right\|_{p'} \le \left\| \frac{J(f;\circ;\rho) - f(\circ)}{\rho^{-\alpha}} - g(\circ) \right\|_p = o(1) \quad (\rho \to \infty)$$

† We shall confine ourselves to singular integrals of Fejér's type. For results concerning the more general integral (3.1.3) we refer to the Problems.

‡ Let us emphasize the importance of the existence of a nonzero constant c. At the end of Sec. 12.5 we shall construct an example of a function $\chi \in NL^1$ for which $\lim_{v\to 0} (\chi^\wedge(v) - 1)/|v|^\alpha = 0$ for every $\alpha > 0$ and which indeed does not possess the saturation property.

by Titchmarsh's inequality. In view of (12.3.8) and Fatou's lemma it follows that $\|c\,|\circ|^\alpha f^\wedge(\circ) - g^\wedge(\circ)\|_{p'} = 0$, and thus the assertion.

Proposition 12.3.2. *Let* χ *satisfy* (12.3.4). *If for* $f \in \mathsf{L}^p$, $1 \le p \le 2$,

$$\|J(f;\circ;\rho) - f(\circ)\|_p = o(\rho^{-\alpha}) \qquad (\rho \to \infty),$$

then $f(x) = 0$ a.e. *Moreover, the nullfunction is the only invariant element of the singular integral* (12.3.1) *in* L^p.

Proof. Since (12.3.6) holds with $g(x) = 0$ a.e., we conclude $|v|^\alpha f^\wedge(v) = 0$, and thus $f^\wedge(v) = 0$ (a.e.). Therefore $f(x) = 0$ a.e. Obviously, the nullfunction is an invariant element of $J(f;x;\rho)$. Conversely, if $f \in \mathsf{L}^p$ is such that $\|J(f;\circ;\rho) - f(\circ)\|_p = 0$ for every $\rho > 0$, then by the convolution theorem $\{\chi^\wedge(v/\rho) - 1\}f^\wedge(v) = 0$ for every $\rho > 0$, and hence as above $f(x) = 0$ a.e.

Theorem 12.3.3. *If* $f \in \mathsf{L}^p$, $1 \le p \le 2$, *and* χ *satisfies* (12.3.5), *then the singular integral* $J(f;x;\rho)$ *is saturated in* L^p *with order* $O(\rho^{-\alpha})$.

Proof. In view of Prop. 12.3.2 we only have to show that the order of approximation

$$\|J(f;\circ;\rho) - f(\circ)\|_p = O(\rho^{-\alpha}) \qquad (\rho \to \infty) \tag{12.3.9}$$

is attained for at least one function $f \in \mathsf{L}^p$ different from the nullfunction. This is given by the following (see also Problem 12.3.8)

Lemma 12.3.4. *Let* $\Phi \in \mathsf{C}_{00}^\infty$ *and* χ *satisfy* (12.3.5). *Then for* $1 \le p \le 2$

$$\lim_{\rho\to\infty} \|\rho^\alpha[J(\Phi;\circ;\rho) - \Phi(\circ)] - c[D_s^{\{\alpha\}}\Phi](\circ)\|_p = 0. \tag{12.3.10}$$

Proof. Since $\Phi \in \mathsf{C}_{00}^\infty$ it follows by Problem 11.2.13 that Φ has strong Riesz derivatives of every order. In particular, $D_s^{\{\alpha\}}\Phi \in \mathsf{L}^p$ for every $1 \le p \le 2$, and $|v|^\alpha\Phi^\wedge(v) = [D_s^{\{\alpha\}}\Phi]^\wedge(v)$. Moreover, $[D_s^{\{\alpha\}}\Phi]^\wedge \in \mathsf{L}^1$ and therefore

$$\begin{aligned}\rho^\alpha[J(\Phi;x;\rho) - \Phi(x)] &= \frac{1}{\sqrt{2\pi}}\int_{-\infty}^{\infty} \frac{\chi^\wedge(v/\rho) - 1}{(|v|\rho^{-1})^\alpha}\,|v|^\alpha\Phi^\wedge(v)\,e^{ixv}\,dv\\ &= \frac{c}{\sqrt{2\pi}}\int_{-\infty}^{\infty} \nu^\vee(v/\rho)[D_s^{\{\alpha\}}\Phi]^\wedge(v)\,e^{ixv}\,dv\\ &= \frac{c}{\sqrt{2\pi}}\int_{-\infty}^{\infty} [D_s^{\{\alpha\}}\Phi](x-u)\,d\nu(\rho u)\end{aligned}$$

by Prop. 5.1.10, 5.3.3. Since ν is normalized, this implies

$$\rho^\alpha[J(\Phi;x;\rho) - \Phi(x)] - c[D_s^{\{\alpha\}}\Phi](x) = \frac{c}{\sqrt{2\pi}}\int_{-\infty}^{\infty}\{[D_s^{\{\alpha\}}\Phi](x-u) - [D_s^{\{\alpha\}}\Phi](x)\}\,d\nu(\rho u),$$

and (12.3.10) follows by Problem 3.1.16.

We here recall Prop. 1.7.4 on positive bounded linear polynomial operators. This suggests that for even and positive functions χ the exponent α of (12.3.4) may be restricted to $0 < \alpha \le 2$. Indeed (cf. also Problem 12.3.2)

Proposition 12.3.5. *If* χ *is an even and positive function satisfying* (12.3.4), *then the exponent* α *is restricted to* $0 < \alpha \le 2$.

Proof. Since $\chi^\wedge$ is an even and real function bounded by 1, it follows that $c < 0$ under (12.3.4). Suppose $\alpha > 2$. Then for $R > 0$

$$\left|\frac{\chi^\wedge(v) - 1}{|v|^\alpha}\right| = \frac{1}{|v|^\alpha}\cdot\sqrt{\frac{2}{\pi}}\int_0^\infty (1 - \cos vu)\chi(u)\,du \tag{12.3.11}$$
$$> \frac{1}{|v|^{\alpha-2}}\cdot\sqrt{\frac{2}{\pi}}\int_0^R \frac{1 - \cos vu}{v^2}\chi(u)\,du.$$

But

$$\lim_{v\to 0}\int_0^R \frac{1 - \cos vu}{v^2}\chi(u)\,du = \frac{1}{2}\int_0^R u^2\chi(u)\,du > 0,$$

so the right-hand side of inequality (12.3.11) tends to infinity as $v \to 0$, a contradiction to (12.3.4). Thus $0 < \alpha \le 2$.

12.3.2 Characterizations of Favard Classes: $p = 1$

The next problem is to characterize the Favard class of the singular integral $J(f; x; \rho)$. To this end we proceed separately for $p = 1$ and $1 < p \le 2$.

Theorem 12.3.6. *Let* $f \in \mathsf{L}^1$.

(i) *If* χ *satisfies* (12.3.4), *then* (12.3.9) *implies the existence of* $\mu \in \mathsf{BV}$ *such that* $c\,|v|^\alpha f^\wedge(v) = \mu^\vee(v)$ *for all* $v \in \mathbb{R}$.

(ii) *If in addition* χ *satisfies* (12.3.5), *then* $f \in \mathsf{V}[\mathsf{L}^1; c\,|v|^\alpha]$ *implies* (12.3.9).

Proof. (i) By the convolution theorem and the hypotheses

$$\left|\frac{\chi^\wedge(v/\rho) - 1}{\rho^{-\alpha}} f^\wedge(v)\right| \le \left\|\frac{J(f; \circ; \rho) - f(\circ)}{\rho^{-\alpha}}\right\|_1 \le M \tag{12.3.12}$$

uniformly for large ρ, and therefore $c\,|v|^\alpha f^\wedge(v)$ defines a bounded continuous function on $\mathbb{R}$.

In order to apply Cramér's representation theorem, let us consider the arithmetic means (cf. the proof of Theorem 10.5.4)

$$\frac{1}{\sqrt{2\pi}}\int_{-R}^R \left(1 - \frac{|v|}{R}\right) c\,|v|^\alpha f^\wedge(v)\,e^{ixv}\,dv \qquad (x \in \mathbb{R}, R > 0).$$

By (12.3.12) and Lebesgue's dominated convergence theorem it follows that

$$\int_{-R}^R \left(1 - \frac{|v|}{R}\right) c\,|v|^\alpha f^\wedge(v)\,e^{ixv}\,dv = \lim_{\rho\to\infty}\int_{-R}^R \left(1 - \frac{|v|}{R}\right)\frac{\chi^\wedge(v/\rho) - 1}{\rho^{-\alpha}} f^\wedge(v)\,e^{ixv}\,dv.$$

Hence by Fatou's lemma and (5.1.18)

$$\left\|\frac{1}{\sqrt{2\pi}}\int_{-R}^R \left(1 - \frac{|v|}{R}\right) c\,|v|^\alpha f^\wedge(v)\,e^{i\circ v}\,dv\right\|_1 \le \liminf_{\rho\to\infty}\left\|\sigma\left(\frac{J(f; \circ; \rho) - f(\circ)}{\rho^{-\alpha}}; \circ; R\right)\right\|_1$$
$$\le \liminf_{\rho\to\infty}\|\rho^\alpha[J(f; \circ; \rho) - f(\circ)]\|_1.$$

Since the latter expression is bounded independently of $R > 0$, (i) follows by an application of Theorem 6.1.5.

(ii) By Prop. 5.3.1 and Theorems 5.1.3, 5.3.5

$$\left[\int_{-\infty}^{\circ} \frac{J(f; u; \rho) - f(u)}{\rho^{-\alpha}}\, du\right]^{\vee}(v) = \left[\frac{J(f; \circ; \rho) - f(\circ)}{\rho^{-\alpha}}\right]^{\wedge}(v) = \frac{\chi^{\wedge}(v/\rho) - 1}{\rho^{-\alpha}} f^{\wedge}(v)$$

$$= \mu^{\vee}(v)\nu^{\vee}(v/\rho) = \left[\frac{1}{\sqrt{2\pi}}\int_{-\infty}^{\infty} \mu(\circ - u)\, d\nu(\rho u)\right]^{\vee}(v),$$

the convolution of μ and ν being understood in the sense of Prop. 0.5.2. Thus by the uniqueness theorem for Fourier–Stieltjes transforms we have for each $\rho > 0$

$$\int_{-\infty}^{x} \frac{J(f; u; \rho) - f(u)}{\rho^{-\alpha}}\, du = \frac{1}{\sqrt{2\pi}}\int_{-\infty}^{\infty} \mu(x - u)\, d\nu(\rho u). \tag{12.3.13}$$

It follows that

$$\left\|\frac{J(f; \circ; \rho) - f(\circ)}{\rho^{-\alpha}}\right\|_1 = \left\|\frac{1}{\sqrt{2\pi}}\int_{-\infty}^{\infty} \mu(\circ - u)\, d\nu(\rho u)\right\|_{\mathsf{BV}} \le \|\mu\|_{\mathsf{BV}}\|\nu\|_{\mathsf{BV}}$$

uniformly for all $\rho > 0$, which completes the proof.

Thus the saturation class for singular integrals (12.3.1) with kernel χ satisfying (12.3.5) may be characterized as the class $\mathsf{V}[\mathsf{L}^1; c\,|v|^{\alpha}]$. Moreover,

Theorem 12.3.7. *Let $f \in \mathsf{L}^1$ and χ satisfy* (12.3.5). *Then the following assertions are equivalent for the singular integral $J(f; x; \rho)$:*

(i) $\|J(f; \circ; \rho) - f(\circ)\|_1 = \mathsf{O}(\rho^{-\alpha}) \qquad (\rho \to \infty)$,

(ii) $f \in \mathsf{V}[\mathsf{L}^1; c\,|v|^{\alpha}]$,

(iii) $\left\|\frac{1}{\sqrt{2\pi}}\int_{-R}^{R}\left(1 - \frac{|v|}{R}\right)c\,|v|^{\alpha} f^{\wedge}(v)\, e^{i\circ v}\, dv\right\|_1 = \mathsf{O}(1) \qquad (R \to \infty)$,

(iv) $f \in \widetilde{\mathsf{W}[\mathsf{L}^1; c\,|v|^{\alpha}]}^{\mathsf{L}^1}$,

(v) *there exists $\mu \in \mathsf{BV}$ such that* (12.3.13) *holds,*

(vi) *there exists $\mu \in \mathsf{BV}$ such that for every $s \in \mathsf{C}_0$*

$$\lim_{\rho\to\infty}\int_{-\infty}^{\infty} s(u)\rho^{\alpha}[J(f; u; \rho) - f(u)]\, du = \int_{-\infty}^{\infty} s(u)\, d\mu(u),$$

(vii) *for $\alpha \ne [\![\alpha]\!]$ ($\alpha = [\![\alpha]\!]$) the $[\![\alpha]\!]$th ($[\![\alpha - 1]\!]$th) strong Riesz derivative of f exists and* ($D^{\{0\}}f = f$)

$\alpha \ne [\![\alpha]\!]$: $\quad \|[R^{\sim}_{[\![\alpha]\!]+1-\alpha} D_s^{\{[\![\alpha]\!]\}} f](\circ + h) - [R^{\sim}_{[\![\alpha]\!]+1-\alpha} D_s^{\{[\![\alpha]\!]\}} f](\circ)\|_1 = \mathsf{O}(|h|) \qquad (h \to 0)$,

$\alpha = [\![\alpha]\!]$: $\quad \|(D_s^{\{[\![\alpha-1]\!]\}} f)^{\sim}(\circ + h) - (D_s^{\{[\![\alpha-1]\!]\}} f)^{\sim}(\circ)\|_1 = \mathsf{O}(|h|) \qquad (h \to 0)$,

(viii) $\left\|\int_{\varepsilon}^{\infty} \frac{\overline{\Delta}_u^{2j} f(\circ)}{u^{1+\alpha}}\, du\right\|_1 = \mathsf{O}(1) \qquad (\varepsilon \to 0+)$

where $j \in \mathbb{N}$ is chosen such that $0 < \alpha < 2j$,

(ix) *there exists* $\mu \in \mathsf{BV}$ *such that for*

$0 < \alpha < 1$: $f(x) = [R_\alpha\, d\mu](x)$ a.e.,

$\alpha \geq 1$: f *belongs to* L^2 *and*

$$\alpha = [\![\alpha]\!]: \qquad [H^{[\alpha]}f](x) = \int_{-\infty}^{x} du_1 \int_{-\infty}^{u_1} du_2 \ldots \int_{-\infty}^{u_{[\alpha]-1}} d\mu(u_{[\alpha]}) \quad \text{a.e.,}$$

$\alpha = [\![\alpha]\!] + \beta$, $0 < \beta < 1$: $R_\beta d\mu$ *belongs to* L^1 *and*

$$[H^{[\alpha]}f](x) = \int_{-\infty}^{x} du_1 \int_{-\infty}^{u_1} du_2 \ldots \int_{-\infty}^{u_{[\alpha]-1}} [R_\beta d\mu](u_{[\alpha]})\, du_{[\alpha]} \quad \text{a.e.,}$$

where H denotes the operation of taking Hilbert transforms.

Proof. The equivalence of (i), (ii), and (v) is given by Theorem 12.3.6 whilst Theorem 6.1.7 and Prop. 10.5.7 imply the equivalence of (ii), (iii), and (iv) (which in particular hold independently of condition (12.3.5)).

Regarding (vi), let $f \in \mathsf{V}[\mathsf{L}^1; c\,|v|^\alpha]$, i.e., there exists $\mu \in \mathsf{BV}$ such that $c\,|v|^\alpha f^\wedge(v) = \mu^\vee(v)$ for all $v \in \mathbb{R}$. As (ii) implies (i), we may define a set of absolutely continuous functions by

$$\mu_\rho(x) = \int_{-\infty}^{x} \frac{J(f; u; \rho) - f(u)}{\rho^{-\alpha}}\, du$$

which have total variation $\|\mu_\rho\|_{\mathsf{BV}} = \|\rho^\alpha[J(f; \circ; \rho) - f(\circ)]\|_1$, uniformly bounded for large ρ. Therefore by the theorems of Helly and Helly–Bray there exists a weakly* convergent subsequence, i.e., there exists a sequence of positive numbers $\{\rho_k\}$ with $\lim_{k\to\infty} \rho_k = \infty$ and $\eta \in \mathsf{BV}$ such that

$$\lim_{k\to\infty} \int_{-\infty}^{\infty} s(u) \frac{J(f; u; \rho_k) - f(u)}{\rho_k^{-\alpha}}\, du = \int_{-\infty}^{\infty} s(u)\, d\eta(u) \tag{12.3.14}$$

for every $s \in \mathsf{C}_0$. Our next aim is to show that μ and η must coincide. For this purpose, if we apply (12.3.14) to the particular functions

$$s_{x,R}(u) = \begin{cases} \dfrac{1}{\sqrt{2\pi}}\left(1 - \dfrac{|u|}{R}\right) e^{-ixu}, & |u| \leq R \\ 0, & |u| > R \end{cases} \qquad (x \in \mathbb{R},\ R > 0), \tag{12.3.15}$$

then it follows by the Parseval formulae (5.3.11), (5.1.5), by Problem 5.1.2, and Lebesgue's dominated convergence theorem that

$$\begin{aligned}
\frac{2}{\pi R}\int_{-\infty}^{\infty} \frac{\sin^2(R(x-u)/2)}{(x-u)^2}\eta^\vee(u)\, du &= \frac{1}{\sqrt{2\pi}}\int_{-R}^{R}\left(1 - \frac{|u|}{R}\right) e^{-ixu}\, d\eta(u) \\
&= \lim_{k\to\infty} \frac{1}{\sqrt{2\pi}}\int_{-R}^{R}\left(1 - \frac{|u|}{R}\right) e^{-ixu} \frac{J(f; u; \rho_k) - f(u)}{\rho_k^{-\alpha}}\, du \\
&= \lim_{k\to\infty} \frac{2}{\pi R}\int_{-\infty}^{\infty} \frac{\sin^2(R(x-u)/2)}{(x-u)^2} \frac{\chi^\wedge(u/\rho_k) - 1}{\rho_k^{-\alpha}} f^\wedge(u)\, du \\
&= \lim_{k\to\infty} \frac{2}{\pi R}\int_{-\infty}^{\infty} \frac{\sin^2(R(x-u)/2)}{(x-u)^2} \nu^\vee(u/\rho_k)\mu^\vee(u)\, du \\
&= \frac{2}{\pi R}\int_{-\infty}^{\infty} \frac{\sin^2(R(x-u)/2)}{(x-u)^2} \mu^\vee(u)\, du.
\end{aligned}$$

Thus Fejér's integral of the function $(\eta^\vee - \mu^\vee) \in \mathsf{C}$ vanishes for all $R > 0$. By Cor. 3.1.10 this gives $\eta^\vee(v) = \mu^\vee(v)$ for all $v \in \mathbb{R}$, and $\eta(x) = \mu(x)$ by the uniqueness theorem for Fourier–Stieltjes transforms. But since μ is the only weak* limit of $\rho^\alpha[J(f; x; \rho) - f(x)]$, this implies the weak* convergence of the functions themselves, i.e., (vi) is established.

Conversely, let (vi) be satisfied. Then for the particular set of functions $s_{x,R}^{\wedge}(-u)$, where $s_{x,R}$ is given by (12.3.15), we have by the same arguments that for each $x \in \mathbb{R}$ and $R > 0$

$$\frac{2}{\pi R}\int_{-\infty}^{\infty} \frac{\sin^2(R(x-u)/2)}{(x-u)^2}\,d\mu(u) = \frac{1}{\sqrt{2\pi}}\int_{-R}^{R}\left(1-\frac{|v|}{R}\right)c\,|v|^{\alpha}f^{\wedge}(v)\,e^{ixv}\,dv.$$

Therefore we have uniformly for all $R > 0$

$$\left\|\frac{1}{\sqrt{2\pi}}\int_{-R}^{R}\left(1-\frac{|v|}{R}\right)c\,|v|^{\alpha}f^{\wedge}(v)\,e^{i\circ v}\,dv\right\|_1 \le \|\mu\|_{\mathsf{BV}}.$$

Thus (vi) implies (iii).

For the proof of the equivalence of (ii), (vii)–(ix) we refer to Theorems 11.2.10–11.2.12, 11.3.6, and Problem 11.5.1.

Let us briefly discuss the problem whether conditions (12.3.4) and (12.3.5) may be weakened. In view of Theorem 12.3.6(i) we have that condition (12.3.4) is sufficient for the inverse part of the saturation theorem to hold whereas the direct part was established under the stronger condition (12.3.5) (cf. Theorem 12.3.6(ii)). But the following result shows that (12.3.5) is indeed *necessary* for the direct part to hold.

Proposition 12.3.8. *Let* $f \in \mathsf{L}^1$, $\chi \in \mathsf{NL}^1$ *and suppose that for some* $\alpha > 0$ *every* $f \in \mathsf{V}[\mathsf{L}^1; |v|^{\alpha}]$ *admits the approximation* (12.3.9). *Then necessarily the representation* (12.3.5) *is valid.*

Proof. Let f_0 be such that $f_0^{\wedge}(v) = (1 + |v|^{\alpha})^{-1}$. In view of Problem 6.3.9, $f_0 \in \mathsf{L}^1$. Since $|v|^{\alpha}f_0^{\wedge}(v) = 1 - f_0^{\wedge}(v)$ belongs to $[\mathsf{BV}]^{\vee}$, the hypothesis implies that f_0 satisfies (12.3.9). Therefore

$$\text{(12.3.16)}\qquad \left\|\frac{\chi^{\wedge}(\circ)-1}{|\circ|^{\alpha}}\,|\rho\circ|^{\alpha}f_0^{\wedge}(\rho\circ)\right\|_{\infty} = \left\|\frac{\chi^{\wedge}(\circ/\rho)-1}{\rho^{-\alpha}}f_0^{\wedge}(\circ)\right\|_{\infty}$$
$$\le \rho^{\alpha}\|J(f_0;\circ;\rho) - f_0(\circ)\|_1 = O(1)$$

and

$$\text{(12.3.17)}\qquad \|\sigma(\rho^{\alpha}[J(f_0;\circ;\rho) - f_0(\circ)];\circ;\rho R)\|_1 \le \rho^{\alpha}\|J(f_0;\circ;\rho) - f_0(\circ)\|_1 = O(1)$$

uniformly for all $\rho > 0$, $R > 0$. On the other hand, by the Parseval formula (5.1.5) and obvious changes of variables

$$\text{(12.3.18)}\qquad \|\sigma(\rho^{\alpha}[J(f_0;\circ;\rho) - f_0(\circ)];\circ;\rho R)\|_1$$
$$= \left\|\frac{1}{\sqrt{2\pi}}\int_{-R}^{R}\left(1-\frac{|v|}{R}\right)\frac{\chi^{\wedge}(v)-1}{|v|^{\alpha}}\,|\rho v|^{\alpha}f_0^{\wedge}(\rho v)\,e^{i\circ v}\,dv\right\|_1.$$

On account of (12.3.16) and $\lim_{|v|\to\infty}|v|^{\alpha}f_0^{\wedge}(v) = 1$, we may apply Lebesgue's dominated convergence theorem to deduce

$$\lim_{\rho\to\infty}\int_{-R}^{R}\left(1-\frac{|v|}{R}\right)\frac{\chi^{\wedge}(v)-1}{|v|^{\alpha}}\,|\rho v|^{\alpha}f_0^{\wedge}(\rho v)\,e^{ixv}\,dv = \int_{-R}^{R}\left(1-\frac{|v|}{R}\right)\frac{\chi^{\wedge}(v)-1}{|v|^{\alpha}}\,e^{ixv}\,dv$$

for every $R > 0$, $x \in \mathbb{R}$. Hence by Fatou's lemma and (12.3.17), (12.3.18)

$$\left\|\frac{1}{\sqrt{2\pi}}\int_{-R}^{R}\left(1-\frac{|v|}{R}\right)\frac{\chi^{\wedge}(v)-1}{|v|^{\alpha}}\,e^{i\circ v}\,dv\right\|_1 = O(1)$$

uniformly for $R > 0$, so that the representation (12.3.5) with $c = 1$ follows by Theorem 6.1.5, since $|v|^{-\alpha}[\chi^{\wedge}(v) - 1]$ is bounded on $\mathbb{R}$ by (12.3.16).

In view of Theorem 6.5.6 and Problem 6.5.5 we may formulate the result of Theorem 12.3.6(ii) and Prop. 12.3.8 as follows:

Corollary 12.3.9. *Let* $\chi \in \mathsf{NL}^1$. *Then* $\mathsf{V}[\mathsf{L}^1; c\,|v|^\alpha] \subset \{f \in \mathsf{L}^1 \mid \|J(f; \circ; \rho) - f(\circ)\|_1 = O(\rho^{-\alpha})\}$ *if and only if* λ *as defined by*

$$\lambda(v) = \begin{cases} \dfrac{\chi^\wedge(v) - 1}{c\,|v|^\alpha}, & v \neq 0 \\ 1, & v = 0 \end{cases} \tag{12.3.19}$$

is a multiplier of type (BV, BV).

12.3.3 Characterizations of Favard Classes: $1 < p \leq 2$

Next we turn to characterizations of Favard classes in L^p-space, $1 < p \leq 2$. The results follow straightforwardly, but instead of carrying over the method of proof of e.g. Theorem 12.3.6 to the present situation which obviously is possible, we shall indicate another way of proof, intimately connected with that of Theorem 6.1.5 and applicable to the case $p = 1$ as well.

Theorem 12.3.10. *Let* $f \in \mathsf{L}^p$, $1 < p \leq 2$.
(i) *If* χ *satisfies* (12.3.4), *then*

$$\|J(f; \circ; \rho) - f(\circ)\|_p = O(\rho^{-\alpha}) \qquad (\rho \to \infty) \tag{12.3.20}$$

implies the existence of $g \in \mathsf{L}^p$ *such that* $c\,|v|^\alpha f^\wedge(v) = g^\wedge(v)$ a.e.
(ii) *If in addition* χ *satisfies* (12.3.5), *then* $f \in \mathsf{V}[\mathsf{L}^p; c\,|v|^\alpha]$ *implies* (12.3.20).

Proof. (i) First we observe that the hypothesis implies by the convolution theorem and Titchmarsh's inequality that

$$\left\| \frac{\chi^\wedge(\circ/\rho) - 1}{\rho^{-\alpha}} f^\wedge(\circ) \right\|_{p'} \leq \left\| \frac{J(f; \circ; \rho) - f(\circ)}{\rho^{-\alpha}} \right\|_p \leq M \tag{12.3.21}$$

uniformly for large ρ. Hence (12.3.4) and Fatou's lemma yield $c\,|v|^\alpha f^\wedge(v) \in \mathsf{L}^{p'}$.†

Let Φ belong to the class F as defined by (6.1.10). Then by the Parseval formula (5.2.27) and Hölder's inequality

$$\begin{aligned} \left| \int_{-\infty}^{\infty} \frac{\chi^\wedge(v/\rho) - 1}{\rho^{-\alpha}} f^\wedge(v) \overline{\Phi^\wedge(v)}\, dv \right| &= \left| \int_{-\infty}^{\infty} \frac{J(f; u; \rho) - f(u)}{\rho^{-\alpha}} \overline{\Phi(u)}\, du \right| \\ &\leq \left\| \frac{J(f; \circ; \rho) - f(\circ)}{\rho^{-\alpha}} \right\|_p \|\Phi\|_{p'} \leq M\, \|\Phi\|_{p'} \end{aligned} \tag{12.3.22}$$

uniformly for large ρ. Furthermore, $\lim_{\rho\to\infty} \rho^\alpha[\chi^\wedge(v/\rho) - 1] f^\wedge(v) = c\,|v|^\alpha f^\wedge(v)$ a.e. by (12.3.8), and since the latter function belongs to $\mathsf{L}^{p'}$, in view of (12.3.21) we may apply Prop. 0.1.11 to obtain

$$\lim_{\rho\to\infty} \int_{-\infty}^{\infty} \frac{\chi^\wedge(v/\rho) - 1}{\rho^{-\alpha}} f^\wedge(v) \overline{\Phi^\wedge(v)}\, dv = \int_{-\infty}^{\infty} c\,|v|^\alpha f^\wedge(v) \overline{\Phi^\wedge(v)}\, dv$$

† If $p = 2$, the assertion follows from this point on by Plancherel's theorem, see also Problem 12.3.10.

for every $\Phi \in \mathsf{F}$, observing that $\Phi \in \mathsf{F}$ implies $\Phi, \Phi^\wedge \in \mathsf{L}^q$ for every $1 \le q \le \infty$. Together with (12.3.22) this yields

$$\left| \int_{-\infty}^{\infty} c\, |v|^\alpha f^\wedge(v) \overline{\Phi^\wedge(v)}\, dv \right| \le M \|\Phi\|_{p'}.$$

Thus the integral on the left defines a bounded linear functional on F considered as a subspace of $\mathsf{L}^{p'}$. Since F is dense in $\mathsf{L}^{p'}$ (cf. Problem 6.1.5), we may extend this functional to a bounded linear one on all of $\mathsf{L}^{p'}$ (cf. Prop. 0.7.1). Therefore by the Riesz representation theorem there exists $g \in \mathsf{L}^p$ such that

$$\int_{-\infty}^{\infty} c\, |v|^\alpha f^\wedge(v) \overline{\Phi^\wedge(v)}\, dv = \int_{-\infty}^{\infty} g(u) \overline{\Phi(u)}\, du = \int_{-\infty}^{\infty} g^\wedge(v) \overline{\Phi^\wedge(v)}\, dv,$$

the latter equality being valid by (5.2.27). Thus for every $\Phi \in \mathsf{F}$

$$\int_{-\infty}^{\infty} [c\, |v|^\alpha f^\wedge(v) - g^\wedge(v)] \overline{\Phi^\wedge(v)}\, dv = 0,$$

and since $[\mathsf{F}]^\wedge$ is dense in L^p as well, part (i) follows by Prop. 0.8.1, 0.8.11.

(ii) Let $g \in \mathsf{L}^p$ be such that $c\, |v|^\alpha f^\wedge(v) = g^\wedge(v)$ a.e. Then by (5.2.27) and (5.3.5) for every $\Phi \in \mathsf{F}$

$$\int_{-\infty}^{\infty} \frac{J(f; u; \rho) - f(u)}{\rho^{-\alpha}} \overline{\Phi(u)}\, du = \int_{-\infty}^{\infty} \frac{\chi^\wedge(v/\rho) - 1}{\rho^{-\alpha}} f^\wedge(v) \overline{\Phi^\wedge(v)}\, dv$$

$$= \int_{-\infty}^{\infty} [g * d\nu(\rho\circ)]^\wedge(v) \overline{\Phi^\wedge(v)}\, dv = \int_{-\infty}^{\infty} (g * d\nu(\rho\circ))(u) \overline{\Phi(u)}\, du.$$

Since $(g * d\nu(\rho\circ))$ belongs to L^p by Prop. 0.5.5 and since F is dense in $\mathsf{L}^{p'}$, it follows that for each $\rho > 0$

$$\frac{J(f; x; \rho) - f(x)}{\rho^{-\alpha}} = \frac{1}{\sqrt{2\pi}} \int_{-\infty}^{\infty} g(x - u)\, d\nu(\rho u) \quad \text{a.e.} \tag{12.3.23}$$

Hence by (0.5.9)

$$\left\| \frac{J(f; \circ; \rho) - f(\circ)}{\rho^{-\alpha}} \right\|_p \le \|g\|_p \|\nu\|_{\mathsf{BV}}$$

uniformly for all $\rho > 0$, and the theorem is completely established.

Thus in L^p-space, $1 < p \le 2$, the saturation class for singular integrals (12.3.1) with kernels χ satisfying (12.3.5) may be characterized as the class $\mathsf{V}[\mathsf{L}^p; c\, |v|^\alpha]$ just as for $p = 1$. Moreover, by the preceding arguments we have immediately

Theorem 12.3.11. *Let $f \in \mathsf{L}^p$, $1 < p \le 2$, and χ satisfy* (12.3.5). *Then the following assertions are equivalent:*

(i) $\|J(f; \circ; \rho) - f(\circ)\|_p = O(\rho^{-\alpha}) \qquad (\rho \to \infty),$

(ii) $f \in \mathsf{W}[\mathsf{L}^p; c\, |v|^\alpha]$,

(iii) $\left\| \frac{1}{\sqrt{2\pi}} \int_{-R}^{R} \left(1 - \frac{|v|}{R}\right) c\, |v|^\alpha f^\wedge(v)\, e^{i\circ v}\, dv \right\|_p = O(1) \qquad (R \to \infty),$

(iv) *there exists* $g \in \mathsf{L}^p$ *such that* (12.3.23) holds,
(v) *there exists* $g \in \mathsf{L}^p$ *such that* $\lim_{\rho\to\infty} \|\rho^\alpha[J(f;\circ;\rho) - f(\circ)] - g(\circ)\|_p = 0$,
(vi) *f has an αth strong Riesz derivative,*
(vii) *there exists* $g \in \mathsf{L}^p$ *such that*

$$\lim_{\varepsilon\to 0+} \left\| \frac{1}{C_{\alpha,2j}} \int_\varepsilon^\infty \frac{\overline{\Delta}_u^{2j} f(\circ)}{u^{1+\alpha}}\, du - g(\circ) \right\|_p = 0,$$

where $j \in \mathbb{N}$ *is chosen such that* $0 < \alpha < 2j$, *and* $C_{\alpha,2j}$ *is a constant given by* (11.3.14); *in this case, f has an αth strong Riesz derivative and* $D_s^{\{\alpha\}} f = g$,
(viii) *there exists* $g \in \mathsf{L}^p$ *such that for*

$0 < \alpha < 1$: $R_{\alpha,p'} g \in \mathsf{L}^p$ *and* $f(x) = [R_{\alpha/p}(R_{\alpha/p'} g)](x)$ a.e.,
$\alpha \geq 1$: $R_{\beta/p'} g$, $R_{\beta/p}(R_{\beta/p'} g)$ *belong to* L^p *and*

$$[H^{[\alpha]} f](x) = \int_{-\infty}^x du_1 \int_{-\infty}^{u_1} du_2 \ldots \int_{-\infty}^{u_{[\alpha]-1}} [R_{\beta/p}(R_{\beta/p'} g)](u_{[\alpha]})\, du_{[\alpha]} \quad \text{a.e.},$$

where $\alpha = [\![\alpha]\!] + \beta$, $0 \leq \beta < 1$, *and* $R_0 g = g$.

For statements (vi)–(viii) we refer to Theorems 11.2.6–11.2.9, 11.3.7, and Problem 11.5.2.

Problems

1. Give an example of a function $\chi \in \mathsf{NL}^1$ which satisfies (12.3.4) but not (12.3.5). (Hint: Sunouchi [8] gives an example due to P. Malliavin)
2. Let $f \in \mathsf{L}^1$ and $\chi \in \mathsf{NL}^1$ be even and positive. Show that if $\|J(f;\circ;\rho) - f(\circ)\|_1 = O(\rho^{-\alpha})$ for some $\alpha > 2$, then $f(x) = 0$ almost everywhere. (Hint: See also Butzer–König [1])
3. Let $\chi \in \mathsf{NL}^1$ be even and positive. Show that (12.3.4) holds for $\alpha = 2$ if and only if the 2nd moment of χ exists. In this case we have $c = -m(\chi; 2)/2$. (Hint: Prop. 5.1.19)
4. Let $f \in \mathsf{L}^p$, $1 \leq p \leq 2$, and χ be an even and positive function satisfying (12.3.4) with $\alpha = 2$. Show that the corresponding singular integral $J(f; x; \rho)$ is saturated in L^p with order $O(\rho^{-2})$ and that the Favard class $\mathsf{F}[\mathsf{L}^p; J(\rho)]$ for e.g. $p = 1$ is characterized by any one of the following statements:
   (i) There exists $\mu \in \mathsf{BV}$ such that $cv^2 f^\wedge(v) = \mu^\vee(v)$ for all $v \in \mathbb{R}$,
   (ii) there exists $\mu \in \mathsf{BV}$ such that $f(x) = \int_{-\infty}^x du_1 \int_{-\infty}^{u_1} d\mu(u_2)$ a.e.,
   (iii) $\|f(\circ + h) + f(\circ - h) - 2f(\circ)\|_1 = O(h^2)$ $\quad (h \to 0)$.
   (Hint: Theorems 12.3.6(i), 10.5.4, and 5.3.15)
5. (a) As an application of the preceding Problem show ($1 \leq p \leq 2$):
   (i) The singular integral of Gauss–Weierstrass is saturated in L^p with order $O(t)$, $t \to 0+$.
   (ii) The singular integral of Jackson–de La Vallée Poussin is saturated in L^p with order $O(\rho^{-2})$, $\rho \to \infty$.
   (iii) The moving averages are saturated in L^p with order $O(h^2)$, $h \to 0+$.
   (iv) The Bessel potentials are saturated in L^p with order $O(\rho^{-2})$, $\rho \to \infty$.
   In each case, the saturation class is characterized as $\mathsf{V}[\mathsf{L}^p; cv^2]$ with $c = -1$; $-3/2$; $-1/6$; $-\alpha/2$, respectively.
   (b) Show that for a twice continuously differentiable $f \in \mathsf{W}[\mathsf{L}^p; |v|^2]$, $1 \leq p \leq 2$, the following limit relations of Voronovskaja type are valid (cf. (3.3.9), Problem 3.4.5):

(i) $$\lim_{t\to 0+} \left\| \frac{W(f;\circ;t) - f(\circ)}{t} - f''(\circ) \right\|_p = 0,$$

(ii) $$\lim_{\rho\to\infty} \left\| \rho^2 [N^*(f;\circ;\rho) - f(\circ)] - \frac{3}{2} f''(\circ) \right\|_p = 0,$$

(iii) $$\lim_{h\to 0+} \left\| \frac{A^r(f;\circ;h) - f(\circ)}{h^2} - \frac{r}{6} f''(\circ) \right\|_p = 0.$$

6. Let $f \in L^p$, $1 \le p \le 2$, and χ satisfy (12.3.5). Show that the following statements are equivalent:
   (i) There exists $g \in L^p$ such that $\lim_{\rho\to\infty} \|\rho^\alpha[J(f;\circ;\rho) - f(\circ)] - cg(\circ)\|_p = 0$,
   (ii) there exists $g \in L^p$ such that $|v|^\alpha f^\wedge(v) = g^\wedge(v)$ (a.e.),
   (iii) there exists $g \in L^p$ such that
   $$\rho^\alpha[J(f;x;\rho) - f(x)] = \frac{c}{\sqrt{2\pi}}\int_{-\infty}^{\infty} g(x-u)\,d\nu(\rho u) \quad \text{a.e.}$$
   In any event, the αth strong Riesz derivative of f exists and $D_s^{\{\alpha\}}f = g$.

As already mentioned, the results of this section may also be established for general singular integrals $I(f;x;\rho)$ as given by (3.1.3). Instead of conditions (12.3.4), (12.3.5) we have (ρ is a positive parameter tending to infinity):

(12.3.24) Given an approximate identity $\{\chi(x;\rho)\}$, let there exist a function $\psi(v)$ defined and continuous on $\mathbb{R}$ with isolated zeros, and a positive function $\varphi(\rho)$ tending monotonely to zero as $\rho \to \infty$ such that
$$\lim_{\rho\to\infty} \frac{\chi^\wedge(v;\rho) - 1}{\varphi(\rho)} = \psi(v).$$

(12.3.25) Given an approximate identity $\{\chi(x;\rho)\}$, let there exist ψ, φ as in (12.3.24) and $\nu_\rho \in BV$ such that
$$\frac{\chi^\wedge(v;\rho) - 1}{\varphi(\rho)} = \psi(v)\nu_\rho^\vee(v)$$
for all $v \in \mathbb{R}$ and $\|\nu_\rho\|_{BV} \le M$ for all $\rho > 0$.

(12.3.26) Given an approximate identity $\{\chi(x;\rho)\}$, let there exist ψ, φ as in (12.3.24) and an approximate identity $\{h(x;\rho)\}$ such that for all $v \in \mathbb{R}$
$$\frac{\chi^\wedge(v;\rho) - 1}{\varphi(\rho)} = \psi(v)h^\wedge(v;\rho).$$

7. Let $f \in L^p$, $1 \le p \le 2$, and $\{\chi(x;\rho)\}$ satisfy (12.3.24). Show that the following conclusions are valid:
   (i) If there exists $g \in L^p$ such that $\lim_{\rho\to\infty} \|(\varphi(\rho))^{-1}[I(f;\circ;\rho) - f(\circ)] - g(\circ)]\|_p = 0$, then $f \in W[L^p;\psi(v)]$.
   (ii) If $\|I(f;\circ;\rho) - f(\circ)\|_p = O(\varphi(\rho))$, then $f \in V[L^p;\psi(v)]$.
   (iii) If $\{\chi(x;\rho)\}$ in addition satisfies (12.3.25), then conversely $f \in V[L^p;\psi(v)]$ implies $\|I(f;\circ;\rho) - f(\circ)\|_p = O(\varphi(\rho))$.

8. (i) Let $f \in L^p$, $1 \le p \le 2$, and $\{\chi(x;\rho)\}$ satisfy (12.3.25). Show that $f * d\nu_\rho \in W[L^p;\psi(v)]$ and $\psi(v)[f * d\nu_\rho]^\wedge(v) = (\varphi(\rho))^{-1}[I(f;\circ;\rho) - f(\circ)]^\wedge(v)$ for every $f \in L^p$. Thus, if $\{\chi(x;\rho)\}$ in addition satisfies (12.3.26), show that $W[L^p;\psi(v)]$ is dense in L^p.
   (ii) Let $f \in L^p$, $1 \le p \le 2$, and $\{\chi(x;\rho)\}$ satisfy (12.3.26). Show that $f \in W[L^p;\psi(v)]$ if and only if there exists $g \in L^p$ so that $\lim_{\rho\to\infty} \|(\varphi(\rho))^{-1}[I(f;\circ;\rho) - f(\circ)] - g(\circ)\|_p = 0$.

9. Let $f \in L^1$ and $\{\chi(x;\rho)\}$ satisfy (12.3.26). Show that the following assertions are equivalent:
   (i) $\|I(f;\circ;\rho) - f(\circ)\|_1 = O(\varphi(\rho))$ $\quad(\rho \to \infty)$,
   (ii) $f \in V[L^1;\psi(v)]$,
   (iii) $\left\| \frac{1}{\sqrt{2\pi}}\int_{-R}^{R}\left(1 - \frac{|v|}{R}\right)\psi(v)f^\wedge(v)\,e^{i\circ v}\,dv \right\|_1 = O(1)$ $\quad(R \to \infty)$,
   (iv) $f \in \widetilde{W[L^1;\psi(v)]}^{L^1}$,
   (v) there exists $\mu \in BV$ such that for each $\rho > 0$

(12.3.27)
$$\frac{I(f;x;\rho) - f(x)}{\varphi(\rho)} = \frac{1}{\sqrt{2\pi}}\int_{-\infty}^{\infty} h(x-u;\rho)\,d\mu(u) \quad \text{a.e.,}$$

(vi) there exists $\mu \in \mathsf{BV}$ such that for every $s \in \mathsf{C}_0$

$$\lim_{\rho\to\infty} \int_{-\infty}^{\infty} s(u)(\varphi(\rho))^{-1}[I(f; u; \rho) - f(u)]\, du = \int_{-\infty}^{\infty} s(u)\, d\mu(u).$$

State and prove analogous results for L^p-space, $1 < p \le 2$ (cf. Theorem 12.3.11(i)–(v)).

10. Given $\chi \in \mathsf{NL}^1$, let there exist constants $c \neq 0$, $\alpha > 0$ such that $|\lambda(v)| \le M$ for all $v \in \mathbb{R}$, λ being defined as in (12.3.19). Show that $f \in \mathsf{V}[\mathsf{L}^2; c\,|v|^\alpha]$ (cf. also Theorem 6.5.5) implies $\|J(f; \circ; \rho) - f(\circ)\|_2 = O(\varphi(\rho))$.
11. Instead of using Cramér's representation theorem prove Theorem 12.3.6 with the aid of Theorem 6.2.1 or 6.2.3. (Hint: See also Butzer–König [1])
12. Let $f \in \mathsf{L}^p$, $1 \le p \le 2$, and χ satisfy (12.3.4). Suppose that $\lambda(v)$ as defined by (12.3.19) is a multiplier of type $(\mathsf{L}^p, \mathsf{L}^p)$. Show that the corresponding singular integral $J(f; x; \rho)$ is saturated in L^p with order $O(\rho^{-\alpha})$ and saturation class $\mathsf{F}[\mathsf{L}^p; J(\rho)]$ characterized as $\mathsf{V}[\mathsf{L}^p; c\,|v|^\alpha]$. (Hint: Theorems 6.5.5–6.5.7) What can be said about the necessity of the multiplier condition?
13. Let the kernel χ of the singular integral $J(f; x; \rho)$ satisfy (12.3.5). Show that for $1 \le p \le 2$ the Banach spaces $\mathsf{V}[\mathsf{L}^p; |v|^\alpha]$, endowed with norm (10.5.11), and $\mathsf{L}^p_{\{J(\rho)\},\rho^{-\alpha}}$, endowed with norm (12.1.8), are equal with equivalent norms. (Hint: Compare with Theorems 12.3.6, 12.3.10)

12.4 Applications to Various Singular Integrals

The object here is to present applications of the general theorems obtained in the preceding section. In all of these, condition (12.3.4) is readily verified. To prove (12.3.5) we shall mainly make use of the actually stronger condition†

(12.4.1) *Given $\chi \in \mathsf{NL}^1$, let there exist constants $c \neq 0$, $\alpha > 0$, and a function $h \in \mathsf{NL}^1$ such that for all $v \neq 0$*

$$\frac{\chi^\wedge(v) - 1}{c\,|v|^\alpha} = h^\wedge(v).$$

Thus, in this section we are concerned with different ways to verify (12.4.1) for a number of examples. Once this is done, the results here follow as immediate consequences of those of Sec. 12.3.

12.4.1 Singular Integral of Fejér

A first application can be made to the singular integral of Fejér as defined by

$$\sigma(f; x; \rho) = \frac{2}{\pi\rho} \int_{-\infty}^{\infty} f(x-u)\, \frac{\sin^2(\rho u/2)}{u^2}\, du \tag{12.4.2}$$

(cf. (3.1.14), Problem 5.1.2). It is of type (3.1.8) with kernel F,

$$F(x) = \frac{1}{\sqrt{2\pi}}\left(\frac{\sin(x/2)}{x/2}\right)^2, \qquad F^\wedge(v) = \begin{cases} 1 - |v|, & |v| \le 1 \\ 0, & |v| \ge 1. \end{cases} \tag{12.4.3}$$

† In the particular instance of the Hilbert space L^2 we may apply Problem 12.3.10, the hypothesis of which is easily verified in the applications.

Since

$$(12.4.4)\qquad \frac{F^\wedge(v)-1}{|v|}=\begin{cases}-1, & |v|\le 1\\ -1/|v|, & |v|\ge 1,\end{cases}$$

it is obvious that the Fejér kernel satisfies (12.3.4) with $\alpha = 1$ and $c = -1$. Regarding (12.4.1) we have by (12.4.4) that $[F^\wedge(v) - 1]/|v|$ is locally absolutely continuous and belongs to L^2, together with its derivative. Hence (12.4.1) follows by Prop. 6.3.10. It is left to Problem 12.4.1 to verify that the corresponding function h is even and given by

$$(12.4.5)\qquad h(x)=\sqrt{\frac{2}{\pi}}\int_x^\infty \frac{\sin u}{u^2}\,du \qquad (x>0).$$

Thus, the integral of Fejér is saturated in L^p, $1 \le p \le 2$, with order $O(\rho^{-1})$, and its Favard class $\mathsf{F}[\mathsf{L}^p; \sigma(\rho)]$ is characterized as $\mathsf{V}[\mathsf{L}^p; -|v|]$. More explicitly, we have

Proposition 12.4.1. *Let $f \in \mathsf{L}^p$, $1 \le p \le 2$.*

(a) *There exists $g \in \mathsf{L}^p$ such that*

$$(12.4.6)\qquad \lim_{\rho\to\infty}\|\rho[\sigma(f;\circ;\rho)-f(\circ)]-g(\circ)\|_p=0$$

if and only if $f \in \mathsf{W}[\mathsf{L}^p; -|v|]$, i.e., if and only if $f^\sim(x)$ is equal a.e. *to a function of AC_{loc} with derivative in L^p. In particular, if*

$$(12.4.7)\qquad \|\sigma(f;\circ;\rho)-f(\circ)\|_p=o(\rho^{-1}) \qquad (\rho\to\infty),$$

then $f(x) = 0$ a.e.

(b) *The following statements are equivalent:*

(i) $\|\sigma(f;\circ;\rho)-f(\circ)\|_p = O(\rho^{-1})$ $\qquad (\rho\to\infty)$,

(ii) *$f^\sim(x)$ is equal* a.e. *to a function* $\begin{cases}\text{in } \mathsf{BV} \text{ if } p=1\\ \text{in } \mathsf{AC}_{\text{loc}} \text{ with derivative in } \mathsf{L}^p \text{ if } 1<p\le 2,\end{cases}$

(iii) $\|f^\sim(\circ+h)-f^\sim(\circ)\|_p = O(|h|)$ $\qquad (h\to 0)$,

(iv) $\|f(\circ+h)-f(\circ)\|_p = O(|h|)$, *provided* $1 < p \le 2$.

For the proof of part (a) we refer to Problems 12.3.6 and 8.3.3, whereas part (b) is a consequence of Theorems 8.3.6, 10.5.6, 12.3.6, and 12.3.10. For further characterizations of the Favard class $\mathsf{F}[\mathsf{L}^p; \sigma(\rho)]$ we refer to Theorems 12.3.7 and 12.3.11.

12.4.2 Generalized Singular Integral of Picard

The next application is to the generalized singular integral of Picard, given by

$$(12.4.8)\qquad C_\kappa(f;x;\rho)=\frac{\rho}{\sqrt{2\pi}}\int_{-\infty}^{\infty} f(x-u)c_\kappa(\rho u)\,du,$$

where $c_\kappa(x)$ is defined by its Fourier transform

$$(12.4.9)\qquad c_\kappa^\wedge(v)=(1+|v|^\kappa)^{-1} \qquad (\kappa>0).$$

In view of Problem 6.4.5, c_κ belongs to NL^1 for every $\kappa > 0$ and, in fact, it is a positive function for $0 < \kappa \le 2$. In particular, $c_2(x) = \sqrt{\pi/2}\exp\{-|x|\}$, and $C_2(f; x; \rho)$ reduces to the (special) singular integral of Picard as considered in Problem 3.2.5.

In view of the identity

$$\frac{c_\kappa^\wedge(v) - 1}{-|v|^\kappa} = c_\kappa^\wedge(v) \tag{12.4.10}$$

it follows that the Picard kernel satisfies (12.3.4) with $\alpha = \kappa$ and $c = -1$. Obviously, (12.4.10) already furnishes the representation (12.4.1).

Proposition 12.4.2. *The generalized singular integral $C_\kappa(f; x; \rho)$ of Picard is saturated in* L^p, $1 \le p \le 2$, *with order* $\mathsf{O}(\rho^{-\kappa})$. *The Favard class* $\mathsf{F}[\mathsf{L}^p; C_\kappa(\rho)]$ *is characterized by any of the nine* [*eight*] *statements of Theorem* 12.3.7 [12.3.11], *respectively. In particular, we have* (*cf. statement* (v) [(iv)]): $f \in \mathsf{F}[\mathsf{L}^p; C_\kappa(\rho)]$ *if and only if there exists* $\mu \in \mathsf{BV}$ *if* $p = 1$ *and* $g \in \mathsf{L}^p$ *if* $1 < p \le 2$ *such that*

$$\rho^\kappa[C_\kappa(f; x; \rho) - f(x)] = \begin{cases} C_\kappa(d\mu; x; \rho) \\ C_\kappa(g; x; \rho) \end{cases} \quad \text{a.e.} \tag{12.4.11}$$

Note that in view of (12.4.10) equation (12.4.1) reduces to a 'recurrence formula' for the generalized singular integral of Picard. Hence the direct part of the saturation theorem follows by the uniform boundedness of the operators $C_\kappa(\rho)$ as defined by the original singular integral.

12.4.3 General Singular Integral of Weierstrass

The general singular integral of Weierstrass is defined by

$$W_\kappa(f; x; t) = \frac{1}{\sqrt{2\pi}} \int_{-\infty}^{\infty} f(x-u)\{t^{-1/\kappa} w_\kappa(t^{-1/\kappa}u)\}\, du, \tag{12.4.12}$$

the kernel w_κ being given by its Fourier transform

$$w_\kappa^\wedge(v) = \exp\{-|v|^\kappa\}. \qquad (\kappa > 0) \tag{12.4.13}$$

In view of Sec. 6.4.1, w_κ belongs to NL^1 for every $\kappa > 0$ and, in fact, it is a positive function for $0 < \kappa \le 2$. Therefore, (12.4.12) defines a singular integral of type (3.1.8) with parameter $\rho = t^{-1/\kappa}$, $t \to 0+$. For $\kappa = 1$ and $\kappa = 2$ the functions w_κ admit explicit representations given by (3.1.39) and (3.1.33); then the integral (12.4.12) is associated with the names of Cauchy–Poisson and Gauss–Weierstrass, respectively.

Now, w_κ satisfies (12.3.4) with $\alpha = \kappa$ and $c = -1$ since

$$\lim_{v\to\infty} \frac{w_\kappa^\wedge(v) - 1}{|v|^\kappa} = -1. \tag{12.4.14}$$

To establish (12.4.1) we could apply Theorem 6.3.11 (cf. Problem 12.4.2). But as in the periodic case of Sec. 12.2.3 we shall prefer to use the identity

$$\frac{e^{-|v|^\kappa} - 1}{-|v|^\kappa} = \int_0^1 e^{-|v|^\kappa \tau}\, d\tau. \tag{12.4.15}$$

This implies that

$$\frac{w_\kappa^\wedge(v) - 1}{-|v|^\kappa} = \int_0^1 [\tau^{-1/\kappa} w_\kappa(\tau^{-1/\kappa} \circ)]^\wedge(v)\, d\tau = \left[\int_0^1 \{\tau^{-1/\kappa} w_\kappa(\tau^{-1/\kappa} \circ)\}\, d\tau\right]^\wedge(v),$$

where the change of the order of integration is justified by Fubini's theorem since

$$\int_0^1 \left\{\frac{1}{\sqrt{2\pi}} \int_{-\infty}^{\infty} |\tau^{-1/\kappa} w_\kappa(\tau^{-1/\kappa}x)|\, dx\right\} d\tau \le \|w_\kappa\|_1.$$

Thus w_κ satisfies (12.4.1) with $h(x) = \int_0^1 \tau^{-1/\kappa} w_\kappa(\tau^{-1/\kappa}x)\, d\tau$.

Proposition 12.4.3. *The general singular integral $W_\kappa(f; x; t)$ of Weierstrass is saturated in L^p, $1 \le p \le 2$, with order $\mathsf{O}(t)$, $t \to 0+$. The Favard class $\mathsf{F}[\mathsf{L}^p; W_\kappa(t)]$ is characterized by any of the nine [eight] statements of Theorem* 12.3.7 [12.3.11], *respectively. In particular (cf. statement* (v) [(iv)]): *$f \in \mathsf{F}[\mathsf{L}^p; W_\kappa(t)]$ if and only if there exists $\mu \in \mathsf{BV}$ if $p = 1$ and $g \in \mathsf{L}^p$ if $1 < p \le 2$ such that*

$$W_\kappa(f; x; t) - f(x) = \int_0^t \begin{Bmatrix} W_\kappa(d\mu; x; \tau) \\ W_\kappa(g; x; \tau) \end{Bmatrix} d\tau \quad \text{a.e.} \tag{12.4.16}$$

Let us mention that the identity (12.4.15) intrinsically depends upon the semi-group property of the general singular integral of Weierstrass. In other words (cf. Sec. 12.2.3 and 13.4.2), the bounded linear operators $W_\kappa(t)$ satisfy the functional equation

$$W_\kappa(t_1)\{W_\kappa(t_2)f\} = W_\kappa(t_1 + t_2)f \qquad (t_1, t_2 > 0) \tag{12.4.17}$$

for every $f \in \mathsf{X}(\mathbb{R})$. This is easily checked by means of Fubini's theorem but all details are left to Sec. 13.4.2. Here we only observe that in view of (12.4.16) the equation (12.4.1) for the present integral again reduces to a certain type of recursive formula. Thus as in Sec. 12.4.2 the direct part of the saturation theorem follows by the uniform boundedness of the operators $W_\kappa(t)$.

At this stage, we recall that for $\kappa = 1$ the integral (12.4.12) reduces to that of Cauchy–Poisson, thus ($t = y$)

$$W_1(f; x; t) = P(f; x, y) = \frac{y}{\pi} \int_{-\infty}^{\infty} \frac{f(x-u)}{u^2 + y^2}\, du.$$

Now, in Chapter 7 we saw that this integral is the unique solution of Dirichlet's problem for the upper half-plane as posed in (7.2.13). We study the behaviour of this solution on the boundary $y = 0$. The question as to the best possible order of approximation to the boundary value f is posed, and we give necessary and sufficient conditions upon f such that this order is attained.

Proposition 12.4.4. *Let $f \in \mathsf{L}^1$.*

(a) *If there is $g \in \mathsf{L}^1$ such that* $\liminf_{y \to 0+} \|y^{-1}[P(f; \circ; y) - f(\circ)] - g(\circ)\|_1 = 0$, *then $f^\sim$ is equal* a.e. *to a function of AC_{loc} with derivative in L^1. In particular,* $\|P(f; \circ; y) - f(\circ)\|_1 = \mathsf{o}(y)$, $y \to 0+$, *implies that f is the nullfunction.*

(b) *For* $0 < \alpha < 1$: $\|P(f; \circ; y) - f(\circ)\|_1 = \mathsf{O}(y^\alpha) \Leftrightarrow f \in \mathsf{Lip}\,(\mathsf{L}^1; \alpha)$,
$\alpha = 1$: $\|P(f; \circ; y) - f(\circ)\|_1 = \mathsf{O}(y) \Leftrightarrow f^\sim$ *is equal* a.e. *to a function in* BV.

This theorem tells us that the difference in L^1-norm of the solution of Dirichlet's problem from the boundary value f is at most $\mathsf{O}(y)$ for small y, unless f is zero a.e., and that the order $\mathsf{O}(y)$ holds if and only if the Hilbert transform $f^\sim$ of f is equivalent to a function in BV. Moreover, the order of approximation is given by $\mathsf{O}(y^\alpha)$ $(0 < \alpha < 1)$ if and only if f is a $\mathsf{Lip}\,(\mathsf{L}^1; \alpha)$-function.

In case $\kappa = 2$, the integral (12.4.12) is that of Gauss–Weierstrass which has been seen to be the solution of the equation of heat-flow in an infinite rod (cf. Sec. 7.2.1). Thus one may interpret the approximation of the initial value $f(x)$ by the solution $W(f; x; t)$ from a physical point of view (cf. Prop. 12.2.2 in the periodic case); for further characterizations of the Favard class we refer to Prop. 12.4.6.

12.4.4 Singular Integral of Bochner–Riesz

So far we have established the important condition (12.4.1) not as much by an application of general representation theorems such as Theorem 6.3.11 as by making the most of the individual properties of the integrals under consideration. Thus we used identity (12.4.15) in the semi-group case. On the other hand, most of the singular integrals which are produced by summation of the Fourier inversion formula do not possess the semi-group property. However, a slight generalization of (12.4.15) will give a simple, but widely applicable test for, at least, (12.3.5) to hold.

In fact, it follows from (12.4.15) that

$$e^{-|v|^\kappa} - 1 = -\kappa \int_0^{|v|} e^{-u^\kappa} u^{\kappa-1}\, du,$$

thus for the kernel w_κ of the general singular integral of Weierstrass we have

$$w_\kappa^\wedge(v) - 1 = \kappa \int_0^{|v|} \{-w_\kappa^\wedge(u)\} u^{\kappa-1}\, du. \tag{12.4.18}$$

This suggests the following

Proposition 12.4.5. *If for an even function $\chi \in \mathsf{NL}^1$ there exists an even $H \in \mathsf{L}^1$ such that for some $\beta > 0$*

$$\chi^\wedge(v) - 1 = \beta \int_0^{|v|} H^\wedge(u) u^{\beta-1}\, du, \tag{12.4.19}$$

then (12.3.5) *is satisfied for $\alpha = \beta$.*

Proof. Since $H^\wedge \in \mathsf{C}_0$, the integral in (12.4.19) exists as a Riemann integral (at least improper) for every $\beta > 0$. Changing variables we obtain for each $v \neq 0$

$$\frac{\chi^\wedge(v) - 1}{|v|^\beta} = \beta \int_0^1 H^\wedge(|v|\, u) u^{\beta-1}\, du = \lim_{n\to\infty} \beta \sum_{k=1}^{n} H^\wedge\Big(|v| \frac{k}{n}\Big)\Big(\frac{k}{n}\Big)^{\beta-1} \frac{1}{n} = \lim_{n\to\infty} R_n^\wedge(v),$$

say. Here $R_n^\wedge(v)$, obviously being the Fourier transform of some $R_n \in \mathsf{L}^1$, denotes the approximate Riemann sum of the integral. It follows that

$$\|R_n\|_1 \le \|H\|_1 \frac{\beta}{n^\beta} \sum_{k=1}^{n} k^{\beta-1} \le 2^\beta \|H\|_1 \tag{12.4.20}$$

uniformly for all $n \in \mathbb{N}$. Theorem 6.2.3 then gives the existence of some $\nu \in \mathsf{BV}$ such that (12.3.5) holds with $\alpha = \beta$.

Turning to applications, we consider the integral of Bochner–Riesz as defined by

$$B_\lambda(f; x; \rho) = \frac{\rho}{\sqrt{2\pi}} \int_{-\infty}^{\infty} f(x-u) b_\lambda(\rho u)\, du, \tag{12.4.21}$$

the kernel b_λ being given by its Fourier transform

$$b_\lambda^\frown(v) = \begin{cases} (1 - v^2)^\lambda, & |v| \le 1 \\ 0, & |v| \ge 1 \end{cases} \qquad (\lambda > 0). \tag{12.4.22}$$

By Problem 6.4.9, b_λ belongs to NL^1 for every $\lambda > 0$ and, in fact, admits the explicit representation

$$b_\lambda(x) = 2^\lambda \Gamma(1 + \lambda)|x|^{-((1/2)+\lambda)} J_{(1/2)+\lambda}(|x|), \tag{12.4.23}$$

where $J_\gamma(x)$ is the Bessel function of order γ.

It follows that

$$\lim_{v \to 0} \frac{b_\lambda^\frown(v) - 1}{v^2} = -\lambda, \tag{12.4.24}$$

and thus b_λ satisfies (12.3.4) with $\alpha = 2$ and $c = -\lambda$.

To establish (12.3.5) we have in view of the identity

$$b_\lambda^\frown(v) - 1 = 2 \int_0^{|v|} [-\lambda b_{\lambda-1}^\frown(u)]u\, du \tag{12.4.25}$$

that b_λ satisfies (12.4.19) with $\beta = 2$ and $H(x) = -\lambda b_{\lambda-1}(x)$. Therefore Prop. 12.4.5 applies for $\lambda > 1$. In fact, for all $\lambda > 0$

Proposition 12.4.6. *Let* $f \in \mathsf{L}^p$, $1 < p \le 2$.

(a) *The approximation*

$$\|B_\lambda(f; \circ; \rho) - f(\circ)\|_p = o(\rho^{-2}) \qquad (\rho \to \infty)$$

implies $f(x) = 0$ a.e.

(b) *The following assertions are equivalent:*

(i) $\|B_\lambda(f; \circ; \rho) - f(\circ)\|_p = O(\rho^{-2}) \qquad (\rho \to \infty)$,

(ii) *there exists* $g \in \mathsf{L}^p$ *such that* $\lim_{\rho\to\infty} \|\rho^2[B_\lambda(f; \circ; \rho) - f(\circ)] - g(\circ)\|_p = 0$,

(iii) *there exists* $g \in \mathsf{L}^p$ *such that* $-\lambda v^2 f^\wedge(v) = g^\wedge(v)$ a.e.,

(iv) f *is equal* a.e. *to a function in* $\mathsf{AC}^1_{\mathrm{loc}}$ *with first and second derivative in* L^p,

(v) *there exists* $g \in \mathsf{L}^p$ *such that*

$$f(x) = (1/\lambda) \int_{-\infty}^{x} du_1 \int_{-\infty}^{u_1} g(u_2)\, du_2 \quad \text{a.e.}$$

(vi) $\|f(\circ + h) - 2f(\circ) + f(\circ - h)\|_p = O(h^2) \qquad (h \to 0)$.

Proof. In view of (12.4.24), part (a) follows for any $\lambda > 0$ by Prop. 12.3.2. Concerning part (b), for $\lambda > 1$ the equivalence of (i), (ii), and (iii) follows by Theorem 12.3.11 since (12.3.5) holds.

Thus let $0 < \lambda \le 1$. Then, according to (12.4.24), (i) implies (iii) by Theorem 12.3.10. If (iii) holds, we use the identity

$$B_\lambda(f; x; \rho) - f(x) = B_{\lambda+1}(f; x; \rho) - f(x) - \frac{1}{\lambda\rho^2} B_\lambda(g; x; \rho) \tag{12.4.26}$$

which follows by comparing the Fourier transforms of both sides. Moreover, (iii) implies $-(\lambda + 1)v^2 f^\wedge(v) = [\lambda^{-1}(\lambda + 1)g]^\wedge(v)$ a.e.; thus (iii) is valid for $\lambda + 1$, too. Since $\lambda + 1 > 1$, $\|B_{\lambda+1}(f; \circ; \rho) - f(\circ)\|_p = O(\rho^{-2})$, and since $\|B_\lambda(g; \circ; \rho)\|_p \le \|g\|_p$, (i) follows by (12.4.26). Regarding (ii) we again use (12.4.26). Indeed, if (iii) holds, then

$$\rho^2[B_\lambda(f; x; \rho) - f(x)] - g(x) = \left\{\rho^2[B_{\lambda+1}(f; x; \rho) - f(x)] - \frac{\lambda + 1}{\lambda} g(x)\right\} - \frac{1}{\lambda}\{B_\lambda(g; x; \rho) - g(x)\}.$$

Since $\lambda + 1 > 1$, the first term on the right tends to zero in L^p-norm as $\rho \to \infty$, whereas $\lim_{\rho\to\infty} \|B_\lambda(g; \circ; \rho) - g(\circ)\|_p = 0$ by Theorem 3.1.6.

Finally, the equivalence of (iii)–(vi) follows by Theorems 5.2.21 and 10.5.4. This completes the proof.

Naturally, there exists an analogous result in L^1-space. Thus the singular integral $B_\lambda(f; x; \rho)$ of Bochner–Riesz is saturated in L^p, $1 \le p \le 2$, with order $O(\rho^{-2})$, and f belongs to the Favard class $\mathsf{F}[\mathsf{L}^p; B_\lambda(\rho)]$ if and only if $\|f(\circ + h) - 2f(\circ) + f(\circ - h)\|_p = O(h^2)$, $h \to 0$.

12.4.5 Riesz Means

Finally we consider the Riesz means

$$R_{\kappa,\lambda}(f; x; \rho) = \frac{1}{\sqrt{2\pi}} \int_{-\rho}^{\rho} \left(1 - \left|\frac{v}{\rho}\right|^\kappa\right)^\lambda f^\wedge(v)\, e^{ixv}\, dv \qquad (\kappa, \lambda > 0) \tag{12.4.27}$$

of the Fourier inversion integrals (5.1.6) and (5.2.19) of $f \in \mathsf{L}^p$, $1 \le p \le 2$. In view of the Parseval formulae (5.1.5) and (5.2.18), $R_{\kappa,\lambda}(f; x; \rho)$ may be rewritten as a singular integral of Fejér's type

$$R_{\kappa,\lambda}(f; x; \rho) = \frac{\rho}{\sqrt{2\pi}} \int_{-\infty}^{\infty} f(x - u) r_{\kappa,\lambda}(\rho u)\, du. \tag{12.4.28}$$

By Problem 6.4.8, $r_{\kappa,\lambda}$ belongs to NL^1 for each $\kappa > 0$, $\lambda > 0$ and has Fourier transform given by

$$r^\wedge_{\kappa,\lambda}(v) = \begin{cases} (1 - |v|^\kappa)^\lambda, & |v| \le 1 \\ 0, & |v| \ge 1. \end{cases} \tag{12.4.29}$$

In fact, $r_{\kappa,\lambda}$ is a positive function for $0 < \kappa \le 1$ and $\lambda \ge 1$. The integral $B_\lambda(f; x; \rho)$ of Bochner–Riesz is the case $\kappa = 2$ of the Riesz means $R_{\kappa,\lambda}(f; x; \rho)$. Since

$$\lim_{v\to 0} \frac{r^\wedge_{\kappa,\lambda}(v) - 1}{|v|^\kappa} = -\lambda, \tag{12.4.30}$$

the kernel $r_{\kappa,\lambda}$ satisfies (12.3.4) with $\alpha = \kappa$ and $c = -\lambda$. To establish (12.4.1) we may apply Theorem 6.3.11. For, if we set

$$\eta_{\kappa,\lambda}(v) = \begin{cases} -\dfrac{r^\wedge_{\kappa,\lambda}(v) - 1}{\lambda|v|^\kappa}, & v \ne 0 \\ 1, & v = 0, \end{cases}$$

then for $\lambda \geq 1$, $\kappa > 0$, $\eta_{\kappa,\lambda}$ and $\eta'_{\kappa,\lambda}$ are locally absolutely continuous on $(0, \infty)$ except that $\eta'_{\kappa,\lambda}$ has a jump for $\lambda = 1$ at $x_1 = 1$: $\eta'_{\kappa,1}(1+) - \eta'_{\kappa,1}(1-) = -\kappa$. Furthermore, $\int_0^\infty v\,|\eta''_{\kappa,\lambda}(v)|\,dv < \infty$ for every $\lambda \geq 1$ and $\kappa > 0$. Thus for these values of κ and λ, $\eta_{\kappa,\lambda}$ is an even and quasi-convex function of class C_0, and so by Theorem 6.3.11 it is representable as the Fourier transform of an even function of class L^1. Moreover, for all $\kappa > 0$, $\lambda > 0$

Proposition 12.4.7. *The Riesz means $R_{\kappa,\lambda}(f; x; \rho)$ are saturated in L^p, $1 \leq p \leq 2$, with order $O(\rho^{-\kappa})$, and the Favard class $\mathsf{F}[\mathsf{L}^p; R_{\kappa,\lambda}(\rho)]$ is characterized as $\mathsf{V}[\mathsf{L}^p; -\lambda\,|v|^\kappa]$.*

Proof. Since (12.3.4) holds for each $\kappa > 0$, $\lambda > 0$, Prop. 12.3.2 and Theorems 12.3.6(i) and 12.3.10(i) apply. Let $\lambda \geq 1$. Then condition (12.4.1) was shown to hold for each $\kappa > 0$, and the proof follows by Theorems 12.3.6(ii) and 12.3.10(ii). If $0 < \lambda < 1$ and $\mu \in \mathsf{BV}$ is such that $-\lambda\,|v|^\kappa f^\wedge(v) = \mu^\vee(v)$ for all $v \in \mathbb{R}$, we use the identity

$$R_{\kappa,\lambda}(f; x; \rho) - f(x) = R_{\kappa,\lambda+1}(f; x; \rho) - f(x) - \frac{1}{\lambda\rho^\kappa} R_{\kappa,\lambda}(d\mu; x; \rho) \tag{12.4.31}$$

and complete the proof as in the case of the integral of Bochner–Riesz.

For further characterizations of the Favard class of the Riesz means we refer to Theorems 12.3.7 and 12.3.11.

Problems

1. Show that the following singular integrals satisfy (12.4.1) with

   (i) Fejér: $\alpha = 1$, $c = -1$,
   $$h(x) = \sqrt{\frac{2}{\pi}} \int_x^\infty \frac{\sin u}{u^2}\,du,$$
   (ii) Cauchy–Poisson: $\alpha = 1$, $c = -1$,
   $$h(x) = \frac{1}{\sqrt{2\pi}} \log \frac{1 + x^2}{x^2},$$
   (iii) Gauss–Weierstrass: $\alpha = 2$, $c = -1$,
   $$h(x) = \sqrt{2}\left\{e^{-x^2/4} - x\int_{x/2}^\infty e^{-u^2}\,du\right\}.$$
   Here h is an even function, and the representations hold for $x > 0$. (Hint: cf. Butzer [7])
2. Show by an application of Theorem 6.3.11 that the kernels of the general singular integral of Weierstrass and the Bessel potentials satisfy (12.4.1).
3. Let $f \in \mathsf{L}^p$, $1 \leq p \leq 2$. Show that if $f^\sim$ is equal a.e. to a function $\Phi \in \mathsf{AC}_{\text{loc}}$ with $\Phi' \in \mathsf{L}^p$, then the following limit relations of Voronovskaja type are valid:

   (i) $\lim_{\rho\to\infty} \|\rho[f(\circ) - \sigma(f; \circ; \rho)] - \Phi'(\circ)\|_p = 0$,

   (ii) $\displaystyle\lim_{y\to 0+} \left\|\frac{f(\circ) - P(f; \circ; y)}{y} - \Phi'(\circ)\right\|_p = 0.$

12.5 Saturation of Higher Order

12.5.1 Singular Integrals on the Real Line

So far we have seen that the most fundamental condition upon the kernel χ such that the corresponding singular integral $J(f; x; \rho)$ may possess saturation phenomena is given by

(12.5.1) $$\lim_{v\to 0} \frac{\chi^{\wedge}(v) - 1}{|v|^{\alpha}} = c_1 \neq 0$$

for some constants $\alpha > 0$, $c_1 \in \mathbb{C}$ (cf. (12.3.4)). If $\chi^{\wedge}$ is a function of $|v|^{\alpha}$, say $\chi^{\wedge}(v) = \eta(|v|^{\alpha})$, then (12.5.1) reduces to

$$\lim_{t\to 0+} \frac{\eta(t) - 1}{t} = c_1 \neq 0,$$

and this means that η has a nonzero right-hand derivative at $t = 0$. This suggests to consider higher order Taylor expansions of η at $t = 0$, or, more generally, Peano expansions at $t = 0$. We are therefore led to discuss condition

(12.5.2) *Given $\chi \in \mathsf{NL}^1$, let there exist constants $\alpha > 0$ and $c_1, c_2, \ldots, c_r \in \mathbb{C}$ with $c_r \neq 0$, $r \in \mathbb{N}$, such that*

$$\lim_{v\to 0} \frac{\chi^{\wedge}(v) - 1 - \sum_{j=1}^{r-1} c_j |v|^{\alpha j}}{|v|^{\alpha r}} = c_r.$$

In connection with (12.3.5) and (12.4.1) we expect that also the following will be needed:

(12.5.3) *Given $\chi \in \mathsf{NL}^1$, let there exist constants $\alpha > 0$ and $c_1, c_2, \ldots, c_r \in \mathbb{C}$, $c_r \neq 0$, and a function $h_r \in \mathsf{NL}^1$ such that*

$$h_r^{\wedge}(v) = \begin{cases} \dfrac{\chi^{\wedge}(v) - 1 - \sum_{j=1}^{r-1} c_j |v|^{\alpha j}}{c_r |v|^{\alpha r}}, & v \neq 0 \\ 1, & v = 0. \end{cases}$$

The interest in these conditions is also motivated by the following considerations. If the kernel χ satisfies the condition

(12.5.4) $$h_1^{\wedge}(v) = \begin{cases} \dfrac{\chi^{\wedge}(v) - 1}{c_1 |v|^{\alpha}}, & v \neq 0 \\ 1, & v = 0 \end{cases}$$

for some $h_1 \in \mathsf{NL}^1$, then for $f \in \mathsf{W}[\mathsf{L}^p; c_1|v|^{\alpha}]$, $1 \leq p \leq 2$, we have by Problem 12.3.6 the representation

(12.5.5) $$\rho^{\alpha}[J(f; x; \rho) - f(x)] = \frac{c_1 \rho}{\sqrt{2\pi}} \int_{-\infty}^{\infty} (D_s^{\{\alpha\}} f)(x - u) h_1(\rho u)\, du \quad \text{a.e.},$$

and therefore by Theorem 3.1.6

(12.5.6) $$\lim_{\rho\to\infty} \|\rho^{\alpha}[J(f; \circ; \rho) - f(\circ)] - c_1 D_s^{\{\alpha\}} f\|_p = 0.$$

On the other hand, by the right side of (12.5.5) there may be defined for all $g \in X(\mathbb{R})$ a new singular integral $J_1(g; x; \rho)$ of Fejér's type, namely

$$J_1(g; x; \rho) = \frac{\rho}{\sqrt{2\pi}} \int_{-\infty}^{\infty} g(x-u)h_1(\rho u)\,du, \tag{12.5.7}$$

which is generated by the kernel h_1. In these terms, the asymptotic approximation (12.5.6) for the original integral $J(f; x; \rho)$ may be interpreted as ordinary convergence of the integral $J_1(f; x; \rho)$. Moreover, we may investigate the degree of convergence of g by the integral $J_1(g; x; \rho)$ and in particular ask for its saturation, thus for *second order saturation* of the original integral $J(f; x; \rho)$. In studying the difference $[h_1^\wedge(v) - 1]$, conditions of type (12.3.4) and (12.4.1) may be posed, thus that there exist constants $\beta > 0$ and $\gamma \neq 0$ such that

$$\lim_{v \to 0} \frac{h_1^\wedge(v) - 1}{|v|^\beta} = \gamma, \tag{12.5.8}$$

and that furthermore there exists $h_2 \in NL^1$ such that

$$h_2^\wedge(v) = \begin{cases} \dfrac{h_1^\wedge(v) - 1}{\gamma\, |v|^\beta}, & v \neq 0 \\ 1, & v = 0. \end{cases} \tag{12.5.9}$$

Going back to (12.5.2) and (12.5.3) for $r = 2$, one expects that one must verify (12.5.8) and (12.5.9) with $\gamma = c_2/c_1$ and $\beta = \alpha$, i.e., that the exponent β of (12.5.9) for h_1 must be equal to the exponent α of (12.5.4) for the original χ. This will do to indicate the connection between the saturation problem for the integrals $J(f; x; \rho)$ and $J_1(f; x; \rho)$.

Theorem 12.5.1. *Let* $f \in L^p$, $1 \leq p \leq 2$, *and* χ *satisfy* (12.5.2).

(i) *If the* $\alpha(r-1)$*th strong Riesz derivative of* f *exists and*

$$\left\| J(f; \circ; \rho) - f(\circ) - \sum_{j=1}^{r-1} c_j\, \rho^{-\alpha j} (D_s^{\{\alpha j\}} f)(\circ) \right\|_p = o(\rho^{-\alpha r}) \qquad (\rho \to \infty), \tag{12.5.10}$$

then $f(x) = 0$ a.e.

(ii) *If the* $\alpha(r-1)$*th strong Riesz derivative of* f *exists and*

$$\left\| J(f; \circ; \rho) - f(\circ) - \sum_{j=1}^{r-1} c_j\, \rho^{-\alpha j} (D_s^{\{\alpha j\}} f)(\circ) \right\|_p = O(\rho^{-\alpha r}) \qquad (\rho \to \infty), \tag{12.5.11}$$

then $f \in V[L^p; c_r\, |v|^{\alpha r}]$.

(iii) *If, in addition,* χ *satisfies* (12.5.3), *then conversely* $f \in V[L^p; c_r\, |v|^{\alpha r}]$ *implies* (12.5.11).

Proof. According to Problem 11.2.9 the existence of the $\alpha(r-1)$th strong Riesz derivative of f implies that the βth strong Riesz derivative of f exists for all $0 < \beta \leq \alpha(r-1)$ and $[D_s^{\{\beta\}} f]^\wedge(v) = |v|^\beta f^\wedge(v)$. Furthermore, setting

$$\nabla_{\alpha r} J(f; x; \rho) = J(f; x; \rho) - f(x) - \sum_{j=1}^{r-1} c_j\, \rho^{-\alpha j} (D_s^{\{\alpha j\}} f)(x), \tag{12.5.12}$$

we have

$$[\nabla_{\alpha r} J(f; \circ; \rho)]^\wedge(v) = \left[\chi^\wedge\left(\frac{v}{\rho}\right) - 1 - \sum_{j=1}^{r-1} c_j \left|\frac{v}{\rho}\right|^{\alpha j}\right] f^\wedge(v). \tag{12.5.13}$$

Thus, if (12.5.10) holds, $\|[\rho^{\alpha r} \nabla_{\alpha r} J(f; \circ; \rho)]^\wedge(\circ)\|_{p'} = o(1)$, $\rho \to \infty$. Therefore (12.5.2) implies $c_r |v|^{\alpha r} f^\wedge(v) = 0$, and (i) follows. Likewise, the proofs of (ii) and (iii) follow along the lines of the proofs of Theorems 12.3.6 and 12.3.10. We only recall the result of Problem 11.2.9 that $f \in \mathsf{V}[\mathsf{L}^p; c_r|v|^{\alpha r}]$, $1 \le p \le 2$, implies the existence of the βth strong Riesz derivative of f for every $0 < \beta < \alpha r$.

As a first example let us consider the generalized singular integral $C_\kappa(f; x; \rho)$ of Picard. In view of (12.4.10) we have $h_1(x) \equiv c_\kappa(x)$. Moreover,

$$\frac{c_\kappa^\wedge(v) - 1 - \sum_{j=1}^{r-1} (-1)^j |v|^{\kappa j}}{(-1)^r |v|^{\kappa r}} = c_\kappa^\wedge(v), \tag{12.5.14}$$

which already establishes (12.5.3) with $\alpha = \kappa$, $c_j = (-1)^j$, and $h_r(x) = c_\kappa(x)$. Therefore in the notation of (12.5.12)

Proposition 12.5.2. *Let $f \in \mathsf{L}^p$, $1 \le p \le 2$, and $\kappa > 0$.*

(a) *The following assertions are equivalent:*

(i) *The $\kappa(r-1)$th strong Riesz derivative of f exists and* $\|\nabla_{\kappa r} C_\kappa(f; \circ; \rho)\|_p = \mathsf{O}(\rho^{-\kappa r})$,

(ii) *the $\kappa(r-1)$th strong Riesz derivative of f exists and $\mu \in \mathsf{BV}$ if $p = 1$ or $g \in \mathsf{L}^p$ if $1 < p \le 2$ such that*

$$\rho^{\kappa r} \nabla_{\kappa r} C_\kappa(f; x; \rho) = \begin{cases} C_\kappa(d\mu; x; \rho) \\ C_\kappa(g; x; \rho) \end{cases} \quad \text{a.e.},$$

(iii) $f \in \mathsf{V}[\mathsf{L}^p; (-1)^r |v|^{\kappa r}]$,

(iv) $$\left\| \frac{1}{C_{\kappa r, 2j}} \int_\varepsilon^\infty \frac{\overline{\Delta}_u^{2j} f(\circ)}{|u|^{1+\kappa r}}\, du \right\|_p = \mathsf{O}(1) \qquad (\varepsilon \to 0+),$$

where $j \in \mathbb{N}$ is chosen such that $0 < \kappa r < 2j$ and $C_{\kappa r, 2j}$ is a constant given by (11.3.14).

(b) *Furthermore, the following statements are equivalent:*

(v) *There exists the $\kappa(r-1)$th strong Riesz derivative of f and $g \in \mathsf{L}^p$ such that* $\lim_{\rho\to\infty} \|\rho^{\kappa r} \nabla_{\kappa r} C_\kappa(f; \circ; \rho) - g(\circ)\|_p = 0$,

(vi) *the κrth strong Riesz derivative of f exists,*

(vii) *there exists $g \in \mathsf{L}^p$ such that $(-1)^r |v|^{\kappa r} f^\wedge(v) = g^\wedge(v)$.*

All of the assertions (i)–(vii) *are equivalent if $p \ne 1$.*

The proof follows by Theorem 12.5.1 in view of (12.5.14) (cf. Sec. 12.4.2).

Whereas for this integral it was very easy to establish condition (12.5.3) since it reduces to the recursive formula (12.5.14), for most of the other examples the following generalization of Prop. 12.4.5 works.

Proposition 12.5.3. *If for an even function $\chi \in \mathsf{NL}^1$ there exists an even $H \in \mathsf{L}^1$ such that for some $\beta > 0$, $r \in \mathbb{N}$*

$$\chi^\wedge(v) - 1 - \sum_{j=1}^{r-1} c_j |v|^{\beta j} = \beta \int_0^{|v|} (|v|^\beta - u^\beta)^{r-1} H^\wedge(u) u^{\beta-1}\, du, \tag{12.5.15}$$

then there exists $\nu_r \in \mathsf{NBV}$ such that condition (12.5.3) *is satisfied for $\alpha = \beta$, $h_r^\wedge(v)$ being replaced by $\nu_r^\vee(v)$.*

Proof. The proof follows as for Prop. 12.4.5. Indeed, changing variables we have by (12.5.15) that for all $v \ne 0$

$$\begin{aligned} \frac{\chi^\wedge(v) - 1 - \sum_{j=1}^{r-1} c_j |v|^{\beta j}}{|v|^{\beta r}} &= \beta \int_0^1 (1 - u^\beta)^{r-1} H^\wedge(|v|\, u) u^{\beta-1}\, du \\ &= \lim_{n\to\infty} \frac{\beta}{n^\beta} \sum_{j=1}^n \left(1 - \left(\frac{j}{n}\right)^\beta\right)^{r-1} H^\wedge\left(|v| \frac{j}{n}\right) j^{\beta-1} \equiv \lim_{n\to\infty} R_n^\wedge(v), \end{aligned}$$

where R_n belongs to L^1 and $R_n^\wedge(v)$ denotes the approximate Riemann sum of the integral. It follows that

$$\|R_n\|_1 \le \|H\|_1 \frac{\beta}{n^\beta} \sum_{j=1}^{n} \left(1 - \left(\frac{j}{n}\right)^\beta\right)^{r-1} j^{\beta-1} \le \|H\|_1 \frac{\beta}{n^\beta} \sum_{j=1}^{n} j^{\beta-1} \le 2^\beta \|H\|_1$$

uniformly for all $n \in \mathbb{N}$. Hence Theorem 6.2.3 gives the existence of some $\nu_r \in \mathsf{NBV}$ such that (12.5.3) holds with $\alpha = \beta$, $h_r^\wedge(v)$ being replaced by $\nu_r^\vee(v)$.

As an application, let us consider the singular integral $W(f; x; t)$ of Gauss–Weierstrass, the particular case $\kappa = 2$ of (12.4.12). Since

$$e^{-v^2} - 1 - \sum_{j=1}^{r-1} \frac{(-1)^j}{j!} v^{2j} = 2\frac{(-1)^r}{(r-1)!} \int_0^{|v|} (v^2 - u^2)^{r-1} e^{-u^2} u \, du, \tag{12.5.16}$$

we see that (12.5.15) is satisfied with $\beta = 2$, $c_j = (-1)^j/j!$ and $H(x) = \{(-1)^r/(r-1)!\}w_2(x)$. Concerning (12.5.12) for $W(f; x; t)$ we have $\alpha = 2$, $\rho = t^{-1/2}$, $t \to 0+$. Observing that $D_s^{\{2j\}}f$ exists if and only if f belongs to $\mathsf{W}[\mathsf{L}^p; (-1)^j(iv)^{2j}]$, and thus by Theorems 5.1.16, 5.2.21 if and only if there exists $\Phi \in \mathsf{AC}_{loc}^{2j-1}$ with $\Phi^{(k)} \in \mathsf{L}^p$, $0 \le k \le 2j$, such that $f(x) = \Phi(x)$ a.e., the expression (12.5.12) for the Gauss–Weierstrass integral is given by

$$\nabla_{2r} W(f; x; t) = W(f; x; t) - f(x) - \sum_{j=1}^{r-1} \frac{t^j}{j!} \Phi^{(2j)}(x). \tag{12.5.17}$$

Hence

Proposition 12.5.4. *For the integral $W(f; x; t)$ of Gauss–Weierstrass of $f \in \mathsf{L}^p$, $1 \le p \le 2$, we have:*

(a) *If $f \in \mathsf{W}[\mathsf{L}^p; |v|^{2r-2}]$ and $\|\nabla_{2r} W(f; \circ; t)\|_p = o(t^r)$, $t \to 0+$, then $f(x) = 0$ a.e.*

(b) *The following assertions are equivalent:*

(i) *f belongs to $\mathsf{W}[\mathsf{L}^p; |v|^{2r-2}]$ and $\|\nabla_{2r} W(f; \circ; t)\|_p = O(t^r)$, $t \to 0+$,*

(ii) *f belongs to $\mathsf{V}[\mathsf{L}^p; |v|^{2r}]$,*

(iii) *f belongs to $\mathsf{W}[\mathsf{L}^p; |v|^{2r-2}]$ and there exists $\mu \in \mathsf{BV}$ if $p = 1$ and $g \in \mathsf{L}^p$ if $1 < p \le 2$ such that*

$$\nabla_{2r} W(f; x; t) = \frac{1}{(r-1)!} \int_0^t (t - \tau)^{r-1} \begin{Bmatrix} W(d\mu; x; \tau) \\ W(g; x; \tau) \end{Bmatrix} d\tau \quad \text{a.e.},$$

(iv) *there exists $\mu \in \mathsf{BV}$ if $p = 1$ and $g \in \mathsf{L}^p$ if $1 < p \le 2$ such that*

$$f(x) = \int_{-\infty}^{x} du_1 \int_{-\infty}^{u_1} du_2 \ldots \int_{-\infty}^{u_{2r-1}} \begin{Bmatrix} d\mu(u_{2r}) \\ g(u_{2r})\, du_{2r} \end{Bmatrix} \quad \text{a.e.},$$

(v) *there exists a 2rth Peano difference of f such that $\|\diamondsuit_h^{2r} f(\circ)\|_p = O(h^{2r})$, $h \to 0$,*

(vi) $\|\Delta_h^{2r} f(\circ)\|_p = O(h^{2r})$ $(h \to 0)$.

The proof is left to Problem 12.5.1.

12.5.2 Periodic Singular Integrals

There is a completely analogous theory of saturation of higher order for periodic singular integrals $I_\rho(f; x)$ as defined by (1.1.3). But if we proceed as above and try to generalize (12.1.1), for example, in the way we generalized (12.3.4) to (12.5.2), then on account of the general dependence of the kernel upon the parameter ρ the coefficients in the respective expansions would be general functions φ_j and ψ_j, $j = 1, 2, \ldots, r$. Although this general approach is possible, the consequence would be a very intricate calculus, difficult to interpret. Therefore we restrict the discussion to the following case which covers many of the examples considered so far:

(12.5.18) *Let ρ be a positive parameter tending to infinity and let the (periodic) kernel $\{\chi_\rho^*(x)\}$ of $I_\rho^*(f; x)$ be generated by $\chi \in \mathsf{NL}^1$ (cf. (3.1.28)):*

$$\chi_\rho^*(x) = \sqrt{2\pi} \sum_{j=-\infty}^{\infty} \rho\chi(\rho(x + 2j\pi)).$$

Then $[\chi_\rho^*]^\wedge(k) = \chi^\wedge(k/\rho)$ by (5.1.58). In view of Sec. 12.3, 12.4 it now seems reasonable to restrict (12.1.1) to (12.3.8), i.e., let there exist $\alpha > 0$ and $c_1 \in \mathbb{C}$, $c_1 \neq 0$, such that for each $k \in \mathbb{Z}$

$$\lim_{\rho\to\infty} \frac{[\chi_\rho^*]^\wedge(k) - 1}{\rho^{-\alpha}} = c_1|k|^\alpha. \tag{12.5.19}$$

For the same reasons let there exist $h_1 \in \mathsf{NL}^1$ such that

$$h_1^\wedge\left(\frac{k}{\rho}\right) = \begin{cases} \dfrac{[\chi_\rho^*]^\wedge(k) - 1}{c_1|k/\rho|^\alpha}, & k \neq 0 \\ 1, & k = 0. \end{cases} \tag{12.5.20}$$

Now we might expand $[\chi_\rho^*]^\wedge$. But this would correspond to the expansion of $\chi^\wedge$ (cf. (12.5.2) and (12.5.3)), and the periodic theory would be reduced to that of the real line, i.e., to Sec. 12.5.1; then one might formulate a general theorem, however this we shall leave to Problem 12.5.2.

But let us continue with the Riesz means $R_{n,\kappa,\lambda}(f; x)$ of the Fourier series of $f \in \mathsf{X}_{2\pi}$ (cf. Problem 6.4.8)

$$R_{n,\kappa,\lambda}(f; x) = \frac{1}{2\pi}\int_{-\pi}^{\pi} f(x - u) r_{n,\kappa,\lambda}(u)\, du \tag{12.5.21}$$

with kernel

$$r_{n,\kappa,\lambda}(x) = \sum_{k=-n}^{n} \left(1 - \left|\frac{k}{n+1}\right|^\kappa\right)^\lambda e^{ikx} \qquad (\kappa, \lambda > 0). \tag{12.5.22}$$

Here, condition (12.1.1) is satisfied with $\varphi(n) = (n + 1)^{-\kappa}$ and $\psi(k) = -\lambda\, |k|^\kappa$ since

$$\lim_{n\to\infty} \frac{r_{n,\kappa,\lambda}^\wedge(k) - 1}{(n + 1)^{-\kappa}} = -\lambda\, |k|^\kappa. \tag{12.5.23}$$

Moreover, the kernel $\{r_{n,\kappa,\lambda}(x)\}$ is generated via (12.5.18) by $r_{\kappa,\lambda}(x)$ (cf. (12.4.29)), thus

$$r_{n,\kappa,\lambda}(x) = \sqrt{2\pi} \sum_{j=-\infty}^{\infty} (n + 1) r_{\kappa,\lambda}((n + 1)(x + 2j\pi)).$$

In view of Sec. 12.4.5, $r_{\kappa,\lambda}$ satisfies (12.4.1), and therefore the representation (12.5.20) holds for $\{r_{n,\kappa,\lambda}(x)\}$. Hence the Riesz means $R_{n,\kappa,\lambda}(f; x)$ are saturated in $\mathsf{X}_{2\pi}$ with order $O(n^{-\kappa})$, and the Favard class $\mathsf{F}[\mathsf{X}_{2\pi}; R_{n,\kappa,\lambda}]$ is characterized as $\mathsf{V}[\mathsf{X}_{2\pi}; -\lambda\, |k|^\kappa]$.

Concerning higher order saturation we note that in view of the Taylor expansion ($r \in \mathbb{N}$, $x > 0$, $\lambda > 0$)

$$(12.5.24)\quad (1-x)^\lambda = \sum_{j=0}^{r-1} (-1)^j \binom{\lambda}{j} x^j + (-1)^r \binom{\lambda}{r} r \int_0^x (x-\tau)^{r-1}(1-\tau)^{\lambda-r}\, d\tau,$$

$r^\wedge_{\kappa,\lambda}(v)$ of (12.4.29) satisfies

$$(12.5.25)\quad r^\wedge_{\kappa,\lambda}(v) - 1 - \sum_{j=1}^{r-1} (-1)^j \binom{\lambda}{j} |v|^{\kappa j} = (-1)^r \binom{\lambda}{r} r\kappa \int_0^{|v|} (|v|^\kappa - u^\kappa)^{r-1} r^\wedge_{\kappa,\lambda-r}(u) u^{\kappa-1}\, du.$$

Thus the representation (12.5.15) holds for $r_{\kappa,\lambda}$ with $\beta = \kappa$, $c_j = (-1)^j \binom{\lambda}{j}$, and $H(x) = (-1)^r r \binom{\lambda}{r} r_{\kappa,\lambda-r}(x)$. Therefore with (cf. Problem 12.5.2)

$$(12.5.26)\quad \nabla_{\kappa r} R_{n,\kappa,\lambda}(f;x) \equiv R_{n,\kappa,\lambda}(f;x) - f(x) - \sum_{j=1}^{r-1} (-1)^j \binom{\lambda}{j} (n+1)^{-\kappa j} f^{\{\kappa j\}}(x)$$

Proposition 12.5.5. *For the Riesz means $R_{n,\kappa,\lambda}(f;x)$ of the Fourier series of $f \in X_{2\pi}$ we have:*

(a) *If $f \in W[X_{2\pi}; |k|^{\kappa(r-1)}]$ and $\|\nabla_{\kappa r} R_{n,\kappa,\lambda}(f;\circ)\|_{X_{2\pi}} = o(n^{-\kappa r})$, then f is constant.*

(b) *The following assertions are equivalent:*

(i) *$f \in W[X_{2\pi}; |k|^{\kappa(r-1)}]$ and $\|\nabla_{\kappa r} R_{n,\kappa,\lambda}(f;\circ)\|_{X_{2\pi}} = O(n^{-\kappa r})$,*

(ii) *$f \in V[X_{2\pi}; |k|^{\kappa r}]$.*

(c) *The following assertions are equivalent:*

(i) *$f \in W[X_{2\pi}; |k|^{\kappa(r-1)}]$ and there exists $g \in X_{2\pi}$ such that*

$$\lim_{n\to\infty} \|(n+1)^{\kappa r} \nabla_{\kappa r} R_{n,\kappa,\lambda}(f;\circ) - g(\circ)\|_{X_{2\pi}} = 0,$$

(ii) *$f \in W[X_{2\pi}; |k|^{\kappa r}]$.*

In case $\lambda = l$, $l \in \mathbb{N}$, a new phenomenon occurs for the Riesz means $R_{n,\kappa,\lambda}(f;x)$. Then $(1-x)^l$ is an algebraic polynomial of degree l, and therefore the expansion (12.5.24) *breaks off* at $r = l + 1$. Indeed,

$$(12.5.27)\quad (1-x)^l = \sum_{j=0}^{l-1} (-1)^j \binom{l}{j} x^j + (-1)^l x^l,$$

and hence (cf. (12.4.29), (12.5.3), (12.5.25))

$$(12.5.28)\quad r^\wedge_{\kappa,l}(v) - 1 - \sum_{j=1}^{l-1} (-1)^j \binom{l}{j} |v|^{\kappa j} = (-1)^l |v|^{\kappa l} h^\wedge_{\kappa,l}(v)$$

for all $v \in \mathbb{R}$, where

$$(12.5.29)\quad h^\wedge_{\kappa,l}(v) = \begin{cases} 1, & |v| \le 1 \\ 1 - (1 - |v|^{-\kappa})^l, & |v| \ge 1. \end{cases}$$

Now it is easy to see that the function defined by (12.5.29) satisfies the conditions of Theorem 6.3.11 for every $\kappa > 0$ and $l \in \mathbb{N}$, and therefore it is actually the Fourier transform of a function $h_{\kappa,l} \in \mathsf{NL}^1$. Since $r^{\wedge}_{\kappa,l}(k/(n+1)) = r^{\wedge}_{n,\kappa,l}(k)$, the periodic kernel $\{r_{n,\kappa,l}(x)\}$ has the representation

$$(12.5.30)\quad r^{\wedge}_{n,\kappa,l}(k) - 1 - \sum_{j=1}^{l-1} (-1)^j \binom{l}{j} \left|\frac{k}{n+1}\right|^{\kappa j} = (-1)^l \left|\frac{k}{n+1}\right|^{\kappa l} [h^*_{n,\kappa,l}]^{\wedge}(k)$$

for every $k \in \mathbb{Z}$, where $\{h^*_{n,\kappa,l}(x)\}$ denotes the periodic approximate identity produced by $h_{\kappa,l}$ via (12.5.18), thus

$$(12.5.31)\qquad h^*_{n,\kappa,l}(x) = \sqrt{2\pi} \sum_{j=-\infty}^{\infty} (n+1) h_{\kappa,l}((n+1)(x+2j\pi)).$$

Furthermore, since $[h^*_{n,\kappa,l}]^{\wedge}(k) = 1$ for $|k| \le n+1$ by (12.5.29), it follows that for each $t_m \in \mathsf{T}_m$, $m \le n+1$, and $x \in \mathbb{R}$

$$(12.5.32)\qquad \frac{1}{2\pi} \int_{-\pi}^{\pi} t_m(x-u) h^*_{n,\kappa,l}(u)\, du = t_m(x).$$

Therefore

Proposition 12.5.6. *Let $f \in \mathsf{X}_{2\pi}$ and $\kappa > 0$, $l \in \mathbb{N}$. If the κlth Riesz derivative of f exists and if $E_n(\mathsf{X}_{2\pi}; f^{\{\kappa l\}})$ denotes the best trigonometric approximation of $f^{\{\kappa l\}}$, then*

$$(12.5.33)\quad \|(-1)^l (n+1)^{\kappa l}\, \nabla_{\kappa l} R_{n,\kappa,l}(f;\circ) - f^{\{\kappa l\}}(\circ)\|_{\mathsf{X}_{2\pi}} = O(E_n(\mathsf{X}_{2\pi}; f^{\{\kappa l\}}))\quad (n \to \infty).$$

Proof. In view of (12.5.30) and Def. 11.5.10 we have

$$(12.5.34)\qquad (-1)^l (n+1)^{\kappa l}\, \nabla_{\kappa l} R_{n,\kappa,l}(f;x) = (f^{\{\kappa l\}} * h^*_{n,\kappa,l})(x).$$

Therefore, if $t^*_n(f^{\{\kappa l\}})$ denotes the trigonometric polynomial of best approximation of degree n for $f^{\{\kappa l\}}$, it follows by (12.5.32) and (3.1.29)(i) that

$$\begin{aligned} &\|(-1)^l (n+1)^{\kappa l} \nabla_{\kappa l} R_{n,\kappa,l}(f,\circ) - f^{\{\kappa l\}}(\circ)\|_{\mathsf{X}_{2\pi}} \\ &\quad\le \|(f^{\{\kappa l\}} - t^*_n(f^{\{\kappa l\}})) * h^*_{n,\kappa,l}\|_{\mathsf{X}_{2\pi}} + \|t^*_n(f^{\{\kappa l\}}) - f^{\{\kappa l\}}\|_{\mathsf{X}_{2\pi}} \\ &\quad\le (\|h_{\kappa,l}\|_{\mathsf{L}^1} + 1)\, \|f^{\{\kappa l\}} - t^*_n(f^{\{\kappa l\}})\|_{\mathsf{X}_{2\pi}}. \end{aligned}$$

This proves (12.5.33).

Now, the right-hand side of (12.5.34) may be extended to all of $\mathsf{X}_{2\pi}$. This defines a new singular integral $I^*_{n,\kappa,l}(f;x)$ of $f \in \mathsf{X}_{2\pi}$, namely

$$(12.5.35)\qquad I^*_{n,\kappa,l}(f;x) = \frac{1}{2\pi} \int_{-\pi}^{\pi} f(x-u) h^*_{n,\kappa,l}(u)\, du,$$

with parameter $n \in \mathbb{N}$, $n \to \infty$ (note that $\kappa > 0$, $l \in \mathbb{N}$ are certain fixed numbers). Since $\{h^*_{n,\kappa,l}(x)\}$ is a (periodic) approximate identity, we have

$$\lim_{n\to\infty} \|I^*_{n,\kappa,l}(f;\circ) - f(\circ)\|_{\mathsf{X}_{2\pi}} = 0$$

for every $f \in X_{2\pi}$. Moreover, it follows by (12.5.32) (as in the proof of (12.5.33)) that for every $f \in X_{2\pi}$

$$\|I^*_{n,\kappa,l}(f; \circ) - f(\circ)\|_{X_{2\pi}} = O(E_n(X_{2\pi}; f)) \qquad (n \to \infty).$$

Since by Corollary 2.2.4, $E_n(X_{2\pi}; f)$ tends to zero arbitrarily rapidly for $n \to \infty$ by taking f sufficiently smooth, it follows that the integral $I^*_{n,\kappa,l}(f; x)$ does not possess the saturation property. This is a consequence of the fact that $[h^*_{n,\kappa,l}]^\wedge(k) = 1$ for $|k| \le n + 1$, which implies that the approximate identity $\{h^*_{n,\kappa,l}(x)\}$ does not satisfy (12.1.1). Indeed, if $k \in \mathbb{Z}$ is arbitrary, then $[h^*_{n,\kappa,l}]^\wedge(k) = 1$ for all $n \ge |k| - 1$ which implies $\psi(k) \equiv 0$ for $k \in \mathbb{Z}$ regardless of $\varphi(n) \ne 0$.

Finally, we observe that $h_{\kappa,l}$ of (12.5.29) furnishes an example of a function which belongs to NL^1 but does not satisfy (12.3.4). Indeed, since $h^\wedge_{\kappa,l}(v) = 1$ for all $|v| \le 1$, we have

$$\lim_{v \to 0} \frac{h^\wedge_{\kappa,l}(v) - 1}{|v|^\alpha} = 0$$

for every $\alpha > 0$.

Problems

1. Prove Prop. 12.5.4. (Hint: See also BUTZER [11])
2. (i) Let χ satisfy (12.5.3) and $\{\chi^*_\rho(x)\}$ be generated via (12.5.18) by χ. Setting
$$\nabla_{\alpha r} I^*_\rho(f; x) = I^*_\rho(f; x) - f(x) - \sum_{j=1}^{r-1} c_j \rho^{-\alpha j} f^{(\alpha j)}(x),$$
show that the assertions of Prop. 12.5.5 hold for $I^*_\rho(f; x)$.
(ii) Give applications to the integrals $W_t(f; x)$ of Weierstrass and $P_r(f; x)$ of Abel–Poisson. (Hint: See also BUTZER–SUNOUCHI [1])
3. Show that if condition (12.5.2) holds for some $r \in \mathbb{N}$, then (12.5.2) holds for every $m \in \mathbb{N}$ with $1 \le m \le r$. What can be said concerning condition (12.5.3)?
4. (i) Show that for the Fejér means of the Fourier series of $f \in \mathsf{W}[\mathsf{C}_{2\pi}; |k|]$ (compare Problem 12.2.12)
$$\|n[f(\circ) - \sigma_n(f; \circ)] - (f^\sim)'(\circ)\|_{\mathsf{C}_{2\pi}} = O(E_n(\mathsf{C}_{2\pi}; (f^\sim)')) \qquad (n \to \infty).$$
(ii) Show that for the typical means $R_{n,2}(f; \circ)$ of the Fourier series of a twice continuously differentiable f
$$\|(n+1)^2\{R_{n,2}(f; \circ) - f(\circ)\} - f''(\circ)\|_{\mathsf{C}_{2\pi}} = O(E_n(\mathsf{C}_{2\pi}; f'')) \qquad (n \to \infty).$$
State and prove analogous results for the singular integral $N^*_n(f; x)$ of Jackson–de La Vallée Poussin. (Hint: See also BERENS–GÖRLICH [1])

12.6 Notes and Remarks

The concept of saturation was first introduced by JEAN FAVARD for summation methods of Fourier series in a lecture in 1947 (cf. [2]). However as early as 1941 ALEXITS [1] had already shown that for the Fejér means $\|\sigma_n(f; \circ) - f(\circ)\|_{\mathsf{C}_{2\pi}} = O(n^{-1})$ if and only if $f^\sim \in \mathsf{Lip}(\mathsf{C}_{2\pi}; 1)$. The proof of the important fact that $\|\sigma_n(f; \circ) - f(\circ)\|_{\mathsf{C}_{2\pi}} = o(n^{-1})$ implies $f(x) = \text{const}$, which in fact establishes the existence of the saturation property, was communicated by ZYGMUND (see also his paper [3]) in a letter to HILLE in July 1940 (cf. HILLE [2, p. 352]). As a matter of fact, HILLE [1] in 1936 had already considered such o-results for particular singular integrals which are semi-group operators. The first important paper largely concerned with saturation is the doctoral dissertation (1949) of MARC ZAMANSKY [1] which pays particular attention to the singular integrals of Fejér and Jackson–de La Vallée Poussin. Although FAVARD does not seem to have been aware of the ALEXITS result, which was followed by a further paper [2] in 1952, he was the first to realize the importance of the concept as such. In 1956 BUTZER [2, 3] showed that semi-groups of operators on Banach spaces possess the saturation behaviour. Applications were given to the singular integrals of Abel–Poisson and Gauss–Weierstrass.

These investigations were followed by a large number of papers concerned with saturation theorems for various particular methods of summation of Fourier series.

It seems that FAVARD [4] was the first to recognize in 1957 that the knowledge of the behaviour of $\chi_\rho^\wedge(k) - 1$ for fixed k and $\rho \to \rho_0$ (cf. condition 12.1.1) suffices to determine saturation for general methods of summation of Fourier series. This paved the way for the *general* approach to the subject, and was developed in 1958/60 by BUCHWALTER† [1], HARŠILADSE [2], and especially by SUNOUCHI–WATARI [1], SUNOUCHI [1]. Furthermore, in 1959/60, BUTZER [5–7] announced an *integral transform method* in order to solve saturation theorems, the setting being that of the theory of singular integrals. Depending upon the nature of the problem in question one would use Fourier transforms, or finite Fourier transforms, or Laplace transforms, or finite Legendre transforms, etc. The choice of the transform may be compared with the choice of a particular transform method in the solution of initial and boundary value problems for partial differential equations (cf. Chapter 7). The advantage of transform methods‡ as a whole lies in the fact that they possess a common methodological basis and allow a unified theory. See the short survey articles by BUTZER [12, 13].

Sec. 12.1. The definition of saturation that has been chosen (cf. Def. 12.0.2) is the one used in BUTZER [3]. It is equivalent to one given by ZAMANSKY [1] (cf. Problem 12.1.1) yet somewhat different to the one used, for example, in BUTZER–BERENS [1, p. 88].

Condition (12.1.1) which is sufficient for the saturation property to exist, may be compared with the definition of the infinitesimal generator (in the transformed state) of semi-group theory as applied to operators of convolution type (cf. (13.4.8)). The Dirichlet singular integral (which is not an approximate identity) does not satisfy (12.1.1). Indeed, it fails to have the saturation property since it furnishes approximations of arbitrarily high order (cf. (2.4.1)). Now on the other hand, the kernel of the integral $I_n(f; x)$ of (12.2.29) satisfies (12.1.1) and (12.2.6) so that $I_n(f; x)$ is saturated in $X_{2\pi}$ with order $O(n^{-1})$, the saturation class $F[X_{2\pi}; I_n]$ being characterized as $V[X_{2\pi}; -(1/2)\,|k|]$. For an example of an approximate identity which does not possess the saturation property we refer to the end of Sec. 12.5.2.

As already remarked, Prop. 12.1.2 for the singular integral of Fejér (or Abel–Poisson) may be found in ZYGMUND [3]. Theorem 12.1.4, a main result of this section, is due to SUNOUCHI–WATARI [1] and HARŠILADSE [2], see also ALEXITS [2] and FAVARD [4]. For weak* compactness arguments of this type we refer to SUNOUCHI [1II] (see also Sec. 10.7). For Cor. 12.1.5 in $C_{2\pi}$-space we recall ZAMANSKY [1]; for $L^p_{2\pi}$-space (and $C_{2\pi}$) see ALEXITS [2], FAVARD [4], and SUNOUCHI–WATARI [1]. Prop. 12.1.6 may be found in BUTZER–GÖRLICH [2, p. 375]. The important rôle of condition (12.1.7) was first noted by KOROVKIN [3]. The fact that the Fejér–Korovkin integral is saturated in $X_{2\pi}$ (Cor. 12.1.7) is shown in BUTZER–GÖRLICH [2, pp. 371, 377], see also KOROVKIN [3].

† BUCHWALTER wrote four noteworthy *Comptes Rendus* notes dealing with saturation theory, unfortunately written in a brief and somewhat too compact style. Since he did not publish longer articles with proofs and details, his papers perhaps did not receive the credit they actually deserved at the time. Often others rediscovered his results and published them in full.

‡ For Laplace transform methods in saturation theory, see BERENS–BUTZER [1], BERENS–WESTPHAL [1]; for Legendre transform methods and surface spherical harmonics see BUTZER [12] and BERENS–BUTZER–PAWELKE [1]; for Mellin transform methods see KOLBE–NESSEL [1, 2].

Sec. 12.2. Concerning Theorem 12.2.1, the assertion $f \in \mathsf{F}[\mathsf{C}_{2\pi}; I_\rho]$ if and only if $f \in \mathsf{Lip}^*(\mathsf{C}_{2\pi}; 2)$ is given in TURECKIĬ [3], where also most of the relevant examples may be found. The Favard class of the (special) integral $W_t(f; x)$ of Weierstrass (cf. Prop. 12.2.2) was first determined in $\mathsf{L}^p_{2\pi}$-space, $1 < p < \infty$, as an application of a general theorem on semi-group operators (cf. Sec. 13.4.2). For an interpretation of the saturation theoretical behaviour of solutions of partial differential equations in terms of their initial or boundary behaviour see e.g. LEIS [1, 2], BUTZER [6, 9, 11], KANIEV [1].

The idea of solving the direct part of the saturation problem using the concept of uniformly bounded multipliers (cf. Theorem 12.2.3) is due to BUCHWALTER [2], but is worked out in SUNOUCHI [3, 6]. Earlier results assumed the uniform convexity or quasi-convexity of the sequences $\{\lambda_\rho(k)\}$ of (12.2.4); see BUCHWALTER [1] and SUNOUCHI [3]. In any case, the essential point is to have suitable sufficient conditions for (12.2.4) to be valid. Here (12.2.6), in connection with the criteria of BEURLING [1] and SZ.-NAGY [3] (cf. Sec. 6.3.3), is widely applicable; see also BERENS–GÖRLICH [1].

Concerning the saturation of the typical means, SZ.-NAGY [3] showed for fractional $\kappa > 0$ that $f \in \mathsf{V}[\mathsf{C}_{2\pi}; -|k|^\kappa]$ implies $\|R_{n,\kappa}(f; \circ) - f(\circ)\|_{\mathsf{C}_{2\pi}} = O(n^{-\kappa})$ while ZAMANSKY [3] and HARŠILADSE [2] established the converse for $\kappa \in \mathbb{N}$. For fractional κ both directions were proved almost simultaneously by ALJANČIĆ [2] and SUNOUCHI–WATARI [1].

The idea of using functional equations in solving saturation problems may perhaps be traced back to the case of semi-group operators. The proof of Prop. 12.2.8 which rests upon the important identity (12.2.15) was carried out by NESSEL [2]; the saturation class itself for $W_{t,\kappa}(f; x)$ may be found in BUTZER–GÖRLICH [2, p. 367]. The recursive formula (12.2.28) was employed by BERENS [3]; for a paper making use of functional equations for solving similar problems we cite BUTZER–PAWELKE [1].

For Problems 12.2.1(ii)–12.2.3 see ZAMANSKY [1], SZ.-NAGY [3, p. 50], SUNOUCHI–WATARI [1], TURECKIĬ [3], and BUTZER–GÖRLICH [2, p. 379].

For further investigations concerned with the Cesàro (C, α), Hölder (H, α), Riesz, Abel, Lambert, Euler, Borel, Riemann, Voronoi, and Nörlund methods of summation of Fourier series we cite ALJANČIĆ [1, 2], BUCHWALTER [1, 2], HARŠILADSE [1, 2], NATANSON [7], TURECKIĬ [1–5]. For instance, if $m \in \mathbb{N}$, ALJANČIĆ [1, 3] showed that the Hölder (H, m)-method has saturation order $O((\log^{m-1} n)/n)$ and class $\mathsf{V}[\mathsf{X}_{2\pi}; |k|]$. However, the methods of proof in the latter papers are often restricted to particular types of summation processes and cannot be generalized to a wider class.

There are also a number of results on 'local saturation'. For instance, if $f \in \mathsf{C}_{2\pi}$ and (a, b) is an arbitrary subinterval of $[-\pi, \pi]$, then $\sigma_n(f; x) - f(x) = O(n^{-1})$ uniformly over (a, b) if and only if $(f^\sim)'$ is bounded over (a, b). See SUNOUCHI [2, 7], MAMEDOV [8], SUZUKI [1], CHEN TIAN-PING [1], IKENO–SUZUKI [1]. For results on saturation concerning strong approximation see FREUD [5].

A first result concerning pointwise saturation (for the Fejér means) was already mentioned in the Notes and Remarks to Sec. 1.7. As precursory material one may regard the lemma of H. A. Schwarz (cf. Problem 10.1.3) as well as its following extension due to DE LA VALLÉE POUSSIN (cf. TITCHMARSH [6, p. 153], BUTZER [10], and the literature cited there): *If* $f \in \mathsf{C}[a, b]$ *is such that* $\lim_{h\to 0} h^{-2}[f(x+h) + f(x-h) - 2f(x)] = g(x)$ *exists finitely for all* $x \in (a, b)$ *with* $g \in \mathsf{L}^1(a, b)$, *then there are constants* α_0, α_1 *such that* $f(x) = \alpha_0 + \alpha_1 x + \int_a^x du_1 \int_a^{u_1} g(u_2)\, du_2$. This leads to the following extension for the Fejér means (cf. BERENS [4, 5]): *Let* $f \in \mathsf{L}^1_{2\pi}$ *be finite in some interval* (a, b) *and* $\lim_{n\to\infty} \sigma_n(f; x) = f(x)$ *for all* $x \in (a, b)$. *If the limit* $\lim_{n\to\infty} n[\sigma_n(f; x) - f(x)] = g(x)$ *exists finitely for all* $x \in (a, b)$ *with* $g \in \mathsf{L}^1(a, b)$, *then there is a constant* α_0 *such that* $f^\sim(x) = \alpha_0 - \int_a^x g(u)\, du$ *for almost all* $x \in (a, b)$. This result may be representative for pointwise saturation theorems obtained recently by several authors. Let us, in addition, mention BAJŠANSKI–BOJANIČ [1], AMEL'KOVIC [1], MÜHLBACH [1, 2], LORENTZ–SCHUMAKER [1], MICCHELLI [1].

We remark that it is possible to deduce theorems on nonoptimal approximation from corresponding saturation theorems. See SUNOUCHI [4, 6], BUTZER–GÖRLICH [2], BUTZER–

SCHERER [1], FREUD [2, 4], and GÖRLICH–STARK [2] and the literature cited there; compare also Sec. 13.3.3.

The saturation problem has furthermore been considered for (i) algebraic approximation, in particular, by the polynomials of Bernstein and Landau (LORENTZ [2], DE LEEUW [1], SUNOUCHI [1II]), (ii) linear summation methods (determined by matrices) of series (which need not necessarily be trigonometric), see the original approach by FAVARD [3, 4], BUCHWALTER [1], ZELLER [2], also ZYGMUND [7I, p. 123], (iii) trigonometric interpolation processes (RUNCK [1], SUNOUCHI [5], KALNI'BOLOCKAJA [1]).

Sec. 12.3. The basic results of this section are to be found in BUTZER [5–7], where the fundamentals on the Fourier transform method for functions defined on the line group are developed. The limiting relation (12.3.4) suffices to prove the inverse part, while the representation of the fundamental quotient of (12.3.5) as a Fourier–Stieltjes transform enables one to prove the direct part of the saturation theorems. The importance of condition (12.3.5) for the method in question was first recognized by H. KÖNIG (see the remarks in BUTZER [7, p. 401]; compare also BUCHWALTER [3]). (It is of interest to note that condition (12.3.5) may also be expressed directly in terms of the kernel χ itself, see KÖNIG [1]). The independent investigation [1] by SUNOUCHI does not contain condition (12.3.5) and is thus mainly concerned with inverse saturation theorems. For Lemma 12.3.4 see KOZIMA–SUNOUCHI [1], for Prop. 12.3.5 see BUTZER–KÖNIG [1]. For the equivalences of (v) [(iv)] with (i) of Theorem 12.3.7 [12.3.11] see BUTZER-NESSEL [2].

The fact that condition (12.3.5) is indeed a necessary one for the direct part to hold (cf. Prop. 12.3.8) was noted by BUTZER–KÖNIG [1]. The proof of Theorem 12.3.10 is modelled after BUCHWALTER [3]. For the integral (12.3.1) does the ratio $[\chi^\wedge(v) - 1]/|v|^\beta$ not only play an intrinsic rôle in the saturation case, say for $\beta = \alpha$, but its behaviour for any $0 < \beta < \alpha$ may just as well be related to nonoptimal approximations; see the fundamental paper of SZ.-NAGY [3] as well as TELJAKOVSKIĬ [2], where $X_{2\pi}$-approximations of the corresponding (periodic) singular integral (3.1.30) of Fejér's type are considered in case it represents a polynomial summation process of Fourier series (i.e. $\chi^\wedge(v)$ has compact support). On the other hand, condition (12.3.5) alone, i.e., the behaviour of $[\chi^\wedge(v) - 1]/|v|^\beta$ in the saturation case $\beta = \alpha$, covers all nonoptimal approximations $O(\rho^{-\beta})$; see Sec. 13.3.3.

Regarding Problem 12.3.5, for saturation of the integrals of Gauss–Weierstrass and Jackson–de La Vallée Poussin see BUTZER [7] and SUNOUCHI [1I]. For extensions of the theory as given in Problems 12.3.7–12.3.9 we may, for example, refer to BUTZER–NESSEL [2]. We may also mention a series of papers of MAMEDOV in which a systematic account on approximation by so-called m-singular integrals (cf. also HARŠILADSE [3]) defined through ($m \in \mathbb{N}$)

$$(-1)^{m+1} \int_{-\infty}^{\infty} \left[\sum_{j=1}^{m} (-1)^{m-j} \binom{m}{j} f(x - ju) \right] \chi(u; \rho)\, du$$

is given. Similar extensions are discussed on the circle group. See e.g. [4, 5, 7] and in particular his monograph [9] (unfortunately written in Azerbaijani as are also many of his papers).

Sec. 12.4. For the material on the generalized integral of Picard as well as on the general integral of Weierstrass, see BERENS–GÖRLICH [1], NESSEL [2]; for saturation of convolution integrals (cf. Sec. 3.6) in the sense of HIRSCHMAN–WIDDER [5] see DITZIAN [1, 2].

Formula (12.4.19) of Proposition 12.4.5, which is a generalization of the semi-group case treated by NESSEL [2] (cf. (12.4.15)), was used by KOZIMA–SUNOUCHI [1] to prove direct parts of saturation theorems for various processes. Obviously, one may replace $H \in \mathsf{L}^1$ by some $\eta \in \mathsf{BV}$, thus $H^\wedge(v)$ by $\eta^\vee(v)$.

In connection with Prop. 12.4.4 there is an extension by KURAN [1], stating that $\|f\|_p - \|P(f; \circ; y)\|_p = o(y)$ implies $f(x) = 0$ a.e. if $f \in \mathsf{L}^p$, $1 < p < \infty$, and f keeps constant sign almost everywhere if $f \in \mathsf{L}^1$. (The results which are actually given for functions in several variables are proved by using the theory of subharmonic functions.)

For the integrals of Bochner–Riesz and Riesz see BERENS–GÖRLICH [1], NESSEL [2], KOZIMA–SUNOUCHI [1], BERENS [3]. The proof of Prop. 12.4.6 and 12.4.7 for $0 < \lambda \leq 1$ is due to H. BERENS (unpublished).

Saturation theory has already been treated in texts on approximation. Apart from the brief accounts in LORENTZ [3, pp. 98–108] and ZYGMUND [7I, pp. 122–124] we can refer to the stimulating presentation of H. S. SHAPIRO [1, 4]; see also MAMEDOV [9].

Sec. 12.5. The approximation of the Gauss–Weierstrass integral $W(f; x; t)$ by Taylor polynomials, particularly the equivalence of parts (i) and (ii) of Prop. 12.5.4(b), was first presented by BUTZER as a conjecture in a lecture at the *Institut Henri Poincaré*, Paris, in Dec. 1958 (see BUTZER [11]). Subsequently, BUTZER–TILLMANN [1, 2] considered approximation and saturation of general semi-group operators $\{S(t)\}$, $t \geq 0$, by Taylor polynomials $\sum_{k=0}^{r-1} (t^k/k!)A^k f$, A being the infinitesimal generator of the semi-group (see also Sec. 13.4.2; for details BUTZER–BERENS [1, Sec. 2.2]). Further papers in their order of appearance are BUCHWALTER [4], LI SJUN-CZIN [1], BUTZER [6, 11], Leis [1], SUNOUCHI [6], BERENS [2], BERENS–GÖRLICH [1]. In BUTZER–SUNOUCHI [1] particular interest is given to asymptotic expansions of the solutions of Dirichlet's problem (7.1.17) for the unit disc and Fourier's problem (7.1.4) for the ring. Condition (12.5.15) for general kernels is given in KOZIMA–SUNOUCHI [1]; it generalizes relation (12.5.16) for the Weierstrass kernel noted in BUTZER [11]. The curious phenomenon stated in Prop. 12.5.6 was observed by BERENS–GÖRLICH [1], for the Fejér means already by LI SJUN-CZIN [1].

13

Saturation on X($\mathbb{R}$)

13.0 Introduction

In the final chapter of this volume we wish to treat: (i) a number of questions, considered particularly in the last three chapters, from a unified source including the L^p-case, $2 < p < \infty$, (ii) the comparison of the error of approximation corresponding to two different processes, and (iii) saturation on arbitrary Banach spaces.

In Chapter 12 we have seen that for the Fejér and Cauchy–Poisson kernel condition (12.3.5) is satisfied with the same exponent ($\alpha = 1$). Thus the saturation classes of the associated convolution integrals in L^p, $1 \le p \le 2$, are both characterized by the same class (apart from a constant factor), namely $V[L^p; c\,|v|\,]$, whereas those of Gauss–Weierstrass and Jackson–de La Vallée Poussin, for example, are characterized by $V[L^p; cv^2]$ (see Problem 12.3.5). This suggests that the order of approximation of the first pair of integrals should be comparable, similarly for the second pair. Indeed, we have for $0 < \alpha \le 1$ ($\rho \to \infty$, $y \to 0+$)

$$\|\sigma(f; \circ; \rho) - f(\circ)\|_p = O(\rho^{-\alpha}) \Leftrightarrow \|P(f; \circ; y) - f(\circ)\|_p = O(y^\alpha).$$

(If $0 < \alpha < 1$, the equivalence follows via the class $\mathrm{Lip}\,(L^p; \alpha)$ by the results of Sec. 3.5.) This leads to a study of so-called *comparison theorems* which relate the order of approximation of a pair of approximate identities *directly* (and not via equivalent function classes such as $V[X; c\,|v|\,]$) so as to cover optimal as well as nonoptimal approximation.

Whereas in the saturation case the quotient $(\chi^\wedge(v) - 1)/|v|^\alpha$ played the key rôle, for the comparison of a pair of convolution integrals we shall see that the quotients

$$\frac{\chi_2^\wedge(v) - 1}{\chi_1^\wedge(v) - 1}, \qquad \frac{\chi_1^\wedge(v) - 1}{\chi_2^\wedge(v) - 1}$$

will do the same, particularly their representations as Fourier–Stieltjes transforms. Thus the notion of *divisibility* in the ring of Fourier–Stieltjes transforms is significant.

For this purpose, following H. S. SHAPIRO [1], it is useful to rewrite the difference $J(f; x; \rho) - f(x)$ (even higher order differences, the operators $R_\varepsilon^{\{\alpha\}}$, the strong Riesz derivative, etc.) in a unified form as an integral

(13.0.1) $$\frac{1}{\sqrt{2\pi}} \int_{-\infty}^{\infty} f(x - \tau u)\, d\sigma(u)$$

with suitable $\sigma \in \mathsf{BV}$ and parameter $\tau > 0$. Thus, let $\chi \in \mathsf{NL}^1$ and $f \in \mathsf{X}(\mathbb{R})$. The difference $J(f; x; \rho) - f(x)$ may be rewritten in the form (13.0.1) with $\tau = 1/\rho$ and σ as defined by

(13.0.2) $$\sigma(x) = \int_{-\infty}^{x} \chi(u)\, du - \delta(x),$$

δ being the Dirac measure of (5.3.3).

Consequently, the error $J(f; x; \rho) - f(x)$ of the approximation of f by the integral $J(f; x; \rho)$ is expressible as

(13.0.3) $$D_\sigma(f; x; \tau) \equiv (1/\sqrt{2\pi}) \int_{-\infty}^{\infty} f(x - \tau u)\, d\sigma(u),$$

σ being defined by (13.0.2), and we are interested in the behaviour of this integral as $\tau \to 0+$.

To this end, it is no longer necessary to restrict attention to functions σ of type (13.0.2). Thus, letting $\sigma \in \mathsf{BV}$ be such that $\int_{-\infty}^{\infty} d\sigma(u) = 0$, i.e. $\sigma \in \mathsf{BV}_0$, we call $\| D_\sigma(f; \circ; \tau)\|_{\mathsf{X}(\mathbb{R})}$ the *σ-deviation* of f in $\mathsf{X}(\mathbb{R})$. Correspondingly,

$$\omega_\sigma(\mathsf{X}(\mathbb{R}); f; t) = \sup_{0 \le \tau \le t} \| D_\sigma(f; \circ; \tau)\|_{\mathsf{X}(\mathbb{R})}$$

is referred to as the *σ-modulus*† of f in $\mathsf{X}(\mathbb{R})$. Noting that $D_\sigma(f; \circ; \tau)$ tends in $\mathsf{X}(\mathbb{R})$-norm to zero‡ as $\tau \to 0+$, a major aim will be to compare (for fixed f) the rate of decrease of the σ-deviation and σ-modulus for different choices of σ.

As a further example, if σ is the binomial measure $\delta_r(x)$ of Problem 5.3.1, then the σ-modulus is the ordinary modulus of continuity of order r of f. In particular, supposing σ to be the dipole measure δ_1, then (13.0.3) becomes $\{f(x + \tau) - f(x)\}$. Thus the statement $\| D_{\delta_1}(f; \circ; \tau)\|_{\mathsf{X}(\mathbb{R})} = O(\tau^\alpha)$ for some $\alpha > 0$ expresses that $f \in \mathsf{Lip}\,(\mathsf{X}(\mathbb{R}); \alpha)$. On the other hand, if σ is defined by (13.0.2), the corresponding statement expresses that $\|J(f; \circ; \rho) - f(\circ)\|_{\mathsf{X}(\mathbb{R})} = O(\rho^{-\alpha})$. Thus, as H. S. SHAPIRO remarks, both 'smoothness' and 'order of approximation' assertions can be expressed in the form (13.0.3) with suitable function σ. From this point of view there is no essential difference between a *direct* and an *inverse* approximation theorem. Both types are merely *comparison* theorems relating the behaviour of $D_{\sigma_1}(f; x; \tau)$ to that of $D_{\sigma_2}(f; x; \tau)$ for a pair of functions σ_1, σ_2.

† This terminology is justified since σ-moduli behave much the same as the ordinary modulus of smoothness.

‡ (13.0.3) is not only defined for $\sigma \in \mathsf{BV}_0$ but also for $\nu \in \mathsf{NBV}$. However, then $f \in \mathsf{X}(\mathbb{R})$ implies $\lim_{\tau \to 0+} \| D_\nu(f; \circ; \tau) - f(\circ)\|_{\mathsf{X}(\mathbb{R})} = 0$ (cf. Problem 13.1.1).

Sec. 13.1 is concerned with the saturation problem for $D_\sigma(f; x; \tau)$ in X($\mathbb{R}$)-space, in particular, in L^p for $2 < p < \infty$. In the latter case, dual arguments are employed in the proofs. Sec. 13.2 is devoted to applications. Whereas those for L^p-space for $2 < p < \infty$ round off previous results begun in the last chapters, at the same time a number of theorems of Chapters 10–12 are reestablished from a unified source. Sec. 13.3 is reserved to a compact account of comparison theory, including reductions to periodic functions. This yields yet further proofs of the fundamental theorems of Jackson and Bernstein on best approximation. Sec. 13.4 treats saturation of strong approximation processes on arbitrary Banach spaces.

13.1 Saturation of $D_\sigma(f; x; \tau)$ in X($\mathbb{R}$), Dual Methods

The saturation problem for the integral (13.0.3) shall be investigated under the following assumption:

(13.1.1) *Given $\sigma \in \mathsf{BV}_0$, let there exist a homogeneous*† *function $\psi(v)$ of order $\alpha > 0$ with $\psi(0) = 0$, $\psi(v) \neq 0$ for $v \neq 0$, and $\nu \in \mathsf{NBV}$ such that $\sigma^\vee(v) = \psi(v)\nu^\vee(v)$ for all $v \in \mathbb{R}$.*

The importance of this condition for the discussion in L^p-space, $1 \le p \le 2$, is revealed by the results of the preceding chapter. In fact, we obtain by the methods of proof of Sec. 12.3

Proposition 13.1.1. *Let $f \in \mathsf{L}^p$, $1 \le p \le 2$, and σ satisfy* (13.1.1).

(a) *If $\|D_\sigma(f; \circ; \tau)\|_p = o(\tau^\alpha)$, $\tau \to 0+$, then $D_\sigma(f; x; \tau) = 0$ a.e. for each $\tau > 0$.*

(b) *The following assertions are equivalent:*

(i) $\|D_\sigma(f; \circ; \tau)\|_p = O(\tau^\alpha)$ $\qquad (\tau \to 0+)$,

(ii) $f \in \mathsf{V}[\mathsf{L}^p; \psi(v)]$.

Proposition 13.1.2. *Let $f \in \mathsf{L}^p$, $1 \le p \le 2$, and σ satisfy* (13.1.1). *The following assertions are equivalent:*

(i) *There exists $g \in \mathsf{L}^p$ such that $\lim_{\tau \to 0+} \|\tau^{-\alpha} D_\sigma(f; \circ; \tau) - g(\circ)\|_p = 0$,*

(ii) *$f \in \mathsf{W}[\mathsf{L}^p; \psi(v)]$,*

(iii) *there exists $g \in \mathsf{L}^p$ such that for each $\tau > 0$*

$$D_\sigma(f; x; \tau) = \frac{\tau^\alpha}{\sqrt{2\pi}} \int_{-\infty}^{\infty} g(x - \tau u)\, d\nu(u) \quad \text{a.e.}$$

The proofs are left to Problem 13.1.2. This solves the saturation problem for $D_\sigma(f; x; \tau)$ in L^p-space, $1 \le p \le 2$.

Concerning extensions to all X($\mathbb{R}$)-spaces, we introduce the following notations:

Definition 13.1.3. *The **adjoint integral** of $D_\sigma(f; x; \tau)$ is given by*

$$D_{\sigma^*}(f; x; \tau) = \frac{1}{\sqrt{2\pi}} \int_{-\infty}^{\infty} f(x - \tau u)\, d\sigma^*(u),$$

where $\sigma^*(u) = -\sigma(-u)$.

† A function $\psi(v)$ defined on $\mathbb{R}$ is called (positive-) *homogeneous* of order $\alpha > 0$ if $\psi(\tau v) = \tau^\alpha \psi(v)$ for every $\tau > 0$ and $v \in \mathbb{R}$.

Definition 13.1.4. *Let $f \in \mathsf{W}[\mathsf{L}^p; \psi(v)]$, $1 \le p \le 2$. If $g \in \mathsf{L}^p$ is such that $\psi(v)f^\wedge(v) = g^\wedge(v)$, then g is called the **ψ-derivative** of f, in notation: $D^\psi f$.*

To illustrate the latter definition let us consider some examples of functions ψ. In case $\psi(v) = |v|^\alpha$, it was the result of Theorems 11.2.6–11.2.9 that $D^\psi f$ is equal to $D_s^{\{\alpha\}}f$, the αth strong Riesz derivative of f. If $\psi(v) = (iv)^r$, $r \in \mathbb{N}$, then $f \in \mathsf{W}[\mathsf{L}^p; (iv)^r]$ implies by Theorem 5.2.21 that f is equal a.e. to some $\Phi \in \mathsf{AC}_{\text{loc}}^{r-1}$ with $\Phi^{(k)} \in \mathsf{L}^p$, $1 \le k \le r$, and $D^\psi f = \Phi^{(r)}$. For the interpretation in case $\psi(v) = \{-i \operatorname{sgn} v\}(iv)^r$ we refer to Problem 8.3.3.

Now, if $\sigma \in \mathsf{BV}_0$, then $\lim_{\tau\to 0+} \|D_{\sigma^*}(f; \circ; \tau)\|_{\mathsf{X}(\mathbb{R})} = 0$ for every $f \in \mathsf{X}(\mathbb{R})$. Moreover, if σ satisfies (13.1.1) and ψ^* is defined by $\psi^*(v) = \psi(-v)$, then (cf. Problems 13.1.3, 13.1.4)

$$D_{\sigma^*}(s; x; \tau) = \frac{\tau^\alpha}{\sqrt{2\pi}} \int_{-\infty}^{\infty} (D^{\psi^*}s)(x - \tau u)\, d\nu^*(u) \tag{13.1.2}$$

for every $s \in \mathsf{W}[\mathsf{L}^p; \psi^*(v)]$, $1 \le p \le 2$, and thus

$$\lim_{\tau\to 0+} \|\tau^{-\alpha} D_{\sigma^*}(s; \circ; \tau) - (D^{\psi^*}s)(\circ)\|_p = 0. \tag{13.1.3}$$

We may now treat the saturation problem for the integral (13.0.3) in L^p-space, $2 < p < \infty$.

Theorem 13.1.5. *Let $f \in \mathsf{L}^p$, $2 < p < \infty$, and σ satisfy* (13.1.1). *The following assertions are equivalent:*

(i) $\|D_\sigma(f; \circ; \tau)\|_p = O(\tau^\alpha)$ $\qquad (\tau \to 0+)$,

(ii) *there exists $g \in \mathsf{L}^p$ such that for every $s \in \mathsf{W}[\mathsf{L}^{p'}; \psi^*(v)]$*

$$\int_{-\infty}^{\infty} f(x)(D^{\psi^*}s)(x)\, dx = \int_{-\infty}^{\infty} g(x)s(x)\, dx, \tag{13.1.4}$$

(iii) *there exists $g \in \mathsf{L}^p$ such that*

$$D_\sigma(f; x; \tau) = \frac{\tau^\alpha}{\sqrt{2\pi}} \int_{-\infty}^{\infty} g(x - \tau u)\, d\nu(u),$$

(iv) *there exists $g \in \mathsf{L}^p$ such that*

$$\lim_{\tau\to 0+} \|\tau^{-\alpha} D_\sigma(f; \circ; \tau) - g(\circ)\|_p = 0.$$

Proof. We recall that p and p' are conjugate numbers, i.e. $p^{-1} + p'^{-1} = 1$. Thus, in the present case, $1 < p' < 2$ and hence the expressions occurring in (13.1.4) are meaningful.

(i) $\Rightarrow$ (ii). For each $s \in \mathsf{W}[\mathsf{L}^{p'}; \psi^*(v)]$ we define the functional (cf. (10.6.13))

$$\Lambda_\tau(f) = \int_{-\infty}^{\infty} f(x)\tau^{-\alpha} D_{\sigma^*}(s; x; \tau)\, dx. \tag{13.1.5}$$

By Hölder's inequality and the theorem of Fubini it follows that

$$\Lambda_\tau(f) = \int_{-\infty}^{\infty} s(x)\tau^{-\alpha} D_\sigma(f; x; \tau)\, dx. \tag{13.1.6}$$

From (13.1.5) we have by (13.1.3) and Hölder's inequality that

$$\lim_{\tau\to 0+} \Lambda_\tau(f) = \int_{-\infty}^{\infty} f(x)(D^{\psi^*}s)(x)\, dx,$$

whereas (13.1.6) implies by (i) and the weak compactness theorem for L^p that there exists a null sequence $\{\tau_j\}$ and $g \in \mathsf{L}^p$ such that

$$\lim_{j\to\infty} \Lambda_{\tau_j}(f) = \int_{-\infty}^{\infty} s(x)g(x)\, dx,$$

which proves (ii).

(ii) $\Rightarrow$ (iii) $\Rightarrow$ (iv) $\Rightarrow$ (i). In view of (13.1.2) we may rewrite the functional (13.1.5) as

$$\Lambda_\tau(f) = \int_{-\infty}^{\infty} f(x) D_{\nu^*}(D^{\psi^*}s; x; \tau)\, dx, \tag{13.1.7}$$

where $\nu^*(x) \equiv -\nu(-x)$. Since by the convolution theorem (cf. Problem 13.1.3)

$$[D_{\nu^*}(D^{\psi^*}s; \circ; \tau)]^\wedge(v) = [D^{\psi^*}s]^\wedge(v)\nu^\vee(-\tau v) = \psi^*(v)\{s^\wedge(v)\nu^\vee(-\tau v)\},$$

it follows that $D_{\nu^*}(s; x; \tau)$ belongs to $\mathsf{W}[\mathsf{L}^{p'}; \psi^*(v)]$ for every $s \in \mathsf{W}[\mathsf{L}^{p'}; \psi^*(v)]$ and

$$[D^{\psi^*}(D_{\nu^*}(s; \circ; \tau))](x) = D_{\nu^*}(D^{\psi^*}s; x; \tau).$$

Therefore we conclude from (13.1.7) and the hypothesis (ii) that

$$\begin{aligned}\Lambda_\tau(f) &= \int_{-\infty}^{\infty} f(x)[D^{\psi^*}(D_{\nu^*}(s; \circ; \tau))](x)\, dx \\ &= \int_{-\infty}^{\infty} g(x) D_{\nu^*}(s; x; \tau)\, dx = \int_{-\infty}^{\infty} s(x) D_\nu(g; x; \tau)\, dx\end{aligned}$$

for every $s \in \mathsf{W}[\mathsf{L}^{p'}; \psi^*(v)]$. Since the latter class is dense in $\mathsf{L}^{p'}$ (cf. Problem 13.1.4), (iii) follows by (13.1.6) and Prop. 0.8.1. Since ν is normalized, this in turn gives (iv) by Problem 13.1.1. Trivially, (iv) implies (i), and the proof is complete.

As an immediate consequence of the above arguments we note (cf. Problem 13.1.2)

Proposition 13.1.6. *Let σ satisfy condition* (13.1.1). *If $f \in \mathsf{L}^p$, $2 < p < \infty$, is such that $\|D_\sigma(f; \circ; \tau)\|_p = o(\tau^\alpha)$, $\tau \to 0+$, then $D_\sigma(f; x; \tau) = 0$ a.e. for each $\tau > 0$.*

This solves the saturation problem for $D_\sigma(f; x; \tau)$ in L^p-space, $2 < p < \infty$. For the counterparts in C we refer to Problems 13.1.5–13.1.7.

Finally, let us remark that in L^p-space, $1 \le p \le 2$, the characterization in terms of the Fourier transform of f (cf. assertion b(ii) of Prop. 13.1.1) was the intermediate station in the derivation of further characterizations of the corresponding Favard classes. In L^p, $2 < p < \infty$, this rôle is now played by assertion (ii) of Theorem 13.1.5. For details we refer to the next section.

Problems

1. (i) Let $\sigma \in \mathsf{BV}_0$, $f \in \mathsf{X}(\mathbb{R})$. Show that $\|D_\sigma(f; \circ; \tau)\|_{(\mathbb{R})}$ is uniformly continuous and bounded for $\tau > 0$, and $\lim_{\tau\to 0+} \|D_\sigma(f; \circ; \tau)\|_{\mathsf{X}(\mathbb{R})} = 0$. (Hint: See Problem 3.1.16)
   (ii) Let $\nu \in \mathsf{NBV}$, $f \in \mathsf{X}(\mathbb{R})$. Show that $\lim_{\tau\to 0+} \|D_\nu(f; \circ; \tau) - f(\circ)\|_{\mathsf{X}(\mathbb{R})} = 0$.
2. Prove Prop. 13.1.1 and 13.1.2. Show that if σ satisfies (13.1.1) and $f \in \mathsf{L}^p$, $1 \le p \le 2$, then $D_\sigma(f; x; \tau) = 0$ a.e. for every $\tau > 0$ implies $f(x) = 0$ a.e. (Hint: Use part (iii) of the next Problem)
3. Let $\sigma \in \mathsf{BV}_0$. (i) Show that $\lim_{\tau\to 0+} \|D_{\sigma^*}(f; \circ; \tau)\|_{\mathsf{X}(\mathbb{R})} = 0$ for every $f \in \mathsf{X}(\mathbb{R})$. (ii) Show that

$$\int_{-\infty}^{\infty} l(x) D_\sigma(f; x; \tau)\, dx = \int_{-\infty}^{\infty} f(x)\, D_{\sigma^*}(l; x; \tau)\, dx$$

   for every $f \in \mathsf{L}^p$, $l \in \mathsf{L}^{p'}$. (iii) If $f \in \mathsf{L}^p$, $1 \le p \le 2$, show that $[D_\sigma(f; \circ; \tau)]^\wedge(v) = f^\wedge(v)\sigma^\vee(\tau v)$, $[D_{\sigma^*}(f; \circ; \tau)]^\wedge(v) = f^\wedge(v)\sigma^\vee(-\tau v)$.
4. Let σ satisfy (13.1.1) with ψ, ν. (i) Show that σ^* satisfies (13.1.1) with $\psi^*(v) = \psi(-v)$, $\nu^*(x) = -\nu(-x)$. Prove (13.1.2), (13.1.3). (ii) Show that $D_\nu(f; x; \tau)$ belongs to $\mathsf{W}[\mathsf{L}^p, \psi(v)]$ for every $f \in \mathsf{L}^p$, $1 \le p \le 2$, and $[D^\psi(D_\nu(f; \circ; \tau))](x) = \tau^{-\alpha} D_\sigma(f; x; \tau)$. (iii) Show that $\mathsf{W}[\mathsf{L}^p; \psi(v)]$ is dense in L^p for $1 \le p \le 2$. (iv) Using (ii), give another proof of the implication (ii) $\Rightarrow$ (iii) of Theorem 13.1.5.
5. Let $f \in \mathsf{C}$ and σ satisfy (13.1.1). Show that the following assertions are equivalent:
   (i) $\|D_\sigma(f; \circ; \tau)\|_{\mathsf{C}} = O(\tau^\alpha)$ $(\tau \to 0+)$,
   (ii) There exists $g \in \mathsf{L}^\infty$ such that for every $s \in \mathsf{W}[\mathsf{L}^1; \psi^*(v)]$

$$\int_{-\infty}^{\infty} f(x)\, (D^{\psi^*}s)(x)\, dx = \int_{-\infty}^{\infty} g(x)s(x)\, dx,$$

   (iii) there exists $g \in \mathsf{L}^\infty$ such that

$$D_\sigma(f; x; \tau) = \frac{\tau^\alpha}{\sqrt{2\pi}} \int_{-\infty}^{\infty} g(x - \tau u)\, d\nu(u).$$

6. Let $f \in \mathsf{C}$ and σ satisfy (13.1.1). Show that the following assertions are equivalent:
   (i) there exists $g \in \mathsf{C}$ such that $\lim_{\tau\to 0+} \|\tau^{-\alpha} D_\sigma(f; \circ; \tau) - g(\circ)\|_{\mathsf{C}} = 0$,
   (ii) there exists $g \in \mathsf{C}$ such that for every $s \in \mathsf{W}[\mathsf{L}^1; \psi^*(v)]$

$$\int_{-\infty}^{\infty} f(x)(D^{\psi^*}s)(x)\, dx = \int_{-\infty}^{\infty} g(x)s(x)\, dx,$$

   (iii) there exists $g \in \mathsf{C}$ such that $D_\sigma(f; x; \tau) = (\tau^\alpha/\sqrt{2\pi}) \int_{-\infty}^{\infty} g(x - \tau u)\, d\nu(u)$ for every $x \in \mathbb{R}$, $\tau > 0$.
7. Let σ satisfy (13.1.1). If $f \in \mathsf{C}$ is such that $\|D_\sigma(f; \circ; \tau)\|_{\mathsf{C}} = o(\tau^\alpha)$, $\tau \to 0+$, then $D_\sigma(f; x; \tau) = 0$ for every $x \in \mathbb{R}$, $\tau > 0$.

13.2 Applications to Approximation in L^p, $2 < p < \infty$

13.2.1 Differences

Let us first consider the case $\sigma = \delta_r$, the binomial measure of Problem 5.3.1. Then $D_{\delta_r}(f; x; h)$ corresponds to the rth right-hand difference $\Delta_h^r f(x)$ of f at x, and an application of the results of Sec. 13.1 would essentially reestablish those of Sec. 10.5, 10.6.

Indeed, since $\delta_r^{\vee}(v) = (e^{iv} - 1)^r$, condition (13.1.1) is satisfied by δ_r with $\psi(v) = (iv)^r$, $\alpha = r$, because

$$(e^{iv} - 1)^r = (iv)^r[m * m * \cdots * m]^{\wedge}(v),$$

the right-hand side being the r-times convolution of the function m of (5.1.27). Thus Prop. 13.1.1 applies which again gives Theorem 10.5.4 by Theorem 5.3.15.

Let $2 < p < \infty$ and $s \in W[L^{p'}; (iv)^r]$ with $(iv)^r s^{\wedge}(v) = l^{\wedge}(v)$, $l \in L^{p'}$. Then $s \in W_{L^{p'}}^r$ by Prop. 5.2.20, i.e., there exists $\Phi \in AC_{loc}^{r-1}$ with $\Phi^{(k)} \in L^{p'}$, $0 \le k \le r$, such that $s(x) = \Phi(x)$ and $\Phi^{(r)}(x) = l(x)$ almost everywhere. Condition (13.1.4) then reads

$$\int_{-\infty}^{\infty} f(x)\Phi^{(r)}(x)\,dx = (-1)^r \int_{-\infty}^{\infty} g(x)\Phi(x)\,dx.$$

By Prop. 10.6.4 this implies $f \in W_{L^p}^r$, and thus Theorem 13.1.5 again gives Theorems 10.6.5 and 10.6.7 in L^p-space, $2 < p < \infty$. Concerning an application of Prop. 13.1.6 this only yields $\Delta_h^r f(x) = 0$ a.e. for each $h > 0$. But by Lemma 10.6.1 and the fact that f belongs to L^p, this actually implies $f(x) = 0$ a.e. (which by the way solves Problem 10.6.18).

13.2.2 Singular Integrals Satisfying (12.3.5)

Let σ be of type (13.0.2), thus $D_\sigma(f; x; \tau) = J(f; x; \rho) - f(x)$, $\rho = \tau^{-1}$. If χ satisfies (12.3.5), then condition (13.1.1) holds for $\sigma^{\vee}(v) = \chi^{\wedge}(v) - 1$ with $\psi(v) = c\,|v|^\alpha = \psi^*(v)$, and an application of Prop. 13.1.1, 13.1.2 would reproduce the results of Sec. 12.3. Moreover,

Proposition 13.2.1. *Let* $f \in L^p$, $2 < p < \infty$, *and* χ *satisfy* (12.3.5).

(a) *If* $\|J(f; \circ; \rho) - f(\circ)\|_p = o(\rho^{-\alpha})$, $\rho \to \infty$, *then* $J(f; x; \rho) = f(x)$ a.e. *for each* $\rho > 0$, *i.e.,* f *is an invariant element of the integral* $J(f; x; \rho)$.

(b) *The following assertions are equivalent:*

(i) $\|J(f; \circ; \rho) - f(\circ)\|_p = O(\rho^{-\alpha})$ $\qquad (\rho \to \infty)$,

(ii) *there exists* $g \in L^p$ *such that for every* $s \in W[L^{p'}; |v|^\alpha]$

$$c\int_{-\infty}^{\infty} f(x)(D_s^{\{\alpha\}}s)(x)\,dx = \int_{-\infty}^{\infty} g(x)s(x)\,dx,$$

(iii) *there exists* $g \in L^p$ *such that for each* $\rho > 0$

$$\rho^\alpha[J(f; x; \rho) - f(x)] = \frac{c}{\sqrt{2\pi}}\int_{-\infty}^{\infty} g(x - u)\,dv(\rho u) \quad \text{a.e.},$$

(iv) *there exists* $g \in L^p$ *such that*

$$\lim_{\rho\to\infty} \|\rho^\alpha[J(f; \circ; \rho) - f(\circ)] - cg(\circ)\|_p = 0.$$

The proof is given by Theorem 13.1.5 and Prop. 13.1.6. The respective result in C is left to Problem 13.2.2. Now one might give applications to the various singular integrals discussed in Sec. 12.3, 12.4 in order to supply saturation theorems in all $X(\mathbb{R})$-spaces. But this is left to the reader (cf. Sec. 13.2.5).

13.2.3 Strong Riesz Derivatives

In order to extend the results of Sec. 11.2 to L^p, $2 < p < \infty$, let us first rewrite Def. 11.2.5 in the setting of integrals of type (13.0.3). Let $0 < \alpha < 1$. Putting

$$\eta_\alpha(x) = \int_{-\infty}^{x} m_{1,1-\alpha}^{\sim}(u)\,du,$$

where $m_{1,1-\alpha}^{\sim}$ is given by (11.2.6), Lemma 11.2.2 shows that η_α belongs to BV_0 for each $0 < \alpha < 1$. Moreover

$$h^{-1}(f * m_{h,1-\alpha}^{\sim})(x) = h^{-\alpha} D_{\eta_\alpha}(f; x; h),$$

thus the αth strong Riesz derivative of f for $0 < \alpha < 1$ is the strong limit (if it exists) of $h^{-\alpha} D_{\eta_\alpha}(f; x; h)$ as $h \to 0$.

Now, η_α satisfies (13.1.1) with $\psi(v) = |v|^\alpha$ since by Lemma 11.2.2

$$\eta_\alpha^{\vee}(v) = [m_{1,1-\alpha}^{\sim}]^{\wedge}(v) = |v|^\alpha \frac{e^{iv} - 1}{iv} = |v|^\alpha m^{\wedge}(v),$$

where $m \in \mathsf{NL}^1$ is given by (5.1.27). We have

Proposition 13.2.2. *Let $f \in \mathsf{L}^p$, $2 < p < \infty$, and $\alpha > 0$. The αth strong Riesz derivative of f exists if and only if there exists $g \in \mathsf{L}^p$ such that for every $s \in \mathsf{W}[\mathsf{L}^{p'}; |v|^\alpha]$*

$$\int_{-\infty}^{\infty} f(x)(D_s^{\{\alpha\}}s)(x)\,dx = \int_{-\infty}^{\infty} g(x)s(x)\,dx. \tag{13.2.1}$$

In this case, $D_s^{\{\alpha\}}f = g$.

Proof. For $0 < \alpha < 1$ the proof follows by the equivalence of assertions (ii) and (iv) of Theorem 13.1.5 as applied to η_α.

Let $\alpha = 1$. We recall (Theorem 8.1.12) that $f \in \mathsf{L}^p$, $2 < p < \infty$, implies that the Hilbert transform $f^\sim$ also belongs to L^p. The first strong Riesz derivative of f has been defined as the limit in norm of $h^{-1}[f^\sim(x + h) - f^\sim(x)]$ for $h \to 0$ (see (11.2.16)). In other words, since $f^\sim(x + h) - f^\sim(x) = D_{\delta_1}(f^\sim; x; h)$, $D_s^{\{1\}}f$ exists if and only if the limit in norm of $h^{-1} D_{\delta_1}(f^\sim; x; h)$ exists. Setting $\xi(v) = -iv$, in view of Theorem 13.1.5, (ii) $\Leftrightarrow$ (iv), this is equivalent to the existence of $g \in \mathsf{L}^p$ such that for every $s \in \mathsf{W}[\mathsf{L}^{p'}; \xi(v)]$

$$\int_{-\infty}^{\infty} f^\sim(x)(D^\xi s)(x)\,dx = \int_{-\infty}^{\infty} g(x)s(x)\,dx. \tag{13.2.2}$$

Since $1 < p' < 2$, we have by Prop. 8.3.1 that $s \in \mathsf{W}[\mathsf{L}^{p'}; \xi(v)]$ if and only if $s \in \mathsf{W}[\mathsf{L}^{p'}; |v|]$. Moreover,

$$[-(D^\xi s)^\sim]^{\wedge}(v) = \{i \operatorname{sgn} v\}(-iv)s^{\wedge}(v) = |v|s^{\wedge}(v) = [D_s^{\{1\}}s]^{\wedge}(v).$$

Therefore by (8.2.1)

$$\int_{-\infty}^{\infty} f^\sim(x)(D^\xi s)(x)\,dx = -\int_{-\infty}^{\infty} f(x)(D^\xi s)^\sim(x)\,dx = \int_{-\infty}^{\infty} f(x)(D_s^{\{1\}}s)(x)\,dx,$$

which together with (13.2.2) proves (13.2.1) for $\alpha = 1$.

For arbitrary $\alpha > 1$, the proof follows in view of the iterative process in defining $D_s^{\{\alpha\}}f$. Indeed, if e.g. $\alpha = 1 + \beta$, $0 < \beta < 1$, we have by the preceding two cases that $D_s^{\{\alpha\}}f$ exists if and only if $D_s^{\{1\}}f$ exists and there is $g \in \mathsf{L}^p$ such that

$$\int_{-\infty}^{\infty} (D_s^{\{1\}}f)(x)(D_s^{\{\beta\}}s)(x)\,dx = \int_{-\infty}^{\infty} g(x)s(x)\,dx \tag{13.2.3}$$

Table 1. Summary Table of School Primary Prevention Approaches

Reference	Subjects $(n)_1$ and grade	Length and type of treatment	Dependent measures	Outcome	Follow-up
Ojemann (1960)	35, 4th grade 35, 5th grade 35, 6th grade	Training in causal thinking; one academic year	1. Problem situation test 2. Causal test (written measure)	Experimental group significantly more causal than control	None
Ojemann & Snider (1964)	206, 4th grade 212, 5th grade	Training in causal thinking; one academic year	1. Problem situation test 2. Observational report, items related to causal behavior	4th grade = 2 of 4 experimental classes significantly more causal than controls 5th grade = 3 of 4 experimental classes significantly more causal than controls	None
Stone, Hinds, & Schmidt (1965)	48, 3rd grade 48, 4th grade 48, 5th grade	Problem-solving skills (information seeking, generation of alternatives, setting goals); 1 week	1. Pre-post test measures of frequency of facts, choices, and solutions to problem situations (oral report)	3rd grade = Facts: significant difference in favor of experimental group; no difference on measures of choices and solutions between control and experimental groups 4th grade = Facts, choices, and solutions: significantly different in favor of experimental group	None
Allen, Chinsky, Larcen, Lochman, & Selinger (1976)	150, 3rd and 4th grades; 24 biweekly 30-minute lessons for 18 weeks	Social problem-solving training	1. Structured Real-life Problem situation 2. Problem Solving Measure (PSM) 3. In-class Assessment Measure (ICM), written quizzes testing learning of problem-solving components	1. Results inconclusive; favor experimental group 2. Significant difference favoring experimental group on PSM 3. No statistical interpretation provided for ICM	None
Shure & Spivack (1979)	131, nursery and kindergarten; daily 20-minute lessons for 3 months	Interpersonal cognitive problem-solving training (alternative solutions, consequential thinking)	1. Preschool Interpersonal Problem Solving Test (oral report measure of alternative solution thinking) 2. What Happens Next Game (oral report measure of consequential thinking) 3. Hahnemann Preschool Behavior Rating Scale	Significant difference favoring experimental groups on all measures	1 year
Shure & Spivack (1980)	219, nursery school	Interpersonal cognitive problem-solving training	Same as 1, 2, 3 above in addition to causal test	Significant difference favoring experimental groups on all measures	1 year

Then $\xi_2 \in \mathsf{L}^1$ and $\xi_2^\wedge(v) = i\,|v|^{-\alpha} \sin v$ by Lemma 11.2.2. Furthermore, $\xi_2(x)/x \in \mathsf{L}^1$ since $\xi_2(x) = O(|x|)$ as $|x| \to 0$. This implies $\xi_2^\wedge(v) = i([\xi_2(\circ)/\circ]^\wedge)'(v)$ by Prop. 5.1.17. Therefore

$$[\xi_2(\circ)/\circ]^\wedge(v+h) - [\xi_2(\circ)/\circ]^\wedge(v) = -i \int_v^{v+h} \xi_2^\wedge(u)\,du.$$

By the lemma of Riemann–Lebesgue the limit for $h \to \infty$ exists and hence

$$[\xi_2(\circ)/\circ]^\wedge(v) = i \int_v^\infty \xi_2^\wedge(u)\,du = -\int_{|v|}^\infty \frac{\sin u}{|u|^\alpha}\,du = -\eta_2(v).$$

Thus $\eta_2 \in [\mathsf{L}^1]^\wedge$ and the case $0 < \alpha < 1$ is complete.

For $\alpha = 1$ the above considerations lead one to consider the limit of $\xi_2(x)/x$ as $\alpha \to 1-$ which is given by (cf. (11.1.19))

$$\lambda_2(x) = \frac{1}{x}[\log|x+1| - \log|x-1|] = \frac{1}{2x}\log\left(\frac{1+x}{1-x}\right)^2.$$

Now $\lambda_2 \in \mathsf{L}^1$, and in view of (see Erdélyi [2I, p. 18(11)])

$$\int_0^\infty \lambda_2(x) \cos vx\,dx = \pi \int_v^\infty \frac{\sin u}{u}\,du$$

for $v > 0$. Since λ_2 is even, this implies $\lambda_2^\wedge(v) = \sqrt{2\pi}\,\eta_2(v)$. Thus we have shown that k_α satisfies (12.4.1) for all $0 < \alpha < 2$.

Now we are ready to apply the results of Sec. 12.3 or Prop. 13.1.1, 13.1.2 to k_α, which would again give Theorems 11.3.3, 11.3.5. For the extension of the latter to L^p, $2 < p < \infty$, we may apply Theorem 13.1.5 or Prop. 13.2.1 to deduce by (13.2.6)

Proposition 13.2.3. *Let* $0 < \alpha < 2$. *For* $f \in \mathsf{L}^p$, $2 < p < \infty$, *the following assertions are equivalent:*

(i) $$\left\| \frac{1}{A_c(-\alpha)} \int_{|u| \ge \varepsilon} \frac{f(\circ - u) - f(\circ)}{|u|^{1+\alpha}}\,du \right\|_p = O(1) \qquad (\varepsilon \to 0+),$$

(ii) *there exists* $g \in \mathsf{L}^p$ *such that for every* $s \in \mathsf{W}[\mathsf{L}^{p'}; |v|^\alpha]$

$$\frac{\alpha}{2} A_c(-\alpha) \int_{-\infty}^\infty f(x)(D_s^{\{\alpha\}}s)(x)\,dx = \int_{-\infty}^\infty g(x)s(x)\,dx,$$

(iii) *there exists* $g \in \mathsf{L}^p$ *such that*

$$\lim_{\varepsilon \to 0+} \left\| \frac{1}{A_c(-\alpha)} \int_{|u| \ge \varepsilon} \frac{f(\circ - u) - f(\circ)}{|u|^{1+\alpha}}\,du - g(\circ) \right\|_p = 0.$$

Concerning the extension of Theorems 11.3.6, 11.3.7 to L^p, $2 < p < \infty$, we note that for given $\alpha > 0$ and $0 < \alpha < 2j$, $j \in \mathbb{N}$, we may by similar computations rewrite the integral (11.3.1) as a singular integral of Fejér's type. The corresponding kernel again satisfies (12.4.1) so that Theorem 13.1.5 furnishes the desired extension (cf. Problem 13.2.4).

13.2.5 Riesz and Fejér Means

Let us finally collect the results so far obtained in order to give a saturation theorem for the Riesz and Fejér means in L^p-space, $2 < p < \infty$. The singular integral of Riesz as given by (12.4.28) satisfies (12.4.1) so that we have by an application of Prop. 13.2.1–13.2.3

Proposition 13.2.4. *The Riesz means $R_{\kappa,\lambda}(f; x; \rho)$ are saturated in L^p, $2 < p < \infty$, with order $O(\rho^{-\kappa})$, $\rho \to \infty$, and the Favard class $F[L^p; R_{\kappa,\lambda}(\rho)]$ is characterized by any one of the following statements:*

(i) *There exists $g \in L^p$ such that for every $s \in W[L^{p'}; |v|^\kappa]$*

$$-\lambda \int_{-\infty}^{\infty} f(x)(D_s^{\{\kappa\}}s)(x)\,dx = \int_{-\infty}^{\infty} g(x)s(x)\,dx,$$

(ii) *there exists $g \in L^p$ such that $\lim_{\rho\to\infty} \|\rho^\kappa[f(\circ) - R_{\kappa,\lambda}(f; \circ; \rho)] - \lambda g(\circ)\|_p = 0$,*

(iii) *f has a κth strong Riesz derivative,*

(iv) *there exists $g \in L^p$ such that*

$$\lim_{\varepsilon\to 0+} \left\| \frac{1}{C_{\kappa,2j}} \int_\varepsilon^\infty \frac{\overline{\Delta}_u^{2j} f(\circ)}{u^{1+\kappa}}\,du - g(\circ) \right\|_p = 0,$$

where $j \in \mathbb{N}$ is chosen so that $0 < \kappa < 2j$, and $C_{\kappa,2j}$ is a constant given by (11.3.14).

In any case, we have $D_s^{\{\kappa\}}f = g$.

For the particular instance $\kappa = 1$ we note (cf. Problems 13.2.1, 13.2.5)

Proposition 13.2.5. *The singular integral $\sigma(f; x; \rho)$ of Fejér is saturated in L^p, $2 < p < \infty$, with order $O(\rho^{-1})$, $\rho \to \infty$, and the Favard class $F[L^p; \sigma(\rho)]$ is characterized as the set of functions f for which $\|f^\sim(\circ + h) - f^\sim(\circ)\|_p = O(|h|)$, $h \to 0$.*

Thus, in view of Problem 8.2.5 the Favard class $F[L^p; \sigma(\rho)]$ for $2 < p < \infty$ is given by Lip $(L^p; 1)$.

Problems

1. Let $f \in L^p$, $2 < p < \infty$, and χ satisfy (12.3.5). Show that f is an invariant element of $J(f; x; \rho)$ if and only if $f(x) = 0$ a.e. In C-space the same is true if and only if $f(x) =$ const.
2. State and prove the analog of Prop. 13.2.1 for functions $f \in C$.
3. (i) Let $\alpha = 1 + \beta$, $0 < \beta < 1$. Show that if (13.2.1) is satisfied, then the αth strong Riesz derivative of f exists. (Hint: Compare the proof of Prop. 10.6.4, (ii) ⇒ (i), as well as (11.2.15))
   (ii) Prove Prop. 13.2.2 for arbitrary values of $\alpha > 0$.
   (iii) Let $f \in L^p$, $2 < p < \infty$, and suppose that for some $\alpha > 0$ there exists $g \in L^p$ such that (13.2.1) holds for every $s \in W[L^{p'}; |v|^\alpha]$. Show that for each $0 < \beta < \alpha$ there exists $g_\beta \in L^p$ such that $\int_{-\infty}^{\infty} f(x)(D_s^{\{\beta\}}s)(x)\,dx = \int_{-\infty}^{\infty} g_\beta(x)s(x)\,dx$ for every $s \in W[L^{p'}; |v|^\beta]$.
   (iv) Let $f \in C$. The αth strong Riesz derivative of f exists if and only if there exists $g \in C$ such that for every $s \in W[L^1; |v|^\alpha]$

$$\int_{-\infty}^{\infty} f(x)(D_s^{\{\alpha\}}s)(x)\,dx = \int_{-\infty}^{\infty} g(x)s(x)\,dx.$$

4. Prove the equivalence of assertions (i) and (iv) of Prop. 13.2.4.
5. Let $f \in L^p$, $2 < p < \infty$. Show that the first strong Riesz derivative of f exists if and only if $\|f^\sim(\circ + h) - f^\sim(\circ)\|_p = O(|h|)$, $h \to 0$.

13.3 Comparison Theorems

13.3.1 Global Divisibility

We start with a very simple comparison theorem based on the *global divisibility* of Fourier–Stieltjes transforms.

Definition 13.3.1. *$\sigma \in \mathsf{BV}$ divides $\xi \in \mathsf{BV}$* ***globally****†, if there exists $\lambda \in \mathsf{BV}$ such that $\xi^\vee(v) = \lambda^\vee(v)\sigma^\vee(v)$ for all $v \in \mathbb{R}$.*

Theorem 13.3.2. *If $\sigma \in \mathsf{BV}$ divides $\xi \in \mathsf{BV}$ globally, then for every $f \in \mathsf{X}(\mathbb{R})$*

$$\|D_\xi(f;\circ;\tau)\|_{\mathsf{X}(\mathbb{R})} \le \|\lambda\|_{\mathsf{BV}}\|D_\sigma(f;\circ;\tau)\|_{\mathsf{X}(\mathbb{R})} \qquad (\tau > 0). \tag{13.3.1}$$

Proof. In view of Theorem 5.3.5 and Prop. 5.3.11, $\xi^\vee(v) = \lambda^\vee(v)\sigma^\vee(v)$ implies $\xi(x) = (\lambda * d\sigma)(x)$, i.e., for every $f \in \mathsf{X}(\mathbb{R})$ and $\tau > 0$ (cf. Problem 5.3.9)

$$\int_{-\infty}^{\infty} f(x - \tau u)\, d\xi(u) = \frac{1}{\sqrt{2\pi}}\int_{-\infty}^{\infty}\int_{-\infty}^{\infty} f(x - \tau u - \tau y)\, d\sigma(u)\, d\lambda(y) \quad \text{(a.e.)}.$$

Hence by the Hölder–Minkowski inequality

$$\left\|\int_{-\infty}^{\infty} f(\circ - \tau u)\, d\xi(u)\right\|_{\mathsf{X}(\mathbb{R})} \le \frac{1}{\sqrt{2\pi}}\int_{-\infty}^{\infty}\left\|\int_{-\infty}^{\infty} f(\circ - \tau u - \tau y)\, d\sigma(u)\right\|_{\mathsf{X}(\mathbb{R})}|d\lambda(y)|$$
$$= \|\lambda\|_{\mathsf{BV}}\left\|\int_{-\infty}^{\infty} f(\circ - \tau u)\, d\sigma(u)\right\|_{\mathsf{X}(\mathbb{R})},$$

which proves (13.3.1).

Theorem 13.3.3. *If $\xi, \sigma_1, \sigma_2, \ldots, \sigma_r \in \mathsf{BV}$ are such that $\xi^\vee(v) = \sum_{j=1}^{r} \lambda_j^\vee(v)\sigma_j^\vee(v)$ for every $v \in \mathbb{R}$ with some $\lambda_1, \ldots, \lambda_r \in \mathsf{BV}$, then for every $f \in \mathsf{X}(\mathbb{R})$*

$$\|D_\xi(f;\circ;\tau)\|_{\mathsf{X}(\mathbb{R})} \le \sum_{j=1}^{r} \|\lambda_j\|_{\mathsf{BV}}\|D_{\sigma_j}(f;\circ;\tau)\|_{\mathsf{X}(\mathbb{R})} \qquad (\tau > 0). \tag{13.3.2}$$

This is an immediate consequence of Theorem 13.3.2.

Corollary 13.3.4. *Let $\chi_1, \chi_2 \in \mathsf{NL}^1$ and suppose that $(\chi_2^\wedge(v) - 1)$ divides $(\chi_1^\wedge(v) - 1)$ globally. Then for every $f \in \mathsf{X}(\mathbb{R})$*

$$\|J_1(f;\circ;\rho) - f(\circ)\|_{\mathsf{X}(\mathbb{R})} \le \|\lambda\|_{\mathsf{BV}}\|J_2(f;\circ;\rho) - f(\circ)\|_{\mathsf{X}(\mathbb{R})} \qquad (\rho > 0), \tag{13.3.3}$$

where $J_1(f;x;\rho)$ and $J_2(f;x;\rho)$ are the singular integrals of Fejér's type corresponding to the kernels χ_1 and χ_2, respectively.

For the proof, set $\xi(x) = \int_{-\infty}^{x} \chi_1(u)\, du - \delta(x)$ and $\sigma(x) = \int_{-\infty}^{x} \chi_2(u)\, du - \delta(x)$, where δ is the Dirac measure (cf. (13.0.2)), and apply Theorem 13.3.2 with $\tau = 1/\rho$.

Definition 13.3.5. *We say that the kernel $\chi_1 \in \mathsf{NL}^1$ is* ***better than*** *the kernel $\chi_2 \in \mathsf{NL}^1$ if there exists a constant M such that for every $f \in \mathsf{X}(\mathbb{R})$ and $\rho > 0$*

$$\|J_1(f;\circ;\rho) - f(\circ)\|_{\mathsf{X}(\mathbb{R})} \le M\,\|J_2(f;\circ;\rho) - f(\circ)\|_{\mathsf{X}(\mathbb{R})}. \tag{13.3.4}$$

Two kernels, each better than the other, are called ***equivalent****.*

Proposition 13.3.6. *Let $\chi_1, \chi_2 \in \mathsf{NL}^1$ be such that $(\chi_2^\wedge(v) - 1)$ divides $(\chi_1^\wedge(v) - \chi_2^\wedge(v))$ globally. Then χ_1 is better than χ_2.*

† This will also be expressed by saying that $\sigma^\vee$ divides $\xi^\vee$ globally.

Suppose there exists $\lambda \in \mathsf{BV}$ such that $\hat{\chi_1}(v) - \hat{\chi_2}(v) = \lambda^\vee(v)(\hat{\chi_2}(v) - 1)$. Then $\hat{\chi_1}(v) - 1 = (\lambda^\vee(v) + \delta^\vee(v))(\hat{\chi_2}(v) - 1)$, and the result follows from Cor. 13.3.4.

Proposition 13.3.7. *The kernel of Fejér is equivalent to that of Cauchy–Poisson, i.e., there exist constants M_1, M_2 such that*

$$(13.3.5)\quad M_1\|P(f;\circ;1/\rho) - f(\circ)\|_{\mathsf{X}(\mathbb{R})} \le \|\sigma(f;\circ;\rho) - f(\circ)\|_{\mathsf{X}(\mathbb{R})} \le M_2\|P(f;\circ;1/\rho) - f(\circ)\|_{\mathsf{X}(\mathbb{R})}$$

for every $f \in \mathsf{X}(\mathbb{R})$ and $\rho > 0$.

Proof. For these kernels we have (cf. Problem 5.1.2)

$$F^\wedge(v) = \begin{cases} 1 - |v|, & |v| \le 1 \\ 0, & |v| > 1 \end{cases}, \quad p^\wedge(v) = e^{-|v|}.$$

Now $h_1(v) \equiv (p^\wedge(v) - F^\wedge(v))/(F^\wedge(v) - 1)$ is an even function, and by an elementary calculation it can be shown that h_1 satisfies the assumptions of Prop. 6.3.10 and thus is the Fourier transform of an L^1-function. Prop. 13.3.6 then gives that p is better than F.

On the other hand, writing $h_2(v) \equiv (F^\wedge(v) - p^\wedge(v))/(p^\wedge(v) - 1)$, then again h_2 is an even function, which by the same reasons is the Fourier transform of an L^1-function. Thus F is better than p, and the proof is complete.

For a selection of further comparison theorems between familiar singular integrals we refer to Problem 13.3.1.

13.3.2 Local Divisibility

If $\sigma^\vee$ does not divide $\xi^\vee$ globally, we cannot in general expect an inequality of type (13.3.1). However, an inequality slightly weaker than (13.3.1) which nevertheless implies (13.3.1) in many cases of interest may be drawn under the weaker hypothesis that σ divides ξ only locally. More precisely,

Definition 13.3.8. $\sigma \in \mathsf{BV}$ *divides* $\xi \in \mathsf{BV}$ ***locally***, *if there exists $\lambda \in \mathsf{BV}$ such that $\xi^\vee(v) = \lambda^\vee(v)\sigma^\vee(v)$ for all v of a neighbourhood of $v = 0$. In this case we also say that $\sigma^\vee$ divides $\xi^\vee$ locally.*

Instead of Theorem 13.3.2 we now have

Theorem 13.3.9. *Let σ be a real-valued† function of* BV *which does not vanish identically. If we suppose that σ divides $\xi \in \mathsf{BV}$ locally, then there is b, $0 < b < 1$, such that for every $f \in \mathsf{X}(\mathbb{R})$ and $\tau > 0$*

$$(13.3.6)\quad \|D_\xi(f;\circ;\tau)\|_{\mathsf{X}(\mathbb{R})} \le \|\lambda\|_{\mathsf{BV}}\|D_\sigma(f;\circ;\tau)\|_{\mathsf{X}(\mathbb{R})} + A\sum_{j=0}^{\infty}\|D_\sigma(f;\circ;Bb^j\tau)\|_{\mathsf{X}(\mathbb{R})},$$

where A, B are constants depending only on ξ and σ.

† If σ is complex-valued, then we have e.g. to suppose that there are two intervals symmetric to the origin where $\sigma^\vee$ does not vanish.

Proof. Since σ real implies $\sigma^\vee(v) = \overline{\sigma^\vee(-v)}$ (cf. Prop. 5.3.1), we observe that in view of Prop. 5.3.2 and 5.3.11 there exist constants $v_0 > 0$ and $b, 0 < b < 1$, such that $\sigma^\vee(v) \neq 0$ on $[b^2v_0, v_0]$. Let $K(v)$ be the continuous function which equals one for $|v| \leq b^2v_0$, zero for $|v| > bv_0$, and is piecewise linear ('trapezoid function'). Then by Prop. 6.3.10 there exists $k \in \mathsf{NL}^1$ such that $K(v) = k^\wedge(v)$ for all $v \in \mathbb{R}$. Let $L(v) = K(v) - K(bv)$. Then $L(v) = l^\wedge(v)$, where $l \in \mathsf{L}^1$ and $l(x) = k(x) - (1/b)k(x/b)$. Now, $L(v) = 0$ except for v in the intervals (b^2v_0, v_0) and $(-v_0, -b^2v_0)$, and on these intervals $\sigma^\vee$ does not vanish by hypothesis since $\sigma^\vee(-v) = \overline{\sigma^\vee(v)}$. Therefore, by the fundamental lemma of Wiener (cf. Sec. 13.5) σ divides L globally, $L(v) = \eta^\vee(v)\sigma^\vee(v)$, say. By Theorem 13.3.2 we then have for every $f \in \mathsf{X}(\mathbb{R})$ and $\tau > 0$

$$\left\| \frac{1}{\sqrt{2\pi}} \int_{-\infty}^{\infty} f(\circ - \tau u)\left[k(u) - \frac{1}{b}k\left(\frac{u}{b}\right)\right] du \right\|_{\mathsf{X}(\mathbb{R})} \leq C\, \|D_\sigma(f; \circ; \tau)\|_{\mathsf{X}(\mathbb{R})},$$

where C is a constant independent of f and τ. Replacing τ by $b\tau$ and setting $bu = t$ gives

$$\left\| \frac{1}{\sqrt{2\pi}} \int_{-\infty}^{\infty} f(\circ - \tau t)\left[\frac{1}{b}k\left(\frac{t}{b}\right) - \frac{1}{b^2}k\left(\frac{t}{b^2}\right)\right] dt \right\|_{\mathsf{X}(\mathbb{R})} \leq C\, \|D_\sigma(f; \circ; b\tau)\|_{\mathsf{X}(\mathbb{R})}.$$

Repeating this process, adding, and using Minkowski's inequality gives ($r \in \mathbb{N}$)

$$\left\| \frac{1}{\sqrt{2\pi}} \int_{-\infty}^{\infty} f(\circ - \tau u)\left[k(u) - \frac{1}{b^r}k\left(\frac{u}{b^r}\right)\right] du \right\|_{\mathsf{X}(\mathbb{R})} \leq C \sum_{j=0}^{\infty} \|D_\sigma(f; \circ; b^j\tau)\|_{\mathsf{X}(\mathbb{R})}.$$

Now, since $k \in \mathsf{NL}^1$, by (3.1.9) for each fixed $\tau > 0$

$$\lim_{\rho\to\infty} \left\| \frac{\rho}{\sqrt{2\pi}} \int_{-\infty}^{\infty} f(\circ - \tau u)k(\rho u)\, du - f(\circ) \right\|_{\mathsf{X}(\mathbb{R})} = 0.$$

Therefore, setting $\rho = b^{-r}$ and letting $r \to \infty$, we have

$$\left\| \frac{1}{\sqrt{2\pi}} \int_{-\infty}^{\infty} f(\circ - \tau u)k(u)\, du - f(\circ) \right\|_{\mathsf{X}(\mathbb{R})} \leq C \sum_{j=0}^{\infty} \|D_\sigma(f; \circ; b^j\tau)\|_{\mathsf{X}(\mathbb{R})}. \tag{13.3.7}$$

Now, by hypothesis, there exists $c > 0$ and $\lambda \in \mathsf{BV}$ such that $\xi^\vee(v) = \lambda^\vee(v)\sigma^\vee(v)$ for all $|v| \leq c$. Hence, since $k(v_0v/c)$ vanishes for $|v| \geq c$, we have for all $v \in \mathbb{R}$

$$\xi^\vee(v) = \lambda^\vee(v)\sigma^\vee(v) + [\lambda^\vee(v)\sigma^\vee(v) - \xi^\vee(v)]\left[K\left(\frac{v_0}{c}v\right) - 1\right].$$

Therefore by Theorem 13.3.3

$$\begin{aligned} \|D_\xi(f; \circ; \tau)\|_{\mathsf{X}(\mathbb{R})} \leq{}& \|\lambda\|_{\mathsf{BV}}\|D_\sigma(f; \circ; \tau)\|_{\mathsf{X}(\mathbb{R})} \\ &+ \|\lambda * d\sigma - \xi\|_{\mathsf{BV}} \left\| \frac{1}{\sqrt{2\pi}} \int_{-\infty}^{\infty} f(\circ - \tau u)\frac{c}{v_0}k\left(\frac{c}{v_0}u\right) du - f(\circ) \right\|_{\mathsf{X}(\mathbb{R})}. \end{aligned}$$

If we set $B = v_0/c$, we obtain in view of (13.3.7)

$$\left\| \frac{1}{\sqrt{2\pi}} \int_{-\infty}^{\infty} f(\circ - \tau u)\frac{c}{v_0}k\left(\frac{c}{v_0}u\right) du - f(\circ) \right\|_{\mathsf{X}(\mathbb{R})} \leq C \sum_{j=0}^{\infty} \|D_\sigma(f; \circ; b^jB\tau)\|_{\mathsf{X}(\mathbb{R})},$$

and thus

$$\|D_\xi(f;\circ;\tau)\|_{X(\mathbb{R})} \leq \|\lambda\|_{BV}\|D_\sigma(f;\circ;\tau)\|_{X(\mathbb{R})} + C\,\|\lambda * d\sigma - \xi\|_{BV}\sum_{j=0}^{\infty}\|D_\sigma(f;\circ;Bb^j\tau)\|_{X(\mathbb{R})},$$

which proves (13.3.6) (and is to be compared with (13.3.1)).

Let us mention that in the important special case where $\sigma(x) = \int_{-\infty}^{x}\chi(u)\,du - \delta(x)$ with $\chi \in \mathsf{NL}^1$, $\sigma^\vee(v)$ does not vanish for large v in view of the Riemann–Lebesgue lemma (regardless of χ being real- or complex-valued).

Also important is the following

Corollary 13.3.10. *If σ is as in Theorem* 13.3.9 *and $\xi \in \mathsf{BV}$ is such that $\xi^\vee(v)$ vanishes in a neighbourhood of the origin, then* (13.3.6) *holds.*

For purposes of application it should be observed that, for a concrete choice of ξ and σ, the various constants occurring can usually be estimated quite explicitly. Also, if σ is fairly smooth which is generally the case in applications, the depth of Wiener's lemma is not required since then the quotient of $K(v) - K(bv)$ by $\sigma^\vee(v)$ is smooth and of compact support. This guarantees by more elementary theorems that it is the Fourier transform of an L^1-function.

As a first application we note the following immediate consequence of Theorem 13.3.9.

Proposition 13.3.11. *Let σ be a real function in* BV *which does not vanish identically. Suppose that σ divides $\xi \in \mathsf{BV}$ locally. If for some $f \in X(\mathbb{R})$ and $\alpha > 0$ $\|D_\sigma(f;\circ;\tau)\|_{X(\mathbb{R})} = O(\tau^\alpha)$, then also $\|D_\xi(f;\circ;\tau)\|_{X(\mathbb{R})} = O(\tau^\alpha)$ as $\tau \to 0+$.*

As an example, choose $\sigma(x) = \delta_2(x)$ and $\xi(x) = \int_{-\infty}^{x} w(u)\,du - \delta(x)$, where δ and δ_2 are defined by (5.3.3) and Problem 5.3.1 and w is the kernel of the singular integral of Gauss–Weierstrass (cf. (3.1.33)). Then

$$\sigma^\vee(v) = \delta_2^\vee(v) = (e^{iv} - 1)^2, \qquad \xi^\vee(v) = w^\wedge(v) - 1 = e^{-v^2} - 1.$$

Now, it is readily checked that $\delta_2^\vee(v)$ divides $\{w^\wedge(v) - 1\}$ at $v = 0$. Indeed, if we set $h_1(v) = (e^{-v^2} - 1)/(e^{iv} - 1)^2$, then, for e.g. $|v| \leq \pi$, $h_1(v)$ is a continuous function. Moreover, h_1' is continuous since $\lim_{v\to 0} h_1'(v) = i$. Let $l(v)$ be an absolutely continuous function which equals $h_1(v)$ on $|v| \leq \pi$ and vanishes for $|v| > 2\pi$. By Prop. 6.3.10 $l \in [\mathsf{L}^1]^\wedge$, and thus h_1 coincides at $v = 0$ with an L^1-Fourier transform, which proves the local divisibility of $\{w^\wedge(v) - 1\}$ by $\delta_2^\vee(v)$. (Note that $\delta_2^\vee(v)$ does not divide $\{w^\wedge(v) - 1\}$ globally.)

On the other hand, $\{w^\wedge(v) - 1\}$ also divides $\delta_2^\vee(v)$ locally (even globally, see Problem 13.3.2). Indeed, if we set $h_2(v) = (e^{iv} - 1)^2/(e^{-v^2} - 1)$, then h_2 is a continuous function. Moreover, h_2' is continuous since $\lim_{v\to 0} h_2'(v) = i$, and local divisibility is checked as above. Hence by Prop. 13.3.11 we have reshown the familiar result (cf. Cor. 3.5.3, Problem 12.3.5)

Corollary 13.3.12. *Let $f \in X(\mathbb{R})$, $0 < \alpha \leq 2$, and $W(f;x;t)$ be the singular integral of Gauss–Weierstrass. Then*

$$\|W(f;\circ;t) - f(\circ)\|_{X(\mathbb{R})} = O(t^{\alpha/2}) \Leftrightarrow \|f(\circ + h) + f(\circ - h) - 2f(\circ)\|_{X(\mathbb{R})} = O(h^\alpha).$$

13.3.3 Special Comparison Theorems with no Divisibility Hypothesis

It is a remarkable fact that for certain functions ξ an estimate for $\|D_\xi(f; \circ; \tau)\|_{X(\mathbb{R})}$ in terms of $\|D_\sigma(f; \circ; \tau)\|_{X(\mathbb{R})}$ is possible with a completely arbitrary choice of σ. To prove our main result we need

Lemma 13.3.13. *Let $\eta(x)$ be defined for $x \geq 0$, and suppose for all x that $0 \leq \eta(x) \leq M$ and*

$$\eta(x) \leq Ax^\alpha + B^{-\beta}\eta(Bx), \tag{13.3.8}$$

where α, β, A, B are positive constants with $B > 1$. Then for each $0 \leq x \leq B^{-2}$

$$\eta(x) \leq \begin{cases} \dfrac{A}{1 - B^{\alpha-\beta}}x^\alpha, & \alpha < \beta \\ (A_1|\log x| + A_2)x^\alpha, & \alpha = \beta \\ A_3 x^\beta, & \alpha > \beta, \end{cases}$$

where A_1, A_2, A_3 are positive constants.

Proof. Replacing x by Bx in (13.3.8) and substituting into (13.3.8) we obtain

$$\eta(x) \leq (1 + B^{\alpha-\beta})Ax^\alpha + B^{-2\beta}\eta(B^2x). \tag{13.3.9}$$

Now we may replace x by B^2x in (13.3.8) and substitute into (13.3.9). If we iterate this process, an obvious induction shows that for every $r \in \mathbb{N}$

$$\eta(x) = \left\{\sum_{j=0}^{r-1} B^{(\alpha-\beta)j}\right\}Ax^\alpha + B^{-r\beta}\eta(B^rx). \tag{13.3.10}$$

We now consider cases.

(i) If $\alpha < \beta$, we may let $r \to \infty$ in (13.3.10) and have

$$\eta(x) \leq \left\{\sum_{j=0}^{\infty} B^{(\alpha-\beta)j}\right\}Ax^\alpha,$$

proving the first part of our assertion.

(ii) If $\alpha = \beta$, we get from (13.3.10)

$$\eta(x) \leq rAx^\alpha + MB^{-r\alpha},$$

and if we determine r so that $(1/Bx) \leq B^r < (1/x)$ which obviously has a solution for each $0 \leq x \leq B^{-2}$, we obtain

$$\eta(x) \leq \frac{A}{\log B}x^\alpha \log\frac{1}{x} + MB^\alpha x^\alpha.$$

(iii) If $\alpha > \beta$, it follows from (13.3.10) that

$$\eta(x) \leq \frac{B^{(\alpha-\beta)r}}{B-1}Ax^\alpha + MB^{-r\beta}.$$

Choosing r as we did in (ii) gives

$$\eta(x) \le \frac{A}{B-1} x^{\beta} + MB^{\beta}x^{\beta},$$

and the proof is complete.

Theorem 13.3.14. *Let σ be a real function of* BV *which does not vanish identically, and let $f \in \mathsf{X}(\mathbb{R})$ be such that for some $\alpha > 0$ $\|D_{\sigma}(f; \circ; \tau)\|_{\mathsf{X}(\mathbb{R})} = O(\tau^{\alpha})$ as $\tau \to 0+$. Let $\xi \in$ BV and suppose there exists a homogeneous function $\psi(v)$ of order $\beta > 0$ with $\psi(0) = 0$, $\psi(v) \ne 0$ for $v \ne 0$ such that both $\psi(v)$ and $\xi^{\vee}(v)/\psi(v)$ coincide in a neighbourhood of $v = 0$ with the Fourier–Stieltjes transform of a* BV*-function. Then*

$$\|D_{\xi}(f; \circ; \tau)\|_{\mathsf{X}(\mathbb{R})} = \begin{cases} O(\tau^{\alpha}), & \alpha < \beta \\ O(\tau^{\alpha}|\log \tau|), & \alpha = \beta \\ O(\tau^{\beta}), & \alpha > \beta. \end{cases} \tag{13.3.11}$$

Proof. Let $\gamma \in$ BV be such that $\psi(v) = \gamma^{\vee}(v)$ in some neighbourhood of $v = 0$. Since ψ is homogeneous, $\{\gamma^{\vee}(v) - 2^{-\beta}\gamma^{\vee}(2v)\}$ vanishes in a neighbourhood of $v = 0$. Therefore by Cor. 13.3.10 and Prop. 13.3.11 there is a constant A such that for all $\tau > 0$

$$\left\| \frac{1}{\sqrt{2\pi}} \int_{-\infty}^{\infty} f(\circ - \tau u)\, d\gamma(u) - 2^{-\beta} \frac{1}{\sqrt{2\pi}} \int_{-\infty}^{\infty} f(\circ - 2\tau u)\, d\gamma(u) \right\|_{\mathsf{X}(\mathbb{R})} \le A\tau^{\alpha}.$$

Thus, if we set

$$\eta(\tau) = \left\| \frac{1}{\sqrt{2\pi}} \int_{-\infty}^{\infty} f(\circ - \tau u)\, d\gamma(u) \right\|_{\mathsf{X}(\mathbb{R})},$$

then $\eta(\tau)$ is defined and bounded for $\tau \ge 0$ and satisfies

$$\eta(\tau) \le A\tau^{\alpha} + 2^{-\beta}\eta(2\tau).$$

An application of Lemma 13.3.13 shows that $\|D_{\gamma}(f; \circ; \tau)\|_{\mathsf{X}(\mathbb{R})}$ admits the estimates of (13.3.11). For the actual proof of (13.3.11) we only have to apply Theorem 13.3.9 with σ replaced by γ.

Proposition 13.3.15. *Let σ be a real function of* BV *which does not vanish identically, and let $f \in \mathsf{X}(\mathbb{R})$ be such that $\|D_{\sigma}(f; \circ; \tau)\|_{\mathsf{X}(\mathbb{R})} = O(\tau^{\alpha})$, $\tau \to 0+$, for some $\alpha > 0$. Then the rth modulus of smoothness $\omega_r(\mathsf{X}(\mathbb{R}); f; t)$ of f is $O(t^{\alpha})$, $O(t^{\alpha}|\log t|)$, or $O(t^{r})$ for $t \to 0+$ according as α is less than, equal to, or greater than r.*

The proof follows by an immediate application of Theorem 13.3.14. Indeed, setting $\psi(v) = (iv)^r$ and $\xi = \delta_r$, we have already seen by Sec. 13.2.1 that $\delta_r^{\vee}(v)/(iv)^r \in [\mathsf{L}^1]^{\wedge}$. Moreover, let l be a twice differentiable function such that $l(v) = (iv)^r$ for $|v| \le 1$ and $l(v) = 0$ for $|v| \ge 2$. Then $l \in [\mathsf{L}^1]^{\wedge}$ by Prop. 6.3.10.

In the important special case that one of the functions σ or ξ is of the form (13.0.2), Theorem 13.3.14 may be used to deduce important results on the approximation of singular integrals of Fejér's type, already familiar to us. Indeed,

Proposition 13.3.16. *Let $\chi \in \mathsf{NL}^1$ be the kernel of the singular integral $J(f; x; \rho)$. If for some $f \in \mathsf{X}(\mathbb{R})$ and $\alpha > 0$*

$$\|J(f; \circ; \rho) - f(\circ)\|_{\mathsf{X}(\mathbb{R})} = O(\rho^{-\alpha}) \qquad (\rho \to \infty),$$

then the rth modulus of smoothness $\omega_r(\mathsf{X}(\mathbb{R}); f; t)$ *of f is* $O(t^\alpha)$, $O(t^\alpha|\log t|)$, *or* $O(t^r)$ *for* $t \to 0+$ *according as* α *is less than, equal to, or greater than r.*

Proposition 13.3.17. *Let* $f \in \mathsf{Lip}_r(\mathsf{X}(\mathbb{R}); \alpha)$, $0 < \alpha \le r$, $r \in \mathbb{N}$. *If* χ *satisfies* (12.3.5) *for some exponent* $\beta > 0$, *then* $\|J(f; \circ; \rho) - f(\circ)\|_{\mathsf{X}(\mathbb{R})}$ *is* $O(\rho^{-\alpha})$, $O(\rho^{-\alpha}\log\rho)$, *or* $O(\rho^{-\beta})$ *according as* α *is less than, equal to, or greater than* β.

In fact, Prop. 13.3.16 follows by Prop. 13.3.15. To prove Prop. 13.3.17 we apply Theorem 13.3.14 with $\sigma = \delta_r$, $\psi(v) = c\,|v|^\beta$ and ξ defined through (13.0.2). By Prop. 6.3.10 it again follows that ψ coincides in a neighbourhood of $v = 0$ with the Fourier–Stieltjes transform of a function of BV, whereas $(\chi^\wedge(v) - 1)/c\,|v|^\beta \in [\mathsf{BV}]^\vee$ by (12.3.5).

Thus, both the *inverse* assertion of Prop. 13.3.16 as well as the *direct* one of Prop. 13.3.17 are obtainable from the same source, namely Theorem 13.3.14. But in contrast to the results on global and local divisibility (cf. Theorem 13.3.2, Prop. 13.3.11 for $\xi = \delta_r$), the occurrence of the log-factor for the critical value $\alpha = \beta$ suggests that the present results are mainly applicable in case of nonoptimal approximation. Thus, for example, Cor. 13.3.12 is reproduced only for $0 < \alpha < 2$.

Let us return to the singular integral $R_{\kappa,\lambda}(f; x; \rho)$ of Riesz. The corresponding kernel satisfies (12.3.5) with $\psi(v) = -\lambda\,|v|^\kappa$ (cf. Sec. 12.4.5). Therefore (cf. Sec. 6.4.2)

Corollary 13.3.18. *Let* $f \in \mathsf{X}(\mathbb{R})$ *and* $\alpha > 0$, $\kappa > 0$, $r \in \mathbb{N}$ *be such that* $0 < \alpha < \kappa$, $0 < \alpha < r$. *Then for the singular integral of Riesz (for arbitrary* $\lambda > 0$*)*

$$\|R_{\kappa,\lambda}(f; \circ; \rho) - f(\circ)\|_{\mathsf{X}(\mathbb{R})} = O(\rho^{-\alpha}) \Leftrightarrow f \in \mathsf{Lip}_r(\mathsf{X}(\mathbb{R}); \alpha).$$

For further applications we refer to the Problems (see also Sec. 13.5).

13.3.4 Applications to Periodic Continuous Functions

Comparison theory in C may be applied to rededuce some of the important results of Chapter 2 in $\mathsf{C}_{2\pi}$. The deduction is based upon the fact that $\mathsf{C}_{2\pi}$ is a subset of C.

Lemma 13.3.19. *Let* $f \in \mathsf{C}_{2\pi}$ *and* $\chi \in \mathsf{NL}^1$ *be such that* $\chi^\wedge(v) = 0$ *for* $|v| \ge 1$. *Then* $J(f; x; n+1)$ *is a trigonometric polynomial of degree n at most, thus is an element of* T_n.

Proof. In view of Prop. 3.1.3 we have $J(f; x; \rho) \in \mathsf{C}_{2\pi}$ and $\|J(f; \circ; \rho)\|_{\mathsf{C}_{2\pi}} \le \|\chi\|_{\mathsf{L}^1}\|f\|_{\mathsf{C}_{2\pi}}$ for every $\rho > 0$. Moreover (cf. (5.1.58)), $[J(f; \circ; n+1)]^\wedge(k) = \chi^\wedge(k/(n+1))f^\wedge(k)$ for every $k \in \mathbb{Z}$. Thus the Fourier coefficients of $J(f; x; n+1)$ vanish for $|k| > n$, and the assertion follows by Problem 4.1.2.

Let us now indicate how to derive a theorem of Jackson's type from Prop. 13.3.11.

Proposition 13.3.20. *If* $f \in \mathsf{Lip}_r(\mathsf{C}_{2\pi}; \alpha)$, *then* $E_n(\mathsf{C}_{2\pi}; f) = O(n^{-\alpha})$, $n \to \infty$.

Proof. Let $L(v)$ be an absolutely continuous function which equals one for $|v| \le 1/2$ and vanishes for $|v| \ge 1$. By Prop. 6.3.10 there exists $l \in \mathsf{NL}^1$ such that $L(v) = l^\wedge(v)$. Furthermore,

$$l_n(x) \equiv \frac{n+1}{\sqrt{2\pi}}\int_{-\infty}^{\infty} f(x-u)\,l((n+1)u)\,du$$

belongs to T_n by Lemma 13.3.19. Since $l^\wedge(v) - 1 = 0$ for $|v| \le 1/2$, $\delta_r^\vee(v) = (e^{iv} - 1)^r$ divides $\{l^\wedge(v) - 1\}$ locally (cf. Cor. 13.3.10). Therefore the hypothesis $\|D_{\delta_r}(f; \circ; \tau)\|_{\mathsf{C}_{2\pi}} = O(\tau^\alpha)$ implies $E_n(\mathsf{C}_{2\pi}; f) \le \|f - l_n\|_{\mathsf{C}_{2\pi}} = O(n^{-\alpha})$ by Prop. 13.3.11, and the proof is complete.

The theorems of Bernstein and Zygmund may be obtained by an application of Prop. 13.3.15. Indeed, to derive Theorems 2.3.3, 2.3.5 in $\mathsf{C}_{2\pi}$-space, suppose that $f \in \mathsf{C}_{2\pi}$ and that for each $n \in \mathbb{N}$ there exists $t_n \in \mathsf{T}_n$ with $\|f - t_n\|_{\mathsf{C}_{2\pi}} \le Mn^{-\alpha}$. Let σ be a non-null real function of BV for which $\sigma^\vee(v)$ vanishes for $|v| \le 1$ (e.g. $F(x)\cos 2x$, F being the Fejér kernel). Writing $t_n(x) = \sum_{k=-n}^{n} c_k e^{ikx}$, we have

$$\int_{-\infty}^{\infty} f(x - \tau u)\, d\sigma(u) = \int_{-\infty}^{\infty} [f(x - \tau u) - t_n(x - \tau u)]\, d\sigma(u) + \sqrt{2\pi} \sum_{k=-n}^{n} c_k \sigma^\vee(k\tau)\, e^{ikx}.$$

Now, given τ (sufficiently small), choose n so that $\tau^{-1} - 1 < n \le \tau^{-1}$. Then the second term on the right vanishes, giving

$$\|D_\sigma(f; \circ; \tau)\|_{\mathsf{C}} \le \|f - t_n\|_{\mathsf{C}_{2\pi}} \frac{1}{\sqrt{2\pi}} \int_{-\infty}^{\infty} |d\sigma| \le M_1 n^{-\alpha} \le M_2 \tau^\alpha,$$

where M_1, M_2 are certain constants. If we take $r = 1$ [$r = 2$] in Prop. 13.3.15, then Theorem 2.3.3 [2.3.5] follows.

Problems

1. (i) Show that the kernels w and N of Gauss–Weierstrass and Jackson–de La Vallée Poussin are equivalent.
   (ii) Let $\kappa > 0$ be fixed. Show that the kernels w_κ and $r_{\kappa,\lambda}$ of Weierstrass and Riesz are equivalent for every $\lambda > 0$.
2. Show that $\{w^\wedge(v) - 1\}$ divides $\delta_2^\vee(v)$ globally. Therefore there exists a constant $M > 0$ such that for every $f \in \mathsf{X}(\mathbb{R})$ and $h > 0$
   $$\|f(\circ + h) + f(\circ - h) - 2f(\circ)\|_{\mathsf{X}(\mathbb{R})} \le M\, \|W(f; \circ; h^2) - f(\circ)\|_{\mathsf{X}(\mathbb{R})}.$$
   State and prove analogous assertions for the singular integrals of Jackson–de La Vallée Poussin and Picard, the integral means and the Bessel potentials.
3. Let $f \in \mathsf{X}(\mathbb{R})$ and $0 < \alpha \le 2$. Show that for the integral means
   $$\|A(f; \circ; h) - f(\circ)\|_{\mathsf{X}(\mathbb{R})} = O(h^\alpha) \Leftrightarrow \|f(\circ + h) + f(\circ - h) - 2f(\circ)\|_{\mathsf{X}(\mathbb{R})} = O(h^\alpha).$$
4. Carry out a similar analysis for the singular integrals of Fejér and Cauchy–Poisson.
5. Let $f \in \mathsf{X}(\mathbb{R})$ and $r, s \in \mathbb{N}$ with $r < s$. Show for $0 < \alpha < r$ that $\omega_r(\mathsf{X}(\mathbb{R}); f; t) = O(t^\alpha)$ if and only if $\omega_s(\mathsf{X}(\mathbb{R}); f; t) = O(t^\alpha)$. Furthermore, show that $\omega_s(\mathsf{X}(\mathbb{R}); f; t) = O(t^r)$ implies $\omega_r(\mathsf{X}(\mathbb{R}); f; t) = O(t^r|\log t|)$.
6. Let $\chi \in \mathsf{NL}^1$ have an integrable nth derivative. Show that if $\|[J(f; \circ; \rho)]^{(n)}\|_{\mathsf{X}(\mathbb{R})} = O(\rho^\alpha)$ for some $f \in \mathsf{X}(\mathbb{R})$ and α with $0 < \alpha < n$, then the rth modulus of smoothness $\omega_r(\mathsf{X}(\mathbb{R}); f; t)$ of f is $O(t^{n-\alpha})$, $O(t^r|\log t|)$, or $O(t^r)$ for $t \to 0+$ according as r is greater than, equal to, or less than $n - \alpha$.
7. If the assumptions of Theorem 13.3.9 are satisfied and $\|D_\sigma(f; \circ; \tau)\|_{\mathsf{X}(\mathbb{R})} \le \varphi(\tau)$, where φ increases with τ, show that $\|D_\xi(f; \circ; \tau)\|_{\mathsf{X}(\mathbb{R})} \le A \int_0^{B\tau} u^{-1}\varphi(u)\, du$ for suitable constants A, B. (Hint: H. S. Shapiro [4, Sec. 5.2.1], where an application to moduli of smoothness is indicated)

13.4 Saturation on Banach Spaces

In this final section we treat approximation theoretical questions for a general class of bounded linear operators $\{T_\rho\}$ defined on a Banach space X. The results may in particular be applied to operators generated by convolution integrals. Here we shall confine ourselves to a brief discussion of the saturation case.

13.4.1 Strong Approximation Processes

Let $\{T_\rho\}$ be a commutative strong approximation process on the Banach space X (cf. Def. 12.0.1). Suppose there exist a closed linear operator B with domain $\mathsf{D}(B)$ dense in X and range in X and a positive number α such that for every $f \in \mathsf{D}(B)$

$$\lim_{\rho \to \infty} \|\rho^\alpha[T_\rho f - f] - Bf\| = 0\dagger. \tag{13.4.1}$$

Then the following theorem states that the process $\{T_\rho\}$ is saturated in X with order $O(\rho^{-\alpha})$, $\rho \to \infty$, and that the Favard class $\mathsf{F}[\mathsf{X}; T_\rho]$ is characterized as the completion of $\mathsf{D}(B)$ relative to X.

Theorem 13.4.1. *Let $f \in \mathsf{X}$, $\{T_\rho\}$ be a commutative strong approximation process on X and B a closed linear operator satisfying* (13.4.1). *Suppose that there exists a* ***regularization process*** *$\{J_n\}$, i.e., a family of bounded linear operators J_n, $n \in \mathbb{N}$, of X into X such that $J_n[\mathsf{X}] \subset \mathsf{D}(B)$ for each $n \in \mathbb{N}$, $\lim_{n\to\infty} \|J_n f - f\| = 0$ for each $f \in \mathsf{X}$ and the operators J_n and T_ρ commute for each $n \in \mathbb{N}$, $\rho > 0$.*

(a) *If $f \in \mathsf{X}$ is such that $\|T_\rho f - f\| = o(\rho^{-\alpha})$, then $f \in \mathsf{D}(B)$ and $Bf = \Theta$.*

(b) *The following statements are equivalent:*

(i) $\|T_\rho f - f\| = O(\rho^{-\alpha})$ $\qquad (\rho \to \infty)$,

(ii) $f \in \widetilde{\mathsf{D}(B)}^{\mathsf{X}}$,

(iii) $f \in \mathsf{D}(B)$, *provided X is reflexive.*

Proof. First of all we observe that f belongs to $\mathsf{D}(B)$ and $Bf = g$ if and only if $\text{s-lim}_{\rho\to\infty} \rho^\alpha[T_\rho f - f]$ exists and is equal to g. Indeed, if $f \in \mathsf{D}(B)$ and $Bf = g$, then by (13.4.1)

$$\lim_{\rho \to \infty} \|\rho^\alpha[T_\rho f - f] - g\| = 0. \tag{13.4.2}$$

Conversely, let $g \in \mathsf{X}$ be such that (13.4.2) holds. For the regularization process $\{J_n\}$ we have on the one hand $\text{s-lim}_{n\to\infty} J_n f = f$, on the other $J_n f \in \mathsf{D}(B)$ for each $n \in \mathbb{N}$, and therefore by (13.4.1) and the commutativity

$$B[J_n f] = \underset{\rho\to\infty}{\text{s-lim}}\, \rho^\alpha[T_\rho[J_n f] - J_n f] = \underset{\rho\to\infty}{\text{s-lim}}\, J_n[\rho^\alpha[T_\rho f - f]].$$

By the continuity of J_n and by (13.4.2) we therefore obtain $B[J_n f] = J_n g$, and thus $\text{s-lim}_{n\to\infty} B[J_n f] = g$. Hence $f \in \mathsf{D}(B)$ and $Bf = g$ by the definition of a closed linear operator.

† Note the equivalent notation $\text{s-lim}_{\rho\to\infty} \rho^\alpha[T_\rho f - f] = Bf$. Here $\|\circ\|$ always denotes the X-norm unless otherwise specified.

(a) Obviously, (13.4.2) is satisfied with $g = \Theta$. Hence $f \in D(B)$ and $Bf = \Theta$.
(b) We recall that $D(B)$ becomes a Banach space under the norm $\|f\|_{D(B)} \equiv \|f\| + \|Bf\|$. By (13.4.1), $\{\rho^\alpha[T_\rho - I]\}$† defines a family of bounded linear operators of $D(B)$ into X such that for each $f \in D(B)$ there exists a constant M_f with $\rho^\alpha\|T_\rho f - f\| \le M_f$ uniformly for all $\rho > 0$. Hence by the uniform boundedness principle there exists a constant M_1 such that

$$\rho^\alpha\|T_\rho f - f\| \le M_1\|f\|_{D(B)} \qquad (f \in D(B), \rho > 0). \tag{13.4.3}$$

By the same reasons there exists a constant M_2 independent of f and n such that

$$\|J_n f\| \le M_2\|f\| \qquad (f \in X, n \in \mathbb{N}). \tag{13.4.4}$$

(ii) $\Rightarrow$ (i). By Def. 10.4.1 there exists a sequence $\{f_n\} \subset D(B)$ such that $\|f_n\|_{D(B)} \le M_3$ and $\lim_{n\to\infty} \|f_n - f\| = 0$. Therefore by (13.4.3)

$$\rho^\alpha\|T_\rho f_n - f_n\| \le M_1\|f_n\|_{D(B)} \le M_1 M_3,$$

uniformly for all $\rho > 0$, $n \in \mathbb{N}$. Hence, letting $n \to \infty$, we obtain (i).
(i) $\Rightarrow$ (ii). For the sequence $\{J_n f\}$ we have on the one hand $\lim_{n\to\infty} \|J_n f - f\| = 0$, on the other hand $J_n f \in D(B)$ for each $n \in \mathbb{N}$, and hence by (13.4.1), (13.4.4)

$$\begin{aligned}\|J_n f\|_{D(B)} &\equiv \|J_n f\| + \|B[J_n f]\| = \|J_n f\| + \lim_{\rho\to\infty} \rho^\alpha\|T_\rho[J_n f] - J_n f\| \\ &\le M_2\|f\| + \lim_{\rho\to\infty} \|J_n[\rho^\alpha[T_\rho f - f]]\| \le M_2(\|f\| + \limsup_{\rho\to\infty} \rho^\alpha\|Tf - f\|).\end{aligned}$$

Thus $\|J_n f\|_{D(B)}$ is bounded for $n \in \mathbb{N}$, and $f \in \widetilde{D(B)}^{X}$.
(ii) $\Leftrightarrow$ (iii). If in particular X is reflexive, then it follows by Prop. 10.4.3 that $f \in D(B)$ if and only if $f \in \widetilde{D(B)}^{X}$, and the proof is complete.

Let us mention that in many cases the strong approximation process itself defines a regularization process of the type needed for the proof.

To give an application let us consider the generalized singular integral $C_\kappa(f; x; \rho)$ of Picard on $X(\mathbb{R})$. For each $\kappa > 0$ the family of operators $C_\kappa(\rho)$ defined through $(C_\kappa(\rho)f)(x) = C_\kappa(f; x; \rho)$ forms a commutative strong approximation process on $X(\mathbb{R})$ (Problem 6.4.5), and it was shown in Sec. 12.4.2 that the kernel c_κ satisfies (12.4.1). Hence in L^p-space, $1 \le p \le 2$, it follows by (12.4.11) (cf. Problem 12.3.6) that there exists $g \in L^p$ with

$$\lim_{\rho\to\infty} \|\rho^\kappa[C_\kappa(f; \circ; \rho) - f(\circ)] - g(\circ)\|_p = 0 \tag{13.4.5}$$

if and only if $f \in W[L^p; -|v|^\kappa]$. In view of Theorems 11.2.6–11.2.9 the operators $C_\kappa(\rho)$ therefore satisfy (13.4.1) with $B_\kappa = -D_s^{\{\kappa\}}$. Furthermore, (13.4.5) holds in L^p,

† I denotes the identity operator.

$2 < p < \infty$, if and only if there exists $g \in \mathsf{L}^p$ such that assertion (ii) of Prop. 13.2.1 is valid. Thus if we set

$$\mathsf{D}(B_\kappa) = \Big\{ f \in \mathsf{L}^p \;\Big|\; -\int_{-\infty}^{\infty} f(x)(D_s^{\{\kappa\}} s)(x)\, dx$$

$$= \int_{-\infty}^{\infty} g(x)s(x)\, dx,\; g \in \mathsf{L}^p, \textit{for every } s \in \mathsf{W}[\mathsf{L}^{p'}; |v|^\kappa] \Big\}, \tag{13.4.6}$$

$$B_\kappa f = g,$$

then (13.4.1) is also satisfied in L^p, $2 < p < \infty$. The case $\mathsf{X}(\mathbb{R}) = \mathsf{C}$ may be treated analogously. Now Theorem 13.4.1 may be applied, the regularization process being given by $C_\kappa(\rho)$ itself (cf. (12.4.10)). Moreover, the concrete form of the (closed) operator B_κ and of $\mathsf{D}(B_\kappa)$ admits further characterizations of the Favard class. This would again lead to Prop. 12.4.2 and 13.2.1 (as applied to $C_\kappa(f; x; \rho)$, cf. also Prop. 13.2.4). These considerations may indicate how methods of Fourier analysis enter into the discussion if the general setting of Theorem 13.4.1 is applied to processes generated by convolution integrals (for an application to the (periodic) singular integral of de La Vallée Poussin see Berens [3, p. 56 ff]).

13.4.2 Semi-Groups of Operators

Let us consider the case when the strong approximation process $\{T_\rho\}$ forms a semi-group of class (C_0).

Definition 13.4.2. *The family $\{S(t)\}$, $t \geq 0$, of bounded linear operators $S(t)$ of X into X is called a (one-parameter)* ***semi-group*** *(of operators) if*

(i) $S(0) = I$ *(I = identity operator),*

(ii) $S(t_1 + t_2) = S(t_1)S(t_2)$ $(t_1, t_2 \geq 0)$.

The semi-group $\{S(t)\}$ is said to be of ***class*** (C_0) *if for each $f \in \mathsf{X}$*

$$\lim_{t\to 0+} \|S(t)f - f\| = 0. \tag{13.4.7}$$

Under these hypotheses, for each $f \in \mathsf{X}$, $S(t)f$ is a strongly continuous vector-valued function defined on $0 \leq t < \infty$ into the Banach space X, and the norm of $S(t)$ is uniformly bounded on every finite interval $0 \leq t \leq t_0$. The semi-group $\{S(t)\}$ of class (C_0) is *uniformly bounded* if there is a constant M such that $\|S(t)f\| \leq M\,\|f\|$ for all $t \geq 0$ and $f \in \mathsf{X}$.

Definition 13.4.3. *The* ***infinitesimal generator*** *A of the semi-group $\{S(t)\}$ of class (C_0) is defined by*

$$Af = \operatorname{s-lim}_{t\to 0+} \frac{S(t)f - f}{t}, \tag{13.4.8}$$

whenever the limit exists, the ***domain*** *$\mathsf{D}(A)$ of A being the set of elements $f \in \mathsf{X}$ for which this limit exists.*

The infinitesimal generator A of such a semi-group is a closed (in general unbounded) linear operator. Its domain $D(A)$ is dense in X and becomes a Banach space under the norm $\|f\|_{D(A)} \equiv \|f\| + \|Af\|$. If $f \in D(A)$, so does $S(t)f$ for each $t \geq 0$ and

$$\frac{d[S(t)f]}{dt} = A[S(t)f] = S(t)[Af],$$

the differentiation being taken in the strong sense. Moreover,

$$S(t)f - f = \int_0^t S(\tau)[Af]\, d\tau \qquad (f \in D(A)). \tag{13.4.9}$$

For these elementary properties we refer to BUTZER–BERENS [1, p. 7 ff].

Theorem 13.4.4. *Let $f \in X$ and $\{S(t)\}$ be a uniformly bounded semi-group of class* (C_0).
(a) *If $f \in X$ is such that*

$$\|S(t)f - f\| = o(t) \qquad (t \to 0+), \tag{13.4.10}$$

then $f \in D(A)$ and $Af = \Theta$, i.e. $S(t)f = f$ for every $t \geq 0$.
(b) *The following statements are equivalent:*
(i) $\|S(t)f - f\| = O(t) \qquad (t \to 0+)$,
(ii) $f \in \widetilde{D(A)}^{X}$,
(iii) $f \in D(A)$, *provided* X *is reflexive.*

Proof. We wish to apply Theorem 13.4.1. Indeed, every uniformly bounded semi-group $\{S(t)\}$ of class (C_0) defines a commutative strong approximation process on X with parameter $\rho = t^{-1}$, $t \to 0+$. It follows by Def. 13.4.3 that

$$\lim_{t \to 0+} \|t^{-1}[S(t)f - f] - Af\| = 0 \tag{13.4.11}$$

for each $f \in D(A)$. Thus $\{S(t)\}$ satisfies (13.4.1) with B given by the infinitesimal generator A, $\rho = t^{-1}$, $t \to 0+$, and $\alpha = 1$. Moreover, if for $f \in X$ and $n \in \mathbb{N}$ we set

$$J_n f = n \int_0^{1/n} S(\tau) f\, d\tau, \tag{13.4.12}$$

then $\{J_n\}$ defines a regularization process. Indeed, J_n defines a bounded linear operator of X into X for each $n \in \mathbb{N}$, and

$$\|J_n f - f\| = \left\| n \int_0^{1/n} [S(\tau)f - f]\, d\tau \right\| \leq \sup_{0 \leq \tau \leq 1/n} \|S(\tau)f - f\|,$$

which tends to zero as $n \to \infty$ by (13.4.7). Since

$$\begin{aligned} S(t)[J_n f] - J_n f &= n \int_0^{1/n} S(t + \tau) f\, d\tau - n \int_0^{1/n} S(\tau) f\, d\tau \\ &= n \int_t^{t+(1/n)} S(\tau) f\, d\tau - n \int_0^{1/n} S(\tau) f\, d\tau \\ &= n \int_{1/n}^{t+(1/n)} S(\tau) f\, d\tau - n \int_0^{t} S(\tau) f\, d\tau \\ &= n\left[S\left(\frac{1}{n}\right) - I\right]\left[\int_0^t S(\tau) f\, d\tau\right], \end{aligned}$$

which implies

$$\underset{t\to 0+}{\text{s-lim}} \frac{S(t)[J_n f] - J_n f}{t} = n\left[S\left(\frac{1}{n}\right) - I\right]f,$$

it follows that $J_n f \in \mathsf{D}(A)$ for each $f \in \mathsf{X}$ and $n \in \mathbb{N}$ with $A[J_n f] = n[S(1/n)f - f]$. Moreover,

$$J_n[S(t)f] \equiv n \int_0^{1/n} S(\tau)[S(t)f]\, d\tau = S(t)\left[n \int_0^{1/n} S(\tau)f\, d\tau\right] \equiv S(t)[J_n f]$$

for each $n \in \mathbb{N}$ and $t \geq 0$. Therefore Theorem 13.4.1 applies. Since by (13.4.9) $f \in \mathsf{D}(A)$ and $Af = \Theta$ is equivalent to the fact that f is an invariant element of the semi-group $\{S(t)\}$, the proof is complete.

As an example, let us consider the general singular integral $W_\kappa(f; x; t)$ of Weierstrass. In Sec. 12.4.3 we solved the saturation problem in the spaces $\mathsf{X}(\mathbb{R}) = \mathsf{L}^p$, $1 \leq p \leq 2$. We shall now treat this problem for general $\mathsf{X}(\mathbb{R})$ as an application of Theorem 13.4.4.

By (12.4.17) and Sec. 6.4.1 it follows that for each $\kappa > 0$ the bounded linear operators $W_\kappa(t)$ of $\mathsf{X}(\mathbb{R})$ into $\mathsf{X}(\mathbb{R})$ as defined by $[W_\kappa(t)f](x) = W_\kappa(f; x; t)$ for each $t > 0$ forms a uniformly bounded semi-group of class (C_0). Therefore Theorem 13.4.4 applies, and denoting the infinitesimal generator of the semi-group $\{W_\kappa(t)\}$ by A_κ, we have

Proposition 13.4.5. *The general convolution integral $W_\kappa(f; x; t)$ of Weierstrass is saturated in $\mathsf{X}(\mathbb{R})$ with order $O(t)$, $t \to 0+$, and its Favard class $\mathsf{F}[\mathsf{X}(\mathbb{R}); W_\kappa(t)]$ is characterized as* $\widetilde{\mathsf{D}(A_\kappa)}^{\mathsf{X}(\mathbb{R})}$.

If we wish to give a more direct characterization of the Favard class, we have to evaluate a concrete analytical representation of the infinitesimal generator A_κ. For this purpose we may use the methods of Fourier analysis as presented in the preceding sections, particularly in Chapter 12 and Sec. 13.1.

Indeed, in view of Sec. 12.4.3 the kernel of $W_\kappa(f; x; t)$ satisfies (12.4.1). Thus, if $\mathsf{X}(\mathbb{R}) = \mathsf{L}^p$, $1 \leq p \leq 2$, then by (12.4.16) (cf. Problem 12.3.6) there exists $g \in \mathsf{L}^p$ such that

$$\lim_{t\to 0+} \|t^{-1}[W_\kappa(f; \circ; t) - f(\circ)] - g(\circ)\|_p = 0 \tag{13.4.13}$$

if and only if $f \in \mathsf{W}[\mathsf{L}^p; -|v|^\kappa]$. Hence for $1 \leq p \leq 2$

$$\mathsf{D}(A_\kappa) = \{f \in \mathsf{L}^p \mid -|v|^\kappa f^\wedge(v) = g^\wedge(v), g \in \mathsf{L}^p\}, \quad A_\kappa f = g. \tag{13.4.14}$$

If $2 < p < \infty$, then (13.4.13) holds if and only if condition (ii) of Prop. 13.2.1 is satisfied. Thus A_κ is equal to B_κ as defined through (13.4.6).

Therefore, if we apply Prop. 13.4.5 in L^p-space, $1 \leq p \leq 2$, we again obtain Prop. 12.4.3. Concerning the application in L^p, $2 < p < \infty$, we have in view of the reflexivity of these spaces

Proposition 13.4.6. *Let $f \in \mathsf{L}^p$, $2 < p < \infty$, and $\kappa > 0$. The following assertions are equivalent for the general singular integral $W_\kappa(f; x; t)$ of Weierstrass:*

(i) $\|W_\kappa(f; \circ; t) - f(\circ)\|_p = O(t) \qquad (t \to 0+)$,

(ii) *there exists $g \in \mathsf{L}^p$ such that for every $s \in \mathsf{W}[\mathsf{L}^{p'}; |v|^\kappa]$*

$$-\int_{-\infty}^{\infty} f(x)(D_s^{\{\kappa\}}s)(x)\,dx = \int_{-\infty}^{\infty} g(x)s(x)\,dx,$$

(iii) *the κth strong Riesz derivative of f exists,*

(iv) *if $j \in \mathbb{N}$ is such that $0 < \kappa < 2j$, then*

$$\left\| \frac{1}{C_{\kappa,2j}} \int_\varepsilon^\infty \frac{\overline{\Delta}_u^{2j} f(\circ)}{u^{1+\kappa}}\,du \right\|_p = O(1) \qquad (\varepsilon \to 0+),$$

where $C_{\kappa,2j}$ is given by (11.3.14).

Obviously, there are also applications of Theorem 13.4.4 to Banach spaces of periodic functions (cf. e.g. Butzer–Berens [1, p. 128 ff] for the (periodic) singular integral of Weierstrass).

In the above, we treated the saturation problem for bounded linear operators on Banach spaces. One may also treat the problem of nonoptimal approximation from this general point of view. In case the operators in question possess the semi-group property, we refer the interested reader to the detailed investigations in Butzer–Berens [1], for other classes of linear operators to Berens [3] and Butzer–Scherer [1–3]. In these monographs the results are presented in the setting of the theory of interpolation or intermediate spaces.

13.5 Notes and Remarks

The material of Sec. 13.3, the second topic of this chapter, is taken from lecture notes [1] and a monograph [4] of H. S. Shapiro. As Shapiro [2] remarks, his contribution 'completes in some essential respects a program outlined by Butzer in a series of papers'. In this way, his work rounds off the material covered in the later chapters of the present book very beautifully. On the occasion of the Oberwolfach and Gatlinburg conferences on *Approximation* in 1963, Jean Favard ([5, 6], see also [3]) proposed an interesting problem on the *comparison of different processes of summation.* He established some first results and showed how the theorem of Banach–Steinhaus may be employed in its solution (see also Haf [1]). This problem seems to be connected with the comparison theorems of H. S. Shapiro who attended the two Favard lectures, but unfortunately did not fit his results into the general setting envisaged by Favard. It would be of great interest to follow up the problem and relate it to the material of Sec. 13.3.

The main references to the third topic here, namely **Sec. 13.4**, are Butzer–Berens [1] and the lecture notes of Berens [3] (see also Kupzov [1]). The purpose of the brief discussion here is to introduce the reader to approximation theory on arbitrary Banach spaces.

Sec. 13.1. The results of this section (for measures of type (13.0.2)) are largely taken from Berens–Nessel [1]. The dual techniques employed there had also been used in Görlich–Nessel [1] (see also Sec. 10.6). This technique is especially widespread in the study of various problems of analysis. Let us just mention WIENER [1], F. RIESZ [3], SCHOENBERG [2], L. SCHWARTZ [1], BUCHWALTER [3], PFLUGER [1], LORENTZ [2], and SUNOUCHI [1II]. Homogeneous functions play an important rôle in much of the work of PEETRE [1–3] and LÖFSTRÖM [1–3].

Condition (13.1.4) may be used to extend the concepts to L^p, $2 < p < \infty$, which are originally given in L^p, $1 \le p \le 2$, by means of the Fourier transforms of the functions involved. Thus, the set of $f \in L^p$, $2 < p < \infty$, satisfying (13.1.4) may again be denoted by $W[L^p; \psi(v)]$ (or $V[L^p; \psi(v)]$), and as an extension of Def. 13.1.4 we may call g of (13.1.4) the *ψ-derivative* of f (for $\psi(v) = |v|^\alpha$ see Prop. 13.2.2). The latter definition is to be compared with the rth (ordinary) distributional derivative of f (cf. Sec. 3.7) where $\psi(v) = (iv)^r$ (cf. Prop. 5.1.15, 5.2.20) and s belongs to C_{00}^∞, thus $D^{\psi*}s = (-1)^r s^{(r)}$ (obviously, C_{00}^∞ is a subset of $W[L^{p'}; (-iv)^r]$, but is nevertheless dense in $L^{p'}$). For corresponding interpretations in C we refer to condition (ii) of Problem 13.1.6.

Sec. 13.2. A number of the applications of this section do not seem to be stated explicitly elsewhere. The idea to rewrite $R_\rho^{\{\alpha\}}f$ as a singular integral of Fejér's type is due to SUNOUCHI [8] where it is also shown that k_α has the property (12.4.1). For the Riesz means see also BERENS–NESSEL [1].

At present, there are (at least) five distinct methods for proving saturation theorems in L^p, $2 < p < \infty$: (i) An elementary distributional method by which GÖRLICH [2] solved the saturation problem in case of a particular approximation process. It depends upon the use of a fundamental identity which reduces to (12.4.16) for the generalized singular integral of Weierstrass. (ii) A general method for solving saturation problems due to KOZIMA–SUNOUCHI [1]. It depends on Lemma 12.3.4 and on some distribution-theoretical arguments which are stated for approximation processes having saturation class $V[L^p; |k|^\alpha]$, $\alpha \in \mathbb{N}$. But, according to a written communication of Prof. SUNOUCHI, there will be a further paper where the arguments will be extended also to fractional α using D'_{L^p}-spaces and a regularization process. (iii) A general method presented by GÖRLICH [3]. It has certain relations to the method of KOZIMA–SUNOUCHI in so far as both commence in the same way and use the multiplier criterion of Prop. 12.4.5. However, GÖRLICH's treatment leads to the spaces of Bessel potentials as Favard classes and uses only elementary facts of distribution theory. (iv) The elementary dual method as presented in Sec. 13.1 which does not use distribution theory explicitly and hence fits precisely into the classical setting of this text. (v) The comparison theory of SHAPIRO as given in Sec. 13.3.

It is a common feature of all methods that they apply to the several variable case as well.

Sec. 13.3. The concepts of global and local divisibility were first made practicable in approximation theory by SHAPIRO [1–3]. Important work in this connection is due to NORBERT WIENER [2]. As a matter of fact, the following fundamental lemma was established by WIENER (cf. PITT [1, pp. 49; 93 ff], KATZNELSON [1, p. 228]): *Let* $\sigma_1, \sigma_2 \in BV$ *and* $\mathbb{E}$ *a compact set of* $\mathbb{R}$. *If* $\sigma_1^\vee(v)$ *vanishes nowhere in* $\mathbb{E}$ *and* $\sigma_2^\vee(v)$ *vanishes everywhere outside of* $\mathbb{E}$, *then there exists* $\tau \in BV$ *such that* $\sigma_2^\vee(v) = \tau^\vee(v)\sigma_1^\vee(v)$ *for all* $v \in \mathbb{R}$. It is used in the proof of Theorem 13.3.9 on local divisibility. Important here is the concept of the Wiener ring W of Fourier–Stieltjes transforms $\sigma^\vee$ where $\sigma \in BV$. W is a commutative ring with unit element, with respect to ordinary multiplication. The norm in W is $\|\sigma^\vee\|_W = (1/\sqrt{2\pi})\int_{-\infty}^{+\infty} |d\sigma|$. For Def. 13.3.5 and certain of its modifications see also FAVARD [3, 5, 6].

To verify the diverse hypotheses of divisibility of measures†, in the global case any of the sufficient conditions of Chapter 6 may be taken into account. Concerning a general criterion on local divisibility, roughly speaking any somewhat smooth function defined in a neighbourhood of the origin is the 'germ' of an element of W.

† Comparison theorems in connection with global divisibility may also be obtained for more general entities than integrals of type (13.0.1). Here we refer to SHAPIRO [5] where the principles of a 'naive' distribution theory are outlined and Jackson-type theorems for periodic functions (in several variables) are derived.

As RAGOZIN pointed out to SHAPIRO [4, Sec. 5.2.2] the converse of Theorem 13.3.2 is also true if $\mathsf{X}(\mathbb{R}) = \mathsf{C}$, i.e., an inequality of type (13.3.1) in this space implies that σ divides ξ globally. This is a consequence of a result of DE LEEUW–MIRKIL [1] (cf. SHAPIRO [4, Sec. 6]). Thus divisibility relations on W belong quite essentially to the study of comparison theorems. It is now instructive to compare via Fourier transforms the Fejér kernel F with the dipole measure δ_1. Neither the ratio $[F^\wedge(v) - 1]/\delta_1^\vee(v)$ nor its reciprocal is in W; in fact both are discontinuous at $v = 0$. Thus, in view of the cited converse, neither a relation of the type $\|\sigma(f; \circ; \rho) - f(\circ)\|_{\mathsf{C}} \le A_1\|f(\circ + \rho^{-1}) - f(\circ)\|_{\mathsf{C}}$ nor a relation of the type $\omega(\mathsf{C}; f; \rho^{-1}) \le A_2\|\sigma(f; \circ; \rho) - f(\circ)\|_{\mathsf{C}}$ is possible. This strongly suggests (although it does not prove) that the saturation class of Fejér's integral neither includes nor is included in $\mathsf{Lip}\,(\mathsf{C}; 1)$. Similar remarks hold in L^1. But note that in L^p, $1 < p < \infty$, $\|\sigma(f; \circ; \rho) - f(\circ)\|_p = \mathsf{O}(\rho^{-1}) \Leftrightarrow f \in \mathsf{Lip}\,(\mathsf{L}^p; 1)$ (cf. Prop. 12.4.1, 13.2.5 and Problem 8.2.5).

Problem 13.3.6 admits important applications to harmonic functions such as theorems of HARDY–LITTLEWOOD in their following extension due to ZYGMUND [2] (half-plane version): *The class of functions $u(x, y)$ which admit a representation*

$$u(x, y) = \frac{y}{\pi}\int_{-\infty}^{\infty} \frac{f(t)}{(x-t)^2 + y^2}\,dt \qquad (y > 0)$$

where f is bounded and $\omega^*(\mathsf{C}; f; \tau) = \mathsf{O}(\tau)$ is identical with the class of functions $u(x, y)$ harmonic and bounded for $y > 0$ for which $u_{xx}(x, y) = \mathsf{O}(y^{-1})$. For details see SHAPIRO [4, Sec. 5.3.3] as well as BERENS [1]. This problem is closely connected with further investigations by BUTZER–BERENS [1] and BUTZER–SCHERER [1].

Concerning the applications to periodic functions of Sec. 13.3.4, see in particular SHAPIRO [2]. The present method of proving the theorems of Bernstein and Zygmund has perhaps some methodological interest in that a Bernstein-type inequality is not used. According to several written communications of HAROLD S. SHAPIRO, results in $\mathsf{L}^p_{2\pi}$ may be derived either by developing a separate comparison theory, especially tailored to $\mathsf{L}^p_{2\pi}$ from the start, or by establishing the results of this section not for $\mathsf{X}(\mathbb{R})$-norm but for (cf. KATZNELSON [1, p. 127])

$$\|f\|_p^* = \sup_t \left(\int_t^{t+2\pi} |f(u)|^p\,du\right)^{1/p}.$$

Indeed, this definition of a norm covers L^p as well as $\mathsf{L}^p_{2\pi}$.

From a general point of view, comparison theory is a stimulating general frame which has to be supplemented with detailed discussion, particularly on saturation theory. The global result covers comparison of the approximation of f by different convolution integrals, although the explicit checking may sometimes be nontrivial. Concerning comparison of the approximation by a convolution integral with Lipschitz conditions, the local result may be used. However, the problem is to find comparable Lipschitz-type conditions for a given convolution integral, particularly if α of (12.3.5) is fractional. In the case of nonoptimal approximation, Theorem 13.3.14 may be applied, but it does not cover the saturation case. Here the global result may be successful provided that the saturation behaviour of a sufficiently general convolution integral is known.

As the reader will have noticed, a number of research problems have been posed implicitly in many of the sections on Notes and Remarks. For further new and unsolved problems connected with the material of this volume the reader may consult the corresponding sections in BUTZER–KOREVAAR [1, p. 179 f], BUTZER–SZ.-NAGY [1, p. 416 f].

List of Symbols

Symbols not Explicitly Defined in Text

o, O — Landau symbols. For example, $a_n = o(b_n)$, $n \to \infty$, means that $\lim_{n\to\infty}(a_n/b_n) = 0$; $a_n = O(b_n)$, $n \to \infty$, means that $|a_n/b_n|$ remains bounded for $n \to \infty$; also $a_n = c_n + O(b_n)$ expresses the fact that there exists a constant $M > 0$ such that $|a_n - c_n|/|b_n| \leqslant M$ for all sufficiently large n.

$[\![\alpha]\!]$ — largest integer $\leqslant \alpha$, $\alpha > 0$.

$\operatorname{sgn} x$ — function with value $x/|x|$ at $x \neq 0$, with value 0 at $x = 0$.

$\{f \in \mathsf{X} \mid \ldots\}$ — the set of all $f \in \mathsf{X}$ such that ... (cf. (4.1.24)).

$\sum_{k=-\infty}^{\prime\infty}$ — the summation index $k = 0$ is omitted (cf. footnote to (9.0.4)).

(...) at the right-hand side of a formula: Assertion holds for *each* ... For example, in (0.7.1) the inner product (f, g) should be taken for *each* $f, g \in \mathsf{L}^2$; in (1.1.1) one has $\int_{-\pi}^{\pi} \chi_\rho(u)\, du = 2\pi$ for *each* $\rho \in \mathbb{A}$; (1.3.3) is valid for each p with $1 \leqslant p \leqslant \infty$ (cf. also p. 3).

General Symbols

Function Spaces

Norms

Differences and Derivatives

Transforms

Certain Operators

General Kernels and Singular (Convolution) Integrals

Particular Kernels

Particular Singular (Convolution) Integrals

Tables of Fourier and Hilbert Transforms

Table 1 Fourier Transforms

$f(x)$		$f^{\wedge}(v) = \dfrac{1}{\sqrt{2\pi}} \displaystyle\int_{-\infty}^{\infty} f(u)\, e^{-ivu}\, du$	(5.1.1)†
Operational Rules			
$\alpha f(x) + \beta g(x)$	$(\alpha, \beta \in \mathbb{C})$	$\alpha f^{\wedge}(v) + \beta g^{\wedge}(v)$	
$f(x+h)$	$(h \in \mathbb{R})$	$e^{ihv} f^{\wedge}(v)$	
$e^{-ihx} f(x)$	$(h \in \mathbb{R})$	$f^{\wedge}(v+h)$	
$\rho f(\rho x)$	$(\rho > 0)$	$f^{\wedge}(v/\rho)$	
$\overline{f(-x)}$		$\overline{f^{\wedge}(v)}$	(Prop. 5.1.1, 5.2.11)
$f(x) = f(-x)$		$f^{\wedge}(v) = f^{\wedge}(-v) = \sqrt{\dfrac{2}{\pi}} \displaystyle\int_0^{\infty} f(u) \cos vu\, du$	
$f(x) = -f(-x)$		$f^{\wedge}(v) = -f^{\wedge}(-v) = -i\sqrt{\dfrac{2}{\pi}} \displaystyle\int_0^{\infty} f(u) \sin vu\, du$	
$f(-x)$		$f^{\wedge}(-v)$	
$f(hx)$	$(h \in \mathbb{R}, h \neq 0)$	$\lvert h\rvert^{-1} f^{\wedge}(v/h)$	(Problems 5.1.1, 5.1.10)
$(f * g)(x)$		$f^{\wedge}(v) g^{\wedge}(v)$	(Theorems 5.1.3, 5.2.12)
$f^{\wedge}(x)$		$f(-v)$	((5.1.6), (5.2.19))
$\left(\dfrac{d}{dx}\right)^r f(x)$		$(iv)^r f^{\wedge}(v)$	(Prop. 5.1.14, 5.2.19)
$x^r f(x)$		$\left(i\dfrac{d}{dv}\right)^r f^{\wedge}(v)$	(Prop. 5.1.17)
$f^{\sim}(x)$		$\{-i \operatorname{sgn} v\} f^{\wedge}(v)$	(Prop. 8.3.1)
$\{-i \operatorname{sgn} x\} f\ \ (x)$		$f^{\sim}(-v)$	(Problem 8.3.2)
Elementary Functions			
$\sqrt{2\pi}\, \kappa_{[-1,0]}(x)$		$\dfrac{e^{iv} - 1}{iv}$	(5.1.29)
$\sqrt{\dfrac{2}{\pi}} \dfrac{\sin ax}{x}$	$(a > 0)$	$1, \quad \lvert v\rvert \neq a$; $\tfrac{1}{2}, \quad \lvert v\rvert = a$	

† This refers to the spot in this volume where the exact interpretation as well as proofs or references may be found.

$f(x)$	$f^\wedge(v) = \frac{1}{\sqrt{2\pi}} \int_{-\infty}^{\infty} f(u)\, e^{-ivu}\, du$	(5.1.1)
Elementary Functions—*continued*		
$\frac{1}{\sqrt{2\pi}} \left[\frac{\sin(x/2)}{x/2}\right]^2$	$1 - \lvert v\rvert, \quad \lvert v\rvert \le 1$ $0, \quad \lvert v\rvert > 1$	
$\frac{3}{\sqrt{8\pi}} \left[\frac{\sin(x/2)}{x/2}\right]^4$	$1 - \frac{3}{2}v^2 + \frac{3}{4}\lvert v\rvert^3, \quad \lvert v\rvert \le 1$ $(2 - \lvert v\rvert)^3/4, \quad 1 < \lvert v\rvert \le 2$ $0, \quad \lvert v\rvert > 2$	(Problem 5.1.2)
$\sqrt{2/\pi}\,(1 + x^2)^{-1}$	$e^{-\lvert v\rvert}$	
$(1/\sqrt{2})\, e^{-x^2/4}$	e^{-v^2}	
$H_n(x) \equiv (-1)^n e^{x^2/2} \frac{d^n}{dx^n}\{e^{-x^2}\}$	$(-i)^n H_n(v)$	
$\sqrt{2\pi}\, e^{-x}, \quad 0 \le x < \infty$ $0, \quad -\infty < x < 0$	$(1 + iv)^{-1}$	(5.1.34)
$\lvert x\rvert^{-s} \quad (0 < \mathrm{Re}\, s < 1)$	$\sqrt{\frac{2}{\pi}} \frac{\Gamma(1-s)}{\lvert v\rvert^{1-s}} \sin\frac{\pi s}{2}$	((11.1.14), Problem 11.1.1)
$x^{-s}\kappa_{(0,\infty)}(x) \quad (0 < \mathrm{Re}\, s < 1)$	$(\Gamma(s)/\sqrt{2\pi})(iv)^{-(1-s)}$	(Problem 11.1.2)
$\sqrt{\frac{2}{\pi}}\, \Gamma(1+\alpha)\, \mathrm{Re}\,(1 - ix)^{-1-\alpha} \quad (\alpha > 0)$	$\lvert v\rvert^{\alpha} e^{-\lvert v\rvert}$	
$\sqrt{\frac{2}{\pi}}\, \Gamma(1+\alpha)\, \mathrm{Im}\,(1 - ix)^{-1-\alpha} \quad (\alpha > 0)$	$\{-i\, \mathrm{sgn}\, v\}\lvert v\rvert^{\alpha} e^{-\lvert v\rvert}$	(Problem 11.2.13)
$\frac{e^{-ihx} - 1}{\lvert x\rvert} \quad (h \in \mathbb{R})$	$\sqrt{\frac{2}{\pi}} \log\left\lvert\frac{v}{h+v}\right\rvert$	(Problem 8.1.5)
$\frac{2^{(2-\kappa)/2}}{\Gamma(\kappa/2)} \lvert x\rvert^{(\kappa-1)/2} K_{(1-\kappa)/2}(\lvert x\rvert) \quad (\kappa > 0)$	$(1 + v^2)^{-\kappa/2}$	(Problem 6.4.6)
$2^{\lambda}\Gamma(1+\lambda) \frac{J_{(1/2)+\lambda}(\lvert x\rvert)}{\lvert x\rvert^{(1/2)+\lambda}} \quad (\lambda > 0)$	$(1 - v^2)^{\lambda}, \quad \lvert v\rvert \le 1$ $0, \quad \lvert v\rvert > 1$	(Problem 6.4.9)
$\frac{1}{\lvert x+h\rvert^{1-\alpha}} - \frac{1}{\lvert x\rvert^{1-\alpha}} \quad (0 < \alpha < 1)$	$\frac{A_c(\alpha)}{\sqrt{2\pi}} \frac{e^{ihv} - 1}{\lvert v\rvert^{\alpha}}$	(Lemma 11.2.2)
$\frac{1}{x} \log\left(\frac{a+x}{a-x}\right)^2 \quad (a > 0)$	$\sqrt{\frac{8}{\pi}} \int_{av}^{\infty} \frac{\sin u}{u}\, du$	(Sec. 13.2.4)
$\sqrt{\frac{2}{\pi}} \int_{\lvert x\rvert}^{\infty} \frac{\sin u}{u^2}\, du$	$1, \quad \lvert v\rvert \le 1$ $\lvert v\rvert^{-1}, \quad \lvert v\rvert > 1$	
$\frac{1}{\sqrt{2\pi}} \log\frac{1+x^2}{x^2}$	$\frac{1 - e^{-\lvert v\rvert}}{\lvert v\rvert}$	
$\sqrt{2}\left(e^{-x^2/4} - \lvert x\rvert \int_{\lvert x\rvert/2}^{\infty} e^{-u^2}\, du\right)$	$\frac{1 - e^{-v^2}}{v^2}$	(Problem 12.4.1)

For a more extensive selection of Fourier transforms the reader is referred to ERDÉLYI [2I].

Table 2 Finite Fourier Transforms

$f(x)$		$f^\wedge(k) = \dfrac{1}{2\pi}\int_{-\pi}^{\pi} f(u)\, e^{-iku}\, du$	(4.1.1)
Operational Rules			
$\alpha f(x) + \beta g(x)$	$(\alpha, \beta \in \mathbb{C})$	$\alpha f^\wedge(k) + \beta g^\wedge(k)$	
$f(x + h)$	$(h \in \mathbb{R})$	$e^{ihk} f^\wedge(k)$	
$e^{-ijx} f(x)$	$(j \in \mathbb{Z})$	$f^\wedge(k + j)$	
$\overline{f(-x)}$		$\overline{f^\wedge(k)}$	(Prop. 4.1.1)
$f(x) = f(-x)$		$f^\wedge(k) = f^\wedge(-k) = \dfrac{1}{\pi}\int_0^\pi f(u) \cos ku\, du$	
$f(x) = -f(-x)$		$f^\wedge(k) = -f^\wedge(-k) = \dfrac{-i}{\pi}\int_0^\pi f(u) \sin ku\, du$	(Problem 4.1.1)
$(f * g)(x)$		$f^\wedge(k) g^\wedge(k)$	(Theorem 4.1.3)
$\left(\dfrac{d}{dx}\right)^r f(x)$		$(ik)^r f^\wedge(k)$	(Prop. 4.1.8)
$f^\sim(x)$		$\{-i \operatorname{sgn} k\} f^\wedge(k)$	(Prop. 9.1.8, 9.2.5)
$\sqrt{2\pi} \sum_{j=-\infty}^{\infty} \rho\chi(\rho(x + 2j\pi))$		$\chi^\wedge(k/\rho)$	(5.1.58)
Elementary Functions			
$\sum_{j=-n}^{n} c_j\, e^{ijx}$		$c_k, \quad \lvert k\rvert \le n$ $0, \quad \lvert k\rvert > n$	
$2\pi\kappa_{[-1,0]}(x)$ on $[-\pi, \pi]$		$\dfrac{e^{ik} - 1}{ik}$	(Problem 4.1.5)
$\dfrac{\sin((2n + 1)x/2)}{\sin(x/2)}$		$1, \quad \lvert k\rvert \le n$ $0, \quad \lvert k\rvert > n$	((1.2.20), (1.2.42))
$\dfrac{1}{n + 1}\left[\dfrac{\sin((n + 1)x/2)}{\sin(x/2)}\right]^2$		$1 - \dfrac{\lvert k\rvert}{n + 1}, \quad \lvert k\rvert \le n$ $0, \quad \lvert k\rvert > n$	((1.2.24), (1.2.43))
$\dfrac{3}{n(2n^2 + 1)}\left[\dfrac{\sin(nx/2)}{\sin(x/2)}\right]^4$		$\dfrac{1}{2n(2n^2 + 1)}\begin{cases} 3\lvert k\rvert^3 - 6nk^2 - 3\lvert k\rvert + 4n^3 + 2n, & \lvert k\rvert \le n \\ -\lvert k\rvert^3 + 6nk^2 - (12n^2 - 1)\lvert k\rvert + 8n^3 - 2n, & n < \lvert k\rvert \le 2n - 2 \\ 0, & \lvert k\rvert > 2n - 2 \end{cases}$	(Problem 1.3.9; cf. Görlich–Stark [2])
$\dfrac{2 + \cos x}{2n^3}\left[\dfrac{\sin(nx/2)}{\sin(x/2)}\right]^4$		$1 - \dfrac{3}{2}\left(\dfrac{k}{n}\right)^2 + \dfrac{3}{4}\left\lvert\dfrac{k}{n}\right\rvert^3, \quad \lvert k\rvert \le n$ $\dfrac{1}{4}\left(2 - \left\lvert\dfrac{k}{n}\right\rvert\right)^3, \quad n < \lvert k\rvert \le 2n - 1$ $0, \quad \lvert k\rvert > 2n - 1$	(Problems 3.1.15, 5.1.2)

$f(x)$	$f\^{}(k) = \dfrac{1}{2\pi}\displaystyle\int_{-\pi}^{\pi} f(u)\, e^{-iku}\, du$	(4.1.1)
Elementary Functions—*continued*		
$\dfrac{1 - r^2}{1 - 2r\cos x + r^2}$	$r^{\lvert k\rvert}$	(1.2.37)
$\displaystyle\prod_{j=1}^{\infty}(1 - e^{-2jt})$ $\times (1 + 2e^{-(2j-1)t}\cos x + e^{-2(2j-1)t})$	e^{-tk^2}	(Problem 1.3.10)
$\dfrac{1}{2}\left[D_n\left(x+\dfrac{\pi}{2n+1}\right) + D_n\left(x-\dfrac{\pi}{2n+1}\right)\right]$	$\cos\dfrac{\lvert k\rvert\pi}{2n+1}$	((1.2.20), (1.3.11))
$\dfrac{2\sin^2(\pi/n)}{n}\,\dfrac{\cos^2(nx/2)}{(\cos(\pi/n) - \cos x)^2}$	$\left(1 - \dfrac{\lvert k\rvert}{n}\right)\cos\dfrac{k\pi}{n} + \dfrac{1}{n}\cot\dfrac{\pi}{n}\sin\dfrac{\lvert k\rvert\pi}{n}$, $\lvert k\rvert \le n-2$ 0, $\lvert k\rvert > n-2$	((1.6.2–4), Sec. 12.1)
$\dfrac{(n!)^2}{(2n)!}\left(2\cos\dfrac{x}{2}\right)^{2n}$	$\dfrac{(n!)^2}{(n-k)!(n+k)!}$	(2.5.13)
$2\displaystyle\sum_{j=1}^{\infty} j^{-\alpha}\cos jx$	$\lvert k\rvert^{-\alpha}$, $k \ne 0$ 0, $k = 0$	(Problem 6.3.1)
$2\displaystyle\sum_{j=1}^{\infty} j^{-\alpha}\sin jx$	$\{-i\,\mathrm{sgn}\,k\}\lvert k\rvert^{-\alpha}$, $k \ne 0$ 0, $k = 0$	(Problem 6.3.3)

Table 3 Hilbert Transforms on the Real Line

$f(x)$		$f^{\sim}(x) = \lim_{\delta\to 0+} \frac{1}{\pi}\int_{\lvert u\rvert \geq \delta} \frac{f(x-u)}{u}\,du$	(8.0.2)
General Formulae			
$\alpha f(x) + \beta g(x)$	$(\alpha, \beta \in \mathbb{C})$	$\alpha f^{\sim}(x) + \beta g^{\sim}(x)$	
$f(x) = f(-x)$		$f^{\sim}(x) = -f^{\sim}(-x)$	
$\overline{f(x)}$		$\overline{f^{\sim}(x)}$	
$f(x+h)$	$(h \in \mathbb{R})$	$f^{\sim}(x+h)$	
$\rho f(\rho x)$	$(\rho > 0)$	$\rho f^{\sim}(\rho x)$	(Problem 8.1.1)
$(f * g)(x)$		$(f^{\sim} * g)(x) = (f * g^{\sim})(x)$	(Prop. 8.2.2–3)
$f^{\sim}(x)$		$-f(x)$	(Prop. 8.2.10)
Elementary Functions			
$\kappa_{[a,b]}(x)$		$\frac{1}{\pi}\log\left\lvert\frac{x-b}{x-a}\right\rvert$	(8.1.8)
$\left[\frac{\sin(x/2)}{x/2}\right]^2$		$2\frac{x - \sin x}{x^2}$	(Problem 8.2.4)
$(1+x^2)^{-1}$		$x(1+x^2)^{-1}$	(Problem 8.1.2)
$\cos x$		$\sin x$	
$\frac{1}{\cos(\pi\alpha/2)}\left[\frac{1}{\lvert x+h\rvert^{1-\alpha}} - \frac{1}{\lvert x\rvert^{1-\alpha}}\right]$		$\frac{1}{\sin(\pi\alpha/2)}\left[\frac{\operatorname{sgn}(x+h)}{\lvert x+h\rvert^{1-\alpha}} - \frac{\operatorname{sgn} x}{\lvert x\rvert^{1-\alpha}}\right]$	(Lemma 11.2.2)
$\operatorname{Re}(1-ix)^{-1-\alpha}$	$(\alpha > 0)$	$\operatorname{Im}(1-ix)^{-1-\alpha}$	(Problem 11.2.13)

Any number of such examples can be written down by starting with a suitably analytic function; compare TITCHMARSH [6, p. 121].

Table 4 Hilbert Transforms of Periodic Functions

$f(x)$		$f^{\sim}(x) = \lim_{\delta\to 0+} \frac{1}{2\pi}\int_{\delta\le\lvert u\rvert\le\pi} f(x-u)\cot\frac{u}{2}\,du$	(9.0.2)
General Formulae			
$\alpha f(x) + \beta g(x)$	$(\alpha, \beta \in \mathbb{C})$	$\alpha f^{\sim}(x) + \beta g^{\sim}(x)$	
$f(x) = f(-x)$		$f^{\sim}(x) = -f^{\sim}(-x)$	
$\overline{f(x)}$		$\overline{f^{\sim}(x)}$	
$f(x+h)$	$(h \in \mathbb{R})$	$f^{\sim}(x+h)$	(Problem 9.1.1)
$(f * g)(x)$		$(f^{\sim} * g)(x) = (f * g^{\sim})(x)$	(Prop. 9.1.12–13)
$f^{\sim}(x)$		$-f(x) + \frac{1}{2\pi}\int_{-\pi}^{\pi} f(u)\,du$	(Prop. 9.1.11, 9.2.5)
Elementary Functions			
$\sum_{k=-n}^{n} c_k e^{ikx}$		$\sum_{k=-n}^{n} \{-i \operatorname{sgn} k\} c_k e^{ikx}$	(Prop. 9.1.6)
$\frac{\sin((2n+1)x/2)}{\sin(x/2)}$		$\cot\frac{x}{2} - \frac{\cos((2n+1)x/2)}{\sin(x/2)}$	
$\frac{1}{(n+1)}\left[\frac{\sin((n+1)x/2)}{\sin(x/2)}\right]^2$		$\cot\frac{x}{2} - \frac{\sin(n+1)x}{2(n+1)\sin^2(x/2)}$	
$\frac{1-r^2}{1-2r\cos x + r^2}$		$\frac{2r\sin x}{1-2r\cos x+r^2}$	(Sec. 1.2.6, Problem 9.1.2)
$\sum_{k=1}^{\infty}\frac{\cos kx}{k^{\alpha}}$		$\sum_{k=1}^{\infty}\frac{\sin kx}{k^{\alpha}}$	(Sec. 11.5)

Bibliography

Abel, N. H.

[1] Solution de quelques problèmes à l'aide intégrales définies. *In:* Oeuvres complètes de Niels Henrik Abel I. (Nouv. Èd.; Ed. L. Sylow–S. Lie) Grøndahl & Søn, Christiania 1881; *pp. 11–27.*

Achieser, N. I.

[1] *The Classical Moment Problem and Some Related Questions in Analysis.* Oliver and Boyd 1965, x + 253 pp. (Orig. Russ. ed. Moscow 1961.)

[2] *Vorlesungen über Approximationstheorie.* Akademie-Verl. 1967, xiii + 412 pp. (Orig. Russ. ed. Moscow 1947.)

Achieser, N. I.–M. G. Krein

[1] *Some Questions in the Theory of Moments.* (Transl. of Math. Monographs **2**) Amer. Math. Soc., Providence, R.I., 1962, v + 265 pp. (Orig. Russ. ed. Kharkov 1938.)

Aissen, M.–I. J. Schoenberg–A. Whitney

[1] On the generating functions of totally positive sequences I. *J. Analyse Math.* **2**, 93–103 (1952).

Akilov, G. P.: *see Kantorovich, L. V.*

Alexits, G.

[1] Sur l'ordre de grandeur de l'approximation d'une fonction par les moyennes de sa série de Fourier (Hungar.). *Mat. Fiz. Lapok* **48**, 410–422 (1941).

[2] Sur l'ordre de grandeur de l'approximation d'une fonction périodique par les sommes de Fejér. *Acta Math. Acad. Sci. Hungar.* **3**, 29–42 (1952).

[3] *Convergence Problems of Orthogonal Series.* Pergamon Press 1961, ix + 350 pp. (Orig. German ed. Berlin 1960.)

Aljančić, S.

[1] Classe de saturation des procédés de sommation de Hölder et de Riesz. *C.R. Acad. Sci. Paris* **246**, 2567–2569 (1958).

[2] Approximation of continuous functions by typical means of their Fourier series. *Proc. Amer. Math. Soc.* **12**, 681–688 (1961).

[3] On logarithmic and Hölder means (Serb.). *Acad. Serbe Sci. Arts. Glas* **260**, 39–46 (1965).

Aljančić, S.–M. Tomić

[1] Transformationen von Fourier–Reihen durch monoton abnehmende Multiplikatoren. *Acta Math. Acad. Sci. Hungar.* **17**, 23–30 (1966).

Amel'kovič, V. G.

[1] A theorem converse to a theorem of Voronovskaja type (Russ.). *Teor. Funkciĭ Funkcional Anal. i Priložen.* **2**, 67–74 (1966).

Andrienko, V. A.

[1] On the approximation of functions by Fejér means (Russ.). *Sibirsk. Mat. Ž.* **9**, 3–12 (1968) = *Siberian Math. J.* **9**, 1–8 (1968).

Anghelutza, T.

[1] Une remarque sur l'intégrale de Poisson. *Bull. Sci. Math. Bibliothéque École Hautes Études = Darboux Bull.* (2) **48**, 138–140 (1924). (See also: *Jahrbuch Fortschritte Math.* **50**, 330 (1924), review by M. Plancherel)

ARONSZAJN, N.–E. GAGLIARDO
[1] Interpolation spaces and interpolation methods. *Ann. Mat. Pura Appl. (4)* **68**, 51–118 (1965).
ARONSZAJN, N.–F. MULLA–P. SZEPTYCKI
[1] On spaces of potentials connected with L^p classes. *Ann. Inst. Fourier (Grenoble)* **13**, 211–306 (1963).
ARONSZAJN, N.–K. T. SMITH
[1] Theory of Bessel potentials I. *Ann. Inst. Fourier (Grenoble)* **11**, 385–475 (1961).
ARSAC, J.
[1] *Transformation de Fourier et théorie des distributions*. Dunod 1961, xv + 347 pp.
ASPLUND, E.–L. BUNGART
[1] *A First Course in Integration*. Holt, Rinehart and Winston 1966, xiii + 489 pp.
AUMANN, G.
[1] Approximation von Funktionen. *In:* Mathematische Hilfsmittel des Ingenieurs III (Ed. R. Sauer–I. Szabó). Springer 1968, xix + 535 pp.; *pp. 320–351.*
BAJŠANSKI, B.–R. BOJANIČ
[1] A note on approximation by Bernstein polynomials. *Bull. Amer. Math. Soc.* **70**, 675–677 (1964).
BALAKRISHNAN, A. V.
[1] Representation of abstract Riesz potentials of the elliptic type. *Bull. Amer. Math. Soc.* **64**, 288–289 (1958).
BANACH, S.
[1] *Théorie des opérations linéaires*. Chelsea Publ. 1955, xii + 259 pp. (Orig. ed. Warsaw 1932.)
BARI, N. K.
[1] *A Treatise on Trigonometric Series I, II*. Pergamon Press 1964, xx + 553 pp./xiv + 508 pp. (Orig. Russ. ed. Moscow 1961.)
BARI, N. K.–S. B. STEČKIN
[1] The best approximation and differential properties of two conjugate functions. *Trudy Moskov. Mat. Obšč.* **5**, 483–522 (1956).
BARLAZ, J.
[1] A corollary of a theorem of Schwarz. Solution of Problem 4906, proposed by P. L. Butzer. *Amer. Math. Monthly* **68**, 672 (1961).
BAUSOV, L. I.
[1] Approximation of functions of class Z_α by positive methods of summation of Fourier series (Russ.). *Uspehi Mat. Nauk* **16**, no. **3 (99)**, 143–149 (1961).
[2] The order of approximation of functions of class Z_α by positive linear polynomial operators (Russ.). *Uspehi Mat. Nauk* **17**, no. **1 (103)**, 149–155 (1962).
[3] On the order of approximation of functions of class Z_α by linear positive operators (Russ.). *Mat. Zametki* **4**, 201–210 (1968) = *Math. Notes* **4** (1968), 612–617 (1969).
BELLMAN, R. E.
[1] *A Brief Introduction to Theta Functions*. Holt, Rinehart and Winston 1961, x + 78 pp.
BELTRAMI, E. J.
[1] Pointwise and norm convergence of distributions. *J. Math. Mech.* **14**, 99–107 (1965).
BELTRAMI, E. J.–M. R. WOHLERS
[1] Distributional boundary value theorems and Hilbert transforms. *Arch. Rational Mech. Anal.* **18**, 304–309 (1965).
[2] *Distributions and the Boundary Values of Analytic Functions*. Academic Press 1966, xiv + 116 pp.
BERENS, H. (*see also Butzer, P. L.*)
[1] Approximationssätze für Halbgruppenoperatoren in intermediären Räumen (Dissertation, T.H. Aachen). *Schr. Math. Inst. Univ. Münster* **32** (1964), 59 pp.
[2] Equivalent representations for the infinitesimal generator of higher orders in semi-group theory. *Nederl. Akad. Wetensch. Proc. Ser. A* **68** = *Indag. Math.* **27**, 497–512 (1965).
[3] *Interpolationsmethoden zur Behandlung von Approximationsprozessen auf Banachräumen.* (Lecture Notes in Math. **64**) Springer 1968, iv + 90 pp.
[4] On the saturation theorem for the Cesaro means of Fourier series. *Acta Math. Acad. Sci. Hungar.* **21**, 95–99 (1970).
[5] On pointwise approximation of Fourier series by typical means. *Tôhoku Math. J.; to appear.*
BERENS, H.–P. L. BUTZER
[1] On the best approximation for singular intregals by Laplace-transform methods. *In:* BUTZER, P. L.–J. KOREVAAR [1]; *pp. 24–42.*

[2] Über die Darstellung vektorwertiger holomorpher Funktionen durch Laplace-Integrale. *Math. Ann.* **158**, 269–283 (1965).

BERENS, H.–P. L. BUTZER–S. PAWELKE

[1] Limitierungsverfahren von Reihen mehrdimensionaler Kugelfunktionen und deren Saturationsverhalten. *Publ. Res. Inst. Math. Sci. Ser. A* **4**, 201–268 (1968).

BERENS, H.–P. L. BUTZER–U. WESTPHAL

[1] Representation of fractional powers of infinitesimal generators of semi-groups. *Bull. Amer. Math. Soc.* **74**, 191–196 (1968).

BERENS, H.–E. GÖRLICH

[1] Über einen Darstellungssatz für Funktionen als Fourierintegrale und Anwendungen in der Fourieranalysis. *Tôhoku Math. J.* (2) **18**, 429–453 (1966).

BERENS, H.–R. J. NESSEL

[1] Contributions to the theory of saturation for singular integrals in several variables V. Saturation in $L_p(E^n)$, $2 < p < \infty$. *Nederl. Akad. Wetensch. Proc. Ser. A* **72** = *Indag. Math.* **31**, 71–76 (1969).

BERENS, H.–U. WESTPHAL

[1] Zur Charakterisierung von Ableitungen nichtganzer Ordnung im Rahmen der Laplacetransformation. *Math. Nachr.* **38**, 115–129 (1968).

[2] A Cauchy problem for a generalized wave equation. *Acta Sci. Math.* (*Szeged*) **29**, 93–106 (1968).

BERGSTRÖM, H.

[1] On some expansions of stable distribution functions. *Ark. Mat.* **2**, 375–378 (1952).

BERNSTEIN, S. N.

[1] Sur l'ordre de la meilleure approximation des fonctions continues par des polynômes de degré donné. *Mém. Cl. Sci. Acad. Roy. Belg.* (2) **4**, 1–104 (1912).

[2] Sur la convergence absolue des séries trigonométriques. *C.R. Acad. Sci. Paris* **158**, 1661–1664 (1914).

[3] Sur un procédé de sommations des séries trigonométriques. *C.R. Acad. Sci. Paris* **191**, 976–979 (1930).

BERRY, A. C.

[1] Necessary and sufficient conditions in the theory of Fourier transforms. *Ann. of Math.* (2) **32**, 830–838 (1931).

BEURLING, A.

[1] Sur les intégrals de Fourier absolument convergentes et leur application à une transformation fonctionelle. *In:* Neuvième Congrès Math. Scand., Helsingfors 1938; *pp. 345–366.*

[2] On two problems concerning linear transformations in Hilbert space. *Acta Math.* **81**, 239–255 (1949).

BEURLING, A.–H. HELSON

[1] Fourier-Stieltjes transforms with bounded powers. *Math. Scand.* **1**, 120–126 (1953).

BOAS, R. P., JR.

[1] *Integrability Theorems for Trigonometric Transforms.* Springer 1967, 65 pp.

BOAS, R. P., JR.–M. KAC

[1] Inequalities for Fourier transforms of positive functions. *Duke Math. J.* **12**, 189–206 (1945).

BOCHNER, S.

[1] A theorem on Fourier–Stieltjes integrals. *Bull. Amer. Math. Soc.* **40**, 271–276 (1934).

[2] Summation of multiple Fourier series by spherical means. *Trans. Amer. Math. Soc.* **40**, 175–207 (1936).

[3] Stable laws of probability and completely monotone functions. *Duke Math. J.* **3**, 726–728 (1937).

[4] Quasi-analytic functions, Laplace operator, positive kernels. *Ann. of Math.* (2) **51**, 68–91 (1950).

[5] Book review of L. SCHWARTZ [1]. *Bull. Amer. Math. Soc.* **58**, 78–85 (1952).

[6] *Harmonic Analysis and the Theory of Probability.* Univ. of California Press 1955, viii + 176 pp.

[7] *Lectures on Fourier Integrals* (with an author's supplement on *Monotonic Functions, Stieltjes Integrals, and Harmonic Analysis*). (Ann. of Math. Stud. **42**) Princeton Univ. Press 1959, viii + 233 pp. (Orig. German ed. Leipzig 1932.)

BOCHNER, S.–K. CHANDRASEKHARAN

[1] *Fourier Transforms.* (Ann. of Math. Stud. **19**) Princeton Univ. Press 1949, ix + 219 pp.

Bohman, H.
[1] On approximation of continuous and of analytic functions. *Ark. Mat.* **2**, 43–56 (1952).
[2] Approximate Fourier analysis of distribution functions. *Ark. Mat.* **4**, 99–157 (1960).
Boĭcun, L. G.: *see Ogieveckiĭ, I. I.*
Bojanič, R.: *see Bajšanski, B.*
Borel, É.
[1] *Leçons sur les fonctions de variables réelles et les développement en séries de polynomes.* Gauthier-Villars 1905, x + 160 pp.
Botts, T. A.: *see McShane, E. J.*
Brainerd, B.–R. E. Edwards
[1] Linear operators which commute with translations I, II. *J. Austral. Math. Soc.* **6**, 289–350 (1966).
Bredihina, E. A.
[1] Some problems in summation of Fourier series of almost periodic functions (Russ.). *Uspehi Mat. Nauk* **15**, no. **5 (95)** 143–150 (1960) = *Amer. Math. Soc. Transl.* (2) **26**, 253–261 (1963).
[2] On simultaneous approximation of almost periodic functions and their derivatives (Russ.). *In:* Ibragimov, I. I. [1]; *pp. 34–39.*
Bremermann, H.
[1] *Distributions, Complex Variables, and Fourier Transforms.* Addison-Wesley 1965, xii + 186 pp.
Brosowski, B.
[1] *Nicht-lineare Tschebyscheff-Approximation.* (Hochschulskripten **808/808a**) Bibliograph. Inst. 1968, 152 pp.
Bruckner, A. M.–Max Weiss
[1] On approximate identities in abstract measure spaces. *Preprint.*
Buchwalter, H.
[1] Saturation de certains procédés de sommation. *C.R. Acad. Sci. Paris* **248**, 909–912 (1959).
[2] Saturation dans un espace normé. *C.R. Acad. Sci. Paris* **250**, 651–653 (1960).
[3] Saturation sur un groupe abélien localement compact. *C.R. Acad. Sci. Paris* **250**, 808–810 (1960).
[4] Saturation et distributions. *C.R. Acad. Sci. Paris* **250**, 3562–3564 (1960).
Bungart, L.: *see Asplund, E.*
Bureau, F. J.
[1] Divergent integrals and partial differential equations. *Comm. Pure Appl. Math.* **8**, 143–202 (1955).
[2] *Problems and Methods in Partial Differential Equations.* Mimeographed Lecture Notes, Duke Univ. 1955/56.
Burkhardt, H.
[1] Trigonometrische Reihen und Integrale. *In:* Encyklopädie der Mathematischen Wissenschaften, Bd. II. 1.2 Analysis; Leipzig: Teubner 1904–1916; *pp. 819–1354.*
Butzer, P. L. (*see also Berens, H.*)
[1] On the singular integral of de la Vallée-Poussin. *Arch. Math.* (*Basel*) **7**, 295–309 (1956).
[2] Sur la théorie des demi-groupes et classes de saturation de certaines intégrales singulières. *C.R. Acad. Sci. Paris* **243**, 1473–1475 (1956).
[3] Über den Grad der Approximation der Identitätsoperators durch Halbgruppen von linearen Operatoren und Anwendungen auf die Theorie der singulären Integrale. *Math. Ann.* **133**, 410–425 (1957).
[4] Zur Frage der Saturationsklassen singulärer Integraloperatoren. *Math. Z.* **70**, 93–112 (1958).
[5] Sur le rôle de la transformation de Fourier dans quelques problèmes d'approximation. *C.R. Acad. Sci. Paris* **249**, 2467–2469 (1959).
[6] Representation and approximation of functions by general singular integrals. *Nederl. Akad. Wetensch. Proc. Ser. A* **63** = *Indag. Math.* **22**, 1–24 (1960).
[7] Fourier transform methods in the theory of approximation. *Arch. Rational Mech. Anal.* **5**, 390–415 (1960).
[8] On some theorems of Hardy, Littlewood and Titchmarsh. *Math. Ann.*, **142**, 259–269 (1961).
[9] On Dirichlet's problem for the half-space and the behaviour of its solution on the boundary. *J. Math. Anal. Appl.* **2**, 86–96 (1961).
[10] Beziehungen zwischen den Riemannschen, Taylorschen und gewöhnlichen Ableitungen reellwertiger Funktionen. *Math. Ann.* **144**, 275–298 (1961).

[11] The dependence of the solution of the equation of heat conduction upon its initial temperature distribution; asymptotic expansions. *Arch. Math.* (*Basel*) **13**, 302–312 (1962).
[12] Integral transform methods in the theory of approximation. *In:* BUTZER, P. L.–J. KOREVAAR [1]; *pp. 12–23.*
[13] Saturation and approximation. *SIAM J. Numer. Anal. Ser. B* **1**, 2–10 (1964).
[14] Semi-groups of bounded linear operators and approximation. *Sém. d'Analyse Harmonique, Univ. de Paris, Fac. des Sci. d'Orsay, Exposé* **15**, 1965/66, pp. 1–15.

BUTZER, P. L.–H. BERENS
[1] *Semi-Groups of Operators and Approximation.* Springer 1967, xi + 318 pp.

BUTZER, P. L.–E. GÖRLICH
[1] Zur Charakterisierung von Saturationsklassen in der Theorie der Fourierreihen. *Tôhoku Math. J.* (2) **17**, 29–54 (1965).
[2] Saturationsklassen und asymptotische Eigenschaften trigonometrischer singulärer Integrale. *In:* Festschrift zur Gedächtnisfeier Karl Weierstraß 1815–1965 (Ed. H. Behnke–K. Kopfermann; Wiss. Abh. Arbeitsgemeinschaft für Forschung des Landes Nordrhein-Westfalen **33**) Westdeutsch. Verl. 1966, 612 pp.; *pp. 339–392.*

BUTZER, P. L.–E. GÖRLICH–K. SCHERER
[1] *Introduction to Interpolation and Approximation.* To appear.

BUTZER, P. L.–J. KEMPER
[1] Operatorkalkül von Approximationsverfahren fastperiodischer Funktionen. *In:* Forschungsber. des Landes Nordrhein-Westfalen **2157**. Westdeutsch. Verl.; *to appear.*

BUTZER, P. L.–H. KÖNIG
[1] An application of Fourier–Stieltjes transforms in approximation theory. *Arch. Rational Mech. Anal.* **5**, 416–419 (1960).

BUTZER, P. L.–J. KOREVAAR (Ed.)
[1] *On Approximation Theory.* (Proc. Conf. Math. Res. Inst. Oberwolfach, Black Forest, 1963; ISNM **5**) Birkhäuser 1964, xvi + 261 pp.

BUTZER, P. L.–W. KOZAKIEWICZ
[1] On the Riemann derivatives for integrable functions. *Canad. J. Math.* **6**, 572–581 (1954).

BUTZER, P. L.–R. J. NESSEL
[1] Über eine Verallgemeinerung eines Satzes von de La Vallée Poussin. *In:* BUTZER, P. L.–J. KOREVAAR [1]; *pp. 45–59.*
[2] Contributions to the theory of saturation for singular integrals in several variables I. General theory. *Nederl. Akad. Wetensch. Proc. Ser. A* **69** = *Indag. Math.* **28**, 515–531 (1966).

BUTZER, P. L.–R. J. NESSEL–K. SCHERER
[1] Trigonometric convolution operators with kernels having alternating signs and their degree of convergence. *Jber. Deutsch. Math.-Verein.* **70**, 86–99 (1967).

BUTZER, P. L.–S. PAWELKE
[1] Ableitungen von trigonometrischen Approximationsprozessen. *Acta Sci. Math.* (*Szeged*) **28**, 173–183 (1967).

BUTZER, P. L.–K. SCHERER
[1] *Approximationsprozesse und Interpolationsmethoden.* (Hochschulskripten **826/826a**) Bibliograph. Inst. 1968, 172 pp.
[2] On the fundamental approximation theorems of D. Jackson, S. N. Bernstein and theorems of M. Zamansky and S. B. Stečkin. *Aequationes Math.* **3**, 170–185 (1969).
[3] Über die Fundamentalsätze der klassischen Approximationstheorie in abstrakten Räumen. *In:* BUTZER, P. L.–B. SZ.-NAGY [1]; *pp. 113–125.*
[4] On fundamental theorems of approximation theory and their dual versions. *J. Approximation Theory* **3**, 87–100 (1970).
[5] Approximation theorems for sequences of commutative operators in Banach spaces. *In:* Proc. of 'Conference on Constructive Function Theory', Varna (Bulgaria) 1970; *to appear.*
[6] Jackson and Bernstein-type inequalities for families of commutative operators in Banach spaces. *J. Approximation Theory, to appear.*

BUTZER, P. L.–E. L. STARK
[1] Wesentliche asymptotische Entwicklungen für Approximationsmaße trigonometrischer singulärer Integrale. *Math. Nachr.* **39**, 223–237 (1969).
[2] On a trigonometric convolution operator with kernel having two zeros of simple multiplicity. *Acta Math. Acad. Sci. Hungar.* **20**, 451–461 (1969).

BUTZER, P. L.–G. SUNOUCHI
[1] Approximation theorems for the solution of Fourier's problem and Dirichlet's problem. *Math. Ann.* **155**, 316–330 (1964).

BUTZER, P. L.–B. SZ.-NAGY (Ed.)

[1] *Abstract Spaces and Approximation.* (Proc. Conf. Math. Res. Inst. Oberwolfach, Black Forest, 1968; ISNM **10**) Birkhäuser 1969, 423 pp.

BUTZER, P. L.–H. G. TILLMANN

[1] An approximation theorem for semi-groups of operators. *Bull. Amer. Math. Soc.* **66**, 191–193 (1960).

[2] Approximation theorems for semi-groups of bounded linear transformations. *Math. Ann.* **140**, 256–262 (1960).

BUTZER, P. L.–W. TREBELS

[1] Hilbert transforms, fractional integration and differentiation. *Bull. Amer. Math. Soc.* **74**, 106–110 (1968).

[2] *Hilberttransformation, gebrochene Integration und Differentiation.* (Forschungsber. des Landes Nordrhein-Westfalen **1889**) Westdeutsch. Verl. 1968, 82 pp.

CALDERÓN, A. P.

[1] Lebesgue spaces of differentiable functions and distributions. *In:* Proceedings of Symposia in Pure Mathematics. **IV**: Partial Differential Equations. Amer. Math. Soc., Providence, R.I., 1961, vi + 169 pp.; *pp. 33–49.*

[2] Singular integrals. *Bull. Amer. Math. Soc.* **72**, 427–465 (1966).

[3] (Ed.) *Proceedings of Symposia in Pure Mathematics.* **X**: *Singular Integrals.* (Proc. Sympos. Pure Math. Amer. Math. Soc., Univ. Chicago, Chicago, Ill., 1966) Amer. Math. Soc., Providence, R.I., 1967, vi + 375 pp.

CALDERÓN, A. P.–A. ZYGMUND

[1] On the existence of certain singular integrals. *Acta Math.* **88**, 85–139 (1952).

[2] Singular integrals and periodic functions. *Studia Math.* **14**, 249–271 (1954).

[3] Local properties of solutions of elliptic partial differential equations. *Studia Math.* **20**, 171–225 (1961).

CARATHEODORY, C.

[1] Über den Variabilitätsbereich der Fourier'schen Konstanten von positiven harmonischen Funktionen. *Rend. Circ. Mat. Palermo* **32**, 193–217 (1911).

CARLEMAN, T.

[1] *L'intégral de Fourier et questions qui s'y rattachent.* Publ. Sci. Inst. Mittag-Leffler **1**, Uppsala 1944, 119 pp.

CARLESON, L.

[1] On convergence and growth of partial sums of Fourier series. *Acta Math.* **116**, 135–157 (1966).

CARLITZ, L.

[1] Note on Lebesgue's constants. *Proc. Amer. Math. Soc.* **12**, 932–935 (1961).

CARSLAW, H. S.

[1] *Introduction to the Theory of Fourier's Series and Integrals and the Mathematical Theory of the Conduction of Heat.* Macmillan 1906, xx + 434 pp.

CARTWRIGHT, M. L.

[1] On the relation between the different types of Abel summation. *Proc. London Math. Soc.* (2) **31**, 81–96 (1930).

CHANDRASEKHARAN, K.–S. MINAKSHISUNDARAM (*see also Bochner, S.*)

[1] *Typical Means.* (Tata Inst. of Fundamental Res., Bombay; Monographs on Math. and Phys. **1**) Oxford Univ. Press 1952, x + 139 pp.

CHAUNDY, T. W.

[1] *The Differential Calculus.* Oxford Clarendon Press 1935, xiv + 459 pp.

CHENEY, E. W.

[1] *Introduction to Approximation Theory.* McGraw-Hill 1966, xii + 259 pp.

CHENEY, E. W.–C. R. HOBBY–P. D. MORRIS–F. SCHURER–D. E. WULBERT

[1] On the minimal property of the Fourier projection. *Bull. Amer. Math. Soc.* **75**, 51–52 (1969).

[2] On the minimal property of the Fourier projection. *Trans. Amer. Math. Soc.* **143**, 249–258 (1969).

CHENEY, E. W.–K. H. PRICE

[1] Minimal projections. *In:* Approximation Theory. (Proc. Int. Symposium Univ. Lancaster, 1969; ed. A. Talbot) Academic Press 1970, viii + 356 pp.; *pp. 261–289.*

CHEN TIAN-PING

[1] Typical means of Fourier series I (Chin.). *Acta Math. Sinica* **16**, 179–193 (1966) = *Chinese Math.–Acta* **8**, 191–204 (1966).

CHURCHILL, R. V.
[1] The operational calculus of Legendre transforms. *J. Math. and Phys.* **33**, 165–177 (1954).
[2] *Operational Mathematics.* McGraw-Hill 1958, ix + 337 pp.
[3] Extensions of operational mathematics. *In:* Proceedings of the Conference on Differential Equations. Univ. of Maryland Book Store, College Park, Md., 1956; *pp. 235–250.*
COOPER, J. L. B.
[1] Positive-definite functions of a real variable. *Proc. London Math. Soc.* (*3*) **10**, 53–66 (1960).
[2] Some problems in the theory of Fourier transforms. *Arch. Rational Mech. Anal.* **14**, 213–216 (1963).
[3] Fourier transforms and inversion formulae for L^p functions. *Proc. London Math. Soc.* (*3*) **14**, 271–298 (1964).
CORDES, H. O.
[1] The algebra of singular integral operators on R^n. *J. Math. Mech.* **14**, 1007–1032 (1965).
COSSAR, J.
[1] On conjugate functions. *Proc. London Math. Soc.* (*2*) **45**, 369–381 (1939).
COURANT, R.–D. HILBERT
[1] *Methods of Mathematical Physics I, II.* Interscience Publ. 1953, xv + 561 pp./1962, xxii + 830 pp. (Orig. German ed. Berlin 1937.)
CRAMÉR, H.
[1] On the representation of a function by certain Fourier-integrals. *Trans. Amer. Math. Soc.* **46**, 190–201 (1939).
[2] *Mathematical Methods of Statistics.* Princeton Univ. Press 1961, xvi + 575 pp.
CURTIS, P. C., JR.
[1] The degree of approximation by positive convolution operators. *Michigan Math. J.* **12**, 155–160 (1965).
DAVIS, H. F.
[1] *Fourier Series and Orthogonal Functions.* Allyn and Bacon 1963, xii + 403 pp.
DAVIS, H. T.
[1] The applications of fractional operators to functional equations. *Amer. J. Math.* **49**, 123–142 (1927).
DAVIS, P. J.
[1] *Interpolation and Approximation.* Blaisdell Publ. 1963, xiv + 393 pp.
DENJOY, A.
[1] Sur l'intégration des coefficients différentiels d'ordre supérieur. *Fund. Math.* **25**, 273–326 (1935).
DITKIN, A. V.–A. P. PRUDNIKOV
[1] Integral Transforms. *In:* Progress in Mathematics. Vol. 4: Mathematical Analysis (Ed. R. V. Gamkrelidze). Plenum Press 1969, viii + 132 pp.; *pp. 1–85.* (Orig. Russ. ed. Moscow 1967.)
DITZIAN, Z.
[1] On inversion of Weierstrass transform and its saturation class. *J. Approximation Theory* **2**, 318–321 (1969).
[2] Saturation class for inversion of convolution transforms. *In:* Proc. of 'Colloquium on Constructive Theory of Functions', Budapest 1969; *to appear.*
DOETSCH, G.
[1] Über das Problem der Wärmeleitung. *Jber. Deutsch. Math.-Verein.* **33**, 45–52 (1925).
[2] Integration von Differentialgleichungen vermittels der endlichen Fourier-Transformationen. *Math. Ann.* **112**, 52–68 (1935).
[3] *Theorie und Anwendung der Laplace-Transformation.* Springer 1937, 436 pp.
[4] *Handbuch der Laplace-Transformation.* Bd. I: *Theorie der Laplace-Transformation;* Bd. II, III: *Anwendungen der Laplace-Transformation.* Birkhäuser 1950–1956, 581/436/300 pp.
[5] Funktionaltransformationen. *In:* Mathematische Hilfsmittel des Ingenieurs I (Ed. R. Sauer–I. Szabó). Springer 1967, xv + 496 pp.; *pp. 232–484.*
DONOGHUE, W. F., JR.
[1] *Distributions and Fourier Transforms.* Academic Press 1969, viii + 315 pp.
DOSS, R.
[1] On the multiplicators of some classes of Fourier transforms. *Proc. London Math. Soc.* (*2*) **50**, 169–195 (1949).
DUFF, G. F. D.–D. NAYLOR.
[1] *Differential Equations of Applied Mathematics.* John Wiley 1966, xi + 423 pp.

DUNFORD, N.–J. T. SCHWARTZ

[1] *Linear Operators*. Vol. I: *General Theory*; Vol. II: *Spectral Theory*. Interscience Publ. 1958, xiv + 858 pp./1963, ix + pp. 859–1923.

DZJADYK, V. K.

[1] On best approximation in the class of periodic functions having a bounded s-th derivative ($0 < s < 1$) (Russ.). *Izv. Akad. Nauk SSSR Ser. Mat.* **17**, 135–162 (1953).

[2] Approximation of functions by positive linear operators and singular integrals (Russ.). *Mat. Sb.* **70 (112)**, 508–517 (1966).

EBERLEIN, W. F.

[1] Characterizations of Fourier–Stieltjes transforms. *Duke Math. J.* **22**, 465–468 (1955).

EDREI, A.

[1] On the generating function of totally positive sequences II. *J. Analyse Math.* **2**, 104–109 (1952).

[2] On the generating function of a double infinite totally positive sequence. *Trans. Amer. Math. Soc.* **74**, 367–383 (1953).

EDWARDS, R. E. (*see also Brainerd, B.*)

[1] *Fourier Series I, II*. Holt, Rinehart and Winston 1967, x + 208 pp./ix + 318 pp.

[2] A class of multipliers. *J. Austral. Math. Soc.* **8**, 584–590 (1968).

[3] *Functional Analysis*. Holt, Rinehart and Winston 1965, xiii + 781 pp.

EFIMOV, A. V.

[1] On the approximation of some classes of continuous functions by Fourier sums and Fejér sums (Russ.). *Izv. Akad. Nauk SSSR Ser. Mat.* **22**, 81–116 (1958).

[2] Approximation of conjugate functions by Fejér means. *Uspehi Mat. Nauk* **14**, no. **1 (85)**, 183–185 (1959).

[3] Linear methods of approximation of certain classes of periodic continuous functions. *Trudy Mat. Inst. Steklov* **62**, 3–47 (1961) = *Amer. Math. Soc. Transl.* (*2*) **28**, 221–268 (1963).

EGERVÁRY, E. v.–O. SZÁSZ

[1] Einige Extremalprobleme im Bereiche der trigonometrischen Polynome. *Math. Z.* **27**, 641–652 (1928).

EISEN, M.

[1] *Introduction to Mathematical Probability Theory*. Prentice-Hall 1969, xxi + 535 pp.

ERDÉLYI, A. et al.

[1] *Higher Transcendental Functions I, II, III*. McGraw-Hill 1953, xxvi + 302 pp./xvii + 396 pp./1955, xvii + 292 pp.

[2] *Tables of Integral Transforms I, II*. McGraw-Hill 1954, xx + 391 pp./xvi + 451 pp.

FAVARD, J.

[1] Sur les meilleures procédés d'approximation de certaines classes des fonctions par des polynômes trigonométriques. *Bull. Sci. Math.* **61**, 209–224, 243–256 (1937).

[2] Sur l'approximation des fonctions d'une variable réelle. *In:* Analyse Harmonique (Colloques Internat. Centre nat. Rech. sci. **15**). Éditions Centre nat. Rech. sci. 1949; *pp. 97–110*.

[3] Sur l'approximation dans les éspaces vectoriels. *Ann. Mat. Pura Appl.* (*4*) **29**, 259–291 (1949).

[4] Sur la saturation des procédés de sommation. *J. Math. Pures Appl.* **36**, 359–372 (1957).

[5] Sur la comparaison des procédés de sommation. *In:* BUTZER, P. L.–J. KOREVAAR [1]; *pp. 4–11*.

[6] On the comparison of the processes of summation. *SIAM J. Numer. Anal. Ser. B* **1**, 38–52 (1964).

FEJÉR, L.

[1] Über trigonometrische Polynome. *J. Reine Angew. Math.* **146**, 53–82 (1916).

[2] Gestaltliches über die Partialsummen und ihre Mittelwerte bei der Fourierreihe und der Potenzreihe. *Z. Angew. Math. Mech.* **13**, 80–88 (1933).

FELLER, W.

[1] On a generalization of Marcel Riesz' potentials and the semi-groups generated by them. *Comm. Sém. Math. Univ. Lund.* [*Medd. Lunds Univ. Mat. Sem.*] *Tome Suppl.* 72–81 (1952).

[2] *An Introduction to Probability Theory and Its Applications I, II*. John Wiley 1950, xii + 419 pp./1966, xviii + 626 pp.

FICHTENHOLZ, G.

[1] Sur l'intégrale de Poisson et quelques questions qui s'y rattachent. *Fund. Math.* **13**, 1–33 (1929).

FIGÀ-TALAMANCA, A.

[1] On the subspaces of L^p invariant under multiplication of transforms by bounded continuous functions. *Rend. Sem. Mat. Univ. Padova*, 176–198 (1965).

[2] Translation invariant operators in L^p. *Duke Math. J.* **32**, 495–502 (1965).
FIGÀ-TALAMANCA, A.–G. I. GAUDRY
[1] Multipliers and sets of uniqueness of L^p. *Preprint.*
FRÉCHET, M.–A. ROSENTHAL
[1] Funktionenfolgen. *In:* Encyklopädie der Mathematischen Wissenschaften. Bd. II.3.2 Analysis; Leipzig: Teubner 1923–1927; *pp. 1136–1187.*
FREUD, G.
[1] Sui procedimenti lineari d'approssimazione. *Atti Accad. Naz. Lincei Rend. Cl. Sci. Fis. Mat. Natur.* (*8*) **26**, 641–643 (1959).
[2] Über positive Zygmundsche Approximationsfolgen. *Magyar Tud. Akad. Mat. Kutató Int. Közl.* **6**, 71–75 (1961).
[3] *Orthogonale Polynome.* Birkhäuser 1969, 294 pp.
[4] On approximation by positive linear methods I, II. *Studia Sci. Math. Hungar.* **2**, 63–66 (1967); **3**, 365–370 (1968).
[5] Über die Sättigungsklasse der starken Approximation durch Teilsummen der Fourierschen Reihe. *Acta Math. Acad. Sci. Hungar.* **20**, 275–279 (1969).
FREUD, G.–S. KNAPOWSKI
[1] On linear processes of approximation I, II, III. *Studia Math.* **23**, 105–112 (1963); **25**, 252–263, 373–383 (1965).
FRIEDMAN, A.
[1] *Generalized Functions and Partial Differential Equations.* Prentice-Hall 1963, xii + 340 pp.
[2] *Partial Differential Equations of Parabolic Type.* Prentice-Hall 1964, xiv + 347 pp.
GAER, M.
[1] *Fractional derivatives and entire functions.* Thesis, Univ. of Illinois 1968.
GAER, M.–L. A. RUBEL
[1] The fractional derivative and entire functions. *Preprint.*
GAGLIARDO, E. (*see also Aronszajn, N.*)
[1] A unified structure in various families of function spaces. Compactness and closure theorems. *In:* Proceedings of the International Symposium on Linear Spaces, Hebrew Univ. Jerusalem 1960; Publ. Israel Acad. Sci. Humanities. Jerusalem Academic Press–Pergamon Press 1961, xi + 456 pp.; *pp. 237–241.*
GANTMACHER, F. R.–M. G. KREIN
[1] *Oszillationsmatrizen, Oszillationskerne und kleine Schwingungen mechanischer Systeme.* Akademie-Verl. 1960, x + 359 pp. (Orig. Russ. ed. Moscow–Leningrad 1950; 1941.)
GARABEDIAN, H. L. (Ed.)
[1] *Approximation of Functions.* (Proc. Sympos. Approximation of Functions, General Motors Res. Laboratories, Warren, Michigan, 1964) Elsevier Publ. 1965, vii + 220 pp.
GARKAVI, A. L.
[1] Approximative properties of subspaces with finite defect in the space of continuous functions (Russ.). *Dokl. Akad. Nauk SSSR* **155**, 513–516 (1964) = *Soviet Math. Dokl.* **5**, 440–443 (1964).
GARNIR, H. G.–M. DE WILDE–J. SCHMETS
[1] *Analyse fonctionnelle.* Tome I: *Théorie générale.* Birkhäuser 1968, x + 562 pp.
GAUDRY, G. I. (*see also Figà-Talamanca, A.*)
[1] Multipliers of type (p, q). *Pacific J. Math.* **18**, 477–488 (1966).
GELFAND, I. M.–G. E. SHILOV
[1] *Generalized Functions.* Vol. I: *Properties and Operations;* Vol. II: *Spaces of Fundamental and Generalized Functions*; Vol. III: *Theory of Differential Equations.* Academic Press 1964, xviii + 423 pp./1968, x + 261 pp./1967, x + 222 pp. (Orig. Russ. ed. Moscow 1958.)
GELFAND, I. M.–N. YA. VILENKIN
[1] *Generalized Functions.* Vol. IV: *Applications of Harmonic Analysis.* Academic Press 1964, xiv + 384 pp. (Orig. Russ. ed. Moscow 1961.)
GERONIMUS, YA. L.
[1] *Polynomials Orthogonal on a Circle and Interval.* Pergamon Press 1960, ix + 210 pp. (Orig. Russ. ed. Moscow 1958.)
[2] Some remarks on the method of coefficients (Russ.). *Dokl. Akad. Nauk SSSR* **166**, 1274–1276 (1966) = *Soviet Math. Dokl.* **7**, 263–265 (1966).
GÖRLICH, E. (*see also Berens, H.; Butzer, P. L.*)
[1] *Saturationssätze und Charakterisierung der Saturationsklassen für Limitierungsverfahren von Fourierreihen.* Staatsexamensarbeit, T.H. Aachen 1964, 137 pp.

[2] Distributional methods in saturation theory. *J. Approximation Theory* **1**, 111–136 (1968).
[3] Saturation theorems and distributional methods. *In:* BUTZER, P. L.–B. SZ.-NAGY [1]; *pp. 218–232.*
[4] Zur Saturation von Summationsverfahren mehrdimensionaler Fourierreihen. *Österreich. Akad. Wiss. Math.-Natur. Kl. S.-B. II* **117**, 171–202 (1968).
[5] Über optimale Approximationsmethoden. *In:* Proc. of 'Conference on Constructive Function Theory', Varna (Bulgaria) 1970; *to appear.*

GÖRLICH, E.–R. J. NESSEL
[1] Über Peano- und Riemann-Ableitungen in der Norm. *Arch. Math. (Basel)* **18**, 399–410 (1967).

GÖRLICH, E.–E. L. STARK
[1] A unified approach to three problems on approximation by positive linear operators. *In:* Proc. of 'Colloquium on Constructive Theory of Functions', Budapest 1969; *to appear.*
[2] Über beste Konstanten und asymptotische Entwicklungen positiver Faltungsintegrale und deren Zusammenhang mit dem Saturationsproblem. *Jber. Deutsch. Math.-Verein.* **72**, 18–61 (1970).

GOES, G.
[1] Multiplikatoren für starke Konvergenz von Fourierreihen I, II. *Studia Math.* **17**, 299–311 (1958).
[2] Complementary spaces of Fourier coefficients, convolutions, and generalized matrix transformations and operators between BK-spaces. *J. Math. Mech.* **10**, 135–157 (1961).
[3] Fourier–Stieltjes transforms of discrete measures; periodic and semiperiodic functions. *Math. Ann.* **174**, 148–156 (1967).

GOLDBERG, R. R.
[1] *Fourier Transforms.* Cambridge Univ. Press 1962, viii + 76 pp.

GOLDBERG, S.
[1] *Unbounded Linear Operators.* McGraw-Hill 1966, viii + 199 pp.

GOLINSKIĬ, B. L.
[1] Approximation of two conjugate functions by means of the Ahiezer and Levitan operators (Russ.). *In:* SMIRNOV, V. I. [1]; *pp. 53–60.*
[2] Approximation on the entire number axis of two functions which are conjugate in the sense of Riesz by integral operators of singular type (Russ.). *Mat. Sb.* **66 (108)**, 3–34 (1965).

GOLOMB, M.
[1] *Lectures on Theory of Functions.* Mimeographed Lecture Notes; Argonne National Laboratory, Appl. Math. Division 1962, iii + 289 pp.

GOOD, I. J.
[1] Analogues of Poisson's summation formula. *Amer. Math. Monthly* **69**, 259–266 (1962).

GRAVES, L. M.
[1] *The Theory of Functions of Real Variables.* McGraw-Hill 1956, xii + 375 pp.

GÜTTINGER, W.
[1] Generalized functions and dispersion relations in physics. *Fortschr. Physik* **14**, 483–602 (1966).

GUO ZHU-RUI
[1] The approximation of a continuous function by the typical mean of its Fourier series I, II (Chin.). *Acta Math. Sinica* **15**, 42–53; 249–273 (1965) = *Chinese Math.-Acta* **6**, 340–352; 561–586 (1965).
[2] On the problem of approximation of a continuous function by linear means of its Fourier series (Chin.). *Acta Math. Sinica* **16**, 328–343 (1966) = *Chinese Math.-Acta* **8**, 348–364 (1966).

HADAMARD, J.
[1] Essai sur l'étude des fonctions données par leur développement de Taylor. *J. Math. Pures Appl. (4)* **8**, 101–186 (1892).

HAF, H.
[1] *Über den Vergleich von Summationsverfahren orthogonaler Reihen.* Diplomarbeit, T.H. Aachen 1966, 74 pp.

HAHN, H.
[1] Über die Darstellung gegebener Funktionen durch singuläre Integrale. *Denkschriften Kais. Akad. Wiss. Wien Math.-Natur. Kl.* **93**, 585–692 (1916).

HALMOS, P. R.
[1] *Measure Theory.* Van Nostrand 1950, xi + 304 pp.

[2] *Introduction to Hilbert Space and the Theory of Spectral Multiplicity*. Chelsea Publ. 1957, 114 pp.
HARDY, G. H.
[1] On Hilbert transforms. *Quart. J. Math. Oxford Ser.* **3**, 102–112 (1932).
[2] *Divergent Series*. Oxford Clarendon Press (1949) 1956, xvi + 396 pp.
HARDY, G. H.–J. E. LITTLEWOOD
[1] Some new properties of Fourier constants. *Math. Ann.* **97**, 159–209 (1926).
[2] Some properties of fractional integrals I, II. *Math. Z.* **27**, 565–606 (1928); **34**, 403–439 (1932).
[3] A convergence criterion for Fourier series. *Math. Z.* **28**, 612–634 (1928).
HARDY, G. H.–J. E. LITTLEWOOD–G. PÓLYA
[1] *Inequalities*. Cambridge Univ. Press (1934) 1952, xii + 324 pp.
HARDY, G. H.–M. RIESZ
[1] *The General Theory of Dirichlet's Series*. Cambridge Univ. Press (1915) 1952, 78 pp.
HARDY, G. H.–W. W. ROGOSINSKI
[1] *Fourier Series*. Cambridge Univ. Press (1944) 1962, 100 pp.
HARŠILADSE, F. I.
[1] On the functions of V. A. Steklov (Grusin.). *Soobšč. Akad. Nauk Gruzin. SSR* **14**, 139–144 (1953).
[2] Saturation classes for some summation processes (Russ.). *Dokl. Akad. Nauk SSSR* **122**, 352–355 (1958).
[3] Approximation of functions by the V. A. Steklov means (Russ.). *Akad. Nauk Gruzin. SSR Trudy Tbiliss. Mat. Inst. Razmadze* **31**, 55–70 (1966).
HAUSDORFF, F.
[1] Momentprobleme für ein endliches Intervall. *Math. Z.* **16**, 220–248 (1923).
HAVINSON, S. YA.
[1] On uniqueness of functions of best approximation in the metric of the space L^1 (Russ.). *Izv. Akad. Nauk SSSR Ser. Mat.* **22**, 243–270 (1958).
HELLWIG, G.
[1] *Partial Differential Equations*. Blaisdell Publ. 1964, xiii + 263 pp. (Orig. German ed. Stuttgart 1960.)
HELMBERG, G.
[1] *Introduction to Spectral Theory in Hilbert Space*. North-Holland 1969, xiii + 346 pp.
HELSON, H. (*see also Beurling, A.*)
[1] *Lectures on Invariant Subspaces*. Academic Press 1964, xi + 130 pp.
HERGLOTZ, G.
[1] Über Potenzreihen mit positivem, reellem Teil im Einheitskreis. *Ber. Verhandl. Sächs. Gesellschaft Wiss. Leipzig Math.-Phys. Kl.* **63**, 501–511 (1911).
HERZ, C. S.
[1] Lipschitz spaces and Bernstein's theorem on absolutely convergent Fourier transforms. *J. Math. Mech.* **18**, 283–323 (1968).
HEWITT, E.
[1] *Topics in Fourier Analysis*. Mimeographed Lecture Notes; Univ. Washington, Seattle, 1959, iv + 210 pp.
HEWITT, E.–I. I. HIRSCHMAN, JR.
[1] A maximum problem in harmonic analysis. *Amer. J. Math.* **76**, 839–852 (1954).
HEWITT, E.–K. A. ROSS
[1] *Abstract Harmonic Analysis*. Vol. I: *Structure of Topological Groups, Integration Theory, Group Representations;* Vol. II: *Structure and Analysis for Compact Groups, Analysis on Locally Compact Abelian Groups*. Springer 1963, viii + 519 pp./1970, ix + 771 pp.
HEWITT, E.–K. STROMBERG
[1] *Real and Abstract Analysis*. Springer 1965, viii + 476 pp.
HEYWOOD, P.
[1] On a modification of the Hilbert transform. *J. London Math. Soc.* **42**, 641–645 (1967).
HILB, E.–M. RIESZ
[1] Neuere Untersuchungen über trigonometrische Reihen. *In:* Encyklopädie der Mathematischen Wissenschaften. Bd. II.3.2 Analysis; Leipzig: Teubner 1923–1927; *pp. 1189–1228*.
HILB, E.–O. SZÁSZ
[1] Allgemeine Reihenentwicklungen. *In:* Encyklopädie der Mathematischen Wissenschaften, Bd. II.3.2 Analysis; Leipzig: Teubner 1923–1927; *pp. 1229–1276*.

HILBERT, D.: *see Courant, R.*
HILDEBRANDT, T. H.
[1] *Introduction to the Theory of Integration.* Academic Press 1963, ix + 385 pp.
HILLE, E.
[1] Notes on linear transformations I. *Trans. Amer. Math. Soc.* **39**, 131–153 (1936).
[2] *Functional Analysis and Semi-Groups.* (Amer. Math. Soc. Colloq. Publ. **31**) Amer. Math. Soc., New York 1948, xii + 528 pp.
[3] On the generation of semi-groups and the theory of conjugate functions. *Kungl. Fysiografiska Sällskapets i Lund Förhandlingar [Proc. Roy. Physiog. Soc. Lund]* **21**, no. **4**, 13 pp. (1952).
[4] *Analytic Function Theory I, II.* Ginn and Comp. 1959, xi + 308 pp./1962, xii + 496 pp.
HILLE, E.–R. S. PHILLIPS
[1] *Functional Analysis and Semi-Groups (rev. ed.).* (Amer. Math. Soc. Colloq. Publ. **31**) Amer. Math. Soc., Providence, R.I., 1957, xii + 808 pp.
HILLE, E.–J. D. TAMARKIN
[1] On the absolute integrability of Fourier transforms. *Fund. Math.* **25**, 329–352 (1935).
HIRSCHMAN, I. I., JR. (*see also Hewitt, E.*)
[1] On multiplier transformations I, II, III. *Duke Math. J.* **26**, 221–242 (1959); **28**, 45–56 (1961); *Proc. Amer. Math. Soc.* **71**, 764–766 (1965).
HIRSCHMAN, I. I., JR.–D. V. WIDDER
[1] An inversion and representation theory for convolution transforms with totally positive kernels. *Proc. Nat. Acad. Sci. U.S.A.* **34**, 152–156 (1948).
[2] Generalized inversion formulas for convolution transforms. *Duke Math. J.* **15**, 659–696 (1948).
[3] The inversion of a general class of convolution transforms. *Trans. Amer. Math. Soc.* **66**, 135–201 (1949).
[4] A representation theory for a general class of convolution transforms. *Trans. Amer. Math. Soc.* **67**, 69–97 (1949).
[5] *The Convolution Transforms.* Princeton Univ. Press 1955, x + 268 pp.
HOBBY, C. R.: *see Cheney, E. W.*
HOBSON, E. W.
[1] On a general convergence theorem, and the theory of the representation of a function by series of normal functions. *Proc. London Math. Soc.* (2) **6**, 349–395 (1908).
[2] *The Theory of Functions of a Real Variable and the Theory of Fourier's Series I, II.* Dover Publ. 1958, xv + 732 pp./x + 780 pp. (First ed. Cambridge 1907.)
HÖRMANDER, L.
[1] Estimates for translation-invariant operators in L^p spaces. *Acta Math.* **104**, 93–140 (1960).
[2] *Linear Partial Differential Operators.* Springer 1963, vii + 287 pp.
HOFF, J. C.
[1] *Approximation with kernels of finite oscillations.* Thesis, Purdue Univ. Lafayette, Indiana 1968, 116 pp. (See also: Approximation with kernels of finite oscillations I. Convergence. *J. Approximation Theory* **3**, 213–228 (1970); II. Degree of approximation, *to appear ibid.*)
HOFFMAN, K.
[1] *Banach Spaces of Analytic Functions.* Prentice-Hall 1962, xiii + 217 pp.
HOLMGREN, H. J.
[1] Om Differentialkalkylen med Indices af hoad Natur som helst. *Kongl. Svenska Vetenskaps-Akademiens Handlingar* (5) **11**, 1–83 (1864).
HORVÁTH, J.
[1] Singular integral operators and spherical harmonics. *Trans. Amer. Math. Soc.* **82**, 52–63 (1956).
HUNT, R. A.
[1] On the convergence of Fourier series. *In:* Orthogonal Expansions and Their Continuous Analogues. (Proc. Conf. Southern Illinois Univ., Edwardsville, Ill., 1967; Ed. D. Tepper Haimo) Southern Illinois Univ. Press; Feffer and Simons 1968; xx + 307 pp.; *pp. 235–255.*
IBRAGIMOV, I. I. (Ed.)
[1] *Studies of Contemporary Problems in the Constructive Theory of Functions* (Russ.) (Proc. Second All-Union Conf. on the Constructive Theory of Functions, Baku 1962.) Acad. Sci. Azerbaijani SSR. Inst. Math. Mech., Izdat. Akad. Nauk Azerbaĭdžan. SSR, Baku 1965, 638 pp.
IKENO, K.–Y. SUZUKI
[1] Some remarks of saturation problems in the local approximation. *Tôhoku Math. J.* (2) **20**, 214–233 (1968).

IVANOV, V. K.
[1] The Cauchy problem for the Laplace equation in an infinite strip (Russ.). *Differentsial'nye Uravneniya* **1**, 131–136 (1965) = *Differential Equations* **1**, 98–102 (1965).
JACKSON, D.
[1] *Über die Genauigkeit der Annäherung stetiger Funktionen durch ganze rationale Funktionen gegebenen Grades und trigonometrische Summen gegebener Ordnung.* Inauguraldissertation und Preischrift, Univ. Göttingen 1911, 98 pp.
[2] *The Theory of Approximation.* (Amer. Math. Soc. Colloq. Publ. **11**) Amer. Math. Soc., New York (1930) 1958, viii + 178 pp.
JOHNEN, H.
[1] Räume stetiger Funktionen und Approximation auf kompakten Mannigfaltigkeiten. *In:* Forschungsber. des Landes Nordrhein-Westfalen **2078**. Westdeutsch. Verl. 1970, 60 pp.; *pp. 3–25.*
KAC, M.: *see Boas, R. P., Jr.*
KACZMARZ, S.–H. STEINHAUS
[1] *Theorie der Orthogonalreihen.* Chelsea Publ. 1951, viii + 296 pp. (Orig. ed. Warsaw 1935.)
KALNI'BOLOCKAJA, L. A.
[1] On the classes of saturation in the theory of summator analogs of integral operators (Russ.). *Vestnik Leningrad. Univ.* **22**, 36–39 (1967).
KANIEV, S.
[1] On the deviation from their boundary values of functions biharmonic in a circle (Russ.). *Dokl. Akad. Nauk SSSR* **153**, 995–998 (1963) = *Soviet Math. Dokl.* **4**, 1758–1761 (1963).
KANTOROVICH, L. V.–G. P. AKILOV
[1] *Functional Analysis in Normed Spaces.* Pergamon Press 1964, xiii + 773 pp. (Orig. Russ. ed. Moscow 1959.)
KARAMATA, J.
[1] Suite de fonctionelles linéaires et facteurs de convergence des séries de Fourier. *J. Math. Pures Appl.* (9) **35**, 87–95 (1956).
KARLIN, S.
[1] Pólya-type distributions II. *Ann. Math. Statist.* **28**, 281–308 (1957).
[2] Total positivity and convexity preserving transformations. *In:* Proceedings of Symposia in Pure Mathematics. Vol. VII: Convexity. Amer. Math. Soc., Providence, R.I., 1963, xv + 516 pp.; *pp. 329–347.*
[3] *Total Positivity I.* Stanford Univ. Press 1968, xi + 576 pp.
KARLIN, S.–A. NOVIKOFF
[1] Generalized convex inequalities. *Pacific J. Math.* **13**, 1251–1279 (1963).
KARLIN, S.–W. J. STUDDEN
[1] *Tchebycheff Systems: With Applications in Analysis and Statistics.* Interscience Publ. 1966, xvii + 586 pp.
KATZNELSON, Y.
[1] *An Introduction to Harmonic Analysis.* John Wiley 1968, xiii + 264 pp.
KEMPER, J.: *see Butzer, P. L.*
KEMPERMAN, J. H. B.
[1] On the regularity of generalized convex functions. *Trans. Amer. Math. Soc.* **135**, 69–93 (1969).
KIRCHBERGER, P.
[1] *Über Tschebyscheffsche Annäherungsmethoden.* Inauguraldissertation, Univ. Göttingen 1902, 96 pp. (See also: *Math. Ann.* **57**, 509–540 (1903).)
KNAPOWSKI, S.: *see Freud, G.*
KNOPP, K.
[1] *Theorie und Anwendung der unendlichen Reihen.* Springer (1921) 1947, xii + 582 pp.
KOBER, H.
[1] A note on Hilbert's operator. *Bull. Amer. Math. Soc.* **48**, 421–426 (1942).
[2] A note on Hilbert transforms. *J. London Math. Soc.* **18**, 66–71 (1943).
[3] A modification of Hilbert transforms, the Weyl integral and functional equations. *J. London Math. Soc.* **42**, 42–50 (1967).
KÖNIG, H. (*see also Butzer, P. L.*)
[1] Einige Eigenschaften der Fourier–Stieltjes Transformation. *Arch. Math.* (*Basel*) **11**, 352–365 (1960).

KOIZUMI, S.
[1] On the singular integrals I–VI. *Proc. Japan Acad.* **34**, 193–198; 235–240; 594–598; 653–656 (1958); **35**, 1–6, 323–328 (1959).
[2] On the Hilbert transform I, II. *J. Fac. Sci. Hokkaido Univ. Ser. I* **14**, 153–224 (1959); **15**, 93–130 (1960).
KOLBE, W.–R. J. NESSEL
[1] Saturation theory in connection with Mellin transform methods. *To appear.*
[2] Simultaneous saturation in connection with Dirichlet's problem for a wedge. *To appear.*
KOREVAAR, J. (*see also Butzer, P. L.*)
[1] *Mathematical Methods.* Vol. I: *Linear Algebra, Normed Spaces, Distributions, Integration.* Academic Press 1968, x + 505 pp.
KOROVKIN, P. P.
[1] On convergence of linear positive operators in the space of continuous functions (Russ.). *Dokl. Akad. Nauk SSSR* **90**, 961–964 (1953).
[2] On the order of the approximation of functions by linear positive operators (Russ.). *Dokl. Akad. Nauk SSSR* **114**, 1158–1161 (1957).
[3] An asymptotic property of positive methods of summation of Fourier series and best approximation of functions of class Z_2 by linear positive polynomial operators (Russ.). *Uspehi Mat. Nauk* **13**, no. **6 (84)**, 99–103 (1958).
[4] Best approximation of functions of class Z_2 by certain linear operators (Russ.). *Dokl. Akad. Nauk SSSR* **127**, 513–515 (1959).
[5] *Linear Operators and Approximation Theory.* Hindustan Publ. 1960, vii + 222 pp. (Orig. Russ. ed. Moscow 1959.)
[6] Asymptotic properties of positive methods for summation of Fourier series (Russ.). *Uspehi Mat. Nauk* **15**, no. **1 (91)**, 207–212 (1960).
[7] Convergent sequences of linear operators (Russ.). *Uspehi Mat. Nauk* **17**, no. **4 (106)**, 147–152 (1962).
[8] On the order of approximation of functions by linear polynomial operators of the class S_m (Russ.). *In:* IBRAGIMOV, I. I. [1]; *pp. 163–166.*
KOZAKIEWICZ, W.: *see Butzer, P. L.*
KOZIMA, M.–G. SUNOUCHI
[1] On the approximation and saturation by general singular integrals. *Tôhoku Math. J.* (2) **20**, 146–169 (1968).
KOVALENKO, A. I.
[1] Certain summation methods for Fourier series (Russ.). *Mat. Sb.* **71 (113)**, 598–616 (1966).
KRABBE, G. L.
[1] A space of multipliers of type $L^p(-\infty, \infty)$. *Pacific J. Math.* **9**, 729–737 (1959).
KRASNOSEL'SKIĬ, M. A.–YA. B. RUTICKIĬ
[1] *Convex Functions and Orlicz Spaces.* Noordhoff 1961, xi + 249 pp. (Orig. Russ. ed. Moscow 1958.)
KRÉE, P.
[1] Sur les multiplicateurs dans FL^p. *C.R. Acad. Sci. Paris* **260**, 4400–4403 (1965).
KREIN, M. G.: *see Achieser, N. I.; Gantmacher, F. R.*
KRIPKE, B. R.–T. J. RIVLIN
[1] Approximation in the metric of $L^1(X, \mu)$. *Trans. Amer. Math. Soc.* **119**, 101–122 (1965).
KUPCOV, N. P.
[1] Direct and inverse theorems of approximation theory and semigroups of operators (Russ.). *Uspehi Mat. Nauk* **23**, no. **4 (142)**, 117–178 (1968) = *Russian Math. Surveys* **23**, no. **4**, 115–177 (1968).
KURAN, Ü.
[1] On the boundary values of Poisson integrals in $R^n \times (0, +\infty)$. *J. London Math. Soc.* **41**, 508–512 (1966).
LANGER, R. E. (Ed.)
[1] *On Numerical Approximation.* (Proc. Sympos., Madison 1958; Publ. no. **1** Math. Res. Center, U.S. Army, Univ. of Wisconsin.) Univ. of Wisconsin Press 1959, x + 462 pp.
LARSEN, R.
[1] *The Multiplier Problem.* (Lecture Notes in Math. **105**) Springer 1969, vii + 284 pp.
LAUWERIER, H. A.
[1] The Hilbert problem for generalized functions. *Arch. Rational Mech. Anal.* **13**, 157–166 (1963).

LEBESGUE, H.
[1] Sur la représentation approchée des fonctions. *Rend. Circ. Mat. Palermo* **26**, 325–328 (1908).
[2] Sur les intégrales singulières. *Ann. Fac. Sci. Univ. Toulouse* (3) **1**, 25–117, 119–128 (1909).
DE LEEUW, K.
[1] On the degree of approximation by Bernstein polynomials. *J. Analyse Math.* **7**, 89–104 (1959).
[2] On the adjoint semi-group and some problems in the theory of approximation. *Math. Z.* **73**, 219–234 (1960).
[3] On L_p multipliers. *Ann. of Math.* **81**, 364–379 (1965).
DE LEEUW, K.–H. MIRKIL
[1] A priori estimates for differential operators in L_∞ norm. *Illinois J. Math.* **8**, 112–124 (1964).
LEIBNIZ, G. W.
[1] *Oeuvres*, Tome III 1859 (4e lettre à Wallis (1725), p. 105).
LEIS, R.
[1] Über das Randverhalten harmonischer Funktionen. *Arch. Rational Mech. Anal.* **7**, 168–180 (1961). Addendum: *ibid.* **8**, 444–445 (1961).
[2] Zur Approximation der Randwerte von Potentialfunktionen. *Arch. Math.* (*Basel*) **16**, 378–387 (1965).
LEVIATAN, D.–M. MÜLLER
[1] Some applications of the gamma-operator. *Arch. Math.* (*Basel*) **20**, 638–647 (1969).
LEVITAN, B. M.
[1] Über die Parsevalsche Gleichung für verallgemeinerte Fouriersche Integrale. *Ann. of Math.* (2) **38**, 784–786 (1937).
LÉVY, P.
[1] Sur la détermination des lois de probabilité par leurs fonctions caractéristiques. *C.R. Acad. Sci. Paris* **175**, 854–856 (1922).
[2] *Calcul des probabilités.* Gauthier-Villars 1925, viii + 352 pp.
LINNIK, YU. V.
[1] *Decomposition of Probability Distributions.* Oliver and Boyd 1964, xii + 242 pp. (Orig. Russ. ed. Moscow 1960.)
LIONS, J. L.
[1] *Équations différentielles opérationelles et problèmes aux limites.* Springer 1961, ix + 292 pp.
LIOUVILLE, J.
[1] Sur le calcul des différentielles à indices quelconques. *J. de l'École Polytéchnique* **13**, 71–162 (1832).
LI SJUN-CZIN
[1] Cesàro summability in Banach space (Chin.). *Acta Math. Sinica* **10**, 41–54 (1960) = *Chinese Math.-Acta* **1**, 40–52 (1960).
LITTLEWOOD, J. E.: *see Hardy, G. H.*
LITTMAN, W.–C. MCCARTHY–N. RIVIÈRE
[1] L^p-multiplier theorems. *Studia Math.* **30**, 193–217 (1968).
[2] The non-existence of L^p estimates for certain translation-invariant operators. *Studia Math.* **30**, 219–229 (1968).
LJUSTERNIK, L. A.–W. I. SOBOLEW
[1] *Elemente der Funktionalanalysis.* Akademie-Verl. 1965, xi + 256 pp. (4. ed. 1968, 375 pp.) (Orig. Russ. ed. Moscow 1951.)
LÖFSTRÖM, J.
[1] Some theorems on interpolation spaces with applications to approximation in L_p. *Math. Ann.* **172**, 176–196 (1967).
[2] On the rate of convergence of difference schemes for parabolic initial-value problems and of singular integrals. *In:* BUTZER, P. L.–B. SZ.-NAGY [1]; *pp. 393–415.*
[3] Besov spaces in the theory of approximation. *Ann. Mat. Pura Appl.; to appear.*
LOÈVE, M.
[1] *Probability Theory.* Van Nostrand (1955) 1960, xvi + 685 pp.
LOOMIS, L. H.
[1] A note on the Hilbert transform. *Bull. Amer. Math. Soc.* **52**, 1082–1086 (1946).
[2] *An Introduction to Abstract Harmonic Analysis.* Van Nostrand 1953, x + 190 pp.
[3] *Harmonic Analysis.* (Mimeographed Lecture Notes; 1965 MAA cooperative summer seminar) Math. Ass. of America, Buffalo, N.Y., 1965, iii + 310 + 60 pp.
LORCH, L.
[1] On approximation by Fejér means to periodic functions satisfying a Lipschitz condition. *Canad. Math. Bull.* **5**, 21–27 (1962).

LORENTZ, G. G.
[1] *Bernstein Polynomials.* Univ. of Toronto Press 1953, x + 130 pp.
[2] Inequalities and the saturation classes of Bernstein polynomials. *In:* BUTZER, P. L.–J. KOREVAAR [1]; *pp. 200–207.*
[3] *Approximation of Functions.* Holt, Rinehart and Winston 1966, ix + 188 pp.
LORENTZ, G. G.–L. L. SCHUMAKER
[1] Saturation of positive operators. *Preprint.*
LORENTZ, G. G.–K. ZELLER
[1] Degree of approximation by monotone polynomials I, II. *J. Approximation Theory* **1**, 501–504 (1968); **2**, 265–269 (1969).
LUKACS, E.
[1] *Characteristic Functions.* (Griffin's Statistical Monographs and Courses **5**) Charles Griffin 1960, 216 pp.
MAIRHUBER, J. C.–I. J. SCHOENBERG–R. E. WILLIAMSON
[1] On variation diminishing transformations on the circle. *Rend. Circ. Mat. Palermo* (*2*) **8**, 241–270 (1959).
MALOZEMOV, V. N.
[1] The generalized differentiation of periodic functions (Russ.). *Vestnik Leningrad Univ.* **20**, 164–167 (1965).
MAMEDOV, R. G.
[1] On the order of the approximation of differentiable functions by linear positive operators (Russ.). *Dokl. Akad. Nauk SSSR* **128**, 674–676 (1959).
[2] Asymptotic value of approximation of periodic functions by means of positive operators (Russ.). *In:* SMIRNOV, V. I. [1]; *pp. 154–158.*
[3] The degree of approximation of functions by linear integral operators at Lebesgue points of given order (Russ.). *Izv. Akad. Nauk Azerbaĭdžan. SSR Ser. Fiz.-Mat. Tehn. Nauk* **1961**, no. 3, 13–21 (1961).
[4] Direct and converse theorems in the theory of approximation of functions by m-singular integrals (Russ.). *Dokl. Akad. Nauk SSSR* **144**, 272–274 (1962) = *Soviet Math. Dokl.* **3**, 711–714 (1962).
[5] On the order of convergence of m-singular integrals of a periodic function at generalized Lebesgue points and in the space $L_p(-\pi, \pi)$ (Azerbaijani). *Izv. Akad. Nauk Azerbaĭdžan. SSR Ser. Fiz.-Mat. Tehn. Nauk* **1962**, no. **2**, 15–26 (1962).
[6] On the asymptotic value of the approximation of repeatedly differentiable functions by linear positive operators (Russ.). *Dokl. Akad. Nauk SSSR* **146**, 1013–1016 (1962) = *Soviet Math. Dokl.* **3**, 1435–1439 (1962).
[7] On the order of convergence of m-singular integrals at generalized Lebesgue points and in the space $L_p(-\infty, \infty)$ (Russ.). *Izv. Akad. Nauk SSSR Ser. Mat.* **27**, 287–304 (1963).
[8] Local saturation of a family of linear positive operators (Russ.). *Dokl. Akad. Nauk SSSR* **156**, 499–502 (1964) = *Soviet Math. Dokl.* **5**, 671–674 (1964).
[9] *Approximation von Funktionen durch lineare Operatoren* (Azerbaijani). Baku 1966, pp. 216.
MARCHAUD, A.
[1] Sur les dérivées et sur les différences des fonctions des variables réelles. *J. Math. Pures Appl.* (*9*) **6**, 337–425 (1927).
MARCINKIEWICZ, J.
[1] Sur les multiplicateurs des séries de Fourier. *Studia Math.* **8**, 78–91 (1939).
MASANI, P.
[1] Wiener's contributions to generalized harmonic analysis, prediction theory and filter theory. *Bull. Amer. Math. Soc.* **72**, no. **1**, Part II (Norbert Wiener 1894–1964), 73–125 (1966).
MATSUOKA, Y.
[1] A refinement of Natanson's estimate concerning the Vallée Poussin's singular integrals for the class $\mathrm{Lip}_1\,\alpha$. *Sci. Rep. Kagoshima Univ.* **9**, 7–10 (1960).
[2] On the degree of approximation of functions by some positive linear operators. *Sci. Rep. Kagoshima Univ.* **9**, 11–16 (1960).
[3] Note on Komleva's theorem. *Sci. Rep. Kagoshima Univ.* **9**, 17–23 (1960).
MCCARTHY, C.: *see Littman, W.*
MCSHANE, E. J.–T. A. BOTTS
[1] *Real Analysis.* Van Nostrand 1959, ix + 272 pp.
MEINARDUS, G.
[1] *Approximation of Functions: Theory and Numerical Methods.* Springer 1967, viii + 198 pp. (Orig. German ed. Berlin 1964.)

MICCHELLI, C. A.
[1] *Saturation classes and iterates of operators*. Thesis, Stanford Univ. 1969.
MIKHLIN, S. G.
[1] On the multipliers of Fourier integrals (Russ.). *Dokl. Akad. Nauk SSSR* **109**, 701–703 (1956).
[2] *Multidimensional Singular Integrals and Integral Equations*. Pergamon Press 1965, xii + 259 pp. (Orig. Russ. ed. Moscow 1962.)
MINAKSHISUNDARAM, S. (*see also Chandrasekharan, K.*)
[1] Studies in Fourier Ansatz and parabolic equations I. *J. Madras Univ.* **14**, 73–142 (1942).
MIRKIL, H.: *see de Leeuw, K.*
MOND, B.: *see Shisha, O.*
MORRIS, P. D.: *see Cheney, E. W.*
MOTZKIN, T. S.
[1] *Beiträge zur Theorie der linearen Ungleichungen*. Dissertation, Univ. Basel 1936 (Jerusalem 1936), 69 pp. (Translation: RAND Corp., Santa Monica, Calif. 1958)
[2] Approximation by curves of a unisolvent family. *Bull. Amer. Math. Soc.* **55**, 789–793 (1949).
MÜHLBACH, G.
[1] *Über das Approximationsverhalten gewisser positiver linearer Operatoren*. Dissertation, T.U. Hannover 1969, 93 pp.
[2] Operatoren vom Bernsteinschen Typ. *J. Approximation Theory* **3**, 274–292 (1970).
MÜLLER, M. (*see also Leviatan, D.*)
[1] *Die Folge der Gammaoperatoren*. Dissertation, T.H. Stuttgart 1967, 87 pp.
[2] Über die Approximation durch Gammaoperatoren. *In:* BUTZER, P. L.–B. SZ.-NAGY [1]; *pp. 290–302.*
MULLA, F.: *see Aronszajn, N.*
MUNROE, M. E.
[1] *Introduction to Measure and Integration*. Addison-Wesley 1953, x + 310 pp.
NATANSON, I. P.
[1] On some estimations connected with singular integral of C. de la Vallée-Poussin (Russ.). *Dokl. Akad. Nauk SSSR* **45**, 290–293 (1944) = *C.R. (Doklady) Acad. Sci. URSS (N.S.)* **45**, 274–277 (1944).
[2] Sur la représentation approchée des fonctions vérifiant la condition de Lipschitz au moyen de l'intégrale de Vallée-Poussin (Russ.). *Dokl. Akad. Nauk SSSR* **54**, 11–14 (1946) = *C. R. (Doklady) Acad. Sci. URSS (N.S.)* **54**, 11–13 (1946).
[3] On the degree of approximation to a continuous function of period 2π by means of its Poisson integral (Russ.). *Dokl. Akad. Nauk SSSR* **72**, 11–14 (1950).
[4] On the accuracy of representation of continuous periodic functions by singular integrals (Russ.). *Dokl. Akad. Nauk SSSR* **73**, 273–276 (1950).
[5] On the approximation of multiply differentiable periodic functions by means of singular integrals (Russ.). *Dokl. Akad. Nauk SSSR* **82**, 337–339 (1952).
[6] *Theorie der Funktionen einer reellen Veränderlichen*. Akademie-Verl. 1961, xii + 590 pp. (Orig. Russ. ed. Moscow 1957.)
[7] Saturation classes in the theory of singular integrals (Russ.). *Dokl. Akad. Nauk SSSR* **158**, 520–523 (1964) = *Soviet Math. Dokl.* **5**, 1257–1260 (1964).
[8] *Constructive Function Theory*. Vol. I: *Uniform Approximation;* Vol. II: *Approximation in the Mean;* Vol. III: *Interpolation and Approximation Quadratures*. Frederick Ungar 1964, ix + 232 pp./1965, 176 pp./176 pp. (Orig. Russ. ed. Moscow 1949.)
NAYLOR, D.: *see Duff, G. F. D.*
NESSEL, R. J. (*see also Berens, H.; Butzer, P. L.; Görlich, E.; Kolbe, W.*)
[1] Über die Darstellung holomorpher Funktionen durch Weierstrass und Weierstrass–Stieltjes–Integrale. *J. Reine Angew. Math.* **218**, 31–50 (1965).
[2] *Das Saturationsproblem für mehrdimensionale singuläre Integrale und seine Lösung mit Hilfe der Fourier-Transformation*. Dissertation, T.H. Aachen 1965, 145 pp.
[3] Contributions to the theory of saturation for singular integrals in several variables II. Applications; III. Radial kernels. *Nederl. Akad. Wetensch. Proc. Ser. A* **70** = *Indag. Math.* **29**, 52–73 (1967).
[4] Über Nikolskii-Konstanten von positiven Approximationsverfahren bezüglich Lipschitz-Klassen. *Jber. Deutsch. Math.-Verein; to appear.*
[5] Nikolskii constants of positive operators for Lipschitz classes. *In:* Proc. of 'Conference on Constructive Function Theory', Varna (Bulgaria) 1970; *to appear.*

NESSEL, R. J.–A. PAWELKE
[1] Über Favardklassen von Summationsprozessen mehrdimensionaler Fourierreihen. *Compositio Math.* **19**, 196–212 (1968).
NESSEL, R. J.–W. TREBELS
[1] Gebrochene Differentiation und Integration und Charakterisierungen von Favardklassen *In:* Proc. of 'Colloquium on Constructive Theory of Functions', Budapest 1969; *to appear.*
NEWCOMB, R. W.
[1] Hilbert transforms–distributional theory. Stanford Electronics Laboratories, Technical Rep. **2250–1** (1962), ii + 11 pp.
NIKOLSKIĬ, S. M.
[1] Sur l'allure asymptotique du reste dans l'approximation au moyen des sommes de Fejér des fonctions vérifiant la condition de Lipschitz (Russ.). *Izv. Akad. Nauk SSSR Ser. Mat.* **4**, 501–508 (1940).
[2] Approximation of periodic functions by trigonometrical polynomials (Russ./Engl. Sum.). *Trudy Mat. Inst. Steklov* **15**, 5–58/59–76 (1945).
[3] Some questions of the approximation of differentiable functions (Russ.). *In:* Comptes Rendus du Premier Congrès des Mathématiciens Hongrois, 1950. Akadémiai Kiadó, Budapest, 1952; *pp. 113–124.*
[4] On linear methods of summation of Fourier series (Russ.). *Izv. Akad. Nauk SSSR Ser. Mat.* **12**, 259–278 (1948).
[5] Estimations of the remainder of Féjèr's sum of periodical functions possessing a bounded derivative (Russ.). *Dokl. Akad. Nauk SSSR* (1941) = *C.R.* (*Doklady*) *Acad. Sci. URSS* (*N.S.*) **31**, 210–214 (1941).
[6] Approximation of functions in the mean by trigonometric polynomials (Russ.). *Izv. Akad. Nauk SSSR Ser. Mat.* **10**, 207–256 (1946).
NOVIKOFF, A.: *see Karlin, S.*
OFFORD, A. C.
[1] On Fourier transforms III. *Trans. Amer. Math. Soc.* **38**, 250–266 (1935).
[2] Note on the uniqueness of the representation of a function by a trigonometric integral. *J. London Math. Soc.* **11**, 171–174 (1936).
[3] On the uniqueness of the representation of a function by a trigonometric integral. *Proc. London Math. Soc.* (*2*) **42**, 422–480 (1937).
OGIEVECKIĬ, I. I.
[1] Integration and differentiation of fractional order of periodic functions and the constructive theory of functions (Russ.). *In:* SMIRNOV, V. I. [1]; *pp. 159–164.*
OGIEVECKIĬ, I. I.–L. G. BOĬCUN
[1] On a theorem of Titchmarsh (Russ.). *Izv. Vysš. Učebn. Zaved. Matematika* **1965**, no. **4 (47)**, 100–103 (1965).
OKIKIOLU, G. O.
[1] Fourier transforms and the operator H_α. *Proc. Cambridge Philos. Soc.* **62**, 73–78 (1966).
OLIVER, H.
[1] The exact Peano derivative. *Trans. Amer. Math. Soc.* **76**, 444–456 (1954).
O'NEIL, M. R.
[1] Les fonctions conjugées et les intégrales fractionnaires de la class $L(\log^+ L)^s$. *C.R. Acad. Sci. Paris Sér. A–B* **263**, A463–A466 (1966).
ORLICZ, W.
[1] Ein Satz über die Erweiterung von linearen Operatoren. *Studia Math.* **5**, 127–140 (1934).
OSLER, T. J.
[1] Leibniz rule for fractional derivatives and an application to infinite series. *SIAM J. Math. Analysis* **1** (1970); *to appear.*
PALEY, R. E. A. C.–N. WIENER
[1] *Fourier Transforms in the Complex Domain.* (Amer. Math. Soc. Colloq. Publ. **19**) Amer. Math. Soc., New York 1934, viii + 184 pp.
PAWELKE, A.: *see Nessel, R. J.*
PAWELKE, S.: *see Berens, H.; Butzer, P. L.*
PEANO, G.
[1] Sulla formula di Taylor. *In:* Opere scelte. Vol. I: Analisi matematica—Calcolo numerico. Edizioni Cremonese 1957, vii + 530 pp.; *pp. 204–209.*
PEETRE, J.
[1] Sur le nombre de paramètres dans la définition de certains espaces d'interpolation. *Ricerche Mat.* **12**, 248–261 (1963).

[2] Espaces d'interpolation, généralisations, applications. *Rend. Sem. Mat. Fis. Milano* **34**, 133–164 (1964).
[3] Applications de la théorie des espaces d'interpolation dans l'analyse harmonique. *Ricerche Mat.* **15**, 3–36 (1966).

PERRON, O.
[1] Über die Approximation stetiger Funktionen durch trigonometrische Polynome. *Math. Z.* **47**, 57–65 (1941).

PETROV, I. M.
[1] Order of approximation of functions belonging to the class Z by some polynomial operators (Russ.). *Uspehi Mat. Nauk* **13**, no. **6 (84)**, 127–131 (1958).
[2] The order of approximation of functions of class Z_α by certain polynomial operators (Russ.). *Izv. Vysš. Učebn. Zaved. Matematika* **1960**, no. **1 (14)**, 188–193 (1960).

PETROVSKY, I. G.
[1] *Lectures on Partial Differential Equations.* Interscience Publ. 1954, x + 245 pp. (Orig. Russ. ed. Moscow 1950.)

PFLUGER, A.
[1] Bemerkungen zur Charakterisierung von Fourier–Stieltjes–Transformationen, Lecture, Conf. 'Harmonische Analysis und Integraltransformationen' dir. by P. L. Butzer, Math. Res. Inst. Oberwolfach, Black Forest, 1965; *unpublished.*

PHELPS, R. R.
[1] Uniqueness of Hahn–Banach extensions and unique best approximation. *Trans. Amer. Math. Soc.* **95**, 238–255 (1960).

PHILLIPS, K. L.
[1] The maximal theorems of Hardy and Littlewood. *Amer. Math. Monthly* **74**, 648–660 (1967).

PHILLIPS, R. S. (*see also Hille, E.*)
[1] On Fourier–Stieltjes integrals. *Trans. Amer. Math. Soc.* **69**, 312–323 (1950).

PINNEY, E.
[1] *Ordinary Difference-Differential Equations.* Univ. of California Press 1958, xii + 262 pp.

PITMAN, E. J. G.
[1] On the derivation of a characteristic function at the origin. *Ann. Math. Statist.* **27**, 1156–1160 (1956).

PITT, H. R.
[1] *Tauberian Theorems.* Oxford Univ. Press 1958, x + 174 pp.

DU PLESSIS, N.
[1] Some theorems about the Riesz fractional integral. *Trans. Amer. Math. Soc.* **80**, 124–134 (1955).
[2] Spherical fractional integrals. *Trans. Amer. Math. Soc.* **84**, 262–272 (1957).

POLLARD, H.
[1] Harmonic analysis of bounded functions. *Duke Math. J.* **20**, 499–512 (1953).

POLLARD, S.
[1] Extension to Stieltjes integrals of a theorem due to Plessner. *J. London Math. Soc.* **2**, 37–41 (1926).

PÓLYA, G.: *see also Hardy, G. H.*

PÓLYA, G.–I. J. SCHOENBERG
[1] Remarks on de La Vallée Poussin means and convex conformal maps of the circle. *Pacific J. Math.* **8**, 295–334 (1958).

PÓLYA, G.–G. SZEGÖ
[1] *Aufgaben und Lehrsätze aus der Analysis I, II.* Springer (1925) 1964, xvi + 338 pp./x + 407 pp.

POST, E. L.
[1] Generalized differentiation. *Trans. Amer. Math. Soc.* **32**, 723–781 (1930).

PRICE, K. H.: *see Cheney, E. W.*

PRIVALOV, I. I.
[1] *Randeigenschaften analytischer Funktionen.* Deutsch. Verl. der Wiss. 1956, viii + 247 pp. (Orig. Russ. ed. Moscow 1950.)

PTÁK, V.
[1] On approximation of continuous functions in the metric $\int_a^b |x(t)|\, dt$. *Czechoslovak Math. J.* **8**, 267–273 (1958).

PYCH, P.
[1] Approximation of functions in L- and C-metrics. *Prace Mat.* **11**, 61–76 (1967).

QUADE, E. S.
[1] Trigonometric approximation in the mean. *Duke Math. J.* **3**, 529–543 (1937).
RAGOZIN, D. L.
[1] Approximation theory of SU(2). *J. Approximation Theory* **1**, 464–475 (1968).
REDHEFFER, R.
[1] A delightful inequality. *Amer. Math. Monthly* **76**, 1153–1154 (1969).
REITER, H.
[1] *Classical Harmonic Analysis and Locally Compact Groups.* Oxford Clarendon Press 1968, xi + 200 pp.
RICE, J. R.
[1] *The Approximation of Functions.* Vol. I: *Linear Theory;* Vol. II: *Nonlinear and Multivariate Theory.* Addison-Wesley 1964, x + 203 pp./1969, xiii + 334 pp.
RIEFFEL, M. A.
[1] Multipliers and tensor products of L^p-spaces of locally compact groups. *Studia Math.* **33**, 71–82 (1969).
RIEMANN, B.
[1] Versuch einer allgemeinen Auffassung der Integration und Differentiation. *In:* Bernhard Riemann's gesammelte mathematische Werke und wissenschaftlicher Nachlass. (2. Aufl.; Ed. H. Weber–R. Dedekind) Leipzig: Teubner 1892; *pp. 353–366.* (Reprinted in: *Collected Works of Bernhard Riemann.* Dover Publ. 1953, 690 pp.)
RIESZ, F.
[1] Sur les polynômes trigonométriques. *C.R. Acad. Sci. Paris* **158**, 1657–1661 (1914).
[2] Sur la formule d'inversion de Fourier. *Acta Sci. Math. (Szeged)* **3**, 235–241 (1927).
[3] Über die Sätze von Stone und Bochner. *Acta Sci. Math. (Szeged)* **6**, 184–198 (1932–1934).
RIESZ, F.–B. SZ.-NAGY
[1] *Functional Analysis.* Frederick Ungar 1955, xii + 468 pp. (Orig. French ed. Budapest 1952.)
RIESZ, M. (*see also Hardy, G. H.; Hilb, E.*)
[1] Sur les fonctions conjugées. *Math. Z.* **27**, 218–244 (1927).
[2] L'intégrale de Riemann–Liouville et le problème de Cauchy. *Acta Math.* **81**, 1–223 (1949).
RIVIÈRE, N.: *see Littman, W.*
RIVLIN, T. J. (*see also Kripke, B. R.*)
[1] *An Introduction to the Approximation of Functions.* Blaisdell Publ. 1969, viii + 150 pp.
RIVLIN, T. J.–H. S. SHAPIRO
[1] A unified approach to certain problems of approximation and minimization. *J. Soc. Indust. Appl. Math.* **9**, 670–699 (1961).
ROGOSINSKI, W. W. (*see also Hardy, G. H.*)
[1] Über die Abschnitte trigonometrischer Reihen. *Math. Ann.* **95**, 110–134 (1925).
[2] Reihensummierung durch Abschnittskoppelungen. *Math. Z.* **25**, 132–149 (1926).
ROONEY, P. G.
[1] On the representation of functions as Fourier transforms. *Canad. J. Math.* **11**, 168–174 (1959).
[2] On the representation of sequences as Fourier coefficients. *Proc. Amer. Math. Soc.* **11**, 762–768 (1960).
ROSENTHAL, A.: *see Fréchet, M.*
ROSS, K. A.: *see Hewitt, E.*
ROULIER, J.
[1] Monotone approximation of certain classes of functions. *J. Approximation Theory* **1**, 319–324 (1968).
ROYDEN, H. L.
[1] *Real Analysis.* Macmillan 1963, xvi + 284 pp.
RUBEL, L. A.: *see Gaer, M.*
RUDIN, W.
[1] Factorization in the group algebra of the real line. *Proc. Nat. Acad. Sci. U.S.A.* **43**, 339–340 (1957).
[2] Representation of functions by convolutions. *J. Math. Mech.* **7**, 103–115 (1958).
[3] *Fourier Analysis on Groups.* Interscience Publ. 1962, ix + 285 pp.
[4] *Real and Complex Analysis.* McGraw-Hill 1966, xi + 412 pp.
RUNCK, P. O.
[1] Über Konvergenzgeschwindigkeit linearer Operatoren in Banach-Räumen. *In:* BUTZER, P. L.–J. KOREVAAR [1]; *pp. 96–106.*

RUTICKIĬ, YA. B.: *see Krasnosel'skiĭ, M. A.*
SAGAN, H.
[1] *Boundary and Eigenvalue Problems in Mathematical Physics.* John Wiley 1961, xviii + 381 pp.
SALEM, R.–A. ZYGMUND
[1] Capacity of sets and Fourier series. *Trans. Amer. Math. Soc.* **59**, 23–41 (1946).
SAMERSKII, A. A.: *see Tychonov, A. N.*
SCHERER, K. (*see also Butzer, P. L.*)
[1] A remark on the characterization of Lip 1 by trigonometric best approximation. *J. Approximation Theory; to appear.*
SCHOENBERG, I. J. (*see also Aissen, M.; Mairhuber, J. C.; Pólya, G.*)
[1] Über variationsvermindernde Transformationen. *Math. Z.* **32**, 321–328 (1930).
[2] A remark on the preceding note by Bochner. *Bull. Amer. Math. Soc.* **40**, 277–278 (1934).
[3] On Pólya frequency functions. II: Variation-diminishing integral operators of the convolution type. *Acta Sci. Math.* (*Szeged*) **12**, 97-106 (1950).
[4] On Pólya frequency functions. I: The totally positive functions and their Laplace transforms. *J. Analyse Math.* **1**, 331–374 (1951).
[5] On smoothing operations and their generating functions. *Bull. Amer. Math. Soc.* **59**, 199–230 (1953).
[6] On the zeros of the generating functions of multiply positive sequences and functions. *Ann. Math.* **62**, 447–471 (1955).
[7] On variation diminishing approximation methods. *In:* LANGER, R. F. [1]; *pp. 249–274.*
SCHOENBERG, I. J.–A. WHITNEY
[1] A theorem on polygons in n dimensions with applications to variation-diminishing and cyclic variation-diminishing linear transformation. *Compositio Math.* **9**, 141–160 (1951).
SCHMETS, J.: *see Garnir, H. G.*
SCHUMAKER, L. L.: *see Lorentz, G. G.*
SCHURER, F. (*see also Cheney, E. W.*)
[1] *On linear positive operators in approximation theory.* Thesis, T.H. Delft 1965, 79 pp.
SCHWARTZ, J. T.: *see Dunford, N.*
SCHWARTZ, L.
[1] *Théorie des distributions I, II.* (Publ. Inst. Math. Univ. Strasbourg **9**, **10**) Hermann 1950, 148 pp./1951, 169 pp.–(Nouv. Èd. 1966, xii + 418 pp.)
[2] Sur les multiplicateurs de FL^p. *Kungl. Fysiografiska Sällskapets i Lund Förhandlingar* [*Proc. Roy. Physiog. Soc. Lund*] **22**, no. **21**, 124–128 (1953).
[3] *Methodes mathematiques de la physique.* I. *Compléments de calcul intégral: Séries et intégrales,* 64 pp.; II. *Théorie élémentaire des distributions,* 36 pp.; III. *Convolution,* 40 pp.; IV. *Séries de Fourier,* 32 pp.; V. *L'intégral de Fourier,* 39 pp.; VI. *Transformations de Laplace,* 25 pp.; VII. *Équations des cordes vibrantes,* 45 pp.; VIII. *Fonctions spéciales,* 24 pp. Les cours de Sorbonne, Paris 1955/56.
SEELEY, R. T.
[1] *An Introduction to Fourier Series and Integrals.* Benjamin 1966, x + 104 pp.
SHAPIRO, H. S. (*see also Rivlin, T. J.*)
[1] *Smoothing and Approximation of Functions.* (Mimeographed Lecture Notes; Matscience Rep. **55**) Inst. Math. Sci., Madras (India) 1967, iii + 109 pp.
[2] Some Tauberian theorems with applications to approximation theory. *Bull. Amer. Math. Soc.* **74**, 500–504 (1968).
[3] A Tauberian theorem related to approximation theory. *Acta Math.* **120**, 279–292 (1968).
[4] *Smoothing and Approximation of Functions.* Van Nostrand 1969, viii + 134 pp.
[5] Approximation by trigonometric polynomials to periodic functions of several variables. *In:* BUTZER, P. L.–B. SZ.-NAGY [1]; *pp. 203–217.*
SHAPIRO, V. L.
[1] Fourier series in several variables. *Bull. Amer. Math. Soc.* **70**, 48–93 (1964).
SHILOV, G. E. (*see also Gelfand, I. M.*)
[1] *Generalized Functions and Partial Differential Equations.* Gordon and Breach 1968, xii + 345 pp. (Orig. Russ. ed. Moscow 1965.)
SHISHA, O.
[1] Monotone approximation. *Pacific J. Math.* **15**, 667–671 (1965).
SHISHA, O.–B. MOND
[1] The degree of convergence of sequences of linear positive operators. *Proc. Nat. Acad. Sci. U.S.A.* **60**, 1196–1200 (1968).

[2] The degree of approximation to periodic functions by linear positive operators. *J. Approximation Theory* **1**, 335–339 (1968).

SINGER, I.

[1] Caractérisation des éléments de meilleure approximation dans une espace de Banach quelconque. *Acta Sci. Math. (Szeged)* **17**, 181–189 (1956).

[2] Sur l'unité de l'élément de meilleure approximation dans des espaces de Banach quelconques (Ruman.). *Acad. R.P. Romîne. Stud. Cerc. Mat.* **8**, 235–244 (1957).

[3] La meilleure approximation interpolative dans les espaces de Banach. *Rev. Roumaine Math. Pures Appl.* **4**, 95–113 (1959).

[4] *Best Approximation in Normed Linear Spaces by Elements of Linear Subspaces*. Springer 1970, 409 pp. (Orig. Ruman. ed. Bucharest 1967.)

SMIRNOV, V. I. (Ed.)

[1] *Investigations in Modern Problems of the Constructive Theory of Functions* (Russ.). (Proc. First Conf. on the Constructive Theory of Functions, Leningrad 1959.) Gosudarstv. Izdat. Fiz.-Mat. Lit., Moscow 1961, 368 pp.

SMITH, K. T.: *see Aronszajn, N.*

SNEDDON, I. N.

[1] *Fourier Transforms*. McGraw-Hill 1951, xii + 542 pp.

[2] Functional Analysis. *In:* Encyclopedia of Physics, Vol. II: Mathematical Methods II (Ed. S. Flügge) Springer 1955, vii + 520 pp.; *pp. 198–348.*

[3] *Elements of Partial Differential Equations*. McGraw-Hill 1957, ix + 327 pp.

SOBOLEW, W. I.: *see Ljusternik, L. A.*

STAKGOLD, I.

[1] *Boundary Value Problems of Mathematical Physics I, II*. Macmillan 1967, viii + 340 pp./1968, viii + 408 pp.

STARK, E. L. (*see also Butzer, P. L.; Görlich, E.*)

[1] Über die Approximationsmaße spezieller singulärer Integrale. *Computing (Arch. Elektron. Rechnen)* **4**, 153–159 (1969).

[2] *Über trigonometrische singuläre Faltungsintegrale mit Kernen endlicher Oszillation*. Dissertation, T.H. Aachen 1970, iv + 85 pp.

[3] On approximation improvement for trigonometric singular integrals by means of finite oscillation kernels with separated zeros. *In:* Proc. of 'Conference on Constructive Function Theory', Varna (Bulgaria) 1970; *to appear.*

[4] Zur vollständigen Entwicklung des Approximationsmaßes für die Fejérschen Mittel. *To appear.*

STEČKIN, S. B. (*see also Bari, N. K.*)

[1] On the best approximation of some classes of periodic functions by trigonometric polynomials (Russ.). *Izv. Akad. Nauk SSSR Ser. Mat.* **20**, 643–648 (1956).

[2] On the best approximation of conjugate functions by trigonometric polynomials (Russ.). *Izv. Akad. Nauk SSSR Ser. Mat.* **20**, 197–206 (1956).

[3] The approximation of periodic functions by Fejér sums (Russ.). *Trudy Mat. Inst. Steklov.* **62**, 48–60 (1961) = *Amer. Math. Soc. Transl.* (2) **28**, 269–282 (1963).

STEČKIN, S. B.–S. A. TELJAKOVSKIĬ

[1] The approximation of differentiable functions by trigonometric polynomials in the L metric. *Trudy Mat. Inst. Steklov* **88**, 20–29 (1967) = *Proc. Steklov Inst. Math.* **88** (1967), 19–29 (1969).

STEIN, E. M.

[1] The characterization of functions arising as potentials I, II. *Bull. Amer. Math. Soc.* **67**, 102–104 (1961); **68**, 577–582 (1962).

[2] On limits of sequences of operators. *Ann. Math.* **74**, 140–170 (1961).

STEIN, E. M.–G. WEISS

[1] An extension of a theorem of Marcinkiewicz and some of its applications. *J. Math. Mech.* **8**, 263–284 (1959).

STEINHAUS, H.: *see Kaczmarz, S.*

STONE, M. H.

[1] A generalized Weierstrass approximation theorem. *In:* Studies in Modern Analysis (Studies in Math. **1**; Ed. R. C. Buck) Amer. Math. Soc., Prentice-Hall 1962, ix + 182 pp.; *pp. 30–87.*

STROMBERG, K.: *see Hewitt, E.*

STUDDEN, W. J.: *see Karlin, S.*

SUNOUCHI, G. (*see also Butzer, P. L.; Kozima, M.*)
[1] On the class of saturation in the theory of approximation I, II. *Tôhoku Math. J.* (2) **12**, 339–344 (1960); **13**, 112–118 (1961).
[2] Local operators on trigonometric series. *Trans. Amer. Math. Soc.* **104**, 457–461 (1962).
[3] Characterization of certain classes of functions. *Tôhoku Math. J.* (2) **14**, 127–134 (1962).
[4] On the saturation and best approximation. *Tôhoku Math. J.* (2) **14**, 212–216 (1962).
[5] Approximation and saturation of functions by arithmetic means of trigonometric interpolating polynomials. *Tôhoku Math. J.* (2) **15**, 162–166 (1963).
[6] Saturation in the theory of best approximation. *In:* BUTZER, P. L.–J. KOREVAAR [1]; *pp. 72–88.*
[7] Saturation in the local approximation. *Tôhoku Math. J.* (2) **17**, 16–28 (1965).
[8] Direct theorems in the theory of approximation. *Acta Math. Acad. Sci. Hungar.* **20**, 409–420 (1969).

SUNOUCHI, G.–C. WATARI
[1] On determination of the class of saturation in the theory of approximation of functions I, II. *Proc. Japan Acad.* **34**, 477–481 (1958); *Tôhoku Math. J.* (2) **11**, 480–488 (1959).

SUZUKI, Y. (*see also Ikeno, K.*)
[1] Saturation of local approximation by linear positive operators. *Tôhoku Math. J.* (2) **17**, 210–221 (1965).

SZÁSZ, O. (*see also Egerváry, E. v.; Hilb, E.*)
[1] Über den Konvergenzexponent der Fourierschen Reihen. *S.-B. Bayer, Akad. Wiss. Math.-Phys. Kl.*, 135–150, 1922.
[2] Über die Fourierschen Reihen gewisser Funktionenklassen. *Math. Ann.* **100**, 530–536 (1928).
[3] Über die Cesàroschen und Rieszschen Mittel Fourierschen Reihen. *Comment. Math. Helv.* **11**, 215–220 (1939).
[4] On the Cesàro and Riesz means of Fourier series. *Compositio Math.* **7**, 112–122 (1939).

SZEGÖ, G. (*see also Pólya, G.*)
[1] Koeffizientenabschätzungen bei ebenen und räumlichen harmonischen Entwicklungen. *Math. Ann.* **96**, 601–632 (1926–27).

SZEPTYCKI, P.: *see Aronszajn, N.*

SZ.-NAGY, B. (*see also Butzer, P. L.; Riesz, F.*)
[1] Approximation der Funktionen durch die arithmetischen Mittel ihrer Fourierschen Reihen (Hungar.). *Mat. Fiz. Lapok* **49**, 123–138 (1942).
[2] Approximation der Funktionen durch die arithmetischen Mittel ihrer Fourierschen Reihen. *Acta Sci. Math. (Szeged)* **11**, 71–84 (1946).
[3] Sur une class générale de procédés de sommation pour les séries de Fourier. *Hungarica Acta Math.* **1**, no. **3**, 14–52 (1948).
[4] Sur l'ordre de l'approximation d'une fonction par son intégrale de Poisson. *Acta Math. Acad. Sci. Hungar.* **1**, 183–188 (1950).
[5] *Introduction to Real Functions and Orthogonal Expansions.* Oxford Univ. Press 1965, xi + 447 pp.
[6] Méthodes de sommation des séries de Fourier I. *Acta Sci. Math. (Szeged)* **12**, 204–210 (1950).
[7] Über gewisse Extremalfragen bei transformierten trigonometrischen Entwicklungen. I: Periodischer Fall; II: Nichtperiodischer Fall. *Ber. Sächs. Akad. Wiss. Leipzig Math.-Phys. Kl.* **90**, 103–134 (1938); **91**, 3–24 (1939).

TABERSKI, R.
[1] On singular integrals. *Ann. Polon. Math.* **4**, 249–268 (1958).
[2] Summability with Korovkin's factors. *Bull. Acad. Polon. Sci. Sér. Sci. Math. Astronom. Phys.* **9**, 385–388 (1961).
[3] Approximation to conjugate function. *Bull. Acad. Polon. Sci. Sér. Sci. Math. Astronom. Phys.* **10**, 255–260 (1962).
[4] Asymptotic formulae for Cesàro singular integrals. *Bull. Acad. Polon. Sci. Sér. Math. Astronom. Phys.* **10**, 637–640 (1962).

TAIBLESON, M. H.
[1] On the theory of Lipschitz spaces of distributions on Euclidean n-space. I. Principal properties; II. Translation invariant operators, duality, and interpolation; III. Smoothness and integrability of Fourier transforms, smoothness of convolution kernels. *J. Math. Mech.* **13**, 407–479 (1964); **14**, 821–839 (1965); **15**, 973–981 (1966).

TAMARKIN, J. D.: *see Hille, E.*

TAYLOR, A. E.
[1] *Introduction to Functional Analysis.* John Wiley 1958, xvi + 423 pp.

[2] *General Theory of Functions and Integration.* Blaisdell Publ. 1965, xvi + 437 pp.
TELJAKOVSKIĬ, S. A. (*see also S. B. Stečkin*)
[1] Approximation to functions differentiable in Weyl's sense by de la Vallée-Poussin sums (Russ.). *Dokl. Akad. Nauk SSSR* **131**, 259–262 (1960) = *Soviet Math. Dokl.* **1**, 240–243 (1960).
[2] On norms of trigonometric polynomials and approximation of differentiable functions by linear averages of their Fourier series I. *Trudy Mat. Inst. Steklov.* **62**, 61–97 (1961) = *Amer. Math. Soc. Transl.* (2) **28**, 283–322 (1963).
[3] Estimations of trigonometric series and polynomials in connection with problems in the theory of approximation of functions. *Mat. Zametki* **1**, 611–623 (1967) = *Math. Notes.*
[4] On approximation by Fejér means to functions satisfying a Lipschitz condition (Russ.). *Ukrain. Mat. Ž.* **21**, 334–342 (1969).
TILLMANN, H. G. (*see also Butzer, P. L.*)
[1] Randverteilungen analytischer Funktionen und Distributionen. *Math. Z.* **59**, 61–83 (1953).
[2] Distributionen als Randverteilungen analytischer Funktionen II. *Math. Z.* **76**, 5–21 (1961).
TIMAN, A. F.
[1] A precise estimate of the remainder in the approximation of differentiable functions by Poisson integrals (Russ.). *Dokl. Akad. Nauk SSSR* **74**, 17–20 (1950).
[2] *Theory of Approximation of Functions of a Real Variable.* Pergamon Press 1963, xii + 613 pp. (Orig. Russ. ed. Moscow 1960.)
TITCHMARSH, E. C.
[1] A contribution to the theory of Fourier transforms. *Proc. London Math. Soc.* (2) **23**, 279–289 (1923).
[2] A theorem on Lebesgue integrals. *J. London Math. Soc.* **2**, 36–37 (1927).
[3] A note on Fourier transforms. *J. London Math. Soc.* **2**, 148–150 (1927).
[4] On conjugate functions. *Proc. London Math. Soc.* (2) **29**, 49–80 (1929).
[5] *The Theory of Functions.* Oxford Univ. Press 1939, x + 454 pp.
[6] *Introduction to the Theory of Fourier Integrals.* Oxford Univ. Press (1937) 1948, viii + 395 pp.
TOEPLITZ, O.
[1] Über die Fourier'sche Entwicklung positiver Funktionen. *Rend. Circ. Mat. Palermo* **32**, 191–192 (1911).
TÖRNIG, W.
[1] Anfangswertprobleme bei gewöhnlichen und partiellen Differentialgleichungen. *In:* Mathematische Hilfsmittel des Ingenieurs II (Ed. R. Sauer–I. Szabó) Springer 1969, xx + 684 pp.; *pp. 1–292.*
TOMIĆ, M.: *see Aljančić, S.*
TRANTER, C. J.
[1] *Integral Transforms in Mathematical Physics.* John Wiley 1951, ix + 133 pp.
TREBELS, W. (*see also Butzer, P. L.; Nessel, R. J.*)
[1] *Charakterisierungen von Saturationsklassen in* $L^1(E_n)$. Dissertation, T.H. Aachen 1969, 93 pp.
[2] Über absolute Summierbarkeit von n-dimensionalen Fourierreihen und Fourierintegralen. *J. Approximation Theory* **2**, 394–399 (1969).
TRÈVES, F.
[1] *Linear Partial Differential Equations with Constant Coefficients.* Gordon and Breach 1966, x + 534 pp.
TRICOMI, F. G.
[1] *Vorlesungen über Orthogonalreihen.* Springer 1955, viii + 264 pp.
TURECKIĬ, A. H.
[1] On the saturation class of Hölder's method of summing Fourier series of continuous periodic functions (Russ.). *Dokl. Akad. Nauk SSSR* **121**, 980–983 (1958).
[2] Saturationsklassen verschiedener Ordnung. Saturationsklasse für die Summationsmethode von Vallée-Poussin (Russ.). *Dokl. Akad. Nauk BSSR* **2**, 395–402 (1958).
[3] On classes of saturation for certain methods of summation of Fourier series of continuous periodic functions (Russ.). *Uspehi Mat. Nauk* **15**, no. **6** (**96**), 149–156 (1960) = *Amer. Math. Soc. Transl.* (2) **26**, 263–272 (1963).
[4] The class of saturation for the 'logarithmic' mean method of summing Fourier series (Russ.). *Dokl. Akad. Nauk BSSR* **4**, 95–100 (1960).
[5] Saturation classes in a space C (Russ.). *Izv. Akad. Nauk SSSR Ser. Mat.* **25**, 411–442 (1961).

TYCHONOV, A. N.–A. A. SAMARSKII
[1] *Equations of Mathematical Physics.* Pergamon Press 1963, xvi + 765 pp. (Russ. ed. Moscow 1953.)
DE LA VALLÉE POUSSIN, C.
[1] Sur l'approximation des fonctions d'une variable réelle et de leurs dérivées par des polynômes et des suites limitées de Fourier. *Acad. Roy. Belg. Bull. Cl. Sci.* **3**, 193–254 (1908).
[2] Note sur l'approximation par un polynôme d'une fonction dont la dérivée est à variation bornée. *Bull. Soc. Math. Belg.* **3**, 403–410 (1908).
[3] *Leçons sur l'approximation des fonctions d'une variable réelle.* Gauthier-Villars (1919) 1952, vi + 151 pp. (Reprinted in: *L'approximation.* Chelsea Publ. 1970, 363 pp.)
VERBLUNSKY, S.
[1] On a class of Cauchy exponential series II. *Proc. London Math. Soc.* (*3*) **12**, 707–724 (1962).
VILENKIN, N. YA.: *see Gelfand, I. M.*
VINOGRADOVA, G. N.
[1] Asymptotic value of approximation of functions by operators of a certain class (Russ.). *Volž. Mat. Sb.* **4**, 24–29 (1966).
DE VORE, R. A.
[1] On Jackson's theorem. *J. Approximation Theory* **1**, 314–318 (1968).
[2] Saturation of positive convolution operators. *Preprint.*
WAINGER, S.
[1] *Special Trigonometric Series in k-Dimensions.* Mem. Amer. Math. Soc. **59** (1965), 102 pp.
WATARI, C.: *see Sunouchi, G.*
WEIL, A.
[1] *L'intégration dans les groupes topologiques et ses applications.* Hermann 1940, 158 pp.
WEINBERGER, H. F.
[1] *Partial Differential Equations with Complex Variables and Transform Methods.* Blaisdell Publ. 1965, ix + 446 pp.
WEINSTEIN, A.
[1] The generalized radiation problem and the Euler–Poisson–Darboux equation. *Summa Brasil. Math.* **3**, 125–147 (1955).
[2] On a singular differential operator. *Ann. Mat. Pura Appl.* (*4*) **49**, 359–366 (1960).
[3] On the wave equation and the equation of Euler–Poisson. *In:* Proceedings of Symposia in Applied Mathematics. V: Wave Motion and Vibration Theory. McGraw-Hill 1954; *pp. 137–147.*
WEISS, G. (*see also Stein, E. M.*)
[1] *Analisis armonico en varias variables. Teoria de los espacios H^p.* (Mimeographed Lecture Notes; Cursos y Seminarios de Matemática **9**) Univ. de Buenos Aires 1960, 139 pp.
[2] Harmonic analysis. *In:* Studies in Real and Complex Analysis (Stud. in Math. **3**; Ed. I. I. Hirschman, Jr.) Math. Ass. of America, Buffalo, N.Y.; Prentice-Hall 1965, v + 213 pp.; *pp. 124–178.*
WEISS, MAX: *see Bruckner, A. M.*
WESTPHAL, U. (*see also Berens, H.*)
[1] Über Potenzen von Erzeugern von Halbgruppenoperatoren. *In:* BUTZER, P. L.–B. SZ.-NAGY [1]; *pp. 82–91.*
[2] Ein Kalkül für gebrochene Potenzen infinitesimaler Erzeuger von Halbgruppen und Gruppen von Operatoren. I: Halbgruppenerzeuger; II: Gruppenerzeuger. *Compositio Math.* **22**, 67–103, 104–136 (1970).
WEYL, H.
[1] Bemerkungen zum Begriff des Differentialquotienten gebrochener Ordnung. *Vierteljschr. Naturforsch. Ges. Zürich* **62**, 296–302 (1917).
WHEEDEN, R. L.
[1] Hypersingular integrals and summability of Fourier integrals and series. *In:* CALDERÓN, A. P. [3]; *pp. 336–369.*
[2] On hypersingular integrals and Lebesgue spaces of differentiable functions. *Trans. Amer. Math. Soc.* **134**, 421–435 (1968).
WHITNEY, A.: *see Aissen, M.; Schoenberg, I. J.*
WHITNEY, H.
[1] Derivatives, difference quotients, and Taylor's formula. *Bull. Amer. Math. Soc.* **40**, 89–94 (1934).
[2] On functions with bounded n^{th} differences. *J. Math. Pures Appl.* **36**, 67–96 (1957).

WIDDER, D. V. (*see also Hirschman, I. I., Jr.*)
[1] *The Laplace Transform.* Princeton Univ. Press 1941, x + 406 pp.
[2] Inversion formula for convolution transforms. *Duke Math. J.* **14**, 217–249 (1947).
[3] Weierstrass transforms of positive functions. *Proc. Nat. Acad. Sci. U.S.A.* **37**, 315–317 (1951).
[4] Necessary and sufficient conditions for the representation of a function by a Weierstrass transform. *Trans. Amer. Math. Soc.* **71**, 430–439 (1951).
WIENER, N. (*see also Paley, R. E. A. C.*)
[1] Laplacians and continuous linear functionals. *Acta Sci. Math. (Szeged)* **3**, 7–16 (1927).
[2] *The Fourier Integral and Certain of its Applications.* Dover Publ. 1933, xi + 201 pp.
DE WILDE, M.: *see Garnir, H. G.*
WILLIAMSON, J. H.
[1] *Lebesgue Integration.* Holt, Rinehart and Winston 1962, viii + 117 pp.
WILLIAMSON, R. E.: *see Mairhuber, J. C.*
WOHLERS, M. R.: *see Beltrami, E. J.*
WULBERT, D. E. (*see also Cheney, E. W.*)
[1] Convergence of operators and Korovkin's theorem. *J. Approximation Theory* **1**, 381–390 (1968).
YOSIDA, K.
[1] *Functional Analysis.* Springer 1965, xi + 458 pp.
ZAANEN, A. C.
[1] *Linear Analysis.* North-Holland 1953, vii + 601 pp.
[2] *An Introduction to the Theory of Integration.* North-Holland (1958) 1969, ix + 254 pp.
ZAMANSKY, M.
[1] Classes de saturation de certains procédés d'approximation des séries de Fourier des fonctions continues et applications à quelques problèmes d'approximation. *Ann. Sci. École Norm. Sup.* (*3*) **66**, 19–93 (1949).
[2] Sur la sommation des séries de Fourier dérivées. *C.R. Acad. Sci. Paris* **231**, 1118–1120 (1950).
[3] Classes de saturation des procédés de sommation des séries de Fourier et applications aux séries trigonométriques. *Ann. Sci. École Norm. Sup.* (*3*) **67**, 161–198 (1950).
ZELLER, K. (*see also Lorentz, G. G.*)
[1] *Theorie der Limitierungsverfahren.* Springer 1959, viii + 242 pp. (2. rev. ed. by Zeller, K.–W. Beekmann; 1970, xii + 314 pp.)
[2] Saturation bei äquivalenten Summierungsverfahren. *Math. Z.* **71**, 109–112 (1959).
ZEMANIAN, A. H.
[1] *Distribution Theory and Transform Analysis. An Introduction to Generalized Functions, with Applications.* McGraw-Hill 1965, xiii + 371 pp.
[2] *Generalized Integral Transformations.* Interscience Publ. 1968, xvi + 300 pp.
ZIEGLER, Z.
[1] Linear approximation and generalized convexity. *J. Approximation Theory* **1**, 420–443 (1968).
ŽUK, V. V.
[1] The approximation of periodic functions by linear approximation methods (Russ.). *Dokl. Akad. Nauk SSSR* **179**, 1038–1041 (1968) = *Soviet Math. Dokl.* **9**, 489–493 (1968).
ZYGMUND, A. (*see also Calderón, A. P.; Salem, R.*)
[1] Sur la sommation des séries trigonométriques conjuguées aux séries de Fourier. *Bull. Int. Acad. Polon. Sci. Lettres Cl. Sci. Math. Natur. Ser. A Sci. Math.* 251–258 (1924).
[2] Smooth functions. *Duke Math. J.* **12**, 47–76 (1945).
[3] The approximation of functions by typical means of their Fourier series. *Duke Math. J.* **12**, 695–704 (1945).
[4] A remark on characteristic functions. *Ann. Math. Statist.* **18**, 272–276 (1947).
[5] *Trigonometric Series.* Dover Publ. 1955, vii + 329 pp. (Orig. ed. Warsaw 1935.)
[6] On singular integrals. *Rend. Mat. e Appl.* (*5*) **16**, 468–505 (1957).
[7] *Trigonometric Series I, II.* Cambridge Univ. Press 1959, xii + 383 pp./vii + 354 pp. (2. ed. 1968; one vol. xiii + 383 pp. + vii + 364 pp.)

Index

For further information see also the table of **CONTENTS** and the **LIST OF SYMBOLS** (*LoS*). The material of the sections on **NOTES AND REMARKS** is not completely listed. For properties and theorems concerning particular kernels and singular integrals see item *singular integrals, particular*. The list of theorems under item *theorem* is by no means complete. The association of a theorem with a mathematician's name is often due to purposes of orientation. Page numbers following ℝ indicate properties referring to the real line.

Printed in the United States
by Baker & Taylor Publisher Services